Calculus

LIPMAN BERS

Columbia University

HOLT, RINEHART AND WINSTON, INC.

Calculus

New York Chicago San Francisco Atlanta Dallas Montreal Toronto London Sydney

Copyright © 1969 by Holt, Rinehart and Winston, Inc.
A preliminary edition of this book was published in two volumes in 1967.
All Rights Reserved

SBN 03-065240-5

Library of Congress Catalog Card Number: 69-11994
Printed in the United States of America

1 2 3 4 5 6 7 8 9

To My Mother

This book is intended for students taking calculus for the first time. The needs of such students, whatever their reason for studying mathematics, are essentially the same. A first course ought to give a grasp of the main ideas of calculus and an ability to use its language and its techniques with ease and understanding. Therefore a first course must be a course in essentials. Calculus is the art of setting up and solving differential equations; this is how it originated and this is what it is about. Our book is based on this point of view without, of course, attempting to give anything approaching a theory of differential equations.

Like every other subject, calculus is learned best with due regard to its history. References to the history of mathematics are therefore made at various places in the book, and historical considerations have influenced the choice of material. Yet, I hope this is a modern calculus book, since it is written by a modern working mathematician. Wherever possible, examples from recent scientific developments are used. But no attempt is made to use artificially modern language; traditional notation and terminology are preserved, with only a few exceptions.

Applications are emphasized throughout, primarily applications to mechanics, since they played a dominant part in the history of calculus. I believe that these applications are as important for students who will go into social and life sciences or into pure mathematics as they are for students who will apply calculus to physical sciences and engineering.

Intuitive reasoning is used and stressed throughout the book, not as a substitute for, but rather as a guide to, rigorous thinking.

Rigor in mathematics means, first of all, honesty and clarity. The purpose of rigor in a beginning course is to make the concepts easily understood and used. The technique of making definitions and of proving theorems is of little concern to us at this point. All definitions and statements in our book are correct and precise. In most chapters, however, certain key theorems are explained and used without proof. Proofs are given in appendixes (and, in the case of a few more difficult results, in the Supplement). The sections containing the proofs make the book logically self-contained, but these sections will and ought to be omitted from formal instruction in most classes. Other appendixes contain additional material which may or may not be included in a first course.

In order to use rigor as an aid rather than as an impediment to understanding, it was necessary to rethink certain standard approaches. The expert will easily note where we deviate from tradition.

Finally, I attempted to write a readable book, and I hope that teachers will find it possible to give substantial reading assignments, thus gaining classroom time for explanations, discussions, and examples. Students should be warned, though, that a mathematical book must be read slowly and, if possible, with pencil in hand. The reader should verify the calculations and supply the steps omitted in the printed text. This may require an effort, but it is the best way to learn mathematics.

The book presupposes no special preparation beyond the ability to perform ordinary algebraic operations and some rudimentary knowledge of geometry. Though the text centers on calculus of one variable, analytic geometry and calculus of several variables are also developed as much as is needed for a first course and as can reasonably be taught without developing substantial amounts of linear algebra simultaneously.

The exercises, of which there are over 3000, serve to develop skills and to strengthen understanding. Few of them, if any, should present difficulties to a student who read the corresponding parts in the text and looked at the examples given there. Answers are provided for all even-numbered exercises, and worked-out solutions for several hundred exercises marked with the symbol ▶. The problems at the end of each chapter, on the other hand, are often challenging and in some cases lead beyond the content of the corresponding chapter.

Besides being useful and powerful, calculus is very beautiful. The aesthetic appreciation of mathematics may require a strenuous apprenticeship, but the success or failure of a mathematical text should be measured, ultimately, by how far it succeeds in conveying to the reader the fundamental beauty of the subject.

Suggestions for Use

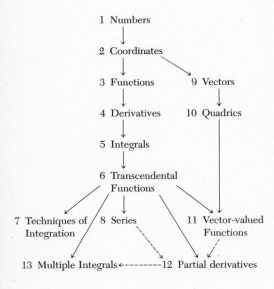

1 Numbers

2 Coordinates

3 Functions   9 Vectors

4 Derivatives   10 Quadrics

5 Integrals

6 Transcendental
  Functions

7 Techniques of / 8 Series   11 Vector-valued
  Integration   Functions

13 Multiple Integrals ←--------12 Partial derivatives

The whole text could be covered in either three or four terms, depending upon the preparation of the students and the number of lectures per week. The material covered, and the way it is organized, is seen from the Table of Contents; the main logical connections between the chapters are shown in the adjoining diagram. The book has been designed so as to provide maximum flexibility for the instructor (or reader). We make only a few comments.

The first two chapters (Numbers, Coordinates) deal with material familiar to many students; these chapters could be reviewed briefly or assigned for independent reading, except perhaps for some work on inequalities in Chapter 1,§3.

The core of calculus is presented in Chapters 3 to 6 (Functions, Derivatives, Integrals, Transcendental Functions). Chapter 3 establishes the basic language and main facts about functions, continuity, and limits. Some teachers may prefer to postpone certain parts of this chapter until it is needed later (for instance, one could take up Chapter 3.5 to 3.7 immediately before Chapter 4,§§2.5–6, and Chapter 3,§§2.4–5 immediately before Chapter 7,§2.) The mean value theorem makes its first appearance in Chapter 8,§1. This section could be studied immediately after Chapter 4.

After Chapter 6, several paths are open. One can go on either to techniques of integration (Chapter 7), or to Taylor's Theorem and infinite series (Chapter 8), or to vectors and vector-valued functions (Chapters 9, 10, and 11), or finally to functions of several variables (Chapters 12 and 13).

Also, Chapters 9 and 10 can be studied immediately after Chapter 2; these three chapters constitute a course in analytic geometry. Instructors who do not intend to cover calculus of several variables may want to omit solid analytic geometry (Chapter 9,§§4 and 5). Of course, students who have already studied analytic geometry, will know most of this material.

The two chapters on several variables are largely independent of each other, except that the definitions in Chapter 12,§1 are used everywhere, and §§2.1 to 6 of Chapter 12 are needed for Chapter 13,§2. Analytic geometry and vectors are used occasionally in these chapters, the latter only as a kind of shorthand.

There are two kinds of appendixes. Some contain optional material at essentially the same level of sophistication as the main text. Others contain more difficult proofs. We repeat that such sections, as well as the Supplement, are intended only for students who are interested in the logical structure of calculus. On the other hand, in so-called honors courses, the emphasis may be put on the theoretical material in the appendixes and in the Supplement.

Acknowledgments

It is a pleasure to thank the colleagues, too numerous to be named, who read the manuscript, tried out the preliminary edition, made suggestions, and discovered mistakes. I was very fortunate to have had the cooperation of Professor J. E. D'Atri, who prepared the exercises for Chapters 1–11, according to my specifications, and displayed originality and taste; of Professor R. Hersch, who did the same for the last two chapters; and of Professor S. Anastasio who supplied the answers and worked-out solutions. I profited greatly by their judgment and taste. I am indebted to the illustrator, Earl Kvam, who combines artistic ability with an understanding of mathematics. I also thank Mrs. Charles Skelley, who prepared a typescript of exceptional quality. During the preparation of the book, I learned to rely on the good will and competence of many members of Holt, Rinehart and Winston; special thanks are due to Chester W. Krone, Jr., Milton S. Mautner, Mrs. Margaret Ong Tsao, Miss Gloria C. Oden, and Mrs. Elaine W. Wetterau.

Finally, I would like to acknowledge the inspiration I received from two highly unconventional books on calculus: a historical account written by the outstanding mathematician O. Toeplitz (now available in an English edition, Chicago University Press) and the text by the distinguished physicist Ya. Zeldovich (available, at this time, only in Russian). Several examples in Chapter 6, §6 are from that book.

New Rochelle, New York   Lipman Bers
January 1969

Contents

1/Numbers

As a preparation for studying calculus, we discuss here rational and irrational numbers, and the rules for operating with them. The material presented in this introductory chapter will be familiar to many readers; such readers should review it rapidly. Since the section on inequalities (§3) deals with techniques that are used quite often, the reader will benefit from doing some of the exercises in this section.

The appendixes to this and to the other chapters can be omitted without the loss of continuity. The first appendix (§6) deals with induction and with the completeness of the system of real numbers. These are theoretical results which will be used only in the appendixes of other chapters that deal with the proofs of some theorems. The theoretical discussion of numbers is continued in §8. Another appendix (§7) treats the accuracy of approximate calculations.

The invention of calculus in the seventeenth century was a turning point in the history of human thought. This turning point made modern science possible. Calculus is the foundation of many branches of contemporary mathematics, and many applications of mathematics to other sciences involve calculus. Two developments prepared the way for the invention of calculus: the gradual extension of the concept of number and the fusion of geometry with algebra. An account of calculus should begin, therefore, with a discussion of numbers.

§1 Rational Numbers. The Need for Extending the Concept of Number

1.1 Positive fractions

The simplest numbers are the positive integers used for counting, 1, 2, 3, 4, $\cdots$, and the positive fractions used for measuring, $\frac{1}{2}$, $\frac{17}{3}$, $\frac{4}{5}$, $\cdots$. Positive integers are themselves fractions ($1 = \frac{1}{1}$, $2 = \frac{2}{1}$, and so forth), and every positive fraction is the ratio of two positive integers. The positive integers and fractions are called **positive rational numbers.**

We should, of course, distinguish the numbers themselves from the symbols used to denote them. Thus, instead of the Arabic numerals 1, 2, 3, 4, $\cdots$, we could use Roman numerals I, II, III, IV, $\cdots$ or some self-explanatory symbols like *, **, ***, ****, $\cdots$.

Positive rational numbers and the arithmetic operations made on them can be represented geometrically. To do this, we must choose a fixed straight segment as a unit of length.

Let the unit segment be divided into n congruent segments (n a positive integer). This can be done by compass and straightedge. Each "small" segment is said to have length $\frac{1}{n}$. A segment is said to have length $\frac{m}{n}$, m and n positive integers, if it can be divided into m congruent segments, each of which has length $\frac{1}{n}$. The length of a segment AB will be denoted by $\overline{AB}$.

Let a and b be positive rational numbers, and let O and A be two points such that $\overline{OA} = a$. Let B be a point on the line through O and A such that A lies between O and B and $\overline{AB} = b$. Then the sum $a + b$, a positive rational number, is the length of OB (see Fig. 1.1). The **commutative law of addition,** $a + b = b + a$, means that the length of the segment OB is the same measured from O to B or from B to O. The geometric meaning of the **associative law of addition** $(a + b) + c = a + (b + c)$, is obvious from Fig. 1.2. If a is greater than b, the **difference** $x = a - b$, that is, the positive rational number x such that $b + x = a$, can also be found geometrically (Fig. 1.3).

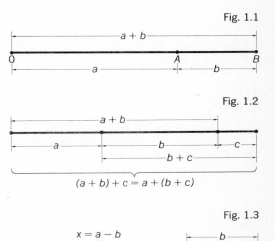

Fig. 1.1

Fig. 1.2

$(a + b) + c = a + (b + c)$

Fig. 1.3

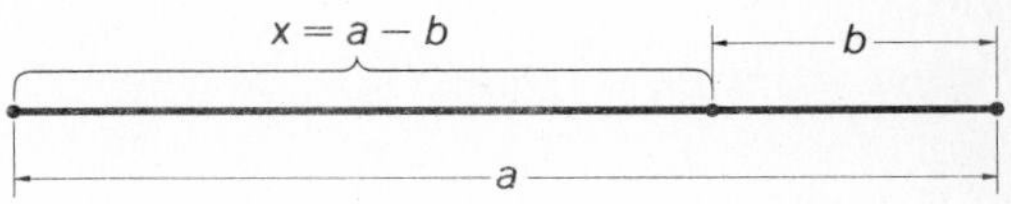

The **product** ab is again a positive rational number, and so is the **quotient** $x = \dfrac{a}{b}$, that is, the number x such that $bx = a$. (From now on, we often write a/b instead of $\dfrac{a}{b}$; this fraction, called a shilling fraction, is frequently used in printed mathematical literature. The geometric construction of segments of length ab or of length a/b is discussed in an appendix to this chapter (see §8.1).

The product ab may also be interpreted as the **area** of a rectangle of length a and width b. Such a rectangle can be cut up into pieces and reassembled so as to form a rectangle of length ab and width 1. The **commutative law of multiplication**, $ab = ba$, means that it does not matter which of the two dimensions of a rectangle we call "length" and which "width." The geometric meaning of the **distributive law**, $a(b + c) = ab + ac$, is shown in Fig. 1.4.

The volume of a right parallelepiped ("box") of length a, width b, and height c is $(ab)c$ ($=$ area of base times height). Such a box may be cut up into pieces and reassembled so as to form a box of length $(ab)c$, width 1, and height 1. The **associative law of multiplication**, $(ab)c = a(bc)$, follows by noting that a box of length a, width b, and height c may also be said to have length b, width c, and height a.

1.2 Rational numbers

Positive fractions were already known in the ancient civilizations of Babylon and Egypt; Hindus are believed to have invented the number **zero. Negative** rational numbers, -1, $-\frac{3}{5}$, and so on, finally, were introduced in Italy during the Renaissance. Positive and negative integers and fractions, and zero, form the system of **rational** numbers.

Rational numbers can be represented by points on a straight line. We choose a point on the line (the **origin**), which we label 0, and another point which we label 1. The distance from 0 to 1 is chosen as the unit of length. A positive rational number a is represented by a point on the line whose distance from 0 is a and which lies on the same side of 0 as 1. The point on the other side of 0 and the same distance from it represents the number $(-a)$.

It is traditional to draw the line horizontally and to choose 1 to the right of 0, as in Fig. 1.5. Note that the terms "horizontal" and "right" cannot be defined mathematically. We shall, nevertheless, use the expressions "to the right" and "to the left," by which we mean, respectively, "in the direction from 0 to 1," and "in the direction from 1 to 0."

Fig. 1.4

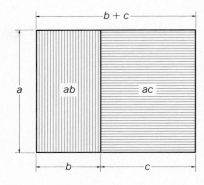

Fig. 1.5

1.3 Incommensurable segments

Rational numbers were invented for measuring, especially for the measurement of length. By means of fractions, the process of measuring can sometimes be reduced to the more elementary process of counting. But this is not always possible. There exist **incommensurable segments,** that is, two segments such that if one is chosen as the unit of length, the length of the other cannot be expressed by any rational number. The simplest example is the side of a square and its diagonal. Let the side have length 1. If the length of the diagonal were a rational number x, we should have $x^2 = 2$. This follows from the Pythagorean Theorem, or simply by noting that the area of the "big" square in Fig. 1.6 is twice that of the "small" square. But *there is no rational number whose square is* 2. Let us recall the famous proof of this fact.

We recall first that every integer is either even, that is, divisible by 2, or odd, that is, not divisible by 2. We remark next that *if m is an integer and m^2 is even, then m is also even.*

Indeed, if m is odd, then $m + 1$ is even and hence $m + 1 = 2n$, n an integer. Thus $m = 2n - 1$ and $m^2 = (2n - 1)(2n - 1) = 4n^2 - 4n + 1 = 4(n^2 - n) + 1 =$ an even integer $+ 1 =$ an odd integer.

Now let x be a rational number. Then $x = p/q$, p and q integers. We can assume that p and q are not both even, for we can cancel any 2 that appears as a factor in both numerator and denominator. Assume next that $x^2 = 2$. We show that this assumption leads to a contradiction and is therefore untenable.

If $x^2 = 2$, then $p^2/q^2 = 2$ or $p^2 = 2q^2$. Hence p^2 is even. By the remark made above, p is even. Hence $p = 2r$, r a positive integer, and $p^2 = 4r^2$. But $p^2 = 2q^2$. Hence $4r^2 = 2q^2$ or $q^2 = 2r^2$. Therefore q^2 is even and, by the remark, q is even. Thus both p and q are even, which is contrary to assumption.

The existence of incommensurable segments was discovered 2500 years ago by Greek mathematicians. There is a legend that the man who announced this discovery was punished by the gods for revealing an imperfection of the universe. This shows how incommensurability shocked the Greek thinkers; it seemed to destroy not only the proofs but even the meaning of simple geometric theorems about ratios of lengths.

The difficulty was overcome in a brilliant way by Eudoxos, who created a rigorous theory of proportions. In modern mathematics the same problem is resolved differently—by introducing a new kind of numbers, real numbers, such that no matter which unit of length we use every segment has a length expressible as a real number.

Fig. 1.6

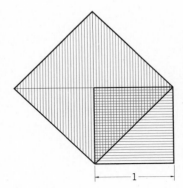

PYTHAGORAS of Samos (6th century B.C.) was the semi-legendary founder of a philosophical school. This school was also a religious community which became entangled in politics.

The Pythagoreans are credited with many mathematical discoveries, including the theorem about the square of the hypotenuse (which, however, was known centuries earlier), and the existence of incommensurable segments. They observed the dependence of musical tones on ratios of lengths of strings, which may have led them to the proposition that "all is number." If we interpret this to mean that nature can be described mathematically, this was one of the most successful guesses in the history of thought.

EUDOXOS of Cnidus (c. 403–355 B.C.) lived for a while in Athens where he was connected with Plato's Academy. The theory of proportions is only one of his brilliant achievements. Eudoxos also gave rigorous proofs for various theorems on volumes, and in astronomy he originated the method of describing the visible motion of planets as a superposition of circular motions.

The need for extending the system of rational numbers was explained in the preceding section. The first step toward this extension is a more thorough examination of the rational numbers themselves.

2.1   Variables

There are infinitely many rational numbers and the only way of making statements about all of them is to use variables, that is, symbols that may be replaced by numbers. We agree that letters stand for rational numbers. Thus the statement "$a + 0 = a$ for all a" means that $0 + 0 = 0$, $1 + 0 = 1$, $-\frac{5}{7} + 0 = -\frac{5}{7}$, and so forth. Also, "$a = b$" means that a and b are the same number, and "$a \neq b$" means that a and b are different numbers. Throughout this and the next two sections we shall often say "number" instead of "rational number."

In a strictly deductive presentation of mathematics, rational numbers are defined in terms of positive integers. The positive integers themselves can be defined by axioms or reduced to more primitive concepts of set theory. Here, however, we take the rational numbers for granted. We shall state explicitly some of their properties.

2.2   Field postulates

It is remarkable that all of the familiar rules of algebra can be derived from 12 simple postulates.

1. FOR ANY a AND b, THERE IS A UNIQUE NUMBER CALLED **THE SUM** OF a AND b AND DENOTED BY $a + b$.
2. $a + b = b + a$ FOR ALL a AND b.
3. $(a + b) + c = a + (b + c)$ FOR ALL a, b, AND c.
4. THERE IS A NUMBER CALLED **ZERO** AND DENOTED BY 0, SUCH THAT $0 + a = a$ FOR ALL a.
5. FOR ANY a, THERE IS A UNIQUE NUMBER CALLED ITS **NEGATIVE** AND DENOTED BY $(-a)$, SUCH THAT $a + (-a) = 0$.
6. FOR ANY a AND b, THERE IS A UNIQUE NUMBER CALLED **THE PRODUCT** OF a AND b AND DENOTED BY ab.
7. $ab = ba$ FOR ALL a AND b.
8. $(ab)c = a(bc)$ FOR ALL a, b, AND c.
9. $a(b + c) = ab + ac$ FOR ALL a, b, AND c.
10. THERE IS A NUMBER CALLED **ONE** AND DENOTED BY 1, SUCH THAT $1 \cdot a = a$ FOR ALL a.
11. $1 \neq 0$.
12. FOR EVERY NUMBER $a \neq 0$, THERE IS A UNIQUE NUMBER CALLED ITS **RECIPROCAL** AND DENOTED BY a^{-1}, SUCH THAT $a(a^{-1}) = 1$.

Postulates **1** to **5** are the postulates of addition; the other postulates refer to multiplication, and Postulate **9** connects addition with multiplication.

Postulates **1** to **12** are called field axioms and mathematicians abbreviate the statement that the rational numbers satisfy these axioms by saying that the rational numbers form a **field.**

Why does one go to the trouble of writing out such almost self-evident statements as **1** to **12** and why does one dignify them by a special name? Because there are many other mathematical systems obeying the same postulates, and it is convenient to know which rules of algebra hold whenever statements **1** to **12** are true. The system of real numbers, for instance, is also a field, that is, it satisfies axioms **1** to **12**. So is the system of so-called complex numbers (which we do not use in this book).

2.3 Consequences of the field postulates

We mention now some properties of rational numbers that can be derived from **1** to **12** and are therefore valid in any field.

We agree to write $a + b + c$ instead of $(a + b) + c$ and abc instead of $(ab)c$, and similarly for sums and products of more than three terms. The associative laws **3** and **8** assure us that this will not lead to confusion. We note that in view of the commutative laws **2** and **7** we may write $a + c + b$ instead of $a + b + c$, bac instead of abc, and so forth. We also agree to write $b - a$ instead of $b + (-a)$ and b/a instead of $b(a^{-1})$. In view of the commutative laws **2** and **7**, we also have $-a + b = b + (-a) = b - a$, $a^{-1}b = ba^{-1} = b/a$. We note that

$$(1) \qquad\qquad a^{-1} = \frac{1}{a} \qquad \text{for } a \neq 0.$$

Using the Postulates **1** to **12**, we can prove that

$$(2) \qquad\qquad 0a = 0 \qquad \text{for all } a.$$

This shows that 0 cannot have a reciprocal and *division by 0 cannot be defined* without violating our postulates.

An important rule is the **cancellation law:**

$$(3) \qquad\qquad \text{If } ab = 0, \text{ then either } a = 0 \text{ or } b = 0.$$

Here "or" is the "nonexclusive or" that is always used in mathematics. The statement "$a = 0$ or $b = 0$" means that the following three cases are possible: (1) $a = 0$, $b \neq 0$, (2) $a \neq 0$, $b = 0$, (3) $a = b = 0$, while Case (4) $a \neq 0$, $b \neq 0$ is impossible.

It follows from the cancellation law that, if a product of several numbers is 0, at least one of the factors is 0.

Finally, we recall that

$$(4) \qquad (-a)b = -(ab), \qquad (-a)(-b) = ab.$$

Readers interested to learn *how* to derive (2), (3), (4) and similar rules from the field axioms are referred to an appendix to this chapter (§8.2). From now on, we shall freely use familiar algebraic rules that are based on Postulates **1** to **12**.

EXERCISES

In Exercises 1 to 8, derive the given statements from Postulates **1** to **12**. In each step of your derivation, indicate exactly which postulate is being used.

1. $2 + 300 = 300 + 2$.
2. $0 + 2 = 2$.
3. $2 + 0 = 2$.
4. $a + 0 = a$.
5. $0 + (0 + 0) = 0$.
► 6. $7 + (-7) = (-6) + 6$.
7. $1 + 1 \neq 1$.
8. $(-76345) + 1 \neq -76345$.

9. Suppose "number" were to denote positive integer instead of rational number. Which of the postulates **1** to **12** would still hold?

10. Suppose "number" were to denote nonzero rational number instead of rational number. Which of the postulates **1** to **12** would still hold? Which would make no sense in this new system?

11. Suppose "number" were to denote integer instead of rational number. Which of the postulates **1** to **12** would still hold?

12. Suppose "number" were to denote positive rational number instead of rational number. Which of the postulates **1** to **12** would still hold?

13. Suppose "number" were to denote a nonpositive rational number instead of a rational number. Which of the postulates **1** to **12** would still hold? Which would make no sense here?

2.4 Integers. Exponents

The field axioms do not mention integers (whole numbers). We define a number to be an **integer** if it is 1 or obtained from 1 by addition and by taking negatives. Thus $1 + 1 = 2$, $2 + 1 = 3$, and so forth, are integers, as are -1, -2, -3, and so forth, and so is $0 = 1 + (-1)$.

Next, one defines

$$a^1 = a, \, a^2 = a \cdot a, \, a^3 = aa^2, \qquad \text{and so forth.}$$

For $a \neq 0$, one sets

$$a^0 = 1$$

and, $a^{-1} = \dfrac{1}{a}$ having been defined, we write

$$a^{-2} = (a^{-1})^2 = \frac{1}{a^2}, \, a^{-3} = \frac{1}{a^3}, \qquad \text{and so forth.}$$

No meaning is assigned to expressions like 0^0, 0^{-1}, 0^{-2}, and so forth.

The familiar **laws of exponents** follow:

$$a^n b^n = (ab)^n, \qquad a^{n+m} = a^n a^m, \qquad (a^n)^m = a^{nm}.$$

Here n and m stand for arbitrary integers and a, b are any numbers, except that the values $a = 0$ and $b = 0$ have to be excluded if the corresponding expressions become meaningless.

Algebraic identities like $(a + b)(a - b) = a^2 - b^2$ or $(a + b)^2 = a^2 + 2ab + b^2$ are also consequences of the field axioms.

EXERCISES

► 14. If $abc = 2$, what is $a^5 b^5 c^5$?

15. If $a^{13} = 4$, what is $(a^{-7})^7 (a^{-6})^{-6}$?

16. Simplify $[(a^{-2})^4]^{-3}$.

17. If $a^{-1} b^{-1} c^{-1} = 2$, what is $(abc)^4$?

18. If n is an integer, what is $(-1)^{2n}$?

19. If n is an integer, what is $(-1)^{2n+1}$?

20. Does $2^5 = 5^2$?

21. Does $(2^3)^2 = 2^{(3^2)}$?

22. Does $(2 + 3)^2 = 2^2 + 3^2$?

23. Does $2^3 3^2 = 2 \cdot 6^2$?

2.5 Geometric progressions

A particularly important identity is the formula for the sum of a **geometric progression:**

$$(5) \qquad \underbrace{1 + q + q^2 + \cdots + q^{n-1}}_{n \text{ terms}} = \frac{1 - q^n}{1 - q}$$

for $q \neq 1$ and $n = 1, 2, 3, 4, \cdots$.

The proof is quite simple. For any q,

$$(1 - q)(1 + q + q^2 + \cdots + q^{n-1})$$
$$= (1 + q + q^2 + \cdots + q^{n-1}) - q(1 + q + q^2 + \cdots q^{n-1})$$
$$= (1 + q + q^2 + \cdots + q^{n-1}) - (q + q^2 + q^3 + \cdots q^n)$$

so that

$$(6) \qquad (1 - q)(1 + q + q^2 + \cdots + q^{n-1}) = 1 - q^n.$$

If $1 - q \neq 0$, then both sides of the last equation may be divided by $1 - q$, that is, multiplied by $1/(1 - q)$, to obtain the desired formula. A closely related identity is

$$(7) \quad A^n - B^n = (A - B)(A^{n-1} + A^{n-2} B + A^{n-3} B^2 + \cdots + B^{n-1})$$

which holds for all numbers A, B, and $n = 2, 3, 4, \cdots$. To prove this, in the case $A \neq 0$, set $q = B/A$ and multiply both sides of (6) by A^n. If $A = 0$, then (7) is obvious.

EXERCISES

24. Compute $1 + \dfrac{1}{2} + \dfrac{1}{4} + \cdots + \dfrac{1}{2^8}$.

25. Compute $1 + 3 + 9 + \cdots + 3^6$.

26. Show that $1 + a^2 + a^4 + \cdots + a^{18} = \dfrac{1 - a^{20}}{1 - a^2}$, for $a \neq 1$ or -1.

27. Find a formula for the sum $1 - b + b^2 - b^3 + \cdots - b^{21}$. Does this hold for all b?

28. Compute $1 + 2 + \dfrac{1}{3} + 4 + \dfrac{1}{9} + \cdots + 2^8 + \dfrac{1}{3^8}$.

29. What is the difference between $1 + \dfrac{1}{3} + \dfrac{1}{3^2} + \cdots + \dfrac{1}{3^{100}}$ and

$$1 + \dfrac{1}{3} + \dfrac{1}{3^2} + \cdots + \dfrac{1}{3^{1000}}?$$

§3 Inequalities

Postulates **1** to **12** do not yet give a sufficiently detailed description of rational numbers, since they do not include the concept of one number's being less than another. The present section deals with this concept.

3.1 Notations

One writes "$x < y$" to say that the number x is less than the number y. This means the same as "$y > x$"(read: y is greater than x). Every **inequality**, that is, a statement that one number is less than another, may be written using either the symbol $<$ or the symbol $>$. An inequality $x < y$ has a simple geometric meaning: the point represented by x (see §1.2) lies to the left of the point represented by y.

The sentence "$x \leq y$" means that either $x < y$ or $x = y$. So $-3 \leq -2$ and $7 \leq 7$, but it is not true that $1 \leq 1/2$. Similarly, "$x \geq y$" means that either $x > y$ or $x = y$. Also, "x is *positive*" and "y is *negative*" is another way of saying that $x > 0$ and $y < 0$, while "*nonnegative*" and "*nonpositive*" may be written as ≥ 0 and ≤ 0, respectively.

EXERCISES

1. Is $7 < 8$? 　　4. Is $-7 > -5$? 　　7. Is $-2 \leq -3/2$?
2. Is $7 \leq 8$? 　　5. Is $1/2 \geq 1/3$? 　　8. Is $-5 < 5$?
3. Is $5 \geq 5$? 　　6. Is $-9 > 1$?

3.2 Order postulates

Facility with inequalities is indispensable for understanding calculus. Fortunately, all rules for working with inequalities follow from a few simple postulates.

13. IF a AND b ARE TWO NUMBERS, THEN ONE OF THE THREE STATEMENTS "$a = b$," "$a < b$," AND "$b < a$" IS TRUE AND THE OTHER TWO ARE FALSE.
14. IF $a < b$ AND $b < c$, THEN $a < c$.
15. IF $a < b$, THEN $a + c < b + c$.
16. IF $a < b$ AND $0 < c$, THEN $ac < bc$.

We abbreviate Postulates **1** to **16** by saying: the rational numbers form an **ordered field**. There are also other ordered fields, for instance, the field of real numbers to be introduced later.

We consider next the following consequences of Postulates **1** to **16**.

(1)　　　　　If $a > 0$, $b > 0$, then $a + b > 0$.
(2)　　　　　If $a > 0$, $b > 0$, then $ab > 0$.
(3)　　　　　If $a > 0$, then $-a < 0$.
(4)　　　　　If $a \neq 0$, then $a^2 > 0$.
(5)　　　　　If $a > 0$, $b < 0$, then $ab < 0$.
(6)　　　　　If $a > 0$, $b > 0$, then $\dfrac{a}{b} > 0$.

(7)　　$\cdots -4 < -3 < -2 < -1 < 0 < 1 < 2 < 3 < 4 < \cdots$.

These statements are quite obvious; nevertheless we shall derive them from Postulates **1** to **16**. Assume that $a > 0$. Then $a + b > 0 + b = b$ (by **15**; here we use $>$ rather than $<$ to get accustomed to both symbols). If, also, $b > 0$, then $a + b > 0$ (by **14**). This proves (1).

If $a > 0$ and $b > 0$, then $ab > 0b = 0$ (by **16**); this proves (2).

If $a > 0$, then $a \neq 0$ (by **13**); hence, either $-a > 0$ or $-a < 0$ (again by **13**). If $-a > 0$, then $a + (-a) > 0$, by (1). This would mean $0 > 0$, which is impossible (by **13**). Hence (3) holds.

If $a \neq 0$, then either $a > 0$ or $a < 0$. If $a > 0$, then $a^2 = aa > 0$ by (2), and if $a < 0$, then $-a > 0$, by (3), and $a^2 = (-a)(-a) > 0$, by (2). This proves (4).

If $b < 0$, then $-b > 0$, by (3), so that, if also $a > 0$, then $a(-b) = -ab > 0$, by (2), and $ab = -(-ab) < 0$, by (3). This proves (5).

If $a > 0$ and $b > 0$, then $\dfrac{a}{b} \neq 0$. For if $\dfrac{a}{b} = 0$, then $0 = \dfrac{a}{b} \cdot b = a$.

We cannot have $\dfrac{a}{b} < 0$, for if $\dfrac{a}{b} < 0$, then $a = \dfrac{a}{b} \cdot b < 0$, by (5).

Therefore, $\dfrac{a}{b} > 0$ and (6) is proved.

Since $1 \neq 0$ (by **11**) and $1^2 = 1$, we have $1 > 0$, by (4). Hence $2 = 1 + 1 > 0 + 1 = 1$ (by **15**). Similarly, $3 = 2 + 1 > 2$, $4 = 3 + 1 > 3$, and so forth. Also $-1 < 0$, by (3), $-2 < -1$, and so on. This proves (7).

It follows from (7) that $1 \neq 2$, $2 \neq 3$, $1 \neq 3$, $1 \neq 4$, and so on. It is interesting to note that, if we use only the field axioms **1** to **12**, we cannot prove that $1 \neq 3$.

EXERCISES

9. If $a \leq b$ and $a \geq b$, what conclusion can you draw?
10. If $a \leq b$ and $a \neq b$, what conclusion can you draw?
11. If $a \leq b$, $b \leq c$, and $c \leq a$, what conclusion can you draw?
12. If $a < b$ and $b \geq c$, what conclusion can you draw?
13. Is $x + 1 > x$ for all x?   15. Is $x + x > x$ for all x? For all $x > 0$?
14. Is $x^2 > x$ for all x?   ▶ 16. Is $(x^2 + 1)^{-1} < 1$ for all x?

3.3 Solving inequalities

It follows from **15** that *a true inequality remains true if we transpose a term.* In other words, if $x + y < z$, then $x < z - y$. For, if $x + y < z$, then $x + y + (-y) < z + (-y)$ or $x < z - y$.

It follows from **16** that *a true inequality remains true if both sides are multiplied by the same positive number.*

We show next that *a true inequality remains true if we multiply both sides by the same negative number and reverse the direction of the inequality.*

(8)   If $a < b$ and $c < 0$, then $ac > bc$.

Proof. If $c < 0$, then $-c > 0$ so that $a(-c) < b(-c)$ (by **16**). This means that $-ac < -bc$ or $-ac + ac < -bc + ac$, by **15**. Hence $ac - bc > 0$ and, again by **15**, $(ac - bc) + bc > 0 + bc$ or $ac > bc$.

For inequalities involving $\geq$, the same rules hold.

As an illustration we *solve* the inequality

(9) $3 - 2x < 4x - 5$.

"To solve" means to find all numbers x for which the statement is true. The steps are similar to those used in solving the equation $3 - 2x = 4x - 5$. Inequality (9) is equivalent to the inequality $-2x - 4x < -5 - 3$ (obtained by transposing terms), and to $-6x < -8$, and to $x > 8/6 = 4/3$ (obtained by multiplying by $-1/6$ and reversing the direction). This is the solution: inequality (9) means that $x > 4/3$. The solution is a *set of numbers:* all numbers greater than $4/3$.

►*Examples*   *1.* Solve the double inequality

$$x - 6 < 2x - 5 \leq x - 3.$$

This means that $x - 6 < 2x - 5$ and $2x - 5 \leq x - 3$. Treating these two inequalities by the method used above, we see that (10) means that $-x < 1$ and $x \leq 2$. The solution set of (10) consists therefore of all x such that $-1 < x \leq 2$. (It is shown in Fig. 1.7.)

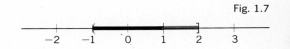

Fig. 1.7

   2. Solve the double inequality

$$x + 10 < 2x - 5 \leq x - 3.$$

The inequality means that $x + 10 < 2x - 5$ and $2x - 5 \leq x - 3$, that is, $-x < -15$ and $x \leq 2$. This is not true for any x (so the solution set is empty).

   3. For which values of x is $\dfrac{2x + 1}{x + 1} > 3$? (We must, of course, exclude the value $x = -1$, since division by 0 makes no sense.)

   The inequality means that

either $x + 1 > 0$ and $2x + 1 > 3(x + 1)$,

or $x + 1 < 0$ and $2x + 1 < 3(x + 1)$.

But $2x + 1 > 3(x + 1)$ means that $2x + 1 > 3x + 3$, that is, $-x > 2$, that is, $x < -2$. And if $x < -2$, then $x + 1 < -1$. Hence the first case is impossible. On the other hand, $2x + 1 < 3(x + 1)$ means that $x > -2$. The solution reads: $x + 1 < 0$ and $x > -2$; in other words: $-2 < x < -1$ (see Fig. 1.8).

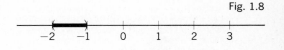

Fig. 1.8

   4. Solve the inequality $x^3 < x$.

   We write this as $x^3 - x < 0$ or $x(x - 1)(x + 1) < 0$. The inequality means that either all three factors, x, $x - 1$, and $x + 1$, are negative, or two are positive and one negative. But for all x we have $x - 1 < x < x + 1$. So the solution reads: either $x + 1 < 0$ or $x - 1 < 0$ and $x > 0$. The solution set consists of all x such that $x < -1$ and of all x such that $0 < x < 1$ (see Fig. 1.9).◄

Fig. 1.9

EXERCISES

Solve the following inequalities.

17. $x - 10 > 2 - 2x$.

18. $x + 6 \leq 5 - 3x$.

19. $6x + 5 \geq x - 5$.

20. $7x - 1 \leq 2x + 1$.

21. $2x + 11 > 10 - 6x$.

22. $\dfrac{5 + x}{5 - x} \leq 2$.

23. $\dfrac{3 + x}{4 - x} \leq 3$.

▶ 24. $\dfrac{6 - 2x}{3 + x} > 2$.

25. $\dfrac{3x - 5}{2x + 4} > 1$.

26. $\dfrac{3x + 8}{x - 1} \geq -2$.

27. $-5 < x - 4 < 2 - x$.

28. $0 < 3x + 6 \leq 1 - 2x$.

29. $x - 2 \leq 2x - 3 \leq x + 2$.

30. $4x + 2 > 5x + 3 > 4x + 4$.

31. $x - 1 < 2x + 5 \leq -x - 10$.

32. $6 - 2x \leq 3x + 1 \leq 9 - 5x$.

33. $1 - 6x \geq 2x + 5 > -3x$.

▶ 34. $5x - 2 \leq 10x + 8 < 2x - 8$.

35. $4 - 3x < 2x + 3 < 3x - 4$.

36. $2x - 3 < 5x + 3 < 2x + 3$.

37. $0 < \dfrac{2x - 1}{x - 1} < 1$.

38. $1 < \dfrac{3x - 1}{x - 3} < 2$.

39. $2 \geq \dfrac{3x + 1}{x} > \dfrac{1}{x}$.

40. $\dfrac{2}{x - 2} < \dfrac{x + 2}{x - 2} < 1$.

41. $\dfrac{2x - 1}{x + 4} < \dfrac{x}{x + 4} \leq \dfrac{x + 1}{x + 4}$.

42. $x(x - 1) < 0$.

43. $x^4 < x^2$.

▶ 44. $x^2 + 3x > -2$.

45. $2x^2 - 2 \leq x^2 - x$.

46. $(x^2 - 1)(x + 4) < 0$.

3.4 Archimedean postulate

Although the next postulate we list sounds completely obvious, it is very important.

17. FOR EVERY NUMBER a, THERE IS AN INTEGER k SUCH THAT $a < k$.

We call this the **Archimedean postulate.** Archimedes, the greatest mathematician of antiquity, stated an equivalent axiom in geometric terms; it was also actually used by Greek mathematicians who lived long before Archimedes. The Archimedean postulate has two corollaries (or consequences).

Corollary 1. IF $a \leq \dfrac{1}{n}$ FOR ALL POSITIVE INTEGERS n, THEN $a \leq 0$.

Proof. Suppose $a > 0$. Then $1/a > 0$. By **17** there is an integer, call it s, such that $1/a < s$. Also $s > 0$, for otherwise we would have $1/a < 0$. Multiplying the inequality $s > 1/a$ by the positive number a/s, we obtain $a > 1/s$. Hence it is not true that $a \leq 1/n$ for all positive integers n.

ARCHIMEDES of Syracuse (287–212 B.C.) spent most of his life in his native city, except for a stay in Alexandria where he studied in the famed Museum, the center of Hellenistic civilization. Many of his works are preserved. Archimedes was a profound mathematician, and also the founder of statics: he discovered the law of floating bodies and the principle of the lever.

When the Romans besieged Syracuse, Archimedes' inventions were used in defending the city; yet reports by ancient writers about his war machines are probably exaggerated. Archimedes was killed by a Roman soldier during the sack of Syracuse.

Corollary 2.   IF $A > 1$ AND $a \leq A^{-n}$ FOR ALL POSITIVE INTEGERS n, THEN $a \leq 0$.

Proof.   Since $A > 1$, we have $A = 1 + b$, $b > 0$. Hence

$$A^n = \underbrace{(1 + b)(1 + b) \cdots (1 + b)}_{n \text{ factors}} = 1 + nb + \text{other} \atop \text{positive numbers} > nb.$$

Thus if $a \leq A^{-n}$, then $a \leq 1/nb$ and $ab \leq 1/n$. If this holds for all positive integers n, then $ab \leq 0$ by Corollary 1. Hence either $ab = 0$ and $a = 0$, by the cancellation law, or $ab < 0$ and $a < 0$.

EXERCISES

47. If $a < \dfrac{1}{\sqrt{n}}$ for all positive integers n, what can you conclude about a?

48. For which numbers x is it true that $-(10^{10})^{-j} \leq x \leq (10^{10})^{-j}$ for $j = 1, 2, 3, \cdots$?

49. For which numbers x is it true that $1 - 2^{-j} \leq x$ for $j = 1, 2, 3, \cdots$?

▶ 50. For which numbers x is it true that $1 - 10^{-j} \leq x \leq 1 + 1/10^j$ for $j = 1, 2, 3, \cdots$?

51. For which numbers x is it true that $3^{-j} \leq x \leq 2^{-j}$ for $j = 1, 2, 3, \cdots$?

52. For which numbers x is it true that $x \leq 4(n + 1)^{-2}$ for all positive integers n?

3.5 Absolute values

The **absolute value** of a rational number a is a number denoted by $|a|$ and defined by

$$|a| = \begin{cases} a & \text{if } a \geq 0, \\ -a & \text{if } a < 0. \end{cases}$$

Thus $|-5| = 5$, $\left|\dfrac{3}{2}\right| = \dfrac{3}{2}$. The same definition applies to any ordered field.

Clearly,

$$|a| > 0 \text{ if } a \neq 0, \qquad |0| = 0$$

and

$$|ab| = |a||b|, \quad |a^{-1}| = |a|^{-1}, \quad \left|\frac{a}{b}\right| = \frac{|a|}{|b|}.$$

The absolute value has a simple geometric interpretation. If numbers are represented by points on the line, then $|a|$ is the **distance** from the point a to 0 (see Fig. 1.10). Therefore,

Fig. 1.10

Fig. 1.11

$$|a| < \alpha \text{ means that } -\alpha < a < \alpha$$

(as is seen from Fig. 1.11). Also $|a - b|$ is the **distance between the points** a and b, as is seen from the various cases shown in Fig. 1.12. When we say that two numbers are *close* to each other, we mean that the absolute value of their difference is small. Thus $-11/10$ is closer to -1 than 5 is to 6 for $|-11/10 - (-1)| = 1/10$, while $|5 - 6| = 1$.

Fig. 1.12

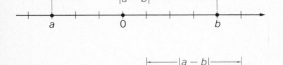

It is geometrically evident that the distance between two points (on a line) is not greater than the sum of their distances from a third point. If we write down this statement for the case in which the two points represent the numbers a and b, and the third point is the origin O, we obtain a simple but very important property of absolute values, the so-called **triangle inequality:**

$$(10) \qquad\qquad |a + b| \leq |a| + |b|.$$

We shall give an analytic proof of the triangle inequality ("analytic" means "based exclusively on properties of numbers").

If either $a = 0$ or $b = 0$, (10) is obvious. If a and b are positive, so is $a + b$ and we have $a = |a|$, $b = |b|$, and $|a + b| = a + b$. If a and b are negative, so is $a + b$, and we have $a = -|a|$, $b = -|b|$, $a + b = -|a + b|$. In both cases, (10) holds with the equality sign. If $a > 0$ and $b < 0$, then $a = |a|$, $b = -|b|$, and $a + b = |a| - |b|$. Hence either $|a + b| = |a| - |b|$ or $|a + b| = -(|a| - |b|) = |b| - |a|$. Since $|a| - |b| < |a| + |b|$ and $|b| - |a| < |a| + |b|$, we see that (10) holds with the strict inequality sign $<$. The case $a < 0$, $b > 0$ is analogous.

Consequences of the triangle inequality are

$$|a + b + c| \leq |a| + |b| + |c|$$

and similar inequalities for more terms.

▶*Example*   If $a \leq x \leq b$ and $a \leq y \leq b$, then $|y - x| \leq b - a$.

Geometric Proof.   Represent the numbers by points on a line. Both x and y lie between the points a and b. Therefore the distance between them, that is, $|y - x|$, is not greater than the distance between the end-points, $|b - a|$.

Analytic Proof.   Assume first that $x \leq y$. Then $a \leq x \leq y \leq b$. Add $-x$ to each term of this inequality. This yields $a - x \leq 0 \leq y - x \leq b - x$. Since $-x \leq -a$, we have $0 \leq y - x \leq b - a$, and $|y - x| \leq b - a$. If $y \leq x$, a similar argument gives the same result. The reader

will note that the geometric argument is much easier. The analytic
argument should be used only to reinforce our geometric intuition.◄

EXERCISES

53. What is $|-3|$?
54. What is $|4 - 6 - 8|$?
55. What is $|(-3)^2|$?
56. Which is larger: $|2 - 3|$ or $|2| + |-3|$?
57. Show that $|x^2| = |x|^2$.
58. Is it true that $|x^5| \neq |x|^5$ if $x < 0$?
59. For which numbers x is it true that $|x - 1| \leq 2^{-j}$ for $j = 1, 2, 3, \cdots$?
60. Show that $|a + b + c| \leq |a| + |b| + |c|$.
61. Show that $|a - b| \geq |a| - |b|$.
► 62. Show that $|a - b| \geq ||a| - |b||$.
63. Show that $|a - b - c| \geq |a| - |b| - |c|$.
64. Show that $|a - b| < r$ if and only if $b - r < a < b + r$, where $r > 0$.
65. Is -10 closer to -22 or to 3?
66. Is -4 closer to 0 or to -7?
67. Is -78 closer to 68 than -14 is to -100?
68. Which is smaller: the distance between -10 and 13 or between -36 and -54?
69. Match each set in the left-hand column below with an equal set from the right-hand column.

(i) The set of numbers whose distance from 3 is less than 2.

(ii) The set of numbers whose distance from -1 is greater than 3.

(iii) The set of numbers whose distance from -2 is less than 4 and greater than 1.

(iv) The set of numbers whose distance from -1 is less than their distance from 1.

(v) The set of numbers whose distance from 0 is greater than their distance from -2.

(vi) The set of numbers whose distance from 1 is greater than the distance from a to 1.

(vii) The set of numbers whose distance from a is less than their distance from -1 and greater than the distance from a to -1.

(1) The set of x such that $|x - 1| > |a - 1|$.

(2) The set of x such that $|x - 1| < |x + 1|$.

(3) The set of x such that $|x - 3| < 2$.

(4) The set of x such that $|x - 1| > 3$.

(5) The set of x such that $|a + 1| < |x - a| < |x + 1|$.

(6) The set of x such that $|x + 1| > 3$.

(7) The set of x such that $|x + 1| < |x - 1|$.

(8) The set of x such that $|x| > |x + 2|$.

(9) The set of x such that $1 < |x + 2| < 4$.

(10) The set of x such that $x < |a| < x - 1$.

3.6 Rational numbers defined by axioms

Postulates 1 to 17 are not yet sufficient to describe rational numbers completely. If we want to define rational numbers by axioms, we must add one more postulate to 1 to 17.

FOR EVERY RATIONAL NUMBER a, THERE IS AN INTEGER $m \neq 0$ SUCH THAT ma IS AN INTEGER.

This postulate asserts that every rational number is a *ratio*, that is, the quotient of two integers. This explains the term "rational."

§4 Decimal Representation of Rational Numbers

4.1 Positional number systems

There have been three major advances in the art of computing. The third, the development of electronic computers, has occurred in our lifetime. The second, the invention of logarithms, took place during the sixteenth century. The first advance, without which the other two would have been impossible, was the development of the **positional number system.**

The number system we use is called decimal, or to *base ten*, since in this system every positive integer is represented as a sum of powers of ten (*decem* in Latin). Thus $203 = 2 \cdot 10^2 + 0 \cdot 10 + 3 \cdot 10^0$. This system came to Europe from the Islamic world, after the Crusades, therefore the name Arabic numerals. But the Arabs learned the positional system from the Hindus, who received it from the Hellenistic civilization, and the Greek astronomers borrowed their number system from the Babylonians, who were using it as early as 1500 B.C.

A positional number system has several advantages over the Roman system or the Hebrew and Greek method of using the letters of the alphabet as names of numbers. Using only a few digits (in our system $0, 1, \cdots, 9$) one can name any number, and by remembering the addition table and multiplication table for these digits, the sum and product of any two integers can be computed by a simple mechanical procedure.

The choice of a base is, of course, arbitrary, and the popularity of ten is an anatomical accident (*digit* means *finger*). British currency, the system of weights and measures used in the English-speaking world, and the French name for 80, *quatre-vingt* $= 4 \cdot 20$, show that men experimented with other bases. The base of the Babylonian system, also used by the Greek astronomer Ptolemy, was 60. This sexagesimal system is preserved in our way of measuring time (1 hour $= 60$ minutes; 1 minute $= 60$ seconds). The binary system has the smallest possible integral base, namely 2. It is used by electronic computing machines,

PTOLEMY (C. 100–187 A.D.) was a mathematician, astronomer, and geographer. His astronomical treatise, known by its Arabic name, *Almagest*, summarized the whole of Greek astronomy including observational data, mathematical methods (in particular spherical trigonometry), and the so-called Ptolemaic world system with the immovable earth at the center of the universe.

since the two digits 0 and 1 can be represented by the on-off positions of a switch. In principle, however, one base is as good as another, and we restrict ourselves to the familiar decimal notation.

4.2 Decimal fractions

Some fractions can also be represented in the decimal notation. One needs only one additional symbol, the **decimal point.** The rule is: if a is a nonnegative integer and α_1, α_2, $\cdots$, α_m are m digits, each of which is one of the numbers 0, 1, 2, $\cdots$, 9, then

$$(1) \qquad a.\alpha_1\alpha_2 \cdots \alpha_m = a + \frac{\alpha_1}{10} + \frac{\alpha_2}{10^2} + \cdots + \frac{\alpha_m}{10^m}.$$

By convention, a may be omitted if $a = 0$. We call such an expression a **terminating decimal.** Its value is not changed if one writes any number of zeros on the right. Thus $2 = 2.0 = 2.00$, $3/2 = 1.5 = 1.50$.

Arithmetic operations on terminating decimals are as easy as the corresponding operations on integers, provided one observes certain well-known rules concerning the position of the decimal point.

Which rational numbers can be represented by terminating decimals? If we bring the fractions into the common denominator, 10^m, we see that $a.\alpha_1 \cdots \alpha_m = (\text{integer})/10^m = (\text{integer})/(\text{product of 2s and 5s})$. Conversely, if $x = p/q$, where p and q are integers and q is a product of 2s and 5s, we may multiply numerator and denominator by the same number, so as to obtain x in the form: integer divided by a power of 10, and this can be written as a terminating decimal. For instance,

$$(2) \qquad \frac{29}{25} = \frac{29 \cdot 4}{25 \cdot 4} = \frac{116}{100} = 1.16.$$

In other words, a rational number can be written as a terminating decimal if and only if it can be written p/q, where q is a product of 2s and 5s.

The number $\frac{1}{3}$ cannot be written as a terminating decimal. Instead one writes

$$(3) \qquad \frac{1}{3} = .33333 \cdots \qquad (\text{and so on, all 3s}).$$

This simple but puzzling equation contains some important ideas, which it will be worthwhile to investigate more closely.

4.3 Nonterminating decimal fractions

We associate with *every* positive rational number a *nonterminating decimal*. Let x be the number considered and let a be the largest integer less than x. (That there is such an integer is obvious and follows from the Archimedean postulate.) We note that

$$a < x \le a + 1,$$

and we agree to write a to the left of the decimal point. (If $x = \dfrac{15}{7}$, for instance, then $a = 2$, for $2 < \dfrac{15}{7} \le 3$.) Next, we consider the 9 numbers

$$a = a + \frac{0}{10}, a + \frac{1}{10}, a + \frac{2}{10}, \cdots, a + \frac{9}{10}$$

and find the largest of these that is less than x. Let it be $a + \dfrac{\alpha_1}{10}$. We note that

$$a + \frac{\alpha_1}{10} < x \le a + \frac{\alpha_1}{10} + \frac{1}{10}$$

or, in decimal notation,

$$a.\alpha_1 < x \le a.\alpha_1 + 10^{-1},$$

and we agree to write α_1 to the right of the decimal point. (If $x = \dfrac{15}{7}$, then $\alpha_1 = 1$ for $2.1 < \dfrac{15}{7} \le 2.2$.) Next we find, in the same way, a digit α_2, which will be one of the numbers $0, 1, \cdots, 9$, such that

$$a.\alpha_1\alpha_2 < x \le a.\alpha_1\alpha_2 + 10^{-2}$$

and write α_2 to the right of α_1. (If $x = \dfrac{15}{7}$, then $\alpha_2 = 4$, since $2.14 < \dfrac{15}{7} \le 2.15$.) Continuing in this manner, we find an infinite sequence of digits $\alpha_1, \alpha_2, \alpha_3, \cdots$ such that

$$(4) \qquad \begin{array}{c} a.\alpha_1\alpha_2 \cdots \alpha_m < x \le a.\alpha_1\alpha_2 \cdots \alpha_m + 10^{-m} \\ \text{for } m = 0, 1, 2, 3, \cdots. \end{array}$$

It can never happen that all α's beginning with, say α_k, are 0. For, if it were so, we would have $a.\alpha_1\alpha_2 \cdots \alpha_m = a.\alpha_1\alpha_2 \cdots \alpha_k$ for $m > k$; hence, by (4)

$$a.\alpha_1 \cdots \alpha_k < x \le a.\alpha_1 \cdots \alpha_k + 10^{-m}$$

for $m = k + 1,\ k + 2,\ \cdots$; hence $0 < x - a.\alpha_1 \cdots \alpha_k < 10^{-m}$ for all positive integers m, and, by Corollary 2 of the Archimedean postulate (see §3.4), $0 < x - a.\alpha_1 \cdots \alpha_k \le 0$. But then $0 < 0$, which is absurd. Thus $a.\alpha_1\alpha_2 \cdots$ is always a nonterminating decimal.

It can never happen that two different positive rational numbers, x and y, lead to the same decimal $a.\alpha_1\alpha_2 \cdots$. For if it were so, we would have the inequalities (4) and the same inequalities for y, that is,

(5) $a.\alpha_1 \cdots \alpha_m < y \le a.\alpha_1 \cdots a_m + 10^{-m}$ $m = 0, 1, 2, \cdots$.

From (4) and (5), we can conclude (see the example in §3.5) that $|y - x| \le 10^{-m}$ for $m = 1, 2 \cdots$, so that $|y - x| = 0$ by Corollary 2 of the Archimedean postulate, and $x = y$.

To summarize: knowing x, we can compute $a.\alpha_1\alpha_2 \cdots$, and the knowledge of $a.\alpha_1\alpha_2 \cdots$ determines x. We can therefore use $a.\alpha_1\alpha_2 \cdots$ as a name for x. That is, we can write

$$x = a.\alpha_1\alpha_2\alpha_3 \cdots.$$

This equation means simply that all inequalities (4) hold; $a.\alpha_1\alpha_2\alpha_3 \cdots$ is called the **nonterminating decimal expansion** of x. The negative of x is then written as $-x = -a.\alpha_1\alpha_2\alpha_3 \cdots$. No new notation is introduced for 0.

The nonterminating decimal expansion of a number is a string of inequalities that describe the number with greater and greater accuracy; all inequalities taken together determine the number completely. In particular, equation (3) means that $0 < 1/3 \le 1,\ .3 < 1/3 \le .4,\ .33 < 1/3 \le .34,\ .333 < 1/3 \le .334$, and so forth.

4.4 Repeating decimals

We show now how the nonterminating decimal expansion of a rational number can be computed.

The number 29/25 can be represented as a terminating decimal, since $25 = 5^2$. We found the terminating decimal above; see (2). It is also possible, and usually easier, to obtain the same decimal by *long division:*

$$
\begin{array}{r}
1.16 \\
25{\overline{\smash{\big)}\,29.00}} \\
\underline{25.} \\
40 \\
\underline{25} \\
150 \\
\underline{150}
\end{array}
\qquad \text{so that } \frac{29}{25} = 1.16.
$$

[The calculation is, of course, essentially the same as in (2).] The non-terminating decimal expansion of the same number is

(6) $$\frac{29}{25} = 1.15999 \cdots \qquad \text{(all 9s),}$$

which one writes as

$$\frac{29}{25} = 1.15\overline{9},$$

the bar denoting that 9 is to be repeated indefinitely. To verify (6) we note that

$$1.15\overbrace{99 \cdots 9}^{n \text{ times}} < 1.16 = 1.15\overbrace{99 \cdots 9}^{n \text{ times}} + 10^{-n-2} \text{ for } n = 1, 2, \cdots.$$

This string of inequalities is the same as equation (6).

The same argument applies to any positive rational number $x = p/q$, p and q integers, which can be written as a terminating decimal. Long division gives this terminating decimal. The nonterminating decimal expansion is obtained by diminishing the last digit by 1 and following it by an infinite string of 9s. For instance, $1/2 = .5 = .4999 \cdots = .4\overline{9}$, $3 = 2.999 \cdots = 2.\overline{9}$.

If we apply long division to a rational number $x = p/q$ which cannot be represented by a terminating decimal, the process never stops and we obtain the nonterminating decimal expansion of x. To verify this, it is best to consider a particular example, say the number $23/18$. The first three steps of long division read:

$$
\begin{array}{r}
1.27 \\
18{\overline{\smash{\big)}\,23.000000 \cdots}} \\
\underline{18} \\
50 \\
\underline{36} \\
140 \\
\underline{126} \\
14
\end{array}
$$

The remainder is $14 \cdot 10^{-2} = .14$. Thus

$$23 = (1.27) \cdot 18 + 14 \cdot 10^{-2} \qquad \text{or} \qquad \frac{23}{18} = 1.27 + \frac{14}{18} \cdot 10^{-2} < 1.28.$$

Therefore $1.27 < 23/18 < 1.28$, which shows that 1.27 gives the first three digits of the decimal expansion of $23/18$. It is clear that the same argument holds in every case.

It turns out that the nonterminating decimal expansion of a rational number (different from zero) is always *repeating*. The same digit, or the same string of digits, is repeated infinitely often; the repetition may start either immediately after the decimal point, or after several digits behind the decimal point. For instance,

$$\frac{23}{99} = .232323 \cdots \text{ (all 23s)} = .\overline{23},$$

$$\frac{203}{165} = 1.2303030 \cdots \text{ (all 30s)} = 1.2\overline{30}.$$

Conversely, every repeating decimal is the decimal expansion of some rational number. We shall recall the proof of this curious fact in an appendix (§8.3). It plays no part, however, in actual calculations performed on decimal fractions.

4.5 Calculating with decimals

In calculating with decimals, either by hand or on a computing machine, one cannot use infinitely many digits. It is always necessary to **round-off,** that is, to replace every nonterminating decimal by a terminating one. The simplest way is to choose a nonnegative integer m and to retain in each number x only the first m digits after the decimal point. This amounts to replacing every number x by a terminating decimal X such that $|X - x| \leq 10^{-m}$. Rounding-off may also be needed if we use for x its terminating decimal representation, provided that it has one.

How large an error do we commit if in computing $x + y$, $x - y$, xy, and x/y we retain only m digits after the decimal point? We expect that the error will be small if m is large and that by choosing m large enough we can compute the sum, difference, product, and quotient of two rational numbers *with any desired degree of accuracy.* This is indeed so, as we shall prove in an appendix (§7.2).

We note that in describing how one actually computes with decimals, we never once used the fact that all decimals considered came from rational numbers, that is, were repeating ones. So why not also consider nonterminating, nonrepeating decimals? Indeed, there are such decimals and their existence suggests one way of defining real numbers—as nonterminating decimals, repeating or nonrepeating. Moreover, this is how real numbers were first introduced—mathematicians worked with them long before the concept was formalized and analyzed.

EXERCISES

Let x, y, z be rational numbers whose nonterminating decimal expansions begin as follows: $x = 1.0234107\cdots$, $y = 1.0235106\cdots$, $z = 1.0235106\cdots$. About each of the statements made below, indicate whether it is true or false, or whether its truth cannot be ascertained from the information given.

1. $1.02 < x \leq 1.03$.
2. $x < 1.03$.
3. $x > 1.0234107$.
4. $x < y$.
5. $x + y < 2.048$.
6. $x + y < 2.046$.
7. $x + y > 2.04692$.
8. $x < z$.
9. $y = z$.
10. $yz > 1.04$.
11. $x = 1.0234108$.
12. $x = 1.0234109$.

Find the first 4 digits, after the decimal point, of the nonterminating decimal expansion of the following rational numbers.

13. $2/3$.
14. $3/2$.
15. $10,000$.
16. $777/33$.
17. $777/333$.
18. $29/25$.
19. $250/29$.
20. $1/100$.

§5 Real Numbers

5.1 Defining real numbers

There are various equivalent definitions of real numbers. We use the following:

A POSITIVE REAL NUMBER IS A NONTERMINATING DECIMAL $a.\alpha_1\alpha_2\alpha_3 \cdots$.

Here a is a positive integer or zero (and we agree to omit a if $a = 0$), each α is one of the digits 0, 1, $\cdots$, 9, and infinitely many α's are not zero. The last condition means that the decimal does not end with a string of zeros.

It may happen that the nonterminating decimal $a.\alpha_1\alpha_2 \cdots$ represents a rational number. (This will be so if the decimal is repeating.) We agree to *include* the rational numbers among the real numbers. Thus $1/3$ is another name for the real number $.333 \cdots = .\bar{3}$ and 1.5 is another name for the real number $1.4999 \cdots = 1.4\bar{9}$.

Let $x = a.\alpha_1\alpha_2\alpha_3 \cdots$ and $y = b.\beta_1\beta_2\beta_3 \cdots$ be two positive real numbers. The statement "$x = y$" means, of course, that $a = b$, $\alpha_1 = \beta_1$,

$\alpha_2 = \beta_2$, and so forth. We *define* "$x < y$" to mean that either $a < b$, or $a = b$ and $\alpha_1 < \beta_1$, or $a = b$, $\alpha_1 = \beta_1$, and $\alpha_2 < \beta_2$, and so on. Thus x is less than y if the first digit of x which is different from the corresponding digit of y is less than that digit, for example: $2.30782 \cdots$ $< 2.30899 \cdots$. The definition just given is *consistent* with the properties of rationals. For instance, $7/3 < 12/5$ whether we think of $7/3$ and $12/5$ as rational numbers or as the real numbers $2.333 \cdots = 2.\overline{3}$ and $2.3999 \cdots = 2.3\overline{9}$.

Also, our definition is such that, for every positive real number x, the statement

$$x = a.\alpha_1\alpha_2 \cdots \qquad \text{(a nonterminating decimal)}$$

is equivalent to the string of inequalities

$$a.\alpha_1 \cdots \alpha_m < x \leq a.\alpha_1\alpha_2 \cdots \alpha_m + 10^{-m}, \qquad m = 1, 2, 3, \cdots,$$

as it was for rational numbers.

A negative real number is defined as a nonterminating decimal preceded by the sign $-$ (minus). If the decimal is repeating, we identify this real number with the corresponding negative rational number. We define "$-a.\alpha_1\alpha_2 \cdots < -b.\beta_1\beta_2 \cdots$" to mean that $b.\beta_1\beta_2 \cdots < a.\alpha_1\alpha_2 \cdots$. This is again consistent with the properties of rational numbers. Finally, we include 0 among the real numbers and we agree that $-a.\alpha_1\alpha_2 \cdots < 0$ and $0 < a.\alpha_1\alpha_2 \cdots$.

The positive and negative real numbers and zero form the **set of real numbers**. The set of real numbers includes the set of rational numbers as a subset. A real number that is not rational is called **irrational**. For instance, $2.1010010001000010 \cdots$ is irrational, since this nonterminating decimal is nonrepeating.

The reader will easily verify that the real numbers satisfy the order postulates **13** and **14** of §3.2 and **17** of §3.4.

EXERCISES

1. Given that $x = .45637070070007 \cdots$ and $y = .4563\overline{707}$, which of the two numbers is larger?

2. Find a rational number between $.10203040506070809010011012\cdots$ and $.112131415161718191101111121 \cdots$.

3. Arrange the numbers $x = .20406080100120 \cdots$, $y = .2040\overline{60}$, and $z = .2040\overline{6}$ in increasing order.

4. Arrange the numbers $x = .2939\overline{49}$, $y = .\overline{293949}$, and $z = 293949596979 \cdots$ in increasing order.

5. Find an irrational number between $.\overline{00112}$ and $.0011200112001\overline{1}$. (You may assume that a nonrepeating decimal represents an irrational number.)

5.2 Calculating with real numbers

Suppose we want to add two real numbers, say two positive real numbers, $x = a.\alpha_1\alpha_2\alpha_3 \cdots$ and $y = b.\beta_1\beta_2\beta_3 \cdots$. The natural way to do this is to "round-off" both numbers by dropping all digits after, say, the mth, and then adding the resulting rational numbers $a.\alpha_1\alpha_2 \cdots \alpha_m$ and $b.\beta_1\beta_2 \cdots \beta_m$. We get a rational number which, we hope, is close to the "true" sum of x and y, and will be as close to the true sum as we want, provided we choose m large enough. The same applies to multiplication: we expect that, if m is large enough, then the product of $a.\alpha_1 \cdots \alpha_m$ and $b.\beta_1 \cdots \beta_m$ will be as close as we like to the "true" product of x and y.

But what is the "true" sum and the "true" product of two real numbers? Can they be defined in a precise manner? The answer is yes.

Proposition A   ONE CAN DEFINE THE SUM AND PRODUCT OF ANY TWO REAL NUMBERS IN SUCH A WAY THAT THE DEFINITION IS CONSISTENT WITH THE PROPERTIES OF RATIONAL NUMBERS AND ALL POSTULATES 1 TO 17 REMAIN VALID.

The proof of Proposition A, while not really difficult, is somewhat delicate and long and we will not carry it out here. The reader is advised to accept Proposition A on faith, for the time being. In doing so, he will be following excellent precedents. The founders of calculus and the great mathematicians who developed and applied it for centuries considered it self-evident that one could work with real numbers by following the usual rules of algebra. The need for proof, and the proof itself, date from the middle of the nineteenth century when a rigorous theory of real numbers was developed by Dedekind and by Cantor.

Proposition A is often abbreviated as follows: The real numbers form an **ordered Archimedean field.** If we admit the truth of this proposition, we may freely use *all* consequences of the field and order axioms derived above. We also may use the method outlined above for computing sums and products (and also differences and quotients) of real numbers with any desired degree of accuracy, by rounding-off. This will be verified in an appendix (§7) where we shall also derive formulas for estimating the accuracy.

5.3 Geometric interpretation. The number line

The field and order properties of real numbers were accepted without question by many generations of mathematicians because they have a natural and simple geometric interpretation.

RICHARD DEDEKIND (1831–1916) was primarily an algebraist, one of the originators of so-called "abstract" algebra.

GEORG CANTOR (1845–1918) single-handedly created the theory of infinite sets, perhaps the most revolutionary development in the history of mathematics since calculus. His work provided a common language for most of mathematics. It also led to hitherto unsuspected paradoxes and to logical difficulties, not yet fully resolved. The development of modern mathematical logic is a response to the challenge of Cantor's set theory.

Cantor studied in Berlin and then taught at a mediocre provincial university. His life was troubled by bitter scientific controversies and bouts with insanity.

Consider once more the straight line on which we marked off the points 0 and 1. We can imagine that we marked off on it all points corresponding to rational numbers (Fig. 1.13). There are infinitely many such "rational points." But there are also infinitely many points on the line that correspond to *no* rational number. This is so for every point P such that the segment OP is not commensurable with the segment 01.

On the other hand, we can associate with *every* point P a real number. We consider first the case when P lies to the right of 0.

The point P is "given." This means that, if Q is any other point on the line, distinct from P, we know whether Q lies to the left or to the right of P. We find the largest integer a such that the point a lies to the left of P. Then $a \geq 0$ and P does not lie to the right of $a + 1$. Next, let a_1 be the largest of the numbers $0, 1, \cdots, 9$ such that P lies to the right of the point $a.a_1 = a + \dfrac{\alpha_1}{10}$. Then P lies either to the left of $a.a_1 + \dfrac{1}{10}$ or coincides with $a.a_1 + \dfrac{1}{10}$. Next we find the largest α_2 among $0, 1, \cdots, 9$, such that P lies to the right of $a.a_1 a_2$, and so on. Thus we obtain an infinite sequence of digits $a.a_1 a_2 a_3 \cdots$ such that P is to the right of $a.a_1 \cdots a_m$ but not to the right of $a.a_1 \cdots a_m + 10^{-m}$, for $m = 0$, $1, 2, \cdots$. The process is illustrated in Fig. 1.14, where $a.a_1 a_2 a_3 \cdots = 1.305 \cdots$.

If P was a rational point to begin with, that is, a point corresponding to a rational number x, then $a.a_1 a_2 a_3 \cdots$ is the nonterminating decimal expansion of x. This becomes clear if we compare what we just did with the method used in §4.3.

We now make three statements that we accept as *geometric axioms*, since they conform to our geometric intuition.

1. The decimal $a.a_1 a_2 \cdots$ corresponding to point P (to the right of 0) is *nonterminating*. (Our geometric intuition tells us that P cannot lie to the right of a rational point, say c, without lying also to the right of a rational point $c + 10^{-m}$, for some integer $m > 0$.)

We call the real number $a.a_1 a_2 \cdots$ the **coordinate** of P.

2. Two *distinct* points P and Q to the right of 0 cannot have the same coordinate.

3. *Each* real number $x > 0$ is the coordinate of some point on the line to the right of 0.

Next, we give to O the coordinate 0, and we assign to a point Q to the left of 0 the coordinate $-x$, x being the coordinate of the point P to the right of 0 such that the segments OQ and OP are congruent.

Now we have associated to each point on the line a unique real number as its coordinate, and each real number is the coordinate of a unique point on the line. One expresses this by saying: *we have*

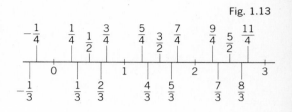

Fig. 1.13

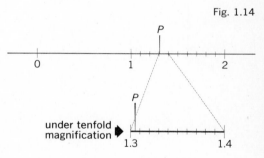

Fig. 1.14

under tenfold magnification ▶

established a one-to-one correspondence between real numbers and points on the line. This correspondence depends, of course, on the choice of the two points with coordinates 0 and 1.

We shall often use the expressions: real number x, the point with the coordinate x, and the point x, interchangeably.

The line on which we marked off the points 0 and 1 is often called the **number line.** It is an idealized measuring tape, which one can use to measure the length of *any* segment. Indeed, let AB be a segment. There is a unique point, call it P, on the number line to the right of 0 such that AB is congruent to the segment OP. The coordinate of P is, by definition, the length of AB. This length is a rational or irrational number according to whether AB is or is not commensurable with the unit segment 01.

Now we can construct sums and products of real numbers geometrically, as we did in §1 for rational numbers. This gives us assurance that the real numbers obey the usual rules of algebra.

Fig. 1.15

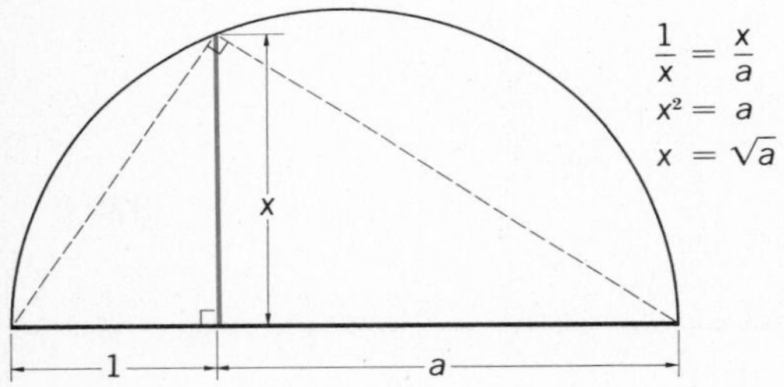

$$\frac{1}{x} = \frac{x}{a}$$
$$x^2 = a$$
$$x = \sqrt{a}$$

5.4 **Roots**

Let c be a positive real number. Then there is a unique real positive number x such that $x^2 = c$. The validity of this statement is obvious geometrically, since there is a ruler and compass construction (sketched in Fig. 1.15) which, given the unit segment and a segment of length c, yields a segment of length x such that $x^2 = c$. The number x is denoted by $\sqrt{c}$ or $\sqrt[2]{c}$.

A stronger statement is also true.

Theorem 1   FOR EVERY REAL POSITIVE NUMBER c AND EVERY INTEGER $n > 1$, THERE IS A UNIQUE POSITIVE REAL NUMBER x SUCH THAT $x^n = c$.

We call this x the nth **root** of c and write $x = \sqrt[n]{c}$. We also set $\sqrt[n]{0} = 0$ and, if n is odd, $\sqrt[n]{-c} = -\sqrt[n]{c}$. Indeed, $0^n = 0$ and, for odd n, $(-\sqrt[n]{c})^n = -c$.

It is clear that the theorem on the existence of roots cannot be true in the field of rational numbers. The number $\sqrt{2}$, for instance, is irrational (compare §1.3).

We shall prove the theorem by exhibiting a method for computing $\sqrt[n]{c}$ (though not necessarily the most efficient method).

Lemma   LET x AND y BE POSITIVE NUMBERS, AND n A POSITIVE INTEGER. THEN

(1)   IF $x < y$, THEN $x^n < y^n$

AND

(2)   IF $x^n < y^n$, THEN $x < y$.

Proof. If $0 < x < y$, then $x^2 < xy$ and $xy < y^2$ (by **16**) and hence $x^2 < y^2$ (by **14**). Again, $x^3 < y^2x$ and $y^2x < y^3$ (by **16**) and hence $x^3 < y^3$. Continuing in this way, we obtain (1). Now, (1) implies (2). For, if it is not true that $x < y$, then either $x = y$ and $x^n = y^n$ or $y < x$ and, by what we just proved, $y^n < x^n$.

The lemma implies at once that, for a given n and $c > 0$, there is at most one positive x with $x^n = c$. The lemma also shows how to compute $\sqrt[n]{c}$, say $\sqrt[3]{13}$, assuming that there is such a number. We must find a nonnegative integer a and digits $\alpha_1, \alpha_2, \alpha_3, \cdots$ each of which should be one of the numbers $0, 1, \cdots, 9$ such that $\sqrt[n]{c} = a.\alpha_1\alpha_2\alpha_3 \cdots$. This means, as we know, that

$$a.\alpha_1\alpha_2 \cdots \alpha_k < \sqrt[n]{c} \leq a.\alpha_1\alpha_2 \cdots \alpha_k + 10^{-k} \qquad \text{for } k = 0, 1, 2, \cdots.$$

By the lemma this is equivalent to

(3) $(a.\alpha_1\alpha_2 \cdots \alpha_k)^n < c \leq (a.\alpha_1\alpha_2 \cdots \alpha_k + 10^{-k})^n \quad k = 0, 1, 2, \cdots.$

Thus we find first a as the largest integer such that $a^n < c$, then α_1 as the largest number among $0, 1, \cdots, 9$ such that $(a.\alpha_1)^n < c$, then α_2 as the largest among the numbers $0, 1, \cdots, 9$ such that $(a.\alpha_1\alpha_2)^n < c$, and so forth.

In our example, $a = 2$, since $2^3 = 8$ and $3^3 = 27$; $\alpha_1 = 3$, since $2.3^3 = 12.167$ and $2.4^3 = 13.824$; $\alpha_2 = 5$, since $2.35^3 = 12.977 \cdots$ and $2.36^3 = 13.144 \cdots$. Thus $\sqrt[3]{13} = 2.35 \cdots$.

We have now obtained a number $x = a.\alpha_1\alpha_2\alpha_3 \cdots$. We have $a.\alpha_1\alpha_2 \cdots \alpha_k < x \leq a.\alpha_1\alpha_2 \cdots \alpha_k + 10^{-k}$ for $k = 0, 1, 2, \cdots$ and therefore, by the lemma,

(4) $(a.\alpha_1 \cdots \alpha_k)^n < x^n \leq (a.\alpha_1 \cdots \alpha_k + 10^{-k})^n, \qquad k = 0, 1, 2, \cdots.$

Comparing it with (3), we see that, for all k, the numbers c and x^n both lie in the same interval of length,

$$L_k = (a.\alpha_1 \cdots \alpha_k + 10^{-k})^n - (a.\alpha_1 \cdots \alpha_k)^n.$$

Therefore (compare the example in §3.5)

$$|x^n - c| \leq L_k.$$

In order to compute L_k we make use of the identity (7) in §2.5:

$$A^n - B^n = (A - B)(A^{n-1} + A^{n-2}B + A^{n-3}B^2 + \cdots + B^{n-1}).$$

We set $A = a.\alpha_1\alpha_2 \cdots \alpha_k + 10^{-k}$ and $B = a.\alpha_1\alpha_2 \cdots \alpha_k$. Then $A - B = 10^{-k}$ and $B < A \le a + 1$. Hence $L_k = A^n - B^n$, and

$$L_k = 10^{-k}(A^{n-1} + A^{n-2}B + A^{n-3}B^2 + A^{n-4}B^3 + \cdots + B^{n-1})$$
$$\le 10^{-k}n(a + 1)^{n-1}.$$

We conclude that
$$\frac{|x^n - c|}{n(a + 1)^{n-1}} \le 10^{-k}$$

for all integers k. This implies (by Corollary 2 to the Archimedean postulate; see §3.4) that $|x^n - c| = 0$ and $x^n = c$.

We found one solution of the equation $x^n = c$. The lemma implies that there can be no other positive solution. The theorem is completely proved.

EXERCISES

In the following exercises, count the digits before and after the decimal point, that is, 211.35 has 5 digits.

6. Compute the first 3 digits in the decimal representation of $\sqrt[3]{7}$.
7. Compute the first 3 digits in the decimal representation of $\sqrt[3]{7.9}$.
8. Compute the first 3 digits in the decimal representation of $(-26)^{1/3}$.
9. Compute the first 2 digits in the decimal representation of $\sqrt[4]{80}$.
► 10. Given that $x = 1.01001000100001 \cdots$, compute the first 4 digits in the decimal representation of $\sqrt[3]{x}$.
11. Given that $x = 1.11213141516 \cdots$, compute the first 4 digits in the decimal representation of $\sqrt[3]{x}$.
12. Compute the first 4 digits in the decimal representation of $\sqrt[4]{1.1}$.
13. Compute the first 4 digits in the decimal representation of $\sqrt[3]{1003}$.

5.5 Fractional exponents

One can now define fractional exponents, that is, one can assign a meaning to the expression

$$c^r$$

where c is a positive number and r a rational number. The definition reads:

$$c^{p/q} = \sqrt[q]{c^p} = (\sqrt[q]{c})^p.$$

where p and q are integers, $q > 0$, and the fraction p/q is reduced to lowest terms. If c is a negative number, c^r is defined whenever the rational number r, when reduced to lowest terms, has an odd denominator. In this case the definition given above can be used. For $c = 0$, finally, one sets $c^r = 0$ for all positive rational r.

The definition is chosen so as to make the laws of exponents stated in §2.4 valid for fractional exponents, whenever meaningful. The verification is easy and need not be carried out here.

▶**Example** $(-1)^{1/3} = \sqrt[3]{-1} = -\sqrt[3]{1} = -1$. But it would be wrong to write $(-1)^{1/3} = (-1)^{2/6} = \sqrt[6]{(-1)^2} = \sqrt[6]{1} = 1$. ◀

EXERCISES

Simplify the following expressions.

14. $[\sqrt[4]{(x^{1/3})^3}]^4$.   17. $(\sqrt{x^{1/3}x^{5/3}})^3$.

15. $\sqrt{(-x^{4/5})^5 y^8}$.   18. $x^2 \sqrt[4]{(x^{-8/3})^3 y^8}$.

16. $[(x^{3/2})^4 x^2]^{1/4}$.

5.6 Intervals

This is a good place to introduce some terminology that will be used throughout the book.

Let a and b be numbers such that $a < b$. (From now on, "number" means real number.) The set of all numbers x such that $a < x < b$ is called the **open interval** (a, b). "Open" means that a and b are not included in the set. The interval (a, b) consists of all points on the segment of the number line with endpoints a and b except for the endpoints themselves. The set of all x such that $a \le x \le b$ is called the **closed interval** $[a, b]$; it consists of all points of the above segment, endpoints included. The set of all x such that $a \le x \le b$ is called the **half-open interval** $[a, b)$. The notation $(a, b]$ is self-explanatory. In all cases $b - a$ is called the **length** of the interval, and the point $(a + b)/2$ is called the **center** or **midpoint** of the interval.

All the intervals considered above are called **finite**.

The set of all x such that $x > a$ is called the **infinite interval** $(a, + \infty)$. Similarly, $[a, + \infty)$ is the set of all x with $x \ge a, (- \infty, b)$ the set of all x with $x < b, (- \infty, b]$ the set of all x with $x \le b$. The whole number line, finally, is denoted by $(- \infty, + \infty)$.

The symbols $+ \infty, - \infty$ are read: plus infinity and minus infinity. These are *not* names of numbers.

Theorem 2   EVERY INTERVAL CONTAINS INFINITELY MANY RATIONAL NUMBERS AND INFINITELY MANY IRRATIONAL NUMBERS.

This is not at all surprising. A formal proof will be found in an appendix (§8.4).

EXERCISES

19. Does 5 lie in the (open) interval $(5, 6)$?
20. Does 5 lie in the (closed) interval $[5, 15]$?
21. Does -1 lie in the (half-open) interval $[-2, -1)$?
22. Does 0 lie in the (closed) interval $[0, 2]$?
23. Does 4 lie in the (half-open) interval $[-1, 5)$?
24. What is the center of the interval $(-3, 1)$?
25. What is the closed interval whose center is at $\frac{1}{2}$ and whose length is 2?
26. Find an interval that contains $1/10^{-j} + 1$ for $j = -1, 0, 1, 2, 3, \cdots$ but does not contain 1.

Appendixes to Chapter 1

§6 Completeness of the Real Numbers. Induction

The rational numbers and the real numbers both satisfy Postulates 1 to 17—and so do other systems of numbers. In this section, we discuss a property that distinguishes the system of real numbers from other sets also satisfying 1 to 17. We combine this discussion with that of mathematical induction.

6.1 Bounds

Let S be a **set** (collection) of real numbers. We say that "x is an **element** of S" or "x belongs to S" or "x is in S" to indicate that x is a member of the set. (One often abbreviates this statement by writing "$x \epsilon S$.") A set that contains no members is called **empty**.

A number c is called an **upper bound** for S if every x in S satisfies $x \leq c$. One also calls such a c an upper bound of S, or one says that S has c for an upper bound. Whether c belongs to S or not is irrelevant.

A set S is called **bounded from above** if it has an upper bound; otherwise S is called **unbounded from above**.

A number c is called a **lower bound** for S if $c \leq x$ for all x in S. If a lower bound exists, S is called **bounded from below**. A set is called **bounded** if it is bounded from above and from below. (The empty set is bounded, by convention.)

An upper bound for S, which is also an element of S, is called the greatest or largest element of S. A lower bound for S, which is also an element of S, is called the smallest or the least element of S. It is clear that S can have at the most one greatest element and at the most one smallest element.

A set of numbers may be considered a set of points on the number line. An upper (lower) bound for S is a point such that no element of S lies to its right (to its left).

►*Examples* The interval $(-10, +\infty)$ is bounded from below, but not from above. The interval $(-100, 100)$ is bounded. Any finite set, that is, a set with finitely many elements, is bounded.

If S consists of 1, 2, and -5, then 2 is an upper bound; so are 2.0001 and 10^{10}. If S is the set of all negative rationals, then 0 is an upper bound. So is every positive number. If S is the set of all integers, S has no upper bound.

Zero is the least element in the set of all nonnegative rational numbers. The interval $(-2, 3)$ has no least element and no greatest element. ◄

EXERCISES

1. Is 1 an upper bound for the interval $(-3, 1]$?
2. Is 12 an upper bound for the interval $(-1, 0)$?
3. Is 3 a lower bound for the interval $(3, +\infty)$?
4. Is 2 a lower bound for the interval $(-\infty, 2]$?
5. Is the set of all integers divisible by 3 bounded? Bounded from above? Bounded from below?
► 6. Is the set of all integral perfect squares bounded? Bounded from above? Bounded from below?
7. Find the least elements and the largest element of the set

$$\left\{ \frac{3}{1}, \frac{3}{4}, \frac{3}{5}, \frac{4}{1}, \frac{4}{3}, \frac{4}{5}, \frac{5}{1}, \frac{5}{3}, \frac{5}{4} \right\}.$$

8. Does the interval $(0, 2)$ have a least element? A largest element?

9. Does the infinite set $\left\{ \frac{1}{2}, \frac{2}{3}, \frac{3}{4}, \frac{4}{5}, \cdots \right\}$ have a least element? A largest element? Is it bounded?

10. Does the set of rational numbers in the interval $[0, 1)$ have a least element? A largest element? Is $\sqrt{2}$ an upper bound for this set?

6.2 Principle of least integer

Before discussing the property of real numbers mentioned before, it will be useful to have a look at a somewhat analogous property of integers.

I. LET S BE A NONEMPTY SET OF INTEGERS. IF S IS BOUNDED FROM BELOW, S CONTAINS A LEAST ELEMENT.

The statement is almost self-evident. It is assumed, perhaps in a different wording, in every axiomatic treatment of integers. We call **I** the **principle of least integer.** The principle has an immediate corollary.

I′. LET S BE A NONEMPTY SET OF INTEGERS. IF S IS BOUNDED FROM ABOVE, S CONTAINS A LARGEST ELEMENT.

Proof. Let T be the set of all negatives of the elements of S. Thus x is in T if and only if $-x$ is in S. The set T is a nonempty set of integers. Let c be an upper bound of S. Then $y \leq c$ for all y in S. Hence $-c \leq -y$ for all y in S, hence $-c$ is a lower bound for T. By **I**, T has a least element b. Then $-b$ is the greatest element of S.

6.3 Mathematical induction

The principle of least integer is the basis of proofs by **mathematical induction.** Let us explain it with an example; we choose one that will be of use to us later.

We want to establish the formula for the sum of squares of the first m positive integers. The formula reads:

$$(1) \qquad 1^2 + 2^2 + 3^2 + \cdots + m^2 = \frac{m^3}{3} + \frac{m^2}{2} + \frac{m}{6}.$$

We claim that it is true for all positive integers m. If the claim is false, there is some positive integer for which the relation (1) is false: the set of positive integers m for which (1) is false is not empty. There is then the smallest integer, call it r, for which (1) is false (by the principle of least integer, since positive integers are bounded from below). Either $r = 1$ (Case 1) or $r > 1$ (Case 2). In Case 2, we may set $r = k + 1$, where $k > 0$ is an integer.

We observe now that Case 1 cannot occur. In other words: the relation to be established holds for $m = 1$. Indeed, for $m = 1$, relation (1) reads:

$$1 = \frac{1}{3} + \frac{1}{2} + \frac{1}{6}.$$

Next, Case 2 cannot occur either. For if it does, we have

$$(2) \qquad 1^2 + 2^2 + \cdots + k^2 + (k + 1)^2 \neq \frac{(k + 1)^3}{3} + \frac{(k + 1)^2}{2} + \frac{k + 1}{6}$$

and

$$(3) \qquad 1^2 + 2^2 + \cdots + k^2 = \frac{k^3}{3} + \frac{k^2}{2} + \frac{k}{6}.$$

Adding $(k + 1)^2$ to both sides of (3), however, we obtain the correct equation,

$$1^2 + 2^2 + \cdots + k^2 + (k + 1)^2 = \frac{k^3}{3} + \frac{k^2}{2} + \frac{k}{6} + (k + 1)^2,$$

and after some simple manipulations

$$1^2 + 2^2 + \cdots + k^2 + (k + 1)^2 = \frac{(k + 1)^3}{3} + \frac{(k + 1)^2}{2} + \frac{k + 1}{6}.$$

This result contradicts (2).

We conclude that (1) is true for all positive integers m.

The example just given is typical. If we want to prove that a certain statement about all positive integers is true, we proceed as follows.

1. *First step:* we show that the statement is true for the integer 1.

2. *Second step:* we show that *if* the statement were true for any integer $k > 0$, *then* it would also be true for $k + 1$.

Strictly speaking, the method of mathematical induction must be used whenever one proves a theorem about all positive integers. If the situation is very simple, however, it is customary (and legitimate) to replace the formal induction proof by a vague "and so on." We did this, for instance, in §5.3, deriving the statement: if $0 < x < y$ and $m > 0$ is an integer, then $x^m < y^m$.

Let us show how a formal induction proof would look in this case.

1. *First step in induction proof:* For $m = 1$, the statement reads: if $0 < x < y$, then $x < y$. This is true.

2. *Second step:* Assume the statement to be true for some fixed integer $k > 0$. Hence, if $0 < x < y$, then $x^k < y^k$. But then $x^{k+1} = x^k x < y^k x$ and $y^k x < y^k y = y^{k+1}$, by **16**, and therefore $x^{k+1} < y^{k+1}$, by **14**. Thus the statement holds for $k + 1$.

This completes the proof.

EXERCISES

In Exercises 11 to 20, the proofs should be by mathematical induction.

11. Prove that $1 + 2 + 3 + \cdots + n = \dfrac{n(n + 1)}{2}$ for every positive integer n.

12. Prove that $1 + q + q^2 + \cdots + q^{n-1} = \dfrac{1 - q^n}{1 - q}$ for every positive integer n and any number $q \neq 1$. Note that this was proved directly in §2.

13. Prove that $1 + 3 + 5 + \cdots + (2n - 1) = n^2$ for every positive integer n.

►14. Prove that $1 \cdot 2 + 2 \cdot 3 + 3 \cdot 4 + \cdots + n(n + 1) = \dfrac{n(n + 1)(n + 2)}{3}$ for every positive integer n.

15. Prove that $\dfrac{1}{1 \cdot 2} + \dfrac{1}{2 \cdot 3} + \dfrac{1}{3 \cdot 4} + \cdots + \dfrac{1}{n(n + 1)} = \dfrac{n}{n + 1}$ for every positive integer n.

16. Prove that $3n^2 \geq 2n + 1$ for every positive integer n.

17. Prove that $3^n \geq 3n$ for every positive integer n.

►18. Prove that $4^n \geq n^2$ for every positive integer n.

19. Prove that $n^3 \geq 3n + 3$ for $n = 3, 4, 5, \cdots$.

20. Prove that $n^3 > n^2 + 3$ for $n = 2, 3, 4, \cdots$.

6.4 Least upper bound principle

Statements **I** and **I′** need not be true for sets of real numbers. Thus, if S is the set of all negative numbers, it has no largest element. But the set of all upper bounds for this S has a least element, namely 0. This situation is typical. We state now the so-called **principle of least upper bound** or **completeness postulate** for real numbers. It is to be added to Postulates **1** to **17**.

18. A NONEMPTY SET OF REAL NUMBERS THAT IS BOUNDED FROM ABOVE HAS A LEAST UPPER BOUND.

This means that, if S is the set considered, and there are numbers that are greater than, or equal to, all members of S, then there is a smallest such number. This least upper bound may, or may not, belong to the set S. If all members of S are rational, the least upper bound may be irrational.

An easy corollary of **18** is

18′. A NONEMPTY SET OF REAL NUMBERS THAT IS BOUNDED FROM BELOW HAS A GREATEST LOWER BOUND.

The reader is urged to derive **18′** from **18** following the pattern used above to derive **I′** from **I**.

Here is the geometric interpretation of **18**. Let S be a nonempty set of points on the number line. Let a be a point such that no point in S is to the right of a. Then there is a point b such that (1) no point of S is to the right of b and (2) if c is any point to the left of b, there is a point in S to the right of c.

That real numbers satisfy Postulates **1** to **17** is intuitively obvious, but the proof is not easy. The meaning of **18** may be difficult to grasp and is far from obvious. The proof, however, is not difficult. Before presenting it, we emphasize that throughout this book Postulate **18** will be used only a few times, but always at crucial points in the argument. Just as mathematical induction is present, explicitly or implicitly, in all theorems about all integers, so the least upper bound principle is present in all significant theorems about real numbers.

Proof of Postulate 18. The proof is a generalization of the argument used in §5.4 to compute roots. Let S be a nonempty set of real numbers and let c be an upper bound for S. We shall compute the least upper bound of S.

By the Archimedean postulate **17**, there is an integer k such that $c \leq k$. This k is also an upper bound of S. The set of all integers that are upper bounds for S is itself bounded from below and has (by **I**) a smallest element. Call it $a + 1$. The integer a is not an upper bound for S, but $a + 1$ is. We consider the numbers $a + (1/10)$, $a + (2/10), \cdots, a + (9/10)$ and find a digit α_1 (among $0, 1, \cdots, 9$) such that $a + .\alpha_1$ is not an upper bound for S, but $a + .\alpha_1 + .1$ is. Continuing in this way, we obtain a real number

$$x = a + .\alpha_1\alpha_2\alpha_3 \cdots$$

such that: for all $k = 1, 2, 3, \cdots$

(4) $a + .\alpha_1\alpha_2 \cdots \alpha_k$ is not an upper bound for S

and

(5) $a + .\alpha_1 \cdots \alpha_k + 10^{-k}$ is an upper bound for S.

(We could write $x = a.\alpha_1\alpha_2 \cdots$ if we know that $a \geq 0$, which it need not be.)

Let y be any number belonging to S. Statement (5) implies that $y \leq a + .\alpha_1 \cdots \alpha_k + 10^{-k}$ (for all $k = 2, 3, \cdots$). Since $a + .\alpha_1 \cdots \alpha_k \leq x$, we have $y \leq x + 10^{-k}$ or $y - x \leq 10^{-k}$, $k = 2, 3, \cdots$. This means that $y - x \leq 0$, by Corollary 2 to **17** (see §3.4) so that, y being an arbitrary element of S, x is *an upper bound* for S.

Assume next that w is an upper bound for S. By (4) we have $w \geq a + \alpha_1\alpha_2 \cdots \alpha_k$ for $k = 2, 3, \cdots$. For otherwise, $a + .\alpha_1\alpha_2 \cdots \alpha_k$ would be an upper bound for S. Hence $w \geq x - 10^{-k}$ or $x - w \leq 10^{-k}$, for all positive integers k. This means that $x - w \leq 0$ or $x \leq w$. Hence x is *the smallest upper bound* for S.

(In the preceding proof we used the principle of least integer. It can be shown that this principle is a consequence of **18** and **17**. On the other hand, **17** is a consequence of **18** and **I**. Also, by more carefully rephrasing our definition of integers (see §2), we could derive **I**. We do not prove these statements here, since the proofs would be of no use to us later.)

EXERCISES

21. Let S be the set of all rational numbers x with $x^2 + 4 < 6$. Find the least upper bound of S.

▶ 22. Let S be the set of all rational numbers z with $2z^3 - 1 < 15$. Find the least upper bound of S.

23. Let S be the set $\left\{ \dfrac{1}{2}, \dfrac{1}{3}, \dfrac{1}{5}, \dfrac{1}{7}, \dfrac{1}{11}, \dfrac{1}{13}, \cdots \right\}$. Find the greatest lower bound of S.

24. Let S be the set of all real numbers whose decimal expansion starts with $.12 \cdots$. Find the least upper bound and greatest lower bound of S.

25. Let S be the set of all rational numbers x with $x^{1/3} + 2 < 4$. Find the least upper bound of S.

26. Let S be the set of all irrational numbers t with $1 < t^3 + 1 \leq 3$. Find the least upper bound and greatest lower bound of S.

27. Let S be the set of all irrational numbers v with $v^2 + v < 2$. Find the least upper bound and greatest lower bound of S.

6.5 Axioms for real numbers

In §5 we asked the reader to accept Proposition A on faith. We now offer him another option. He can treat Postulates **1** to **18** as **axioms for the real number system,** that is, as assumed statements whose logical consequences we are investigating. It is, by the way, customary to abbreviate Postulates **1** to **18** by saying that **the real numbers form a complete, Archimedean, ordered field.**

The whole stupendous structure of mathematical analysis rests on Axioms **1** to **18**. So does most of the new mathematics that is being discovered or invented now. (See §8 for further discussion.)

§7 Approximate Calculations

7.1 An estimate for the accuracy

Let X, Y and x, y be numbers such that X is close to x and Y is close to y. We expect then that $X + Y$ is close to $x + y$ and $X - Y$ to $x - y$. We also expect that XY is close to xy and, if $Y \neq 0$, $y \neq 0$, that X/Y is close to x/y. Using Postulates **1** to **16**, we can verify that this is so and make a precise statement.

Theorem 1   SUPPOSE THAT

(1) $$|X - x| \leq a, \qquad |Y - y| \leq b.$$

THEN

(2) $$|X + Y - (x + y)| \leq a + b, \qquad |X - Y - (x - y)| \leq a + b,$$

(3) $$|XY - xy| \leq |x|b + |y|a + ab$$

AND, IF

(4) $$Y \neq 0, y \neq 0, \qquad b \leq \tfrac{1}{2}|y|,$$

THEN ALSO

(5) $$\left| \frac{X}{Y} - \frac{x}{y} \right| \leq \frac{2(|x|b + |y|a)}{|y|^2} = \frac{2|x|}{|y|^2}b + \frac{2}{|y|}a.$$

Before proving the theorem, we point out its significance. Suppose we want to compute $x + y$ or $x - y$, or xy or x/y, and we know the numbers x and y only approximately, say with errors not exceeding a and b, respectively. This means that we know numbers X and Y satisfying (1). Such a situation arises, for instance, if x and y are physical quantities. No measurement can give us the precise values of x and y; we must be satisfied with approximate values X and Y. But, even if we know the precise values of x and y, we might want to round them off in order to save computational labor, and we might have to do so if we use a computer.

Our theorem tells us how large an error we commit if we compute, instead of the desired numbers, the numbers $X + Y$, $X - Y$, XY, X/Y. Indeed, the theorem implies that, in order to compute the sum, difference, product, and quotient of two numbers with an error not exceeding some positive number, however small, it is enough to know the two numbers with a certain accuracy.

Thus, if we want to know $x + y$ or $x - y$ with an error of not more than $1/100$, the inequalities (2) tell us that it suffices to know x and y each with an accuracy of at least $1/200$.

The situation is somewhat more complicated for products and quotients. In order to know xy with an error not exceeding, say, $1/100$, it suffices, according to (3), to know x with an error not exceeding a and y with an error not exceeding b, such that

$$|y|a \leq \frac{1}{300}, \qquad |x|b \leq \frac{1}{300}, \qquad ab < \frac{1}{300}.$$

If we know, for instance, that $|x| \leq 10$, $|y| \leq 100$, it suffices to have $a \leq 1/30{,}000$, $b \leq 1/3000$.

In order to know x/y with an accuracy of $1/100$, it suffices that the permissible errors a and b satisfy $b < 1/2\, |y|$ and

$$\frac{2b|x|}{|y|^2} \le \frac{1}{200}, \qquad \frac{2a}{|y|} \le \frac{1}{200},$$

as is seen from (4) and (5). If we know, for instance, that $|x| \le 100$, $|y| \le 100$, and $|y| \ge 5$, it suffices to have

$$a \le \frac{5}{2 \cdot 200} = \frac{1}{80}, \qquad b \le \frac{25}{2 \cdot 100 \cdot 200} = \frac{1}{64}.$$

For the sake of definiteness, we talked about specific numbers, but it is clear that the argument is general. By choosing a and b small enough, the right-hand side of (3) and (5) can be made smaller than any preassigned positive number.

Now we prove the theorem. Without much comment, we shall use the rules for inequalities and absolute values derived in §3. By hypothesis $X = x + \alpha$, $Y = y + \beta$ with $|\alpha| \le a$, $|\beta| \le b$. Hence, using the triangle inequality [see inequality (10) in §3.5]

$$|(X + Y) - (x + y)| = |(x + \alpha + y + \beta) - (x + y)|$$
$$= |\alpha + \beta| \le |\alpha| + |\beta| \le a + b,$$

which proves the first inequality (2). The second is established similarly.

Next, using the triangle inequality, we find that

$$|XY - xy| = |(x + \alpha)(y + \beta) - xy| = |x\beta + y\alpha + \alpha\beta| \le |x\beta| + |y\alpha| + |\alpha\beta|$$
$$= |x||\beta| + |y||\alpha| + |\alpha||\beta| \le |x|b + |y|a + ab,$$

which proves (3).

Now we assume (4). Then $|y| = |(y + \beta) + (-\beta)| \le |y + \beta| + |\beta|$ and $|y + \beta| \ge |y| - |\beta| \ge |y| - 1/2\,|y| = 1/2\,|y|$ and hence $|y(y + \beta)| = |y||y + \beta| \ge 1/2\,|y|^2$. Therefore,

$$\frac{1}{|y(y + \beta)|} \le \frac{2}{|y|^2}.$$

Using this, we see that

$$\left| \frac{X}{Y} - \frac{x}{y} \right| = \left| \frac{x + \alpha}{y + \beta} - \frac{x}{y} \right| = \frac{|(x + \alpha)y - (y + \beta)x|}{|y(y + \beta)|} = \frac{|xy + y\alpha - yx - \beta x|}{|y(y + \beta)|}$$
$$\le \frac{|y||\alpha| + |x||\beta|}{|y(y + \beta)|} \le \frac{2|y||\alpha| + 2|x||\beta|}{|y|^2} \le \frac{2|y|a + 2|x|b}{|y|^2}$$

which proves (5).

EXERCISES

1. Find a so that whenever X is within a of 2 and Y is within a of 6, then $X + Y$ is within $1/10$ of 8.

2. Find b so that whenever X is within .1 of 5 and Y is within b of 2, then $X - Y$ is within .15 of 3.

3. Find a so that, whenever U, V, W, X, Y, and Z are each within a of 3, then $U + V + W + X + Y + Z$ is within .25 of 18.

► 4. Find a and b so that, whenever X is within a of 5 and Y is within b of 8, then XY is within 1 of 40.

5. How close should X be to 1 in order to guarantee that X^2 is within .01 of 1?

6. Find a so that, whenever X is within a of 10 and Y is within a of 4, then XY is within $1/4$ of 40.

7. How close should X be to 3 in order to guarantee that XY is within .25 of 6 for any Y such that $1.95 < Y < 2.05$?

8. How close should Y be to 8 in order to guarantee that Y/X is within 1.1 of 4 for any X such that $1.75 \leq X \leq 2.25$?

9. Suppose x and y are fixed numbers with $|x| < 100$, $1 < |y| < 100$. Find a so that, whenever X is within a of x and Y is within $3a$ of y, then X/Y is within $1/2$ of x/y.

10. If X is within .05 of x, $1 \leq x \leq 2$, and Y is within .001 of 1, estimate how close X/Y is to x.

11. Find a and b so that, whenever X is within a of 4, Y is within .01 of 8, and Z is within b of 2, then $X + Y/Z$ is within .011 of 6.

12. Suppose x and y are fixed numbers with $1 < x < 10, 3 < y < 5$. Find a and b so that, whenever X is within a of x, Y is within a of y, and Z is within b of 1, then XYZ is within .1 of xy.

7.2 Rounding-off decimals

We apply Theorem 1 to the case in which x and y are given as decimal fractions and X and Y are obtained by the simplest method of rounding-off, that is, by dropping all digits after the decimal point, after the mth. Then (as we already observed in §4.5), $|X - x| \leq 10^{-m}$, $|Y - y| \leq 10^{-m}$. The following statement follows at once from Theorem 1.

Theorem 2   IF IN COMPUTING $x + y, x - y, xy$ AND x/y WE RETAIN ONLY m DIGITS AFTER THE DECIMAL POINT IN THE DECIMAL EXPANSIONS OF x AND y, THEN

$$(6) \qquad |\text{THE ERROR IN } x + y \text{ OR IN } x - y| \leq 2(10^{-m}),$$

$$(7) \qquad |\text{THE ERROR IN COMPUTING } xy| \leq 10^{-m}|x| + 10^{-m}|y| + 10^{-2m},$$

AND IF $2(10^{-m}) \leq |y|$, THEN

$$(8) \qquad \left| \text{THE ERROR IN COMPUTING } \frac{x}{y} \right| \leq \frac{2(|x| + |y|)10^{-m}}{|y|^2}.$$

These formulas show that, as asserted in §4.5, we can, by choosing m large enough, compute the sum, difference, product, and quotient of two rational numbers *with any desired degree of accuracy*. How large m must be chosen depends only on the size of the error we are willing to permit, and, in the case of products and quotients, also on the sizes of the given numbers.

We remark that, while we established that in calculating with decimals one may achieve any desired degree of accuracy while retaining only finitely many digits, we did not indicate the most efficient way of doing so.

In computing the product xy, for instance, it may not be wise to round-off both x and y to the same number of digits. It is better first to write x and y as powers of ten multiplied by factors of absolute value less than one: $x = 10^p r$, $y = 10^q s$ with $|r| < 1$, $|s| < 1$, and then round-off r and s to the same number of digits.

The way in which we rounded-off, simply dropping all digits after the mth, is also not efficient. We may diminish errors if in rounding-off, say to two digits, we replace $.2345 \cdots$ by $.23$ but $.2378 \cdots$ by $.24$.

Actual calculations performed on modern computers may involve hundreds of millions of additions and multiplications. The control over round-off errors in such calculations is a difficult and fascinating task involving refined techniques from calculus and other mathematical disciplines.

EXERCISES

13. Compute $22.\overline{41} + 10.13\overline{2}$ with an error not exceeding 2×10^{-4}.
14. Compute $3.163\overline{87} - 1.0\overline{25}$ with an error not exceeding $.0001$.
15. Compute $1.167\overline{3210} - 1.0\overline{672}$ with an error not exceeding 10^{-4}.
16. Compute $2.\overline{86} + .\overline{998} - .9832\overline{21}$ with an error not exceeding $.001$.
17. Compute $(8.\overline{3})(.99\overline{88})$ with an error not exceeding $.01$.
▶ 18. Compute $(.\overline{717})(.18\overline{19})$ with an error not exceeding $.0001$.
19. Compute $(30.\overline{41})(61.0\overline{25})$ with an error not exceeding 1.
20. Compute $3.\overline{6}/2.\overline{15}$ with an error not exceeding 10^{-2}.
21. Compute $.5/.000\overline{2}$ with an error not exceeding 10.
22. Compute $998/1.\overline{01}$ with an error not exceeding 1.

§8 Rational and Irrational Numbers

In this section we prove and amplify some statements made in Chapter 1.

8.1 Geometric construction of products and quotients

In §1.1 we mentioned that, if a fixed segment is chosen as a unit of length, then the product ab of the lengths of two segments, and the quotient a/b of these two lengths, can be constructed geometrically as a length. This construction is based on the following theorem.

LET ABC BE A TRIANGLE, AND LET D BE A POINT ON THE SEGMENT AB AND E A POINT ON THE SEGMENT CB SUCH THAT DE IS PARALLEL TO AC (SEE FIG. 1.16). THEN

$$\frac{\overline{BA}}{\overline{BD}} = \frac{\overline{BC}}{\overline{BE}} = \frac{\overline{AC}}{\overline{DE}}.$$

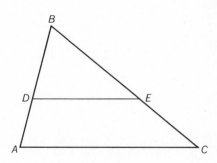

Fig. 1.16

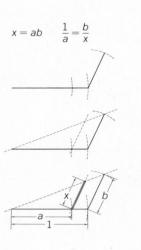

$$x = ab \qquad \frac{1}{a} = \frac{b}{x}$$

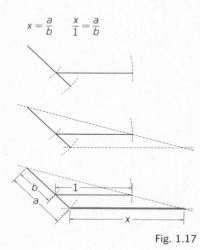

$$x = \frac{a}{b} \qquad \frac{x}{1} = \frac{a}{b}$$

Fig. 1.17

The actual construction of ab and a/b is shown in Fig. 1.17, for the case $1 > a > b$. In order to find $x = ab$, we construct a triangle like the one in Fig. 1.16 with $\overline{BA} = 1$, $\overline{BD} = a$, and $\overline{AC} = b$. Then $x = \overline{DE}$ is the desired length; indeed, we have $1/a = b/x$.

In order to find $x = a/b$, we construct a triangle like the one in Fig. 1.16 with $\overline{BA} = a$, $\overline{BD} = b$ and $\overline{DE} = 1$. Then $x = \overline{AC}$ is the desired length; indeed, we have $x/1 = a/b$.

The basic theorem on similar triangles stated above illustrates the difficulty presented by incommensurability. It is easily proved if the segments BD and DA are commensurable. We indicate the proof in Fig. 1.18, where $\overline{BA}/\overline{BD} = 5/3$, all parallelograms, and all shaded triangles are congruent. But even the meaning of the theorem becomes difficult to grasp if the two segments are not commensurable.

The introduction of real numbers resolves the difficulty. The theorem of similar triangles now has a clear meaning even if the segments involved are not commensurable. It can indeed be proved. The same is true of the elementary area theorems. The area of a rectangle of length a and width b is ab. This means: the rectangle can be cut up into pieces and reassembled so as to form a rectangle of length 1 and width ab. This can be proved whether the numbers a, b are rational or not.

The proofs of these theorems in the "incommensurable case" are not simple and we shall not carry them out here. As a matter of fact, having introduced real numbers, we can bypass geometric proofs altogether and reduce all geometric theorems to statements about numbers. This will be discussed further in Chapter 2.

It is nevertheless instructive to recall how the incommensurability difficulty was overcome by the Greeks. The question is: if PQ, RS, P_1Q_1 and R_1S_1 are four segments, which need not be commensurable, what does it mean to say that

$$\frac{\overline{PQ}}{\overline{RS}} = \frac{\overline{P_1Q_1}}{\overline{R_1S_1}} ?$$

The key observation is as follows. Let m and n be positive integers. Whether PQ and RS are commensurable or not, the statement $\dfrac{\overline{PQ}}{\overline{RS}} > \dfrac{m}{n}$ has a simple geometric meaning: a segment n times as long as PQ is longer than a segment m times as long as RS; in symbols: $n\overline{PQ} > m\overline{RS}$. Now Eudoxos calls two ratios of lengths equal if there are *no* integers m and n such that one ratio is greater than m/n and the other ratio is not.

The same definition applies to ratios of other geometric quantities: areas, volumes, and angles.

The reader will note how close Eudoxos' definition is to our use of real numbers. We "know" a real number x if we know for which rational numbers r the inequality $x > r$ is true; this is the information conveyed by the decimal expansion of x. Eudoxos "knew" the ratio of two geometric quantities α and β if he knew for which integers m and n the quantity $m\alpha$ exceeded $n\beta$.

8.2 Deductions from the field postulates

In §2.3, we stated that all familiar rules of algebra can be derived from the postulates **1** to **12**. The derivations are not difficult; one must only be careful to justify each step by referring to the postulates or some consequence of the postulates already established. We illustrate this by giving a few derivations.

▶ *Examples 1.* $-0 = 0$, $-(-a) = a$ for all a.

The first statement is true because -0 is the number x such that $0 + x = 0$, and we know that $0 + 0 = 0$ by **4**. The second statement means that $(-a) + a = 0$; this is so by **5** and **2**.

2. Subtraction. $a + x = b$ if and only if $x = b - a$.

Proof. If $a + x = b$, then $(a + x) + (-a) = b + (-a)$. But $b + (-a) = b - a$, by definition. Also, by **2, 3, 4**, and **5**, we have $(a + x) + (-a) = (x + a) + (-a) = x + (a + (a)) = x + 0 = 0 + x = x$. Hence $x = b - a$.

Assume next that $x = b - a = b + (-a)$. Then, using **2, 3, 4**, and **5**, we have $a + x = a + (b + (-a)) = a + ((-a) + b) = (a + (-a)) + b = 0 + b = b$.

3. Division. Let $a \neq 0$. Then $ax = b$ if and only if $x = ba^{-1}$ $(= b/a$, by definition).

Proof. If $ax = b$, then $(ax)^{-1} = ba^{-1}$. But by **7, 8, 10**, and **12**, we have $(ax)a^{-1} = (xa)a^{-1} = x(aa^{-1}) = x \cdot 1 = 1 \cdot x = x$. Hence $x = ba^{-1}$.

Assume next that $x = ba^{-1}$. Then $ax = a(ba^{-1}) = a(a^{-1}b) = (aa^{-1})b = 1 \cdot b = b$ where we used **7, 8, 10**, and **12**.

4. For all a, $0a = a$.

Proof. The distributive law **9** applied to the case $c = 0$ yields $a(b + 0) = ab + a0 = ab + 0a$, by **7**. But $a(b + 0) = a(0 + b) = ab$, by **2** and **4**. Hence $ab + 0a = ab$. Thus, by Example 2, $0a = ab - ab = ab + (-ab) = 0$ by **5**.

5. Cancellation law: If $ab = 0$, then either $a = 0$ or $b = 0$.

Proof. Assume that $a \neq 0$, $b \neq 0$. We must show that $ab \neq 0$. Set $c = b^{-1}a^{-1}$. Then, using Postulates **8, 10**, and **12**, we have $(ab)c = a(bc) = a(b(b^{-1}a^{-1})) = a((bb^{-1})a^{-1}) = a(1 \cdot a^{-1}) = aa^{-1} = 1 \neq 0$, by **11**. This shows that $ab \neq 0$. For, if ab were 0, we should have $(ab/c = 0)$ by Example 4.

6. $(-a)b = -(ab)$, $(-a)(-b) = ab$.

Proof. The first statement means that $ab + ((-a)b) = 0$. It is so, because by **9** and **7** we have $ab + ((-a)b) = (a + (-a))b = 0b = 0$, the last step being a consequence of Example 4. To prove the second statement, we use the result just established and Example 1. We have $(-a)(-b) = -(a(-b)) = -((-b)a) = -(-ba) = ba = ab$. ◀

EXERCISES

Verify the following statements, valid for all a, b, c, d, using the field postulates.

1. $(a + b) + (c + d) = (d + a) + (c + b)$.
▶ 2. $(a + b)(c + d) = (ac + bd) + (bc + ad)$.
3. If $a \neq 0$ and $a = aa$, then $a = 1$.
4. $(2 + 3)(a + 1 + b) = [(2 + 3) + (2 + 3)b] + (2 + 3)a$.
5. $(ab + ac)d = a(bd + cd)$.

6. $\dfrac{1}{1} = 1$.

7. $\dfrac{a}{1} = a$.

8. $\dfrac{1}{a^{-1}} = a$ for $a \neq 0$.

9. $-\dfrac{1}{a} = \dfrac{1}{-a}$ for $a \neq 0$.

10. $\dfrac{a}{-1} = -a$.

11. $\dfrac{-a}{-b} = \dfrac{a}{b}$ for $b \neq 0$.

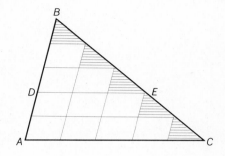
Fig. 1.18

▶ 12. $\dfrac{1}{ab} = \dfrac{1}{a}\dfrac{1}{b}$ for $a \neq 0$, $b \neq 0$.

16. $(a + b)^2 = a^2 + 2ab + b^2$.

17. $(a + b)(a - b) = a^2 - b^2$.

18. $(-a)^5 = -a^5$.

13. $\dfrac{ab}{ac} = \dfrac{b}{c}$ for $a \neq 0$, $c \neq 0$.

19. $(-a)^4 = a^4$.

20. $(a - b)^3 = a^3 - 3a^2b + 3ab^2 - b^3$.

14. $\dfrac{a + b}{c} = \dfrac{a}{c} + \dfrac{b}{c}$ for $c \neq 0$.

15. $\dfrac{a}{b} + \dfrac{c}{d} = \dfrac{ad + bc}{bd}$ for $b \neq 0$, $d \neq 0$.

8.3 Periodicity of rational decimals

We show now why the nonterminating decimal expansion of a positive rational number is always repeating, as stated in §4.4. If the number has a terminating decimal representation, then the nonterminating one ends with a string of 9s. We consider next a number, say 23/18, which cannot be written as a terminating decimal. Its decimal representation is obtained by long division, which begins as follows.

$$
\begin{array}{r}
1.27 \\
18\overline{)23.000000\,\cdots} \\
\underline{18} \\
50 \\
\underline{36} \\
140 \\
\underline{126} \\
14
\end{array}
$$

After three steps the remainder is $14 \cdot 10^{-2}$. The previous remainder was $14 \cdot 10^{-1}$, that is, the same apart from a power of ten. This is no accident. We could have foreseen, without carrying out any computations, that we must eventually get the same remainder again, apart from a power of ten. For, every remainder we can get must be of the form "r times a power of ten," where r is a positive integer less than 18, that is, one of the numbers 1, 2, $\cdots$, 17. After at most 17 steps we must get the same r twice.

The argument is general. If p and q are positive integers, $q > 1$, and we carry out the division $p \div q$, the remainders we get are of the form, "a number r times a power of ten," where r is one of the numbers 1, 2, $\cdots$, $q - 1$. Hence the division process either terminates or leads, after at most $q - 1$ steps, to the same value of r.

We return to our example 23/18. We stopped at the remainder $14 \cdot 10^{-2}$. We have already encountered the remainder $14 \cdot 10^{-1}$, obtaining the digit 7. Hence, if we continue the division, we shall again obtain the digit 7. Earlier, this digit led to the remainder $14 \cdot 10^{-2}$. The next remainder will therefore be $14 \cdot 10^{-3}$, the next digit again 7, and so on. Thus

$$\frac{23}{18} = 1.2777 \cdots \text{(all 7s)} = 1.2\overline{7}.$$

Once more the argument is general. Repeated remainders lead to repeated digits or repeated sequences of digits. Therefore, *every positive rational number can be represented, uniquely, by a nonterminating repeating decimal.* ("Repeating" means that repetition starts either immediately after the decimal point or after finitely many places to the right of the decimal point.)

The converse statement is also true, as we said in §4.4: *Every nonterminating repeating decimal represents a rational number.* We verify this statement, and the rule for converting a repeating decimal into a common fraction, by an example. It will be obvious that our method would work in any given case.

Assume first that $a = 0$ and that repetition starts immediately after the decimal point. In this case the rule reads:

$$.\overline{\alpha_1 \alpha_2 \cdots \alpha_s} = \frac{\alpha_1 \alpha_2 \cdots \alpha_s}{9\,9 \cdots 9}$$

with s consecutive 9s in the denominator. We have, for instance,

$$(1) \qquad\qquad\qquad .\overline{7} = \frac{7}{9}.$$

To verify this we compute, using the formula in §2.5 for the geometric progression, the number $.77 \cdots 7$ with n 7s behind the decimal point. It equals

$$7 \cdot 10^{-1} + 7 \cdot 10^{-2} + \cdots + 7 \cdot 10^{-n} = (.7)[1 + (.1) + (.1)^2 + \cdots + (.1)^{n-1}]$$

$$= \frac{7}{10} \frac{1 - (.1)^n}{1 - (.1)} = \frac{7}{9} - \frac{7}{9} 10^{-n},$$

so that
$$\overbrace{.77 \cdots 7}^{n} < \frac{7}{9} < \overbrace{.77 \cdots 7}^{n} + 10^{-n}.$$

This means that the first n digits in the decimal expansion of $7/9$ are all 7s. And this holds for all n. Hence (1) is true.

Another example is $.\overline{312} = 312/999 = 104/333$. For we have

$$\overbrace{.312312 \cdots 312}^{3n \text{ terms}} = 312 \cdot 1000^{-1} + 312 - 1000^{-2} + \cdots + 312 \cdot 1000^{-n}$$

$$= .312[1 + (.001) + (.001)^2 + \cdots + (.001)^{n-1}]$$

$$= \frac{312}{1000} \frac{1 - (.001)^n}{1 - (.001)} = \frac{312}{999} - \frac{312}{999} 10^{-3n}.$$

Thus we have, for every n,

$$\overbrace{.312\ 312 \cdots 312}^{3n} < \frac{312}{999} < \overbrace{.312\ 312 \cdots 312}^{3n} + 10^{-3n},$$

so that the first $3n$ digits in the decimal expansion of $312/999 = 104/333$ are $.312\ 312 \cdots 312$.

The case where $a \neq 0$ or the repetition does not start immediately after the decimal point is now easy. We have, for instance,

$$2.00\overline{7} = 2 + .00\overline{7} = 2 + \frac{1}{100}\,(.\overline{7}) = 2 + \frac{1}{100}\,\frac{7}{9} = \frac{1807}{900},$$

as the reader may verify.

EXERCISES

In Exercises 21 to 26, represent the given numbers both as terminating and as nonterminating decimals.

21. 3.   23. 1/1000.   25. 8/50.
22. 11/2.   24. 33/5.   26. 11/80.

In Exercises 27 to 36, represent the given numbers as decimals.

27. 30/11.   31. 2/7.   35. 25/26.
28. 7/60.   32. 299/54.   36. 30/31.
29. 3/22.   33. 45/74.
30. 223/111.   34. 99/101.

In Exercises 37 to 45, represent the given numbers as common fractions.

37. $.\overline{4}$.   40. $.\overline{41}$.   43. $.0\overline{198}$.
▶ 38. $.8\overline{1}$.   41. $1.1\overline{25}$.   44. $1.\overline{0011}$.
39. $3.\overline{9}$.   42. $.08\overline{1}$.   45. $1.00\overline{11}$.

8.4 Density

Now we prove Theorem 2 in §5.6 which asserts that *every interval contains infinitely many rational numbers and infinitely many irrational numbers.* One also expresses this property of numbers by saying that the rational numbers are dense in the number line, and so are the irrational numbers.

Every interval contains an open finite interval, say (a, b). Let x denote its midpoint, $x = (a + b)/2$. Let m be a positive integer such that $10^{-m} < b - x$; there is such an integer, by Corollary 2 to 17 (see §3.4). If y is a number with $|x - y| < 10^{-m}$, then y lies in (a, b). Using this remark, we shall find in (a, b) a rational number c and an irrational number d. More precisely, let c be the number obtained by replacing all digits in the decimal expansion of x, after the mth, by 0. Then c is rational. Also, let d be the number obtained by replacing the digits mentioned above by the nonrepeating sequence $101101110111101 \cdots$. Then d is irrational. (There is another way of defining d: if x is irrational, set $d = x$; if x is rational, set $d = x + 10^{-m-1}\sqrt{2}$.)

The process can be continued. We can find a rational number c_1 between a and c, another rational number c_2 between a and c_1, and so forth. Also, we can find an irrational number d_1 between a and d, another such number between a and d_1, and so forth. This proves the assertion.

8.5 Sums and products of real numbers

In §5.2, we stated that one *can* define addition and multiplication of real numbers so as to satisfy the postulates **1** to **18**. For the sake of completeness, we indicate very briefly *how* this can be done.

We assume that the rational numbers are known, and that the real numbers and the relation $<$ between real numbers are defined as in §5.1. The postulates **13**, **14**, and **17** follow easily. Also, and this is most important, we can prove **18** (see §6.4). We use the least upper bound principle **18** in defining addition and multiplication.

Let α and β be positive real numbers. Let A be the set of all rational numbers a with $0 < a < \alpha$, B the set of all rational numbers b with $0 < b < \beta$, S the set of all rational numbers $a + b$ with a in A and b in B, and P the set of all rational numbers ab with a in A and b in B. These sets are bounded. Now one proves that if α and β are rational, then the least upper bound of S is $\alpha + \beta$ and that of P is $\alpha\beta$. If α and β are not both rational, then one *defines* the sum $\alpha + \beta$ and the product $\alpha\beta$ as the least upper bounds of S and P, respectively.

Next, one proves that $1 \cdot \alpha = \alpha$ for all α, where 1 is the real number $.\overline{9}$, that for $0 < \alpha < \beta$ there is a unique real number x with $\alpha + x = \beta$, and that for $0 < \alpha$, $0 < \beta$ there is a unique real number y with $\alpha y = \beta$. (One obtains x as the least upper bound of the set of all positive rational c such that $a + c < \beta$ for all a in A, and one obtains y as the least upper bound of the set of all positive rational d with $ad < \beta$ for all a in A.) It is now not too difficult to define the sum and product of any two real numbers and to verify all remaining postulates.

8.6 A different viewpoint

Readers interested in the foundations of mathematics should be told that some mathematicians (called *intuitionists* or *constructivists*) reject the theory of real numbers summarized in Postulates **1** to **18**. They contend that a statement about a number x is meaningful only if it contains an actual recipe for calculating this number. The proof for the existence of roots (see §5.4) meets this condition. But the proof for the existence of a least upper bound for any bounded set, given in §6.4, does not. For it contains no prescription for deciding, at each stage, whether a rational number A is or is not an upper bound for the set S. But, says a "classical" mathematician, one thing is obvious: either A is a bound, or it is not! This time-honored "principle of the excluded middle," however, is not accepted by the intuitionists.

To develop calculus in a way satisfactory to constructivists is possible, but more laborious. The constructivist school of thought numbered among its adherents some of the greatest mathematicians, for instance, Kronecker, Poincaré, Brouwer, and Weyl.

LEOPOLD KRONECKER (1823–1891), a distinguished algebraist, summarized his attitude toward the foundations of mathematics in the epigram: "The good Lord made positive integers, everything else is the handiwork of men."

HENRI POINCARÉ (1857–1912). For several decades, Poincaré's restless intellect dominated mathematics and mathematical physics. He created, or transformed, several branches of mathematics, and he arrived at the equations of special relativity, independently of Einstein. Poincaré also wrote several popular books about science that earned him a seat in the literary Académie Francaise.

L. E. J. BROUWER (1881–1967) did some of the fundamental work in topology, a modern branch of geometry. Then the Dutch mathematician turned to the foundations of mathematics and became the leader of the "intuitionists."

HERMANN WEYL (1885–1955) was one of the few truly universal mathematicians of our time; his interests ranged from philosophy to physics and comprised all of mathematics. One of his many famous books (*Symmetry*, Princeton, 1952) can even be enjoyed by the general reader.

Weyl left his native Germany after Hitler came to power and accepted a chair at the Institute for Advanced Study in Princeton, N. J.

Problems

1. An integer $p > 1$ which is not divisible by a positive integer other than 1 or p itself is called a **prime.** Thus 2, 3, 5, 7, 11, 13, and so forth, are primes. The two following theorems are proved in every book on elementary number theory: there are *infinitely many primes,* and every integer $n > 1$ can be written as a product of primes *in one way only,* except for the order of the factors. The latter statement is called the **fundamental theorem of arithmetic.** (Example: $12 = 2 \cdot 2 \cdot 3 = 2 \cdot 3 \cdot 2 = 3 \cdot 2 \cdot 2$; there is no other way of writing 12 as a product of primes.)

    Using the fundamental theorem of arithmetic, show that, if k and n are positive integers and k is not the nth power of another integer, then $\sqrt[n]{k}$ is irrational.

2. Let a, b, c, d be rational numbers. Show that if $a + b\sqrt{2} = c + d\sqrt{2}$, then $a = c$ and $b = d$.

3. Show that, for any two rational numbers a and b, not both 0, one can find another pair of rational numbers, c and d, such that $(a + b\sqrt{2})(c + d\sqrt{2}) = 1$.

4. The set of all numbers of the form $a + b\sqrt{2}$, where a and b are rational numbers, forms a field—indeed, an ordered Archimedean field. Prove this by verifying Postulates **1 to 17.**

5. Does the statement in Problem 4 remain true if $\sqrt{2}$ is replaced by $\sqrt{3}$? By $\sqrt{5}$? By $\sqrt{4}$? By $\sqrt{6}$?

6. Using the results of Problems 1 to 5, prove that given any integer r, which is not a square of another integer, the set of all real numbers of the form $a + b\sqrt{r}$, a and b rational, is an ordered Archimedean field. [This field is usually denoted by $\mathbf{Q}(\sqrt{r})$.]

7. Let $n > 1$ be an integer. Consider the set $\mathbf{Z}_n$ consisting of the integers 0, 1, 2, $\cdots$, $n - 1$. Define the sum of the two elements in $\mathbf{Z}_n$ to be the remainder obtained by dividing the "ordinary" sum by n; define the product of two elements in the same way. (Example: if $n = 3$, then the "new" sum of 2 and 1 is 0, for $2 + 1 = 3$ is divisible by 3, the "new" product of 2 and 1 is 2, the "new" sum of 2 and 2 is 1 for $2 + 2 = 4 = 3 \cdot 1 + 1$, and so forth.)

    Prepare multiplication and addition tables for $\mathbf{Z}_2$, $\mathbf{Z}_3$, and $\mathbf{Z}_4$. Check which of the field Postulates **1 to 12** hold in these sets.

8. Show that Postulates **1 to 11** hold in $\mathbf{Z}_n$ for every n.

9. Show that $\mathbf{Z}_5$ is a field and $\mathbf{Z}_6$ is not.

10. Show that $\mathbf{Z}_p$ is a field if p is a prime.

11. Show that $\mathbf{Z}_n$ is not a field if n is not a prime; more specifically, show that the cancellation law does not hold in $\mathbf{Z}_n$.

12. None of the fields $\mathbf{Z}_p$, p a prime, can be ordered. Prove this. (This statement means that it is not possible to introduce a relation "$<$" between the elements of $\mathbf{Z}_p$ so as to satisfy Postulates **1 to 16.**)

13. There exists a field with exactly 4 elements. Call the elements 0, 1, α, and β. Find the addition and multiplication tables for this field.

14. Is the number .1234567891011121314151617181920212223 24 ⋯ rational? Give a reason for your answer.

15. Give an example of a number x such that adding 10^{-10000} to it changes all digits in the decimal expansion of x, from the decimal point to the 10,000th digit.

16. Does there exist an irrational number whose decimal expansion uses only 1 digit? Only 2 digits? (Example: the decimal expansion $1/4 = .24999 \cdots = .24\overline{9}$ uses only the 3 digits 2, 4, and 9.)

17. Does every interval on the number line contain a rational number that can be written as a terminating decimal fraction?

18. Does every interval on the number line contain a number whose decimal expansion can be written using only the digit 3?

19. Suppose we were using not base ten but some other base of a positional number system, say base k. Then we could again represent numbers by non-terminating "k-ary" fractions. Would it still be true that rational numbers, and only rational numbers, would be represented by repeating fractions?

20. An interval I_1 with endpoints a and b is said to contain the interval I_2 with endpoints c and d if $a \leq c < d \leq b$. A sequence of intervals, I_1, I_2, I_3, $\cdots$, is said to be **nested** if every interval I_j contains the next I_{j+1}. The **principle of nested intervals** asserts that, for every such sequence of *closed* nested intervals, there is a point (that is, a real number) that lies in all intervals of the sequence. Prove the principle of nested intervals, using the least upper bound principle. (It is also possible, though harder, to derive the least upper bound principle from the principle of nested intervals.)

21. Show by an example that the statement of the principle of nested intervals need not be true for *open* intervals.

22. Construct a nested sequence of closed intervals with rational endpoints such that no rational point lies in all intervals of the sequence.

23. Suppose that all real numbers are divided into two classes, the left class L and the right class R, such that (1) there are numbers in both classes, (2) every number in L is less than every number in R. Assuming the least upper bound principle, prove **Dedekind's principle:** either L has a largest element, or R has a smallest element. (It is also possible, though harder, to derive the least upper bound principle from Dedekind's principle.)

24. Show that it is possible to divide all rational numbers into two classes L and R, having the same properties as above, so that L has no largest and R no smallest element. In other words, give an example to show that Dedekind's principle does not hold in the field of rational numbers.

2/Coordinates

This is another introductory chapter. It contains the elements of analytic geometry and may be omitted or gone over very rapidly by a reader familiar with this material. The development of analytic geometry is continued in the chapters on vectors (Chapter 9) and on quadrics (Chapter 10). These chapters may be read immediately after Chapter 2.

Two appendixes are included: one presents one of the first calculus theorems in history, Archimedes' calculation of the area of a parabolic sector; the other deals with the two ways of laying the foundations of geometry, the axiomatic method and the analytic method.

While the concept of real numbers developed gradually and was first fully understood only long after calculus had become a flourishing mathematical discipline, the second prerequisite of calculus, analytic geometry, was invented at a definite time, and independently, by two men: Descartes, a philosopher, and Fermat, a judge. Both are among the best mathematicians of all time.

The basic idea of analytic geometry is simple: one can specify the position of a point in the plane by two numbers and thus translate any statement about points into a statement about numbers.

1.1 Cartesian coordinate systems

We draw two perpendicular lines in the plane and call one the **horizontal axis,** the other the **vertical axis.** The intersection point O of the axes is called the **origin.** We give each of the axes a direction; the direction of the horizontal axis is called "right" and that of the vertical is called "up"; by left we mean the direction opposite to right, by down the direction opposite to up. We also choose a **unit of length.** (The customary way of drawing the axes is shown in Fig. 2.1.) The two axes and the unit of length are called a **Cartesian coordinate system** in the plane. (Cartesius is the Latinized form of the name Descartes.)

Let P be a point in the plane; we associate with it two numbers, called its **coordinates.** The first coordinate is zero if P lies on the vertical axis; the distance from P to the vertical axis if P is to the right of the vertical axis; and the negative of this distance if P is to the left of the vertical axis. The second coordinate of P is zero if P lies on the horizontal axis, and the distance from P to the horizontal axis (or the negative of that distance) if P lies above the horizontal axis (or below it). Some points and their coordinates are shown in Fig. 2.2.

We often denote the first coordinate of a point by the letter x, and the second by y. Accordingly, we call the horizontal axis the x **axis,** and the vertical axis, the y **axis.** Note that each of the axes is a number line in the sense of Chapter 1, §5.3.

Here is another way of interpreting the coordinates. The point (a, b) is reached if we first go a units in the x direction and then b units in the y direction. To go a units in the x direction means: not to move at all if $a = 0$, to move the distance $|a|$ in the direction of the x axis if $a > 0$, to move the distance $|a|$ in the opposite direction if $a < 0$. An analogous explanation holds for the y direction.

A point P is determined by its coordinates, and for any two numbers

§1 Coordinates of a Point. Distance Formula

Fig. 2.1

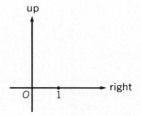

RENÉ DESCARTES (1596–1650) is the father of modern Western thought. He based his philosophy on independent reasoning and research, rather than on reliance on authorities. His philosophy also contains an attempt to explain all of natural phenomena, including plant life and animal life, in strictly mechanical terms. Descartes' only mathematical book, *Geometry*, appeared in 1635.

As a young man, Descartes traveled extensively and was a soldier. Then he settled in Holland, which he left only shortly before his death to go to Stockholm as tutor to Queen Christina.

PIERRE DE FERMAT (1601–1665) was a Counsellor at the Parlement of Toulouse, a judicial position that gave him time for mathematics. He was the cofounder of analytic geometry (with Descartes) and of the theory of probability (with Pascal), one of the forerunners of calculus, and a great number theorist. Fermat never published his discoveries.

One of the best-known unsolved mathematical problems concerns the proof, or disproof, of "Fermat's last theorem": if $n \geqslant 3$ is an integer and x and y are rational numbers, then $x^n + y^n \neq 1$. "I have found a truly remarkable proof," wrote Fermat on the margin of a book, "but have no room to reproduce it here." (Recent work, utilizing electronic computers, shows that Fermat's theorem is true for all $n < 100,000$.)

Fig. 2.2

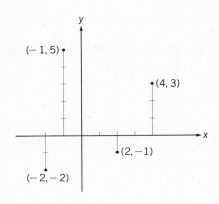

Fig. 2.3

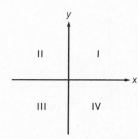

a and b there is a point P having a as the first coordinate and b as the second. We can use the ordered pair of coordinates (a, b) as a name for the point P. We do so in the sequel.

When we say that the coordinates of a point form an **ordered pair,** we mean that the order of the two numbers in the pair is important. Thus $(2, 3)$ and $(3, 2)$ are two distinct ordered pairs. They are names of two distinct points.

All points to the right of the y axis have a positive first coordinate. The set of all these points is called "the set $x > 0$," or the **right half-plane.** Similarly, the set $x < 0$ is the **left half-plane** (all points to the left of the y axis), the set $y > 0$ is the **upper half-plane** (all points above $y = 0$), and the set $y < 0$ is the **lower half-plane** (all points below $y = 0$).

The intersection of the upper and right half-planes is the set of points with positive first and positive second coordinates. It is called the **first quadrant.** The second, third, and fourth quadrants, respectively, are the sets: $x < 0$ and $y > 0$; $x < 0$ and $y < 0$; and $x > 0$ and $y < 0$. They are shown in Fig. 2.3.

EXERCISES

Draw a coordinate system and mark the points with the following coordinates.

1. $(2, 3)$.   3. $(0, 4)$.   5. $(-2, 3)$.   7. $(2, -3)$.
2. $(-\frac{1}{2}, 1)$.   4. $(-1, 1)$.   6. $(3, -2)$.   8. $(-\frac{3}{2}, 0)$.

In doing the following exercises, make a drawing for each (preferably on graph paper).

9. Find the coordinates of all points in the third quadrant that are at a distance 5 from the origin and a distance 3 from the x axis.
10. Find the coordinates of all points in the left half-plane that are at a distance 3 from the x axis and a distance 2 from the y axis.
11. Find the coordinates of all points in the upper half-plane that are at a distance 7 from the x axis and a distance 4 from the y axis.
12. Find the coordinates of all points in the second quadrant that are at a distance 5 from the origin and a distance 5 from the line $y = -1$.
13. Find the coordinates of all points in the right half-plane that are at a distance 2 from the line $y = 3$ and a distance 4 from the line $x = 5$.
14. Find the coordinates of all points that are at a distance 13 from the point $(1, 0)$ and a distance 5 from the x axis.
15. Find the coordinates of all points in the fourth quadrant that are at a distance 8 from the line $y = \frac{1}{2}$ and a distance $\frac{1}{2}$ from the line $x = 8$.
▶ 16. Find the coordinates of all points on the line $x = 10$ that are at a distance 5 from the point $(7, -1)$.

1.2 Translating coordinates

By choosing a Cartesian coordinate system, one establishes a one-to-one correspondence between points in the plane and ordered pairs of real numbers. This means: to each point there corresponds a unique pair of numbers, and to every ordered pair of numbers there corresponds a unique point.

But the choice of the coordinate system is arbitrary. The same point will, of course, have different coordinates in different coordinate systems. We consider here only the simplest case of this situation.

We choose first one coordinate system and then another "new" system, with a different origin, but with the axes parallel to the old axes and having the same directions. One says in this case that the new coordinate system is obtained from the old one by a **translation** (see Fig. 2.4).

Let a point P have coordinates (x, y) in the old system and coordinates (X, Y) in the new system, and let (a, b) be the coordinates of the new origin in the old system. Then

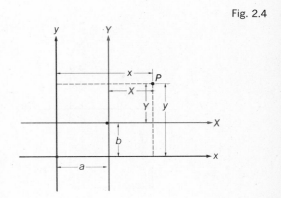

Fig. 2.4

$$(1) \qquad x = a + X \qquad \text{and} \qquad y = b + Y.$$

This can be read off Fig. 2.4 for the case when the new origin lies in the first quadrant and P in the (new) first quadrant. The reader can convince himself that the formulas also hold in all other cases. One of these other cases is shown in Fig. 2.5.

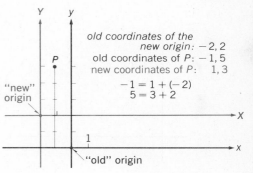

Fig. 2.5

old coordinates of the
new origin: $-2, 2$
old coordinates of P: $-1, 5$
new coordinates of P: $\ 1, 3$

$$-1 = 1 + (-2)$$
$$5 = 3 + 2$$

EXERCISES

In each of the following exercises, the new coordinate system is obtained from the old one by a translation. Illustrate each exercise by a sketch.

17. If the coordinates of the new origin in the old system are $(0, 3)$ and the coordinates of a point P in the old system are $(-1, 4)$, what are the coordinates of P in the new system?

18. If the coordinates of the new origin in the old system are $(-1, 5)$ and the coordinates of a point P in the old system are $(0, 8)$, what are the coordinates of P in the new system?

19. If the coordinates of a point P in the old system are $(1, -2)$ and in the new system are $(3, 1)$, what are the coordinates of the new origin in the old system?

►20. If the coordinates of the old origin in the new system are $(-3, 2)$, what are the coordinates of the new origin in the old system?

21. If the coordinates of the new origin in the old system are $(-1, -4)$ and the coordinates of a point P in the new system are $(-6, 2)$, what are the coordinates of P in the old system?

22. If the coordinates of the new origin in the old system are $(3, -2)$ and the coordinates of a point P in the new system are $(\frac{1}{2}, -1)$, what are the coordinates of P in the old system?

Fig. 2.6

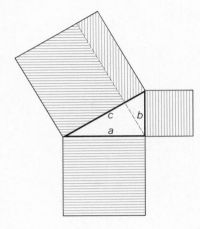

Fig. 2.7

1.3 Distance formula

We recall now the famous *Pythagorean Theorem*. In a right triangle the two squares erected on the legs can be cut up and reassembled so as to form the square erected on the hypotenuse (see Fig. 2.6). This is, by the way, one of the oldest results in mathematics. The discovery was made by the Babylonians; the general statement and proof are due to the Greeks.

For the Greeks, the Pythagorean Theorem was a statement about areas. Having real numbers, we of today can restate it as a proposition about numbers. If a and b are the lengths of two legs of a right triangle and c is the length of the hypotenuse, then $a^2 + b^2 = c^2$.

We use the Pythagorean Theorem to establish an important formula.

Theorem 1   THE DISTANCE d BETWEEN THE POINTS (x_1, y_1) AND (x_2, y_2) IS

$$(2) \qquad d = \sqrt{(x_1 - x_2)^2 + (y_1 - y_2)^2}.$$

To prove this distance formula, we consider first a special case: one point is the origin $(0, 0)$. The other point may then be denoted by (x, y), and we must show that the distance from $(0, 0)$ to $(x, y) = \sqrt{x^2 + y^2}$. This is the same as

$$(\text{distance from } (0, 0) \text{ to } (x, y))^2 = |x|^2 + |y|^2.$$

This is certainly so if $(x, y) = (0, 0)$, or if either $x = 0$ or $y = 0$. In all other cases, our assertion is the Pythagorean Theorem; see Fig. 2.7.

Now we consider the general case. We translate the coordinate system so that the new origin is the point (x_1, y_1). Then the new coordinates of the second point are $(x_2 - x_1, y_2 - y_1)$, as is seen by using (1) with $(a, b) = (x_1, y_1)$ and $(x, y) = (x_2, y_2)$. By the result just proved, the distance between the two points is

$$\sqrt{(x_2 - x_1)^2 + (y_2 - y_1)^2},$$

as asserted.

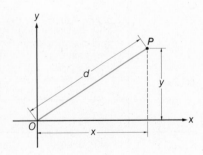

1.4 Coordinates in space

From the outset, we shall have little occasion to use solid geometry, but we shall describe briefly how a Cartesian coordinate system is set up in space.

We again select a unit of length and a point in space called the origin. Through this point we draw three mutually perpendicular lines and give each of them a direction. The lines are called the x **axis**, the y **axis**, and the z **axis**, respectively, and they are usually drawn as in Fig. 2.8. Each pair of axes determines a **coordinate plane** containing them.

The position of a point P in space is completely described by the ordered triplet (a, b, c), its coordinates. P is the point reached by going from the origin a units in the x direction, b units in the y direction, and c units in the z direction (see Fig. 2.8).

The set $x = a$, that is, the set of all points with x coordinate (first coordinate) equal to a, is a plane parallel to the yz plane and thus perpendicular to the x axis. It coincides with the yz plane if $a = 0$; otherwise it has distance $|a|$ from that plane. The plane $x = a$ lies on the side of the yz plane toward which the x axis is pointing if $a > 0$, on the opposite side if $a < 0$. Similarly, the sets $y = b$ (points with y coordinate b) and $z = c$ (points with z coordinate c) are planes perpendicular to the y axis and z axis, respectively. The point (a, b, c) is the intersection of the three planes $x = a$, $y = b$, and $z = c$.

The three coordinate planes divide the space into 8 **octants**. The one containing points with all three coordinates positive is called the first octant.

The distance d of a point P with coordinates (a, b, c) from the origin is

$$(3) \qquad d = \sqrt{a^2 + b^2 + c^2}.$$

The proof, for the case when P lies in the first octant, is indicated in Fig. 2.9. Let s denote the distance from O to P_z; P_z is the point in the xy plane with coordinates (a, b). By the distance formula for the plane,

$$s^2 = a^2 + b^2.$$

Since OP is the hypotenuse of a right triangle with legs OP_z and P_zP and the length of P_zP is c, the Pythagorean Theorem gives

$$d^2 = s^2 + c^2.$$

Hence $d^2 = a^2 + b^2 + c^2$ and (3) follows.

From (3) one obtains, as in the case of the plane, the formula

$$(4) \qquad d = \sqrt{(a_1 - a_2)^2 + (b_1 - b_2)^2 + (c_1 - c_2)^2}$$

for the distance between two points in space, (a_1, b_1, c_1) and (a_2, b_2, c_2).

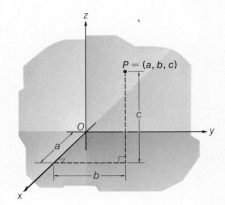

Fig. 2.8

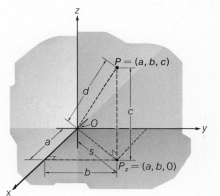

Fig. 2.9

In the following exercises, find the distance between the points with the given coordinates.

23. $(0, 1)$ and $(1, 0)$.
24. $(-1, -2)$ and $(3, 1)$.
25. $(-2, 0)$ and $(-1, 1)$.
26. $(0, 1, 1)$ and $(-2, 1, 4)$.

27. $(1, -6, 0)$ and $(2, 3, 1)$.
28. $(2, 1/2)$ and $(3, -1/2)$.
29. $(3, -1)$ and $(-1, 3)$.
▶ 30. $(1, 3, 2)$ and $(2, 1, 3)$.

§2 Straight Lines

2.1 Linear equations

We assume in this and in the following sections that a Cartesian coordinate system has been chosen. The translation of geometric statements into algebraic formulas, that is, into statements about numbers, is based on the distance formula and on another result, which we now state.

Theorem 1   LET A, B, C BE NUMBERS, SUCH THAT A AND B ARE NOT BOTH ZERO. THEN THE SET OF ALL POINTS WHOSE COORDINATES SATISFY THE EQUATION

$$(1) \qquad\qquad Ax + By + C = 0$$

IS A (STRAIGHT) LINE. EVERY LINE IS THE SET OF ALL POINTS SATISFYING AN EQUATION OF THIS FORM.

Proof.   We note first that a vertical line, that is, either the y axis or a line parallel to it, is the set of all points (x, y) satisfying the equation $x = a$, for some fixed number. (The number a is 0 if the line in question is the y axis. Otherwise, $|a|$ is the distance from the y axis to the line, and a is positive or negative, according to whether the line is to the right or to the left of the y axis.) We observe that the equation $x = a$ is of the form (1), with $A = 1$, $B = 0$, and $C = -a$. Also, a horizontal line is the set of all points (x, y) with $y = b$, b being a fixed number; this equation is also of the form (1), with $A = 0$, $B = 1$, and $C = -b$.

Let l be a line passing through the origin that does not coincide with either axis. Let P be the intersection point of l with the line $x = 1$; this point has coordinates $(1, m)$ with $m > 0$ or $m < 0$, according to whether l passes through the first or the fourth quadrant (see Fig. 2.10, where $m > 0$). Let Q be any point on l distinct from O and P; we denote its coordinates by (x, y) and note that if $m > 0$ then x and y have the same sign, and if $m < 0$ then x and y have opposite signs. Let R be the point $(x, 0)$. Then the triangles $O1P$ and ORQ are similar and therefore

Fig. 2.10

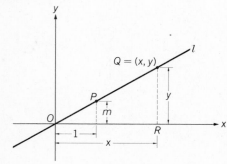

$$\frac{|m|}{1} = \frac{|y|}{|x|} \qquad \text{or} \qquad |y| = |m||x|.$$

Recalling what we said about the signs of x and y, we conclude that $y = mx$ whether $m > 0$ or $m < 0$.

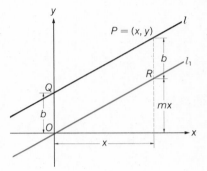

Fig. 2.11

Further, all points (x, y) with $y = mx$ lie on l. For if a point (x, y) does not lie on l, it lies on another line, call it l_1, which passes through the origin. This other line intersects the vertical line $x = 1$ at a point $(1, m_1)$ with $m_1 \neq m$. By the same reasoning as before, we have $y = m_1 x$ and therefore $y \neq mx$. We conclude that the line l is the set of *all* points (x, y) with $y = mx$. This equation is of the form (1) with $A = m$, $B = -1$, and $C = 0$.

Finally, let l be a line that is neither vertical nor horizontal and does not pass through the origin. Then l intersects the y axis at a point Q with coordinates $(0, b)$, $b \neq 0$ (see Fig. 2.11, where $b > 0$). Draw the line l_1 through the origin, parallel to l. Note that l lies above l_1 if $b > 0$ and below l_1 if $b < 0$. We just saw that there is a number m such that l_1 is the set of all points (x, y) with $y = mx$. Let P be some point on l distinct from Q, and R a point on l_1 such that the line PR is vertical. Then $ORPQ$ is a parallelogram and the segments OQ and RP are congruent. The points P and R have the same first coordinate, and the second coordinate of P = (second coordinate of R) + b. If P and R have first coordinate x, then the second coordinate of R is mx, and that of P is $mx + b$. We conclude: l is the set of points (x, y) such that $y = mx + b$. This equation is of the form (1) with $A = m$, $B = -1$, $C = b$.

So far we have shown that the points (x, y) on any straight line satisfy an equation of the form (1).

Now we begin with a given equation of the form (1), A and B not both zero. If $B = 0$, then $A \neq 0$ and the equation says that $x = -(C/A)$; the set of points (x, y) satisfying this equation (the "solution set") is a vertical line. If $B \neq 0$, the equation says: $y = -(A/B)x - (C/B)$ or, setting $-(A/B) = m$, $-(C/B) = b$,

$$(2) \qquad\qquad y = mx + b.$$

If $m = 0$, the solution set is a horizontal line. If $m \neq 0$ and $b = 0$, then, by what we saw before, the solution set is the line passing through O and the point $(1, m)$. If $b \neq 0$, then, by what we saw before, the solution set is the line passing through the point $(0, b)$ and parallel to the line l_1 which passes through O and $(1, m)$.

Theorem 1 is now completely proved. The argument shows that, given a line l, we can write down equation (1) satisfied by all its points. The equation is determined uniquely except, of course, that it may be multiplied by some number different from zero.

An equation of the form (1) is called **linear.** Our theorem explains the choice of this name.

▶*Examples*   *1.* Draw the line whose equation reads $2x + 3y - 1 = 0$.

Solution.   In order to draw the line, it suffices to know two distinct points on the line. We ask: is there a point on the line with $x = 0$? For $x = 0$, the equation gives $3y - 1 = 0$ or $y = \frac{1}{3}$. Thus the point $(0, \frac{1}{3})$ is on our line. Similarly, for $x = 5$ the equation reads $10 + 3y - 1 = 0$ or $y = -3$. Thus the point $(5, -3)$ lies on one line. It is now easy to draw it.

   2. Does the point $(-3, 2)$ lie on the line $5x + 4y - 7 = 0$?

Answer.   No, since $5(-3) + 4 \cdot 2 - 7 = -14 \neq 0$.

   3. Find the intersection point of the lines $2x - 3y = 0$ and $3x + 2y + 13 = 0$.

Answer.   A point (x, y) lies on both lines if its coordinates satisfy both equations. Solving the two equations simultaneously, we obtain $x = -3, y = -2$. The desired intersection point is therefore $(-3, -2)$. The reader is asked to verify this by drawing the two lines.◀

EXERCISES

1. Draw the line whose equation is $5x - 2y + 1 = 0$. Draw the line whose equation is $5x - 2y + 2 = 0$. Describe the collection of lines $5x - 2y + c = 0$, where c may be any real number.
2. Draw the line $y = (1/2)x + 1$. On the same set of axes, draw the line $y = -2x + 1$. Where do these lines intersect? At what angle do they meet?
3. Draw the line $y = -(1/3)x - 2$. On the same set of axes, draw the line $y = 3x - 2$. Where do these lines intersect? At what angle do they meet?
4. Show that the equation $y - 2 = 3(x - 1)$ is the equation of a line. Is the point $(1, 2)$ on this line?
5. Show that the equation $x/3 + y/2 = 1$ is the equation of a line. Draw the line. Where does it intersect the x axis? The y axis?
6. Find the intersection point of the lines $4x - 3y - 2 = 0$ and $x - 5y + 10 = 0$.
7. Show, without drawing, that the lines $114x - 2y - 26 = 0$ and $y = 57x + 57$ have no point in common.
8. Since the x axis and the y axis are lines, they have equations of the form (1). Find them.

2.2 Slope

A nonvertical line is the set of all points (x, y) satisfying an equation of the form $y = mx + b$, as we have seen above. [Compare equation (2).]

The number m is called the **slope** of the line. This number is uniquely determined by the line and the coordinate system. If the line is given by an equation $Ax + By + C = 0$, the slope is given by

$$(3) \qquad\qquad m = -\frac{A}{B}$$

($B \neq 0$, because the line is not vertical). This is seen by "solving" the equation $Ax + By + C = 0$ for y.

Suppose the line, given by $y = mx + b$, contains the points (x_1, y_1) and (x_2, y_2). Then

$$y_1 = mx_1 + b, \qquad y_2 = mx_2 + b.$$

Subtracting one equation from the other, we obtain

$$(4) \qquad\qquad y_2 - y_1 = m(x_2 - x_1).$$

Since we deal with a nonvertical line, $x_1 \neq x_2$. Hence

$$(5) \qquad\qquad m = \frac{y_2 - y_1}{x_2 - x_1}.$$

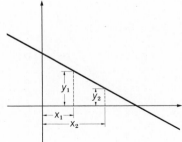

Fig. 2.12

These formulas show the geometric meaning of the slope. If $m = 0$, the line is horizontal; any two points on it have the same y coordinate. If $m > 0$, the line rises as we move from left to right, and the y coordinate of a point on the line increases m times as fast as the x coordinate. This is seen from (4). Thus small positive slope means a gentle rise, large positive slope, a steep rise. If $m < 0$, the line is descending as we move on it from left to right; the y coordinate decreases m times as fast as the x coordinate increases (see Fig. 2.12).

We do not define the slope of a vertical line.

EXERCISES

9. If the point Q is 3 units above and 2 units to the right of the point P, what is the slope of the line running through P and Q?

10. If the point Q is 6 units above and 1 unit to the left of the point P, what is the slope of the line running through P and Q?

11. If the point P is 2 units below and 10 units to the right of the point Q, what is the slope of the line running through P and Q?

12. The equation of line l_1 is $3y + 2x = 6$ and that of line l_2 is $5x - 2 = 2y$. Which line descends as we move from left to right?

13. The equation of line l_1 is $6y - 4x - 2 = 0$, that of line l_2 is $2y - 40x + 7 = 0$, and that of l_3 is $18y - 17x + 51 = 0$. Which line rises most steeply? Which line rises most gently?

2.3 Parallel and perpendicular lines

Theorem 2 TWO DISTINCT LINES ARE PARALLEL IF AND ONLY IF BOTH ARE
EITHER VERTICAL OR HAVE THE SAME SLOPE.

The proof is contained, implicitly, in the proof of Theorem 1. We saw
there that the slope of a nonvertical line l is equal to the slope of a line l_1
parallel to l and passing through the origin. This shows that two nonvertical lines are parallel if and only if their slopes are equal. On the
other hand, it is obvious that a vertical line is not parallel to a nonvertical one.

Theorem 3 TWO LINES ARE PERPENDICULAR IF AND ONLY IF EITHER ONE
IS HORIZONTAL AND THE OTHER VERTICAL, OR THEIR SLOPES, m_1 AND m_2,
SATISFY

$$(6) \qquad\qquad m_1 m_2 = -1.$$

Proof. Let l_1, l_2 be two lines. Let $\widehat{l}_1$, and $\widehat{l}_2$ (read "l-one hat," "l-two
hat") be two lines through the origin such that $\widehat{l}_1$ is parallel to l_1 and $\widehat{l}_2$
to l_2 (Fig. 2.13). Then the lines l_1, l_2 are perpendicular if and only if
$\widehat{l}_1$, $\widehat{l}_2$ are perpendicular. It is therefore enough to prove the theorem for
the two lines passing through the origin.

A horizontal line is perpendicular to a vertical one, and vice versa.
We consider next two lines that are neither vertical nor horizontal. Then
both have slopes, m_1 and m_2, and $m_1 \neq 0$, $m_2 \neq 0$. The equations of
our lines are

$$y = m_1 x, \qquad y = m_2 x.$$

If $m_1 > 0$, $m_2 > 0$, both lines pass through the first quadrant. Then
they are not perpendicular (see Fig. 2.14) and $m_1 m_2 > 0$ so that
(6) does not hold. If $m_1 < 0$, $m_2 < 0$, both lines pass through the second
quadrant. Again the lines are not perpendicular and $m_1 m_2 > 0$.

Assume next that m_1 and m_2 have opposite signs, say $m_1 > 0$ and
$m_2 < 0$, as in Fig. 2.15. The two lines are perpendicular if and only if
the angles α and β are complementary (add up to a right angle). This is
so if and only if the two right triangles are similar, and they are similar
if and only if $|m_1|/1 = 1/|m_2|$, which is the same as $|m_1 m_2| = 1$. Since
m_1 and m_2 have opposite signs, this is equivalent to (6).

EXERCISES

14. Find the equation of the line through $(0, 0)$ which is parallel to the line whose
    equation is $2x - 3y + 7 = 0$.

Fig. 2.13

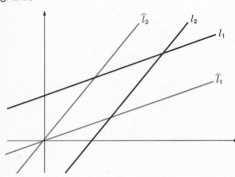

Fig. 2.14
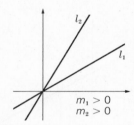

Fig. 2.15
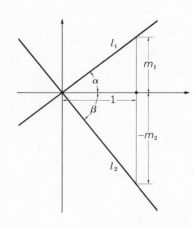

15. Find the equation of the line through $(0, 0)$ which is perpendicular to the line whose equation is $2x - 3y + 7 = 0$.

16. For which values of α is $\alpha x + 3y - 5 = 0$ the equation of a line perpendicular to the line whose equation is $x - y + 9 = 0$?

17. For which values of u is $2ux - 3y + 6 = 0$ the equation of a line parallel to the line whose equation is $6x + 4y - 3 = 0$?

18. For which values of t are $tx - 2y + 8 = 0$ and $3tx + 6y + 4 = 0$ the equations of perpendicular lines?

19. Find the equation of the line through $(-1, 2)$ which is parallel to the line whose equation is $4x + 12y + 3 = 0$.

▶ 20. Find the equation of the line through $(6, -3)$ which is perpendicular to the line whose equation is $x - 3y + 12 = 0$.

21. Find the equation of the line through $(1/2, -1)$ which is perpendicular to the line whose equation is $18x + 12y - 1 = 0$.

2.4 Equations of lines

We list some convenient ways of writing equations of nonvertical lines. If l intersects the x axis at the point $(a, 0)$, a is called the x **intercept** of l. Similarly, if l intersects the y axis, the intersection point has coordinates $(0, b)$; b is the y **intercept** of l.

A. The equation of a line with slope m and y intercept b is

$$y = mx + b.$$

Proof. For $x = 0$, the equation yields $y = b$; thus $(0, b)$ is on the line. And the line with this equation is parallel to the desired line, and thus is the desired line.

B. The equation of a line with slope m and containing the point (x_1, y_1) is

(7) $$y - y_1 = m(x - x_1).$$

Proof. Equation (7) is equivalent to $y = mx + b$ with $b = y_1 - mx_1$. Hence it is the equation of a line with slope m. One checks easily that the point (x_1, y_1) satisfies (7).

C. The equation of a line with slope m and x intercept a is

$$y = m(x - a).$$

This is a special case of (B), namely $(x_1, y_1) = (a, 0)$.

D. The equation of a line passing through two distinct points (x_1, y_1) and (x_2, y_2) is

(8) $$(y - y_2)(x_1 - x_2) = (y_1 - y_2)(x - x_2).$$

Proof.   Equation (8) is a linear equation in the variables x and y, since x_1, y_1, x_2, y_2 are fixed numbers. Hence the solution set of (8) is a line. One checks easily that (x_1, y_1) and (x_2, y_2) satisfy (8).

E. The equation of a line with x intercept a and y intercept b, $a \neq 0$, $b \neq 0$, is

$$(9) \qquad \frac{x}{a} + \frac{y}{b} = 1.$$

Proof.   Equation (9) is a linear equation satisfied by $(a, 0)$ and $(0, b)$.

There is no particular need to memorize these formulas, since whenever they are needed they can easily be derived from the **slope-intercept form** (2) or the general form (1).

▶ *Example*   Find the equation of a line through the points $(3, 4)$ and $(4, 3)$.

First Method.   Using D, we obtain $(y - 3)(3 - 4) = (4 - 3)(x - 4)$ or, simplifying, $x + y - 7 = 0$.

Second Method.   The desired equation is of the form $Ax + By + C$ and is to be satisfied by $(3, 4)$ and $(4, 3)$. Thus we must have

$$3A + 4B + C = 0, \qquad 4A + 3B + C = 0.$$

Treat C as given, and solve these two equations simultaneously for A and B. This yields $A = B = -C/7$. The equation of the line reads

$$-\frac{C}{7}x - \frac{C}{7}y + C = 0.$$

The value of C is immaterial, as long as $C \neq 0$. Choosing $C = -7$, we obtain the same equation as above. ◀

EXERCISES

22. Find the equation of the line with slope 4 passing through the point $(-2, -3)$.
23. Find the equation of the line with slope $-1/2$ passing through the point $(1, 7)$.
24. Find the equation of the line passing through the points $(-6, 2)$ and $(3, -2)$.
25. Find the equation of the line passing through the points $(0, 2)$ and $(-24, 0)$.
26. Find the equation of the line with slope $3/4$ passing through the point $(100, 30)$.
27. Find the equation of the line with slope 6 whose y intercept is 2.
▶ 28. Find the x intercept and y intercept of the line whose equation is $2x/3 - y/4 = 1$.
29. Find the equation of the line whose x intercept is -2 and whose y intercept is $3/5$.

30. What is the slope of a line which does not pass through the origin and whose x intercept equals its y intercept?
31. Find the equation of the line passing through the points $(36, -10)$ and $(34, -4)$.
32. The line $y = mx + b$ passes through $(1, 2)$ and $(3, 5)$. Find m and b.
33. Find the equation of the perpendicular bisector of the segment joining $(3, -1)$ and $(-2, -2)$.
34. Find the equation of the line that cuts off from the third quadrant an isosceles triangle of area 4.
35. Find the point on the line $2x - y - 1 = 0$ that is equidistant from the points $(3, 2)$ and $(2, 3)$.
36. Find the equation of the line through the origin that is parallel to the line $3x - 2y - 1 = 0$.
37. The points $X = (3, -2)$, $Y = (4, 1)$, and $Z = (-3, 5)$ are the vertices of a triangle. Find the equation of the line through Y perpendicular to the side XZ.
38. Show that the altitudes of triangle XYZ in Exercise 37 intersect in a point.

3.1 The equation of a circle

A **circle** is the set of all points that have the same distance, say r, from a fixed point, say Q. The point Q is called the **center** of the circle and r the **radius**; the radius is always a positive number. The circle with center at the origin and radius 1 is called the **unit circle**.

Let the center Q have coordinates (a, b). The distance of a point (x, y) from Q is given by

$$\sqrt{(x - a)^2 + (y - b)^2}.$$

Hence a circle with center (a, b) and radius $r > 0$ is the solution set of the equation

(1) $$(x - a)^2 + (y - b)^2 = r^2.$$

A point (x, y) is said to lie interior (or exterior) to the circle if its distance from the center (a, b) is less than the radius r (is greater than r). Therefore, the set of points interior to the circle is the solution set of the inequality $(x - a)^2 + (y - b)^2 < r^2$, while the set of exterior points is given by $(x - a)^2 + (y - b)^2 > r^2$.

Theorem 1 LET A, B, AND C BE NUMBERS. THE SOLUTION SET OF THE EQUATION

(2) $$x^2 + y^2 + Ax + By + C = 0$$

IS EITHER A CIRCLE, OR A POINT, OR THE EMPTY SET.

Proof. We use the method of **completing the square**, one of the oldest devices in mathematics (it was known in Babylon). We have

$$x^2 + Ax = x^2 + 2\left(\frac{A}{2}x\right) = x^2 + 2\left(\frac{A}{2}x\right) + \frac{A^2}{4} - \frac{A^2}{4}$$

$$= \left(x + \frac{A}{2}\right)^2 - \frac{A^2}{4},$$

$$y^2 + By = \left(y + \frac{B}{2}\right)^2 - \frac{B^2}{4},$$

so that equation (2) may be written as

$$\left(x + \frac{A}{2}\right)^2 - \frac{A^2}{4} + \left(y + \frac{B}{2}\right)^2 - \frac{B^2}{4} + C = 0$$

or

$$(3) \qquad \left(x + \frac{A}{2}\right)^2 + \left(y + \frac{B}{2}\right)^2 = D,$$

where $D = (A^2 + B^2)/4 - C$. If $D > 0$, set $r = \sqrt{D}$. Then (3) is the equation of a circle with center $(-A/2, -B/2)$ and radius r. If $D = 0$, equation (3) says that the sum of two nonnegative numbers, $(x + A/2)^2 + (y + B/2)^2$, is zero. Hence both numbers must be zero. The solution set contains the single point $x = -A/2$, $y = -B/2$. If $D < 0$, finally, (3) cannot hold for any pair of real numbers; the solution set is empty.

▶*Example*   Describe the solution set of

$$2x^2 + 2y^2 - 2x + 4y - 10 = 0.$$

Solution.   We first divide both sides of the equation by 2 and obtain

$$x^2 + y^2 - x + 2y - 5 = 0.$$

Now we complete the square:

$$x^2 - x = x^2 - 2 \times \frac{1}{2} = x^2 - 2 \times \frac{1}{2} + \left(\frac{1}{2}\right)^2 - \left(\frac{1}{2}\right)^2$$

$$= \left(x - \frac{1}{2}\right)^2 - \frac{1}{4},$$

$$y^2 + 2y = y^2 + 2y + 1^2 - 1^2 = (y + 1)^2 - 1,$$

so that our equation becomes

$$\left(x - \frac{1}{2}\right)^2 + (y + 1)^2 - \frac{1}{4} - 1 - 5 = 0$$

or
$$\left(x - \frac{1}{2}\right)^2 + (y + 1)^2 = \frac{25}{4}.$$

The solution set is a circle with center $(\frac{1}{2}, -1)$ and radius $\frac{5}{2}$. ◄

EXERCISES

1. Find the equation of the circle of radius 2 whose center is at $(0, 3)$.
2. Find the equation of the circle of radius 3 whose center is at $(-1, 4)$.
3. Find the equation of the circle whose center is at $(1, 6)$ and which passes through $(-2, 2)$.
4. Find the equation of the circle whose center is at $(-3, 2)$ and which passes through $(-2, 1)$.
5. Find the equation of the circle of radius 4 that passes through $(-3, 0)$ and $(5, 0)$.
6. Find the equation of the circle of radius 3 whose center is at $(-10, 12)$.
7. Describe the solution set of $x^2 + y^2 + 2x + 4y + 4 = 0$.
8. Describe the solution set of $x^2 + y^2 + 4x - 4y + 8 = 0$.
9. Describe the solution set of $x^2 + y^2 - x - y + 1 = 0$.
10. Describe the solution set of $x^2 + y^2 + 6x - 2y - 6 = 0$.
11. Describe the solution set of $2x^2 + 2y^2 + 2x - 2y - 1 = 0$.
12. Describe the solution set of $4x^2 + 4y^2 + 8x - 4y + 5 = 0$.

3.2 Tangents

We now give two examples of algebraic proofs of geometric theorems.

Theorem 2   THE INTERSECTION OF A LINE l AND A CIRCLE C IS EITHER EMPTY, OR CONSISTS OF ONE POINT, OR CONSISTS OF TWO POINTS.

Proof.   We choose the coordinate system so that the circle C is the unit circle $x^2 + y^2 = 1$ (this is achieved by taking the center of the circle as the origin and the radius as the unit of length) and the line l is horizontal (this is achieved by choosing the x axis as parallel to l). Then the equation of the line is $y = b$. A point common to C and l must have coordinates (x, b) with $x^2 + b^2 = 1$ or

(4)
$$x^2 = 1 - b^2.$$

If $|b| > 1$, then $1 - b^2 < 0$, so that no x satisfies (4); there are no points common to C and l. If $|b| = 1$, that is, $b = 1$ or $b = -1$, then (4) means that $x = 0$. There is exactly one intersection point, namely $(0, b)$.

Fig. 2.16

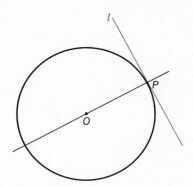

If $|b| < 1$, then $1 - b^2 > 0$ and (4) means that either $x = \sqrt{1 - b^2}$ or $x = -\sqrt{1 - b^2}$. There are exactly two intersection points, $(\sqrt{1 - b^2}, b)$ and $(-\sqrt{1 - b^2}, b)$. The theorem is proved.

A **diameter** of a circle is that segment of a line through the center of the circle whose endpoints lie on the circle. A **tangent** to a circle is a line that has exactly one point in common with the circle (see Fig. 2.16).

Theorem 3   LET P BE A POINT ON A CIRCLE C, AND l A LINE THROUGH P. THEN l IS TANGENT TO C IF AND ONLY IF IT IS PERPENDICULAR TO THE DIAMETER THROUGH P.

Proof.   We choose the coordinate system as in the preceding proof. Since P lies on C, the equation of l is $y = b$ with $|b| \leq 1$. If l is a tangent line, then, by the preceding argument, either $b = 1$ or $b = -1$. This means: either $P = (0, 1)$ or $P = (0, -1)$. In both cases OP is vertical, hence perpendicular to l. If l is not a tangent line, then, by the preceding argument, $|b| < 1$. Hence P is either the point $(\sqrt{1 - b^2}, b)$ or the point $(-\sqrt{1 - b^2}, b)$. The line QP has either slope $b/\sqrt{1 - b^2}$ or $-b/\sqrt{1 - b^2}$. It is not vertical and thus not perpendicular to l.

Note that the proofs of Theorems 2 and 3 were very easy, only because we chose the coordinate system conveniently.

We shall not discuss theorems about the areas or the lengths of circles as yet. These matters belong to calculus.

EXERCISES

13. A circle whose center is at $(1, 2)$ passes through the point $(-1, 1)$. What is the slope of the line tangent to this circle at $(-1, \cdot 1)$?

14. A line meets a circle in the point $(0, 1)$ and no other. If the center of the circle is at $(-1, 3)$, what is the slope of the line?

15. The equation of a circle is $x^2 + y^2 - 10x = 28$. Find the equation of the tangent line that meets the circle in the point $(3, 7)$.

16. The equation of a circle is $x^2 + y^2 + 4x - 8y - 5 = 0$. What is the equation of the line tangent to the circle at $(1, 0)$?

17. The equation of a circle is $x^2 + y^2 - 2x - 24 = 0$. At what point will the line tangent to the circle at $(1, 5)$ intersect the line tangent to the circle at $(4, -4)$?

► 18. If the line $y = x$ is tangent to a circle at $(3, 3)$ and the line $y = 2x$ passes through the center of the circle, what is the equation of the circle?

19. Find the equations of all lines that are tangent to the circle $x^2 + y^2 = 1$ and pass through the point $(3, 1)$.

20. Find the equations of all lines that are tangent to the circle $x^2 + y^2 - 2x = 0$ and pass through the point $(4, 0)$.

The full power of analytic geometry can be seen only if one considers more complicated curves than lines or circles. In this section, we discuss the parabola, a curve discovered by the Greek geometers. (The study of the parabola will be continued in Chapter 10. There we shall also consider hyperbolas and ellipses.)

4.1 Definition of the parabola

Let l be a line and F a point not on l. The **parabola** with **directrix** l and **focus** F is defined as the set of all points equidistant from l and F (Fig. 2.17). This means: a point P is on the parabola whenever

(1)   distance from P to F = distance from P to l.

We want to represent the parabola as the solution set of an equation. The equation will depend on the choice of a coordinate system. We try to find a coordinate system that leads to a simple equation.

Drop a perpendicular from F to l; it meets l at a point Q. We choose the midpoint of FQ as the origin, and we choose the horizontal direction parallel to l. Then F is the point $(0, p)$ with $p \neq 0$ and l the line $y = -p$ (see Fig. 2.18). Let P be the point (x, y). Then

$$\text{distance from } P \text{ to } F = \sqrt{x^2 + (y - p)^2}$$

by the distance formula in §1.3. Also,

$$\text{distance from } P \text{ to } l = |y + p|$$

as is seen from Fig. 2.18. [One can also reason as follows. The point Q with coordinates $(x, -p)$ lies on l, and the line through P and Q is perpendicular to l. Therefore the distance from P to l is the distance from P to Q, and this is, by the distance formula, $\sqrt{(y + p)^2} = |y + p|$.] We conclude that equation (1) says that

$$\sqrt{x^2 + (y - p)^2} = |y + p|,$$

which means (since two nonnegative numbers are equal if and only if their squares are equal)

$$x^2 + (y - p)^2 = (y + p)^2.$$

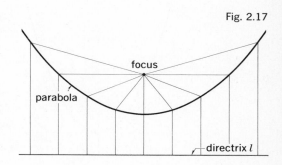

Fig. 2.17

focus

parabola

directrix l

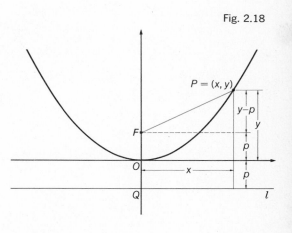

Fig. 2.18

$P = (x, y)$

$y-p$

y

F

p

O

x

p

Q

l

or $$x^2 + y^2 - 2py + p^2 = y^2 + 2py + p^2$$

or

(2) $$x^2 = 4py.$$

Thus we have proved the following theorem:

Theorem 1   THE SOLUTION SET OF $x^2 = 4py$, FOR ANY FIXED $p \neq 0$, IS THE PARABOLA WITH FOCUS $(0, p)$ AND DIRECTRIX $y = -p$.

Had we chosen the unit of length as half of the distance between the focus and the directrix, and the vertical direction in such a way that the focus lay in the upper half-plane, we would have $p = \frac{1}{4}$ and equation (2) would read simply

(3) $$y = x^2.$$

In studying geometric properties of parabolas, we may restrict ourselves to this equation.

The parabola (3) is shown in Fig. 2.19. The diagram was obtained by computing many points (x, x^2) and finding their position in the coordinate plane.

Any other parabola will have the same shape as this one, though it may have a different location and a different size. In this respect, parabolas are like circles. Given any circle, there is a coordinate system in which the equation of this circle reads $x^2 + y^2 = 1$ (namely, the system for which the center of the circle is the origin and the radius of the circle is the unit of length). Given any parabola, there is a coordinate system in which the equation of this parabola reads $y = x^2$. One expresses this by saying: all circles are similar to each other; all parabolas are similar to each other. Analogously, all squares are similar to each other, whereas two triangles need not be similar.

4.2 Properties of the parabola

The line through the focus perpendicular to the directrix is called the **axis** of the parabola. The intersection point of the parabola with its axis is called the **vertex** of the parabola. In the case of the parabola $y = x^2$, the axis is the y axis and the vertex is the origin.

The vertex is closer to the directrix than any other point of a parabola.

In proving this, we may assume that the parabola has the equation $y = x^2$, since this can be achieved by a proper choice of the coordinate system. Then the vertex has coordinates $(0, 0)$ and the directrix is the line $y = -1/4$. The distance from the vertex to the directrix is $1/4$, and the

Fig. 2.19

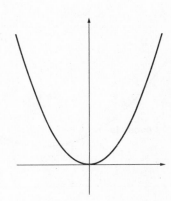

distance from any point (x, y) on the parabola to the directrix is $y + 1/4$; since $y = x^2$, y is never negative.

If a point (x, y) satisfies $y = x^2$, so does the point $(-x, y)$. But the points (x, y) and $(-x, y)$ are symmetric with respect to the y axis. This means that the line joining these points is perpendicular to the y axis and the two points are equidistant from the y axis. Thus:

A parabola is symmetric with respect to its axis.

The **latus rectum** of a parabola is defined to be the finite segment of the line through the focus and parallel to the directrix that is inside the parabola. More precisely, the parabola cuts this line in two points and hence divides it into three segments, two of which are infinite in extent. The remaining finite segment is the latus rectum (see Fig. 2.20). For the parabola $y = x^2$, the focus is at $(0, 1/4)$ and the line through the focus parallel to the directrix is the line $y = 1/4$. It meets the parabola at points where $1/4 = x^2$. Thus the endpoints of the latus rectum are $(-1/2, 1/4)$ and $(1/2, 1/4)$; the length of the latus rectum is 1.

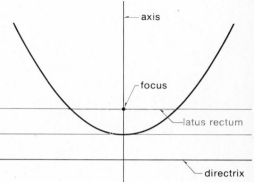

EXERCISES

In Exercises 1 to 6, make a sketch of each parabola and show the vertex, focus, directrix, and axis.

1. Find the focus and directrix of the parabola whose equation is $x^2 = 8y$.
2. Find the focus and directrix of the parabola whose equation is $8x^2 = y$.
3. Find the equation of the parabola whose directrix is the line $y = -3$ and whose focus is the point $(2, 1)$. Find the vertex and axis of symmetry of this parabola. [Hint: First try deriving the equation directly as in the text, that is, compute the distance from the point (x, y) to the focus and the distance from the point (x, y) to the directrix, and set these equal. Simplify the resulting equation. Then try to derive the equation from the one already established in the text by using a new coordinate system and the formulas for changing coordinates. Naturally, you should be able to obtain the same equation by either method.]
4. Find the equation of the parabola whose directrix is the line $y = 1$ and whose focus is the point $(-2, -4)$. Find the vertex and axis of symmetry of this parabola.
5. Find the equation of the parabola whose directrix is the line $x = 2$ and whose focus is the point $(1, 1)$. Find the vertex and axis of symmetry of this parabola.
6. Find the equation of the parabola whose directrix is the line $x = 0$ and whose focus is the point $(-4, 2)$. Find the vertex and axis of symmetry of this parabola.
7. Find the endpoints and the length of the latus rectum of the parabola $x^2 = 16y$. Sketch the parabola and the latus rectum.
8. Find the endpoints and the length of the latus rectum of the parabola $x^2 = 4py$.
9. The directrix of a parabola is the line $y = 1$ and the focus is $(2, 9)$. Find the endpoints of the latus rectum. Sketch the parabola and the latus rectum.
10. The directrix of a parabola is the line $x = 2$ and the focus is $(-1, 1)$. Find the endpoints of the latus rectum. Sketch the parabola and the latus rectum.

4.3 Parabolas with vertical axes

Theorem 2   LET A, B, C BE NUMBERS AND $B \neq 0$. THE SOLUTION SET OF THE EQUATION

(4) $$x^2 + Ax + By + C = 0$$

IS A PARABOLA WITH A VERTICAL AXIS.

Proof.   Completing the square, we write (4) in the form

$$\left(x + \frac{A}{2}\right)^2 - \frac{A^2}{4} + By + C = 0$$

or

(5) $$\left(x + \frac{A}{2}\right)^2 = -By - C + \frac{A^2}{4} = -B\left(y - \frac{A^2 - 4C}{4B}\right).$$

We introduced a new coordinate system obtained from our original system by translation, so that the new coordinates X, Y are connected with the old coordinates x, y by the equations

$$x + \frac{A}{2} = X, \qquad y - \frac{A^2 - 4C}{4B} = Y.$$

This means, of course, that the new origin has the coordinates

$$\left(-\frac{A}{2}, \frac{A^2 - 4C}{4B}\right)$$

in the old system. We also set $-B = 4p$. Equation (5) states

$$X^2 = 4pY,$$

which we know to represent a parabola.

▶*Example*   Find the focus and directrix of the parabola

$$x^2 + 4x + 2y - 1 = 0.$$

Solution.   We write the equation as

$$(x + 2)^2 - 4 + 2y - 1 = 0$$

or
$$-2(y - \tfrac{5}{2}) = (x + 2)^2.$$

Setting $\qquad x + 2 = X \qquad$ and $\qquad y - \frac{5}{2} = Y,$

we find that this becomes $-2Y = X^2$. By Theorem 1, this is the equation of a parabola with focus $X = 0$, $Y = -\frac{1}{2}$, and directrix $Y = \frac{1}{2}$. We conclude that in the original x, y coordinate system the focus of our parabola is at $(-2, 2)$ and the directrix is the line $y = 3$. The parabola is shown in Fig. 2.21. ◄

Fig. 2.21

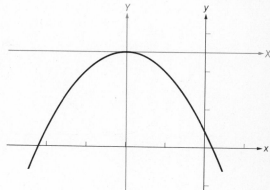

EXERCISES

Find the focus and directrix of each of the parabolas whose equation is given below. Make a sketch of each parabola and show the vertex, the focus, the directrix, and the axis.

11. $x^2 - 6x - 7 - 8y = 0$.
12. $2x^2 + 4x + 5 + y = 0$.
13. $x^2 - 14x + 1 = 16y$.
14. $4x^2 - 4x - 7 = 16y$.
15. $4x^2 - 4x + 1 = y$.
16. $x^2 + 8x + 17 = -4y$.

4.4 Tangents

We consider now the intersection of a parabola with a straight line.

Theorem 3 A STRAIGHT LINE INTERSECTS A PARABOLA, IN NO POINT, ONE POINT, OR TWO POINTS.

Proof. We may assume that the coordinate system is chosen so that the equation of the parabola is $y = x^2$.

Let us consider first a vertical line, that is, a line parallel to the axis of the parabola. Such a line has the equation $x = a$, a being some number. It intersects the parabola at (a, a^2) and at no other point.

If the nonvertical line $y = mx + b$ intersects $y = x^2$, the x coordinate of the intersection point satisfies the quadratic equation

$$mx + b = x^2.$$

Completing the square, we write it as

(6) $$\left(x - \frac{m}{2}\right)^2 = b + \frac{m^2}{4}.$$

There are no intersections if the right-hand side is negative, two if it is positive, and one if it is zero. This proves the theorem.

A line that meets the parabola at exactly one point and is not parallel to the axis is called **tangent** to the parabola.

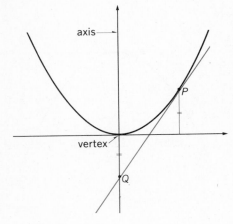

Fig. 2.22

Theorem 4 THROUGH EACH POINT OF THE PARABOLA THERE PASSES EXACTLY ONE TANGENT. IF THE PARABOLA CONSIDERED IS $y = x^2$ AND THE POINT IS (x, x^2), THE TANGENT HAS SLOPE $2x$.

Proof. The tangent is a line $y = mx + b$ which intersects the parabola at exactly one point. By the previous proof, the coordinate x of the intersection point satisfies $x - (m/2) = 0$. [See equation (6).] Hence the slope, m, equals $2x$.

Theorem 5 LET P BE A POINT ON THE PARABOLA, DISTINCT FROM THE VERTEX. THE TANGENT TO THE PARABOLA THROUGH P INTERSECTS THE AXIS AT A POINT Q SUCH THAT P AND Q ARE EQUIDISTANT FROM THE LINE THROUGH THE VERTEX AND PERPENDICULAR TO THE AXIS (FIG. 2.22).

Proof. We may assume that the coordinate system is chosen so that the parabola is given by $y = x^2$. Let P have coordinates (α, α^2) with $\alpha \neq 0$. The slope of the tangent to the parabola at P is 2α. Referring to §2.4B, we see that the equation of the tangent reads

$$y - \alpha^2 = 2\alpha(x - \alpha).$$

For $x = 0$, we obtain $y = -\alpha^2$. Thus $Q = (0, -\alpha^2)$. It is clear that P and Q are equidistant from the x axis.

A word about the definition of the tangent. The tangent to a circle through a point P on the circle has been defined as the line that intersects the circle at P and only at P. Such a definition would not work for the parabola. For each point on the parabola, there are two lines that meet the parabola only at this point: a line parallel to the axis and one other line (the "tangent"). We feel that this other line is the "true" tangent; it "just touches" the parabola, while the line parallel to the axis actually goes through the parabola.

Theorems 3 and 5 were known to the Greeks. The problem of finding tangents to other curves was one of the stimuli that led to the development of calculus.

▶*Example* Find a point Q on the parabola $y = x^2$ such that the tangent through Q is perpendicular to the tangent through the point $(2, 4)$.

Solution. The tangent through $(2, 4)$ has slope 4; that through the point $Q = (x, x^2)$ has slope $2x$. We must have $4(2x) = -1$. Thus $x = -1/8$ and $Q = (-1/8, 1/64)$. ◀

EXERCISES

17. Find the equation of the tangent to the parabola $y = x^2$ at the point $(-2, 4)$.
18. At what point will the line tangent to the parabola $y = x^2$ at $(1/4, 1/16)$ intersect the extension of the latus rectum? Sketch the parabola, the latus rectum, and the tangent.
19. What is the point of intersection of the lines tangent to the parabola $y = x^2$ at $(2, 4)$ and at $(4, 16)$?
► 20. Find a point Q on the parabola $y = x^2$ so that the line tangent to the parabola at Q is perpendicular to the line tangent to the parabola at $(-10, 100)$.
21. The line tangent to the parabola $y = x^2$ at a point Q intersects the extension of the latus rectum when $x = 1$. Find all possible coordinates of Q.
22. Find the slopes of all tangents to the parabola $y = x^2$ that pass through the point $(4, -9)$. Sketch the parabola and the tangents.

Appendixes to Chapter 2

§5 Squaring the Parabola

5.1 Archimedes' Theorem

The problem of tangents was one of the stimuli in the development of calculus. Another stimulus was the problem of areas: how to calculate the area of a region with a curved boundary. The main result of calculus is that these two, seemingly very different, problems are intimately connected. (This connection will become apparent in Chapter 5.)

The Greeks tried to solve a very difficult problem involving areas. They wanted to "square a circle," that is, to construct by ruler and compass a square whose area was precisely equal to that of a given circle. This cannot be done; the impossibility proof was not given until the nineteenth century. But Archimedes, the greatest mathematician of antiquity, succeeded in squaring a region bounded by an arc of the parabola. His famous theorem can be stated as follows.

Consider a rectangle such that its two opposite vertices are the vertex and a second point of a parabola, and such that one side lies on the axis of the parabola. The area of this rectangle is divided by the parabola in the ratio 2/1 (Fig. 2.23).

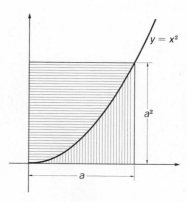

Fig. 2.23

5.2 Archimedes' proof

We give one of Archimedes' proofs, in modern notation. Let the parabola be given by the equation $y = x^2$ and let the vertices of the rectangle be $(0, 0)$ and (a, a^2) with $a > 0$ (Fig. 2.23). The area of the rectangle is $a \cdot a^2 = a^3$. Let A be the area of the part of the rectangle below the parabola. We must show that

(1)
$$A = \frac{1}{3} a^3.$$

It should be noted that at this point we do not ask what is meant by the area of a figure bounded by a curve, but rely upon our intuitive notion of area.

The proof of (1) uses the formula

(2)
$$1^2 + 2^2 + 3^2 + \cdots + n^2 = \frac{1}{3} n^3 + \frac{1}{2} n^2 + \frac{1}{6} n,$$

which was proved in Chapter 1, (see §6.3). Another proof of this formula will be found later in this section.

Let n be a positive integer. We divide the interval $(0, a)$ into n equal segments, by the points $0, \frac{a}{n}, \frac{2a}{n}, \cdots, \frac{(n-1)a}{n}, a$. (See Fig. 2.24, where $n = 6$.) Each segment has length a/n. Over each segment we erect a "tall" rectangle whose upper right corner lies on the parabola and a "short" rectangle whose upper left vertex lies on the parabola; there is no short rectangle over the segment $\left(0, \frac{a}{n}\right)$.

The heights of the tall rectangles are

$$\left(\frac{a}{n}\right)^2, \quad \left(\frac{2a}{n}\right)^2, \quad \left(\frac{3a}{n}\right)^2, \quad \cdots, \quad \left(\frac{(n-1)a}{n}\right)^2, \quad \left(\frac{na}{n}\right)^2 = a^2,$$

so that the sum of their areas is

$$\frac{a}{n}\left(\frac{a}{n}\right)^2 + \frac{a}{n}\left(\frac{2a}{n}\right)^2 + \frac{a}{n}\left(\frac{3a}{n}\right)^2 + \cdots + \frac{a}{n}\left(\frac{na}{n}\right)^2$$
$$= \frac{a^3}{n^3}(1^2 + 2^2 + 3^2 + \cdots + n^2) = \frac{a^3}{n^3}\left(\frac{n^3}{3} + \frac{n^2}{2} + \frac{n}{6}\right),$$

which is the same as

(3)
$$\frac{a^3}{3} + \frac{a^3}{2n} + \frac{a^3}{6n^2}.$$

The area of each of the short rectangles is the area of the corresponding tall rectangle minus the area of a small rectangle (heavily shaded in Fig. 2.24). Each of these small rectangles has width a/n. If they are placed on top of each other, we see that their heights add up to a^2. Their total area therefore is a^3/n.

Hence the sum of the areas of the short rectangles is the number (3) minus a^3/n or

(4)
$$\frac{a^3}{3} + \frac{a^3}{2n} + \frac{a^3}{6n^2} - \frac{a^3}{n} = \frac{a^3}{3} - \frac{a^3}{2n} + \frac{a^3}{6n^2}.$$

Obviously, A is less than (3) and greater than (4). Hence

$$\frac{a^3}{3} - \frac{a^3}{2n} + \frac{a^3}{6n^2} < A < \frac{a^3}{3} + \frac{a^3}{2n} + \frac{a^3}{6n^2}.$$

Fig. 2.24

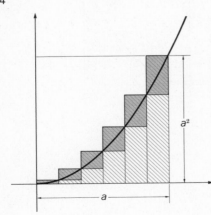

Since
$$\frac{1}{2n} + \frac{1}{6n^2} = \frac{3n+1}{6n^2} < \frac{6n}{6n^2} = \frac{1}{n}$$

and
$$-\frac{1}{2n} + \frac{1}{6n^2} = \frac{-3n+1}{6n^2} > \frac{-6n}{6n^2} = \frac{-1}{n},$$

we have that
$$\frac{a^3}{3} - \frac{a^3}{n} < A < \frac{a^3}{3} + \frac{a^3}{n}.$$

Subtracting $a^3/3$ from each term and then multiplying by $1/a^3$, we obtain

$$-\frac{1}{n} < \frac{A}{a^3} - \frac{1}{3} < \frac{1}{n}.$$

This holds for all $n = 2, 3, 4, \cdots$. Therefore $A/a^3 = 1/3$, which is precisely (1).

Archimedes' proof is one of the first calculus results in history. We will see later how the same problem, when treated by a general method, becomes much simpler.

5.3 Sum of squares

We now give another proof of (2). For every number t, we have

$$(1 + t)^3 - t^3 = 3t^2 + 3t + 1.$$

We write this for $t = 1, 2, \cdots, n - 1, n$:

$$2^3 - 1^3 = 3 \cdot 1^2 + 3 \cdot 1 + 1$$
$$3^3 - 2^3 = 3 \cdot 2^2 + 3 \cdot 2 + 1$$
$$4^3 - 3^3 = 3 \cdot 3^2 + 3 \cdot 3 + 1$$
$$\cdots \quad \cdots \quad \cdots \quad \cdots \quad \cdots$$
$$\cdots \quad \cdots \quad \cdots \quad \cdots \quad \cdots$$
$$(n + 1)^3 - n^3 = 3 \cdot n^2 + 3n + 1.$$

Adding all n formulas, we have

$$(n + 1)^3 - 1 = 3(1^2 + 2^2 + \cdots + n^2) + 3(1 + 2 + \cdots + n) + n.$$

Replace here $1 + 2 + \cdots + n$ by

$$\frac{n(n + 1)}{2},$$

solve the resulting equation for $1^2 + 2^2 + \cdots + n^2$, and simplify. This yields (2).

EXERCISES

1. Suppose the vertices of a rectangle are at $(0, 0)$, $(2, 0)$, $(0, 4)$, and $(2, 4)$. Find the area of each of the two regions into which the rectangle is divided by the parabola $y = x^2$.
2. Find the area of the region bounded by the parabola $y = x^2$ and the latus rectum.

3. Find the area of the rectangle whose base is the interval $[2, 3]$ and whose upper left-hand corner lies on the parabola $y = x^2$.

4. Find the area of the rectangle whose base is the interval $[2, 3]$ and whose upper right-hand corner lies on the parabola $y = x^2$.

5. Over each of the intervals $[0, 1/2]$, $[1/2, 1]$, $[1, 3/2]$, $[3/2, 2]$ construct a rectangle whose upper right-hand corner lies on the parabola $y = x^2$. Find the sum of the areas of these rectangles. Compare your computations step by step with those given for general a and n in the text, and show that they agree.

6. Over each of the intervals $[0, 1/2]$, $[1/2, 1]$, $[1, 3/2]$, $[3/2, 2]$ construct a rectangle whose upper left-hand corner lies on the parabola $y = x^2$. (The height of the rectangle over the first interval is zero and so it can be neglected.) Find the sum of the areas of these rectangles. Compare your computations step by step with those given for general a and n in the text and show that they agree.

7. Find the area of the region in the first quadrant bounded by the parabola $y = x^2$, the x axis, and the line $x = 2$ and compare this with the figures obtained in Exercises 5 and 6. Verify that the difference of these figures is a^3/n and that the area indeed does lie between them.

§6 Geometry and Numbers

6.1 Axiomatic geometry

The word geometry means, literally, measuring earth, that is, surveying. Geometry was originally a practical art and, as far as it was a science, experimental. Geometry as a **deductive science** is a creation of the Greeks. Thales, traditionally considered the first Greek philosopher, is also credited with the first proof of a geometric theorem. In deductive geometry certain simple statements about points and lines (and, in the case of solid geometry, also about planes) are assumed as **axioms;** all other statements are **theorems,** that is, logical consequences of the axioms.

Plato, whose philosophy was influenced by Greek mathematics, emphasized that deductive geometry deals with abstract or ideal, rather than physical, figures. The line of the geometer is not the line drawn on paper, a straight rod, or a ray of light. Drawings are only a device to reinforce our memory and stimulate our imagination. Modern mathematicians accept Plato's statement, though perhaps not in his original sense. What matters is that no concepts or properties except those explicitly stated in the axioms are to be used in the proofs.

This does not mean, of course, that geometry has nothing to do with the physical world. The axioms are designed so as to give a description of a part of physical reality. The abstract character of the definitions and proofs ensures the validity of the theorems whenever the axioms are assumed. The history of science shows that, by becoming more abstract, mathematics becomes more, rather than less, useful as a tool for understanding nature.

The strictly logical construction of geometry from axioms is the aim of Euclid's *Elements*, a work that summarized the results of a century of Greek mathematics and became the second most influential book of our civilization. It is no disparagement of the stupendous achievement of Euclid and his predecessors to observe that, judged by modern standards of rigor, they did not achieve their aim.

THALES (6th century B.C.). The reports about Thales' mathematical and astronomical achievements are considered to lack historical credibility.

PLATO (429–348 B.C.), a pupil of Socrates and founder of the idealistic philosophy, was also the inventor of universities. He established the Academy, a school that endured nearly a thousand years.

This great thinker and writer extolled mathematics as a road to true knowledge, and he assigned to it a prominent place in the curriculum of the Academy. Though Plato himself was not a mathematician, and there is no evidence that he was fully abreast of the mathematical achievements of his time, the Platonic tradition played its part in the creation of modern mathematical science.

EUCLID (c. 365–300 B.C.). Until recently, most high-school texts of geometry were more or less watered-down versions of the *Elements*. But Euclid's book was not intended for elementary instruction. It is a sophisticated treatise dealing with geometry, as well as number theory, and describes contributions by Euclid's predecessors as well as his own work.

A system of axioms really sufficient for Euclidean geometry was created relatively recently. Hilbert's *Foundation of Geometry,* which contains such a system, appeared in 1899.

6.2 Analytic geometry

We are not going to list Hilbert's axioms here. There is no need to do so, since they can be summarized in one sentence: *Euclidean geometry is analytic geometry based on a Cartesian coordinate system.* Let us make this more precise.

We assume that we know real numbers, either because we constructed them from positive integers, say as nonterminating decimals, or because we accepted Postulates 1 to 18 of the preceding chapter. We define the **plane** to be the set of all ordered pairs of numbers (x, y). Every such pair is called a **point.** The **distance** between two points, (x_1, y_1), and (x_2, y_2), is defined to be the number $\sqrt{(x_1 - x_2)^2 + (y_1 - y_2)^2}$. A **straight line** is, by definition, the set of all points (x, y) that satisfy some equation of the form $Ax + By + C = 0$, where A, B, and C are numbers such that $A^2 + B^2 \neq 0$. (We write $A^2 + B^2 \neq 0$ to show that A and B are not both zero.)

Having made these conventions, we can translate every axiom of Hilbert's list, as far as it pertains to the plane, into a statement about numbers, and it turns out that each of these statements can be proved from Postulates 1 to 18.

Conversely, starting with Hilbert's axioms, we can construct a Cartesian coordinate system, describe points by coordinates, derive the distance formula, and so forth. This is essentially what we did in §§1 to 4, except that we used geometry informally.

The same procedure applies to geometry of space, that is, solid geometry. If we assume solid geometry, we can prove the following: the set of points (x, y, z) in space whose Cartesian coordinates satisfy a linear equation

$$Ax + By + Cz + D = 0,$$

where not all three numbers A, B, and C are zero, is a **plane,** and every plane can be so represented. (The actual proof is postponed until Chapter 9, §4, since the result will not be used earlier.) This gives the clue to a purely analytic foundation of solid geometry. We define space as the set of all ordered triples (x, y, z) of numbers. These triples are called points. The distance between two points is defined by the formula (4) of §1. The set of all points satisfying a linear equation $Ax + By + Cz + D = 0$, with $A^2 + B^2 + C^2 \neq 0$, is a **plane.** A line in space is the intersection of two distinct planes that have a point in common.

Again it turns out that all of Hilbert's axioms for solid geometry can be derived from the Postulates 1 to 18 about real numbers.

6.3 Axioms derived from algebra

We give two samples of geometric axioms derived from properties of numbers.

I. *Let P and Q be two distinct points. Then there is a line containing them.*

Two distinct points means two distinct ordered pairs of numbers (x_1, y_1) and (x_2, y_2). By "a line containing them" we mean the solution set of a linear equation, satisfied by (x_1, y_1) and (x_2, y_2). We have written down such an equation in §2 [equation (8)].

II. *Two distinct lines have at most one point in common.*

This axiom means that, if the solution sets of two linear equations

$$(1) \qquad A_1x + B_1y + C_1 = 0 \qquad \text{and} \qquad A_2x + B_2y + C_2 = 0$$

are distinct, there is at most one point (x, y) satisfying both. It is assumed that $A_1{}^2 + B_1{}^2 \neq 0$, $A_2{}^2 + B_2{}^2 \neq 0$. Now if (x, y) satisfies both equations, it also satisfies the equations obtained (a) by multiplying the first equation by B_2, the second by B_1, and subtracting; and (b) by multiplying the first by A_2, the second by A_1, and subtracting. These equations read:

$$(2) \quad (A_1B_2 - A_2B_1)x = -B_2C_1 + B_1C_2, \qquad (A_1B_2 - A_2B_1)y = A_2C_1 - A_1C_2.$$

If $A_1B_2 - A_2B_1 \neq 0$, then we obtain

$$(3) \qquad x = \frac{-B_2C_1 + B_1C_2}{A_1B_2 - A_2B_1}, \qquad y = \frac{A_2C_1 - A_1C_2}{A_1B_2 - A_2B_1}$$

as the only possible common solution of (1). It is indeed a common solution, as can be ascertained by substituting these values of x, y into (1).

Next, if $A_1B_2 - A_2B_1 = 0$ and there is some point (x, y) satisfying both equations (2), then $B_2C_1 - B_1C_2 = A_2C_1 - A_1C_2 = 0$. If $A_1 \neq 0$, we have

$$B_2 = \frac{A_2B_1}{A_1}, \qquad C_2 = \frac{A_2C_1}{A_1},$$

which means that $A_2 \neq 0$, for we cannot have $A_2 = B_2 = 0$. We see that the second equation (1) reads

$$A_2\left(x + \frac{B_1}{A_1}y + \frac{C_1}{A_1}\right) = 0$$

or

$$\frac{A_2}{A_1}(A_1x + B_1y + C_1) = 0,$$

and we conclude that the two equations (1) have the same solution set. The same conclusion holds if $A_1B_2 - A_2B_1 = 0$ and $B_1 \neq 0$, as the reader is asked to verify.

6.4 Spaces of more than three dimensions

The arithmetic interpretation of geometry outlined above leads at once to the concept of n-**dimensional** space, where n may be any positive integer. A point of this space is an n-tuple of real numbers $(x_1, x_2, \cdots, x_n)$. The distance between two points, say $(\alpha_1, \cdots, \alpha_n)$ and $(\beta_1, \cdots, \beta_n)$, is defined as

$$\sqrt{(\alpha_1 - \beta_1)^2 + \cdots + (\alpha_n - \beta_n)^2}.$$

A $(n - 1)$-dimensional **hyperplane** is the solution set of a linear equation $A_1x_1 + \cdots + A_nx_n + D = 0$, not all A's zero. The set of points equidistant from a fixed point is called an $(n - 1)$-dimensional sphere.

Proceeding in this way one can develop an n-dimensional geometry. This geometry contains none of the mystery with which philosophers and science-fiction writers surrounded the "fourth dimension." Every statement in n-dimensional geometry is simply a statement about numbers. The geometric language helps, by the analogy with the "visible" cases $n = 1, 2, 3$, to ask interesting questions and to guess at the correct answers.

The primary motivation for developing n-dimensional geometry was the desire of mathematicians to generalize and develop concepts in their natural setting. But n-dimensional spaces, as well as more complicated spaces of infinitely many dimensions, are not at all mere playthings of pure mathematicians; they are also indispensable working tools of modern physicists.

We shall continue to use geometrical language and geometrical arguments throughout this book. But we only use arguments that can easily be proved as statements about real numbers. There will be a few exceptions; they will be pointed out.

EXERCISES

1. Does the point $(0, 1, 3, -2)$ lie on the hyperplane $3x_1 + 2x_2 - x_3 + x_4 + 3 = 0$?
2. Does the point $(2, 0, -1, 4)$ lie on the hyperplane $3x_1 - x_2 + x_4 - 2 = 0$?
3. Does the point $(0, 0, 0, 1, 0)$ lie on the hyperplane $5x_1 + x_2 - x_3 + 6x_4 - 7x_5 - 6 = 0$?
4. What is the distance between $(0, 0, 0, 0)$ and $(2, -1, 3, 1)$? Notice that it would not make any sense to ask for the distance between the points $(0, 0, 0)$ and $(2, -1, 3, 1)$. Why not?
5. What is the distance between $(0, 1, 0, 1)$ and $(1, 1, 2, -3)$?
6. The hyperplane $x_4 = 2$ intersects the 3-dimensional sphere

$$x_1{}^2 + x_2{}^2 + x_3{}^2 + x_4{}^2 = 9$$

in an ordinary (2-dimensional) sphere. Find the radius and center of this sphere.

Problems

1. Show that, for every choice of the number b, the equation $x^2 + y^2 - 2by = 1$ is the equation of a circle passing through the points $(1, 0)$ and $(-1, 0)$. Where is the center of this circle?
2. Write the equation of all circles passing through the points (a, b) and (c, d). (Assume these points to be distinct.)
3. Give an analytic proof of the theorem that through any 3 points in the plane which do not lie on one line there passes a unique circle.
4. Generalizing the considerations of §3, show that the solution set of an equation

$$x^2 + y^2 + z^2 + Ax + By + Cz + D = 0$$

is either a sphere, or a point, or an empty set.

5. Give an analytic proof of the theorem that the intersection of a sphere with a plane is either empty, or a point, or a circle. (Hint: Introduce a conveniently located coordinate system.)

6. Given 4 points in space that do not lie on one plane, there is a unique sphere through these points. Prove this analytically. (Hint: Introduce a coordinate system in which one point is the origin, another lies on the x axis, and a third lies on the plane $z = 0$.)

7. Find a parabola with a vertical axis that passes through the points $(1, 3)$, $(2, -3)$, and $(4, 6)$.

8. Prove that given 3 distinct points, P, Q, and R, in the plane and a line l, there is a unique parabola passing through P, Q, and R, whose axis is parallel to l, provided that none of the segments PQ, QR, and RP, is parallel to l. (Hint: Introduce a conveniently located coordinate system.)

9. Consider a parabola with vertical axis, vertex O, and focus F. Suppose that a line through F intersects the parabola at 2 distinct points, P and Q. The segment PQ and an arc of the parabola bound a region R. Let A be the area of R. Write a formula expressing A in terms of the two lengths, OF and PQ.

10. The theorems stated in Problems 3 and 6 have a generalization in 4-space, and also in n-space, for every n. State these generalizations.

3/Functions

The study of calculus begins in this chapter. Here we develop the language of functions and limits. Since some readers may find it tedious to learn the fundamentals of a new language, some parts may be skipped and returned to later as they are needed.

There are three appendixes. The first (§5) deals with one-sided limits, infinite limits, and limits at infinity, and may be omitted at first reading. The other two contain proofs of some theorems that are stated, but not proved, in the main text. Since these theorems appear to be geometrically obvious, the reader may accept them on faith for the time being.

Mathematical analysis, the part of mathematics that includes calculus and all its ramifications, is dominated by the concept of continuity and the related concept of limits. These concepts were cast in their present form in the nineteenth century, primarily by Cauchy and Weierstrass. Since they were unavailable to the founders of calculus, the original calculus had a flavor of mystery. Mathematicians used the powerful new tool for solving problems and gaining insight without being able to justify their steps.

In presenting calculus to the modern reader, however, there is no need to be mysterious. We shall freely use the concept of continuity from the very beginning. To explain this concept, we must first define what is meant by a function.

§1 Functions and Graphs

1.1 Functions

A **function,** in mathematics, is a rule that assigns to each element of a set an element of the same or of another set. In the first half of this book we deal mostly with functions that assign to each real number, from some set of numbers, another real number. Such functions are called **real-valued functions of a real variable,** and will be called simply functions from now on. The set of numbers to which other numbers are assigned is called the **domain of definition** of a function. For functions we shall consider that this will be an interval or a collection of intervals. In particular, the domain of definition may be the whole number line. We shall not specify the domain of definition, unless this is necessary. Often it will be obvious from the context.

The rule that assigns one number to another may be of any nature. For example, we may want to assign to every number its square. We write this function in the form

$$x \mapsto x^2$$

(read: x goes into x^2). Here x is a **variable;** it represents an unspecified real number. The rule assigns a number to every number; the domain of definition is the whole number line.

Often one uses a variable, say y, to denote the number that is assigned to another number by our rule. In this case one writes

$$x \mapsto y = x^2,$$

or more briefly, $$y = x^2,$$

AUGUSTIN LOUIS CAUCHY (1789–1857). His indefatigable productivity, his inventiveness, and the breadth of his mathematical interests are similar to those of the great mathematicians of the eighteenth century. But Cauchy was a nineteenth-century mathematician and paid great attention to the logical foundation of mathematical analysis.

Cauchy had strong religious and political convictions. After the revolution of 1830, he followed the Bourbons to exile; when he returned to France, he would not accept a university position until the requirement of a loyalty oath to the government was waived.

KARL WEIERSTRASS (1815–1897) began his career as a secondary-school teacher. His profound papers, some of which appeared in such unlikely places as a high school graduation program, earned him a professorship in Berlin. From then on, his lectures and seminars had an immense impact on mathematics. The tendency to reduce all mathematics to numbers, and the insistence on complete rigor, are a result, in part, of Weierstrass' influence.

and one calls x (representing the number to which we assign another) the **independent variable** and y (which represents the number assigned to the first one) the **dependent variable.**

The function $x \mapsto x^2$ assigns to 2 the number 4, to -7 the number 49, and so forth. We say the function equals 4 at 2, or, the function takes on the value 49 at the point -7.

The symbols used for the variables are, of course, irrelevant. The formulas

$$t \mapsto t^2, \qquad \beta = \alpha^2, \qquad w \mapsto u = w^2, \qquad y \mapsto x = y^2$$

all represent the same function. In each case the rule is: assign to each number its square.

The function

$$x \mapsto \frac{1}{x}$$

assigns to each number, except 0, of course, its reciprocal. The domain of definition consists of the intervals $(-\infty, 0)$ and $(0, +\infty)$. The same function may be written as

$$y = \frac{1}{x}.$$

The rule defining the function may be more complicated, for instance,

$$x \mapsto \sqrt{1 + x} + \sqrt[4]{2 - x} + x^{10}.$$

This function is defined for x such that $1 + x \geq 0$ and $2 - x \geq 0$, that is, for $-1 \leq x \leq 2$.

Here we used a **convention:** if a function is defined by a formula, then, in the absence of specific instructions to the contrary, the domain of definition is assumed to be the largest set on which the formula makes sense.

Originally, mathematicians considered only functions defined by formulas. This turned out to be inconvenient, indeed an impediment to the development and applicability of mathematics. Therefore, we do not demand that the rule defining a function be given by a single formula. For instance,

$$x \mapsto \begin{cases} -x & \text{if } x \leq 0 \\ 1 & \text{if } 0 < x < 3 \\ \sqrt{x} & \text{if } 3 \leq x \end{cases}$$

is a well-defined function, defined for all x. An even weirder function is

$$y = \begin{cases} 0 & \text{if } x \text{ is rational} \\ 1 & \text{if } x \text{ is irrational,} \end{cases}$$

again defined for all x.

1.2 Variables denoting functions

Just as we use variables to denote unspecified numbers, we also use variables to denote unspecified functions. When we say "let f be a function," f represents some particular, but not specified, rule assigning numbers to other numbers. If we want f to represent some specified function, say the function $x \mapsto x^2$, we write $f(x) = x^2$. The symbol $f(x)$ is read "f of x" or "the value of f at x"; it represents the number the rule f assigns to the number x. Thus, if $f(x) = 2x^3$, then $f(0) = 0$, $f(-1) = -2$, $f(t) = 2t^3$, $f(2y) = 2(2y)^3 = 16y^3$, and so on.

If f and g are functions, the statement $f = g$ means that f and g are the same function, which means that f and g are defined on the same set, and $f(x) = g(x)$ for every x in this set. Instead of writing $f = g$, we also write $f(x) \equiv g(x)$.

Strictly speaking, one should distinguish between the function f and the number $f(x)$ which the rule f assigns to a particular number x. In practice, confusion seldom arises. One speaks, for instance, about the function x^2, that is, the rule $x \mapsto x^2$.

With every number, say a, there is associated a function $x \mapsto a$ that assigns to every number x the number a. This function is usually denoted by the same symbol as the number a. It is called a **constant,** or a constant function.

▶ *Examples*   *1.* What value does the function

$$u \mapsto \frac{2u + 1}{u^2 + 1}$$

assign to the number 3?

Answer.   If $u = 3$, then $(2u + 1)/(u^2 + 1) = (2 \cdot 3 + 1)/(3^2 + 1) = 7/10$. Our function assigns to the number 3 the number 7/10.

2. If $f(t) = (2t + 1)/(t^2 + 1)$, what is $f(3)$?

Answer.   This is the same function as above, only the names of the variables have been changed. We have $f(3) = 7/10$.

3. For which values of x is the function $x \mapsto \sqrt{x - 1} + 1/\sqrt{2 - x}$ defined?

Answer. In order to be able to compute $\sqrt{x - 1}$, the number $x - 1$ must be positive or zero. This gives the condition $x \geq 1$. Similarly, $\sqrt{2 - x}$ makes sense if $2 - x \geq 0$, that is, if $x \leq 2$. But in order to compute $1/\sqrt{2 - x}$, we must be sure that the denominator is not zero; thus the value $x = 2$ must be excluded. We collect all conditions obtained: $x \geq 1$, $x \leq 2$, $x \neq 2$. Conclusion: our function is defined for $1 \leq x < 2$.

4. For which values of t is the function $h(t) = \sqrt{\sqrt{t} - 1}$ defined?

Answer. Clearly $h(t)$ is defined whenever (1) $\sqrt{t} - 1$ is defined and (2) $\sqrt{t} - 1 \geq 0$. The first condition demands that $t \geq 0$. The second condition demands that $\sqrt{t} \geq 1$, that is, $t \geq 1$. Conclusion: $h(t)$ is defined for $t \geq 1$.◄

EXERCISES

1. What value does the function $x \mapsto 2x^2 - 1$ assign to the number 7? To the number $3/2$? To the number -4?
2. What value does the function $u \mapsto (u + 1)/(u^2 + 1)$ take on at the point -1? At the point $1/4$? At the point $\sqrt{2}$?
3. What does the function $y \mapsto 4y^3 + y^2 + 1$ equal at 2? At -2? At 0?
4. What is the value of the function $h \mapsto h^{1/3} + h$ when $h = -8$? When $h = 27$? When $h = 1/125$?
5. What value does the function $\alpha \mapsto (\alpha^2 + 1)(\alpha^3 - 2\alpha)$ assign to the number -1? To the number .3? To the number .02?
6. Consider the function $y = 3x^{-2} + x$ (that is, $x \mapsto 3x^{-2} + x$). What value does this function assign to the number -3? To the number 10? To the number $-\sqrt{2}$?
7. Consider the function $y = 2u^{-2/3} + (u + 1)^2$. What is the value of y when $u = 8$? When $u = .027$? When $u = -1$?
8. Consider the function $f(x) = 1/(x + 1) - x^3$ (that is, f is the name of the function $x \mapsto (1/(x + 1)) - x^3$. What is $f(2)$? $f(1/2)$? $f(-1/3)$?
9. Let $h(t) = (-t)^{-3/2} + (t - 1)^2$. What is $h(-4)$? $h(-9/4)$? $h((-8)^{1/3})$?
► 10. For which values of x is the function $x \mapsto 1/(x - 1) + 1/(x + 2) + \sqrt[4]{3x - 2}$ defined?
11. For which values of u is the function $f(u) = \dfrac{1}{\sqrt{4u - 1}} - \sqrt{1 - u^2}$ defined?
12. What is the domain of definition of the function $y = \dfrac{1}{r^3 - 1} + \sqrt{(r + 1)(r + 2)}$?
13. For which values of θ is the function $\theta \mapsto \dfrac{\sqrt{\theta^2 - 9}}{\sqrt{4 - (\theta - 8)^2}}$ defined?
14. What is the domain of definition of the function $f(z) = \sqrt{z - \sqrt{z^2 - z - 2}}$?
15. For which values of v is the function $\zeta(v) = \sqrt{|v| - \sqrt{|v| - v^2}}$ defined?

1.3 Sums, differences, products, and quotients of functions

Let f and g be two functions. The sum $f + g$, the difference $f - g$, the product fg, and the quotient f/g are the functions defined by

$$x \mapsto f(x) + g(x),$$
$$x \mapsto f(x) - g(x),$$
$$x \mapsto f(x) \cdot g(x)$$

and

$$x \mapsto \frac{f(x)}{g(x)},$$

respectively. These functions are defined for all those x for which both f and g are defined, except that the quotient f/g is not defined at points x such that $g(x) = 0$. The commutative, associative, and distributive laws hold for functions, because they hold for numbers. Thus $f + g = g + f$, $fg = gf$, $(f + g) + h = f + (g + h)$, $(fg)h = f(gh)$, and $f(g + h) = fg + fh$.

▶**Example** Let $f(x) = 2x^2 + 3$, $g(x) = \sqrt{x - 1}$. Then $f + g$, $f - g$, and fg are defined for all x such that $x \geq 1$, and f/g is defined for $x > 1$. For instance, $(fg)(5) = f(5)g(5) = 53 \cdot 2 = 106$. Also $2f$ is the function $x \mapsto 4x^2 + 6$; $-g = (-1)g$ is the function $x \mapsto -\sqrt{x - 1}$; $f - 3$ is the function $x \mapsto 2x^2$.◀

1.4 Composition of functions

If f and g are two functions, the **composite** function, denoted by $f \circ g$, is the function

$$x \mapsto f(g(x)).$$

It is defined for numbers x such that $g(x)$ is defined and the number $g(x)$ is in the domain of definition of f. The function $f \circ g$ is the rule: apply first g, then f. It is, in general, distinct from the function $g \circ f$ (apply first f, then g).

▶**Example** Let f and g be as in the previous example. Then $f \circ g$ is the function $x \mapsto 2g(x)^2 + 3$. Since $g(x) = \sqrt{x - 1}$, we see that $f \circ g$ is the function $x \mapsto 2(\sqrt{x - 1})^2 + 3 = 2x + 1$ for $x \geq 1$ and is not defined for $x \leq 1$. The function $g \circ f$ is the function $x \mapsto \sqrt{f(x) - 1}$. Since $f(x) = 2x^2 + 3$, we see that $g \circ f$ is the function $x \mapsto \sqrt{2x^2 + 2}$ and is defined for all x. In particular, $f(g(1)) = 3$ and $g(f(1)) = 2$.◀

EXERCISES

16. Let $f(x) = x - 5$, $g(x) = x + 5$. What value does $f + g$ assign to the number 4? What value does fg assign to the number 2? What value does $f \circ g$ assign to the number -1? What value does $g \circ f$ assign to the number 10?

17. Let $f(x) = 2x - 1$, $g(x) = x^2$. What are the following: $(f - g)(3)$, $(f/g)(3)$, $(f/g)(-2)$, $(f \circ g)(3)$, and $(g \circ f)(5/2)$?

18. Let $h(s) = s^2 - s + 1$, $L(s) = s/(s - 1)$. What are the following: $(2h - 3L)(4)$, $(h \circ h)(1)$, $(h \circ L \circ h)(2)$, and $(L \circ L \circ h)(-1)$?

19. Let $\alpha(t) = \sqrt{t^2 - 5}$, $\beta(t) = \sqrt{t^2 + 5}$. What are the following: $(\alpha \circ \beta)(2)$, $(\alpha \circ \beta)(-2)$, $(\beta \circ \alpha)(4)$, and $(\beta \circ \alpha)(-4)$?

▶ 20. Let $F(u) = u^2 + 2u - 3$, $K(u) = 1/(u - 1)$. What are the following: $(F + K)(5)$, $((F + K) \circ F)(2)$, $(F \circ (FK))(-1)$, and $(F + (K \circ K))(3)$?

21. Let $f(x) = 2x + 5$, $g(x) = y^2 + 1$, $h(x) = 1/x$. For each function in the left-hand column, find a function equal to it from the right-hand column.

(1) $f + g - h$.	(a) $w \mapsto (2w + 5)(w^2 + 1) + (2/w)$.
(2) $fg + 2h$.	(b) $w \mapsto 1/(w^2 + 1)$.
(3) $f \circ h$.	(c) $w \mapsto (2/w + 5)^2 + 1$.
(4) $h \circ g$.	(d) $w \mapsto 1/(w^2 + 2w + 6)$.
(5) $g \circ f \circ h$.	(e) $w \mapsto (w^2 + 1)^4 + 1$.
(6) $h \circ f \circ g$.	(f) $w \mapsto (2w + 5)/(10w + 27)$.
(7) $h \circ h \circ f \circ f$.	(g) $w \mapsto w^2 + 2w + 6 - (1/w)$.
(8) $h \circ f \circ h \circ f$.	(h) $w \mapsto 4w + 15$.
(9) $h \circ (f + g)$.	(i) $w \mapsto 1/(2w^2 + 7)$.
(10) $g \circ (gg)$.	(j) $w \mapsto (2/w) + 5$.

22. Let $R(x) = 1 + \sqrt{x}$, $S(x) = 1/(1 + x)$. Express $R + S$, RS, $R \circ S$, $S \circ R$ directly, using z as the independent variable.

23. Let $A(s) = (s + 1)^{2/3}$, $B(t) = t^3 - 2$. What are the following: $(A \circ B)(v)$, $(B \circ A)(w)$, and $(A \circ (B + 2))(x)$?

24. Let $f(x) = x^2$, $g(x) = x + 1$, $h(x) = 1/x$. Find an expression for the function $s \mapsto (s + 3)^{-4}$, using only f, g, h, and composition (do not use addition or multiplication). Find an expression for the function $t \mapsto t^3 + 2t^2 + 2t$, $t \neq 0$, using only f, g, h, composition, and multiplication.

25. Where is $f \circ g$ defined if $f(x) = \sqrt{x + 3}$ and $g(y) = 1 + y^2$?

26. Where is $\sigma \circ \tau$ defined if $\sigma(u) = 1/\sqrt{1 - u}$ and $\tau(v) = v^3$?

27. What is the domain of $G \circ K$ if $G(s) = \sqrt{s^2 - 4}$, $K(y) = \sqrt{y - 1}$?

28. For which values of u is $(g \circ f)(u)$ defined if $g(x) = \sqrt{9 - (x - 1)^2}$ and $f(z) = \sqrt{25 - z^2}$?

29. Where is $h \circ \zeta$ defined if $h(y) = (|y| - 12)^{1/6}$ and $\zeta(x) = x + 4$?

30. Where is $A \circ B$ defined if $A(s) = s^2$ and $B(t) = \sqrt{1 - t^2}$?

31. If $x \mapsto f(x)$ is defined for $x \leq 0$ and for $1 < x < 5$ and if $g(y) = 5/(y^2 - 4)$, where is $f \circ g$ defined?

1.5 Graphs

Fig. 3.1

In principle, the simplest way of representing a function f is by a **table,** listing in one column all numbers x for which f is defined, and in the second the value $f(x)$ assigned to each x. A partial table for the function $x \mapsto y = x^2$ looks as follows.

x	y
0	0
.1	.01
.2	.04
.3	.09
.4	.16

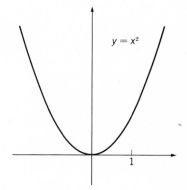

We can never write down all the entries in a table, since our functions are defined on infinite sets. It is possible and useful, however, to think of a function as a set of all entries in its table, that is, the set of all ordered pairs $(x, f(x))$, one for each number x for which the function is defined.

Fig. 3.2

The **graph** of a function f is the set of all points in the plane with coordinates $(x, f(x))$. The graph is thus a complete table of the function: all information about the function is contained therein and, if we draw the graph, or a part of it, we can "see" the function.

Here are a few examples. The graph of $y = x^2$ is, as we already know, a **parabola** (Fig. 3.1). That of $y = 1/x$ is shown in Fig. 3.2. This graph is called an **equilateral hyperbola.** The graph of $y = x^3$, shown in Fig. 3.3, is called a **cubic parabola.** The graph of

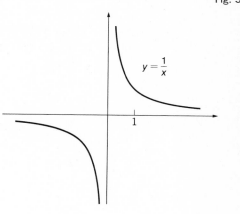

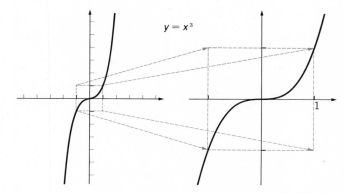

Fig. 3.3

$$f(x) = \begin{cases} x + 1 & \text{for } x < 0 \\ x^2 & \text{for } 0 \leq x \leq 1 \\ 2/x & \text{for } 1 < x, \end{cases}$$

Fig. 3.4

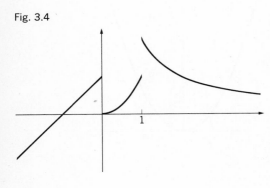

shown in Fig. 3.4, consists of three pieces.

For functions studied in calculus, the points of the graph are not scattered all over the plane but form what one calls a **curve** or several curves, as it was in the four examples just considered.

Not every set of points in the plane is the graph of a function. *A necessary and sufficient condition is that every vertical line should meet the set in at most one point.* If so, the set is the graph of the following function: if the vertical line $x = a$ does not meet the set, $f(a)$ is not defined; if $x = a$ meets the set at the point (a, b), then $f(a) = b$. (See Fig. 3.5.)

Fig. 3.5 illustrates this statement. The curve to the left satisfies our condition and is the graph of a function. The function is defined in an interval consisting of points on the x axis directly above or directly under the curve, like the point a. (The number a_1 is not in the domain of definition.)

On the other hand, the curve to the right violates our condition; there are vertical lines that meet the curve at two points. The curve is not the graph of a function. Indeed, if it were, what value would we assign to the number a_2?

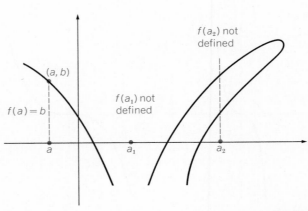

Fig. 3.5

▶ *Example*   The unit circle (center at the origin, radius 1) is not the graph of a function; the condition stated above is violated. The upper semicircle, the set of points (x, y) such that $x^2 + y^2 = 1$ and $y \geq 0$ is the graph of the function $x \mapsto \sqrt{1 - x^2}$. Similarly, the lower semicircle is the graph of $x \mapsto -\sqrt{1 - x^2}$. Both functions are defined for $-1 \leq x \leq 1$.◀

The shape of the graph reflects the properties of the function. We shall consider some examples.

1.6 Monotone functions

A function is called **increasing** if

$$f(x_1) < f(x_2) \qquad \text{whenever } x_1 < x_2.$$

This means that, given any two points on the graph, the one to the right is above the one to the left. The graph rises as one traverses it from left to right. An example is $f(x) = x^3$ graphed in Fig. 3.3.

A function is called **nondecreasing** if

$$f(x_1) \leq f(x_2) \qquad \text{for } x_1 < x_2.$$

This means that, given two points on the graph, the one to the right is not below the one to the left. The graph does not descend. An example is: $f(x) = 0$ for $x \leq 0$, $f(x) = x$ for $x > 0$ (Fig. 3.6).

Similarly, f is called **decreasing** if

$$f(x_1) > f(x_2) \qquad \text{for } x_1 < x_2$$

(the graph descends), or **nonincreasing** if

$$f(x_1) \geq f(x_2) \qquad \text{for } x_1 < x_2$$

(the graph does not rise).

A function is called **strictly monotone** if it is either increasing or decreasing, **monotone** if it is either nonincreasing or nondecreasing.

The function $f(x) = x^2$ is neither. It decreases for $x < 0$ and increases for $x > 0$. This is an example of a **piecewise monotone function**. A function is called piecewise monotone if every finite interval on which it is defined can be divided into finitely many intervals, on each of which the function is monotone. Another example is the function $x \mapsto \sin x$, which we shall discuss in Chapter 6.

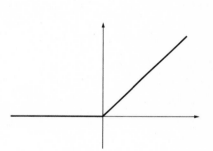

Fig. 3.6

Fig. 3.7

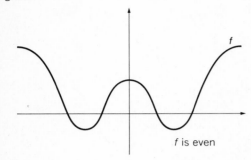

f is even

Fig. 3.8

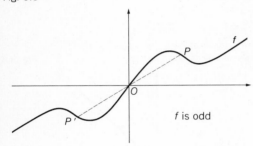

f is odd

1.7 Even and odd functions

A function f is called **even** if

$$f(-x) = f(x),$$

odd if

$$f(-x) = -f(x)$$

for all x for which $f(x)$ is defined; in both cases it is assumed that $f(x)$ is defined whenever $f(-x)$ is. For instance, x^2 is even and x^3 odd.

The graph of an even function is **symmetric about the y axis**. (See Fig. 3.7.) The graph of an odd function is **symmetric about the origin**. (This means that, if a point P lies on the graph, so does the point P' on the line joining P to O, and such that O is the midpoint of the segment PP'; see Fig. 3.8.)

Note that in general a function will be **neither even nor odd**; the function graphed in Fig. 3.6 is such a function.

EXERCISES

In Exercises 32 to 37, decide whether the given function is even, odd, or neither.

32. $x \mapsto x^4 + 2$.

33. $s \mapsto (s^{1/3} + 5)/s$.

34. $f(u) = (u^2 + 1)^3 - u^4$.

35. $y = t^3 + t^2$.

▶ 36. $g(z) = \sqrt{z^3 + 1}$.

37. $v \mapsto \sqrt{v^2 + 1}/|v|$.

38. Can a function be both odd and even? If so, try to determine all such functions, or at least give an example of one.

In Exercises 39 to 44, decide whether the given function is increasing, nonincreasing, nondecreasing, decreasing, or none of these. Naturally, an increasing function is automatically nondecreasing; in this case, give the stronger of the two answers.

39. $t \mapsto t^5$.

40. $f(x) = 1/\sqrt{x}$.

41. $z = x^3 + x^2$.

42. $g(y) = |y| + y$.

43. $h(u) = |u|/u$.

▶ 44. $k(z) = z + 1$ if $z \le 0$; z^2 if $z > 0$.

In Exercises 45 to 48, each of the functions is a piecewise monotone function. In each case, divide the domain of the function into intervals where it is monotone and, for each such interval, decide whether the function is increasing, nondecreasing, decreasing, or nonincreasing.

45. $A(x) = (x - 1)^4$.

46. $\phi(y) = y^2 + y$.

47. $t \to t^2$ if $t \le 0$; $1 - t$ if $t > 0$.

48. $g(h) = 1/h$ if $h < 0$; $(h - 1)^2$ if $0 \le h \le 2$; $1/h^3$ if $h > 2$.

In Exercises 49 to 56, decide whether the function whose graph is given is even, odd, or neither. If the function is piecewise monotone, indicate on the horizontal axis the intervals where it is monotone. For each such interval, indicate whether the function is increasing, nondecreasing, decreasing, or nonincreasing.

49.

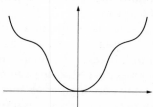

53.

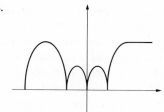

50.

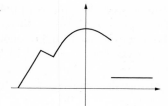

54.

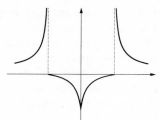

51.

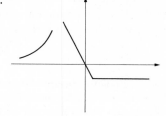

55.

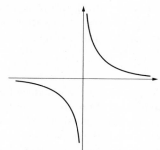

52.

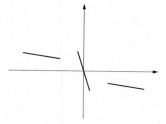

56.

1.8 Applications of functions

In the application of mathematics, functions are used to represent relationships between measurable, observable quantities.

Suppose we measure the temperature at some spot during a certain period of time. These measurements can be plotted as a curve; this curve is the graph of a function. One says in this case: the temperature is a function of time. But one ought to remember that we obtain a mathematical function, that is, a rule assigning one number to another, only after we agree on the **units of measurement** and on **reference points**. If temperature is measured in degrees Fahrenheit and time in minutes, time zero being noon, we obtain one function. If temperature is measured in degrees centigrade and time in seconds, time zero being 2 P.M., the same measurements will be represented by a different function. It is, of course, easy to convert the first function into the second.

Similar remarks apply when the numbers (variables) involved in the definition of a function measure distances, areas, volumes, weights, pressures, intensities, velocities, accelerations, costs, prices, and so forth. A particular case is of special significance.

Given a function $x \mapsto y = f(x)$, we may think of x as standing for time and y as being the coordinate of a point that moves along a number line. A constant function, for instance, represents a point at rest, while an increasing one represents motion in the positive direction. This kinematic interpretation of a function will turn out to be very important.

EXERCISES

57. The distance of a moving particle from a fixed base point is measured during a certain interval of time (that is, distance is being treated as a function of time). If distance is measured in feet and time in seconds (from the start of the interval), one gets the function $t \mapsto t^2 + 3t$. What function does one get if distance is measured in inches and time in minutes (again from the start of the interval)?

58. A house is being moved 1200 ft along a straight road starting at noon. If the distance from the house to the starting point is measured in feet, and the time from noon is measured in minutes, one gets the function

$$\text{distance} = f(t) = \begin{cases} t & \text{for } 0 \leq t < 600 \\ 600 & \text{for } 600 \leq t \leq 660 \\ 2t - 720 & \text{for } 660 < t \leq 960. \end{cases}$$

What function does one get if distance from the house to the terminal point is measured in yards and time from noon is measured in hours?

59. The temperature of an incandescent wire 1 meter long is measured at each point. If the temperature is measured in degrees centigrade and distance from

one end is measured in centimeters, one gets the function

$$\text{temperature} = \phi(s) = \begin{cases} s^2 + 100 & \text{for } 0 \le s \le 50 \\ (100 - s)^2 & \text{for } 50 \le s \le 100. \end{cases}$$

What function does one get if temperature is measured in degrees Fahrenheit and distance (from the same end) is measured in meters? Remember that, if a temperature is C on the centigrade scale and F on the Fahrenheit scale, $F = (9C/5) + 32$.

60. A coke oven is being fired up and the temperature is measured during a certain interval of time. If temperature is measured in degrees Fahrenheit and time (from the start of the interval) is measured in minutes, one gets the function

$$\text{temperature} = h(t) = \begin{cases} \frac{1}{3}t^2 & \text{for } 0 \le t \le 30 \\ 10t & \text{for } 30 \le t \le 120 \\ 1200 & \text{for } 120 \le t. \end{cases}$$

What function does one get if temperature is measured in degrees centigrade and time (from the start of the interval) in hours?

61. The velocity of a car is measured during a certain interval of time. If velocity is measured in feet per minute and time in seconds, one gets the function

$$\text{velocity} = v(t) = \begin{cases} 60t & \text{for } 0 \le t \le 15 \\ 900 & \text{for } 15 \le t. \end{cases}$$

What function does one get if velocity is measured in miles per hour and time in minutes?

The definition of a function as given in §1.1 is very general. In practice, one uses mostly certain special classes of functions. Three such classes will be described in the present section.

2.1 Linear functions

The simplest functions and some of the most important ones are **linear functions** $f(x) = mx + b$. The graph of a linear function is a line. (See Chapter 2, §2.) Conversely, every nonvertical line is the graph of a linear function. The slope of the graph is also called the slope of the linear function. The function is increasing if the slope m is positive, and the slope measures how fast the function is increasing. If $m = 0$, the function is constant; if $m < 0$, the function decreases. The **identity function** $x \mapsto x$ has slope 1.

For any function f, a number α such that $f(\alpha) = 0$ is called a **root,** or a **zero,** of f. A nonconstant linear function has exactly one zero (since $mx + b = 0$ for $x = -b/m$ and for no other value of x).

2.2 Quadratic functions

Let a, b, and c be fixed numbers with $a \neq 0$. The function

$$y = f(x) = ax^2 + bx + c$$

is called a **quadratic function** with *leading coefficient a*. The graph of a quadratic function is a *parabola* with vertical axis (see Chapter 2, §4.3).

Completing the square, we write the function in the form

$$y = f(x) = a\left\{\left(x + \frac{b}{2a}\right)^2 + \frac{c}{a} - \frac{b^2}{4a^2}\right\}$$

or $\qquad y = a\left(x + \frac{b}{2a}\right)^2 + \frac{D}{4a}$ $\qquad$ where $D = 4ac - b^2$.

Assume that $a > 0$. Then the smallest value of y corresponds to $x = -b/2a$. Hence the point

$$\left(-\frac{b}{2a}, \frac{D}{4a}\right)$$

is the vertex of the parabola. If $D > 0$, the vertex and the whole curve lie in the upper half-plane; there are no roots. If $D = 0$, the vertex lies on the x axis. In this case the function may be written as

$$f(x) = a(x - r)^2 \qquad \text{where } r = -\frac{b}{2a}.$$

Thus f is the square of a linear function $\sqrt{a}(x - r)$. The root r is called a **double root**. If $D < 0$, finally, the vertex lies in the lower half-plane. The parabola intersects the x axis at two distinct points $x = r_1$ and $x = r_2$, where

$$r_1 = \frac{-b + \sqrt{-D}}{2a}, \qquad r_2 = \frac{-b - \sqrt{-D}}{2a},$$

and the function may be written as a product of two distinct linear functions

$$f(x) = a(x - r_1)(x - r_2),$$

which the reader is asked to verify. The various cases are shown in Fig. 3.9.

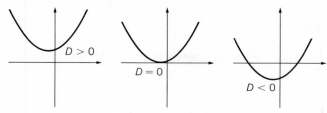

Fig. 3.9

The reader may also verify that, in the case $a < 0$, we have the same result: our function has two distinct roots, one double root, or no roots when $D < 0$, $D = 0$, or $D > 0$, respectively.

EXERCISES

1. For which values of α does $x \mapsto x^2 - 2\alpha x + 10$ have two distinct roots?
2. For which values of s does $z \mapsto z^2 + z + 4s$ have no roots?
3. For which values of a does $f(y) = ay^2 + 2y + 1$ have one double root?
▶ 4. For which values of z does the graph of $x \mapsto x^2 + 3zx + 4$ intersect the graph of $x \mapsto 1$ in two distinct points?
5. Suppose $x \mapsto x^2 + rx + s$ and $y \mapsto y^2 + rx - s + 1/2$ each have only one double root. Determine r and s as far as possible.
6. For which values of u does $s \mapsto s^2 + 2u^2s + u$ have two distinct roots?
7. Write $x \mapsto x^2 + 3x + 1$ as the product of two linear functions.
8. Write $s \mapsto 2s^2 - 4s + 1$ as the product of two linear functions.

2.3 Polynomials

Linear and quadratic functions are special cases of **polynomials**. A polynomial function is one of the form

$$f(x) = a_0 x^n + a_1 x^{n-1} + a_2 x^{n-2} + \cdots + a_{n-1}x + a_n,$$

where n is a nonnegative integer and a_0, a_1, $\cdots$, a_n are fixed numbers, with $a_0 \neq 0$. One calls n the **degree** of the polynomial, a_0, a_1, $\cdots$, a_n the **coefficients**, and a_0 the **leading coefficient**. A polynomial of degree 0 is a constant, one of degree 1, a linear function, and one of degree 2, a quadratic function.

We shall review some properties of polynomials which are probably familiar to the reader.

Theorem 1   IF f AND g ARE POLYNOMIALS, SO ARE $f + g$, fg, AND $f \circ g$.

The statement is obvious. Instead of giving a formal proof, we ask the reader to consider a few examples.

2.4 Long division

We recall that the process of long division may be applied to polynomials. Suppose, for instance, that we want to divide $f(x) = 4x^4 - 3x + 15$ by $g(x) = 2x^2 + 2x + 5$. The calculation reads

$$
\begin{array}{r}
2x^2 - 2x\ \ - 3 \\[2pt]
2x^2 + 2x + 5 \overline{)\,4x^4 - 3x\ \ + 15\ \ } \\[2pt]
\underline{4x^4 + 4x^3 + 10x^2\ \ } \\[2pt]
-\,4x^3 - 10x^2 - 3x + 15 \\[2pt]
\underline{-\,4x^3 - \ 4x^2 - 10x\ \ } \\[2pt]
-\ 6x^2 + \ 7x + 15 \\[2pt]
\underline{-\ 6x^2 - \ 6x - 15\ } \\[2pt]
13x + 30;
\end{array}
$$

we stop here, since the degree of the remainder (namely, 1) is less than that of the divisor (namely, 2). The result of the calculation reads

$$4x^4 - 3x + 15 = (2x^2 + 2x + 5)(2x^2 - 2x - 3) + (13x + 30)$$

$$\underset{\text{dividend}}{} \qquad \underset{\text{divisor}}{} \qquad \underset{\text{quotient}}{} \qquad \underset{\text{remainder}}{}$$

This process can be applied in all cases and yields Theorem 2.

Theorem 2   LET $f(x)$ AND $g(x)$ BE POLYNOMIALS OF DEGREES n AND m, RESPECTIVELY, WITH $n \geq m > 0$. THEN THERE IS A POLYNOMIAL $Q(x)$, THE QUOTIENT, OF DEGREE $n - m$, AND A POLYNOMIAL $R(x)$, THE REMAINDER, OF DEGREE LESS THAN m, SUCH THAT

$$f(x) = g(x)Q(x) + R(x).$$

(IF $R \equiv 0$, ONE SAYS THAT f IS DIVISIBLE BY g, OR THAT g DIVIDES f.)

An important special case occurs if $n > 1$ and $g(x) = x - r$, r being some number.

Theorem 3   IF A POLYNOMIAL $f(x)$ OF DEGREE $n > 1$ IS DIVIDED BY THE LINEAR FUNCTION $x - r$, THE REMAINDER IS THE NUMBER $f(r)$.

Proof.   By Theorem 2 applied to the case $n > 1$, $m = 1$, there is a polynomial $Q(x)$ of degree $n - 1$ and a polynomial of degree 0, that is, a constant R, such that

$$f(x) = (x - r)Q(x) + R$$

for every number x. For $x = r$, we get $f(r) = 0 \cdot Q(r) + R = R$.

2.5 Roots of polynomials

Theorem 3 yields information on the **roots of polynomials.** Let $f(x)$ be a polynomial of degree $n > 1$ and r_1 a root of f. This means that $f(r_1) = 0$. By Theorem 3, the linear function $x - r_1$ divides $f(x)$. Thus $f(x) = (x - r_1)Q_1(x)$, where Q_1 is a polynomial of degree $n - 1$. Let Q_1 have a root r_2. Then we have, as before, $Q_1(x) = (x - r_2)Q_2(x)$ with Q_2 of degree $n - 2$. If Q_2 has a root r_3, then $Q_2(x) = (x - r_3)Q_3(x)$, Q_3 of degree $n - 3$. This process must stop after n steps; it may stop earlier if we obtain a polynomial Q without roots. (Here we do not admit so-called complex numbers as roots.) Thus one obtains

$$f(x) = (x - r_1)(x - r_2) \cdots (x - r_k)Q(x),$$

where $Q(x)$ is a polynomial of degree $n - k$ without roots. (If f itself has no roots, $k = 0$ and the linear factors are absent, so that $f = Q$.)

Of course, the numbers $r_1, \cdots, r_k$ need not be distinct. If r is a number such that exactly m of the numbers $r_1, \cdots, r_k$ are equal to r, one says that r is an m-fold root of f, or a **root of multiplicity** m. In other words, r is a root of $f(x)$ of multiplicity m if

$$f(x) = (x - r)^m P(x),$$

where $P(x)$ is a polynomial such that $P(r) \neq 0$.

Thus we have established the following theorem.

Theorem 4   A POLYNOMIAL OF DEGREE n HAS AT MOST n (REAL) ROOTS, COUNTING THEIR MULTIPLICITIES.

▶**Example**   Consider the polynomial

$$f(x) = 2x^8 - 10x^7 + 8x^6 + 8x^5 + 6x^4 + 18x^3.$$

Clearly, 0 is a triple root. We want to find the other roots. Division by $2x^3$ yields the polynomial $x^5 - 5x^4 + 4x^3 + 4x^2 + 3x + 9$. By trial and error we see that -1 is a root. Division by $x + 1$ yields the polynomial $x^4 - 6x^3 + 10x^2 - 6x + 9$; -1 is not a root of the polynomial, but 3 is. Division by $x - 3$ yields $x^3 - 3x^2 + x - 3$. This polynomial again has the root 3; dividing it by $x - 3$, we obtain $x^2 + 1$, a polynomial without roots. Hence

$$f(x) = 2x^3(x + 1)(x - 3)^2(x^2 + 1).$$

The roots are 0 with multiplicity 3, 3 with multiplicity 2, and -1 with multiplicity 1.◀

EXERCISES

In Exercises 9 to 16, perform the indicated divisions, writing your answer in the form (dividend) = (divisor)(quotient) + (remainder).

9. Divide $f(x) = 3x^2 + x^2 - x + 2$ by $g(x) = x^2 + x - 2$.
10. Divide $h(s) = s^5 - 2s^4 + s^3 - 6s^2 + s + 8$ by $k(s) = s^2 + 1$.
11. Divide $y \mapsto y^4 + y - 1$ by $y \mapsto y^3 + 1$.
12. Divide $A(t) = t^5 + 2t^4 - t - 2$ by $B(t) = t + 2$.
13. Divide $f(x) = 2x^5$ by $g(x) = 3x^2 + x$.
14. Divide $\alpha(z) = z^3 + z^2 - 1$ by $\beta(z) = z - c$, where c is a constant.
15. Divide $x \mapsto x^3 + 2x - 5$ by $x \mapsto 7$.
16. Divide $h(u) = u^2 + 2u - 3$ by $k(u) = u^2 - u - 1$.

In Exercises 17 to 21, find the value of the polynomial at the given point r by long division.

17. $f(x) = x^3 + x - 1$, $r = 2$.
18. $g(x) = 2x^4 - x^2 + 2x - 1$, $r = -1$.
19. $s \mapsto 4 - 2s + s^2 - s^3$, $r = -2$.
20. $h(y) = 2y^3 + 2y - 1$, $r = 1/2$.
21. $z \mapsto z^3/3 - 2z^2 + z + 6$, $r = 3$.
22. Find a polynomial that has a double root at 1, single roots at -1 and 0, and no others.
23. Find a polynomial whose roots are 2 with multiplicity 3 and $\pm\sqrt{2}$, each with multiplicity 1.
24. Find a polynomial that has a double root at -2, a double root at 3, a triple root at $1/2$, and no others.
25. The polynomial $f(x) = x^4 + x^3 - 7x^2 - 13x - 6$ has a double root at -1; find the other roots.
▶ 26. The polynomial $y \mapsto y^5 - 7y^3 + 2y^2 + 12y - 8$ has a double root at -2 and a single root at 2; find the other roots.
27. Find all the real roots (and their multiplicities) of $(x - 2)^{10}(x^{10} + 2)$.
28. Find all the real roots (and their multiplicities) of $(x + 3)^3(x - 1)^2(x^4 - 1)^5$.

2.6 Rational functions

A **rational function** is a function of the form

$$\phi(x) = \frac{f(x)}{g(x)}$$

where f and g are polynomials, $g(x) \not\equiv 0$. A polynomial is also a rational function [corresponding to $g(x) \equiv 1$.] The function ϕ is defined for all values of x, except for the roots of the denominator g.

The validity of the following statement is obvious.

Theorem 5   LET f AND g BE RATIONAL FUNCTIONS. THEN $f + g$, fg AND $f \circ g$ ARE RATIONAL FUNCTIONS, AND IF $g(x) \not\equiv 0$, SO IS f/g.

The relation between polynomials and rational functions has a formal similarity to that between integers and rational numbers. The rational numbers form a field, that is, they satisfy Postulates **1** to **12** of Chapter 1, §2. The integers do not form a field, since the quotient of two integers needs to be an integer; but they satisfy Postulates **1** to **11**. We abbreviate this by saying that the integers form a **ring.** Similarly, the *rational functions form a field;* they satisfy all Postulates **1** to **12**, with the word "number" replaced by "function." The polynomials, on the other hand, form a ring. (Note that in this interpretation the parts of 0 and 1 are played by the constant functions 0 and 1.) Every rational function is a quotient of polynomials, just as every rational number is a quotient of integers.

EXERCISES

29. Given $f(x) = (1 + x)/(1 - x)$ and $g(x) = 1 + x^2$, find $f + g$, $f \circ g$, and $g \circ f$. Show that these are rational functions.
30. Given $R(y) = 1/(1 + y - y^2)$ and $S(y) = y/(1 + y)$, find $R - S$, R/S, and $S \circ R$. Show that these are rational functions.
31. Given $A(t) = (t^2 + 2)/t$ and $B(t) = t/(t - 3)$, find $A + 2B$, AB, and $A \circ B$. Show that these are rational functions.
32. Given $\alpha(z) = (z^2 + z - 3)/(z + 2)$ and $\beta(z) = 1/z^2$, find $\alpha\beta + \beta$, α/β, and $\alpha \circ \beta$. Show that these are rational functions.
33. Given $f(x) = (x^2 - x)/(x + 3)$, $g(x) = 1/(x + 1)$, and $h(x) = x^2/(x + 1)$, find $[g \circ (g \circ h)] + f$. Show that this is a rational function.

2.7 Radical functions

A rational function may be defined as a function obtained by applying the **rational operations** (addition, multiplication, subtraction, division) to the constants and the identity function $x \mapsto x$. A **radical function** is a function obtained by applying the rational operations *and composition* to the constants and the functions $x \mapsto x^{1/n}$, n a positive integer. In other words, a radical function is one that is defined by a formula involving the symbol (say, x) for the independent variable, numbers, the signs $+$, $-$, $.$, $/$, and rational exponents. The domain of definition of such a function is an interval or several intervals.

▶*Example*   The function

$$f(x) = 2x^{-1/2} + (1 - x^{1/2})^{1/4}$$

is a radical function. It is defined if $x \geq 0$, $x^{1/2} > 0$, and $1 - x^{1/2} \geq 0$, that is, for $0 < x \leq 1$. ◀

The following statement is essentially a repetition of the definition.

Theorem 6   IF f AND g ARE RADICAL FUNCTIONS, SO ARE $f + g$, fg, f/g, AND $f \circ g$, PROVIDED THESE FUNCTIONS ARE DEFINED.

Observe that if, for instance, $f(x) = \sqrt{x - 2}$ and $g(x) = -\sqrt{1 - x}$, then we cannot define $f + g$, fg, f/g, or $f \circ g$.

We note that the set of radical functions contains the field of rational functions as a subset.

EXERCISES

34. Given $f(x) = (1 + \sqrt{x})/x$ and $g(x) = x/\sqrt{x - 1}$, find $f + g$, f/g, and $f \circ g$. Find $(f \circ g)$ (4).
35. Given $h(y) = 2y^{-2/3} + y^2$ and $k(y) = (y^3 + 2)/\sqrt{y - 1}$, find $h - k$, hk, and $k \circ h$. Find $(k \circ h)$ (9).
▶ 36. Suppose $f(x) = \sqrt{1 + \sqrt{x}}$, $g(x) = 1/\sqrt{x}$, and $h(x) = 4/x$. Find the function $f \circ g \circ h \circ f$. Find the value of this function at 225.
37. Suppose $A(z) = \sqrt{z}$, $B(z) = \sqrt{-z}$, $C(z) = 1/z$. Is $A + B + C$ defined? A/B? $A \circ B$?

2.8 Extending the definition of a rational function

We stated before that a rational function, that is, a quotient of two polynomials, is defined wherever the denominator is not zero. But it may happen that the definition of such a function can also be extended, *in a natural way,* to some points that are roots of the denominator.

Consider, for instance, the very simple function

$$f(x) = \frac{x^2 - 1}{x - 1}.$$

The function is not defined for $x = 1$, since substituting $x = 1$ into the formula gives the meaningless expression $\frac{0}{0}$. But we observe that we can also write

$$f(x) = \frac{(x + 1)(x - 1)}{x - 1}$$

or, canceling $(x - 1)$, $f(x) = x + 1$. This function is also defined for $x = 1$; there it takes on the value 2.

In working with rational functions, we often assume that every linear factor that appears both in the numerator and in the denominator has been canceled and that the domain of definition of the function has been changed accordingly.

For instance, the function

$$(1) \qquad \phi(x) = \frac{x^3 - 2x^2 + x}{x^2 - 1}$$

is originally defined for $x \neq 1, -1$. But $x^3 - 2x^2 + x = x(x - 1)^2$ and $x^2 - 1 = (x - 1)(x + 1)$. Hence we would like to write

$$(2) \qquad \phi(x) = \frac{x(x - 1)}{x + 1}$$

and consider the function as defined also for $x = 1$, with $\phi(1) = 0$.

2.9 Limits of rational functions

Let us have a critical look at the previous discussion. We argued:

$$(3) \qquad \frac{x^3 - 2x^2 + x}{x^2 - 1} = \frac{x(x - 1)^2}{(x + 1)(x - 1)} = \frac{x(x - 1)}{x + 1}.$$

There can be no doubt as to the legitimacy of the first step; no matter which number we substitute for x, the first and second fractions either both represent the same number, or are both not defined. The second step, however, dividing the numerator and the denominator by $x - 1$, is meaningful only if $x - 1$ stands for a number other than zero, that is, if x is not 1. Have we any right to say that

$$x \mapsto \frac{x^3 - 2x^2 + x}{x^2 - 1} \qquad \text{and} \qquad x \mapsto \frac{x(x - 1)}{x + 1}$$

are the same function? For values of x different from -1 and 1, we do. For the value $x = 1$, no. All we can say is that we *want* to modify the definition of the first function, so as to have it take on the value of 0 at $x = 1$, where it was previously not defined.

This way of extending the definition of our function is *natural*, while any other way of defining the value of the function at 1 would be unnatural. For, if we consider a value x that is not 1 but close to 1, then all three fractions (3) are equal to each other. In the last fraction the numerator is close to 0, whereas the denominator is close to 2. The whole fraction is therefore close to 0. It is also clear (and can be made precise) that the closer x gets to 1, the closer the common value of the three fractions (3) gets to 0. Indeed, the fractions (3) will be as close to 0 as we want, provided we take for x numbers that are distinct from, but sufficiently close to, 1.

Fig. 3.10

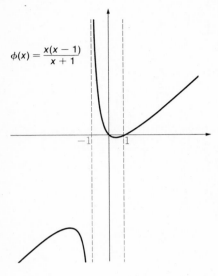

$$\phi(x) = \frac{x(x-1)}{x+1}$$

Fig. 3.11

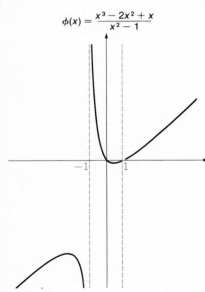

$$\phi(x) = \frac{x^3 - 2x^2 + x}{x^2 - 1}$$

We see therefore that our way of extending the definition of $\phi(x)$ to $x = 1$ is natural in the following sense: as the independent variable x "approaches 1," the function value $\phi(x)$ "approaches 0." We abbreviate this statement by writing

$$\lim_{x \to 1} \phi(x) = 0.$$

(This is read: the **limit** of $\phi(x)$, as x approaches 1, is 0.)

Another way of looking at this is as follows. Suppose we graph the function (2). We obtain the curve shown in Fig. 3.10; it has no breaks except at $x = -1$. Suppose we graph the function (1); we obtain the same curve, except that the point $(1, 0)$ is missing. (See Fig. 3.11.) If we define the value $\phi(1)$ to be any number different from 0, the graph of the function would have a break at $x = 1$.

2.10 Limits of other functions

The method of extending the definition of a function $f(x)$ to a point x_0 where it has not yet been defined, by finding the number the dependent variable "approaches" as the independent variable "approaches" the value x_0, is not restricted to rational functions. Consider, for instance, the function

$$f(x) = \frac{\sqrt{|x|} - 16}{\sqrt[4]{|x|} - 4}.$$

This function is not defined for $x = 256$. However, we have $\sqrt{|x|} - 16 = (\sqrt[4]{|x|} - 4)(\sqrt[4]{|x|} + 4)$, so that for $x \neq 256$ we may write

$$f(x) = \sqrt[4]{|x|} + 4.$$

It is now easy to see that

$$\lim_{x \to 256} f(x) = 8;$$

the natural way of defining $f(256)$ is to set it equal to 8.

One should not believe that the method always works—it does not. The function

$$\phi(x) = \frac{x^3 - 2x^2 + x}{x^2 - 1},$$

considered before and graphed in Fig. 3.11, is not defined for $x = -1$, and whereas we are free to choose any number we want and call it

$\phi(-1)$, there is no natural way of making this choice. As x approaches -1, the function $\phi(x)$ takes on arbitrarily large positive and negative values. There is no number that the function "approaches." We say in this case that the function has no limit at $x = -1$.

The limit concept is crucial in calculus. We have given only a very brief and informal introduction to this concept. We shall consider it again later, and we shall base our discussion on the closely related concept of continuity.

EXERCISES

In Exercises 38 to 47, determine any points where the given function is undefined and, if possible, extend the definition of the function to these points so that the graph is continuous.

38. $x \mapsto (x-1)^2/(x^2-1)$.
39. $x \mapsto (2x^2 + x - 3)/(2x^2 - x - 1)$.
40. $x \mapsto (x^2 - x)/[(x-1)^2x]$.
41. $x \mapsto (x^{2/3} - 4)/(2x^{2/3} - 3x^{1/3} - 2)$.
42. $x \mapsto (x-2)/(x^2-4)$.
43. $x \mapsto (x^{2/3} - 5x^{1/3} + 4)/[(x^{2/3} - 1)(x^{2/3} - 4)]$.
▶ 44. $x \mapsto (x^2 - 6x + 9)/[(x^2 - 9)(x - 3)]$.
45. $x \mapsto (x^3 + 3x^2 - x - 3)/(x^3 + x^2 - 2x)$.
46. $x \mapsto (x-2)(x-1)^3/[x(x^2 + x - 2)^2]$.
47. $x \mapsto x^2/(\sqrt{x^2 + 1} - 1)$.

3.1 Examples

An important property that a function may or may not have is continuity. Roughly speaking, *a function is* **continuous** *if its graph has no breaks.* Before giving a precise definition, we look at some examples.

Consider the functions whose graphs are shown in Figs. 3.12 to 3.17. We note that the graphs in Figs. 3.12 and 3.13 each *consist of one piece.*

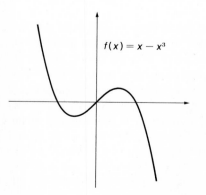

$f(x) = x - x^3$

Fig. 3.12

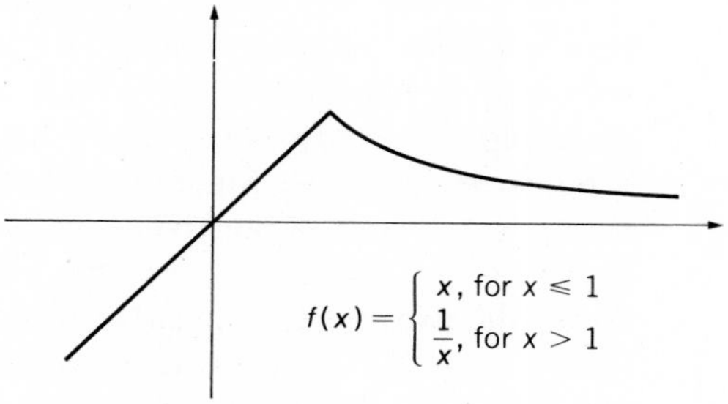

$f(x) = \begin{cases} x, & \text{for } x \le 1 \\ \dfrac{1}{x}, & \text{for } x > 1 \end{cases}$

Fig. 3.13

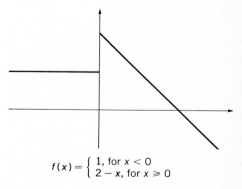

$f(x) = \begin{cases} 1, & \text{for } x < 0 \\ 2 - x, & \text{for } x \ge 0 \end{cases}$

Fig. 3.14

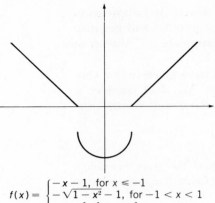

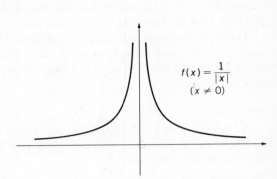

$$f(x) = \begin{cases} -x - 1, & \text{for } x \leqslant -1 \\ -\sqrt{1 - x^2} - 1, & \text{for } -1 < x < 1 \\ x - 1, & \text{for } x \geqslant 1 \end{cases}$$

Fig. 3.15

$$f(x) = \frac{1}{|x|}$$
$$(x \neq 0)$$

Fig. 3.16

Fig. 3.17

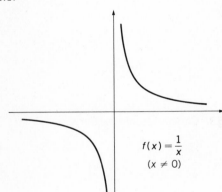

$$f(x) = \frac{1}{x}$$
$$(x \neq 0)$$

Fig. 3.18

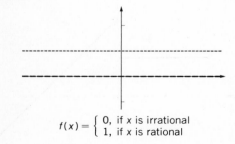

$$f(x) = \begin{cases} 0, & \text{if } x \text{ is irrational} \\ 1, & \text{if } x \text{ is rational} \end{cases}$$

The graphs of the other functions have breaks. For the function in Fig. 3.14, the break occurs at $x = 0$, for the one in Fig. 3.15 at $x = -1$ and $x = 1$. The functions in Figs. 3.16 and 3.17 have breaks at $x = 0$. These functions are not defined at $x = 0$ and for small values of $|x|$ the number $y = f(x)$ becomes very large in absolute value. The first two functions are called continuous. Functions like those in Figs. 3.14, 3.15, 3.16, and 3.17 are called **discontinuous**. Another example of a discontinuous function is the function whose graph is sketched (or rather, hinted at) in Fig. 3.18; its graph has a break at every point.

If we interpret x as time and $y = f(x)$ as the position of a moving point, the first two functions represent what one commonly calls continuous motions. The function in Fig. 3.14 represents the motion of a point that is at rest until the time 0, then abruptly jumps to a new position, and then moves continuously. In the motion represented by the function, there are two sudden jumps, at time -1 and 1. It is difficult or even impossible to visualize the other functions as representing motion of a point.

There is another way of describing the difference between the first two functions and the others. In Cases 1 and 2, *if we consider two values of x, say x_1 and x_2, which are close to each other, the corresponding values of y, $f(x_1)$ and $f(x_2)$, are also close to each other.* This is not so for the other functions. For the one in Fig. 3.14, for instance, if x_1 is negative and x_2 positive, then $|f(x_2) - f(x_1)|$ is close to 1 when x_2 is close to x_1. For the function in Fig. 3.16, if x_1 is positive and small, there is always another positive number x_2, such that $0 < x_2 < x_1$ and $|f(x_2) - f(x_1)|$ is as large as we please. The same is true for Fig. 3.15 and Fig. 3.17. Finally, for Fig. 3.15 and Fig. 3.18, $|f(x_2) - f(x_1)| = 1$ whenever x_2 is rational and x_1 is irrational, no matter how small the difference $|x_2 - x_1|$ is.

Also, we note that the functions in Figs. 3.14 to 3.17 are "discontinuous" only at certain points ($x = 0$ for Figs. 3.14, 3.16 and 3.17 and $x = -1$ and $x = 1$ for Fig. 3.15) and continuous elsewhere, while the function in Fig. 3.18 is discontinuous everywhere.

Continuous functions are of paramount importance in mathematics and its applications. In order to make valid general statements about such functions, we must give a precise definition. This will be done in the next subsection. We shall state precisely what we mean by saying "the function $f(x)$ is continuous at a point x_0." In doing this, we shall be guided by the examples above. However, for most functions ordinarily encountered in calculus, one needs no formal definition in order to tell where these functions are continuous and where they are discontinuous.

EXERCISES

In Exercises 1 to 6, indicate on the horizontal axis where the function whose graph is given is discontinuous.

1.

4.

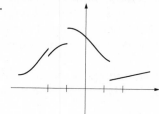

2.

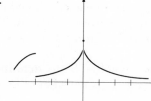

5.

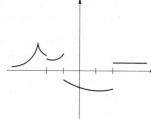

3.

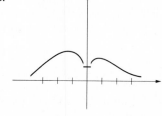

6.

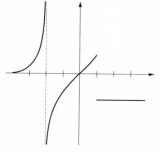

7. Give a function $x \mapsto f(x)$ that is defined for all x and is continuous everywhere, except at -1, 2, and 3. Try to define this function by a simple formula.

8. Give a function $s \mapsto g(s)$ that is defined for all s and is continuous everywhere, except at $-1/2$, 0, and 8.

9. Give a function $y = h(z)$ that is defined for all z except $z = 0$ and is continuous everywhere in its domain, except at -1 and $+1$.

10. Give a function $x \mapsto f(x)$ that is defined and continuous for all x, and equals $x \mapsto x^2$ for $x \leq 0$ but does not equal 4 at 2.

3.2 Definition of continuity

Let x_0 be a point on the number line, that is, a real number, and let $x \mapsto f(x)$ be a function defined for $x = x_0$. We want to give a precise meaning to the statement: if x is close to x_0, then $f(x)$ is close to x_0. In order to achieve this, we first agree that "**near** x_0" should mean "in some interval with x_0 as midpoint," or, which is the same, "for all x satisfying an inequality $|x - x_0| < \delta$, where δ is some positive number." (Why do we not insist that "near x_0" should mean "in some *small* interval with x_0 as midpoint?" Because what numbers are to be considered "small" depends on the circumstances, and cannot be determined once and for all.)

We now give the basic **definition of continuity**. *A function $f(x)$ defined for $x = x_0$ is called continuous at x_0 if it has the two properties:*

(I) *if A is any number such that $f(x_0) < A$, then $f(x) < A$ for all x near x_0 (at which f is defined);*

(II) *if B is any number such that $f(x_0) > B$, then $f(x) > B$ for all x near x_0 (at which f is defined).*

A function $f(x)$ defined in an interval is called continuous in that interval if it is continuous at every point of the interval.

This definition expresses the geometric meaning of continuity: *the graph has no breaks*. If a point of the graph is at $(x_0, f(x_0))$, the graph cannot immediately jump to a point above or below the line $y = f(x_0)$. We illustrate this in Figs. 3.19 and 3.20.

From our definition of continuity, one can conclude: *a small change in x produces a small change in y*. Indeed, let f be continuous at x_0, and let ϵ be a positive number, however small. Then $f(x_0) < f(x_0) + \epsilon$ and $f(x_0) > f(x_0) - \epsilon$. Therefore we have (using I with $A = f(x_0) + \epsilon$ and II with $B = f(x_0) - \epsilon$) that near x_0

$$f(x_0) - \epsilon < f(x) < f(x_0) + \epsilon \qquad \text{or} \qquad |f(x_0) - f(x)| < \epsilon.$$

Near x_0 means: in some interval with x_0 as midpoint. Denote the length of such an interval by 2δ; the interval is the set of all x with $|x - x_0| < \delta$.

Fig. 3.19

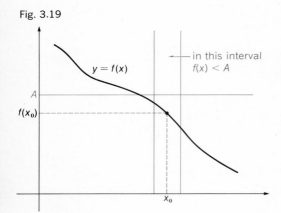

Fig. 3.20

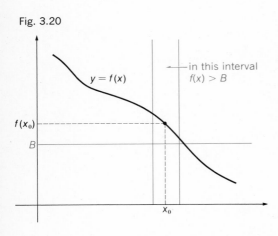

In other words: *given a positive ϵ, we can find a positive δ such that, for all x with $|x - x_0| < \delta$, we have $|f(x) - f(x_0)| < \epsilon$.*

This statement is easily seen to be equivalent to Conditions I and II and is called the **$\epsilon - \delta$ definition of continuity.** Its geometric meaning is shown in Fig. 3.21: if we want y to change by less than the distance ϵ, it suffices to restrict x to change by less than δ. The function is continuous at x_0 if one can find an appropriate distance δ for every given distance ϵ. (There is no need to find the largest δ that will do for a given ϵ; the distance δ shown in Fig. 3.21, for instance, could be increased.)

We shall give below some examples and exercises on using the definition. In all cases the use of the definition will confirm our intuitive idea of continuity as the absence of abrupt changes. Yet the reader may be dismayed that even simple functions require relatively complicated arguments. Fortunately, we shall soon learn more efficient ways of proving that a function is continuous.

▶*Examples*   The first two examples elucidate the technical meaning of the term "near."

1. Is the function $x \mapsto 1 - x^2$ positive near 0?

Answer.   Yes, since there is an interval with 0 as midpoint in which the function is positive. Such an interval is, for instance, the interval $-1 < x < 1$.

Is the function $1 - 10^{10} x^2$ positive near 0? Yes, it is positive for $-.00001 < x < .00001$.

2. If f and g are defined near x_0, so is $f + g$. For if f is defined for all x such that $|x - x_0| < a$ (with $a > 0$) and g for all x such that $|x - x_0| < b$ (with $b > 0$), then $f + g$ is defined in the interval $|x - x_0| < \delta$, where δ is the smaller of the numbers a and b.

3. A constant function is continuous everywhere. For if a constant function is $< A$ (or $> B$) at some point, it is so everywhere.

4. The identity function $f(x) = x$ is continuous everywhere. For if, for some fixed x_0, $f(x_0) < A$, then $x_0 < A$ and, setting $\delta = A - x_0$ we have $x < A$ for $|x - x_0| < \delta$. Thus Condition I is verified. The reader may verify for himself Condition II.

5. The function $f(x) = 1/|x|$ (see Fig. 3.16) is continuous at all points other than zero.

We shall verify Condition I at a point $x_0 > 0$, and leave the rest of the proof to the reader. Let A be a number such that $f(x_0) < A$, that is, $1/|x_0| < A$. We must show that

$$\frac{1}{|x|} < A \qquad \text{near } x_0,$$

Fig. 3.21

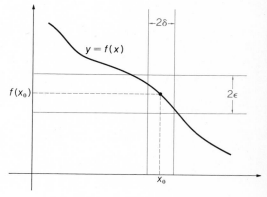

for those x for which our function is defined, that is, for $x \neq 0$. Since for such x we have $|x| > 0$, the statement to be established becomes

$$|x| > \frac{1}{A} \qquad \text{near } x_0,$$

which is the same as

$$|x| - |x_0| > \frac{1}{A} - |x_0| \qquad \text{near } x_0.$$

Set $\delta = |x_0| - 1/A$. Then $\delta > 0$, and we must show that

$$|x| - |x_0| > -\delta \qquad \text{near } x_0.$$

But this is certainly so; indeed we have

$$|x_0| = |x - (x - x_0)| \leq |x| + |x - x_0|$$

so that $\qquad\qquad\qquad |x| - |x_0| \geq -|x - x_0|$

and $\qquad\qquad\quad |x| - |x_0| > -\delta \qquad \text{for } |x - x_0| < \delta.$

6. The function in Fig. 3.14, that is, the function $f(x) = 1$ for $x < 0$, $f(x) = 2 - x$ for $x \geq 0$, is not continuous at 0. Condition I is satisfied, but Condition II is not. Indeed $f(0) = 2 > 3/2$, but every interval with 0 as midpoint contains negative points x at which $f(x) = 1$ and hence $f(x) < 3/2$. Of course, we could have used any other B, rather than $B = 3/2$, provided we choose a B that lies between 1 and 2.

7. The function f such that $f(x) = 0$ if x is rational, $f(x) = 1$ if x is irrational, is not continuous at any point.

Proof. Let x_0 be rational. Then $f(x_0) = 0 < 1/2$. But every interval with x_0 as midpoint contains irrational points x at which $f = 1$ and thus not less than $1/2$. Condition I is violated. One sees similarly that Condition II is violated at every irrational point x_0.◄

EXERCISES

11. Let $f(x) = x^2 + 1$. Is $f < 2$ near 0? If so, find the largest α such that $f(x) < 2$ for $|x| < \alpha$. (Naturally, α must be positive. Conceivably, in some examples, we could take $\alpha = \infty$, meaning that the inequality always holds.)
12. Let $g(x) = 1/x^2$. Is $g > 1/4$ near 1? If so, find the largest δ such that $g(x) > 1/4$ for $|x - 1| < \delta$.

13. Let $A(t) = \sqrt{t+1}$. Is $A < 1$ near 3? If so, find the largest ϵ such that $A(t) < 3$ for $|t - 3| < \epsilon$.
14. Let $L(z) = z^2$ for $z \le 1$, $z + 1$ for $z > 1$. Is $L < 3/2$ near 1? If so, find the largest β such that $L(z) < 3/2$ for $|z - 1| < \beta$.
15. Let $H(u) = u$ for $u \le 0$, $-2u$ for $u > 0$. Is $H > -1$ near 0? If so, find the largest γ such that $H(u) > -1$ for $|u| < \gamma$.
► 16. Let $f(x) = x^2 + 1$ and consider any fixed number $A > 1$. Is $f < A$ near 0? If so, find the largest α such that $f(x) < A$ for $|x| < \alpha$. How did you use the condition $A > 1$? Why would it be naïve to ask if $f < A$ near 0 without this condition?
17. Let $f(x) = x^2 + 1$ and consider any fixed number $\epsilon > 0$. Is $f < 1 + \epsilon$ near 0? If so, find the largest α such that $f(x) < 1 + \epsilon$ for $|x| < \alpha$.
18. Let $F(y) = 2y + 3$ and consider any fixed number $B < 5$. Is $F > B$ near 1? If so, find the largest β such that $F(y) > B$ for $|y - 1| < \beta$.
19. Let $A(t) = (t - 4)^{1/3} + 2$ and consider any fixed number $\gamma > 0$. Is $A < 2 + \gamma$ near 4? If so, find the largest ϵ such that $A(t) < 2 + \gamma$ for $|t - 4| < \epsilon$.
20. Let $g(x) = 1/x^2$ and consider any fixed number $B < 1$. Is $g > B$ near 1? If so, find the largest δ such that $g(x) > B$ for $|x - 1| < \delta$.
21. Let $f(u) = u^3$ for $u \le 0$, $u^2 + 1$ for $u > 0$. Is $f < A$ near 0 for every $A > 0$? If not, find the largest A so that $f \not< A$ near 0. (The symbol $\not<$ means not $<$. For numbers, $a \not< b$ is equivalent to $a \ge b$. However, for functions, $g \not< h$ is not equivalent to $g \ge h$.)
22. Let $G(z) = z^3 + 2$ for $z \le 1$, $2z$ for $z > 1$. Is $G < A$ near 1 for every $A > 3$? Is $G > B$ near 1 for every $B < 3$? In each case, if your answer is yes, find the largest δ so that the inequality holds whenever $|z - 1| < \delta$. If your answer is no, find an A or B for which the inequality fails near 1.
23. Prove that the function $h(x) = 2x + 1$ is continuous at 3 (that is, verify Properties I and II).
24. Prove that the function $g(x) = (x^2 - 1)/3$ is continuous at 0.
25. Prove that the function $f(y) = 3y - 2$ for $y \le 0$, $y + 1$ for $y > 0$, is not continuous at 0.
► 26. Prove that the function $h(t) = t^2 + 1$ for $t < -1$, $-2t$ for $t \ge -1$, is continuous at -1.
27. Prove that the function $K(u) = u^3 - 1$ for $u \le 1$, $2u$ for $u > 1$, is not continuous at 1.

The $\epsilon - \delta$ definition of continuity is very common and becomes increasingly useful in more advanced topics in calculus. The following exercises illustrate the type of computation involved in verifying continuity from this definition.

28. Let $f(x) = 9x - 5$. Find a $\delta > 0$ so that $|f(x) - 4| < 1/10$ for $|x - 1| < \delta$.
29. Let $g(y) = 2y^2 + 3$. Find an $\alpha > 0$ so that $|g(y) - g(0)| < 1/2$ for $|y| < \alpha$.
30. Let $h(t) = t^2 - t$. Find a $\lambda > 0$ so that $|h(t) - h(1)| < 1/5$ for $|t - 1| < \lambda$.
31. Let $f(u) = u + 3$ for $u \le 2$, $3u - 1$ for $u > 2$. Find a $\zeta > 0$ so that $|f(u) - f(2)| < 1/3$ for $|u - 2| < \zeta$.
► 32. Let $f(x) = (2x + 3)/5$. Using the $\epsilon - \delta$ definition, prove that f is continuous at 1. (Hint: Start with the phrase "let $\epsilon > 0$ be given.")

3.3 Continuity of sums, differences, products, and quotients

In practice we usually recognize that a function is continuous not by verifying the definition but by using certain general rules which we proceed to state.

Theorem 1   ASSUME THAT THE FUNCTIONS f AND g ARE CONTINUOUS AT x_0. THEN SO ARE THE FUNCTIONS $f + g$, $f - g$, fg, AND, IF $g(x_0) \neq 0$, ALSO THE FUNCTION f/g.

The truth of the theorem is intuitively clear if we think of continuity as meaning that a small change in x results in a small change in y. A formal proof of Theorem 1 is given in the appendix, §6.

3.4 Continuity of rational functions

Theorem 2   A POLYNOMIAL IS CONTINUOUS EVERYWHERE.

Proof.   We already saw that the identity function $x \mapsto x$ is continuous. Therefore, by Theorem 1, so are the functions $x^2 = xx$, $x^3 = x^2x$, and so forth. Thus every function x^n, n a positive integer, is continuous. We already know that every constant is continuous. Therefore, by Theorem 1, terms like cx^n (c a constant, n a nonnegative integer) are continuous. A polynomial is a sum of such terms and hence is continuous, again by Theorem 1.

Theorem 3   A RATIONAL FUNCTION IS CONTINUOUS AT ALL POINTS AT WHICH IT IS DEFINED.

Proof.   A rational function may be written as f/g, where f and g are polynomials without common roots. By Theorems 1 and 2, it is continuous at every point that is not a root of the denominator g.

3.5 Continuity of composite functions

Theorem 4   LET THE FUNCTION f BE CONTINUOUS AT x_0 AND SET $y_0 = f(x_0)$. LET g BE A FUNCTION CONTINUOUS AT y_0. THEN THE COMPOSITE FUNCTION $g \circ f$ IS CONTINUOUS AT x_0. [Recall that $g \circ f$ is the function $x \mapsto g(f(x))$.]

Let us denote the composite function by $h(x)$ so that $h(x) = g(f(x))$. Let x_1 be close to x_0. Then $y_1 = f(x_1)$ is close to y_0, since f is continuous at y_0. And $h(x_1) = g(f(x_1)) = g(y_1)$ is close to $h(x_0) = g(y_0)$, since g is continuous at y_0. This argument shows why we must expect h to be continuous. A formal proof will be given in the Appendix (see §6).

EXERCISES

In Exercises 33 to 39, it is assumed that $\phi(x)$ and $\psi(x)$ are continuous functions defined for all values of x. Each of the functions listed below is continuous. Indicate how we know this is so. (Hint: Use Theorems 1, 2, 3, and 4.)

33. $f(x) = \phi(x) + \psi(x) + \phi(x)\psi(x)$.   37. $f(x) = \phi(\psi(x^2))$.

34. $f(x) = \phi(x^2 + 2x)$.   38. $f(x) = x^2 + \phi(1 + \psi(x^2))$.

35. $f(x) = \phi(x)/(1 + x^2)$.   39. $f(x) = x\phi(2 + \psi(x)^2)$.

36. $f(x) = \psi(x)/(1 + \phi(x)^4)$.

3.6 Intermediate values

A very important property of continuous functions is stated in Theorem 5.

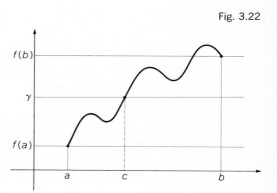

Fig. 3.22

Theorem 5 (The Intermediate Value Theorem)   LET $f(x)$ BE CONTINUOUS FOR $a \leq x \leq b$ AND ASSUME THAT $f(a) \neq f(b)$. IF γ IS ANY NUMBER STRICTLY BETWEEN $f(a)$ AND $f(b)$, THEN THERE IS A POINT c, $a < c < b$, SUCH THAT $f(c) = \gamma$.

The validity of the theorem is geometrically evident. Since f is continuous, the graph consists of one piece. This curve joins the points $(a, f(a))$ and $(b, f(b))$, one of which lies below the line $y = \gamma$ and one above this line. Hence the curve must cross the line $y = \gamma$ somewhere (Fig. 3.22). There is at least one c with $a < c < b$ and $f(c) = \gamma$. An analytic proof of Theorem 5 will be given in the appendix (§7).

Theorem 5 is used to locate zeros of continuous functions. If such a function is positive at some point a and negative at a point b, then there is a root of the function between a and b. Here is an application.

Theorem 6   A POLYNOMIAL OF ODD DEGREE HAS A (REAL) ROOT.

Proof.   Let n be an odd positive integer and f a polynomial of degree n. We may assume that the leading coefficient of f is 1. (If not, we may divide f by its leading coefficient without changing the roots.) Thus

$$f(x) = x^n + a_1 x^{n-1} + a_2 x^{n-2} + \cdots + a_{n-1}x + a_n,$$

where $a_1, a_2, \cdots, a_n$ are some numbers. For $x \neq 0$, we have

$$f(x) = x^n\left(1 + \frac{a_1}{x} + \frac{a_2}{x^2} + \cdots + \frac{a_{n-1}}{x^{n-1}} + \frac{a_n}{x^n}\right).$$

Since n is odd, $x^n > 0$ for $x > 0$ and $x^n < 0$ for $x < 0$. If $|x|$ is large enough, each of the terms a_1/x, a_2/x^2, $\cdots$, a_n/x^n will be less than $1/n$ in absolute value. For such an x, by the triangle inequality,

$$\left| \frac{a_1}{x} + \frac{a_2}{x^2} + \cdots + \frac{a_n}{x^n} \right| \le \left| \frac{a_1}{x} \right| + \left| \frac{a_2}{x^2} \right| + \cdots + \left| \frac{a_n}{x^n} \right|$$

$$< \frac{1}{n} + \frac{1}{n} + \cdots + \frac{1}{n} = 1,$$

so that

$$1 + \frac{a_1}{x} + \cdots + \frac{a_n}{x^n} \ge 1 - \left| \frac{a_1}{x} + \cdots + \frac{a_n}{x^n} \right| > 0$$

Therefore $f(x) > 0$ for $|x|$ large and $x > 0$, and $f(x) < 0$ for $|x|$ large and $x < 0$. By Theorem 2, the function $f(x)$ is continuous. Hence, by the Intermediate Value Theorem, there is a number c with $f(c) = 0$.

Remark   The theorem would have been false for polynomials of even degrees. For instance, $x^2 + 1$ has no real roots.

EXERCISES

40. Suppose f is a function that is continuous in the interval $[-1, 1]$, and $f(-1) = 2$, $f(0) = -1$, and $f(1) = 3$. What is the minimum number of roots that f can have in this interval? Draw a graph fitting the given data with exactly this number of roots. Draw another with two more roots.

41. Suppose g is a function that is continuous in the interval $[-2, 3]$, and $g(-2) = 1/2$, $g(-1) = -1$, $g(0) = 2$, $g(1) = 2$, $g(2) = -2$, and $g(3) = 4$. What is the minimum number of roots that g can have in this interval? Draw a graph fitting the given data with exactly this number of roots.

42. Graph a function continuous on the interval $[-1, 2]$ and positive at both endpoints that has exactly two roots in this interval.

43. Define a function on the interval $[-1, 1]$, continuous on $[-1, 1]$ except at 0, negative at -1, and positive at 1, but which has no roots. (First try to sketch a graph.) Does this contradict the Intermediate Value Theorem?

► 44. Use the method of proving Theorem 6 (on real roots of polynomials of odd degree) to find a number K such that the polynomial $f(x) = x^9 - 100x^4 + 3x^3 - 2$ has a real root within the interval $(-K, +K)$.

45. Find a number K such that the polynomial $x^5 - 18x^3 + 333$ has a root r with $|r| < K$.

3.7 Inverse functions

Let $f(x)$ be defined and continuous for $a \le x \le b$ and assume that $\alpha = f(a) < \beta = f(b)$; see Fig. 3.23. By the Intermediate Value Theorem,

Fig. 3.23

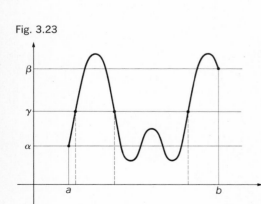

the function f takes on every value γ between α and β somewhere in the interval (a, b). Of course, it may take on this value at several points, even at infinitely many points. Also, f may take on the value α and β not only at the endpoints of the interval but also at inner points (see, for instance, the function graphed in Fig. 3.23.)

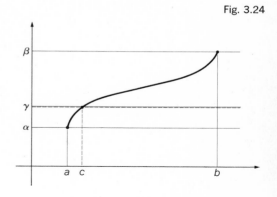

Fig. 3.24

But if $f(x)$ is an increasing function, the situation is different. For such a function, if $x_1 < x_2$, then also $f(x_1) < f(x_2)$. Hence, f takes on the value α only at a, the value β only at b, and every value γ between α and β at exactly one point between a and b (see Fig. 3.24). This means that f establishes a one-to-one correspondence between the closed interval $[a, b]$ on the x axis and the closed interval $[\alpha, \beta]$ on the y axis. The rule

$$\text{``}y \mapsto \text{that } x \text{ with } a \le x \le b \text{ for which } f(x) = y\text{''}$$

is a well-defined function, defined for all y, $\alpha \le y \le \beta$. This function is called **inverse** to f. If we denote it by the letter g, we have

$$g(f(x)) = x \text{ for all } x, a \le x \le b.$$

Fig. 3.25

For what is $g(f(x))$? The unique number to which f assigns the value $f(x)$; this is x. A similar argument (reminiscent of the question "Who is buried in Grant's Tomb?") shows that

$$f(g(y)) = y \text{ for all } y, \alpha \le y \le \beta,$$

so that f is the function inverse to g. It is evident that g is an increasing function.

A similar argument applies to continuous decreasing functions, except that the inverse of such a function is decreasing.

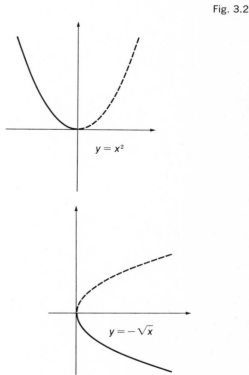

▶ **Examples** **1.** The function $f(x) = x^2$ considered in any interval to the right of 0, $0 \le x \le b$, is continuous and increasing, and takes on non-negative values. The inverse function $g(y)$ is defined as that nonnegative number whose square is equal to y. In other words, $g(y) = \sqrt{y}$.

2. $x \mapsto x^2$ is a decreasing continuous function in any interval $a \le x \le 0$ to the left of 0. The inverse function is now $y \mapsto -\sqrt{y}$. (See Fig. 3.25.)

3. The function $x \mapsto x^3$ is continuous and increasing for all x. The inverse function is $y \mapsto y^{1/3}$. ◀

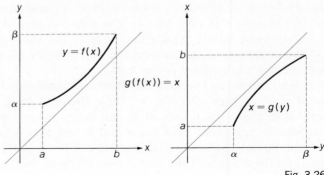

Fig. 3.26

The preceding considerations contain the proof of the first half of the following theorem.

Theorem 7 (The Inverse Function Theorem) LET f BE A CONTINUOUS STRICTLY MONOTONE (INCREASING OR DECREASING) FUNCTION DEFINED IN THE CLOSED INTERVAL $[a, b]$. SET $\alpha = f(a)$, $\beta = f(b)$. THEN (1) THERE EXISTS A STRICTLY MONOTONE (INCREASING OR DECREASING) FUNCTION g, CALLED INVERSE TO f, DEFINED IN THE CLOSED INTERVAL WITH ENDPOINTS α AND β, SUCH THAT $g(f(x)) = x$, $f(g(y)) = y$, AND (2) THIS INVERSE FUNCTION IS CONTINUOUS.

An analytic proof of Statement 2 will be found in the appendix, §6.3. Geometrically the statement becomes obvious if one notes that the graph of g is obtained from the graph of f by interchanging the x and y axes, that is, by reflecting ("flipping") the graph about the bisector of the first quadrant (see Fig. 3.26). The graph of f consists of one piece if f is continuous. This property of the graph is not destroyed by "flipping." Hence the graph of g consists of one piece, so that g is continuous.

EXERCISES

Each of the functions in Exercises 46 to 49 has an inverse on the domain indicated. Find the inverse function.

► 46. $f(t) = (3t + 1)/t$, $t > 0$.   48. $h(x) = (1 + \sqrt{x})^3$, $x > 0$.
47. $g(y) = (1 - y^3)/y^3$, $y > 1$.   49. $f(z) = 1/(1 - z^3)$, $z > 1$.

In Exercises 50 and 51, decide if the given function has an inverse.

50. (a) $x \mapsto x^2$ defined for all x, (b) $x \mapsto x^2$ defined for $x \geq 0$, (c) $x \mapsto x^2$ defined for $x \leq 0$.

51. (a) $x \mapsto x^2 + x - 1$ defined for all x, (b) $x \mapsto x^2 + x - 1$ defined for $x \geq -1/2$, (c) $x \mapsto x^2 + x - 1$ defined for $x \leq -1/2$.

In Exercises 52 to 54, decide if the function whose graph is given has an inverse, and if so, graph it (make a rough sketch only).

52.

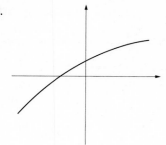

54.

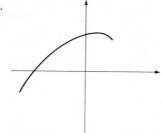

53.

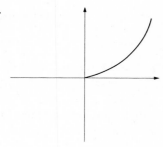

In Exercises 55 to 60 for each point marked (on the horizontal axis) in the domain of the function whose graph is given, find the largest interval containing that point (and contained in the domain) in which the function is strictly monotone. On each such interval, the function (more precisely, the "restriction" of the function) will have an inverse. Graph the inverse (make a rough sketch only).

55.

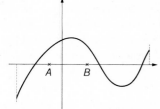

57.

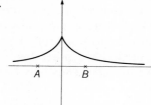

56.

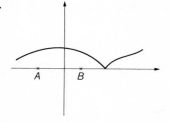

58.

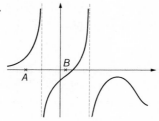

59. 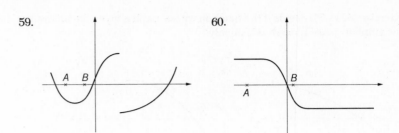 60.

3.8 Continuity of radical functions

We can now establish the continuity of the third class of functions defined in §2.

Theorem 8 A RADICAL FUNCTION IS CONTINUOUS AT ALL POINTS AT WHICH IT IS DEFINED.

Proof. The functions $x^{1/2}$, $x^{1/4}$, $x^{1/6}$, $x^{1/8}$, $\cdots$ are defined for $x \geq 0$. They are inverse to the continuous functions x^2, x^4, x^6, x^8, $\cdots$ which are increasing for $x \geq 0$. Hence the functions $x^{1/2}$, $x^{1/4}$, and so forth, are continuous (by Theorem 7).

The functions $x^{1/3}$, $x^{1/5}$, $x^{1/7}$, $\cdots$ are defined for all x. They are inverse to the continuous increasing functions x^3, x^5, x^7, $\cdots$ and hence continuous (again by Theorem 7).

Next, $x^{p/q}$, where p and q are integers, $q > 0$, is the composition of two continuous functions, $x^{1/q}$ and x^p [for $x^{p/q} = (x^{1/q})^p$]. Hence this function is continuous, wherever defined (by Theorem 4).

Finally, a radical function is continuous, wherever defined, since it is obtained from continuous functions (rational powers) by rational operations and composition, processes that lead from continuous functions to continuous functions (Theorems 1 and 4).

Remark The reader might have noticed that, using the existence of inverse functions asserted by Theorem 7, we could have proved the existence of the nth root of a positive real number, independently of the considerations of Chapter 1, §5.4.

§4 Limits

In this section the concept of limits, which was touched upon in §2, will be developed systematically. The concept is a natural outgrowth of the concept of continuity.

4.1 Definition of limits

Let $f(x)$ be a function defined near x_0, except possibly not at x_0 itself. We look for a number α such that, if we define or redefine $f(x_0)$ to be α, then f becomes continuous at x_0. If there is such an α, we call it the **limit** of f at x_0 and we write

$$\lim_{x \to x_0} f(x) = \alpha$$

(read: the limit of f of x as x approaches x_0 is α)

or
$$f(x) \to \alpha \qquad \text{as } x \to x_0$$

[read: $f(x)$ approaches α as x approaches x_0]. For typographical convenience, one sometimes writes $\lim_{x \to x_0} f(x)$.

If $f(x_0)$ was defined to begin with, and the function was continuous at x_0, no redefinition is needed. Thus:

(1) $\lim_{x \to x_0} f(x) = f(x_0)$ means that f is defined near and continuous at x_0.

If there is no way to define $f(x_0)$ so as to make the function continuous at x_0, we say that f has **no limit** at x_0, or that the limit of f at x_0 does not exist.

Fig. 3.27

▶**Examples**   **1.** $\lim_{x \to 3} x^3 = 27$, for $x \mapsto x^3$ is continuous at 3 and $3^3 = 27$.

2. Set $f(x) = x^3$ for $x \neq 3$ and $f(3) = 26$. Then $\lim_{x \to 3} f(x) = 27$ for the same reason as above.

3. Assume that $f(x) = x$ for $x < 0$ and $f(x) = x^2$ for $x > 0$ (see Fig. 3.27). Then $\lim_{x \to x_0} f(x) = 0$, since $f(0) = 0$ is the choice that makes the function continuous at $x = 0$.

4. Assume that $f(x) = 2$ for $x \leq 0$ and $f(x) = 1/2 + x$ for $x > 0$ (see Fig. 3.28). Then f has no limit at $x = 0$, since there is no way of redefining $f(0)$ so as to make the function continuous at that point. ◀

We remark that a function *cannot have two different limits at a point.* Indeed, assume that $f(x)$ has at x_0 the limit α and also the limit $\beta > \alpha$. This means that if we assign to $f(x)$ the value α, we obtain a function, call it $f_1(x)$, which is continuous at x_0, and if we assign to $f(x_0)$ the value $\beta > \alpha$ we obtain another function, call it $f_2(x)$, which is also continuous at x_0. Thus $f_1(x_0) = \alpha$, $f_2(x_0) = \beta$, $f_1(x) = f_2(x)$ for $x \neq x_0$, and both f_1 and f_2 are continuous at x_0. But this is absurd. For the func-

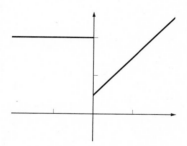
Fig. 3.28

tion $g(x) = f_2(x) - f_1(x)$ is continuous at x_0 (by Theorem 1 of §3.3). If γ is a number such that $\beta - \alpha > \gamma > 0$, we have $g(x_0) > \gamma$, but since $g(x) = 0$ for $x \neq x_0$, it is not true that $g(x) > \gamma$ near x_0. Thus g violates Condition II in the definition of continuity.

We repeat the **definition of limit.** The statement "$\lim_{x \to x_0} f(x) = \alpha$" means that (1) *the function f is defined at all points near x_0 (except possibly it may be undefined at x_0), and* (2) *if we define or redefine $f(x_0) = \alpha$, then f is continuous at x_0.*

Another way of saying that $\lim_{x \to x_0} f(x) = \alpha$ is as follows: the function $f(x)$ is defined for x near x_0, except perhaps at x_0, and for every number $\epsilon > 0$ there is a number $\delta > 0$ such that for all x with $0 < |x - x_0| < \delta$ we have $|f(x) - \alpha| < \epsilon$.

This $\epsilon - \delta$ **definition of limit** is equivalent to the original definition, since if it is satisfied and we set $f(x_0) = \alpha$, we obtain a function continuous at x_0. This definition interprets the statement "$f(x) \to \alpha$ as $x \to x_0$" to mean that $f(x)$ is as close as we like to α, provided x is sufficiently close to, but distinct from, x_0.

The reader will notice that the use of the term "limit" in §2 conforms to the present definition.

EXERCISES

In Exercises 1 to 15, determine if the indicated limits exist, and if so, evaluate them.

1. $\lim_{x \to 3} f(x)$ if $f(x) = (x^3 - 7x)/(x + 5)$.
▶ 2. $\lim_{t \to 4} h(t)$ if $h(t) = (\sqrt{t} - 1)/(t^2 + 1)$.
3. $\lim_{y \to -1} (y^2 + y - 1)/(y - 1)$.
4. $\lim_{u \to 1/2} (u^2 + 1)/(1 + \sqrt{2u + 8})$.
5. $\lim_{h \to 0} (h^2 + 2h - 1)$.
6. $\lim_{h \to 0} (h^3 + 2h^2 - h)/h$.
7. $\lim_{u \to 1} (u^2 - 1)/(u - 1)$.
8. $\lim_{z \to 2} (z^2 + z - 6)/(z^2 - 4)$.
9. limit of $r \mapsto (4r^3 - 3r + 1)/(4r^3 - 4r^2 + r)$ as r approaches $1/2$.
▶ 10. $\lim_{x \to 1} (x\sqrt{x} - x + \sqrt{x} - 1)/(x - 1)$.
11. $\lim_{x \to 2} f(x)$ if $f(x) = x^2 + 4$ for $x < 2$, x^3 for $x > 2$.
12. $\lim_{x \to 0} |x|/x$.
13. $\lim_{x \to 0} g(x)$ if $g(x) = x - 2$ for $x < 0$, $x^3 - 3$ for $x > 0$.
▶ 14. $\lim_{y \to 1/3} H(y)$ if $H(y) = y^2 + y$ for $y < 1/3$, 5 for $y = 1/3$, $12y^3$ for $y > 1/3$.
15. $\lim_{t \to -2} K(t)$ if $K(t) = t + 4$ for $t < -2$, 0 for $t = -2$, t^2 for $t > -2$.

4.2 Computing limits

We now state the basic rules for computing limits.

Theorem 1 (Limits of Sums, Products, and Quotients)   ASSUME THAT $f(x)$ AND $g(x)$ ARE FUNCTIONS THAT HAVE LIMITS AT A POINT x_0. THEN (ALL LIMITS BEING TAKEN AS $x \to x_0$)

$$\lim (f + g) = \lim f + \lim g,$$
$$\lim (fg) = (\lim f)(\lim g),$$
$$\lim \frac{f}{g} = \frac{\lim f}{\lim g} \quad \text{if } \lim g \neq 0.$$

In words: *the limit of a sum (product, quotient) is the sum (product, quotient) of limits.*

This statement follows at once from Theorem 1 in §3.3 and the definition of limits. We prove only the statement about sums, the argument for the other rules being similar. Set

$$\lim_{x \to x_0} f(x) = \alpha, \qquad \lim_{x \to x_0} g(x) = \beta.$$

This means that if we define or redefine $f(x_0) = \alpha$ and $g(x_0) = \beta$, the functions f and g become continuous at x_0. Then the function $f + g$, with the value $\alpha + \beta$ at x_0, is continuous at x_0 (Theorem 1, §3.3) so that $\lim_{x \to x_0} [f(x) + g(x)] = \alpha + \beta$. But this is the statement to be established.

A special case of the theorem is the rule

$$\lim (cf) = c \lim f \quad (c \text{ a constant}).$$

For c is continuous. Since $f - g = f + (-g) = f + (-1)g$, it follows also that

$$\lim (f - g) = \lim f - \lim g.$$

It is clear that the rules may be applied to sums, differences, products, and quotients of more than two terms. For instance, if the functions f, g, and h have at x_0 the limits α, β, and γ, then the function $f + g - h$ has at x_0 the limit $\alpha + \beta - \gamma$.

▶*Example*   We see easily that

$$\lim_{x \to 0} \left(\frac{|x|^a}{x} \right) = 0 \text{ if } a > 1.$$

Therefore, by Theorem 1, we have

$$\lim_{x\to 0}\frac{|x|^{1.3}(1-x)}{x(2x+3)}=\lim_{x\to 0}\left(\frac{|x|^{1.3}}{x}\right)\left(\lim_{x\to 0}\frac{1-x}{2x+3}\right)=0\cdot\left(\frac{1}{3}\right)=0.\blacktriangleleft$$

Theorem 2 (Limits of Composite Functions)   IF

$$\lim_{x\to x_0}f(x)=y_0\qquad\text{AND}\qquad\lim_{y\to y_0}g(y)=\alpha,$$

THEN
$$\lim_{x\to x_0}g(f(x))=\alpha.$$

The statement follows at once from the theorem on the continuity of composite functions (Theorem 4, §3.5).

▶*Example*   In order to compute

$$\lim_{x\to 1}\sqrt[3]{\frac{27x^3+4x-4}{x^{10}+4x^2+3x}}$$

we note that the function $y\mapsto y^{1/3}$ is continuous everywhere, so that the desired limit is

$$\lim_{x\to 1}\sqrt[3]{\frac{27x^3+4x-4}{x^{10}+4x^2+3x}}=\sqrt[3]{\frac{\lim_{x\to 1}(27x^3+4x-4)}{\lim_{x\to 1}(x^{10}+4x^2+3x)}}=\sqrt[3]{\frac{27}{8}}=\frac{3}{2}.\blacktriangleleft$$

EXERCISES

▶ 16. If $\lim_{x\to 0}f(x)=12$ and $\lim_{x\to 0}g(x)=3$, find $\lim_{x\to 0}\sqrt{fg}(x)$.

17. If $\lim_{x\to 1/2}f(x)=1/3$ and $\lim_{x\to 1/2}g(x)=30$, find $\lim_{x\to 1/2}(fg-f-g)(x)$.

18. If $\lim_{x\to 1}f(x)=2$ and $\lim_{x\to 2}g(x)=8$, find $\lim_{x\to 1}(g\circ f)(x)$.

19. If $\lim_{x\to 1}f(x)=9$ and $\lim_{x\to 3}g(x)=5$, find $\lim_{x\to 1}(g\circ\sqrt{f})(x)$.

20. If $\lim_{x\to 2}f(x)=4$ and $\lim_{x\to 4}g(x)=36$, find $\lim_{x\to 2}\sqrt{g\circ f}(x)$.

21. If $\lim_{x\to 1}f(x)=30$ and $\lim_{x\to 1}g(x)=1/2$, find $\lim_{x\to 1}(f+g)/(f-g)(x)$.

22. If $\lim_{x\to 2}f(x)=3$, $\lim_{x\to 2}g(x)=1/3$, and $\lim_{x\to 3}g(x)=2$, find $\lim_{x\to 2}(f+(g\circ f)/g)(x)$.

4.3 Extensions

There are certain extensions of the language of limits which we shall have occasion to use. We shall explain these informally by some examples.
Consider first the function

$$f(x)=\frac{|x|}{x+x^3}=\frac{|x|}{x}\frac{1}{1+x^2};$$

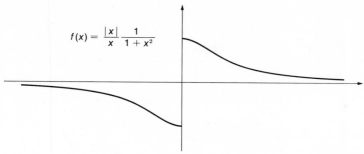

$$f(x) = \frac{|x|}{x}\frac{1}{1+x^2}$$

Fig. 3.29

it is defined for all $x \neq 0$ and is graphed in Fig. 3.29. Note that $|x|/x = 1$ for $x > 0$ and $|x|/x = -1$ for $x < 0$. The function $f(x)$ has no limit at $x = 0$, but for x positive and sufficiently small $f(x)$ is as close to $1/2$ as we want, and for x negative and of sufficiently small absolute value $f(x)$ as close to $-1/2$ as we want. One says in this case that f has at 0 the **one-sided limits:** $1/2$ from the right and $-1/2$ from the left, and one writes

$$\lim_{x \to 0^+} f(x) = \frac{1}{2}, \qquad \lim_{x \to 0^-} f(x) = -\frac{1}{2}.$$

(The symbols $x \to 0^+$ and $x \to 0^-$ are read: as x approaches 0 from the right, as x approaches 0 from the left.)

The same function, $f(x)$, has the property that $f(x)$ becomes as close as we want to 0 if x is large enough. One expresses this by saying that $f(x)$ has the limit 0 as x approaches $+\infty$ (read: plus infinity) and one writes

$$\lim_{x \to +\infty} f(x) = 0.$$

We also have

$$\lim_{x \to -\infty} f(x) = 0;$$

the meaning of this statement should be obvious. The function $f(x)$ is said to have limits **at infinity.**

Consider next the function

$$g(x) = \frac{x^2}{2(1+x)}$$

Fig. 3.30

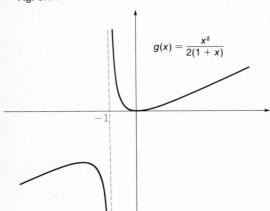

$$g(x) = \frac{x^2}{2(1 + x)}$$

defined for all $x \neq -1$. Its graph is shown in Fig. 3.30. No number can be called a one-sided limit of this function at $x = -1$; no number can be called a limit of this function at infinity. Rather, we claim that as x approaches -1 from the right, or as x approaches $+\infty$, $g(x)$ becomes as large as we want. One expresses this by writing

$$\lim_{x \to -1^+} g(x) = +\infty, \qquad \lim_{x \to +\infty} = +\infty.$$

Similarly, in self-explanatory notation,

$$\lim_{x \to -1^-} g(x) = -\infty, \qquad \lim_{x \to -\infty} g(x) = -\infty.$$

One says that $g(x)$ has one-sided **infinite limits** at $x = -1$ and also infinite limits at infinity.

Let us verify our claim. Consider first an x near -1, and set $x = -1 + t$, so that $|t|$ is small. For $x \neq -1$, we can write

$$g(x) = \frac{(-1 + t)^2}{2t} = \frac{1}{2t} - 1 + \frac{t}{2}.$$

The number $1/2t$ is positive if $x > -1$ (that is, if $t > 0$), negative if $x < -1$ (if $t < 0$); its absolute value can be made as large as we want by making $|t| = |x - (-1)|$ small. The same is true for $g(x)$, since the number $-1 + t/2$ is close to -1 if $|t|$ is small.

In order to see what happens to $g(x)$ when $|x|$ becomes large, we note that, for $x \neq 0, 1$,

$$g(x) = x \frac{x}{2(1 - x)} = x \frac{1}{2 - 2/x}.$$

If $|x|$ is large, then $2/x$ is close to 0, $2 - 2/x$ is close to 2, and the fraction $1/(2 - 2/x)$ is close to $1/2$. Therefore $g(x)$ is, for large $|x|$, of the same sign as x, and $|g(x)|$ can be made as large as we want by making $|x|$ large. This is what we claimed.

The symbols $+\infty$ and $-\infty$ are very handy, but the reader should remember that *they are not names of numbers,* or of anything else, but simply a form of shorthand used in certain sentences.

One should also remember that the statement "f has a limit at x_0" always means that there is a *number* α such that $\lim_{x \to x_0} f(x) = \alpha$. If we want to say that either there is a number α that is the limit of f at x_0, or $\lim_{x \to x_0} f(x) = +\infty$, or $\lim_{x \to x_0} f(x) = -\infty$, we say that "$f$ has at x_0 a finite or infinite limit." The same convention applies to one-sided limits and limits at infinity.

Formal definitions of the "limits" described above, and rules for computing them, will be found in the appendix, §5.

EXERCISES

In Exercises 23 to 27, state which statements are correct.

23. (a) $\lim_{x \to 0^+} x^2 = 0$, (b) $\lim_{x \to +\infty} x^2 = +\infty$, (c) $\lim_{x \to -\infty} x^2 = -\infty$.

24. (a) $\lim_{x \to 0^+} |x|^3/x^3 = 1$, (b) $\lim_{x \to 0^-} |x|^3/x^3 = 0$, (c) $\lim_{x \to 0^{-1}} |x|^3/x^3 = -1$.

25. (a) $\lim_{x \to +\infty} (1 + x)/(1 - x) = 1$, (b) $\lim_{x \to -\infty} (1 + x)/(1 - x) = -1$,
    (c) $\lim_{x \to -\infty} (1 + x)/(1 - x) = -\infty$.

▶ 26. (a) $\lim_{x \to +\infty} x^3/(1 + x^2) = +\infty$, (b) $\lim_{x \to -\infty} x^3/(1 + x^2) = -\infty$,
    (c) $\lim_{x \to 2^+} x^3/(1 + x^2) = 8/5$.

27. (a) $\lim_{x \to 1^+} x^2/(1 - x) = -\infty$, (b) $\lim_{x \to 1^+} x^2/(1 - x) = +\infty$,
    (c) $\lim_{x \to 1^-} x^2/(1 - x) = +\infty$.

—————————————————————————— Appendixes to Chapter 3

§5 One-sided Limits, Infinite Limits, and Limits at Infinity

This section contains a precise discussion of the various ramifications of the limit concept described above.

5.1 One-sided limits

The **definition of one-sided limits** is very similar to the definition of limit. The statement

$$\lim_{x \to x_0^+} f(x) = \alpha$$

[read: the limit of $f(x)$, as x approaches x_0 from the right, is α] or

$$f(x) \to \alpha \qquad \text{as } x \to x_0^+$$

[read: $f(x)$ approaches α as x approaches x_0 from the right] means that (1) $f(x)$ is defined in some interval with x_0 as the left endpoint (that is, for $x_0 < x < b$) and may or may not be defined at x_0, and (2) if we define or redefine $f(x_0)$ to be α, then f, when considered as defined for $x_0 \leq x < b$, is continuous at x_0. (If there is such an α, it is unique; this follows as discussed in §4.1.)

The definition of the limit of f as x approaches x_0 from the left, that is, the meaning of the statement

$$\lim_{x \to x_0^-} f(x) = \beta$$

is completely analogous. (Sometimes one writes $x \uparrow x_0$ instead of $x \to x_0^-$ and $x \downarrow x_0$ instead of $x \to x_0^+$.)

Suppose that f is defined near x_0, except perhaps at x_0 itself, and that both one-sided limits

$$\lim_{x \to x_0^+} f(x) = \alpha \qquad \text{and} \qquad \lim_{x \to x_0^-} f(x) = \beta$$

exist. If $\alpha = \beta$, then, as one sees at once,

$$\lim_{x \to x_0} f(x) = \alpha.$$

If $\alpha \neq \beta$, then there is no way of making f continuous at x_0, and one says that f has a **jump discontinuity** at x_0. A look at Fig. 3.29 will explain the name.

▶ *Examples* *1.* For the function $f(x) = 1$ for $x < 0$, and $f(x) = 2 - x$ for $x > 0$, $f(0)$ not defined, we have $f(x) \to 2$ as $x \to 2^+$, $f(x) \to 1$ as $x \to 2^-$. Indeed, the function $2 - x$ is continuous at 0 and $2 - 0 = 2$, and the constant function 1 is also continuous at 0. The function has a jump discontinuity at $x = 0$.

2. Set $g(t) = 1/t$ for $t < 0$, $g(0) = 4$, $g(t) = 7 + t^2$ for $t > 0$. Then $\lim_{t \to 0^+} g(t) = 7$, and g has no limit as t approaches 0 from the left.

3. For every integer x, let $[x]$ denote the largest integer not exceeding x. Thus $[-5] = -5$, $[-5.3] = -6$, $[1/2] = 0$, and so forth. Let n be any integer; then $\lim_{x \to n^-} [x] = n - 1$, $\lim_{x \to n^+} [x] = n$. All discontinuities of the function $[x]$ are jump discontinuities. ◀

EXERCISES

In Exercises 1 to 8 determine if the indicated one-sided limits exist and, if they do, evaluate them.

1. $\lim_{x \to 1^+} f(x)$ if $f(x) = x^3 - 1$ for $x < 1$; $1/2$ for $x = 1$; x^2 for $x > 1$.
2. $\lim_{x \to (1/2)^-} h(x)$ if $h(x) = (x^2 + 1)^2$ for $x < 0$; $2x - 8$ for $x > 0$.
3. $\lim_{t \to 1^-} g(t)$ if $g(t) = (t^2 + 1)(t - 1)^{-1}$ for $t < 1$; $t^2 - 1$ for $t \geq 1$.
4. $\lim_{u \to 3^+} A(u)$ if $A(u) = u + 1$ for $u \leq 3$; $u^2 - 1$ for $u > 3$.
5. (a) $\lim_{v \to 1^+} \sqrt{1 - v^2}$, (b) $\lim_{v \to 1^-} \sqrt{1 - v^2}$.
6. $\lim_{v \to -2^-} (v + 2)(4 - v^2)^{1/2}$.
7. $\lim_{x \to -3^+} (x^2 - (9 - x^2)^{1/2})^{1/2}$.
8. $\lim_{y \to 0^-} f(y)$ if $f(y) = y^2$ for $y < -.0001$; $y + 1$ for $-.0001 \leq y < 0$; $\sqrt{2y - y^2}$ for $0 \leq y < 4$.
9. Sketch the graph of $G(s) = 1$ for $s \geq 1$; $1/2$ for $1 > s \geq 1/2$, $1/4$ for $1/2 > s \geq 1/4$, $1/6$ for $1/4 > s \geq 1/8$, $\cdots$, $1/2^n$ for $1/2^{n-1} > s > 1/2^n$ and $n = 1, 2, 3, \cdots$; 1 for $s = 0$; 0 for $s < 0$.
   Determine if $\lim_{s \to 0^+} G(s)$ and $\lim_{s \to 0^-} G(s)$ exist and, if so, evaluate them. What conclusion can you draw about the behavior of G at 0?
10. Sketch the graph of $H(t) = 1$ for $t \geq 1$; $2t - 1$ for $1 > t \geq 1/2$, $4t - 1$ for $1/2 > t \geq 1/4$; $8t - 1$ for $1/4 > t \geq 1/8$, $\cdots$, $2^n t - 1$ for $1/2^{n-1} > t \geq 1/2^n$ and $n = 1, 2, 3, \cdots$; 1 for $t = 0$; 0 for $t < 0$.

Determine if $\lim_{t \to 0^+} H(t)$ and $\lim_{t \to 0^-} H(t)$ exist and, if so, evaluate them. What conclusion can you draw about the behavior of H at 0?

5.2 Infinite limits

The function $f(x) = 1/|x|$ is not defined at $x = 0$ and has no limit there, and the discontinuity is not a jump. Rather, for every number A no matter how large, we have $f > A$ near 0, except at 0 itself. This is the same as saying: $f > 0$ near 0, except at 0, and for every positive number ϵ, no matter how small, $1/f < \epsilon$ near 0, except at 0.

This suggests the following **definition of infinite limit.** The statements

$$\lim_{x \to x_0} f(x) = +\infty$$

(read: the limit of f at x_0 is plus infinity)

or

$$f \to +\infty \qquad \text{as } x \to x_0$$

(read: f approaches plus infinity as x approaches x_0, or f becomes positive infinite at x_0) mean that (1) $f(x) > 0$ near x_0 (except perhaps at x_0), and (2) $\lim_{x \to x_0} 1/f(x) = 0$.

In a similar way, we define the meaning of the statements

$$\lim_{x \to x_0^-} f(x) = -\infty, \qquad \lim_{x \to x_0^+} f(x) = +\infty,$$

and so forth. We recall once more (see §4.3) that the symbols $+\infty$, $-\infty$ are not names of numbers.

▶ **Examples** **1.** $\lim_{x \to 0} 1/x^4 = +\infty$, because $x^4 > 0$ if $x \neq 0$, x^4 is continuous at 0, and $1/f(x) = x^4$, and $0^4 = 0$.

2. $\lim_{x \to 0^+} 1/x = +\infty$, $\lim_{x \to 0^-} 1/x = -\infty$. Indeed, $1/x > 0$ if $x > 0$; $1/x < 0$ if $x < 0$ and $\lim_{x \to 0} x = 0$.

3. $\lim_{x \to 1^+} (1 + x)/(1 - x) = -\infty$. For if $x > 1$, then $1 + x > 0$, $1 - x < 0$, so that $(1 + x)/(1 - x) < 0$, and $(1 - x)/(1 + x)$ is continuous at $x = 1$ and has there the value 0. ◀

EXERCISES

In Exercises 11 to 15, determine if the given function has a limit (meaning finite limit), an infinite limit, or neither, at the indicated point. In the first two cases, evaluate the limit (we use this somewhat misleading but common phrase even for infinite limits).

11. (a) $\lim_{x \to 2^+} (x^2 + 1)/(x - 2)$, (b) $\lim_{x \to 2^-} (x^2 + 1)/(x - 2)$,
   (c) $\lim_{x \to 2} (x^2 + 1)/(x - 2)$.
▶ 12. (a) $\lim_{x \to 1^+} (x + 1)/(|x - 1|)$, (b) $\lim_{x \to 1^-} (x + 1)/(|x - 1|)$,
   (c) $\lim_{x \to 1} (x + 1)/(|x - 1|)$.
13. (a) $\lim_{x \to 0^+} 1/\sqrt{x}$, (b) $\lim_{x \to 2^-} (4 - x^2)/(\sqrt{4 - x^2})$.

14. (a) $\lim_{t \to -(2/3)^+} (t^3 + 1)/(2t + 3)$, (b) $\lim_{t \to -(2/3)^-} (t^3 + 1)/(2t + 3)$,
    (c) $\lim_{t \to -2/3} (t^3 + 1)/(2t + 3)$.
15. (a) $\lim_{u \to 1^+} (u^5 + 2u)/(u^{5/3} - u)$, (b) $\lim_{u \to 1^-} (u^5 + 2u)/(u^{5/3} - u)$,
    (c) $\lim_{u \to 1} (u^5 + 2u)/(u^{5/3} - u)$.
16. Sketch the graph of $f(z) = 1$ for $z \geq 1$; -2 for $1 > z \geq 1/2$, 4 for $1/2 > z \geq 1/4$, -8 for $1/4 > z \geq 1/8$, $\cdots$, $(-1)2^n$ for $1/2^{n-1} > z \geq 1/2^n$ and $n = 1, 2, 3, \cdots$.

   Determine if $f(z)$ has a limit, an infinite limit, or neither, as z approaches 0 from the right. In the first two cases, evaluate the limit.

5.3 Limits at infinity

The language of limits can also be used to describe the behavior of a function $f(x)$ for large values of x. Let us agree that the expression "**near** $+\infty$" should mean "for all numbers x greater than some positive number." We shall say that $\lim_{x \to +\infty} f(x) = \alpha$ if (1) f is defined near $+\infty$, and (2) if $\alpha < A$, then $f < A$ near $+\infty$ and if $\alpha > B$, then $f > B$ near $+\infty$. In this case, we also write: $f \to \alpha$ as $x \to +\infty$.

Similarly, "**near** $-\infty$" means "for all x smaller than some negative number." It is hardly necessary to state explicitly what we mean by the statement $\lim_{x \to -\infty} f(x) = \beta$.

We now observe that, if $f(x)$ is defined near $+\infty$ (that is, for all x greater than some number), the function $t \mapsto f(1/t)$ is defined for all positive t near 0 (that is, for all positive t less than some number). If $f(x) < A$ (or $f(x) > B$) for all x near $+\infty$, then $f(1/t) < A$ (or $f(1/t) > B$) for all positive t near 0. Hence we have the following equivalent **definition of limits at infinity:**

$$\lim_{x \to +\infty} f(x) = \alpha \qquad \text{means that} \qquad \lim_{t \to 0^+} f\left(\frac{1}{t}\right) = \alpha.$$

Similarly,

$$\lim_{x \to -\infty} f(x) = \beta \qquad \text{means that} \qquad \lim_{t \to 0^-} f\left(\frac{1}{t}\right) = \beta.$$

If both limits exist and are equal, we write

$$\lim_{x \to \infty} f(x) = \alpha, \qquad \text{which means} \qquad \lim_{t \to 0} f\left(\frac{1}{t}\right) = \alpha.$$

We agree to use the same definitions if α and β are not numbers but one of the symbols $+\infty$ or $-\infty$.

▶ **Examples** **1.** $\lim_{x \to \infty} (1/x) = 0$. For (replacing x by $1/t$) $\lim_{t \to 0} t = 0$.

2. We have

$$\lim_{x \to +\infty} \frac{1 + 2x}{1 - 3x} = -\frac{2}{3}.$$

For this means (if we replace x by $1/t$) that

$$\lim_{t \to 0^+} \frac{1 + 2/t}{1 - 3/t} = -\frac{2}{3}.$$

But

$$\frac{1 + 2/t}{1 - 3/t} = \frac{t + 2}{t - 3};$$

this function is continuous at $t = 0$; its value there is $-2/3$.

3. $\lim_{x \to +\infty} x^n = +\infty$ for every positive integer n. For $\lim_{t \to 0^+} 1/t^n = +\infty$, since $t^n > 0$ for $t > 0$, t^n is continuous at $t = 0$, and $0^n = 0$.

4. $\lim_{x \to -\infty} x^n = +\infty$ for every even positive integer n and $\lim_{n \to -\infty} x^n = -\infty$ for every odd positive integer n. (Use the same argument as in 3.)◄

EXERCISES

17. (a) $\lim_{x \to +\infty} (1 + 6x)/(-2 + x) = ?$ (Fill in with a number, $+\infty$, $-\infty$, or say that the function has neither a finite nor an infinite limit.)
   (b) $\lim_{x \to -\infty} (1 + 6x)/(-2 + x) = ?$ (c) $\lim_{x \to \infty} (1 + 6x)/(-2 + x) = ?$

► 18. (a) $\lim_{y \to +\infty} (1 + y + 2y^2)/(10 + 5y - 4y^2) = ?$
   (b) $\lim_{y \to -\infty} (1 + y + 2y^2)/(10 + 5y - 4y^2) = ?$
   (c) $\lim_{y \to \infty} (1 + y + 2y^2)/(10 + 5y - 4y^2) = ?$

19. (a) $\lim_{t \to +\infty} 2t^3/(6 + t + 2t + t^3) = ?$ (b) $\lim_{t \to -\infty} 2t^3/(6 + t + 2t + t^3) = ?$
   (c) $\lim_{t \to \infty} 2t^3/(6 + t + 2t + t^3) = ?$

20. Let $x \mapsto p(x)$ be a polynomial of degree n and leading coefficient r. Let $x \mapsto q(x)$ be a polynomial (not identically zero) of degree n and leading coefficient s. (a) $\lim_{x \to +\infty} p(x)/q(x) = ?$ (b) $\lim_{x \to -\infty} p(x)/q(x) = ?$
   (c) $\lim_{x \to \infty} p(x)/q(x) = ?$

21. (a) $\lim_{u \to +\infty} 3/(u^3 + u - 1) = ?$ (b) $\lim_{u \to -\infty} 3/(u^3 + u - 1) = ?$
   (c) $\lim_{u \to \infty} 3/(u^3 + u - 1) = ?$

22. (a) $\lim_{z \to +\infty} (100z^3 + z + 1)/z^4 = ?$ (b) $\lim_{z \to -\infty} (100z^3 + z + 1)/z^4 = ?$
   (c) $\lim_{z \to \infty} (100z^3 + z + 1)/z^4 = ?$

23. (a) $\lim_{v \to +\infty} (v^{100} + v^{99})/(v^{101} - v^{100}) = ?$
   (b) $\lim_{v \to -\infty} (v^{100} + v^{99})/(v^{101} - v^{100}) = ?$
   (c) $\lim_{v \to \infty} (v^{100} + v^{99})/(v^{101} - v^{100}) = ?$

► 24. Let $x \mapsto p(x)$ be a polynomial of degree n. Let $x \mapsto q(x)$ be a polynomial (not identically zero) of degree $m > n$. (a) $\lim_{x \to +\infty} p(x)/q(x) = ?$
   (b) $\lim_{x \to -\infty} p(x)/q(x) = ?$ (c) $\lim_{x \to \infty} p(x)/q(x) = ?$

25. (a) $\lim_{v \to +\infty} (v^3 + 1)/(v^2 - 1) = ?$ (b) $\lim_{v \to -\infty} (v^3 + 1)/(v^2 - 1) = ?$
   (c) $\lim_{v \to \infty} (v^3 + 1)/(v^2 - 1) = ?$

26. (a) $\lim_{r \to +\infty} (2r^3 + r^2 - 1)/(r + 5) = ?$ (b) $\lim_{r \to -\infty} (2r^3 + r^2 - 1)/(r + 5) = ?$
   (c) $\lim_{r \to \infty} (2r^3 + r^2 - 1)/(r + 5) = ?$

27. (a) $\lim_{y \to +\infty} (y^4 - 2y^2 + y - 1)/(y - 1) = ?$
   (b) $\lim_{y \to -\infty} (y^4 - 2y^2 + y - 1)/(y - 1) = ?$
   (c) $\lim_{y \to \infty} (y^4 - 2y^2 + y - 1)/(y - 1) = ?$

28. Let $x \mapsto p(x)$ be a polynomial of degree n. Let $x \mapsto q(x)$ be a polynomial (not identically zero) of degree $m < n$. (a) $\lim_{x \to +\infty} p(x)/q(x) = ?$
   (b) $\lim_{x \to -\infty} p(x)/q(x) = ?$ (c) $\lim_{x \to \infty} p(x)/q(x) = ?$ (Hint: Consider separately the cases $n - m$ even and $n - m$ odd.)

5.4 Computing infinite limits

In §4.2 we showed (see Theorem 1) that the limit of the sum, difference, product, and quotient of two functions (having limits at a point) is the sum, difference, product, or quotient of the limits of the two functions (in the case of quotient, we must also assume that the limit of the divisor is not 0). *The same is true, as one sees at once, for one-sided limits.*

What about infinite limits? It turns out that the same rule for computing limits can be applied, provided that the "calculation" of the limit can be carried out by using one of the following rules (in the following formulas, α stands for a number):

$$
\begin{aligned}
&+\infty + \alpha = \alpha + (+\infty) = +\infty + (+\infty) = +\infty, \\
&-\infty + \alpha = \alpha + (-\infty) = -\infty + (-\infty) = -\infty, \\
&\quad \alpha(+\infty) = (+\infty)\alpha = (-\alpha)(-\infty) = (-\infty)(-\alpha) = +\infty \text{ if } \alpha > 0, \\
&\quad \alpha(-\infty) = (-\infty)\alpha = (-\alpha)(+\infty) = (+\infty)(-\alpha) = -\infty \text{ if } \alpha > 0, \\
&(+\infty)(+\infty) = (-\infty)(-\infty) = +\infty, \\
&(+\infty)(-\infty) = (-\infty)(+\infty) = -\infty, \\
&\quad \frac{\alpha}{+\infty} = \frac{\alpha}{-\infty} = 0.
\end{aligned}
$$

(*)

We must prove that each of these rules gives a correct result. It is easy to see this, by using a little common sense. The first of the above formulas, for instance, asserts essentially that, if one function becomes very large positive, as the independent variable approaches some point, and another function is continuous at this point, then their sum also becomes very large positive. A formal proof can be given along the same lines; we have only to talk not about "very large" but about "larger than any given number." The reader is asked to give similar "common sense proofs" for the other statements (*). A formal proof of one of these statements will be given in §6.4.

Note that there are cases not covered by our rules. There is no "formula" for expressions like $0/0$, or $0 \cdot (+\infty)$, or $+\infty + (-\infty)$. There is a good reason for this. If, for instance, two functions $f(x)$ and $g(x)$ both have the limit 0 at a point, there is no general way of predicting how the quotient $h(x) = f(x)/g(x)$ will behave at this point. Consider, for instance, the case $f(x) = x^2$, $g(x) = x^n$, n being an integer. The quotient $h(x) = x^{2-n}$ may have at 0 the limit 0 (if $n = 1$), or 1 (if $n = 2$), or $+\infty$ (if $n = 4$), or it may happen that the left limit at 0 is $-\infty$ and the right limit is $+\infty$ (if $n = 3$).

The applicability of the rules (*) to the case of one-sided limits or limits at infinity requires no further comment.

It is also evident that *the rule for computing limits of composite functions (stated in §4.2 as Theorem 2) applies to one-sided limits, limits at infinity, and infinite limits.*

▶ *Examples 1.* We want to compute $\lim_{x \to +\infty} (x^2 - 3x)$. We know that $\lim_{x \to +\infty} x^2 = +\infty$ (Example 3 in §5.3) and it follows from rule (*) that $\lim_{x \to +\infty} (-3x) = (-3)(+\infty) = -\infty$. But, using this information, we cannot compute $\lim_{x \to +\infty} (x^2 - 3x)$, since no meaning is assigned to the expression $+\infty + (-\infty)$. Instead we use rule (*) as follows:

$$\lim_{x \to +\infty} (x^2 - 3x) = \lim_{x \to +\infty} x(x - 3) = \lim_{x \to +\infty} x \lim_{x \to +\infty} (x - 3)$$
$$= (+\infty)(+\infty - 3) = (+\infty)(+\infty) = +\infty.$$

2. We want to compute the limit as $x \to 1$ of

$$f(x) = \frac{1 + x^3}{1 - 2x + x^2}.$$

We cannot use the meaningless expression $2/0$. Instead we note that

$$\frac{1 + x^3}{1 - 2x + x^2} = (1 + x^3) \frac{1}{(1 - x)^2}.$$

Since $(1 - x)^2 > 0$ for $x \neq 1$, we have

$$\lim_{x \to 1} \frac{1}{(1 - x)^2} = +\infty.$$

Hence

$$\lim_{x \to 1} f(x) = 2(+\infty) = +\infty.$$

3. In a similar way one sees that if

$$f(x) = \frac{1 + x^3}{1 - 3x + 3x^2 - x^3},$$

then

$$\lim_{x \to 1^+} f(x) = -\infty, \qquad \lim_{x \to 1^-} f(x) = +\infty.$$

Indeed,

$$f(x) = \frac{1 + x^3}{(1 - x)^3}$$

and

$$\lim_{x \to 1^+} \frac{1}{(1 - x)^3} = -\infty, \qquad \lim_{x \to 1^-} \frac{1}{(1 - x)^3} = +\infty. \blacktriangleleft$$

EXERCISES

Although many of the following exercises can be done directly, the use of some combination of the rules stated above will usually be simpler.

29. $\lim_{x \to 0} (x^3 + 2x - 5)(x^8 + 2x^7 + x^3 - 1) = ?$

30. $\lim_{z \to 1} (z^{11} - 2z + 3)^5 (z^7 + z^2 - 1)^{-8} = ?$

31. $\lim_{y \to 3^-} (9 - y^2)^{-1/2} (y^3 + 2)^{10} = ?$

▶ 32. $\lim_{x \to 1^-} (2x + 1)^{100}/(x^2 + x - 2) = ?$

33. $\lim_{x \to 0^+} (|x|/x)^3 (x^2 + \sqrt{x + 4})^4 = ?$

34. $\lim_{t \to +\infty} (t^2 + 1)^5 (\sqrt{t} - 1)^3 [(t^2 + 1)/(2t^2 - 5)]^2 = ?$

35. $\lim_{u \to -\infty} \frac{(u^3 + 2u - 1)^5}{(u^2 + u - 6)^4} = ?$

36. $\lim\limits_{z \to \infty} \sqrt{\dfrac{(z^3 - 2z + 1)^2}{(2z^2 + z - 1)^3}} = ?$

37. $\lim\limits_{w \to -\infty} \left[\dfrac{(w^3 + 2w)^4}{w^5 + w^4 - w^2} \right]^{1/3} = ?$

38. $\lim\limits_{x \to +\infty} \dfrac{1/x - 1/x^2}{1/x^3 - 1/x^4} = ?$

Certain expressions involving $+\infty$ and $-\infty$ are not "computable," because any attempt to define a computational rule would be inconsistent with the theorems of the past sections. The following exercises will illustrate this.

39. Let $f(x) = 2x$, $g(x) = -x$. Compute $\lim_{x \to +\infty} f(x)$, $\lim_{x \to +\infty} g(x)$, $\lim_{x \to +\infty} (f + g)(x)$.

40. Let $f(x) = 2x$, $h(x) = -3x$. Compute $\lim_{x \to +\infty} f(x)$, $\lim_{x \to +\infty} h(x)$, $\lim_{x \to +\infty} (f + h)(x)$.

41. Let $A(t) = t^{-2}$, $B(t) = t^{-1}$. Compute $\lim_{t \to 0^+} A(t)$, $\lim_{t \to 0^+} B(t)$, $\lim_{t \to 0^+} (A/B)(t)$, $\lim_{t \to 0^+} (B/A)(t)$.

§6 Proofs of Some Continuity Theorems

In this section we prove some of the theorems stated in §3 without proof, and also the rule for computing infinite limits stated in §5.4. The proofs are essentially exercises in applying the definition of continuity. The proof of the Intermediate Value Theorem, which is based on the fundamental properties of the real numbers, is postponed until the next section.

In the theorems to be proved, the conclusion is always that some function is continuous, that is, has Properties I and II stated in the definition of continuity. In every case we shall verify only Property I. The reader is asked to carry out the proof of Property II, which follows the same pattern.

6.1 Sums, differences, products, and quotients

We begin by establishing Theorem 1 in §3.3, which asserts that if two functions, f and g, are continuous at a point x_0, so are $f + g$, $f - g$, fg, and, assuming that $g(x_0) \neq 0$, also f/g.

The proof is divided into several steps.

1. Let f be a function continuous at x_0, and let c be a constant. Then $f + c$ is continuous at x_0.

Proof. Assume that $f(x_0) + c < A$. Then $f(x_0) < A - c$ and, f being continuous, $f < A - c$ near x_0, so that $f + c < A$ near x_0. Thus $f + c$ has Property I.

2. Let f and c be as in Step (1). Then the function cf is continuous at x_0.

Proof. If $c = 0$, the function $cf(x)$ is the constant 0 and hence continuous everywhere. We assume now that $c > 0$. If $cf(x_0) < A$, then $f(x_0) < A/c$. Since f is continuous, we have that $f < A/c$ near x_0. Hence $cf < A$ near x_0. Thus cf has Property I. The case $c < 0$ can be treated similarly.

3. Let f and g be continuous at x_0 and assume that $f(x_0) = g(x_0) = 0$. Then $f + g$ is continuous at x_0.

Proof. Assume that $f(x_0) + g(x_0) < A$. Then $A > 0$ and therefore $f(x_0) < A/2$, $g(x_0) < A/2$. Since f and g are continuous, $f < A/2$ and $g < A/2$ near x_0. Hence $f + g < A/2 + A/2 = A$ near x_0. Thus $f + g$ has Property I.

4. If f and g are continuous at x_0, so is $f + g$.

Proof. Set $f_1(x) = f(x) - f(x_0)$, $g_1(x) = g(x) - g(x_0)$. Then f_1 and g_1 are continuous at x_0, by (1), and $f_1(x_0) = g_1(x_0) = 0$. Hence the function $\phi = f_1 + g_1$ is continuous at x_0, by (3). But $f(x) + g(x) = \phi(x) + c$, where $c = f(x_0) + g(x_0)$. Hence $f + g$ is continuous at x_0, by (1).

5. If f and g are continuous at x_0, so is $f - g$.

Proof. The function $-g = (-1)g$ is continuous at x_0, by (2). The function $f - g = f + (-g)$ is continuous at x_0, by (4).

6. Let f and g be as in (3). Then fg is continuous at x_0.

Proof. Assume that $f(x_0)g(x_0) < A$. Then $A > 0$. Hence $f(x_0) = 0 < A$ and $f(x_0) = 0 > -A$ and, since f is continuous, $f < A$ and $f > -A$ near x_0. In other words, $|f| < A$ near x_0.
Also, $g(x_0) < 1$ and $g(x_0) > -1$ and, since g is continuous, $g < 1$ and $g > -1$ near x_0. In other words, $|g| < 1$ near x_0.
We conclude that $|fg| = |f||g| < A \cdot 1 < A$ near x_0. This implies that $fg < A$ near x_0. The function fg has Property I.

7. If f and g are continuous at x_0, so is fg.

Proof. Set $f_1(x) = f(x) - f(x_0)$, $g_1(x) = g(x) - g(x_0)$, and $\phi = f_1g_1$. We have that $f(x)g(x) = \phi(x) + f(x_0)g(x) + g(x_0)f(x) - f(x_0)g(x_0)$. Now f_1 and g_1 are continuous at x_0, by (1). So are the functions $f(x_0)g(x)$ and $g(x_0)f(x)$, by (2). So is the function $\phi(x)$, by (6). So is, finally, the function $f + g$ [by (4)], since it is a sum of continuous functions.

8. Assume that f is continuous at x_0 and $f(x_0) = 1$. Then $1/f$ is continuous at x_0.

Proof. Assume that $1/f(x_0) < A$. This means that $A > 1$. Hence $f(x_0) > 1/A$ and, since f is continuous, $f > 1/A$ near x_0. Then $1/f < A$ near x_0. The function $1/f$ has Property I.

9. If f is continuous at x_0, $f(x_0) \neq 0$, then $1/f$ is continuous at x_0.

Proof. The function $[1/f(x_0)]f(x)$ is continuous at x_0 by (2), and its value at x_0 is 1. Hence its reciprocal, the function $f(x_0)/f(x)$, is continuous at x_0 by (8), and so is the function $1/f(x) = [1/f(x_0)][f(x_0)/f(x)]$, again by (2).

10. Let f and g be continuous at x_0, and $g(x_0) \neq 0$. Then f/g is continuous at x_0.

Proof. The function $1/g$ is continuous at x_0, by (9). So is $f/g = (1/g)f$, by (7).

Theorem 1 follows from Steps (4), (5), (7), and (10).

1. Verify that Property II is satisfied by the function $f + c$ in Step (1).
2. Verify that Property I is satisfied by the function cf in Step 2 for $c < 0$.
3. Verify that Property II is satisfied by the function cf in Step (2).
4. Verify that Property II is satisfied by the function $f + g$ in Step (3).
5. Verify that Property II is satisfied by the function fg in Step (6).
6. Verify that Property II is satisfied by the function $1/f$ in Step (8).

6.2 Composed functions

We prove next that the composite of two continuous functions is continuous (Theorem 4 in §3.5).

We are given two functions f and g; f is continuous at x_0, $f(x_0) = y_0$, and g is continuous at y_0. We must show that $\phi = g \circ f$ is continuous at x_0.

Assume that $\phi(x_0) < A$. This means that $g(f(x_0)) < A$, that is, $g(y_0) < A$. Since g is continuous, $g < A$ near y_0. This means that there is a positive number δ such that $g(y) < A$ for any y with $y_0 - \delta < y < y_0 + \delta$.

Now, $f(x_0) = y_0 < y_0 + \delta$ and $f(x_0) = y_0 > y_0 - \delta$. Since f is continuous, $f < y_0 + \delta$ and $f > y_0 - \delta$ near x_0. Therefore, by what was said before, $g(f(x)) < A$ for x near x_0. Property I holds.

7. Verify that Property II is satisfied by the function $\phi = g \circ f$.

6.3 Inverse functions

Now we establish the continuity of the inverse function of a continuous strictly monotone function. This is the part of Theorem 7 in §3.7 which we did not prove there. We are not concerned here with establishing the existence of the inverse function; this follows, as we noted in §3, from the Intermediate Value Theorem.

Thus we assume that we are given two strictly monotone functions, f and g, which are inverse to each other. Let f be defined for $a \leq x \leq b$. We assume that f is continuous. The function g is such that $g(f(x)) = x$ for all x, $a \leq x \leq b$. We must show that g is continuous.

We assume, for the sake of definiteness, that f is increasing and we set $\alpha = f(a)$, $\beta = f(b)$. Then $g(y)$ is defined for $\alpha \leq y \leq \beta$ and is increasing, and $g(\alpha) = a$, $g(\beta) = b$.

Let y_0 be a number such that $\alpha \leq y_0 \leq \beta$. Let A be a number such that $g(y_0) < A$. We must show that $g < A$ near y_0. We may assume that $A \leq b$; otherwise there is nothing to prove, since g takes on no values greater than b.

Set $g(y_0) = A_0$. Then, by the definition of g, $a \leq A_0 \leq b$ and $f(A_0) = y_0$. Set $f(A) = y_1$. Then $y_1 \leq \beta$. Since $A_0 < A$ and f is increasing, we have $y_0 < y_1$. If y is a number such that $y_0 < y < y_1$, then, by the intermediate value theorem, $y = f(C)$ for some C. Since f is increasing, $C < A$ as $f(C) < f(A)$. But if $y = f(C)$, then $g(y) = C$. Hence $g(y) < A$ for $y_0 < y < y_1$. On the other hand, since g is increasing, $g(y) < A$ for $y \leq y_0$. Thus $g < A$ near y_0. Property I is verified.

► 8. Verify that Property II is satisfied by the function g.

6.4 Computing infinite limits

Now we shall prove one case of the rules (*) for computing infinite limits (see §5.4). These rules cover various cases. In each case the proof follows the same pattern. We consider only the case when (all limits being taken as $x \to x_0$)

(1)
$$\lim f = \alpha > 0, \qquad \lim g = +\infty.$$

In this case the rules (*), applied to sums, products, and quotients, yield the statements

(2)
$$\lim (f + g) = +\infty,$$

(3)
$$\lim (fg) = +\infty,$$

(4)
$$\lim \frac{f}{g} = 0.$$

We must show that this is indeed so. By (1), we have that $f > 0$ and $g > 0$ near x_0. Hence $f + g > 0$ near x_0. Also, by (1), $\lim 1/g = 0$. Hence, by repeated application of Theorem 1, §4.2,

$$\lim \frac{1}{f + g} = \lim \frac{1}{g[1 + (f/g)]} = \lim \frac{1}{g} \lim \frac{1}{1 + f \cdot (1/g)}$$

$$= \lim \frac{1}{g} \frac{1}{\lim [1 + f \cdot (1/g)]}$$

$$= \lim \frac{1}{g} \frac{1}{1 + (\lim f)(\lim (1/g))} = 0 \frac{1}{1 + \alpha \cdot 0} = 0,$$

which means the same as (2).

Next, $fg >$ near x_0 and, by Theorem 1, §4.2,

$$\lim \frac{1}{fg} = \lim \frac{1}{f} \lim \frac{1}{g} = \frac{1}{\lim f} \lim \frac{1}{g} = \frac{1}{\alpha} \cdot 0 = 0,$$

which means the same as (3).

Finally, again by Theorem 1, §4.2,

$$\lim \frac{f}{g} = \lim \left(f \frac{1}{g} \right) = (\lim f) \left(\lim \frac{1}{g} \right) = \alpha \cdot 0 = 0,$$

so that (4) holds, too.

9. Prove that, if at some point x_0 we have $\lim f = +\infty$ and $\lim g = +\infty$, then also $\lim (f + g) = \lim fg = +\infty$.

10. Prove that, if at some point x_0 we have $\lim f = +\infty$ and $\lim g = -\infty$, then $\lim fg = -\infty$.

§7 Proof of the Intermediate Value Theorem

Let $f(x)$ be defined and continuous for $a \leq x \leq b$. Set $\alpha = f(a)$, $\beta = f(b)$, and let $\alpha \neq \beta$. We assume, for the sake of definiteness, that $\alpha < \beta$. Let γ be a number such that $\alpha < \gamma < \beta$. Then there is a number c in the interval (a, b) with $f(c) = \gamma$.

This statement is the Intermediate Value Theorem (Theorem 5 in §3.6). The proof involves directly the least upper bound principle (Postulate **18** in Chapter 1).

7.1 The proof

Let S denote the set of all numbers x such that $a \leq x \leq b$ and $f(x) < \gamma$. The set S is not empty, since it contains the number a; it is bounded from above, because b is an upper bound for S. Hence there is a number, call it c, which is the least upper bound for S. This c is the smallest number with the property: if $f(x) < \gamma$, then $x \leq c$. Clearly $a \leq c$ (for a belongs to S) and $c \leq b$ (for b is an upper bound for S).

Three cases are conceivable: $f(c) < \gamma$, $f(c) > \gamma$, and $f(c) = \gamma$. We shall prove that the first two cases lead to contradictions. Therefore the third must take place, so that $f(c) = \gamma$ and the theorem is proved.

Assume that $f(c) < \gamma$. Then $f(c) < \beta$; hence $c \neq b$ and $c < b$. Since f is continuous at c, we have $f(x) < \gamma$ near c (by Condition I of continuity). Therefore there are points to the right of c where $f < \gamma$, but all such points belong to S. Since c is an upper bound for S, such points cannot lie to the right of c. This is a contradiction.

Assume next that $f(c) > \gamma$. Thus $f(c) > \alpha$; hence $c \neq a$ and $c > a$. Since f is continuous at c, we have $f(x) > \gamma$ near c (by Condition II of continuity). This means that there is a number c_1 such that $a < c_1 < c$ and $f(x) > \gamma$ for every x with $c_1 < x \leq c$. Since no point of S (the set where $f < \gamma$) lies to the right of c, no point of S lies to the right of c_1. Thus c_1 is an upper bound for S, and is smaller than c. But c was the smallest upper bound for S. This is a contradiction.

1. Prove the following theorem: Let $f(x)$ be continuous for $-\infty < x < +\infty$ and assume that $\lim_{x \to -\infty} f(x) = \alpha$, $\lim_{x \to +\infty} f(x) = \beta > \alpha$ (α and β numbers). If γ is any number strictly between α and β, then there is a point c, $-\infty < c < +\infty$, such that $f(c) = \gamma$. (Of course, this theorem also holds if $\beta < \alpha$.)

2. Reformulate and prove the theorem in Exercise 1 if $\alpha = -\infty$ and $\beta = +\infty$.

7.2 The need for proofs

In §3.6 we concluded that the Intermediate Value Theorem held by appealing to our geometric intuition. The graph of a continuous function consists of one piece; if it has a point below a line $y = c$ and a point above this line, there must be a point on the graph at which the graph crosses the line $y = c$. What can be simpler than this argument? Why complicate matters by appealing to the least upper bound principle and by giving a rather involved indirect proof?

The same question can be asked about many "geometrically obvious" theorems for which mathematicians nevertheless give analytic proofs.

The answer is: Our geometric intuition is an invaluable guide, but, unfortunately, not an infallible one. There are statements that seem geometrically obvious, but are still false.

The analytic proofs do not replace intuition; they reinforce it. Only our intuition and imagination, applied to particular cases, can suggest general theorems. Only a rigorous proof can assure us that intuition did not lead us astray.

Problems

1. Let $x \mapsto g(x)$ denote the function such that $g(x) = 1$ if x is rational, $g(x) = -1$ if x is irrational. Set $f(x) = xg(x)$. At which points is $f(x)$ continuous?
2. Let $f(x)$ and $g(x)$ be two continuous functions defined in an interval I. At every point x of this interval, set $h(x) = \max(f(x), g(x))$ and $k(x) = \min(f(x), g(x))$, where $\max(a, b)$ denotes the larger of the two numbers a and b, and $\min(a, b)$ the smaller. Show that h and k are continuous functions.
3. Under the hypotheses of Problem 3, assume that the functions f and g are nondecreasing. Is h necessarily nondecreasing? Is k?
4. If $f(x)$ and $g(x)$ are two odd functions, and p and q are two positive integers, is $\phi(x) = f(x)^p g(x)^q$ even, odd, or neither?
5. Is the composite of two monotone functions always monotone?
6. Is the composite of two odd functions even or odd? Is the composite of an even and an odd function even or odd?
7. Show that, if the function $x \mapsto f(x)$ is continuous, then the function $x \mapsto |f(x)|$ is also continuous, but the converse statement need not be true.
8. Find a function $F(x)$ that is not continuous at any point, and such that $F(x)^2$ is continuous everywhere.
9. Let $f(x)$ be a rational function and assume that both statements

$$\text{``} \lim_{x \to +\infty} f(x) = +\infty \text{''} \quad \text{and} \quad \text{``} \lim_{x \to +\infty} f(x) = -\infty \text{''}$$

are false. Show that in this case the finite limit $\lim_{x \to +\infty} f(x)$ exists.
10. Let $f(x)$ be a rational function, not the constant function 0. Show that there exists a number A such that either $f(x) > 0$ for $x > A$ or $f(x) < 0$ for $x > A$.
11. We mentioned in §2.6 that the rational functions form a field. For two elements of this field, f and g, define the statement "$f < g$" to mean

that $f(x) < g(x)$ for all x that are larger than some number A (A may depend on f and g). Prove that this definition makes the field of rational functions into an ordered field, that is, that Postulates **13** to **16** of Chapter 1 are valid.

12. Show that the field of rational function is not Archimedean, that is, that Postulate **17** is not valid. (Note that every integer, as every other real number, is an element of the field of rational functions, since it is a constant function.)

13. Is the least upper bound principle valid in the field of rational functions?

14. For every rational function $f(x)$, there exists an integer n (which may be positive, negative, or zero) such that the finite limit $\lim_{x \to +\infty} (x^n f(x))$ exists. Prove this statement.

15. For every radical function $f(x)$ which is defined for all $x > A$, there is a rational number r such that the finite limit $\lim_{x \to +\infty} (x^r f(x))$ exists. Verify this statement using a few examples. Then try to prove it. (Hint: Every radical function is obtained from rational functions by finitely many applications of the following operations: (1) addition of two functions, (2) multiplication of two functions, (3) taking the reciprocal of a function, and (4) raising a function to a rational power. By the result of Problem 14, the statement to be proved is true for rational functions. Hence it suffices to prove that, if we apply any of the operations (1) to (4) to functions having the desired property, we obtain a function having the same property.)

16. Define a function that is continuous for all values of the independent variable and which you are sure is not a radical function.

17. Suppose that a function is continuous in an interval and takes on only rational values. What can you say about this function?

18. Let $f(x)$ be defined for $0 < x < 1$. Suppose we know that for all such x, and for all positive integers n, the number $10^n f(x)$ is never an integer. What can you say about the function f?

19. Can an odd function have an inverse? If it does, is the inverse odd? Ask and answer the same question for even functions.

20. Find three functions, $f(x)$, $g(x)$, and $h(x)$ such that for $x \to +\infty$, we have $f \to +\infty$, $g \to +\infty$, $h \to +\infty$, $f - g \to -\infty$, $f - h \to -3$, $f/g \to 1/2$, $f/h \to 1$.

4/Derivatives

In this chapter, we develop the first of the two basic concepts of calculus—the derivative. (The second basic concept, the integral, will be discussed in Chapter 5.) The reader should strive to acquire both an intuitive understanding of the derivative concept and the ability to compute and use derivatives with ease and accuracy. The various applications of the derivative discussed in this chapter serve both purposes.

The appendixes contain a section on one-sided and infinite derivatives (§6) which may be omitted at first reading, and two sections (§§7 and 8) with proofs of several theorems on derivatives. These theorems are fortunately geometrically obvious; therefore the proofs may be omitted by a reader willing to trust his intuition.

The concept of derivative originated from two seemingly unrelated problems: how to draw a tangent line to a curve and how to compute the velocity of a moving body. We shall consider both problems.

1.1 The derivative, the slope of the tangent line

In the preceding chapter, we considered continuous functions, those whose graphs "have no breaks." Now we consider continuous functions whose graphs "have no corners." An example of a graph with corners is shown in Fig. 4.1; the graph in Fig. 4.2 has no corners. Through every point of the latter graph passes a unique line, the tangent to the curve, which "just touches" the curve.

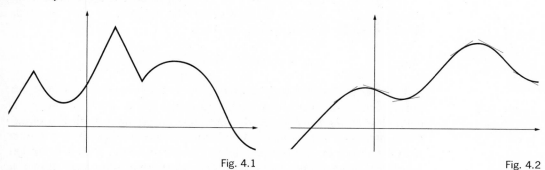

Fig. 4.1   Fig. 4.2

Consider a function $x \mapsto f(x)$. *The derivative of f at x_0 is the slope of the line tangent to the graph of f at the point $(x_0, f(x_0))$.* This, essentially, is the description given by Leibniz in his first published paper on calculus (1684). This paper contains a drawing not unlike our Fig. 4.3. The derivative of f at a point x_0 will be denoted by $f'(x_0)$. (Read: f prime of x_0.)

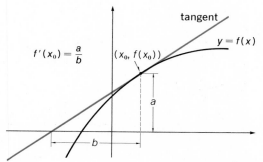

$$f'(x_0) = \frac{a}{b}$$

tangent

$y = f(x)$

$(x_0, f(x_0))$

a

b

Fig. 4.3

GOTTFRIED WILHELM LEIBNIZ (1646–1716) spent most of his life in the service of a German ducal family (one of his employers became George I of England). He is equally famous as a philosopher and as a mathematician. He was also a lawyer, diplomat, theologian, historian, geologist, economist, librarian, and linguist. Leibniz founded the Berlin Academy of Sciences and one of the first scientific journals. In this journal, he published the first account of calculus. Leibniz developed calculus later than Newton, but independently of him.

A central theme in Leibniz' thought was the search for a "universal method" that would reduce all reasoning to calculations. In this sense, Leibniz is a precursor of mathematical logic. He also invented determinants and designed a computing machine.

Fig. 4.4

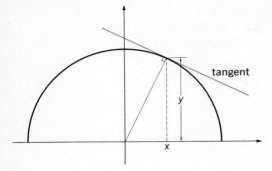

For some functions we can at once write down the derivative.

The graph of a linear function $x \mapsto mx + b$ is a straight line; it is considered to be its own tangent. Hence

(1)   if $f(x) = mx + b$, then $f'(x) = m$.

The graph of the function $x \mapsto \sqrt{1 - x^2}$ is the upper half of the circle with center at the origin and radius 1 (Fig. 4.4). The tangent to this circle passing through the point (x, y), $y = \sqrt{1 - x^2}$ is perpendicular to the line joining $(0, 0)$ and (x, y) and therefore has slope $-x/y$ (see Chapter 2, §3.2). Therefore

Fig. 4.5

(2)   if $f(x) = \sqrt{1 - x^2}$, then $f'(x) = \dfrac{-x}{\sqrt{1 - x^2}}$.

The graph of the function $x \mapsto x^2$ is a parabola (Fig. 4.5), and we already know (Chapter 2, §4.4) that the tangent to this parabola passing through (x, x^2) has slope $2x$. Thus

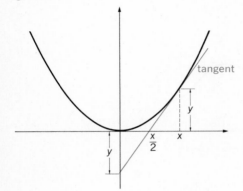

(3)   if $f(x) = x^2$, then $f'(x) = 2x$.

In order to use Leibniz's definition for other functions, however, we must define what we mean by the tangent. For a circle, a tangent is defined as a line having exactly one point in common with the circle. For a parabola, a tangent is a line having exactly one point in common with the parabola and not parallel to the axis. Clearly, such a definition would not work in general. Consider, for example, the graph shown in Fig. 4.6. There are several lines (for instance, the lines labeled 1, 2, and 3) passing through the point P on the curve and having only this point in common with the curve. None of these fits the description of the tangent

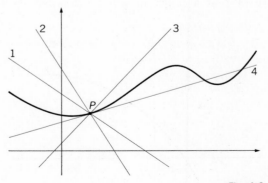

Fig. 4.6

as the line that "just touches" the curve at P. The line labeled 4 does, but this line intersects the curve at other points also.

EXERCISES

In doing Exercises 1 to 5, use your intuitive understanding of the word "tangent."

1. For each point marked on the given graph, draw the tangent to the graph and measure its slope.

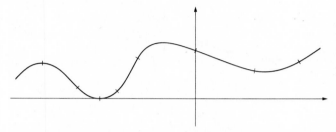

2. Draw the graph of a function $x \mapsto f(x)$, defined and continuous for $-4 \leq x \leq 4$, such that $f(-3) = 0$, $f'(-3) = 1$, $f(0) = 1$, $f'(0) = 0$, $f(3) = -3$, and $f'(3) = -2$.
3. Draw the graph of a function $v = g(u)$, defined and continuous for $-1/2 \leq u \leq 9/2$, such that $g(0) = 0$, $g'(1) = 0$, $g(2) = 1$, $g'(2) = 1$, $g(3) = -1$, and $g'(4) = -1/2$.
4. Draw the graph of a function $x \mapsto h(x)$, defined and continuous for $-5/4 \leq x \leq 7/2$ such that $h(-1) = 0$, h is increasing in the interval $[-1, 0]$, $h'(0) = 0$, $h(2) = -1$, and $h'(3) = 3/2$.
5. Draw the graph of a function $y = g(x)$, defined and continuous for $-3 \leq x \leq 5/2$, such that $g(-2) = 1$, $g'(-1/2) = -1$, $g(0) = 0$, the graph of g does not have a tangent at the point $(0, 0)$ and (a) g is odd, (b) g is even.

1.2 Drawing the tangent line

Our task is therefore to give a precise meaning to the expression: a line just touches the curve at a point. We proceed, as one often does in mathematics, by first assuming that we know what is meant by a tangent to a graph and by asking ourselves how we can compute the slope of this line. Then we shall make the method of computing into a definition.

We are given a function $x \mapsto f(x)$ and a point $P = (x_0, f(x_0))$; we want to compute the slope m of the tangent l to the graph of f at the point P (Fig. 4.7). The difficulty is that we know only one point, namely, P, on l, while we need two points on a line in order to compute the slope (see Chapter 2, §2.2). The key observation now is that, if we pick another point, Q, on the curve, which is close to P, then the slope m_1 of the line l_1 joining P and Q will be close to m. But m_1 can be computed. For, if the coordinates of Q are $(x_1, f(x_1))$, then

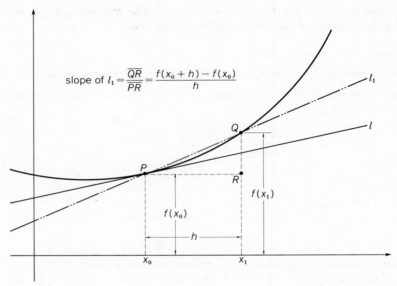

$$\text{slope of } l_1 = \frac{\overline{QR}}{\overline{PR}} = \frac{f(x_0 + h) - f(x_0)}{h}$$

Fig. 4.7

$$m_1 = \frac{f(x_1) - f(x_0)}{x_1 - x_0}.$$

It is convenient to set $h = x_1 - x_0$. Then,

$$m_1 = \frac{f(x_0 + h) - f(x_0)}{h}.$$

This number is called the **difference quotient** of f at x_0 for the difference h. For Q close to P, it is an approximation to m.

The point Q being close to P means that $|h|$ is small. We expect that m_1 is close to m if $|h|$ is small, and that the smaller $|h|$ is, the closer m_1 is to m. Let us test this for the parabola $f(x) = x^2$ and the point $P = (1, 1)$, where we know that $m = 2$. First we compute m_1 for various values of h of small absolute value:

h	$f(1 + h) = (1 + h)^2$	$m_1 = \dfrac{(1 + h)^2 - 1}{h}$
.1	1.21	2.1
−.1	0.81	1.9
.01	1.0201	2.01
−.01	0.9801	1.99
.001	1.002001	2.001
−.001	0.998001	1.999

The results confirm our expectation. But we can do better than that. If $f(x) = x^2$, $x_0 = 1$, we have

$$m_1 = \frac{f(1 + h) - f(1)}{h} = \frac{(1 + 2h + h^2) - 1}{h} = 2 + h \quad \text{(when } h \neq 0\text{)}.$$

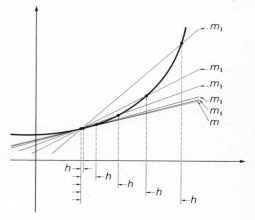

Fig. 4.8

We observe that m_1, as a function of h, has a limit as $h \to 0$ (since $2 + h$ is a continuous function of h). This limit is 2, which is precisely the value of the slope m.

Therefore, we expect that, whenever the graph of a function f has a tangent l at the point $(x_0, f(x_0))$, we can find the slope of l by first computing the difference quotient m_1 of f at x_0 for a variable difference h, and then finding the limit of m_1 at $h = 0$. The process is illustrated in Fig. 4.8.

EXERCISES

6. Given $f(x) = x^2 - 4$, tabulate the values of the difference quotient $(f(2 + h) - f(2))/h$ for $h = \pm 1, \pm 1/2$, and $\pm 1/4$.
7. Given $g(t) = 2t^2 - t + 1$, tabulate the values of the difference quotient $(g(t_0 + \lambda) - g(t_0))/\lambda$ at $t_0 = 1$ for $\lambda = \pm 1, \pm .1$, and $\pm .01$. (Leave the answers in decimal form.)
8. Given $h(z) = z/(z + 1)$, tabulate the values of the difference quotient $(h(z_0 + \delta) - h(z_0))/\delta$ at $z_0 = 0$ for $\delta = \pm 1, \pm .1$, and $\pm .01$.
9. Given $f(x) = 1000x^3 - 900x^2 - 10x + 1$, tabulate the values of the difference quotient $(f(x_0 + h) - f(x_0))/h$ at $x_0 = 0$ for $h = \pm 1, \pm .1$, and $\pm .01$.

1.3 Definition of the derivative

The preceding considerations suggest the following **definition of the derivative.** Let f be a function defined near a point x_0. The derivative of f at x_0 is the number

$$(4) \qquad f'(x_0) = \lim_{h \to 0} \frac{f(x_0 + h) - f(x_0)}{h},$$

provided that this limit exists.

We note that (setting $x_1 = x_0 + h$) we can also write

$$(4') \qquad f'(x_0) = \lim_{x_1 \to x_0} \frac{f(x_1) - f(x_0)}{x_1 - x_0}.$$

The quotient

$$(5) \qquad \frac{f(x_1) - f(x_0)}{x_1 - x_0} = \frac{f(x_0 + h) - f(x_0)}{h}$$

is called, as we already noted, the **difference quotient** (of f at x_0 for the difference h). It is the slope of the line joining the point $(x_0, f(x_0))$ and $(x_1, f(x_1))$ with $x_1 = x_0 + h$, $h \neq 0$. (Such a line is called a chord of the curve.)

Fig. 4.9

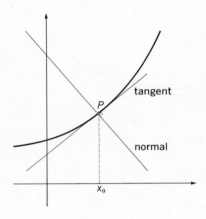

1.4 Tangents and normals

The derivative being defined, we can give the **definition of the tangent.** The tangent to the graph of $x \mapsto f(x)$ at a point x_0 is the line with the equation

$$(6) \qquad\qquad y - f(x_0) = f'(x_0)(x - x_0),$$

that is, the line through $(x_0, f(x_0))$ with slope $f'(x_0)$.

The line through $(x_0, f(x_0))$ perpendicular to the tangent is called the **normal** to the graph at this point (see Fig. 4.9). Hence (see Chapter 2, §2.3) the equation of the normal is

$$(7) \qquad\qquad y - f(x_0) = -\frac{1}{f'(x_0)}(x - x_0)$$

if $f'(x_0) \neq 0$, and $x = x_0$ if $f'(x_0) = 0$.

Of course, it may happen that a function does not have a derivative at a point x_0. In this case, there is no tangent to the graph at the point $(x_0, f(x_0))$.

1.5 Computing the derivative from the definition

The actual calculation of a derivative for a given function f consists in writing down the difference quotient

$$\frac{f(x_0 + h) - f(x_0)}{h}$$

(for x_0, a fixed number or a symbol representing an unspecified fixed number, and for a variable h representing a number with small but positive absolute value) and transforming it, by legitimate algebraic operations, into a form from which the limit (as $h \to 0$) can be read off.

One must remember that x_0 is fixed throughout the calculation.

▶ *Examples* *1.* Compute $f'(2)$ for $f(x) = x^3 - x$.

Answer. We have

$$f'(2) = \lim_{h \to 0} \frac{f(2 + h) - f(2)}{h} = \lim_{h \to 0} \frac{[(2 + h)^3 - (2 + h)] - (2^3 - 2)}{h}$$

$$= \lim_{h \to 0} \frac{(8 + 12h + 6h^2 + h^3 - 2 - h) - (8 - 2)}{h}$$

$$= \lim_{h \to 0} \frac{11h + 6h^2 + h^3}{h} = \lim_{h \to 0}(11 + 6h + h^2) = 11.$$

2. Compute the equation of the tangent and of the normal to the curve $y = \sqrt{x}$ at the point $(4, 2)$.

Answer. First we compute the derivative of the function $f(x) = \sqrt{x}$ at $x = 4$. We have

$$\frac{f(4 + h) - f(4)}{h} = \frac{\sqrt{4 + h} - 2}{h}$$

In order to compute the limit of this function of h as $h \to 0$, we multiply the numerator and denominator of the last fraction by $\sqrt{4 + h} + 2$. This yields

$$\frac{f(4 + h) - f(4)}{h} = \frac{(\sqrt{4 + h} - 2)(\sqrt{4 + h} + 2)}{h(\sqrt{4 + h} + 2)} = \frac{1}{(\sqrt{4 + h} + 2)} \cdot$$

This is a radical function of h, defined and continuous at $h = 0$; its value at $h = 0$ is $\frac{1}{4}$. Hence

$$f'(4) = \lim_{h \to 0} \frac{\sqrt{4 + h} - 2}{h} = \frac{1}{4}.$$

Noting equations (6) and (7) above, we conclude that the equation of the desired tangent is

$$y - 2 = \frac{1}{4}(x - 4) \qquad \text{or} \qquad x - 4y + 4 = 0,$$

and the equation of the desired normal is

$$y - 2 = -4(x - 4) \qquad \text{or} \qquad 4x + y - 18 = 0. \blacktriangleleft$$

EXERCISES

In Exercises 10 to 21, the computation of the derivative should be based directly on the definition (as in the foregoing examples).

10. If $f(x) = x^2 - x - 1$, find $f'(1)$.
11. Find the derivative of $g(s) = (s + 1)/(s - 1)$ at $s = 0$.
12. Find the slope of the line tangent to the graph of $H(x) = x^2/(x + 1)$ at $x = 1$.
13. Find the equation of the line tangent to the graph of $y = x + (1/x)$ at the point $(1, 2)$.
14. Find the equation of the line normal to the graph of $z = x^3 - 1$ at the point $(2, 7)$.
15. If $g(t) = \sqrt{t^2 + 4}$, find $g'(3)$.
► 16. Find the equation of the line tangent to the graph of $s = t/(t^2 - 1)$ at the point $(2, 2/3)$.
17. Find the equation of the line normal to the graph of $v = 2x + (3/x^2)$ at $x = 1$.
18. If $h(y) = (y + 1)^4 - y$, find $h'(-1)$.
19. Find the equation of the line tangent to the graph of $x = u^3 + u - 1$ at $u = 2$.
20. Find where the line tangent to the graph of $K(s) = (s + 2)^5(s - 1)^2$ at the point $(-2, 0)$ intersects the vertical axis.
21. Find $f'(1)$ if $f(x) = x^2$ for $x \leq 1$, $2x - 1$ for $x > 1$.

1.6 The derivative, a new function

Strictly speaking, one should distinguish between the derivative $f'(x_0)$ of a function $x \mapsto f(x)$ at some point x_0, and the rule $x \mapsto f'(x)$ which associates with every relevant number x the value $f'(x)$ of the derivative at this point. The derivative computed at some point is a number, the slope of the tangent at the point considered. The rule $x \mapsto f'(x)$ is a new function, which is sometimes called the derived function of f.

It is customary, however, to use the term "derivative" in both cases. Thus, if $f(x) = x^3$, we say "the derivative of this function is the function $x \mapsto 3x^2$" and "the derivative of $f(x)$ at $x = 2$ is the number 12." The process of computing the derivative is called **differentiation**.

►*Examples 1.* Set $f(x) = mx + b$, and compute the derivative at some point x.

Answer. We have

$$f(x) = mx + b, f(x + h) = m(x + h) + b,$$

and $$\frac{f(x + h) - f(x)}{h} = \frac{(mx + mh + b) - (mx + b)}{h} = m.$$

The difference quotient is a constant. Hence

$$f'(x) = \lim_{h \to 0} \frac{f(x + h) - f(x)}{h} = \lim_{h \to 0} m = m,$$

confirming the statement in §1.1.

2. Compute $f'(x)$ for $f(x) = 1/x$ (and, of course, $x \neq 0$).

Answer. We have

$$\frac{f(x + h) - f(x)}{h} = \frac{1}{h}\left(\frac{1}{x + h} - \frac{1}{x}\right) = \frac{1}{h} \cdot \frac{x - (x + h)}{x(x + h)}$$

$$= -\frac{1}{x(x + h)}.$$

The last expression defines a continuous function of h, at $h = 0$ (since $x \neq 0$). Hence $f'(x) = -1/x^2$.

3. Compute $f'(x)$ for $f(x) = \sqrt{x}$ (and $x > 0$).

Answer. We have

$$\frac{f(x + h) - f(x)}{h} = \frac{\sqrt{x + h} - \sqrt{x}}{h}.$$

It is not clear that this can be extended to a continuous function of h, at $h = 0$. However, the following device helps (compare Example 2 in §1.5).

$$\frac{\sqrt{x + h} - \sqrt{x}}{h} = \frac{\sqrt{x + h} - \sqrt{x}}{h} \frac{\sqrt{x + h} + \sqrt{x}}{\sqrt{x + h} + \sqrt{x}}$$

$$= \frac{(\sqrt{x + h})^2 - (\sqrt{x})^2}{h(\sqrt{x + h} + \sqrt{x})} = \frac{1}{\sqrt{x + h} + \sqrt{x}}.$$

This is a radical function of h, defined and continuous at $h = 0$. Therefore

$$f'(x) = \lim_{h \to 0} \frac{1}{\sqrt{x + h} + \sqrt{x}} = \frac{1}{\sqrt{x} + \sqrt{x}} = \frac{1}{2\sqrt{x}}.$$

4. Compute $f'(x)$ for $f(x) = \sqrt{1 - x^2}$ and for some x, $-1 < x < 1$.

Answer. We use the same device as before. We have

$$\frac{f(x + h) - f(x)}{h} = \frac{\sqrt{1 - (x + h)^2} - \sqrt{1 - x^2}}{h}$$

$$= \frac{\sqrt{1 - (x + h)^2} - \sqrt{1 - x^2}}{h} \frac{\sqrt{1 - (x + h)^2} + \sqrt{1 - x^2}}{\sqrt{1 - (x + h)^2} + \sqrt{1 - x^2}}$$

$$= \frac{(1 - (x + h)^2) - (1 - x^2)}{h(\sqrt{1 - (x + h)^2} + \sqrt{1 - x^2})} = \frac{1 - x^2 - 2xh - h^2 - (1 - x^2)}{h(\sqrt{1 - (x + h)^2} + \sqrt{1 - x^2})}$$

$$= \frac{-2x - h}{\sqrt{1 - (x + h)^2} + \sqrt{1 - x^2}}.$$

This is a continuous function of h at $h = 0$. Setting $h = 0$, we obtain

$$f'(x) = \frac{-x}{\sqrt{1 - x^2}}$$

confirming the statement in §1.1.◄

EXERCISES

In Exercises 22 to 29, the computation of the derivative should be based directly on the definition (as in the foregoing examples).

► 22. If $g(x) = 1/(2x - 3)$, find $g'(x)$.
23. If $f(x) = x^2 - (1/x)$, find $f'(x)$.
24. If $H(x) = 1/\sqrt{x}$, find $H'(x)$.
25. Find the derivative of $s \mapsto s^3$.
26. Find the derivative of $G(x) = (2x - 3)/(3x + 2)$.
27. Find the derivative of $t \mapsto (1/t) + (1/t^2)$.
28. Differentiate (that is, find the derivative of) the function $z \mapsto \sqrt{z} + (1/\sqrt{z})$.
29. Differentiate the function $h \mapsto 2h^2 - 3h + 5$.

1.7 The Leibniz notation

The notation $f'(x)$ for the derivative of a function $y = f(x)$ was introduced by Lagrange. Leibniz denoted the same derivative by

$$\frac{df(x)}{dx} \quad \text{or} \quad \frac{df}{dx} \quad \text{or} \quad \frac{dy}{dx}$$

(read "df over dx" or "dy over dx"). Thus we have, from the examples above,

JOSEPH LOUIS LAGRANGE (1736–1813) was a leading mathematician of the eighteenth century. One of his books, the famous *Analytical Mechanics*, is a systematic development of mechanics by the methods of calculus; not a single diagram appears in the book—everything is done by formulas.

Lagrange was under 20 when he became professor of mathematics at the artillery school in his native Turin. Later he was Euler's successor in the Berlin Academy. From 1786 on, Lagrange lived in Paris and after the French revolution he taught at the newly established École Normale and École Polytechnique.

$$\frac{d(mx + b)}{dx} = m, \qquad \frac{d(x^2)}{dx} = 2x, \qquad \frac{d(x^3 - x)}{dx} = 3x^2 - 1,$$

$$\frac{d(1/x)}{dx} = -\frac{1}{x^2}, \qquad \frac{d\sqrt{x}}{dx} = \frac{1}{2\sqrt{x}}, \qquad \frac{d\sqrt{1 - x^2}}{dx} = -\frac{x}{\sqrt{1 - x^2}}.$$

The Leibniz notation may be explained by a mathematical myth (represented in Fig. 4.10).

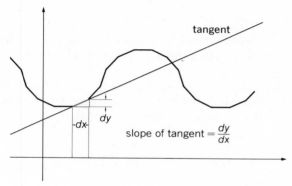

Fig. 4.10

A curve, so the myth goes, consists of infinitely many "infinitesimal" (infinitely small) straight segments. A tangent to the curve is a line containing such an infinitesimal segment. Now let the curve be the graph of a function $y = f(x)$. To compute the slope of the tangent, at the point (x, y), we increase x by an infinitesimal amount dx; this corresponds to a change of $y = f(x)$ by an infinitesimal amount dy such that $f(x + dx) = y + dy$. The piece of the graph between the infinitely close points (x, y) and $(x + dx, y + dy)$ is an infinitesimal segment; its slope is

$$\frac{(y + dy) - y}{(x + dx) - x} = \frac{dy}{dx}.$$

This is how mathematicians thought about calculus before the clarification of the limit concept. They were, of course, aware of the logical deficiencies of this approach, but they followed d'Alembert's advice, "Proceed and faith shall be given to you."

Today, we do not think of dy and dx as names of infinitely small quantities or of dy/dx as a fraction, but we still use Leibniz's notation. First of all, it is highly suggestive. The symbol dy/dx is a name for the derivative; its form reminds us that the derivative is the limit of the difference quotient (5), which is indeed a fraction with the increment of x in the denominator and the corresponding increment of $y = f(x)$ in the

JEAN LE ROND D'ALEMBERT (1717–1783) was a highly influential scientist and mathematician. He belonged to the group of French Enlightenment philosophers who published the *Encyclopédie*, and he was permanent secretary of the Paris Academy of Sciences. D'Alembert is remembered mostly for his work in mechanics.

numerator. One often denotes the increment of x by Δx. (It is important to remember that Δx is a single symbol, not the product of Δ and x. This symbol is pronounced "delta x." Previously, we denoted this increment by the letter h.) Let us now denote the increment of the function $y = f(x)$ by Δy. Thus $\Delta y = f(x + \Delta x) - f(x)$, and the difference quotient may be written as $\Delta y / \Delta x$. Hence we have

$$\frac{dy}{dx} = \lim_{\Delta x \to 0} \frac{\Delta y}{\Delta x}.$$

This is another "justification" of the Leibniz notation. The principal justification, however, is that the notation is very convenient, as we shall see later on many occasions.

Unfortunately, the Leibniz notation does not indicate the point at which the derivative is computed. To overcome this disadvantage, we indicate this point by a subscript. Thus we write

$$\frac{dx^2}{dx} = 2x, \qquad \left(\frac{dx^2}{dx}\right)_{x=2} = 4.$$

We shall use this convention whenever convenient.

Other notations for the derivative, which we shall not use, are Df or $(Df)(x)$.

EXERCISES

In Exercises 30 to 37, the differentiations of the derivative should be based directly on the definition or on an example or exercise worked out previously.

30. If $y = u^2$, what is $(dy/du)_{u=\sqrt{3}}$?
31. If $u = y^2$, what is $(du/dy)_{y=1/3}$?
32. If $r = \sqrt{2s - 1}$, what is dr/ds?
33. Find $\dfrac{d(y^2 - y)}{dy}$.
34. Find $\dfrac{d\sqrt{1 - u}}{du}$.
35. Find $\left(\dfrac{d(2v^3 + 3v^2)}{dv}\right)_{v=1}$.
► 36. If $s = 1/(t^2 + t + 1)$, find $(ds/dt)_{t=-1}$.
37. Find dz^4/dz.

1.8 The Linear Approximation Theorem

The mathematical myth told above contains a kernel of truth. To see this, we reformulate the definition of the derivative.

Theorem 1   LET f BE A FUNCTION DEFINED NEAR x_0. THE FUNCTION HAS
A DERIVATIVE m AT x_0 IF AND ONLY IF

$$(8) \qquad f(x) = f(x_0) + m(x - x_0) + r(x)(x - x_0),$$

WHERE

$$(9) \qquad \lim_{x \to x_0} r(x) = 0$$

OR, WHICH IS THE SAME,

$$(9') \qquad r(x) \text{ IS CONTINUOUS AT } x_0 \text{ AND } r(x_0) = 0.$$

Proof.   If (8) and (9') are true, then

$$\frac{f(x) - f(x_0)}{x - x_0} = m + r(x),$$

and the limit of this expression at $x = x_0$ is m. Conversely, if $f'(x_0) = m$, and if we define $r(x)$, for $x \neq x_0$ and close to x_0, by the above equation, then r is continuous and satisfies (9), and (8) holds.

To understand the meaning of the theorem, note that

$$x \mapsto f(x_0) + m(x - x_0), \text{ where } m = f'(x_0),$$

is the linear function whose graph is the line tangent to the graph of f at the point $(x_0, f(x_0))$. The number

$$(10) \qquad r(x)(x - x_0)$$

is the error we commit if, in computing the value $f(x)$, for some x close to x_0, we replace the graph of the function by the tangent at x_0. This meaning of the term (10) is obtained from Fig. 4.11. The error will be small when $x - x_0$ is small, and it will be much smaller than $x - x_0$, as equation (9) shows. For, if $x - x_0$ is small, the error is $(x - x_0)$ times the small number $r(x)$. In this sense one can replace a small arc of a curve by its tangent.

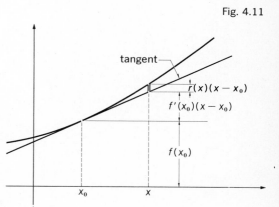

Fig. 4.11

tangent

$r(x)(x - x_0)$

$f'(x_0)(x - x_0)$

$f(x_0)$

x_0 x

▶*Example*   Represent the function $f(x) = x^3 - 2x - 1$ near $x = 2$ as a linear function plus a small error term.

Answer.   We have $f(2) = 3$ and we compute, by the method of §1.5, that $f'(2) = 10$. By Theorem 1 the linear function approximating $f(x)$ near $x = 2$ is the function

$$x \mapsto 3 + 10(x - 2) = 10x - 17.$$

The error term is

$$(x^3 - 2x - 1) - (10x - 17) = x^3 - 12x + 16.$$

We want to write this error term as $(x - 2)$ times a factor that is small near $x = 2$. Long division gives

$$x^3 - 12x + 16 = (x - 2)(x^2 + 2x - 8).$$

We note that the polynomial $x^2 - 2x - 8$, which corresponds to $r(x)$ in Theorem 1, is indeed 0 for $x = 2$. Summarizing, we obtain

$$x^3 - 2x - 1 = \underbrace{10x - 17}_{\substack{\text{linear} \\ \text{function}}} + \underbrace{(x - 2)(x^2 + 2x - 8)}_{\text{error term} = (x - 2)r(x)}$$

and $r(2) = 0$. We can also write

$$x^3 - 2x - 1 = 10x - 17 + (x - 2)^2(x + 4)$$

since $x^2 + 2x - 8 = (x - 2)(x + 4)$.

We observe that, for $x = 2.1$, the error term is $(.1)^2(6.1) = .061$ and for $x = 2.01$ the error term is $(.01)^2(6.01) = .000601.$ ◀

EXERCISES

In Exercises 38 to 45, represent the given function in the form (8) near the given point, identifying which terms in the expression correspond to m and r (of course, you may have to modify notation, for instance, change x to z or f to h). Verify that r is continuous at the given point and vanishes there.

38. $f(x) = 2x^2 - x$ near the point $x_0 = 3$.
39. $g(s) = 1 - 3s + s^2$ near the point $s_0 = -1$.
40. $y \mapsto 2y - 1$ near the point $y_0 = 1/2$.
41. $H(z) = -2$ near the point $z_0 = 10$.
► 42. $t \mapsto 1/t$ near the point $t_0 = 1$.
43. $f(u) = 10 - 2/3\,u$ near the point $u_0 = -3$.
44. $h(v) = v^3$ near the point $v_0 = 1$.
45. $Q(s) = s/(s - 1)$ near the point $s_0 = 0$.

1.9 The use of derivatives in computing function values

Theorem 1 suggests a method for computing **approximately** the value $f(x)$, for x close to a number x_0 for which we know $f(x_0)$ and $f'(x_0)$; we disregard the error term and write

(11) $f(x) \approx f(x_0) + f'(x_0)(x - x_0).$

The sign $\approx$ is read: approximately equal. In a later chapter we shall consider the approximation (11) in greater detail (see Chapter 8, §1).

▶*Examples* *1.* We want to compute 1.023^2. The value 1.023 is close to 1 and we are looking for $f(1.023) = f(1 + .023)$, where $f(x) = x^2$. We know that $f(1) = 1$ and $f'(1) = 2$. Hence (11) yields

$$1.023^2 \approx 1 + 2(.023) = 1.046.$$

Similarly, since $.98 = 1 - .02 = 1 + (-.02)$

$$(.98)^2 \approx 1 + 2(-.02) = 1 - .04 = .96.$$

In general, $(1 + h)^2 \approx 1 + 2h$ if $|h|$ is small.

But $(1 + h)^2 = 1 + 2h + h^2$. In this case the error committed is h^2, hence much smaller than $|h|$ if $|h|$ is small.

2. We know that for $f(x) = \sqrt{x}$, $f'(x) = 1/2\sqrt{x}$ (see §1.7). Hence $f(1) = 1$, $f'(1) = 1/2$, and

$$\sqrt{1 + h} \approx 1 + \frac{1}{2}h   \text{if } |h| \text{ is small.}$$

For instance, $\sqrt{1.03} \approx 1.015$ and $\sqrt{.98} \approx .99$. Similarly, since $f(100) = 10$ and $f'(100) = 1/20 = .05$,

$$\sqrt{100 + h} \approx 10 + (.05)h   \text{if } |h| \text{ is small.}$$

For instance, $\sqrt{101} \approx 10 + .05 = 10.05$, $\sqrt{97} \approx 10 - (.05)3 = 10 - .15 = 9.85$.

3. Using

$$\left(\frac{d(1/x)}{dx}\right)_{x=1} = -1,$$

(see Example 2 in §1.6), we obtain

$$\frac{1}{1 + h} \approx 1 - h   \text{if } |h| \text{ is small.}$$

For instance, $1/1.02 \approx .98$ and $1/.99 \approx 1.01$.◀

EXERCISES

In Exercises 46 to 55, use the approximation (11).

46. Compute $(3.05)^2$ approximately.
47. Compute $\sqrt{49.1}$ approximately.
48. Compute $\sqrt{35.99}$ approximately.
49. Compute $(1.988)^2$ approximately.
50. Compute $(1.08)^2$ approximately.
51. Find an approximation to $(1 + h)^3$ for $|h|$ small.
52. Compute $1/\sqrt{25.025}$ approximately.
53. Find an approximation to $1/(\sqrt{100 + h})$ for $|h|$ small.
54. If $g(s) = (s + 1)/(s - 1)$, compute $g(.025)$ approximately.
55. Compute $(2.95)^4$ approximately.

1.10 Continuity of differentiable functions

A function that has a derivative is called **differentiable**. In discussing the derivative of a function, we have assumed the function to be continuous. This is natural; we do not expect to have a tangent where the graph breaks. The following result justifies our assumption.

Theorem 2   IF f HAS A DERIVATIVE AT x_0, THEN f IS CONTINUOUS AT x_0.

Proof. If $f'(x_0) = m$, then equation (8) holds. In it, f is represented as a sum of three functions [a constant, a linear function, and the function $r(x)(x - x_0)$], each of which is continuous at x_0. A sum of continuous functions is continuous (by Theorem 1, in Chapter 3, §3.3).

1.11 Rectilinear motion

The problem of tangents was only one of the two questions that led to the concept of the derivative. The other was the description of motion. This problem dominated the work of Newton, who developed calculus at the same time that he created the science of mechanics.

We consider a **particle** moving along a straight line (**rectilinear motion**). By a particle we mean a body whose dimensions can be disregarded in the problem considered, so that it can be treated as a mathematical point. For instance, in dealing with the motion of the earth around the sun, the earth can be treated as a point.

We choose a point O on the straight line along which our particle moves and we designate one direction as positive. We also choose a unit of length, a unit of time, and an instant from which we count time. Then the particle moves along a number line and the motion defines a function

ISAAC NEWTON (1642–1727) was educated at Cambridge University and later taught there. During the Great Plague of 1664–65, Newton, who had just received his B.A., withdrew to his native hamlet of Woolsthorpe. In these two years, he discovered that white light can be decomposed into rays of different colors, developed calculus, formulated the law of universal gravitation, and then derived from it Kepler's laws of planetary motion. These great discoveries were published much later. For instance, *The Mathematical Principles of Natural Philosophy*, containing Newtonian mechanics, appeared in 1687; in this book, calculus is not used.

Newton's intellectual interests were not restricted to physics and mathematics, the two fields in which he was never surpassed. He left many manuscripts dealing with theology and alchemy. He was also a successful Master of the Mint (for which he was knighted), and several times represented his university in Parliament.

$$t \mapsto s(t),$$

which associates to every value of the time t the coordinates of the particle at that time. The graph of this function can be thought of as the picture of the motion. It should be remembered, of course, that the graph of the motion depends on the units of length and time. The graph of the same motion will look quite different if we first measure distance in feet and time in hours, and then use centimeters and seconds.

1.12 Uniform motion

The motion is called **uniform** if the distance traversed by the moving point (particle) is proportional to time elapsed. Assume that this is so and let α be a number defined as follows: the absolute value $|\alpha|$ is the distance traversed by the point during one time unit; α is positive or negative according to whether the particle moves in the positive or negative direction. We say that our uniformly moving particle has the **velocity**

$$\alpha \, \frac{\text{length units}}{\text{time units}},$$

for instance, α(cm/sec) (read "α centimeters per second"), and it has the **speed** $|\alpha|$(length units/time units). Assume also that, at time $t = 0$, our point is at the distance β from 0. After t time units, its distance from β will be αt and its distance from 0 will be $\beta + \alpha t$. Thus the motion is represented by the function

$$t \mapsto s(t) = \alpha t + \beta.$$

We see that *the graph of a uniform motion is a straight line and the slope of this straight line is the velocity.*

1.13 Velocity in nonuniform motion

Consider next the motion of a particle that is not uniform, for instance, the motion of a car along a straight highway. What is the velocity of the particle at a given time instant? A possible answer is: "the number read on the speedometer." In order to understand motion, however (in particular, in order to build a speedometer), we must reduce the concept of velocity to the more primitive concepts of distance ("the number read on a measuring tape") and time ("the number read on a clock").

Suppose that, at the time t, the particle is located at the point

$s = s(t)$, and at the time $t + h$ it is located at $s_1 = s(t + h)$. This means that, in h time units, the particle traversed $s_1 - s$ length units. We cannot conclude, of course, that its velocity at time t was

$$(12) \qquad \frac{s_1 - s}{h} = \frac{s(t + h) - s(t)}{h}$$

length units per time units, since the particle could have, for instance, moved very fast during the first $h/2$ time units and very slowly during the remaining time. But our experience with motion suggests that, if h is small, the "average velocity" (12) is close to the "true" or "instantaneous" velocity of the particle at the time t. We also feel that, the smaller h, the better the agreement between the "average velocity during the time h" and the instantaneous velocity. We therefore define the velocity $v(t)$ of the moving point at the time t as that unique number (of length units per time unit) to which the average velocity during the time interval from t to $t + h$ is arbitrarily close, provided that h is small enough. In other words,

$$v(t) = \lim_{h \to 0} \frac{s(t + h) - s(t)}{h} = s'(t)$$

or, in Leibniz's notation, $v = ds/dt$.

Thus the *velocity is the derivative of the distance traveled with respect to time, or, which is the same, the slope of the tangent to the graph of the function (time)* $\mapsto$ *(distance traveled)*.

Applying this definition to the uniform motion $s = \alpha t + \beta$, we obtain

$$v = \frac{d(\alpha t + \beta)}{dt} = \alpha,$$

as we should.

Given any differentiable function whatsoever, we can think of it either as describing a curve (geometric image) or as describing a motion (kinematic image), and we may think of the derivative either as the slope of the tangent or as the instantaneous velocity. The two descriptions complement each other.

EXERCISES

In Exercises 56 to 62, the given functions describe the motion of a particle moving in a straight line.

56. If $s = t^3 - t$, where the distance s is measured in feet and the time t is measured in seconds, what is the velocity when $t = 10$?

57. If $s = \sqrt{t}$, where the distance s is measured in centimeters and the time t is measured in seconds, what is the velocity when $t = 100$?

58. If $s = 100 - 60t$, where the distance s is measured in miles and the time t is measured in hours, what is the velocity when $t = 1$?

59. If $s = 1/(1 - t)$, $0 < t < 1$, where the distance s is measured in inches and the time t is measured in minutes, what is the velocity when $t = .75$?

60. If the function relating time and distance is $t \mapsto 1$, where distance is measured in centimeters and time is measured in minutes, what is the velocity when $t = 1.5$?

61. If $r = t^2 - t$, where the distance r is measured in feet and the time t is measured in seconds, what is the velocity when $t = .25$? When $t = .5$? When $t = .75$?

62. If $r = 3x^2 - 2x$, where r is distance and x is time, during what time interval is the particle moving in the positive direction?

1.14 Free fall

The simplest example of a nonuniform rectilinear motion is that of a body released at a certain height and permitted to fall to the ground. The discovery of the nonuniformity of this motion was an important event in the history of science.

The ancient Greeks were excellent mathematicians and also developed a sophisticated statics (that part of mechanics which deals with bodies at rest). But their ideas of dynamics (the mechanics of moving bodies) were very naïve. Aristotle taught that the motion of falling bodies is uniform, and that the heavier the body, the faster it falls. Since Aristotle's authority acquired an almost religious character, his statement was unchallenged until the sixteenth century. The true law of falling bodies was discovered by Galileo, who thereby founded modern physics. Since his work involved an analysis of the idea of velocity, he became one of the precursors of calculus.

Galileo's law asserts that *the distance traversed by a freely falling body released from rest is proportional to the square of the time it has been falling, the factor of proportionality being the same for all bodies.* Let us formulate these statements mathematically.

We consider the direction of free fall as negative vertical, choose the point at which the body is released as our reference point O, and count the time from the moment of release. According to Galileo, the motion of the falling body is described by the equation

$$(13) \qquad\qquad s = -\beta t^2.$$

The graph of this motion is a parabola. The value of β is the same for all bodies; of course, it depends on the units used. If one uses feet and

ARISTOTLE (384–322 B.C.) was a member of Plato's Academy for 20 years before founding his own school, the Lyceum. For a while, he was tutor to the future Alexander the Great. Aristotle did not contribute to mathematics proper, but he set the pattern for formal logic, as he did for many other subjects (like political science, literary criticism, and descriptive biology). The first significant advance in logic beyond Aristotle occurred in the nineteenth century, with the appearance of symbolic or mathematical logic.

A reconciliation between the Platonic-Pythagorean mathematical world view and the Aristotelian emphasis on observing actual phenomena occurred only when calculus made mathematical science possible.

GALILEO GALILEI (1564–1642). The founder of modern science, who said that the book of nature is written in mathematical symbols, was a mathematician, a physicist, an astronomer, a brilliant polemicist, and a talented writer. He formulated the principle of inertia and the laws of motion of the pendulum, of falling bodies, and of projectiles. He built some of the first telescopes and used them to discover the moons of Jupiter and Saturn, the phases of Venus, the rotation of the sun, and the ruggedness of the lunar landscape.

Galileo was professor at Pisa and at Padua, and later "chief philosopher and mathematician" at the court of the Grand Duke of Florence. At the age of 70, Galileo was tried by the Inquisition for his *Dialogue on the Two Great World Systems.* He was forced to renounce the Copernican system and spent the rest of his life under house arrest. Yet he continued to work and to write.

seconds, β is (approximately) 16; for centimeters and seconds β is (approximately) 490. We put a minus sign in (13) so that β would be positive.

Let us compute the velocity at time t. We have

$$v = \frac{ds}{dt} = \frac{d(-\beta t^2)}{dt} = -2\beta t.$$

Fig. 4.12

Thus the *speed* (absolute value of the velocity) is proportional to the time elapsed since the release of the body.

A particle released from rest and sliding down a smooth inclined plane (see Fig. 4.12) obeys essentially the same law. If s now denotes the length measured along the path from the point of release, and if the upward direction is again defined as positive, Galileo's law reads

(14)
$$s = -\frac{\beta m}{\sqrt{1 + m^2}}\, t^2,$$

where β is the same number as before and m is the slope of the inclined plane. The graph of the motion is again a parabola. The velocity is now

$$v = \frac{ds}{dt} = -\frac{2\beta m}{\sqrt{1 + m^2}}\, t.$$

Again the velocity is proportional to time, but if m is small, the speed increases slowly. This was very fortunate for Galileo since, with the primitive observational tools available to him (he had no clock), he was unable to measure fast motions accurately.

One should remember that the law (13) of falling bodies neglects the effect of air resistance, just as the law (14) neglects the friction between the body and the plane on which it slides.

EXERCISES

63. A ball is dropped from a height of 100 ft. What is its velocity after 2 sec?
64. A stone is dropped from a height of 10 cm. What is its velocity after .1 sec?
65. A book is dropped from a height of 81 ft. In how many seconds will the book strike the ground? What is its velocity at that instant? (More precisely, what is the limit of the velocity as t approaches t_0 from the left, where t_0 is the instant of impact?)
▶ 66. A brick is dropped from a building. How far has it fallen when its velocity is -128 ft/sec?
67. A ball rolls down an inclined plane that has a slope of 1/2. If distance is measured in centimeters, what is the velocity after 2 sec?
68. A particle sliding down an inclined plane achieves a velocity of -16 ft/sec 1 sec after it is released. What is the slope of the plane?

69. A ball slides down an inclined plane that has a slope of 1/4. If distance is measured in centimeters, what is the velocity after 4 sec?
70. A particle sliding down a 3-ft-long inclined plane achieves a velocity of −8 ft/sec when it reaches the bottom. What is the slope of the plane?

1.15 Rate of change

Velocity is an example of a **rate of change.** Whenever a function $x \mapsto y$ describes the relation between two quantities, the derivative dy/dx is called the rate of change of y with respect to x. Thus the velocity is the rate of change of distance with respect to time.

▶ *Examples* **1.** The rate of change of the weight W of a cube of uniform material with respect to the length a of the side of the cube is $3\alpha a^2$, where α is the weight of a cube with side of unit length. For $W = \alpha a^3$, and $dW/da = 3\alpha a^2$.

Recalling the linear approximation theorem (Theorem 1 above) we may interpret this as follows: given a cube of side length a and specific weight α, if we increase the sides by a small amount h, then the weight will increase by approximately $3\alpha a^2 h$.

2. In economics, the rate of change of a quantity q is usually called "marginal q." For instance, if $q(x)$ is the cost of producing x pieces of some product, then $q'(x)$ is called **marginal cost.** In practice x is a very large number, so that 1 is small compared to x, and $q'(x)$ is approximately $q(x + 1) - q(x)$. Therefore economists also define marginal cost as the cost of producing "one more piece."

3. The rate of change of the velocity with respect to time is called acceleration. We shall have many occasions to deal with this important concept later. We mention here that, for a body in free fall, the acceleration, that is, the time derivative of the velocity $-2\beta t$, is the constant -2β.◀

EXERCISES

In these exercises we assume, without proof, some simple geometric formulas.

71. Find the rate of change of the area of a disk (interior of a circle) with respect to its radius.
72. Find the rate of change of the surface area of a cube with respect to the length of a side.
73. The motion of a particle moving in a straight line is described by $s = t^3 - t$, where the distance s is measured in feet and the time t is measured in seconds. What is the acceleration when the velocity is 11 ft/sec?
74. Let h denote the height and r the radius of a cylindrical can. If the can is 1 in. high, what is the rate of change of volume with respect to the radius when the radius is 1/2 in.?

75. Let h denote the height and r the radius of a cylindrical can. If the radius of the can is 1/2 in., what is the rate of change of volume with respect to height when the height is 1 in.?

76. Suppose that the cost q of producing x pieces of some product is given by the formula $q = 10,000 + 22x + x^2/12,000,000$, where q is measured in dollars. What is the marginal cost when $x = 120,000$?

§2 Differentiation

The process of finding the derivative of a function is called **differentiation.** In differentiating functions, we hardly ever go back to the general definition. Instead, we use a set of rules by which the differentiation of functions commonly encountered in calculus is reduced to a mechanical process. We proceed to derive these rules.

2.1 The basic rules

Theorem 1   LET f AND g BE TWO FUNCTIONS THAT HAVE DERIVATIVES AT x_0 AND LET c BE A CONSTANT. THEN, AT x_0:

(1) $$c' = 0,$$

(2) $$(cf)' = cf',$$

(3) $$(f + g)' = f' + g',$$

(4) $$(f - g)' = f' - g',$$

(5) $$(fg)' = f'g + fg',$$

AND, IF $g(x_0) \neq 0$,

(6) $$\left(\frac{1}{g}\right)' = -\frac{g'}{g^2},$$

(7) $$\left(\frac{f}{g}\right)' = \frac{f'g - fg'}{g^2}.$$

Before proving the theorem, we rewrite the statement in the Leibniz notation:

(1′) $$\frac{dc}{dx} = 0 \qquad (c \text{ a constant}),$$

(2′) $$\frac{dcf}{dx} = c\frac{df}{dx} \qquad (c \text{ a constant}),$$

(3′) $$\frac{d(f + g)}{dx} = \frac{df}{dx} + \frac{dg}{dx},$$

(4') $$\frac{d(f-g)}{dx} = \frac{df}{dx} - \frac{dg}{dx},$$

(5') $$\frac{d(fg)}{dx} = \frac{df}{dx}g + f\frac{dg}{dx},$$

(6') $$\frac{d}{dx}\frac{1}{g} = -\frac{1}{g^2}\frac{dg}{dx},$$

(7') $$\frac{d}{dx}\frac{f}{g} = \frac{1}{g^2}\left(\frac{df}{dx}g - f\frac{dg}{dx}\right).$$

One advantage of the Leibniz notation is that the simple rules (2'), (3'), and (4') look like rules for fractions. These rules read, in words: a constant may be factored out in front of the differentiation sign; the derivative of a sum is the sum of the derivatives; and the derivative of a difference is the difference of the derivatives.

The rules (5), (6), and (7) for the derivative of a product (derivative of the first factor times the second factor plus the first factor times the derivative of the second), the derivative of a reciprocal (minus the derivative divided by the square of the function), and the derivative of a quotient (the derivative of the numerator times the denominator minus the numerator times the derivative of the denominator, this whole quantity divided by the square of the denominator) are more complicated and the Leibniz notation does not help in remembering them.

Theorems (1) to (7) are among the few rules in calculus that *must be memorized*.

The rules just stated may be applied to more than two functions. As an example, if the functions f, g, and h have the derivatives f', g', and h', then $(2f - 3g + h)' = 2f' - 3g' + h'$, and $(fgh)' = f'gh + fg'h + fgh'$; for $(fgh)' = (fg)'h + fgh' = (f'g + fg')h + fgh'$.

EXERCISES

1. Let $f(x) = 2x^3$ and $g(x) = (1/2)x + 1$. Compute $f'(x)$, $g'(x)$, $(f + g)'(x)$, and $(fg)'(x)$ directly from the definition. Verify that $(f + g)' = f' + g'$ and $(fg)' = f'g + fg'$.

2. Let $h(s) = 2s + 1$ and $k(s) = 3s^2 - 5$. Compute $h'(s)$, $k'(s)$, $(k - h)'(s)$, and $(k/h)'(s)$ directly from the definition. Verify that $(k - h)' = k' - h'$ and $(k/h)' = (k'h - kh')/h^2$.

3. Let $P(x) = x^2 - x - 1$ and $Q(x) = x + 1$. Compute $P'(x)$, $Q'(x)$, $(2P - 3Q)'(x)$, $(PQ)'(x)$ directly from the definition. Verify that Theorem 1 is satisfied.

4. Let $A(t) = t^2 + t$ and $B(t) = t^2 - t$. Compute $A'(t)$, $B'(t)$, $(\frac{1}{2}A - \frac{2}{3}B)'(t)$, and $(A/B)'(t)$ directly from the definition. Verify that Theorem 1 is satisfied.

5. Let $f(x) = x^2 - 2x$ and $g(x) = x$. Compute $f'(x)$, $g'(x)$, $(f + g^2)'(x)$, and $(g/f)'(x)$ directly from the definition. Verify that Theorem 1 is satisfied.

▶ 6. Suppose $f(0) = 1$, $f'(0) = 2$, $g(0) = 1/2$, and $g'(0) = -3$. Using the rules in Theorem 1, find $(fg)'(0)$ and $(g/f)'(0)$.

7. Suppose $A(3) = -1$, $B(3) = -4$, $A'(3) = 2$, and $B'(3) = 5$. Using the rules in Theorem 1, find $(2A - B)'(3)$, $(AB)'(3)$, and $(A/B)'(3)$.

8. Suppose $\phi(10) = -\frac{1}{2}$, $\psi(10) = 6$, $\phi'(10) = \frac{1}{3}$, $\psi'(10) = 8$. Using the rules in Theorem 1, find $(5\phi + \phi\psi)'(10)$ and $(3\phi\psi - 10\psi)'(10)$.

9. Suppose $f(x) = x^2$, $g(x) = x$, and $h(x) = 1/x$. Compute $f'(x)$, $g'(x)$, $h'(x)$, $(f + g + h)'(x)$, and $(fgh)'(x)$ directly from the definition. Verify that $(f + g + h)' = f' + g' + h'$ and $(fgh)' = f'gh + fg'h + fgh'$.

10. Let $A(x) = x - 1$, $B(x) = x + 1$, and $C(x) = x + 2$. Compute $A'(x)$, $B'(x)$, $C'(x)$, $(2A - B + 3C)'(x)$, $(ABC)'(x)$ directly from the definition. Verify that the extension of Theorem 1 to three functions is satisfied.

11. Suppose $f(1) = -2$, $f'(1) = 3$, $g(1) = -5$, $g'(1) = 1$, $h(1) = 2$, and $h'(1) = 4$. Using the rules in Theorem 1, find $(fgh)'(1)$, $(fg/h)'(1)$, $(f/gh)'(1)$, and $(g/f - h)'(1)$.

12. Suppose $A(2) = 1$, $B(2) = 10$, $C(2) = -2$, $A'(2) = \frac{1}{2}$, $B'(2) = 3$, and $C'(2) = 4$. Using the rules in Theorem 1, find $(AB/C)'(2)$, $(A^2/(B + C))'(2)$, and $((A + B)/(B + C))'(2)$.

2.2 Outline of proof

We already know that $c' = 0$, so that (1) holds. This is a special case of an example in §1.1, and it is geometrically obvious (a horizontal line has slope 0).

Next we shall prove the sum rule (3), the product rule (5), and the rule (6). We consider first the case in which the functions f and g are linear. Then (compare §1.8)

$$f(x) = a + \alpha(x - x_0) \qquad \text{where } a = f(x_0), \ \alpha = f'(x_0)$$
$$g(x) = b + \beta(x - x_0) \qquad \text{where } b = g(x_0), \ \beta = g'(x_0).$$

Now

$$f(x) + g(x) = a + b + (\alpha + \beta)(x - x_0).$$

This is again a linear function. Its slope (that is, the derivative of $f + g$) is $\alpha + \beta$. This is $f'(x_0) + g'(x_0)$ as claimed in (3).

Also,

$$f(x)g(x) = [a + \alpha(x - x_0)][b + \beta(x - x_0)]$$
$$= ab + (\alpha b + a\beta)(x - x_0) + \alpha\beta(x - x_0)^2.$$

Thus fg is the sum of a linear function with slope $\alpha b + a\beta$ and an error term $\alpha\beta(x - x_0)^2 = (x - x_0)r(x)$, where $r(x) = \alpha\beta(x - x_0)$ is a continuous function with $r(x_0) = 0$. By the linear approximation theorem (Theorem 1 in §1.8) we conclude that the derivative of fg at x_0 is $\alpha b + a\beta = f'g + fg'$ as claimed in (5).

Next we assume that $g(x_0) = b \neq 0$, set

$$\phi(x) = \frac{1}{g(x)} = \frac{1}{b + \beta(x - x_0)}$$

and compute $\phi'(x_0)$ using the definition of the derivative. We have

$$\phi(x_0) = \frac{1}{b}, \; \phi(x_0 + h) = \frac{1}{b + \beta(x_0 + h - x_0)} = \frac{1}{b + \beta h}$$

and $\dfrac{\phi(x_0 + h) - \phi(x_0)}{h} = \dfrac{1/(b + \beta h) - 1/b}{h} = -\dfrac{\beta}{b(b + \beta h)}.$

This is a continuous function of h; the value at $h = 0$ is $\phi'(x_0) = -\beta/b^2$. Thus $(1/g)' = -g'/g^2$ as claimed in (6).

The argument just given can be repeated, with little change, for functions that are not linear. Indeed, by the linear approximation theorem we have

$$f(x) = a + \alpha(x - x_0) + r_1(x)(x - x_0) \qquad \text{where } a = f(x_0), \; \alpha = f'(x_0),$$
$$g(x) = b + \beta(x - x_0) + r_2(x)(x - x_0) \qquad \text{where } b = g(x_0), \; \beta = g'(x_0)$$

and $r_1(x)$, $r_2(x)$ are continuous functions with

$$r_1(x_0) = r_2(x_0) = 0.$$

In order to compute the derivatives of $f + g$, fg, and $1/g$ near x_0, we must consider only values of x close to x_0. For such x, the error terms $r_1(x)(x - x_0)$, $r_2(x)(x - x_0)$ are *very* small and will not affect the final result.

Let us carry out the argument for the sum rule (3). We have

$$f(x) + g(x) = a + b + (\alpha + \beta)(x - x_0) + [r_1(x) + r_2(x)](x - x_0).$$

Hence $f + g$ is a linear function of slope $\alpha + \beta$ plus a term $(x - x_0)r(x)$ where $r(x) = r_1(x) + r_2(x)$ is continuous and $r(x) = 0$. By the linear approximation theorem, we conclude that the derivative of $f + g$ at x_0 is $\alpha + \beta = f'(x_0) + g'(x_0)$.

The complete proofs of (5) and (6) are similar but longer; they will be found in an appendix, §§6.1 and 6.2.

Using (3) and (5), we derive relation (2), that is, the statement $(cf)' = cf'$, by writing

$$(cf)' = c'f + cf' = 0f + cf' = cf'.$$

Using (2) and (3), we derive relation (4) by writing

$$(f - g)' = (f + (-1)g)' = f' + ((-1)g)' = f' + (-1)g' = f' - g'.$$

Of course, one can also prove (2) and (4) directly; the reader may try to do so.

Using (5) and (6), we now prove the quotient rule $(f/g)' = (f'g - fg')/g^2$ by writing

$$\left(\frac{f}{g}\right)' = \left(f\frac{1}{g}\right)' = f'\frac{1}{g} + f\left(\frac{1}{g}\right)' = f'\frac{1}{g} + f\left(-\frac{g'}{g^2}\right) = \frac{f'}{g} - \frac{fg'}{g^2}$$
$$= \frac{f'g - fg'}{g^2}.$$

This is precisely (7), and this derivation completes the proof of Theorem 1.

Theorem 1 can also be proved by a rather straightforward application of the definition of the derivative and the rules for limits proved in Chapter 3, §4.2. To prove that $(f+g)'=f'+g'$, we set $\phi(x)=f(x)+g(x)$. Then

$$\phi'(x_0) = \lim_{h \to 0} \frac{\phi(x_0 + h) - \phi(x_0)}{h}$$

(by the definition of the derivative)

$$= \lim_{h \to 0} \frac{[f(x_0 + h) + g(x_0 + h)] - [f(x_0) + g(x_0)]}{h}$$

$$= \lim_{h \to 0} \left(\frac{f(x_0 + h) - f(x_0)}{h} + \frac{g(x_0 + h) - g(x_0)}{h}\right)$$

$$= \lim_{h \to 0} \frac{f(x_0 + h) - f(x_0)}{h} + \lim_{h \to 0} \frac{g(x_0 + h) - g(x_0)}{h}$$

(since the limit of a sum is the sum of limits)

$$= f'(x_0) + g'(x_0)$$

(by the definition of the derivative).

Proofs of (5) and (6) along the same lines will be found in the Appendix, §§6.1 and 6.2.

2.3 Differentiating powers

Theorem 2   FOR EVERY POSITIVE INTEGER n, IF $f(x) = x^n$, THEN $f'(x) = nx^{n-1}$, THAT IS,

(8)
$$\frac{dx^n}{dx} = nx^{n-1}.$$

For $n = 1$, this means that the derivative of $x \mapsto x$ is 1. This is a special case of an example in §1.1, which is geometrically obvious (the graph of the identity function has slope 1). For $n > 1$, we give three different proofs.

I. We use the product rule (5) and the fact that $x' = 1$. Thus

$(x^2)' = (xx)' = x'x + xx' = x + x = 2x,$
$(x^3)' = (xx^2)' = x'x^2 + x(x^2)' = x^2 + x(2x) = x^2 + 2x^2 = 3x^2,$
$(x^4)' = (xx^3)' = x'x^3 + x(x^3)' = x^3 + x(3x^2) = x^3 + 3x^3 = 4x^3,$

and so forth. (To make the argument formally rigorous, we would have to use mathematical induction.)

II. Set $f(x) = x^n$, n an integer > 1. Then

$$f'(0) = \lim_{h \to 0} \frac{h^n - 0}{h} = \lim_{h \to 0} h^{n-1} = 0,$$

which is (8) for $x = 0$. For $x \neq 0$, we write the difference quotient at x as

$$\frac{(x + h)^n - x^n}{h} = \frac{x^n(1 + (h/x))^n - x^n}{x(h/x)} = \frac{x^n}{x} \frac{(1 + (h/x))^n - 1}{h/x}$$

$$= x^{n-1} \frac{(1 + (h/x))^n - 1}{(1 + (h/x)) - 1} \quad \text{\small (using the formula for a geometric progression with } q = 1 + (h/x))$$

$$= x^{n-1} \left\{ 1 + \left(1 + \frac{h}{x}\right) + \left(1 + \frac{h}{x}\right)^2 + \cdots + \left(1 + \frac{h}{x}\right)^{n-1} \right\}.$$

The difference quotient is a polynomial in h, and hence a continuous function of h; its value at $h = 0$ is nx^{n-1}. Thus, $f'(x) = nx^{n-1}$, as asserted.

III. Since $(x + h)^n = \overbrace{(x + h)(x + h) \cdots (x + h)}^{n \text{ factors}} =$ (sum of products, each of which contains one factor, either x or h, from each of the n terms) $= x^n + nx^{n-1}h + h^2 \cdot$ (some polynomial in h), we have

$$\frac{(x + h)^n - x^n}{h} = nx^{n-1} + h \cdot \text{some polynomial in } h.$$

This is a continuous function of h, and its value at $h = 0$, that is, dx^n/dx, is nx^{n-1}.

Theorem 3   THE FORMULA (8), THAT IS,

$$\frac{dx^n}{dx} = nx^{n-1},$$

HOLDS FOR ALL INTEGERS n.

Proof.   For $n > 0$, this is the statement of Theorem 2. For $n = 0$, we must show that the derivative of $x^0 = 1$ is 0. This is so, by rule (1) of Theorem 1. If n is a negative integer, then $m = -n$ is a positive integer. Using rule (6) of Theorem 1 and Theorem 2, we have

$$\frac{dx^n}{dx} = \frac{dx^{-m}}{dx} = \frac{d(1/x^m)}{dx} = -\frac{1}{x^{2m}}\frac{dx^m}{dx} = -\frac{1}{x^{2m}}mx^{m-1}$$

$$= -mx^{-m-1} = nx^{n-1},$$

as asserted.

2.4 Derivatives of polynomials and rational functions

Consider a polynomial function, say,

$$y = 3x^5 + 2x^3 - 6x + 1.$$

Using (1), (2), (3), (4), and (8), we compute its derivatives as follows

$$y' = (3x^5 + 2x^3 - 6x + 1)' = (3x^5)' + (2x^3)' - (6x)' + 1'$$
$$= 3(x^5)' + 2(x^3)' - 6(x)' + 1' = 3(5x^4) + 2(3x^2) - 6 + 0$$
$$= 15x^4 + 6x^2 - 6.$$

The same argument shows that, in general,

$$(a_0x^n + a_1x^{n-1} + \cdots + a_{n-1}x + a_n)'$$
$$= na_0x^{n-1} + (n-1)a_1x^{n-2} + \cdots + a_{n-1}.$$

This establishes

Theorem 4   THE DERIVATIVE OF A POLYNOMIAL OF DEGREE $n > 0$ IS A POLYNOMIAL OF DEGREE $n - 1$.

Theorem 5   A RATIONAL FUNCTION HAS A DERIVATIVE WHEREVER THE FUNCTION IS DEFINED; THE DERIVATIVE IS AGAIN A RATIONAL FUNCTION.

Proof. Let f be a rational function; then $f = p/q$, where p and q are polynomials without common roots, and f is defined wherever $q \neq 0$. At such points we have, by (7), that $f' = (p'q - pq')/q^2$. By Theorem 4, p' and q' are polynomials. So are $p'q - pq'$ and q^2 by Theorem 1 in Chapter 3, §2.3. Hence f' is a quotient of polynomials, and hence rational.

▶ *Example* Compute the derivative of $y = (3x^2 + 5x - 7)/(x^{10} + 1)$ at $x = 1$.

Solution. We have

$$\frac{dy}{dx} = \frac{(3x^2 + 5x - 7)'(x^{10} + 1) - (3x^2 + 5x - 7)(x^{10} + 1)'}{(x^{10} + 1)^2}$$

$$= \frac{(6x + 5)(x^{10} + 1) - (3x^2 + 5x - 7)10x^9}{(x^{10} + 1)^2};$$

it is optional whether or not to simplify. At any rate,

$$\left(\frac{dy}{dx}\right)_{x=1} = \frac{11 \cdot 2 - 1 \cdot 10}{2^2} = 3. ◄$$

EXERCISES

In Exercises 13 to 32, compute the derivatives using the rules that have been proved so far.

13. If $f(x) = x^5 - 2x^3 + 1$, find $f'(x)$.
14. If $z = (y^{11} - 3y^7 + 1)/(y - 1)$, find dz/dy.
15. If $g(t) = (t^3 + 3)(t^7 + 3t^4 - t^2 + 1)$, find $g'(t)$.
16. Differentiate the function $u \mapsto (u^2 - 2u + 3)/(u^2 + 2u - 3)$.
17. If $h(w) = (w^3 - 2w)/(w^2 - w - 1)$, find $h'(1)$.
▶ 18. Find the equation of the line tangent to the graph of $y = (x^2 + 1)/(x^3 + 1)$ at the point $(0, 1)$.
19. Find $\dfrac{d}{dx} \dfrac{1 - 2x - x^2}{1 + x^2}$.
20. Differentiate the function $t \mapsto (t^3 + 2t^2 - 4)(t^2 - 2t + 3)/(t^3 - 1)$.
21. Find the equation of the line normal to the graph of $z = (t^3 + t)/(t - 1)$ at the point $(2, 10)$.
22. If $s = (u^4 + u^3 - 2u + 1)(u^3 - 2u^2 + u + 6)$, find $(ds/du)_{u=2}$.
23. Find the equation of the line tangent to the graph of

$$s = \frac{t^2 - 5t + 10}{3t^3 - 1}$$

at $t = 0$.
24. Find the derivative of $(t^3 - 2t + 6)(t^5 - 3t^3 - 1)$.

25. Differentiate the function $r \mapsto (r^2 - 2r)(r^{10} + 9r + 5)/(r^5 - 2r^2 + 2)$.
26. Differentiate the function

$$w \mapsto (5w^3 + w^2 - 1)(w^2 - 10w + 6) + \frac{w - 1}{w + 1}.$$

27. Find the equation of the line tangent to the graph of $x = (y^4 - 2)/(y^5 + 1)$ at the point $(0, -2)$.
28. If $x = (z^{15} - 3z^{12} + z^6 - 1)(z^6 + 3z + 1)(z^3 + 5)$, find dx/dz.
29. If $A(s) = (s + 1)/(s^4 - 3s^3 + s^2 - 1)$, find $A'(1)$.
30. Find the equation of the line tangent to the graph of the function $z = (1 - 3y + y^2)(1 + y^2)$ at the point $(0, 1)$.
31. Differentiate the function $x \mapsto [(x^3 - 3x)/(x^2 - 1)] + [(x^5 - 3)/(x^4 + 2)]$.
32. If $f(x) = (x^5 - 2x^4 + 3x^3 - x + 1)^2$, find $f'(x)$.

2.5 The chain rule

We come now to the very important **chain rule** for differentiating composite functions.

Theorem 6 IF THE FUNCTION $u = f(x)$ HAS AT x_0 THE DERIVATIVE $f'(x_0)$ AND THE FUNCTION $y = g(u)$ HAS AT $u_0 = f(x_0)$ THE DERIVATIVE $g'(f(x_0))$, THEN THE COMPOSED FUNCTION $\phi = g \circ f$, THAT IS, THE FUNCTION $y = g(f(x))$, HAS AT x_0 THE DERIVATIVE

(9)
$$\phi'(x_0) = g'(f(x_0))f'(x_0).$$

(The derivative of a composition is the product of the derivatives.)

The formulation of this theorem in the Leibniz notation is more elegant:

(9')
$$\frac{dy}{dx} = \frac{dy}{du}\frac{du}{dx}.$$

One is tempted to say, "We simply cancel du," but this makes no sense, since dy/du and du/dx are not fractions and du is not a number. But formula (9') is easy to remember and convenient to use.

Before proving Theorem 6, we shall show how to apply it, choosing a very simple example. We want to differentiate $x \mapsto (1 + x^3)^2$. Since $(1 + x^3)^2 = 1 + 2x^3 + x^6$, the derivative is, of course, $6x^2 + 6x^5$. In order to apply the chain rule, we set $f(x) = 1 + x^3$ and $g(u) = u^2$. Then $\phi(x) = g(f(x)) = (1 + x^3)^2$. Since $f'(x) = 3x^2$ and $g'(u) = 2u$ we have, by (9),

$$\phi'(x) = g'(f(x))f'(x) = 2(1 + x^3)3x^2 = 6x^2 + 6x^5,$$

as expected. Using the Leibniz notation, we find that the same calculation reads

$$\frac{d(1 + x^3)^2}{dx} = \frac{d(1 + x^3)^2}{d(1 + x^3)} \frac{d(1 + x^3)}{dx} = (\text{set } u = 1 + x^3)$$

$$= \frac{du^2}{du} \frac{d(1 + x^3)}{dx} = (2u)(3x^2) = 2(1 + x^3)3x^2 = 6x^2 + 6x^5.$$

With a little experience, one can dispense with the step "set $u = 1 + x^3$" and treat $1 + x^3$ as one symbol:

$$\frac{d(1 + x^3)^2}{dx} = \frac{d(1 + x^3)^2}{d(1 + x^3)} \frac{d(1 + x^3)}{dx} = 2(1 + x^3)3x^2.$$

(With even more experience, it will not be necessary to write the term $d(1 + x^3)$ at all, but one will be able to differentiate from "outside" to "inside" routinely.)

The chain rule may, of course, be applied repeatedly. For instance, we showed in §1.6 that

$$\frac{d\sqrt{x}}{dx} = \frac{1}{2\sqrt{x}}.$$

Therefore

$$\frac{d\sqrt{1 + \sqrt{1 + x^2}}}{dx}$$

$$= \frac{d\sqrt{1 + \sqrt{1 + x^2}}}{d(1 + \sqrt{1 + x^2})} \frac{d(1 + \sqrt{1 + x^2})}{d\sqrt{1 + x^2}} \frac{d\sqrt{1 + x^2}}{d(1 + x^2)} \frac{d(1 + x^2)}{dx}$$

$$= (\text{set } u = 1 + \sqrt{1 + x^2}, v = \sqrt{1 + x^2}, w = 1 + x^2)$$

$$= \frac{d\sqrt{u}}{du} \frac{d(1 + v)}{dv} \frac{d\sqrt{w}}{dw} \frac{d(1 + x^2)}{dx} = \frac{1}{2\sqrt{u}} \cdot 1 \cdot \frac{1}{2\sqrt{w}} (2x)$$

$$= \frac{x}{2\sqrt{(1 + \sqrt{1 + x^2})} \sqrt{1 + x^2}}.$$

Let us now prove the chain rule, assuming that both functions f and g are linear. We have $f(x) = a + \alpha x$, $g(u) = b + \beta u$ and $\phi(x) = (g \circ f)(x)$ $= g(f(x)) = b + \beta(a + \alpha x) = (b + \beta a) + (\beta\alpha)x$. This is again a linear function. Its slope, the derivative of ϕ, is $\beta\alpha$, that is, the product of the derivatives g' and f'.

We have already remarked in §2.2 that, when one computes derivatives, one "may" replace functions by their linear approximations. The complete proof of the chain rule, which will be found in an appendix (§6.3), follows the same lines as the proof given above for linear functions.

▶**Examples**   **1.** Differentiate $\sqrt{1 + ax^2}$, where a is a constant (and, of course, $1 + ax^2 > 0$). Using Theorem 6, we have

$$\frac{d\sqrt{1 + ax^2}}{dx} = \frac{d\sqrt{1 + ax^2}}{d(1 + ax^2)} \frac{d(1 + ax^2)}{dx} = \frac{1}{2\sqrt{1 + ax^2}} 2ax = \frac{ax}{\sqrt{1 + ax^2}}.$$

For $a = -1$, we obtain equation (2) in §1.

2. Let $f(x) = (1 + (1 + x)^{100})^{1000}$. Compute $f'(0)$ and $f'(-1)$. Using the chain rule, we have

$$\frac{df}{dx} = \frac{d(1 + (1 + x)^{100})^{1000}}{d(1 + (1 + x)^{100})} \frac{d(1 + (1 + x)^{100})}{d(1 + x)^{100}} \frac{d(1 + x)^{100}}{d(1 + x)} \frac{d(1 + x)}{dx}$$

$$= 1000(1 + (1 + x)^{100})^{999} \cdot 1 \cdot 100(1 + x)^{99} \cdot 1$$

$$= 10^5[1 + (1 + x)^{100}]^{999}(1 + x)^{99}.$$

Therefore

$$f'(0) = \left(\frac{df}{dx}\right)_{x=0} = 10^5 2^{999} \quad \text{and} \quad f'(-1) = \left(\frac{df}{dx}\right)_{x=-1} = 0.$$

3. Knowing that

$$\frac{d(1/x)}{dx} = -\frac{1}{x^2}$$

(which we proved directly in §1) and the chain rule, we can establish equation (6). Indeed, if $g(x) \neq 0$, we have

$$\frac{d}{dx}\frac{1}{g} = \frac{d(1/g)}{dg}\frac{dg}{dx} = -\frac{1}{g^2}\frac{dg}{dx} = -\frac{g'}{g^2}. \blacktriangleleft$$

EXERCISES

In Exercises 33 to 52, compute the derivatives using the rules that have been proved so far.

33. If $f(x) = (x^5 - 3x^3 + 2x^2 + x - 1)^{100}$, find $f'(x)$.
34. If $g(z) = (z^6 + 3z^2 + z - 3)^{12}$, find $g'(z)$.
35. If $z = \sqrt{y^4 - y^2 + 5}$, find dz/dy.
36. Differentiate the function $x \mapsto x\sqrt{x^5 - 2x^4 + x}$.
37. Suppose that $g(s) = (s^3 - s + \sqrt{s})^6$; find $g'(s)$.
▶ 38. Suppose that $f(t) = \sqrt{t^3 + 2t} - \sqrt{t}$; find $f'(t)$.

39. If $h(t) = ((t^3 - 2t^2 + t - 1)/(t^4 + t^2 + 1))^{50}$, find $h'(t)$.
40. Differentiate the function $y \mapsto (y^6 + y^3 - 2y + 1)^{15}(y^2 + 1)^{10}$.
41. Suppose that $x = (y^5 + 2y - 3)^8(y^2 + y - 3)^5$. Find dx/dy.
42. Suppose that $f = g \circ h$, $h(2) = 6$, $h'(2) = -10$, and $g'(6) = -\frac{1}{2}$. Find $f'(2)$.
43. Suppose that $A = B \circ C$, $C(0) = -2$, $C'(0) = 1/2$, $B(0) = 3$, $B'(0) = 5$, $B(-2) = 10$, and $B'(-2) = 6$. Find $A'(0)$.
44. Suppose that $\phi(12) = -3$, $\phi'(12) = 10$, $\psi(12) = 2$, $\psi'(12) = \frac{1}{2}$, $\psi(-3) = 12$, and $\psi'(-3) = 4$. Find $(\psi \circ \phi)'(12)$ and $(\phi \circ \psi)'(-3)$.
45. If $g(t) = [(t^3 - 6t^2 - 3)^{10} + t^2 - 1]^3$, find $g'(t)$.
46. Suppose that $z = \sqrt{(y^2 + 1)^3 - 1}$. Find dz/dy.
47. Find $d(t^3 - \sqrt{1 + t^2})^5/dt$.
48. Differentiate the function $u \mapsto (u^3 + (u^2 - 3)^6)^8(u^3 - 1)$.
49. Differentiate the function $s \mapsto [s^2 + \sqrt{s^2 + 1}]^4(s + 1)$.
50. Suppose that $F(0) = 1$, $F'(0) = -2$, $G(1) = 3$, $G'(1) = -5$, and $H'(3) = 4$. Find $(H \circ G \circ F)'(0)$.
51. If $x = (t^2 - 3)^6(t^3 - t)^4(t + 1)^{1/2}$, find dx/dt.
52. If $z = \sqrt{s + \sqrt{s + \sqrt{s}}}$, find dz/ds.

2.6 Derivatives of inverse functions

We recall (see Chapter 3, §3.7) that two functions $x \mapsto f(x)$ and $y \mapsto g(y)$ are called inverse to each other if $g(f(x)) = x$ and $f(g(y)) = y$. Assume that we are given two such functions, and we know that both have derivatives. Then we may apply the chain rule and conclude that

$$g'(f(x))f'(x) = \frac{dx}{dx} = 1.$$

This shows that both functions can be differentiable only if neither of the derivatives is zero. It turns out that, conversely, if one of the functions has a derivative that is not zero, the other also has a nonzero derivative.

Theorem 7   LET $y = f(x)$ AND $x = g(y)$ BE STRICTLY MONOTONE FUNCTIONS INVERSE TO EACH OTHER. ASSUME THAT f HAS AT x_0 THE DERIVATIVE $f'(x_0) \neq 0$, AND SET $y_0 = f(x_0)$. THEN g HAS AT y_0 THE DERIVATIVE

(10) $$g'(y_0) = \frac{1}{f'(x_0)}.$$

Again this rule looks more elegant in the Leibniz notation:

(10') $$\frac{dx}{dy} = \frac{1}{dy/dx}.$$

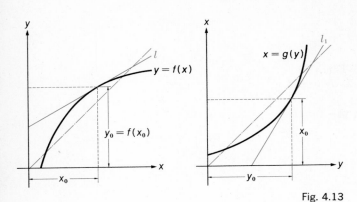

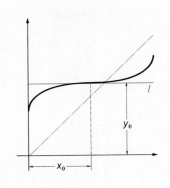

Fig. 4.13

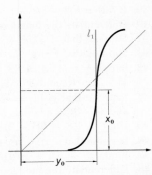

Fig. 4.14

The statement of Theorem 7 is easily verified geometrically. Let l be the tangent to the graph of $x \mapsto f(x) = y$ at the point (x_0, y_0). Then l is the graph of a linear function

$$x \mapsto mx + b = y, \qquad m = f'(x_0).$$

The graph of the inverse function $y \mapsto g(y) = x$ is obtained by reflecting the graph of f about the bisector of the first quadrant, that is, the line $y = x$ (see Fig. 4.13). If we also reflect l, we obtain a line l_1 which is tangent to the graph of g at (y_0, x_0). Now l_1 is the graph of the linear function inverse to $y = mx + b$, that is, of the function

$$x = \frac{1}{m} y - \frac{b}{m}.$$

Therefore $\qquad\qquad g'(y_0) = \dfrac{1}{m} = \dfrac{1}{f'(x_0)}.$

The argument also shows what happens if $f'(x_0) = 0$. Then l is horizontal and l_1 vertical, and g has no derivatives at y_0 (see Fig. 4.14).

An analytic proof of Theorem 7 will be found in the Appendix, §6.4.

EXERCISES

In Exercises 53 to 66, compute the derivatives using the rules that have been proved so far.

53. Let $y = f(x)$ and $x = g(y)$ be strictly monotone functions, inverse to each other. If $f(0) = 3$ and $f'(0) = -\frac{1}{4}$, find $g'(3)$.

54. Let $x \mapsto h(x)$ and $x \mapsto k(x)$ be strictly monotone functions, inverse to each other. If $h(-1) = 2$, $h'(-1) = \frac{1}{3}$, and $h'(2) = -3$, find $k'(2)$. (Note: The

concept of inverse functions does not depend on what symbols, if any, are used to denote the dependent and independent variables. Sometimes it is convenient to reverse the symbols for dependent and independent variables in the inverse function, especially if the functions are derived from some physical situation, but this is not necessary if one uses the proper notation.)

55. Let ϕ and ψ be strictly monotone functions, inverse to each other. If $\phi(0) = 1$ and $\phi'(0) = -2/3$, find $\psi'(1)$.

56. The function $f(x) = x^3 + 2x$ is increasing. If g is the function inverse to f, find $g'(0)$ and $g'(3)$.

57. The function $y = 3 - x - 2x^3$ is decreasing. Find $(dx/dy)_{y=3}$ and $(dx/dy)_{y=0}$. (Note: If one uses the pure Leibniz notation as in the exercise above, one must reverse the symbols for dependent and independent variables in the inverse function. In fact, that is the only way one can denote the inverse function.)

58. The function $s \mapsto s^5 + 3s^3 - 1$ is increasing. If f is the inverse function, find $f'(-1)$ and $f'(3)$.

59. The function $x \mapsto \sqrt{2x - 1}$ is increasing. Find the derivative of the inverse function at 3.

60. The function $r \mapsto r + \sqrt{r}$ is increasing. Find the derivative of the inverse function at 2.

61. (a) The function $h(z) = z^2 - z + 1$ is increasing for $z > \frac{1}{2}$. If k is the function inverse to h in this interval, find $k'(3)$ and $k'(7)$.
    (b) The function $h(z) = z^2 - z + 1$ is decreasing for $z < \frac{1}{2}$. If l is the function inverse to h in this interval, find $l'(3)$ and $l'(7)$.

▶ 62. (a) The function $y = f(x) = x^4 + 1$ is increasing for $x > 0$. If $x = g(y)$ is the function inverse to f in this interval, find $(dg/dy)_{y=2}$ and $(dg/dy)_{y=82}$.
    (b) The function $y = f(x) = x^4 + 1$ is decreasing for $x < 0$. If $x = h(y)$ is the function inverse to f in this interval, find $(dh/dy)_{y=2}$ and $(dh/dy)_{y=82}$.
    (Note: The pure Leibniz notation dx/dy is not really adequate in this situation, because it does not specify whether dg/dy or dh/dy is intended.)

63. (a) The function $f(z) = z^3 - 9z$ is increasing for $z < -\sqrt{3}$. If g is the function inverse to f in this interval, find $g'(0)$.
    (b) The function $f(z) = z^3 - 9z$ is decreasing for $-\sqrt{3} < z < \sqrt{3}$. If h is the function inverse to f in this interval, find $h'(0)$.
    (c) The function $f(z) = z^3 - 9z$ is increasing for $z > \sqrt{3}$. If k is the function inverse to f in this interval, find $k'(0)$.

64. The function $y = f(x) = 1 - x^3 - x^5$ is decreasing. If $x = g(y)$ is the function inverse to f, find $g' \circ f$.

65. (a) The function $f(s) = s^4 - 4s$ is increasing for $s > 1$. If g is the function inverse to f in this interval, find $g' \circ f$.
    (b) The function $f(s) = s^4 - 4s$ is decreasing for $s < 1$. If h is the function inverse to f in this interval, find $h' \circ f$.

66. (a) The function $P(s) = s^3 - 2s^2 + s$ is increasing for $s < \frac{1}{3}$. If f is the function inverse to P in this interval, find $f' \circ P$.
    (b) The function $P(s) = s^3 - 2s^2 + s$ is decreasing for $\frac{1}{3} < s < 1$. If g is the function inverse to P in this interval, find $g' \circ P$.
    (c) The function $P(s) = s^3 - 2s^2 + s$ is increasing for $s > 1$. If h is the function inverse to P in this interval, find $h' \circ P$.

2.7 Differentiating powers with rational exponents

We are now in a position to establish the following theorem.

Theorem 8   LET r BE A RATIONAL NUMBER. THEN

$$\frac{dx^r}{dx} = r x^{r-1}$$

(FOR ALL x FOR WHICH x^{r-1} AND x^r ARE DEFINED).

Proof.   If r is an integer, this is Theorem 3 of this section. We consider first the case in which $r = 1/n$, n being a positive integer. We shall show that

$$\frac{dx^{1/n}}{dx} = \frac{1}{n} x^{(1/n)-1},$$

that is,

$$\frac{d \sqrt[n]{x}}{dx} = \frac{\sqrt[n]{x}}{nx}.$$

For $n = 2$, this is the formula for differentiating $\sqrt{x}$, which we have already proved directly and used repeatedly.

In order to prove the above statement, we note that the functions

$$y = x^{1/n} \qquad \text{and} \qquad x = y^n$$

are inverse to each other. By Theorems 3 and 7, we have

$$\frac{dx^{1/n}}{dx} = \frac{dy}{dx} = \frac{1}{dx/dy} = \frac{1}{dy^n/dy} = \frac{1}{ny^{n-1}}$$

$$= \frac{1}{n(x^{1/n})^{n-1}} = \frac{1}{nx^{1-(1/n)}} = \frac{1}{n} x^{(1/n)-1},$$

as asserted.

Assume next that $r = m/n$, where m is an integer and n a positive integer. By Theorem 3, Theorem 7, and the result just proved, we have

$$\frac{dx^r}{dx} = \frac{d(x^{1/n})^m}{dx} = \frac{d(x^{1/n})^m}{d(x^{1/n})} \frac{dx^{1/n}}{dx}$$

$$= m(x^{1/n})^{m-1} \frac{1}{n} x^{(1/n)-1} = \frac{m}{n} x^{(m/n)-(1/n)+(1/n)-1} = r x^{r-1}.$$

Theorem 8 is proved.

2.8 Derivatives of radical functions

The differentiation rules proved above suffice to differentiate any radical function.

Theorem 9 A RADICAL FUNCTION HAS A DERIVATIVE WHEREVER IT IS DEFINED, EXCEPT PERHAPS AT FINITELY MANY POINTS. THE DERIVATIVE IS AGAIN A RADICAL FUNCTION.

The proof becomes obvious from the examples given below.

►*Examples* *1.* Differentiate $f(x) = \sqrt[3]{1 + x^2}/\sqrt[4]{x}$. We have

$$f'(x) = \frac{d}{dx}\left(\frac{(1 + x^2)^{1/3}}{x^{1/4}}\right) = \frac{d}{dx}((1 + x^2)^{1/3}(x)^{-1/4})$$

$$= \frac{d(1 + x^2)^{1/3}}{dx} x^{-1/4} + (1 + x^2)^{1/3} \frac{d(x^{-1/4})}{dx} \qquad \text{[by formula (5)]}$$

$$= \frac{d(1 + x^2)^{1/3}}{d(1 + x^2)} \frac{d(1 + x^2)}{dx} x^{-1/4} + (1 + x^2)^{1/3} \frac{dx^{-1/4}}{dx}$$

$$\hspace{8cm} \text{(by Theorem 6)}$$

$$= \frac{1}{3}(1 + x^2)^{-2/3}(2x)(x^{-1/4}) + (1 + x^2)^{1/3}\left(-\frac{1}{4}\right)x^{-5/4}$$

$$\hspace{8cm} \text{(by Theorems 1 and 8)}$$

$$= \frac{2}{3}(1 + x^2)^{-2/3}x^{3/4} - \frac{1}{4}(1 + x^2)^{1/3}x^{-5/4}.$$

If we want, we may "simplify" this to

$$f'(x) = (1 + x^2)^{-2/3}x^{3/4}\left(\frac{2}{3} - \frac{1}{4}(1 + x^2)x^{-2}\right)$$

$$= \frac{\sqrt[4]{x^3}}{\sqrt[3]{(1 + x^2)^2}}\left(\frac{2}{3} - \frac{1 + x^2}{4x^2}\right) = \frac{(5x^2 - 3)\sqrt[4]{x^3}}{12x^2 \sqrt[3]{(1 + x^2)^2}}.$$

The simplification is, however, not significant. The formula for $f'(x)$ is valid for $x > 0$, that is, in the whole domain of definition of $f(x)$.

2. Differentiate $f(x) = (1 + x^3)^{4/5}$. We have

$$f'(x) = \frac{d(1 + x^3)^{4/5}}{dx} = \frac{d(1 + x^3)^{4/5}}{d(1 + x^3)} \frac{d(1 + x^3)}{dx} = \frac{4}{5}(1 + x^3)^{-1/5} 3x^2.$$

This makes sense only for $1 + x^3 \neq 0$, that is, for $x \neq -1$. The function f, however, is defined also for $x = -1$.

3. Differentiate $f(x) = \sqrt{\sqrt[3]{x} + \sqrt[5]{x}}$. We have $f(x) = (x^{1/3} + x^{1/5})^{1/2}$. Hence

$$f'(x) = \frac{d(x^{1/3} + x^{1/5})^{1/2}}{d(x^{1/2} + x^{1/5})} \frac{d(x^{1/3} + x^{1/5})}{dx}$$

$$= \frac{1}{2}(x^{1/3} + x^{1/5})^{-1/2}\left(\frac{1}{3}x^{-2/3} + \frac{1}{5}x^{-4/5}\right).$$

Here $f(0) = 0$ is defined, while $f'(0)$ is not. ◄

EXERCISES

In Exercises 67 to 76, compute the derivatives using the rules that have been proved so far.

67. If $f(s) = s^{2/3} + 3s^{1/4} - 2s^{-3}$, find $f'(s)$.
68. If $y = z^{5/2} - (z^2 + 1)^{2/3}$, find dy/dz.
69. Find $d(v^{3/2} + (v^2 - v + 1)^{-2/3})\, dv$.
70. Differentiate the function $u \mapsto \dfrac{u^{1/3} - 2u^{1/2}}{u^{2/5} + 3u}$.
71. If $s = (z^3 + 2z - 1)^{3/4} + z^{-2/5}$, find ds/dz.
72. If $h(x) = (x + (1 - x^3)^{1/2})^{3/2}$, find $h'(x)$.
73. If $g(t) = (t^2 + (t^3 - 1)^{1/5})(t^{1/2} - 1)^4$, find $g'(t)$.
74. If $k(z) = \dfrac{\sqrt{z} + 2}{(z^3 - 1)^{1/3}}$, find $k'(z)$.
75. Differentiate the function $y \mapsto ((4 - 2y^{1/3})^{1/4} + y)^{3/2}$.
76. If $g(w) = (w^{2/3} + 1)(w^{1/2} - w^{1/4})^{1/2}$, find $g'(w)$.

2.9 Nondifferentiable functions

We defined the derivative of a function f at a point x_0 as the limit of the difference quotient

$$\frac{f(x_0 + h) - f(x_0)}{h}$$

as $h \to 0$, *provided this limit exists*. Of course, the limit may fail to exist; the graph need not have a tangent. If $f'(x_0)$ does exist, f is said to be **differentiable** at x_0. We already know (Theorem 2 in §1) that every differentiable function is continuous.

Not every continuous function is differentiable, however. The simplest example is the function $f(x) = |x|$; its graph, shown in Fig. 4.15, has a **corner** at $x = 0$. This function is said to have, at $x = 0$, a derivative

Fig. 4.15

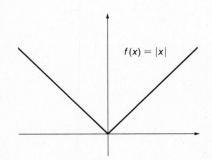

$f(x) = |x|$

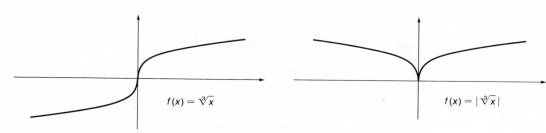

Fig. 4.16   Fig. 4.17

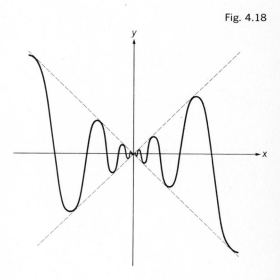

Fig. 4.18

from the right (equal to 1) and a derivative from the left (equal to -1). One writes $f'(0^+) = 1$, $f'(0^{-1}) = -1$. Similarly, the function graphed in Fig. 3.6 (Chap. 3, §1.6) has at $x = 0$ the one-sided derivatives 0 (from the left) and 1 (from the right).

The graph of the function $f(x) = \sqrt[3]{x}$ (see Fig. 4.16) has at $x = 0$ a vertical tangent; one says that $f'(0) = +\infty$. The graph of $f(x) = |\sqrt[3]{x}|$ shown in Fig. 4.17 also has a vertical tangent at $x = 0$. One says, however, for obvious geometric reasons, that for this function $f'(0^-) = -\infty$, $f'(0^+) = +\infty$. One also calls the point 0 a **cusp** of the graph.

Precise definitions of one-sided and infinite derivatives, and rules for computing them, will be found in the Appendix, §7.

A continuous function may fail to be differentiable in a complicated way. An example is the function graphed in Fig. 4.18, which equals 0 at 0. The graph of this function oscillates infinitely often between the lines $y = x$ and $y = -x$. Consider the difference quotient at $x = 0$, that is, $f(h)/h$. There are arbitrarily small values of h for which the difference quotient is 1, as well as arbitrarily small values of h for which the difference quotient is -1. Therefore the difference quotient has no limit as $h \to 0$. There are even no one-sided or infinite limits.

In all examples described thus far, the functions considered failed to have derivatives only at certain points. It would be natural to guess that a continuous function must be differentiable at most points. In this case, however, intuition leads us astray. There are continuous functions that are *nowhere* differentiable. To construct such functions is not easy, and it is almost impossible to visualize the graph of such a function. We shall not pursue this matter further, since calculus deals with functions having derivatives.

3.1 Derivatives of derivatives

Consider a differentiable function $x \mapsto y = f(x)$ and its derivative $x \mapsto dy/dx = f'(x)$. This rule, which associates with x the slope of the tangent of the curve $y = f(x)$ at the point x, is itself a function. We can compute its derivative if it has one. The derivative of the derivative of f is denoted by $f''(x)$ and is called the **second derivative** of f; f' itself is sometimes called the **first derivative.**

This process may be continued. The derivative of $f''(x)$, that is, the derivative of the second derivative or the derivative of the derivative of the derivative, is denoted by $f'''(x)$ and is called the **third derivative** of f. Similarly, the **fourth derivative** of f is defined as the derivative of f''' and could be denoted by f''''. It is customary, however, to denote the fourth derivative by $f^{(4)}$, the fifth by $f^{(5)}$, and so on.

The kth derivative of f, that is, $f^{(k)}(x)$, is also called the **derivative of order** k. The Leibniz notation for the higher derivatives is as follows:

$$f''(x) = \frac{d^2f}{dx^2} = \frac{d^2y}{dx^2}$$

(read "d square of f over dx square" or "d square of y over dx square"),

$$f'''(x) = \frac{d^3f}{dx^3} = \frac{d^3y}{dx^3}, f^{(4)}(x) = \frac{d^4f}{dx^4} = \frac{d^4y}{dx^4}, \cdots, f^{(k)}(x) = \frac{d^kf}{dx^k} = \frac{d^ky}{dx^k}.$$

This notation can be motivated by noting that we obtain the derivative $f'(x)$ of a function $f(x)$ by applying to f the differentiation process. Since we write f' as df/dx, we interpret the symbol d/dx as representing the differentiation "operator" and write $(d/dx)f = df/dx$. The second derivative f'' is obtained by applying the differentiation process twice:

$$f'' = \frac{d}{dx}\frac{d}{dx}f \quad \text{or} \quad f'' = \left(\frac{d}{dx}\right)^2 f,$$

and, if we treat d and dx as numbers, this may be written

$$f'' = \frac{d^2}{(dx)^2}f = \frac{d^2f}{dx^2}.$$

Similarly,
$$f''' = \frac{d}{dx}\frac{d}{dx}\frac{d}{dx}f = \left(\frac{d}{dx}\right)^3 f = \frac{d^3f}{dx^3}$$

and so on, for $f^{(4)}, f^{(5)}, \cdots$.

Although this reasoning sounds contrived, experience shows that the Leibniz notation is very convenient.

A function f is called k **times differentiable** in an interval if at every point of this interval there exist the derivatives $f'(x)$, $f''(x)$, $\cdots$, $f^{(k)}(x)$. Of particular importance are functions that have **derivatives of all orders.** Polynomials have this property. So do rational functions at all points at which they are defined, and radical functions, except at certain points.

▶ **Examples** **1.** If $f(x) = 2 - \frac{1}{2}x^3 + x^5$, then $f'(x) = -\frac{3}{2}x^2 + 5x^4$, and $f''(x) = -3x + 20x^3$, $f'''(x) = -3 + 60x^2$, $f^{(4)}(x) = 120x$, $f^{(5)}(x) = 120$, $f^{(6)}(x) \equiv 0$, and, of course, $f^{(7)}(x) = f^{(8)}(x) = \cdots \equiv 0$.

2. If $y = x^{7/3}$, then

$$\frac{dy}{dx} = \frac{7}{3}x^{4/3}, \frac{d^2y}{dx^2} = \frac{28}{9}x^{1/3} \text{ for all } x, \frac{d^3y}{dx^3} = \frac{28}{27}x^{-2/3} \text{ for } x \neq 0.$$

In the interval $-\infty < x < +\infty$ the function $x^{7/3}$ is twice differentiable but not 3 times differentiable. In the intervals $-\infty < x < 0$ and $0 < x < +\infty$, the same function has derivatives of all orders. ◀

EXERCISES

1. If $y = x^3 - x^2 + 2x + 1$, find d^2y/dx^2.
2. If $f(t) = t^4 - 2t^{3/2} + (3/t)$, find $f^{(4)}(t)$.
3. Find the third derivative of the function $z \mapsto z^{5/2} - 2z^{2/3} + z$.
4. If $h(s) = (s^2 - 1)^5$, find $h''(s)$.
5. Find the equation of the line tangent to the graph of the derivative of $y = \sqrt{x^2 - 3}$ at $x = 2$.
6. Find the second derivative of the function $u \mapsto (1 - \sqrt{u})^5$.
7. If $f(x) = (2x + 1)^{5/2}$, find $f'''(x)$. Does the graph of f'' have any cusps or vertical tangents?
8. If $v = (y + 2)^5(2y - 3)^6$, find dv/dy, d^2v/dy^2, d^3v/dy^3.
9. If $z = t^4(1 - 2t)^5$, what is the rate of change of dz/dt with respect to t?
10. Find the third derivative of the function $w \mapsto w^3(2w - 1)^4$.
11. If $h(u) = u/(u - 1)$, find $h'''(u)$.
12. If $x = \sqrt{1 + \sqrt{y}}$, find d^2x/dy^2.
13. Suppose that $k(x) = x^3 h(x)$ and $h(1) = 3$, $h'(1) = 1/2$, $h''(1) = 4$. Find $k''(1)$.
14. Suppose that $A(z) = (z - 1)^{1/2} B(z)$ and $B(5) = -1$, $B'(5) = 1/3$, $B''(5) = 2$, $B'''(5) = 10$. Find $A'''(5)$.
15. Suppose that $f(0) = 2$, $f'(0) = -4$, $f''(0) = 1/5$, $g(0) = 10$, $g'(0) = 2$, $g''(0) = -1$. Find $(fg)''(0)$.
▶ 16. Suppose that $G(x) = H(x)^3$ and $H(1) = 2$, $H'(1) = 1/3$, $H''(1) = 1/4$. Find $G''(1)$.
17. Suppose that $\phi(s) = \sqrt{1 - \psi(s)}$ and $\psi(-2) = -3$, $\psi'(-2) = 3$, $\psi''(-2) = 5$. Find $\phi''(-2)$.
18. Suppose that $A(u) = B(u^3 + 1)$ and $B'(1) = 1/6$, $B''(1) = -2$, $B'(2) = 3$, $B''(2) = 1/4$. Find $A''(1)$.
19. Suppose that $f(s) = sg(\sqrt{s})$ and $g'(3) = 6$, $g''(3) = -1$, $g'''(3) = 2$. Find $f'''(9)$.
20. Suppose that $G(0) = 1$, $G'(0) = 2$, $G''(0) = 1/2$, $H(0) = 0$, $H'(0) = -1$, $H'(1) = 6$, $H''(1) = 10$. Find $(H \circ G)''(0)$.
21. If $f(t) = t^{-1}$, find $f^{(k)}(t)$ for $k = 1, 2, 3, \cdots$.
22. If $g(u) = (u + 1)^{-1/2}$, find $g^{(k)}(u)$ for $k = 1, 2, 3, \cdots$.
23. If $A(x) = (1 - 2x)^{2/3}$, find $A^{(k)}(x)$ for $k = 1, 2, 3, \cdots$.
24. If $h(z) = z^2 + z + \sqrt{z}$, find $h^{(k)}(z)$ for $k = 1, 2, 3, \cdots$.

3.2 Acceleration

The meaning of the second derivative becomes particularly clear if we think of a function as describing motion along a straight line. In conformity with our previous notation, we write our function as

$$t \mapsto s(t),$$

where t denotes time and s the distance traveled. The first derivative $v(t) = s'(t)$ is the instantaneous velocity (see §1.13).

Let h be a small number; the difference $v(t + h) - v(t)$ is the change in the velocity during the time interval from t to $t + h$; this change is, of course, to be measured in units of velocity, that is, in units of length per unit of time (for instance, in cm/sec). The fraction

$$\frac{v(t + h) - v(t)}{h}$$

is the average change in the velocity during the time interval $[t, t + h]$, or the **average acceleration.** It is measured in units of length per unit of time per unit of time, for instance, in cm/sec² (read "centimeters per second per second"). As we take smaller and smaller values of h, the fraction comes closer and closer to the derivative

$$a(t) = \frac{dv}{dt} = v'(t).$$

We call $a(t)$ the **instantaneous acceleration** and note that

$$a(t) = s''(t) = \frac{d^2s}{dt^2};$$

the acceleration is the second derivative of the distance traveled with respect to the time elapsed.

Let us now consider a freely falling body (see §1.14). According to Galileo, we have $s = -\beta t^2$. Hence

$$\frac{d^2s}{dt^2} = -2\beta.$$

Thus Galileo's law reads: *The acceleration of all freely falling bodies has the same constant value.* One usually denotes the quantity 2β by g, so that the above equation is written as

(1)
$$\frac{d^2s}{dt^2} = -g.$$

The numerical value of g depends on the units used; in particular,

$$g = 32\,\frac{\text{ft}}{\text{sec}^2} = 980\,\frac{\text{cm}}{\text{sec}^2}, \qquad \text{approximately.}$$

3.3 Newton's law of motion

Having defined acceleration, we are now able to state one of the most important laws of physics, Newton's law of motion. This law involves two more basic concepts beside the acceleration: **force** and **mass.**

The idea of force originates in the subjective feeling of effort that we experience when we change the velocity of a body, for instance, when we throw a stone. In physics, force is defined as that which causes acceleration. In the case of rectilinear motion, we assume that force can be expressed by a number.

It is a common experience that the same effort produces different accelerations in different bodies. We therefore assign to each particle a positive number, its mass, which measures its resistance to force. Experiments show that this resistance to acceleration is a property of the particle and does not depend on its velocity or on the force applied (this statement must be revised for very fast moving bodies; see §3.4). Experiments also show that the mass of a body consisting of two parts is equal to the sum of the masses of the parts. We therefore think of the mass of a body as measuring the *quantity of matter* contained in it. In order to measure mass we must, of course, choose a unit; in scientific computations one commonly uses grams. We shall therefore mostly use the CGS system of units (centimeters, grams, seconds) in our examples.

We now state the law of motion for a particle of mass m moving along a straight line. At every time instant the acceleration a and the force F acting on the particle are connected by the relation

(2)
$$ma = F.$$

This shows that force is measured in units of "mass times acceleration," for instance, in gram · cm/sec² (gram centimeter per second per second). The force is positive if $a > 0$, that is, if it pushes the body in the positive direction. Since the acceleration is the derivative of the velocity, Newton's law may be also written as

(2′)
$$m\frac{dv}{dt} = F.$$

The force that makes a body fall to the ground is called its **weight.** According to Galileo, a freely falling body always has the same acceleration g. Denoting the weight of our body by W, we have, therefore,

$$(3) \qquad\qquad mg = W.$$

Thus the weight of the body depends only on its mass. The masses of two bodies can be compared by comparing their weights, that is, by putting them on a scale. The weight of a body of mass 1 gram is a convenient measure of force, called 1 gram weight.

EXERCISES

25. A moving particle whose mass is .3 gram has an acceleration of 10 cm/sec^2 at a certain instant. What is the force acting on the particle at this instant?

26. A body whose mass is 10 grams is pushed across a level frictionless table in a straight line. If the velocity of the body is given by $v = (t^2/10) + t$ for $0 \le t \le 10$, where v is measured in cm/sec and t in seconds, what is the force acting on the body when $t = 1$? When $t = 2$? When $t = 10$?

27. A projectile whose mass is 30 grams is shot horizontally into a viscous fluid. Assume that the path is a straight line. If the distance traveled by the projectile is given by $s = 100 - (100 - t)^4$ for $0 \le t \le 100$, where the distance s is measured in centimeters and the time t is measured in seconds, with what force is the fluid resisting the projectile when $t = 10$? When $t = 50$?

28. A body whose mass is 500 mass-pounds is put on a hydraulic lift and pushed straight up. What force must the lift exert on the body in order to give it a constant acceleration of $\frac{1}{2}$ ft/sec^2?

3.4 Einstein's law of motion

In the beginning of this century, Einstein discovered that Newtonian physics was not applicable to very high speeds. In the *theory of relativity,* the differential equation of rectilinear motion of a particle, under the influence of a force F, is not (2) but reads

$$(4) \qquad\qquad m_0 \frac{d}{dt} \frac{v}{\sqrt{1 - (v^2/c^2)}} = F,$$

where m_0 is the mass of the particle measured at rest and c is the speed of light in vacuum ($c = 3 \cdot 10^{10}$ cm/sec). When Einstein proposed this law, its significance was only theoretical, since nobody observed bodies moving with velocities v comparable to the velocity of light. And if v/c is very small, the number $\sqrt{1 - (v^2/c^2)}$ is so close to 1 that (4) may be replaced by Newton's law (2′). In modern accelerators, however, some subatomic particles reach speeds up to $(.99)c$.

ALBERT EINSTEIN (1879–1955) was working for the Swiss Patent Office in 1905, when he published three papers, each of which initiated a new physical theory. One dealt with Brownian motion; another contained the first application of Planck's quantum hypothesis to a subatomic process. Through this and other papers, Einstein became one of the fathers of quantum theory—yet he never fully accepted this theory. The third paper contained the Special Relativity Theory. This was followed in 1916 by the General Relativity Theory, and then by a stubborn search, extending over decades, for a unified theory describing both gravitation and electromagnetism.

Einstein came to the United States as a refugee from Hitler's Germany in 1933 and settled in Princeton, New Jersey. In 1939, he signed a letter alerting President Roosevelt to the possibility of an atomic bomb. Throughout his life, Einstein maintained a strong interest in social causes, particularly in world peace, the establishment of a Jewish homeland in Palestine, and civil liberties.

Let us carry out the differentiation in (4) to obtain the acceleration $a = dv/dt$. We have (using the rules in §2, including the chain rule)

$$\frac{d}{dt}\frac{v}{\sqrt{1 - (v^2/c^2)}} = \left\{\frac{d}{dv}\frac{v}{\sqrt{1 - (v^2/c^2)}}\right\}\frac{dv}{dt}.$$

This equals

$$a\frac{d}{dv}v\left(1 - \frac{v^2}{c^2}\right)^{-1/2} = a\left\{\left(1 - \frac{v^2}{c^2}\right)^{-1/2} + v\frac{d}{dv}\left(1 - \frac{v^2}{c^2}\right)^{-1/2}\right\}$$

$$= a\left\{\left(1 - \frac{v^2}{c^2}\right)^{-1/2} + v\left(-\frac{1}{2}\right)\left(1 - \frac{v^2}{c^2}\right)^{-3/2}\left(-\frac{2v}{c^2}\right)\right\}$$

$$= a\left\{\left(1 - \frac{v^2}{c^2}\right)^{-1/2} + \frac{v^2}{c^2}\left(1 - \frac{v^2}{c^2}\right)^{-3/2}\right\}$$

$$= \frac{a}{\left(1 - \frac{v^2}{c^2}\right)^{3/2}}.$$

Thus we may write (4) in the form

$$\frac{m_0 a}{\sqrt{1 - (v^2/c^2)^3}} = F.$$

This shows that the ratio F/a depends on the velocity of the body rather than on its mass alone. The greater the velocity, the greater the force needed to produce a given acceleration.

EXERCISES

In Exercises 29 to 32, use the relativistic equations of motion unless otherwise specified. All units are in the CGS system and c denotes the speed of light.

29. A particle whose rest mass is .001 gram has at a certain instant a velocity of .9c and an acceleration of 1 cm/sec². What is the force acting on the particle at this instant? Compare with the force that would be predicted by the Newtonian law of motion.

30. A particle whose rest mass is 1 gram has at a certain instant a velocity of .1c and an acceleration of 10 cm/sec². What is the force acting on the particle at this instant? Compare with the force that would be predicted by the Newtonian law of motion.

31. A particle whose rest mass is .003 gram is excited, so that its velocity is given by $v = tc/(t + 100)$, $t \geq 0$. When is the velocity equal to .5c? What is the force acting on the particle at this instant? When is the velocity equal to .9c? What is the force acting on the particle at this instant? Compare with the forces that would be predicted by the Newtonian law of motion.

▶ 32. To produce an acceleration of 5 cm/sec² in the motion of a certain particle, a force of 10 gram · cm/sec² is required. If the velocity of the particle is .8c, what is its rest mass? Compare with the mass that would be predicted by the Newtonian law of motion.

3.5 Differentiability assumptions in physics

A tacit assumption underlies Newton's (and Einstein's) law of motion—the position of a moving particle is described by a twice differentiable function of time $s(t)$, so that we can talk about the velocity $s'(t)$ and acceleration $s''(t)$. This assumption is amply justified in many cases, but there are some situations in which it is untenable.

A microscopic particle floating in a liquid collides with the much smaller molecules of the liquid, and these innumerable collisions produce an erratic motion of the particle, a phenomenon first noticed by the botanist Brown. In the mathematical theory of *Brownian motion*, the position of the particle is described by a continuous but nondifferentiable function. The particle has therefore no definite instantaneous velocity.

About forty years ago, physicists realized that subatomic particles, like electrons, protons, and neutrons, do not obey the laws of classical physics of Galileo and Newton. In *quantum mechanics*, the theory of motion of such particles, the velocity is not the derivative of the position, at least not in the same sense as in classical physics. For when we think of velocity as a derivative, we assume that we can measure it by measuring the position $s(t)$ at different times, plotting the curve, drawing the tangent and measuring its slope. The more accurately we know the position, therefore, the more accurately we know the velocity. But a basic principle of quantum theory asserts that, in any experiment, a greater precision in measuring one of the two quantities, position and velocity, results in a smaller precision in measuring the other. If we measure the position of a particle with the accuracy α and its velocity with the accuracy β, then the Heisenberg *uncertainty principle* states that $\alpha\beta \geq h/m$, where m is the mass of the particle and h is a universal constant (Planck's constant).

This does not mean that calculus is not applicable in quantum theory. On the contrary, the mathematics of quantum mechanics is rooted in calculus.

ROBERT BROWN (1773–1858).

WERNER HEISENBERG (b. 1906) is one of the inventors of quantum mechanics. He published his uncertainty principle in 1926.

MAX PLANCK (1858–1947) proposed in 1900 the "quantum hypothesis," according to which energy can be emitted by electromagnetic radiation only in quantities $N h\nu$, where N is an integer, ν the frequency of radiation, and h a universal constant.

§4 Signs of Derivatives. Maxima and Minima

Having learned how to compute derivatives, we ask what information about the function can be obtained from knowing whether the derivative is positive or negative. This section contains the answer to this question, and some of its ramifications.

4.1 A preliminary result

We state first a simple result which asserts that, if at some point the tangent points up, the curve rises there, and if the tangent points down, the curve falls.

Theorem 1 THE FOLLOWING CONCLUSIONS HOLD:

IF	THEN FOR $x < x_0$ AND NEAR x_0	AND FOR $x > x_0$ AND NEAR x_0
$f'(x_0) > 0$	$f(x) < f(x_0)$	$f(x_0) < f(x)$
$f'(x_0) < 0$	$f(x) > f(x_0)$	$f(x_0) > f(x)$

Proof. We prove only the first row; the second is established similarly. The derivative $f'(x_0)$ is the limit of the difference quotient

$$\frac{f(x_0 + h) - f(x_0)}{h}$$

as $h \to 0$. The derivative is positive. Therefore the difference quotient is positive for sufficiently small h. This means that, if a point $x = x_0 + h$ is close and to the right of x_0, so that h is positive and small, the numerator in (1) is also positive and so $f(x) = f(x_0 + h) > f(x_0)$. Similarly, if $x_0 + h$ is a point close and to the left of x_0, then h is small in absolute value and negative, so that $f(x_0 + h) < f(x_0)$.

The next result refers not to a single point but to a whole interval.

4.2 Functions with derivatives of one sign

Theorem 2 LET $f(x)$ BE CONTINUOUS IN A CLOSED INTERVAL $a \leq x \leq b$. IF AT ALL POINTS x WITH $a < x < b$, THE DERIVATIVE EXISTS AND

IF $f' > 0$,	THEN f IS INCREASING IN THE INTERVAL.
IF $f' < 0$,	THEN f IS DECREASING IN THE INTERVAL.

This is one of the fundamental results in calculus. We recall that f is increasing if for any two points in the interval, $x_1 < x_2$, we have $f(x_1) < f(x_2)$. The statement of the theorem is quite believable: if the tangent points up at every point, the curve rises. (See Fig. 4.19.) The

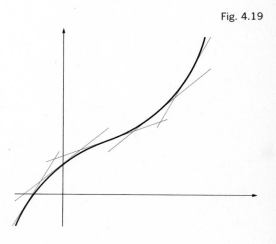

Fig. 4.19

rigorous proof of Theorem 2 is, however, somewhat delicate. We give it in the Appendix (§8.1). For the time being, we may accept the theorem as geometrically obvious.

▶**Examples**   **1.** Consider the functions

$$f(x) = x^3, g(x) = x^{1/3}, h(x) = x \text{ for } x \leq 0, = 2x \text{ for } x > 0.$$

All functions are continuous everywhere. For $x \neq 0$, all derivatives exist and are positive. But $f'(0) = 0$, while g and h are not differentiable at $x = 0$. Still, all three functions are everywhere increasing (see Figs. 4.20, 4.21, and 4.22). This shows that $f' > 0$ is a sufficient condition for f to be increasing but not a necessary one.

2. Find the intervals of increase and decrease of the function $f(t) = (1 + t)/(1 + t^2)$.
Solution.   We have

$$f'(t) = \frac{(1 + t^2) - (1 + t)2t}{(1 + t^2)^2} = \frac{1 - 2t - t^2}{(1 + t^2)^2} = \frac{2 - (t + 1)^2}{(1 + t^2)^2}.$$

We note that $(1 + t^2)^2$ is positive for all t. Hence $f'(t) > 0$ if $2 - (t + 1)^2 > 0$, that is, if $|t + 1| < \sqrt{2}$, that is, if $-1 - \sqrt{2} < t < -1 + \sqrt{2}$. Also $f'(t) = 0$ for $t = -1 + \sqrt{2}$ and $-1 - \sqrt{2}$, and $f'(t) < 0$ for all other t. We conclude that f is increasing for

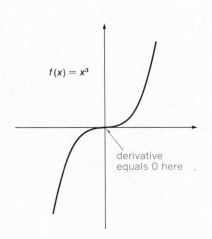

$f(x) = x^3$

derivative
equals 0 here

Fig. 4.20

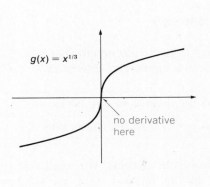

$g(x) = x^{1/3}$

no derivative
here

Fig. 4.21

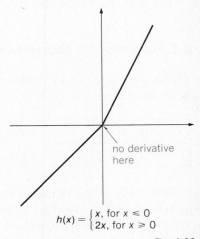

no derivative
here

$h(x) = \begin{cases} x, \text{ for } x \leq 0 \\ 2x, \text{ for } x \geq 0 \end{cases}$

Fig. 4.22

$-1 - \sqrt{2} < t < -1 + \sqrt{2}$, decreasing for $t < -1 - \sqrt{2}$ and for $t > -1 + \sqrt{2}$.

3. Find the intervals of increase or decrease of the function $g(x) = 3x^{1/3} - 5x^{1/5}$.

Solution. We have

$$g'(x) = x^{-2/3} - x^{-4/5} = x^{-4/5}(x^{2/15} - 1)$$

for $x \neq 0$, while there is no derivative at $x = 0$. Since $x^{-4/5} = (x^{-1/5})^4 > 0$ for $x \neq 0$, and $x^{2/15} - 1 > 0$ for $|x| > 1$, $x^{2/15} - 1 < 0$ for $|x| < 1$, we conclude that g increases for $x \leq -1$, decreases for $-1 \leq x \leq 1$, and increases for $x \geq 1$.

4. Consider the function $f(x) = 1/x$ for $x \neq 0$, $f(0) = 0$. We have that $f' < 0$, except at $x = 0$ where there is no derivative. But f is not decreasing for all x since, for instance, $f(-1) = -1 < 1 = f(1)$. This does not contradict Theorem 2, since f is not continuous at $x = 0$. (Of course, f is decreasing for $x < 0$ and for $x > 0$.)◄

4.3 Functions with nonnegative and nonpositive derivatives

We note, for later reference, a consequence of Theorem 2.

Theorem 3   LET $f(x)$ BE CONTINUOUS IN A CLOSED INTERVAL $a \leq x \leq b$. IF AT ALL POINTS x WITH $a < x < b$, THE DERIVATIVE EXISTS AND

IF $f' \geqslant 0$,	THEN f IS NONDECREASING IN THE INTERVAL.
IF $f' \leqslant 0$,	THEN f IS NONINCREASING IN THE INTERVAL.

We prove only the first statement. The second may be proved similarly, or it may be established by applying the first statement to the function $(-f)$. Examples illustrating the theorem are shown in Figs. 4.20 and 4.23.

Proof of Theorem 3.   We must show that, if $x_1 < x_2$, then $f(x_1) \leq f(x_2)$. Let n be a positive integer, and set $g(x) = f(x) + (x/n)$. At every point where $f' \geq 0$, we have $g' = f' + (1/n) > 0$, since the sum of a nonnegative number and a positive number is positive. Let x_1 and x_2 be any two points in the interval considered with $x_1 < x_2$. By Theorem 2 we have $g(x_1) < g(x_2)$. This means that $f(x_1) + (x_1/n) < f(x_2) + (x_2/n)$, which is the same as $f(x_1) - f(x_2) < (x_2 - x_1)/n$. Since this holds for $n = 1, 2, 3, \cdots$, we conclude $f(x_1) - f(x_2) \leq 0$, so that $f(x_1) \leq f(x_2)$, as asserted. (Note that here we used the Archimedean postulate.)

Fig. 4.23

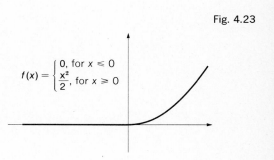

$$f(x) = \begin{cases} 0, & \text{for } x \leq 0 \\ \dfrac{x^2}{2}, & \text{for } x \geq 0 \end{cases}$$

EXERCISES

In Exercises 1 to 8, find the intervals of increase and decrease of the given functions. It is understood that these intervals must be contained in the domain of definition of the function and should be maximal (as large as possible).

1. $f(x) = \sqrt{1 + x^3}$.

2. $f(t) = (t - 2)/(t + 1)$.

3. $s \mapsto (s^2/2) + 3s^{1/3}$.

4. $g(x) = x^3 - 2x^2 + x - 1$.

5. $h(z) = z^2/\sqrt{z^2 + 1}$.

6. $\phi(s) = s^4 - 2s^2 + 4$.

7. $g(y) = (y/3) + y^{2/3} - 8y^{1/3}$.

8. $f(x) = x^2(x - 1)^{1/3}$.

4.4 Local maxima and minima

One says that a function f defined near x_0 has a **local maximum** at this point if $f(x) \leq f(x_0)$ for all x near x_0. The point x_0 is then called a local maximum point, and the number $f(x_0)$ is called a local maximum. If $f(x) < f(x_0)$ for $x \neq x_0$ and x near x_0, we say that the maximum is **strict**. The various possibilities are shown in Fig. 4.24.

The function f is said to have a **local minimum** at x_0 if $f(x) \geq f(x_0)$ for x near x_0, a **strict** local minimum if $f(x) > f(x_0)$ for $x \neq x_0$ and x near x_0 (see Fig. 4.25). In this case, x_0 is called a local minimum point and $f(x_0)$ is called a local minimum.

The reader will note that if a function is constant in an interval, that is, if the graph has a horizontal segment, then each point of this interval, except the endpoints, is a local maximum point and also a local minimum point, according to our definition.

If $f'(x_0) > 0$ or $f'(x_0) < 0$, then f cannot have a local maximum or local minimum at x_0; this follows from Theorem 1. In geometric language: at a peak or a trough of a curve, the tangent is horizontal, if there is a tangent. We formulate this as a theorem.

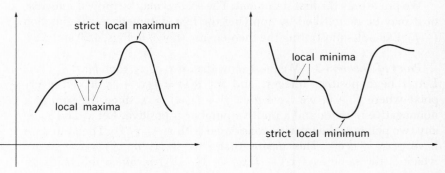

Fig. 4.24   Fig. 4.25

Theorem 4 IF f HAS A LOCAL MAXIMUM OR MINIMUM AT x_0, AND f IS DIFFERENTIABLE AT x_0, THEN $f'(x_0) = 0$.

One will not think, however, that if $f(x_0) = 0$, then f must have a maximum or minimum at x_0 (see Fig. 4.26). The points at which the derivative of a function is zero are called **critical points**. A local maximum or minimum point of a differentiable function is always a critical point, but a critical point need not be a local maximum point or minimum point.

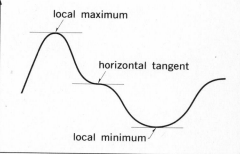

Fig. 4.26

Let f be defined for $a \leq x \leq b$, and let x_0 be a point with $a < x_0 < b$. The remarks summarized in the following table are self-evident.

if in $[a, x_0]$ f is	and in $[x_0, b]$ f is	then f has at x_0
decreasing	increasing	a strict local minimum
nonincreasing	nondecreasing	a local minimum
increasing	decreasing	a strict local maximum
nondecreasing	nonincreasing	a local maximum

As an application, we prove the following result.

Theorem 5 LET f BE DIFFERENTIABLE NEAR x_0.

IF	AND IF	THEN f HAS at x_0 A STRICT LOCAL
$f'(x_0) = 0$	$f''(x_0) > 0$	MINIMUM
$f'(x_0) = 0$	$f''(x_0) < 0$	MAXIMUM

We prove only the first statement. The proof of the second is similar. Also, the second follows by applying the first to the function $(-f)$.

Assume that $f'(x_0) = 0$ and $f''(x_0) > 0$. Recall that $f'' = (f')'$. Applying Theorem 1 to f', we conclude that for $x < x_0$ and near x_0, we have $f'(x) < f'(x_0) = 0$, and that for $x > x_0$ and near x_0, we have $f'(x) > f'(x_0) = 0$. Therefore f is decreasing in an interval with x_0 as the right endpoint, by Theorem 2, and for the same reason f is increasing in an interval with x_0 as the left endpoint. Now we see from the preceding table that f has a strict local minimum at x_0.

▶*Examples 1.* Find the local maxima and minima of the function $f(x) = 4x^4 - x^2 + 5$.

Solution. We first find the critical points of f, that is, the points at which $f' = 0$. Next we determine the sign of f'' at all critical points.

(This is called *applying the second derivative tests.*) In our case $f'(x) = 16x^3 - 2x = 2x(8x^2 - 1)$ for all x. The critical points are $x = 0$, $x = 1/\sqrt{8}$, and $x = -1/\sqrt{8}$. Also, $f''(x) = 48x^2 - 2$. Hence $f''(0) = -2 < 0$, $f''(1/\sqrt{8}) = f''(-1/\sqrt{8}) = 4 > 0$. Therefore f has a strict local maximum at $x = 0$, and strict local minima at $x = 1/\sqrt{8}$, $-1/\sqrt{8}$, by Theorem 5. The local maximum is $f(0) = 5$. The local minima are $f(1/\sqrt{8}) = f(-1/\sqrt{8}) = 79/16$.

2. The function $f(x) = -|x|$ has a strict local maximum at $x = 0$ but no derivative at this point. This local maximum point could not have been detected by the derivative test. ◄

EXERCISES

In the following exercises, find all local minima and local maxima of the given function.

9. $f(x) = x^3 + 3x^2 - 3x + 4$.

10. $g(x) = 2x^3 - 15x^2 - 36x + 1$.

11. $f(x) = 2x^5 - 10x - 7$.

► 12. $f(x) = x^{100} + x^{1000}$.

13. $\phi(x) = 6x^5 + 10x^3 + 15x^{-4}$.

14. $u \mapsto |1 + u^2|$.

15. $x \mapsto |x^3 + 3x^2 + 6x|$.

16. $f(y) = |1 + y|$.

17. $x \mapsto (2 + x^2)/(1 + x^2)$.

18. $g(t) = \sqrt{1 + t^4}$.

4.5 Piecewise monotone functions

We recall that a function, defined in some finite interval, is called **piecewise monotone** if the interval of definition can be divided into finitely many subintervals in each of which the function is monotone (see Fig. 4.27).

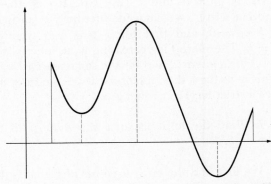

Fig. 4.27

Theorem 6   IF A FUNCTION $f(x)$ DEFINED IN AN INTERVAL HAS A CON-
TINUOUS DERIVATIVE $f'(x)$ THAT HAS ONLY FINITELY MANY ROOTS, THE
FUNCTION f IS PIECEWISE MONOTONE.

Fig. 4.28

Proof.   Between two points at which $f'(x)$ takes on a positive and a
negative value lies a root of the derivative; this follows from the intermediate value theorem. It follows that, between two successive roots of
the derivative, either the derivative is everywhere positive or everywhere
negative. Therefore, by Theorem 2, $f(x)$ is either increasing or decreasing between two successive roots of $f'(x)$.

It follows from Theorem 6 that polynomial, rational, and radical
functions are piecewise monotone. So are most functions commonly
encountered in calculus. A continuous function that is not piecewise
monotone is shown in Fig. 4.28.

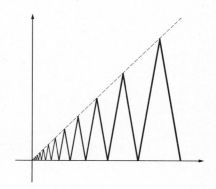

4.6 Absolute maxima and minima

Let $f(x)$ be a function defined in an interval, and let x_0 be a point in
this interval. If $f(x) \leq f(x_0)$ for all other points in the interval, one says
that f has at x_0 an **absolute maximum** (for the interval) and one calls the
value $f(x_0)$ **the maximum** of f in the interval [one often writes $f(x_0) =$
max f]. Similarly, the function f has at x_0 an **absolute minimum** if
$f(x_0) \leq f(x)$ for all x in the interval; the value of $f(x_0)$ is then called **the
minimum** of f [$f(x_0) = $ min f]. A function defined in an interval need
not have in this interval a maximum or a minimum.

▶*Examples*   *1.* The continuous function $f(x) = x$ has neither a maximum
nor a minimum in the open interval $0 < x < 1$. In the closed interval
$0 \leq x \leq 1$, the same function has the maximum 1 and the minimum 0.

2. Set $f(x) = x^2$ for $0 < x < 1$, $f(0) = f(1) = \frac{1}{2}$. The function f is
defined but not continuous in the closed interval $0 \leq x \leq 1$. In this
interval, it has neither a maximum nor a minimum.

3. The continuous function $f(x) = 1/\sqrt{1 - x^2}$ defined in the open
interval $-1 < x < 1$ has a minimum 1 at $x = 0$ but no maximum. ◀

EXERCISES

19. Draw the graph of a function defined and continuous in the interval $[-2, 2]$
    which has an absolute maximum at 3 points and a strict local minimum at 2
    other points. If this is impossible, say so.
20. Draw the graph of a function defined and continuous in the interval $[0, 5]$
    which has an absolute maximum at 1 and 3, a strict local minimum at 2, and
    no absolute minimum. If this is impossible, say so.

21. Draw the graph of a function defined and continuous in the interval $[-2, 2]$ which has an absolute maximum at 1, a cusp at 1, and an absolute minimum at -1 and $3/2$. If this is impossible, say so.

22. On the horizontal axis, indicate where the function whose graph is given below has a local maximum or minimum, or an absolute maximum or minimum. Also indicate where the derivative is 0 and where the derivative fails to exist.

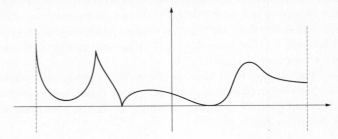

4.7 Existence of maxima and minima

A continuous function defined in a finite closed interval always has a maximum and a minimum in that interval. This statement is not difficult to believe, but the proof is not quite easy. We shall present the proof in the Supplement at the end of the book. For the time being, we shall make no use of this result, since a much simpler theorem is sufficient for our purposes.

Theorem 7   A FUNCTION DEFINED, CONTINUOUS, AND PIECEWISE MONOTONE ON A CLOSED (FINITE) INTERVAL HAS AN ABSOLUTE MAXIMUM AND AN ABSOLUTE MINIMUM.

Proof.   If f is continuous and monotone for $\alpha \le x \le \beta$, it is clear that f takes on its largest and smallest values at the endpoints α and β. If f is continuous and piecewise monotone for $a \le x \le b$, there are finitely many points $\alpha_1, \alpha_2, \cdots, \alpha_k$ such that

$$a < \alpha_1 < \alpha_2 < \cdots < \alpha_k < b$$

and f is monotone in each of the intervals $[a, \alpha_1], [\alpha_1, \alpha_2], \cdots, [\alpha_k, b]$. Therefore the largest value of f is the largest of the numbers

$$f(a), f(\alpha_1), f(\alpha_2), \cdots, f(\alpha_k), f(b),$$

and the smallest value of f is the smallest of these numbers.

4.8 Finding maxima and minima

One of the first applications of calculus was in finding maxima and minima of functions. One usually deals with functions f that satisfy the conditions of Theorems 6 and 7. Then max f and min f are found by comparing the local maxima and minima of f and the values of f at the endpoints of the interval considered, or simply by computing f at the endpoints and at all roots of f'.

▶*Examples* *1.* Find the largest and smallest values of $f(x) = (.1)(x^3 - 12x + 10)$ for $-3 \le x \le 3$.

Solution. We have $f'(x) = (.1)(3x^2 - 12) = (.3)(x + 2)(x - 2)$. The roots of f' are -2 and 2. Since $f(-3) = 1.9, f(-2) = 2.6, f(2) = -.6$, and $f(3) = .1$, we conclude that the maximum of f is $f(-2) = 2.6$ and the minimum is $f(2) = -.6$. (See Fig. 4.29.)

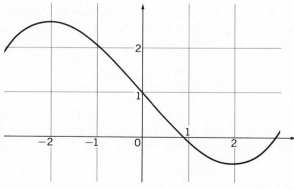

Fig. 4.29

2. Find the maximum and minimum of $f(x) = \sqrt{x} - 2\sqrt[4]{x}$ for $0 \le x \le 100$.

Solution. We have

$$f'(x) = \tfrac{1}{2}x^{-1/2} - 2 \cdot \tfrac{1}{4}x^{-3/4} = \tfrac{1}{2}x^{-3/4}(x^{1/4} - 1).$$

We note that f is not differentiable at $x = 0$. The only root of f' is $x = 1$. We note that $f(0) = 0, f(1) = -1$, and $f(100) = 10 - 2\sqrt{10} > 0$. The answer is: max $f = 10 - 2\sqrt{10}$, min $f = -1$.

3. Find the maximum and minimum of $f(x) = 1/\sqrt{1 - x^2}$, $-1 < x < 1$.

Solution. Instead of working with f, we look at $g(x) = f(x)^2 = 1/(1 - x^2)$. For, since $f \geq 0$, and $u \mapsto u^2$ is an increasing function for $u \geq 0$, the functions f and g attain their maxima and minima at the same point.

Now, $g'(x) = 2x/(1 - x^2)^2$. Thus $g' = 0$ if and only if $x = 0$. The minimum of g is attained at $x = 0$; g has no maximum. Hence f has the minimum 1, and no maximum. The function f was defined in an open interval; there was no need to look at the endpoints.

It would be even simpler to consider the function $h(x) = 1/f(x)^2 = 1 - x^2$ and to note that the maximum of h is the reciprocal of the minimum of f, and the minimum of h, provided it is not zero, is the reciprocal of the maximum of f. It is clear that in the interval considered the function h has the maximum 1 and no minimum. ◄

Maximum and minimum problems often arise as (or are disguised as) geometric or arithmetic theorems or as "practical" problems. In such a case, some preliminary work is needed to reformulate the problem into one of finding the largest or smallest value of a function in an interval. We illustrate the procedure by several examples.

► *Examples* **1.** Prove the theorem: Among all rectangles of given diagonal the square has the largest area.

Fig. 4.30

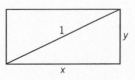

Solution. Let x and y be the sides of the rectangle (we mean, of course, the lengths of the sides, but such permissible inaccuracies in language cannot lead to confusion). The area A is given by $A = xy$. We may choose the diagonal as unit segment. Then (see Fig. 4.30) $x^2 + y^2 = 1$ and $y = \sqrt{1 - x^2}$, so that $A = x\sqrt{1 - x^2}$. Only the values $0 \leq x \leq 1$ are of interest. We must find the maximum of A in $[0, 1]$. We may consider instead $f(x) = A^2 = x^2(1 - x^2) = x^2 - x^4$. Indeed, A is positive so that A and $f = A^2$ will attain their maxima at the same point. Now, $f'(x) = 2x - 4x^3 = 2x(1 - 2x^2)$. The roots of f' are 0 and $1/\sqrt{2}$. Now $f(0) = 0$, $f(1/\sqrt{2}) = 1/2 - 1/4 = 1/4$, $f(1) = 0$. The maximum is attained for $x = 1/\sqrt{2}$. Then $y = \sqrt{1 - x^2} = 1/\sqrt{2}$; the rectangle is a square.

2. Show that, for $x \leq 0$, we have $\sqrt{1 + x} \geq -1 + x/2$.

Elementary Proof. The statement is equivalent to the true inequality

$$1 + x \leq \left(1 + \frac{x}{2}\right)^2 = 1 + x + \frac{x^2}{4}.$$

Proof by Calculus. Set $f(x) = (1 + x)^{1/2} - 1 - (x/2)$. We have

$$f'(x) = \frac{1}{2}(1 + x)^{-1/2} - \frac{1}{2} = \frac{1}{2}\left(\frac{1}{\sqrt{1 + x}} - 1\right).$$

Hence $f'(x) < 0$ and f is decreasing for $x > 0$, and $f'(x) > 0$ and f is increasing for $-1 \leq x < 0$. But $f(0) = 0$. Hence $f(x) \leq 0$ for $-1 \leq x$, and this is our assertion.

3. Show that, for $x > 0$ and a rational number $r > 2$, we have

$$(1 + x)^r > 1 + rx + \frac{r(r - 1)x^2}{2}.$$

Proof. Set $\quad f(x) = (1 + x)^r - 1 - rx - \frac{r(r - 1)x^2}{2}.$

We have

$$f'(x) = r(1 + x)^{r-1} - r - r(r - 1)x,$$
$$f''(x) = r(r - 1)(1 + x)^{r-2} - r(r - 1) = r(r - 1)[(1 + x)^{r-2} - 1].$$

Since $r > 2$, we have $f''(x) > 0$ for $x > 0$. Therefore $f'(x)$ increases. Since $f'(0) = 0$, $f'(x) > 0$ for $x > 0$. Therefore $f(x)$ increases. Since $f(0) = 0$, $f(x) > 0$ for $x > 0$. This is the assertion.

4. What is the most advantageous shape of a tin can? (The can is a right circular cylinder of radius r and height h. The amount of tin is measured by the total area: $A = \text{base} + \text{lateral surface} + \text{top} = \pi r^2 + 2\pi rh + \pi r^2 = 2\pi(r^2 + rh)$. The volume is $V = \pi r^2 h$. We want to make V as large as possible, for a given A.)

Since $A = 2\pi r^2 + 2\pi rh$, we have

$$h = \frac{A - 2\pi r^2}{2\pi r} \quad \text{and} \quad V = \pi r^2 h = \frac{1}{2}(Ar - 2\pi r^3) = \frac{1}{2}r(A - 2\pi r^2).$$

We are only interested in values of r for which V is not negative, that is, for

$$0 \leq r \leq \sqrt{\frac{A}{2\pi}}.$$

We need the point in this interval at which V has its maximum. At the endpoints we have $V = 0$. Next,

$$\frac{dV}{dr} = \frac{1}{2}(A - 6\pi r^2), \qquad \frac{d^2V}{dr^2} = -6\pi r < 0.$$

The first derivative is zero at precisely one point; the second derivative is negative there. This is a local maximum point, and V has its absolute maximum there. This maximum is attained when $A - 6\pi r^2 = 0$, that is, when $2\pi r^2 + 2\pi rh - 6\pi r^2 = 0$ or $h = 2r$.

The answer is: The diameter must equal the height.

(We have used here certain elementary results about areas and volumes. A discussion of these formulas will be given later.)

5. Find the largest area of a rectangle with vertices at the origin of a Cartesian coordinate system, on the x axis, on the y axis, and on the parabola $y = 4 - x^2$ (see Fig. 4.31).

Let the vertex on the parabola be $(x, 4 - x^2)$. The area of the rectangle is $A = (4 - x^2)x = 4x - x^3$. The interval in question is $[0, 2]$. Since $dA/dx = 4 - 3x^2$, $d^2A/dx^2 = -6x$, the maximum occurs for $x = 2/\sqrt{3}$ and the desired largest area is $16/3\sqrt{3}$.◄

Fig. 4.31

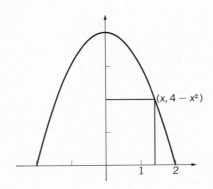

$(x, 4 - x^2)$

EXERCISES

23. (a) Find the maximum and minimum of the function $f(x) = (x^2 + 5)/(x + 2)$ for $0 \le x \le 10$.

(b) Find the maximum and minimum, if they exist, of the function $f(x) = (x^2 + 5)/(x + 2)$ for $0 \le x$.

24. Find the maximum and minimum of the function $f(x) = x - x^{2/3}$ for $-1 \le x \le 1$. Where does f attain its maximum? Its minimum?

25. Find the maximum and minimum of $\phi(y) = (2 + y - y^2)/(2 - y + y^2)$ for $-4 \le y \le 4$.

26. Find the maximum and minimum of the function $s \mapsto s\sqrt{1 - s^2}$.

27. Find all points where the function $g(u) = u^2 \sqrt{1 + u}$, $-1 \le u \le 1/2$, attains its maximum and all points at which it attains its minimum.

28. Let $f(y) = 3y^4 - 4y^3 - 12y^2 + 12$. Find the intervals of increase and decrease of f. Find all points, if any, where f has a local maximum or minimum. Find all points, if any, where f has an absolute maximum or minimum.

29. Let $h(z) = z^3/(z^4 + 27)$. Find the intervals of increase and decrease of h. Find all points, if any, where h has a local maximum or minimum. Find all points, if any, where h has an absolute maximum or minimum.

30. Show that the function $x \mapsto 3 + 3x - x^3$ is positive for $x \le 2$.

31. If x and y are nonnegative numbers and $x^2 + y^2 = 1$, what is the smallest value that $2x^3 + y^3$ can attain? If k is a positive constant and the same conditions hold, what is the smallest value that $kx^3 + y^3$ can attain?

► 32. What dimensions does a closed box with a square base have if its volume is V and the smallest amount of material is in it?

33. What dimensions does an open box with a square base have if its volume is V and the smallest amount of material is in it?

34. An artist decides to paint a picture consisting of a red rectangle surrounded by a white border. If the red rectangle is to have an area of 12 sq in. and the border is to be 1 in. wide along each side and 2 in. wide along the top and bottom, what dimensions should the picture have in order to have the smallest total area?

35. A sports field consists of a rectangular region with a semicircular region adjoined at each of two opposing sides. If the perimeter is to be 1000 ft, find the area of the largest possible field.

36. An open box is made from a rectangular piece of tin, 5 in. by 8 in., by cutting out equal squares at each corner and then folding up the remaining flaps. What size squares should be cut out so that the box will have maximum volume?

37. A country is designing a new flag, which is to consist of an orange rectangular region divided by a pink stripe. The perimeter of the entire flag is to be 14 ft and the orange part is to have an area of 9 sq ft. One faction wants the stripe to be horizontal; another wants the stripe to be vertical. They both agree that the stripe must be as wide as possible. Who wins? (By convention, the standard position for a rectangular flag is with the longest edge horizontal.)

38. What is the largest size rectangle that can be inscribed in a semicircle of radius 1 so that 2 vertices lie on the diameter?

39. A man walking in a forest is 5 miles from an infinitely long straight road and 13 miles from a house that is on the road. Suppose the man can walk at a rate of 3 miles per hour (mph) in the forest and 5 mph on the road. If he walks in a straight line to the road and then along the road to the house, in how short a time can he reach the house?

40. (a) A farmer wants to clear a field in the shape of a rectangular plot with a semicircular plot adjoined to one side of the rectangle. The rectangular plot is to be planted in hay, which will yield a profit at 5 cents per sq ft; and the semicircular plot is to be planted in rye, which will yield a profit at 6 cents per sq ft. If the perimeter of the field is to be 800 ft, how should the farmer lay out the field to earn the most money?

(b) Suppose the price of rye increases so that a field planted in rye would yield 10 cents per sq ft. How should the farmer lay out the field now?

41. A wire is to be cut into 2 pieces. One piece will be bent into a square, the other piece into a circle. If the total area enclosed by the 2 pieces is to be 16 sq in., what is the shortest length of wire that can be used? What is the longest length of wire that can be used?

42. A farmer has a herd of 100 cows, each weighing 500 lb. It costs 50 cents a day to keep 1 cow. The cows are gaining weight at the rate of 6 lb a day. The market price for cows is now $1 per lb and is falling by 1 cent a day. How long should the farmer wait to sell the cows in order to earn the maximum profit? How much has he gained by waiting? (Note: Assume that cows gain weight uniformly during each day, the cost of keeping a cow is distributed uniformly throughout the day, and so forth.)

43. A line is drawn from the point P with coordinates $(3, 0)$ to the curve given by $y = x^2$, $0 \leq x \leq 3$, intersecting the curve at the point Q. How should the line be drawn if the triangle bounded by the line, the x axis (the horizontal line through P), and the vertical line through Q is to have maximal area?

▶ 44. What is the shortest distance from the point $(1, 0)$ to the curve given by $y = \sqrt{x^2 + 6x + 10}$?

45. Find the shortest and the longest distances from the point $(-2, 1)$ to the curve given by $y = 1 + \sqrt{18 - 2x^2}$.

46. Find the trapezoid of largest area that can be inscribed in a semicircle of radius r, the lower base being a diameter.

47. The illumination provided by a point light source is directly proportional to the intensity of the source and inversely proportional to the square of the distance from the source. The illumination provided by several light sources is the sum of the illuminations provided by each one. Suppose there are 2 light beacons 1 mile apart. If the intensity of the first beacon is 8 times the intensity of the second, what point on the straight-line segment joining the beacons receives the least illumination? Generalize this result to the case in which the first beacon is k^3 times stronger than the second.

48. A hay fever sufferer discovers that the amount of pollen that reaches him from a given source is directly proportional to the strength of the source and inversely proportional to the distance from the source. Unfortunately, he is forced to live somewhere on a straight line joining 2 pollen sources that are 1 mile apart. If one source is 4 times stronger than the other, where should the man live to suffer the least discomfort? Generalize this result to the case in which one source is k^2 times stronger than the other.

49. The amount of gas produced by a coke oven operating at $1000°F$ is 100 cubic feet per minute (cu ft/min) and will increase by 1/5 cu ft/min for each degree rise in temperature up to $1500°F$. Above $1500°F$, the amount of gas produced will increase by 1/4 cu ft/min for each degree rise in temperature. It costs $1000 + (T - 1000/10)^2$ cents to operate the oven for 1 hr at a temperature of $T°F$ $(T \geq 1000)$. If gas can be sold at 1 cent per cu ft, what is the most profitable temperature to run the oven?

50. Previously, before improvements in gas retrieval at high temperatures were introduced, the amount of gas produced by the oven in Exercise 49 would only increase by 1/10 cu ft/min for each degree rise in temperature above $1500°F$. If other factors were the same, what would have been the most profitable temperature to run the oven?

51. A child who is 1 yd tall is standing 8 yd from a light pole. What is the shortest possible distance from the light to the tip of the child's shadow?

52. A 3-ft-wide corridor intersects a hallway at right angles. A 24-ft-long pole is being pushed along the floor down the corridor. How wide must the hallway be so that the pole will go around the corner?

Fig. 4.32

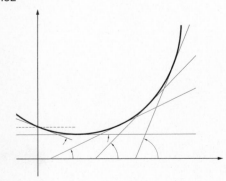

4.9 Convexity

So far we have discussed mainly the sign of the first derivative; that of the second derivative also has a simple geometric meaning. If $f'' > 0$ in an interval, f' is increasing. This means that the slope of the tangent to the graph increases as we go along the curve from left to right; the tangent turns counterclockwise (see Fig. 4.32). The graph is "bending upward" and "bulging downward." Such a curve, and such a function, are called **convex functions** and their graphs are called **convex down-**

ward or **concave upward.** It is seen from Fig. 4.33 that a convex graph lies "below its chords and above its tangents."

Similarly, if $f'' < 0$, then f' decreases, the tangent turns in the clockwise direction, and the graph lies "above its chords and below its tangents" (see Fig. 4.34). Such a function, and its graph, are called **concave functions** and their graphs are called **concave downward** or **convex upward.**

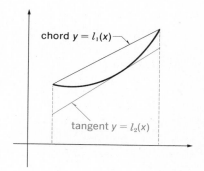

Fig. 4.33

We formulate these observations as the following theorem.

Theorem 8 LET $f(x)$ BE TWICE DIFFERENTIABLE FOR $a \leq x \leq b$. LET $x \mapsto l_1(x)$ BE A LINEAR FUNCTION SUCH THAT $l_1(a) = f(a)$, $l_1(b) = f(b)$. (THE GRAPH OF l_1 IS A CHORD OF THE GRAPH OF f.) LET x_0 BE A POINT SUCH THAT $a < x_0 < b$, AND $x \mapsto l_2(x)$ A LINEAR FUNCTION WITH $l_2(x_0) = f(x_0)$, $l'_2(x_0) = f'(x_0)$. (THE GRAPH OF l_2 IS A TANGENT TO THE GRAPH OF f.) UNDER THESE CIRCUMSTANCES

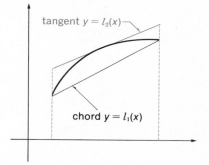

Fig. 4.34

IF FOR $a < x < b$	THEN FOR $a < x < b$	AND FOR $a < x < b$, $x \neq x_0$
$f'' > 0$	$f < l_1$	$f > l_2$
$f'' < 0$	$f > l_1$	$f < l_2$

An analytic proof will be given in the Appendix (§8.2).

Remark Let f be twice differentiable near x_0. If $f''(x_0) = 0$ and f'' changes sign at x_0, one says that $(x_0, f(x_0))$ is a **point of inflection** of the graph (see Fig. 4.35). At such a point the direction of bending changes.

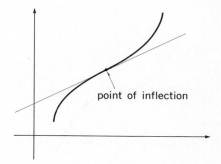

Fig. 4.35

4.10 Sketching graphs

The information collected in this section leads to a procedure for sketching quite accurately a graph of a function $f(x)$, while computing only relatively few points. The procedure, applicable to many functions defined by formulas, involves the following steps:

1. Determine whether the function is even, odd, or neither. If the function is even or odd, only values $x \geq 0$ need to be considered.

2. Find the intervals where the function is defined. Determine the (one-sided) limits of the function at $+\infty$, $-\infty$, and at the endpoints of the intervals of definition. If f has a finite limit c for $x \to +\infty$ or $x \to -\infty$, draw the line $y = c$ as an aid in sketching the curve. Such a line is called a **horizontal asymptote.** If f becomes infinite at $x = \alpha$, that is, if f has infinite one-sided limits at $x = \alpha$, draw the line $x = \alpha$ as an aid to drawing the curve. Such a line is called a **vertical asymptote.**

(We used here the language of one-sided and infinite limits, introduced informally in Chap. 3, §4.3. The precise definitions and the methods of computing these limits are discussed in appendix to that chapter (§5). For the benefit of readers who did not read the appendix, we give rather detailed explanations in the examples below.)

3. Locate the roots of f, if possible. (Mark these points on the graph.)

4. Compute f' and locate the roots of f', if possible. Determine the intervals of decrease and increase of f; for each root α of f' determine whether it is a local maximum, minimum, or neither, and compute $f(\alpha)$. Draw the points $(\alpha, f(\alpha))$ and the horizontal tangent through them. (Also note the points where f' becomes infinite or where only one-sided derivatives exist.)

5. Compute f''; locate the roots of f'', if possible, and determine the intervals of convexity and concavity of f. Determine which roots β of f'' are points of inflection; for each such β compute $f(\beta)$ and $f'(\beta)$. Draw the points $(\beta, f(\beta))$ and the tangents, with slope $f'(\beta)$, through them.

6. Sketch the graph of f, using the results of Steps 1 to 5. Better accuracy can be obtained by computing additional points and slopes.

We note that in any given case not all of Steps 1 to 5 are needed or possible.

▶ *Examples*   In the examples that follow, we shall write $f(+\infty), f(-\infty), f(x_0^+)$, and $f(x_0^-)$ as abbreviations for the symbols for one-sided limits, $\lim_{x \to +\infty} f(x), \lim_{x \to +\infty} f(x)$, and so forth.

1. $f(x) = -(x^3/6) - (x/6) + 2$. The function is a polynomial, defined everywhere. For $x \neq 0$, we have $f(x) = -(x^3/6)((1 + 1/x^2) - (12/x^7))$. If $|x|$ is large enough, then the term $((1 + 1/x^2) - (12/x^3))$ is close to 1, hence $f(x) < 0$ for $x > 0$ and $f(x) > 0$ for $x < 0$. The absolute value of $f(x)$ will be as large as we like if $|x|$ is large enough. This means that $f(x) \to -\infty$ as $x \to +\infty$ and $f(x) \to +\infty$ as $x \to -\infty$. We note this by writing: $f(-\infty) = +\infty$ and $f(+\infty) = -\infty$. Also $f'(x) = -(x^2/2) - 1/6 < 0$; f is decreasing. We note that $f(0) = 2$ and $f'(0) = -1/6$. Next, $f''(x) = -x$; hence f is convex for $x < 0$, concave for $x > 0$. There is an inflection point at $x = 0$. Clearly, a root of f lies between 0 and $+\infty$. We have $f(1) = 5/3, f'(1) = -2/3, f(2) = 1/3, f'(2) = -(13/6), f(3) = -3$, and $f'(3) = -28/9$. Thus f has a root between 2 and 3.

The graph is sketched in Fig. 4.36.

Fig. 4.36

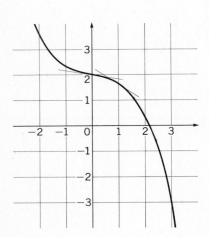

2. $f(x) = (x - 1)/(x - 2)$. We have $f'(x) = -1/(x - 2)^2$ and $f'(x) < 0$ for all $x \neq 2$; $f''(x) = 2/(x - 2)^3$ and $f''(x) < 0$ for $x < 2$, $f''(x) > 0$ for $x > 2$. The function is defined for $x \neq 2$, decreasing and concave for $x < 2$, decreasing and convex for $x > 2$. We proceed to compute the one-sided limits at $2^-, 2^+, +\infty$ and $-\infty$.

If x is close to 2, then $x - 1$ is close to 1. If x is close to 2, then $x - 2$ is close to 0, so that $f(x)$ is as large as we like in absolute value provided x is sufficiently close to 2. Also $f(x) > 0$ for $x > 2$ and $f(x) < 0$ for $x < 2$ and x close to 2. Thus $f(x) \to +\infty$ as $x \to 2^+$ (that is, as x approaches 2 from the right), and $x \to -\infty$ as $x \to 2^-$ (as x approaches 2 from the left). We write $f(2^-) = -\infty$, $f(2^+) = +\infty$. There is a vertical asymptote $x = 2$. (In Fig. 4.37 and in other figures, asymptotes are shown as dotted lines.)

For $x \neq 0$, we have

$$f(x) = \frac{1 - (1/x)}{1 - (2/x)}.$$

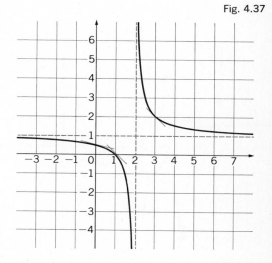

Fig. 4.37

If x is sufficiently large in absolute value, numerator and denominator will be as close as we like to 1. Hence $f(x) \to 1$ as $x \to +\infty$ and as $x \to -\infty$, or $f(-\infty) = f(+\infty) = 1$. There is a horizontal asymptote $y = 1$. Also, $f(0) = 1/2$, $f'(0) = -1/4$, $f(1) = 0$, $f'(1) = -1$, $f(3) = 2$, $f'(3) = -1$. The graph is sketched in Fig. 4.37.

3. $f(x) = x^2/(1 - x^2)$. This function is even; it is defined for $|x| \neq 1$. For $x = 0$, we have

$$f(x) = \frac{-1}{1 - (1/x^2)}$$

which shows that $f(x)$ is as close as we want to -1 for $|x|$ sufficiently large. Thus $f(+\infty) = -1$. We have a horizontal asymptote $y = -1$.

Next we note that

$$f(x) = \frac{x^2}{1 + x} \frac{1}{1 - x}.$$

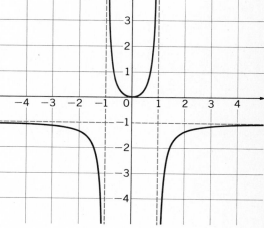

Fig. 4.38

For x close to 1, the first fraction is close to $1/2$. The second fraction has a large absolute value and is positive or negative according to whether $x < 1$ or $x > 1$. We conclude that $f(1^-) = +\infty$, and $f(1^+) = -\infty$. The line $x = 1$ is a vertical asymptote.

Also,

$$f'(x) = \frac{2x}{(1 - x^2)^2}, \qquad f''(x) = \frac{2(1 + 2x^2 - 3x^4)}{(1 - x^2)^4}.$$

We have $f(0) = f'(0) = 0$ and $f''(0) = 2$, hence a local minimum at $x = 0$. There are no inflection points. The graph is shown in Fig. 4.38. Since the function is even, the graph is symmetric about the y axis.

Fig. 4.39

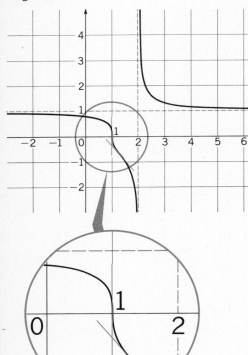

4. $f(x) = \sqrt[3]{(x - 1)/(x - 2)} = (x - 1)^{1/3}(x - 2)^{-1/3}$. The function is defined for $x \neq 2$. As in Example 2, or simply making use of the results in Example 2, we see that $f(-\infty) = f(\infty) = 1$, $f(1) = 0$, $f(2^-) = -\infty$, and $f(2^+) = +\infty$. There are two asymptotes: $x = 2$ and $y = 1$. Also,

$$f'(x) = \frac{1}{3}(x - 1)^{-2/3}(x - 2)^{-1/3} - \frac{1}{3}(x - 2)^{-4/3}(x - 1)^{1/3}$$

$$= \frac{1}{3}(x - 1)^{-2/3}(x - 2)^{-4/3}[(x - 2) - (x - 1)]$$

$$= \frac{-1}{3\sqrt[3]{(x - 1)^2(x - 2)^4}}.$$

One sees that f' becomes infinite at $x = 1$; thus there is a vertical tangent at $x = 1$. For $x \neq 1, 2$, we have $f'(x) < 0$. Next,

$$f''(x) = \frac{2}{9}(x - 1)^{-5/3}(x - 2)^{-4/3} + \frac{4}{9}(x - 2)^{-7/3}(x - 1)^{-2/3}$$

$$= \frac{2}{3}(x - 1)^{-5/3}(x - 2)^{-7/3}\left[\frac{1}{3}(x - 2) + \frac{2}{3}(x - 1)\right]$$

$$= \frac{2(x - (4/3))}{3\sqrt[3]{(x - 1)^5(x - 2)^7}}.$$

We observe that $f'' > 0$ and the graph is convex for $1 < x < 4/3$ and for $x > 2$. For $x < 1$ and for $4/3 < x < 2$, we have $f'' < 0$ and the graph is concave. There is an inflection point at $x = 4/3$. We compute $f(4/3) = -.15 \cdots$, $f'(4/3) = -.22 \cdots$. The graph is sketched in Fig. 4.39.◄

EXERCISES

Sketch the graphs of the functions given in Exercises 53 to 82. Be sure to find any breaks in the curve (points of discontinuity); the intervals of increase and decrease; any points where there is a horizontal tangent, vertical tangent, cusp, or corner (more extreme behavior is possible but unlikely in these examples); any local or absolute maxima or minima; the intervals of concavity and convexity; any points of inflection; and so forth. In sketching these graphs, it may occasionally be convenient to use different units of length along the two coordinate axes (see the answer to Exercise 54).

53. $f(x) = x^3 + 3x^2 - 24x + 20$.

54. $f(x) = x^3 - 12x + 11$.

55. $g(x) = x^4 - 18x^2 + 17$.

56. $k(x) = 3x^4 + 8x^3 - 6x^2 - 24x - 13$.

57. $F(x) = x^4 - 4x^3 + 10$.

► 58. $f(x) = |x|/(x + 1)$.

59. $x \mapsto (x^5/5) - (7/2)x^4 + 15x^3$.

60. $y = (x^2 - 7x + 10)/(x - 1)$.

61. $y = (x - 1)/(x + 1)^2$.

62. $f(t) = (t + 1)^2/(t^2 + 1)$.

63. $f(x) = x/(x^2 - 1)$.

64. $h(s) = (2 - s + s^2)/(2 + s - s^2)$.

65. $t \mapsto t\sqrt{t} + 5t + 3\sqrt{t}.$
66. $t \mapsto t\sqrt{t} + 4t - 3\sqrt{t}.$
67. $f(u) = u/\sqrt{u+1}.$
68. $g(x) = x/(x-1)^2.$
69. $y = \sqrt{u^2 + 2u + 2}.$
70. $f(x) = \sqrt{(4-x)/(4+x)}.$
71. $g(z) = z\sqrt{z+3}.$
72. $h(x) = x^2\sqrt{x+5}.$
73. $g(x) = x/\sqrt{x^2+1}.$

▶ 74. $y = (8/x^3) - (6/x).$
75. $y = x + (1/x).$
76. $g(x) = (x^2 + 3)/\sqrt{x^2 + 1}.$
77. $x \mapsto 2x^{1/3} - x^{2/3}.$
78. $h(x) = 3x^{5/3} - 5x.$
79. $y = (x-1)^{2/3} + (x+1)^{2/3}.$
80. $y = (x-1)^{1/3} + (x+1)^{1/3}.$
81. $F(z) = (z-1)^2 z^{1/3}.$
82. $g(x) = (x^2 + x + 1)^{1/3}.$

5.1 Definition of primitive functions

Let $f(x)$ and $F(x)$ be two functions. If $F'(x) = f(x)$ in some interval, we call F a **primitive function** or **antiderivative** of f. For instance, the function $F(x) = 1/x$ is a primitive function of $f(x) = -(1/x^2)$, since we know that

$$\frac{d}{dx}\left(\frac{1}{x}\right) = -\frac{1}{x^2}.$$

We say "a primitive function" and not "the primitive function" because, for example, $F_1(x) = (1/x) + 73$ is also a primitive function of $-(1/x^2)$.

In general, if $F(x)$ is a primitive function of $f(x)$, and c is any constant, then $F(x) + c$ is also a primitive function of $f(x)$. This is also obvious geometrically. The graph of the function $y = F(x) + c$ is obtained by translating the graph of $F(x)$ in the y direction by the amount c (up, if $c > 0$; down, if $c < 0$). This does not change the slope of the tangent (see Fig. 4.40).

Fig. 4.40

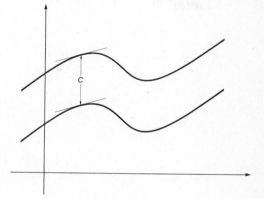

5.2 Uniqueness

If we know one primitive function $F(x)$ of $f(x)$, we know many others, namely, all functions of the form $F(x) + c$, c a constant. And there are no others. To prove this statement, we make use of the following theorem.

Theorem 1   IF $g'(x) = 0$ FOR ALL x IN AN INTERVAL, THEN $g(x)$ IS CONSTANT IN THIS INTERVAL.

The theorem asserts that a curve with a tangent horizontal everywhere is a horizontal straight line. Our geometric intuition certainly tells us that this is so. The statement is also obvious if we think of our function as describing the motion of a point. If the velocity is always zero, the point does not move. Of course, a proof is still needed. It follows easily from Theorem 3 in §4.3.

Let $x_2 > x_1$ be two points in our interval. Since $g' \geq 0$, g is not decreasing; we have $g(x_1) \leq g(x_2)$. Since $g' \leq 0$, g is not increasing; we have $g(x_1) \geq g(x_2)$. Therefore $g(x_1) = g(x_2)$. The function takes on the same value at any two points.

Theorem 2   IF $F(x)$ AND $F_1(x)$ ARE PRIMITIVES OF THE SAME FUNCTION, THEN $F_1(x) - F(x)$ IS A CONSTANT FUNCTION; THAT IS, THERE IS A NUMBER c SUCH THAT $F_1(x) = F(x) + c$.

Proof.   Let $F(x)$ and $F_1(x)$ be primitive functions of $f(x)$ and set $g(x) = F_1(x) - F(x)$. Then $g'(x) = F_1'(x) - F'(x) = f(x) - f(x) \equiv 0$ and, by Theorem 1, $g(x) \equiv c$, as asserted.

5.3 A simple differential equation

As an application, let us find all functions $\phi(x)$ whose second derivative vanishes identically (for all values of x); in other words, let us solve the differential equation:

$$(1) \qquad\qquad\qquad \phi''(x) = 0.$$

This means that
$$\frac{d\phi'(x)}{dx} = 0,$$

so that, by Theorem 1, there exists a number α such that

$$\phi'(x) = \alpha.$$

Hence ϕ is a primitive function of the constant α. One primitive function is αx; every other is of the form $\alpha x + \beta$, β being some other number. Thus ϕ is a linear function. Conversely, every linear function has an identically vanishing second derivative.

5.4 Finding primitives

It is natural to ask whether every continuous function has a primitive function, that is, whether every continuous function is a derivative of another function. The answer is in the affirmative, as we will see in the next chapter.

At the moment we can only guess primitive functions. Every differentiation rule is at the same time a rule for finding primitive functions. Some results that follow from the differentiation rules of §2 are tabulated in the following table.

	Every primitive function of	*is of the form* $(C = \text{a constant})$
(a)	$cf'(x),\ c = \text{a constant}$	$cf(x) + C$
(b)	$f'(x) + g'(x)$	$f(x) + g(x) + C$
(c)	$f'(x)g(x) + f(x)g'(x)$	$f(x)g(x) + C$
(d)	$-\dfrac{f'(x)}{f(x)^2}$	$\dfrac{1}{f(x)} + C$
(e)	$x^\alpha,\ \alpha \text{ rational},\ \alpha \neq -1$	$\dfrac{x^{\alpha+1}}{\alpha + 1} + C$

▶**Examples** **1.** A primitive function of $2x^3 + x + 1$ is the function $\frac{1}{2}x^4 + \frac{1}{2}x^2 + x + c$ [by (a) and (b)]. This result, like any result on primitive functions, can be checked by differentiation.

2. In general, a primitive function of a polynomial of degree n, $a_0 + a_1 x + a_2 x^2 + \cdots + a_n x^n$, is the polynomial

$$c + a_0 x + \frac{a_1}{2} x^2 + \frac{a_2}{3} x^3 + \cdots + \frac{a_n}{n+1} x^{n+1}$$

of degree $n + 1$.

3. Find a primitive function of $f(x) = \dfrac{x^2}{(x^3 + 1)^2}$.

Solution. We note that $(x^3 + 1)' = 3x^2$. Therefore,

$$\left(\frac{1}{x^3 + 1}\right)' = \frac{-3x^2}{(x^3 + 1)^2}$$

and, by (a),

$$-\frac{1}{3}\frac{1}{x^3 + 1} + C$$

is a primitive function of f.

4. The primitive function of $\dfrac{x}{\sqrt{1 + x^2}}$ is $\sqrt{1 + x^2} + C$.

5. Find all solutions of the differential equation $f'''(x) = 0$.

Solution. Since $f''' = (f'')' = 0$ everywhere, $f''(x)$ is a constant; call it α. Hence $(f')' = \alpha$; $f'(x)$ is a primitive function of α. We conclude that $f'(x) = \alpha x + \beta$, where β is another constant. Now f is a primitive function of $\alpha x + \beta$. One such primitive function is $\frac{1}{2}\alpha x^2 + \beta x$. Hence $f(x) = \frac{1}{2}\alpha x^2 + \beta x + \gamma$, with α, β, γ constants. In other words, f is a quadratic polynomial.

6. Find a function f such that $f'(x) = \sqrt{x}$ and $f(1) = 2$.

Since $f'(x) = x^{1/2}$ we have, by (e), that $f(x) = \frac{2}{3}x^{3/2} + C$. The condition $f(1) = 2$ yields $C = \frac{4}{3}$. Thus $f(x) = \frac{1}{3}(2x\sqrt{x} + 4)$.◄

EXERCISES

1. Find all functions ϕ with $\phi'(x) = 1/2$.
2. Find all functions ϕ with $\phi'(x) = x^2$.
3. Find all functions ϕ with $\phi''(x) = -1$.
4. Find all primitive functions of $x \mapsto (x + 1)^9$.
5. Find all primitive functions of $s \mapsto s^4 + 4s^3 + 4s - 2$.
6. Find all primitive functions of $f(u) = u^{1/3} + 3u^{1/2}$.
7. Find all primitive functions of $f(z) = z/\sqrt{z^2 + 1}$.
8. Find all primitive functions of $G(x) = (\sqrt{x} + 1)^5/2\sqrt{x}$.
9. Find all primitive functions of $z \mapsto (z^3 + z^2 - 2z + 3)^{10}(9z^2 + 6z - 6)$.
10. Find all functions f with $f''(s) = 12s^2 + s - 1$.
11. Find all solutions of the differential equation $f''(x) = \sqrt{x}$.
12. Find a function g such that $g'(s) = s^3 + (1/s^2)$ and $g(1) = 1$.
13. Find a function h such that $h'(t) = t(t^2 + 1)^9$ and $h(0) = 1/10$.
► 14. Find a function A such that $A''(v) = 3v^2$, $A'(0) = 1$, and $A(0) = -1$.
15. Find a function F such that $F''(s) = 6s + 1$, $F(0) = 2$, and $F(1) = 0$.

5.5 Inertia

We now apply the concept of primitive functions to Newton's law of motion. For a particle of mass m moving on a rectilinear path, this law reads

$$ma = F,$$

where F is the force applied and a the acceleration. Since a is the first derivative of the velocity $v = v(t)$ and the second derivative of the distance, $s = s(t)$, the above formula may also be written either as

$$(2) \qquad\qquad m\frac{dv}{dt} = F$$

or

$$m\frac{d^2s}{dt^2} = F.$$

In order to determine the motion from this **differential equation,** we must know the force. We shall consider some simple cases.

Assume first that there is *no* force acting on the body. Then $F = 0$ and (2) becomes $m(dv/dt) = 0$ or $dv/dt = 0$. By Theorem 1, we have that v is constant. This is a special case of the **law of inertia** stated by Galileo. *A body acted on by no force remains at rest or in a state of uniform motion along a straight line.*

5.6 Vertical motion under the influence of gravity

Assume next that the body moves vertically under the influence of its weight. Then $F = $ weight $ = -mg$, as we saw in §3.3. Here g is a constant, the same for all bodies. The equation of motion can be written as

$$m\frac{d^2s}{dt^2} = -mg \qquad \text{or} \qquad m\frac{dv}{dt} = -mg$$

or, upon canceling m,

$$\frac{d^2s}{dt^2} = -g \qquad \text{or} \qquad \frac{dv}{dt} = -g.$$

This is, of course, merely Galileo's law. But this time we shall apply it not only to a body freely released from 0, but to a body thrown upward or downward with velocity v_0 from the height s_0 at the time $t = 0$.

Let $v(t)$ be the velocity at the time t. Then $v(t)$ is a primitive function of the constant function $-g$. Hence $v(t) = -gt + \alpha$, where α is some constant. To find this constant, we note that $v(0) = \alpha$. But we denoted the velocity imparted to our body at $t = 0$ by v_0. Hence $\alpha = v_0$ and

$$v(t) = -gt + v_0.$$

The velocity is seen to be composed of two parts: the initial velocity v_0, which would be the steady velocity of the body according to the law of inertia if there were no force acting, and the velocity $-gt$ resulting from the acceleration caused by gravity. (One can say that the body simultaneously executes two motions: a uniform motion with velocity v_0 prescribed by the law of inertia and the falling motion with velocity $-gt$ prescribed by Galileo's law.)

What is the height $s(t)$ of the body at the time t? Since $s'(t) = v(t)$, the function $s(t)$ is a primitive function of $-gt + v_0$. It is therefore of the form $s(t) = -\frac{1}{2}gt^2 + v_0t + \alpha$, where α is a constant. To find α, note that $s(0) = \alpha$. Thus $\alpha = s_0$, the height (above a reference point) of the body at the time 0. Therefore

$$s(t) = -\frac{1}{2}gt^2 + v_0 t + s_0.$$

The graph of the motion is a parabola, or rather the part of it corresponding to positive values of t.

Let us assume, for the sake of definiteness, that $v_0 > 0$ (the body is thrown upward at $t = 0$), and that $s_0 = 0$ (at $t = 0$ the height of the body above the reference point is zero). We also assume that the reference point is at ground level. The motion is now described by

$$s(t) = -\frac{1}{2}gt^2 + v_0 t.$$

At which time will the body hit the ground? We must solve the equation $s(t) = 0$ which gives $t = 0$ (the initial time) and

$$t = \frac{2v_0}{g}.$$

What is the largest height reached by our body? We must find the absolute maximum of $s(t)$ in the interval $[0, 2v_0/g]$. According to the rules explained in §4, we find the point at which $s'(t) = 0$. Since $s'(t) = -gt + v_0$, this derivative is zero for

$$t = \frac{v_0}{g}.$$

For this t, $s''(t) = -g < 0$ and s takes on the value

$$\frac{v_0{}^2}{2g}.$$

This is therefore the maximum height achieved. At the instant it is attained, the velocity is, of course, zero, since the body reverses the direction of motion. This observation explains again why the derivative (= velocity) is zero at a maximum point.

It is interesting to compute the velocity with which the body hits the ground, that is, the value of $v(t) = s'(t) = -gt + v_0$ at $t = 2v_0/g$. Substitution yields the value $(-v_0)$. The final speed is therefore equal to the initial speed. Later, we shall recognize in this statement a special case of the law of conservation of energy.

EXERCISES

16. A particle is shot straight up with an initial velocity of 12 ft/sec. Assuming that the particle is shot from ground level, what is the maximum height achieved by the particle? How long after being shot will the particle hit the ground?

17. A stone is thrown straight down from a height of 160 ft with an initial velocity of 8 ft/sec. How long does it take the stone to reach the ground and what is its velocity then?

▶18. A man on the ground is throwing a ball straight up. If he wants the ball to reach a height of at least 50 ft, what is the minimal velocity he must impart to the ball?

19. A stone is dropped from a height of 72 ft. If a second stone is to be thrown down 1 sec later, what velocity must be imparted to it so that the two stones hit the ground at the same time?

20. A stone is thrown straight up and hits the ground after t seconds. How far up did it go?

21. A particle with mass 2 grams is at rest at time $t = 0$. A constant force is applied. If the velocity at the end of 4 sec is 8 cm/sec, find the force.

22. A car moves along a 10-mile-long test track, starting at one end at 10 mph. If the time t is measured in hours from the start of the test run, the acceleration of the car is found to be $a = 4 - 12t$ (where t runs through the interval of time when the test is being conducted). Where is the car when $t = \frac{1}{2}$?

23. A car moves along a 10-mile-long test track. When it reaches the end of the track, it runs in reverse back to the starting point. Time is measured in hours from the start of the test run. The car returns to the starting point at time $t = 2$ and its acceleration is found to be $a = -120(t - 1)^2$. Where is the car when $t = \frac{1}{2}$? What is the velocity of the car at the start of the test run?

5.7 Relativistic motion under a constant force

Consider now a particle of rest mass m_0 that moves according to Einstein's equation (4) of §3.4, under the influence of a constant force $F > 0$ and beginning at rest ($v = 0$ at $t = 0$). Then

$$\frac{v}{\sqrt{1 - (v^2/c^2)}}$$

is the primitive function of the constant function F/m_0. Also, this primitive function has the value 0 at $t = 0$. Therefore

$$\frac{v}{\sqrt{1 - (v^2/c^2)}} = \frac{Ft}{m_0}.$$

We square both sides and obtain that

$$\frac{v^2}{1 - (v^2/c^2)} = \frac{F^2 t^2}{m_0{}^2}.$$

Solving for v^2, we get

$$v^2 = \frac{F^2 t^2/m_0{}^2}{F^2 t^2/m_0{}^2 c^2} = c^2 \frac{F^2 t^2}{m_0{}^2 c^2 + F^2 t^2}$$

or, since $v = v(t)$ is positive,

$$v(t) = c \frac{Ft}{\sqrt{m_0{}^2 c^2 + F^2 t^2}}.$$

The denominator in the fraction is always larger than the numerator. This shows that, for all t, $v(t) < c$. *The body can never reach the speed of light.* (But the speed of the body will, if the motion continues indefinitely, come as close to the speed of light as we want. In other words, $\lim_{t \to +\infty} v(t) = c$.)

EXERCISES

In Exercises 24 and 25, use the relativistic equations of motion. Use the value $c = 3.10^{10}$ cm/sec for the speed of light.

24. A particle with rest mass .02 gram is at rest at time $t = 0$. A constant force of 10^{10} gram $\cdot$ cm/sec^2 is applied. What is the velocity after 10 sec?

25. A particle with rest mass .1 gram has velocity $c/2$ at time $t = 0$. A constant force is applied. If the velocity at the end of 10 sec is $3c/4$, find the force.

5.8 The area under the parabola

As another application of primitive functions, we shall compute the area of the region bounded by the parabola $y = x^2$, the x axis, and the line $x = t > 0$. Denote this area by $A(t)$; it is shown in Fig. 4.41. In an appendix to Chapter 2 (§5), we treated this problem by a method of Archimedes and proved that

(4)
$$A(t) = \frac{1}{3} t^3.$$

We shall now obtain the same result using derivatives. (At this point, we accept the notion of area as intuitively obvious.)

Fig. 4.41

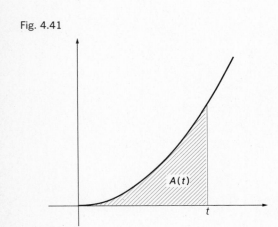

$A(t)$

For a fixed t_0, we compute the derivative $A'(t_0)$. Let $h > 0$ be a small number. Then $A(t_0 + h) - A(t_0)$ is the area shown in Fig. 4.42. It is less than the area $h \cdot (t_0 + h)^2$ of the "taller" rectangle and greater than the area $h \cdot t_0^2$ of the "shorter" rectangle. Thus

$$ht_0^2 < A(t_0 + h) - A(t_0) < h(t_0 + h)^2.$$

Dividing these inequalities by h, we obtain

$$(5) \qquad t_0^2 < \frac{A(t_0 + h) - A(t_0)}{h} < t_0^2 + 2t_0 h + h^2.$$

For a negative h of small absolute value we obtain, from Fig. 4.43, the inequalities

$$(6) \qquad t_0^2 + 2t_0 h + h^2 < \frac{A(t_0 + h) - A(t_0)}{h} < t_0^2.$$

The inequalities (5) and (6) show that

$$\lim_{h \to 0} \frac{A(t_0 + h) - A(t_0)}{h} = t_0^2.$$

Hence $A'(t_0) = t_0^2$. This is so for every t_0. Therefore $A(t)$ is a primitive function of t^2. This means that $A(t) = \frac{1}{3}t^3 + \alpha$, where α is some constant. For $t = 0$, we get $A(0) = \alpha$. But $A(0) = 0$, so that $\alpha = 0$ and we obtain (4).

Archimedes' discovery was one of the high points of Greek mathematics. It remained an isolated achievement until nineteen centuries later when Cavalieri showed, by a method similar to that of Archimedes, that the area bounded by the curve $y = x^n$, the x axis, and the line $x = t$ is equal to

$$\frac{t^{n+1}}{n + 1}.$$

In fact, Cavalieri could prove it only for $n = 3, 4, \cdots, 9$. Today such problems, and even much more complicated problems, are easily solved by anybody who masters the rudiments of calculus. The key to the matter is the connection between two seemingly unrelated problems, measuring areas and drawing tangents. The above example illustrates this connection, which we develop systematically in the next chapter.

Fig. 4.42

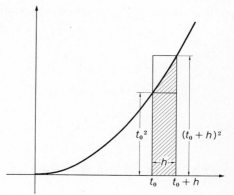

Fig. 4.43

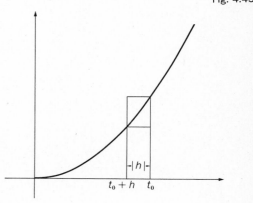

BONAVENTURA CAVALIERI (1598–1647) was a pupil of Galileo. He belongs to the group of brilliant mathematicians who anticipated many ideas and results of calculus before its formulation by Newton and Leibniz.

EXERCISES

26. Use the method of this section to compute the area of the region bounded by the line $y = x$, $x \geq 0$, the x axis, and the line $x = t > 0$.
27. Use the method of this section to compute the area of the region bounded by the curve $y = x^3$, $x \geq 0$, the x axis, and the line $x = t > 0$.
28. Use the method of this section to compute the area of the region bounded by the curve $y = \sqrt{x}$, the x axis, and the line $x = t > 0$.

Appendixes to Chapter 4

§6 Proofs of Differentiation Rules

In this section we complete the proofs of the differentiation rules of §2.

6.1 The product rule

Let $f(x)$ and $g(x)$ be defined near x_0 and let $f(x)$ and $g(x)$ be differentiable at x_0. Then the product $f(x)g(x)$ has a derivative at x_0 given by the product rule $(fg)' = f'g + fg'$. In §2.2, we proved this for the case in which f and g are linear. We show next that, as claimed there, the argument is easily extended to the general case.

By the linear approximation theorem (Theorem 1 in §1.8), we can write

$$(1) \qquad f(x) = a + \alpha(x - x_0) + r_1(x)(x - x_0),$$

$$(2) \qquad g(x) = b + \beta(x - x_0) + r_2(x)(x - x_0)$$

where

$$(3) \qquad a = f(x_0), \ \alpha = f'(x_0), \ b = g(x_0), \ \beta = g'(x_0)$$

and $r_1(x)$, $r_2(x)$ are continuous functions with

$$(4) \qquad r_1(x_0) = r_2(x_0) = 0.$$

Hence

$$f(x)g(x) = [a + \alpha(x - x_0) + r_1(x)(x - x_0)][b + \beta(x - x_0) + r_2(x)(x - x_1)]$$
$$= ab + (\alpha b + a\beta)(x - x_0) + r(x)(x - x_0)$$

where $\quad r(x) = ar_2(x) + br_1(x) + [\alpha r_2(x) + \beta r_1(x) + r_1(x)r_2(x)](x - x_0).$

Clearly, $r(x)$ is continuous at x_0 and $r(x_0) = 0$. The linear approximation theorem assures us that fg has at x_0 the derivative $a\beta + \alpha b = f'g + fg'$.

We give another proof of the product rule, based on the rules for computing limits.

Set
$$\phi(x) = f(x)g(x).$$

Then
$$\frac{\phi(x_0 + h) - \phi(x_0)}{h} = \frac{f(x_0 + h)g(x_0 + h) - f(x_0)g(x_0)}{h}.$$

We use a simple and useful device: the numerator to the right is not changed if we subtract and add the number $f(x_0)g(x_0 + h)$. Hence

$$\phi'(x_0) = \lim_{h \to 0} \frac{\phi(x_0 + h) - \phi(x_0)}{h}$$

$$= \lim_{h \to 0} \frac{f(x_0 + h)g(x_0 + h) - f(x_0)g(x_0 + h) + f(x_0)g(x_0 + h) - f(x_0)g(x_0)}{h}$$

$$= \lim_{h \to 0} \left(\frac{f(x_0 + h) - f(x_0)}{h} g(x_0 + h) + f(x_0) \frac{g(x_0 + h) - g(x_0)}{h} \right)$$

$$= \lim_{h \to 0} \frac{f(x_0 + h) - f(x_0)}{h} \lim_{h \to 0} g(x_0 + h) + \lim_{h \to 0} f(x_0) \lim_{h \to 0} \frac{g(x_0 + h) - g(x_0)}{h}$$

$$= f'(x_0)g(x_0) + f(x_0)g'(x_0).$$

Here we used the rules for limits of Chapter 3, §4 and Theorem 2 of §1, which tell us that $g(x_0 + h)$ is a continuous function of h at $h = 0$. We also used the fact that $g(x_0)$ is a constant function of h. We have shown that $\phi' = f'g + fg'$ at x_0; this is what we wanted to prove.

6.2 The derivative of a reciprocal

We shall prove that $(1/g)' = -g'/g^2$ at a point x_0 assuming that g is differentiable and different from 0 at x_0. We may represent $g(x)$ in the form (2) with $b \neq 0$. Then $g(x) \neq 0$ for x near x_0 and the function

$$\phi(x) = \frac{1}{g(x)}$$

is defined near x_0. We have $\phi(x_0) = 1/b$ and we want to show that $\phi'(x_0) = -\beta/b^2$. This means: if $r(x)$ is defined by the equation

$$\phi(x) = \frac{1}{b} - \frac{\beta}{b^2}(x - x_0) + r(x)(x - x_0),$$

then $r(x)$ is continuous near x_0 and $r(x_0) = 0$. Substituting the expression (2) for $g(x)$, we obtain that

$$r(x) = \frac{[b - \beta(x - x_0)]r_2(x) + \beta^2(x - x_0)}{b^2[b + \beta(x - x_0) + r_2(x)(x - x_0)]}.$$

This r has the required properties.

We also give an alternative proof:

$$\phi'(x_0) = \lim_{h \to 0} \frac{\phi(x_0 + h) - \phi(x_0)}{h} = \lim_{h \to 0} \frac{\dfrac{1}{g(x_0 + h)} - \dfrac{1}{g(x_0)}}{h}$$

$$= \lim_{h \to 0} \frac{g(x_0) - g(x_0 + h)}{g(x_0 + h)g(x_0)h} = \lim_{h \to 0} \frac{-\dfrac{g(x_0 + h) - g(x_0)}{h}}{g(x_0 + h)g(x_0)}$$

$$= \frac{-\lim_{h \to 0} \dfrac{g(x_0 + h) - g(x_0)}{h}}{\lim_{h \to 0} g(x_0 + h) \; \lim_{h \to 0} g(x_0)} = -\frac{g'(x_0)}{g(x_0)^2}.$$

The reader is asked to justify each step in this calculation.

EXERCISES

1. In §2.2 the differentiation rules $(cf)' = cf'$ and $(f - g)' = f' - g'$ have been derived from other differentiation rules. Give direct proof of these rules using the linear approximation theorem.
2. Prove the rules $(cf)' = cf'$, $(f - g)' = f' - g'$ using the definition of the derivative and the rules for computing limits.
3. Prove the rule for computing the derivative of a quotient using the definition of the derivative and the rules for computing limits.

6.3 Proof of the chain rule

The chain rule (Theorem 6 in §2.5) asserts that if the function $x \mapsto f(x)$ is differentiable at x_0 and the function $u \mapsto g(u)$ is differentiable at $u_0 = f(x_0)$, then the composite function $\phi(x) = (g \circ f)(x) = g(f(x))$ is differentiable at x_0 and

$$\phi'(x_0) = g'(f(x_0))f'(x_0).$$

By hypothesis, and by the linear approximation theorem (Theorem 1 of §1.8), we have

$$f(x) = u_0 + \alpha(x - x_0) + r_1(x)(x - x_0), \quad g(u) = C + D(u - u_0) + r_2(u)(u - u_0),$$

with r_1 continuous at x_0, r_2 continuous at u_0, and

$$r_1(x_0) = 0, \quad r_2(u_0) = 0.$$

Also $\qquad\qquad f'(x_0) = \beta, \quad g(u_0) = C, \quad g'(u_0) = D.$

Substituting the expression for f into that for g, we obtain

$$\phi(x) = g(f(x)) = C + D[f(x) - u_0] + r_2(f(x))[f(x) - u_0]$$
$$= C + D[\alpha(x - x_0) + r_1(x)(x - x_0)] + r_2(f(x))[\alpha(x - x_0) + r_1(x)(x - x_0)]$$
$$= C + D\alpha(x - x_0) + r(x)(x - x_0)$$

where

$$r(x) = Dr_1(x) + \alpha r_2(f(x)) + r_2(f(x))r_1(x).$$

By the theorem on continuity of composite functions (Theorem 4 in Chapter 3, §3), $x \mapsto r_2(f(x))$ is continuous at x_0. So, therefore, is $r(x)$, by Theorem 1 in Chapter 3, §3. Also, $r(x_0) = 0$. Therefore, $\phi'(x_0) = D\alpha = g'(u_0)f'(x_0) = g'(f(x_0))f'(x_0)$, again by Theorem 1 of §1. The chain rule is proved.

6.4 The derivative of an inverse function

Now we give an analytic proof of Theorem 7 of §2.6 about the derivative of an inverse function. The theorem states:

If $y = f(x)$ and $x = g(y)$ are strictly monotone functions inverse to each other, and the function f has at $x = x_0$ a derivative, and $f'(x_0) \neq 0$, then g has a derivative at $y = y_0 = f(x_0)$, and $g'(y_0) = 1/f'(x_0)$.

To prove this, we note that there is a function $h \mapsto \phi(h)$ defined near $h = 0$, such that

$$y_0 + h = f(x_0 + \phi(h)).$$

Indeed, the function

$$\phi(h) = g(y_0 + h) - x_0 = g(y_0 + h) - g(y_0)$$

has the required property. It is seen at once that this function is continuous and that $\phi(0) = 0$. Now, the difference quotient of the function g at y_0 may be written as

$$\frac{g(y_0 + h) - g(y_0)}{h} = \frac{g(f(x_0 + \phi(h))) - g(f(x_0))}{f(x_0 + \phi(h)) - f(x_0)}$$

$$= \frac{x_0 + \phi(h) - x_0}{f(x_0 + \phi(h)) - f(x_0)} = \frac{1}{\dfrac{f(x_0 + \phi(h)) - f(x_0)}{\phi(h)}}.$$

The limit of this expression at $h = 0$ is the reciprocal of the limit of $[f(x_0 + \phi(h)) - f(x_0)]/\phi(h)$ at $h = 0$, which is (see Chapter 3, §4.2, Theorem 2) the reciprocal of the limit of $[f(x_0 + k) - f(x_0)]/k$ at $k = 0$, that is, $1/f'(x_0)$, as asserted.

§7 One-sided Derivatives. Infinite Derivatives. Differentiable and Nondifferentiable Functions

This section deals with various ramifications of the concept of the derivative which have been mentioned informally in §2.9. It may be omitted at first reading. We make use of the notations introduced in an appendix to Chapter 3 (see §5 of that chapter).

7.1 One-sided derivatives

Let $f(x)$ be a function; the **right derivative** of f at x_0 is the limit

$$f'(x_0{}^+) = \lim_{h \to 0^+} \frac{f(x_0 + h) - f(x_0)}{h}$$

whenever it exists. This derivative may exist for a function which is defined only at x_0 and at points near, and to the right of, x_0. The **left derivative** at x_0 is defined similarly:

$$f'(x_0{}^-) = \lim_{h \to 0^-} \frac{f(x_0 + h) - f(x_0)}{h},$$

if this limit exists.

▶ *Examples*   **1.** If $f(x) = x^{3/2}$, then $f'(0^+) = 0$. For, if $h > 0$, then

$$\frac{f(0 + h) - f(0)}{h} = \frac{h^{3/2} - 0}{h} = h^{1/2}$$

and $h^{1/2}$ is continuous from the right at 0 and equal to 0 at this point. (There is, of course, no left derivative, since the function $f(x) = \sqrt{x^3}$ is defined only for $x \geq 0$.)

2. Set $f(x) = 2x$ for $x < 0$, $f(x) = x^2$ for $x \geq 0$. Then $f'(0^-) = 2$ (since the function $2x$ at $x = 0$ has the derivative 2) and $f'(0^+) = 0$ (since the function x^2 at $x = 0$ has the derivative 0). ◀

Let f be a function such that $f'(x_0{}^+) = \alpha$, $f'(x_0{}^-) = \beta$. Then, if $\alpha = \beta$, f is differentiable at x_0 and $f'(x_0) = \alpha$ as one sees easily. If $\alpha \neq \beta$, then f is not differentiable at x_0. We say, in this case, that the graph of f has a **corner** at the point $(x_0, f(x_0))$. A typical case is shown in Fig. 4.44.

Fig. 4.44

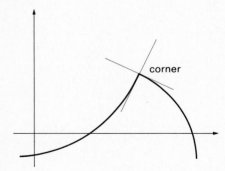

corner

Theorem 1   IF A FUNCTION HAS A RIGHT (OR LEFT) DERIVATIVE AT A POINT x_0, THEN IT IS CONTINUOUS IN AN INTERVAL WITH THIS POINT AS A LEFT (OR RIGHT) ENDPOINT. MORE PRECISELY, SUCH A FUNCTION IS A SUM OF A CONSTANT, A LINEAR FUNCTION, AND A FUNCTION $(x - x_0)r(x)$ WHERE $r(x)$ HAS AT x_0 THE LIMIT 0 FROM THE RIGHT (OR LEFT).

THE RULES FOR COMPUTING DERIVATIVES OF SUMS, DIFFERENCES, PRODUCTS, AND QUOTIENTS (THEOREM 1 OF §2); THE CHAIN RULE (THEOREM 6 OF §2); AND THE RULE FOR COMPUTING THE DERIVATIVE OF AN INVERSE FUNCTION (THEOREM 7 OF §2) REMAIN VALID FOR ONE-SIDED DERIVATIVES.

The first statement is proved almost exactly as the corresponding result (Theorems 1 and 2) in §1. The remaining statements are proved as the corresponding results in §2.

EXERCISES

1. If $g(s) = s^3$ for $s \leq 1$ and s^2 for $s > 1$, find $g'(1^-)$ and $g'(2^+)$.
2. If $f(x) = (1 - x)^{3/2}$, find $f'(1^-)$.
3. Let $h(t) = (t^2 - 4)^{5/2}$. Determine which of the one-sided derivatives $h'(-2^-)$, $h'(-2^+)$, $h'(2^-)$, and $h'(2^+)$ exist; evaluate any that do.
▶ 4. If $f(y) = y + 1$ for $y \leq 3$ and $y^2 - 2y + 1$ for $y > 3$, find $f'(3^-)$ and $f'(3^+)$.
5. If $f(x) = x$ for $x \leq 0$ and $x^{5/2}$ for $x > 0$, find $f'(0^-)$ and $f'(0^+)$.
6. If $\phi(u) = (u - 1)^{2/3}$ for $u \leq 0$ and $(u + 1)^{2/3}$ for $u > 0$, find $\phi'(0^-)$ and $\phi'(0^+)$.
7. If $h(s) = s^2 - 1$ for $s \leq 1$ and s for $s > 1$, find $h'(1^-)$ and $h'(1^+)$.
8. Draw the graph of a function $f(x)$, continuous for $|x| \leq 1$, which satisfies the conditions $f(0) = 0$, $f'(0^-) = 0$, and $f'(0^+) = 1$.
9. Draw the graph of a function $g(x)$, continuous for $-(3/2) \leq x \leq (5/4)$, which satisfies the conditions $g(-1) = 0$, $g'(0^-) = -1$, $g'(0^+) = 2$, and $g(1) = 1$.
10. Draw the graph of a function $h(t)$, defined and continuous for $|t| \leq 1$, which satisfies the conditions $h'(-1^+) = \frac{1}{2}$, $h(0) = 0$, $h'(0) = 0$, $h'(\frac{1}{2}^-) = -1$, $h'(\frac{1}{2}^+) = 2$, and $h'(1^-) = -\frac{1}{2}$.

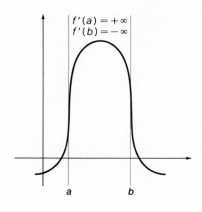

Fig. 4.45

$f'(a) = +\infty$
$f'(b) = -\infty$

a b

7.2 Infinite derivatives

A continuous function f may fail to have a derivative at a point x_0, because the difference quotient at x_0 can be made larger than any number provided we take h small enough. This means

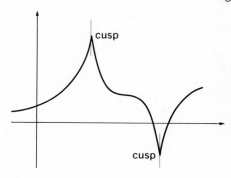

Fig. 4.46

cusp

cusp

$$\lim_{h \to 0} \frac{f(x_0 + h) - f(x_0)}{h} = +\infty.$$

In this case, we write $f'(x_0) = +\infty$. If

$$\lim_{h \to 0^+} \frac{f(x_0 + h) - f(x_0)}{h} = +\infty,$$

one says that $f'(x_0^+) = +\infty$. It is now clear what is meant by statements like $f'(x_0) = -\infty$, $f'(x_0^+) = -\infty$, $f'(x_0^-) = +\infty$, and $f'(x_0^-) = -\infty$. In all of these cases, f is not differentiable at x_0. If, at x_0, $f' = +\infty$ or $f' = -\infty$, we say that the graph of f has a *vertical tangent* (see Fig. 4.45). If, at x_0, one of two one-sided derivatives is $+\infty$ and the other $-\infty$, the graph of f is said to have a *cusp*, with a vertical tangent (see Fig. 4.46).

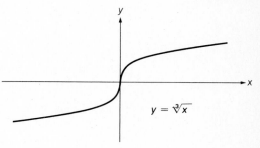

Fig. 4.47

y

$y = \sqrt[3]{x}$

x

▶*Examples* *1.* Let $f(x) = x^{1/3}$ (Fig. 4.47). Then $f'(0) = +\infty$, for

$$\frac{f(h) - f(0)}{h} = \frac{h^{1/3}}{h} = h^{-2/3} = \frac{1}{\sqrt[3]{h^2}}.$$

This is nonnegative for $h \neq 0$ and $\lim_{h \to 0} \sqrt[3]{h^2} = 0$.

Fig. 4.48

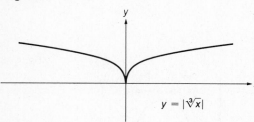

$y = |\sqrt[3]{x}|$

2. Set $f(x) = |x^{1/3}|$. Then, by a similar argument, $f'(0^+) = +\infty$, $f'(0^-) = -\infty$ (cusp). See Fig. 4.48.

3. The tangent to the circle $x^2 + y^2 = 1$ at $(1, 0)$ and also at $(-1, 0)$ is vertical, so that we expect, for $f(x) = \sqrt{1 - x^2}$, that $f'(-1^+) = +\infty$, $f'(1^-) = -\infty$. This is indeed so, for we have (for $h > 0$)

$$\frac{f(-1 + h) - f(-1)}{h} = \frac{\sqrt{1 - (-1 + h)^2}}{\sqrt{h}} = \frac{\sqrt{2h - h^2}}{h} = \frac{\sqrt{h(2 - h)}}{h}$$

$$= \frac{\sqrt{2 - h}}{\sqrt{h}} \to +\infty \text{ for } h \to 0^+,$$

and (for $h < 0$)

$$\frac{f(1 + h) - f(1)}{h} = \frac{\sqrt{1 - (1 + h)^2}}{h} = \frac{\sqrt{-2h - h^2}}{h}$$

$$= \frac{\sqrt{-h(2 + h)}}{h} = \frac{\sqrt{-h}\sqrt{2 + h}}{-(-h)} = -\frac{\sqrt{2 + h}}{\sqrt{-h}} \to -\infty \text{ as } h \to 0^-. \blacktriangleleft$$

7.3 Limits of derivatives

In computing derivatives (including one-sided derivatives and infinite derivatives), the following rule may be useful.

Theorem 2 ASSUME THAT THE FUNCTION $f(x)$ HAS A DERIVATIVE $f'(x)$ FOR ALL x NEAR A POINT x_0, EXCEPT PERHAPS AT x_0 ITSELF. ASSUME THAT THE FINITE LIMIT

$$\lim_{x \to x_0} f(x)$$

EXISTS, AND THAT THE FINITE OR INFINITE LIMIT

$$\alpha = \lim_{x \to x_0} f'(x)$$

EXISTS. THEN $f'(x_0) = \alpha.$

THE CORRESPONDING STATEMENT INVOLVING ONE-SIDED LIMITS AND ONE-SIDED DERIVATIVES IS ALSO TRUE.

The proof of this result will be given in the next section (§8.3).

▶ *Examples* Consider the function $f(x) = x^{1/3}$. For $x \neq 0$, we have the (finite) derivative $\frac{1}{3}x^{-2/3}$. Therefore

$$\lim_{x \to 0} f'(x) = +\infty.$$

Also, the finite limit $\lim_{x \to 0} (x^{1/3})$ exists, and equals 0. Therefore $f'(0) = +\infty. \blacktriangleleft$

EXERCISES

In Exercises 11 to 17, describe the behavior of the derivative of each function as completely as possible; that is, first determine where the derivative exists and then evaluate it at these points. Next, determine where the derivative is infinite and evaluate it as $+\infty$ or $-\infty$ at these points. Finally, determine any additional points where only a one-sided derivative exists or is infinite and evaluate the appropriate one-sided derivative(s) at these points.

11. $f(x) = x^{2/3}$.
12. $f(t) = \sqrt{t^2 - 4}$.
13. $s \mapsto (s - 1)^{3/7}$.
14. $k(x) = x^{1/3} + x^{2/3}$.
15. $g(u) = (8u - 1)^{2/3}$ for $u \leq 1/8$; $(8u - 1)^2$ for $u > 1/8$.
16. $\phi(y) = (y + 1)^{3/5}$ for $y \leq 0$; $\frac{3}{2}(y - 1)^{2/5} - \frac{1}{2}$ for $y > 0$.
17. $A(t) = \sqrt{t^{1/3} + 1}$.
18. Draw the graph of a function $f(x)$, defined and continuous for $-1 \leq x \leq 2$, which satisfies the conditions $f(-1) = 1$, $f'(-1^+) = -\infty$, $f(1) = 2$, $f'(1^-) = -1$, and $f'(1^+) = +\infty$.
19. Draw the graph of a function $g(t)$, defined and continuous for $-3 \leq t \leq 9/8$, which satisfies the conditions $g'(-2^-) = -\frac{1}{2}$, $g'(-2^+) = 1$, $g(0) = 0$, and $g'(1) = +\infty$.
20. Draw the graph of a function $h(u)$, defined and continuous for $|u| \leq 1$, which satisfies the conditions $g'(-1^+) = +\infty$, $g'(-\frac{1}{2}) = -\infty$, $g'(0^-) = -1$, and $g'(0^+) = +1$.

§8 Proof of Some Theorems about Derivatives

This section contains proofs of several theorems stated above without proof.

8.1 Functions with a positive derivative

We prove first the basic Theorem 2 of §4.2: *a function with a positive derivative in an interval is increasing; a function with a negative derivative is decreasing.* We shall prove only the first statement (the first row in the table), leaving the proof of the second statement to the reader.

The proof depends on Theorem 1 of §4.1 [if $f'(x_0) > 0$, then $f(x_0) < f(x)$ for x close to, and to the right of, x_0; and $f(x) < f(x_0)$ for x close to, and to the left of, x_0] and on the least upper bound principle.

We are given that $f(x)$ is continuous for $a \leq x \leq b$ and $f'(x) > 0$ for $a < x < b$. We show first that the function f is increasing in the open interval. Let x_1 and x_2 be points such that $a < x_1 < x_2 < b$. We shall prove that $f(x_1) < f(x_2)$.

Let S be the set of all numbers x such that $x_1 \leq x < x_2$ and $f(x) \geq f(x_2)$. We must show that S is empty.

Assume that S is not empty. We shall show that this leads to a contradiction. The set S is bounded from above, since x_2 is an upper bound. Hence S has a least upper bound; call it y. Clearly, $x_1 \leq y \leq x_2$.

Since $f'(x_2) > 0$, we have (by Theorem 1 of §4.1) that $f(x) < f(x_2)$ for x near, and to the left of, x_2. Hence there is a number $x_3 < x_2$ such that $f(x) < f(x_2)$ for all x such that $x_3 < x < x_2$. This x_3 is an upper bound for S. Therefore $y \leq x_3 < x_2$.

Assume that $f(y) \geq f(x_2)$. Since $f'(y) > 0$, we have, by Theorem 1, that there are points to the right of y where $f > f(y) \geq f(x_2)$. These points belong to S. Hence y is not an upper bound for S. This is a contradiction.

Thus $f(y) < f(x_2)$. Hence $x_1 < y$; for, if $y = x_1$, we should have $f(x_1) < f(x_2)$ and $f(x) < f(x_2)$ for all x with $x_1 < x < x_2$, and S would be empty.

But $f'(y) > 0$. Hence, by Theorem 1, $f < f(y) < f(x_2)$ near, and to the left of, y. Thus there is a number y_1 such that $x_1 < y_1 < y$ and $f(x) < f(x_2)$ for all x with $y_1 < x \leq y$. Since also $f < f(x_2)$ to the right of y, y_1 is an upper bound for S. Hence y is not the least upper bound. This is a contradiction.

We conclude that S is empty. In particular, x_1 is not in S, so that $f(x_1) < f(x_2)$, as asserted.

By hypothesis, f is continuous at a and at b, and we still must show that, for any fixed x_0 with $a < x_0 < b$, we have $f(a) < f(x_0) < f(b)$.

Assume that $f(a) > f(x_0)$. Then, by the definition of continuity, $f(x) > f(x_0)$ for x near a. Hence there is an x_1 with $a < x_1 < x_0$ and $f(x_1) > f(x_0)$. This is impossible by what was proved above.

Thus $f(a) \leq f(x_0)$. By the same argument $f(a) \leq f(x)$ for $a < x < x_0$. Assume that $f(a) = f(x_0)$. Then, for $a < x < x_0$, we have $f(x_0) = f(a) \leq f(x)$ and, by what was proved before, $f(x_0) > f(x)$. This is a contradiction.

Therefore $f(a) < f(x_0)$. One proves in the same way that $f(x_0) < f(b)$.

Remark. One often proves Theorem 2 from the so-called Mean Value Theorem. We discuss this approach in the Supplement.

EXERCISES

1. Prove the second row of Theorem 1.
2. Complete the proof given above by showing that $f(x_0) < f(b)$.

8.2 Functions with a positive second derivative

We prove next Theorem 8 of §4.9. This theorem contains two statements (the two rows in the table). We prove only the first statement dealing with a function $f(x)$, which has a positive second derivative $f''(x)$ in the interval $a \leq x \leq b$.

We consider two linear functions, $x \mapsto l_1(x)$ and $x \mapsto l_2(x)$; the graph of the first passes through the points $(a, f(a))$ and $(b, f(b))$; the graph of the second is the tangent line through the graph of f at a point $(x_0, f(x_0))$, x_0 being some number with $a < x_0 < b$. We must show that

$$l_2(x) \leq f(x) \leq l_1(x)$$

for $a \leq x \leq b$, with equality holding only for $x = a$ and b in the case of l_1 and only for $x = x_0$ in the case of l_2.

To establish this, set

$$g_1(x) = f(x) - l_1(x), \qquad g_2(x) = f(x) - l_2(x).$$

Since l_1 and l_2 are linear, $l_1''(x) = l_2''(x) = 0$, and

$$g_1'' = g_2'' = f'' > 0.$$

By Theorem 2 of §4.2, we have that $g_1'(x)$ increases since $(g_1')' = g_1'' > 0$. By hypothesis, $g_1(a) = g_1(b) = 0$. Therefore $g_1(x)$ is neither an increasing nor a decreasing function in $[a, b]$. By Theorem 2, the derivative g_1' must change sign. Since g_1' increases and is continuous, it has exactly one root in (a, b), call it x_1. Thus $a < x_1 < b$ and $g_1'(x) < 0$ for $a < x < x_1$, $g_1'(x) > 0$ for $x_1 < x < b$. By Theorem 2, we have that g_1 decreases in $[a, x_1]$, increases in $[x_1, b]$. Therefore $g_1(x) < 0$ for $a < x \leq x_1$ and $x_1 \leq x < b$. Thus $f(x) < l_1(x)$ for $a < x < b$.

Next, g_2' is also an increasing function. Since $g_2'(x_0) = 0$ by hypothesis, $g_2'(x) < 0$ for $a \leq x < x_0$ and $g_2'(x) > 0$ for $x_0 < x < b$. Therefore $g_2(x)$ decreases in $[a, x_0]$, and increases in $[x_0, b]$. But $g_2(x_0) = 0$, by hypothesis. Therefore $g_2(x) > 0$ for $a \leq x < x_0$ and $x_0 < x \leq b$. Thus $l_2(x) < f(x)$ for $a \leq x \leq b$, $x \neq x_0$.

EXERCISE

3. Prove the second row of Theorem 8.

8.3 Limits of derivatives

Finally, we prove Theorem 2 of §7.3, which asserts that, if $\lim_{x \to x_0} f(x)$ exists (as a finite limit) and $\lim_{x \to x_0} f'(x)$ exists (as a finite *or* infinite limit), then $f'(x_0) = \lim_{x \to x_0} f'(x)$.

To simplify writing, we assume that $x_0 = 0$. We also assume that $f(0)$ has been defined so as to make f continuous at 0, and that $f(0) = 0$. The latter assumption involves no loss of generality. If $f(0) \neq 0$, we may consider instead of f the function $f_1(x) = f(x) - f(0)$, which has the same derivative as f.

We consider first the case in which $f'(x)$ has a finite limit at 0. We may assume that this limit is 0. For, if it were $\alpha \neq 0$, we could consider instead of f the function $f_1(x) = f(x) - \alpha x$, whose derivative is $f'(x) - \alpha$; this derivative has the limit 0 at 0.

Thus we know that

$$(1) \qquad\qquad f(0) = 0, \; \lim_{x \to 0} f'(x) = 0$$

and we must show that $f'(0) = 0.$

Let ϵ be a given positive number. It will suffice to show that the difference quotient $f(x)/x$ lies between ϵ and $-\epsilon$ for all $x \neq 0$ and near 0. Now, in view of the second equation in (1), we know that $-\epsilon < f'(x) < \epsilon$ for $x \neq 0$ and near 0. For such x, the functions $\epsilon x + f(x)$ and $\epsilon x - f(x)$ are increasing, since their derivatives, $\epsilon + f'(x)$ and $\epsilon - f'(x)$ are positive. Both functions are 0 at 0. Hence they are positive for $x > 0$ and negative for $x < 0$. For positive x, we divide the two inequalities, $\epsilon x + f(x) > 0$ and $\epsilon x - f(x) > 0$, by x. For $x < 0$, we divide the two opposite inequalities by the positive number $-x$. In both cases, we conclude that $-\epsilon < f(x)/x < \epsilon$, as asserted.

We consider now the case in which the limit of $f'(x)$ at 0 is infinite. It will suffice to treat the case

$$(2) \qquad\qquad\qquad\qquad \lim_{x \to 0} f'(x) = +\infty.$$

We must show that $\qquad\qquad\qquad f'(0) = +\infty.$

Let ϵ be a given positive number. It will suffice to show that $f(x)/x > 1/\epsilon$ for $x \neq 0$ and near 0. Now (2) shows that $f'(x) > 1/\epsilon$ for x near 0 and $\neq 0$. For such x, we have that the function $f(x) - x/\epsilon$ is increasing, since its derivative is $f'(x) - 1/\epsilon > 0$. This function is 0 at 0. Hence $f(x) - x/\epsilon > 0$ for $x > 0$ and $f(x) - x/\epsilon < 0$ for $x < 0$. We conclude that, for $x \neq 0$ and sufficiently close to 0, the inequality $f(x)/x > 1/\epsilon$ holds. This completes the proof.

Theorem 2 of §7.3 applies also to one-sided derivatives; the proof is the same.

EXERCISE

4. Prove the statement of Theorem 2 for the case in which $f'(x)$ has the limit $-\infty$ at $x = 0$.

Problems

1. Let $f(x)$ denote the function that is 1 for rational x and 0 for irrational x. Set $g(x) = x^r f(x)$, where r is a positive rational number. Where is this function continuous? For which values of x does $f'(x)$ exist? What is it?

2. What is wrong with the following "proof" of the theorem for differentiating inverse functions?

   Assume that $f(x)$ is differentiable and has a positive derivative $f'(x)$. Let $g(y)$ be the function inverse to f. Then $g(f(x)) = x$. By the chain rule, $g'(f(x))f'(x) = 1$. Therefore $g'(f(x)) = 1/f'(x)$.

3. Prove the following extension of the rule for differentiating a product. If $f(x)$ and $g(x)$ are defined and continuous near x_0, if the (finite) derivative $f'(x_0)$ exists, and if $g'(x_0) = +\infty$, then the function $\phi(x) = f(x)g(x)$ satisfies: $\phi'(x_0) = +\infty$ if $f(x_0) > 0$ and $\phi'(x_0) = -\infty$ if $f(x_0) < 0$. Also, show on examples that, if $f(x_0) = 0$, then it may happen that $\phi(x)$ has a finite derivative at x_0; and it may happen that $\phi(x)$ has an infinite derivative at x_0.

4. State and prove an extension of the chain rule for the case in which one of the functions being composed has an infinite derivative and the other a finite one.

5. Let r be a positive rational number, and set $f(x) = x^r$. What is the largest integer k such that f has at $x = 0$ a (finite) derivative of order k?

6. Suppose the functions $f(x)$ and $g(x)$ have derivatives of all orders. Write down a formula for $(fg)''$ and for $(fg)'''$. Then guess at the formula for $(fg)^{(4)}$; verify this formula. Continue, if you are ambitious.

7. Define a continuous function that is not piecewise monotone. (Hint: Let the function be defined for $x > 0$, and let it be linear in each of the intervals: $(1, +\infty)$, $(1/2, 1)$, $(1/3, 1/2)$, $(1/4, 1/3)$, and so forth.

8. Suppose that $f(x)$ has a continuous derivative for all values of x and $f(0) = 0$, $|f'(x)| < 1$ for all x. Show that $|f(x)| < |x|$ for all x. Make a sketch to illustrate the result. (Hint: Consider the functions $x - f(x)$ and $f(x) - x$; ask whether they are monotone.)

9. Suppose that $f(x)$ has a continuous second derivative, $f(0) = 0$, $f'(0) = 0$, $|f''(x)| < 1$ for all x. Show that $|f(x)| < \frac{1}{2}x^2$ for all x.

10. Show that the converse to Theorem 3 of §4 is true: if a function is not decreasing and has a derivative, its derivative is nonnegative.

11. Show that, if at $x = 0$ we have $f' = f'' = 0$ and $f''' \neq 0$, then the graph of $f(x)$ has an inflection point at $x = 0$.

12. Show that, if at $x = 0$ we have $f' = f'' = f''' = 0$ and $f^{(4)} = a \neq 0$, then $f(x)$ has at $x = 0$ a local minimum or a local maximum according to whether $a > 0$ or $a < 0$.

13. Generalize the results of Problems 11 and 12 by completing the following statement. If a function $f(x)$ is such that all its derivatives of orders 1, 2, $\cdots$, $k - 1$ are 0 at $x = 0$, and the kth derivative has at $x = 0$ a value $a \neq 0$, then $\cdots$. Try to prove this theorem.

14. Prove that every function whose fourth derivative is the constant 0 is a polynomial of degree at most 3.

15. Let n be a fixed positive integer. For what functions $f(x)$ is it true that $f^{(n)}(x) = 0$ for all x?

16. Suppose two functions have the same second derivative. Prove that their difference is a linear function.

17. Find all functions whose third derivative is x.

18. Suppose that $f(x)$ is continuous for all x and that, in every open interval between two successive integers, the function $f(x)$ is linear. Prove that $f(x)$ has a primitive function and show how to compute it.

19. Prove or disprove the statement: a primitive function of a positive function defined in an interval has at most one root. What if we replace the word "positive" by "nonnegative"? By "negative"?

20. Suppose $f(x)$ and $g(x)$ have primitive functions $F(x)$ and $G(x)$, respectively. Find a primitive function of $f(x)G(x) + F(x)g(x)$.

21. Suppose that $f(x)$ has the primitive function $F(x)$. Find a primitive function for $h(x) = F(x)^k f(x)$, where k is an integer, $k \neq 1$.

22. Suppose that $g(x)$ has the primitive function $G(x)$. Find a primitive function for $h(x) = G(x^k)x^{k-1}$, where k is any integer.

23. Is there a function $f(x)$ such that $\lim_{x \to +\infty} f(x) = +\infty$ and $\lim_{x \to +\infty} f'(x) = 0$? Such that $\lim_{x \to +\infty} f(x) = 0$ and $\lim_{x \to +\infty} f'(x) = +\infty$?

24. What can you say about a function whose second derivative is odd? Even?

5/Integrals

This chapter deals with the second basic concept in calculus—the integral. We define integrals and derive their main properties geometrically; then we show how a purely analytic definition could be given. The fundamental theorem about the connection between integrals and derivatives is explained in §2. In §3, we show how this theorem is used to compute integrals, and in §4 we describe the main geometric applications of integrals.

There are five appendixes: one dealing with physical applications, another with a useful extension of the concept of integral, and the other three containing proofs of theorems accepted before as geometrically obvious. These sections are of interest mainly to a reader who insists on a completely rigorous development.

1.1 The integral of a nonnegative function

Consider a continuous function $y = f(x)$ which takes on no negative values, so that its graph lies above the x axis, though it may touch it at some points (Fig. 5.1). Let a and b be two numbers such that our function is defined for $a \leq x \leq b$. The curve $y = f(x)$ and the lines $x = a$, $x = b$, and $y = 0$ bound a certain region of the plane, the *region under the curve* $y = f(x)$ *from a to b*. This region has a definite area A.

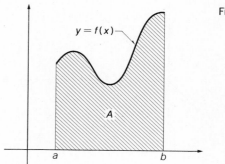

Fig. 5.1

Is this last statement true? Our intuition and our experience tell us it is. If asked to compute the area, we could, for instance, cover the plane with a mesh of small squares and count the number of squares within our region (Fig. 5.2). This would not give us the area precisely, since some squares would lie partly inside and partly outside our region. But we expect that, by making the mesh sufficiently fine, we could compute A with any desired degree of accuracy. That there is indeed a number A which can be so computed is a mathematical statement that requires a proof, and we shall discuss this matter later. For the moment, we take the concept of area for granted.

The number A depends on the function f and on the numbers a and b. Following Leibniz, we write

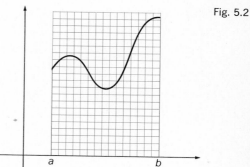

Fig. 5.2

$$A = \int_a^b f(x)\,dx.$$

The right-hand side is read "the **integral** from a to b of the function $f(x)$ with respect to x" or "the integral from a to b of $f(x)\,dx$." The symbol $\int$ is the **integral sign**, the function f the **integrand**, and the interval (a, b) the **interval of integration**. The numbers a and b are usually called the lower and upper limits of integration. (The "limits of integration" have nothing to do with the term "limit" as defined in Chapter 3.)

The Leibniz notation can be justified by another mathematical myth (see Chapter 4, §1.7). The region under the curve is composed, of infinitely many infinitely thin rectangles (Fig. 5.3). Each rectangle is erected over a point x in the interval $[a, b]$; it has height $f(x)$ and an infinitely small width dx; its area is therefore $f(x)\,dx$. The total area A is the sum of all these areas. The symbol $\int$ means "sum"; it is, in fact, an elongated letter S.

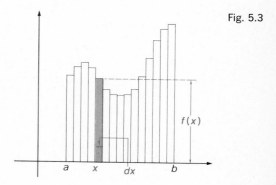

Fig. 5.3

This myth is not literally true, but it contains a kernel of truth. We shall see later what it is. Meanwhile, we use the Leibniz notation, because it is both traditional and convenient.

Fig. 5.4

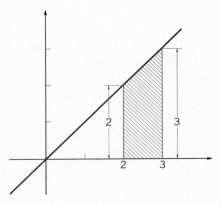

Here are some examples. The symbol

$$\int_2^3 x \, dx$$

represents the area indicated on Fig. 5.4. The region in question is a trapezoid, so that we know how to compute its area. Thus

$$\int_2^3 x \, dx = \frac{5}{2}.$$

The Archimedes result about the area under the parabola, which we derived in an appendix to Chapter 2, §5, and again in Chapter 4, §6, can be written as

$$\int_0^t x^2 \, dx = \frac{t^3}{3}.$$

The variable x occurring in these formulas (the **variable of integration**) is a so-called **dummy variable**. It serves only to identify the form of the function which we "integrate," that is, for which we compute the area. The dummy variable may be replaced by any other symbol. For instance,

$$\int_0^1 y^2 \, dy = \frac{1}{3} \qquad \text{and} \qquad \int_0^1 \alpha^2 \, d\alpha = \frac{1}{3}.$$

1.2 The integral of a function that takes on negative values

We now extend the definition of the integral to functions that take on negative values. *If $a < b$, the number*

$$\int_a^b f(x) \, dx$$

is the total area of the region bounded by the curve $y = f(x)$, the x axis, and the lines $x = a$ and $x = b$, the areas lying above the x axis being counted as positive and those below the axis as negative.

Thus, if f is given by the graph in Fig. 5.5, the number $\int_a^b f(x) \, dx$ equals the sum of the areas I and III minus the sum of the areas II and IV.

As an example we compute $\int_{-2}^1 2x \, dx$ (see Fig. 5.6). The triangle under the x axis has area 4; the one above the axis has area 1. Hence

$$\int_{-2}^1 2x \, dx = -3.$$

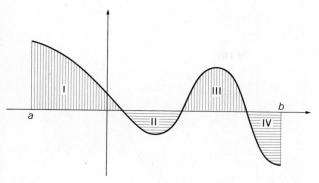

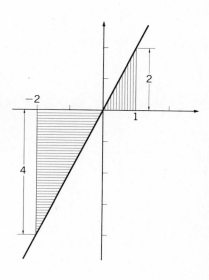

Fig. 5.6

Fig. 5.5

EXERCISES

In Exercises 1 to 10, evaluate the given integrals. Sketch the graph of the integrand and shade the relevant area. (Hint: In Exercises 3, 4, 6, and 10, use the result about the area under the parabola quoted in §1.1.)

1. $\int_1^4 3z\, dz.$

2. $\int_{-1}^{1/2} (1 - s)\, ds.$

3. $\int_0^{2/3} t^2\, dt.$

▶ 4. $\int_0^1 (1 - y^2)\, dy.$

5. $\int_{-1}^1 |u|\, du.$

6. $\int_{-a}^a z^2\, d_3$ where $a = 2^{-2/3}.$

7. $\int_{-2}^1 h(t)\, dt$ if $h(t) = 1$ for $t \leq 0$; $t + 1$ for $t > 0.$

8. $\int_1^4 \phi(r)\, dr$ if $\phi(r) = 2 - r$ for $r \leq 2$; $2r - 4$ for $2 < r \leq 3$; 2 for $r > 3.$

9. $\int_{-1}^3 f(v)\, dv$ if $f(v) = -2$ for $v \leq 0$; $v - 2$ for $0 < v \leq 2$; $2v - 4$ for $v > 2.$

10. $\int_{-4}^4 g(y)\, dy$ if $g(y) = 1$ for $y \leq -1$; y^2 for $-1 < y < 1$; $3 - 2y$ for $y \geq 1.$

Fig. 5.7

1.3 Three basic properties

A special but important case is that of a **constant** function, $f(x) = \alpha$, for all x in the interval considered. Clearly,

$$(1) \qquad \int_a^b \alpha\, dx = (b - a)\alpha,$$

for the rectangle width $b - a$, and height $|\alpha|$ has area $|\alpha|(b - a)$. If $\alpha = 0$, there is no actual rectangle and zero is the only value that makes sense for the area.

The formula is illustrated in Fig. 5.7.

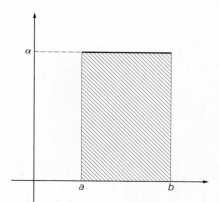

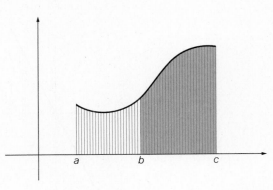

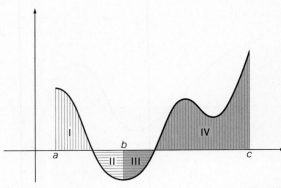

Fig. 5.8   Fig. 5.9

Fig. 5.10

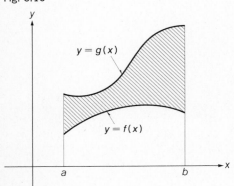

Fig. 5.11

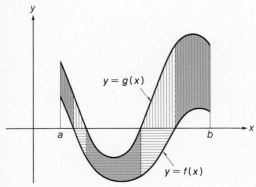

A basic and intuitively obvious property of areas states: if $a < b < c$, then

$$(2) \qquad \int_a^b f(x)\,dx + \int_b^c f(x)\,dx = \int_a^c f(x)\,dx.$$

This formula is illustrated in Fig. 5.8: the total shaded area is the sum of the two areas shaded in black and white and in color. We call the property just derived the **additivity** of the integral, with respect to the interval of integration.

Figure 5.8 shows a positive function, but by considering the functions shown in Fig. 5.9 the reader will convince himself that the argument is general. In this figure, $\int_a^b f(x)\,dx = \mathrm{I} - \mathrm{II}$, $\int_b^c f(x)\,dx = -\mathrm{III} + \mathrm{IV}$, $\int_a^c f(x)\,dx = \mathrm{I} + \mathrm{IV} - (\mathrm{II} + \mathrm{III})$, so that (2) holds.

Another intuitively obvious property of the area is expressed in the following rule, called the **monotonicity** of the integral:

(3) if $a < b$ and $f(x) \le g(x)$ for $a < x < b$, then $\int_a^b f(x)\,dx \le \int_a^b g(x)\,dx$.

In words: "the larger function has the larger integral (over the same interval, from left to right)." This statement is illustrated in Fig. 5.10. The area under f is contained in that under g. A glance at Fig. 5.11 will convince the reader that our statement also holds for functions that take on negative values.

1.4 Piecewise continuous, bounded functions

The additivity property (2) suggests a way of defining integrals of certain functions with discontinuities.

A function $x \mapsto f(x)$ defined in an interval $a < x < b$ is called

bounded in that interval if there is a number M such that $|f(x)| \leq M$ for all x with $a < x < b$. If $f(x)$ is also continuous, we expect to be able to define unambiguously the area $\int_a^b f(x)\,dx$.

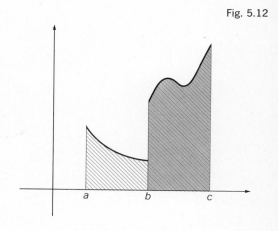

Fig. 5.12

Next, if $f(x)$ is bounded in a finite interval (a, c), and if there is a number b between a and c such that $f(x)$ is continuous for $a < x < b$ and also for $b < x < c$, then we can define $\int_a^c f(x)\,dx$, by additivity, as the sum $\int_a^b f(x)\,dx + \int_b^c f(x)\,dx$ (see Fig. 5.12). The same method works, of course, if the interval (a, c) can be divided, not into two, but into several intervals in each of which the function $f(x)$ is bounded and continuous.

A function f that is continuous in an interval, except perhaps at finitely many points, is called **piecewise continuous** in that interval. (A function defined in an infinite interval is called piecewise continuous if it is so in every finite subinterval.) Most functions encountered in calculus are piecewise continuous. Needless to say, a continuous function is also considered to be piecewise continuous.

We conclude that the number $\int_a^b f(x)\,dx$ ought to be well defined whenever (a, b) is a finite interval and the function $f(x)$ is bounded and piecewise continuous in (a, b).

We shall show that the properties (1), (2), and (3) of integrals permit us to compute approximately the value of an integral of a given function. First we consider a simple special case.

1.5 Step functions

A function $g(x)$ defined for $a < x < b$ is called a **step function** if the interval (a, b) can be divided into finitely many subintervals in each of which $g(x)$ is constant. For example,

Fig. 5.13

$$g(x) = \begin{cases} 6 \text{ for } -2 < x \leq -1.5 \\ 2 \text{ for } -1.5 < x \leq 0 \\ 3 \text{ for } 0 < x \leq 1 \\ 4 \text{ for } 1 < x \leq 2.5 \\ -2 \text{ for } 2.5 < x \leq 4 \end{cases}$$

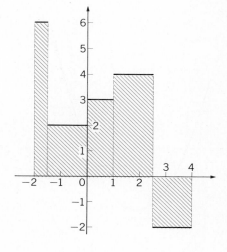

is a step function defined in the interval $(-2, 4)$. Its subintervals of constancy are $(-2, -1.5)$, $(-1.5, 0)$, $(0, 1)$, $(1, 2.5)$, and $(2.5, 4)$; the values g takes on at the endpoints of the subintervals are of no interest. The graph of the function g is shown in Fig. 5.13; its shape explains the name "step function."

A step function is, of course, bounded and piecewise continuous, and its integral over every finite interval can be computed at once, either using additivity and the formula for the integral of a constant or, even simpler, using the definition of the integral as an area. For the example given above, for instance,

$$\int_{-2}^{4} g(x)\,dx = \int_{-2}^{-1.5} 6\,dx + \int_{-1.5}^{0} 2\,dx + \int_{0}^{1} 3\,dx + \int_{1}^{2.5} 4\,dx + \int_{2.5}^{4}(-2)\,dx$$

$$= 6 \cdot \frac{1}{2} + 2 \cdot \frac{3}{2} + 3 \cdot 1 + 4 \cdot \frac{3}{2} + (-2) \cdot \frac{3}{2} = 12,$$

as can also be seen in Fig. 5.13.

1.6 Computing the integral

Fig. 5.14

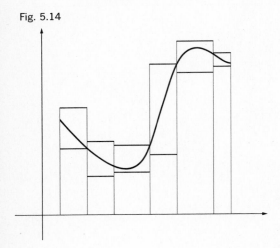

Now let $f(x)$ be any bounded piecewise continuous function defined in a finite interval (a, b), and let $\phi(x)$ and $\psi(x)$ be two step functions such that

$$\phi(x) \leq f(x) \leq \psi(x)$$

for $a < x < b$. Then, by the monotonicity property (3)

$$\int_{a}^{b} \phi(x)\,dx \leq \int_{a}^{b} f(x)\,dx \leq \int_{a}^{b} \psi(x)\,dx.$$

The left and right integrals are computable, and knowing them we also know $\int_{a}^{b} f(x)\,dx$ with an error not exceeding $\int_{a}^{b} \psi(x)\,dx - \int_{a}^{b} \phi(x)\,dx$. The geometric meaning of the statement just made is shown in Fig. 5.14; the area under the curve is greater than the sum of the areas of the "short" rectangles and less than that of the "tall" ones. Figure 5.15 illustrates the situation for a function that takes on both positive and negative values.

Fig. 5.15

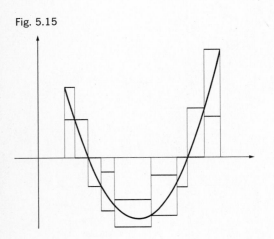

When f is given, it is easy to find step functions ϕ and ψ such that $\phi \leq f \leq \psi$. We simply divide the interval (a, b) into subintervals and in each of the subintervals we find values for ϕ and ψ satisfying our condition. It turns out that in this way *the value of $\int_{a}^{b} f(x)\,dx$ can be computed with any desired degree of accuracy.*

1.7 Error estimate for a monotone function

Let us verify the last statement for an *increasing* function $f(x)$. We divide the interval (a, b) into N equal subintervals each of length $(b - a)/N$. In each of these subintervals, we choose the largest possible value for the step function ϕ, namely, the value of f at the left endpoint of the subinterval. Similarly, we choose for ψ the smallest possible value in each subinterval, the values of f at the right endpoint (see Fig. 5.16). Now, what is the difference $\int_a^b \psi(x)\,dx - \int_a^b \phi(x)\,dx$? Clearly, it is the sum of the N shaded areas in Fig. 5.17. The sum equals $(b-a)/N$ times the sum of the heights, which is $f(b) - f(a)$. Thus we know $\int_a^b f(x)\,dx$ with an error not exceeding

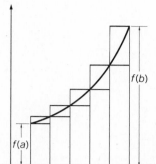

Fig. 5.16

$$\int_a^b \psi(x)\,dx - \int_a^b \phi(x)\,dx = \frac{1}{N}(b-a)[f(b) - f(a)].$$

This number can be made arbitrarily small by choosing N large enough.

▶ **Example** Compute

$$\int_0^{.5}(1 + x^2)\,dx$$

Fig. 5.17

with an error not exceeding $(.03)$.

We use the method just outlined. Since $b - a = (.5) - 0 = .5$ and $f(b) - f(a) = 1.25 - 1 = .25$, we have $(b-a)[f(b) - f(a)] = .125$, so that we need an N with $.125/N < .03$. We choose $N = 5$, so that $.125/N = .025 < .03$. From the following table,

x	0	.1	.2	.3	.4	.5
$f(x)$	1	1.01	1.04	1.09	1.16	1.25

we see that the values of ϕ in the 5 subintervals are 1, 1.01, 1.04, 1.09, and 1.16, while the values of ψ are 1.01, 1.04, 1.09, 1.16, and 1.25. The length of each subinterval is $.5/5 = .1$. Therefore

$$\int_0^{.5}\phi\,dx = (.1)(1 + 1.01 + 1.04 + 1.09 + 1.16) = .530$$

and

$$\int_0^{.5}\psi\,dx = (.1)(1.01 + 1.04 + 1.09 + 1.16 + 1.25) = .555.$$

Thus

$$.530 \le \int_0^{.5}(1 + x^2)\,dx \le .555.$$

In the following section, we shall learn how to compute this particular integral precisely; we shall find that

$$\int_0^{.5}(1 + x^2)\,dx = .541\overline{6}. ◀$$

We have shown above how to compute the integral of an increasing function to any preassigned degree of accuracy. We remark that the

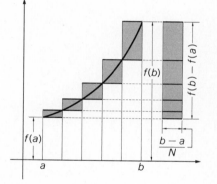

same proof works for a nondecreasing function. Also, it was only for the sake of convenience that we divided the interval (a, b) into N *equal* subintervals; what matters is only that the subintervals be small enough. The same argument goes through for a nonincreasing function. In view of the additivity property (2), we conclude that the integral of any **piecewise monotone** function is computable, using step functions, to any prescribed degree of accuracy. This takes care of all functions commonly encountered in applications of calculus.

In an appendix to this chapter (§7), we shall show that the method of computing integrals by "trapping" a function between two step functions works equally well for functions that are not piecewise monotone.

EXERCISES

11. Find $\int_1^2 g(t)\, dt$ approximately if g is the function whose graph is given below.

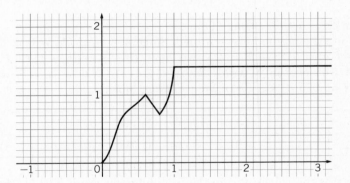

Find a bound for the error.

12. Find $\int_{-1}^3 f(y)\, dy$ approximately if f is the function whose graph is given below.

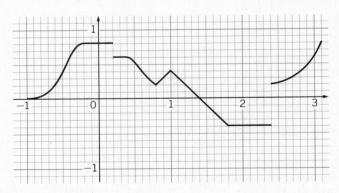

Find a bound for the error.

13. Find $\int_{1/2}^{5/2} h(v)\, dv$ approximately if h is the function whose graph is given below.

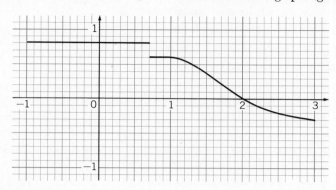

    Find a bound for the error.

14. Find $\int_{-1/2} \phi(t)\, dt$ approximately if ϕ is the function whose graph is given below.

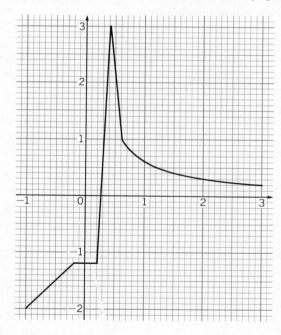

    Find a bound for the error.

In Exercises 15 to 25, compute the given integrals to within the desired degree of accuracy by using step functions. Wherever possible, compute also the exact value.

15. $\int_0^1 (x^2 + 1)\, dx$, with an error not exceeding .1.

16. $\int_{.1}^{.2} t^2 \, dt$, with an error not exceeding .002.

17. $\int_0^{.3} (y^2 + y) \, dy$, with an error not exceeding .03.

18. $\int_{.2}^{.3} (1 - y^2) \, dy$, with an error not exceeding .002.

19. $\int_{-1}^{0} (t^2 + 2) \, dt$, with an error not exceeding .05.

▶ 20. $\int_1^2 dx/x$, with an error not exceeding 1/4.

21. $\int_0^{.36} \sqrt{v} \, dv$, with an error not exceeding .15.

22. $\int_{1/2}^{3/2} [2x/(2x + 1)] \, dx$, with an error not exceeding 1/10.

23. $\int_0^2 f(x) \, dx$, with an error not exceeding 2.5, if $f(x) = -x$ for $0 < x \le 1$ and $f(x) = 1 + x^2$ for $x > 1$.

24. $\int_{-1}^{1} f(t) \, dt$, with an error not exceeding 2.5 if $f(t) = t^3$ for $t \le -9$ and $f(t) = (t + 10)^2$ for $g < t$.

25. $\int_{-1}^{3/2} f(z) \, dz$, with an error not exceeding 1 if $f(z) = 1/(z + 2)$ for $z \le 1$ and $f(z) = z^3$ for $z > 1$.

1.8 Analytic definition of integrals

Thus far, we have proceeded as if we knew all about areas, and we have used this knowledge to define integrals. Now we reverse our point of view. We shall give an analytic definition of the integral, that is, a definition based only on properties of numbers.

It is very easy to give such a definition for step functions: if a step function $\phi(x)$ defined in a finite interval (a, b) has k intervals of constancy, of length $l_1, l_2, \cdots, l_k$, and if ϕ takes on in these intervals the values $\alpha_1, \alpha_2, \cdots, \alpha_k$, then

$$\int_a^b \phi(x) \, dx = \alpha_1 l_1 + \alpha_2 l_2 + \cdots \alpha_k l_k.$$

We use step functions for defining the integrals of other functions. Our previous discussion suggests that the following statement is true.

Theorem 1 (Existence of Integrals)   LET $f(x)$ BE A BOUNDED, PIECEWISE CONTINUOUS FUNCTION DEFINED IN THE FINITE INTERVAL (a, b). THEN THERE IS A NUMBER A, AND ONLY ONE SUCH NUMBER, WITH THE PROPERTY: IF $\phi(x)$ AND $\psi(x)$ ARE STEP FUNCTIONS AND $\phi(x) \le f(x) \le \psi(x)$ FOR $a < x < b$, THEN

$$\int_a^b \phi(x) \, dx \le A \le \int_a^b \psi(x) \, dx.$$

Definition   This number A is denoted by the symbol

$$\int_a^b f(x) \, dx$$

*and is called the **integral** of $f(x)$ over (a, b).*

The proof of Theorem 1 will be given in the Appendix (§7). The reader is invited to accept this theorem without proof for the time being. The same applies to the next statement.

Theorem 2 (Basic Properties of the Integral) LET (a, b) BE A FINITE INTERVAL. THEN

(1) $\quad \int_a^b \alpha \, dx = \alpha(b - a) \quad$ FOR EVERY CONSTANT α,

(2) $\quad \int_a^c f(x) \, dx + \int_c^b f(x) \, dx = \int_a^b f(x) \, dx \quad$ FOR ANY c WITH $a < c < b$,

(3) $\quad \int_a^b f(x) \, dx \leq \int_a^b g(x) \, dx \quad$ IF $f(x) \leq g(x)$ FOR $a < x < b$,

WHERE f AND g ARE ANY TWO BOUNDED PIECEWISE, CONTINUOUS FUNCTIONS.

This theorem merely summarizes the geometrically obvious properties of areas that we already noted. An analytic proof will be found in the Appendix (§7).

1.9 Extending the notations

It is convenient to extend our notations. The number $\int_a^b f(x) \, dx$ has been defined, thus far, only for $a < b$. We want to define it also for $a = b$, so that the additivity rule (2) can remain valid. Therefore we want this to be true:

$$\int_a^a f(x) \, dx + \int_a^c f(x) \, dx = \int_a^c f(x) \, dx.$$

This can be achieved only by defining

(4) $$\int_a^a f(x) \, dx = 0.$$

We also want to define $\int_a^b f(x) \, dx$ for $b < a$. Again, we want (2) to remain valid. Hence we must have

$$\int_a^b f(x) \, dx + \int_b^a f(x) \, dx = \int_a^a f(x) \, dx = 0.$$

so that we must define

(5) $$\int_a^b f(x) \, dx = -\int_b^a f(x) \, dx.$$

From now on, we may use (2) for any three numbers a, b, and c. (The reader should verify this, say, by considering some examples.)

EXERCISES

Compute the following integrals:

26. $\int_0^{-1} 2\,dx.$

27. $\int_{-1}^{-3} x\,dx.$

28. $\int_5^5 (x^{100} - x)\,dx.$

29. $\int_4^2 (-4)\,dx.$

30. $\int_2^{\frac{1}{2}} x^2\,dx.$

31. $\int_{-2}^2 x^4\,dx + \int_2^{-2} x^4\,dx.$

32. Find a number a such that $\int_{2a}^{a-1} 2\,dx = 5.$

33. Find a number b such that $\int_b^{b-2} x\,dx = 6.$

34. Let $f(x)$ be bounded continuous and nonnegative for $a < x < b$. Show that $\int_a^b f(x)\,dx > 0.$

35. Let $f(x)$ be as in Exercise 34 and let c be a number such that $a < c < b$. Show that $\int_a^b f(x)\,dx \geq \int_a^c f(x)\,dx.$

36. Show that $2 < \int_{-1}^1 (1 + x^{22})\,dx < 4.$

§2 The Fundamental Theorem of Calculus

2.1 First part of the fundamental theorem

Let $f(t)$ be a function defined for $a < t < b$, and assumed to be bounded and piecewise continuous. Then, for every number x between a and b, we can compute the number $G(x) = \int_a^x f(t)\,dt$, the area under the curve $t \mapsto f(t)$ from a to x. Thus we obtain a new function, $x \mapsto G(x)$, defined for $a \leq x \leq b$; clearly $G(a) = 0$. We ask: does this new function have a derivative $G'(x)$, and if so, what is this derivative?

We begin by considering some examples. Let $f(t) = (t/2) + 1$ and $a = 0$; see Fig. 5.18. We read off this figure that $G(x) = \int_0^x [(t/2) + 1]\,dt = (x^2/4) + x$; the graph of $G(x)$ is shown in Fig. 5.19. Differentiating $G(x)$, we obtain $G'(x) = (x/2) + 1$ or $G'(x) = f(x)$. We are back to the function with which we started.

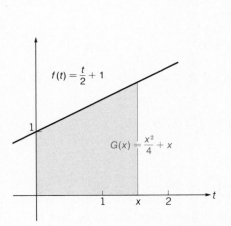

Fig. 5.18

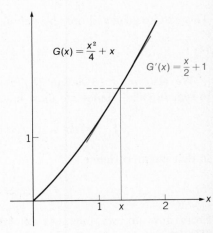

Fig. 5.19

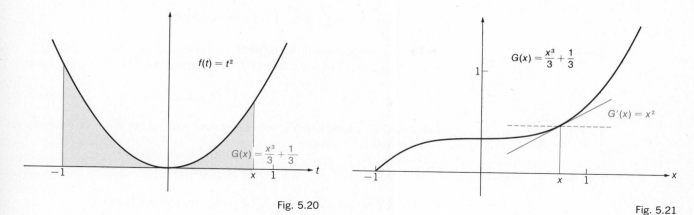

Fig. 5.20

Fig. 5.21

Next, let $f(t) = t^2$ and $a = -1$. Now $G(x) = \int_{-1}^{x} t^2 \, dt = (x^3/3) + 1/3$. Let us verify this for $x > 0$. We have, by the additivity property (2), $G(x) = \int_{-1}^{0} t^2 \, dt + \int_{0}^{x} t^2 \, dt$; this is the sum of the two areas under the parabola shown in Fig. 5.20. By the result in Chapter 4, §5.8, we know that the first area is $1/3$ and the second is $x^3/3$. The graph of G is shown in Fig. 5.21. We differentiate $G(x)$ and obtain $G'(x) = x^2$. Again $G'(x) = f(x)$.

Before stating the general theorem, we consider one more example. Let $f(t) = 2t$ for $0 \leq t \leq 1$ and $f(t) = 1$ for $t > 1$; the graph of f is shown in Fig. 5.22. Set $G(x) = \int_{0}^{x} f(t) \, dt =$ the area under f from 0 to x. From the graph, we see that $G(x) = x^2$ for $0 \leq x \leq 1$ and $G(x) = 1 + (x - 1) = x$ for $x > 1$. The function G is graphed in Fig. 5.23. One sees easily that, if $0 \leq x < 1$, then $G'(x) = 2x = f(x)$, and if $1 < x$, then $G'(x) = 1 = f(x)$. At $x = 1$, the function G has no derivative. The function f has a discontinuity at this point, and the graph of G has a corner.

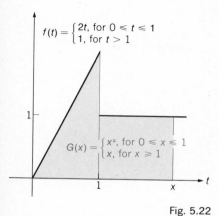

Fig. 5.22

Fig. 5.23

The examples suggest the following statement.

Theorem 1 (Fundamental Theorem of Calculus, First Part) LET $f(t)$, $a < t < b$, BE A BOUNDED PIECEWISE, CONTINUOUS FUNCTION, AND SET

$$G(x) = \int_a^x f(t)\, dt.$$

THEN $G(x)$ IS A CONTINUOUS FUNCTION, AND AT ALL POINTS x AT WHICH f IS CONTINUOUS, WE HAVE

$$G'(x) = f(x).$$

This last equality is the essential point. It can be written as

$$\frac{d}{dx} \int_a^x f(t)\, dt = f(x) \qquad \text{if } f \text{ is continuous at } x.$$

(If $x = a$ or $x = b$, the derivative may be a one-sided derivative.)

The reasoning behind the fundamental theorem is the same as that used in Chapter 4, §6, in computing the area under the parabola.

We assume that $f > 0$, for the sake of simplicity. Let x_0 be a point in the interval considered and h a small positive number. Then $G(x_0 + h)$ is the area under the curve from a to $x_0 + h$, $G(x_0)$ is the area from a to x_0

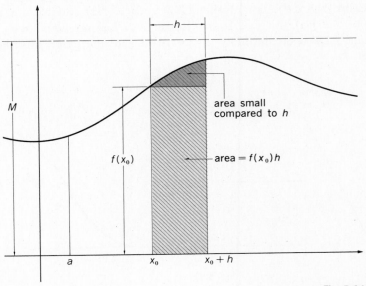

Fig. 5.24

and the difference $G(x_0 + h) - G(x_0)$ is the whole shaded area in Fig. 5.24. It is not greater than Mh if $|f(t)| < M$ everywhere. Hence $|G(x_0 + h) - G(x_0)|$ is small when h is small. A similar argument holds for negative h. Thus G is continuous at x_0. This holds even if f is not continuous at x_0.

Next, if f is continuous at x_0, then $G(x_0 + h) - G(x)$ is approximately the area of the shaded rectangle, that is, $hf(x_0)$. Thus

$$\frac{G(x_0 + h) - G(x_0)}{h} = f(x_0)$$

plus something as small as we like, if h is small enough. A similar argument holds for negative h. Hence $G'(x_0) = f(x_0)$.

The formal proof given in the appendix to this chapter (§7) is an elaboration and verification of this intuitive argument.

The discovery of calculus is usually ascribed to Newton and Leibniz, and there was an ugly and senseless priority dispute among their followers. But it is wrong to assign the sole credit for such a momentous achievement to two men, even to giants like Leibniz and Newton. The essential content of the crucial fundamental theorem of calculus was stated for the first time by Newton's teacher, Barrow.

ISAAC BARROW (1630–1677) was a theologian and a mathematician. In 1669, he resigned his mathematics chair at Cambridge in favor of his onetime pupil Newton.

EXERCISES

In Exercises 1 to 5, verify that Theorem 1 is satisfied for the given functions.

1. $G(x) = \int_{-2}^{x} h(t)\, dt$, where $h(t) = 1$ for $t \leq 0$; $t + 1$ for $t > 0$.
2. $G(x) = \int_{2}^{x} \phi(r)\, dr$, where $\phi(r) = 2 - r$ for $r \leq 2$; $2r$ for $r > 2$.
3. $H(y) = \int_{-1}^{y} f(v)\, dv$, where $f(v) = -2$ for $v \leq 0$; v for $0 < v \leq 2$; $2v$ for $v > 2$.
4. $K(v) = \int_{-4}^{v} g(u)\, du$, where $g(u) = 4$ for $u \leq -1$; 2 for $-1 < u \leq 0$; u^2 for $u > 0$.
5. $F(z) = \int_{1}^{z} h(t)\, dt$, where $h(t) = 1 - 2t$ for $t \leq 2$; 1 for $2 < t \leq 3$; $3t - 8$ for $t > 3$.

In Exercises 6 to 8. let $f(t)$ be a bounded, piecewise continuous function defined for $a < t < b$. Assume Theorem 1.

6. Let $H(x) = \int_{x}^{b} f(t)\, dt$. Show that H is continuous for $a \leq x \leq b$, and, at all points x in $[a, b]$ at which f is continuous, $H'(x) = -f(x)$. [Hint: Use the additivity property of the integral, (2) of §1.]
7. Let c be any point in $[a, b]$. Let $K(x) = \int_{c}^{x} f(t)\, dt$. Show that K is continuous for $a \leq x \leq b$ and, at all points in x in $[a, b]$ at which f is continuous, $K'(x) = f(x)$. [Hint: Use Exercise 6 and equation (5) in §1.]
8. Let $\beta(x)$, $l \leq x \leq m$, be a differentiable function whose values lie in $[a, b]$. Let $L(x) = \int_{c}^{\beta(x)} f(t)\, dt$. Show that L is continuous for $l \leq x \leq m$ and, at all points x in $[l, m]$ such that f is continuous at $\beta(x)$, $L'(x) = \beta'(x)f(\beta(x))$. [Hint: Write $L(x)$ as a composite function involving $K(x)$ and use Exercise 7.]

2.2 Second part of the fundamental theorem

There is another part of the fundamental theorem, and it is this second part that we shall constantly use.

Theorem 2 (Fundamental Theorem of Calculus, Second Part) LET $F(x)$ BE A CONTINUOUS FUNCTION DEFINED FOR $a \leq x \leq b$. LET $f(x)$ BE A BOUNDED PIECEWISE, CONTINUOUS FUNCTION AND ASSUME THAT

$$F'(x) = f(x)$$

EXCEPT PERHAPS AT FINITELY MANY POINTS. THEN

$$\int_a^b f(t)\, dt = F(b) - F(a).$$

Proof. We assume first that f is continuous and $F'(x) = f(x)$ for all x. Set

$$G(x) = \int_a^x f(t)\, dt;$$

we must show that $G(b) = F(b) - F(a)$. Set $H(x) = G(x) - F(x) + F(a)$. We must show that $H(b) = 0$. We shall show that $H(x) \equiv 0$. Indeed, $G'(x) = f(x)$ by Theorem 1, so that $H'(x) = f(x) - f(x) = 0$ at all points x. By the basic Theorem 1 of Chapter 4, §5.2, we conclude that $H(x)$ is a constant. But $H(a) = G(a) - F(a) + F(a) = \int^a f(t)\, dt = 0$. Hence $H(x) \equiv 0$.

The same argument holds if f is only piecewise continuous and $F' = f$ does not hold at finitely many points. Then the interval $[a, b]$ can be divided into finitely many intervals in each of which $H'(x) \equiv 0$ and hence H is constant. But since H is known to be continuous over $[a, b]$, it is constant over the whole interval. This completes the proof.

For any function $F(x)$, it is customary to write

$$F(x)\Big|_a^b = F(b) - F(a).$$

With this notation, the second part of the fundamental theorem may be written thus:

$$\int_a^b F'(x)\, dx = F(x)\Big|_a^b,$$

provided $F'(x)$ exists, is bounded, and is piecewise continuous. This formula permits us to compute the area under a curve $y = f(x)$ if we know a primitive function for f.

▶ *Examples* *1.* We return to the example treated in §1. A primitive function for $1 + x^2$ is $x + (x^3/3)$. Therefore

$$\int_0^{1/2}(1 + x^2)\, dx = \left(x + \frac{x^3}{3}\right)\Big|_0^{1/2} = \frac{1}{2} + \frac{1}{24} = \frac{13}{24} = .541\overline{6}.$$

2. As another application, let us solve the problem considered by Cavalieri (see Chapter 4, §6), that is, let us compute the area under the curve $y = x^n$ (where n is some positive integer) from 0 to $t > 0$. We have that x^n is the derivative of $x^{n+1}/(n + 1)$. Therefore

$$\int_0^t x^n\, dx = \frac{x^{n+1}}{n + 1}\Big|_0^t = \frac{t^{n+1}}{n + 1}.$$

Note the ease with which we obtained the result for all n.

3. The formula just obtained is a special case of a general rule:

$$\int_a^b(a_0 + a_1 x + \cdots + a_n x^n)\, dx = \left(a_0 x + \frac{a_1 x^2}{2} + \cdots + \frac{a_n x^{n+1}}{n + 1}\right)\Big|_a^b.$$

This relation follows at once from the fundamental theorem and the differentiation rules. It enables us to compute the area under the graph of any polynomial. ◀

EXERCISES

In Exercises 9 to 23, compute the integrals by using the fundamental theorem.

9. $\int_{-2}^3 (1 - y^2)\, dy$. Compare with Exercise 18 in §1.
10. $\int_{-1}^0 (t^2 + 2)\, dt$. Compare with Exercise 19 in §1.
11. $\int_0^{36} \sqrt{v}\, dv$. Compare with Exercise 21 in §1.
12. $\int_{10}^{20} dt/t^2$.
13. $\int_{-2}^1 (2x^3 + x + x^{-3})\, dx$.
14. Find the area bounded by the graph of $t \mapsto t^{-2/3}$, the horizontal axis, and the lines $t = 8$ and $t = 27$.
15. $\int_0^1 (u + 1)^5\, du$.
▶ 16. Find the area bounded by the graph of $z \mapsto \sqrt{z - 1}$, the horizontal axis, and the lines $z = 2$ and $z = 5$.
17. $\int_{-1}^2 (2s - 1)^{2/3}\, ds$.
18. Find the area bounded by the graph of $u \mapsto 1/\sqrt{u + 3}$, the horizontal axis, and the lines $u = 1$ and $u = 6$.
19. $\int_0^3 s(s^2 - 2)^5\, ds$.
▶ 20. Find the area bounded by the graph of $v \mapsto v\sqrt{v^2 + 9}$, the horizontal axis, and the line $v = 4$.
21. $\int_1^2 [s^2/(s^3 + 2)^2]\, ds$.
22. $\int_{1/4}^2 f(x)\, dx$ if $f(x) = \sqrt{x} + (1/x^2)$ for $0 < x \le 1$; $2x^{2/3}$ for $x > 1$.
23. $\int_{-3}^5 g(t)\, dt$ if $g(t) = (1 - t)^{3/2}$ for $t < 0$; 1 for $t = 0$; $(t + 4)^{1/2}$ for $t > 0$.

2.3 Indefinite integrals

The fundamental theorem answers the question raised in Chapter 4, §6. *Every continuous function is a derivative of another function, that is, it has a primitive function.* More precisely, the continuous function $f(x)$ is the derivative of the function

$$F(x) = \int_a^x f(t)\, dt,$$

the area under the curve f from some fixed point to x. For this reason, a primitive function of f is often called the **indefinite integral** of f. Instead of the relation $F'(x) = f(x)$, one often writes

$$\int f(x)\, dx = F(x) + C$$

without limits of integration but with an undetermined constant C (the **constant of integration**). The constant reminds us that F is determined by f only up to a constant.

It is important to remember that, if we are given the function f, by a formula or by a table of values or as a graph, the values of $F(x)$ can actually be computed.

▶*Example*   If $r \neq -1$ is a rational number, then

$$\int x^r\, dx = \frac{x^{r+1}}{r+1} + C. \blacktriangleleft$$

EXERCISES

24. If $F'(t) = t - 1$ and $F(0) = 4$, what is $F(6)$?
25. If $G''(u) = u^2 - 3u$ and $G(0) = 1$, $G(1) = 7/12$, what is $G(-1)$?
26. Find a primitive that vanishes at 1 for the function $u \mapsto 2u^{2/3} - u^5$.
27. Find a primitive that attains the value 1 at 2 for the function $z \mapsto z(z^2 - 5)^6$.
28. Find a primitive that takes on the value 0 at 1 for the function $t \mapsto \sqrt{t}(t^{3/2} - 1)^4$.
29. If $x^3 + 2x = \int_0^x f(t)\, dt$, what is $f(3)$?
30. If $\frac{1}{5}(x^{10} - 1) = \int_0^{\alpha(x)} (t + 1)^4\, dt$, what is $\alpha(x)$?
31. Graph the functions $x \mapsto x^2$, $x \mapsto 2x + 1$, and $x \mapsto (x + 1)^2 = x^2 + 2x + 1$ for $x \geq 0$. Compute $\int_0^a x^2\, dx$, $\int_0^a (2x + 1)\, dx$, and $\int_0^a (x^2 + 2x + 1)\, dx$ and show that the sum of the first two integrals is always equal to the third.
▶ 32. If $f(x)$ is a primitive that attains the value -1 at 0 for the function $x \mapsto \int_0^x (s^3 - 1)\, ds$, what is $f(-1)$?

2.4 Inertial navigation

As an application, we discuss **inertial navigation.** Imagine a closed windowless box moving along a straight line. An observer located in the box has to determine the distance traveled by means of measurements and computation performed inside the box. (Thus a speedometer which must be connected to wheels that touch the ground is not permitted.)

If the box moves with constant speed, the observer inside will be unaware of the motion and will be unable to detect it by any mechanical experiment. This is the *classical principle of relativity* which was known to Galileo. Uniform motion cannot be detected even by experiments inside the box involving electromagnetic phenomena, such as the propagation of light. This is *Einstein's principle of relativity,* the basis of the so-called special relativity theory.

But anybody who has ridden in a moving vehicle knows that changes in speed, that is, acceleration, can be detected without looking out of a window. This is not merely a subjective feeling. A spring placed in the direction of motion, with one end attached to a wall (Fig. 5.25), will contract during (positive) acceleration and expand during deceleration (negative acceleration). By measuring the length of the spring, our observer will be able to determine the acceleration $a(t)$ of the box at time t.

We recall that the acceleration is the derivative of the instantaneous velocity $v(t)$, that is, $a(t) = v'(t)$. Hence $v(t)$ is a primitive function of $a(t)$. Assume that at the initial instant $(t = 0)$ the box is at rest. Then the value of $v(t)$ at $t = 0$ is zero. Hence, by the fundamental theorem,

$$v(t) = \int_0^t a(\tau)\, d\tau.$$

(Here τ is, as we know, a dummy variable; it could be replaced by any other symbol or letter, except for v, a, and t, which are used to denote other quantities.) Our observer can therefore compute, for any instant t,

position during uniform motion

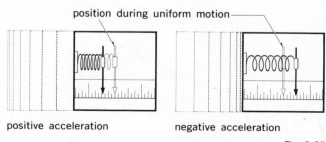

positive acceleration   negative acceleration

Fig. 5.25

the instantaneous velocity of the box, say, by drawing the curve $a(t)$ and measuring the area under the curve.

Next, the velocity $v(t)$ is the derivative of the distance $s(t)$ traveled from the time $t = 0$. Thus $s(0) = 0$ and $s'(t) = v(t)$. The fundamental theorem implies that

$$s(t) = \int_0^t v(\tau) \, d\tau.$$

In this way, the observer can determine his position at any time t.

The situation just described is far from fanciful. It occurs in atomic submarines, which remain submerged for months at a time, and also in rockets. The principle of inertial navigation is always the same: the accelerations can be measured and then the velocities and position are computed by integration. Of course, rockets and submarines do not move along straight lines only, and accelerations in various directions must be measured by complicated and highly sensitive instruments. The integrations, that is, the calculations of areas, are in practice performed by high speed electronic computers.

EXERCISES

33. A rocket is at rest at time $t = 0$. Measurements made inside the rocket show that it experiences an acceleration $a(t) = t^{1/2} + 1$ for $t \geq 0$, where time t is measured in seconds and acceleration a is measured in ft/sec². What is the velocity at time $t = 64$? How far is the rocket from the starting point at time $t = 64$?

34. Suppose that the rocket in Exercise 33 experienced $a(t) = t^{1/6} + 1$. What would the velocity be at time $t = 64$? How far would the rocket have been from the starting point at time $t = 64$?

35. Suppose that the rocket in Exercise 33 experienced an acceleration which was given graphically by the following drawing. Estimate the velocity at time

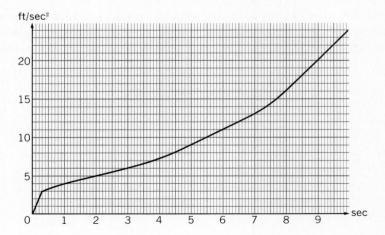

$t = 1$, $t = 2$, $t = 3$, $\cdots$, $t = 10$ and use this information to sketch the graph of the velocity for $0 \leq t \leq 10$. From your sketch, estimate how far the rocket would have been from the starting point at time $t = 10$.

3.1 Numerical and formal integration

Let f be a given continuous function. To integrate f means to compute its integral, either some **definite integral,** that is, $\int_a^b f(x)\,dx$ or an **indefinite integral,** that is, a primitive function F such that $F'(x) \equiv f(x)$. We know that the integral can be computed, and that there are two ways of doing it: numerical integration and formal integration.

Numerical integration is the computing of the number $\int_a^b f(x)\,dx$ using the properties of the integral stated in §1. For instance, we may use two step functions ϕ and ψ such that $\phi \leq f \leq \psi$ and compute the two integrals $\int_a^b \phi(x)\,dx$ and $\int_a^b \psi(x)\,dx$. This is, of course, easy. Then we know that the desired value $\int_a^b f(x)\,dx$ lies between $\int_a^b \phi(x)\,dx$ and $\int_a^b \psi(x)\,dx$. Later (see Chapter 7), we shall discuss more efficient numerical procedures. Since we know how to compute definite integrals, we are also able to find primitive functions. For, if a is some fixed number, then the function $x \mapsto F(x) = \int_a^x f(t)\,dt$ is a primitive function for f.

Numerical integration always works. The other method, **formal integration,** works only in some cases, but, when it does, it is fast and elegant. The method consists of finding among "known" functions a primitive function F for f. A "known" function is simply a function that has been defined and given a name on some previous occasion, and for which we have either a good table or a method of computing a table. If we know a primitive function F, we can compute $\int_a^b f(x)\,dx$, since this definite integral equals $F(b) - F(a)$.

In finding primitive functions one uses certain integration rules. These will be discussed in the present section. For the sake of simplicity, we assume that all functions considered are bounded and continuous in the intervals over which we integrate. It will be clear how to extend the integration rules to piecewise continuous functions.

3.2 Basic integration rules

In view of the fundamental theorem of calculus, every rule for differentiation can be rewritten as a rule for integration. Of particular importance is the rule:

(1)
$$\int_a^b [f(x) + g(x)]\,dx = \int_a^b f(x)\,dx + \int_a^b g(x)\,dx.$$

In words: *the integral of a sum is the sum of the integrals.* To prove this, let $F(x)$ and $G(x)$ be primitive functions of $f(x)$ and $g(x)$, respectively $(F' = f;\ G' = g)$. Then $F + G$ is a primitive function of $f + g$, by the rule for differentiating a sum. The left side of (1) is therefore equal to $[F(b) + G(b)] - [F(a) + G(a)]$, while the right side equals $[F(b) - F(a)] + [G(b) - G(a)]$, which is the same.

By the same token, the rule for differentiating differences implies that

$$(2) \qquad \int_a^b [f(x) - g(x)]\, dx = \int_a^b f(x)\, dx - \int_a^b g(x)\, dx;$$

the integral of the difference is the difference of the integrals.

It is clear that rules (1) and (2) can be applied to more than two functions. For instance,

$$\int_a^b [f(x) + g(x) + h(x)]\, dx = \int_a^b f(x)\, dx + \int_a^b g(x)\, dx + \int_a^b h(x)\, dx.$$

The rule for differentiating a constant times a function implies that

$$(3) \qquad \int_a^b cf(x)\, dx = c \int_a^b f(x)\, dx \qquad (c \text{ a constant})$$

or, in words: *a constant in front of an integrand may be taken out in front of the integral sign.* The proof of (3) is left to the reader.

One sometimes writes the same results for indefinite integrals, that is, for primitive functions,

$$\int [f(x) + g(x)]\, dx = \int f(x)\, dx + \int g(x)\, dx,$$

$$\int [f(x) - g(x)]\, dx = \int f(x)\, dx - \int g(x)\, dx,$$

$$\int cf(x)\, dx = c \int f(x)\, dx.$$

The formulas just written differ only in form from the lines (a) and (b) in the table in Chapter 4, §5.4.

Since

$$(4) \qquad \int_a^b x^\alpha\, dx = \frac{x^{\alpha+1}}{\alpha + 1}\Bigg|_a^b$$

for any rational number $\alpha \neq -1$ (in view of Theorem 8 in Chapter 4, §2), we can integrate every function that is a sum of terms of the form cx^α ($\alpha \neq -1$ and rational, c arbitrary).

▶ *Example*

$$\int_1^2 \left(3\sqrt[3]{x} + 4\sqrt[4]{x} - \frac{1}{x^2}\right) dx = 3\int_1^2 x^{1/3}\, dx + 4\int_1^2 x^{1/4}\, dx - \int_1^2 x^{-2}\, dx$$

$$= \frac{3}{4/3}\, x^{4/3}\Big|_1^2 + \frac{4}{5/4}\, x^{5/4}\Big|_1^2 - \frac{x^{-1}}{-1}\Big|_1^2$$

$$= \frac{9}{4}(\sqrt[3]{16} - 1) + \frac{16}{5}(\sqrt[4]{32} - 1) + \left(\frac{1}{2} - 1\right).\blacktriangleleft$$

EXERCISES

Compute the following definite and indefinite integrals.

1. $\int (3x^{10} + 5x^7)\, dx.$

2. $\int_0^1 (z^4 - z^3 + z^2 - z + 1)\, dz.$

3. $\int_{-1}^1 (100t^{99} + 99t^{100})\, dt.$

4. $\int (u^6 - 7u^8 - u^2)\, du.$

5. $\int \sqrt{z^7}\, dz.$

6. $\int_{1/4}^1 [z^2 + \sqrt{z} - z^{-2}]\, dz.$

7. $\int [y^{2/3} - 3y^{1/2} + 3y^{-3/5}]\, dy.$

8. $\int_0^1 (\sqrt{t^3} + \sqrt[3]{t^2})\, dt.$

9. $\int (1 + t)^5\, dt.$

10. $\int_0^2 [(x - 1)^4 + (x + 1)^7]\, dx.$

11. $\int_1^2 [s^3 + (s - 1)^3 + (s + 2)^3]\, ds.$

12. $\int [(z + 1)^3 - (2z + 3)^4]\, dz.$

13. $\int z\left(\frac{z^2}{2} + 1\right) dz.$

14. $\int_{-1}^1 (x^3 + 1)^2\, 3x^2\, dx.$

15. $\int (z^2 + \sqrt{z - 1})\, dz.$

16. $\int [5(s^3 - s + 2) + 3s^4(s^5 + 1)^7]\, ds.$

17. $\int [\sqrt{2t + 1} + \sqrt{3t - 1}]\, dt.$

18. $\int_{-1}^0 [t\sqrt{t^2 + 1} + t^{1/3}]\, dt.$

19. $\int \left[y\sqrt{y} + \frac{1}{\sqrt{y}} - \frac{1}{(y + 1)^2}\right] dy.$

20. $\int f(x)\, dx$ where $f(x) = \int_0^x (t^2 - 1)\, dt.$

21. $\int_a^b [f(x) + g(x)]\, dx$ where $\int_a^b f(x)\, dx = 3,\ \int_a^b g(x)\, dx = -4.$

22. $\int_{-1}^1 [2\phi(x) - 3\psi(x) + x]\, dx$ where $\int_{-1}^1 \phi(x)\, dx = 4$ and $\psi(x) = \int_{-1}^x t^2\, dt.$

23. $\int_{-1}^{1/2} [f(x) + g(x)]\, dx$ if $f(x) = \sqrt{2x + 3}$ and $g(x) = x^2$ for $x \le 0,\ x$ for $x > 0.$

24. $\int_0^1 [f(z) + g(z)]\, dz$ if $f(z) = \int_0^z t^3\, dt$ and $g(z) = \int_0^1 zt\, dt.$

25. $\int_0^1 [\phi'(x) - 3\psi'(x) + g(x)]\, dx$ where $\phi(1) = \psi(1) = 2,\ \phi(0) = 1,\ \psi(0) = 0$ and $g(x) = \int_0^x (1 - t)\, dt.$

26. $\int_0^1 (\alpha + x)\, dx$ where $\alpha = \int_0^1 (2 + t)\, dt.$

27. $\int_0^1 \{1 + x + \int_{-1}^0 t^2\, dt\}\, dx.$

▶ 28. $\int \{x + \int_{-1}^1 t^3\, dt\}\, dx.$

29. $\int [f(x) - 2g(x)]\, dx$ where $f(x) = g(x) = x$ for $x < 0$ and $f(x) = x^2,\ g(x) = x^3$ for $x > 0.$

30. $\int_0^x [t + \int_0^t z\, dz]\, dt.$

3.3 Integration by parts

The rule for differentiating products

$$\frac{d\,[\,f(x)g(x)\,]}{dx} = f'(x)g(x) + f(x)g'(x),$$

(see Chapter 2, §2) leads to a method of evaluating integrals known as **integration by parts.** By (1), we have

$$\int_a^b \frac{d\,[\,f(x)g(x)\,]}{dx}\,dx = \int_a^b f'(x)g(x)\,dx + \int_a^b f(x)g'(x)\,dx$$

and, by the fundamental theorem of calculus,

$$f(x)g(x)\Big|_a^b = \int_a^b f'(x)g(x)\,dx + \int_a^b f(x)g'(x)\,dx$$

or

$$(5) \qquad \int_a^b f(x)g'(x)\,dx = f(x)g(x)\Big|_a^b - \int_a^b f'(x)g(x)\,dx.$$

In order to apply this rule to computing an integral

$$\int_a^b \phi(x)\,dx,$$

one should be able first to write $\phi(x)$ in the form $f(x)g'(x)$ and one must know a primitive function for $f'(x)g(x)$. Here is an example.

We want to compute

$$\int_{-1}^{1} (1 - x)^{100} x\,dx.$$

Set $f(x) = x$, $g(x) = -(1/101)(1 - x)^{101}$; then $(1 - x)^{100}x = f(x)g'(x)$ and $f'(x)g(x) = -(1/101)(1 - x)^{101}$. We note that $-(1/101)(1 - x)^{101}$ is the derivative of the function $(1/101 \cdot 102)(1 - x)^{102}$. Now, by (5),

$$\int_{-1}^{1} (1 - x)^{100} x\,dx = -\frac{x(1 - x)^{101}}{101}\Big|_{-1}^{1} + \frac{1}{101}\int_{-1}^{1} (1 - x)^{101}\,dx$$

$$= -\frac{2^{101}}{101} + \frac{1}{101}\int_{-1}^{1} (1 - x)^{101}\,dx = -\frac{2^{101}}{101} - \frac{(1 - x)^{102}}{101 \cdot 102}\Big|_{-1}^{1}$$

$$= -\frac{2^{101}}{101} + \frac{2^{102}}{101 \cdot 102} = -\frac{2^{101}}{101}\left(1 - \frac{2}{102}\right).$$

The rule for integration by parts can be remembered and applied by using an extension of the Leibniz notation, which is described in the following subsection.

3.4 The language of differentials

Noting that

$$(6) \qquad \frac{df(x)}{dx} = f'(x),$$

we write formally,

$$(7) \qquad df(x) = f'(x)\,dx.$$

"Formally" means: (7) is merely an abbreviation of (6). One calls the symbol "$df(x)$" the **differential** of $f(x)$, and one calls "dx" the differential of x. At this stage, however, we do not attach any meaning to these symbols beyond that just explained. With this notation, the fundamental theorem can be written as

$$d \int_a^x f(t)\,dt = f(x)\,dx$$

and

$$\int_a^b df(x) = f(x)\Big|_a^b.$$

The rules for differentiating sums and products by constants may be written as

$$d(f + g) = df + dg, \qquad d(cf) = c\,df.$$

The product rule reads

$$d(fg) = g\,df + f\,dg,$$

and the rule for integrating by parts

$$\int_a^b f\,dg = fg\Big|_a^b - \int_a^b g\,df.$$

We rewrite the example considered above in the new notation.

$$\int_{-1}^{1} (1 - x)^{100} x \, dx = \int_{-1}^{1} x \, d\left(-\frac{1}{101}(1 - x)^{101}\right)$$

$$= -\frac{x(1 - x)^{101}}{101}\bigg]_{-1}^{1} - \int_{-1}^{1}\left(-\frac{1}{101}(1 - x)^{101}\right) dx$$

$$= -\frac{2^{101}}{101} + \frac{1}{101}\int_{-1}^{1}(1 - x)^{101} \, dx$$

$$= -\frac{2^{101}}{101} - \frac{1}{101}\frac{1}{102}\int_{-1}^{1} d(1 - x)^{102}$$

$$= -\frac{2^{101}}{101} - \frac{1}{101}\frac{1}{102}(1 - x)^{102}\bigg|_{-1}^{1}$$

$$= -\frac{2^{101}}{101} + \frac{2^{102}}{101 \cdot 102}.$$

Sometimes, integration by parts must be used a few times in succession. Here is an example:

$$\int_{0}^{a} (1 + x)^{1/3} x^2 \, dx = \int_{0}^{a} x^2 \, d\left(\frac{3}{4}(1 + x)^{4/3}\right)$$

$$= \frac{3}{4}\int_{0}^{a} x^2 \, d(1 + x)^{4/3} = \frac{3}{4} x^2 (1 + x)^{4/3}\bigg|_{0}^{a} - \frac{3}{4}\int_{0}^{a}(1 + x)^{4/3} \, dx^2$$

$$= \frac{3}{4} a^2 (1 + a)^{4/3} - \frac{3}{4}\int_{0}^{a}(1 + x)^{4/3} \, 2x \, dx$$

$$= \frac{3}{4} a^2 (1 + a)^{4/3} - \frac{3}{2}\int_{0}^{a} x \, d\left(\frac{3}{7}(1 + x)^{7/3}\right)$$

$$= \frac{3}{4} a^2 (1 + a)^{4/3} - \frac{9}{14}\int_{0}^{a} x \, d(1 + x)^{7/3}$$

$$= \frac{3}{4} a^2 (1 + a)^{4/3} - \frac{9}{14} x(1 + x)^{7/3}\bigg|_{0}^{a} + \frac{9}{14}\int_{0}^{a}(1 + x)^{7/3} \, dx$$

$$= \frac{3}{4} a^2 (1 + a)^{4/3} - \frac{9}{14} a(1 + a)^{7/3} + \frac{9}{14}\int_{0}^{a}\frac{3}{10} d(1 + x)^{10/3}$$

$$= \frac{3}{4} a^2 (1 + a)^{4/3} - \frac{9}{14} a(1 + a)^{7/3} + \frac{27}{140}(1 + x)^{10/3}\bigg|_{0}^{a}$$

$$= \frac{3}{4} a^2 (1 + a)^{4/3} - \frac{9}{14} a(1 + a)^{7/3} + \frac{27}{140}(1 + a)^{10/3} - \frac{27}{140}.$$

The example just given shows how to use integration by parts for finding primitive functions. Indeed, our calculation shows that

$$\frac{3}{4} x^2 (1 + x)^{4/3} - \frac{9}{14} x(1 + x)^{7/3} + \frac{27}{140}(1 + x)^{10/3}$$

is a primitive function of $x^2(1 + x)^{1/3}$. This can be verified by differentiation.

EXERCISES

31. If $d(5x^2 + x^3) = f(x)\,dx$, what is $f(2)$?

▶ 32. If $d(2\sqrt{x} + x^4 - 1) = f(x)\,dx$, what is $f(1)$?

33. If $d[x(x + 1)^5] = g(x)\,dx$, what is $g(-1)$?

34. If $d(t^2/(t - 1)) = h(t)\,dt$, what is $h(2)$?

35. If $d(u/\sqrt{u^2 + 1}) = f(u)\,du$, what is $f(u)$?

36. If $d\int_1^s \sqrt{x^2 + 1}\,dx = h(s)\,ds$, what is $h(1)$?

37. Suppose f, g, and h are differentiable functions of x. Which of the following always hold?
    (a) $d(fgh) = (df)(dg)(dh)$.
    (b) $d(fgh) = [d(fgh)]h + fg\,dh$.
    (c) $d(fgh) = fgh[(1/f)\,df + (1/g)\,dg + (1/h)\,dh]$ wherever f, g, and h are nonzero.
    (d) $d(fgh) = (g/h)\,df + (f/h)\,dg + (f/g)\,dh$ wherever f, g, and h are nonzero.

38. Find a function f such that $df = (x^{3/2} - 2(x + 1)^3)\,dx$.

39. If $x^2\,df = (x^3 - 1)\,dx$ and $f(1) = 0$, what is $f(2)$?

40. If $d(f + g) = x\,dx$, $d(f - g) = (1 - x)\,dx$, and $f(0) = g(0) = 1$, what are $f(x)$ and $g(x)$?

41. If $d(xf) = (2 + 3\sqrt{x})\,dx$ and f' is defined at 0, what is $f(0)$?

42. If $f(x) = x^3 - 1$ and $g(x) = 3x - 1$, for what values of x does $df = dg$?

In Exercises 43 to 57 evaluate the given integrals. It will be useful to use the technique of integration by parts. In each case in which an indefinite integral is asked for, check your answer by differentiation.

43. $\int_1^2 x(3 - 2x)^9\,dx$.

44. $\int x\sqrt{x + 1}\,dx$.

45. $\int_0^1 x(2x - 1)^5\,dx$.

▶ 46. $\int (2z - 3)\sqrt{z - 3}\,dz$.

47. $\int_{-1}^3 [(3x + 1)/(x + 2)^3]\,dx$.

48. $\int [(3 - 2u)/(1 - u)^{3/2}]\,du$.

49. $\int_0^1 (9t + 1)^{-1/3}(3t + 2)\,dt$.

50. $\int (x + 1)^2 x^{1/3}\,dx$.

51. $\int_0^2 (x^2 + 1)(x - 1)^{2/3}\,dx$.

52. $\int [x^2/\sqrt{x + 4}]\,dx$.

53. $\int [(v + 1)^2/(v - 1)^4]\,dv$.

54. $\int [(s^2 + s)/(s + 4)^5]\,ds$.

55. $\int_0^1 (y - 1)^2\,\sqrt{y}\,dy$.

56. $\int (3t^2 - 1)(2t + 3)^{10}\,dt$.

57. $\int (z + 1)^{1/3}(z - 1)^3\,dz$.

3.5 Changing variables

If we rewrite the chain rule (see Theorem 6 in Chapter 4, §2) as an integration rule, we obtain the basic **substitution rule** for changing the variable of integration. *If $x = \phi(t)$ is a function with a continuous derivative $\phi'(t)$ defined for $\alpha \leq t \leq \beta$, and if $\phi(\alpha) = a$, $\phi(\beta) = b$, then*

(8)
$$\int_a^b f(x)\, dx = \int_\alpha^\beta f(\phi(t))\phi'(t)\, dt$$

or

(8')
$$\int_{\phi(\alpha)}^{\phi(\beta)} f(x)\, dx = \int_\alpha^\beta f(\phi(t))\phi'(t)\, dt.$$

Proof. Let $F(x)$ be a primitive function for $f(x)$ so that $F'(x) = f(x)$ and, by the fundamental theorem,

$$F(b) - F(a) = \int_a^b f(x)\, dx.$$

Set $G(t) = F(\phi(t))$. By the chain rule, $G'(t) = F'(\phi(t))\phi'(t) = f(\phi(t))\phi'(t)$. Hence, by the fundamental theorem,

$$F(b) - F(a) = F(\phi(\beta)) - F(\phi(\alpha)) = G(\beta) - G(\alpha) = \int_\alpha^\beta G'(t)\, dt$$
$$= \int_\alpha^\beta f(\phi(t))\phi'(t)\, dt.$$

Comparing the two expressions for $F(b) - F(a)$, we see that the substitution rule (8) holds. We remark explicitly that the function ϕ in (8) need not be monotone.

Here is an application. We want to compute

$$\int_\alpha^\beta \frac{\phi'(t)}{\phi(t)^2}\, dt.$$

Setting $f(x) = x^{-2}$, we have, by (8),

$$\int_\alpha^\beta \frac{\phi'(t)}{\phi(t)^2}\, dt = \int_\alpha^\beta f(\phi(t))\phi'(t)\, dt = \int_{\phi(\alpha)}^{\phi(\beta)} f(x)\, dx \qquad \text{by (8')}$$

$$= \int_{\phi(\alpha)}^{\phi(\beta)} x^{-2}\, dx = -\frac{1}{x}\Big|_{\phi(\alpha)}^{\phi(\beta)} = \frac{1}{\phi(\alpha)} - \frac{1}{\phi(\beta)}.$$

The resulting formula

(9)
$$\int_\alpha^\beta \frac{\phi'(t)}{\phi(t)^2}\, dt = \frac{1}{\phi(\alpha)} - \frac{1}{\phi(\beta)}$$

could have been obtained directly, from the rule for differentiating the reciprocal of a function.

We rewrite the substitution rule interchanging the roles of x and t:

$$(9) \qquad \int_a^b g(\psi(x))\psi'(x)\,dx = \int_{\psi(a)}^{\psi(b)} g(t)\,dt;$$

here $\psi(x)$ is a function with a continuous derivative defined for $a \le x \le b$.

In applying the substitution rule, the main difficulty is to recognize that a function to be integrated, say, a function of x, can be written in the form $g(\psi(x))\psi'(x)$ where g is a function for which we know a primitive function. The Leibniz notation may be helpful in making the right guess.

We note that, using the Leibniz notation, we could have guessed the substitution rule, say, in the form (8). Let us write the integral to be evaluated as

$$\int_{x=a}^{x=b} f(x)\,dx$$

in order to emphasize that we deal with a function of x. If

$$x = \phi(t),$$

then

$$\frac{dx}{dt} = \phi'(t)$$

and "therefore"

$$dx = \phi'(t)\,dt.$$

Also, if $t = \alpha$, then $x = a$; if $t = \beta$, then $x = b$. "Substituting," we get

$$\int_{t=\alpha}^{t=\beta} f(\phi(t))\phi'(t)\,dt = \int_{x=a}^{x=b} f(x)\,dx,$$

as expected.

This is certainly *not* a proof. We acted as if dx and dt were numbers, while only the expression dx/dt has a meaning. But the formal procedure leads to the correct result and it is advisable to imitate it when applying the substitution rule. We shall illustrate this with an example that also shows how the substitution rule is used to find primitive functions.

We want to evaluate

$$\int_0^1 \frac{dx}{\sqrt[3]{1+5x}} = \int_0^1 (1+5x)^{-1/3}\,dx.$$

We do not know a primitive function of $(1+5x)^{-1/3}$, but we note that $(3/2)t^{2/3}$ is a primitive function of $t^{-1/3}$. We therefore try the substitution

$$1 + 5x = t.$$

We have

$$\frac{dt}{dx} = 5, \qquad dx = \frac{1}{5}\,dt,$$

and $\qquad\qquad t = 1$ for $x = 0$, $\qquad t = 6$ for $x = 1$.

Hence

$$\int_0^1 (1 + 5x)^{-1/3} \, dx = \int_1^6 t^{-1/3} \frac{1}{5} \, dt = \frac{1}{5} \frac{3}{2} t^{-2/3} \Big|_1^6$$

$$= \frac{3}{10}(6)^{2/3} - \frac{3}{10} = \frac{3}{10}(\sqrt[3]{36} - 1).$$

Looking over the calculation, we see that a primitive function for $(1 + 5x)^{-1/3}$ is $(3/10)t^{-2/3}$, that is, $(3/10)(1 + 5x)^{2/3}$. Thus

$$\frac{d}{dx}\left(\frac{3}{10}(1 + 5x)^{2/3}\right) = (1 + 5x)^{-1/3}.$$

This can, of course, be verified by differentiation.

Needless to say, not every substitution leads to a simplification. When it pays to make a substitution, and which one to use, are matters of experience.

3.6 Integrals of even and odd functions

We give one more application of the substitution rule. Recall that a function $f(x)$ is called *even* if $f(-x) = f(x)$, *odd* if $f(-x) = -f(x)$.

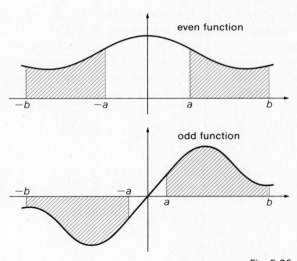

Fig. 5.26

A function is even if and only if its graph is symmetric with respect to the x axis, odd if and only if the graph is symmetric with respect to the origin (see Chapter 3, §1). This shows (compare Fig. 5.26) that

$$(10) \qquad \int_a^b f(x)\, dx = \int_{-b}^{-a} f(x)\, dx \qquad \text{if } f \text{ is even,}$$

$$(11) \qquad \int_a^b f(x)\, dx = -\int_{-b}^{-a} f(x)\, dx \qquad \text{if } f \text{ is odd.}$$

Let us prove (10) analytically. We use the substitution $x = -t$, $dx = -dt$, $t = -a$ for $x = a$, $t = -b$ for $x = b$. Then, for f even,

$$\int_a^b f(x)\, dx = \int_{-a}^{-b} f(-t)(-dt) = -\int_{-a}^{-b} f(t)\, dt = \int_{-b}^{-a} f(t)\, dt.$$

The same substitution proves (11), as the reader is asked to verify.

EXERCISES

In Exercises 58 to 79, evaluate the given integrals. In each case look for a suitable substitution. Where an indefinite integral is asked for, check your answer by differentiation.

58. $\int_1^2 (2x - 3)\, dx.$

59. $\int \dfrac{dt}{\sqrt{3 + 5t}}.$

60. $\int \dfrac{(x - 1)^3}{(1 + (x - 1)^4)^3}\, dx.$

61. $\int_{1/2}^1 (1 - 2y)^3 \sqrt{1 + (1 - 2y)^4}\, dy.$

62. $\int_0^{\sqrt{5}} z\sqrt{z^2 + 4}\, dz.$

63. $\int_0^1 \dfrac{u^2\, du}{(2 - u^3)^3}.$

64. $\int \dfrac{z + 1}{(z^2 + 2z + 3)^{2/3}}\, dz.$

65. $\int_{-1}^{\sqrt{2}} \dfrac{x\, dx}{(x^2 + 1)^3}.$

66. $\int \left(2t - \dfrac{1}{t^2}\right)\left(t^2 + \dfrac{1}{t}\right)^{1/2} dt.$

67. $\int x^{-1/3}(x^{2/3} + 1)^{1/3}\, dx.$

▶ 68. $\int_0^1 y^{1/2}(y^{3/2} + 1)^{1/2}\, dy.$

69. $\int \dfrac{\sqrt{1 + \sqrt{y}}}{\sqrt{y}}\, dy.$

70. $\int_1^8 x^3 \sqrt{1 + x^4}\, dx.$

71. $\int (y^{-3/5} + y^{1/5})^{1/3}\, dy.$

(Hint: Try some algebraic manipulation of the integrand before deciding on a suitable substitution.)

72. $\int_0^{5/9} \dfrac{dy}{\sqrt{1 - y}\,(1 + \sqrt{1 - y})^2}.$

73. $\int \dfrac{v(1 + \sqrt{v^2 + 4})^3}{\sqrt{v^2 + 4}}\, dv.$

▶ 74. $\int_0^1 \dfrac{z}{(z + 1)^4}\, dz.$

75. $\int \dfrac{t^2 - t}{(2t - 1)^4}\, dt.$

76. $\int_1^{\sqrt{2}} \dfrac{x^3 + 6x}{(x^2 + 2)^3}\, dx.$

77. $\int_0^9 \sqrt{1 + \sqrt{y}}\, dy.$

78. $\int \dfrac{(t^{1/3} - 1)^6}{t^{1/3}}\, dt.$

79. $\int \dfrac{dx}{x^{1/3}(x^{1/3} - 1)^{3/2}}.$

3.7 Remark on improper integrals

There are cases in which the symbol $\int_a^b f(x)\,dx$ denotes a definite number, even though the function f is not bounded or the interval (a, b) is not finite, so that the definition given in §1 does not apply. We shall discuss such improper integrals in an appendix (§6).

§4 Area. Volume. Length.

In this section, we describe some applications of integrals to geometry, assuming throughout that a unit of length has been chosen.

4.1 Computing areas

We begin by considering areas of plane regions. The area of a region bounded by the x axis, two lines $x = a$ and $x = b$, and the graph of a nonnegative function $y = f(x)$ is, as we know,

$$\int_a^b f(x)\,dx.$$

There are two ways of interpreting the statement just made. We may feel that we know what area is, and regard the statement above as a definition of the integral, or we may consider the integral to have been defined analytically, as in §1.8, and take the statement above as the *definition of area* for a region below the graph of a function. (We take the second point of view.)

Now we shall consider areas of more general regions. It will turn out that these areas can also be expressed as integrals. First, we assume that we know what area is and are concerned only with computing it. Later we shall transform the method of computing into a definition.

Fig. 5.27

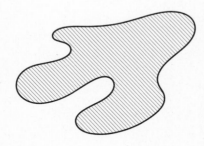

Let R be a "reasonable" region in the plane, that is, a set of points lying inside a smooth closed curve (Fig. 5.27) or between several such curves (Fig. 5.28); a precise definition need not concern us at this point. We want to compute A, the area of R.

Fig. 5.28

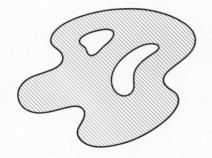

Choose a straight line L, and on it a reference point O and a positive direction. We call this direction "to the right." For every number t, let M_t denote the straight line perpendicular to L which intersects L at the positive or negative distance t from O. We assume, for the sake of simplicity, that our region R lies entirely between the lines M_a and M_b with $a < b$ (see Fig. 5.29). Let us denote by $A(t)$ the area of that part of R which lies to the left of M_t. Then $A(a) = 0$, since all of R lies to the right of M_a; and $A(b) = A$, since all of R lies to the left of M_b. Therefore $A = A(b) - A(a)$ and, by the fundamental theorem,

$$A = \int_a^b A'(t)\,dt,$$

if the function $A(t)$ has a bounded piecewise, continuous derivative. Since $A(t) = 0$ for $t < a$ and $A(t) = A$, a constant, for $t > b$, we certainly have $A'(t) = 0$ for $t < a$ and $t > b$, so that we may write, without specifying the limits of integration,

$$A = \int_{-\infty}^{+\infty} A'(t)\, dt.$$

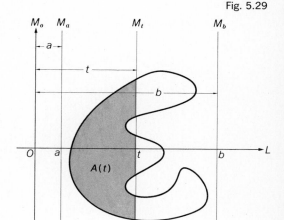

Fig. 5.29

But what is $A'(t)$?

The **intersection** of R and the line M_t is the set of points lying in R and on M_t. (In Fig. 5.29, it consists of three disjoint segments.) The **length** $l(t)$ of the intersection is defined whenever the intersection consists of one segment or of several disjoint segments and points: $l(t)$ is the sum of the lengths of the segments. The length $l(t)$ is 0 if $l(t)$ is empty or consists of finitely many points.

Let h be a small positive number. The quantity $A(t + h) - A(t)$ is the area of the part of R between M_t and M_{t+h}. If the intersection of R and M_t consists of one or several segments, then the area $A(t + h) - A(t)$ differs little from the area of one or several rectangles of total height $l(t)$ and width h, as is seen from Figs. 5.30 and 5.31. Hence $A(t + h) - A(t)$ is approximately $l(t)h$, so that

$$\frac{A(t + h) - A(t)}{h} = l(t) \qquad \text{approximately.}$$

This suggests that $A'(t) = l(t)$ and therefore

(1) $$A = \int_{-\infty}^{+\infty} l(t)\, dt.$$

We call (1) the **area formula.**

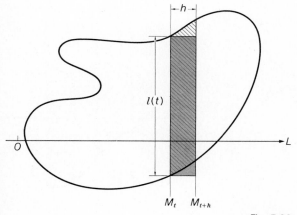

Fig. 5.30

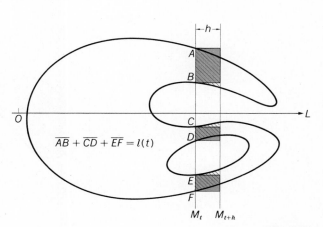

$$\overline{AB} + \overline{CD} + \overline{EF} = l(t)$$

Fig. 5.31

Fig. 5.32

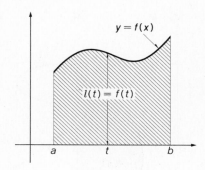

4.2 Defining areas

Now we change our point of view and give a **definition of area.** We simply make (1) into a definition. (This is legitimate since in §1 we defined the integral using only properties of numbers.) A set R of points in the plane will be called a region with area A if the area formula (1) is applicable. This means that there is a line L such that the integral (1) is defined.

Let us show that the area formula gives the expected result in simple cases. Suppose R is the region bounded by the x axis, the lines $x = a$ and $x = b > a$, and by the graph of a continuous positive function $y = f(x)$; see Fig. 5.32. For L we choose the x axis. Then $l(t) = f(t)$ for $a \leq t \leq b$ and $l(t) = 0$ for $t < a$ and $t > b$. The area formula yields

$$A = \int_{-\infty}^{+\infty} l(t)\, dt = \int_a^b f(t)\, dt,$$

as expected.

Another example is the area between two graphs. Let the functions $f(x)$ and $g(x)$ be bounded and piecewise continuous for $a \leq x \leq b$, and assume that $g(x) \leq f(x)$. Let R denote the region between the graphs of these functions and the lines $x = a$, $x = b$, and let A denote the area of R. Then

$$A = \int_a^b f(x)\, dx - \int_a^b g(x)\, dx$$

as we see from Fig. 5.33 (where f and g are both positive) or from Fig. 5.34 (where f is positive and g negative), by using our intuitive ideas about area. This can be written as

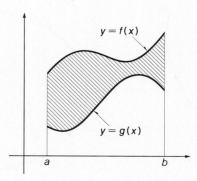

Fig. 5.33

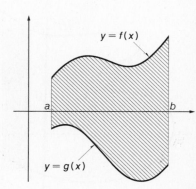

Fig. 5.34

$$A = \int_a^b [f(x) - g(x)]\, dx.$$

If we choose for L the x axis, we have $l(t) = f(t) - g(t)$ for $a < t < b$, $l(t) = 0$ for $t < a$ and $t > b$. Hence the relation above is equivalent to the area formula.

In order to justify the definition of area given by the area formula (1), we would have to show that it leads to the familiar additivity property of areas, that is, that the area of a region composed of two nonoverlapping regions R_1 and R_2 is equal to the sum of the areas of R_1 and R_2. Also, we would have to show that, if we compute the area of the same region R using two different lines L, we get the same result. Both statements can be proved, using properties of numbers. But it is easier to do this in the context of calculus of several variables; for this reason we do not present the proofs here but in Chapter 13.

As an illustration, we compute in two different ways the area bounded by the parabola $y = x^2$ and the lines $x = 0$ and $y = 1$ [Fig. 5.35(a)]. If we choose for L the x axis [Fig. 5.35(b)], then

$$l(t) = 1 - t^2 \text{ for } 0 \leq t \leq 1, \qquad l(t) = 0 \text{ for } t < 0,\, 1 < t,$$

and the area equals

$$A = \int_0^1 (1 - t^2)\, dt = t - \frac{t^3}{3}\Big|_0^1 = 1 - \frac{1}{3} = \frac{2}{3}.$$

If we take for L the y axis [Fig. 5.35(c)], then

$$l(t) = \sqrt{t} \text{ for } 0 \leq t \leq 1, \qquad l(t) = 0 \text{ for } t < 0,\, t > 1,$$

and

$$A = \int_0^1 \sqrt{t}\, dt = \frac{2}{3}\sqrt{t^3}\Big|_0^1 = \frac{2}{3},$$

as before. This is not surprising, but nevertheless reassuring.

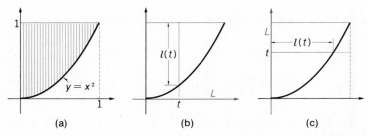

(a)   (b)   (c)

Fig. 5.35

It follows from the area formula that two regions that give rise to the same function $l(t)$ have the same area. This is the famous **Cavalieri's principle**. Its geometric interpretation is shown in Fig. 5.36. The two regions in this figure have the same area.

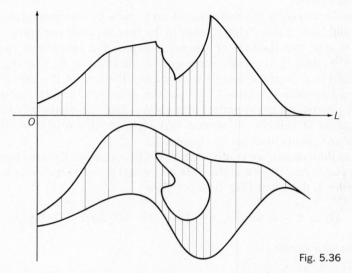

Fig. 5.36

There are very complicated regions to which the area formula is not applicable, but they do not occur in the usual applications of calculus.

▶**Examples**   **1.** Find the area A of the region enclosed between the curve $y = x^3$ and the line $y = x$ (see Fig. 5.37).

Fig. 5.37

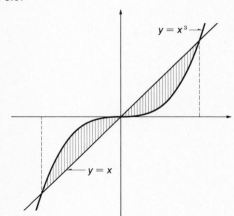

Solution.   The region consists of two pieces. We choose for L the x axis. Then

$$l(x) = \begin{cases} 0 & \text{for } x < -1, \\ x^3 - x & \text{for } -1 \leq x \leq 0, \\ x - x^3 & \text{for } 0 < x \leq 1, \\ 0 & \text{for } x > 1, \end{cases}$$

as is seen from the figure. Hence, by the area formula,

$$A = \int_{-1}^{1} l(x)\, dx = \int_{-1}^{0} (x^3 - x)\, dx + \int_{0}^{1} (x - x^3)\, dx$$

$$= \left(\frac{x^4}{4} - \frac{x^2}{2} \right)\Big|_{-1}^{0} + \left(\frac{x^2}{2} - \frac{x^4}{4} \right)\Big|_{0}^{1} = -\left(\frac{1}{4} - \frac{1}{2} \right) + \left(\frac{1}{2} - \frac{1}{4} \right) = \frac{1}{2}.$$

2. Find the area enclosed between the parabola $y = x^2$, the y axis, and the tangent to this parabola, at the point $(1, 1)$; see Fig. 5.38.

First solution. The equation of the tangent is $y = 2x - 1$. Hence, if L is the x axis, we have $l(x) = x^2 - 2x + 1$ for $0 \leq x \leq 1$ and $l(x) = 0$ for $x < 0$ and $x > 1$. By the area formula,

$$A = \int_0^1 (x^2 - 2x + 1)\, dx = \frac{x^3}{3} - x^2 + x \Big|_0^1 = \frac{1}{3}.$$

Second solution. The area under the parabola from 0 to 1 is $1/3$. The area of the triangle I is $1/4$, and so is the area of triangle II. Therefore $A = (1/3 - 1/4) + 1/4 = 1/3$. ◀

Fig. 5.38

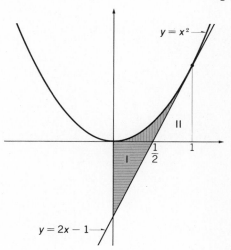

EXERCISES

In Exercises 1 to 15, find the area of the region enclosed by the given set of curves. Make a sketch of the region in question before applying the area formula. In sketching these graphs, it may occasionally be convenient to use different units of length along the two coordinate axes (see the answer to Exercise 14).

1. $y = x^{1/3}$, $x = 0$, and $y = 1$.
2. $y = x^2$ and $y = x + 2$.
3. $y = x^3$ and $y = x^2$.
4. $y = x^{1/3}$ and $y = x^{1/5}$.
5. $y = \sqrt{x}$, $y = x - 2$, and $x = 0$.
6. $y = -x^2$ and $y = x^4 - 20$.
7. $y^2 = x$ and $y = 2 - x$.
8. $y = x^2$, $y = (x - 1)^2$, and $y = 1/9$.
9. $y = x$, $y = x^2 + 1$, $x = 0$, and $y = 2$.
10. $x + y = 1$, $x + y = -1$, $x - y = 1$, and $x - y = -1$.
11. $y = x^3$, $y = 0$, and the tangent to $y = x^3$ at the point $(1, 1)$.
12. $y = \sqrt{x}$, $y = 1/4\, x$, and the normal to $y = \sqrt{x}$ at the point $(1, 1)$.
13. $y^2 = x$ and $y = \sqrt{x}\,(x^2 - 1)$.
▶ 14. $y = x^3 - 12x$ and $y = x^2$.
15. $y = x^3 - x$ and the tangent to $y = x^3 - x$ at $x = -1$. (Hint: Remember that the tangent and the curve will certainly intersect at $x = -1$.)

Fig. 5.39

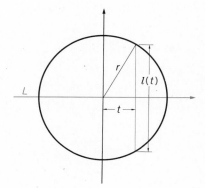

4.3 Area of a circle. The number π

As another example we compute the area interior to the circle

$$x^2 + y^2 = r^2.$$

We have (compare Fig. 5.39)

$$l(t) = 2\sqrt{r^2 - t^2} \qquad \text{for } -r < t < +r$$

while $l(t) = 0$ for $|t| \geq r$. The area is therefore

$$A = \int_{-r}^{r} 2\sqrt{r^2 - t^2}\, dt = 2\int_{-r}^{0} \sqrt{r^2 - t^2}\, dt + 2\int_{0}^{r} \sqrt{r^2 - t^2}\, dt.$$

In the first integral, we make the substitution

$$t = -rx, dt = -r\, dx, \qquad t = -r \text{ for } x = 1, \qquad t = 0 \text{ for } x = 0.$$

In the second integral, we make the substitution

$$t = rx, dt = r\, dx, \qquad t = 0 \text{ for } x = 0, \qquad t = r \text{ for } x = 1.$$

This yields

$$(2) \qquad A = 2\int_{1}^{0} \sqrt{r^2 - r^2 x^2}\,(-r)\, dx + 2\int_{0}^{1} \sqrt{r^2 - r^2 x^2}\, r\, dx$$

$$= -2r^2 \int_{1}^{0} \sqrt{1 - x^2}\, dx + 2r^2 \int_{0}^{1} \sqrt{1 - x^2}\, dx = 4r^2 \int_{0}^{1} \sqrt{1 - x^2}\, dx.$$

We *define* the number π to be the area of a circular disk of radius 1. Thus we have

$$(3) \qquad\qquad\qquad \pi = 4\int_{0}^{1} \sqrt{1 - x^2}\, dx.$$

Note that, since the integral can be defined analytically, the number π is defined by the above formula, without reference to geometry. That π is also the ratio of the length of the circumference to that of the diameter will be proved later.

We can now interpret the result (2) as follows: *the area of a disk of radius r is $A = \pi r^2$.*

We note that the function $y = \sqrt{1 - x^2}$ is decreasing for $0 < x < 1$, since $dy/dx = -x/\sqrt{1 - x^2} < 0$ in this interval. Hence π can be computed with any desired degree of accuracy by evaluating the integral in (3) by the method outlined in §1. Better methods will be discussed in Chapter 8.

The number π is irrational; its decimal expansion

$$\pi = 3.1415926536 \cdots$$

never terminates and never becomes periodic. We shall not prove this result. With the aid of electronic computers it is quite easy to calculate π to thousands of digits. It was not so before the advent of automatic computers and before the discovery of logarithms. Archimedes established that

$$3\frac{10}{71} < \pi < 3\frac{10}{70},$$

and when Ludolph (around 1600) computed 35 digits for π this achievement so impressed mathematicians that π was named "Ludolph's number."

4.4 Computing volumes

We consider next the calculation of volumes. Let R be a plane region with area A. At each point of R, erect a segment of length H, perpendicular to the plane of R; let all segments lie on the same side of R. The solid body formed by all these segments is called the right cylinder with base R and height H (see Fig. 5.40). In elementary geometry, one is taught that the volume of such a cylinder is $V = AH$, at least in the case in which R is a polygon or a circular disk. We shall show how one can arrive at this formula, using the formula for the volume of a right parallelepiped (box), our intuitive concept of volume, and the basic ideas of calculus.

We use a Cartesian coordinate system in space (see Chapter 2, §1.4). Let R be a region in the x, y plane with area A. We assume it to be bounded by a smooth curve. The right cylinder with base R and height H is the set C of all points (x, y, z) such that (x, y) is a point in R and $0 \leq z \leq H$ (see Fig. 5.41). Let us assume that R lies between the lines $x = a$ and $x = b$, $b > a$. Let $V(t)$ denote the volume of the part

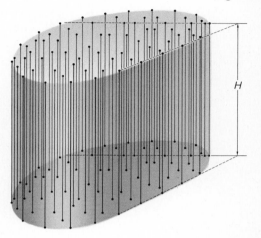

Fig. 5.40

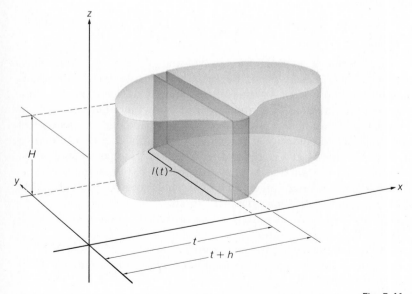

Fig. 5.41

of C that lies to the left of the plane $x = t$. Then $V(a) = 0$, and $V(b) = V$ is the total volume of C. If the function $V(t)$ has a bounded piecewise, continuous derivative $V'(t)$, we have $V = \int_a^b V'(t)\, dt$, which we can also write as

$$V = \int_{-\infty}^{+\infty} V'(t)\, dt,$$

since clearly $V'(t) = 0$ for $t < a$ and for $t > b$.

Let h be a small positive number. Then $V(t + h) - V(t)$ is the volume of the slice of C contained between the planes $x = t$ and $x = t + h$. In Fig. 5.41, we drew the case in which the line $x = t$ intersects R in one segment of length $l(t)$. The slice of C between $x = t$ and $x = t + h$ is nearly a parallelepiped of base area $h \cdot l(t)$ and height H. The volume of this slice is approximately $hl(t)H$. Hence

$$\frac{V(t + h) - V(t)}{h} = Hl(t) \qquad \text{approximately.}$$

This suggests that $\qquad\qquad V'(t) = Hl(t)$

and therefore, $\qquad V = \int_{-\infty}^{+\infty} Hl(t)\, dt = H \int_{-\infty}^{+\infty} l(t)\, dt.$

Since $\int_{-\infty}^{+\infty} l(t)\, dt = A$ by the area formula, we have that $V = AH$.

In particular, if the base R is a circular disk of radius r, we obtain the familiar formula

$$V = \pi r^2 H.$$

We now derive a formula for the volume of more general solids, proceeding as with areas. First, we use freely our geometric intuition: we assume we know what volume means and try to compute it. Then we transform the method of computing into a precise definition.

Consider a "reasonable" solid body B in space; let V denote its volume. In order to compute V, we choose a straight line L, a reference point O on the line, and a positive direction. For the sake of simplicity, we assume that the intersection of any plane P perpendicular to L with B consists of a single piece. Let P_t be such a plane, t denoting the (positive or negative) distance from O to the intersection point of the plane and the line L. We denote the area of the intersection of P_t and B by $A(t)$ (see Fig. 5.42).

Assume that B lies between the planes P_a and P_b, $a < b$. Let $V(t)$ denote the volume of the part of B contained between P_a and P_t. Then $V(a) = 0$ and $V(b) = V$.

The difference $V(t + h) - V(t)$ is the volume of a thin slice that is

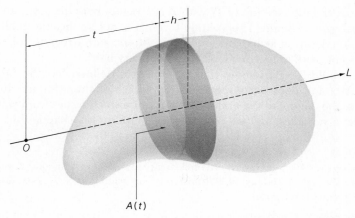

Fig. 5.42

nearly a right cylinder of height h and base area $A(t)$ (see Fig. 5.42). Thus $V(t + h) - V(t)$ is close to $A(t)h$, the fraction

$$\frac{V(t + h) - V(t)}{h} \approx A(t),$$

and we may expect that

$$V'(t) = \lim_{h \to 0} \frac{V(t + h) - V(t)}{h} = A(t).$$

If so, the fundamental theorem of calculus gives the formula $V = \int_a^b A(t)\, dt$.

We note that $A(t) = 0$ for $t < a$ and for $t > b$; hence

(4)
$$V = \int_{-\infty}^{+\infty} A(t)\, dt.$$

We call (4) the **volume formula**. This formula implies **Cavalieri's principle for solids:** if two bodies give rise to the same function $A(t)$, they have the same volume.

4.5 Definition of volume

Up to now, we have regarded the concept of volume as known and we have arrived at equation (4) by informed guessing. Now we change the point of view and *define the volume by the volume formula.* The formula is applicable to sets B in space such that, if the line L is chosen properly, the intersection of P_t with B is a region with area $A(t)$ and the integral (4) is defined. The volume formula is not applicable to all conceivable regions in space, but one does not encounter regions for which it does not work unless one looks for them very hard.

We would have to show that volume just defined has the familiar properties and does not depend on the choice of the reference line L. This can be done, but the easiest proof requires techniques from calculus of several variables, and we postpone the matter until Chapter 13.

EXERCISES

In Exercises 16 to 20, we use the following notation. The symbol $\{(x, y): \cdots\}$ denotes the set of points (x, y) satisfying the condition $\cdots$. For instance, $\{(x, y): y = 0\}$ is the x axis.

▶ 16. The base of a solid is the triangle $\{(x, y): 0 \le x \le 1, 0 \le y \le x\}$ in the x, y plane. Each cross section perpendicular to the x axis is a square. Compute the volume of the solid.

17. The base of a solid is the triangle $\{(x, y): 0 \le x \le 1, 0 \le y \le x\}$ in the x, y plane. Each cross section perpendicular to the y axis is a semicircle. Compute the volume of the solid. Assume that the area of a circle of radius r is πr^2.

18. The base of a solid is the square $\{(x, y): 0 \le x \le 1, 0 \le y \le 1\}$ in the plane $z = 0$. Each cross section perpendicular to the x axis is a triangle. If the upper vertices of the triangular cross sections all lie on the straight line joining the point $x = 0$, $y = 1/2$, $z = 2$ to the point $x = 1$, $y = 1/2$, $z = 1$, compute the volume of the solid.

19. The base of a solid is the circle $\{(x, y): x^2 + y^2 \le 1\}$. Each cross section perpendicular to the x axis is a square. Compute the volume of the solid.

20. The base of a solid is the triangle $\{(x, y): -1 \le x \le 1, |x| \le y \le 1\}$ in the plane $z = 0$. Each cross section perpendicular to the y axis is a triangle. If the upper vertices of the triangular cross sections lie on the quarter circle

$$\{(y, z): y^2 + z^2 = 1, y \ge 0, z \ge 0\}$$

in the plane $x = 0$, compute the volume of the solid.

21. Compute the volume of a sphere of radius R using the volume formula. Assume that the area of a circle of radius r is πr^2.

22. Compute the volume of a right circular cylinder of radius R and height H, using the volume formula and taking for L a line perpendicular to the axis of the cylinder.

23. Compute the volume of a right circular cone whose height is H and which has a base of radius R, using the volume formula. Assume that the area of a circle of radius r is πr^2.

4.6 Solids of revolution

Consider a plane region R under the graph of a function $y = f(x)$ from a to b (Fig. 5.43); we assume f to be nonnegative. When R is rotated about the x axis by 360° (one full revolution), it sweeps out a solid B called a **solid of revolution.**

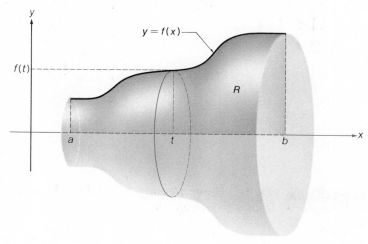

Fig. 5.43

In order to compute the volume V of B, we choose for L the x axis. Then the intersection of the plane P_t with B is a circular disk of radius $f(t)$, if $a < t < b$. Hence $A(t) = \pi f(t)^2$ for $a < t < b$ and, of course, $A(t) = 0$ for $t < a$ and $t > b$. The general volume formula now yields the formula

$$(5) \qquad\qquad V = \pi \int_a^b f(t)^2 \, dt$$

for the volume of a body of revolution. It is clear that the formula may be applied also to bodies of revolution for which the axis of rotation is not the x axis.

Assume, for instance, that R is the region bounded by a curve $x = f(y)$, $a \le y \le b$, the y axis, and the lines $y = a$, $y = b$, and that B is the body swept out by R when it is rotated about the y axis by 300° (such a body is considered in Example 2 hereafter). The volume of B is again given by (5). Indeed the volume does not depend on the names of the axes.

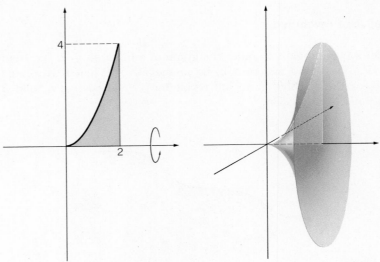

Fig. 5.44

►*Examples* *1.* For the volume obtained by rotating the area under the parabola $y = x^2$ from 0 to 2 about the x axis (Fig. 5.44), $a = 0$, $b = 2$, $f(t) = t^2$, and

$$V = \pi \int_0^2 t^4 \, dt = \pi \cdot \frac{t^5}{5}\Big|_0^2 = \frac{32}{5}\, \pi.$$

2. Find the volume obtained by rotating the region bounded by the arc of the parabola $y = x^2$ from $x = 0$ to $x = 2$, the y axis, and the line $y = 4$ about the y axis. (See Fig. 5.45.)

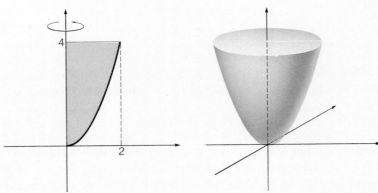

Fig. 5.45

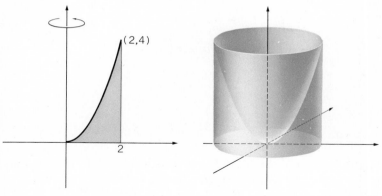

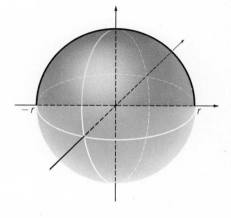

Fig. 5.46

Fig. 5.47

We have $a = 0$, $b = 4$, $f(t) = \sqrt{t}$. Therefore, by (5),

$$V = \pi \int_0^4 t\, dt = \pi \frac{t^2}{2}\Big|_0^4 = 8\pi.$$

3. Find the volume obtained by rotating the area under the parabola $y = x^2$ from 0 to 2 about the y axis. (See Fig. 5.46.)

First Solution.   We use equation (4) with the y axis as the reference line L. For $0 < t < 4$, the intersection P_t with our volume is a circular annulus with radii 2 and $\sqrt{t}$. (See Fig. 5.47.) Hence, for such t, $A(t) = \pi 2^2 - \pi t = \pi(4 - t)$. Also, $A(t) = 0$ for $t < 0$ and $t > 4$. Hence

$$V = \int_0^4 A(t)\, dt = \int_0^4 \pi(4 - t)\, dt = \pi\left(4t - \frac{t^2}{2}\right)\Big|_0^4 = 8\pi.$$

Fig. 5.48

Second Solution.   The volume considered, together with the volume computed in Example 2, form a circular cylinder of base radius 2 and height 4. The volume of this cylinder is 16π. Hence the volume is $16\pi - 8\pi = 8\pi.$◀

4.7 Spheres and circular cones

The sphere of radius r is obtained by rotating the region under $y = \sqrt{r^2 - x^2}$ from $-r$ to r about the x axis (see Fig. 5.48). The **volume of the sphere** can be computed from equation (5), setting

$$a = -r, \qquad b = r, \qquad f(t) = \sqrt{r^2 - t^2}.$$

Fig. 5.49

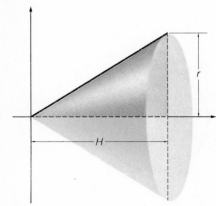

Fig. 5.50

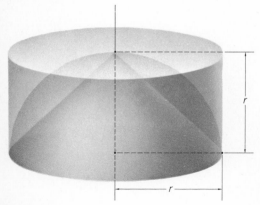

DEMOCRITUS of Abdera (ca. 400–370 B.C.) founded Greek "atomism," probably together with the philosopher Leucippus. He contributed to most fields of human knowledge, but of his many writings, only a few fragments have reached us.

The atomist had the bold idea of explaining the world by the motion of unchangeable, indivisible atoms through empty space.

It equals

$$V = \pi \int_{-r}^{r} (r^2 - t^2)\, dt = \pi \left(r^2 t - \frac{t^3}{3} \right)\Big|_{-r}^{r}$$

or

(6)
$$V = \frac{4}{3}\, \pi r^3.$$

The right circular cone of radius r and height H is obtained by rotating the region under the straight line $y = (r/H)x$ from $x = 0$ to $x = H$ about the x axis (see Fig. 5.49). From (5) we obtain, for the **volume of the cone,**

$$V = \pi \int_{0}^{H} \frac{r^2}{H^2}\, t^2\, dt = \frac{\pi r^2}{H^2} \left(\frac{t^3}{3} \right)\Big|_{0}^{H}$$

or

(7)
$$V = \frac{\pi}{3}\, r^2 H.$$

Consider now a circular cylinder whose radius is equal to its height, a hemisphere inscribed in the cylinder, and a cone inscribed in the hemisphere (Fig. 5.50). It follows from the formulas just derived that the ratios of the three volumes are like $3:2:1$. This beautiful relationship occurs in a work of Archimedes.

The ancient Greeks always expressed formulas for lengths, areas, and volumes as statements about ratios of "like" quantities. They rightly counted the formulas for the volumes of spheres and circular cones among the glories of their mathematics. These formulas were first discovered by the "materialist" philosopher Democritus, but a rigorous proof was given only by the mathematician Eudoxus, a friend of the "idealist" Plato. (The words "idealist" and "materialist" are used in their philosophical sense; they do not imply a moral superiority of Plato over Democritus.) The ideas of Plato and of his onetime pupil Aristotle triumphed in Greek thought and dominated Western culture for centuries, but no works by Democritus have been preserved. Modern mathematicians learned of Democritus' discovery only when a manuscript of a long-lost book by Archimedes was accidentally found in 1906. From this book, we also know that Democritus' method was based on a mathematical myth not unlike the one we retold in §1, and his method was somewhat akin to calculus.

EXERCISES

24. Find the volume of the solid formed by rotating the region under the graph of $y = 1 - x^2$ from $x = 0$ to $x = 1$ around the x axis.
25. Find the volume of the solid formed by rotating the region under the graph of $y = 1 - x^2$ from $x = 0$ to $x = 1$ around the y axis.
26. Find the volume of the solid formed by rotating the region under the graph of $y = x^3$ from $x = 1$ to $x = 2$ around the x axis.
27. Find the volume of the solid formed by rotating the region under the graph of $y = (2x + 1)^{1/3}$ from $x = 0$ to $x = 13$ around the y axis.
28. Find the volume of the solid formed by rotating the region under the graph of $y = 1/x$ from $x = 1$ to $x = a > 1$ around the x axis. (What happens to this volume as $a \to \infty$?)
29. Find the volume of the solid formed by rotating the region bounded by the curves $y = \sqrt{x}$, $x = 0$, and $y = 1$ around the x axis.
30. Find the volume of the solid formed by rotating the region bounded by the curves $y = \sqrt{x}$, $x = 0$, and $y = 1$ around the y axis.
31. Find the volume of the solid formed by rotating the region bounded by the curves $y = \sqrt{1 - x^2}$, $y = x$, and $y = 0$ around the x axis. (Note that the curve $y = \sqrt{1 - x^2}$ is just the upper half of the unit circle $x^2 + y^2 = 1$.)
32. Find the volume of the solid formed by rotating the region bounded by the curves $y = \sqrt{1 - x^2}$, $y = x$, and $y = 0$ around the y axis.
33. Find the volume of the solid formed by rotating the region bounded by the curves $y = x^2 + 1$, $x = 0$, and the tangent to $y = x^2 + 1$ at $x = 1$ around the x axis.
► 34. Find the volume of the solid formed by rotating the region bounded by the curves $y = x^{2/3}$, $x = 0$, and the tangent to $y = x^{2/3}$ at $x = 1$ around the y axis.
35. Find the volume of the solid formed by rotating the region bounded by the curves $y = 4 - x^2$ and $y = 3$ around the line $y = -1$.
36. Find the volume of the solid formed by rotating the region bounded by the curves $y = 1/(x + 1)$, $x = 0$, $x = 1$, and $y = 0$ around the line $x = -1$.
37. Find the volume of the solid formed by rotating the region bounded by the curves $y = x^3$ and $y = x$ around the line $x = -2$.
38. Find the volume of the solid formed by rotating the region bounded by the curves $y = x^2$ and $y = 1$ around the line $y = 2$.
39. Find the volume of the solid formed by rotating the region bounded by the curves $y = x^2$, $x = -1$, $x = 0$, and $y = 4$ around the line $x = -2$.
40. Find the volume of the solid formed by rotating the region inside the circle $(x - 4)^2 + y^2 = 1$ [circle with center at $(4, 0)$ and radius 1] around the y axis. The solid formed in this way is called a solid torus.
41. Find the volume inside the cylinder $x^2 + y^2 = 1$, $0 \le z \le 1$ and between the plane $z = 0$ and the plane $z = x$ (that is, the plane that meets the base plane $z = 0$ in the y axis at an angle of $45°$).
42. Find the volume inside the sphere $x^2 + y^2 + z^2 = 1$ and above the plane $z = 1/2$.
43. Two cylindrical pipes, each 2 in. in diameter, intersect at right angles (the axes of the pipes lie in the same plane). Find the volume of the intersection.

4.8 Length

The calculating of lengths of curves also leads to integrals, as we shall see presently. We assume first that we know what is meant by the length of a curve and use this knowledge to make an educated guess about a method for computing it. Then we define length by the formula obtained. This is, of course, the procedure we also followed for defining areas and volumes.

Consider the graph of a smooth function $y = f(x)$, $a \leq x \leq b$, and denote its length by l. Let $l(t)$ denote the length of that part of the curve which corresponds to the values of x between a and t. Thus $l(a) = 0$, $l(b) = l$, and if $l(t)$ is a "nice" function, then by the Fundamental Theorem

$$l = \int_a^b l'(t)\,dt.$$

To find $l'(t)$, let h be a small positive number and let us consider the difference quotient

$$\frac{l(t+h) - l(t)}{h}.$$

The numerator is the length of the small arc of the curve lying above the interval $[t, t+h]$. We now make an assumption: if we replace the curve, for values of x between t and $t + h$, by its tangent at the point $(t, f(t))$, the resulting error in computing the difference quotient will be very small (see Fig. 5.51). By replacing the curve by its tangent we mean that we replace the length of the arc PQ in this figure by the length of the segment PQ', which equals

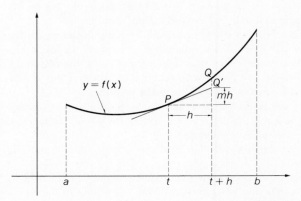

Fig. 5.51

$$\sqrt{h^2 + (mh)^2} = h\sqrt{1 + m^2}$$

as is seen in Fig. 5.51. But $m = f'(t)$ by the definition of the derivative. Thus the ratio $[l(t + h) - l(t)]/h$ is approximately

$$\frac{h\sqrt{1 + m^2}}{h} = \sqrt{1 + f'(t)^2}.$$

We surmise that $l'(t) = \sqrt{1 + f'(t)^2}$ and obtain

(8) $$l = \int_a^b \sqrt{1 + f'(t)^2}\, dt.$$

We call this equation the **length formula.**

Now we change our point of view and *define the length* of the graph of the function $y = f(x)$, $a < x < b$, by the length formula. The formula is applicable whenever the function considered has a piecewise continuous derivative.

If a curve is not the graph of a function (like the curve shown in Fig. 5.52) but can be decomposed into several arcs that are graphs of functions (like the arcs PQ, QR, and RS in Fig. 5.52), we define the length of the curve as the sum of the lengths of these arcs. It remains to be shown that our definition of length does not depend on the position of the coordinate system. This will be done in Chapter 11, §3.10.

▶**Examples**   *1.* Compute the length of the straight segment joining the points $(-2, 4)$ and $(3, 5)$.

First Solution.   By the distance formula

$$l = \sqrt{(-2 - 3)^2 + (4 - 5)^2} = \sqrt{26}.$$

Second Solution.   The equation of the straight line joining the two points is (see Chapter 2, §3)

$$(y - 5)(-2 - 3) = (4 - 5)(x - 3)$$

or $$y = \frac{1}{5}x + \frac{22}{5}.$$

By the length formula,

$$l = \int_{-2}^3 \sqrt{1 + \left(\frac{dy}{dx}\right)^2}\, dx = \int_{-2}^3 \sqrt{1 + \frac{1}{25}}\, dx = \frac{\sqrt{26}}{5} \int_{-2}^3 dx = \sqrt{26}$$

as it should.

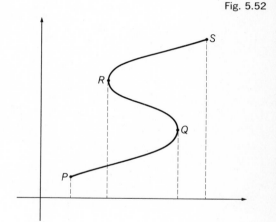

Fig. 5.52

2. Find the length of the graph $x \mapsto y = x^{3/2}$ between $x = 1$ and $x = 2$.

Solution. By the length formula

$$l = \int_1^2 \sqrt{1 + \left(\frac{dy}{dx}\right)^2}\, dx = \int_1^2 \sqrt{1 + \frac{9}{4} x}\, dx$$

$$= \int_{13/4}^{22/4} t^{1/2} \frac{4}{9}\, dt \quad \left(\text{setting } 1 + \frac{9}{4} x = t\right)$$

$$= \frac{4}{9} \left(\frac{2}{3} t^{3/2}\right)\Big|_{13/4}^{22/4} = \frac{(22^{3/2} - 13^{3/2})}{27}. \quad \blacktriangleleft$$

EXERCISES

In Exercises 44 to 53, use the arc-length formula to find the length of the given graphs between the given points.

44. $f(x) = (x + 1)^{3/2}$ between $x = 3$ and $x = 8$.
45. $g(x) = x^{2/3} - 1$ between $x = 8$ and $x = 27$.

Note: The functions in Exercises 44 and 45 are inverse functions, and hence the curves whose lengths are to be computed are "the same." More precisely, one curve is obtained from the other by a reflection about the line $y = x$.

46. $f(x) = \int_1^x \sqrt{t^2 - 1}\, dt$ between $x = 1$ and $x = 3$.
47. $f(x)$ between $x = 1$ and $x = a (a > 1)$, where $y = f(x)$ is any solution of the differential equation $dy/dx = (x^4 - 1)^{1/2}$.
48. $f(x) = (1/3)(x^2 + 2)^{3/2}$ between $x = \sqrt{2}$ and $x = \sqrt{7}$.
49. $f(x) = (2/3)x^{3/2} - (1/2)x^{1/2}$ between $x = 1$ and $x = 4$.
50. $f(x) = (5/6)x^{6/5} - (5/8)x^{4/5}$ between $x = 1$ and $x = 32$.
51. $f(x) = (5/48)(4x^{4/5} + 1)^{3/2}$ between $x = 1/32$ and $x = 1$.
52. $f(x) = x^4/4 + x^{-2}/8$ between $x = 1$ and $x = 2$.
53. $f(x) = \int_1^x [t + (1/t) + 1]^{1/2}\, dt$ between $x = 1$ and $x = 4$.
54. Let $P = (a, b)$ and $Q(c, d)$ be two distinct points in the plane; assume for simplicity that $a \neq c$ and $b \neq d$. Find the equation of the line through P and Q and use the arc-length formula to find the distance from P to Q. Show that your answer agrees with that given by the usual distance formula.

4.9 Circular sectors. Circumference

We shall use the length formula to establish two theorems of elementary geometry: *The area of a circular sector is half the product of its radius and the length of its arc. The circumference (that is, the length) of a circle is $2\pi r$, where r is the radius and π the number defined by the definite integral*

$$\frac{\pi}{4} = \int_0^1 \sqrt{1 - x^2}\, dx,$$

see §4.3. (Since we already know that the area of a circular disk is πr^2 and a disk can be decomposed into sectors, the second theorem follows from the first.)

Fig. 5.53

To simplify writing, we consider the case of unit radius. We must show that (i) *the area of a circular sector of radius 1 equals half the length of its arc, and* (ii) *the circumference of a circle of radius 1 is 2π.*

Let the center of the circle be chosen as the origin O, let a and b be two numbers such that $-1 < a < b < 1$ and let P and Q be points on the upper semicircle with first coordinates a and b. Then $P = (a, \sqrt{1 - a^2})$, $Q = (b, \sqrt{1 - b^2})$; see Fig. 5.53. The upper semicircle is the graph of the function

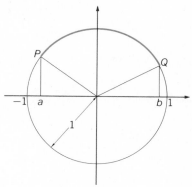

$$x \mapsto y = \sqrt{1 - x^2}.$$

In order to apply the length formula (8), we compute

$$\frac{dy}{dx} = -\frac{x}{\sqrt{1 - x^2}}, \quad 1 + \left(\frac{dy}{dx}\right)^2 = 1 + \frac{x^2}{1 - x^2} = \frac{1}{1 - x^2}$$

$$\sqrt{1 + \left(\frac{dy}{dx}\right)^2} = \frac{1}{\sqrt{1 - x^2}}.$$

Therefore

(9)
$$\text{length of circular arc } PQ = \int_a^b \frac{dx}{\sqrt{1 - x^2}}.$$

It is clear that the same integral gives the length of the arc of the lower semicircle between $x = a$ and $x = b$.

Does formula (9) remain valid if $a = -1$ and/or $b = 1$? For $a = -1$ or for $b = 1$, the definition of the integral given in §1 is not applicable, since the integrand $(1 - x^2)^{-1/2}$ is not bounded in the interval $(-1, 1)$; it becomes arbitrarily large as x approaches -1 or 1. But it is geometrically obvious that, for a fixed a, the length of the circular arc PQ is a continuous function of b defined also for $b = 1$, and that, for a fixed b, the length PQ is a continuous function of a defined also for $a = -1$. (We shall soon verify this analytically.) Therefore (9) holds also for $a = -1$ or for $b = 1$ if we interpret, for instance, the integral $\int_a^1 (1 - x^2)^{-1/2}\, dx$ as the limit of $\int_a^b (1 - x^2)^{-1/2}\, dx$ as $b \to 1$ from the left. (This is an example of an improper integral; compare §6.)

Fig. 5.54

Next we compute the area of the sector OPQ. Consider first the case $-1 < a < 0 < b < 1$; see Fig. 5.54. Let I, II, III be the areas indicated in the figure. Then I + II + III is the area under the curve $y = \sqrt{1 - x^2}$ from $x = a$ to $x = b$, hence

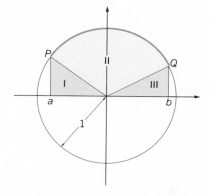

$$\text{I} + \text{II} + \text{III} = \int_a^b \sqrt{1 - x^2}\, dx.$$

Also, $\qquad$ $\mathrm{I} = -\dfrac{1}{2}a\sqrt{1-a^2}, \mathrm{II} = \dfrac{1}{2}b\sqrt{1-b^2},$

(since $a < 0$ and $b > 0$). The area of the sector is III. Thus

Fig. 5.55

(10) $\quad\begin{array}{l}\text{area of}\\ \text{sector } OPQ\end{array} = \displaystyle\int_a^b \sqrt{1-x^2}\,dx - \dfrac{1}{2}b\sqrt{1-b^2} + \dfrac{1}{2}a\sqrt{1-a^2}.$

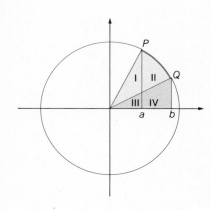

The same formula holds in the case $0 < a < b < 1$ shown in Fig. 5.55. Now the integral in (10) equals $\mathrm{II} + \mathrm{IV}$, and $(1/2)a\sqrt{1-a^2} = \mathrm{I} + \mathrm{III}$, $(1/2)b\sqrt{1-b^2} = \mathrm{III} + \mathrm{IV}$. The right-hand side of (10) equals $(\mathrm{II} + \mathrm{IV}) - (\mathrm{III} + \mathrm{IV}) + (\mathrm{I} + \mathrm{III}) = \mathrm{II} + \mathrm{I}$; this is the area of the sector. The reader will easily convince himself that (10) holds also if $a = -1$ or $a = 0$ and if $b = 0$ or $b = 1$.

The geometric statement (i) which we want to establish reads

(11) $\quad \displaystyle\int_a^b \sqrt{1-x^2}\,dx - \dfrac{1}{2}b\sqrt{1-b^2} + \dfrac{1}{2}a\sqrt{1-a^2} = \dfrac{1}{2}\int_a^b \dfrac{dx}{\sqrt{1-x^2}}.$

(We note that the left-hand side depends continuously on a and on b for $|a| \leq 1$, $|b| \leq 1$. Thus, once we prove (11), we shall have verified the statement we made above about the limits of (9) for $a \to -1^+$ and $b \to 1^-$.)

We rewrite (11) in the form

$$\int_a^b \sqrt{1-x^2}\,dx - \frac{1}{2}b\sqrt{1-b^2} + \frac{1}{2}a\sqrt{1-a^2} - \frac{1}{2}\int_a^b \frac{dx}{\sqrt{1-x^2}} = 0,$$

keep a fixed, and consider the left-hand side as a function of b. The derivative of this function is, by the fundamental theorem,

$$\frac{d}{db}\left(\int_a^b \sqrt{1-x^2}\,dx - \frac{1}{2}b\sqrt{1-b^2} + \frac{1}{2}a\sqrt{1-a^2} - \frac{1}{2}\int_a^b \frac{dx}{\sqrt{1-x^2}} \right)$$

$$= \sqrt{1-b^2} - \frac{1}{2}\sqrt{1-b^2} + \frac{1}{2}\frac{b^2}{\sqrt{1-b^2}} + 0 - \frac{1}{2}\frac{1}{\sqrt{1-b^2}} = 0.$$

The function is constant. But, for $b = a$, it equals 0. Hence (11) holds for all b.

For $a = 0$, $b = 1$ equation (11) becomes

$$\int_0^1 \frac{dx}{\sqrt{1-x^2}} = 2\int_0^1 \sqrt{1-x^2}\,dx$$

or, recalling the definition of π,

$$\int_0^1 \frac{dx}{\sqrt{1 - x^2}} = \frac{\pi}{2}.$$

Similarly, one obtains from (11), for $a = -1, b = 1$,

$$\int_{-1}^1 \frac{dx}{\sqrt{1 - x^2}} = \pi.$$

Since the integral above is half the unit circumference, this proves the geometric statement (ii): the unit circumference has length 2π.

EXERCISES

Exercises 55 to 60 refer to a circle of radius r with center O.

55. Let $L(a, b)$ denote the length of the arc of the upper semicircle from the point $P = (a, \sqrt{r^2 - a^2})$ to the point $Q = (b, \sqrt{r^2 - b^2})$, with $-r \leq a < b \leq r$. Let $A(a, b)$ denote the area of the sector OPQ. Write the two theorems stated in the beginning of this section as equations involving L and A.
56. Write $L(a, b)$ as an integral.
57. Find a formula for $A(a, b)$, analogous to relation (10).
58. Prove that $A(a, b) = (1/2)rL(a, b)$.
59. Show that $L(0, r) = (1/2)\pi r$.
60. Compute $L(-r, r)$.

Appendixes to Chapter 5

§5 Energy

5.1 Forces depending on position

We now apply integrals to mechanics. We consider a particle moving on a straight line, so that its position is described by a single number. We assume that the force F acting on the particle (in the direction of the line along which the particle moves) depends only on its location. This means that

$$s \mapsto F(s)$$

is a given function.

Fig. 5.56

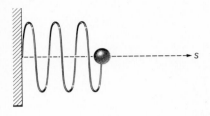

An example is the weight of a vertically moving particle of mass m. In this case

$$F(s) = -mg,$$

where g is the acceleration of gravity and the upward direction is considered as positive. This force is a constant.

Another example is a particle attached to a **spring** (Fig. 5.56). We choose $s = 0$ as the point at which the spring is relaxed and exerts no force on the particle. Thus $F(0) = 0$. If the particle is at a point $s \neq 0$, the spring will exert a **restoring force** on the particle, since it will tend to contract if extended and to expand if compressed. Thus $F(s)$ will be negative for $s > 0$ and positive for $s < 0$ (compare Fig. 5.57). The precise shape of the curve $F = F(s)$ depends on the spring. The simplest case of a **linear restoring force**

$$F = -ks$$

Fig. 5.57

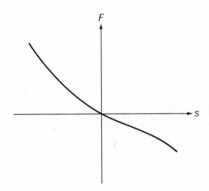

(see Fig. 5.58) is of particular importance. Most springs will exhibit approximately such a restoring force for small deflections. Indeed, if the curve $F = F(s)$ has a tangent at $s = 0$ and we consider only small values of s, we can replace the curve by its tangent without committing a significant error. A spring with a linear restoring force is called **elastic** and the value k is called the **spring constant**. Since F is measured in units of force (say gram cm/sec^2) and s in units of length (say cm), k is measured in units of mass per unit of time per unit of time (say gram/sec^2). Thus the numerical value of k depends on the units used.

EXERCISES

1. An elastic spring has a spring constant of 3 grams/sec^2. What restoring force will the spring exert on a body attached to one end if the spring is stretched 2 cm from its relaxed position?

2. When a $\frac{1}{2}$-gram mass is hung from a vertical elastic spring, the spring is stretched 5 cm from its relaxed position. What is the spring constant? (Neglect the weight of the spring.)

3. A 2-gram mass is hung from a vertical elastic spring whose spring constant is 100 gram/sec^2. How far is the spring stretched from its relaxed position? (Neglect the weight of the spring.)

4. Suppose that a spring has a restoring force given by $F(s) = 5s - 16s^3 + s^5$. For small deflections, we may treat it as an elastic spring with constant k. What is k?

Fig. 5.58

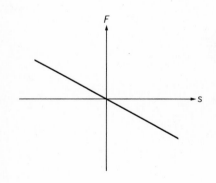

5.2 Work

We call the quantity

$$W = -\int_{s_0}^{s} F(x)\,dx$$

the **work against the force** F needed to move the particle from s_0 to s. This is a *definition;* it cannot be proved, but it can be motivated. Let us do so.

It is a common experience that in moving a particle against a force acting on it, say in lifting a heavy stone, effort is expended. Even before we define work, we have a feeling that it takes twice as much work to lift two equal weights as to lift one, and that the work needed to lift a stone to the height, say, of 3 ft is three times that needed to lift it 1 ft. We therefore define work against a constant force as the product of the *force times distance*. The work is positive if the object is moved in the direction opposite to that of the force, negative otherwise.

Consider now the work done against a nonconstant force in moving the particle from s_0 to, say, x: denote it by $W(x)$. We still have to define $W(x)$, but we certainly want to have

$$W(s_0) = 0,$$

since no work is involved in not moving the particle at all. It is also natural to require that

$$W(x + h) - W(x) = w$$

be the work done in moving the particle from the position x to the position $x + h$. Assume now that h is small. Over the small interval $(x, x + h)$, the force F will differ only a little from its value at x. We treat it as a constant and see that

$$w = -F(x)h \qquad \text{approximately.}$$

The minus sign is correct. If $F(x)$ is negative (say, the force pushes to the left) and h is positive (the particle is moved to the right), we want w to be positive. From the preceding equations, we conclude that

$$\frac{W(x + h) - W(x)}{h} = -F(x) \qquad \text{approximately.}$$

This suggests the requirement

$$W'(x) = -F(x),$$

and, by the condition $W(s_0) = 0$ and the Fundamental Theorem of Calculus, we obtain

(1) $$W(s) = -\int_{s_0}^{s} F(x)\, dx.$$

Work is measured in units of force times distance, say gram $\cdot$ cm^2/sec^2.

5.3 Potential energy

Let us fix s_0 ("the zero level") arbitrarily. Then $W(s)$ is called the **potential energy** of the particle located at s. In the case of a body subject to weight $F = -mg$, it is convenient to choose for $s_0 = 0$ the ground level. Then the potential energy of a particle at height s is

(2) $$W(s) = -\int_{0}^{s} (-mg)\, dx = mgs,$$

which is weight times height.

In the case of a spring, we also take $s_0 = 0$ as the zero level. If the tip of the spring is at the point s, the potential energy is

$$W(s) = - \int_0^s F(x) \, dx.$$

The reader should check that $W(s) > 0$ for $s < 0$ and for $s > 0$. For the linear restoring force $F = -ks$, we have

(3) $$W(s) = \frac{1}{2} ks^2.$$

One of the everyday meanings of the word "energy" is "the ability to do work." This explains the name "potential energy" for $W(s)$. A stone lifted to the height s will do work (fall) if released. So will a stretched spring.

EXERCISES

5. Two positively charged particles with charge 1 statcoulomb repel each other with a force equal to $1/r^2$, where r is the distance between the particles measured in centimeters. One particle is held fixed and the second is moved directly toward the first from a point 100 cm away to a point 10 cm away. How much work is done?

6. How much work is done in lifting a body whose mass is 10 grams from a height of 100 cm to a height of 200 cm?

7. How much work is done in stretching the end of an elastic spring 3 cm from its relaxed position if the spring constant is 15 grams/sec²?

8. What is the potential energy of a stone of weight 10 lb at a height of 50 ft?

9. What is the potential energy of an elastic spring if it is compressed 1 cm from its relaxed position and the spring constant is 4.5 grams/sec²?

10. How far is it necessary to compress an elastic spring whose spring constant is 3 grams/sec² in order that the compressed spring have a potential energy of 13.5 gram · cm/sec²?

5.4 Kinetic energy. Conservation of energy

A moving particle can do work. For instance, a stone that falls into water produces waves. We want therefore to define the energy of motion or **kinetic energy** K of a moving particle. Since K should be measured in units of work, that is, in gram · cm²/sec², mass measured in grams and velocity in cm/sec, the proper definition is $K = \alpha m v^2$, where α is a numerical constant. It turns out that we should choose $\alpha = 1/2$ and set

(4) $$K = \frac{1}{2} mv^2.$$

Why 1/2 is the "right" constant will be seen presently.

If a particle moves along a straight line under the influence of a force that depends only on its position, then the sum

$$E = W + K$$

of its potential and kinetic energies remains constant during the motion.

We call E the total energy of the particle. The fact that E is a constant is a special case of the **law of conservation of energy.** It is a mathematical consequence of Newton's law of motion and of the definitions of W and K. In following the proof, the reader should observe that it would have failed had we used a number different from $1/2$ in defining the kinetic energy K.

We have that E, W, and K are functions of time t. In particular,

$$K = \frac{1}{2} mv^2, \qquad v = \frac{ds}{dt},$$

so that, by the chain rule,

$$\frac{dK}{dt} = \frac{1}{2} m \frac{dv^2}{dt} = \frac{1}{2} m \frac{dv^2}{dv} \frac{dv}{dt} = mv \frac{d^2s}{d^2t}$$

or

$$\frac{dK}{dt} = mva,$$

where a is the acceleration. On the other hand, the potential energy at time t is the potential energy at the point $s(t)$, that is,

$$W(s(t)) = - \int_{s_0}^{s(t)} F(x)\, dx.$$

By the Fundamental Theorem of Calculus,

$$\frac{dW}{ds} = -F(s),$$

so that, by the chain rule,

$$\frac{dW}{dt} = \frac{dW}{ds} \frac{ds}{dt} = -Fv.$$

Thus

$$\frac{dE}{dt} = \frac{dW}{dt} + \frac{dK}{dt} = mva - Fv = (ma - F)v.$$

But, by Newton's law of motion,

$$ma = F.$$

Hence

$$\frac{dE}{dt} = 0 \qquad \text{for all } t \text{ considered,}$$

and E is constant, by Theorem 1 of Chapter 4, §5, as asserted.

5.5 Free fall

Let us apply the energy principle just proved to free fall. Here W is given by (2) so that the total energy is

$$E = \frac{1}{2}\, mv^2 + mgs.$$

This quantity is constant during motion. If a particle is released from height s_0, its velocity, at the instant of release, is 0. Hence

$$\frac{1}{2}\, mv^2 + mgs = mgs_0$$

during the motion. When the particle hits the ground, $s = 0$. Thus the terminal velocity, call it v_1, is computed from the equation

$$\frac{1}{2}\, mv_1^2 = mgs_0,$$

so

(5) $$v_1 = \sqrt{2gs_0}.$$

If a particle is thrown upward from the ground with speed v_0, its potential energy at the initial instant is 0, and the total energy is therefore $E = (\frac{1}{2})mv_0^2$. When the particle reaches ground level again, its kinetic energy is $\frac{1}{2}mv_1^2$, v_1 being the terminal velocity, and its potential energy is again 0. Thus $\frac{1}{2}mv_0^2 = \frac{1}{2}mv_1^2$, and since $v_0 > 0$ and $v_1 < 0$,

(6) $$v_1 = -v_0,$$

as we already observed in Chapter 4, §5.

The results (5) and (6) can also be derived from the general formula describing the position s at the time t of a freely falling particle which had the position s_0 and the velocity v_0 at the time $t = 0$. This formula (equation (3) in Chapter 4, §5) reads

$$s(t) = -\frac{1}{2}\, gt^2 + v_0 t + s_0.$$

It is also easy to derive directly from this formula the validity of the energy principle. In the following subsections, we shall apply the law of conservation of energy to motions for which we are as yet unable to compute the function $s(t)$.

EXERCISES

11. What velocity must a 1/2-lb snowball have in order to have the same kinetic energy as a 3000-lb car traveling at 10 mph? (Note: Pounds are a unit of force, not a unit of mass.)

12. From how high must a 2-lb flower pot be dropped in order to have the same kinetic energy as a 3000-lb car traveling at 10 mph?

13. A particle of mass m is moving in a straight line. If distance s from some fixed point on the line is measured in centimeters and time t is measured in seconds, then the position of the particle is given by $s = t^{3/2} + t + 1$ for $t \geq 0$. What is the kinetic energy of the particle when $t = 49$?

▶ 14. Let $K(t)$ denote the kinetic energy at time t of a body of mass m falling freely under the influence of gravity. Show that $K''(t)$ is constant and find its value.

15. A particle of unit mass is moving in a straight line under the influence of a force whose action at each point of the line is independent of position. Suppose the force acting on the moving particle at time t is found to be $6t$. Suppose also that the particle is at rest at time $t = 0$.
   (a) Find the velocity of the particle.
   (b) Find the kinetic energy of the particle directly from the definition.
   (c) Find the position of the particle (take the position at $t = 0$ as a reference point).
   (d) Find the force at each point s of the line.
   (e) Find the potential energy of the particle (take the position at $t = 0$ as the zero level) directly from the definition.
   (f) Verify that the law of conservation of energy holds.

16. A particle of unit mass is moving in a straight line under the influence of a force whose action at each point of the line is independent of time. The distance of the particle from some fixed point on the line is given by $s = t^{5/2}$, $t \geq 0$, where t denotes time.
   (a) Find the kinetic energy of the particle directly from the definition.
   (b) Find the force at each point s of the line.
   (c) Find the potential energy of the particle (take the base point as zero level) directly from the definition.
   (d) Verify that the law of conservation of energy holds.

5.6 Gravitation

Newton's law of universal gravitation states that, between any two particles with masses M and m and situated at a distance r from each other, there is an attracting force of magnitude

$$\frac{\gamma M m}{r^2}.$$

Here γ is a constant (**gravitational constant**), which is the same for all bodies. Since force is measured in units of mass times length per unit of time squared (say gram $\cdot$ cm/sec^2) and the expression Mm/r^2 is measured in units of mass squared per length squared (say gram/cm^2), γ must be measured in units of length cubed per unit of mass per unit of time squared (say cm^3/gram $\cdot$ sec^2). Thus the numerical value of γ depends on the units used. In applying Newton's law to a spherical body, like the sun or a planet, one may consider the total mass concentrated at the center. (This last statement is a mathematical theorem, which we shall not prove here, but only in Chapter 13.)

Fig. 5.59

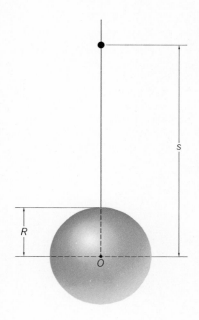

We shall show later how the law of gravitation follows from observed facts about planetary motions. Let us now apply it to the case where one "particle" is the earth and the other some body moving along a straight line passing through the center of the earth (Fig. 5.59). We denote this center by O and the distance of the moving particle from O by s. Let M be the mass of the earth and let m be the mass of the moving particle. Finally, let R denote the radius of the earth.

By the law of gravity, the force F exerted by the earth on the particle considered is

$$(7) \qquad F = -\frac{\gamma Mm}{s^2};$$

it is negative, since it pulls the body toward the earth. In everyday life, we deal with bodies whose distance x from the surface of the earth is very small compared to the radius R of the earth. For such a body, we have

$$s = R + x$$

and

$$F = -\frac{\gamma Mm}{(R + x)^2}.$$

If x/R is small, the number $1/(R + x)^2$ differs very little from $1/R^2$. In fact, for $x > 0$,

$$(8) \qquad \frac{1}{R^2} - \frac{1}{(R + x)^2} = \frac{2Rx + x^2}{(R + x)^2 R^2} < \frac{2x}{R^3} + \frac{x^2}{R^4}.$$

An approximate value of R, determined by geodesic measurements, is

$$R = 6.37 \cdot 10^8 \text{ cm} = 6370 \text{ km} = 3959 \text{ miles}.$$

If x is less than $1/1000$ of the earth radius, that is, less than 3.9 miles, the difference (8) is less than $1/10^{19}$ cm^2. We commit, therefore, an exceedingly small error if, for bodies close to the surface of the earth, we treat F as a constant and write

$$F = -\frac{\gamma Mm}{R^2}.$$

Setting

$$(9) \qquad \frac{\gamma M}{R^2} = g,$$

we obtain, from the law of motion,

$$-mg = ma \qquad (a = \text{acceleration}).$$

This is Galileo's law of constant acceleration $a = -g$.

(As a matter of fact, the value of g, the acceleration of gravity at the surface of the earth, is not the same everywhere. The dependence of g on geographic latitude was discovered in 1672. Newton concluded from this that the earth is not a perfect sphere; the equator is farther from the earth's center than are the poles. Very precise measurements disclose variations in the value of g due to the non-uniformity in the crust of the earth. This fact is used in scientific prospecting to locate the probable whereabouts of oil deposits, minerals, and the like.)

5.7 Escape velocity

We return to a body whose distance from the earth's surface is large enough to warrant the use of equation (7). Its law of motion is

$$ma = -\frac{\gamma Mm}{s^2}$$

or

(10)
$$\frac{d^2s}{dt^2} = -\frac{\gamma M}{s^2}.$$

At the present, we are unable to solve this differential equation, that is, to compute from it the function $s(t)$. However, since F depends only on s, the energy principle is applicable. We compute the potential energy of the body at the distance s from the center O, choosing $s = R$ (the earth's surface) as the zero level. By (1) and (7) we have

(11)
$$W = -\int_R^s \frac{-\gamma Mm}{x^2}\, dx = \gamma Mm \int_R^s \frac{dx}{x^2} = \gamma Mm\left(\frac{1}{R} - \frac{1}{s}\right).$$

Let us use this to compute the initial velocity v_0 which must be given to a body, say, a rocket, in order that it reach the distance A from O before beginning to fall back. At the initial instance, we have

$$K = \frac{1}{2}mv_0^2, \qquad W = 0.$$

Assume that when the body reaches height A its velocity becomes zero, so that it begins to fall back toward the earth. At this instant we have $s = A$, $v = 0$, and hence

$$K = 0, \qquad W = \gamma Mm\left(\frac{1}{R} - \frac{1}{A}\right).$$

But the sum $K + W$ remains the same throughout the motion. Hence

$$\frac{1}{2}mv_0^2 = \gamma Mm\left(\frac{1}{R} - \frac{1}{A}\right)$$

or

$$v_0 = \sqrt{2\gamma M\left(\frac{1}{R} - \frac{1}{A}\right)}.$$

In view of (9), we have that $\gamma M = gR^2$. Thus

(12) $$v_0 = \sqrt{2g\left(R - \frac{R^2}{A}\right)}.$$

Note that v_0 does not depend on m.

As A grows without bound, v_0 approaches the value

(13) $$V = \sqrt{2gR}.$$

A body projected with a speed greater than, or equal to, V will *never* return; one therefore calls V the *escape velocity*. Substitution of the value of R given above and of the value $g = 980$ cm/sec^2 yields

$$V = 1.17 \cdot 10^6 \text{ cm/sec.}$$

Of course, our calculation neglected the effect of air resistance and the gravitational effect of other heavenly bodies. But the value we obtained for V is not too far from the actual escape velocity from earth.

EXERCISES

In Exercises 17 to 21, the value of the gravitational constant γ should be taken as 6.67×10^{-8} cm^3/gram $\cdot$ sec^2.

17. The radius of the moon is approximately 1.74×10^8 cm; its mass is approximately 7.35×10^{25} grams. Compute the escape velocity of a particle from the moon.

▶ 18. The radius of the planet Jupiter is approximately 70.96×10^8 cm; its mass is approximately 1.88×10^{30} grams. Compute the escape velocity of a particle from Jupiter.

19. The radius of the sun is approximately 692.32×10^8 cm; its mass is approximately 1.97×10^{33} grams. Compute the escape velocity of a particle from the sun.

20. Let v_0 denote the escape velocity of a particle from a large spherical body of mass M and radius R. Suppose R remains constant while M varies; for example, if meteors continually fall from space and are embedded in the sphere, the mass will increase while the radius remains effectively unchanged. Find dv_0/dM.

21. In Exercise 20, suppose M remains constant while R varies. (An example of a spherical astronomical body with changing radius is a pulsating star, although in the case of a star the mass is not constant either.) Find dv_0/dR.

5.8 Energy and mass in relativity theory

Let us now consider a particle that moves so fast that one should use Einstein's equation of motion [see equation (4) in Chapter 4, §3], that is,

$$m_0 \frac{d}{dt} \frac{v}{\sqrt{1 - (v^2/c^2)}} = F,$$

where c is the velocity of light and m_0 the mass of the particle measured at rest (the so-called rest mass). In Chapter 4, §3, we showed that the Einstein equation of motion can also be written in the form

$$(14) \qquad \frac{m_0 a}{\sqrt{(1 - (v^2/c^2)^3}} = F \qquad \text{where } a = \frac{dv}{dt}.$$

The energy of the moving particle is now *defined* by

$$(15) \qquad K = \frac{m_0 c^2}{\sqrt{1 - (v^2/c^2)}}.$$

This definition is justified by noting that the sum $E = K + W$ again remains constant during the motion. [Here W is the potential energy defined as before by (1).] Indeed, we have [by (14)]:

$$\frac{dE}{dt} = \frac{dK}{dt} + \frac{dW}{dt} = \text{(applying the chain rule)} \frac{dK}{dv} \frac{dv}{dt} + \frac{dW}{ds} \frac{ds}{dt}$$

$$= \frac{m_0 c^2 \left(\frac{-1}{2}\right)\left(-\frac{2v}{c^2}\right)}{\sqrt{\left(1 - \frac{v^2}{c^2}\right)^3}} \frac{dv}{dt} + (-F)v = \frac{m_0 v}{\sqrt{\left(1 - \frac{v^2}{c^2}\right)^3}} a - Fv = 0.$$

There are good physical reasons to call the quantity

$$(16) \qquad m = \frac{m_0}{\sqrt{1 - (v^2/c^2)}}$$

the mass of the moving particle. Comparing (15) and (16), we conclude that an increase (or decrease) in energy produces an increase (or decrease) in mass according to the relation

$$(17) \qquad E = mc^2.$$

We gave, of course, not a proof, but only an example of this celebrated relation discovered by Einstein.

§6 Improper Integrals

In this section we describe a useful extension of the concept of the integral. We already used this extension, rather informally, in §4.9.

6.1 Unbounded integrands

The number

$$\int_a^b f(x)\, dx$$

has been defined thus far only for bounded (and piecewise continuous) functions f and for finite intervals (a, b). In some cases, however, there is a natural way of assigning a meaning to the integral also when the function $f(x)$ becomes arbitrarily large for x close to a or b, and also for infinite intervals (a, b). We shall use such **improper integrals** only in the case of nonnegative functions. Instead of giving a general definition, we consider some typical examples.

The graph of the function

$$y = \frac{1}{\sqrt{x}}$$

defined for $x > 0$ is shown in Fig. 5.60. If ϵ is a small positive number, we have

$$\int_\epsilon^1 \frac{dx}{\sqrt{x}} = 2\sqrt{x} \Big|_\epsilon^1 = 2 - 2\sqrt{\epsilon}.$$

Since $\sqrt{\epsilon}$ is a continuous function of ϵ for $0 \le \epsilon$, and $\sqrt{0} = 0$, we have

(1)
$$\lim_{\epsilon \to 0^+} \int_\epsilon^1 \frac{dx}{\sqrt{x}} = 2.$$

(Recall that "$\epsilon \to 0^+$" means "as ϵ approaches zero through positive values.") It is natural to say that the area enclosed between the y axis, the line $x = 1$, and our curve has the value 2, even though the region considered extends arbitrarily far upward. Therefore we write

$$\int_0^1 \frac{dx}{\sqrt{x}} = 2.$$

This is to be understood as an abbreviated form of equation (1). One also says that the improper integral $\int_0^1 x^{-1/2}\, dx$ **converges**, and that it converges to the value 2.

Consider next the function

$$y = \frac{1}{x^2}$$

(defined for $x \ne 0$). We know that $(-1/x)$ is a primitive function. If ϵ is a small positive number

$$\int_\epsilon^1 \frac{dx}{x^2} = -\frac{1}{x} \Big|_\epsilon^1 = \frac{1}{\epsilon} - 1.$$

This quantity will be arbitrarily large positive if ϵ is sufficiently small; the statement just made is abbreviated by writing

$$\lim_{\epsilon \to 0^+} \int_\epsilon^1 \frac{dx}{x^2} = +\infty.$$

Fig. 5.60

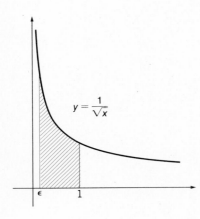

$$y = \frac{1}{\sqrt{x}}$$

(Recall that "∞" is read "infinity;" it is *not* the name of a number.) We will therefore not ascribe any numerical value to the symbol

$$\int_0^1 \frac{dx}{x^2}.$$

We also say that the improper integral $\int_0^1 x^{-2}\, dx$ **diverges.**

6.2 Infinite intervals of integration

We compute next the area under the curve $y = 1/x^2$ from 1 to a large positive number a. We have that

$$\int_1^a \frac{dx}{x^2} = -\frac{1}{x}\Big|_1^a = 1 - \frac{1}{a}.$$

Therefore

$$\lim_{a \to +\infty} \int_1^a \frac{dx}{x^2} = 1$$

or, in an abbreviated notation,

(2) $$\int_1^{+\infty} \frac{dx}{x^2} = 1.$$

We also say that the improper integral $\int_1^{+\infty} x^{-2}\, dx$ converges (to 1). Similarly,

(3) $$\int_{-\infty}^{-1} \frac{dx}{x^2} = 1,$$

which means that

$$\lim_{a \to -\infty} \int_a^{-1} \frac{dx}{x^2} = 1.$$

On the other hand, the improper integral

$$\int_1^{+\infty} \frac{dx}{\sqrt{x}}$$

diverges, so that no numerical value is attached to the symbol above. To verify this, we note that, for $a > 0$, we have

$$\int_1^a \frac{dx}{\sqrt{x}} = 2\sqrt{x}\,\Big|_1^a = 2\sqrt{a} - 2$$

and the limit of this expression for $a \to +\infty$ is $+\infty$.

6.3 Other examples

In the examples above, the "impropriety" always occurred at one endpoint of the interval of integration. This need not be so in all cases. Here is an example.

The meaning of the equation

$$\int_{-1}^{1} \frac{dx}{\sqrt{|x|}} = 4$$

is seen from Fig. 5.61. It is an abbreviation of the statement: if ϵ_1 and ϵ_2 are small positive numbers, then

$$\int_{-1}^{-\epsilon_2} \frac{dx}{\sqrt{|x|}} + \int_{\epsilon_1}^{1} \frac{dx}{\sqrt{|x|}}$$

is close to 4, as close as we like provided ϵ_1 and ϵ_2 are small enough. Another way of writing this is as follows: The improper intervals

$$\int_{-1}^{0} \frac{dx}{\sqrt{|x|}} , \int_{0}^{1} \frac{dx}{\sqrt{|x|}}$$

converges, and

$$\int_{-1}^{1} \frac{dx}{\sqrt{|x|}} = \int_{-1}^{0} \frac{dx}{\sqrt{|x|}} + \int_{0}^{1} \frac{dx}{\sqrt{|x|}} = 2 + 2 = 4.$$

In §4.9, we showed that

(4) $$\int_{0}^{1} \frac{dx}{\sqrt{1 - x^2}} = \frac{\pi}{2} ;$$

this is an improper integral; the equation just written is an abbreviation for the statement

$$\lim_{b \to 1^-} \int_{0}^{b} \frac{dx}{\sqrt{1 - x^2}} = \frac{\pi}{2}.$$

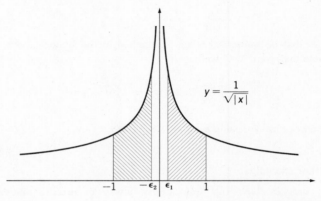

$$y = \frac{1}{\sqrt{|x|}}$$

Fig. 5.61

The rules for integration derived in §3 remain valid for integrals of piecewise continuous functions and also for improper integrals. The proof is almost obvious and need not be spelled out.

EXERCISES

In Exercises 1 to 10, determine if the given improper integrals exist (that is, can be assigned a finite value) and evaluate them if they do. (Hint: Use integration by parts if you cannot obtain the answer at once.)

1. $\int_0^1 \dfrac{dx}{x^{1/3}}$.

2. $\int_1^\infty \dfrac{dx}{x^{3/2}}$.

3. $\int_{-1}^0 \dfrac{dt}{t^{4/3}}$.

4. $\int_{-1}^0 (s+1)^{-5/4}\,ds$.

5. $\int_0^\infty \dfrac{ds}{\sqrt{s+1}}$.

6. $\int_2^\infty \dfrac{u\,du}{(u^2+1)^3}$.

7. $\int_0^2 \dfrac{dy}{(1-y)^{2/3}}$.

▶ 8. $\int_0^{\sqrt{2}} \dfrac{v\,dv}{(v^2-1)^{4/5}}$.

9. $\int_{-1}^1 \dfrac{dy}{(2y+1)^2}$.

10. $\int_0^\infty \dfrac{x\,dx}{(x+1)^3}$.

6.4 Comparison Theorem

There exists a useful criterion for the convergence of improper integrals for nonnegative functions.

Theorem 1 (Comparison Test for Improper Integrals) LET $f(x)$ AND $g(x)$ BE TWO FUNCTIONS DEFINED FOR $a < x < b$, WHERE a IS A NUMBER AND b EITHER A NUMBER OR THE SYMBOL $+\infty$. ASSUME THAT, FOR EVERY c SUCH THAT $a < c < b$, THE FUNCTIONS f AND g ARE BOUNDED AND PIECEWISE CONTINUOUS IN (a, c), AND THAT

(5) $$0 \le f(x) \le g(x) \qquad \text{FOR } a < x < b.$$

THEN, IF THE INTEGRAL

$$\int_a^b g(x)\,dx$$

CONVERGES, THE INTEGRAL

$$\int_a^b f(x)\,dx$$

ALSO CONVERGES.

The theorem covers two cases: $b = +\infty$ and $b < +\infty$, and the functions f and g become infinite as $x \to a^+$. The validity of the theorem is geometrically obvious. Condition (5) implies that the graph of f lies between the x axis and the graph of g. Therefore, if the area under the graph of g (from a to b) is finite, so is the area under the graph of f. An analytic proof is given in another appendix (§9).

There is an analogous statement for the case in which the "impropriety" occurs at the left-hand endpoint.

As an example, we show that the improper integral

$$\int_{-\infty}^{+\infty}\frac{dx}{1+x^2}$$

converges. We note first that

$$\int_{-\infty}^{+\infty}\frac{dx}{1+x^2} = \int_{-\infty}^{-1}\frac{dx}{1+x^2} + \int_{-1}^{1}\frac{dx}{1+x^2} + \int_{1}^{+\infty}\frac{dx}{1+x^2}.$$

The left-hand side of this equation will have a meaning if all three terms on the right do. About the middle term we need not worry, since the function $1/(1+x^2)$ is continuous. Do the first and last terms have meaning? In other words, do these two integrals converge? They do, by the comparison test, since

$$\frac{1}{1+x^2} < \frac{1}{x^2}$$

and we know that the integrals (2) and (3) have definite finite values (see Fig. 5.62). In the next subsection, we shall obtain the remarkable result

(6)
$$\int_{-\infty}^{+\infty}\frac{dx}{1+x^2} = \pi,$$

where π is the ratio of the length of a circle to that of its diameter.

It is also geometrically obvious, and follows from Theorem 1, that, under the hypotheses of the theorem, if the integral $\int_a^b f(x)\, dx$ diverges, then the integral $\int_a^b g(x)\, dx$ also diverges.

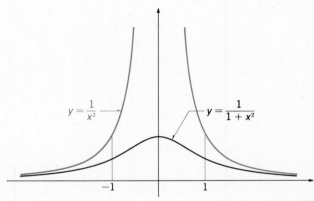

Fig. 5.62

EXERCISES

In Exercises 11 to 20, use one of the tests discussed in this section to determine if the given improper integrals converge or diverge.

11. $\int_1^\infty \dfrac{dx}{x^3 + 1}$.

▶ 16. $\int_0^1 \dfrac{\sqrt{x}}{(x - 1)^{2/3}}\, dx$.

12. $\int_4^\infty \dfrac{dx}{\sqrt{x} - 1}$.

17. $\int_{-1}^1 \dfrac{(t + 2)^{1/3}}{(4t - 1)^2}\, dt$.

13. $\int_0^1 \dfrac{dt}{t + \sqrt{t}}$.

18. $\int_0^2 \dfrac{z + 1}{(z^2 - 4)^{4/3}}\, dz$.

14. $\int_1^\infty \dfrac{dt}{(2t - 1)^2 + 1}$.

19. $\int_1^\infty \dfrac{dy}{(1 + y^2)^2}$.

15. $\int_0^1 \dfrac{\sqrt{v + 1}}{v^3}\, dv$.

20. $\int_{1/4}^1 \dfrac{dz}{(\sqrt{z} - 1)^2}$.

6.5 Some integral formulas involving π

In order to prove relation (6), we note first that the function $t = 1/x$ is decreasing for $x > 0$. Choose a number $N > 0$ and apply to the integral

$$\int_{1/N}^1 \frac{dx}{1 + x^2}$$

the substitution

$$x = \frac{1}{t}, \qquad dx = -\frac{dt}{t^2}, \qquad t = N \text{ for } x = \frac{1}{N}, \qquad t = 1 \text{ for } x = 1.$$

We obtain

$$\int_{1/N}^1 \frac{dx}{1 + x^2} = \int_N^1 \frac{1}{1 + (1/t)^2}\left(-\frac{1}{t^2}\right) dt = \int_1^N \frac{dt}{1 + t^2}$$

or, since both t and x are dummy variables,

$$\int_{1/N}^1 \frac{dx}{1 + x^2} = \int_1^N \frac{dx}{1 + x^2}.$$

The first integral is, for large N, as close as we like to $\int_0^1 dx/(1 + x^2)$. The same is therefore true for the second integral. Hence

(7) $$\int_0^1 \frac{dx}{1 + x^2} = \int_1^{+\infty} \frac{dx}{1 + x^2}.$$

We see now that we could have been more courageous; we could have applied at once to the first integral above the substitution $x = 1/t$, $dx = -(dt/t^2)$, "$t = +\infty$" for $x = 0$, $t = 1$ for $x = 1$. Since the function $1/(1 + x^2)$ is even, we have (see §3.6)

$$\int_{-1}^0 \frac{dx}{1 + x^2} = \int_0^1 \frac{dx}{1 + x^2}, \qquad \int_{-\infty}^{-1} \frac{dx}{1 + x^2} = \int_1^{+\infty} \frac{dx}{1 + x^2}.$$

Therefore

(8)
$$\int_{-\infty}^{-1} \frac{dx}{1+x^2} = \int_{-1}^{0} \frac{dx}{1+x^2} = \int_{0}^{1} \frac{dx}{1+x^2} = \int_{1}^{+\infty} \frac{dx}{1+x^2}$$

and since
$$\int_{-\infty}^{+\infty} \frac{dx}{1+x^2} = \int_{-\infty}^{-1} + \int_{-1}^{0} + \int_{0}^{1} + \int_{1}^{+\infty} \frac{dx}{1+x^2},$$

we have
$$\int_{-\infty}^{+\infty} \frac{dx}{1+x^2} = 4 \int_{0}^{1} \frac{dx}{1+x^2}.$$

Thus we must show that

(9)
$$\int_{0}^{1} \frac{dx}{1+x^2} = \frac{\pi}{4}.$$

To do this, we apply to the integral in (4) the substitution

$$x = \frac{2t}{1+t^2}$$

and note that $x = 0$ for $t = 0,$ $x = 1$ for $t = 1.$

Also
$$\frac{dx}{dt} = \frac{(1+t^2)2 - 2t(2t)}{(1+t^2)^2} = \frac{2(1-t^2)}{(1+t^2)^2}.$$

This shows that x is a function of t increasing for $0 < t < 1$. Finally,

$$\sqrt{1-x^2} = \sqrt{1 - \frac{4t^2}{(1+t^2)^2}} = \frac{1-t^2}{1+t^2}.$$

We obtain

$$\frac{\pi}{2} = \int_{0}^{1} \frac{dx}{\sqrt{1-x^2}} = \int_{0}^{1} \frac{1+t^2}{1-t^2} \frac{2(1-t^2)}{(1+t^2)^2} dt = 2 \int_{0}^{1} \frac{dt}{1+t^2},$$

which proves (9). The geometric meaning of (6) and (9) is shown in Fig. 5.63.

The manipulations we carried out in this and the preceding paragraphs are of

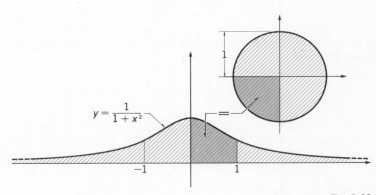

Fig. 5.63

the kind that made Euler say: "My pen is more intelligent than I." We remark, however, that in Chapter 6 the relations (5) and (8) will appear less mysterious than they seem now.

LEONHARD EULER (1707–1789) was born in Basel, Switzerland, and studied there under Johann Bernoulli. He spent most of his life as a member of the Academies in Berlin and in St. Petersburg (now Leningrad). Euler was not only the most productive mathematician of all time, but one of the best; he wrote on all mathematical subjects and on some nonmathematical ones as well. During the last 13 years of his life, he worked in total blindness.

Euler also wrote textbooks of calculus of which all subsequent texts are derivatives. Euler's collected works are still in the process of publication; over 70 large volumes have appeared thus far.

EXERCISES

21. Find the volume of the solid formed by rotating the region under the graph of $y = 1/\sqrt{1 + x^2}$ from $x = 0$ to $x = 1$ around the x axis.

22. Let V_a denote the volume of the solid formed by rotating the region under the graph of $y = 1/\sqrt{1 + x^2}$ from $x = 1$ to $x = a (a > 1)$ around the x axis. Find $\lim_{a \to \infty} V_a$ if it exists. This can be thought of as the volume of the solid formed by rotating the region under the graph for $1 \leq x < \infty$ around the x axis.

23. Determine the value of $\int_{-\infty}^{\infty} dx/(1 + \pi x^2)$ if it exists.

24. Determine the value of $\int_1^{\infty} u \, du/(1 + u^4)$ if it exists. (Hint: Try the substitution $u^2 = x$.)

25. Determine the value of $\int_0^1 \sqrt{u} \, du/\sqrt{1 - u}$ if it exists. (Hint: Try the substitution $u = x^2$.)

§7 Proof of the Fundamental Theorem

In §2, we gave only a plausability argument for the first part of the fundamental theorem of calculus. Now we give a formal proof.

Let (a, b) be a finite interval and let $f(x)$ be a bounded piecewise continuous function defined for $a < x < b$. We set

$$G(x) = \int_a^x f(t) \, dt$$

and must show that $G(x)$ is continuous for $a \leq x \leq b$, and that $G'(x_0) = f(x_0)$ if f is continuous at x_0.

7.1 Continuity proof

Let x_0 and x be two distinct points in the interval $[a, b]$. By the additivity of the integral

$$\int_a^x f(t) \, dt = \int_a^{x_0} f(t) \, dt + \int_{x_0}^x f(t) \, dt,$$

so that

$$G(x) - G(x_0) = \int_{x_0}^x f(t) \, dt.$$

Since f is bounded, there is a number M such that $|f(t)| \leq M$ for all t considered, that is,

$$-M \leq f(t) \leq M.$$

Assume that $x > x_0$. By Theorem 2 in §1, we have

$$\int_{x_0}^x (-M) \, dt \leq \int_{x_0}^x f(t) \, dt \leq \int_{x_0}^x M \, dt$$

or
$$-M(x - x_0) \leq G(x) - G(x_0) \leq M(x - x_0)$$

which is the same as

$$|G(x) - G(x_0)| \leq M|x - x_0|.$$

One proves in the same way that the inequality holds also if $x < x_0$. This shows that G is continuous at x_0: if x lies in an interval of length 2δ with x_0 as midpoint, $G(x)$ differs from $G(x_0)$ by at most $2M\delta$. (We note explicitly that x_0 could have been one of the endpoints a or b.)

7.2 Computing the derivative

Assume now that x_0 is a point in (a, b) and f is continuous at x_0. Let ϵ be any positive number. Then, by the definition of continuity,

$$f(x_0) - \epsilon \leq f(t) \leq f(x_0) + \epsilon$$

for t near x_0, that is, for $x_0 - \delta < t < x_0 + \delta$, where δ is a sufficiently small positive number. Let x be so close to x_0 that $|x - x_0| < \delta$. Then the above inequalities hold for t between x_0 and x. By the monotonicity property of the integral, we have

$$\int_{x_0}^{x} [f(x_0) - \epsilon]\, dt \leq \int_{x_0}^{x} f(t)\, dt \leq \int_{x_0}^{x} [f(x_0) + \epsilon]\, dt \qquad \text{if } x > x_0$$

and
$$\int_{x}^{x_0} [f(x_0) - \epsilon]\, dt \leq \int_{x}^{x_0} f(t)\, dt \leq \int_{x}^{x_0} [f(x_0) + \epsilon]\, dt \qquad \text{if } x < x_0.$$

Evaluating the integrals of constants, we obtain

$$[f(x_0) - \epsilon](x - x_0) \leq \int_{x_0}^{x} f(t)\, dt \leq [f(x_0) + \epsilon](x - x_0) \qquad \text{if } x > x_0$$

and
$$[f(x_0) - \epsilon](x_0 - x) \leq \int_{x}^{x_0} f(t)\, dt \leq [f(x_0) + \epsilon](x_0 - x) \qquad \text{if } x < x_0.$$

Dividing the first inequality by the positive number $x - x_0$, we obtain

$$f(x_0) - \epsilon \leq \frac{1}{x - x_0} \int_{x_0}^{x} f(t)\, dt \leq f(x_0) + \epsilon.$$

The same result is obtained by dividing the second inequality by $x_0 - x$, since

$$\frac{1}{x_0 - x} \int_{x}^{x_0} f(t)\, dt = \frac{1}{x - x_0} \int_{x_0}^{x} f(t)\, dt.$$

Thus, in all cases,

$$f(x_0) - \epsilon \leq \frac{G(x) - G(x_0)}{x - x_0} \leq f(x_0) + \epsilon$$

if $x \neq x_0$ and $|x - x_0| < \delta$. More precisely, we have shown that, given any $\epsilon > 0$, there is a $\delta > 0$ such that the above holds. This means that

$$\lim_{x \to x_0} \frac{G(x) - G(x_0)}{x - x_0} = f(x_0)$$

or $G'(x_0) = f(x_0)$, as asserted. A similar argument applies to the case in which x_0 is one of the endpoints of (a, b).

§8 Existence of Integrals

In this section, we prove the existence and the three basic properties of integrals, that is, Theorems 1 and 2 of §1. The proofs are simple but rather long; they depend on the least upper bound principle.

The definition of the integral discussed below was first given, in a different form, by Riemann in 1854.

8.1 Integrals of step functions

Throughout this section (a, b) denotes a finite interval, and all functions considered are assumed to be defined in this interval and to be bounded. If f and g are two functions, the inequality $f \leq g$ means that

$$f(x) \leq g(x) \qquad \text{for all } x, \qquad a < x < b.$$

The letters ϕ, ψ, χ will always denote step functions.

We recall that if ϕ is a step function defined in (a, b), then (a, b) can be divided into a finite number of subintervals (**intervals of constancy**) in each of which $\phi(x)$ has a constant value. The values of ϕ at the endpoints of the intervals of constancy will be of no importance in what follows.

Let ϕ be a step function defined in (a, b), let it have n intervals of constancy of lengths $l_1, l_2, \cdots, l_n$, and let it take on the values $v_1, v_2, \cdots, v_n$ in those intervals. We recall that the integral $\int_a^b \phi(x)\, dx$ is defined as

$$(1) \qquad \int_a^b \phi(x)\, dx = v_1 l_1 + v_2 l_2 + \cdots + v_n l_n.$$

This is precisely "the area between the curve $y = \phi(x)$ and the x axis from a to b, areas under the x axis being counted as negative" (see Fig. 5.64). Since this is a sum of areas of rectangles, the definition is unambiguous.

Remark Given a step function with n intervals of constancy, we may, if we choose, divide some of them into several smaller subintervals and call each of those an interval of constancy. (See Figs. 5.64 and 5.65, where the same function ϕ is considered as having first 5 and then 8 intervals of constancy.) One sees easily that this does not change the value of the integral. Simply observe what happens when one interval of constancy is divided into two.

BERNHARD RIEMANN (1826–1866) was educated at Goettingen University and later taught there. At the age of 33, he became a full professor and no longer had to worry about how to support himself. At 40, he died of tuberculosis.

Riemann's collected papers form a small volume of about 500 pages; about half of these contain posthumous works. Yet, a large part of modern research in mathematics and mathematical physics consists of using and developing the methods and ideas of Riemann.

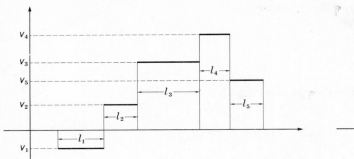

Fig. 5.64

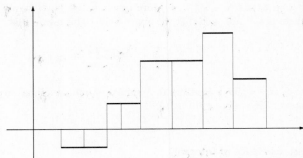

Fig. 5.65

Similarly, if ϕ takes on the same value in one or more adjacent intervals of constancy, we may, if we choose, combine them into a single interval of constancy. Again, this does not change the value of the integral.

Lemma A THE FOLLOWING STATEMENTS ARE TRUE FOR STEP FUNCTIONS:

(A1)   IF $\phi(x) = \alpha$, α A CONSTANT, THEN $\int_a^b \phi(x)\, dx = \alpha(b - a)$,

(A2)   IF $a < c < b$, THEN $\int_a^c \phi(x)\, dx + \int_c^b \phi(x)\, dx = \int_a^b \phi(x)\, dx$,

(A3)   IF $\phi \leq \psi$, THEN $\int_a^b \phi(x)\, dx \leq \int_a^b \psi(x)\, dx$.

Proof (A1). The function ϕ may be considered to have a single interval of constancy (by the remark made above). The length of this interval is $b - a$. Hence

$$\int_a^b \phi(x)\, dx = \alpha(b - a)$$

by the definition of the integral of a step function.

(A2). We may assume that c is an endpoint of an interval of constancy of ϕ (by the remark made above). Suppose that (a, c) consists of the first k intervals of constancy and (c, b) of the remaining $n - k$. Then

$$\int_a^c \phi(x)\, dx = v_1 l_1 + \cdots + v_k l_k, \int_c^b \phi(x)\, dx = v_{k+1} l_{k+1} + \cdots + v_n l_n.$$

Comparing this with (1), we see that the desired conclusion holds.

(A3). We may assume that ϕ and ψ have the same intervals of constancy (by the remark made above). Then

(2) $$\int_a^b \psi(x)\, dx = w_1 l_1 + w_2 l_2 + \cdots + w_n l_n$$

where $w_1, w_2, \cdots, w_n$ are the values of ψ in the first, second, $\cdots$, nth interval of constancy. If $\phi \leq \psi$, then $v_1 \leq w_1$, $v_2 \leq w_2$, $\cdots$, $v_n \leq w_n$ and therefore $v_1 l_1 \leq w_1 l_1$, $v_2 l_2 \leq w_2 l_2$, $\cdots$, $v_n l_n \leq w_n l_n$. Comparing (1) and (2), we see that the desired inequality holds.

8.2 Upper and lower integrals

Let f be a bounded function and let M be a number such that $|f| \leq M$. Note that, if ϕ is a step function such that $\phi \leq M$, then by (A1) and (A3) we have that $\int_a^b \phi(x)\,dx \leq M(b-a)$. Hence, if we consider all step functions ϕ such that $\phi \leq f$, the set of numbers $\int_a^b \phi(x)\,dx$ is bounded from above. By the least upper bound principle, the set has a least upper bound. We call this number the **lower integral** of f over (a, b) and we denote it by $\underline{\int_a^b} f(x)\,dx$. Thus

$$\underline{\int_a^b} f(x)\,dx = \text{the least upper bound of } \int_a^b \phi(x)\,dx \text{ for all step functions } \phi \leq f.$$

One defines similarly the **upper integral** of f:

$$\overline{\int_a^b} f(x)\,dx = \text{greatest lower bound of } \int_a^b \psi(x)\,dx \text{ for all step functions } \psi \geq f.$$

The definition is legitimate, since the set of numbers $\int_a^b \psi(x)\,dx$ for all $\psi \geq f$ is bounded from below. For if $\psi \geq f$, then $\psi \geq -M$ and hence $\int_a^b \psi(x)\,dx \geq -M(b-a)$, by (A1) and (A3).

We note that lower and upper integrals are defined for every bounded function. For step functions, we get nothing new.

Lemma B   IF χ IS A STEP FUNCTION, THEN

$$\underline{\int_a^b} \chi(x)\,dx = \int_a^b \chi(x)\,dx = \overline{\int_a^b} \chi(x)\,dx.$$

Proof.   Let S be the set of all numbers $\int_a^b \phi(x)\,dx$ where $\phi \leq \chi$ and let I be the least upper bound of S. Since χ is a step function and $\chi \leq \chi$, the number $\int_a^b \chi(x)\,dx$ belongs to S. Therefore

$$\int_a^b \chi(x)\,dx \leq I.$$

On the other hand, if $\phi \leq \chi$, then $\int_a^b \phi(x)\,dx \leq \int_a^b \chi(x)\,dx$ by (A3), so that $\int_a^b \chi(x)\,dx$ is an upper bound for S. Therefore, I being the smallest upper bound,

$$I \leq \int_a^b \chi(x)\,dx.$$

Since $I = \underline{\int_a^b} \chi(x)\,dx$ by definition, we have proved half of the lemma. The other half can be proved similarly.

Lemma C   FOR EVERY BOUNDED FUNCTION f AND FOR EVERY NUMBER $\epsilon > 0$, THERE ARE STEP FUNCTIONS ϕ AND ψ SUCH THAT

$$\phi \leq f \text{ AND } \int_a^b \phi(x)\,dx \leq \underline{\int_a^b} f(x)\,dx \leq \int_a^b \phi(x)\,dx + \epsilon,$$

$$\psi \geq f \text{ AND } \int_a^b \psi(x)\,dx \geq \overline{\int_a^b} f(x)\,dx \geq \int_a^b \psi(x)\,dx - \epsilon.$$

Proof. The lemma asserts that the lower (upper) integral of f can be approximated arbitrarily closely by the integral of a step function that lies below (above) f. We only prove the existence of ϕ; that of ψ can be proved similarly.

Let S be the set of all numbers $\int_a^b \phi(x)\, dx$ for all $\phi \leq f$. Set $\underline{\int_a^b} f(x)\, dx = I$. Since I is the least upper bound of S, we have $\int_a^b \phi(x)\, dx \leq I$ whenever $\phi \leq f$. Let $\epsilon > 0$ be given. If for all $\phi \leq f$ we had $\int_a^b \phi(x)\, dx < I - \epsilon$, $I - \epsilon$ would be an upper bound for S, a smaller bound than I. This is impossible. Hence there is a ϕ with $I \leq \int_a^b \phi(x)\, dx + \epsilon$, that is a ϕ with the required properties.

We remark that, in general, the lower and upper integrals of a function are different. For instance, if $f(x) = 0$ for x rational, and $f(x) = 1$ for x irrational, then $\underline{\int_0^1} f(x)\, dx = 0$, $\overline{\int_0^1} f(x)\, dx = 1$.

Lemma D   FOR EVERY BOUNDED FUNCTION f,

$$\underline{\int_a^b} f(x)\, dx \leq \overline{\int_a^b} f(x)\, dx.$$

Proof.   Let ϕ and ψ be as in Lemma C. By (A3), we have $\int_a^b \phi(x)\, dx \leq \int_a^b \psi(x)\, dx$. Hence, by Lemma C,

$$\underline{\int_a^b} f(x)\, dx \leq \int_a^b \phi(x)\, dx + \epsilon \leq \int_a^b \psi(x)\, dx + \epsilon \leq \overline{\int_a^b} f(x)\, dx + 2\epsilon.$$

Since this is so for every positive ϵ, the desired inequality follows.

8.3 Properties of upper and lower integrals

We can now show that upper and lower integrals have the three characteristic properties of integrals.

Lemma E   LET f AND g BE BOUNDED FUNCTIONS. THEN

(E1)   IF $f(x) = \alpha$, A CONSTANT, THEN $\displaystyle\underline{\int_a^b} f(x)\, dx = \overline{\int_a^b} f(x)\, dx = \alpha(b - a)$,

(E2)   IF $a < c < b$, THEN $\displaystyle\underline{\int_a^c} f(x)\, dx + \underline{\int_c^b} f(x)\, dx = \underline{\int_a^b} f(x)\, dx$,

$$\overline{\int_a^c} f(x)\, dx + \overline{\int_c^b} f(x)\, dx = \overline{\int_a^b} f(x)\, dx,$$

(E3)   IF $f \leq g$, THEN $\displaystyle\underline{\int_a^b} f(x)\, dx \leq \underline{\int_a^b} g(x)\, dx,\ \overline{\int_a^b} f(x)\, dx \leq \overline{\int_a^b} g(x)\, dx.$

Proof.   (E1) follows from Lemma B and (A1).

(E2): Let $\epsilon > 0$ be a given number. By Lemma C, there exist step functions ϕ_1 and ϕ_2 defined in (a, c) and (c, b), respectively, such that $\phi_1(x) \leq f(x)$ for $a < x < c$, $\phi_2(x) \leq f(x)$ for $c < x < b$, and

$$\underline{\int_a^c} f(x)\, dx \leq \int_a^c \phi_1(x)\, dx + \epsilon, \ \underline{\int_c^b} f(x)\, dx \leq \int_c^b \phi_2(x)\, dx + \epsilon.$$

Define the step function $\phi(x)$ by the conditions: $\phi(x) = \phi_1(x)$ for $a < x < c$, $\phi_2(x)$ for $c < x < b$. Then ϕ is defined in (a, b) and $\phi \le f$. Using the above inequalities, (A2), and the definition of the lower integral, we have

$$\underline{\int_a^c} f(x)\, dx + \underline{\int_c^b} f(x)\, dx \le \int_a^c \phi_1(x)\, dx + \int_c^b \phi_2(x)\, dx + 2\epsilon$$

$$= \int_a^c \phi(x)\, dx + \int_c^b \phi(x)\, dx + 2\epsilon = \int_a^b \phi(x)\, dx + 2\epsilon$$

$$\le \underline{\int_a^b} f(x)\, dx + 2\epsilon.$$

Since ϵ could be any positive number, this means that

$$\underline{\int_a^c} f(x)\, dx + \underline{\int_c^b} f(x)\, dx \le \underline{\int_a^b} f(x)\, dx.$$

On the other hand, there is again, by Lemma C, a $\phi_3 \le f$ with

$$\underline{\int_a^b} f(x)\, dx \le \int_a^b \phi_3(x)\, dx + \epsilon.$$

Using (A2), we find that this becomes

$$\underline{\int_a^b} f(x)\, dx \le \int_a^c \phi_3(x)\, dx + \int_c^b \phi_3(x)\, dx + \epsilon \le \underline{\int_a^c} f(x)\, dx + \underline{\int_c^b} f(x)\, dx + \epsilon,$$

and, since ϵ is arbitrary,

$$\underline{\int_a^b} f(x)\, dx \le \underline{\int_a^c} f(x)\, dx + \underline{\int_c^b} f(x)\, dx.$$

Together with the reverse inequality already established, this proves statement (E2) for $\underline{\int}$. The statement for $\overline{\int}$ can be proved similarly.

(E3): Let $\epsilon > 0$ be given. By Lemma C, there is a step function $\phi \le f$ with

$$\underline{\int_a^b} f(x)\, dx \le \int_a^b \phi(x)\, dx + \epsilon.$$

But since $f \le g$, we have $\phi \le g$, and hence

$$\int_a^b \phi(x)\, dx \le \underline{\int_a^b} g(x)\, dx,$$

so that

$$\underline{\int_a^b} f(x)\, dx \le \underline{\int_a^b} g(x)\, dx + \epsilon.$$

Since ϵ is arbitrary, we get the desired inequality for $\underline{\int}$. The inequality for $\overline{\int}$ can be proved similarly.

We recall now the proof, given in §7, of the first part of the fundamental theorem of calculus (Theorem 1 in §2). The proof used *only* the formula for the integral of a constant and the additivity and monotonicity properties of the integral. By Lemma E, these properties hold for lower and upper integrals. Therefore the proof given in §7 can be repeated for lower and upper integrals, and we obtain:

Lemma F   LET $f(t)$ BE A BOUNDED FUNCTION DEFINED FOR $a < t < b$. FOR EVERY x, $a < x \leq b$, SET

$$\underline{G}(x) = \underline{\int_a^x} f(t) \, dt, \; \overline{G}(x) = \overline{\int_a^x} f(t) \, dt,$$

AND SET

$$\underline{G}(a) = \overline{G}(a) = 0.$$

THEN THE FUNCTIONS $\underline{G}(x)$ AND $\overline{G}(x)$ ARE CONTINUOUS FOR $a \leq x \leq b$, AND AT EVERY POINT x AT WHICH f IS CONTINUOUS, WE HAVE

$$\underline{G}'(x) = \overline{G}'(x) = f(x).$$

8.4 The Riemann integral

A bounded function f is called **Riemann integrable** over an interval (a, b) if

$$\underline{\int_a^b} f(x) \, dx = \overline{\int_a^b} f(x) \, dx.$$

For instance, a step function is Riemann integrable, by Lemma B. If f is Riemann integrable, the common value of its lower and upper integrals is called the **Riemann integral** of f and is denoted by $\int_a^b f(x) \, dx$.

Theorem 1   LET f BE BOUNDED AND PIECEWISE CONTINUOUS IN AN INTERVAL (a, b). THEN f IS RIEMANN INTEGRABLE OVER THE INTERVAL.

Proof.   Let $\underline{G}(x)$ and $\overline{G}(x)$ be defined as in Lemma F, and set

$$H(x) = \overline{G}(x) - \underline{G}(x).$$

Then $H(x)$ is continuous everywhere, $H(a) = 0$, and $H'(x) = 0$ for all x, except perhaps the finitely many points at which f is discontinuous. By Theorem 1 of Chapter 4, §5, the function H is constant. Hence $H(b) = H(a) = 0$, so that

$$\overline{G}(b) = \overline{\int_a^b} f(x) \, dx = \underline{G}(b) = \underline{\int_a^b} f(x) \, dx.$$

Lemma G   LET f BE A BOUNDED FUNCTION DEFINED IN THE INTERVAL (a, b). THEN f IS RIEMANN INTEGRABLE IF AND ONLY IF THERE IS A UNIQUE NUMBER A SUCH THAT, IF ϕ AND ψ ARE TWO STEP FUNCTIONS AND $\phi \leq f \leq \psi$, THEN

$$(3) \qquad\qquad \int_a^b \psi(x) \, dx \leq A \leq \int_a^b \phi(x) \, dx.$$

Proof.   A number A has the required property if and only if it is a lower bound for the integrals of all step functions that are $\geq f$, and an upper bound for the integrals of all step functions that are $\leq f$. In other words, A has the required property if and only if

$$(4) \qquad\qquad \underline{\int_a^b} f(x) \, dx \leq A \leq \overline{\int_a^b} f(x) \, dx.$$

If f is Riemann integrable, there is only one number A that satisfies the above inequality, namely, $A = \int_a^b f(x)\, dx$. If f is not Riemann integrable, then it follows from Lemma D that there are infinitely many numbers A satisfying (4).

Lemma G shows that the Riemann integral and the integral as defined in §1 are one and the same thing.

In view of Theorem 1 just stated, and Lemma E, the existence of integrals of bounded piecewise continuous functions (that is, Theorem 1 of §1), and the three properties of integrals (Theorem 2 of §1) are established from the properties of real numbers, without appealing to geometric intuition.

8.5 Computing integrals using step functions

Theorem 2   LET f BE A BOUNDED FUNCTION DEFINED IN THE INTERVAL (a, b). THE FUNCTION f IS RIEMANN INTEGRABLE IF AND ONLY IF, FOR EVERY NUMBER $\epsilon > 0$, ONE CAN FIND TWO STEP FUNCTIONS ϕ AND ψ SUCH THAT

(5)
$$\phi \le f \le \psi, \; 0 \le \int_a^b \psi(x)\, dx - \int_a^b \phi(x)\, dx \le \epsilon.$$

Proof.   Let ϕ and ψ have the properties stated. Since $\underline{\int_a^b} f(x)\, dx \ge \int_a^b \phi(x)\, dx$ and $\overline{\int_a^b} f(x)\, dx \le \int_a^b \psi(x)\, dx$, we see that the upper and lower integrals of f differ by at most ϵ. If this is so for every positive ϵ, the upper and lower integrals are the same number and f is Riemann integrable.

On the other hand, if f is Riemann integrable, the existence of ϕ and ψ having the desired property follows at once from Lemma C.

Theorem 2 contains the statement made without proof in §1. The integral can be computed as accurately as desired, using step functions.

8.6 Monotone functions

Theorem 3   EVERY BOUNDED MONOTONE FUNCTION IS RIEMANN INTEGRABLE.

Proof.   The argument in §1.7 shows that such a function satisfies the condition of Theorem 2.

Since a monotone function can have infinitely many discontinuity points, this theorem shows that there are Riemann integrable functions that are not piecewise continuous.

EXERCISES

1. Prove the other half of Lemma B.
2. Prove the existence of ψ in Lemma C.
3. Prove statement (E2) in Lemma E for $\overline{\int}$.
4. Prove statement (E3) in Lemma E for $\underline{\int}$.
5. Prove the statement made in the text concerning the upper and lower integrals of the function $f(x)$, which is 0 or 1 according to whether x is rational or irrational.
6. Give an example of a bounded monotone function that has infinitely many jump discontinuities in the interval $(0, 1)$.

§9 Existence of Improper Integrals

In this section, we prove the comparison theorem for improper integrals (Theorem 1 of §6). The proof is based on a result that is itself quite important.

9.1 Limits of monotone functions

Theorem A (On Limits of Monotone Functions)   LET $x \mapsto f(x)$ BE A NONDECREAS-ING FUNCTION DEFINED IN THE INTERVAL (a, b), WHERE a IS A NUMBER AND b EITHER A NUMBER OR THE SYMBOL $+\infty$. ASSUME THAT THERE IS A NUMBER M SUCH THAT $f(x) \leq M$ FOR $a < x < b$. THEN THE FINITE LIMIT $\alpha = \lim_{x \to b^-} f(x)$ EXISTS (AND $\alpha \leq M$).

In other words, a nondecreasing function bounded from above has a limit at the right endpoint of the interval in which it is defined. (If $b = +\infty$, then b^- denotes the same symbol.)

The proof of the theorem depends on the least upper bound principle. By hypothesis, the set S of all numbers $f(x)$, where x lies between a and b, is bounded from above. Indeed, M is an upper bound. Let α be the least upper bound of this set (so that $\alpha \leq M$). We assert that α is the limit from the left of f at $x = b$. Hence we must show that two statements are true: (1) if A is a number such that $\alpha < A$, then $f(x) < A$ for $a < x < b$; and (2) if B is a number such that $\alpha > B$, then there is a number d, $a < d < b$, such that $f(x) > B$ for $d < x < b$.

Statement 1 is true for the simple reason that, if $A > \alpha$, then $f(x) < A$ for all x such that $a < x < b$, since $f(x) < \alpha$ for all such x.

The proof of Statement 2 is indirect. Assume that the statement is false. Then there is a number $B < \alpha$ such that, for every number d between a and b, there is a number d_1 with $d < d_1 < b$ and with $f(d_1) \leq B$. Since the function f is nondecreasing, we also have $f(d) \leq B$. But d was any number between a and b. We conclude that $f(x) \leq B$ for $a < x < b$. But this is impossible; α was the smallest number greater than, or equal to, all values of $f(x)$. Hence Statement 2 is true.

9.2 Proof of the comparison test

The proof of the comparison test for improper integrals is now easy. Let (a, b) be as before, and let f and g be two functions defined in this interval. We are given that, for every c between a and b, both functions are bounded and piecewise continuous in the interval (a, c), that

(1) $$0 \leq f(x) \leq g(x) \qquad \text{for } a < x < b,$$

and that the integral

(2) $$\int_a^b g(x)\, dx$$

exists (perhaps as an improper integral). We must show that the integral

$$(3) \qquad \int_a^b f(x)\,dx$$

also exists.

For every y between a and b, set

$$F(y) = \int_a^y f(x)\,dx, \qquad G(y) = \int_a^y g(x)\,dx.$$

Both functions are clearly nondecreasing. Indeed, if $y_1 < y_2$, then

$$F(y_2) - F(y_1) = \int_{y_1}^{y_2} f(x)\,dx \geq 0$$

by (1) and the monotonicity property of integrals; the same reasoning applies to G. Since the integral (2) exists, the function $G(y)$ has a finite limit as $y \to b^-$. Therefore there is a number M such that $G(y) \leq M$ for $a < y < b$. By the monotonicity of the integral and by (1), we have that $F(y) \leq G(y)$ for $a < y < b$. Therefore $F(y) \leq M$ for such y. Therefore, by Theorem A, the function $F(y)$ has a finite limit as $y \to b^-$. This is the same as to say that the integral (3) exists.

EXERCISES

1. Formulate and prove an analogue of Theorem A for a nonincreasing function defined in (a, b) and bounded from below.
2. Formulate and prove the analogue of Theorem A, which asserts that a function defined in an interval (a, b) has, under certain conditions, a limit at the left endpoint. (The left endpoint may be $-\infty$.)
3. In the proof of the comparison test, we used the fact that, if $G(y)$ is a nondecreasing function defined in (a, b), and the finite limit $\lim_{y \to b^-} G(y)$ exists, then G is bounded from above. Give a complete proof of this statement.

Problems

In Problems 1 to 17, all functions are assumed to be bounded and piecewise continuous in every interval over which they are being integrated.

1. Suppose that $f(x)$ is continuous for $0 < x < 1$, and takes on no negative values. Prove that $\int_0^1 f(x)\,dx > 0$ unless $f(x)$ is the constant function 0.
2. Show that $\left|\int_a^b f(x)\,dx\right| \leq M\,|b - a|$ if $|f(x)| \leq M$ for x between a and b. Verify that, unlike the monotonicity rule (3) in §1, the present statement is not tied to the assumption $a < b$.
3. Let $f(x)$ be continuous and bounded for $a < x < b$. Assume that $f(x)$ does not take on the value m. Prove that either $\int_a^b f(x)\,dx > m(b - a)$ or $\int_a^b f(x)\,dx < m(b - a)$. (Hint: Use the intermediate value theorem and the monotonicity rule for integrals.) Show that the corresponding statement for piecewise continuous functions is false.

4. Prove that, if $f(x)$ is bounded and continuous for $a < x < b$, then there is a number x_0 such that

$$a < x_0 < b \qquad \text{and} \qquad f(x_0) = \int_a^b f(x)\, dx.$$

(This is a form of the so-called mean value theorem for integrals.)

5. Prove the following **mean value theorems for integrals.** Let $p(x)$ and $q(x)$ be bounded continuous functions defined for $a < x < b$, and assume that for such x we have $p(x) > 0$. Then there is a number x_0 such that

$$a < x_0 < b \qquad \text{and} \qquad q(x_0) \int_a^b p(x)\, dx = \int_a^b p(x)q(x)\, dx.$$

6. Prove the following extension of the fundamental theorem of calculus. If $F(x) = \int_a^x f(t)\, dt$ for $a \le x \le b$, and, if at a point x_0 with $a < x_0 < b$ the finite one-sided limit $m = \lim_{x \to x_0^-} f(x)$ exists, then the one-sided derivative $F'(x_0^-)$ exists and equals m.

7. For a continuous function $f(t)$, set

$$F(x) = \int_0^x (x - t)f(t)\, dt.$$

Prove that F is twice differentiable and $F''(x) = f(x)$.

8. For a continuous function $f(t)$, set

$$F(x) = \frac{1}{2} \int_0^x (x - t)^2 f(t)\, dt.$$

Prove that F is three times differentiable and $F'''(x) = f(x)$.

9. Generalizing Problems 7 and 8, prove by mathematical induction that, if $f(t)$ is a continuous function, then the function

$$F(x) = \frac{1}{1 \cdot 2 \cdot 3 \cdots (n - 1)} \int_0^x (x - t)^{n-1} f(t)\, dt$$

is n times differentiable and $F^{(n)}(x) = f(x)$. Also find the values of $F(0)$, $F'(0)$, $F''(0)$, $\cdots F^{(n-1)}(0)$.

10. It follows from the fundamental theorem, without any calculations that, given a bounded continuous function $f(x)$ and a positive integer k, there is a function $G(x)$ such that $G^{(k)}(x) = f(x)$. Use this remark in proving the statements in Problems 7, 8, and 9. Integrate by parts.

11. How must the statements in Problems 7, 8, and 9 be modified if the function f is assumed to be bounded and piecewise continuous?

12. Prove **Schwarz' inequality** for integrals:

$$\left| \int_a^b f(x)g(x)\, dx \right| \le \sqrt{\int_a^b f(x)^2\, dx \int_a^b g(x)^2\, dx}.$$

[Hint: Let u be a number. Consider the integral

HERMAN AMADEUS SCHWARZ (1843–1921) was a very original nineteenth-century mathematician.

$$\int_a^b [uf(x) + g(x)]^2 \, dx$$

and ask whether it can ever be negative.]

13. Show that, for any two functions f and g,

$$\sqrt{\int_a^b [f(x) + g(x)]^2 \, dx} \leq \sqrt{\int_a^b f(x)^2 \, dx} + \sqrt{\int_a^b g(x)^2 \, dx}.$$

(Hint: Use Schwarz' inequality.)

14. We know that a differentiable function is continuous. If we know something about the derivative, we may be able to say "how continuous" the function is. The present problem is an example. Let $f'(x)$ be continuous in an interval, and assume that, for every x in this interval, $|f'(x)| \leq K$. Prove that, for any two points in the interval, x_1 and x_2, we have

(a) $$|f(x_1) - f(x_2)| \leq K \, |x_1 - x_2|.$$

15. Only constants are "more continuous" than functions with a bounded derivative. More precisely, if a function $f(x)$ satisfies, in some interval, the inequality

$$|f(x_1) - f(x_2)| \leq K \, |x_1 - x_2|^k, \qquad k > 1$$

for any two points x_1 and x_2 in the interval, then f is a constant. Prove this statement. (Hint: You do not need integrals but only derivatives to solve this problem.)

16. Prove that, for $a < b$ and any function f, we have

$$\left| \int_a^b f(x) \, dx \right| \leq \sqrt{(b - a) \int_a^b f(x)^2 \, dx}.$$

(Hint: Use Schwarz' inequality.)

17. Assume that $f(x)$ has a derivative in the interval (a, b) and that $\int_a^b f'(x)^2 \, dx \leq M$. Show that, for any two points x and x_2 in the interval a, we have

(b) $$|f(x_1) - f(x_2)| \leq \sqrt{M \, |x_1 - x_2|}.$$

This is another example of saying something about "how continuous" a function is from some information about the derivative. (Hint: Use the result of Problem 16.)

18. Find a continuous function that satisfies Inequality (b) of Problem 17, for some M, but satisfies no inequality of the form (a) in Problem 14.

19. In the text, improper integrals have been considered only for nonnegative functions. But one can consider such integrals also for functions that take on both positive and negative values. Prove that, if $f(x)$ is bounded and piecewise continuous in every finite interval (a, b), and if the improper integral

$$\int_a^{+\infty} |f(x)| \, dx$$

converges, then the improper integral

$$\int_a^{+\infty} f(x)\, dx = \lim_{b \to +\infty} \int_a^b f(x)\, dx$$

also exists, and $\left| \int_a^{+\infty} f(x)\, dx \right| \leq \int_a^{+\infty} |f(x)|\, dx.$

[Hint: Write $f(x) = f_1(x) - f_2(x)$, where $f_1(x) = f(x)$ if $f(x) \geq 0$ and $f_1(x) = 0$ if $f(x) < 0$.]

20. Let $x \mapsto f(x)$, $a \leq x \leq b$, be a positive function with a continuous derivative. It is geometrically evident that the length of its graph is at least the distance between the points $(a, f(a))$ and $(b, f(b))$. Give an analytic proof of this fact.

21. In the text, the two important formulas,

(i) $\int_a^b Cf(x)\, dx = C \int_a^b f(x)\, dx,$ C a constant, and

(ii) $\int_a^b f(x)\, dx + \int_a^b g(x)\, dx = \int_a^b [f(x) + g(x)]\, dx$

have been proved, for bounded piecewise continuous functions, using the fundamental theorem of calculus. It is easy to see directly that (i) and (ii) hold for step functions. Using this fact, prove (i) and (ii) for all Riemann integrable functions, without using the fundamental theorem of calculus.

22. Let f be a bounded function, not necessarily Riemann integrable, and let C be a constant. Show that

$$\overline{\int_a^b} Cf(x)\, dx = C \, \overline{\int_a^b} f(x)\, dx \qquad \text{if } C > 0$$

and $$\overline{\int_a^b} Cf(x)\, dx = C \, \underline{\int_a^b} f(x)\, dx \qquad \text{if } C < 0.$$

23. Find two bounded functions $f(x)$ and $g(x)$ defined in $(0, 1)$ such that

$$\underline{\int_0^1} f(x)\, dx = \underline{\int_0^1} g(x)\, dx = 0,$$

but $$\underline{\int_0^1} [f(x) + g(x)]\, dx = 1.$$

6/Transcendental Functions

This chapter is devoted to the most important transcendental (neither rational nor algebraic) functions used in mathematics and in its applications. These elementary transcendental functions include trigonometric functions, inverse trigonometric functions, logarithmic functions, and exponential functions. We study them by the methods of calculus, without assuming any previous knowledge.

There are three appendixes. The first deals with various applications of the exponential functions, the second considers hyperbolic functions, and the third contains examples of some nonelementary functions.

A function $x \mapsto f(x)$ is called **algebraic** if, given the number x, the number $f(x)$ can be computed by performing a finite number of additions, multiplications, subtractions, and divisions, and by solving a finite number of algebraic equations, that is, finding roots of polynomials. *All polynomials, rational functions, and radical functions are algebraic.*

Consider a typical example, such as the function

$$x \mapsto y = \left(\frac{3x^{1/2} - 2}{5x^{1/2} + x^{1/3}} \right)^{3/7}.$$

Suppose $x = 4$. To find y, we first solve the algebraic equations $t^2 = 4$ and $t^3 = 4$ in order to find $\sqrt{4} = 2$ and $\sqrt[3]{4}$. Then we compute the number

$$\frac{3 \cdot 2 - 2}{5 \cdot 2 + \sqrt[3]{4}} = \frac{4}{10 + \sqrt[3]{4}}$$

of which y is the $(3/7)$th power; this involves multiplication, addition, subtraction, and division. Finally, we find y as the solution of the algebraic equation $y^7 = [4/(10 + \sqrt[3]{4})]^3$.

A function that is not algebraic is called **transcendental.** The simplest and most important transcendental functions are the so-called *elementary* transcendental functions: logarithmic functions, the power function $x \mapsto a^x$, and the trigonometric functions and their inverses. These functions are familiar to the reader; they are usually taught in high school. Since the easiest and most illuminating approach to them is through calculus, however, we do not make use of any previous knowledge of logarithms or trigonometry and will develop these subjects from the beginning.

To prove that the functions mentioned above are really transcendental would require, at this stage, complicated arguments. With more techniques available, the proofs become very simple. For this reason, we ask the reader to accept on faith the statement that the functions we claim to be transcendental *are* transcendental.

EXERCISES

1. Describe the algebraic operations involved in finding y for a given x if $y = \sqrt[3]{1 + x^4} + \sqrt{1 + x^2}$. Compute y if $x = 2\sqrt{2}$.
2. Describe the algebraic operations involved in finding y for a given x if $y = \sqrt{x^4 + 4} + \sqrt[3]{x^2 - x + 1}$. Compute y if $x = 1$.

3. Describe the algebraic operations involved in finding u for a given t if

$$u = \frac{\sqrt[3]{t^2 + 1}}{\sqrt{t^4 + 1} + \sqrt[4]{t^6 + 1}}.$$

4. Describe the algebraic operations involved in finding s for a given ϕ if

$$s = \frac{1}{\sqrt{1 + (1/\sqrt{\phi + 1})}}.$$

Find s if $\phi = 8$.

5. Describe the algebraic operations involved in finding z for a given y if

$$z = \left(\frac{y^{-2/3} + y}{y^{2/3} - y^{-1}}\right)^{1/3}.$$

Compute z if $y = 8$.

6. Describe the algebraic operations involved in finding x for a given u if

$$x = \frac{\sqrt{u} + \sqrt[3]{u} + 1}{\sqrt[3]{u^2 + u + 1}}.$$

1.2 Periodic functions

Let $p \neq 0$ be a number and $x \mapsto f(x)$ a function. One says that f has **period** p, or is **periodic** with period p, if

$$f(x + p) \equiv f(x).$$

A graph of a periodic function is shown in Fig. 6.1.

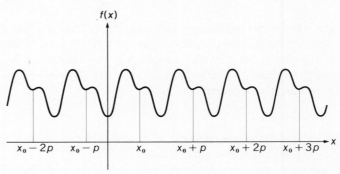

Fig. 6.1

If f has periods p and q, then $p + q$ is also a period, for $f(x + p + q) = f(x + p) = f(x)$. It follows that, if f has period p, then all integral multiples of p, that is, p, $-p$, $2p$, $-2p$, $3p$, $-3p$, and so forth, are periods of f. The smallest positive period of f, if there is such a number, is often called **the period** (or, better, the **primitive period**) of f.

Periodic functions are important in science and technology, because so many natural and artificial phenomena are periodic with respect to time: planetary motions, tides, sound waves, electromagnetic radiation, breathing, movements of clocks and other mechanisms, and so forth.

We record some simple properties of periodic functions.

1. If $f(x)$ has period p and $g(x) = f(\alpha x)$, then g has period p/α (we assume, of course, that $\alpha \neq 0$).

Proof. $g(x + (p/\alpha)) = f[\alpha(x + (p/\alpha))] = f(\alpha x + p) = f(\alpha x) = g(x)$.

2. If $f(x)$ has period p and is differentiable, then $f'(x)$ also has period p.

Proof. We have $f(x + p + h) - f(x + p) = f(x + h) - f(x)$. Divide this equation by h and take the limit as $h \to 0$. This yields

$$f'(x + p) = f'(x).$$

3. If a function $f(x)$ is defined and continuous in the closed interval $[x_0, x_0 + p]$, and if $f(x_0) = f(x_0 + p)$, then one can define $f(x)$ for all x outside this interval, so as to make f periodic with period p and continuous. This can be done in one way only.

Proof.   Every x can be written, as $x = np + y$, where n is an integer and $x_0 \leq y \leq x_0 + p$. We know $f(y)$ and, for periodicity, requires $f(x) = f(y)$. The condition $f(x_0) = f(x_0 + p)$ ensures that the resulting function is continuous everywhere. The situation is illustrated in Fig. 6.2.

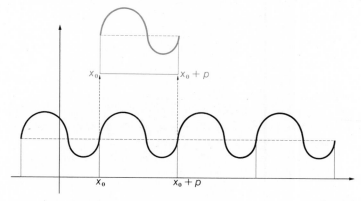

Fig. 6.2

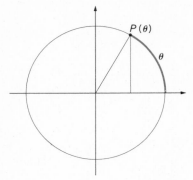

Fig. 6.3

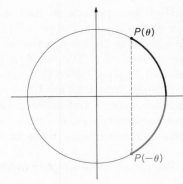

Fig. 6.5

Fig. 6.6

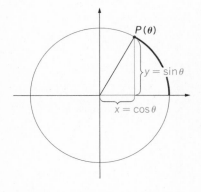

Fig. 6.4

1.3 Geometric definition of sines and cosines

The most important periodic functions are the **trigonometric** functions, in particular the sine and the cosine. We recall the geometric definition of these functions.

Consider the unit circle $x^2 + y^2 = 1$. Let θ be a nonnegative number, and let $P(\theta)$ be *"the point on the unit circle that is reached by moving along the circle in the positive (counterclockwise) direction, starting at the point $(1, 0)$ and covering θ length units."* More precisely: if $\theta = 0$, let $P(0)$ be the point $(1, 0)$; if $0 < \theta \le \pi$, let $P(\theta)$ be that point on the upper semicircle for which the arc of the semicircle from $P(0)$ to $P(\theta)$ has length θ, as in Fig. 6.3 (that there is exactly one such point is geometrically obvious and will be proved later analytically); if $\pi < \theta \le 2\pi$, let $P(\theta)$ be that point on the lower semicircle for which the length of the arc between $P(\pi) = (-1, 0)$ and $P(\theta)$ is $\theta - \pi$, as in Fig. 6.4; if $\theta > 2\pi$, write θ in the form $\theta = 2\pi n + \phi$, n a positive integer and $0 \le \phi < 2\pi$, and set $P(\theta) = P(\phi)$. Thus we have

$$(1) \qquad\qquad P(\theta + 2\pi) = P(\theta).$$

We also define $P(-\theta)$ to be the point that is symmetrical to $P(\theta)$ with respect to the x axis; this definition is justified by Fig. 6.5. One calls $P(-\theta)$ *"the point reached by moving θ length units along the unit circle in the negative (clockwise) direction, starting from the point $(1, 0)$."* Note that relation (1) holds also if θ is negative.

Now let θ be any number and let the point $P(\theta)$ have coordinates (x, y). One defines (see Fig. 6.6):

$$\cos \theta = x, \qquad \sin \theta = y.$$

Since the point $P(\theta)$ is always on the unit circle, we have

(2) $$(\cos\theta)^2 + (\sin\theta)^2 = 1.$$

Also

(3) $$\sin\theta > 0 \text{ for } 0 < \theta < \pi.$$

The definition implies at once that *sine and cosine are periodic with period* 2π; in view of (1), we have

(4) $$\cos(\theta + 2\pi) = \cos\theta, \sin(\theta + 2\pi) = \sin\theta.$$

Also, $P(0) = (1, 0)$, $P(\pi/2) = (0, 1)$ and $P(\pi) = (-1, 0)$, so that

(5) $$\cos 0 = \sin\frac{\pi}{2} = 1, \sin 0 = \cos\frac{\pi}{2} = \sin\pi = 0, \cos\pi = -1$$

Since $P(\theta)$ and $P(-\theta)$ are symmetrically situated with respect to the x axis, we have

(6) $$\cos(-\theta) = \cos\theta, \qquad \sin(-\theta) = -\sin\theta,$$

so that the function *cosine is even* and the function *sine is odd*.

It is geometrically evident (and will be proved later) that the functions sine and cosine are continuous, that cosine is decreasing in the interval $(0, \pi)$ and that $\sin\theta$ is increasing in the interval $[0, \pi/2]$ and decreasing in the interval $[\pi/2, 0]$. The graphs of $\sin\theta$ and $\cos\theta$ in the interval $[0, \pi]$ are shown in Figs. 6.7 and 6.8. Using (6), one can easily draw the graphs for the interval $[-\pi, \pi]$; see Fig. 6.9. It is important to remember the general shape of these curves; note in particular where the functions sine and cosine are positive, negative, decreasing, increasing. A bigger

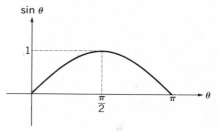

Fig. 6.7

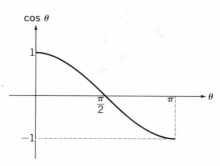

Fig. 6.8

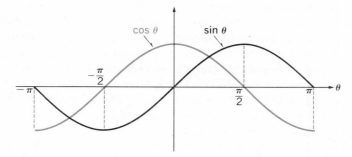

Fig. 6.9

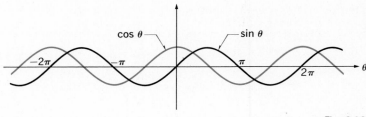

Fig. 6.10

part of the graph is drawn in Fig. 6.10. One sees that 2π is the primitive period of sine and cosine.

EXERCISES

Determine if the following functions are periodic and, if so, find their primitive periods.

7. $x \mapsto \sin 3x$.   9. $x \mapsto (1 + \cos^3 x) \sin (x/8)$.
8. $x \mapsto \cos x + \cos 2x$.   10. $x \mapsto \sin (x/\pi)$.

1.4 Angles

We have defined sines and cosines of numbers. In elementary mathematics, one deals with sines and cosines of angles. Let us establish the connection between these concepts.

A point O on a line l divides the line into two **rays**; O is called the origin of the two rays. An **angle** is a pair of rays with a common origin. The origin is called the **vertex** of the angle, and the rays are called the **legs.** To describe an angle, it suffices to name the vertex O and to name two points, say P and Q, distinct from O, one on each leg (see Fig. 6.11). We call the angle so described $\angle POQ$ or $\angle QOP$. At this point, we deal only with undirected angles, that is, we do not prescribe the order of the legs.

Given an angle with distinct legs, say $\angle POQ$, draw a circle of radius 1 about O and note the points, say A and B, at which the legs intersect the circle. These points divide the circle into two arcs; θ, the length of the smaller arc (see Fig. 6.11), is called the **measure of the angle.** If the two arcs are equal in length, it does not matter which we choose; in this case, the measure is π and the two legs are two rays of one line. If the two legs of an angle coincide, its measure is 0, by definition. The measure of an undirected angle is always a number θ such that $0 \leq \theta \leq \pi$.

If $\theta = \pi/2$, the angle is called **right.** Two angles with the same measure are called **congruent.** Often one calls two angles "equal" instead of "congruent," and one often says "the angle POQ equals θ, or is θ" in-

Fig. 6.11

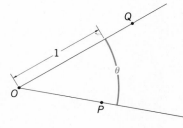

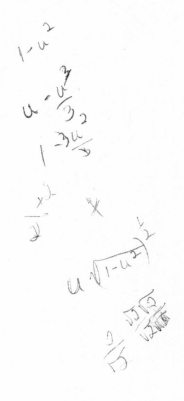

stead of saying that "the measure of ∠POQ is θ." Such "abuses of language" are harmless if they do not lead to confusion.

Given an angle ∠POQ of measure θ, we define sin (∠POQ) to mean sin θ and cos (∠POQ) to mean cos θ. The same convention applies to other functions.

Fig. 6.12

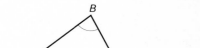
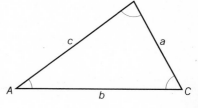

1.5 Triangles

Given a triangle with vertices A, B, and C, we shall, for the sake of brevity, denote the lengths of the sides BC, CA, and AB, by a, b, and c, respectively, and the measures of the angles ∠BAC, ∠CBA, and ∠BCA by A, B, and C (see Fig. 6.12). In particular, if the angle at C is a right angle, ($C = \pi/2$, see Fig. 6.13), we have

Fig. 6.13

$$(7) \qquad \sin A = \frac{a}{c}, \qquad \cos A = \frac{b}{c}.$$

To see this, introduce a coordinate system with the origin at A such that $c = 1$ and C lies on the positive x axis. Then B has the coordinates $(\cos A, \sin A)$ and the assertion follows. Formulas (7) are the "high school definition" of sine and cosine.

Note now that, in the triangle considered, $A + B = \pi/2$, and that $\sin B = b/c = \cos A$. This shows that

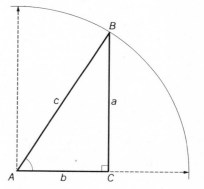

$$(8) \qquad \sin \theta = \cos \left(\frac{\pi}{2} - \theta \right).$$

[We proved this for $0 < \theta < \pi/2$; relations (5) show that (8) holds also for $\theta = 0$ and $\theta = \pi/2$; we shall see later that (8) holds for all θ.]

1.6 Degrees and radians

Trigonometric functions are used in surveying for computing plane triangles, and in astronomy and navigation for computing spherical triangles, that is, triangles formed by arcs of large circles on a sphere. The astronomical applications predate all others. Tables of the sine function were compiled as early as the second century. B.C. The oldest preserved sine tables are contained in Ptolemy's work, known under its Arabic name *Almagest*. The word sine, *sinus* in Latin, is a corruption of an Arabic word meaning "chord."

It is customary to divide the full circle into 360 equal parts, called **degrees,** and to divide each degree into 60 **minutes** ($1° = 60'$), each minute into 60 **seconds** ($1' = 60''$). This tradition goes back to the sexagesimal system of the Babylonians. In everyday life and in most technological and scientific applications, angles are measured by degrees, rather than by numbers. Degrees are used in practically all trigonometric tables.

In mathematics, it is almost imperative to use the "natural" way of measuring angles by numbers. The simplest way of reconciling the two methods is to agree, once and for all, that $360°$ (360 degrees) is another name for the number 2π, $1°$ (one degree) is another name for $\dfrac{\pi}{180}$, and in general $k°$ (k degrees) another name for $\dfrac{k\pi}{180}$. Thus

$$(9) \qquad k° = \frac{k\pi}{180} \qquad \text{and} \qquad x = \left(\frac{180x}{\pi}\right)°.$$

In particular,

$$1 = \left(\frac{180}{\pi}\right)° = 57°\ 17'\ 45'' \qquad \text{approximately.}$$

An angle of measure 1 is often called a **radian.**

▶ **Examples** **1.** What is $\cos(-90°)$?

Answer. Since $90° = \dfrac{90 \cdot \pi}{180} = \dfrac{\pi}{2}$, we have $\cos(-90°) = \cos 90° = \cos \dfrac{\pi}{2} = 0$.

2. $\sin 30° = \dfrac{1}{2}$, $\sin 60° = \dfrac{\sqrt{3}}{2}$, $\sin 45° = \dfrac{\sqrt{2}}{2}$.

The first two relations are easily read off, using the Pythagorean theorem, from an equilateral triangle (see Fig. 6.14), the last from an isosceles right triangle (Fig. 6.15).

3. What is $\sin 360030°$?

Answer. Using the periodicity of the sine function with period $2\pi = 360°$ and the previous example, we obtain

$$\sin 360030° = \sin(1000 \cdot 360° + 30°) = 1/2. ◀$$

EXERCISES

11. Use equation (7) to compute $\cos(\pi/4)$ and $\sin(\pi/4)$.
12. Use equation (7) to compute cos and sin for $\pi/3$ and $\pi/6$.

Fig. 6.14

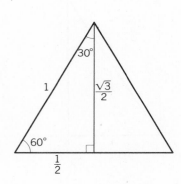

Fig. 6.15

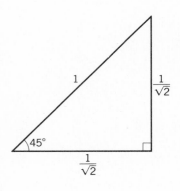

In Exercises 13 to 17, the measure of an angle is given in degrees. Convert these to the equivalent numerical (radian) expression.

13. $10°$.   15. $24°$.   17. $90°/\pi$.
14. $15°$.   16. $36°$.

In Exercises 18 to 22, the measure of an angle is given numerically (in radians). Convert these to the equivalent expression in degrees.

18. $\pi/9$.   22. $1/3$.
19. $\pi/10$.   23. Evaluate $\sin(2925°)$.
20. $\pi/30$.   24. Evaluate $\cos(2925°)$.
21. $\pi/8$.

1.7 Inverse sines and cosines

Our next task is to express sines and cosines by integrals. It turns out that this is possible if one considers not the functions $\cos\theta$ and $\sin\theta$ themselves but their inverse functions.

Of course, the functions sine and cosine are not monotone. In order to define an inverse function of, say, $x = \cos\theta$, we must select an interval on which this function is either increasing or decreasing. We choose the interval $[0, \pi]$. The function inverse to $\cos\theta$ on this interval is called **arc cosine** (or inverse cosine) and is denoted by arc cos. Thus

$$\theta = \text{arc cos } x \qquad \text{means that} \qquad x = \cos\theta,\, 0 \le \theta \le \pi$$

or

$$\text{arc cos }(\cos\theta) = \theta \text{ for } 0 \le \theta \le \pi, \qquad \cos(\text{arc cos } x) = x.$$

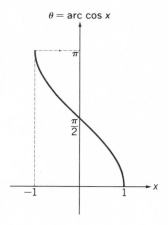

Fig. 6.16

The function arc cos x is defined only for $-1 \le x \le 1$, since $|\cos\theta|$ is never greater than 1. The graph of the inverse cosine is shown in Fig. 6.16; it is obtained by "flipping" the graph in Fig. 6.8.

We recall now (see §1.3) that, for $0 \le \theta \le \pi$, $\cos\theta$ and $\sin\theta$ are the coordinates of a point (x, y) with $x^2 + y^2 = 1$ and $y \ge 0$, such that the length of the arc of the upper semicircle from (x, y) to $(1, 0)$ (the heavy arc in Fig. 6.17) is θ. But according to the formulas in Chapter 5, §4.9, the length of such an arc is

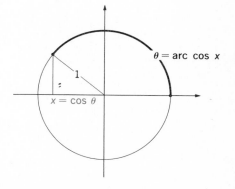

Fig. 6.17

(10)
$$\theta = \int_x^1 \frac{du}{\sqrt{1 - u^2}}.$$

It is also equal to twice the area of the circular sector with this arc, that is,

(10′)
$$\theta = x\sqrt{1 - x^2} + 2\int_x^1 \sqrt{1 - u^2}\, du.$$

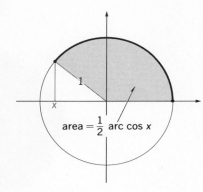

Fig. 6.18

area $= \frac{1}{2}$ arc cos x

(The two expressions for θ are equal, since both represent continuous functions of x which are 0 at $x = 1$, and which have the same derivative. The first expression is shorter; the second has two advantages: it involves no "improper" integral and it could have been obtained without the length formula.)

Since (10) and (10′) mean that $x = \cos\theta$, we have $\theta = \text{arc cos } x$. Thus

$$(11) \qquad \text{arc cos } x = \int_x^1 \frac{du}{\sqrt{1-u^2}} \qquad \text{for } |x| \leq 1,$$

and

$$(11') \quad \text{arc cos } x = x\sqrt{1-x^2} + 2\int_x^1 \sqrt{1-u^2}\,du \qquad \text{for } |x| \leq 1.$$

The geometric meaning of this last formula is shown in Fig. 6.18.

Differentiating both sides of (11) or of (11′), we obtain

$$(12) \qquad \frac{d \text{ arc cos } x}{dx} = -\frac{1}{\sqrt{1-x^2}} \qquad \text{for } -1 < x < 1$$

which shows once more that the arc cosine is a decreasing function. It is quite remarkable that the derivative of the transcendental function arc cosine is a radical function.

Next we look for a function inverse to the sine. We note from the graph that the function $\theta \mapsto \sin\theta$ is increasing for $-\pi/2 \leq \theta \leq \pi/2$. In the interval $[-\pi/2, \pi/2]$ therefore, the sine function has an inverse, called the **arc sine**. The inverse is, of course, defined in the interval $[-1, 1]$; its graph is shown in Fig. 6.19. We note that

$$\theta = \text{arc sin } x \qquad \text{means that} \qquad x = \sin\theta, \ -\frac{\pi}{2} \leq \theta \leq \frac{\pi}{2}.$$

Fig. 6.19

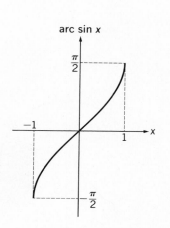

arc sin x

In order to find a formula for the arc sine, we consider first a number x such that $0 \leq x \leq 1$. Then (see Fig. 6.16) $0 \leq \text{arc cos } x \leq \pi/2$, so that, by equation (8) in §1.5, we have

$$\sin\left(\frac{\pi}{2} - \text{arc cos } x\right) = \cos(\text{arc cos } x) = x.$$

Hence

$$(13) \qquad \text{arc sin } x = \frac{\pi}{2} - \text{arc cos } x,$$

and since

$$\frac{\pi}{2} - \text{arc cos } x = \text{arc cos } 0 - \text{arc cos } x = \int_0^1 \frac{du}{\sqrt{1-u^2}} - \int_x^1 \frac{du}{\sqrt{1-u^2}}$$

$$= \int_0^1 \frac{du}{\sqrt{1-u^2}} + \int_1^x \frac{du}{\sqrt{1-u^2}} = \int_0^x \frac{du}{\sqrt{1-u^2}},$$

we obtain that

(14)
$$\text{arc sin } x = \int_0^x \frac{du}{\sqrt{1-u^2}}.$$

We proved this for $0 \le x \le 1$, but the formula holds also for $-1 \le x < 0$. Indeed, since $\sin \theta$ is odd, arc sin x must be odd, and the right-hand side of (14) is easily seen to be an odd function of x. Hence (14) holds for $-1 \le x \le 1$, and so does (13) which is equivalent to (14). The differentiation rule

(15)
$$\frac{d \text{ arc sin } x}{dx} = \frac{1}{\sqrt{1-x^2}} \qquad \text{for } -1 < x < 1$$

follows at once from (14):

EXERCISES

In Exercises 25 to 30, find the indicated numbers.

25. arc cos $(1/2)$.

26. arc sin $(1/2)$.

27. arc cos $(1/\sqrt{2})$.

28. arc sin $(-3\sqrt{2})$.

29. arc cos $(-1/2)$.

30. sin (arc sin $(1/27)$).

In Exercises 31 to 34, find $f'(x)$.

31. $f(x) = $ arc sin $(2x)$, $|x| < 1/2$.

32. $f(x) = (\text{arc cos } x)^3$, $|x| < 1$.

33. $f(x) = $ arc sin $(\sin 2x)$.

34. $f(x) = \sin [\text{arc sin } (2x)]$.

35. Find $\dfrac{d}{dx}$ (arc cos $\sqrt{1-x)^2}$, $0 < x < 1$.

▶ 36. Find $\dfrac{d}{du}$ [(arc cos u)(arc cos $2u$)].

37. Differentiate $t \mapsto \sqrt{1-t^2}$ arc cos t.

38. Find $\displaystyle\int \frac{du}{\sqrt{1-16u^2}}$.

39. Find $\displaystyle\int \sqrt{\frac{\text{arc cos } z}{1-z^2}} \, dz$.

40. Find $\displaystyle\int \frac{8u^{3/2}}{\sqrt{1-u^5}} \, du$.

41. Find $\int_{-1}^0$ arc cos $x \, dx$. (Hint: Use integration by parts and be watchful for an improper integral.)

42. Find $\int \dfrac{dx}{\sqrt{a - bx^2}}$, where a and b are positive constants.

$$\left(\text{Hint: Note that } \sqrt{a - bx^2} = \sqrt{a}\,\sqrt{1 - \frac{b}{a}x^2} = \sqrt{a}\,\sqrt{1 - \left(\frac{\sqrt{bx}}{\sqrt{a}}\right)^2}.\right)$$

43. Compute $\displaystyle\int_0^{1/2} \frac{dx}{\sqrt{1 - x^2}}$.

44. Compute $\displaystyle\int_{1/2}^{\sqrt{2}/2} \frac{dz}{\sqrt{1 - z^2}}$.

45. Compute $\displaystyle\int_{\sqrt{2}/2}^{\sqrt{3}/2} \frac{\text{arc cos } v}{\sqrt{1 - v^2}}\, dv$.

46. Compute $\displaystyle\int_{1/4}^{1/2} \frac{1}{\sqrt{1 - t}}\, \frac{dt}{\sqrt{t}}$.

(Hint: Use a substitution $t = x^2$.)

1.8 Recapitulation

We summarize the results of our discussion thus far.

Theorem 1   THERE EXIST UNIQUELY DETERMINED CONTINUOUS FUNCTIONS $\theta \mapsto \cos\theta, \theta \mapsto \sin\theta$, DEFINED FOR ALL VALUES OF θ, WITH THE FOLLOWING PROPERTIES. (i) THE FUNCTION $\cos\theta$ IS EVEN, HAS PERIOD 2π, AND IN THE INTERVAL $0 \leq \theta \leq \pi$ IT IS INVERSE TO THE FUNCTION

$$(16) \qquad\qquad \text{arc cos } x = \int_x^1 \frac{du}{\sqrt{1 - u^2}}$$

(SO THAT $x = \cos\theta$ MEANS THAT $\theta = \text{arc cos } x$). (ii) THE FUNCTION $\sin\theta$ IS ODD, HAS PERIOD 2π, AND

$$(17) \qquad\qquad \sin\theta = \sqrt{1 - (\cos\theta)^2} \qquad \text{for } 0 \leq \theta \leq \pi.$$

(iii) EVERY POINT (x, y) WITH $x^2 + y^2 = 1$ CAN BE WRITTEN IN THE FORM $x = \cos\theta, y = \sin\theta$, AND THIS REPRESENTATION IS UNIQUE IF WE REQUIRE THAT $0 \leq \theta < 2\pi$.

In arriving at this theorem, we have accepted, in §1.3, certain statements as geometrically obvious. This is easily remedied. The formula (17) defining arc cos x shows that the function $x \mapsto \text{arc cos } x$ is continuous and decreasing for $-1 \leq x \leq 1$, and that arc cos $(-1) = \pi$, arc cos $1 = 0$. We can now use properties (i) and (ii) of the theorem as *definitions* of $\cos\theta$ and $\sin\theta$, and prove the continuity of these functions, using Theorem 7 in Chapter 3, §3.7 on the continuity of inverse functions.

The formulas for arc cosine and arc sine show that one could have arrived at the trigonometric functions without any help from geometry, simply by studying the inverse to a primitive function of the radical function $1/\sqrt{1 - x^2}$.

1.9 Differentiating sine and cosine

Theorem 2   WE HAVE

$$(18) \qquad \frac{d \cos \theta}{d\theta} = -\sin \theta, \qquad \frac{d \sin \theta}{d\theta} = \cos \theta.$$

Proof.   Consider first the interval $0 < \theta < \pi$. By (12) and the theorem on differentiating inverse functions (Theorem 7 in Chapter 4, §2.1) we have, setting $\theta = \text{arc cos } x, x = \cos \theta$

$$\frac{d \cos \theta}{d\theta} = \frac{1}{d\theta/d \cos \theta} = \frac{1}{d \text{ arc cos } x/dx} = \frac{1}{-1/\sqrt{1 - x^2}}$$

$$= -\sqrt{1 - x^2} = -\sqrt{1 - (\cos \theta)^2} = -\sin \theta,$$

as asserted. Also, by (17) and the chain rule,

$$\frac{d \sin \theta}{d\theta} = \frac{d\sqrt{1 - (\cos \theta)^2}}{d\theta} = \frac{d\sqrt{1 - (\cos \theta)^2}}{d \cos \theta} \frac{d \cos \theta}{d\theta}$$

$$= \frac{1}{2} \frac{1}{\sqrt{1 - (\cos \theta)^2}} (-2 \cos \theta)(-\sin \theta) = \frac{\cos \theta \sin \theta}{\sin \theta} = \cos \theta.$$

Hence (18) holds in $(0, \pi)$.

For θ in $(-\pi, 0)$ the number $-\theta$ lies between 0 and π, and we have, by (6) and the chain rule,

$$\frac{d \cos \theta}{d\theta} = \frac{d \cos (-\theta)}{d\theta} = -\frac{d \cos (-\theta)}{d(-\theta)} = -(-\sin (-\theta)) = -\sin \theta,$$

$$\frac{d \sin \theta}{d\theta} = -\frac{d \sin (-\theta)}{d\theta} = \frac{d \sin (-\theta)}{d(-\theta)} = \cos (-\theta) = \cos \theta.$$

In view of the periodicity, it follows [see Property (2) of periodic functions in §1.2] that (18) holds for all values of θ, except perhaps for $\theta = 0, \pi, -\pi, 2\pi, -2\pi, \cdots$, and so forth. But since cosine and sine are continuous everywhere, (18) holds also at these points. Indeed, let $F(\theta)$ be a primitive function for $\sin \theta$ and set $G(\theta) = F(\theta) + \cos \theta$. Then

$G(\theta)$ is continuous everywhere and we have $G'(\theta) = F'(\theta) - \sin \theta =$ $\sin \theta - \sin \theta = 0$, except perhaps for $\theta = n\pi$, n being an integer. But then $G(\theta)$ must be a constant, so that $G'(\theta) = 0$ and therefore $(\cos \theta)' = -\sin \theta$ everywhere. The second statement (18) is proved similarly.

▶ *Examples*   We note two important special cases.

   1. Show that

$$\lim_{\alpha \to 0} \frac{\sin \alpha}{\alpha} = 1,$$

(so that for small values of $|\alpha|$, the sine of α is close to α).

   Proof.   Since $\sin 0 = 0$, we have

$$\lim_{\alpha \to 0} \frac{\sin \alpha}{\alpha} = \lim_{\alpha \to 0} \frac{\sin (0 + \alpha) - \sin 0}{\alpha} = \left(\frac{d \sin \theta}{d\theta}\right)_{\theta=0} = \cos 0 = 1.$$

   2. Show that

$$\lim_{\alpha \to 0} \frac{\cos \alpha - 1}{\alpha} = 0.$$

   Proof.   One shows, as above, that the limit in question is the derivative of $\cos \theta$ at $\theta = 0$, that is, $\sin 0$, and thus 0. ◀

Remark   It is, unfortunately, customary to write $\cos^2 \theta$ instead of $(\cos \theta)^2$, $\sin^3 \theta$ instead of $(\sin \theta)^3$, and so forth, and similarly for other trigonometric functions. We shall adhere to this tradition. For instance, we write $\cos^2 \theta + \sin^2 \theta = 1$ instead of $(\cos \theta)^2 + (\sin \theta)^2 = 1$.

EXERCISES

47. Differentiate $t \mapsto \sin (2t + 1)$.

48. Find $\dfrac{d}{dx} \cos x^2$.

49. Find $\dfrac{d}{dx} \cos^2 x$.

50. Find $\dfrac{d}{dx} \cos \cos x$.

51. Differentiate $z \mapsto \sin (\cos z)$.

52. Find $\dfrac{dy}{d\theta}$ if $y = \cos \theta \sin \theta$.

53. Find $\dfrac{d}{du} \sin (\text{arc cos } u)$.

▶54. Find $\dfrac{du}{dz}$ if $u = \cos^2 (2z + 1) \sin (2z + 1)^2$.

55. Find $\dfrac{d^2}{d\theta^2} \cos^3 (2\theta)$.

56. Find $\dfrac{d^2y}{ds^2}$ if $y = \text{arc cos } (s^2)$.

57. Find $\int_0^{9\pi/2} \sin 3t \, dt$.

58. Find $\int_{3\pi}^{5\pi/2} \cos \dfrac{\theta}{2} \, d\theta$.

59. Find $\int (\cos x^2) \, x \, dx$.

▶60. Find $\int (\cos \sqrt{x}) \dfrac{dx}{\sqrt{x}}$.

61. Find $\int (\sin^2 \theta) \cos \theta \, d\theta$.

62. Find $\int \left(\dfrac{1}{\cos \theta}\right)^3 \sin \theta \, d\theta$.

63. Find $\int x \cos x \, dx$. (Hint: Use integration by parts.)

64. Find a solution of the differential equation $\dfrac{d^2y}{dx^2} = \cos \dfrac{x}{2}$.

65. Find a solution of the differential equation $\dfrac{d^2y}{dx^2} - 2\dfrac{dy}{dx} = 3 \cos x$. (Hint: Set $y = a \cos x + b \sin x$, where a and b are constants, and substitute into the differential equation.)

66. Find $\lim\limits_{h \to 0} \dfrac{1}{h} \cos \left(\dfrac{\pi}{2} + h\right)$.

67. Find $\lim\limits_{t \to 0} \dfrac{\sin (\pi + t)}{t}$.

68. Find $\lim\limits_{\lambda \to 0} \dfrac{\sin^2 \lambda}{\lambda}$.

69. Find $\lim\limits_{\theta \to 0} \dfrac{\cos \theta \sin \theta}{\theta}$.

1.10 The differential equation of sine and cosine

While the applications of trigonometric functions to geometry continue to be very significant, the importance of the sine and cosine functions in mathematics and science derives primarily from the differential equations they satisfy.

Since, by Theorem 2, $(\cos x)' = -\sin x$ and $(\sin x)' = \cos x$, we have that $(\cos x)'' = -\cos x$. Similarly, $(\sin x)'' = -\sin x$, for $(\sin x)'' =$

$(\cos x)' = -\sin x$. Thus both functions $x \mapsto \cos x$ and $x \mapsto \sin x$ satisfy the differential equation $f''(x) + f(x) = 0$. We proceed to find all solutions of this **second order differential equation.**

(The equation is said to be of second order, since it involves a second derivative. We now denote the independent variable by x rather than by θ; this is, of course, of no importance. We chose this notation, since the geometric origin of the functions sine and cosine now recedes into the background.)

Theorem 3   LET a AND b BE TWO NUMBERS. IF

(19) $$f(x) = a \cos x + b \sin x,$$

THEN

(20) $$f''(x) + f(x) = 0$$

AND

(21) $$f(0) = a, \qquad f'(0) = b.$$

CONVERSELY, IF THE FUNCTION $f(x)$ SATISFIES EQUATION (20) AND THE CONDITIONS (21), THEN (19) HOLDS.

In the second part of the theorem, it is assumed that $f(x)$ is defined for all x or, at least, in an interval containing $x = 0$. We note three important special cases: If $f'' + f \equiv 0$ and $f(0) = f'(0) = 0$, then $f \equiv 0$. If $f'' + f \equiv 0$ and $f(0) = 1$, $f'(0) = 0$, then $f(x) \equiv \cos x$. If $f'' + f \equiv 0$ and $f(0) = 0$, $f'(0) = 1$, then $f(x) \equiv \sin x$.

Proof of Theorem 3.   If $f(x) = a \cos x + b \sin x$, then one computes easily that $f''(x) + f(x) = 0$ and $f(0) = a, f'(0) = b$.

We establish first a special case of the second statement. Assume that $f(x)$ satisfies $f'' + f = 0$ and that $f(0) = f'(0) = 0$. We want to show that $f(x) \equiv 0$. To do this, consider the function

$$\phi(x) = f(x)^2 + f'(x)^2$$

Then $\phi(0) = 0$ and

$$\phi'(x) = 2f(x)f'(x) + 2f'(x)f''(x) = 2f'(x)(f''(x) + f(x)) \equiv 0.$$

Hence ϕ is constant, and since $\phi(0) = 0$ we have $\phi(x) \equiv 0$. Since $f(x)^2$ and $f'(x)^2$ cannot be negative, we have $f(x)^2 \equiv 0$ and $f(x) \equiv 0$ as asserted.

Now let $f(x)$ be a given function satisfying $f'' + f = 0$ with $f(0) = a$

and $f'(0) = b$. We set $g(x) = f(x) - a \cos x - b \sin x$ and must show that $g(x) \equiv 0$. But we have that $g(0) = f(0) - a = 0$, and

$$g'(x) = f'(x) + a \sin x - b \cos x, \qquad g'(0) = f'(0) - b = 0,$$
$$g''(x) = f''(x) + a \cos x + b \sin x = -f(x) + a \cos x + b \sin x = -g(x).$$

Thus $g''(x) + g(x) \equiv 0$ and $g(0) = g'(0) = 0$. By what was proved before, $g(x) \equiv 0$. This completes the argument.

The content of Theorem 3 is sometimes expressed by saying that $f(x) = a \cos x + b \sin x$ is the **general solution** of the differential equation $f''(x) + f(x) = 0$. This means that, for every choice of constants a and b, the function $a \cos x + b \sin x$ satisfies the equation, and every function satisfying the equation can be written as $a \cos x + b \sin x$ for some choice of a and b.

The theorem also asserts that a solution $f(x)$ of $f'' + f = 0$ is uniquely determined by the so-called **initial conditions,** that is, by the values of $f(0)$ and $f'(0)$, and these values may be prescribed. One can also assign the values of f and its derivative f' at any point x_0. For if $f(x)$ is a solution satisfying the conditions $f(0) = a$, $f'(0) = b$, then $f_1(x) = f(x - x_0)$ is a solution satisfying $f_1(x_0) = a$, $f'_1(x_0) = b$.

EXERCISES

In Exercises 70 to 75, find a solution of the differential equation (20) satisfying the indicated conditions.

70. $f(0) = f'(0) = 1$.
71. $f(0) = 2f'(0) = 8$.
72. $f(\pi) = 1, f'(\pi) = -1$.

73. $f(\pi/2) = 0, f'(\pi/2) = 1$.
74. $f(-\pi) = f'(-\pi) = 3$.
75. $f(3\pi) = f'(3\pi) = 3\pi$.

1.11 Addition theorem

All properties of the functions sine and cosine can be obtained from Theorem 3. An important example is the so-called **addition theorem.**

Theorem 4   WE HAVE

$$(22) \qquad \begin{aligned} \sin (x + y) &= \sin x \cos y + \cos x \sin y, \\ \cos (x + y) &= \cos x \cos y - \sin x \sin y. \end{aligned}$$

Proof.   For a fixed y, set $f(x) = \sin (x + y)$. Differentiate this function twice, remembering that y is to be treated as a constant. We obtain $f'(x) = \cos (x + y)$ and $f''(x) = -\sin (x + y)$, so that $f''(x) + f(x) = 0$ for all x. By Theorem 3, we have that

$$f(x) = a \cos x + b \sin x \qquad \text{where } a = f(0) \text{ and } b = f'(0).$$

But $f(0) = \sin y$ and $f'(0) = \cos y$. Substituting these values of a and b, we obtain the first relation (22). The second relation can be proved similarly, or can be obtained by differentiating both sides of the first with respect to x, for a fixed value of y.

The addition theorem has many interesting consequences. For $x = y$, we obtain the **doubling theorem**

$$\sin 2x = 2 \sin x \cos x, \qquad \cos 2x = \cos^2 x - \sin^2 x.$$

The second relation can also be written in the form

$$\cos 2x = 2 \cos^2 x - 1 = 1 - 2 \sin^2 x.$$

For $y = \pi/2$ and for $y = \pi$, we get the relation

$$(23) \qquad \sin\left(x + \frac{\pi}{2}\right) = \cos x, \qquad \cos\left(x + \frac{\pi}{2}\right) = -\sin x$$

$$\sin(x + \pi) = -\sin x, \qquad \cos(x + \pi) = -\cos x.$$

The important formulas

$$\sin\left(\frac{\pi}{2} - x\right) = \cos x, \qquad \cos\left(\frac{\pi}{2} - x\right) = \sin x$$

also follow from the addition theorems; the details are left to the reader. (Earlier we obtained this for $0 \leq x \leq \pi/2$ only.) This relation shows that the graph of the cosine can be obtained from that of the sine by translating the latter to the left by the distance $\pi/2$.

▶ **Examples   1.** Express $\sin\dfrac{\theta}{2}$ and $\cos\dfrac{\theta}{2}$ in terms of $\sin\theta$ and $\cos\theta$, assuming that $0 \leq \theta \leq \pi$.

Answer.   Using the doubling theorem with $2x = \theta$, we see that

$$\cos\theta = \cos^2\frac{\theta}{2} - \sin^2\frac{\theta}{2}.$$

Also $$1 = \cos^2\frac{\theta}{2} + \sin^2\frac{\theta}{2}.$$

If we add and subtract these two equations, we obtain

$$1 - \cos \theta = 2 \sin^2 \frac{\theta}{2}, \qquad 1 + \cos \theta = 2 \cos^2 \frac{\theta}{2}.$$

Dividing by 2, and noting that $\sin \frac{\theta}{2}$ and $\cos \frac{\theta}{2}$ must be nonnegative,

we get

$$\sin \frac{\theta}{2} = \sqrt{\frac{1 - \cos \theta}{2}}, \qquad \cos \frac{\theta}{2} = \sqrt{\frac{1 + \cos \theta}{2}}.$$

2. Express $\cos 3x$ and $\sin 3x$ in terms of $\cos x$ and $\sin x$.

Solution.   We have, using the addition formulas and the doubling theorems,

$$\cos 3x = \cos 2x \cos x - \sin 2x \sin x$$
$$= \cos^3 x - \cos x \sin^2 x - 2 \sin^2 x \cos x$$
$$= \cos^3 x - 3 \cos x \sin^2 x.$$

One shows similarly that

$$\sin 3x = 3 \cos^2 x \sin x - \sin^3 x.$$

3. Prove the formulas

$$\cos x + \cos y = 2 \cos \frac{x + y}{2} \cos \frac{x - y}{2},$$

$$\sin x + \sin y = 2 \sin \frac{x + y}{2} \sin \frac{x - y}{2}.$$

Answer.   We prove only the first relation; the second is established similarly. By the addition theorem,

$$\cos (\alpha + \beta) = \cos \alpha \cos \beta - \sin \alpha \sin \beta$$

and, since $\alpha - \beta = \alpha + (-\beta)$,

$$\cos (\alpha - \beta) = \cos \alpha \cos \beta + \sin \alpha \sin \beta.$$

Adding the two displayed equations, we obtain

$$\cos(\alpha + \beta) + \cos(\alpha - \beta) = 2\cos\alpha\cos\beta.$$

Setting $\alpha + \beta = x$, $\alpha - \beta = y$, we have $\alpha = \dfrac{x + y}{2}$ and $\beta = \dfrac{x - y}{2}$.
The desired relation follows.◄

EXERCISES

76. Find a formula for $\sin 4\theta$ in terms of $\cos\theta$ and $\sin\theta$.
77. Find a formula for $\cos 5x$ in terms of $\cos x$ and $\sin x$.
78. Find a formula for $\cos(\theta/4)$ in terms of $\cos\theta$ and $\sin\theta$, assuming $0 \le \theta \le 2\pi$.
79. Find a formula for $\cos(3t/2)$ in terms of $\cos t$ and $\sin t$, assuming $0 \le t \le \pi$.
80. Find $\int \sin^2 x\, dx$. (Hint: Use the doubling theorem.)
81. Find $\int \cos^2 3x\, dx$.
82. Find $\int \cos^2 x \sin^2 x\, dx$. (Hint: Use the doubling theorem repeatedly.)
83. Find $\int \cos^4 2z\, dz$. (Hint: Use the doubling theorem repeatedly.)
84. Find $\int_{\pi/4}^{\pi/3} \sin\theta \sin(\theta/2)\, d\theta$. (Hint: Use the half-angle equations in Example 2.)
85. Find $\int \cos 5u \cos 3u\, du$. (Hint: Solve $(x + y)/2 = 5u$, $(x - y)/2 = 3u$ for x and y and use the equations in Example 3.)

In Exercises 86 to 91, sketch the given functions. In particular, find the points where the tangent is horizontal or vertical, points of inflection, and maxima and minima (absolute and local). Also show where the function is decreasing, where it is increasing, and where it is concave up and where it is concave down.

86. $\theta \mapsto 2\sin(\theta/2)$. Use only Fig. 6.9.
87. $\theta \mapsto 3\sin 2\theta$. Use only Fig. 6.9.
88. $\theta \mapsto \frac{1}{4}\cos 4\theta$. Use only Fig. 6.9.
89. $x \mapsto x + \sin x$ for $-\pi \le x \le \pi$.
90. $x \mapsto \cos^2 x$ for $-\pi \le x \le \pi$.
91. $x \mapsto \sqrt{\cos x}$ for $-\pi/2 \le x \le \pi/2$.

1.12 A second order differential equation

Theorem 3 has a simple but important generalization.

Theorem 5   LET $\omega > 0$ BE A GIVEN NUMBER. THE GENERAL SOLUTION OF THE DIFFERENTIAL EQUATION

$$(24) \qquad\qquad f''(x) + \omega^2 f(x) = 0$$

IS

$$(25) \qquad\qquad f(x) = a\cos\omega x + b\sin\omega x.$$

One can prove this in the same way as we proved Theorem 3. It is perhaps even simpler to reduce Theorem 5 to Theorem 3. Set $F(t) = f(t/\omega)$. Then $f(x) = F(\omega x)$, $f'(x) = \omega F'(\omega x)$, and $f''(x) = \omega^2 F''(\omega x)$. Hence the equation $f'' + \omega^2 f = 0$ is equivalent to $F'' + F = 0$. We know that $F'' + F = 0$ if and only if $F = a \cos x + b \sin x$. Therefore $f'' + \omega^2 f = 0$ if and only if $f = a \cos \omega x + b \sin \omega x$.

The constants a and b in (19) can be determined if one knows the values of f and f' at some point. For instance, if $f(0) = A$ and $f'(0) = B$, then $a = A$ and $b = B/\omega$.

We remark that the general solution of (24) is periodic with period $2\pi/\omega$ [see Property (1) of periodic functions in §1.2].

1.13 Motion of an elastic spring

We apply Theorem 5 to the motion of a particle attached to an **elastic spring**. Such a spring is shown in Fig. 6.20. It is assumed that the spring has an equilibrium position, and that if it is displaced from that position by an amount s, there is a restoring force F acting on the particle that tends to restore the spring to its original position. (We may assume that $s > 0$ if the spring is stretched, and $s < 0$ if the spring is compressed.) The spring is called elastic if the restoring force is a linear function of the displacement

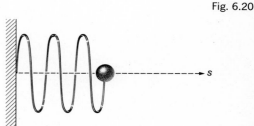
Fig. 6.20

$$F = -ks$$

where k is a positive constant (**spring constant**). Practically all springs may be considered as elastic, provided one considers only small displacements. (This has already been discussed in Chapter 5, §5.)

Let m be the mass of the particle attached to the spring. If we neglect the mass of the spring itself, and if we observe that ds/dt, the time derivative of s, is the velocity, and d^2s/dt^2 is the acceleration, Newton's law of motion reads

(26) $$m\frac{d^2s}{dt^2} = -ks.$$

We rewrite this as $$\frac{d^2s}{dt^2} + \omega^2 s = 0,$$

where $$\omega = \sqrt{\frac{k}{m}}.$$

Equation (26) is identical with equation (24), except for the names of the quantities involved. Using Theorem 5, we conclude that the function

$t \mapsto s(t)$ describing the motion is of the form

(27) $$s(t) = a \cos \omega t + b \sin \omega t.$$

In particular, the motion is **periodic** with period T, where

(28) $$T = \frac{2\pi}{\omega} = 2\pi \sqrt{\frac{m}{k}}.$$

Such a perpetual motion is possible, because we neglect friction.

In order to know the motion completely, we must know the numbers a and b in (27). This means that we must know the position $s(t)$ and velocity $s'(t)$ of our spring at some fixed time t_0, say, at the initial time $t = 0$. Once these *initial conditions* are given, we can compute the constants a and b, and hence also the position $s(t)$ and the velocity $s'(t)$ of the spring for any time t.

We claim that $s(t)$ can also be written in the form

(29) $$s(t) = A \cos [\omega(t - t_0)],$$

where $A > 0$ and t_0 are constants depending on the initial conditions. To see this, start with the expression (27) for $s(t)$. If $a = b = 0$, set $A = 0$. If $a^2 + b^2 > 0$, let $A = \sqrt{a^2 + b^2}$. By Theorem 1, there is a number α such that $a/A = \cos \alpha$, $b/A = \sin \alpha$. Set $t_0 = \alpha\omega$. Then we have

$$s(t) = a \cos \omega t + b \sin \omega t = A \cos \omega t_0 \cos \omega t + A \sin \omega t_0 \sin \omega t$$

$$= A \cos (\omega t - \omega t_0) = A \cos [\omega(t - t_0)]$$

by the addition theorem.

Since the cosine oscillates between -1 and 1, the number A gives the maximal deviation of the spring from the equilibrium position; A is called the **amplitude** of the oscillation. The number t_0 describes the time at which this maximal deviation takes place; it is called the **phase difference**. Between the times t_0 and $t_0 + T$ the spring carries out a complete cycle.

Amplitude and phase difference depend on how the motion was started at $t = 0$; the period T, however, depends only on the spring.

1.14 Harmonic oscillations

When we strike a tuning fork, we always hear the same tone, no matter how hard or how lightly we strike it. Similarly, hitting the same piano key once lightly and once full force will produce sounds of different intensities but of the same pitch. These and many similar phenomena can be understood by considering a vibrating body as an elastic spring. This may be done because the differential equations describing the motion of a vibrating tuning fork, a vibrating piano string, and so on, are approximately the same as the equation of motion of an elastic spring, that is, equation (26). In other words, there is a number s describing the deviation of the vibrating body from the position of equilibrium, a constant k that measures the **stiffness** of the body, and a constant m that measures the **inertia** of the body. These three quantities and the time t are connected by equation (26). Therefore s as a function of time is given by (27) or by (29).

The amplitude and the phase difference depend on the initial conditions; the period is given by the formula (28) and depends only on the stiffness and the inertia of the vibrating body. So does the number

$$\nu = \frac{1}{T},$$

the **frequency** of the oscillation; it is the number of cycles per unit time. In sound (audible oscillations of air) the frequency determines the **pitch.** In our case

$$(30) \qquad \qquad \nu = \frac{1}{2\pi} \sqrt{\frac{k}{m}}.$$

The frequency is always the same. It does not depend on the amplitude A (which in sound determines the intensity).

A motion governed by the same differential equation as the elastic spring is called a **harmonic oscillation.** Harmonic oscillations occur not only in elastic bodies but in many other situations, for instance, in a radio transmitter or inside an atom. The constancy of pitch is their most important property. It explains why a wrist watch keeps correct time and why an atom emits light of a definite frequency. The surprising fact that the number π appears in the formula for the frequency suggests a connection between circles and harmonic oscillations. This connection will be pursued in Chapter 11.

EXERCISES

In Exercises 92 to 97, a particle of mass m is attached to one end of a horizontal elastic spring whose spring constant is k. The other end of the spring is held fixed and the particle is set vibrating along the axis of the spring. The displacement of the particle from the rest position is denoted by $s = s(t)$ with $s > 0$ when the spring is stretched.

92. Suppose that $m = 5$ grams and $s = 2 \cos (t/3)$. Find the spring constant.
93. Suppose that $m = 2$ grams and $k = 8$ grams/sec². Suppose also that, at time $t = 0$, the spring is stretched 1 cm from its rest position, and that the speed of the particle is 4 cm/sec and is increasing. Find $s(t)$. Find the amplitude.
94. Suppose that $m = 1.5$ grams and $k = 4.5$ grams/sec². Suppose also that at times $t = 0$ and $t = \pi \sqrt{3}/6$, the spring is compressed 1 cm from its rest position. Find an expression for the velocity of the particle.
95. Suppose that $m = .2$ gram and $k = 1.8$ grams/sec². Suppose also that at time $t = 0$, the spring is compressed 2 cm from its rest position. Find the velocity of the particle at time $t = \pi/6$. For what other values of t can you compute the velocity?
96. Suppose that $m = 1$ gram and $k = .25$ gram/sec². Suppose also that at time $t = 0$, the spring is stretched 1 cm from its rest position, and at time $t = \pi/2$ the velocity of the particle is $(\sqrt{2}/2)$ cm/sec. Find $s(t)$. Find the amplitude.
97. Suppose that the period of the motion is $(2\pi/15)$ sec. Suppose also that at time $t = 0$, the velocity of the particle is -15 cm/sec, and at time $t = \pi/6$ the velocity of the particle is 3 cm/sec. Find $s(t)$. Find the amplitude.

§2   The Tangent and the Arc Tangent

In the preceding section, we noted that one could have defined the functions sine and cosine using a primitive function of $\dfrac{1}{\sqrt{1 - x^2}}$. In this section we shall show how these functions could have been obtained from a primitive function of $\dfrac{1}{1 + x^2}$. At the same time, we shall study a third important trigonometric function.

2.1   The tangent

We begin by defining the function $\theta \mapsto \tan \theta$ called the **tangent**:

$$(1) \qquad \tan \theta = \frac{\sin \theta}{\cos \theta}.$$

This function is defined and continuous for all values of θ with $\cos \theta \neq 0$, that is, for $\theta \neq \pi/2 + n\pi$, $n = 0, \pm 1, \pm 2, \cdots$.

It is regrettable that the word "tangent" is used to denote two different things, the line that "just touches" a curve and the function defined above. But both names are so traditional that they cannot be changed.

The tangent is an *odd* function, since

$$(2) \qquad \tan (-\theta) = \frac{\sin (-\theta)}{\cos (-\theta)} = \frac{-\sin \theta}{\cos \theta} = -\tan \theta.$$

The tangent is *periodic* with *period* π, since, noting equation (23),

(3) $\qquad \tan(\theta + \pi) = \dfrac{\sin(\theta + \pi)}{\cos(\theta + \pi)} = \dfrac{-\sin\theta}{-\cos\theta} = \tan\theta.$

The tangent is strictly increasing in the interval $(-\pi/2, \pi/2)$ and therefore also in every interval $(n\pi - \pi/2, n\pi + \pi/2)$, $n = 0, \pm1, \pm2, \cdots$. Indeed in $(-\pi/2, \pi/2)$ the tangent has a positive derivative.

$$(\tan\theta)' = \left(\frac{\sin\theta}{\cos\theta}\right)' = \frac{(\sin\theta)'\cos\theta - \sin\theta\,(\cos\theta)'}{\cos^2\theta} = \frac{\cos^2\theta + \sin^2\theta}{\cos^2\theta}.$$

This can be written either as

(4) $$\frac{d\tan\theta}{d\theta} = \frac{1}{\cos^2\theta}$$

or as

(5) $$\frac{d\tan\theta}{d\theta} = 1 + \tan^2\theta.$$

Since $\sin(\pi/2) = 1$, $\cos(\pi/2) = 0$ and $\tan\theta$ is positive and increasing for $0 < \theta < \pi/2$, $\tan\theta$ becomes arbitrarily large as θ approaches $\pi/2$ from the left. One expresses this by writing

(6) $$\lim_{\theta \to (\pi/2)^-} \tan\theta = +\infty.$$

One sees similarly that $\tan\theta$ becomes negative and arbitrarily large in absolute value as θ approaches $-\pi/2$ from the left:

(7) $$\lim_{\theta \to (-\pi/2)^+} \tan\theta = -\infty.$$

Since $\tan\theta$ has period π, we conclude that

(8) $$\lim_{\theta \to (\pi/2 + n\pi)^-} \tan\theta = +\infty,$$

$$\lim_{\theta \to (\pi/2 + n\pi)^-} \tan\theta = -\infty, \, n = 0, \pm1, \pm2, \cdots.$$

The graph of the function $\tan\theta$ is shown in Fig. 6.21. Note that π is the primitive period of the tangent.

Remark   At present, we have no convenient name for a primitive function for $\theta \mapsto \tan\theta$. We return to this question in §4.7.

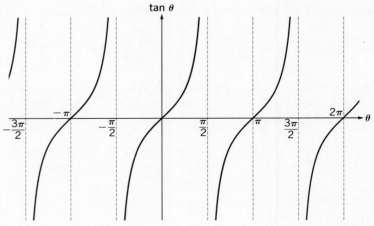

Fig. 6.21

EXERCISES

1. Find $\dfrac{d}{d\theta} \tan (2\theta + \pi)$ at $\theta = 22\frac{1}{2}°$.   ► 4. Find $\dfrac{d^3 \tan x}{dx^3}$ at $x = 9\pi/4$.

2. Differentiate $u \mapsto \tan (u^2)$.   5. Find $\dfrac{d^2}{d\theta^2} \sin (\tan \theta)$.

3. Find $\dfrac{d}{dz} \tan^2 z$.   6. Find $\dfrac{d^2 z}{du^2}$ if $z = u \tan u$.

2.2 Addition theorem

The function tan obeys an **addition theorem**

(9) $$\tan (\alpha + \beta) = \frac{\tan \alpha + \tan \beta}{1 - \tan \alpha \tan \beta}$$

(assuming that neither α, nor β, nor $\alpha + \beta$ is $\pi/2 + n\pi$). To prove this, we use the definition (1) of tan and the addition theorems of sin and cos. We have

$$\tan (\alpha + \beta) = \frac{\sin (\alpha + \beta)}{\cos (\alpha + \beta)} = \frac{\sin \alpha \cos \beta + \cos \alpha \sin \beta}{\cos \alpha \cos \beta - \sin \alpha \sin \beta}$$

$$= \frac{\cos \alpha \cos \beta \left(\dfrac{\sin \alpha}{\cos \alpha} + \dfrac{\sin \beta}{\cos \beta} \right)}{\cos \alpha \cos \beta \left(1 - \dfrac{\sin \alpha \sin \beta}{\cos \alpha \cos \beta} \right)} = \frac{\tan \alpha + \tan \beta}{1 - \tan \alpha \tan \beta},$$

as asserted.

For $\alpha = \beta$, we obtain the **doubling formula**

$$(10) \qquad \tan 2\alpha = \frac{2 \tan \alpha}{1 - \tan^2 \alpha}.$$

2.3 The functions sin θ and cos θ as rational functions of tan ($\theta/2$)

We show next that the functions $\cos \theta$ and $\sin \theta$ can be written as rational functions of $\tan (\theta/2)$.

Theorem 1 FOR $-\pi < \theta < \pi$, WE HAVE

$$(11) \qquad \sin \theta = \frac{2 \tan (\theta/2)}{1 + \tan^2 (\theta/2)},$$

$$(12) \qquad \cos \theta = \frac{1 - \tan^2 (\theta/2)}{1 + \tan^2 (\theta/2)}.$$

Remark Since both sides of (11) and (12) are periodic with period 2π, it follows that these relations hold for all values of θ for which the right sides are defined, that is, except for $\theta = n\pi$, n an integer. At the exceptional points, the limits of the right sides are 0 for (11) and 1 or -1 for (12).

First proof. Setting $\alpha = \theta/2$ in (10), we obtain the relation

$$\tan \theta = \frac{2 \tan (\theta/2)}{1 - \tan^2 (\theta/2)}.$$

We set

$$(13) \qquad t = \tan \frac{\theta}{2}$$

and rewrite the above formula in the form

$$(14) \qquad \frac{\sin \theta}{\cos \theta} = \frac{2t}{1 - t^2}.$$

Squaring both sides and adding 1 to each side, we find that

$$1 + \left(\frac{\sin \theta}{\cos \theta}\right)^2 = \frac{\cos^2 \theta + \sin^2 \theta}{\cos^2 \theta} = \frac{1}{\cos^2 \theta}$$

$$= 1 + \left(\frac{2t}{1 - t^2}\right)^2 = 1 + \frac{4t^2}{1 - 2t^2 + t^4} = \left(\frac{1 + t^2}{1 - t^2}\right)^2$$

so that

$$(15) \qquad \cos^2 \theta = \left(\frac{1 - t^2}{1 + t^2}\right)^2.$$

If $-\pi < \theta \leq -\pi/2$, then $\cos \theta \leq 0$. Then also $-\pi/2 < \theta/2 \leq -\pi/4$ and therefore $\tan (\theta/2) \leq -1$, that is, $t \leq -1$, $t^2 \geq 1$, $1 - t^2 \leq 0$, and $(1 - t^2)/(1 + t^2) \leq 0$. One verifies in the same way that $\cos \theta$ and $(1 - t^2)/(1 + t^2)$ have the same sign for θ in the intervals $[-\pi/2, 0]$, $[0, \pi/2]$, $[\pi/2, \pi]$. Therefore we may conclude from (15) that

$$(16) \qquad \cos \theta = \frac{1 - t^2}{1 + t^2}.$$

Hence, using (14), we also have

$$(17) \qquad \sin \theta = \frac{2t}{1 + t^2}.$$

Formulas (16) and (17) are equivalent to (11) and (12).

   Second proof.   Let the function $t(\theta)$ be defined by (13). Then

$$t(0) = \tan 0 = 0$$

and $\qquad t'(\theta) = \frac{1}{2}\left(1 + \tan^2 \frac{\theta}{2}\right) = \frac{1}{2}(1 + t^2)$

as is seen from (5). Now set

$$f(\theta) = \frac{2t}{1 + t^2}, \qquad g(\theta) = \frac{1 - t^2}{1 + t^2}.$$

Then $\qquad f'(\theta) = \frac{2(1 + t^2) - (2t)(2t)}{(1 + t^2)^2} \frac{dt}{d\theta}$

$$= \frac{2(1 - t^2)}{(1 + t^2)^2} \frac{1}{2}(1 + t^2) = \frac{1 - t^2}{1 + t^2} = g(\theta).$$

A similar computation shows that

$$g'(\theta) = -f(\theta).$$

Therefore $f''(\theta) = g'(\theta) = -f(\theta)$ and $g''(\theta) = -f'(\theta) = -g(\theta)$. Also $f(0) = 0$ and $g(0) = 1$ so that $f'(0) = g(0) = 1$ and $g'(0) = -f(0) = 0$. Now we conclude, by Theorem 3 in §1.10, that $f(\theta) = \sin \theta$, $g(\theta) = \cos \theta$.

EXERCISES

7. Show that $\tan x = \dfrac{\sin 2x}{1 + \cos 2x}$.

8. Find a formula for $\tan 3\theta$ in terms of $\tan \theta$.

9. Show that $\tan 2\theta = \dfrac{\sin 3\theta + \sin \theta}{\cos 3\theta + \cos \theta}$.

10. Express $\sin x$ and $\cos x$ as rational functions of the function $\tan (x/4)$.

11. Using the formulas from Theorem 1 and the periodicity and oddness of tan, show that the sine is an odd function, the cosine an even one, and that both have period 2π.

12. Let (x, y) be a point of the unit circle, not the point $(-1, 0)$. Show that there is a real number t such that $x = \dfrac{1 - t^2}{1 + t^2}$, $y = \dfrac{2t}{1 + t^2}$. Is this t uniquely determined?

13. Show that $\sin x$ and $\cos x$ can be expressed as radical functions of $\tan x$.

14. Find $\displaystyle\int_0^{\pi/3} \dfrac{\tan \theta}{1 + \tan^2 \theta} \, d\theta$.

15. Find $\displaystyle\int_0^{\pi/4} \dfrac{1 - \tan^2 x}{1 + \tan^2 x} \, dx$.

2.4 The arc tangent

Since the function $u = \tan \theta$ is strictly increasing and has a positive derivative in the interval $-\pi/2 < \theta < \pi/2$, it has a differentiable inverse function (see Theorem 7 in Chapter 3, §3.7, and Theorem 7 in Chapter 4, §2.6). We call this function the **arc tangent** or inverse tangent of u, and we denote it by arc tan u. In view of (6) and (7), the equation $\tan \theta = u$ has a solution θ in the interval $-\pi/2 < \theta < \pi/2$ for every number u. The function $u \mapsto$ arc tan u is therefore defined for all values of u. The graph of the arc tan function (see Fig. 6.22) is obtained by "flipping"

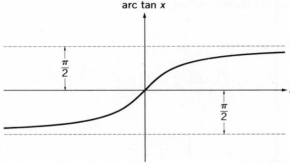

Fig. 6.22

the central branch of the graph in Fig. 6.21. Since $\tan 0 = 0$, we have

(18) $\text{arc } \tan 0 = 0.$

Next, we compute the derivative of the arc tan. We have that $\text{arc } \tan (\tan \theta) = \theta$ for $-\pi/2 < \theta < \pi/2$, so that

$$\left(\frac{d \text{ arc } \tan u}{du}\right)_{u=\tan \theta} = \frac{1}{d \tan \theta/d\theta} = \frac{1}{1 + \tan^2 \theta}$$

and

(19) $\dfrac{d \text{ arc } \tan u}{du} = \dfrac{1}{1 + u^2}.$

From (12), (19), and the fundamental theorem of calculus, we obtain

Theorem 2   THE FUNCTION

$$\theta \mapsto \tan \theta = \frac{\sin \theta}{\cos \theta}$$

CONSIDERED IN THE INTERVAL $-\pi/2 < \theta < \pi/2$ HAS THE INVERSE FUNCTION

(20) $u \mapsto \text{arc } \tan u = \displaystyle\int_0^u \frac{dt}{1 + t^2}.$

Theorem 2 shows the possibility of another, purely analytic, definition of the trigonometric functions. Indeed, we could have defined the function $\tan \theta$, without using the sine and cosine function, as the function inverse to the function (20). Then we could have defined sine and cosine by the relations (11) and (12).

Theorem 2 has an interesting corollary.

Corollary.   *We have*

(21) $\displaystyle\int_{-\infty}^{-1} \frac{dt}{1 + t^2} = \int_1^0 \frac{dt}{1 + t^2} = \int_0^1 \frac{dt}{1 + t^2} = \int_1^{+\infty} \frac{dt}{1 + t^2} = \frac{\pi}{4}$

and

(22) $\displaystyle\int_{-\infty}^{+\infty} \frac{dt}{1 + t^2} = \pi.$

Proof. We have

$$\tan 45° = \tan \frac{\pi}{4} = 1, \tag{23}$$

and therefore

$$\int_1^A \frac{dt}{1 + t^2} = \text{arc tan } u \Big|_1^A = \text{arc tan } A - \text{arc tan } 1 = \text{arc tan } A - \frac{\pi}{4}.$$

Since, by (6) and Theorem 1,

$$\lim_{A \to +\infty} \text{arc tan } A = \frac{\pi}{2},$$

we have

$$\int_1^{+\infty} \frac{dt}{1 + t^2} = \lim_{A \to +\infty} \int_1^A \frac{dt}{1 + t^2} = \lim_{A \to +\infty} \left(\text{arc tan } A - \frac{\pi}{4} \right)$$
$$= \frac{\pi}{2} - \frac{\pi}{4} = \frac{\pi}{4}.$$

The other relations (21) are derived similarly. Relation (22) follows, since the integral in (22) is the sum of the four integrals in (21).

In an appendix to Chapter 5 (see §6.5) we have already derived (21) and (22), without using trigonometric functions. The present derivation is less mysterious.

EXERCISES

16. Show that $\int_0^1 \frac{dt}{1 + t^2} = \frac{\pi}{4}$, using the arc tangent function.

17. Find $\frac{d}{dx} (\text{arc tan } x^3)^2$.

18. Differentiate $z \mapsto z \text{ arc tan } \sqrt{z}$.

19. Find $\frac{d}{dt} \text{arc tan } (\sqrt{\text{arc tan } t})$.

20. Find $\frac{d}{dz} \text{arc tan } \frac{z}{\sqrt{z - 1}}$.

21. Find $f''(s)$ if $f(s) = \left(\text{arc tan } \frac{1}{s} \right)^2$.

▶ 22. Find $\frac{d^2}{dx^2} \left(x \text{ arc tan } \frac{1}{\sqrt{x}} \right)$.

23. Find $g'(x)$ if $g(x) = \text{arc tan } x + \text{arc tan } \frac{1}{x}$. What can you conclude about g?

24. Suppose that $f(u) = \arctan \dfrac{u^2}{\sqrt{3}}$. In what intervals is f increasing? In what intervals is f concave upward?

25. Show that $\sin(2 \arctan u) = \dfrac{2u}{1 + u^2}$.

26. Show that $\cos(2 \arctan u) = \dfrac{1 - u^2}{1 + u^2}$.

27. Show that $\sin(\arctan u) = \dfrac{u}{\sqrt{1 + u^2}}$. (Hint: Use Exercise 25 and the half-angle formula for sine. Treat $u \geq 0$ and $u < 0$ separately.)

28. Show that $\cos(\arctan u) = \dfrac{1}{\sqrt{1 + u^2}}$.

29. Show that $\tan(\arccos u) = \dfrac{\sqrt{1 - u^2}}{u}$. [Hints: (1) Show that both sides have the same derivative, and next show that both sides are equal when $u = 1$. (2) Write $\tan = \dfrac{\sin}{\cos} = \pm \dfrac{\sqrt{1 - \cos^2}}{\cos}$ and use the relation $\cos(\arccos u) = u$. Show that the minus sign never holds.]

2.5 Using the arc tangent for integration

Theorem 2 is important in integrating rational functions. The function $1/(1 + x^2)$ is rational, but it is not the derivative of a rational function or even of an algebraic one. Indeed, it is the derivative of the transcendental function arc tan x. Using the arc tangent, we can write down explicitly indefinite integrals for many functions. At the moment, however, we are unable to write down a convenient formula for a primitive of the function arc tangent. We shall discuss this later (see §4.7).

▶ *Example* Compute

$$\int_2^3 \frac{x\,dx}{1 + x^4}.$$

Solution. We note the differentiation formula (19) and the fact that x is the derivative of $x^2/2$. Hence

$$\int_2^3 \frac{x\,dx}{1 + x^4} = \frac{1}{2}\int_2^3 \frac{d(x^2)}{1 + (x^2)^2} = \frac{1}{2}\int_{x=2}^{x=3} d\arctan(x^2) = \frac{1}{2}\arctan(x^2)\Big|_{x=2}^{x=3}$$

$$= \frac{1}{2}\arctan 9 - \frac{1}{2}\arctan 4 = .13 \qquad \text{approximately.} ◀$$

Remark To find values of the arc tangent, we use a table of trigonometric functions. If the table lists trigonometric functions of degrees, one must transform degrees to radians.

EXERCISES

30. Find $\int \dfrac{dx}{1 + (2x - 1)^2}$.

31. Find $\int \dfrac{dz}{z^2 + 2z + 2}$. [Hint: Note that $z^2 + 2z + 2 = 1 + (z + 1)^2$.]

32. Find $\int \dfrac{dx}{2x^2 - 6x + 5}$. $\left[\text{Hint: Note that } 2x^2 - 6x + 5 = \dfrac{1}{2}(1 + (2x - 3)^2).\right]$

33. Find $\int \dfrac{\sqrt{u}\, du}{1 + u^3}$.

▶ 34. Find $\int_1^2 \dfrac{t^2\, dt}{1 + 4t^6}$ approximately. Use the tables of values of arc tangent.

35. Find $\int \dfrac{x^{1/4}\, dx}{4 + x^{5/2}}$.

36. Find $\int \dfrac{x\, dx}{x^4 + 2x^2 + 2}$. [Hint: Note that $x^4 + 2x^2 + 2 = 1 + (x^2 + 1)^2$.]

37. Find $\int \dfrac{du}{a + bu^2}$, where a and b are any positive constants. $\Bigg[\text{Hint: Note that}$

$$a + bu^2 = a\left(1 + \frac{b}{a}u^2\right) = a\left(1 + \frac{\sqrt{b}}{\sqrt{a}}u\right)^2.\Bigg]$$

38. Find all primitive functions of $g(x) = \dfrac{\arctan x}{1 + x^2}$.

39. Find all primitive functions of $h(t) = \dfrac{(\arctan t)^2 + (\arctan t)^{-2}}{1 + t^2}$.

40. Find a solution of the differential equation $\dfrac{dz}{du} = \dfrac{u^{-3}}{1 + u^{-4}}$ such that $\lim_{u \to \infty} z = 1$.

41. Find a solution of the differential equation $\dfrac{d^2x}{dt^2} = \dfrac{-t}{(1 + t^2)^2}$ such that $x = \dfrac{\pi}{8}$ when $t = 0$ and $x = \pi$ when $t = 1$.

3.1 Three more trigonometric functions

In addition to the three trigonometric functions sine, cosine, and tangent, one sometimes uses three more: the **secant, cosecant,** and **cotangent,** defined by

(1) $\sec x = \dfrac{1}{\cos x}, \qquad \csc x = \dfrac{1}{\sin x}, \qquad \cot x = \dfrac{\cos x}{\sin x}.$

Their graphs are shown in Figs. 6.23, 6.24, and 6.25. Note that

(2) $\tan x \cot x = 1.$

Fig. 6.23

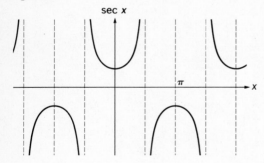

Fig. 6.24

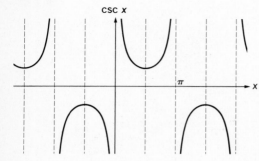

Fig. 6.25

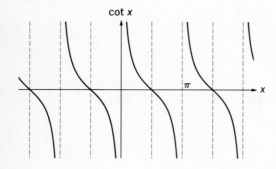

The functions cosecant and cotangent are odd, while the secant is even. The functions secant and cosecant have period 2π, the function cot period π. It is clear that sec x is not defined if $x = n + \pi/2$, n an integer; csc x and cot x are not defined if $x = n\pi$.

We note the differentiation formulas

(3) $$\frac{d \sec x}{dx} = \frac{\sin x}{(\cos x)^2} = \tan x \sec x,$$

$$\frac{d \csc x}{dx} = -\frac{\cos x}{(\sin x)^2} = -\cot x \csc x,$$

(4) $$\frac{d \cot x}{dx} = -1 - (\cot x)^2 = -\frac{1}{(\sin x)^2} = -(\csc x)^2,$$

and

(5) $$\frac{d \tan x}{dx} = (\sec x)^2.$$

It is convenient to remember the relations

(6) $$\sin\left(\frac{\pi}{2} - x\right) = \cos x, \quad \sec\left(\frac{\pi}{2} - x\right) = \csc x, \quad \tan\left(\frac{\pi}{2} - x\right) = \cot x$$

(the first of which we already noted in §1). The proof of all these statements, from the definitions (1), are quite easy and are left to the reader.

▶ *Examples*   *1.*   Given a *right triangle* with vertices A, B, and C, the angle at C being right, and using the notations introduced in §1, we have (see Fig. 6.26)

$$\sin A = \frac{a}{c}, \qquad \cos A = \frac{b}{c}, \qquad \tan A = \frac{a}{b},$$

$$\sec A = \frac{c}{b}, \qquad \csc A = \frac{c}{a}, \qquad \cot A = \frac{b}{a}.$$

Indeed, the first two relations we already know; the others follow easily.

2. The *addition theorem* for the cotangent reads

$$\cot(x + y) = \frac{\cot x \cot y - 1}{\cot x + \cot y}.$$

This can be proved either from (2) using the addition theorem for the tangent, or from (1) using the addition theorem for sine and cosine.

3. The identity

Fig. 6.26

$$\sec^2 \theta + \csc^2 \theta = 4 \csc^2 2\theta$$

follows from the definition of secant and cosecant, and from the identities $\sin 2\theta = 2 \sin \theta \cos \theta$ and $\sin^2 \theta + \cos^2 \theta = 1.$ ◀

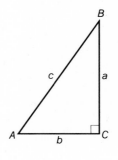

EXERCISES

1. Find $\dfrac{d}{dx} \tan (\sec x)$.

2. Find $\dfrac{d}{dz} \sqrt{\sec z}$.

3. Differentiate $u \mapsto \sec (\text{arc tan } u)$.

4. Find $\dfrac{d}{d\theta} \cot^2 \sqrt{\theta}$.

5. Find $\dfrac{d^2}{dy^2} \csc y$.

6. Find $\int_0^{\pi/4} \sec^2 x \, dx$.
7. Find $\int \sec^2 x \tan x \, dx$.
▶ 8. Find $\int \sec^2 x \tan^2 x \, dx$.
9. Find $\int \tan^2 x \, dx$. (Hint: Note that $\tan^2 x = \sec^2 x - 1$.)
10. Sketch the graph of $y = \sec^2 x$ for $-\pi \leq x \leq \pi$.
11. Sketch the graph of $y = \cot (x/2)$ for $-\pi \leq x \leq \pi$.
12. Sketch the graph of $y = \csc (2x + \pi)$ for $-\pi \leq x \leq \pi$.

3.2 Inverse trigonometric functions

We have already discussed three of the so-called **inverse trigonometric functions,** the inverse cosine or arc cosine, the inverse sine or arc sine, and the inverse tangent or arc tangent. We shall discuss briefly the remaining three functions.

The functions (1) are not monotone. In order to define inverse functions, we must choose for each of the functions (1) an interval (or a collection of intervals) on which it is monotone. This involves a certain amount of arbitrariness and is sometimes done differently by different authors.

We define the **arc cotangent, arc secant,** and **arc cosecant** as the functions inverse to the cotangent, secant, and cosecant, respectively, in the intervals $(0, \pi)$, $(0, \pi)$, and $(-\pi/2, \pi/2)$. A look at the graph of $\cot \theta$ (see Fig. 6.26) indicates that arc cot x is defined everywhere. Its analytic definition reads

$$\text{arc cot } (\cot \theta) = \theta \qquad \text{for } 0 < \theta < \pi.$$

Noting (4) and the relation cot (arc cot x) $= x$, we obtain

$$\frac{d \text{ arc cot } x}{dx} = \frac{1}{dx/d \text{ arc cot } x} = \frac{1}{d \cot (\text{arc cot } x)/d \text{ arc cot } x}$$

$$= -\frac{1}{1 + \cot^2 (\text{arc cot } x)} = -\frac{1}{1 + x^2}$$

so that

(7)
$$\frac{d \text{ arc cot } x}{dx} = -\frac{1}{1 + x^2}.$$

Also

(8)
$$\text{arc cot } 0 = \frac{\pi}{2}.$$

The functions arc sec x and arc csc x are defined for $|x| \geq 1$. We have

$$\text{arc sec } (\sec \theta) = \theta \qquad \text{for } 0 < \theta < \pi,$$
$$\text{arc csc } (\csc \theta) = \theta \qquad \text{for } -\frac{\pi}{2} < \theta < \frac{\pi}{2},$$
$$\sec (\text{arc sec } x) = \csc (\text{arc csc } x) = x \qquad \text{for } |x| \geq 1$$

and

(9)
$$\text{arc sec } 1 = 0, \qquad \text{arc sec } (-1) = \pi,$$
$$\text{arc csc } 1 = \frac{\pi}{2}, \qquad \text{arc csc } (-1) = -\frac{\pi}{2}.$$

Noting (3), we have

$$\frac{d \text{ arc sec } x}{dx} = \frac{1}{dx/d \text{ arc sec } x} = \frac{1}{d \sec (\text{arc sec } x)/d \text{ arc sec } x}$$

$$= \frac{1}{\sin (\text{arc sec } x)/\cos^2 (\text{arc sec } x)} = \frac{\cos^2 (\text{arc sec } x)}{\sqrt{1 - \cos^2 (\text{arc sec } x)}}$$

$$= \frac{1}{x^2 \sqrt{1 - (1/x^2)}} = \frac{1}{|x| \sqrt{x^2 - 1}}$$

or

(10)
$$\frac{d \text{ arc sec } x}{dx} = \frac{1}{|x| \sqrt{x^2 - 1}}.$$

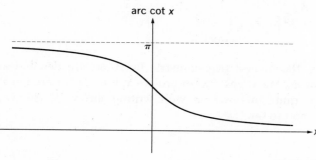

Fig. 6.27

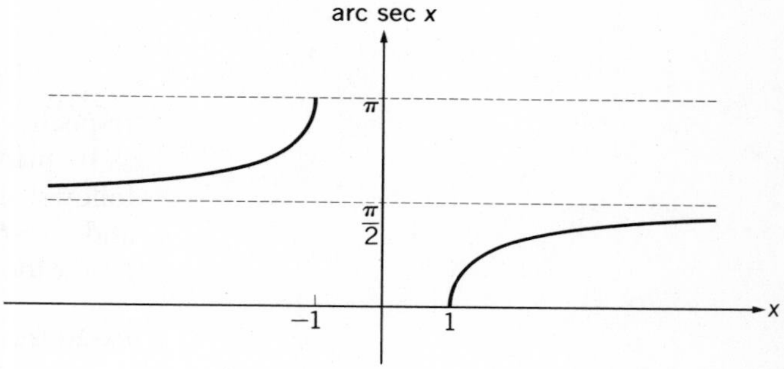

Fig. 6.28

One proves similarly that

(11) $$\frac{d \text{ arc csc } x}{dx} = -\frac{1}{|x| \sqrt{x^2 - 1}}.$$

The graphs of the functions arc cotangent, arc secant, and arc cosecant are shown in Figs. 6.27, 6.28, 6.29. Our definitions are such that

(12) $$\text{arc sec } x = \text{arc cos}\left(\frac{1}{x}\right), \qquad \text{arc csc } x = \text{arc sin}\left(\frac{1}{x}\right)$$

for $|x| \geq 1$. For if $0 < \theta < \pi$ and $\sec \theta = x$, then $\cos \theta = 1/x$, which proves the first relation. The second is proved similarly.

Remarks   Of the six trigonometric functions, it is advisable to use only sin x, cos x, and tan x. Of the six inverse trigonometric functions, it is advisable to use only arc tan x, arc sin x, and arc sec x, the primitive functions of

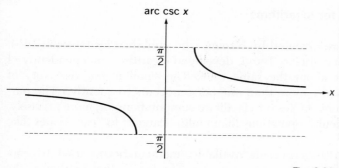

Fig. 6.29

$$\frac{1}{1 + x^2}, \qquad \frac{1}{\sqrt{1 - x^2}}, \qquad \frac{1}{|x|\sqrt{x^2 - 1}},$$

respectively.

In many books, the inverse trigonometric functions are denoted as follows: $\sin^{-1} x$ for arc sin x, $\cos^{-1} x$ for arc cos x, $\tan^{-1} x$ for arc tan x, and so on. This is quite inconsistent with writing $\sin^2 x$ for $(\sin x)^2$, $\tan^3 x$ for $(\tan x)^3$, and so on.

EXERCISES

13. Find $\dfrac{d}{du}$ arc sin $\sqrt{u + 1}$.

17. Find $\dfrac{d}{dz}$ tan $\sqrt{\text{arc sec } z}$.

14. Find $\dfrac{d}{dt}$ arc sec t^2.

18. Find $\dfrac{d}{du}$ arc $\sin^{3/2}(u^{2/3})$.

15. Differentiate $\theta \mapsto \dfrac{\text{arc csc } \theta}{\text{csc } \theta}$.

19. Find $\displaystyle\int \dfrac{x \, dx}{\sqrt{1 - x^4}}$.

16. Find $\dfrac{dy}{dx}$ if $y = x^2$ arc sec $\sqrt{x}$.

20. Find $\displaystyle\int_0^{\sqrt{2}/2} \dfrac{\text{arc sin } u}{\sqrt{1 - u^2}} \, du$.

21. Find $\displaystyle\int \dfrac{dz}{z\sqrt{z^4 - 1}}$. (Hint: Multiply numerator and denominator of the integrand by z.)

▶ 22. Find $\displaystyle\int_1^{\sqrt{2}} x$ arc sec $x \, dx$. (Hint: Use integration by parts and be watchful for an improper integral.)

23. Find $\displaystyle\int_{1/2}^1 \dfrac{1}{\sqrt{x}}$ arc sin $\sqrt{x} \, dx$. (Hint: Use integration by parts and be watchful for an improper integral.)

24. Find $\int$ arc sec $\sqrt{x} \, dx$. (Hint: Use integration by parts.)

25. Prove the second relation (12).

26. Show that, for $x \neq 0$, we have arc tan $x =$ arc cot $(1/x) + \alpha$, α a constant. Find α.

§4 Logarithms

4.1 The need for logarithms

Logarithms were invented by Napier, who published his tables in 1614. (Another mathematician, Bürgi, developed logarithms independently of Napier, and at about the same time. The simultaneous discovery of important ideas by more than one person occurs frequently.) The purpose of logarithmic tables is to facilitate computations, or, more precisely, to reduce "difficult" operations (like multiplications) to "easy" ones (like additions).

Before logarithms became available, mathematicians used trigonometric functions for the same purpose. The key to that method is the addition theorem (see §1.11)

JOHN NAPIER (1550–1617), a Scots noble, lived and worked on his ancestral estate near Edinburgh. He was an advocate of the Protestant cause and the author of a theological treatise.

Beside inventing logarithms, Napier also constructed another device to facilitate calculations, the Napier rods.

JOST BÜRGI (1552–1632) was a Swiss watchmaker, mathematician and inventor. He published his tables in 1620 but the promised book of instructions and explanations never appeared.

$$\cos (\alpha + \beta) = \cos \alpha \cos \beta - \sin \alpha \sin \beta.$$

Replacing β by $(-\beta)$, we obtain

$$\cos (\alpha - \beta) = \cos \alpha \cos \beta + \sin \alpha \sin \beta.$$

Adding the two formulas, we see that

$$2 \cos \alpha \cos \beta = \cos (\alpha + \beta) + \cos (\alpha - \beta).$$

Suppose we want, using this identity, to multiply two numbers a and b; we assume that $0 < a < 1$ and $0 < b < 1$. From trigonometric tables, we find numbers (or "angles") α and β such that $a = \cos \alpha$ and $b = \cos \beta$. Then we compute $\alpha + \beta$ and $\alpha - \beta$, look up in the tables the values of $\cos (\alpha + \beta)$ and $\cos (\alpha - \beta)$, add these values, and take one half of the sum. This is the desired product $ab = \cos \alpha \cos \beta$.

Of course, whenever one works with tables, the result will be only approximate. For instance, given an a, we shall in general not find in the tables an α such that $\cos \alpha = a$, but only an α such that $\cos \alpha$ is close to a. The same is true for calculations with logarithmic tables. The more extensive the tables, though, the more accurate will be the result.

Logarithmic tables enable one to perform not only multiplication, but also division, raising to powers and extracting roots, with comparative ease. Soon after the appearance of Napier's first tables, a better set was computed by Napier jointly with Briggs. Much more elaborate tables became available later. For over three centuries, all extensive calculations were performed by logarithms. There is hardly a scientific discovery or a technological advance that did not use Napier's invention, either directly or indirectly.

HENRY BRIGGS (1561–1631) was a professor of mathematics at London University and later at Oxford.

Only very recently, with the advent of the modern automatic computer, have logarithmic (and other) tables become somewhat obsolete. An electronic computer performs multiplications and divisions directly, though in the binary system, and with incredible speed. And when a computer needs a logarithm, or a cosine of a number, it does not look up a table, but finds it by performing the necessary calculation. (Yet computers will not replace slide rules. A slide rule is simply a table of logarithms in which certain numbers are represented by lengths.)

The lasting value, however, of Napier's work lies in the important functions to which it led.

EXERCISES

In Exercises 1 to 4, perform the indicated multiplications by using cosines.

1. $.873 \times .802$.
2. $.921 \times .758$.
3. $.90475 \times .75836$.
4. $(.85771)^2$.

4.2 The search for logarithms

The purpose of logarithmic tables is to reduce multiplication to addition. How can this be done in the simplest possible way? We want to associate with every positive number, say x, another number, called its **logarithm,** so that multiplying two numbers should amount to adding their logarithms. Let us, for the time being, denote the logarithm of x by $\phi(x)$. We want the logarithm of a product to be the sum of logarithms. Hence we want the function $x \mapsto \phi(x)$ to be such that

$$(1) \qquad\qquad\qquad \phi(xy) = \phi(x) + \phi(y).$$

We call this the **functional equation** of logarithms. Applying this to $x = y = 1$, we obtain that $\phi(1) = 2\phi(1)$ or

$$(2) \qquad\qquad\qquad \phi(1) = 0.$$

We do not yet know, of course, that there is a function ϕ satisfying (1). Let us assume that there is, and let us further assume that the function is differentiable. It turns out, then, that we can find its derivative. For if $\phi'(x)$ exists, then

$$\phi'(x) = \lim_{h \to 0} \frac{\phi(x + h) - \phi(x)}{h} \qquad \text{by definition}$$

$$= \lim_{h \to 0} \frac{\phi\left[x\left(1 + \dfrac{h}{x}\right)\right] - \phi(x)}{h} \qquad \text{since } x > 0$$

$$= \lim_{h \to 0} \frac{\phi(x) + \phi\left(1 + \dfrac{h}{x}\right) - \phi(x)}{h} \qquad \text{by the functional equation (1)}$$

$$= \lim_{h \to 0} \frac{\phi\left(1 + \dfrac{h}{x}\right)}{h} = \lim_{h \to 0} \frac{1}{x} \frac{\phi\left(1 + \dfrac{h}{x}\right)}{\dfrac{h}{x}}$$

$$= \frac{1}{x} \lim_{h \to 0} \frac{\phi\left(1 + \dfrac{h}{x}\right)}{\dfrac{h}{x}} \qquad \text{since } \frac{1}{x} \text{ is here a fixed number}$$

$$= \frac{1}{x} \lim_{k \to 0} \frac{\phi(1 + k)}{k} \qquad \begin{array}{l}\text{setting } \dfrac{h}{x} = k, \text{ noting that } \lim_{h \to 0} \left(\dfrac{h}{x}\right) = 0 \\ \text{and using Theorem 4 in Chapter 3, §3.}\end{array}$$

Denoting the number

$$\lim_{k \to 0} \frac{\phi(1 + k)}{k}$$

by c, we have

(3)
$$\phi'(x) = \frac{c}{x}.$$

We have shown that, if a differentiable function $x \mapsto \phi(x)$ satisfying (1) exists, its derivative is c/x.

The choice of the constant c is up to us. For, if ϕ is some function satisfying (1) and α any fixed number, then $\psi(x) = \alpha\phi(x)$ also satisfies the functional equation of logarithms; indeed

$$\psi(xy) = \alpha\phi(xy) = \alpha(\phi(x) + \phi(y)) = \psi(x) + \psi(y).$$

And if $\phi'(x) = c/x$, then $\psi'(x) = \alpha c/x$.

The simplest and most natural choice for c is 1. Therefore the function $x \mapsto \phi(x)$ with $\phi(1) = 0$ and $\phi'(x) = 1/x$ is called the **natural logarithm**. This function is denoted by the abbreviation ln in most engineering books and in many elementary texts, and by log in mathematical and scientific literature. We shall follow the latter convention.

Napier worked before calculus was developed. Yet he seems to have carried out an analysis similar to the one given above, though in different terminology. He recognized that Property (1) of a function $\phi(x)$ implies that the rate of change of $\phi(x)$ is proportional to $1/x$, and he chose 1 as the factor of proportionality. Indeed, his 1614 tables are in effect tables of natural logarithms.

4.3 Definition of natural logarithms

The preceding discussion suggests the following definition of the natural logarithmic function (or simply the **logarithmic function**, if there is no danger of confusion): $x \mapsto \log x$ is the function satisfying the conditions

(4)
$$\log 1 = 0, \qquad \frac{d \log x}{dx} = \frac{1}{x} \qquad \text{for } x > 0.$$

That such a function exists and is unique follows from the fundamental theorem of calculus; we have the explicit formula

(5)
$$\log x = \int_1^x \frac{dt}{t} \qquad \text{for } x > 0.$$

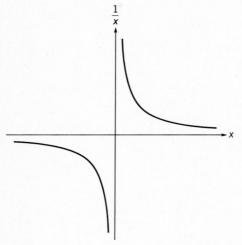

Fig. 6.30

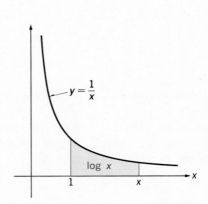

Fig. 6.31

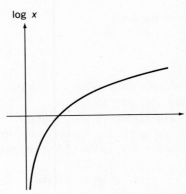

Fig. 6.32

The graph of the function $y = 1/x$, shown in Fig. 6.30, is called the **equilateral hyperbola;** it consists of two branches. The natural logarithm of a number $x > 0$ is the area under this hyperbola, from 1 to x (see Fig. 6.31). The values of log x are graphed in Fig. 6.32.

EXERCISES

In each of the exercises below, indicate for which values of the independent variable the function considered is defined. (Remember that log x is defined only for $x > 0$.)

5. Find $\dfrac{d}{du} \log (1 + u^2)$.　　　　　　7. Find $\dfrac{d}{dz} \log \log \log z$.

6. Find $\dfrac{d}{dt} \log \log t$.　　　　　　　　8. Find $\dfrac{du}{dy}$ if $u = (\log y)^3$.

9. Find the derivative of $u \mapsto$ arc tan $(\log u)$.
10. Find $f'(x)$ if $f(x) = \log (\text{arc tan } x)$.
11. Differentiate $y = \log (\sin t)$.
▶ 12. Differentiate $y \mapsto (\log y)(\text{arc tan } y^2)$.
13. Differentiate $y = \sin (\log t)$.
14. Find the derivative of $v \mapsto u = \cos \log \log v$.

4.4 Properties of logarithms

Now we shall show that the logarithmic function actually has the required property (the logarithm of a product is the sum of the logarithms of the factors).

Theorem 1   THE FUNCTION $x \mapsto \log x$ IS CONTINUOUS, INCREASING, AND CONCAVE DOWNWARD. IT SATISFIES THE FUNCTIONAL EQUATION

$$(6) \qquad\qquad \log (xy) = \log x + \log y$$

AND ALSO HAS THE FOLLOWING PROPERTIES:

$$(7) \qquad\qquad \log \frac{x}{y} = \log x - \log y,$$

$$(8) \qquad\qquad \log x^r = r \log x,$$

FOR EVERY RATIONAL NUMBER r, AND

$$(9) \qquad\qquad \lim_{x \to +\infty} \log x = +\infty, \qquad \lim_{x \to 0^+} \log x = -\infty.$$

Proof.   The function $\log x$ is continuous, because it has a derivative $(1/x)$, increasing because the derivative is positive ($\log x$ is defined only for $x > 0$), concave downward because the second derivative $(-1/x^2)$ is negative. We give two arguments for the functional equation (6).

First Proof.   Set, for a fixed y,

$$f(x) = \log (xy) - \log x - \log y.$$

We must show that $f(x) \equiv 0$. Since

$$f(1) = \log y - \log 1 - \log y = -\log 1 = 0,$$

it will suffice to show that $f(x)$ is constant, that is, that $f'(x) \equiv 0$. But, remembering that y is to be treated as a constant, we have

$$f'(x) = \frac{d \log (xy)}{d(xy)} \frac{d(xy)}{dx} - \frac{d \log x}{dx} - \frac{d \log y}{dx}$$

$$= \frac{1}{xy} y - \frac{1}{x} - 0 = 0,$$

as expected.

Second Proof.   We have

$$\log (xy) = \int_1^{xy} \frac{dt}{t} = \int_1^x \frac{dt}{t} + \int_x^{xy} \frac{dt}{t}.$$

In the second integral, we make the substitution $t = xs$, so that $dt = x \, ds$, $s = 1$ for $t = x$, $s = y$ for $t = xy$; this yields

$$\int_1^x \frac{dt}{t} + \int_1^y \frac{x \, ds}{xs} = \int_1^x \frac{dt}{t} + \int_1^y \frac{ds}{s} = \log x + \log y,$$

as asserted.

We continue the proof of Theorem 1. For $y = \dfrac{1}{x}$, the functional equation yields $\log \left(x \cdot \dfrac{1}{x} \right) = \log 1 = 0 = \log x + \log \dfrac{1}{x}$. Thus,

$$(10) \qquad\qquad\qquad \log \frac{1}{x} = - \log x$$

and $\log \dfrac{x}{y} = \log \left(x \, \dfrac{1}{y} \right) = \log x + \log \dfrac{1}{y} = \log x - \log y$. This proves (7).

Next, if n is a positive integer, then

$$\log x^n = \log \overbrace{(x \cdot x \cdots x)}^{n \text{ times}} = \overbrace{\log x + \log x + \cdots + \log x}^{n \text{ times}}$$

or

$$(11) \qquad\qquad\qquad \log x^n = n \log x.$$

The same is true if $n = 0$, since $x^0 = 1$ and $\log 1 = 0$.

If n is a negative integer, then $n = -m$ where n is positive and

$$\log x^n = \log x^{-m} = \log \frac{1}{x^m} = -\log x^m = -m \log x = n \log x.$$

This implies that (10) holds for all integers. Since $(\sqrt[n]{x})^n = x$, we have

$$\log x = \log (\sqrt[n]{x})^n = n \log \sqrt[n]{x}$$

or

$$(12) \qquad\qquad\qquad \log \sqrt[n]{x} = \frac{1}{n} \log x.$$

We conclude that, for p a positive integer and q any integer,

$$\log x^{q/p} = \log (x^{1/p})^q = q \log x^{1/p} = q \log \sqrt[p]{x} = q \left(\frac{1}{p} \log x \right) = \frac{q}{p} \log x.$$

Hence (8) is established.

Since $\log x$ is an increasing function and $\log 1 = 0$, we have that

(13) $\qquad \log x > 0 \qquad$ if and only if $x > 1$.

In particular, $\log 2 > 0$. Also, for $x > 2^n$,

$$\log x > \log(2^n) = n \log 2.$$

This shows that $\lim_{x \to +\infty} \log x = +\infty$.

Finally, we show that $\lim_{x \to 0^+} \log x = -\infty$. Indeed, $\log x = -\log(1/x)$ and $\lim_{x \to 0^+} (1/x) = +\infty$. The proof of Theorem 1 is now complete.

4.5 The number e

In view of (9), the function $\log x$ takes on arbitrarily large and arbitrarily small (that is, arbitrarily large negative) values. It follows from the Intermediate Value Theorem (Chapter 3, §3.6) that, FOR EVERY NUMBER α THERE IS A POSITIVE NUMBER x SUCH THAT $\log x = \alpha$; this x is unique, since the log is an increasing function.

In particular, there is a number e such that

(14) $\qquad \log e = 1.$

This means: the area under the hyperbola $xy = 1$ from 1 to e is 1 (see Figs. 6.33 and 6.34). The number e is irrational, as we will prove later (see Chapter 8, §2.9); its approximate value is

$$e = 2.718 \cdots.$$

The importance of the number e, and its surprising connection with the number π, will become apparent from what follows.

EXERCISES

In Exercises 15 to 20, perform the indicated calculations by using natural logarithms.

15. 4.53×3.09.
16. $(8.56)^{1/3}$.
17. $6.12 \div 5.59$.
18. $\sqrt{4.12 \times .138}$.
19. $.389 \times 12.2$.
20. $(.621 \div .382)^{1/4}$.

4.6 Logarithms to different bases

We noted above that, if we know one function satisfying the functional equation of logarithms, we can obtain other such functions by multiplying the given one by constants. We make use of this remark in defining, for

Fig. 6.33

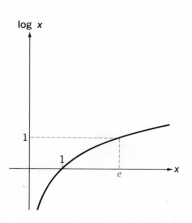

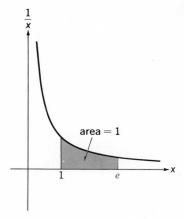

Fig. 6.34

every positive number $a \neq 1$, **logarithms to base** a. If x is a positive number, then $\log_a x$ (read "logarithm of x to base a") is the number

$$(15) \qquad \log_a x = \frac{\log x}{\log a}.$$

It follows from the definition that

$$(16) \qquad \log_a 1 = 0$$

and

$$(17) \qquad \frac{d \log_a x}{dx} = \frac{1}{\log a} \frac{1}{x}.$$

Also, by Theorem 1,

$$(18) \qquad \log_a (xy) = \log_a x + \log_a y, \qquad \log_a \frac{x}{y} = \log_a x - \log_a y$$

and

$$(19) \qquad \log_a (x^r) = r \log_a x$$

(for r rational). For every positive number u, there is a unique positive number x such that $\log_a x = u$.

Since $\log e = 1$, we have, using (15),

$$(20) \qquad \log_e x = \frac{\log x}{\log e} = \log x;$$

therefore e is called the **base of natural logarithms**.

For every $a > 0$, a table of logarithms to base a can be used for computation. To find xy (or x/y or x^r), one finds $\alpha = \log_a x$, $\beta = \log_a y$, and then the number whose logarithm to base a is $\alpha + \beta$ (or $\alpha - \beta$ or $r\alpha$). But if one computes in the decimal system, then logarithms to base 10 are most convenient. For instance, we have

$$\log_{10} 22.340 = 1 + \log_{10} 2.2340$$

(since $10 \cdot 2.234 = 22.34$ and $\log_{10} 10 = 1$),

$$\log .02234 = -2 + \log 2.2340$$

(since $\log_{10} 10^{-2} = -2$), and so forth.

The number $\log_{10} x$ is called the **common logarithm** of x. Most tables, beginning with those computed by Napier and Briggs, list common logarithms. (Authors who denote the natural logarithms by ln usually reserve the notation log for common logarithms.) The function $x \mapsto \log_{10} x$ is graphed in Fig. 6.35.

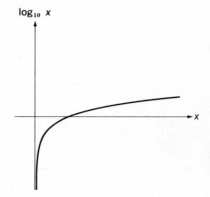

Fig. 6.35

$\log_{10} x$

EXERCISES

In Exercises 21 to 28, perform the indicated calculations by using logarithms to base 10.

21. 10.8×2.91.
22. $108 \div .0365$.
23. $(225)^{1/3}$.
24. 1.29×32.8.

25. 1.86×35.2.
26. $(.0178 \div .138)^{1/3}$.
27. $(327)^5$.
28. $(1.89 \div 1.87)^4$.

4.7 Using logarithms for integration

The function $x \mapsto 1/x$ is the simplest rational function that is not the derivative of a rational or of an algebraic function. In the interval $(0, +\infty)$ this function has a primitive function $\log x$. If x is negative, $-x$ is positive, so that the function $x \mapsto \log(-x) = \log |x|$ is defined for $x < 0$. We have, by the chain rule and by (4),

$$\frac{d \log(-x)}{dx} = \frac{d \log(-x)}{d(-x)} \frac{d(-x)}{dx} = \frac{1}{(-x)} \cdot (-1) = \frac{1}{x}.$$

Thus

$$(21) \qquad \frac{d \log |x|}{dx} = \frac{1}{x} \qquad \text{for } x \neq 0.$$

Using the logarithmic function, we can write down explicitly the indefinite integrals (primitive functions) of many functions. The rule to remember reads:

$$(22) \qquad \int \frac{\phi'(x)}{\phi(x)} \, dx = \log |\phi(x)| + C \qquad C \text{ constant of integration}$$

if the function $x \mapsto \phi(x)$ is not zero in the interval considered. If $\phi(x)$ is positive, we have simply

$$(23) \qquad \int \frac{\phi'(x)}{\phi(x)} \, dx = \log \phi(x) + C.$$

To verify these statements, we note that by the chain rule and by (4) we have

$$\frac{d \log |\phi(x)|}{dx} = \frac{\phi'(x)}{\phi(x)} \qquad \text{if } \phi(x) \neq 0.$$

We illustrate the use of logarithms in finding integrals by several examples. In particular, we are now able to write down explicit formulas for primitive functions of the functions arc tangent, tangent, cotangent, and others.

▶ *Examples*   **1.** Find a primitive function of $f(x) = \dfrac{x^{99}}{1 + x^{100}}$.

Solution.   We note that the numerator in the expression for f is "almost" the derivative of the denominator. Indeed, if $\phi(x) = 1 + x^{100}$, then $\phi'(x) = 100x^{99}$, and $\dfrac{\phi'(x)}{100\phi(x)} = \dfrac{x^{99}}{1 + x^{100}}$. Therefore we write

$$\int \frac{x^{99}\,dx}{1 + x^{100}} = \frac{1}{100}\int \frac{100x^{99}\,dx}{1 + x^{100}} = \frac{1}{100}\int \frac{1}{1 + x^{100}} \frac{d}{dx}(1 + x^{100})\,dx$$

$$= \frac{1}{100}\log(1 + x^{100}) + C.$$

We can also write this in the form

$$\frac{1}{100}\int \frac{100x^{99}}{1 + x^{100}}\,dx = \frac{1}{100}\int \frac{d(1 + x^{100})}{1 + x^{100}} = \frac{1}{100}\int d\log(1 + x^{100})$$

$$= \frac{1}{100}\log(1 + x^{100}) + C.$$

2. Find a primitive function for $\theta \mapsto \tan \theta$.

Solution.   Since $\tan \theta = \dfrac{\sin \theta}{\cos \theta}$ and $d\cos \theta = -\sin \theta\,d\theta$, we have

$$\int \tan \theta\,d\theta = -\log|\cos \theta| + C.$$

3. Integrate the function $\cot \theta$.

Solution.   An argument similar to the one used above yields

$$\int \cot \theta\,d\theta = \log|\sin \theta| + C.$$

4. Find $\int$ arc tan $x\,dx$.

Solution.   We use integration by parts and obtain

$$\int \text{arc tan } x \, dx = x \text{ arc tan } x - \int x \, d \text{ arc tan } x$$

$$= x \text{ arc tan } x - \int \frac{x \, dx}{1 + x^2} = x \text{ arc tan } x - \frac{1}{2} \int \frac{d(x^2)}{1 + x^2}$$

$$= x \text{ arc tan } x - \frac{1}{2} \int \frac{d(1 + x^2)}{1 + x^2} = x \text{ arc tan } x -$$

$$\frac{1}{2} \int d \log (1 + x^2) = x \text{ arc tan } x - \frac{1}{2} \log (1 + x^2) + C.$$

5. Find $\int \log x \, dx$.

Solution. We again use integration by parts:

$$\int \log x \, dx = x \log x - \int x \frac{d \log x}{dx} \, dx$$

$$= x \log x - \int dx = x \log x - x + C.$$

6. Find a primitive function for $x \mapsto x \log x$.

Solution. Use, again, integration by parts.

$$\int x \log x \, dx = \int (\log x) x \, dx = \int (\log x) \, d \frac{x^2}{2}$$

$$= \frac{x^2}{2} \log x - \int \frac{x^2}{2} \frac{d \log x}{dx} \, dx$$

$$= \frac{x^2}{2} \log x - \frac{1}{2} \int x \, dx = \frac{x^2}{2} \log x - \frac{x^2}{4} + C.$$

7. Find $\int \frac{dx}{x \log x}$.

Solution. We have

$$\int \frac{dx}{x \log x} = \int \frac{d \log x/dx}{\log x} \, dx = \log |\log x| + C. \blacktriangleleft$$

EXERCISES

29. Find $f'(x)$ if $f(x) = \log (x^3)$.
30. Find $h'(u)$ if $h(u) = (\log (u^2)^{1/3}$.
31. Find $\frac{d^2 y}{dx^2}$ if $y = \frac{1}{\log x}$.
32. Find $g''(s)$ if $g(s) = \log \log \log s$.

33. If $f(x)$ is a positive function such that $\log f(x) = x \log x$ and $f(2) = 4$, find $f'(2)$ approximately. Use the table of values of $\log$.

34. Where will the graph of the function $y = \dfrac{1 + \log |x|}{x}$ have a horizontal tangent?

35. Find $\int \dfrac{2x^3 + x}{x^4 + x^2}\, dx.$

36. Find all primitive functions of $s \mapsto \left(\dfrac{1}{\sqrt{s}}\right)\left(\dfrac{1}{1 + \sqrt{s}}\right).$

37. Find $\int \dfrac{1/y^2}{1 + (1/y)}\, dy.$

38. Find all primitive functions of $u \mapsto \dfrac{u + u^{-3}}{u^2 + u^{-2}}.$

39. Find all primitive functions of $z \mapsto \left(\dfrac{z}{z^2 + 1}\right)\left(\dfrac{1}{\log (z^2 + 1)}\right).$

40. Find $\int \dfrac{dt}{t \log (1/t)}.$

41. Find $\int \dfrac{\log u}{u}\, du.$

42. Find $\int \dfrac{\sqrt{\log z}}{z}\, dz.$

43. Find all primitive functions of $g(t) = \sqrt{t} \log t.$

44. Find a solution for the differential equation $\dfrac{dy}{dx} = x^3 \log \sqrt{x}$ that vanishes when $x = 1$.

45. Find $\int \sec x\, dx$. (Hint: Write $\sec x = \sec x \dfrac{\sec x + \tan x}{\sec x + \tan x}$ and compare the derivative of the denominator of this expression with the numerator.)

46. Find $\int \sec^2 x \log (\sec x)\, dx$. (Hint: Use integration by parts.)

47. Find the length of the graph of $y = \log (\cos x)$ from $x = 0$ to $x = \dfrac{\pi}{4}$. (Hint: Use the relation $\tan^2 x + 1 = \sec^2 x$.)

48. Find $\int \dfrac{\sin u}{\cos^3 u}\, du.$ $\left(\text{Hint: Note that } \dfrac{\sin u}{\cos^3 u} = \tan u \dfrac{1}{\cos^2 u}.\right)$

49. Find $\int (\tan x + \tan^3 x)\, dx.$

50. Find $\int \dfrac{u}{\cos^2 u^2}\, du.$

51. Find $\int \dfrac{x}{\cos^2 x}\, dx$. (Hint: Use integration by parts and Example 2.)

52. Find $\int \dfrac{1 + \tan^2 \log u}{u}\, du.$

53. Find a solution that vanishes when $t = 1$ and takes on the value $1 - \log 2$ when $t = 2$ for the differential equation $\dfrac{d^2 v}{dt^2} = \dfrac{1}{t^2}.$

54. Find $\int x \log (x^2 + 1)\, dx.$

55. Find $\int \frac{1}{t} \log \log t \, dt$.

56. Find $\int_0^1 \frac{u^{1/2}}{1 + u^{3/2}} \, du$.

57. Suppose $f(x) = \frac{x^2}{2} - \frac{1}{4} \log x$. Find the length of the graph of f from $x = 1$ to $x = 2$.

▶ 58. Find $\int \frac{dx}{x^2 + x}$. $\left(\text{Hint: Find constants } a \text{ and } b, \text{ so that } \frac{1}{x^2 + x} = \frac{a}{x} + \frac{b}{x + 1}.\right)$

59. Find $\int \frac{dx}{x^3 + x}$. $\left(\text{Hint: Find constants } a, b, \text{ and } c, \text{ so that } \frac{1}{x^3 + x} = \frac{a}{x} + \frac{bx + c}{x^2 + 1}.\right)$

60. Find $\int \frac{dz}{(1 + z^2)(1 + z)}$. $\left(\text{Hint: Find constants } a, b, \text{ and } c, \text{ so that } \frac{1}{(1 + z^2)(1 + z)} = \frac{a}{1 + z} + \frac{bz + c}{1 + z^2}.\right)$

5.1 Irrational exponents

The relation

$$(1) \qquad\qquad \log a^r = r \log a$$

has been proved in §4.4 for rational numbers r. Is it valid also for irrational r? For instance, is it true that $\log 2^{\sqrt{2}} = \sqrt{2} \log 2$? As it stands, this question makes no sense. We have not yet defined the meaning of a^r; in particular $2^{\sqrt{2}}$ is, at this stage of our presentation, not a name of a number. But nothing prevents us from defining powers with irrational exponents by the requirement that the relation analogous to (1) be true. For instance, we define $2^{\sqrt{2}}$ to be that number whose natural logarithm is $\sqrt{2} \log 2$.

In general, we agree that, for $a > 0$ and any number x, the symbol a^x denotes the unique number whose natural logarithm is $x \log a$. Thus

$$(2) \qquad\qquad \log a^x = x \log a.$$

This is a **definition** of a^x. If x is a rational number, then a^x can, of course, be computed differently; but in this case we already know that (2) holds. We have made relation (1) true for all r; it is easy to see that now

$$(3) \qquad\qquad \log_a (x^r) = r \log_a x$$

is also true for all r.

For $a = e$, relation (3) becomes

$$(4) \qquad\qquad\qquad \log e^x = x$$

(since $\log e = 1$). Let u be any positive number; if we apply (4) for $x = \log u$, we obtain $\log (e^{\log u}) = \log u$. Since two different numbers cannot have the same logarithm, we conclude that

$$(5) \qquad\qquad\qquad e^{\log u} = u.$$

We also have, for every $a > 0$,

$$(6) \qquad\qquad\qquad a^x = e^{x \log a}.$$

Indeed, $\log a^x = x \log a$ by (2) and $\log e^{x \log a} = x \log a$ by (4). Finally, for every $a > 0$,

$$(7) \qquad\qquad \log_a (a^x) = x, \qquad a^{\log_a u} = u.$$

Indeed, $\log_a a^x = \dfrac{\log a^x}{\log a} = \dfrac{x \log a}{\log a} = x$, which proves the first relation. The second now follows by observing that both sides have the same logarithm to base a.

The following proposition is merely a restatement of relations (4), (5), and (7).

Theorem 1   FOR EVERY FIXED $a > 0$, THE FUNCTIONS

$$u \mapsto \log_a u \text{ and } x \mapsto a^x$$

ARE INVERSE TO EACH OTHER. IN PARTICULAR, THE FUNCTIONS

$$u \mapsto \log u \text{ and } x \mapsto e^x$$

ARE INVERSE TO EACH OTHER.

In Theorem 1, we have recaptured the "high school definition" of logarithms: the logarithm of u to base a is the power to which a must be raised to obtain u. Why did we not begin our discussion with this definition? Because we would have had to define the meaning of a^x for x irrational, and would have had to prove that the equation $a^x = u$ has, for given a and u, a unique solution. Both steps could be accomplished without calculus, but would take much longer.

EXERCISES

1. What is $\log_{z^2} z^{5/2}$? (Assume $z > 0$.)
2. What is $\log_{1/u^2} u^{3/5}$? (Assume $u > 0$.)
3. What is $\log_{v^2+1} (v^4 + 2v^2 + 1)$?
4. What is $2^{4\log_2 3}$?
5. What is $\log_2 (x^{3\log x^2})$?
6. For what values of x is $\log_x (x + 6) = 2$?
7. For what values of x is $\log_{4/x} (x^2 - 6) = 2$?
8. For what values of x is $\log_x [(x^2/2) + 2] < 2$? (Hint: Consider the cases $x > 1$ and $x < 1$ separately.)

5.2 Laws of exponents

Theorem 2   FOR a AND b POSITIVE, AND FOR ALL x AND y, WE HAVE

(8) $$a^{x+y} = a^x a^y,$$

(9) $$a^{x-y} = \frac{a^x}{a^y},$$

(10) $$(a^x)^y = a^{xy},$$

(11) $$(ab)^x = a^x b^x.$$

These are the familiar laws of exponents, which we already know for the case in which x and y are rational numbers (see Chapter 1, §5.5). We now must prove them for all cases. This is quite easy; we merely verify that both sides of each equation have the same natural logarithm. We carry out the proof for (11) and ask the reader to carry out the remaining part of the argument. We have, using (2),

$$\log (ab)^x = x \log (ab) = x (\log a + \log b) = x \log a + x \log b$$
$$= \log a^x + \log b^x = \log (a^x b^x)$$

and therefore $(ab)^x = a^x b^x$.

5.3 Exponential functions

Having defined powers with arbitrary (rational or irrational) exponents, we can study functions like $x \mapsto a^x$, where a is some fixed positive number. The graphs of the functions $x \mapsto e^x$ and $x \mapsto 10^x$ (obtained by "flipping" the graphs in Figs. 6.32 and 6.35 of §4) are shown in Fig. 6.36 and

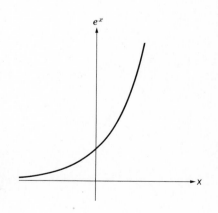

Fig. 6.36

Fig. 6.37

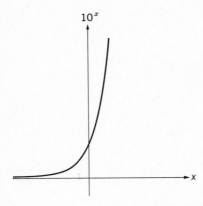

Fig. 6.37. The function e^x is often written $\exp x$,

$$\exp x = e^x$$

and is called the **exponential function;** it is one of the most important functions in mathematics.

A special case of (8) is

(12) $$e^{x+y} = e^x e^y.$$

This is called the **addition theorem** or the **functional equation** of the exponential function.

5.4 Differentiation rules

Theorem 3   WE HAVE

(13) $$\frac{de^x}{dx} = e^x,$$

AND, FOR $a > 0$,

(14) $$\frac{da^x}{dx} = (\log a)a^x.$$

Proof.   Using Theorem 1 and the theorem on differentiating inverse functions (Theorem 7 in Chapter 4, §2.6) we have, setting $e^x = u$,

$$\frac{de^x}{dx} = \frac{du}{dx} = \frac{1}{dx/du} = \frac{1}{d\log u/du} = \frac{1}{1/u} = u = e^x.$$

Relation (14) could be proved similarly. It is simpler to use (6), (13), and the chain rule:

$$\frac{da^x}{dx} = \frac{de^{x\log a}}{dx} = \frac{de^{x\log a}}{d(x\log a)}\frac{d(x\log a)}{dx} = e^{x\log a}(\log a) = (\log a)a^x,$$

as asserted.

In Chapter 4, §2.7, we derived the relation $dx^\alpha/dx = \alpha x^{\alpha-1}$ for rational α. We can now show that this restriction is unessential for $x > 0$.

Theorem 4   FOR EVERY NUMBER α AND FOR $x > 0$

(15) $$\frac{dx^\alpha}{dx} = \alpha x^{\alpha-1}.$$

Proof.   We have

$$\frac{dx^{\alpha}}{dx} = \frac{de^{\alpha \log x}}{dx} = \frac{de^{\alpha \log x}}{d(\alpha \log x)} \frac{d(\alpha \log x)}{dx} = e^{\alpha \log x} \frac{\alpha}{x} = \alpha \frac{x^{a}}{x} = \alpha x^{\alpha - 1},$$

as asserted.

EXERCISES

9. Find $f'(2)$ if $f(x) = e^{2x}$.

10. Find $f'(3)$ if $f(t) = 5^{t}$.

11. Differentiate $\sin (e^{x})$.

12. Differentiate $e^{\sin x}$.

13. Find the fifth derivative of e^{2x}.

14. Differentiate $e^{\log x}$.

15. Find $\int e^{x} \, dx$.

16. Find $\int e^{-2x} \, dx$.

17. Find $\int x^{\sqrt{2}} \, dx$.

18. For which values of x is $e^{x^{2}}$ an increasing function?

5.5 Logarithmic differentiation

If $x \mapsto \phi(x)$ is a positive function, we can form the function $x \mapsto \log \phi(x)$. If ϕ is differentiable, then by the chain rule

$$\frac{d \log \phi(x)}{dx} = \frac{d \log \phi}{d\phi} \frac{d\phi}{dx} = \frac{1}{\phi(x)} \phi'(x)$$

or

$$(16) \qquad\qquad \phi'(x) = \phi(x) \frac{d \log \phi(x)}{dx}.$$

The use of this formula, together with the properties of the log function, sometimes facilitate differentiation. This way of finding ϕ' is called **logarithmic differentiation.** One need not memorize the equation above, but should simply remember that one usually achieves the desired result by writing $\phi(x) = e^{\psi(x)}$, where $\psi(x) = \log \phi(x)$.

▶**Examples   1.** Find the derivative of $x \mapsto x^{x}$ (for $x > 0$).

Solution.   Set $y = x^{x}$. Then, by (16),

$$\frac{dy}{dx} = y \frac{d \log y}{dx} = y \frac{d \log x^{x}}{dx} = y \frac{d(x \log x)}{dx}$$

$$= y\left(\log x + x\frac{1}{x}\right) = x^{x}(\log x + 1).$$

Second Solution. $x^{x} = e^{x \log x}$; therefore

$$\frac{dx^x}{dx} = \frac{de^{x \log x}}{dx} = e^{x \log x} \frac{d(x \log x)}{dx}$$

$$= e^{x \log x} \left(\log x + x \cdot \frac{1}{x} \right) = x^x (\log x + 1),$$

as before.

Third Solution. Set $y = x^x$; then $\log y = x \log x$. Differentiation yields

$$\frac{1}{y} y' = \log x + \frac{x}{x} \qquad \text{or} \qquad y' = y(\log x + 1) = x^x(\log x + 1).$$

2. Find the derivative of $y = \left(\dfrac{1 + x^{10}}{2 + \cos x} \right)^{1/10}$ at $x = 0$ and at $x = \pi$.

Answer. We use logarithmic differentiation.

$$\log y = \frac{1}{10} \Big(\log (1 + x^{10}) - \log (2 + \cos x) \Big),$$

$$\frac{d \log y}{dx} = \frac{1}{10} \frac{10x^9}{1 + x^{10}} - \frac{1}{10} \frac{(-\sin x)}{2 + \cos x},$$

$$\frac{dy}{dx} = y \frac{d \log y}{dx} = \left(\frac{1 + x^{10}}{2 + \cos x} \right)^{1/10} \left(\frac{x^9}{1 + x^{10}} + \frac{\sin x}{20 + 10 \cos x} \right),$$

and $\qquad \left(\dfrac{dy}{dx} \right)_{x=0} = 0, \left(\dfrac{dy}{dx} \right)_{x=\pi} = \left(\dfrac{\pi}{(1 + \pi^{10})^{1/10}} \right)^9.$

3. Differentiate $f(x) = e^{e^{e^x}}$.

Answer. By the chain rule

$$f'(x) = e^{e^{e^x}} e^{e^x} e^x.$$

Or, by logarithmic differentiation,

$$f'(x) = e^{e^{e^x}} \frac{d \log e^{e^{e^x}}}{dx} = e^{e^{e^x}} \frac{de^{e^x}}{dx} = e^{e^{e^x}} \cdot e^{e^x} \frac{d \log e^{e^x}}{dx}$$

$$= e^{e^{e^x}} \cdot e^{e^x} \cdot \frac{de^x}{dx} = e^{e^{e^x}} \cdot e^{e^x} \cdot e^x. \blacktriangleleft$$

EXERCISES

In the following exercises, an expression of the form a^{bc} should be interpreted as $a^{(bc)}$, that is, as a raised to the power bc. In general, this is different from $(a^b)^c = a^{bc}$. Thus, compare $2^{32} = 2^{(32)} = 2^9 = 512$ and $(2^3)^2 = 8^2 = 64$.

19. Find $\dfrac{d \log (1 + e^x)}{dx}$.

20. Find $\dfrac{d \log (x + e^{x^2})}{dx}$.

21. Find $\dfrac{dy}{dx}$ if $y = \cos (e^x)$.

22. Differentiate $u \mapsto e^{\sqrt{\log u}}$.

(27.) Verify that $y = e^{1/x}$ is a solution of the differential equation

$$x^3 \frac{d^2 y}{dx^2} + x \frac{dy}{dx} - 2y = 0.$$

28. Find $\dfrac{d}{dx} x^{x^2}$.

29. Differentiate $v \mapsto (v^2 + 1)^v$.

▶ 30. Find $\dfrac{dz}{dx}$ if $z = (x^2 + 1)^{\log x}$.

31. Differentiate $\theta \mapsto (\cos \theta)^{\sin \theta}$.

(32) Differentiate $x \mapsto \tan (x^{\sqrt{x}})$.

33. Find $\int (\cos \theta) e^{\sin \theta} \, d\theta$.

23. Differentiate $u \mapsto 2^u 3^{1/u}$.

24. Differentiate $z \mapsto \arctan e^z$.

25. Find $\dfrac{d^2 e^{\sin u}}{du^2}$.

26. Find $f''(t)$ if $f(t) = t e^{\sqrt{t}}$.

34. Find $\int \dfrac{e^{\sqrt{x}}}{\sqrt{x}} \, dx$.

(35.) Find $\int \dfrac{\pi^{1/x}}{x^2} \, dx$.

36. Find $\int e^{\theta} \cos (e^{\theta}) \, d\theta$.

37. Find $\int_0^{\sqrt{2}} 2^{x^4+1} x^3 \, dx$.

38. Find $\int_1^8 \dfrac{4z^{1/3}}{z^{2/3}} \, dz$.

39. Find $\int x e^x \, dx$. (Hint: Use integration by parts.)

40. Find $\int x^2 e^x \, dx$. (Hint: Use integration by parts and Exercise 39.)

41. Sketch the graph of the function $x \mapsto x^2 e^{2x}$ for $-4 \le x \le 1$.

42. Sketch the graph of the function $x \mapsto x e^{x^2/3}$ for $-3 \le x \le 1$.

43. Sketch the graph of the function $x \mapsto \sqrt{x} e^{\sqrt{x}}$ for $0 \le x \le 4$.

44. Sketch the graph of the function $x \mapsto x^3 e^{3x}$ for $-3 \le x \le 1$.

5.6 The differential equation of the exponential function

The exponential function $x \mapsto e^x$ is equal to its own derivative. It is, in fact, completely determined by this property and by the condition that it takes on the value 1 at $x = 0$.

Theorem 5 THERE IS PRECISELY ONE FUNCTION $x \mapsto f(x)$ SATISFYING THE DIFFERENTIAL EQUATION

$$(17) \qquad\qquad f'(x) = f(x)$$

AND THE CONDITION

$$(18) \qquad\qquad f(0) = 1;$$

THIS IS THE FUNCTION $f(x) = e^x$.

Proof. By Theorem 3, the function e^x has the required properties. We must show that there is no other function satisfying (17) and (18). Assume that $f(x)$ satisfies these relations and set $g(x) = e^{-x}f(x)$. Then $g(0) = 1$ and $g'(x) = -e^{-x}f(x) + e^{-x}f(x) = 0$ for all x. Hence $g(x) \equiv 1$, so that $f(x) \equiv e^x$.

We establish next a slight generalization of Theorem 5. This generalization is the source of most applications of the exponential function.

Theorem 6   LET A AND α BE NUMBERS. THE FUNCTION

$$(19) \qquad f(x) = Ae^{\alpha x}$$

IS A SOLUTION OF THE DIFFERENTIAL EQUATION

$$(20) \qquad f'(x) = \alpha f(x)$$

SATISFYING THE "INITIAL CONDITION"

$$(21) \qquad f(0) = A;$$

IT IS THE ONLY FUNCTION SATISFYING (20) AND (21).

Proof. The first statement is verified by a direct calculation. Assume now that f satisfies (20) and (21). Define $g(x) = f(x)e^{-\alpha x}$. Then $g'(x) = f'(x)e^{-\alpha x} - \alpha f(x)e^{-\alpha x} = e^{-\alpha x}(f'(x) - \alpha f(x)) = 0$. Hence g is constant. Since $g(x) = g(0) = f(0) = A$, this constant is A. Relation (19) follows.

The content of Theorem 6 can be stated as follows: *the general solution of the differential equation $f' + \alpha f = 0$ is $Ae^{-\alpha x}$.*

All properties of the exponential function can be obtained from Theorem 6. As an example, we give a new proof of the addition theorem: $e^{x+y} = e^x e^y$.

Set $f(x) = e^{x+y}$, where y is a fixed number. By the chain rule, we have

$$\frac{de^{x+y}}{dx} = \frac{de^{x+y}}{d(x+y)} \frac{d(x+y)}{dx} = e^{x+y},$$

so that $f'(x) = f(x)$. Also, $f(0) = e^{0+y} = e^y$. Thus, by Theorem 6, $f(x) = e^y e^x$, as asserted. (This proof of the addition theorem of the exponential function should be compared with the proof of the addition theorem of the functions sine and cosine given in §1.11.)

EXERCISES

45. Find a function $f(x)$ such that $f'(x) = \frac{1}{2}f(x)$ and $f(0) = 4$.
▶ 46. Find a function $f(x)$ such that $f'(x) = 4f(x)$ and $f(1) = e$.
47. Find a function $f(x)$ such that $3f'(x) = -2f(x)$ and $f(3) = e$.
48. Find a function $f(x)$ such that $f'(x)/f(x) = \frac{2}{5}$ and $f'(1) = e$.

In Exercises 49 to 52, assume that a solution of the form $f(x) = e^{g(x)}$ exists and try to determine $g(x)$. Then verify that $e^{g(x)}$ actually is a solution. Be sure to include a constant of integration in your computations.

49. Find a function $f(x)$ such that $f'(x) = 2xf(x)$ and $f(0) = 1$.
50. Find a function $f(x)$ such that $f'(x) = 3x^2f(x)$ and $f(0) = 2$.
51. Find a function $f(x)$ such that $2\sqrt{x}f'(x) = f(x)$ and $f(4) = 1$.
52. Find a function $f(x)$ such that $x^2f'(x) = -f(x)$ and $\lim_{x \to \infty} f(x) = e$.

5.7 Exponential growth

In applying the exponential function, one should remember that, if $\alpha > 0$, then the function $e^{\alpha x}$ is **rapidly increasing.** For instance, for $\alpha = 1$, e^2 is slightly greater than 7 and e^{10} exceeds 22,000. In general, if x increases "in arithmetic progression," that is, if one considers an increasing sequence of values,

$$x = a, \; x = a + b, \; x = a + 2b, \; x = a + 3b, \; x = a + 4b, \; \cdots,$$

the corresponding values of $e^{\alpha x}$ form a geometric progression

$$y_0, \; y_0q, \; y_0q^2, \; y_0q^3, \; y_0q^4, \cdots, \qquad \text{where } y_0 = e^{\alpha a}, \; q = e^{\alpha b}.$$

Since we assume $b > 0$, we have $q > 1$. People long ago noted how fast the terms of a geometric progression with $q > 1$ increase. (Recall the oriental legend about the reward requested by the inventor of chess: 1 grain of wheat on the first square of the board, 2 on the second, 4 on the third, 8 on the fourth, and so on, up to 2^{63} on the last square.)

Let us now discuss the growth of the exponential function in a more precise way.

Theorem 7   IF α IS ANY NUMBER, THEN

(22) $$\lim_{x \to +\infty} \frac{e^x}{x^\alpha} = +\infty.$$

One expresses this by saying that "e^x goes to infinity faster than any power of x, as $x \mapsto +\infty$." The theorem is of interest only for $\alpha > 0$.

If $\alpha > 0$, then $\lim_{x \to +\infty} x^\alpha = +\infty$. Hence the denominator in (22) will be very large for large positive x. The theorem asserts that e^x will eventually be even larger. The word "eventually" is essential. For instance, if we choose $\alpha = 10^{10}/\log 10$, then $x^\alpha > e^x$ for $x = 10^{10}$. Indeed, we have $(10^{10})^\alpha = 10^{10\alpha} = e^{10\alpha \log 10} = e^{10^{11}} > e^{10^{10}}$. But even for this tremendous α, the number e^x will eventually, as x grows, become much greater than x^α.

Proof of Theorem 7. We consider first the function $\phi(t) = e^t - 1 - t$. Since $\phi'(t) = e^t - 1$ and therefore $\phi'(t) > 0$ for $t > 0$, we have that $\phi(t)$ is an increasing function of t for $t > 0$. Therefore $\phi(t) > \phi(0) = 0$ so that $e^t > 1 + t$, and

$$(23) \qquad \frac{e^t}{t} > 1 + \frac{1}{t} > 1 \qquad \text{for } t > 0.$$

Let a number $\alpha > 0$ be given. We choose an integer $N > \alpha$ and apply (23) to $t = x/N$. We obtain that, for $x > 0$,

$$\frac{e^{x/N}}{x/N} > 1 \qquad \text{or} \qquad \frac{e^{x/N}}{x} > \frac{1}{N}$$

so that $\qquad \dfrac{e^x}{x^N} > \dfrac{1}{N^N} \qquad$ and $\qquad \dfrac{e^x}{x^\alpha} = \dfrac{e^x}{x^N} x^{N-\alpha} > \dfrac{x^{N-\alpha}}{N^N}.$

Since $\lim_{x \to \infty} x^{N-\alpha} = +\infty$, assertion (22) follows.

We record a consequence of Theorem 7.

Corollary. FOR EVERY $\alpha > 0$

$$(24) \qquad \lim_{x \to +\infty} \frac{x^\alpha}{\log x} = +\infty.$$

One expresses this by saying: "the function $\log x$ goes to infinity slower than any power of x, as $x \to +\infty$."

Proof of the Corollary. Set $\log x = y$. Then $y \to +\infty$ for $x \to +\infty$ and

$$\frac{x^\alpha}{\log x} = \frac{e^{\alpha y}}{y} = \left(\frac{e^y}{y^{1/\alpha}}\right)^\alpha,$$

so that (24) follows from (23).

EXERCISES

53. Let α be any positive number. Sketch the function $x \mapsto \dfrac{e^x}{x^\alpha}$ for $x > 0$. In particular, determine the intervals of increase and decrease, and locate any maxima or minima, either local or absolute.

54. Evaluate $\dfrac{e^x}{x^2}$ for $x = 1, 2, 3,$ and 4.

55. Evaluate $\dfrac{e^x}{x^3}$ for $x = 1, 3, 5,$ and 7.

56. Evaluate $\dfrac{x^2}{\log x}$ for $x = 1, 2, 3,$ and 4.

57. Sketch the function $x \mapsto \dfrac{x^2}{\log x}$ for $x > 0$. In particular, determine the intervals of increase and decrease, and locate any maxima or minima, either local or absolute.

58. Find a value x_0 so that $\dfrac{e^x}{x^4} > 2^5$ for $x > x_0$. $\left(\text{Hint: Let } N = 5 \text{ and use the inequality } \dfrac{e^x}{x^\alpha} > \dfrac{x^{N-\alpha}}{N^N}.\right)$

59. Find a value x_0 so that $\dfrac{e^x}{x^{3/2}} > 10^4$ for $x > x_0$.

60. Find a value x_0 so that $\dfrac{x^2}{\log x} > 10^4$ for $x > x_0$. $\left[\text{Hint: Let } y = \log x \text{ and use the relation } \dfrac{x^2}{\log x} = \left(\dfrac{e^y}{y^{1/2}}\right)^2.\right]$

5.8 Continuously compounded interest

The discovery of the exponential function was anticipated in a question asked by Jacob Bernoulli: Suppose a borrower pays a lender a certain annual interest, but compounds the interest *continuously;* what will he owe after one year, for each unit lent?

Assume that the interest rate is k percent and set $\alpha = k/100$. If the interest is computed at the end of the year, the borrower owes $1 + \alpha$ times the original amount. We assume, for the sake of simplicity, that this amount was unity. If the interest is compounded semiannually, then the borrower owes $1 + \dfrac{\alpha}{2}$ after half a year and therefore

$$\left(1 + \frac{\alpha}{2}\right)\left(1 + \frac{\alpha}{2}\right) = \left(1 + \frac{\alpha}{2}\right)^2$$

after one year. If the interest is compounded quarterly, the amount owed

BERNOULLI. This Basel (Switzerland) family produced several outstanding mathematicians in three generations. The most famous were Jacob (1654–1705), his brother, Johann (1667–1748), and Johann's son Daniel (1700–1784). The two older Bernoullis were the first scholars to understand, use, and advance Leibniz's calculus. Jacob Bernoulli is also the author of the first treatise on the mathematical theory of probability. Daniel is remembered primarily for his work in fluid mechanics.

The Bernoullis maintained close scientific contacts with each other and occasionally quarrelled bitterly about priorities.

increases by a factor of $1 + \frac{\alpha}{4}$ each quarter and it increases by the factor $\left(1 + \frac{\alpha}{4}\right)^4$ after a year. Compounding monthly yields $\left(1 + \frac{\alpha}{12}\right)^{12}$. In general, if we divide the year into n equal periods and compound the interest at the end of each period, a debt of 1 will grow during the year into

$$(25) \qquad \left(1 + \frac{\alpha}{n}\right)^n.$$

One has the feeling, that, for a very large n, the number (25) will be close to the true answer, that is, to the result of the continual compounding of interest.

We now look at the problem from the point of view of calculus. If the interest is compounded continuously, the amount owed will be an increasing function of the time t elapsed; call this function $f(t)$. Of course, we measure time in years. Since the debt was 1 at $t = 0$, we have

$$(26) \qquad f(0) = 1.$$

We assume f to be differentiable and try to compute $f'(t)$. Let h be a small positive number. Consider some instant t_0 and a very short time interval from t_0 to $t_0 + h$. During this short interval, $f(t)$ will change very little. If we neglect this change, we may say that during this short time interval the amount $f(t_0)$, which would have earned $\alpha f(t_0)$ interest in a year, increased by $h\alpha f(t_0)$. Hence $f(t_0 + h) = f(t_0) + h\alpha f(t_0)$ approximately and

$$\frac{f(t_0 + h) - f(t_0)}{h} = \alpha f(t_0) + \text{a small error};$$

we guess the error to be arbitrarily small for h sufficiently small. If so, then

$$(27) \qquad f'(t) = \alpha f(t).$$

Noting Theorem 6, we have

$$(28) \qquad f(t) = e^{\alpha t},$$

and, for $t = 1$, $f(1) = e^{\alpha}$. We conclude that the answer to Bernoulli's question is e^{α}.

The procedure by which we arrived at the "solution" is typical of that generally used in applications of calculus. It consisted of a "fuzzy" part and a "sharp" part. In getting to the differential equation (27), we freely used our intuition. We certainly did not "prove" that $f(t)$ satisfies (27). How could we? We do not even possess a clear definition of "continuous compounding of interest." Rather, we are inclined to consider the differential equation as a substitute for such a definition.

Yet, once we arrived at the differential equation, the fuzzy part of the argument was completed. The next step, that is, obtaining (28), was accomplished by a purely mathematical and completely rigorous method, the application of Theorem 6.

If we dealt with a natural process, we could next check how successful our "fuzzy" guessing was, by comparing the solution of the differential equation with observations. Here we cannot do it, since Bernoulli's question concerns a purely fanciful situation. But we can perform an indirect check. If our reasoning was sound, then the number $\left(1 + \dfrac{\alpha}{n}\right)^n$ should be, for very large n, very close to e^α. This is indeed so.

5.9 Exponentials and logarithms as limits

Theorem 8   FOR ANY NUMBER α,

$$(29) \qquad \lim_{n \to +\infty} \left(1 + \frac{\alpha}{n}\right)^n = e^\alpha.$$

We guessed at this theorem in the preceding paragraph. There n was restricted to integral values and α was positive. Neither restriction is needed.

Before proceeding with the proof, we note that $\left(\text{setting } n = \dfrac{1}{h}\right)$ we can rewrite (29) in the form

$$\lim_{h \to 0^+} (1 + \alpha h)^{1/h} = e^\alpha.$$

We shall prove more; the restriction to positive h is not needed. One has

$$(30) \qquad \lim_{h \to 0} (1 + \alpha h)^{1/h} = e^\alpha.$$

Proof.   Since $\log x$ is a continuous function, and so is its inverse e^x, we see (compare Theorem 2 in Chapter 3, §4.2) that, instead of proving (30), it suffices to show that

$$\lim_{h \to 0} \log (1 + \alpha h)^{1/h} = \log e^\alpha = \alpha.$$

But $\quad \log{(1 + \alpha h)^{1/h}} = \dfrac{1}{h}\log{(1 + \alpha h)} = \dfrac{1}{h}\{\log{(1 + \alpha h)} - \log{1}\}$

$$= \alpha\,\frac{\log{(1 + \alpha h)} - \log{1}}{\alpha h}$$

and $\quad \lim_{h \to 0}\alpha\,\dfrac{\log{(1 + \alpha h)} - \log{1}}{\alpha h} = \alpha\lim_{h \to 0}\dfrac{\log{(1 + \alpha h)} - \log{1}}{\alpha h}$

$$= \alpha\lim_{k \to 0}\frac{\log{(1 + k)} - \log{1}}{k} = \alpha\left(\frac{d\log{x}}{dx}\right)_{x=1} = \alpha\left(\frac{1}{x}\right)_{x=1} = \alpha$$

as asserted.

We note the special case $\alpha = 1$:

$$\lim_{n \to +\infty}\left(1 + \frac{1}{n}\right)^{n} = e.$$

This is sometimes used to define e (then one must prove that the limit exists). We also can take either (29) or (30) as a definition of the exponential function. There is a similar definition of the natural logarithm.

Theorem 9   FOR $a > 0$,

(31) $$\lim_{h \to 0}\frac{a^h - 1}{h} = \log{a}.$$

Proof.   We have

$$\lim_{h \to 0}\frac{a^h - 1}{h} = \lim_{h \to 0}\frac{a^h - a^0}{h} = \left(\frac{da^x}{dx}\right)_{x=0} = (\log{a})a^0 = \log{a},$$

as asserted.

EXERCISES

61. Find $\displaystyle\lim_{x \to +\infty}\left(1 + \frac{1}{x}\right)^{x+2}$.

62. Find $\displaystyle\lim_{x \to +\infty}\left(1 + \frac{1}{x + 1}\right)^{x}$.

63. Find $\lim_{x \to 0}(1 + 4x)^{3x}$.

▶ 64. Find $\lim_{x \to 0}(1 - x)^{(x+1)/x}$.

65. Find $\displaystyle\lim_{u \to 0}\frac{2^{2u} - 2^{u+1} + 1}{u^2}$.

5.10 Population explosion

There are real situations that imitate Bernoulli's notion that capital continuously increases at a rate proportional to itself. An example is the **growth of a bacteria population.** Bacteria are indestructible unless killed, and a given population of N bacteria will, after a very short time interval of length h, increase approximately by αNh, where α is some positive constant. Thus $\dfrac{N(t + h) - N(t)}{h} = \alpha N(t)$, approximately. We assume therefore that $N(t)$, the number of bacteria at the time t, satisfies the differential equation

$$(32) \qquad\qquad N'(t) = \alpha N(t),$$

and is therefore (see Theorem 6) given by the formula

$$N(t) = N_0 e^{\alpha t}, \qquad N_0 = N(0).$$

The rapid growth of $e^{\alpha t}$ as $t \to +\infty$ describes a "population explosion." The example just considered illustrates an important point. The number of bacteria is an integer. Also, $N(t)$ is a step function which cannot, strictly speaking, be regarded as a continuous function. But N is a very large integer, and the appearance of a few new bacteria change it by a relatively insignificant amount. We therefore obtain a sufficiently accurate picture of reality by treating $N(t)$ *as if* it were a continuous and even differentiable function.

A differential equation describing the growth of a human population would have to be more complicated than (32). At the very least, it would have to contain two constants, the birth rate α and the death rate β. If we assume that in a population of N there occur, during a short time interval h, $\alpha h N$ births and $\beta h N$ deaths, we obtain the differential equation

$$\frac{dN}{dt} = \alpha N - \beta N$$

[since, at the time $t + h$, the population is approximately $N(t) + \alpha h N(t) - \beta h N(t)$]. This can be written as

$$(33) \qquad\qquad N'(t) = (\alpha - \beta)N(t)$$

and has the solution

$$(34) \qquad\qquad N = N_0 e^{(\alpha - \beta)t}, \qquad N_0 = N(0).$$

The shape of the population curve described by (34) depends on the sign of $\alpha - \beta$. If $\alpha = \beta$, the population is constant. If $\alpha < \beta$, we have extinction, since $\lim_{t \to +\infty} N(t) = 0$. If the birth rate α exceeds the death rate β, the total population will increase, maybe slowly at first, but necessarily very fast after a while.

While a human population explosion is a serious danger, the assumptions underlying the differential equation (33) are, of course, too simplified to give anything approaching a true picture. For instance, the numbers of deaths and births depend on both the size of the population and on the age distribution within it, and the death and birth rates are not constant.

EXERCISE

66. Suppose that, in an experiment, a culture of 100 bacteria is allowed to reproduce under favorable conditions. Twelve hours later, the culture is found to have 500 bacteria. How many bacteria will there be two days after the start of the experiment?

5.11 Radioactive disintegration

As an example of exponential decay, we consider a radioactive element—for instance, radium. Given N atoms of such an element, we find that, after a short time interval h, approximately $\alpha N h$ of those will disintegrate. Therefore the amount $F(t)$ of the radioactive material present at time t (measured by mass, or by the number of atoms, or in any other way) obeys the differential equation

$$(35) \qquad \frac{dF}{dt} = -\alpha F.$$

Hence

$$(36) \qquad F(t) = F_0 e^{-\alpha t}, \qquad F_0 = F(0).$$

The time t after which the original amount of a radioactive substance will be reduced by $\frac{1}{2}$ is called the **half life** of the substance. The half life is found from the equation $e^{-\alpha t} = \frac{1}{2}$. Solving this equation for t, we obtain

$$(37) \qquad t = \frac{\log 2}{\alpha}.$$

The half life t depends only on the substance considered. For radium, it is 1656 years; for the isotopes radium A and radium C', it is $t = 2\,\text{min}$

and 45 sec and $t = 1.4 \times 10^{-6}$ sec. The various isotopes of uranium have half lives of the order of 10^9 years.

The universality of the decay rule (36) shows that the laws of atomic and nuclear physics are quite different from the more familiar laws describing objects of ordinary size. Given N atoms of polonium, we find that about half of them will disintegrate in 140 days (for this element $t = 140$ days, approximately). Suppose we divide the N atoms into several piles, and observe them separately. Half of *each* will disintegrate in 140 days. This fact shows that there is no way to predict when any individual atom will decay.

Other applications of the exponential function (to atomic fission, to atomic reactors, to radioactive equilibrium, and to barometric pressure) are considered in an appendix to this chapter (§6).

EXERCISES

67. Suppose a radioactive material has a half life of 1 year. How long will it take for 10 grams of this material to decay to 1 gram?
68. If 10 percent of a certain radioactive material decays in 5 days, what is the half life of this material?

5.12 Hyperbolic functions

Since the function $t \mapsto e^t$ equals its derivative and the function $t \mapsto e^{-t}$ is the negative of its derivative, each of these functions is a solution of the second order differential equation

Fig. 6.38

$$(38) \qquad\qquad f''(t) - f(t) = 0.$$

This equation differs from the one satisfied by sines and cosines ($f'' + f = 0$) "only" in a sign. This "slight" difference has spectacular consequences. The graphs of e^t or e^{-t} look quite different from the sine curve. Yet there are striking formal similarities between the trigonometric and the exponential functions. They become particularly apparent when one considers e^t and e^{-t} in the following combinations:

$$(39) \qquad \sinh t = \frac{e^t - e^{-t}}{2}, \quad \cosh t = \frac{e^t + e^{-t}}{2}.$$

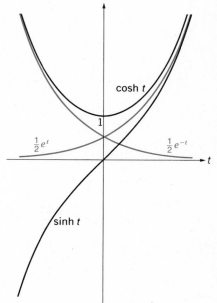

The functions $t \mapsto \sinh t$ and $t \mapsto \cosh t$ are called the **hyperbolic sine** and **hyperbolic cosine,** respectively.

The graphs of the hyperbolic sine and cosine are shown in Fig. 6.38, together with the graphs of the functions $\frac{1}{2}e^t$, $\frac{1}{2}e^{-t}$. We note that

$$(40) \qquad\qquad \sinh 0 = 0, \quad \cosh 0 = 1,$$

(41) $$\sinh(-t) = -\sinh t, \qquad \cosh(-t) = \cosh t,$$

(42) $$\cosh^2 t - \sinh^2 t = 1.$$

(We write $\cosh^2 t$ instead of $(\cosh t)^2$, and similarly for other hyperbolic functions.) These relations follow at once from the definition (42) and remind one of the corresponding relations for sines and cosines. On the other hand, the functions $\sinh$ and $\cosh$ are not periodic. Also

(43) $$\lim_{t \to +\infty} \frac{\sinh t}{\frac{1}{2}e^t} = \lim_{t \to +\infty} \frac{\cosh t}{\frac{1}{2}e^t} = 1.$$

Indeed,

$$\frac{\sinh t}{\frac{1}{2}e^t} = \frac{e^t - e^{-t}}{e^t} = 1 - e^{-2t}, \quad \frac{\cosh t}{\frac{1}{2}e^t} = 1 + e^{-2t}, \quad \lim_{t \to +\infty} e^{-2t} = 0.$$

For large t, therefore, the functions $\sinh t$ and $\cosh t$ are, essentially, $\frac{1}{2}e^t$. We should remember that

(44) $$\cosh t \geq 1 \qquad \text{for all } t.$$

Indeed, $\cosh t - 1 = \frac{1}{2}(e^t + e^{-t} - 2) = \frac{1}{2}(e^{t/2} - e^{-t/2})^2 \geq 0$. Therefore the **hyperbolic tangent** defined by

(45) $$\tanh t = \frac{\sinh t}{\cosh t} = \frac{e^t - e^{-t}}{e^t + e^{-t}}$$

is, unlike the ordinary tangent, continuous for all values of the independent variable. The graph of the hyperbolic tangent is shown in Fig. 6.39. The hyperbolic tangent has limit 1 at $+\infty$ and limit -1 at $-\infty$.

We have, in analogy with the trigonometric functions, the differentiation formulas

Fig. 6.39

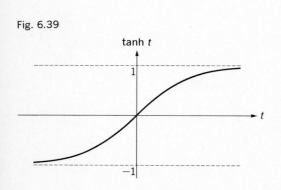

(46) $$\frac{d \sinh t}{dt} = \cosh t, \qquad \frac{d \cosh t}{dt} = \sinh t, \qquad \frac{d \tanh t}{dt} = \frac{1}{\cosh^2 t}.$$

(Again, note the difference in sign; the derivative of $\cosh$ is $\sinh$, not $-\sinh$.)

The proofs follow directly from (39). For instance, to prove the first relation (46), we note that

$$\frac{d \sinh t}{dt} = \frac{d(\frac{1}{2}e^t - \frac{1}{2}e^{-t})}{dt} = \frac{1}{2}e^t + \frac{1}{2}e^{-t} = \cosh t.$$

The reader should supply the other proofs.

There are three more hyperbolic functions and six so-called inverse hyperbolic functions. We describe them and indicate the reason for the name "hyperbolic" in an appendix to this chapter (§7).

EXERCISES

69. Show that $\dfrac{d \cosh t}{dt} = \sinh t$.

70. Show that $\dfrac{d \tanh t}{dt} = \dfrac{1}{\cosh^2 t}$.

71. Show that $\sinh (x + y) = \sinh x \cosh y + \cosh x \sinh y$.
72. State and prove the addition theorems for cosh and tanh.
73. Simplify $\cosh \log x$ and $\sinh \log x$.

► 74. Find $\dfrac{d \cosh^{1/2} z^2}{dz}$.

75. Find $\dfrac{dx \sinh^2 x}{dx}$.

76. Express $\dfrac{d}{dx} \log \sinh x$ in terms of the exponential function.

77. Find $\dfrac{d}{du} (\cosh u \sinh u)$.

78. Find $\dfrac{d^2 z}{dt^2}$ if $z = \cosh^3 (2t)$.

79. Find $\int \dfrac{\sinh u}{\cosh u} \, du$.
80. Find $\int \cosh^2 t \, dt$.
81. Find $\int \cosh^3 z \sinh z \, dz$.
82. Find $\int t \sinh t \, dt$.
83. Find $f'(x)$ if $f(x) = \tanh \tan x$.
84. Find $f'(x)$ if $f(x) = \sinh x^2 - \cosh \sqrt{x}$.
85. Prove that $\lim_{t \to -\infty} \tanh t = -1$, $\lim_{t \to ++\infty} \tanh t = 1$.

Appendixes to Chapter 6

§6 Other Applications of the Exponential Function

In this section, we continue the discussion of the differential equation of the exponential function. In particular, we consider the so-called nonhomogeneous equation [equation (1) below]. This equation will be applied to atomic fission and to radioactive equilibrium.

6.1 The general solution of a nonhomogeneous equation

Consider the differential equation

(1) $$f'(x) - \alpha f(x) = \phi(x),$$

where ϕ is a given function. One can solve this equation, that is, find all functions f satisfying (1) by a device that is also useful for other differential equations. If $\phi = 0$, we know that the solution is $Ae^{\alpha x}$, where A is a constant (see Theorem 6 in §5.6). We try to solve (1) by writing the desired solution f in the form

$$f(x) = A(x)e^{\alpha x},$$

where $A(x)$ is a function to be determined. (If there is a solution, it certainly can be so written.) Substituting this expression for f into (1), we obtain

$$f'(x) - \alpha f(x) = A'(x)e^{\alpha x} + A(x)\alpha e^{\alpha x} - \alpha A(x)e^{\alpha x} = \phi(x)$$

or $$A'(x) = e^{-\alpha x}\phi(x).$$

Hence, by the fundamental theorem of calculus,

$$A(x) = \int_0^x e^{-\alpha \xi}\phi(\xi)\,d\xi + A,$$

where A is a constant, the value of $A(x)$ at $x = 0$. Thus

(2) $$f(x) = Ae^{\alpha x} + e^{\alpha x}\int_0^x e^{-\alpha \xi}\phi(\xi)\,d\xi.$$

We formulate the result as a theorem.

Theorem 1   THE (ONLY) SOLUTION OF THE DIFFERENTIAL EQUATION (1) SUBJECT TO THE INITIAL CONDITION $f(0) = A$ IS GIVEN BY (2).

Note that Theorem 5 in §5.6 is a special case, obtained by setting $\phi(x) \equiv 0$.

EXERCISES

1. Find a function $f(x)$ such that $f'(x) - 2f(x) = x$ and $f(0) = \frac{1}{2}$.
2. Find a function $f(x)$ such that $f'(x) + f(x) = e^{2x}$ and $f(0) = -1$.
3. Find a function $f(x)$ such that $f'(x) - \frac{1}{2}f(x) = e^{x/2}$ and $f(0) = 2$.
4. Find a function $f(x)$ such that $f'(x) + f(x) = \cos(\pi e^x)$ and $f(0) = e$.
5. Find a function $f(x)$ such that $f'(x) + 3f(x) = xe^x$ and $f(0) = 1/e$.

6.2 Atomic fission

Atomic **fission**, the breakup of the nucleus into several parts, can occur in certain elements, such as U-235 (the uranium isotope of atomic weight 235), when a nucleus of an atom is hit by a neutron, an uncharged subatomic particle. Stray neutrons are always present in a material. Some will hit atomic nuclei, producing fission; some of the newly produced neutrons will in turn cause fission of other atoms. At the same time, some neutrons will escape from the material. (In an atomic reactor there is also an independent source of neutrons.) When neutrons are produced by fission, large amounts of energy are liberated, mainly in the form of heat. We shall derive a differential equation describing what happens. It is, of course, very simplified. (Among other things we neglect in what follows is the use of "moderators," materials that absorb neutrons and slow down the process.)

Consider a ball of radius r and of uniform density ρ, made of a fissionable material. If N is the number of neutrons at a time t, then, after a very short time interval of length h, a certain number of new neutrons will be produced by fission. This number is proportional to N and to h, and therefore equals αNh. The positive proportionality factor α depends on the material used and on the density, but not on the shape or size of the piece of material. This is so, since each individual process of atomic fission takes place in a volume whose dimensions are insignificant compared to the "ordinary" dimensions of a piece of metal.

During the same very short time interval h, a certain number of neutrons will escape. This number is βNh, since it is proportional to the number N of neutrons present and to the time elapsed. It is reasonable to assume that the proportionality factor β depends on the radius r, for only neutrons already close to the surface of the ball will have a chance to escape during the short interval h. Therefore β will be proportional to the ratio of the surface area of the ball to its volume. This ratio is

$$\frac{4\pi r^2}{(4/3)\pi r^3} = \frac{3}{r}.$$

Thus

$$\beta = \frac{\beta_0}{r},$$

where the positive constant β_0 depends on the material and the density.

The number of neutrons after the time interval h is therefore

$$N + \alpha Nh - \beta Nh.$$

The differential equation of N as a function of time t reads

$$\frac{dN}{dt} = (\alpha - \beta)N$$

and has the solution

(3) $$N = N_0 e^{(\alpha - \beta)t}.$$

If t is measured in seconds, then the dimension of the constant rates α and β is $1/\text{sec}$; the dimension of β_0 is therefore that of a velocity, namely, cm/sec if the radius of the ball is given in centimeters.

The value of the radius r for which $\alpha = \beta = \beta_0/r$ is called the **critical radius**; it is given by $r_{\text{cr}} = \beta_0/\alpha$. The corresponding mass of fissionable material, $m_{\text{cr}} = (4/3)\pi r_{\text{cr}}^3 \rho$, is known as the **critical mass**.

The amount of energy liberated during fission is proportional to the number of neutrons produced. If this number increases very fast, enormous amounts of energy are liberated in a very short time. This will occur whenever the mass exceeds the critical mass. For if $r > r_{cr}$, then $\alpha - \beta > 0$ and N increases "exponentially." This is how an **atomic explosion** takes place.

6.3 Atomic reactors

Consider next the case in which there is also a neutron source that introduces into the ball q neutrons per unit time. Now, if there are N neutrons at time t, the number at time $t + h$, h small, will be

$$N + \alpha Nh - \beta Nh + qh,$$

approximately, and we obtain the differential equation

$$(4) \qquad \frac{dN}{dt} = (\alpha - \beta)N + q.$$

This is a special case of equation (1), the "given function" ϕ being the constant q. By Theorem 1, we have

$$(5) \qquad N = N_0 e^{(\alpha-\beta)t} + e^{(\alpha-\beta)t} \int_0^t e^{-(\alpha-\beta)\xi} q \, d\xi.$$

Since $\qquad \displaystyle\int_0^t e^{-(\alpha-\beta)\xi} q \, d\xi = -\left.\frac{q e^{-(\alpha-\beta)\xi}}{\alpha - \beta}\right|_0^t = -\frac{\alpha}{\alpha - \beta}\{e^{-(\alpha-\beta)t} - 1\},$

we obtain

$$(6) \qquad N = N_0 e^{(\alpha-\beta)t} + \frac{q}{\beta - \alpha}\{1 - e^{(\alpha-\beta)t}\}.$$

We assumed here that $\alpha - \beta \neq 0$, that is, $r \neq r_{cr}$. If $\alpha - \beta = 0$, we obtain directly from (4) that $N = N_0 + qt$.

If $\alpha > \beta$ (supercritical mass), we see from (6) that N grows exponentially; in this case, we have an explosion. If $\alpha < \beta$ (subcritical mass), then the quantity $e^{(\alpha-\beta)t}$ in (6) rapidly becomes very close to zero and we obtain

$$(7) \qquad N = \frac{q}{\beta - \alpha}, \qquad \text{approximately.}$$

This means that the number of neutrons is kept steady and we have a source of heat. This is the principle of an **atomic reactor.** (Note that in order that N be large for q small, $\beta - \alpha$ must be small. Thus one needs a subcritical mass that is close to the critical one. This requires delicate controls.)

6.4 Two radioactive materials

As another application of Theorem 1, we consider a radioactive element with half life, $(\log 2)/\alpha$. Given N atoms of such an element, we find that, after a short time interval h, approximately αNh of those will disintegrate. We showed in §5.11 that the amount $F(t)$ of the radioactive material present at time t is $F(t) = F(0)e^{-\alpha t}$.

When a radioactive substance disintegrates, it is transformed into another element, usually also radioactive. We assume therefore that substance I with half life $(\log 2)/\alpha$ disintegrates into a substance II with different half life $(\log 2)/\beta$. Denote the amounts of the materials I and II present at the time t by $F(t)$ and $G(t)$, respectively, and assume that

$$(8) \qquad\qquad F(0) = F_0, \quad G(0) = 0.$$

The function F is already computed. During a short time interval from t to $t + h$, the amount $G(t)$ will increase by $\alpha F(t)h$ as a result of the disintegration of I and will decrease by $\beta G(t)h$ in view of the disintegration of II. Hence we have $G(t + h) - G(t) = -\beta G(t)h + \alpha F(t)h$. The differential equation for G reads

$$G'(t) = -\beta G(t) + \alpha F(t),$$

that is,

$$(9) \qquad\qquad G'(t) + \beta G(t) = \alpha F_0 e^{-\alpha t}.$$

This is a special case of the equation considered in Theorem 1. We have therefore

$$G(t) = e^{-\beta t} \int_0^t e^{\beta \xi} \alpha F_0 e^{-\alpha \xi}\, d\xi = \alpha F_0 e^{-\beta t} \int_0^t e^{(\beta - \alpha)\xi}\, d\xi$$

$$= \alpha F_0 e^{-\beta t} \frac{e^{(\beta - \alpha)\xi}}{\beta - \alpha}\bigg|_{\xi=0}^{\xi=t} = \alpha F_0 e^{-\beta t} \frac{1}{\beta - \alpha}\{e^{(\beta - \alpha)t} - 1\}$$

or

$$(10) \qquad\qquad G(t) = \frac{\alpha F_0}{\beta - \alpha}(e^{-\alpha t} - e^{-\beta t}).$$

It is easy to check by differentiation that this G satisfies (9); of course, $G(0) = 0$.

We shall discuss two extreme cases of special interest. Assume first that I is short-lived, while the half life of II is large. Then α is very large compared to β. After a comparatively short time, $e^{-\alpha t}$ will become close to 0, and $G(t)$ will be approximately $[\alpha F_0/(\alpha - \beta)]e^{-\beta t}$. Roughly speaking, substance I has completely disintegrated, producing the amount $\alpha F_0/(\alpha - \beta)$ of II, and this amount is disintegrating according to the decay law of II.

More interesting is the case of a long-lived "mother substance" I and a short-lived "daughter substance" II. Now α is small compared to β, so that $\beta - \alpha$ is approximately β.

We consider first the beginning of the process, when t is very small. For such values of t, we commit only a small error if we replace $G(t)$ by a linear function that equals $G(0) = 0$ at $t = 0$, and whose slope is $G'(0)$ (see Theorem 1 in Chapter 4, §1.8). The derivative of $e^{-\alpha t} - e^{-\beta t}$ at $t = 0$ is $(-\alpha e^{-\alpha t} + \beta e^{-\beta t})_{t=0} = \beta - \alpha$. Thus we have, for small t,

$$(11) \qquad\qquad G(t) = \alpha F_0 t, \qquad \text{approximately.}$$

Assume now that a considerable time t has elapsed. Then the term $e^{-\beta t}$ in (10) will be very small and may be neglected in comparison with $e^{-\alpha t}$. If we replace the denominator in (10) by β, we obtain that

(12) $$G(t) = \frac{\alpha}{\beta} F_0 e^{-\alpha t}, \qquad \text{approximately.}$$

We note that $F_0 e^{-\alpha t} = F(t)$ and conclude that

(13) $$G(t) = \frac{\alpha}{\beta} F(t).$$

Thus, while the amount of I is slowly decreasing; the amount of II is, approximately, a *fixed* small fraction of that of I. That fraction is α/β, that is, the ratio (half life of II/half life of I). One says in this case that the substances I and II are in **radioactive equilibrium.**

►*Example* The law of decay for radium is $G(t) = G(0)e^{-t/2400}$, t being measured in years. The present average concentration of radium in the earth is 10^{-12}. (This means that, of 10^{12} atoms in our planet, about one is a radium atom.) We want to find the concentration of radium a million years ago. If we obtain it from the equation $10^{-12} = G(0)e^{-10^6/2400}$, we get $G(0) = 10^{-12}e^{10^6/2400} = 65$, an absurdity, since the answer must be less than 1. The solution of the seeming paradox is that the radium found today is not what has remained from the radium present on earth 10^6 years ago. Radium is a descendant of the slower decaying uranium 238 (though not an immediate descendant) and is in a state of radioactive equilibrium with it. The present concentration of uranium 238 is $3 \cdot 10^{-6}$, which is 3,000,000 times that of radium; 3,000,000 is about the ratio of the half life of uranium to that of radium.◄

6.5 The barometric formula

Not all applications of the exponential function to physical problems involve growth and decay processes. As an example, we shall derive the **barometric altitude formula.**

A barometer measures air pressure, p. We consider p as a function of the height y above the ground. The basic physical assumptions are: (1) pressure p and density ρ are connected by the equation

(14) $$p = a\rho,$$

where a is a constant, and (2) the air pressure at a point is equal to the weight of the column of air of cross section 1 extending from the point upward and extending "to infinity."

We proceed to derive a differential equation for the function $p(y)$.

Let h be a small positive number. By assumption (2) the difference $p(y + h) - p(y)$ is the weight of a cylinder of air located between y and $y + h$, of cross section 1 (see Fig. 6.40). The volume of this cylinder is h. Since h

Fig. 6.40

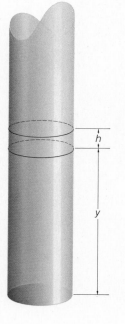

is small, we can assume that, in the cylinder considered, the density of air has the constant value

$$\rho = \frac{1}{a}p(y).$$

The mass of air contained in the cylinder is therefore

$$\rho h = \frac{h}{a}p(y).$$

Its weight is $-\dfrac{gh}{a}p(y)$, and we have

$$\frac{p(y+h)-p(y)}{h} = -bp(y), \qquad \text{approximately,}$$

where $b = g/a$. The function p therefore satisfies the differential equation

$$p'(y) = -bp(y).$$

This may be written as

$$\frac{p'(y)}{p(y)} = -b \qquad \text{or} \qquad \frac{d\log p(y)}{dy} = -b.$$

Therefore $\log p(y) = -by + c$, c a constant, and setting $y = 0$ we see that the ground pressure $p(0)$ is given by $\log p(0) = c$. Thus

(15)
$$y = \frac{1}{b}\log\frac{p(0)}{p(y)}.$$

This formula permits one to determine the altitude y by measuring pressure. We can also write it in the form

(16)
$$p(y) = p(0)e^{-by}.$$

Note that the constant a in (14) depends on the temperature. Since we tacitly assumed that temperature does not depend on altitude, the formula (16) is only approximately true.

EXERCISES

6. Suppose that, for a given fissionable material, we have $\alpha = 2 \cdot 10^8/\text{sec}$ and suppose the critical radius is 50 cm. If a ball of this material has radius 40 cm and contains N_0 free neutrons at time $t = 0$, how many neutrons does it contain at time $t = 1$ sec?

7. Suppose that, for a given fissionable material, we have $\alpha = 2 \cdot 10^8/\text{sec}$ and $\beta_0 = 4 \cdot 10^{10}/\text{sec}$. Find the critical radius r_{cr}. Now find the smallest value of r

so that a ball of this material of radius r will lose no more than $\frac{1}{2}$ of its free neutrons after 1 sec. Find the ratio r/r_{cr}.

▶ 8. Suppose that a ball of the material of Exercise 7 is formed with radius $r = \frac{1}{2}r_{cr}$. Suppose that there is a neutron source which introduces 100 neutrons per sec into the ball. If the ball contains $N(t)$ free neutrons at time t and $N(0) = N_0$, determine $N(t)$ and $\lim_{t \to +\infty} N(t)$.

In Exercises 9 to 11, we consider a radioactive substance I with half life $\log 2/\alpha$ which disintegrates into a radioactive substance II with half life $\log 2/\beta$. We suppose that, at any time t, the amount of substance I present is $F(t)$, and the amount of substance II present is $G(t)$. We assume $F(0) = F_0$ and $G(0) = 0$.

9. Suppose that $\beta = 100\alpha$. Let $\tilde{G}(t)$ denote the approximation $\frac{\alpha}{\beta} F_0 e^{-\alpha t}$ [see equation (12)]. Show that there is a t_0 so that $\dfrac{\tilde{G}(t)}{G(t)} > 1$ for $t < t_0$, $\dfrac{\tilde{G}(t_0)}{G(t_0)} = 1$, and $1 > \dfrac{\tilde{G}(t)}{G(t)} > \dfrac{99}{100}$ for $t > t_0$ to find t_0. Note that this shows something about the percentage error involved in using $\tilde{G}(t)$ in place of $G(t)$.

10. Equation (10) holds only under the assumption that $\alpha \neq \beta$. Find an equation for $G(t)$ if $\alpha = \beta$.

11. Determine when the function $G(t)$ is increasing, when decreasing, and if and when it has any maxima or minima, either absolute or local. Consider the cases $a = \beta$, $a > \beta$, and $\alpha < \beta$ separately.

12. Suppose that we have a ball of fissionable material with $\beta = 2\alpha$ and that there is a neutron source which, for any time t, has introduced a total of qt^2 neutrons into the ball in the time interval $[0, t]$. If $N(t)$ denotes the number of free neutrons in the ball at time t, find a differential equation satisfied by $N(t)$ and solve this equation by using Theorem 1.

13. Suppose that the barometric formula reads: $\rho = \dfrac{1}{\alpha} p(y) + e^{-\alpha y}$ for some constant α. Find a differential equation satisfied by $p(y)$ and solve this equation by using Theorem 1.

14. Suppose one has a rectangular water tank with a base 1 ft square. There is a valve in the base, and water will flow out of the base at a rate directly proportional to the height of the water. Finally, suppose that water is being poured into the tank so that, at any time t, $g(t)$ cu ft of water has been added to the tank in the time interval $[0, t]$. If $y(t)$ denotes the height of the water at any time t and $y(0) = y_0$, find a differential equation satisfied by $y(t)$ and solve this equation by using Theorem 1. [Note: If the water is being poured in at a constant rate of b cu ft per min, and if t is measured in minutes, then $g(t) = bt$.]

§7 Hyperbolic Functions

In this section, we consider in greater detail the hyperbolic functions defined in §5.12. The two basic hyperbolic functions are

(1) $$\sinh t = \frac{e^t - e^{-t}}{2}, \qquad \cosh t = \frac{e^t + e^{-t}}{2}.$$

We recall that both functions are defined for all values of t, that $\sinh 0 = 0$, $\cosh 0 = 1$, that sinh is odd and cosh is even, and that $\cosh^2 t - \sinh^2 t = 1$.

7.1 Differential equations

We could have identified the hyperbolic sine and cosine as solutions of certain differential equations. (a) WE HAVE

$$(2) \qquad \frac{d \sinh t}{dt} = \cosh t, \qquad \frac{d \cosh t}{dt} = \sinh t.$$

This follows at once from the definitions, as we have already noted in §5. The following statement is analogous to Theorem 3 in §1.10.

(b) THE GENERAL SOLUTION OF THE EQUATION

$$(3) \qquad f''(t) - f(t) = 0$$

IS

$$(4) \qquad f(t) = a \cosh t + b \sinh t.$$

Proof. (Note how the proof differs from that of Theorem 3 in §1.10.) We can verify directly that, for any choice of numbers a and b, the function (4) satisfies (3). Now let $f(t)$ be a given solution of (3). We set $a = f(0)$, $b = f'(0)$, and

$$g(t) = f(t) - a \cosh t - b \sinh t.$$

We must show that $g(t) \equiv 0$. One computes easily that $g''(t) \equiv g(t)$ and $g(0) = 0$, $g'(0) = 0$. Set $\phi(t) = g(t) - g'(t)$. Then $\phi(0) = g(0) - g'(0) = 0$ and $\phi'(t) = g'(t) - g''(t) = g'(t) - g(t) = -\phi(t)$. Applying Theorem 6 of §5.6, we conclude that $\phi(t) \equiv 0$. But this means that $g(t) = g'(t)$. Since $g(0) = 0$, we have $g(t) \equiv 0$, again by Theorem 6 of §5.6.

(c) LET $\omega > 0$ BE A GIVEN NUMBER. THE GENERAL SOLUTION OF THE EQUATION

$$f''(t) - \omega^2 f(t) = 0$$

IS

$$f(t) = a \cosh \omega t + b \sinh \omega t.$$

This follows from (4) in the same way as Theorem 5 in §1.12 follows from Theorem 3 in §1.10. The details are left to the reader.

EXERCISES

1. Find the solution of $f''(t) - 4f(t) = 0$ such that $f(0) = 8$ and $f'(0) = -4$.
2. Find the solution of $2f''(t) - 9f(t) = 0$ such that $f(0) = 2\pi$ and $f'(0) = \pi/2$.
3. Find the solution of $f''(t) - f(t) = 0$ such that $f(0) = 1$ and $f(1) = e$.
4. Find the solution of $f'''(t) - f'(t) = 0$ such that $f(0) = 1$, $f'(0) = 1$, $f''(0) = 2$.
   [Hint: First solve for $f'(t)$.]

Fig. 6.41

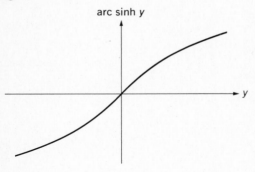

arc sinh y

7.2 The inverse hyperbolic sine

Since the hyperbolic sine is increasing (its derivative is always positive) and has the limits $-\infty$ at $-\infty$ and $+\infty$ at $+\infty$, the function $t \mapsto \sinh t$ has an inverse function which we shall denote by arc sinh (the notation $\sinh^{-1}$ is also used, see the remark in §3.2). The graph of the function $y \mapsto$ arc sinh y is shown in Fig. 6.41. To differentiate arc sinh y, we set $t =$ arc sinh y, that is, $y = \sinh t$ and write

$$\frac{d \text{ arc sinh } y}{dy} = \frac{dt}{dy} = \frac{1}{dy/dt} = \frac{1}{d \sinh t/dt} = \frac{1}{\cosh t} = \frac{1}{\sqrt{1 + \sinh^2 t}} = \frac{1}{\sqrt{1 + y^2}}$$

or

$$(5) \qquad \frac{d \text{ arc sinh } y}{dy} = \frac{1}{\sqrt{1 + y^2}},$$

and since arc sinh $0 = 0$,

$$(6) \qquad \text{arc sinh } y = \int_0^y \frac{du}{\sqrt{1 + u^2}}.$$

This is similar to the corresponding formulas for arc sine and arc cosine (see §1.7). But there is also, as one would expect, a close relation between arc sinh, the inverse function to the hyperbolic sine, and log, the inverse of the exponential function. To derive it, we note that if $t =$ arc sinh y, then $y = \sinh t$, that is,

$$2y = e^t - e^{-t}$$

or, multiplying both sides by e^t and transposing,

$$e^{2t} - 2ye^t - 1 = 0,$$

so that, by the quadratic formula,

$$e^t = y + \sqrt{1 + y^2},$$

since the minus sign before the square root would give a negative value for e^t. Therefore

$$t = \log\left(y + \sqrt{1 + y^2}\right)$$

or

$$(7) \qquad \text{arc sinh } y = \log\left(y + \sqrt{1 + y^2}\right).$$

7.3 The inverse hyperbolic cosine

The function $t \mapsto \cosh t$ is increasing for $t \geq 0$. In this interval, it has an inverse function, we call it $x \mapsto$ arc cosh x. Thus

$$(8) \qquad \cosh(\text{arc cosh } x) = x \text{ for } x \geq 1, \quad \text{arc cosh}(\cosh t) = |t|.$$

The graph of arc hyperbolic cosine is shown in Fig. 6.42.

Let us express arc cosh x by a logarithm. If $x = \cosh t$, then

$$2x = e^t + e^{-t},$$

or

$$e^{2t} - 2xe^t + 1 = 0.$$

and therefore, remembering that we want to have $t > 0$ and $e^t > 1$,

$$e^t = x + \sqrt{x^2 - 1}$$

(a minus would give $e^t < 1$). Hence

$$t = \log\left(x + \sqrt{x^2 - 1}\right)$$

or

(9) $$\text{arc cosh } x = \log\left(x + \sqrt{x^2 - 1}\right).$$

Now, by the chain rule,

$$\frac{d \text{ arc cosh } x}{dx} = \frac{1}{x + \sqrt{x^2 - 1}}\left(1 + \frac{x}{\sqrt{x^2 - 1}}\right)$$

or

(10) $$\frac{d \text{ arc cosh } x}{dx} = \frac{1}{\sqrt{x^2 - 1}}.$$

Since arc cosh $1 = 0$,

(11) $$\text{arc cosh } x = \int_1^x \frac{du}{\sqrt{u^2 - 1}}.$$

EXERCISES

5. Find $\dfrac{d \text{ arc sinh } \sqrt{z}}{dz}$.

6. Find $\dfrac{d \text{ arc cosh } (\sec \theta)}{d\theta}$ for $0 \le \theta < \dfrac{\pi}{2}$.

7. Find $\dfrac{dy}{dx}$ if $y = (\text{arc cos } x)(\text{arc sinh } 2x)$.

8. Find $\dfrac{dz}{dt}$ if $z = \sqrt{4t^2 - 1} \text{ arc cosh } 2t$.

9. Find $\dfrac{d^2u}{dv^2}$ if $u = v \text{ arc sinh } v^2$.

10. Find $\int \dfrac{t^2\,dt}{\sqrt{1 + t^6}}$.   12. Find $\int \dfrac{e^x\,dx}{\sqrt{e^{2x} - 1}}$.

11. Find $\int \sqrt{\dfrac{z}{1 + z^3}}\,dz$.   13. Find $\int \dfrac{du}{\sqrt{u^2 + 2u}}$.

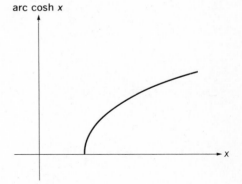

Fig. 6.42

arc cosh x

14. Find $\int$ arc cosh $z\ dz$. (Hint: Use integration by parts.)

15. Find $\int \dfrac{\text{arc sinh } u}{\sqrt{1 + u^2}}\ du$.

16. Find $\int \sqrt{\dfrac{\text{arc cosh } x}{x^2 - 1}}\ dx$.

7.4 Geometric interpretation of hyperbolic function

We now give a geometric interpretation of the functions hyperbolic cosine and hyperbolic sine, analogous to the geometric significance of the sine and cosine functions. This will also explain the term "hyperbolic."

(d) EVERY POINT (x, y) WITH

$$(12) \qquad\qquad x^2 - y^2 = 1$$

AND $x > 0$ CAN BE REPRESENTED UNIQUELY AS $x = \cosh t$, $y = \sinh t$.

Fig. 6.43

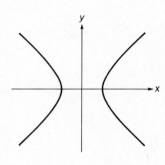

Proof. From y, we determine t as $t = $ arc sinh y; if $x = \cosh t$, then $x = \sqrt{1 + y^2}$.

The set of points (x, y) satisfying (12) is shown in Fig. 6.43. It is an **equilateral hyperbola** (as we shall verify in Chapter 10; at the moment this is not too important). We are concerned only with the right branch of the curve. Let P be a point on this branch. Then

$$P = (\cosh t, \sinh t).$$

We claim that t is twice the area of the sector bounded by the x axis, the segment OP, and the arc of the curve from P to $(1, 0)$ (see Fig. 6.44) (or minus this area if $t < 0$).

It will suffice to prove the statement for $t \geq 0$. Let the area in question be $A(t)$. Clearly $A(0) = 0$ and, if $t > 0$, then $A(t)$ is the area of the right triangle in Fig. 6.44 minus the area under the curve from $x = 1$ to $x = \cosh t$. Thus

Fig. 6.44

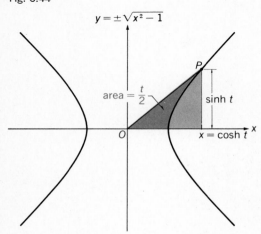

$$A(t) = \tfrac{1}{2} \cosh t \sinh t - \int_1^{\cosh t} \sqrt{x^2 - 1}\ dx$$

$$= \tfrac{1}{4} \sinh 2t - \int_1^{\cosh t} \sqrt{x^2 - 1}\ dx.$$

Therefore

$$A'(t) = \tfrac{1}{4}(\cosh 2t)2 - (\sinh t)\sqrt{\cosh^2 t - 1}$$

$$= \tfrac{1}{2}\cosh^2 t + \tfrac{1}{2}\sinh^2 t - \sinh^2 t = \tfrac{1}{2}(\cos h^2 t - \sinh^2 t) = \tfrac{1}{2}.$$

Therefore, $A(t) = (t/2) + $ const. Since $A(0) = 0$, we have $2A(t) = t$, as asserted.

The result just proved should be compared with the corresponding statement about a point $P = (\cos t, \sin t)$ on the unit circle, see §3. If $0 < t < \pi$, then t can be considered as either the length of the circular arc from $(0, 1)$ to P or twice the area of the circular sector bounded by this arc, the x axis, and the segment OP. *Only* the second interpretation, shown in Fig. 6.45, generalizes to the case of hyperbolic functions.

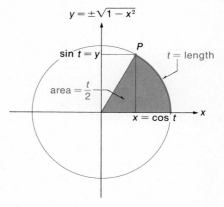

Fig. 6.45

7.5 The inverse hyperbolic tangent

We recall that the **hyperbolic tangent** is defined by

$$\tanh t = \frac{\sinh t}{\cosh t} = \frac{e^t - e^{-t}}{e^t + e^{-t}},$$

and has limit 1 at $+\infty$, limit -1 at $-\infty$; it has the derivative

$$(13) \qquad \frac{d \tanh t}{dt} = \frac{1}{\cosh^2 t},$$

which is positive everywhere. Thus the inverse function arc tanh t is defined for $|t| < 1$, by the condition

$$\tanh (\text{arc tanh } t) = t.$$

We verify in the usual way that

$$(14) \qquad \frac{d \text{ arc tanh } x}{dx} = \frac{1}{1 - x^2} \qquad |x| < 1.$$

Fig. 6.46

Since arc tanh $0 = 0$, we have

$$\text{arc tanh } x = \int_0^x \frac{du}{1 - u^2} = \int_0^x \left(\frac{1}{2} \frac{1}{1 + u} + \frac{1}{2} \frac{1}{1 - u} \right) du$$
$$= \frac{1}{2} (\log (1 + u) - \log (1 - u)) \Big|_0^x = \frac{1}{2} \log \frac{1 + x}{1 - x}$$

so that

$$(15) \qquad \text{arc tanh } x = \frac{1}{2} \log \frac{1 + x}{1 - x}.$$

The function arc tanh t is graphed in Fig. 6.46.

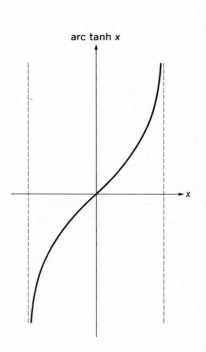

arc tanh x

7.6 Hyperbolic sine and cosine expressed by the hyperbolic tangent

Here is the analogue of Theorem 1 of §2.3.

(e) THE FUNCTIONS $\sinh t$, $\cosh t$, AND $\tanh t$ CAN be WRITTEN AS RATIONAL FUNCTIONS OF $\tanh (t/2)$.

(16)

$$\sinh t = \frac{2 \tanh t/2}{1 - \tanh^2 t/2}, \qquad \cosh t = \frac{1 + \tanh^2 t/2}{1 - \tanh^2 t/2}, \qquad \tanh t = \frac{2 \tanh t/2}{1 + \tanh^2 t/2}.$$

Fig. 6.47

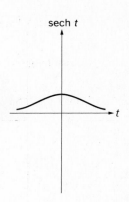

sech t

The proof follows by replacing tanh $(t/2)$ by $(e^{t/2} - e^{-t/2})/(e^{t/2} + e^{-t/2})$ and carrying out the obvious simplifications. The details are left to the reader.

7.7 Other hyperbolic functions and their inverses

It remains to discuss briefly the **hyperbolic cotangent**, the **secant**, and the **cosecant**. They are defined as follows:

$$\text{(17)} \qquad \text{sech } t = \frac{1}{\cosh t} = \frac{2}{e^t + e^{-t}}, \qquad \text{csch } t = \frac{1}{\sinh t} = \frac{2}{e^t - e^{-t}},$$

$$\coth t = \frac{1}{\tanh t} = \frac{e^t + e^{-t}}{e^t - e^{-t}},$$

the last two functions being defined only for $t \neq 0$. The graphs are shown in Figs. 6.47, 6.48, and 6.49. Note that coth t has limit 1 at $\pm\infty$, while sech t and csch t have limit 0 there. The reader will easily verify that

Fig. 6.48

csch t

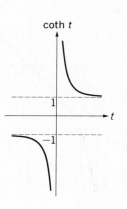

$$\text{(18)} \quad \frac{d \text{ sech } t}{dt} = -\frac{\tanh t}{\cosh t}, \qquad \frac{d \text{ csch } t}{dt} = -\frac{\coth t}{\sinh t}, \qquad \frac{d \coth t}{dt} = -\frac{1}{\sinh^2 t}.$$

A look at the graphs of the functions shows how to define the inverse functions. These are: arc coth x (defined for $|x| > 1$), arc sech x (defined for $0 < x \le 1$ and taking on only nonnegative values), and arc csch x (defined for $x \neq 0$). We have

$$\text{(19)} \qquad \text{arc coth } x = \text{arc tanh } \frac{1}{x};$$

for, if $x = \coth t$, then $(1/x) = \tanh t$. Similarly,

$$\text{(20)} \qquad \text{arc sech } x = \text{arc cosh } \frac{1}{x}, \qquad \text{arc csch } x = \text{arc sinh } \frac{1}{x}.$$

Therefore we have, using our previous results,

Fig. 6.49

coth t

$$\text{(21)} \quad \frac{d \text{ arc sech } x}{dx} = -\frac{1}{x\sqrt{1 - x^2}}, \qquad \frac{d \text{ arc csch } x}{dx} = -\frac{1}{|x|\sqrt{1 + x^2}},$$

$$\frac{d \text{ arc coth } x}{dx} = \frac{1}{1 - x^2}$$

and

$$\text{(22)} \quad \text{arc sech } x = \log \frac{1 + \sqrt{1 - x^2}}{x}, \qquad \text{arc csch } x = \log \left(\frac{1}{x} + \frac{\sqrt{1 + x^2}}{|x|} \right),$$

$$\text{arc coth } x = \frac{1}{2} \log \frac{x + 1}{x - 1};$$

the formulas involving arc sech are valid for $0 < x < 1$, those involving arc csch for $x \neq 0$, and those involving arc coth for $|x| > 1$. The details are left to the reader. Graphs of the functions just discussed are in Figs. 6.50, 6.51, and 6.52.

The preceding discussion revealed a strong formal analogy between exponential functions, written in the "hyperbolic form," and the trigonometric functions. This will become even more pronounced in Chapter 8. However, the real reason for this formal similarity becomes apparent only when one introduces complex numbers and studies functions of complex variables.

EXERCISES

17. Prove the first relation (16).
18. Prove the second and third relations (16).
19. Show that $\cosh \alpha \cosh \beta = \frac{1}{2} \{\cosh (\alpha + \beta) + \cosh (\alpha - \beta)\}$.
20. Derive a formula for $\sinh \alpha \sinh \beta$ in terms of hyperbolic functions of $\alpha + \beta$ and $\alpha - \beta$.
21. Derive a formula for $\sinh \alpha \cosh \beta$ in terms of hyperbolic functions of $\alpha + \beta$ and $\alpha - \beta$.
22. Simplify $e^{\text{arc tanh } x^2}$ (that is, express as a radical function of x).
23. Simplify $e^{\text{arc cosh } x^2}$ (that is, express as a radical function of x).
24. Verify equations (18).
25. Find $\int \cosh 3z \cosh 5z \, dz$.

26. Find $\displaystyle\int \frac{\tanh u}{1 - \tanh^2 u} \, du$.

27. Find $\displaystyle\int \frac{dx}{x\sqrt{1 - 9x^2}}$.

28. Find $\displaystyle\int \frac{x \, dx}{1 - x^4}$.

29. Prove relations (21).
► 30. Prove relations (22).

§8 Some Nonelementary Functions

8.1 Elementary and nonelementary functions

The elementary transcendental functions discussed in the preceding sections were discovered in various ways. Trigonometric functions originated in astronomy. Logarithms were introduced in order to facilitate numerical computations. The function $x \mapsto a^x$ is a natural generalization of the operation of raising a number to a rational power. The reader has noticed, however, that all these functions could have been obtained in a unified way by trying to integrate rational functions. All the functions considered above can be obtained from the primitive functions of $1/x$ and of $1/(1 + x^2)$ that is, from the functions log and arc tan. Indeed, the inverse function to the log is the exponential function e^x. Every function $x \mapsto a^x$ can be written as $e^{\alpha x}$, where $\alpha = \log a$. The hyperbolic functions are simple combinations of exponentials. Inverting the arc tan, we obtain the function $\tan \theta$. All other trigonometric functions are obtainable from the tangent by rational operations.

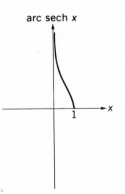

arc sech x

Fig. 6.50

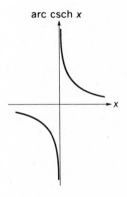

arc csch x

Fig. 6.51

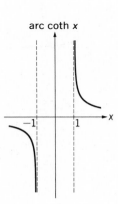

arc coth x

Fig. 6.52

We could expect to obtain other interesting functions by looking for primitives of other rational functions, but this is not so. It turns out that, having at our disposal the functions log and tangent, we can write an explicit formula for the indefinite integral of any rational function. This remarkable fact will be discussed in Chapter 7, §2.

Indefinite integrals of radical and other algebraic functions are, in general, transcendental and are not elementary functions. Exceptions are indefinite integrals of square roots of quadratic functions. They lead to the same functions we studied before. In fact, we can define the sine and the cosine as inverses of indefinite integrals of $1/\sqrt{1 - x^2}$, and one can define the hyperbolic sine and cosine as inverses of indefinite integrals of the functions $1/\sqrt{x^2 + 1}$ and $1/\sqrt{x^2 - 1}$, respectively.

Integrals of square roots of polynomials of third or fourth degrees are called elliptic integrals. A typical example is the function

$$x \mapsto \int_0^x \frac{du}{\sqrt{1 - u^4}}.$$

By inverting it, one obtains a so-called elliptic function.

There are also many important, nonelementary functions that are defined in completely different ways. We shall now discuss three examples; all three functions will be defined using the exponential function.

8.2 The normal curve

The function

(1) $$x \mapsto e^{-x^2}$$

(graphed in Fig. 6.53) plays an important part in mathematical statistics.

The area between the graph of e^{-x^2} and the x axis is finite, that is, the improper integral $\int_{-\infty}^{+\infty} e^{-x^2} \, dx$ exists. To prove this, we note the criterion stated in §6.4 of Chapter 5. It is enough to prove that $e^{-x^2} < |x|^{-4}$ for large x. Now, according to Theorem 7 in §5.7, we have that $y^{-2}e^y > 1$ for large y, so that $e^{-y} < y^{-2}$ and hence, setting $y = x^2$, we obtain the desired inequality. It turns out that

(2) $$\int_{-\infty}^{+\infty} e^{-x^2} = \sqrt{\pi}.$$

This is a surprising relation between e and π. The simplest proof of this relation involves functions of two variables (see Chapter 13, §7.3).

The graph of the function e^{-x^2} is *bell-shaped;* by this we mean that it has a single maximum, at $x = 0$, and two inflection points, for $x = -1/\sqrt{2}$ and $x = 1/\sqrt{2}$. Also the graph is symmetric about the y axis. Using the function e^{-x^2}, we construct a function depending on two parameters, a number m and a positive number σ:

(3) $$x \mapsto \frac{1}{\sigma\sqrt{2\pi}} e^{-\frac{(x - m)^2}{2\sigma^2}}$$

Fig. 6.53

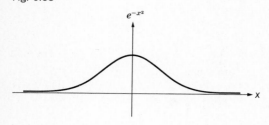

Its graph has the following properties: (a) it is symmetric with respect to the line $x = m$; (b) the total area under it is 1; and (c) it is bell-shaped and its two inflection points are $x = m + \sigma$ and $x = m - \sigma$. The last property shows that the graph of (3) is peaked for σ small and flat for σ large.

The proof of (a) is by direct calculation. If $x_1 = m + \xi$ and $x_2 = m - \xi$, rule (3) assigns the same values to x_1 and x_2. To prove (b), we set $(x - m)/\sqrt{2}\sigma = t$ and compute

Fig. 6.54

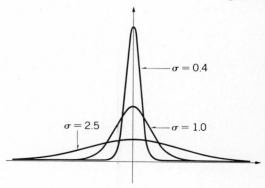

$$\int_{-\infty}^{+\infty} \frac{1}{\sigma\sqrt{2\pi}} e^{-\frac{(x-m)^2}{2\sigma^2}} dx = \int_{-\infty}^{+\infty} \frac{1}{\sigma\sqrt{2\pi}} e^{-t^2} \sqrt{2}\sigma\, dt$$

$$= \frac{1}{\sqrt{\pi}} \int_{-\infty}^{+\infty} e^{-t^2}\, dt = 1 \qquad \text{by (2).}$$

To prove (c), we compute the derivative of (3):

$$-\frac{x - m}{\sigma^3\sqrt{2\pi}} e^{-\frac{(x-m)^2}{2\sigma^2}},$$

Fig. 6.55

and the second derivative

$$-\frac{1}{\sigma^3\sqrt{2\pi}} e^{-\frac{(x-m)^2}{2\sigma^2}} + \frac{(x-m)^2}{\sigma^5\sqrt{2\pi}} e^{-\frac{(x-m)^2}{2\sigma^2}} = \frac{1}{\sigma^3\sqrt{2\pi}} \left\{ \left(\frac{x-m}{\sigma}\right)^2 - 1 \right\} e^{-\frac{(x-m)^2}{2\sigma^2}}.$$

Clearly, this second derivative is positive for $|x - m| > \sigma$, and negative for $|x - m| < \sigma$.

The graph of (3) is called a **normal curve** with mean m and **standard deviation** σ. Normal curves with $m = 0$ and $\sigma = 1, \frac{1}{2}, \frac{1}{4}$ are shown in Fig. 6.54.

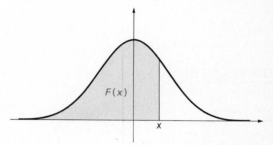

8.3 The Gauss Error Law

The normal distribution function with mean m and standard deviation σ is the function

Fig. 6.56

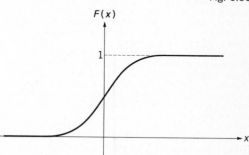

$$(4) \qquad F(x) = \frac{1}{\sigma\sqrt{2\pi}} \int_{-\infty}^{x} e^{-\frac{(t-m)^2}{2\sigma^2}}\, dt.$$

Thus $F(x)$ is the area under the normal curve from $-\infty$ to x (see Fig. 6.55). The function $F(x)$, whose graph for $m = 0$, $\sigma = 1$ appears in Fig. 6.56, is nonelementary. We know no way of writing it in terms of known functions without using an integral sign or its equivalent.

We have

$$(5) \qquad \lim_{x \to -\infty} F(x) = 0, \qquad \lim_{x \to +\infty} F(x) = 1.$$

Also

$$(6) \qquad F(m) = \frac{1}{2},$$

since half of the total area lies to the left of line $x = m$. By a change of the variable of integration, similar to the one used in proving (b) above, we can show that

$$(7) \qquad F(m - \sigma) = \frac{1}{\sqrt{2\pi}} \int_{-\infty}^{-1} e^{-t^2/2} \, dt, \qquad F(m + \sigma) = \frac{1}{\sqrt{2\pi}} \int_{-\infty}^{1} e^{-t^2/2} \, dt,$$

and numerical calculation shows that

$$F(m - \sigma) = .159, \qquad F(m + \sigma) = .841, \qquad \text{approximately.}$$

Thus the areas to the left of the line $x = m - \sigma$ and to the right of $x = m + \sigma$ are each about 16 percent of the total area, while about 68 percent of the total area lies between the lines $x = m - \sigma$, and $x = m + \sigma$.

Suppose we measure a quantity, say, the weight of some object, many times, using the same measuring device each time. Let the true value of the quantity measured be a and let ξ be the number read off on our device. We cannot expect, of course, to obtain $\xi = a$, precisely. But if the device is "good" or "unbiased," we shall get, on the average, as many low readings $(\xi < a)$ as high readings $(\xi > a)$. Moreover, *the fraction of times we shall get a reading $\xi < x$ will be, approximately, equal to $F(x)$, where $F(x)$ is the normal distribution function with mean a and some standard deviation σ.* The smaller σ, the more sensitive is our device, for a larger fraction of all measurements ξ is close to a.

This statement about the distribution of the results of measurements is known as the **Gauss Error Law** and also **Laplace's Law**. Its validity is much wider than may appear at first glance. Whenever we make many measurements, the results of which are determined by many independent random events, we may expect to encounter a normal distribution. This applies, for instance, to the distribution of IQ's in a large population, to the distribution of heights among a large number of men of the same age, and so forth. We formulated the Error Law as a "fuzzy" empirical rule. In the mathematical theory of probability, there is a corresponding "sharp" theorem. We cannot discuss this here.

8.4 The gamma function

We give now an example of a nonelementary transcendental function defined by a definite improper integral. This is the **gamma function** invented by Euler:

$$(8) \qquad x \mapsto \Gamma(x) = \int_0^\infty e^{-t} t^{x-1} \, dt \qquad x > 0.$$

(Note that the dummy variable of integration is t and not x. Note also that we write, as is customary, ∞ instead of $+\infty$.)

The integral in (8) makes sense, for near the upper limit $(t = +\infty)$, $t^{-x-1}e^t > 1$ by Theorem 7 in §5.7, and therefore $e^{-t}t^{x-1} < t^{-2}$. Near the lower limit $(t = 0)$, we note that $e^{-t}t^{x-1} < t^{x-1}$ and we assumed that $x > 0$. For $x \geq 1$, the integral is, of course, not improper at $t = 0$.

KARL FRIEDRICH GAUSS (1777–1855) was the son of a poor laborer. He recalled, half jokingly, that he was able to do sums before he could talk. A perceptive grammar school teacher noticed Gauss' genius. He secured the patronage that enabled Gauss to continue his education. The young Gauss hesitated between philology and mathematics; he made his choice after his first big discovery—the ruler and compass construction of a regular polygon with 17 sides.

While he was still alive, Gauss was recognized as an equal of Archimedes and Newton. The fame of Goettingen as a mathematical center dates from Gauss, who was professor there and director of the Goettingen observatory. Gauss' interests embraced all of mathematics, astronomy, physics, and geodesy. This aloof, awe–inspiring scholar published somewhat reluctantly, and his notebooks revealed that he anticipated many discoveries made later by other mathematicians.

PIERRE-SIMON LAPLACE (1749–1827). His two great works are the many-volume *Celestial Mechanics* and the *Analytic Theory of Probability*. The importance of these books lies primarily in the many new methods they contain, methods that play an important part in contemporary mathematics.

Laplace was of peasant origin. He was active in politics and was showered with honors by Napoleon (whom he served for a short while as a cabinet minister) and by the restored Bourbons (who made him a marquis).

We have

$$\Gamma(1) = \lim_{N \to +\infty} \int_0^N e^{-t}\, dt = \lim_{N \to +\infty} (-e^{-t})\Big|_0^N = \lim_{N \to +\infty} (-e^{-N} + 1) = 1.$$

Knowing that the improper integral defining $\Gamma(1)$ exists, we can write the above calculation in an abbreviated form:

$$(9) \qquad \Gamma(1) = \int_0^\infty e^{-t}\, dt = -e^{-t}\Big|_0^{+\infty} = -0 + 1 = 1.$$

The relation

$$(10) \qquad \Gamma\!\left(\frac{1}{2}\right) = \sqrt{\pi}$$

can be verified using the substitution rule; we set $t = y^2$, so that $dt = 2y\, dy = 2t^{1/2}\, dy$ and $t^{-1/2}\, dt = 2\, dy$ and obtain

$$\Gamma\!\left(\frac{1}{2}\right) = \int_0^\infty e^{-t} t^{-1/2}\, dt = 2\int_0^\infty e^{-y^2}\, dy = \int_{-\infty}^{+\infty} e^{-y^2}\, dy = \sqrt{\pi}.$$

Here we used the fact that e^{-y^2} is even and equation (2).

We derive next the **functional equation:**

$$(11) \qquad \Gamma(x) = (x - 1)\Gamma(x - 1) \qquad \text{for } x > 1.$$

The proof reads, in an abbreviated form, as follows.

$$\Gamma(x) = \int_0^\infty e^{-t} t^{x-1}\, dt = -\int_0^\infty t^{x-1}\, de^{-t}$$

$$= -t^{x-1} e^{-t}\Big|_0^{+\infty} + \int_0^\infty e^{-t}\, dt^{x-1} = (x - 1)\int_0^\infty e^{-t} t^{x-2}\, dt$$

so that

$$\Gamma(x) = (x - 1)\int_0^\infty e^{-t} t^{x-2}\, dt = (x - 1)\Gamma(x - 1).$$

We recall now the well-known notation:

$$0! = 1,\ 1! = 1,\ 2! = 1 \cdot 2,\ 3! = 1 \cdot 2 \cdot 3,\ 4! = 1 \cdot 2 \cdot 3 \cdot 4, \qquad \text{and so forth}$$

($n!$ is read "n **factorial**"). Since $\Gamma(1) = 1$, we have by (11) that $\Gamma(2) = 1 \cdot \Gamma(1) = 1$, $\Gamma(3) = 2\Gamma(2) = 2 = 2 \cdot 1$, $\Gamma(4) = 3\Gamma(2) = 3 \cdot 2 \cdot 1$, and in general

$$(12) \qquad \Gamma(n) = (n - 1)! \qquad \text{for } n = 1, 2, 3, \cdots.$$

These are only some of the many remarkable properties of the gamma function.

8.5 A function with all derivatives zero at one point

We shall use Theorem 7 in §5.7 to exhibit a function that has derivatives of all orders and all of whose derivatives are 0 at one point. Of course, a constant function has this property; but the function we construct will *not* be a constant.

Theorem 1   LET n BE ANY INTEGER. SET

$$(13) \qquad f_n(x) = \begin{cases} x^n e^{-1/x^2} & \text{if } x \neq 0, \\ 0 & \text{if } x = 0. \end{cases}$$

THEN (a) THE FUNCTION f_n IS CONTINUOUS EVERYWHERE; (b) THE FUNCTION f_n HAS DERIVATIVES OF ALL ORDERS; AND (c) ALL OF THESE DERIVATIVES ARE 0 AT $x = 0$.

Proof.   For $x \neq 0$, the continuity is obvious. The derivative of $f_n(x)$ can be computed formally:

$$f'_n(x) = nx^{n-1}e^{-1/x^2} + x^n \frac{de^{-1/x^2}}{dx} = nx^{n-1}e^{-1/x^2} + x^n e^{-1/x^2}\frac{2}{x^3}$$

$$= nx^{n-1}e^{-1/x^2} + 2x^{n-3}e^{-1/x^2}.$$

Therefore, for $x \neq 0$,

$$(14) \qquad f'_n(x) = nf_{n-1}(x) + 2f_{n-3}(x).$$

To show that f_n is continuous at $x = 0$, we must show that $\lim_{x \to 0} f_n(x) = 0$, which is the same as $\lim_{x \to 0} |f_n(x)| = 0$. We use Theorem 7 in §5.7 and the rules for computing limits stated in Chapter 3, §4.2 and §4.3. Also, we set $x^{-2} = t$ so that $|x| = t^{-1/2}$. We obtain

$$\lim_{x \to 0} |f_n(x)| = \lim_{x \to 0}(|x|^n e^{-1/x^2}) = \lim_{t \to +\infty} t^{-n/2}e^{-t} = \frac{1}{\lim_{t \to +\infty} t^{n/2}e^t} = \frac{1}{\infty} = 0.$$

Thus (a) is proved.

Now we know that the right-hand side of (14) is also continuous for $x = 0$, and has the value 0 there. This follows by applying (a) to f_{n-1} and f_{n-3}. We conclude (using Theorem 2 in Chapter 4, §1.10) that equation (14) holds also for $x = 0$, and that $f'_n(0) = 0$. Applying (14) and the result just proved once more, we have

$$\begin{aligned} f''_n(x) &= nf'_{n-1}(x) + 2f'_{n-3}(x) \\ &= n\{(n-1)f_{n-2}(x) + 2f_{n-4}(x)\} + 2\{(n-3)f_{n-4}(x) + 2f_{n-6}(x)\} \\ &= n(n-1)f_{n-2}(x) + (4n-6)f_{n-4}(x) + 4f_{n-6}(x). \end{aligned}$$

Hence f_n has a second derivative that is 0 at $x = 0$. Now we use the same procedure to obtain f'''_n and to prove that $f'''_n(x) = 0$, and so forth.

The graph of the function $y = f_0(x) = e^{-1/x^2}$ is shown in Fig. 6.57.

Fig. 6.57

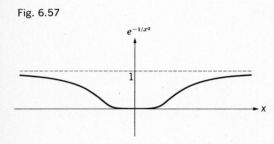

Problems

1. Let $f(x)$ be a continuous function with period p. Show that the number $\int_{x_0}^{x_0+p} f(x)\,dx$ does not depend on x_0.

2. Let $f(x)$ be a continuous periodic function with period p. Let $F(x)$ be an indefinite integral of $f(x)$. Under which circumstances is $F(x)$ periodic with period p? (After you find the condition, verify that it is satisfied for $f(x) = \sin x$, $p = 2\pi$.)

3. Let $f(x)$ and $g(x)$ be periodic, with periods p and q, respectively. Show that if p and q are rational numbers, then the function $h(x) = f(x) + g(x)$ is periodic.

4. Suppose $f(x)$ is periodic and monotone. What can you say about f?

5. Suppose that $f(x)$ is defined and continuous for all x, and that every number $p = 1, \frac{1}{2}, \frac{1}{3}, \cdots$ is a period of f. Prove that f is a constant.

6. Using the geometric definition of sines and cosines, prove that the inequality $0 < |\sin \theta|\cos \theta < |\theta| < |\tan \theta|$ holds for $0 < |\theta| < \pi/2$. (Hint: Let P be the point with the coordinates $(\cos \theta, \sin \theta)$, Q the point with the coordinates $(1, \tan \theta)$, P_1 the point $(\cos \theta, 0)$, and Q_1 the point $(1, 0)$. Express the areas of the triangles OPP_1 and OQQ_1 and of the circular sector OPQ_1 in terms of θ, $\sin \theta$, $\cos \theta$, and $\tan \theta$.)

7. Without using any information about the derivatives of sine and cosine, prove that $\lim_{\theta \to 0} \frac{\sin \theta}{\theta} = 1$ and $\lim_{\theta \to 0} \frac{\cos \theta - 1}{\theta} = 0$. (Hint: Use the preceding problem.)

8. Assume the addition theorems for sine and cosine, and prove that the functions $\sin \theta$ and $\cos \theta$ have the derivatives $\cos \theta$ and $-\sin \theta$, respectively. (Hint: Use the result of the preceding problem.) Note that the addition theorems can be proved geometrically. The above derivation of the differentiation formulas is the traditional one.

9. Assume that $f'(x) = g(x)$ and $g'(x) = -f(x)$ for all x near $x = 0$, and that $f(0) = 0$, $g(0) = 1$. Show that $f(x) = \sin x$, $g(x) = \cos x$.

10. Prove that the functions sine and cosine are not radical functions by showing that no nonconstant periodic function can be radical. (Hint: Use the result of Problem 15 in Chapter 3.)

11. The functional equation of the arc tan function reads:

$$\text{arc tan } x + \text{arc tan } y = \text{arc tan } \frac{x + y}{1 - xy}.$$

Prove this, using the addition theorem for tangent.

12. Give a direct proof of the functional equation for the function arc tan x, using the definition: $x \mapsto \text{arc tan } x$ is that primitive function of $x \mapsto \frac{1}{1 + x^2}$ which is 0 at $x = 0$.

13. Express the total energy of a vibrating elastic spring as a function of amplitude and velocity at the instant when the deviation from equilibrium is 0. Also, express the kinetic energy as a function of time. (Concerning the definition of energy, see Chapter 5, §6.5.)

14. State and prove the functional equation of the function arc cot, analogous to that of the arc tan (see Problem 11).

15. Prove that $\log x$ is not a radical function. (Hint: Use the result of Problem 15 in Chapter 3.)

16. Let $x \mapsto f(x)$ be a positive differentiable function and $x \mapsto g(x)$ a differentiable function. Find the derivative of the function $x \mapsto F(x) = f(x)^{g(x)}$.

17. Prove Theorem 5 in §5.6 (characterization of the exponential function by its differential equation and initial condition) without assuming any previous knowledge of the exponential function. (Hint: First show that equation $f'(x) = f(x)$ implies that $f(-x)f(x)$ is a constant. Then conclude that $f(x)$, a solution of this equation satisfying $f(0) = 1$, must be (1) positive and (2) strictly monotone. Write $y = f(x)$, consider x as a function of y, compute its derivative, apply the fundamental theorem of calculus, and conclude that $f(x)$ is inverse to the function log.)

18. Give a direct proof of Theorem 5 in §5.6. (Hint: Follow the suggestion given above for solving Problem 17.)

19. Given two functions $f(x)$ and $g(x)$ such that $\lim_{x \to +\infty} f(x) = \lim_{x \to +\infty} g(x) = +\infty$, one says that g grows faster than f if $\lim_{x \to +\infty} [g(x)/f(x)] = +\infty$. Find functions that grow faster than e^x, faster than e^{x^2}, faster than e^{e^x}, and faster than a given function $\phi(x)$.

20. Find functions that grow faster than e^x but slower than xe^x, faster than any polynomial but slower than $e^{\sqrt{x}}$, faster than $\log x$ but slower than $\sqrt{x}$, and slower than $\log \log x$. (One says that f grows slower than g if g grows faster than f.)

21. Define two continuous functions $f(x)$ and $g(x)$ such that both grow faster than e^x, the quotient $f(x)/g(x)$ has no limit as $x \mapsto +\infty$, and neither $f(x)$ nor $g(x)$ grows faster than the other.

22. Prove that $e^t > 1 + t + \frac{t^2}{2}$ for $t > 0$.

23. Guess at the approximate values of the following numbers:

$$a = (1.00000001)^{10000000}, \qquad b = (.99999999)^{10000000},$$
$$c = (1.00000001)^{100000}, \qquad d = (.99999999)^{10000000000}.$$

Give reasons for your guess.

24. Prove that the function $t \mapsto \left(1 + \frac{1}{t}\right)^t$ is increasing for large t. How large?

25. The device that we used in §6.1 to solve the nonhomogeneous equation $f'(x) - f(x) = \phi(x)$, ϕ a given function, is called the "variation of constants." Apply the same idea to solving the nonhomogeneous equation

$$f''(x) + f(x) = \phi(x) \qquad \phi \text{ a given function.}$$

(Hint: For $\phi(x) \equiv 0$, we know that the general solution is $f(x) = a \cos x + b \sin x$, a and b constants. Hence try to solve the given equation by setting $f(x) = a(x) \cos x + b(x) \sin x$, where a and b are functions to be determined. Try to impose conditions on these functions that would permit you to find them by integration.)

26. Find the general solution of the differential equation

$$f''(x) + f(x) = \cos \omega x.$$

(Hint: Use the method of the preceding problem.) This equation describes the motion of a harmonic oscillator with frequency $1/2\pi$ subject to an external periodic force with frequency $\omega/2\pi$. Note what happens in the case of "resonance," that is, when $\omega = 1$ and the frequency of the disturbing force equals the frequency of the oscillator.

27. The addition theorems of the hyperbolic sine and cosine read

$$\sinh (x + y) = \sinh x \cosh y + \cosh x \sinh y,$$
$$\cosh (x + y) = \cosh x \cosh y + \sinh x \sinh y.$$

Derive them from Theorem (b) in §7.1.

28. Find the general solution of the nonhomogeneous differential equation

$$f''(x) - f(x) = \phi(x) \qquad \phi \text{ a given function}$$

by following the hint given for solving Problem 26.

29. Show that the nonhomogeneous differential equation in the preceding problem is equivalent to a system of two first order equations, namely,

$$g'(x) + g(x) = \phi(x), \qquad f'(x) - f(x) = g(x).$$

Use this observation to give a new proof for Theorem (b) in §7.1, and also to obtain a new expression for the general solution of the nonhomogeneous equation.

30. Show that the differential equation $f''(x) - \omega^2 f(x) = \phi(x)$ is equivalent to a system of two first order differential equations.

31. Using the geometric interpretation of hyperbolic functions, find and prove an inequality, analogous to that in Problem 6 above, for the functions $\sinh t$, t, and $\tanh t$.

32. Use the result of the previous problem and the addition theorems for the hyperbolic sine and cosine, in order to prove that the derivatives of $\sinh t$ and $\cosh t$ are $\cosh t$ and $\sinh t$, respectively. (Hint: Note Problems 7 and 8 above.)

7/Techniques of Integration

This is an essentially technical chapter. It deals with methods for computing integrals and primitive functions. In accordance with modern developments, we put numerical integration ahead of so-called "formal" integration.

The appendix contains a summary of integration formulas.

1.1 Formal differentiation

The basic processes of calculus are differentiation and integration or, geometrically speaking, finding tangents and computing areas. The fundamental theorem of calculus discloses the surprising fact that the two processes are inverse to each other. Nevertheless, there are significant differences between the two. This can be seen when one tries to differentiate or integrate a given function.

A function $x \mapsto f(x)$ may be "given" in various ways. It can be defined by an explicit formula involving a finite number of rational operations, extractions of roots, and substitutions into certain "known" or "elementary" functions. We usually mean by these the elementary transcendental functions that we studied in Chapter 6.

Finding the derivative of such a function is a purely mechanical process, provided of course that we know the derivatives of the "elementary" functions. No matter how complicated the formula, if we systematically apply the differentiation rules we have learned, after finitely many steps we shall obtain an explicit formula for the derivative $f'(x)$. It is even possible to program a computer to perform differentiations. In the language of computer experts, differentiating functions defined by explicit formulas involves a finite algorithm.

It is quite different with integration. We know, by the fundamental theorem of calculus, that every continuous function $f(x)$ has a primitive function $F(x)$ but if a continuous function $x \mapsto f(x)$ is defined by an explicit formula, there is no reason at all why a primitive function $F(x)$ should be defined by such a formula. In general, it will not. Forming primitive functions of simple "known" functions can lead to interesting new functions. Recall, in this connection, how log and arc tan can be defined as integrals.

The exceptional cases in which a function $f(x)$ can be **integrated in closed form,** that is, has a primitive function (indefinite integral) $F(x)$ given by an explicit formula, are nevertheless quite important. One may be able to find a formula for $F(x)$ using the integration rules of Chapter 5, §3, especially change of variables (substitution) and integration by parts. This process is called **formal integration.** In almost all cases, it involves trial and error. There is no "finite algorithm" for carrying out formal integration, and this task cannot be delegated to a computer. Formal integration will be discussed in the following section of this chapter.

1.2 Numerical differentiation

In applying mathematics, we often encounter functions $x \mapsto f(x)$ defined in ways other than by an explicit formula. A function may be "given" numerically, that is, we may be presented with a table of certain values of the function. This is so, for instance, when $f(x)$ is the result of an experiment or measurement. Sometimes a measuring device produces the graph of the function, that is, an approximate table of values. Or, we may be given an infinite process by which values of the function can be computed with any required degree of accuracy. In a certain sense, all elementary functions are defined in this way. And when $f(x)$ is defined by an explicit formula, what is this formula but the description of a method for computing the number $f(x)$ for a given number x?

Finding the derivative of a numerically given function $f(x)$ is, in general, an unpleasant task. It may be meaningless, since even if we know that $f(x)$ is continuous, it may not have a derivative. If we know that a derivative exists, computing it at some point x_0 is difficult. We can compute the "difference quotient"

$$(1) \qquad \frac{f(x_0 + h) - f(x_0)}{h}$$

for a small h, and consider it an approximation to $f'(x_0)$. In order to estimate how good the approximation is, we need additional information about the function. If we know that it has a second derivative $f''(x)$ and that $|f''(x)| \leq M_2$ for x between x_0 and $x_0 + h$, then the error committed is at most $\frac{1}{2}M_2h^2$ (this follows from Taylor's Theorem which we shall discuss in the next chapter). Thus, to get a high accuracy, we need a very small h, that is, we need a fine table for the function. But if $|h|$ is small, then a small error in computing the values $f(x_0 + h)$ and $f(x_0)$ will produce a large error in the quotient (1).

We say no more about numerical differentiation, since this process is hardly ever used. Some modern texts on numerical analysis do not mention it at all.

1.3 Numerical integration

Integration of numerically given functions, on the other hand, is a basic technique that is used quite often. Even if a function $f(x)$ is given by an explicit formula, the only way of computing

$$(2) \qquad \int_a^b f(x) \, dx$$

may be to treat $f(x)$ as a numerically given function and to use **numerical integration.** The only way to compute a primitive function $F(x)$ may consist in tabulating F by computing

$$\int_a^x f(t)\,dt$$

for various values of x. We assume in what follows that $f(x)$ is continuous for $a \le x \le b$, so that the integral (2) is well defined.

We divide the interval into n equal subintervals, each of length

$$h = \frac{b-a}{n}.$$

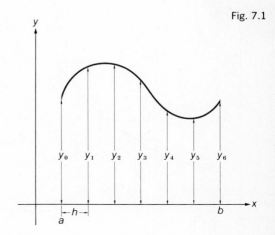

Fig. 7.1

The endpoints of these small intervals are

(3) $\qquad x_0 = a,\ x_1 = a + h,\ x_2 = a + 2h,\ \cdots,\ x_n = a + nh = b$

(see Fig. 7.1, where $n = 6$). We assume that the values

(4) $\qquad y_0 = f(x_0),\ y_1 = f(x_1),\ y_2 = f(x_2),\ \cdots,\ y_n = f(x_n)$

are known, and we want to find a formula for an approximate value of the integral (2) in terms of the $2n + 2$ numbers $x_0, x_1, \cdots, x_n, y_0, y_1, y_2, \cdots y_n$.

An approximate integration formula is often called a **quadrature formula.** No such formula can be useful for *all* conceivable continuous functions. For instance, if a quadrature formula is at all reasonable, it should give the value 0 for the case in which all function values (4) are 0. But there are continuous functions with $f(x_j) = 0$ for $j = 0,\ 1,\ \cdots,\ n$ and a large value for the integral $\int_a^b f(x)\,dx$; such a function is shown in Fig. 7.2.

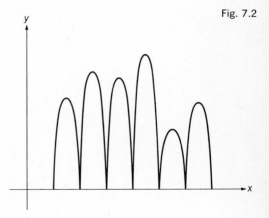

Fig. 7.2

1.4 Trapezoidal rule

In Chapter 5, §1.6, we have already discussed one approximate integration formula, for a monotone function $f(x)$. It amounts to replacing the area under the curve $x \mapsto f(x)$ by the sum of areas of n rectangles of width h and heights $y_0 = f(x_0),\ y_1 = f(x_1),\ \cdots,\ y_{n-1} = f(x_{n-1})$; see Fig. 7.3. Each rectangle has area hy_j (which is negative if $y_j < 0$). Hence the approximate formula reads

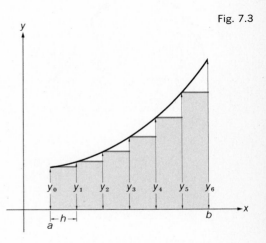

Fig. 7.3

(5)
$$\int_a^b f(x)\,dx \approx hy_0 + hy_1 + \cdots + hy_{n-1}$$
$$= \frac{b-a}{n}(y_0 + y_1 + y_2 + \cdots + y_{n-1}).$$

Fig. 7.4

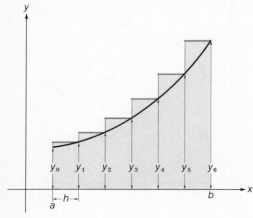

We could also use the rectangles with heights y_1, y_2, $\cdots$, y_n. Then we get (see Fig. 7.4)

(6)
$$\int_a^b f(x)\, dx \approx hy_1 + hy_2 + \cdots + hy_{n-1}$$
$$= \frac{b-a}{n}(y_1 + y_2 + \cdots + y_n).$$

For a monotone function, the true value of the integral lies between the two approximate values (5) and (6). This suggests that a better quadrature formula may be obtained by using the arithmetic mean of (5) and (6). Adding (5) and (6) and dividing by 2, we obtain the approximate formula

(7)
$$\int_a^b f(x)\, dx \approx \frac{b-a}{2n}(y_0 + 2y_1 + 2y_2 + \cdots + 2y_{n-1} + y_n)$$
$$= \frac{h}{2}(y_0 + 2y_1 + 2y_2 + \cdots + 2y_{n-1} + y_n).$$

This is known as the **trapezoidal rule.**

There is another way of arriving at this formula, which also explains its name. Let $\phi(x)$ denote a continuous function which coincides with $f(x)$ for $x = x_0$, x_1, x_2, $\cdots$, x_n, but is linear between two consecutive values x_j and x_{j+1}; this function is graphed in Fig. 7.5 [in this figure $f(x)$ is not monotone]. We want to use $\int_a^b \phi(x)\, dx$ as an approximation to $\int_a^b f(x)\, dx$. But $\int_a^b \phi(x)\, dx$ is the sum of n integrals:

Fig. 7.5

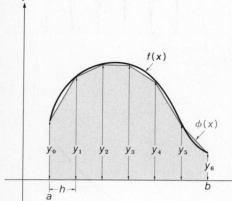

$$\int_a^b \phi(x)\, dx = \int_{x_0}^{x_1} \phi(x)\, dx + \int_{x_1}^{x_2} \phi(x)\, dx + \cdots + \int_{x_{n-1}}^{x_n} \phi(x)\, dx.$$

Since $\int_{x_0}^{x_1} \phi(x)\, dx$ is the area of a trapezoid,

$$\int_{x_0}^{x_1} \phi(x)\, dx = \frac{h}{2}(y_0 + y_1).$$

[It would be possible, though pedantic, to obtain this by integrating the linear function $\phi(x)$.] By the same token,

$$\int_{x_1}^{x_2} \phi(x)\, dx = \frac{h}{2}(y_1 + y_2),$$

. . . .

$$\int_{x_{n-1}}^{x_n} \phi(x)\, dx = \frac{h}{2}(y_{n-1} + y_n).$$

Adding these equations, we obtain that

$$\int_a^b \phi(x)\, dx = \frac{h}{2}(y_0 + 2y_1 + 2y_2 + \cdots + 2y_{n-1} + y_n).$$

The right-hand side here is the same as in (7).

1.5 Error estimate for the trapezoidal rule

Now let E denote the absolute value of the **error** committed in evaluating the integral (2) by the trapezoidal rule, that is, set

$$E = \left| \int_a^b f(x)\, dx - \frac{h}{2}(y_0 + 2y_1 + 2y_2 + \cdots + 2y_{n-1} + y_n) \right|.$$

The precise value of E depends, of course, on the function f, on a and b, and on n.

The following **estimates** *hold:*

(8) $E \leq \dfrac{|f(b) - f(a)|(b - a)}{n} = |f(b) - f(a)|h \quad$ *if $f(x)$ is monotone,*

(9) $E \leq \dfrac{(b - a)^3 M_2}{12n^2} = \dfrac{1}{12}(b - a)M_2 h^2 \quad$ *if $|f''(x)| \leq M_2$.*

(Here and hereafter the conditions imposed on f are supposed to be valid for all x, $a \leq x \leq b$.)

Let us prove (8), assuming that $f(x)$ is monotone. Then the true value of the integral lies between (5) and (6), as is seen from Figs. 7.3 and 7.4. So does the approximate value, given by the trapezoidal rule, since this value is the arithmetic mean between (5) and (6). Hence the absolute value of the error is not greater than the absolute value of the difference,

$$\left| \frac{b - a}{n}(y_1 + y_2 + \cdots + y_n) - \frac{b - a}{n}(y_0 + y_1 + \cdots + y_{n-1}) \right|.$$

But this absolute value is

$$\left| \frac{b - a}{n}(y_n - y_0) \right| = \frac{b - a}{n}|f(b) - f(a)|.$$

The other estimate will be proved in the last appendix to Chapter 8.

EXERCISES

Find an approximate value for the following integrals using the trapezoidal rule for the given value of n. Estimate the error both by using the estimate in (8) and the estimate in (9), wherever possible. First write your answer exactly as a sum and then approximate this as a decimal, using the tables only if necessary. Naturally, the number of decimal places you compute will in part be determined by your error estimate. (You may use the relation $\log 2 < .75$, which follows from Exercise 4, whenever needed.)

1. $\int_0^1 \dfrac{dx}{x^2 + 1}$ with $n = 4$.   6. $\int_0^2 \dfrac{4^x}{x + 1}\, dx$ with $n = 4$.

2. $\int_0^2 (x^4 + x)\, dx$ with $n = 6$.   7. $\int_0^3 x\sqrt{x + 1}\, dx$ with $n = 6$.

3. $\int_{\pi/2}^{\pi/2} \dfrac{dx}{2 + \sin x}$ with $n = 4$.   8. $\int_0^3 \dfrac{x}{x + 1}\, dx$ with $n = 6$.

4. $\int_1^2 \dfrac{dx}{x}$ with $n = 6$.   9. $\int_0^3 \sqrt{4x^2 + 2x}\, dx$ with $n = 6$.

5. $\int_0^1 16^x\, dx$ with $n = 4$.   10. $\int_{-1/2}^{3/2} (4x)^{1/3}\sqrt{4x + 5}\, dx$ with $n = 4$.

1.6 Simpson's rule

There is an equally simple quadrature formula which, in most cases, gives a better approximation. It is known as the **parabolic rule**, or **Simpson's rule.**

We assume now that n is even, so that $n = 2m$, and take as an approximation to $\int_a^b f(x)\, dx$ the number

$$\int_a^b \phi(x)\, dx,$$

where $\phi(x)$ coincides with $f(x)$ for $x = x_0, x_1, x_2, \cdots, x_n$ and $\phi(x)$ equals a quadratic polynomial in each of the m intervals of length $2h$: $[x_0, x_2]$, $[x_2, x_4], \cdots, [x_{2m-2}, x_{2m}]$. The graph of $\phi(x)$, shown in Fig. 7.6, consists of parabolic arcs.

Let us compute $\int_{x_0}^{x_2} \phi(x)\, dx$. In the interval $[x_0, x_2]$, the function $\phi(x)$ is a quadratic polynomial, $\phi(x) = Ax^2 + Bx + C$. We can substitute for x the expression $(x - x_1) + x_1$, collect terms, and write

$$\phi(x) = \alpha(x - x_1)^2 + \beta(x - x_1) + \gamma.$$

The constants α, β, and γ are found from the conditions

$$\phi(x_0) = f(x_1 - h) = f(x_0) = y_0,$$
$$\phi(x_1) = f(x_1) = y_1,$$
$$\phi(x_2) = f(x_1 + h) = f(x_2) = y_2.$$

THOMAS SIMPSON (1710–1761) was a silkweaver by trade. His mathematical books earned him a professorship at the military college in Woolwich.

Fig. 7.6

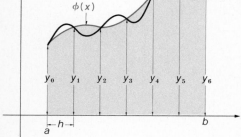
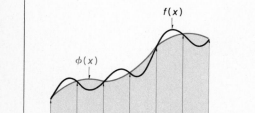

The second relation gives

$$\gamma = y_1;$$

the first and third can now be written as

$$\alpha h^2 - \beta h = y_0 - y_1, \qquad \alpha h^2 + \beta h = y_2 - y_1.$$

Adding and subtracting the two equations, we get

$$\alpha = \frac{y_0 + y_2 - 2y_1}{2h^2}, \qquad \beta = \frac{y_2 - y_0}{2h}.$$

Thus $\phi(x) = \dfrac{y_0 + y_2 - 2y_1}{2h^2}(x - x_1)^2 + \dfrac{y_2 - y_0}{2h}(x - x_1) + y_1$

for $x_0 \le x \le x_2$. This polynomial has the primitive function

$$\frac{y_0 + y_2 - 2y_1}{6h^2}(x - x_1)^3 + \frac{y_2 - y_0}{4h}(x - x_1)^2 + y_1 x.$$

Substituting $x = x_2 = x_1 + h$ and $x = x_0 = x_1 - h$ and subtracting, we obtain

$$\int_{x_0}^{x_2} \phi(x)\, dx = \frac{y_0 + y_2 - 2y_1}{6h^2}[h^3 - (-h)^3]$$

$$+ \frac{y_2 - y_1}{4h}[h^2 - (-h)^2] + y_1[h - (-h)]$$

$$= \frac{y_0 + y_2 - 2y_1}{3}h + 2y_1 h = \left(\frac{1}{3}y_0 + \frac{4}{3}y_1 + \frac{1}{3}y_2\right)h.$$

Thus $\qquad \displaystyle\int_{x_0}^{x_2} \phi\, dx = \left(\frac{1}{3}y_0 + \frac{4}{3}y_1 + \frac{1}{3}y_2\right)h$

and, by the same token,

$$\int_{x_2}^{x_4} \phi\, dx = \left(\frac{1}{3}y_2 + \frac{4}{3}y_3 + \frac{1}{3}y_4\right)h,$$

$$\int_{x_4}^{x_6} \phi\, dx = \left(\frac{1}{3}y_4 + \frac{4}{3}y_5 + \frac{1}{3}y_6\right)h,$$

$$\cdots\cdots\cdots\cdots\cdots\cdots\cdots\cdots\cdots\cdots\cdots$$

$$\int_{x_{2m-2}}^{x_{2m}} \phi\, dx = \left(\frac{1}{3}y_{2m-2} + \frac{4}{3}y_{2m-1} + \frac{1}{3}y_{2m}\right)h.$$

If we add all these equations, we obtain Simpson's rule in the form

$$\int_a^b f(x)\, dx \approx \frac{b-a}{n}\left(\frac{1}{3}y_0 + \frac{4}{3}y_1 + \frac{2}{3}y_2 + \frac{4}{3}y_3 + \frac{2}{3}y_4 + \cdots\right.$$

(10)
$$\left. + \frac{2}{3}y_{n-2} + \frac{4}{3}y_{n-1} + \frac{1}{3}y_n\right)$$

$$= \frac{h}{3}(y_0 + 4y_1 + 2y_2 + 4y_3 + 2y_4 + \cdots + 2y_{n-2} + 4y_{n-1} + y_n).$$

It is almost as easy to use as the trapezoidal rule.

1.7 Error estimate for Simpson's rule

If E denotes the absolute error committed in using Simpson's rule,

$$E = \left| \int_a^b f(x)\, dx - \frac{h}{3}(y_0 + 4y_1 + 2y_2 + \cdots + 4y_{n-1} + y_n) \right|,$$

*then, for $f(x)$ having a continuous fourth derivative, we have the **estimate***

(11) $$E \le \frac{(b-a)^5 M_4}{180n^4} = \frac{1}{180}(b-a)M_4 h^4 \qquad \text{if } |f^{(4)}(x)| \le M_4.$$

Compare this with (9); the error term for the trapezoidal rule is of the order h^2, that for Simpson's rule of the order h^4. (This result shows that $E = 0$ if the function is a polynomial of degree at most 3.) The proof of (11) will be given in the last appendix to Chapter 8.

▶*Examples   1.* Compute

$$\int_0^6 x^2\, dx \left(= \frac{x^3}{3}\Big|_0^6 = 72\right),$$

using the trapezoidal rule and also Simpson's rule, with 6 subintervals. Compare the actual error with the error estimates given above.

The calculations are shown in the following table. Here j denotes the subscript of the point considered and v denotes the "weight factor." Thus for the trapezoidal rule $v = 1$, if $j = 0$ or n, $v = 2$ for all other j. For Simpson's rule $v = 1$, if $j = 1$ and n, and for other j we have $v = 2$ or $v = 4$, according to whether j is odd or even.

j	x_j	$y_j = x_j^2$	Trapezoidal rule		Simpson's rule	
			v_j	$v_j y_j$	v_j	$v_j y_j$
0	0	0	1	0	1	0
1	1	1	2	2	4	4
2	2	4	2	8	2	8
3	3	9	2	18	4	36
4	4	16	2	32	2	32
5	5	25	2	50	4	100
6	6	36	1	36	1	36
				146		216

The result of addition must be multiplied by $h/2$ (for the trapezoidal rule) or by $h/3$ (for Simpson's rule). In our case, $h = 1$. Hence

$$\int_0^6 x^2 \, dx \approx \begin{cases} \dfrac{1}{2} \cdot 146 = 73 & \text{trapezoidal rule,} \\[2mm] \dfrac{1}{3} \cdot 216 = 72 & \text{Simpson's rule.} \end{cases}$$

For the trapezoidal rule, the absolute error E is 1. Estimates (8) and (9) give, with $n = 6$, $a = 0$, $b = 6$, $f(b) - f(a) = 36$, and $M_2 = 2$, the inequalities $E \le 36$ and $E \le 1$, respectively.

For Simpson's rule we have, of course, $E = 0$ [see (11)] and our result is precise.

2. Compute

$$\log 3 = \int_1^3 \frac{dx}{x}$$

by the trapezoidal rule and by Simpson's rule, using $n = 4$. Compare the error with the error estimates.

The calculations are:

j	x_j	$y_j = \dfrac{1}{x_j}$	Trapezoidal rule		Simpson's rule	
			v_j	$v_j y_j$	v_j	$v_j y_j$
0	1.0	1.0000	1	1.0000	1	1.0000
1	1.5	.6667	2	1.3334	4	2.6668
2	2.0	.5000	2	1.0000	2	1.0000
3	2.5	.4000	2	.8000	4	1.6000
4	3.0	.3333	1	.3333	1	.3333
				4.4667		6.6001

Since $h = \frac{1}{2}$, we have

$$\log 3 \approx \begin{cases} \dfrac{1}{2} \cdot \dfrac{1}{2} \cdot (4.4667) = 1.1167 & \text{trapezoidal rule,} \\[2ex] \dfrac{1}{3} \cdot \dfrac{1}{2} \cdot (6.6001) = 1.1000 & \text{Simpson's rule.} \end{cases}$$

Now for $f(x) = 1/x$ we have $f''(x) = 2/x^3$, $f^{(4)}(x) = 24/x^5$. For $1 \le x \le 3$ we may apply (9) with $M_2 = 2$, and (11) with $M_4 = 24$. This gives (for $b - a = 2$, $n = 4$)

$$E \le \begin{cases} \dfrac{1}{12} = .0833 \cdots & \text{trapezoidal rule,} \\[2ex] \dfrac{1}{60} = .0166 \cdots & \text{Simpson's rule.} \end{cases}$$

Actually, $\log 3 = 1.0986 \cdots$, and therefore

$$E = \begin{cases} .0181 \cdots & \text{trapezoidal rule,} \\ .0014 \cdots & \text{Simpson's rule.} \end{cases}$$

Both errors are well below the theoretical bounds, and Simpson's rule gives a more accurate result. ◄

EXERCISES

Find an approximate value for the following integrals using Simpson's rule for the given value of n. Estimate the error by using the estimate in (11).

11. $\displaystyle\int_0^1 \frac{dx}{x^2 + 1}$ with $n = 4$. Compare with the result of Exercise 1.

12. $\int_0^2 (x^4 + x)\, dx$ with $n = 6$. Compare with the result of Exercise 2. What result do you obtain if $n = 2$?

13. $\displaystyle\int_{-\pi/2}^{\pi/2} \frac{dx}{2 + \sin x}$ with $n = 4$. Compare with the result of Exercise 3.

14. $\displaystyle\int_1^2 \frac{dx}{x}$ with $n = 6$. Compare with the result of Exercise 4.

15. $\int_0^1 16^x\, dx$ with $n = 4$. Compare with the result of Exercise 5.

Compute each of the following integrals exactly. Then find approximate values, first by using the trapezoidal rule and then by using Simpson's rule, each for the given value of n. Estimate the error term of each approximation by using (9) and (11). Compare with the actual error. Use the tables.

16. $\displaystyle\int_0^1 \frac{dx}{2x + 1}$ for $n = 6$.   19. $\int_{2/3}^2 \sqrt{x}\, dx$ for $n = 4$.

17. $\displaystyle\int_0^1 \frac{dx}{1 + x^2}$ for $n = 2$.   20. $\int_0^4 x^5\, dx$ for $n = 4$.

18. $\int_0^{4/5} e^x\, dx$ for $n = 4$.

1.8 Truncation error and round-off error

We have discussed only two quadrature formulas and have estimated only the error due to replacing the integral by a weighted sum. This error is called the **truncation error.** In actual calculations, there is almost always another source of error. The values y_i cannot be computed precisely, but only with a certain accuracy, and in carrying out the arithmetical operations one must round off, that is, throw away digits. The two examples given above have been chosen so as to make the **round-off error** negligible. But in practice there will be a round-off error and one must be able to control it.

These matters, however, as well as more sophisticated quadrature formulas, are best learned together with the technique of automatic computing.

We already stated (in Chapter 6) that, having defined the transcendental functions $x \mapsto \log |x|$ and $x \mapsto \arctan x$, which have derivatives $1/x$ and $1/(1 + x^2)$, respectively, we are in a position to write down *an explicit formula for the indefinite integral of every rational function.*

The proof of this remarkable fact is based on two results. First of all, there are certain special rational functions, called **partial fractions,** that can all be integrated in closed form. Secondly, an arbitrary rational function can be written as a sum of a polynomial and of partial fractions.

§2 Integration of Rational Functions

2.1 Partial fractions

A partial fraction is a rational function of the form

$$f(x) = \frac{g(x)}{h(x)^n},$$

where n is a positive integer and either (1) g is a constant and h a nonconstant linear function, or (2) g is constant or a linear function, and h is a nonconstant quadratic function without (real) roots.

For instance,

$$\frac{2}{x - 3}, \quad \frac{6}{(x + 100)^{100}}, \quad \frac{7 + 3x}{2x^2 + 2x + 10}, \quad \frac{3 - 4x}{(x^2 + 2x + 10)^3}$$

are partial fractions. We do *not* call

$$\frac{1}{(x^2 - 3x + 2)^2}$$

a partial fraction, since the denominator has roots $x = 2$ and $x = 1$. Our first task is to learn how to integrate partial fractions.

2.2 Powers of linear functions in the denominator

Integration of the first type of partial fractions, with a power of a linear function in the denominator, presents no difficulty whatsoever. We need only apply the method of substitution. Given an integrand of the form $\alpha/(ax + b)^n$, where $\alpha \neq 0$, $a \neq 0$, we set $ax + b = t$ so that $adx = dt$ and obtain

$$\int \frac{\alpha \, dx}{(ax + b)^n} = \int \frac{\alpha(1/a) \, dt}{t^n} = \frac{\alpha}{a} \int t^{-n} \, dt$$

$$= \begin{cases} \dfrac{\alpha}{a(1 - n)(ax + b)^{n-1}} + C \text{ if } n \neq 1, \\[2ex] \dfrac{\alpha}{a} \log |ax + b| + C \text{ if } n = 1. \end{cases}$$

The primitive function is therefore either a rational function or a logarithmic function.

One should remember the method, not the result. It works also if n is not an integer but any rational number.

►*Examples* *1.* Compute

$$\int_3^5 \frac{2 \, dx}{3x - 6}.$$

We set $3x - 6 = t$, $3 \, dx = dt$, and obtain

$$\int_3^5 \frac{2 \, dx}{3x - 6} = \int_3^9 \frac{(2/3) \, dt}{t} = \frac{2}{3} \int_3^9 \frac{dt}{t} = \frac{2}{3} \log t \Big|_3^9 = \frac{2}{3} (\log 9 - \log 3)$$

$$= \frac{2}{3} \log 3 \approx \frac{2}{3} (1.0986) = .7324.$$

2. Evaluate $\displaystyle\int_{-10}^{-9} \frac{dx}{(2x + 21)^2}.$ We have

$$\int_{-10}^{-9} \frac{dx}{(2x + 21)^2} = \int_{x=-10}^{x=-9} \frac{(1/2) \, d(2x + 21)}{(2x + 21)^2}$$

$$= \int_{t=1}^{t=3} \frac{(1/2) \, dt}{t^2} = -\frac{1}{2} \frac{1}{t} \Big|_1^3 = \frac{1}{3}.$$

3. Find a primitive function of $(2x - 5)^{-6/7}$. We have

$$\int (2x - 5)^{-6/7}\, dx = \int (2x - 5)^{-6/7} \frac{1}{2}\, d(2x - 5)$$

$$= \frac{1}{2} \int t^{-6/7}\, dt = \frac{7}{2} t^{1/7} + C = \frac{7}{2}(2x + 5)^{1/7} + C.$$

4. Find $\displaystyle\int_0^2 \frac{dx}{(2 - x)^5}$.

Answer. The integrand is not bounded in the interval considered. The integral does not exist even as an improper integral. For if ϵ is a number such that $0 < \epsilon < 2$, we have, setting $2 - x = t$, $dx = -dt$,

$$\int_0^{2-\epsilon} \frac{dx}{(2 - x)^5} = -\int_2^\epsilon \frac{dt}{t^5} = \int_\epsilon^2 \frac{dt}{t^5} = -\frac{1}{4t^4}\Big|_\epsilon^2 = \frac{1}{4\epsilon^4} - \frac{1}{64}.$$

This number becomes arbitrarily large as ϵ becomes small. ◄

EXERCISES

Perform the following integrations:

1. $\displaystyle\int_1^3 \frac{1}{(2x - 1)^2}\, dx.$

2. $\displaystyle\int_0^1 \frac{1}{(2 - x)^3}\, dx.$

3. $\displaystyle\int_2^3 \frac{1}{3x - 5}\, dx.$

4. $\displaystyle\int (3x - 8)^{2/5}\, dx.$

5. $\displaystyle\int (3 - 2x)^{-2/3}\, dx.$

6. $\displaystyle\int_1^{6/5} \frac{1}{(5x - 4)^6}\, dx.$

7. $\displaystyle\int \frac{1}{(3x - 7)^5}\, dx.$

8. $\displaystyle\int \frac{1}{\sqrt{4 - 3x}}\, dx.$

9. $\displaystyle\int_0^2 \frac{1}{(2x + 4)^{1/3}}\, dx.$

10. $\displaystyle\int \frac{1}{8 - 7x}\, dx.$

2.3 Powers of quadratic functions in the denominator. First case

We discuss next how to find the primitive function of a partial fraction with a power of a quadratic polynomial, without (real) roots in the denominator. We look first at a fraction with a constant denominator, that is, we try to compute

(1) $$\int \frac{\alpha\, dx}{(ax^2 + bx + c)^n}, \qquad \text{where } 4ac - b^2 > 0.$$

The first step consists in reducing this to an indefinite integral of the form

(2) $$\int \frac{dx}{(1 + x^2)^n}.$$

This step is carried out as follows. We complete the square and write

$$(ax^2 + bx + c)^n = a^n\left(x^2 + \frac{b}{a}x + \frac{c}{a}\right)^n$$

$$= a^n\left[\left(x + \frac{b}{2a}\right)^2 + \frac{c}{a} - \frac{b^2}{4a^2}\right]^n = a^n\left[\left(x + \frac{b}{2a}\right)^2 + \frac{4ac - b^2}{4a^2}\right]^n.$$

Now set

$$q = \frac{\sqrt{4ac - b^2}}{2a} \qquad \text{and} \qquad \frac{x + (b/2a)}{q} = t.$$

We obtain that

$$(ax^2 + bx + c)^n = a^n q^{2n}(t^2 + 1)^n, \qquad dx = q\,dt,$$

and $$\int \frac{\alpha\,dx}{(ax^2 + bx + c)^n} = \frac{\alpha}{a^n q^{2n-1}} \int \frac{dt}{(1 + t^2)^n}.$$

Again, the method, rather than the result, ought to be remembered.

► **Example** Compute $\int_0^3 \dfrac{dx}{4x^2 + 6x + 9}$. We have

$$4x^2 + 6x + 9 = 4\left(x^2 + \frac{3}{2}x + \frac{9}{4}\right) = 4\left[\left(x + \frac{3}{4}\right)^2 + \frac{27}{16}\right]$$

$$= 4 \cdot \frac{27}{16}\left[\left(\frac{x + \frac{3}{4}}{\sqrt{\frac{27}{16}}}\right)^2 + 1\right] = \frac{27}{4}\left[\left(\frac{4\left(x + \frac{3}{4}\right)}{\sqrt{27}}\right)^2 + 1\right] = \frac{27}{4}(1 + t^2)$$

where we set

$$\frac{4(x + \frac{3}{4})}{\sqrt{27}} = t,$$

so that

$$dx = \frac{\sqrt{27}}{4}\,dt, \quad \text{and} \quad t = \frac{1}{\sqrt{3}} \text{ for } x = 0, \quad t = \frac{5}{\sqrt{3}} \text{ for } x = 3.$$

We obtain

$$\int_0^3 \frac{dx}{4x^2 + 6x + 9} = \frac{4}{27}\,\frac{\sqrt{27}}{4} \int_{1/\sqrt{3}}^{5/\sqrt{3}} \frac{dt}{1 + t^2}$$

$$= \frac{1}{\sqrt{27}} \left\{ \text{arc tan } \frac{5}{\sqrt{3}} - \text{arc tan } \frac{1}{\sqrt{3}} \right\}.\blacktriangleleft$$

EXERCISES

Perform the following integrations:

11. $\int_0^1 \frac{dx}{x^2 + x + 1}$.

14. $\int_0^{\sqrt{3}} \frac{dx}{x^2 + \sqrt{3}x + 1}$.

12. $\int \frac{dx}{x^2 - x + 2}$.

15. $\int \frac{dx}{5x^2 - 2x + 1}$.

13. $\int \frac{dx}{2x^2 + 2x + 1}$.

16. $\int_0^1 \frac{dx}{4x^2 + 9}$.

2.4 A recursion formula

Next we must learn how to evaluate the integral (2). If $n = 1$, we obtain, except for a constant of integration, the arc tangent function. If $n > 1$, we write $n = m + 1$ and derive a **recursion formula** expressing the integral (2) with $n = m + 1$ by a similar integral with $n = m$. Repeating this process m times, we arrive at the integral (2) with $n = 1$ and thus at the arc tangent.

The recursion formula is obtained by noting that

$$\frac{d\,\dfrac{1}{(1 + x^2)^m}}{dx} = -\frac{2mx}{(1 + x^2)^{m+1}}$$

so that

(3)
$$\frac{dx}{(1 + x^2)^{m+1}} = -\frac{1}{2mx}\,d\,\frac{1}{(1 + x^2)^m}.$$

On the other hand,

$$\frac{1}{(1 + x^2)^m} = \frac{1 + x^2}{(1 + x^2)^{m+1}} = \frac{1}{(1 + x^2)^{m+1}} + \frac{x^2}{(1 + x^2)^{m+1}}.$$

Therefore, using (3) and integration by parts,

$$\int \frac{dx}{(1 + x^2)^{m+1}} = \int \frac{dx}{(1 + x^2)^m} - \int \frac{x^2\, dx}{(1 + x^2)^{m+1}}$$

$$= \int \frac{dx}{(1 + x^2)^m} - \int x^2 \left(-\frac{1}{2mx}\, d\, \frac{1}{(1 + x^2)^m} \right)$$

$$= \int \frac{dx}{(1 + x^2)^m} + \frac{1}{2m} \int x d\, \frac{1}{(1 + x^2)^m}$$

$$= \int \frac{dx}{(1 + x^2)^m} + \frac{1}{2m}\, \frac{x}{(1 + x^2)^m} - \frac{1}{2m} \int \frac{1}{(1 + x^2)^m}\, dx$$

$$= \frac{1}{2m}\, \frac{x}{(1 + x^2)^m} + \left(1 - \frac{1}{2m} \right) \int \frac{dx}{(1 + x^2)^m}.$$

Thus

$$(4) \qquad \int \frac{dx}{(1 + x^2)^{m+1}} = \frac{1}{2m}\, \frac{x}{(1 + x^2)^m} + \left(1 - \frac{1}{2m} \right) \int \frac{dx}{(1 + x^2)^m}.$$

This is the desired reduction formula. In terms of definite integrals, it reads

$$\int_a^b \frac{dx}{(1 + x^2)^{m+1}} = \frac{1}{2m}\, \frac{x}{(1 + x^2)^m} \bigg|_a^b + \left(1 - \frac{1}{2m} \right) \int_a^b \frac{dx}{(1 + x^2)^m}.$$

The formula is valid also if m is *not* an integer, as long as $m \neq 0$.

The derivation just given involves a trick. Once the recursion formula (4) is found, however, the verification is direct and easy; this involves only differentiations. The formula asserts that

$$\frac{1}{(1 + x^2)^{m+1}} - \left(1 - \frac{1}{2m} \right) \frac{1}{(1 + x^2)^m} = \frac{d}{dx}\, \frac{1}{2m}\, \frac{x}{(1 + x^2)^m},$$

and this is so.

It is up to the reader whether he prefers to remember that there is a recursion formula like (4), which can be looked up in a calculus text or a table of integrals; or whether to remember the method of obtaining the recursion formula, or whether to remember the formula itself.

▶ **Examples** **1.** Compute $\int_{-1}^0 \dfrac{dx}{(x^2 + 2x + 2)^3}$. For $t = x + 1, dt = dx$, we have,

$$\int_{-1}^0 \frac{dx}{(x^2 + 2x + 2)^3} = \int_{-1}^0 \frac{dx}{((x + 1)^2 + 1)^3} = \int_0^1 \frac{dt}{(1 + t^2)^3},$$

$$= \frac{1}{4} \frac{t}{(1 + t^2)^2} \Big|_0^1 + \frac{3}{4} \int_0^1 \frac{dt}{(1 + t^2)^2}$$

$$= \frac{1}{4} \frac{t}{(1 + t^2)^2} \Big|_0^1 + \frac{3}{4} \cdot \frac{1}{2} \frac{t}{1 + t^2} \Big|_0^1 + \frac{3}{4} \cdot \frac{1}{2} \int_0^1 \frac{dt}{1 + t^2}$$

$$= \frac{1}{4} \Big(\frac{1}{4} - 0 \Big) + \frac{3}{8} \Big(\frac{1}{2} - 0 \Big) + \frac{3}{8} \cdot \frac{\pi}{4} = \frac{1}{4} + \frac{3\pi}{32}.$$

2. Find a primitive function of $(x^2 + 2x + 2)^{-3}$. We repeat the computation made above using indefinite integrals and obtain

$$\int \frac{dx}{(x^2 + 2x + 2)^3} = \int \frac{dt}{(1 + t^2)^3} = \cdots$$

$$= \frac{1}{4} \frac{t}{(1 + t^2)^2} + \frac{3}{8} \frac{t}{1 + t^2} + \frac{3}{8} \text{ arc tan } t + C$$

$$= \frac{1}{4} \frac{x + 1}{(x^2 + 2x + 2)^2} + \frac{3}{8} \frac{x + 1}{x^2 + 2x + 2} + \frac{3}{8} \text{ arc tan } (x + 1) + C. \blacktriangleleft$$

EXERCISES

Perform the following integrations:

17. $\int_0^{1/2} \frac{dx}{(4x^2 + 1)^2}$.

21. $\int_0^{1/3} \frac{dx}{(9x^2 + 1)^4}$.

► 18. $\int \frac{dx}{(x^2 + 4)^3}$.

22. $\int_0^2 \frac{dx}{(5x^2 - 2x + 1)^2}$.

19. $\int_{-2}^0 \frac{dx}{(x^2 + 6x + 10)^2}$.

23. $\int \frac{dx}{(x^2 + \sqrt{3}x + 1)^3}$.

20. $\int \frac{dx}{(2x^2 + 2x + 1)^2}$.

24. $\int_{1/2}^1 \frac{dx}{(4x^2 + 4x + 5)^3}$.

2.5 Powers of quadratic functions in the denominator. Second case

Now we must consider integrals of the form

$$\int \frac{\alpha x + \beta}{(ax^2 + bx + c)^n} dx, \qquad \text{where } \alpha \neq 0, 4ac - b^2 > 0.$$

We write (again using a "trick")

$$\int \frac{\alpha x + \beta}{(ax^2 + bx + c)^n} dx = \int \frac{\frac{\alpha}{2a}(2ax + b) + \beta - \frac{\alpha b}{2a}}{(ax^2 + 2bx + c)^n} dx$$

$$= \frac{\alpha}{2a} \int \frac{(2ax + b)\, dx}{(ax^2 + bx + c)^n} + \Big(\beta - \frac{\alpha b}{2a} \Big) \int \frac{dx}{(ax^2 + 2bx + c)^n}.$$

We learned above how to evaluate the second integral. The first can be treated by substitution. We set $t = ax^2 + 2bx + c$. Then

$$\int \frac{(2ax + b)\, dx}{(ax^2 + bx + c)^n} = \int \frac{d(ax^2 + bx + c)}{(ax^2 + bx + c)^n} = \int \frac{dt}{t^n}$$

$$= \begin{cases} \log |t| + C & \text{if } n = 1 \\ \dfrac{1}{(1 - n)t^{n-1}} + C & \text{if } n \neq 1 \end{cases}$$

$$= \begin{cases} \log |ax^2 + bx + c| + C & \text{if } n = 1 \\ \dfrac{1}{(1 - n)(ax^2 + bx + c)^{n-1}} + C & \text{if } n \neq 1. \end{cases}$$

We have now established that:

A *primitive function of any partial fraction is the sum of a rational function, a logarithmic function, and an arc tangent function.*

Of course, in any given case, not all three types need be present. [By a logarithmic function we mean, in this connection, a function of the form $x \mapsto \alpha \log |ax + b|$ or $x \mapsto \alpha \log |ax^2 + bx + c|$. By an arc tangent function we mean a function of the form $x \mapsto \alpha$ arc tan $(ax + b)$.]

▶*Examples* 1. Evaluate $\int_0^2 \dfrac{(x + 3)\, dx}{x^2 + 2x + 2}$. We have

$$\int_0^2 \frac{(x + 3)\, dx}{x^2 + 2x + 2} = \int_0^2 \frac{\frac{1}{2}(2x + 2) + 2}{x^2 + 2x + 2}\, dx$$

$$= \frac{1}{2} \int_0^2 \frac{(2x + 2)\, dx}{x^2 + 2x + 2} + 2 \int_0^2 \frac{dx}{x^2 + 2x + 2}$$

$$\frac{1}{2} \int_{x=0}^{x=2} \frac{d(x^2 + 2x + 2)}{x^2 + 2x + 2} + 2 \int_{x=0}^{x=2} \frac{d(x + 1)}{(x + 1)^2 + 1}$$

$$= \frac{1}{2} \int_2^{10} \frac{dt}{t} + 2 \int_1^3 \frac{ds}{s^2 + 1}$$

(since if $x = 0$, then $x^2 + 2x + 2 = 2$, $x + 1 = 1$ and if $x = 2$, then $x^2 + 2x + 2 = 10$, $x + 1 = 3$)

$$= \frac{1}{2} \log t \Big|_2^{10} + 2 \text{ arc tan } s \Big|_1^3 \approx 1.7.$$

2. Find the primitive function of $\dfrac{x + 3}{(2x^2 + 18)^2}$. We have

$$\int \frac{(x+3)\,dx}{(2x^2+18)^2} = \int \frac{\frac{1}{4}(4x)+3}{(2x^2+18)^2}\,dx$$

$$= \frac{1}{4}\int \frac{4x\,dx}{(2x^2+18)^2} + 3\int \frac{dx}{(2x^2+18)^2}$$

$$= \frac{1}{4}\int \frac{d(2x^2+18)}{(2x^2+18)^2} + \frac{3}{4}\int \frac{dx}{(x^2+9)^2},$$

setting $t = 2x^2 + 18$,

$$= \frac{1}{4}\int \frac{dt}{t^2} + \frac{3}{4}\int \frac{dx}{\left[9\left(\left(\frac{x}{3}\right)^2+1\right)\right]^2},$$

setting $s = \frac{x}{3}$,

$$= \frac{1}{4}\int \frac{dt}{t^2} + \frac{9}{4}\int \frac{ds}{[9(s^2+1)]^2}$$

$$= \frac{1}{4}\int \frac{dt}{t^2} + \frac{1}{36}\int \frac{ds}{(s^2+1)^2},$$

by (4),

$$= \frac{1}{4}\int \frac{dt}{t^2} + \frac{1}{36}\left\{\frac{1}{2}\frac{s}{s^2+1} + \frac{1}{2}\int \frac{ds}{s^2+1}\right\}$$

$$= \frac{-1}{4}\cdot\frac{1}{t} + \frac{1}{72}\cdot\frac{s}{s^2+1} + \frac{1}{72}\arctan s + C$$

$$= \frac{-1}{8x^2+72} + \frac{x}{24x^2+216} + \frac{1}{72}\arctan\frac{x}{3} + C. \blacktriangleleft$$

EXERCISES

Perform the following integrations; check by differentiation.

25. $\displaystyle\int_0^{1/2} \frac{3x+2}{4x^2+1}\,dx.$

26. $\displaystyle\int \frac{3x-1}{x^2+4x+5}\,dx.$

27. $\displaystyle\int_0^2 \frac{2x+1}{(5x^2-2x+1)^2}\,dx.$ See Exercise 22.

28. $\displaystyle\int \frac{1-3x}{4x^2+3x+1}\,dx.$

29. $\displaystyle\int \frac{x-2}{(2x^2+2x+1)^2}\,dx.$ See Exercise 20.

30. $\displaystyle\int_{1/2}^1 \frac{x+3}{(4x^2+4x+5)^3}\,dx.$ See Exercise 24.

31. $\displaystyle\int \frac{x}{(9x^2+6x+2)^3}\,dx.$

32. $\displaystyle\int \frac{4x-3}{3x^2+3x+1}\,dx.$

33. $\displaystyle\int_1^2 \frac{x}{2x^2-4x+3}\,dx.$

► 34. $\displaystyle\int \frac{2-x}{(x^2-2x+2)^3}\,dx.$

2.6 Main theorem

The key to integrating all rational functions in closed form is the following proposition.

Theorem 1 EVERY RATIONAL FUNCTION THAT IS NOT A POLYNOMIAL CAN BE WRITTEN, UNIQUELY, AS THE SUM OF A POLYNOMIAL AND ONE OR SEVERAL PARTIAL FRACTIONS.

Assuming Theorem 1, and noting the results of §§2.2 to 2.5, we obtain at once the following theorem.

Theorem 2 A PRIMITIVE FUNCTION OF A RATIONAL FUNCTION IS ALWAYS A SUM OF A RATIONAL FUNCTION, LOGARITHMIC FUNCTIONS, AND ARC TANGENT FUNCTIONS.

In order to find an explicit primitive function for a rational function, one must decompose the function into partial fractions. This can be done in practice only for relatively simple cases. Thus it may be necessary to use numerical integration even for rational functions, and the main significance of Theorem 2 is theoretical.

The proof of Theorem 1 will not be given in this book. This proof involves the so-called **fundamental theorem of algebra,** which asserts that every nonconstant polynomial has roots, provided that one has extended the real number system and also admits so-called complex numbers. The "real" form of the theorem reads:

EVERY POLYNOMIAL (OF POSITIVE DEGREE, WITH REAL COEFFICIENTS) CAN BE WRITTEN AS A PRODUCT OF LINEAR AND QUADRATIC POLYNOMIALS.

There are many proofs of the fundamental theorem of algebra; most of them are due to Gauss, the greatest mathematician of the 19th century. The truth of the theorem had been suspected, however, long before Gauss. Indeed, during the 16th century, Italian mathematicians found explicit formulas (usually associated with the name of Cardano) for the roots of third and fourth degree equations thus generalizing the age-old formula for solving quadratic equations. But the search for an explicit formula for solving all fifth degree equations was unsuccessful, and finally Abel (at the age of 22) showed that there can be no such formula.

An even more remarkable discovery was made by Galois (who died before reaching the age of 21). Galois found that, given any specific equation of degree $n \geq 5$, one cannot, except under very special circumstances, obtain the zeros (roots) from the coefficients by performing rational operations and by extracting roots.

HIERONIMO CARDANO (1501–1576), a colorful Renaissance character and a many-sided scientist, published in 1545 a method for solving cubic equations. He acknowledged that a special case of this method had been communicated to him, under pledge of secrecy, by Tartaglia. A bitter polemic followed between Tartaglia and Cardano's pupil, Ferrari.

Ferrari himself was a talented mathematician, who discovered that solving fourth degree equations can be reduced to solving cubics.

NIELS HENRIK ABEL (1802–1829). His work on fifth degree equations is only one of his great achievements. Abel contributed to the rigorous theory of infinite series, and his discovery of "elliptic" and other transcendental functions initiated a new era in mathematical analysis.

At 16, Abel began, under the influence of a perceptive teacher, to read the works of Newton, Euler, and Lagrange; after a few years he started to make his discoveries. Meanwhile his father died and Abel became responsible for a large family; he was poor for the rest of his life. A Norwegian government subsidy enabled him to visit Germany and France, but the leading mathematicians there failed to recognize his genius. Abel died of tuberculosis at the age of 27.

ÉVARISTE GALOIS (1811–1832) was persecuted by his government for radicalism, and by some of his teachers for impertinence. He was twice denied admission to the prestigious École Polytechnique, was expelled from the École Normale, and he was in jail for several months. On May 30, 1832, Galois fought a duel and received a fatal wound. The night before, he composed a farewell letter to a friend, describing his discoveries. This letter contains the "Galois theory," one of the foundation stones of modern algebra.

2.7 Decomposition into partial fractions

Let $x \mapsto f(x)$ be a given rational function. Then $f(x) = \phi(x)/\psi(x)$, where ϕ and ψ are polynomials. We may assume that f is not a polynomial, since we already know how to integrate polynomials. Hence we assume that the degree of the polynomial $\psi(x)$ is not zero and that ϕ is not divisible by ψ.

If the degree of ϕ is not less than that of ψ, we apply long division and obtain

$$\phi(x) = q(x)\psi(x) + r(x),$$

where q and r are polynomials and the degree of r is less than that of ψ. Thus

$$f(x) = \frac{\phi(x)}{\psi(x)} = q(x) + \frac{r(x)}{\psi(x)}.$$

In order to integrate f, it suffices to integrate r/ψ, since we know how to integrate q. Thus we may assume, to begin with, that the degree of ϕ is less than that of ψ, and that the two polynomials have no common linear or quadratic factors.

The first and most difficult step in representing ϕ/ψ as a sum of partial fractions is to represent the denominator ψ as a product of a constant A and factors, each of which is either a power of a linear function $x - \alpha$, or a power of a quadratic function $x^2 + \beta x + \gamma$ with no (real) roots. In other words, we must write

(5)
$$\psi(x) =$$
$$A(x - \alpha_1)^{\nu_1} \cdots (x - \alpha_k)^{\nu_k}(x^2 + \beta_1 x + \gamma_1)^{\mu_1} \cdots (x^2 + \beta_l x + \gamma_l)^{\mu_l},$$

where the ν's and μ's are positive integers, the α's are distinct numbers, and the quadratic functions $x^2 - \beta x - \gamma$ are all distinct. The α's are the real roots of ψ and ν_i is the multiplicity of α_i. Also $\nu_i + \cdots + \nu_k + 2\mu_1 + \cdots + 2\mu_l$ is the degree of ψ. That ψ can be represented, uniquely, in this form is a consequence of the fundamental theorem of algebra, as we noted above. Unfortunately, there is no simple way of finding the decomposition of a given polynomial ψ into linear and quadratic factors. In general, the numbers α, β, and γ can be found only approximately.

2.8 Indeterminate coefficients

We assume that the denominator ψ is already given in the form (5). We may also assume that $A = 1$, since we can divide both numerator and denominator by A. Now we must find partial fractions whose sum is $\phi(x)/\psi(x)$. Only powers of the linear and quadratic factors in (5), with exponents not greater than those in (5), can occur as denominators. Once this is realized, finding the actual representation of ϕ/ψ as a sum of partial fractions is reduced to solving a system of simultaneous linear equations. This **method of indeterminate coefficients** is best explained by examples.

Consider the rational function

$$\frac{x^4 + x^2 + 2x + 1}{x^4(x + 1)^2}$$

The factors x and $x + 1$ in the denominator occur with exponents 4 and 2, respectively. Therefore, if we can write the given rational function as a sum of partial fractions, we may encounter the denominators x^4, x^3, x^2, x, $(x + 1)^2$ and $x + 1$, and no others. Since the factors x and $x + 1$ are linear (and not quadratic), the numerators of the partial fractions will be constants (and not linear functions). Thus Theorem 1 tells us that there exist numbers A, B, C, D, E, and F such that

$$\frac{x^4 + x^2 + 2x + 1}{x^4(x + 1)^2} = \frac{A}{x^4} + \frac{B}{x^3} + \frac{C}{x^2} + \frac{D}{x} + \frac{E}{(x + 1)^2} + \frac{F}{(x + 1)}.$$

(We do not yet know what the numbers A, B, C, D, E, and F are; hence the name "indeterminate coefficients.") If we add the six partial fractions, we obtain

$$\frac{A(x+1)^2 + Bx(x+1)^2 + Cx^2(x+1)^2 + Dx^3(x+1)^2 + Ex^4 + Fx^4(x+1)}{x^4(x+1)^2}.$$

The numerator equals

$$(D + F)x^5 + (C + 2D + E)x^4 + (B + 2C + D)x^3 +$$
$$(A + 2B + C)x^2 + (2A + B + F)x + A.$$

We want this numerator to be $x^4 + x^2 + 2x + 1$, that is, we require that

$$D + F = 0$$
$$C + 2D + E = 1$$
$$B + 2C + D = 0$$
$$A + 2B + C = 1$$
$$2A + B + F = 2$$
$$A = 1.$$

Solving this system of equations, we obtain $A = 1$, $B = C = D = 0$, $E = 1$, $F = 0$. Thus

$$\frac{x^4 + x^2 + 2x + 1}{x^4(x + 1)^2} = \frac{1}{x^4} + \frac{1}{(x + 1)^2}.$$

Consider next

$$\frac{2x^3}{x^4 - 1}.$$

The denominator can be written as

$$x^4 - 1 = (x^2 - 1)(x^2 + 1) = (x - 1)(x + 1)(x^2 + 1).$$

Each factor in the denominator occurs in the first power; the last factor is a quadratic function without real roots. Therefore, according to Theorem 1, there are numbers A, B, C, and D such that

$$\frac{2x^3}{x^4 - 1} = \frac{2x^3}{(x - 1)(x + 1)(x^2 + 1)} = \frac{A}{x - 1} + \frac{B}{x + 1} + \frac{Cx + D}{x^2 + 1}$$

$$= \frac{A(x + 1)(x^2 + 1) + B(x - 1)(x^2 + 1) + (Cx + D)(x - 1)(x + 1)}{x^4 - 1}.$$

The numerator is

$$A(x^3 + x^2 + x + 1) + B(x^3 - x^2 + x - 1) + (Cx + D)(x^2 - 1)$$
$$= (A + B + C)x^3 + (A - B + D)x^2 + (A + B - C)x + A - B - D.$$

We want the numerator to be $2x^3$, so that we need

$$A + B + C = 2,$$
$$A - B + D = 0,$$
$$A + B - C = 0,$$
$$A - B - D = 0.$$

Solving this system of equations, we obtain $A = B = \frac{1}{2}$, $C = 1$, $D = 0$. Thus

$$(6) \qquad \frac{2x^3}{x^4 - 1} = \frac{\frac{1}{2}}{x - 1} + \frac{\frac{1}{2}}{x + 1} + \frac{x}{x^2 + 1}$$

is the desired representation.

As a final example, we treat

$$\frac{x^4 + 2x^2 + x + 1}{(x^2 + 1)^3}.$$

Here the denominator is the third power of a quadratic function without real roots. Hence, if the given function is represented as a sum of partial functions, the following denominators may occur: $(x^2 + 1)^3$, $(x^2 + 1)^2$, and $x^2 + 1$; the corresponding numerators will be linear functions, which may happen to be constants. Hence, by Theorem 1, these are numbers A, B, C, D, E, and F such that

$$\frac{x^4 + 2x^2 + x + 1}{(x^2 + 1)^3} = \frac{Ax + B}{(x^2 + 1)^3} + \frac{Cx + D}{(x^2 + 1)^2} + \frac{Ex + F}{x^2 + 1}.$$

Adding the partial fractions, we get the numerator

$$Ax + B + (Cx + D)(x^2 + 1) + (Ex + F)(x^2 + 1)^2$$
$$= Ax + B + (Cx + D)(x^2 + 1) + (Ex + F)(x^4 + 2x^2 + 1)$$
$$= Ex^5 + Fx^4 + (C + 2E)x^3 + (D + 2F)x^2 + (A + C + E)x$$
$$\qquad\qquad\qquad\qquad\qquad\qquad\qquad + (B + D + F).$$

This should be $x^4 + 3x^2 + 1$. Hence

$$E = 0,$$
$$F = 1,$$
$$C + 2E = 0,$$
$$D + 2F = 2,$$
$$A + C + E = 1,$$
$$B + D + F = 1.$$

Solving this system, we obtain $A = 1$, $B = C = D = E = 0$, $F = 1$. Thus

$$\frac{x^4 + 2x^2 + x + 1}{(x^2 + 1)^3} = \frac{x}{(x^2 + 1)^3} + \frac{1}{x^2 + 1}.$$

These examples show how simple the method is and how tedious it often becomes.

We have now learned how to integrate any rational function.

▶*Examples*   *1.* Find a primitive function of $\dfrac{x^7 + x^3}{x^4 - 1}$.

Solution.   Since the degree of the numerator exceeds that of the denominator, we use long division and obtain

$$x^7 + x^3 = x^3(x^4 - 1) + 2x^3.$$

Therefore, using (6),

$$\int \frac{x^7 + x^3}{x^4 - 1} \, dx = \int \left(x^3 + \frac{2x^3}{x^4 - 1} \right) dx$$

$$= \int \left(x^3 + \frac{1}{2(x-1)} + \frac{1}{2(x+1)} + \frac{x}{x^2+1} \right) dx$$

$$= \int x^3 \, dx + \frac{1}{2} \int \frac{dx}{x-1} + \frac{1}{2} \int \frac{dx}{x+1} + \int \frac{x \, dx}{x^2+1}$$

$$= \int x^3 \, dx + \frac{1}{2} \int \frac{d(x-1)}{x-1} + \frac{1}{2} \int \frac{d(x+1)}{x+1} + \frac{1}{2} \int \frac{d(x^2+1)}{x^2+1}$$

$$= \frac{x^4}{4} + \frac{1}{2} \log |x-1| + \frac{1}{2} \log |x+1| + \frac{1}{2} \log |x^2+1| + \text{const.}$$

In this particular case, it was not necessary to use partial fractions at all, for

$$\int \frac{2x^3 \, dx}{x^4 - 1} = \frac{1}{2} \int \frac{d(x^4 - 1)}{x^4 - 1} = \frac{1}{2} \log |x^4 - 1| + C.$$

The reader should check that this answer is the same as the one obtained before.

2. Evaluate $\int_1^2 \frac{3x^2 + 2x + 5}{x^3 + x^2 + x + 1} \, dx$.

Solution.   The denominator can be factored into $(x^2 + 1)(x + 1)$. Hence

$$\frac{3x^2 + 2x + 5}{x^3 + x^2 + x + 1} = \frac{A}{x+1} + \frac{Bx + C}{x^2 + 1}$$

with as yet undetermined coefficients A, B, and C. Adding the partial fractions gives the denominator

$$A(x^2 + 1) + (Bx + C)(x + 1) = (A + B)x^2 + (B + C)x + A + C.$$

This equals $3x^2 + 2x + 5$ if

$$A + B = 3, \qquad B + C = 2, \qquad A + C = 5,$$

that is, for $A = 3$, $B = 0$, $C = 2$. Thus

$$\int_1^2 \frac{3x^2 + 2x + 5}{x^3 + x^2 + x + 1}\, dx = \int_1^2 \left(\frac{3}{x + 1} + \frac{2}{x^2 + 1}\right) dx$$

$$= 3 \log |x + 1| \Big|_1^2 + 2 \text{ arc tan } x \Big|_1^2 \approx .66.$$

This problem could have been done differently. Noting that

$$d(x^3 + x^2 + x + 1) = 3x^2 + 2x + 1,$$

we set

$$\int_1^2 \frac{3x^2 + 2x + 5}{x^3 + x^2 + x + 1}\, dx$$

$$= \int_1^2 \frac{3x^2 + 2x + 1}{x^3 + x^2 + x + 1}\, dx + 4 \int_1^2 \frac{dx}{x^3 + x^2 + x + 1}.$$

Then, using partial fractions,

$$\frac{1}{x^3 + x^2 + x + 1} = \frac{A}{x + 1} + \frac{Bx + C}{x^2 + 1}$$

and $A = C = \frac{1}{2}$, $B = -\frac{1}{2}$. Thus we have

$$\int_1^2 \frac{d(x^3 + x^2 + x + 1)}{x^3 + x^2 + x + 1} + 4 \int_1^2 \left(\frac{1}{2(x + 1)} + \frac{1 - x}{2(x^2 + 1)}\right) dx$$

$$= \log |x^3 + x^2 + x + 1| \Big|_1^2 + 2 \log |x + 1| \Big|_1^2$$

$$+ 2 \text{ arc tan } x \Big|_1^2 - \log |x^2 + 1| \Big|_1^2$$

$$= 3 \log |x + 1| \Big|_1^2 + 2 \text{ arc tan } x \Big|_1^2,$$

as before. ◀

EXERCISES

Perform the following integrations:

35. $\int \dfrac{x^2 - 2}{x^3 - 4x}\, dx.$

36. $\int_1^{\sqrt{3}} \dfrac{x^4 + 3x^2 + x + 1}{x^3 + x}\, dx.$

37. $\int_0^2 \dfrac{x^3 + 2x^2 + 5x + 1}{x^2 + 2x + 1}\, dx.$

38. $\int \dfrac{3x^2 - 6x + 2}{x^3 - 2x^2 + x}\, dx.$

39. $\int \dfrac{x^3 + 4x^2 + 5x - 2}{x^4 - 1}\, dx.$

40. $\int_0^{\sqrt{3}} \dfrac{x^3 + x^2 + x + 2}{(x^2 + 1)(x^2 + 2)}\, dx.$

41. $\int \dfrac{2x^2 + 4x - 1}{(x^2 + 2x + 2)(x - 1)}\, dx.$

48. $\int_0^1 \dfrac{2x^3 + x}{x^2 - x + 1}\, dx.$

▶ 42. $\int \dfrac{x^4 + 3x^3 + 4x^2 + 4x + 1}{(x^2 + x + 1)^3}\, dx.$

49. $\int \dfrac{4x^4 + 8x^3 + 9x^2 + 4x}{(x + 1)(2x^2 + 2x + 1)^2}\, dx.$

43. $\int \dfrac{2x^2 + 2x - 1}{x^3 - 1}\, dx.$

50. $\int \dfrac{x^2 + x - 3}{x^3 - 2x^2 - x + 2}\, dx.$

44. $\int \dfrac{2x^3 + 7x^2 + 4x - 3}{x^3 + 3x^2 + 3x}\, dx.$

51. $\int \dfrac{x^3 - x^2 + 2x}{x^4 + x^2 + 1}\, dx.$

45. $\int_3^4 \dfrac{5x^2 - 4x + 4}{(x^2 - 4)^2}\, dx.$

52. $\int \dfrac{2x^3 - 3x^2 + 2x}{(x^2 + 1)^3}\, dx.$

46. $\int \dfrac{2x^4 - x^3 + 2}{x^2(x^2 + 1)^2}\, dx.$

53. $\int_{\sqrt{3}}^3 \dfrac{x^3 + 4x^2 + 7x + 12}{(x^2 + x)(x^2 + 3)}\, dx.$

47. $\int_0^2 \dfrac{x^3 - x^2 + 3x - 1}{(x^2 + 4)(x^2 + x + 1)}\, dx.$

54. $\int \dfrac{1 - 4x + 8x^2 - 8x^3}{x^4(2x^2 - 2x + 1)^2}\, dx.$

In this section, we enumerate various classes of integrals that can be reduced, by an appropriate substitution, to an integral of a rational function, and hence evaluated in closed form. In each case, the proof will provide a method of carrying out this evaluation. But often it is simpler to compute the integral by some special method rather than by proceeding with the rationalization process. We shall illustrate this by several examples.

In studying this section, the reader should strive to develop the ability to recognize integrals that can be rationalized. Once it is known that a given integral can be computed in closed form (that is, by a formula involving elementary functions), one can perform the integration in various ways. In particular, one can consult a table of integrals (for instance, the one in the *Rinehart Mathematical Tables*[*]). It is advisable always to check the result of integration by differentiating.

§3 Rationalizable Integrals

3.1 Notations

In order to describe rationalizable integrals, we use the following conventions. The symbol $R(u)$ will always denote a rational function of the variable u. The symbol $R(u, v)$ will always denote a rational function of two variables, u and v. This means that $R(u, v)$ is a quotient of two polynomials in u and v. Note that if $x \mapsto f(x)$ and $x \mapsto g(x)$ are rational functions, then $x \mapsto R(f(x), g(x))$ is a rational function of the variable x. For instance, suppose that

$$R(u, v) = \frac{u^2 - v^2}{2uv}.$$

[*] HAROLD D. LARSEN, *Rinehart Mathematical Tables, Formulas and Curves.* New York: Holt, Rinehart and Winston, Inc., 1962, pp. 234–266.

Then $R(2x^3, 4x^2) = \dfrac{1}{4}x - \dfrac{1}{x}$, $R(\cos t, \sin t) = \cot 2t$.

In what follows, we shall not indicate where the functions considered are defined, since this will be clear from the context. If we consider the function $\sqrt{1 - x^2}$, for instance, it is clear that x stands for a number such that $|x| \leq 1$.

3.2 Rational functions of the exponential function

We begin with a particularly simple case of a rationalizable integral.

(I) AN INTEGRAL OF THE FORM $\int R(e^x)\, dx$ CAN BE RATIONALIZED BY THE SUBSTITUTION

$$y = e^x.$$

Proof.   We have $x = \log y$, therefore $dx = dy/y$ and

$$\int R(e^x)\, dx = \int R(y)\frac{1}{y}\, dy.$$

This is an integral of a rational function.

►*Examples*   *1.* $\displaystyle \int \sinh x\, dx = \frac{1}{2}\int (e^x - e^{-x})\, dx = \frac{1}{2}\int \left(y - \frac{1}{y}\right)\frac{dy}{y}$

$[\text{where } y = e^x] = \dfrac{1}{2}\int dy - \dfrac{1}{2}\int \dfrac{dy}{y^2} = \dfrac{1}{2}y + \dfrac{1}{2y} + C = \dfrac{e^x}{2} + \dfrac{e^{-x}}{2} + C$

$= \cosh x + C$, as expected.

2. We have

$$\int \tanh x\, dx = \int \frac{e^x - e^{-x}}{e^x + e^{-x}}\, dx = \int \frac{y - \dfrac{1}{y}}{y + \dfrac{1}{y}}\, \frac{dy}{y} \text{ [where } y = e^x]$$

$$= \int \frac{y^2 - 1}{y^3 + y}\, dy = \int \frac{y^2 - 1}{y(1 + y^2)}\, dy = \int \left(\frac{A}{y} + \frac{B + Cy}{1 + y^2}\right) dy,$$

which, evaluating the indeterminate coefficients, becomes

$$\int \left(-\frac{1}{y} + \frac{2y}{1 + y^2}\right) dy = -\int \frac{dy}{y} + \int \frac{d(1 + y^2)}{(1 + y^2)}$$

$$= -\log|y| + \log|1 + y^2| + C$$

$$= \log\left|\frac{1 + y^2}{y}\right| + C = \log\left|\frac{1}{y} + y\right| + C$$

$$= \log|e^x + e^{-x}| + C = \log\cosh x + C',$$

where C' is another constant. The reader should justify the last step.

Of course, we could have proceeded more simply by noting that

$$\int \tanh x \, dx = \int \frac{\sinh x}{\cosh x} \, dx = \int \frac{d\cosh x}{\cosh x}.$$

This shows that following the general procedure may not be the shortest way to the answer. ◄

EXERCISES

Perform the following integrations:

1. $\int \dfrac{e^{2x}}{e^{2x} - 1} \, dx.$

2. $\int_0^\infty \dfrac{e^x}{e^{2x} + 1} \, dx.$

3. $\int \dfrac{e^x}{(e^{2x} - e^x + 1)^2} \, dx.$

4. $\int \dfrac{2e^{2x} + e^x + 1}{e^{2x} + 1} \, dx.$

5. $\int \dfrac{2\cosh x + 3\sinh x}{3\cosh x + 5\sinh x} \, dx.$

6. $\int \dfrac{1 + \sinh x}{1 + \cosh x} \, dx.$

3.3 Roots of linear functions

(II) AN INTEGRAL $\int R(x, \sqrt[q]{\alpha x + \beta}) \, dx$, $q > 1$ AN INTEGER, $\alpha \neq 0$, CAN BE RATIONALIZED BY THE SUBSTITUTION

$$y = \sqrt[q]{\alpha x + \beta}.$$

Proof. We have

$$\alpha x + \beta = y^q, \qquad x = \frac{y^q}{\alpha} - \frac{\beta}{\alpha},$$

hence

$$dx = \frac{q}{\alpha} y^{q-1} \, dy$$

and

$$\int R(x, \sqrt[q]{\alpha x + \beta}) \, dx = \int R\left(\frac{y^q}{\alpha} - \frac{\beta}{\alpha}, y\right) \frac{q}{\alpha} y^{q-1} \, dy;$$

this is an integral of a rational function.

▶ **Examples** **1.** Evaluate $\int_{3/2}^{2} \sqrt[3]{2x - 3}x\, dx$.

Solution. We set

$$y = \sqrt[3]{2x - 3}, \qquad y^3 = 2x - 3, \qquad x = \frac{1}{2}y^3 + \frac{3}{2},$$

so that $$dx = \frac{3}{2}y^2\, dy$$

and

$$\int_{3/2}^{2} \sqrt[3]{2x - 3}\, dx = \int_{0}^{1} y\left(\frac{1}{2}y^3 + \frac{3}{2}\right)\frac{3}{2}y^2\, dy$$

$$= \int_{0}^{1}\left(\frac{3}{4}y^6 + \frac{9}{4}y^3\right) dy = \frac{3}{28}y^7 + \frac{9}{16}y^4\bigg|_{0}^{1} = \frac{75}{112}.$$

2. Evaluate $\int \dfrac{dx}{1 + \sqrt{x}}$.

Solution. We set $y = \sqrt{x}$, $y^2 = x$, $dx = 2y\, dy$ and obtain

$$\int \frac{dx}{1 + \sqrt{x}} = \int \frac{2y\, dy}{1 + y} = \int\left(2 + \frac{-2}{1 + y}\right) dy = 2y - 2\log(1 + y) + C$$

$$= 2\sqrt{x} - 2\log(1 + \sqrt{x}) + C.$$

We check by differentiating:

$$\frac{d}{dx}\left[2x^{1/2} - 2\log(1 + x^{1/2})\right] = x^{-1/2} - \frac{x^{-1/2}}{1 + x^{1/2}} = \frac{1}{1 + x^{1/2}},$$

as expected. ◀

EXERCISES

Perform the following integrations:

7. $\int_{0}^{1} x(2x - 1)^{1/3}\, dx$.

8. $\int \dfrac{x + \sqrt{x + 1}}{x + 2}\, dx$.

9. $\int \dfrac{x^{2/3} - 2x^{1/3} + 1}{x + 1}\, dx$.

10. $\int \dfrac{x^2}{(x + 4)^{3/2}}\, dx$.

11. $\int \dfrac{2x + 3\sqrt{x + 1}}{2x - 3\sqrt{x + 1}}\, dx$.

12. $\int \dfrac{x\, dx}{x + (x + 1)^{3/5} + 1}$.

3.4 Roots of fractional linear functions

(III) AN INTEGRAL $\int R\left(x, \sqrt[q]{\dfrac{\alpha x + \beta}{\gamma x + \delta}}\right) dx$, $q > 1$ AN INTEGER, AND $\alpha\delta - \gamma\beta \neq 0$, CAN BE RATIONALIZED BY THE SUBSTITUTION

$$(1) \qquad y = \sqrt[q]{\dfrac{\alpha x + \beta}{\gamma x + \delta}}.$$

Before proving this, we make two remarks. 1. The present result contains (II) as a special case. 2. The condition $\alpha\delta - \beta\gamma \neq 0$ means that $(\alpha x + \beta)/(\gamma x + \delta)$ is not a constant.

Proof of (III). If y is defined by (1), we have

$$y^q = \dfrac{\alpha x + \beta}{\gamma x + \delta}.$$

After multiplying both sides by $\gamma x + \delta$, solve for x. This yields

$$x = \dfrac{-\delta y^q + \beta}{\gamma y^q - \alpha}, \qquad dx = \dfrac{\alpha\delta - \beta\gamma}{(\gamma y^q - \alpha)^2} q y^{q-1}\, dy$$

and

$$\int R\left(x, \sqrt[q]{\dfrac{\alpha x + \beta}{\gamma x + \delta}}\right) dx = \int R\left(\dfrac{-\delta y^q + \beta}{\gamma y^q - \alpha}, y\right) \dfrac{(\alpha\delta - \beta\gamma) q y^{q-1}}{(\alpha y^q - \alpha)^2}\, dy;$$

this is an integral of a rational function.

▶**Examples** *1.* Evaluate $\int \sqrt{\dfrac{1 - x}{1 + x}}\, dx$.

Answer. We set

$$y^2 = \dfrac{1 - x}{1 + x} \text{ so that } y^2 + y^2 x = 1 - x, \ x = \dfrac{1 - y^2}{1 + y^2}, \ dx = \dfrac{-4y}{(1 + y^2)^2}\, dy,$$

and obtain

$$\int \sqrt{\dfrac{1 - x}{1 + x}}\, dx = \int y \dfrac{-4y}{(1 + y^2)^2}\, dy = \int \left(\dfrac{-4}{(1 + y^2)} + \dfrac{4}{(1 + y^2)^2}\right) dy.$$

In order to evaluate this integral, we apply the recursion formula (4) derived in §2.4. We obtain

$$\int \left(\frac{-4}{1+y^2}\right) dy + 4\left\{\frac{1}{2}\frac{y}{1+y^2} + \frac{1}{2}\int \frac{dy}{1+y^2}\right\}$$

$$= -2\int \frac{dy}{1+y^2} + 2\frac{y}{1+y^2} = -2 \arctan y + 2\frac{y}{1+y^2} + C$$

$$= -2 \arctan \sqrt{\frac{1-x}{1+x}} + (1+x)\sqrt{\frac{1-x}{1+x}} + C$$

$$= -2 \arctan \sqrt{\frac{1-x}{1+x}} + \sqrt{1-x^2} + C.$$

The reader should check by differentiation.

2. Compute $\int_0^1 \frac{1}{(1+x)^2} \sqrt[3]{\frac{1-x}{1+x}}\, dx$.

Answer. Set

$$y = \sqrt[3]{\frac{1-x}{1+x}}.$$

Then

$$x = \frac{1-y^3}{1+y^3}, \qquad dx = -\frac{6y^2}{(1+y^3)^2}\, dy, \quad \text{and} \quad 1+x = \frac{2}{1+y^3},$$

so that $\int_0^1 \frac{1}{(1+x)^2} \sqrt[3]{\frac{1-x}{1+x}}\, dx = -\int_1^0 \frac{(1+y^3)^2}{4} y \frac{6y^2}{(1+y^3)^2}\, dy$

$$= \frac{3}{2}\int_0^1 y^3\, dy = \frac{3}{2}\cdot\frac{1}{4} y^4 \Big|_0^1 = \frac{3}{8}.$$

3. Compute $\int_1^{+\infty} \frac{1}{(1+x^2)} \sqrt[3]{\frac{1-x}{1+x}}\, dx$.

Answer. The substitution used above shows that this improper integral equals

$$-\frac{3}{2}\int_0^{-1} y^3\, dy = -\frac{3}{8}.$$

The reader should check the details.◄

EXERCISES

Perform the following integrations:

13. $\int \dfrac{\sqrt{4 + 3x}}{4 - 3x}\, dx.$

14. $\int \dfrac{2 + \sqrt{\dfrac{4x - 1}{x}}}{2 - \sqrt{\dfrac{4x - 1}{x}}}\, dx.$

15. $\int_0^1 \dfrac{x}{2x + 1} \sqrt{\dfrac{x}{2x + 1}}\, dx.$

16. $\int \dfrac{\left(\dfrac{x}{1 - x}\right)^{3/2}}{x + \left(\dfrac{x}{1 - x}\right)^{1/2}}\, dx.$

17. $\int \dfrac{dx}{(1 - x)^2 + (1 + x)^2 \left(\dfrac{1 - x}{1 + x}\right)^{4/3}}.$

18. $\int \sqrt{\dfrac{4 + 3x}{4 - 3x}}\, dx.$

3.5 Rational functions of trigonometric functions

We come now to a basic result.

(IV) AN INTEGRAL OF THE FORM $\int R(\cos x, \sin x)\, dx$ CAN BE RATIONALIZED BY THE SUBSTITUTION

$$(2) \qquad\qquad y = \tan \frac{x}{2}.$$

Proof. We use the result established in Chapter 6, §2.3:

$$\sin x = \frac{2 \tan \dfrac{x}{2}}{1 + \tan^2 \dfrac{x}{2}}, \qquad \cos x = \frac{1 - \tan^2 \dfrac{x}{2}}{1 + \tan^2 \dfrac{x}{2}}.$$

Thus
$$\sin x = \frac{2y}{1 + y^2}, \qquad \cos x = \frac{1 - y^2}{1 + y^2}.$$

Also, by (2) we have $\dfrac{x}{2} = \text{arc tan } y$, $x = 2 \text{ arc tan } y$. Hence

$$dx = \frac{2}{1 + y^2}\, dy,$$

and
$$\int R(\cos x, \sin x)\, dx = \int R\!\left(\frac{1 - y^2}{1 + y^2}, \frac{2y}{1 + y^2}\right) \frac{2\, dy}{1 + y^2};$$

this is an integral of a rational function.

This method permits us to calculate integrals of functions formed rationally from $\cos x$, $\sin x$, $\sec x = 1/\cos x$, $\csc x = 1/\sin x$, $\tan x = \sin x/\cos x$, $\cot x = \cos x/\sin x$. But it often pays to use some special device, rather than the general method.

►*Examples* **1.** Compute $\int \sec x\, dx$.

Answer. We set $y = \tan \frac{1}{2}x$ and obtain

$$\int \sec x\, dx = \int \frac{dx}{\cos x} = \int \frac{1 + y^2}{1 - y^2}\, \frac{2\, dy}{1 + y^2} = \int \frac{2}{1 - y^2}\, dy$$

$$= \int \left(\frac{1}{1 + y} + \frac{1}{1 - y} \right) dy = \log |1 + y| - \log |1 - y| + C$$

$$= \log \frac{1 + y}{1 - y}\, dy = \log \left| \frac{1 + \tan \dfrac{x}{2}}{1 - \tan \dfrac{x}{2}} \right| + C.$$

We check by differentiation and obtain

$$\frac{d}{dx} \log \left| \frac{1 + \tan \dfrac{x}{2}}{1 - \tan \dfrac{x}{2}} \right| = \frac{1 - \tan \dfrac{x}{2}}{1 + \tan \dfrac{x}{2}} \frac{2}{\left(1 - \tan \dfrac{x}{2}\right)^2} \frac{1}{\cos^2 \dfrac{x}{2}} \frac{1}{2}$$

$$= \frac{1}{\left(1 - \tan^2 \dfrac{x}{2}\right) \cos^2 \dfrac{x}{2}} = \frac{1}{\cos^2 \dfrac{x}{2} - \sin^2 \dfrac{x}{2}} = \frac{1}{\cos x} = \sec x,$$

as expected. Since

$$\frac{1 + \tan \dfrac{x}{2}}{1 - \tan \dfrac{x}{2}} = \frac{\cos \dfrac{x}{2} + \sin \dfrac{x}{2}}{\cos \dfrac{x}{2} - \sin \dfrac{x}{2}} = \frac{\cos \dfrac{x}{2} + \sin \dfrac{x}{2}}{\cos \dfrac{x}{2} - \sin \dfrac{x}{2}} \cdot \frac{\cos \dfrac{x}{2} + \sin \dfrac{x}{2}}{\cos \dfrac{x}{2} + \sin \dfrac{x}{2}}$$

$$= \frac{1 + 2 \cos \dfrac{x}{2} \sin \dfrac{x}{2}}{\cos^2 \dfrac{x}{2} - \sin^2 \dfrac{x}{2}} = \frac{1 + \sin x}{\cos x} = \sec x + \tan x,$$

we have

(3) $$\int \sec x\, dx = \log |\sec x + \tan x| + C.$$

This is the formula that appears in most tables of integrals.

2. Compute $\int \csc x \, dx$.

Answer. We could obtain this integral in closed form using the same method as above. It is simpler, however, to use (3) and to write

$$\int \csc x \, dx = \int \sec \left(\frac{\pi}{2} - x \right) dx = - \int \sec \left(\frac{\pi}{2} - x \right) d \left(\frac{\pi}{2} - x \right)$$

$$= -\log \left| \sec \left(\frac{\pi}{2} - x \right) + \tan \left(\frac{\pi}{2} - x \right) \right| C = -\log |\csc x + \cot x| + C.$$

This can also be written in the form

(4) $$\int \csc x \, dx = \log |\csc x - \cot x| + C,$$

as the reader would verify.

3. Evaluate $\int \sin^2 x \, dx$.

Answer. Instead of using the general method, we note that

$$\int \cos^2 x \, dx + \int \sin^2 x \, dx = \int 1 \, dx = x + C_1,$$

$$\int \cos^2 x \, dx - \int \sin^2 x \, dx = \int \cos 2x \, dx = \frac{1}{2} \sin 2x + C_2.$$

Adding and subtracting the two equations, we obtain

(5) $$\int \cos^2 x \, dx = \frac{1}{2} x + \frac{1}{4} \sin 2x + C,$$

(6) $$\int \sin^2 x \, dx = \frac{1}{2} x - \frac{1}{4} \sin 2x + C.$$

One could also have obtained this result as follows

$$\int \sin^2 x \, dx = \int \frac{1 - \cos 2x}{2} \, dx = \frac{1}{2} x - \frac{1}{4} \sin 2x + C.$$

4. Compute $\int \dfrac{dx}{\sin x + \cos x}$.

Answer. Instead of using the general method, note that

$$\sin x + \cos x = \sqrt{2} \left(\frac{1}{\sqrt{2}} \sin x + \frac{1}{\sqrt{2}} \cos x \right)$$

$$= \sqrt{2} \left(\sin x \cos \frac{\pi}{4} + \cos x \sin \frac{\pi}{4} \right) = \sqrt{2} \sin \left(x + \frac{\pi}{4} \right).$$

Therefore, using Example 2,

$$\int \frac{dx}{\sin x + \cos x} = \frac{1}{\sqrt{2}} \int \frac{dx}{\sin\left(x + \frac{\pi}{4}\right)}$$

$$= -\frac{1}{\sqrt{2}} \log \left| \csc\left(x + \frac{\pi}{4}\right) + \cos\left(x + \frac{\pi}{4}\right) \right| + C. \blacktriangleleft$$

EXERCISES

Perform the following integrations. (The reader may want to look for a special device rather than (or in addition to) using the general method.)

19. $\int \dfrac{2 - \cos x}{2 + \cos x}\, dx.$

20. $\int \dfrac{1 - \sin x}{1 + \sin x}\, dx.$

21. $\int \dfrac{\cos^3 x}{1 + \cos^2 x}\, dx.$

22. $\int \dfrac{2 \cos x - 5 \sin x}{3 \cos x + 4 \sin x}\, dx.$

23. $\int \dfrac{dx}{\sin x + \sqrt{3} \cos x}.$

24. $\int \dfrac{\sec x}{1 + \sin x}\, dx.$

25. $\int \dfrac{\csc x}{3 + 4 \tan x}\, dx.$

26. $\int \dfrac{\sec x \csc x}{3 + 5 \cos x}\, dx.$

27. $\int \dfrac{dx}{4 \cos^2 x + 3 \sin^2 x}.$

28. $\int \dfrac{\sec x}{1 - \sec^2 x}\, dx.$

29. $\int \dfrac{\sec x\, dx}{1 - 4 \sec^2 x}.$

30. $\int \dfrac{dx}{9 \cos^2 x - 16 \sin^2 x}.$

31. Show that

$$\int \csc x\, dx = \log\left|\tan \tfrac{1}{2} x\right| + C$$

(a) by the general method of this subsection, (b) by direct differentiation, and (c) by using the result of Example 2.

32. Show how the method of Example 4 can be used to compute

$$\int \frac{dx}{a \cos x + b \sin x}$$

where a and b are numbers and that $a^2 + b^2 \neq 0$.

33. Show how the method of Example 4 can be used to compute

$$\int \frac{\alpha + \tan x}{1 - \alpha \tan x}\, dx$$

where α is any number.

3.6 Rational functions of hyperbolic functions

(V) AN INTEGRAL OF THE FORM $\int R(\cosh x, \sinh x)\, dx$ CAN BE RATIONALIZED, FOR INSTANCE, BY THE SUBSTITUTION

$$(7) \qquad\qquad y = \tanh \frac{x}{2}.$$

This is a direct analogue of (IV). To prove (V), note that, assuming (7), we have (see Chapter 6, §7.6, or verify directly)

$$\sinh x = \frac{2y}{1 - y^2}, \qquad \cosh x = \frac{1 + y^2}{1 - y^2}, \qquad dx = \frac{2\, dy}{1 - y^2}.$$

Thus $\displaystyle \int R(\cosh x, \sinh x)\, dx = \int R\!\left(\frac{1 + y^2}{1 - y^2}, \frac{2y}{1 + y^2}\right) \frac{2\, dy}{1 - y^2};$

this is an integral of a rational function.

There is, of course, no necessity to use the substitution (7). One can note that $\cosh x = \frac{1}{2}(e^x + e^{-x})$, $\sinh x = \frac{1}{2}(e^x - e^{-x})$ and use the substitution $y = e^x$ [see (I) above]. Which method is more advantageous depends upon the integral considered.

EXERCISES

Perform the following integrations:

34. $\displaystyle \int \frac{dx}{\cosh x + \sinh x}.$

35. $\displaystyle \int \frac{\tanh x}{1 + \cosh x}\, dx.$

36. $\displaystyle \int \frac{2 + 2 \cosh x - \sinh x}{2 + \cosh x + \sinh x}\, dx.$

Rationalize each of the following integrals.

37. $\displaystyle \int \frac{\cosh x}{(1 + \cosh x)^2}\, dx.$

38. $\displaystyle \int \frac{\sinh x\, dx}{(4 + 3 \sinh x)(1 + 2 \cosh x)}.$

3.7 Integrals involving $\sqrt{1 - x^2}$

(VI) AN INTEGRAL OF THE FORM $\int R(x, \sqrt{1 - x^2})\, dx$ CAN BE RATIONALIZED.

Proof. We indicate two methods of rationalizing this integral.

(a) Use the *trigonometric substitution*

$$x = \cos y, \qquad \sqrt{1 - x^2} = \sin y, \qquad dx = -\sin y\, dy \qquad 0 \le y \le \pi.$$

We obtain $\qquad \int R(x, \sqrt{1-x^2})\,dx = -\int R(\cos y, \sin y) \sin y\,dy;$

we are back to Case IV.

(b) Write

$$\sqrt{1-x^2} = (1+x)\sqrt{\frac{1-x}{1+x}},$$

so that $\quad \int R(x, \sqrt{1-x^2})\,dx = \int R\!\left(x, (1+x)\sqrt{\frac{1-x}{1+x}}\right)dx;$

we are back to Case III.

▶ *Example*   Compute $\int \sqrt{1-x^2}\,dx$.

Answer.   We use Method (a) and obtain

$$\int \sqrt{1-x^2}\,dx = \int \sin y\,(-\sin y\,dy) = -\int \sin^2 y\,dy.$$

Now, using relation (6), we have

$$\int \sqrt{1-x^2}\,dx = -\frac{1}{2}y + \frac{1}{4}\sin 2y + C$$

$$= -\frac{1}{2}\arccos x + \frac{1}{2}\sin y \cos y + C$$

$$= -\frac{1}{2}\arccos x + x\sqrt{1-x^2} + C.$$

In this case, Method (b) requires more complicated calculations. ◀

EXERCISES

Perform the following integrations:

39. $\int \dfrac{x^2}{\sqrt{1-x^2}}\,dx$. Use Method (a).

40. $\int_0^{\sqrt{2}/2} x^3 \sqrt{1-x^2}\,dx$. Use Method (a).

41. $\int_{1/2}^1 \dfrac{(1-x^2)^{3/2}}{x^2}\,dx$. Use Method (a).

$$\frac{\sin^4 x}{\cos^2 x} = \frac{(1-\cos^2 x)^2}{\cos^2 x} = \sec^2 x - 2 + \cos^2 x.$$

42. $\int \dfrac{x^3}{(1-x^2)^{3/2}}\,dx$. Use Method (a).

43. $\int \dfrac{\sqrt{1-x^2}}{x(1+x)}\,dx$. Use Method (b).

► 44. $\int x^3(1 - x^2)^{1/4}\, dx$. The integrand in this case is a *radical* function of x and $\sqrt{1 - x^2}$, *not* a rational function. Method (a) still works.

45. $\int_1^{\sqrt{2}} x^3 \sqrt{1 - x^2}\, dx$. Use Method (a).

46. $\int \dfrac{\sqrt{1 - x^2}}{x^2}\, dx$. Use Method (a).

3.8 Integrals involving $\sqrt{x^2 - 1}$

(VII) AN INTEGRAL OF THE FORM $\int R(x, \sqrt{x^2 - 1})\, dx$ CAN BE RATIONALIZED.

Proof. We indicate three methods of rationalization.

(a) Use the *trigonometric substitution*

$$x = \sec y = \frac{1}{\cos y}$$

so that $\qquad \sqrt{x^2 - 1} = \tan y, \qquad dx = \tan y \sec y\, dy.$

We obtain

$$\int R(x, \sqrt{x^2 - 1})\, dx = \int R(\sec y, \tan y) \tan y \sec y\, dy.$$

This is an integral of Type IV.

(b) Use the *hyperbolic substitution*

$$x = \cosh y, \qquad \sqrt{x^2 - 1} = \sinh y, \qquad dx = \sinh y\, dy.$$

Then $\qquad \int R(x, \sqrt{x^2 - 1})\, dx = \int R(\cosh y, \sinh y) \sinh y\, dy.$

This is an integral of Type V or I.

(c) Write

$$\int R(x, \sqrt{x^2 - 1})\, dx = \int R\left(x, (x + 1)\sqrt{\frac{x - 1}{x + 1}}\right) dx.$$

This is an integral of Type III.

► *Example* Compute $\displaystyle\int_1^{\sqrt{2}} \frac{dx}{x^2 \sqrt{x^2 - 1}}.$

Answer. We use Method (a):

$$\int_1^{\sqrt{2}} \frac{dx}{x^2\sqrt{x^2-1}} = \int_0^{\pi/4} \frac{\tan y \sec y \, dy}{\sec^2 y \tan y}$$

$$= \int_0^{\pi/4} \cos y \, dy = \sin y \Big|_0^{\pi/4} = \frac{1}{\sqrt{2}}. \blacktriangleleft$$

EXERCISES

Perform the following integrations.

47. $\int \dfrac{dx}{x^3\sqrt{x^2-1}}$. Use Method (a) and rewrite the new integrand in terms of sines and cosines.

48. $\int \dfrac{x^3 \, dx}{\sqrt{x^2-1}}$. Use Method (a) and rewrite the new integrand in terms of sines and cosines and use the main method of §3.4.

49. $\int \dfrac{dx}{(x^2-1)^{3/2}}$. Use Method (a) and rewrite the new integrand in terms of sines and cosines.

50. $\int \sqrt{x^2-1} \, dx$. Use Method (b). Alternatively, use Method (a) and rewrite the new integrand in terms of sines and cosines. Then use the main method of §3.4.

51. $\int_1^{\sqrt{2}} x^3\sqrt{x^2-1} \, dx$. Use Method (a) and the method of Exercise 29.

52. $\int \dfrac{\sqrt{x^2-1}}{x} \, dx$. Use Method (a).

3.9 Integrals involving $\sqrt{1+x^2}$

(VIII) AN INTEGRAL OF THE FORM $\int R(x, \sqrt{1+x^2}) \, dx$ CAN BE RATIONALIZED.

Proof. We indicate two methods.

(a) Use the *trigonometric substitution*

$$x = \tan y, \qquad \sqrt{1+x^2} = \sec y, \qquad dx = \sec^2 y \, dy.$$

We obtain

$$\int R(x, \sqrt{1+x^2}) \, dx = \int R(\tan y, \sec y) \sec^2 y \, dy.$$

This is an integral of Type IV.

(b) Use the *hyperbolic substitution*

$$x = \sinh y, \qquad \sqrt{1+x^2} = \cosh y, \qquad dx = \cosh y \, dy.$$

Thus $\qquad \int R(x, \sqrt{1 + x^2})\,dx = \int R(\sinh y, \cosh y)\cosh y\,dy,$

an integral of Type V or I.

EXERCISES

Perform the following integrations:

53. $\int \sqrt{x^2 + 1}\,dx$. Use Method (b). Alternatively, use Method (a) and rewrite the new integrand in terms of sines and cosines. Then use the main method of §3.4.

► 54. $\int_1^{\sqrt{3}} \dfrac{\sqrt{x^2 + 1}}{x^4}\,dx$. Use Method (a) and rewrite the new integrand in terms of sines and cosines.

55. $\int \dfrac{x^2}{(x^2 + 1)^{3/2}}\,dx$. Use Method (a).

56. $\int \dfrac{dx}{\sqrt{x^2 + 1}}$. Use Method (a) or (b).

57. $\int \dfrac{x^2}{\sqrt{x^2 + 1}}\,dx$. Use Method (b).

58. $\int \dfrac{dx}{(x^2 + 1)^{3/2}}$.

3.10 Square roots of quadratic functions

We are now able to make a rather general statement.

(IX) ANY INTEGRAL OF THE FORM $\int R(x, \sqrt{\alpha x^2 + 2\beta x + \gamma})\,dx$ CAN BE RATIONALIZED.

We assume $\alpha \neq 0$; otherwise we are in Case II and our assertion is true. Set

$$\alpha x^2 + 2\beta x + \gamma = \alpha\left(x + \frac{\beta}{\alpha}\right)^2 + \frac{\alpha\gamma - \beta^2}{\alpha}.$$

Assume that $\alpha\gamma - \beta^2 > 0$: then $\sqrt{\alpha x^2 + 2\beta x + \gamma}$ equals

$$\sqrt{\alpha\left(x + \frac{\beta}{\alpha}\right)^2 + \frac{\alpha\gamma - \beta^2}{\alpha}} = \sqrt{\frac{\alpha\gamma - \beta^2}{\alpha}}\sqrt{\left(\frac{\alpha x + \beta}{\sqrt{\alpha\gamma - \beta^2}}\right)^2 + 1}.$$

Thus $\alpha > 0$, for otherwise the square root is meaningless. Setting

$$\xi = \frac{\alpha x + \beta}{\sqrt{\alpha\gamma - \beta^2}}, \qquad d\xi = \frac{\alpha\,dx}{\sqrt{\alpha\gamma - \beta^2}},$$

we transform our integral into one of the form (VIII).

Assume next that $\alpha\gamma - \beta^2 < 0$. We set

$$\xi = \frac{\alpha x + \beta}{\sqrt{\beta^2 - \alpha\gamma}}, \qquad d\xi = \frac{\alpha\, dx}{\sqrt{\beta^2 - \alpha\gamma}}$$

and note that $\sqrt{\alpha x^2 + 2\beta x + \gamma}$ equals

$$\sqrt{\alpha\left(x + \frac{\beta^2}{\alpha}\right) + \frac{\alpha\gamma - \beta^2}{\alpha}} = \sqrt{\frac{\beta^2 - \alpha\gamma}{\alpha}\left[\left(\frac{\alpha x + \beta}{\beta^2 - \alpha\gamma}\right)^2 - 1\right]}$$

$$= \sqrt{\frac{\beta^2 - \alpha\gamma}{\alpha}(\xi^2 - 1)}.$$

Hence our integral becomes one of Type VI or VII, according to whether $\alpha < 0$ or $\alpha > 0$.

The case $\alpha\gamma - \beta^2 = 0$, finally, is of no interest. In this case, we must have $\alpha > 0$. Thus

$$\sqrt{\alpha x^2 + 2\beta x + \gamma} = \sqrt{\alpha}\left(x + \frac{\beta}{\alpha}\right)$$

and the integrand is rational.

Assertion IX is now completely proved.

▶*Example* Compute $\displaystyle\int \frac{dx}{\sqrt{-4x^2 + 8x - 3}}$.

Answer. We have

$$-4x^2 + 8x - 3 = -4(x - 1)^2 + 1 = 1 - (2x - 2)^2.$$

Therefore we set $\xi = 2x - 2$, $d\xi = 2\,dx$ and obtain

$$\int \frac{dx}{\sqrt{-4x^2 + 8x - 3}} = \frac{1}{2}\int \frac{d\xi}{\sqrt{1 - \xi^2}} = \frac{1}{2}\,\text{arc sin } \xi + C$$

$$= \frac{1}{2}\,\text{arc sin }(2x - 2) + C.\blacktriangleleft$$

EXERCISES

Transform each of the following integrals into an integral of a form considered previously.

59. $\int x^3 \sqrt{9 - 4x^2}\, dx.$ 60. $\displaystyle\int \frac{x^3\, dx}{\sqrt{4x^2 - 3}}.$

61. $\int \dfrac{1 - 4x + 4x^2}{\sqrt{2 - 4x + 4x^2}}\, dx.$

66. $\int \dfrac{\sqrt{x^2 + x}}{4x + 2}\, dx.$

▶ 62. $\int \dfrac{dx}{(3x^2 + 8x + 5)^{3/2}}.$

67. $\int \dfrac{4x^2 - 4x + 1}{(2x^2 - 2x + 5)^{3/2}}\, dx.$

63. $\int \dfrac{4 + 4x + x^2}{\sqrt{-3 - 4x - x^2}}\, dx.$

68. $\int \dfrac{dx}{\sqrt{9x^2 - 30x + 9}}$

64. $\int \dfrac{dx}{\sqrt{2x^2 - 6x + 5}}.$

69. $\int \dfrac{1 - x}{(2 + \sqrt{3 + 6x - 9x^2})^2}\, dx.$

65. $\int \dfrac{(2x - x^2)^{3/2}}{x^2 - 2x + 1}\, dx.$

Reduction formulas are used to evaluate certain integrals not easily amenable to the methods just described. We encountered an example in §2.4, in evaluating an integral of the form

$$I_n = \int \dfrac{dx}{(1 + x^2)^n}, \qquad n = 1, 2, 3, \cdots.$$

The integrand depends on the integer n. For $n = 1$, we know the indefinite integral; it is the function arc tan $x + C$. Integration by parts gave us a relation between I_n and I_{n-1}. Using this *reduction formula* $n - 1$ times, we compute I_n. The examples considered below will show that the situation just described is typical.

For the convenience of the reader, we recall the formula for integration by parts (see Chapter 5, §3.4)

$$\int u \, dv = uv - \int v \, du.$$

§4 Reduction Formulas

4.1 Integrating $x^n e^x$

We begin with a simple case:

$$\int x^n e^x \, dx.$$

Integration by parts gives

$$\int x^n e^x \, dx = \int x^n \, d(e^x) = x^n e^x - \int e^x \, d(x^n) = x^n e^x - n \int x^{n-1} e^x \, dx.$$

Thus

(1) $$\int x^n e^x \, dx = x^n e^x - n \int x^{n-1} e^x \, dx.$$

This is the desired reduction formula.

▶**Example**   Find a primitive function for $x \mapsto x^3 e^x$.

Answer.   Applying (1), we have

$$\int x^3 e^x \, dx = x^3 e^x - 3 \int x^2 e^x \, dx, \qquad \int x^2 e^x \, dx = x^2 e^x - 2 \int x e^x \, dx,$$

$$\int x e^x \, dx = x e^x - \int e^x \, dx = x e^x - e^x + C.$$

Thus

$$\int x^3 \, e^x \, dx = x^3 e^x - 3x^2 e^x + 3 \cdot 2 x e^x - 3 \cdot 2 e^x + C'$$
$$= (x^3 - 3x^2 + 6x - 6)e^x + C'. \blacktriangleleft$$

EXERCISES

Perform the following integrations:

1. $\int x^4 e^x \, dx$.

2. $\int (3x^4 - 2x^2) e^x \, dx$.

3. $\int x^3 e^{3x} \, dx$.

4. $\int (2x^3 - x + 1) e^{x/2} \, dx$.

5. $\int (3x^3 - 2x^2 + x - 1) e^{2x} \, dx$.

6. $\int x^2 (e^{2x} + e^{3x}) \, dx$.

7. $\int (2x^3 + x)(e^{x/2} - 2e^{2x}) \, dx$.

8. $\int x^5 e^{x^2} \, dx$. Substitute $u = x^2$.

4.2 Integrals involving $x^\alpha (\log x)^n$

We want to compute

$$\int x^\alpha (\log x)^n \, dx,$$

where $\alpha \neq -1$ is some number. We note that $dx^{\alpha+1} = (\alpha + 1) x^\alpha \, dx$, so that

$$\int x^\alpha (\log x)^n \, dx = \frac{1}{\alpha + 1} \int (\log x)^n \, d(x^{\alpha+1})$$

$$= \frac{1}{\alpha + 1} x^{\alpha+1} (\log x)^n - \frac{1}{\alpha + 1} \int x^{\alpha+1} \, d(\log x)^n$$

$$= \frac{1}{\alpha + 1} x^{\alpha+1} (\log x)^n - \frac{n}{\alpha + 1} \int x^{\alpha+1} (\log x)^{n-1} \frac{1}{x} \, dx.$$

Thus

(2) $$\int x^\alpha (\log x)^n \, dx = \frac{x^{\alpha+1} (\log x)^n}{\alpha + 1} - \frac{n}{\alpha + 1} \int x^\alpha (\log x)^{n-1} \, dx.$$

This is the desired reduction formula.

The case $\alpha = -1$ can be handled directly:

$$\int \frac{(\log x)^n}{x}\, dx = \int (\log x)^n\, d(\log x) = \frac{(\log x)^{n+1}}{n+1} + C.$$

EXERCISES

Perform the following integrations:

9. $\int_1^e \sqrt{x}\log x\, dx.$

13. $\int (x^{1/3} + x^{1/2})(\log x)^2\, dx.$

▶ 10. $\int x^{1/3}\log^2 x\, dx.$

14. $\int \dfrac{\log^2 (2x)}{x^2}\, dx.$

11. $\int \dfrac{\log^3 x}{x}\, dx.$

15. $\int (x^{2/3} + 3x^{3/5})(\log \tfrac{1}{2}x)^2\, dx.$

12. $\int \dfrac{\log^3 x}{\sqrt{x}}\, dx.$

16. $\int (2x)^2[\log (3x)]^3\, dx.$

4.3 Integrals of $x^n \cos x$ and $x^n \sin x$

Consider next the two integrals

$$\int x^n \cos x\, dx, \quad \int x^n \sin x\, dx.$$

We obtain at once a reduction formula connecting the two; indeed,

$$\int x^n \cos x\, dx = \int x^n\, d(\sin x) = x^n \sin x - \int \sin x\, d(x^n)$$

or

(3) $$\int x^n \cos x\, dx = x^n \sin x - n \int x^{n-1} \sin x\, dx.$$

One proves similarly that

(4) $$\int x^n \sin x\, dx = -x^n \cos x + n \int x^{n-1} \cos x\, dx.$$

▶ *Example*   Compute $\int_0^{\pi/2} x^3 \sin x\, dx.$

Answer.   Using (3) and (4), we have

$$\int_0^{\pi/2} x^3 \sin x\, dx = -x^3 \cos x \,\Big|_0^{\pi/2} + 3 \int_0^{\pi/2} x^2 \cos x\, dx$$

$$= -x^3 \cos x \,\Big|_0^{\pi/2} + 3x^2 \sin x \,\Big|_0^{\pi/2} - 3\cdot 2 \int_0^{\pi/2} x \sin x\, dx$$

$$= (-x^3 \cos x + 3x^2 \sin x + 6x \cos x) \,\Big|_0^{\pi/2} - 6 \int_0^{\pi/2} \cos x\, dx$$

$$= (-x^3 \cos x + 3x^2 \sin x + 6x \cos x - 6 \sin x) \,\Big|_0^{\pi/2} = \frac{3\pi^2}{4} - 6. \blacktriangleleft$$

EXERCISES

Perform the following integrations:

17. $\int_0^{\pi/2} x^2 \cos x \, dx$.

18. $\int_{\pi/3}^{\pi/2} x^2 \sin x \, dx$.

19. $\int_0^{\pi/4} (3x^2 - 2x + 1) \cos x \, dx$.

20. $\int x^2(\cos x + \sin x) \, dx$.

21. $\int x^3 \sin (2x) \, dx$.

22. $\int x^4 \sin x \, dx$.

23. $\int_0^{\pi/9} (2x)^2 \cos (3x) \, dx$.

24. $\int_0^{\pi/2} x^{99}(x \cos x + 100 \sin x) \, dx$.

25. $\int x^{3/2} \cos \sqrt{x} \, \dfrac{dx}{\sqrt{x}}$. (Substitute $u = \sqrt{x}$.)

26. $\int (x - \tfrac{1}{2}\pi)^4 \cos x \, dx$.

4.4 Products of powers of sines and cosines

Our next aim is to compute integrals of the form

$$(5) \qquad\qquad \int \cos^n x \sin^m x \, dx.$$

To obtain an appropriate reduction formula, we use a "trick." For any two numbers, a and b, we have

$$d[\cos^a x \sin^b x] = [-a \cos^{a-1} x \sin^{b+1} x + b \cos^{a+1} x \sin^{b-1} x] \, dx.$$

Now write $\cos^{a+1} x = \cos^{a-1} x \cos^2 x = \cos^{a-1} x(1 - \sin^2 x)$. Thus

$$d[\cos^a x \sin^b x] = [-(a + b) \cos^{a-1} x \sin^{b+1} x + b \cos^{a-1} x \sin^{b-1} x] \, dx.$$

Taking the indefinite integral of both sides, we obtain

$$\cos^a x \sin^b x = -(a + b)\int \cos^{a-1} x \sin^{b+1} x \, dx + b \int \cos^{a-1} x \sin^{b-1} x \, dx.$$

Now set $a - 1 = n$, $b + 1 = m$. The above relation yields

$$(6) \quad \int \cos^n x \sin^m x \, dx = -\frac{1}{n + m} \cos^{n+1} x \sin^{m-1} x$$
$$+ \frac{m - 1}{n + m}\int \cos^n x \sin^{m-2} x \, dx.$$

This is the desired reduction formula.

A special case is obtained by setting $n = 0$; we have

$$(7) \qquad \int \sin^m x \, dx = -\frac{1}{m} \cos x \sin^{m-1} x + \frac{m - 1}{m}\int \sin^{m-2} x \, dx.$$

Using this formula, we can reduce the calculation of $\int \sin^m x \, dx$ to the case $m = 0$ (if m is even) or to $m = 1$ (if m is odd); these two integrals are, of course, easily computable.

The corresponding reduction formula for integrating $\cos^m x$ reads:

(8) $$\int \cos^m x \, dx = \frac{1}{m} \cos^{m-1} x \sin x + \frac{m-1}{m} \int \cos^{m-2} x \, dx.$$

This is easily obtained from (7) by using the relation $\cos(90° - x) = \sin x$. The reader should carry out the calculation. Using this reduction formula, we can find a primitive function of $\cos^m x$ for every integer $m > 0$. Now we see that the reduction formula (6) is sufficient for calculating the integral (5), since it reduces this integral either to

$$\int \cos^n x \, dx$$

(if m is even) or to

$$\int \cos^n x \sin x \, dx = -\frac{1}{n+1} \cos^{n+1} x + C$$

(if m is odd).

▶ **Examples** **1.** Compute $\int_0^{\pi/2} \sin^{2k} x \, dx$, $k \geq 0$ an integer.

Answer. Relation (7) with $m = 2k$ gives

(9) $$\int_0^{\pi/2} \sin^{2k} x \, dx = \frac{2k-1}{2k} \int_0^{\pi/2} \sin^{2k-2} x \, dx$$

(since $\cos x \sin^{2k-1} x$ takes on the value 0 at $x = 0$ and at $x = \pi/2$). On the other hand, for $k = 0$ the desired integral is $\pi/2$. Repeated application of the reduction formula (9) therefore gives

$$\int_0^{\pi/2} \sin^{2k} x \, dx = \frac{2k-1}{2k} \cdot \frac{2k-3}{2k-2} \cdot \frac{2k-5}{2k-4} \cdots \frac{1}{2} \cdot \frac{\pi}{2}.$$

2. Compute $\int_0^{\pi/2} \sin^{2k+1} x \, dx$, $k \geq 0$ an integer.

Answer. First of all, we have for $k = 0$, $\int_0^{\pi/2} \sin x \, dx = \cos x \big|_0^{\pi/2} = 1$. Also, Relation (7) yields, for $m = 2k + 1$, the reduction formula

(10) $$\int_0^{\pi/2} \sin^{2k+1} x \, dx = \frac{2k}{2k+1} \int_0^{\pi/2} \sin^{2k-1} x \, dx.$$

Repeated application of this reduction formula gives

$$\int_0^{\pi/2} \sin^{2k+1} x \, dx = \frac{2k}{2k+1} \cdot \frac{2k-2}{2k-1} \cdot \frac{2k-4}{2k-3} \cdots \frac{2}{3}.$$

3. Prove that $\displaystyle\lim_{k\to\infty} \frac{\int_0^{\pi/2} \sin^{2k} x \, dx}{\int_0^{\pi/2} \sin^{2k+1} x \, dx} = 1.$

Proof. In the interval $(0, \pi/2)$, the function $\sin x$ is positive and less than 1. Therefore

$$0 < \int_0^{\pi/2} \sin^{2k+1} x \, dx \le \int_0^{\pi/2} \sin^{2k} x \, dx \le \int_0^{\pi/2} \sin^{2k-1} x \, dx.$$

Dividing by $\int_0^{\pi/2} \sin^{2k+1} x \, dx$, we obtain

$$1 < \frac{\int_0^{\pi/2} \sin^{2k} x \, dx}{\int_0^{\pi/2} \sin^{2k+1} x \, dx} \le \frac{\int_0^{\pi/2} \sin^{2k-1} x \, dx}{\int_0^{\pi/2} \sin^{2k+1} x \, dx}.$$

But the last fraction is, by the reduction formula (10), equal to

$$\frac{2k+1}{2k} = 1 + \frac{1}{2k}$$

and hence as close as we like to 1, for sufficiently large k. This proves the assertion. ◀

Remark The preceding examples contain the proof of a remarkable formula for the number π discovered by Wallis. To obtain this, we set $I_n = \int_0^{\pi/2} \sin^n x \, dx$ and write the results of Examples 1 and 2 in the form

$$\frac{\pi}{2} = \frac{2}{1} \cdot \frac{4}{3} \cdot \frac{6}{5} \cdot \frac{8}{7} \cdots \frac{2k-2}{2k-3} \cdot \frac{2k}{2k-1} I_{2k},$$

$$1 = \frac{3}{2} \cdot \frac{5}{4} \cdot \frac{7}{6} \cdots \frac{2k-1}{2k-2} \cdot \frac{2k+1}{2k} I_{2k+1}.$$

Dividing the first equation by the second, we obtain

$$\frac{\pi}{2} = \frac{2}{1} \cdot \frac{2}{3} \cdot \frac{4}{3} \cdot \frac{4}{5} \cdot \frac{6}{5} \cdot \frac{6}{7} \cdots \frac{2k}{2k-1} \cdot \frac{2k}{2k+1} \frac{I_{2k}}{I_{2k+1}}.$$

Noting the result of Example 3, we have

$$\frac{\pi}{2} = \lim_{k\to\infty} \frac{2}{1} \cdot \frac{2}{3} \cdot \frac{4}{3} \cdot \frac{4}{5} \cdot \frac{6}{5} \cdot \frac{6}{7} \cdots \frac{2k}{2k-1} \cdot \frac{2k}{2k+1}.$$

JOHN WALLIS (1616–1703) was professor of mathematics at Oxford and one of the precursors of calculus.

One usually writes this, in a self-explanatory notation,

$$\frac{\pi}{2} = \frac{2}{1} \cdot \frac{2}{3} \cdot \frac{4}{3} \cdot \frac{4}{5} \cdot \frac{6}{5} \cdot \frac{6}{7} \cdot \frac{8}{7} \cdot \frac{8}{9} \cdots.$$

This is **Wallis' infinite product** for $\pi/2$.

EXERCISES

Perform the following integrations using the reduction formula (6).

27. $\int \sin^3 x \, dx$.

28. $\int \cos^5 x \, dx$.

29. $\int \sin^4 x \, dx$. [Remark: An alternative reduction is to write $\sin^4 x \, dx = (\sin^2 x)^2 \, dx = \frac{1}{8}(1 - \cos 2x)^2 \, d(2x)$, and so on.]

30. $\int \cos^2 x \sin^2 x \, dx$. [Remark: An alternative reduction is to write $\cos^2 x \sin^2 x \, dx = \frac{1}{8}(1 - \cos 2x)(1 + \cos 2x) \, d(2x) = \frac{1}{8} \sin^2 2x \, d(2x)$, and so on.]

31. $\int \cos^4 x \sin^3 x \, dx$.   34. $\int \cos^3 (x + 90°) \sin^2 x \, dx$.

▶ 32. $\int_{\pi/4}^{\pi/3} \cos^3 x \sin^3 x \, dx$.   35. $\int \cos^3 (2x + 90°) \cos^2 (2x) \, dx$.

33. $\int \cos^3 x \sin^2 x \, dx$.   36. $\int_0^{\pi/2} \cos^3 (3x) \sin^5 (3x) \, dx$.

4.5 Products of sines and cosines

We conclude this section by mentioning a type of trigonometric integral that can be evaluated in closed form, but by a somewhat different method from the one used up to now.

A typical example is the integral

$$\int \cos mx \cos nx \, dx,$$

where m and n are two distinct numbers. To evaluate it, we make use of the trigonometric identity (see Chapter 6, §4.1)

$$\cos \alpha \cos \beta = \frac{1}{2} [\cos (\alpha + \beta) + \cos (\alpha - \beta)].$$

This yields $\cos mx \cos nx = \frac{1}{2}[\cos (m + n)x + \cos (m - n)x]$. Therefore

$$\int \cos mx \cos nx \, dx = \frac{\sin (m + n)x}{2(m + n)} + \frac{\sin (m - n)x}{2(m - n)} + C.$$

We observe that, if m and n are integers, then we could, by repeated application of the addition theorems, write $\cos mx \cos nx$ as a rational function of $\cos x$ and $\sin x$. In this case, we could conclude, from Statement IV in §3.5, that the integral under consideration could be evaluated in closed form. A similar reasoning would hold if the ratio m/n were rational. At any rate, the method by which we arrived at the result is more convenient than the general method of (IV). It is also the only method that works when m/n is irrational.

Integrals of the form

$$\int \sin mx \sin nx \, dx, \qquad \int \cos mx \sin nx \, dx$$

can be evaluated similarly, by making use of the trigonometric identities

$$\sin \alpha \sin \beta = \frac{1}{2}[\cos(\alpha - \beta) - \cos(\alpha + \beta)]$$

and

$$\cos \alpha \sin \beta = \frac{1}{2}[\sin(\alpha + \beta) - \sin(\alpha - \beta)].$$

The reader is asked to derive these identities from the addition theorems. [It is also possible to reduce the two latter integrals to the case considered first by using the relation $\cos(90° - x) = \sin x$.]

It is clear that the method just described permits one to evaluate in closed form a primitive function of a product of more than two functions, each of which is a sine or a cosine of a multiple of x.

▶**Example**   We have

$\cos x \cos 2x \cos 3x$

$$= \frac{1}{2}[\cos 3x + \cos x]\cos 3x$$

$$= \frac{1}{2}\cos^2 3x + \frac{1}{2}\cos x \cos 3x = \frac{1}{2}\cos 3x + \frac{1}{4}\cos 4x + \frac{1}{4}\cos 2x$$

$$= \left(\frac{1}{4} + \frac{1}{4}\cos 6x\right) + \left(\frac{1}{4}\cos 4x + \frac{1}{4}\cos 2x\right),$$

so that

$$\int \cos x \cos 2x \cos 3x \, dx = \frac{1}{4}x + \frac{1}{8}\sin 2x + \frac{1}{16}\sin 4x + \frac{1}{24}\sin 6x + C. ◀$$

EXERCISES

Perform the following integrations:

37. $\int_0^{\pi/8} \cos 3x \cos x \, dx.$ 39. $\int \sin 5x \sin 6x \, dx.$

38. $\int_0^{\pi/8} \cos 5x \sin 3x \, dx.$ 40. $\int \cos 2x \sin 3x \sin 4x \, dx.$

41. $\int \cos 5x \cos 3x \cos x \, dx.$

42. $\int \cos^2 3x \sin 2x \, dx.$

43. $\int_0^{\pi/6} \cos 2x \sin 3x \sin x \, dx.$ ·

44. $\int (2 \cos 3x + 3 \sin x) \cos 2x \, dx.$

45. $\int (\sin 3x \sin 2x)^2 \, dx.$

46. $\int \cos x \cos 2x \sin 3x \sin 4x \, dx.$

Appendix to Chapter 7

§5 A Collection of Integration Formulas

We collect here the most important integration rules and formulas. To simplify writing, *we consistently omit constants of integration* (which is customarily done in tables of integrals).

Formulas involving hyperbolic functions and their inverses have been omitted. These functions can be expressed in terms of exponential and logarithmic functions.

5.1 Basic formulas

Let u and v be functions of x, and let du and dv be abbreviations for $u'(x) \, dx$ and $v'(x) \, dx$, respectively. With these notations, the **fundamental theorem of calculus** reads

$$(1) \qquad \int du = u$$

and the **substitution rule** (Chapter 5, §3.5) may be written as

$$(2) \qquad \int f'(u) \, du = f(u).$$

We have

$$(3) \qquad \int ku \, dx = k \int u \, dx, \ k \text{ a constant,}$$

$$(4) \qquad \int (u + v) \, dx = \int u \, dx + \int v \, dx,$$

$$(5) \qquad \int u \, dv = uv - \int u \, du.$$

The last formula describes **integration by parts** (Chapter 5, §§3.3, 3.4).

5.2 Powers. Rational functions.

$$(6) \qquad \int x^m \, dx = \frac{x^{m+1}}{m+1} \qquad \text{if } m \neq -1,$$

$$(7) \qquad \int \frac{dx}{x} = \log |x|,$$

(8) $\int \dfrac{dx}{(1 + x^2)^{m+1}} = \dfrac{1}{2m} \dfrac{x}{(1 + x^2)^m} + \left(1 - \dfrac{1}{2m}\right) \int \dfrac{dx}{(1 + x^2)^m}$ if $m \neq 0$.

This is a "reduction formula." If $m > 0$ is an integer, repeated application of (8) leads to the integral with $m = 0$; the latter is given by

(9) $$\int \dfrac{dx}{1 + x^2} = \arctan x.$$

Every rational function can be integrated using formulas (6), (7), (8), *and* (9); compare §2.

There are two reduction formulas analogous to (8):

(10) $\int \dfrac{dx}{(1 - x^2)^{m+1}} = \dfrac{1}{2m} \dfrac{x}{(1 - x^2)^m} + \left(1 - \dfrac{1}{2m}\right) \int \dfrac{du}{(1 - x^2)^m}$ if $m \neq 0$,

(11) $\int \dfrac{dx}{(x^2 - 1)^{m+1}} = -\dfrac{1}{2m} \dfrac{x}{(x^2 - 1)^m} - \left(1 - \dfrac{1}{2m}\right) \int \dfrac{dx}{(x^2 - 1)^m}$ if $m \neq 0$.

They are of interest if m is an integer or a half-integer. For $m = 0$, we have

(12) $$\int \dfrac{dx}{1 - x^2} = \dfrac{1}{2} \log \left| \dfrac{1 + x}{1 - x} \right|.$$

5.3 Some radical functions

(13) $$\int \dfrac{dx}{\sqrt{1 - x^2}} = \arcsin x,$$

(14) $$\int \sqrt{1 - x^2}\, dx = \dfrac{1}{2} x \sqrt{1 - x^2} + \dfrac{1}{2} \arcsin x,$$

(15) $$\int \dfrac{dx}{\sqrt{1 + x^2}} = \log (x + \sqrt{1 + x^2}),$$

(16) $$\int \sqrt{1 + x^2}\, dx = \dfrac{1}{2} x \sqrt{1 + x^2} + \dfrac{1}{2} \log (x + \sqrt{1 + x^2}),$$

(17) $$\int \dfrac{dx}{\sqrt{x^2 - 1}} = \log |x + \sqrt{x^2 - 1}|,$$

(18) $$\int \sqrt{x^2 - 1}\, dx = \dfrac{1}{2} x \sqrt{x^2 - 1} - \dfrac{1}{2} \log |x + \sqrt{x^2 - 1}|.$$

Compare §5.8 below.

5.4 Trigonometric functions. Basic formulas

(19) $$\int \sin x\, dx = -\cos x,$$

(20) $$\int \cos x\, dx = \sin x,$$

(21) $\quad \int \tan x \, dx = -\log|\cos x|,$

(22) $\quad \int \sec x \, dx = \log|\sec x + \tan x|,$

(23) $\quad \int \csc x \, dx = \log|\csc x - \cot x|,$

(24) $\quad \int \cot x \, dx = \log|\sin x|,$

(25) $\quad \int \sec^2 x \, dx = \tan x,$

(26) $\quad \int \csc^2 x \, dx = -\cot x,$

(27) $\quad \int \sec x \tan x \, dx = \sec x,$

(28) $\quad \int \csc x \cot x \, dx = -\csc x.$

5.5 Powers and products of trigonometric functions

(29) $\quad \int \sin ax \sin bx \, dx = \dfrac{\sin(a-b)x}{2(a-b)} - \dfrac{\sin(a+b)x}{2(a+b)},$

(30) $\quad \int \cos ax \cos bx \, dx = \dfrac{\sin(a-b)x}{2(a-b)} + \dfrac{\sin(a+b)x}{2(a+b)},$

(31) $\quad \int \sin ax \cos bx \, dx = -\dfrac{\cos(a-b)x}{2(a-b)} - \dfrac{\cos(a+b)x}{2(a+b)},$

provided $a^2 - b^2 \neq 0$.

(32) $\quad \int x^k \sin x \, ux = -x^k \cos x + k \int x^{k-1} \cos x \, dx,$

(33) $\quad \int x^k \cos x \, dx = x^k \sin x - k \int x^{k-1} \sin x \, dx,$

(34) $\quad \int \sin^k x \, dx = -\dfrac{\sin^{k-1} x \cos x}{k} + \dfrac{k-1}{k} \int \sin^{k-2} x \, dx,$

(35) $\quad \int \cos^k x \, dx = \dfrac{\cos^{k-1} x \sin x}{k} + \dfrac{k-1}{k} \int \cos^{k-2} x \, dx,$

(36) $\quad \int \tan^k x \, dx = \dfrac{\tan^{k-1} x}{k-1} - \int \tan^{k-2} x \, dx,$

if $k \neq -1$. The reduction formulas (34) and (35) are special cases of the following formula:

(37) $\int \sin^m x \cos^n x \, dx = -\dfrac{\sin^{m-1} x \cos^{n+1} x}{m + n} + \dfrac{m - 1}{m + n} \int \sin^{m-2} x \cos^n x \, dx$

$\qquad\qquad\qquad = \dfrac{\sin^{m+1} x \cos^{n-1} x}{m + n} + \dfrac{n - 1}{m + n} \int \sin^m x \cos^{n-2} x \, dx$

provided $m \neq -n$. Compare also §5.8 below.

5.6 Inverse trigonometric functions

(38) $\qquad\qquad\qquad \int \text{arc sin } x \, dx = x \arcsin x + \sqrt{1 - x^2},$

(39) $\qquad\qquad\qquad \int \text{arc cos } x \, dx = x \arccos x - \sqrt{1 - x^2},$

(40) $\qquad\qquad\qquad \int \text{arc tan } x \, dx = x \arctan x - \dfrac{1}{2} \log (1 + x^2),$

(41) $\qquad\qquad\qquad \int \text{arc cot } x \, dx = x \text{ arc cot } x + \dfrac{1}{2} \log (1 + x^2).$

(42) $\qquad\qquad\qquad \int \text{arc sec } x \, dx = x \text{ arc sec } x - \log (x + \sqrt{x^2 - 1}),$

(43) $\qquad\qquad\qquad \int \text{arc csc } x \, dx = x \text{ arc csc } x + \log (x + \sqrt{x^2 - 1}),$

In formulas (42) and (43), it is assumed that $x > 1$.

5.7 Exponential and logarithmic functions

(44) $\qquad\qquad\qquad \int e^x \, dx = e^x,$

(45) $\qquad\qquad\qquad \int x^k e^x \, dx = x^k e^x - k \int x^{k-1} e^x \, dx,$

(46) $\qquad\qquad\qquad \int e^x \sin ax \, dx = \dfrac{e^x}{1 + a^2} (\sin ax - a \cos ax),$

(47) $\qquad\qquad\qquad \int e^x \cos ax \, dx = \dfrac{e^x}{1 + a^2} (\cos ax + a \sin ax),$

(48) $\qquad\qquad\qquad \int \log x \, dx = x(\log x - 1),$

(49) $\qquad\qquad\qquad \int (\log x)^k \, dx = x(\log x)^k - k \int (\log x)^{k-1} \, dx,$

(50) $\qquad\qquad\qquad \int x^k \log x \, dx = x^{k+1} \left\{ \dfrac{\log x}{k + 1} - \dfrac{1}{(k + 1)^2} \right\}$

for $k \neq -1$. For $k = -1$, we have

(51) $\qquad\qquad\qquad \int \dfrac{\log x}{x} \, dx = \dfrac{1}{2} (\log x)^2.$

Integrals involving a^x and $\log_a x$, for some $a > 0$, are treated by setting $a^x = e^{x \log a}$ and $\log_a x = \log x / \log a$.

5.8 Rationalizable integrals

In this subsection $R(u)$ and $R(u, v)$, denote unspecified rational functions. The following integrals can be *rationalized* (transformed into integrals of rational functions) by the indicated substitutions.

(52) $$\int R(e^x)\, dx; \qquad \text{set } y = e^x.$$

(53) $$\int R\!\left(x, \sqrt[q]{\frac{\alpha x + \beta}{\gamma x + \delta}}\right) dx; \qquad \text{set } y = \sqrt[q]{\frac{\alpha x + \beta}{\gamma x + \delta}}.$$

(Here α, β, γ, δ are numbers such that $\alpha\delta - \beta\gamma \neq 0$ and $q > 1$ is an integer.)

(54) $$\int R(\sin x, \cos x)\, dx; \qquad \text{set } y = \tan\frac{x}{2}.$$

[This amounts to setting $\sin x = 2y/(1 + y^2)$, $\cos x = (1 - y^2)/(1 + y^2)$, and $dx = 2dy/(1 + y)$.]

The following integrals can be reduced to the form (54) by the indicated substitutions.

(55) $$\int R(x, \sqrt{1 - x^2})\, dx; \qquad \text{set } x = \cos y,$$

(56) $$\int R(x, \sqrt{1 + x^2})\, dx; \qquad \text{set } x = \tan y,$$

(57) $$\int R(x, \sqrt{x^2 - 1})\, dx; \qquad \text{set } x = \sec y.$$

(Concerning the last three integrals, see also §§3.7, 3.8, and 3.9.)

Every integral of the form

(58) $$\int R(x, \sqrt{\alpha x^2 + 2\beta x + \gamma})\, dx$$

can be reduced to integrals of the form (53), with $q = 2$, (55), (56), or (57) (compare §3.10).

Problems

1. Show how to find a primitive function of $x^m f(x)$ where $m > 1$ is an integer and $f(x)$ an inverse trigonometric function.
2. Derive the formulas in §5.3 by using the method described in §5.8.
3. Derive formulas analogous to those in §§5.4 and 5.5 for hyperbolic functions.
4. Find indefinite integrals of the functions arc sec x and arc csc x in the interval $x < -1$.

8/Series

This chapter has two main themes, the Mean Value Theorem and infinite series.

The Mean Value Theorem expresses a geometric property of differentiable functions. It has many applications, among them Taylor's formula for writing any function with sufficiently many derivatives as the sum of a polynomial and a small remainder term.

Functions for which the remainder can be made as small as desired, by choosing the degree of the polynomial sufficiently high are called analytic. Most of the important functions used in calculus have this property.

In order to study analytic functions, we first investigate (§§3 and 4) infinite sequences and series.

There are four appendixes. One (§6) deals with l'Hospital's rule for evaluating limits. The other three contain proofs of some theorems stated in the text: §7 deals with sequences and series, §8 with power series and §9 with the error estimates in numerical integration.

Statement of the Mean Value Theorem

At the very beginning of our discussion of calculus, as soon as we defined the derivative, we arrived at an approximate method for computing the value of a function $f(x)$ close to a point x_0 at which we knew $f(x_0)$ and $f'(x_0)$ (see Chapter 4, §1.8). The formula read

$$(1) \qquad f(x) \approx f(x_0) + f'(x_0)(x - x_0),$$

where "$\approx$" stands for "approximately equal." Now we take up this subject once more. We shall give a precise meaning to the words "approximately equal" by estimating the error committed when using (1). Calculation of this error has far-reaching consequences. We begin by showing that a formula that differs very little from (1) is precisely true.

Consider a graph of a function $x \mapsto f(x)$ defined for $a \leq x \leq b$ and assume that $f'(x)$ exists and is continuous. The graph is a smooth curve joining the points $P = (a, f(a))$ and $Q = (b, f(b))$. It seems geometrically obvious that there must be points on this graph, between P and Q, at which the tangent to the curve is parallel to the chord $\overrightarrow{PQ}$. (See Fig. 8.1.)

Let us formulate the statement "there is a tangent parallel to the chord" analytically. The slope of the chord is, of course,

$$\frac{f(b) - f(a)}{b - a},$$

and the slope of a tangent at a point $(\xi, f(\xi))$ is $f'(\xi)$. The tangent and the slope are parallel if their slopes are equal. Thus we expect the following to hold.

Theorem 1 (*Mean Value Theorem*)   IF THE FUNCTION $x \mapsto f(x)$ IS CONTINUOUS FOR $a \leq x \leq b$, AND HAS A CONTINUOUS DERIVATIVE FOR $a < x < b$, THEN THERE IS A NUMBER ξ SUCH THAT

$$a < \xi < b$$

AND

$$(2) \qquad \frac{f(b) - f(a)}{b - a} = f'(\xi).$$

Let us set $a = x_0$, $b = x$; then we can rewrite (2) as $f(x) - f(x_0) = f'(\xi)(x - x_0)$, or

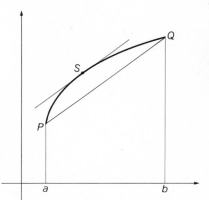

Fig. 8.1

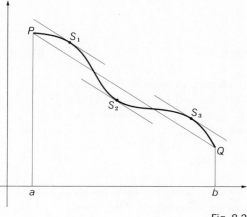

Fig. 8.2

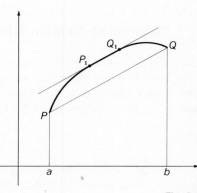

Fig. 8.3

(3) $$f(x) = f(x_0) + f'(\xi)(x - x_0).$$

This looks very much like (1), except that we have an $=$ sign instead of the $\approx$ sign and $f'(\xi)$ instead of $f'(x_0)$, where ξ is a point between x_0 and x.

There may be, of course, more than one point on the arc $y = f(x)$, $a \leq x \leq b$, at which the tangent is parallel to the chord. In Fig. 8.1, there is one and only one such point, namely, S; in Fig. 8.2, there are three points, S_1, S_2, and S_3, with the desired property; in Fig. 8.3, all points between P_1 and Q_1 will do. We note that, if the derivative $f'(x)$ fails to exist at a single point, there need be no tangent parallel to the chord; such a case is shown in Fig. 8.4. Thus the condition that f has a derivative at all points between a and b is essential for Theorem 1.

Fig. 8.4

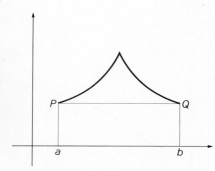

1.2 Rolle's Theorem

An important special case of the Mean Value Theorem (Theorem 1) is Rolle's Theorem: if $f(a) = f(b)$, then there is a number ξ between a and b with $f'(\xi) = 0$. Rolle's Theorem is sometimes stated as follows:

BETWEEN TWO ZEROS OF A FUNCTION, THERE IS A ZERO OF THE DERIVATIVE.

We prove Rolle's Theorem first. If, for all x, $a < x < b$, $f'(x) \neq 0$, then either $f'(x) > 0$ in the whole interval, or $f'(x) < 0$ in the whole interval. Indeed, $f'(x)$ is continuous and if it takes on a positive and a negative value, it must also take on the value 0 [see the Intermediate

MICHEL ROLLE (1652–1719) was a French mathematician. He stated the theorem that bears his name as a rule for locating roots of polynomials.

Value Theorem (Theorem 5) in Chapter 3, §3.6]. If $f'(x) > 0$ for all x, then f is increasing [see Theorem 2 in Chapter 4, §4.2] and $f(b) > f(a)$. If $f'(x) < 0$ for all x, then f is decreasing and $f(b) < f(a)$. Neither is possible. Thus our assertion that $f'(\xi) = 0$ for some ξ between a and b is proved.

1.3 Proof of the Mean Value Theorem

The general form of Theorem 1 follows easily from Rolle's Theorem. Indeed, define the function $\phi(x)$ by

(4) $$\phi(x) = \frac{f(b) - f(a)}{b - a}(x - a) - [f(x) - f(a)].$$

Then ϕ is continuous where f is, and ϕ has the continuous derivative

$$\phi'(x) = \frac{f(b) - f(a)}{b - a} - f'(x).$$

Also we verify at once that $\phi(a) = \phi(b) = 0$. Therefore Rolle's Theorem applies: there is a number ξ between a and b with $\phi'(\xi) = 0$. But the relation $\phi'(\xi) = 0$ is the same as (2).

Remark   One can prove Rolle's Theorem, and hence also the Mean Value Theorem, *without* assuming the continuity of the derivative. (See the Supplement, §2.3, for this proof.) The form stated above, however, is sufficient for most applications.

EXERCISES

1. If $f(x) = 1 - 3x$, find all numbers ξ strictly between 1 and 4 such that
$$f'(\xi) = \frac{f(4) - f(1)}{4 - 1}.$$

2. If $f(x) = x^2 + 1$, find all numbers ξ strictly between 1 and 2 such that
$$f'(\xi) = \frac{f(2) - f(1)}{2 - 1}.$$

3. If $f(x) = \log x$, find all numbers ξ strictly between e^2 and e^3 such that
$$f'(\xi) = \frac{f(e^3) - f(e^2)}{e^3 - e^2}.$$

4. If $f(x) = x^3 - x$, find all numbers ξ strictly between 0 and 1 such that
$$f'(\xi) = f(1) - f(0).$$

5. If $f(x) = \sqrt{5}x^3 + x^2$, find all numbers ξ strictly between -1 and 1 such that
$$f'(\xi) = \frac{f(1) - f(-1)}{1 - (-1)}.$$

▶ 6. If $f(x) = x^{10} + x^4 + 1$, find all numbers ξ strictly between -1 and 1 such that
$$f'(\xi) = \frac{f(1) - f(-1)}{1 - (-1)}.$$

7. If $f(x) = x^3 + x^2 - 2x$, find all numbers ξ strictly between 0 and 1 such that
$$f'(\xi) = f(1) - f(0).$$

8. If $f(x) = x^{5/3} + x^{4/3}$, find all numbers ξ strictly between -8 and 8 such that
$$f'(\xi) = \frac{f(8) - f(-8)}{8 - (-8)}.$$

9. If $f(x) = \sqrt{x^2 + 9}$, find all numbers ξ strictly between 0 and 4 such that
$$f'(\xi) = \frac{f(4) - f(0)}{4}.$$

10. If $f(x) = e^x$, find all numbers ξ strictly between 1 and 2 such that $f'(\xi) = \frac{f(2) - f(1)}{2 - 1}.$

1.4 Generalized Mean Value Theorem

There is an easy but extremely useful generalization of the Mean Value Theorem.

Theorem 2 (Generalized Mean Value Theorem)   LET $F(x)$ AND $G(x)$ BE CONTINUOUS FUNCTIONS DEFINED FOR $a \leq x \leq b$, AND ASSUME THAT THE DERIVATIVES $F'(x)$ AND $G'(x)$ EXIST AND ARE CONTINUOUS, AND $G'(x) \neq 0$, FOR $a < x < b$. THEN THERE IS A NUMBER ξ SUCH THAT

$$a < \xi < b$$

AND

(5)
$$\frac{F(b) - F(a)}{G(b) - G(a)} = \frac{F'(\xi)}{G'(\xi)}.$$

One can dispense with the condition $G' \neq 0$, if one agrees to interpret (5) to mean that $[F(b) - F(a)]G'(\xi) = [G(b) - G(a)]F'(\xi)$.

Proof.   Define the new function

$$f(x) = [F(b) - F(a)][G(x) - G(a)] - [G(b) - G(a)][F(x) - F(a)].$$

It is continuous for $a \leq x \leq b$, and if $a < x < b$, then

$$f'(x) = [F(b) - F(a)]G'(x) - [G(b) - G(a)]F'(x).$$

Also $f(a) = f(b) = 0$. By Rolle's Theorem, there is a number ξ between a and b with $f'(\xi) = 0$. Thus $[F(b) - F(a)]G'(\xi) = [G(b) - G(a)]F'(\xi)$. This is the same as (5), provided $G'(\xi) \neq 0$ and $G(b) - G(a) \neq 0$.

Corollary. *If* $F(a) = G(a) = 0$, *then for every* $b \neq a$ *there is a* ξ *between a and b with*

(6)
$$\frac{F(b)}{G(b)} = \frac{F'(\xi)}{G'(\xi)}.$$

It is assumed that $F'(t)$, $G'(t)$ exist and are continuous for $a < t < b$, and $G'(t) \neq 0$. This corollary follows at once from the theorem. (In the corollary we may have $b < a$.)

EXERCISES

11. If $F(x) = 2x + 1$ and $G(x) = 3x - 4$, find all numbers ξ strictly between 1 and 3 such that $\dfrac{F(3) - F(2)}{G(3) - G(2)} = \dfrac{F'(\xi)}{G'(\xi)}$.

12. If $F(x) = x^3$ and $G(x) = 2 - x$, find all numbers ξ strictly between 0 and 9 such that $\dfrac{F(9) - F(0)}{G(9) - G(0)} = \dfrac{F'(\xi)}{G'(\xi)}$.

13. If $F(x) = 1/x$ and $G(x) = x^2$, find all numbers ξ strictly between 1 and 2 such that $\dfrac{F(2) - F(1)}{G(2) - G(1)} = \dfrac{F'(\xi)}{G'(\xi)}$.

14. If $F(x) = \sin x$ and $G(x) = \cos x$, find all numbers ξ strictly between $\pi/4$ and $3\pi/4$ such that $\dfrac{F(3\pi/4) - F(\pi/4)}{G(3\pi/4) - G(\pi/4)} = \dfrac{F'(\xi)}{G'(\xi)}$.

15. If $F(x) = \log x$ and $G(x) = 1/x$, find all numbers ξ strictly between 1 and e such that $\dfrac{F(e) - F(1)}{G(e) - G(1)} = \dfrac{F'(\xi)}{G'(\xi)}$.

16. If $F(x) = (x^4/4) - x^3 + x^2$ and $G(x) = x^2$, find all numbers ξ strictly between 0 and 2 such that $\dfrac{F(2) - F(0)}{G(2) - G(0)} = \dfrac{F'(\xi)}{G'(\xi)}$.

17. If $F(x) = \sqrt{x + 9}$ and $G(x) = \sqrt{x}$, find all numbers ξ strictly between 0 and 16 such that $\dfrac{F(16) - F(0)}{G(16) - G(0)} = \dfrac{F'(\xi)}{G'(\xi)}$.

18. If $F(x) = x^4 + 4x$ and $G(x) = x^2 + 2x$, find all numbers ξ strictly between -1 and 2 such that $\dfrac{F(2) - F(-1)}{G(2) - G(-1)} = \dfrac{F'(\xi)}{G'(\xi)}$.

19. If $F(x) = \sqrt{x^2 + 9}$ and $G(x) = x^2 + 1$, find all numbers ξ strictly between 0 and 4 such that $\dfrac{F(4) - F(0)}{G(4) - G(0)} = \dfrac{F'(\xi)}{G'(\xi)}$.

20. If $F(x) = x^3$ and $G(x) = x^3 + 3x^2 + 3x$, find all numbers ξ strictly between 0 and 3 such that $\dfrac{F(3) - F(0)}{G(3) - G(0)} = \dfrac{F'(\xi)}{G'(\xi)}$.

1.5 The error in linear approximation

We now have the tools for estimating the error in the approximate formula (1). The use of this formula amounts to replacing the graph of f

Fig. 8.5

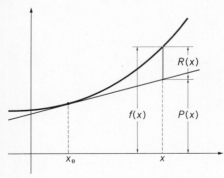

by its tangent at $(x_0, f(x_0))$. This tangent is the graph of the linear function

$$x \mapsto P(x) = f(x_0) + f'(x_0)(x - x_0),$$

which has at x_0 the value and slope of $f(x)$; see Fig. 8.5. Let $R(x)$ denote the error committed if we compute the value of $f(x)$ using P instead of f, that is, let $R(x) = f(x) - P(x)$. Then we have

(7) $$f(x) = f(x_0) + f'(x_0)(x - x_0) + R(x).$$

Therefore $$\frac{R(x)}{x - x_0} = \frac{f(x) - f(x_0)}{x - x_0} - f'(x_0),$$

and since $$\lim_{x \to x_0} \frac{f(x) - f(x_0)}{x - x_0} = f'(x_0)$$

we have that

(8) $$\lim_{x \to x_0} \frac{R(x)}{x - x_0} = 0.$$

This means that $|R(x)|$ is small if x is close to x_0, and even becomes arbitrarily small compared to $|x - x_0|$ (compare Theorem 1 in Chapter 4, §1.8). But statement (8) does not tell us *how* small $R(x)$ is for a given $x \neq x_0$.

In order to answer the latter question, we need additional information on the function $f(x)$. We shall assume that f has a continuous *second* derivative.

We consider the remainder $R(x)$ for a fixed x_0 and a variable x. Clearly

$$R(x_0) = 0$$

[this is seen by substituting $x = x_0$ in (7)]. Also, differentiating both sides of (7), and noting that x_0 is fixed, we get

$$f'(x) = f'(x_0) + R'(x)$$

so that $$R'(x_0) = 0.$$

Another differentiation gives

$$f''(x) = R''(x).$$

Now we apply the corollary to Theorem 2 with $F(t) = R(t)$, $G(t) = (t - x_0)^2$, $a = x_0$, $b = x$. There exists a number τ between x_0 and x such that

$$\frac{R(x)}{(x - x_0)^2} = \frac{R'(\tau)}{2(\tau - x_0)}.$$

Applying the corollary to Theorem 2 once more, this time to $F(t) = R'(t)$ and $G(t) = 2(t - x_0)$, we conclude that there exists a ξ between x_0 and τ with

$$\frac{R'(\tau)}{2(\tau - x_0)} = \frac{R''(\xi)}{2} = \frac{f''(\xi)}{2}.$$

Thus

$$\frac{R(x)}{(x - x_0)^2} = \frac{f''(\xi)}{2}$$

or

$$(9) \qquad R(x) = \frac{f''(\xi)(x - x_0)^2}{2}.$$

In particular, if

$$|f''(t)| \leq M \qquad \text{for } t \text{ between } x \text{ and } x_0$$

then

$$(10) \qquad |R(x)| \leq \frac{M}{2}(x - x_0)^2.$$

Thus we proved that

$$(11) \qquad f(x) = f(x_0) + f'(x_0)(x - x_0) + R(x),$$

where the remainder $R(x)$ is given by (9) for some ξ between x and x_0, and satisfies inequality (10).

▶ **Example** What error is committed in computing $\sqrt{1 + h}$, for small $|h|$, by the approximate formula $\sqrt{1 + h} = 1 + \frac{1}{2}h$?

Answer. The approximate formula is a special case of (1) with $f(x) = \sqrt{x}$, $f'(x) = 1/2\sqrt{x}$, $x_0 = 1$, and $x = 1 + h$. We have

$$f''(x) = -\frac{1}{4x\sqrt{x}} = -\frac{1}{4(1 + h)\sqrt{1 + h}},$$

and, in order to apply the inequality (10) for the error R, we must find a number M such that $|f''(x)| \leq M$ for small $|h|$.

We note that $|f''(x)| = \frac{1}{4}$ if $h = 0$. If $h > 0$, then $(1 + h)\sqrt{1 + h} > 1$ and $|f''(x)| < \frac{1}{4}$. Also, $|f''(x)|$ is an increasing function of h. For $h = -\frac{3}{4}$, we have $1 + h = \frac{1}{4}$, $\sqrt{1 + h} = \frac{1}{2}$ and hence $|f''(x)| = 2$. Thus $|f''(x)| \leq 2$ for $|h| < \frac{3}{4}$. Applying (10) with $M = 2$, we obtain $|R| \leq h^2$. For small h (more precisely, for $|h| < \frac{3}{4}$) the error is not greater than h^2.

For instance, $\sqrt{1.02} = 1.01$ and $\sqrt{98} = .99$ with an accuracy of .0004. Similarly, $\sqrt{1.03} = 1.015$ and $\sqrt{.97} = .985$ with an error not exceeding .0009.

Another application of the Mean Value Theorem is l'Hospital's rule for evaluating limits. We discuss it in an appendix (see §6). ◄

EXERCISES

For each of Exercises 21 to 30, find an approximate value for the given expression and estimate the error. Use the linear approximation of this section.

21. $\sqrt{50}$.　　　25. $(1.1)^{3/5}$.　　　29. $e^{-.01}$.
22. $\sqrt{.8}$.　　　26. $\log(.98)$.　　　30. $\tan(.24\pi)$.
23. $(.025)^{1/3}$.　　　27. $\cos(.01)$.
24. $(999)^{1/3}$.　　　28. $\arctan(.01)$.

§2　Taylor's Theorem

2.1　Parabolic approximation

Let $f(x)$ be a function defined near the point x_0. In the preceding section, we derived the Mean Value Theorem

$$(1) \qquad f(x) = f(x_0) + f'(\xi)(x - x_0)$$

and the formula

$$(2) \qquad f(x) = f(x_0) + f'(x_0)(x - x_0) + \frac{1}{2}f''(\xi)(x - x_0)^2.$$

It is assumed that the derivatives occurring in these relations are continuous. The number ξ is a suitably chosen number between x and x_0, not necessarily the same in both formulas.

The next step in the discussion is almost self-evident. We consider a function $f(x)$ for which we know the values of

$$f(x_0), f'(x_0), \text{ and } f''(x_0).$$

We want to compute, approximately, the value of $f(x)$ at a point near x_0 by using not a constant, as in (1), and not a linear function as in (2), but a quadratic polynomial $P(x)$. This polynomial we choose so that

$$(3) \qquad P(x_0) = f(x_0), P'(x_0) = f'(x_0), P''(x_0) = f''(x_0).$$

In other words: the parabola $y = P(x)$ should pass through the point $(x_0, f(x_0))$ and should, at this point, have the same slope and the same derivative of the slope (one also says: the same curvature, cf. Chapter 11, §3.15) as the graph of $y = f(x)$. We expect that the error committed,

$$R(x) = f(x) - P(x),$$

will be, for x close to x_0, significantly smaller than the error in the linear approximation considered before. The situation is depicted in Fig. 8.6.

First we must find $P(x)$. It is easy to guess that

$$P(x) = f(x_0) + f'(x_0)(x - x_0) + \frac{1}{2} f''(x_0)(x - x_0)^2.$$

Indeed, this $P(x)$ is a quadratic polynomial in x (x_0 is kept fixed) and one computes at once that (3) holds. In order to determine the remainder $R(x)$, we note that

$$R(x_0) = R'(x_0) = R''(x_0) = 0,$$

since P has been chosen to satisfy condition (3), and that

(4) $$R'''(x) = f'''(x),$$

since P is a quadratic polynomial and therefore $P'''(x) \equiv 0$. Now we shall apply several times the corollary to the Generalized Mean Value Theorem (see §1.4). We obtain

$$\frac{R(x)}{(x - x_0)^3} = \frac{R'(t_1)}{3(t_1 - x_0)^2} \qquad \text{for some } t_1 \text{ between } x_0 \text{ and } x,$$

$$\frac{R'(t_1)}{3(t_1 - x_0)^2} = \frac{R''(t_2)}{3 \cdot 2(t_2 - x_0)} \qquad \text{for some } t_2 \text{ between } x_0 \text{ and } t_1,$$

$$\frac{R''(t_2)}{3 \cdot 2(t_2 - x_0)} = \frac{R'''(\xi)}{3 \cdot 2} \qquad \text{for some } \xi \text{ between } x_0 \text{ and } t_2.$$

Noting (4), we combine these results into

$$\frac{R(x)}{(x - x_0)^3} = \frac{f'''(\xi)}{3 \cdot 2} \qquad \text{for some } \xi \text{ between } x \text{ and } x_0.$$

This may be written as

(5) $$R(x) = \frac{f'''(\xi)}{1 \cdot 2 \cdot 3} (x - x_0)^3.$$

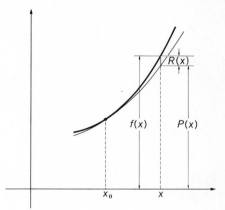

Fig. 8.6

Thus we obtain:

$$(6) \quad f(x) = f(x_0) + f'(x_0)(x - x_0) + \frac{1}{1 \cdot 2} f''(x_0)(x - x_0)^2 + R(x),$$

where R is given by (5), for a suitable point ξ between x_0 and x. If we also know that

$$|f'''(t)| \leq M$$

for all t between x and x_0, then

$$(7) \qquad |R| \leq \frac{M|x - x_0|^3}{6}.$$

▶**Example** We use the notations of the example in §1.5 and note that, for $f(x) = \sqrt{x}$, we have $f'''(x) = \frac{3}{8}x^{-5/2}$. We compute easily that $|f'''(x)| < 12$ for $\frac{1}{4} \leq x \leq \frac{5}{4}$, that is, for $|h| = |1 - x| \leq \frac{3}{4}$. Thus we obtain the relation

$$\sqrt{1 + h} = 1 + \frac{1}{2} h - \frac{1}{8} h^2 + R, \qquad |R| \leq 2h^3$$

valid for $|h| < \frac{3}{4}$. [Set $f(x) = \sqrt{x}$, $x_0 = 1$, $x = 1 + h$ and $M = 12$ in (6) and in (7).]

For instance, we have $\sqrt{1.2} = 1.1 - \frac{1}{8}(.04) = 1.095$ and $\sqrt{.98} = .99 - \frac{1}{8}(.04) = .985$ with an error not exceeding $2(.2)^3 = .016$. Also $\sqrt{1.03} = 1 + \frac{1}{2}(.03) - \frac{1}{8}(.0009) = 1.0149$ with an error not exceeding $2(.03)^3 < .00006$. ◀

EXERCISES

In each of Exercises 1 to 5, first find all values ξ strictly between x_1 and x_0 such that $f(x_1) = f(x_0) + f'(x_0)(x_1 - x_0) + \frac{1}{2}f''(\xi)(x_1 - x_0)^2$. Then find all values η strictly between x_1 and x_0 such that

$$f(x_1) = f(x_0) + f'(x_0)(x_1 - x_0) + \frac{1}{2}f''(x_0)(x_1 - x_0)^2 + \frac{1}{6}f'''(\eta)(x_1 - x_0)^3.$$

1. $f(x) = x^3$, $x_0 = 1$, $x_1 = \frac{1}{2}$.
2. $f(x) = 1/x$, $x_0 = 1$, $x_1 = \frac{1}{8}$.
3. $f(x) = e^x$, $x_0 = 0$, $x_1 = 1$.
4. $f(x) = \log x$, $x_0 = 1$, $x_1 = e$.
5. $f(x) = x^4 - x^2$, $x_0 = 1$, $x_1 = 2$.

For each of Exercises 6 to 10, find an approximate value for the given expression and estimate the error. Use the quadratic approximation of this section. Compare your answers with those for Exercises 21, 23, 25, 27, and 29 in §1.

6. $\sqrt{50}$.
7. $(.025)^{1/3}$.
8. $(1.1)^{3/5}$.
9. $\cos(.01)$.
▶10. $e^{-.01}$.

2.2 Taylor polynomials

The three relations (1), (2), and (6) are answers to three special cases of the following question. Let $x \mapsto f(x)$ be a function; suppose that we know the values of f and its first n derivatives f', f'', $\cdots$, $f^{(n)}$ at a point x_0. Now can we compute, approximately, the value of $f(x)$ at a point x close to x_0?

At the first glance, this question seems very special and somewhat contrived. It turns out, however, that it leads to some of the most important topics in mathematics.

The answer to the question will involve two steps. First we shall construct a polynomial $P(x)$, of degree n, which has at $x = x_0$ the same value and the same first, second, third, $\cdots$, nth derivatives as the function f. There is, as we shall see, only one such polynomial. We call it the nth **Taylor polynomial** of $f(x)$ at x_0. We expect that, for x not too far from x_0, the value of $P(x)$ is close to that of $f(x)$.

The second step will consist in studying the **remainder**

BROOKS TAYLOR (1685–1731) published the series which bears his name in a book that appeared in 1715. Taylor for a time was secretary of the Royal Society.

$$(8) \qquad R(x) = f(x) - P(x).$$

We begin by finding P.

Theorem 1 GIVEN $n + 2$ NUMBERS $x_0, \alpha_0, \alpha_1, \cdots, \alpha_n$, THERE IS EXACTLY ONE POLYNOMIAL $P(x)$ OF DEGREE AT MOST n THAT SATISFIES THE CONDITIONS

$$(9) \quad P(x_0) = \alpha_0, \qquad P'(x_0) = \alpha_1, \qquad P''(x_0) = \alpha_2, \cdots, P^{(n)}(x_0) = \alpha_n.$$

THIS IS THE POLYNOMIAL

$$(10) \quad P(x) = \alpha_0 + \frac{\alpha_1}{1}(x - x_0) + \frac{\alpha_2}{1 \cdot 2}(x - x_0)^2 + \frac{\alpha_3}{1 \cdot 2 \cdot 3}(x - x_0)^3$$
$$+ \cdots + \frac{\alpha^n}{1 \cdot 2 \cdot 3 \cdots n}(x - x_0)^n.$$

Proof. $P(x)$ is certainly a polynomial and its degree is at most n. (It may be less than n, since it may happen that $\alpha_n = 0$.) We note next that, for all positive integers j and k,

$$\frac{d^k(x - x_0)^j}{dx^k} = j(j - 1)(j - 2) \cdots (j - k + 1)(x - x_0)^{j-k} \quad \text{if } k < j,$$

$$\frac{d^j(x - x_0)^j}{dx^j} = 1 \cdot 2 \cdot 3 \cdots j, \qquad \frac{d^k(x - x_0)^j}{dx^k} = 0 \quad \text{if } k > j.$$

(The reader may convince himself that this is so by computing some special case, like $k = 4$ and $k = 5$. A formal proof would involve mathematical induction. But the picture is so clear that a formal proof would be an unnecessary pedantry. We shall encounter such situations several times in this chapter.)

Now we write down the derivatives of (10), using the above relations. We have

$$P'(x) = \alpha_1 + \alpha_2(x - x_0) + \frac{\alpha_3}{1 \cdot 2}(x - x_0)^2$$
$$+ \cdots + \frac{\alpha_n}{1 \cdot 2 \cdots (n-1)}(x - x_0)^{n-1},$$

$$P''(x) = \alpha_2 + \alpha_3(x - x_0) + \frac{\alpha_4}{1 \cdot 2}(x - x_0)^2$$
$$+ \cdots + \frac{\alpha_n}{1 \cdot 2 \cdots (n-2)}(x - x_0)^{n-2},$$

and so on, the kth derivative $(k < n)$ being

$$P^{(k)}(x) = \alpha_k + \alpha_{k+1}(x - x_0) + \cdots + \frac{\alpha_n}{1 \cdot 2 \cdots (n-k)}(x - x_0)^{n-k}$$

and the nth

$$P^{(n)}(x) = \alpha_n.$$

If we substitute $x = x_0$ into these expressions, we obtain (9).

Suppose that there is another polynomial of degree at most n, call it $\hat{P}(x)$, which also satisfies (9). We must show that $\hat{P} \equiv P$.

Set $Q(x) = P(x) - \hat{P}(x)$. Then Q is a polynomial of degree at most n, so that the nth derivative $Q^{(n)}(x)$ is a constant. But $Q^{(n)}(x_0) = P^{(n)}(x_0) - \hat{P}^{(n)}(x_0) = \alpha_n - \alpha_n = 0$. Hence $Q^{(n)}(x) \equiv 0$ and the degree of $Q(x)$ is at most $n - 1$. Therefore $Q^{(n-1)}(x)$ is a constant. But $Q^{(n-1)}(x_0) = P^{(n-1)}(x_0) - \hat{P}^{(n-1)}(x_0) = \alpha_{n-1} - \alpha_{n-1} = 0$. Hence $Q^{(n-1)}(x) \equiv 0$ and the degree of $Q(x)$ is at most $n - 2$. Continuing this way, we finally arrive at the conclusion that the degree of $Q(x)$ is 0, that is, that $Q(x)$ is constant. Since $Q(x_0) = \hat{P}(x_0) - P(x_0) = \alpha_0 - \alpha_0 = 0$, we have $Q(x) \equiv 0$.

Corollary. *If the function $f(x)$ has n derivatives, that is, derivatives of order $1, 2, \cdots, n$ at x_0, then the nth Taylor polynomial of f at x_0 is*

$$P(x) = f(x_0) + \frac{f'(x_0)}{1}(x - x_0) + \frac{f''(x_0)}{1 \cdot 2}(x - x_0)^2$$

(11)

$$+ \frac{f'''(x_0)}{1 \cdot 2 \cdot 3}(x - x_0)^3 + \cdots + \frac{f^{(n)}(x_0)}{1 \cdot 2 \cdot 3 \cdots n}(x - x_0)^n.$$

We note some properties of Taylor polynomials. First, the $(n + 1)$st Taylor polynomial is obtained from the nth by adding one term, namely,

$$\frac{f^{(n+1)}(x_0)}{1 \cdot 2 \cdot 3 \cdots (n + 1)} (x - x_0)^{n+1}.$$

Next, let $P(x)$ be the Taylor polynomial of degree n of $f(x)$ at x_0. Then $P'(x)$ is the Taylor polynomial of $f'(x)$ at x_0, of degree $n - 1$. Indeed P' is a polynomial of degree $n - 1$. Since $P^{(j)}(x_0) = f^{(j)}(x_0)$ for $j = 1, \cdots, n$, we see that the values of P' and f' as well as those of their respective derivatives up to the order $n - 1$ coincide at x_0.

One proves similarly that $\int_{x_0}^{x} P(t)\, dt$ is the Taylor polynomial of degree $n + 1$ of the function $\int_{x_0}^{x} f(t)\, dt$, again at x_0. Also $cP(x)$, c a constant, is the Taylor polynomial of $cf(x)$. If $x_0 = 0$, then $P(-x)$ is the Taylor polynomial of $f(-x)$, at $-x_0$, $P(x^k)$ is the Taylor polynomial of degree nk of $f(x^k)$, and so forth.

The coefficients of the Taylor polynomial of a function f at a point x_0 are called the **Taylor coefficients** of f at x_0; the coefficient of $(x - x_0)^j$ is called the jth coefficient.

▶**Examples**   **1.** Find the fourth Taylor polynomial, at $x_0 = 0$, for

$$f(x) = \sqrt{1 + x}$$

Answer.   We have $f'(x) = \frac{1}{2}(1 + x)^{-1/2}$, $f''(x) = -\frac{1}{4}(1 + x)^{-3/2}$, $f'''(x) = \frac{3}{8}(1 + x)^{-5/2}$, $f^{(4)}(x) = -\frac{15}{16}(1 + x)^{-7/2}$. Therefore $f(0) = 1$, $f'(0) = \frac{1}{2}, f''(0) = -\frac{1}{4}, f'''(0) = \frac{3}{8}, f^{(4)}(0) = -\frac{15}{16}$. The desired polynomial is

$$1 + \frac{1}{2} x - \frac{1}{1 \cdot 2} \cdot \frac{1}{4} x^2 + \frac{1}{1 \cdot 2 \cdot 3} \cdot \frac{3}{8} x^3 - \frac{1}{1 \cdot 2 \cdot 3 \cdot 4} \cdot \frac{15}{16} x^4$$

or

$$1 + \frac{x}{2} - \frac{x^2}{8} + \frac{x^3}{16} - \frac{5x^4}{128}.$$

2. Find the fourth Taylor coefficient of the function $f(x) = \sqrt{1 + x}$ at $x_0 = -\frac{1}{2}$.

Answer.   We saw that $f^{(4)}(x) = -\frac{15}{16}(1 + x)^{-7/2}$. Thus $f^{(4)}(x_0) = -\frac{15}{16}(\frac{1}{2})^{-7/2} = -15/\sqrt{2}$, and the fourth Taylor coefficient is

$$\frac{1}{1 \cdot 2 \cdot 3 \cdot 4} f^{(4)}(x_0) = \frac{-5}{8\sqrt{2}}.$$

3. Find the fourth Taylor polynomial of the function $f(x) = (1 + x)^{-3/2}$ at $x = 0$.

Answer. We note that $(1 + x)^{3/2} = 1 + \frac{3}{2} \int_0^x \sqrt{1 + t}\, dt$. Using the result of Example 1, we conclude that the desired polynomial is

$$1 + \frac{3}{2} \int_0^x \left(1 + \frac{t}{2} - \frac{t^2}{8} + \frac{t^3}{16}\right) dt = 1 + \frac{3x}{2} + \frac{3x^2}{8} - \frac{3x^3}{48} + \frac{3x^4}{128}.$$

Of course, this can also be obtained directly.

4. The function $f(x) = x^{11/3}$ has at $x = 0$ the third Taylor polynomial $P(x) \equiv 0$, since $f(0) = f'(0) = f''(0) = f'''(0) = 0$. There is *no* fourth Taylor polynomial, since the function has no fourth derivative at $x = 0$. ◄

EXERCISES

11. Find the third Taylor polynomial of the function $x \mapsto 1/(x + 1)$ at $x_0 = 0$.
12. Find the fourth Taylor polynomial of the function $x \mapsto (x + 1)^{4/3}$ at $x_0 = 0$.
13. Find the fifth Taylor polynomial of the function $x \mapsto \cos x$ at $x_0 = 0$.
14. Find the third Taylor polynomial of the function $x \mapsto \log (x^2 + 1)$ at $x_0 = 0$.
15. Find the fourth Taylor polynomial of the function $x \mapsto \int_0^x \log (u^2 + 1)\, du$ at $x_0 = 0$.
16. Find the fourth Taylor polynomial of the function $x \mapsto x^{13/3}$ at $x_0 = 0$.
17. Find the third Taylor polynomial of the function $x \mapsto x^4 - x^3 + 2x^2 - 1$ at $x_0 = 1$.
18. Find the fourth Taylor polynomial of the function $x \mapsto \log^2 x$ at $x_0 = 1$.
19. Find the sixth Taylor polynomial of the function $x \mapsto \cos x$ at $x_0 = \pi/2$.
20. Find the tenth Taylor polynomial of the function $x \mapsto e^{x^5}$ at $x_0 = 0$.
21. Find the thirty-first Taylor polynomial of the function $x \mapsto \int_0^x \arctan u^{10}\, du$ at $x_0 = 0$.
22. Find the twelfth Taylor polynomial of the function $x \mapsto (x^4 + 1)^{4/3}$ at $x_0 = 0$.

2.3 Taylor's Theorem

We now consider the "remainder" $R_{n+1}(x) = f(x) - P_n(x)$, where P_n is the nth Taylor polynomial of f at x_0. We assume that $f(x)$ is defined and continuous in an interval I containing x_0 and x, and that it has a continuous $(n + 1)$st derivative in I, except perhaps at the endpoints. We often write $R(x)$ for $R_{n+1}(x)$.

By construction $R(x_0) = R'(x_0) = \cdots = R^{(n)}(x_0) = 0$. Hence repeated application of the corollary to the Generalized Mean Value Theorem (see §1.4) gives

$$\frac{R(x)}{(x - x_0)^{n+1}} = \frac{R'(t_1)}{(n + 1)(t_1 - x_0)^n} \qquad \text{for some } t_1 \text{ between } x_0 \text{ and } x$$

$$= \frac{R''(t_2)}{(n + 1)n(t_2 - x_0)^{n-1}} \qquad \text{for some } t_2 \text{ between } x_0 \text{ and } t_1$$

$$= \frac{R'''(t_3)}{(n+1)n(n-1)(t_3-x_0)^{n-2}} \qquad \text{for some } t_3 \text{ between } x_0 \text{ and } t_2$$

$$= \cdots$$

$$= \frac{R^{(n+1)}(\xi)}{(n+1)n(n-1)\cdots 2 \cdot 1} \qquad \begin{array}{l}\text{for some } \xi \text{ between } x_0 \text{ and } t_n, \\ \text{hence between } x_0 \text{ and } x.\end{array}$$

Since $R^{(n+1)}(x) = f^{(n+1)}(x)$ by construction, we have

$$\frac{R_{n+1}(x)}{(x-x_0)^{n+1}} = \frac{f^{(n+1)}(\xi)}{1 \cdot 2 \cdot 3 \cdots (n+1)}.$$

We summarize our results in the following theorem.

Theorem 2 (Taylor's Theorem) LET $f(x)$ BE DEFINED AND CONTINUOUS IN AN INTERVAL I CONTAINING x AND x_0, AND LET IT HAVE A DERIVATIVE OF ORDER n AT x AND A CONTINUOUS DERIVATIVE OF ORDER $n+1$ AT ALL INNER POINTS OF I. SET

$$(12) \quad \begin{aligned} f(x) = {} & f(x_0) + \frac{f'(x_0)}{1}(x-x_0) + \frac{f''(x_0)}{1 \cdot 2}(x-x_0)^2 \\ & + \frac{f'''(x_0)}{1 \cdot 2 \cdot 3}(x-x_0)^3 + \cdots + \frac{f^{(n)}(x_0)}{1 \cdot 2 \cdot 3 \cdots n}(x-x_0)^n + R_{n+1}(x). \end{aligned}$$

THEN THERE IS A NUMBER ξ BETWEEN x AND x_0 SUCH THAT

$$(13) \qquad R_{n+1}(x) = \frac{f^{(n)}(\xi)}{1 \cdot 2 \cdots (n+1)}(x-x_0)^{n+1}.$$

An obvious consequence of (13) is

$$(14) \quad \text{IF } |f^{(n+1)}| \le M \text{ IN } I, \text{ THEN } |R_{n+1}(x)| \le \frac{M|x-x_0|^{n+1}}{1 \cdot 2 \cdot 3 \cdots (n+1)}.$$

The significance of **Taylor's formula** (12) is that it represents an essentially arbitrary, sufficiently differentiable function $f(x)$ defined near x_0 as a polynomial of degree n plus a correction term that is small of the order of $|x-x_0|^{n+1}$ for x close to x_0.

For $x_0 = 0$, Taylor's formula reads

$$(15) \quad \begin{aligned} f(x) = {} & f(0) + \frac{f'(0)}{1}x + \frac{f''(0)}{1 \cdot 2}x^2 + \cdots + \frac{f^{(n)}(0)}{1 \cdot 2 \cdots n}x^n \\ & + \frac{f^{(n+1)}(\xi)}{1 \cdot 2 \cdots n(n+1)}x^{n+1}. \end{aligned}$$

COLIN MACLAURIN (1698–1746), a professor at Edinburgh, was a former pupil of Newton.

Here ξ is a point between 0 and x. This special case is sometimes called **Maclaurin's formula.**

We note that the interval I may be infinite. Also, the theorem and proof remain valid if x_0 is an endpoint of I; in this case, the derivatives $f'(x_0), \cdots, f^{(n)}(x_0)$ are taken as one-sided derivatives.

EXERCISES

In Exercises 23 to 32, find an estimate (upper bound) for the absolute value of the remainder term $R(x) = f(x) - P(x)$.

23. $f(x) = x^4$, $P(x)$ is the third Taylor polynomial, $x_0 = 0$ and $x = \frac{1}{2}$.
24. $f(x) = x^{13/3}$, $P(x)$ is the third Taylor polynomial, $x_0 = 0$ and $x = \frac{1}{2}$.
25. $f(x) = \arctan x$, $P(x)$ is the third Taylor polynomial, $x_0 = 1$ and $x = \frac{3}{4}$.
26. $f(x) = e^x$, $P(x)$ is the fourth Taylor polynomial, $x_0 = 0$ and $x = 1$.
27. $f(x) = \log x$, $P(x)$ is the fourth Taylor polynomial, $x_0 = 1$ and $x = e$.
28. $f(x) = \sqrt{x+1}$, $P(x)$ is the fourth Taylor polynomial, $x = 0$ and $-\frac{1}{2} \leq x \leq \frac{1}{2}$.
29. $f(x) = 1/x$, $P(x)$ is the fourth Taylor polynomial, $x_0 = 1$ and $\frac{1}{2} \leq x \leq \frac{3}{2}$.
▶ 30. $f(x) = \cos x$, $P(x)$ is the sixth Taylor polynomial, $x_0 = \pi/2$ and $3\pi/8 \leq x \leq 5\pi/8$.
31. $f(x) = \arctan x$, $P(x)$ is the third Taylor polynomial, $x_0 = 0$ and $-\frac{1}{2} \leq x \leq \frac{1}{2}$.
32. $f(x) = e^x$, $P(x)$ is the nth Taylor polynomial, $x_0 = 0$ and $x = 1$. (Note: The answer will depend on n.)

2.4 Binomial theorem

As the first application of Taylor's formula, we consider the function $f(x) = (a + x)^n$, setting $x_0 = 0$. Since $f(x)$ is a polynomial of degree n, $f^{(n+1)}(x) \equiv 0$ and $R = 0$. Also $f(0) = a^n$. Since $f'(x) = n(a + x)^{n-1}$, $f''(x) = n(n - 1)(a + x)^{n-2}$, and in general

$$f^{(j)}(x) = n(n - 1) \cdots (n - j + 1)(a + x)^{n-j} \text{ for } j \leq n,$$

we have
$$f^{(j)}(0) = n(n - 1) \cdots (n - j + 1)a^{n-j}.$$

Substituting into (12) and noting that $R = 0$, we obtain the identity

$$(a + x)^n = a^n + \frac{n}{1} a^{n-1}x + \frac{n(n - 1)}{1 \cdot 2} a^{n-2}x^2 + \frac{n(n - 1)(n - 2)}{1 \cdot 2 \cdot 3} a^{n-3}x^3$$

$$+ \frac{n(n - 1)(n - 2)(n - 3)}{1 \cdot 2 \cdot 3 \cdot 4} a^{n-4}x^4 + \cdots + \frac{n(n - 1)(n - 2) \cdots 2 \cdot 1}{1 \cdot 2 \cdots n} x^n.$$

This is, of course, the well-known **binomial formula.** The coefficient of $a^{n-j}x^j$ is called the **binomial coefficient** $\binom{n}{j}$. We have

$$(16) \quad \binom{n}{0} = 1, \quad \binom{n}{j} = \frac{n(n - 1) \cdots (n - j + 1)}{1 \cdot 2 \cdot 3 \cdots j} \quad \text{for } j = 1, 2, 3, \cdots.$$

The binomial formula can be written as

$$(17) \qquad (a + x)^n = \binom{n}{0}a^n + \binom{n}{1}a^{n-1}x^2 + \binom{n}{2}a^{n-2}x^2 + \cdots + \binom{n}{n}x^n.$$

Of course, this is an algebraic identity and can easily be established directly.

EXERCISES

33. Show that $\dfrac{n}{j}\binom{n-1}{j-1} = \binom{n}{j}$ for $j = 1, \cdots, n$. Use equation (16).

34. Show that $\binom{n}{j} = \binom{n}{n-j}$ for $j = 0, 1, \cdots, n$. Use equation (16) and note that it suffices to consider the case $j \le n - j$.

35. Show that $\binom{n}{j-1} + \binom{n}{j} = \binom{n+1}{j}$ for $j \ge 1$. Use equation (16).

▶ 36. Show that $\binom{n}{0} + \binom{n}{1} + \binom{n}{2} + \cdots + \binom{n}{n} = 2^n$. Use the binomial theorem with $x = 1$ and an appropriate choice for a.

37. Show that $\binom{n}{0} - \binom{n}{1} + \binom{n}{2} - \cdots - \binom{n}{n} = 0$ for $n \ge 1$. Use the binomial theorem with $x = 1$ and an appropriate choice for a.

38. Show that $\binom{m}{1} + 2\binom{m}{2} + 3\binom{m}{3} + \cdots + m\binom{m}{m} = m2^{m-1}$ for $m \ge 2$. Use the result of Exercise 36 with $n = m - 1$ and the result of Exercise 33.

2.5 Binomial coefficients for arbitrary exponents

The above derivation of the binomial formula from Taylor's Theorem is a mere curiosity. Yet it suggests an interesting next step. We consider the function

$$f(x) = (1 + x)^\alpha,$$

where $\alpha \neq 0$ is any real number, not necessarily a positive integer. For $x > -1$, this function has derivatives of all orders:

$$f'(x) = \alpha(1 + x)^{\alpha-1}, \qquad f''(x) = \alpha(\alpha - 1)(1 + x)^{\alpha-2}, \cdots,$$

and, in general,

$$f^{(j)}(x) = \alpha(\alpha - 1)(\alpha - 2) \cdots (\alpha - j + 1)(1 + x)^{\alpha-j},$$

so that

$$f^{(j)}(0) = \alpha(\alpha - 1) \cdots (\alpha - j + 1).$$

Substitution into the Maclaurin formula (15) yields the relation

$$(1 + x)^{\alpha} = 1 + \frac{\alpha}{1} x + \frac{\alpha(\alpha - 1)}{1 \cdot 2} x^2 + \frac{\alpha(\alpha - 1)(\alpha - 2)}{1 \cdot 2 \cdot 3} x^3$$

(18)

$$+ \cdots + \frac{\alpha(\alpha - 1) \cdots (\alpha - n + 1)}{1 \cdot 2 \cdots n} x^n + R,$$

with
$$R = \frac{\alpha(\alpha - 1) \cdots (\alpha - n)}{1 \cdot 2 \cdots (n + 1)} (1 + \xi)^{\alpha - n - 1} x^{n+1},$$

where ξ is a number between 0 and x. We can extend the definition (16) of binomial coefficients by setting, for any α,

(19) $\quad \binom{\alpha}{0} = 1, \binom{\alpha}{j} = \dfrac{\alpha(\alpha - 1)(\alpha - 2) \cdots (\alpha - j + 1)}{1 \cdot 2 \cdot 3 \cdots j}, j = 1, 2, 3, \cdots.$

Then we can write (18) as

(20) $\quad (1 + x)^{\alpha} = 1 + \binom{\alpha}{1} x + \binom{\alpha}{2} x^2 + \cdots + \binom{\alpha}{n} x^n + R$, for $x > -1$,

with

(21) $$R = \binom{\alpha}{n + 1}(1 + \xi)^{\alpha - n - 1} x^{n+1}.$$

▶ *Example*   Compute $\sqrt[3]{1.03}$ using formula (20) with $n = 2$, and estimate the error.

Answer.   We have, setting $\alpha = \frac{1}{3}$ and $x = .03$,

$$\sqrt[3]{1.03} = (1 + .03)^{1/3} = 1 + \tfrac{1}{3}(.03) + \frac{\tfrac{1}{3}(\tfrac{1}{3} - 1)}{1 \cdot 2} (.03)^2$$

$$+ \frac{\tfrac{1}{3}(\tfrac{1}{3} - 1)(\tfrac{1}{3} - 2)}{1 \cdot 2 \cdot 3} (1 + \xi)^{(1/3) - 3} (.03)^3$$

$$= 1 + \frac{.03}{3} - \frac{(.03)^2}{9} + \frac{5(.03)^3}{81} (1 + \xi)^{(1/3) - 3}$$

Here $0 < \xi < .03$, so that $0 < (1 + \xi)^{(1/3) - 3} < 1$ and the last term (that is, R) is at most $5(.000027)/81 = < .000004$. Since $1 + .03/3 - (.03)^2/9 = 1.0099$, we have $1.0099 < \sqrt[3]{1.03} < 1.0099 + .000004$.   ◀

EXERCISES

For each of Exercises 39 to 46, find an approximate value for the given number, and estimate the error. Use formulas (20) and (21) with the given value of n.

39. $\sqrt[3]{.97}$ $(n = 3)$.   43. $(.98)^{1/5}$ $(n = 3)$.
40. $\sqrt[3]{1.02}$ $(n = 3)$.   44. $\sqrt[3]{2}$ $(n = 4)$.
41. $\sqrt[3]{.75}$ $(n = 4)$.   45. $(1.1)^{1.1}$ $(n = 3)$.
42. $\sqrt{1.4}$ $(n = 4)$.   ▶ 46. $(.8)^{.01}$ $(n = 3)$.

2.6 Elementary Taylor formulas

Some of the most important special cases of Taylor's formula can be obtained without using Theorem 1, by "elementary" methods.

The nth Taylor polynomial of the function $1/(1 + x)$ at $x_0 = 0$ is

$$(22) \qquad 1 - x + x^2 - x^3 + \cdots + (-1)^n x^n.$$

(Note that $(-1)^n$ is 1 if n is even, -1 if n is odd.) This can be seen directly from the definition, or from (20) by noting that

$$\binom{-1}{j} = \frac{(-1)(-1-1)(-1-2)\cdots(-1-j+1)}{1\cdot 2\cdot 3\cdots j}$$

$$= \frac{(-1)(-2)\cdots(-j)}{1\cdot 2\cdots j} = (-1)^j.$$

Replacing x by $-x$ in (22), we obtain the nth Taylor polynomial of $1/(1 - x)$ at 0 as

$$(23) \qquad 1 + x + x^2 + \cdots + x^n.$$

There is no need to use Theorem 1 for computing the remainder, since there is the identity

$$(24) \qquad \frac{1}{1-x} = 1 + x + x^2 + \cdots + x^n + \frac{x^{n+1}}{1-x}.$$

Replacing x by $(-x)$, we obtain

$$(25) \qquad \frac{1}{1+x} = 1 - x + x^2 - \cdots + (-1)^n x^n + \frac{(-1)^{n+1} x^{n+1}}{1+x}.$$

We can use these two formulas to verify once more that (22) and (23) are the nth Taylor polynomials of $(1 - x)^{-1}$ and $(1 + x)^{-1}$, respectively. Formula (24) shows, for instance, that the difference

$$\frac{1}{1-x} - (1 + x + x^2 + \cdots + x^n)$$

is 0 at $x = 0$, and so are its derivatives of orders $1, 2, \cdots, n$.

Now we replace x in (25) by x^2 and obtain the identity

$$(26) \qquad \frac{1}{1+x^2} = 1 - x^2 + x^4 - \cdots + (-1)^n x^{2n} + \frac{(-1)^{n+1} x^{2n+2}}{1+x^2}.$$

2.7 Computing logarithms

We obtain interesting cases of Taylor's formula by integrating the identities just obtained from $x = 0$. First we rewrite (25), replacing x by t and n by $n - 1$. We obtain

$$\frac{1}{1 + t} = 1 - t + t^2 - t^3 + \cdots + (-1)^{n-1}t^{n-1} + \frac{(-1)^n t^n}{1 + t},$$

so that

$$\log (1 + x) = \int_0^x \frac{dt}{1 + t}$$

(27)

$$= x - \frac{x^2}{2} + \frac{x^3}{3} - \frac{x^4}{4} + \cdots + (-1)^{n-1}\frac{x^n}{n} + (-1)^n \int_0^x \frac{t^n \, dt}{1 + t}.$$

To estimate the remainder, assume first that $x > 0$. For $0 < t < x$, we have $1/(1 + t) < 1$, and therefore

$$0 < \int_0^x \frac{t^n \, dt}{1 + t} \leq \int_0^x t^n \, dt = \frac{x^{n+1}}{n + 1} \qquad \text{for } x > 0.$$

If $x < 0$, we must assume that $x > -1$. For $x < t < 0$, we have

$$\frac{1}{1 + t} < \frac{1}{1 + x} = \frac{1}{1 - |x|}$$

and

$$\left| \int_0^x \frac{t^n \, dt}{1 + t} \right| \leq \frac{1}{1 - |x|} \left| \int_0^x t^n \, dt \right| = \frac{|x|^{n+1}}{(1 - |x|)(n + 1)}.$$

We conclude that

(28) $$\log (1 + x) = x - \frac{x^2}{2} + \frac{x^3}{3} - \cdots + (-1)^{n-1}\frac{x^n}{n} + R,$$

where $|R| \leq \begin{cases} \dfrac{x^{n+1}}{n + 1} & \text{for } x \geq 0, \\[3mm] \dfrac{1}{1 - |x|} \dfrac{|x|^{n+1}}{n + 1} & -1 < x \leq 0. \end{cases}$

Starting with relation (24) we obtain, in a similar way,

$$(29) \quad -\log \frac{1}{1-x} = x + \frac{x^2}{2} + \frac{x^3}{3} + \cdots + \frac{x^n}{n} + \int_0^x \frac{t^n \, dt}{1-t}.$$

We leave it to the reader to estimate the remainder.

If we write the relations (27) and (29) for n even, that is, for $n = 2m + 2$, and add, we get the relation

$$(30) \qquad \log \frac{1+x}{1-x} = 2x + \frac{2x^3}{3} + \cdots + \frac{2x^{2m+1}}{2m+1} + R,$$

where

$$R = 2 \int_0^x \frac{t^{2m+2} \, dt}{1-t^2}.$$

Assume that $x > 0$. Then the integral above will increase if we replace $1 - t^2$ by $1 - x^2$. The resulting integral is easily computed, and we obtain that

$$(31) \qquad 0 < R \le \frac{2}{1-x^2} \frac{x^{2m+3}}{2m+3} \qquad \text{for } 0 < x < 1.$$

Now we observe that every number $a > 0$ can be written as

$$a = \frac{1+x}{1-x} \qquad \text{with } x = \frac{a-1}{a+1}.$$

This x satisfies $0 < x < 1$ if $a > 1$. Substituting into (30) and (31), we obtain

$$(32) \quad \log a = 2\frac{a-1}{a+1} + \frac{2}{3}\left(\frac{a-1}{a+1}\right)^2 + \cdots + \frac{2}{2m+1}\left(\frac{a-1}{a+1}\right)^{2m+1} + R$$

with

$$(33) \qquad 0 \le R \le \frac{(a+1)^2}{2a} \frac{1}{2m+3}\left(\frac{a-1}{a+1}\right)^{2m+3} \qquad \text{for } a > 1.$$

The error can be made arbitrarily small by choosing m large, but if $a - 1$ is small, even a small m gives good accuracy.

For instance, setting $a = 2$ and $m = 1$ in (32), we get:

$$\log 2 = \frac{2}{3} + \frac{2}{3}\frac{1}{3^3} + R, \qquad 0 < R \le \frac{9}{4}\frac{1}{5}\frac{1}{3^5},$$

so that

$$.692 < \log 2 < .694$$

(actually $\log 2 = .6931 \cdots$). On the other hand, (28) applied to $x = 1$ gives the beautiful relation

(34) $\log 2 = 1 - \dfrac{1}{2} + \dfrac{1}{3} - \dfrac{1}{4} + \cdots + (-1)^n \dfrac{1}{n} + R, \quad |R| \leq \dfrac{1}{n+1}.$

But to compute log 2 with an accuracy of .002 by this formula, we need about 50 terms.

2.8 Computing π

Now we rewrite (26), replacing x by t and n by $n - 1$:

$$\frac{1}{1+t^2} = 1 - t^2 + t^4 - \cdots + (-1)^{n-1} t^{2n-2} + (-1)^n \frac{t^{2n}}{1+t^2}.$$

Integration yields

(35) $\arc \tan x = x - \dfrac{x^3}{3} + \dfrac{x^5}{5} - \dfrac{x^7}{7} + \cdots + (-1)^{n-1} \dfrac{x^{2n-1}}{2n-1} + R,$

where $$R = (-1)^n \int_0^x \frac{t^{2n}\,dt}{1+t^2}$$

and, since $1/(1 + t^2) \leq 1$, we see that

(36) $$|R| \leq \frac{|x|^{2n+1}}{2n+1}.$$

For $x = 1$, arc tan $x = \pi/4$ and we obtain the interesting relation

(37) $\dfrac{\pi}{4} = 1 - \dfrac{1}{3} + \dfrac{1}{5} - \dfrac{1}{7} + \cdots + \dfrac{(-1)^{n-1}}{2n-1} + R, |R| \leq \dfrac{1}{2n+1}.$

This is, however, not an efficient way to compute $\pi/4$; one would need about $n/2$ terms to obtain an accuracy of $1/n$.

The formula (35) for arc tan x gives good results with relatively few terms if x is close to 0. Now if u and v are small, we can use this formula to compute arc tan u and arc tan v, and we obtain at once, from the addition theorem for tan (see Chapter 6, §2.2):

(38) $$\arc \tan \frac{u+v}{1-uv} = \arc \tan u + \arc \tan v.$$

▶ *Example* For $u = \frac{1}{2}$, $v = \frac{1}{3}$, we have $(u + v)/(1 - uv) = 1$ and equation (38) gives

$$\frac{\pi}{4} = \text{arc tan} \frac{1}{2} + \text{arc tan} \frac{1}{3}.$$

For $u = \frac{1}{3}$, $v = \frac{1}{7}$ and for $u = \frac{1}{5}$, $v = \frac{1}{8}$, we obtain

$$\text{arc tan} \frac{1}{2} = \text{arc tan} \frac{1}{3} + \text{arc tan} \frac{1}{7}, \text{arc tan} \frac{1}{3} = \text{arc tan} \frac{1}{5} + \text{arc tan} \frac{1}{8}.$$

Combining these formulas, we obtain

$$\frac{\pi}{4} = 2 \text{ arc tan} \frac{1}{5} + \text{arc tan} \frac{1}{7} + 2 \text{ arc tan} \frac{1}{8}.$$

This, together with (35), gives a good basis for computing π. ◄

EXERCISES

47. Find the nth Taylor polynomial for $1/(1 + x^{10})$ at $x_0 = 0$. What is the remainder?
48. Find the nth Taylor polynomial for $1/(1 - x^8)$ at $x_0 = 0$. What is the remainder?
49. Suppose $P(x)$ is the nth Taylor polynomial for $f(x)$ at $x_0 = 0$ and $f(x)$ has at least $n + 1$ derivatives at 0. Show that $xP(x)$ is the $(n + 1)$st Taylor polynomial for $xf(x)$ at $x_0 = 0$.
50. Find the nth Taylor polynomial for $x^2/(1 + x)$ at $x_0 = 0$. What is the remainder? Use the result of Exercise 49.
51. Find the nth Taylor polynomial for $x^3/(1 + x^5)$ at $x_0 = 0$. What is the remainder? Use the result of Exercise 49.
52. Find the nth Taylor polynomial for $f''(x)$ at $x_0 = 0$ if $f(x) = 1/(1 + x^4)$. What is the remainder?
53. Find the nth Taylor polynomial for $f^{(4)}(x)$ at $x_0 = 0$ if $f(x) = x/(1 - x^3)$. What is the remainder?
54. Estimate the remainder term $\int_0^x \frac{t^n \, dt}{1 - t}$ in equation (29).

For each of Exercises 55 to 64, find an approximate value for the given expression and estimate the error. Use equations (32), (33), (35), and (36), as indicated.

55. $\log \frac{3}{2}$ with $m = 1$.
56. $\log 4$ with $m = 1$.
57. $\log \frac{5}{4}$ with $m = 1$.
58. $\log 3$ with $m = 2$.
59. $\log 9$ with $m = 1$.

60. arc tan $\frac{1}{5}$ with $n = 2$.
61. arc tan $\frac{1}{7}$ with $n = 2$.
62. arc tan $\frac{1}{8}$ with $n = 2$.
63. arc tan 1 with $n = 5$.
64. arc tan $\frac{1}{2}$ with $n = 4$.

2.9 Taylor formula for the exponential function

We return to the general formulas (12) and (13), and apply them to the exponential function $f(x) = e^x$ with $x_0 = 0$. Since all derivatives of e^x

are e^x, and therefore equal to 1 at $x = 0$, we obtain

(39)
$$e^x = 1 + \frac{x}{1} + \frac{x^2}{1 \cdot 2} + \frac{x^3}{1 \cdot 2 \cdot 3} + \cdots + \frac{x^n}{1 \cdot 2 \cdot 3 \cdots n}$$
$$+ \frac{e^\xi x^{n+1}}{1 \cdot 2 \cdot 3 \cdots n(n+1)},$$

where ξ is some number between 0 and x.

In particular, for $x = 1$, we have

(40)
$$e = 1 + 1 + \frac{1}{1 \cdot 2} + \frac{1}{1 \cdot 2 \cdot 3} + \frac{1}{1 \cdot 2 \cdot 3 \cdot 4} + \cdots$$
$$+ \frac{1}{1 \cdot 2 \cdot 3 \cdots n} + \frac{e^\xi}{1 \cdot 2 \cdot 3 \cdots n + 1}, \quad \text{where } 0 < \xi < 1.$$

This can be used to show that *e is irrational*. Indeed, assume that e is rational. Then, since $2 < e < 3$, we must have $e = p/q$, where p and q are integers, $q \geq 2$. Let n be chosen so that $n > q$. We multiply both sides of (40) by $1 \cdot 2 \cdot 3 \cdots n$ and observe that, since q is one of the numbers $2, 3, \cdots, n$, all terms in (40), except perhaps the last, become integers. We conclude that $e^\xi/(n + 1)$ is also an integer. This is impossible, because $0 < e^\xi < e < 3$ and $n + 1 \geq 3$.

2.10 Taylor formulas for sines and cosines

As the last example, we consider the Taylor formulas for the functions $x \mapsto \cos x$ and $x \mapsto \sin x$, with $x_0 = 0$. We note the table:

		Value at $x = 0$		*Value at* $x = 0$
Function	$\cos x$	1	$\sin x$	0
First derivative	$-\sin x$	0	$\cos x$	1
Second derivative	$-\cos x$	-1	$-\sin x$	0
Third derivative	$\sin x$	0	$-\cos x$	-1
Fourth derivative	$\cos x$	1	$\sin x$	0
Fifth derivative	$-\sin x$	0	$\cos x$	1
Sixth derivative	$-\cos x$	-1	$-\sin x$	0

and obtain

$$\cos x = 1 - \frac{x^2}{1 \cdot 2} + \frac{x^4}{1 \cdot 2 \cdot 3 \cdot 4} - \cdots + (-1)^n \frac{x^{2n}}{1 \cdot 2 \cdot 3 \cdots (2n)}$$
$$+ (-1)^{n+1} \frac{(\cos \xi)x^{2n+2}}{1 \cdot 2 \cdot 3 \cdots (2n+2)},$$

where ξ is between 0 and x. Since $|\cos \xi| \leq 1$, we have

(41)
$$\cos x = 1 - \frac{x^2}{1 \cdot 2} + \frac{x^4}{1 \cdot 2 \cdot 3 \cdot 4} - \cdots$$
$$+ (-1)^n \frac{x^{2n}}{1 \cdot 2 \cdot 3 \cdots (2n)} + R, |R| \leq \frac{|x|^{2n+2}}{1 \cdot 2 \cdot 3 \cdots (2n + 2)}.$$

In the same way, we get

(42)
$$\sin x = x - \frac{x^3}{1 \cdot 2 \cdot 3} + \frac{x^5}{1 \cdot 2 \cdot 3 \cdot 4 \cdot 5} - \cdots$$
$$+ (-1)^n \frac{x^{2n+1}}{1 \cdot 2 \cdot 3 \cdots (2n + 1)} + R, |R| \leq \frac{|x|^{2n+3}}{1 \cdot 2 \cdot 3 \cdots (2n + 3)}.$$

EXERCISES

65. Find the nth Taylor polynomial for $x \mapsto e^{x^2}$ at $x_0 = 0$. Estimate the remainder.
66. Find the nth Taylor polynomial for $x \mapsto \sin (x^2)$ at $x_0 = 0$. Estimate the remainder.
67. Find the nth Taylor polynomial for $x \mapsto \cos (x^3)$ at $x_0 = 0$. Estimate the remainder.
68. Find the nth Taylor polynomial for $x \mapsto x \sin (x^3)$ at $x_0 = 0$. Estimate the remainder.

2.11 The sigma notation

From now on, we use an abbreviated notation for sums, which involves the letter Σ (Greek capital "sigma"). Given a rule

$$j \mapsto a_j$$

which associates with every integer j from a given range a number a_j, and given two integers k and $l \geq k$, we agree that the symbol $\sum_{j=k}^{l} a_j$ denotes the sum $a_k + a_{k+1} + \cdots + a_l$. The variable j is, of course, a "dummy variable." For instance,

$$\sum_{j=1}^{n} j = 1 + 2 + 3 + \cdots + n = \frac{1}{2} n(n + 1)$$

and also

$$\sum_{k=1}^{n} k = \sum_{p=1}^{n} p = \sum_{q=0}^{n-1} (q + 1) = 1 + 2 + 3 + \cdots + n = \frac{1}{2} n(n + 1).$$

▶**Examples** **1.** What is $\sum_{j=-2}^{2} (j^2 - 1)$?

Answer. We have

$$\sum_{j=2}^{2} (j^2 - 1)$$
$$= ((-2)^2 - 1) + ((-1)^2 - 1) + (0^2 - 1) + (1^2 - 1) + (2^2 - 1)$$
$$= 3 + 0 + (-1) + 0 + 3 = 5.$$

2. The formula for the sum of a geometric progression reads

$$\sum_{i=0}^{n} q^i = \frac{1 - q^{n+1}}{1 - q}. \blacktriangleleft$$

We also recall the notation

(43) $0! = 1, 1! = 1, 2! = 1 \cdot 2, 3! = 1 \cdot 2 \cdot 3, \cdots.$

($j!$ is read j **factorial**.) The binomial coefficients are

$$\binom{\alpha}{j} = \frac{\alpha(\alpha - 1)(\alpha - 2) \cdots (\alpha - j + 1)}{j!}.$$

If n is a positive integer, then

$$n(n - 1)(n - 2) \cdots (n - j + 1)[(n - j)!]$$
$$= n(n - 1)(n - 2) \cdots (n - j + 1)(n - j)(n - j - 1) \cdots 2 \cdot 1 = n!,$$

so that $n(n - 1)(n - 2) \cdots (n - j + 1) = \dfrac{n!}{(n - j)!}$

and

(44) $\binom{n}{j} = \dfrac{n!}{(n - j)!j!}.$

This explains the symmetry

$$\binom{n}{j} = \binom{n}{n - j}.$$

The binomial formula can be written as

(45) $(a + x)^n = \displaystyle\sum_{j=0}^{n} \binom{n}{j} a^{n-j}x^j.$

Setting $a = x = 1$ and then $a = 1, x = -1$, we see that

$$\sum_{j=0}^{n} \binom{n}{j} = 2n, \qquad \sum_{j=0}^{n} \binom{n}{j}(-1)^j = 0.$$

We rewrite Taylor's general formula in the sigma notation

$$(46) \quad f(x) = \sum_{j=0}^{n} \frac{f^{(j)}(x_0)}{j!}(x - x_0)^j + R_{n+1}, \; R_{n+1} = \frac{f^{(n+1)}(\xi)}{(j+1)!}(x - x_0)^{j+1}.$$

It is assumed here that

$$f^{(0)}(x) = f(x).$$

We adopt this convention in what follows.

EXERCISES

Evaluate the following sums:

69. $\Sigma_{j=0}^{4} j!$

70. $\Sigma_{p=2}^{5} \dfrac{p}{p-1}.$

71. $\Sigma_{k=0}^{4} (-1)^{k+1}(k+1)^3.$

72. $\Sigma_{l=1}^{2} \dfrac{l}{l^2+1}.$

73. $\Sigma_{\beta=0}^{3} 2^{2\beta+1}$

74. $\Sigma_{\xi=0}^{3} (2\xi)!$

75. $\Sigma_{\lambda=1}^{5} \begin{pmatrix} \lambda+1 \\ \lambda-1 \end{pmatrix}.$

76. $\Sigma_{l=1}^{3} \dfrac{(-1)^l}{(3l-2)^2}.$

77. $\Sigma_{k=1}^{5} \dfrac{1 \cdot 3 \cdot 5 \cdots (2k-1)}{2 \cdot 4 \cdot 6 \cdots (2k)}.$

Write the following in sigma notation:

78. $\sqrt{1} + \sqrt{3} + \sqrt{5} + \sqrt{7} + \sqrt{9}.$

79. $\dfrac{1}{2} + \dfrac{1}{6} + \dfrac{1}{10} + \dfrac{1}{14} + \dfrac{1}{18} + \dfrac{1}{22}.$

80. $\dfrac{1}{1} + \dfrac{1 \cdot 3}{1 \cdot 4} + \dfrac{1 \cdot 3 \cdot 5}{1 \cdot 4 \cdot 7} + \dfrac{1 \cdot 3 \cdot 5 \cdot 7}{1 \cdot 4 \cdot 7 \cdot 10}.$

81. $1 - 8 + 27 - 64 + 125 - 216.$

82. $\dfrac{1}{3} + \dfrac{3}{6} + \dfrac{5}{9} + \dfrac{7}{12} + \dfrac{9}{15}.$

83. $\dfrac{1 \cdot 2}{3 \cdot 4} - \dfrac{2 \cdot 3}{4 \cdot 5} + \dfrac{3 \cdot 4}{5 \cdot 6} - \dfrac{4 \cdot 5}{6 \cdot 7} + \dfrac{5 \cdot 6}{7 \cdot 8}.$

84. $\dfrac{1 \cdot 4}{3 \cdot 4} + \dfrac{9 \cdot 16}{5 \cdot 6} + \dfrac{25 \cdot 36}{7 \cdot 8} + \dfrac{49 \cdot 64}{9 \cdot 10} + \dfrac{81 \cdot 100}{11 \cdot 12}.$

85. Evaluate $\Sigma_{k=1}^{2n} \left[\cos\left(\dfrac{k\pi}{n}\right) - \cos\left(\dfrac{(k-1)\pi}{n}\right) \right]$ for any positive integer n. This is an example of what is called a telescoping sum.

86. Show that $\Sigma_{j=0}^{n+1} (-1)^j \begin{pmatrix} n+1 \\ j \end{pmatrix} = (-1)^{n+1} \begin{pmatrix} n \\ n+1 \end{pmatrix}$ for any nonnegative integer n. (Hint: Use the result of Exercise 35 and the principle of the telescoping sum.)

87. Show that $\sum_{\lambda=0}^{n}\binom{n-\lambda}{j-1} = \binom{n+1}{j} - \binom{0}{j}$ for any nonnegative integer n and positive integer j. See hint above.

88. Show that $\sum_{\lambda=0}^{n}\binom{\lambda+k}{k} = \binom{n+k+1}{k+1}$ for any nonnegative integers n and k. See hint above.

89. Show that

$$\sum_{\lambda=0}^{\left[\frac{n}{2}\right]}\binom{n}{2\lambda} = \sum_{\lambda=0}^{\left[\frac{n-1}{2}\right]}\binom{n}{2\lambda+1} = 2^{n-1}$$

for any positive integer n where $[\beta]$, as usual, denotes the largest integer not exceeding β. First prove the left-hand equality using the result of Exercise 37. (You may find it helpful to consider the cases n even and n odd separately.) Then use the left-hand equality and the result of Exercise 36 to prove the right-hand equality.

2.12 A list of Taylor formulas

Now we rewrite some of the Taylor formulas for particular functions in the sigma notation:

$$(47) \qquad\qquad (1+x)^{\alpha} = \sum_{j=0}^{n}\binom{\alpha}{j}x^j + R,$$

where $R = \binom{\alpha}{n+1}(1+\xi)^{\alpha-n-1}x^{n+1}$ \qquad with ξ between 0 and x,

$$(48) \qquad\qquad \log(1+x) = \sum_{j=1}^{n}(-1)^{j+1}\frac{x^j}{j} + R,$$

where $\qquad |R| \leq \begin{cases} \dfrac{x^{n+1}}{(n+1)} & \text{if } x \geq 0 \\[2ex] \dfrac{|x|^{n+1}}{(1-|x|)(n+1)} & \text{if } x < 0, \end{cases}$

and, in particular,

$$(49) \qquad \log 2 = \sum_{j=1}^{n}\frac{(-1)^{j+1}}{j} + R, \qquad |R| \leq \frac{1}{n+1}.$$

Also,

$$(50) \qquad \arctan x = \sum_{j=1}^{n}(-1)^{j+1}\frac{x^{2j-1}}{2j-1} + R, \qquad |R| \leq \frac{|x|^{2n+1}}{2n+1}$$

and, in particular,

$$(51) \qquad \frac{\pi}{4} = \sum_{j=1}^{n} \frac{(-1)^{j+1}}{2j-1} + R, \qquad |R| \leq \frac{1}{2n+1}.$$

Also

$$(52) \qquad e^x = \sum_{j=0}^{n} \frac{x^j}{j!} + R, \qquad |R| \leq \frac{e^x |x|^{n+1}}{(n+1)!},$$

$$(53) \qquad \cos x = \sum_{j=0}^{n} (-1)^j \frac{x^{2j}}{(2j)!} + R, \qquad |R| \leq \frac{|x|^{2n+2}}{(2n+2)!},$$

$$(54) \qquad \sin x = \sum_{j=0}^{n} (-1)^j \frac{x^{2j+1}}{(2j+1)!} + R, \qquad |R| \leq \frac{|x|^{2n+3}}{(2n+3)!}.$$

The reader is asked to check that these are indeed the same expansions we obtained before.

3.1 Adding infinitely many numbers

According to Taylor's Theorem, a function $f(x)$ can be represented, under certain circumstances, as a polynomial plus a small remainder. A polynomial is a sum of finitely many monomials ax^j. It is natural to ask whether one can dispense with the remainder and represent a function precisely as a sum of infinitely many monomials.

But how can infinitely many numbers "add up" to a number? Perhaps the simplest way of visualizing this is by looking at the following example.

Fig. 8.7

Consider a segment of length 2 (see Fig. 8.7). Divide it into two equal segments (of length 1 each). Leave the left segment alone, and divide the right one into two equal segments (of length $\frac{1}{2}$ each). Divide the right segment of length $\frac{1}{2}$ into two equal segments (of length $\frac{1}{4}$ each). Continue this process indefinitely. We obtain a decomposition of the segment of length 2 into segments of length 1, $\frac{1}{2}$, $\frac{1}{4}$, $\frac{1}{8}$, $\frac{1}{16}$, and so forth. "Therefore,"

$$(1) \qquad 1 + \frac{1}{2} + \frac{1}{4} + \frac{1}{8} + \frac{1}{16} + \cdots = 2.$$

This argument was known to the Greeks, and the philosopher Zeno objected to its validity. Zeno is known to us only through his "paradoxes,"

ZENO of Elea (ca. 490 B.C.) was a follower of the Eleatic philosopher Parmenides, who taught that ultimate reality is unchangeable. Zeno's paradoxes may be interpreted as an attempt to show that motion is illusory.

which are preserved in the works of others. One of these paradoxes asserts that a runner cannot complete a race-course, because he would first have to traverse half of the distance, then half of the remaining distance, then half of the then remaining distance, and so on. Thus the runner must traverse infinitely many distances, and this will take forever.

Zeno certainly saw runners arriving at the finishing line, and precisely what he meant by this and other paradoxes is a matter of controversy. But if he wanted to say that one cannot talk about adding infinitely many numbers as if this were a procedure analogous to adding finitely many numbers, he was certainly right. If we were to compute the sum in (1) by carrying out all indicated additions, this would indeed take forever.

Yet we feel that equation (1) is correct. What is its precise meaning?

We can compute, given any positive integer n, the sum of the first n terms on the left-hand side. We obtain, using the formula for summing a geometric series

$$1 + \frac{1}{2} = \frac{3}{2}, \qquad 1 + \frac{1}{2} + \frac{1}{4} = \frac{7}{4}, \qquad 1 + \frac{1}{2} + \frac{1}{4} + \frac{1}{8} = \frac{15}{8},$$

and, in general,

$$\underbrace{1 + \frac{1}{2} + \frac{1}{4} + \cdots + \frac{1}{2^{n-1}}}_{n \text{ terms}} = \frac{1 - (\frac{1}{2})^n}{1 - \frac{1}{2}} = 2 - \frac{1}{2^{n-1}}.$$

Thus *the difference between the sum of the first n terms on the left side of (1) and the number 2 is a number that will be arbitrarily small if n is sufficiently large.* We interpret (1) to be simply an abbreviation for the statement just made.

3.2 The need for precision

A reader may object that the preceding discussion amounts to hair-splitting. "It is true, is it not, that $1 + \frac{1}{2} + \frac{1}{4} + \frac{1}{8} + \cdots = 2$? What difference does it make how you choose to interpret this statement?" To counter such an attitude, we present an example which shows that one cannot always operate with infinite sums as one does with finite ones. In §2.12, we showed [see (49)] that the difference between the finite sum $1 - \frac{1}{2} + \frac{1}{3} - \frac{1}{4} + \frac{1}{5} - \cdots + (-1)^{n-1}\frac{1}{n}$ and $\log 2$ is at most $\frac{1}{n+1}$

in absolute value, that is, as small as we like, provided n is large enough. Therefore

$$(2) \quad 1 - \frac{1}{2} + \frac{1}{3} - \frac{1}{4} + \frac{1}{5} - \frac{1}{6} + \frac{1}{7} - \frac{1}{8} + \frac{1}{9} - \frac{1}{10} + \cdots = \log 2.$$

Consider now the infinite sum

$$(3) \quad 1 + \frac{1}{3} - \frac{1}{2} + \frac{1}{5} + \frac{1}{7} - \frac{1}{4} + \frac{1}{9} + \frac{1}{11} - \frac{1}{6} + \cdots.$$

It consists of the same terms as the sum in (2). We expect therefore that this sum is again $\log 2$. On the other hand, we conclude from (2) that

$$\frac{1}{2} - \frac{1}{4} + \frac{1}{6} - \frac{1}{8} + \frac{1}{10} - \cdots = \frac{1}{2} \log 2.$$

We write (2) and the relation just obtained one above the other,

$$1 - \frac{1}{2} + \frac{1}{3} - \frac{1}{4} + \frac{1}{5} - \frac{1}{6} + \frac{1}{7} - \frac{1}{8} + \frac{1}{9} - \frac{1}{10} + \frac{1}{11} - \cdots = \log 2$$

$$\frac{1}{2} \quad\quad - \frac{1}{4} \quad\quad + \frac{1}{6} \quad\quad - \frac{1}{8} \quad\quad + \frac{1}{10} \quad\quad - \cdots = \frac{1}{2} \log 2$$

and add. This yields

$$(4) \quad 1 + \frac{1}{3} - \frac{1}{2} + \frac{1}{5} + \frac{1}{7} - \frac{1}{4} + \frac{1}{9} + \frac{1}{11} - \frac{1}{6} + \cdots = \frac{3}{2} \log 2.$$

So what is the true value of the infinite sum (3), $\log 2$ or $\frac{3}{2} \log 2$? We can answer this only on the basis of a precise definition. When we give this definition, it will turn out that (4) is a correct equation.

3.3 Infinite sequences

In order to develop a rigorous theory of infinite series, we must first consider **infinite sequences** of numbers. ("Series" and "sequences" are closely related but distinct terms; care should be taken not to confuse them.) An infinite sequence is a rule that associates a number to each integer $1, 2, 3, \cdots$; the number associated with the integer k is called the kth term of the sequence. From now on, we shall say "sequence" instead of "infinite sequence," for the sake of brevity.

Sequences are often denoted by writing down the first few terms; in this case one assumes that the rule of forming the "general" term, that is, the kth term for any k, is clear. For instance, consider the sequences

(5) $\qquad\qquad\qquad\qquad\qquad$ $1, 3, 5, 7, \cdots$

(6) $\qquad\qquad\qquad\qquad\qquad$ $1, 0, 1, 0, 1, 0, \cdots$

(7) $\qquad\qquad\qquad\qquad\qquad$ $1, 1, 2, 3, 5, 8, 13, \cdots.$

In writing $\cdots$, we indicate that the sequence is infinite. Sequence (5) is the sequence of odd integers. The fifth term is 9. The nth term is

$$2n - 1.$$

In the sequence (6) the nth term is 0 or 1 according to whether n is even or odd. We may write this nth term as

$$\frac{1 - (-1)^n}{2}.$$

LEONARDO of Pisa, also known as FIBONACCI (13th century) wrote two mathematical books—one on algebra, the other on geometry. They contain material that Leonardo acquired during his travels in the Orient and, probably, also include some of his original investigations.

Sequence (7), finally, is the famous **Fibonacci sequence,** in which each term is computed by adding the preceding two, the first two terms being 1. Therefore the eighth term in (7) is $8 + 13 = 21$, the ninth is $13 + 21 = 34$, and so forth. The Fibonacci numbers have many curious properties and have been a source of delight to professional and amateur mathematicians for seven centuries.

In talking about sequences, we use variables; the term "the sequence $a_1, a_2, a_3, \cdots$" or "the sequence a_n, $n = 1, 2, 3, \cdots$" "the sequence $\{a_j\}$," or simply "the sequence a_j" denotes a definite but unspecified rule $k \mapsto a_k$ which assigns to every positive integer k the number a_k. If we want to specify the sequence, we write out the rule. For instance,

$$a_n = 2n - 1, \qquad n = 1, 2, \cdots$$

denotes the sequence (5) of odd numbers, whereas

$$a_1 = a_2 = 1, \qquad a_n = a_{n-1} + a_{n-2}, \qquad n = 3, 4, \cdots$$

denotes the Fibonacci sequence. Of course, in the above formulas a and n are "dummy variables." The Fibonacci sequence can be just as well described by writing

$$b_1 = b_2 = 1, \qquad b_j = b_{j-1} + b_{j-2}, \qquad j = 3, 4, \cdots$$

or $\qquad\quad$ $\alpha_k = 1$ for $k = 1, 2, \qquad \alpha_{k+2} + \alpha_{k+1}$ for $k \geq 1.$

Finally, we note that labeling the terms in a sequence by the integers 1, 2, 3, $\cdots$ is only a matter of convenience. We may begin at 0 or at -1 or at 5.

EXERCISES

In Exercises 1 to 9, write the first six terms of the given sequence.

1. $a_n = n^2 - n$, $n = 1, 2, \cdots$.

2. $a_\lambda = (-1)^\lambda (\lambda - 1)^3$, $\lambda = 1, 2, \cdots$.

3. $b_k = (-2)^k$, $k = 0, 1, \cdots$.

4. $k_b = b^2 - b$, $b = -1, 0, \cdots$.

5. $n_j = \dfrac{j}{j^2 + 1}$, $j = -1, 0, \cdots$.

6. $t_x = x^{x-4}$, $x = 1, 2, \cdots$.

7. $x_1 = 1$, $x_2 = 2$, $x_n = 2x_{n-1} - x_{n-2}$, $n = 3, 4, \cdots$.

8. $b_s = 2s + (-1)^s s$, $s = 3, 4, \cdots$.

9. $a_r = \dfrac{2r + 1}{2r}$, $r = 5, 6, \cdots$.

10. Write the first five terms of the sequence: $s_1 = 2$, $s_2 = 4$, $s_n = \left(\dfrac{s_{n-1}}{s_{n-2}}\right)$, $n = 3, 4, \cdots$.

11. If $a_{-1} = 1$, $a_0 = 4$, $a_1 = 9$, $a_2 = 16$, $a_3 = 25$, $a_4 = 36$, $\cdots$, what is a_n?

12. If $b_2 = 1$, $b_3 = 2$, $b_4 = 4$, $b_5 = 8$, $b_6 = 16$, $b_7 = 32$, $\cdots$, what is b_k?

13. What is the "general" term of the sequence 0, 4, 8, 12, 16, 20, $\cdots$? Indicate from where you begin labeling the terms.

14. What is the "general" term of the sequence, 1, -4, 7, -10, 13, -16, $\cdots$? Indicate from where you begin labeling the terms.

15. If $a_k = (-1)^k \dfrac{(k - 1)!}{(k + 1)^2}$, $k = 1, 2, \cdots$, how would you define b_j, $j = -1$, 0, $\cdots$, so that $b_{-1} = a_1$, $b_0 = a_2$, $b_1 = a_3$, $\cdots$?

16. If $x_t = \dfrac{(2t + 2)(2t + 3)}{2t + 4}$, $t = -1, 0, \cdots$, how would you define z_λ, $\lambda = 2$, 3, $\cdots$, so that $z_2 = x_{-1}$, $z_3 = x_0$, $z_4 = x_1$, $\cdots$?

3.4 Convergent sequences. Limits

Consider the four sequences

$$1, \frac{1}{2}, \frac{1}{3}, \frac{1}{4}, \frac{1}{5}, \cdots$$

$$1, -\frac{1}{2}, \frac{1}{3}, -\frac{1}{4}, \frac{1}{5}, \cdots$$

$$1, 0, \frac{1}{10}, 0, \frac{1}{100}, 0, \frac{1}{1000}, \cdots$$

$$\frac{1}{\sqrt{2}}, \frac{1}{\sqrt{3}}, \frac{1}{\sqrt{4}}, \frac{1}{\sqrt{5}}, \frac{1}{\sqrt{6}}, \cdots$$

All four have one property in common. If we go far enough in the sequence, the terms become as small in absolute value as we like. In the first sequence, for instance, all terms after the fifth lie between 0 and $\frac{1}{5}$, all terms after the tenth between 0 and $\frac{1}{10}$, and so on. We describe this property by saying that the four sequences considered have the limit 0, or that they converge to 0, or that they approach 0.

The sequence

$$5.1, \ 5.01, \ 5.001, \ 5.0001, \ \cdots$$

does not converge to 0. But if we go far enough in this sequence, the terms become as close as we want to 5. We say that this sequence has the limit 5 (or converges to 5, or approaches 5).

We now define the terms **convergence** and **limit of a sequence** precisely. It is convenient to use the expression **nearly all** to mean "all but a finite number." For instance, in the sequence

$$-10, \ -9, \ -8, \ -7, \ \cdots$$

nearly all terms are positive (for the eleventh term is 0, and all terms after the eleventh are greater than 0).

Let a_1, a_2, a_3, $\cdots$ be a sequence and let α be a number. We say that the sequence converges to α, or has α as a limit, and we write

$$\lim_{i \to \infty} a_i = \alpha$$

if the following two conditions are satisfied:

I. If A is any number such that $\alpha < A$, then nearly all terms of the sequence are $< A$.

II. If B is any number such that $\alpha > B$, then nearly all terms of the sequence are $> B$.

Let us use the definition to prove that the sequence $1, \frac{1}{2}, \frac{1}{3}, \frac{1}{4}, \frac{1}{5}, \frac{1}{6}, \cdots$ converges to 0, that is, that

$$(8) \qquad \qquad \qquad \lim_{i \to \infty} \frac{1}{i} = 0.$$

First we verify Condition I. Let A be any number such that $A > 0$. We ask: for which i is $1/i < A$? For those positive integers i, clearly, which satisfy $1/A < i$. Let N be an integer such that $N > 1/A$. Then $1/i < A$ for all $i \geq N$. Thus, all but maybe the first $N - 1$ terms of our sequence are $< A$. Condition I is satisfied. Note that the number N which we found depends on A. What matters is that, for every $A > 0$, there is an N with the required property. Now we verify Condition II. This is, in our particular case, very easy. For if B is any number such that $0 > B$, then all terms of our sequence are $> B$.

As another example, we prove that the sequence

$$2 + \frac{1}{2}, 2 - \frac{1}{3}, 2 + \frac{1}{4}, 2 - \frac{1}{5}, \cdots$$

converges to 2, that is, that

$$(9) \qquad \lim_{n \to \infty} \left(2 + (-1)^n \frac{1}{n} \right) = 2.$$

The verification of I follows. Choose a number $A > 2$. Then, if n is odd,

$$a_n = 2 - \frac{1}{n} < 2 < A.$$

If n is even, then

$$a_n = 2 + \frac{1}{n} < A$$

will hold, provided $1/n < A - 2$, that is, for $n > 1/(A - 2)$. Hence, if N is an integer such that $N \geq 1/(A - 2)$, we have $a_n < A$ for all n, except perhaps the first N. For instance, if $A = 2.01$, all but the first 100 terms of our sequence satisfy $a_n < A$, and if $A = 2.0001$ all but the first 10,000 do.

The verification of II is similar and is left to the reader.

There is another, equivalent, form of the definition of the limit of a sequence.

The statement

$$(10) \qquad \lim_{n \to \infty} a_n = \alpha$$

means that, for every positive number ϵ, there is an integer N such that $|a_n - \alpha| < \epsilon$ for $n > N$, that is, for $n = N + 1, N + 2, N + 3, \cdots$, and so forth.

To see that this "ϵ-N-definition" means the same as Conditions I and II, note that every $A > \alpha$ can be written as $\alpha + \epsilon_1$ for some $\epsilon_1 > 0$, that every $B < \alpha$ can be written as $\alpha - \epsilon_2$ for some $\epsilon_2 > 0$, and that the inequality $|a_n - \alpha| < \epsilon$ means that $\alpha - \epsilon < a_n < \alpha + \epsilon$.

The ϵ-N-definition discloses the intuitive meaning of limits. If (9) holds, then the terms of the sequence a_1, a_2, a_3, $\cdots$ that are far enough to the right, that is, have a high-enough subscript, are as close to α as desired.

Let us prove (9), using the ϵ-N-definition. Let ϵ be a given positive number. We ask: for which positive integers n is

$$\left| \left(2 + (-1)^n \frac{1}{n} \right) - 2 \right| < \epsilon?$$

For those n, of course, for which $1/n < \epsilon$, that is, $n > 1/\epsilon$, that is, $n > N$ where N is the smallest integer with $N \geq 1/\epsilon$. This proves (9),

We shall write $\lim a_i = \alpha$ instead of $\lim_{i \to \infty} a_i = \alpha$ when there is no danger of confusion. Many authors also write

$$a_j \to \alpha \text{ as } j \to \infty$$

(read "a_j approaches α as j approaches infinity").

EXERCISES

The application of the definition of convergence of a sequence directly to a concrete problem consists of two stages. First, we must decide if we are trying to prove that the sequence converges or that it does not converge and, in the former case, we must also decide on the value to which we believe it converges. Only after this can we pass to the second stage, namely, the actual formal proof; of course, if our guess in the first stage is incorrect, it will be impossible to complete the second stage. Remember that the process is not complete without the formal proof, but the main thing here is to guess at the correct answer.

In each of Exercises 17 to 28, decide if the given sequence converges or not and, in the former case, decide on what the limit should be.

17. $0, 1, 0, \dfrac{1}{2}, 0, \dfrac{1}{3}, 0, \dfrac{1}{4}, 0, \dfrac{1}{5}, \cdots$.

▶ 18. $0, \dfrac{1}{2}, \dfrac{2}{3}, \dfrac{3}{4}, \dfrac{4}{5}, \dfrac{5}{6}, \dfrac{6}{7}, \cdots$.

19. $-8, 27, -4, 9, -2, 3, -1, 1, -\dfrac{1}{2}, \dfrac{1}{3}, -\dfrac{1}{4}, \dfrac{1}{9}, \cdots$.

20. $1, \dfrac{1}{2}, 1, \dfrac{1}{4}, 1, \dfrac{1}{8}, 1, \dfrac{1}{16}, 1, \dfrac{1}{32}, \cdots$.

21. $1, -3, 0, -2, -\dfrac{1}{2}, -\dfrac{3}{2}, -\dfrac{3}{4}, -\dfrac{5}{4}, -\dfrac{7}{8}, -\dfrac{9}{8}, \cdots$.

▶ 22. $\dfrac{1}{2}, -\dfrac{1}{2}, \dfrac{1}{2}, -\dfrac{1}{2}, \dfrac{1}{2}, -\dfrac{1}{2}, \dfrac{1}{2}, -\dfrac{1}{2}, \cdots$.

23. $1, 1.1, 1.11, 1.111, 1.1111, \cdots$.

24. $11, 101, 1001, 10001, 100001, 1000001, \cdots$.

25. $1.1, 1.01, 1.001, 1.0001, 1.00001, 1.000001, \cdots$.

26. $\dfrac{\pi}{2}, \dfrac{\pi}{3}, \dfrac{\pi}{5}, \dfrac{\pi}{7}, \dfrac{\pi}{11}, \dfrac{\pi}{13}, \dfrac{\pi}{17}, \dfrac{\pi}{19}, \dfrac{\pi}{23}, \dfrac{\pi}{29}, \cdots$.

27. $0, 3, 1, 4, 2, 5, 3, 6, 4, 7, 5, 8, 6, \cdots$.

28. $\dfrac{1}{2}, -\dfrac{1}{2}, \dfrac{1}{4}, -\dfrac{3}{4}, \dfrac{1}{8}, -\dfrac{7}{8}, \dfrac{1}{16}, -\dfrac{15}{16}, \cdots$.

3.5 Properties of limits

It follows from the definition that a sequence can have *only one* limit. (For if we have $\lim a_j = \alpha$ and also $\lim a_j = \beta > \alpha$, then we must have $a_j < (\alpha + \beta)/2$ for nearly all j and $a_j > (\alpha + \beta)/2$ for nearly all j, which is absurd.)

It follows from the definition that, if we have $\lim a_j = \alpha$, and if the sequence $b_1, b_2, b_3, \cdots$ is obtained from $a_1, a_2, a_3, \cdots$ by changing finitely many terms, then $\lim b_j = \alpha$. Also, if $c_1, c_2, c_3, \cdots$ is a **subsequence** of $a_1, a_2, a_3, \cdots$, that is, a sequence obtained from $a_1, a_2, a_3, \cdots$ by dropping some, perhaps infinitely many, terms, then $\lim c_j = \alpha$. (For anything that holds for nearly all a_j is true for nearly all b_j and nearly all c_j.)

Finally, we note that the statements

$$\lim a_j = 0 \text{ and } \lim |a_j| = 0$$

are equivalent. So are the statements

$$\lim a_j = \alpha \text{ and } \lim a_{j+1} = \alpha.$$

The reader will easily supply the reasons.

▶ *Examples*   *1.* $\lim_{n \to \infty} q^n = 0$ if $0 < q < 1$.

To verify this statement, we must show that if ϵ is any positive number, then $q^n < \epsilon$ for nearly all n. (In the case considered, Condition II is satisfied trivially; if $B < 0$, then $q^n > B$ for all n.)

First proof.   Let $\epsilon > 0$ be given. We ask: for which positive integers n is it true that $q^n < \epsilon$? Now, since the log is an increasing function, the inequality $q^n < \epsilon$ may be written as $\log (q^n) < \log \epsilon$, or as $n \log q < \log \epsilon$. Since $0 < q < 1$, we have $\log q < 0$ so that the last inequality may be written as $n > \log \epsilon / \log q$. This is true for all n if $\log \epsilon \geq 0$ and for all large n, that is, for nearly all n if $\log \epsilon < 0$.

Second proof. Write $1/q = 1 + a$. Then $a > 0$ and $(1 + a)^n > 1 + an$. Hence $(1/q)^n = (1 + a)^n > 1 + an$ and $q^n < 1/(1 + an) < 1/an$. Hence we certainly have $q^n < \epsilon$ if $1/(an) < \epsilon$, that is, if $n > 1/(a\epsilon)$, hence for nearly all n.

2. $\lim_{n \to \infty} q^n = 0$ if $|q| < 1$.

Proof. If $q = 0$, the statement is trivial. For $q \neq 0$, $|q| < 1$, it suffices to show that $\lim |q^n| = 0$. But $\lim |q^n| = \lim |q|^n = 0$ by Example 1.

3. For every x, $\lim_{n \to \infty} \dfrac{x^n}{n!} = 0$.

Proof. We may assume that $x > 0$. Let m be a fixed number, such that $m > x$. For $n > m$, we have

$$0 < \frac{x^n}{n!} = \frac{x}{1} \cdot \frac{x}{2} \cdot \frac{x}{3} \cdots \frac{x}{m} \cdot \frac{x}{m+1} \cdot \frac{x}{m+2} \cdots \frac{x}{n}$$

$$= \frac{x^m}{m!} \cdot \frac{x}{m+1} \cdot \frac{x}{m+2} \cdots \frac{x}{n} \leq \frac{x^m}{m!} \underbrace{\frac{x}{m+1} \frac{x}{m+1} \frac{x}{m+1}}_{n - m \text{ terms}}$$

$$= \frac{x^m}{m!} \left(\frac{x}{m+1}\right)^{n-m}.$$

Let $\epsilon > 0$ be a given number. For which n is it true that $\left|\dfrac{x^n}{n}\right| < \epsilon$? Certainly for those n for which $\dfrac{x^m}{m!} \left(\dfrac{x}{m+1}\right)^{n-m} < \epsilon$, that is, for those n for which $\left(\dfrac{x}{m+1}\right)^{n-m} < \dfrac{m!\epsilon}{x^m}$. This is so for nearly all n, by Example 1. (To see this, set $q = \dfrac{x}{m+1}$.)

4. We have, for every x,

$$\lim \frac{x^{2n+2}}{(2n+2)!} = 0, \qquad \lim \frac{x^{2n+3}}{(2n+3)!} = 0.$$

Proof. Note that the sequences $\dfrac{x^4}{4!}, \dfrac{x^6}{6!}, \dfrac{x^8}{8!}, \cdots$ and $\dfrac{x^5}{5!}, \dfrac{x^7}{7!}, \dfrac{x^9}{9!}, \cdots$ are subsequences of $\dfrac{x}{1!}, \dfrac{x^2}{2!}, \dfrac{x^3}{3!}, \dfrac{x^4}{4!}, \dfrac{x^5}{5!}, \cdots$ and apply Example 4. ◄

3.6 Computing limits

Fig. 8.8

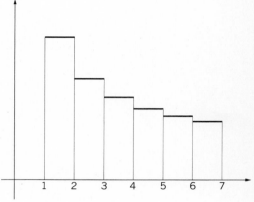

It is possible and useful to reduce the concept of limits of sequences to that of limits of functions. This is easily done. We recall that, for every real number x, the symbol $[x]$ denotes the greatest integer not exceeding x (thus, for instance, $[2] = 2$, $[2.001] = 2$, $[\pi] = 3$). Now let $a_1, a_2, a_3, \cdots$ be a sequence. We associate with it the function

$$x \mapsto a_{[x]}$$

defined for all $x \geq 1$. It is a step function that takes on the value a_1 in the interval $[1, 2)$, the value a_2 in the interval $[2, 3)$, the value a_3 in $[3, 4)$, and so forth. In Fig. 8.8, we show the graph of the function $x \mapsto a_{[x]}$ for the sequence $a_n = 5/\sqrt{n}$, $n = 1, 2, \cdots$.

We can also associate with our sequence the function

$$x \mapsto a_{[|1/x|]}$$

defined for all values of $x \neq 0$. This function takes on the value a_1 for $x \leq -1$ and $x \geq 1_1$, the value a_2 for $-1 < x \leq -\frac{1}{2}$ and $\frac{1}{2} \leq x < 1$, the value a_3 for $-\frac{1}{2} < x \leq \frac{1}{3}$ and $\frac{1}{3} \leq x < \frac{1}{2}$, and so forth. The graph of this function, for $a_n = 5/\sqrt{n}$, is shown in Fig. 8.9.

Recalling the definition (in Chapter 3, §§4.1, 4.3) of the limits of a function, we see that the three statements

$$\lim_{i \to \infty} a_i = \alpha, \qquad \lim_{x \to +\infty} a_{[x]} = \alpha, \qquad \lim_{x \to 0} a_{[|1/x|]} = \alpha$$

are *equivalent*.

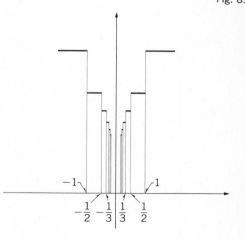
Fig. 8.9

We can now transfer to limits of sequences theorems previously proved about limits of functions (see Chapter 3, §4.2).

Theorem 1　ASSUME THAT

$$\lim a_n = \alpha, \lim b_n = \beta.$$

THEN

$$\lim (a_n + b_n) = \alpha + \beta,$$
$$\lim (ca_n) = c\alpha,$$
$$\lim (a_n b_n) = \alpha\beta,$$

AND

$$\lim \frac{a_n}{b_n} = \frac{\alpha}{\beta} \text{ if } \beta \neq 0.$$

Theorem 2   IF $x \mapsto f(x)$ IS A FUNCTION CONTINUOUS AT $x = \alpha$, AND
IF
$$\lim a_n = \alpha,$$
THEN
$$\lim f(a_n) = f(\alpha).$$

Using these theorems, various limits can be evaluated. Of course, the theorems could have been proved directly, too.

▶ *Example*   Compute $\displaystyle\lim_{n \to \infty} \frac{3n^2 + 2}{5n^2 - 2n + 1}$.

Solution.   The desired limit is

$$\lim \frac{3 + \dfrac{2}{n^2}}{5 - \dfrac{2}{n} + \dfrac{1}{n^2}} = \frac{\lim \left(3 + \dfrac{2}{n^2}\right)}{\lim \left(5 - \dfrac{2}{n} + \dfrac{1}{n^2}\right)} = \frac{\lim 3 + \lim \dfrac{2}{n^2}}{\lim 5 - \lim \dfrac{2}{n} + \lim \dfrac{1}{n^2}}$$

$$= \frac{3 + 0}{5 - 0 + 0} = \frac{3}{5}.$$

We used the fact that

$$\lim \frac{1}{n^2} = 0.$$

This can be seen (a) by the definition, (b) using (8) and Theorem 1 since $\lim \dfrac{1}{n^2} = \lim \dfrac{1}{n} \lim \dfrac{1}{n}$, or (c) using (8) and Theorem 2 with $f(x) = x^2$, since $\lim \dfrac{1}{n^2} = \left(\lim \dfrac{1}{n}\right)^2 = 0^2 = 0.$ ◀

3.7 Divergent sequences

Not every sequence has a limit; a sequence without a limit is called **divergent**. A sequence $a_1, a_2, a_3, \cdots$ is said to *diverge to* $+\infty$, in symbols

$$\lim a_i = +\infty,$$

if, for every number A, nearly all terms a_i are $> A$. This is equivalent to the statement

$$\lim_{x \to +\infty} a_{[x]} = +\infty.$$

Similarly, $$\lim a_i = -\infty$$

means that $$\lim_{x \to +\infty} a_{[x]} = -\infty.$$

This is so if, for every number A, nearly all terms a_i are $< A$. [There is no need to write out explicitly the rules that follow from the results of Chapter 3, §5.2 on infinite limits.]

Of course, a sequence may diverge without having even an infinite limit. For instance, the sequence

$$1, 2, 3, 1, 2, 3, 1, 2, 3, 1, 2, \cdots$$

diverges.

►*Examples*   *1.* For $a > 1$, we have

$$\lim_{n \to \infty} \frac{a^n}{n} = +\infty.$$

Proof.   Since $a > 1$, we may write $a = 1 + b$, $b > 0$. By the binomial theorem we have

$$a^n = (1 + b)^n = \sum_{j=0}^{n} \binom{n}{j} b^j > \binom{n}{2} b^2 = \frac{n(n-1)}{2} b^2.$$

Hence $\frac{a^n}{n} > \frac{b^2}{2}(n - 1)$. This becomes as large as we want.

2. Does the sequence $a_n = \dfrac{3 + 2n^2}{1 - n}$, $n = 1, 2, \cdots$, have a finite limit? An infinite limit?

Solution.   We use the rules of Chapter 3, §5.2, and conclude that

$$\lim a_n = \lim \frac{n^2\left(2 + \dfrac{3}{n^2}\right)}{-n\left(1 - \dfrac{1}{n}\right)} = \lim(-n) \lim \frac{2 + \dfrac{3}{n^2}}{1 - \dfrac{1}{n}} = (-\infty) \cdot 2 = -\infty.$$

The sequence diverges to $-\infty$; it has no finite limit. One can also argue directly:

$$a_n = (-n) \frac{2 + \dfrac{3}{n^2}}{1 - \dfrac{1}{n}}$$

is $(-n)$ times a number that is, for large n, very close to 2. Therefore a_n is, for large n, nearly $-2n$, therefore $\lim_{n \to \infty} a_n = -\infty$. ◄

EXERCISES

29. Prove that the sequence $a_n = \dfrac{n}{n + 1}$, $n = 0, 1, \cdots$, converges to 1.

30. Prove that the sequence $e_j = \dfrac{1}{\sqrt{j}}$, $j = 1, 2, \cdots$, converges to 0.

31. Prove that the sequence $c_k = (-1)^k$, $k = 1, 2, \cdots$, does not converge to 1.

32. Prove that the sequence $w_l = \dfrac{2l}{3l + 1}$, $l = 0, 1, \cdots$, converges to $\dfrac{2}{3}$.

33. Prove that $\lim\limits_{n \to \infty} \sqrt{4 + \dfrac{1}{n}} = 2$.

34. Prove that the sequence $x_u = \left(\dfrac{1000}{u}\right)$, $u = 1, 2, \cdots$, converges to 0.

35. Prove that the sequence $b_k = q^k$, $k = 0, 1, \cdots$, does not converge to 0 for $q > 1$.

36. Prove that the sequence $u_r = \dfrac{1}{r} \sin r$, $r = 1, 2, \cdots$, converges to 0.

37. Prove that $\lim_{n \to \infty} q^{1/n} = 1$ for $0 < q$. (Hint: Consider the cases $q \leq 1$ and $q > 1$ separately.)

38. Prove that $\lim_{n \to \infty} q^{n/(n+1)} = q$. (Hint: Consider the cases $q \leq 1$ and $q > 1$ separately.)

In Exercises 39 to 51, find which of the given sequences converge, which diverge to $+\infty$ or $-\infty$, and which diverge with no infinite limit. In the first case, find the limit. Your answers must be justified either by a proof from the definition or by using the theorems that have been proved.

39. $a_n = \dfrac{2n^3 - n}{3n^3 + n^2 + 1}$, $n = 0, 1, \cdots$.

40. $a_n = \dfrac{1 - 2n^2}{1 + n - n^2}$, $n = 1, 2, \cdots$.

41. $u_n = \dfrac{n^3 - 3n + 4}{n^2 + n}$, $n = 1, 2, \cdots$.

42. $c_k = \dfrac{k^{1/2} + 1}{k - 1}$, $k = 2, 3, \cdots$.

43. $a_k = \dfrac{k^{1/3} + k^{2/3}}{1 + k^{1/2} + k^{3/2}}$, $k = 1, 2, \cdots$.

► 44. $v_k = \dfrac{2k + \sin k}{5k + 1}$, $k = 0, 1, \cdots$.

45. $u_\lambda = \dfrac{\lambda! + 1}{\lambda! - 1}$, $\lambda = 2, 3, \cdots$.

46. $a_n = \dfrac{(n + 3)! - n!}{(n + 4)!}$, $n = 0, 1, \cdots$.

47. $x_n = \dfrac{a^n}{(n-1)^2}$, $n = 2, 3, \cdots$, for $a > 1$.

48. $x_n = na^n$, $n = 0, 1, \cdots$, for $0 < a < 1$.

49. $a_n = \cos\dfrac{n\pi}{7}$, $n = 0, 1, \cdots$.

50. $a_n = \dfrac{n^n}{n!}$, $n = 1, 2, \cdots$.

51. $c_j = \dfrac{8^j - 4^j}{3^j}$, $j = 0, 1, \cdots$.

3.8 Monotone sequences

There is an exceedingly important theorem which asserts that certain sequences are convergent.

A sequence $a_1, a_2, a_3, \cdots$ is called **increasing** if $a_1 < a_2 < a_3 < \cdots$, **nondecreasing** if $a_1 \leq a_2 \leq a_3 \leq \cdots$. Thus, in an increasing sequence, every term is greater than its predecessor, while in a nondecreasing sequence every term is not less than its predecessor. Similarly, $\{a_i\}$ is called **decreasing** if $a_i > a_{i+1}$ for all i, **nonincreasing** if $a_i \geq a_{i+1}$ for all i. A sequence that is either nondecreasing or nonincreasing is called **monotone.**

A sequence $\{a_i\}$ is called **bounded from above** if there is a number M such that $a_i \leq M$ for all i. A sequence $\{a_i\}$ is called **bounded from below** if there is a number m such that $m \leq a_i$ for all i. A **bounded** sequence is one bounded from both above and below.

►*Examples*   *1.* The sequence $a_n = 1 + (-1)^n/n^2$ is neither nonincreasing nor nondecreasing, but it is bounded.

2. The sequence $a_n = n^2 - (1/n)$ is increasing and unbounded.

3. The sequence $a_n = \sin n$ is bounded, but not monotone. ◄

Theorem 3   A BOUNDED NONDECREASING SEQUENCE $a_1, a_2, a_3, \cdots$ HAS A LIMIT; THE LIMIT IS THE SMALLEST NUMBER THAT IS NOT LESS THAN ANY a_j.

The statement is rather obvious geometrically; see Fig. 8.10. The proof will be found in an appendix to this chapter (§7.1). In view of the connection between limits of sequences and limits of functions mentioned in §3.6, Theorem 3 also follows from the theorem on bounded monotone functions (Theorem A) proved in Chapter 5, §9.1.

We note the following almost obvious corollary.

Corollary.   A BOUNDED NONINCREASING SEQUENCE CONVERGES.

Fig. 8.10

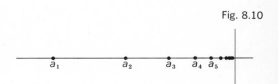

For if $\{a_i\}$ is nonincreasing, the sequence $\{-a_i\}$ is nondecreasing.

We observe that a nondecreasing sequence $\{a_i\}$ is always bounded from below, since $a_1 \leq a_i$ for all i. The term "bounded" in Theorem 3 may therefore be replaced by "bounded from above." Also, if a nondecreasing sequence $\{a_i\}$ is not bounded from above, then, for every number A, there is a number j such that $a_j > A$. Then also, $a_i > A$ for all $i > j$, and therefore $\lim a_i = +\infty$. We may restate Theorem 3 as follows.

Theorem 3'   IF $a_1 \leq a_2 \leq a_3 \leq \cdots$, THEN EITHER $\lim a_i = +\infty$ OR THERE IS A NUMBER α SUCH THAT $\lim a_j = \alpha$. (IN THE LATTER CASE $a_j \leq \alpha$ FOR ALL j, AND α IS THE SMALLEST NUMBER WITH THIS PROPERTY.)

EXERCISES

Determine which of the following sequences are nondecreasing or nonincreasing after a finite number of terms. Among these, determine which are bounded (and so converge) and which are unbounded (and so diverge to $+\infty$ or $-\infty$). In the first case, find bounds. Do not be surprised if it is possible to determine the convergence or divergence of some of the sequences directly, and perhaps more easily, without the aid of Theorem 3.

52. $a_n = \dfrac{2n - 1}{3n + 4}$, $n = 1, 2, \cdots$.

54. $b_k = \dfrac{k^2 + 10k}{k + 10}$, $k = 0, 1, \cdots$.

53. $c_k = \dfrac{k^2 + 1}{k^2 + k}$, $k = 1, 2, \cdots$.

55. $u_n = \dfrac{10^n}{n!}$, $n = 0, 1, \cdots$.

56. $a_n = \dfrac{1 \cdot 3 \cdot 5 \cdots (2n - 1)}{2 \cdot 4 \cdot 6 \cdots (2n)}$, $n = 1, 2, \cdots$.

57. $x_\lambda = (\sin 1)(\sin 2) \cdots (\sin \lambda)$, $\lambda = 1, 2, \cdots$.

58. $y_\lambda = (\sin^2 1)(\sin^2 2) \cdots (\sin^2 \lambda)$, $\lambda = 1, 2, \cdots$.

59. $u_k = 1 + \dfrac{1}{k^2 + k \cos k + 1}$, $k = 1, 2, \cdots$. (Hint: Show that the derivative of $x \mapsto x^2 + x \cos x + 1$ is positive for $x > 1$.)

60. $a_\lambda = \dfrac{1}{\lambda + \sin (\lambda^2)}$, $\lambda = 1, 2, \cdots$.

61. $a_n = \dfrac{n}{\log (n!)}$, $n = 1, 2, \cdots$.

3.9 Euler's constant

As an application of Theorem 3, we shall prove that there exists a number γ such that

(11) $$\lim_{n \to \infty} \left(1 + \frac{1}{2} + \frac{1}{3} + \cdots + \frac{1}{n} - \log n \right) = \gamma.$$

This number is called **Euler's constant,** and

$$\gamma \approx .5772 \cdots.$$

Nobody knows whether γ is rational. Equation (11) is equivalent to the statement

$$\log n = 1 + \frac{1}{2} + \cdots + \frac{1}{n} - \gamma + \epsilon_n, \qquad \lim_{n \to \infty} \epsilon_n = 0.$$

In order to prove the convergence of the sequence

$$a_n = 1 + \frac{1}{2} + \frac{1}{3} + \cdots + \frac{1}{n} - \log n, \qquad n = 1, 2, \cdots,$$

we start with the inequality

$$\frac{1}{2} + \frac{1}{3} + \cdots + \frac{1}{n-1} + \frac{1}{n} < \log n < 1 + \frac{1}{2} + \frac{1}{3} + \cdots + \frac{1}{n-1}.$$

This inequality is obvious from Fig. 8.11, where $n = 4$. It expresses the fact that the area, from $x = 1$ to $x = n$, under the step function $1/[x + 1]$ is less than the area under the function $1/x$, which in turn is less than the area under the step function $1/[x]$. Set

$$b_n = \log n - \left(\frac{1}{2} + \frac{1}{3} + \cdots + \frac{1}{n} \right).$$

Subtracting $\frac{1}{2} + \frac{1}{3} + \cdots + \frac{1}{n}$ from the second inequality above, we see that

$$0 < b_n < 1 - \frac{1}{n} < 1,$$

so that the sequence $\{b_n\}$ is bounded. Since b_n is the sum of the shaded areas in Fig. 8.11, the sequence $\{b_n\}$ is increasing. Hence the finite limit, $\lim b_n$, exists; so does $\lim a_n$, for $a_n = 1 - b_n$.

4.1 Partial sums

After the preparations in §3, we can consider infinite series. An **infinite series** is a sequence of numbers with plus signs between these numbers. The numbers, called the **terms** of the series, may be labeled by inte-

Fig 8.11

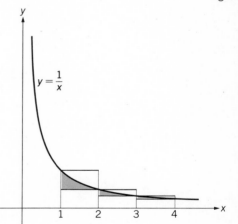

$y = \frac{1}{x}$

§4 Infinite Series

gers 0, 1, 2, $\cdots$; or 1, 2, 3, $\cdots$; or in any other way. If some term is a negative number, say (-6), we agree to write -6 instead of $+(-6)$. Otherwise the same conventions as for sequences are used. When we say "the infinite series $a_0 + a_1 + a_2 + \cdots$," we mean that the terms a_j are definite but unspecified numbers. One also uses for infinite series the sigma notation introduced in §2.11. Thus the series $a_0 + a_1 + a_2 + \cdots$ may also be written as

$$(1) \qquad \sum_{j=0}^{\infty} a_j.$$

For the sake of brevity, we say henceforth "series" instead of "infinite series."

The nth **partial sum** of the series (1) is, by definition, the (finite) sum of the terms up to, and including, the nth, that is, the sum

$$a_0 + a_1 + \cdots + a_n = \sum_{j=0}^{n} a_j.$$

Thus there is an infinite sequence, the sequence of partial sums, associated with every infinite series. Conversely, every sequence is the sequence of partial sums of some series. The terms

$$a_1, a_2, a_3, a_4 \cdots,$$

of an infinite sequence are the partial sums of the infinite series

$$a_1 + (a_2 - a_1) + (a_3 - a_2) + (a_4 - a_3) + \cdots,$$

as the reader will verify.

4.2 Convergent series

The infinite series $a_0 + a_1 + a_2 + \cdots = \sum_{j=0}^{\infty} a_j$ is said to **converge** if the sequence of partial sums converges, that is, if the finite limit

$$(2) \qquad S = \lim_{n \to \infty} \sum_{j=0}^{n} a_j = \lim_{n \to \infty} (a_0 + a_1 + \cdots + a_n)$$

exists. In this case, S is called the **sum of the infinite series,** and we write $a_0 + a_1 + a_2 + \cdots = S$ or

$$\sum_{j=0}^{\infty} a_j = S.$$

Similarly, the statement $a_k + a_{k+1} + a_{k+2} + \cdots = S$ or

$$\sum_{j=k}^{\infty} a_j = S$$

is an abbreviation of the statement

$$\lim_{n \to \infty} (a_k + a_{k+1} + \cdots + a_n) = \lim_{n \to \infty} \sum_{j=k}^{n} a_j = S.$$

Theorem 1   IF AN INFINITE SERIES CONVERGES, SO DOES THE SERIES OBTAINED BY ADDING, DROPPING, OR CHANGING FINITELY MANY TERMS.

Proof.   It suffices to consider what happens if one changes *one* term. Suppose $a_0 + a_1 + a_2 + \cdots = S$. This means that

$$\lim_{n \to \infty} (a_0 + a_1 + \cdots + a_n) = S.$$

Consider now the series $\hat{a}_0 + a_1 + a_2 + \cdots$. Its partial sums are

$$\hat{a}_0 + a_1 + a_2 + \cdots + a_n = a_0 + a_1 + a_2 + \cdots + a_n + (\hat{a}_0 - a_0).$$

Hence, by Theorem 1 of §3.6,

$$\lim_{n \to \infty} (\hat{a}_0 + a_1 + \cdots + a_n) = S + (\hat{a}_0 - a_0).$$

The series $\hat{a}_0 + a_1 + a_2 + \cdots$ converges.

Taylor's Theorem gives us many interesting examples of convergent series. For instance, it follows from equation (48) in §2.12 that

$$(3) \qquad \sum_{j=1}^{\infty} (-1)^{j+1} \frac{x^j}{j} = \log(1 + x) \qquad \text{for } -1 < x \le 1,$$

that is,

$$x - \frac{x^2}{2} + \frac{x^3}{3} - \frac{x^4}{4} + \frac{x^5}{5} - \cdots = \log(1 + x), \qquad -1 < x \le 1.$$

This is so, since $\lim \dfrac{x^{n+1}}{n+1} = 0$ for $0 \le x \le 1$ and $\lim \dfrac{|x|^{n+1}}{(1 - |x|)(n+1)} = 0$ for $-1 < x < 0$. A special case of (3) is the series

$$(4) \qquad 1 - \frac{1}{2} + \frac{1}{3} - \frac{1}{4} + \cdots = \log 2,$$

which we already noted in §3.2.

Similarly, we obtain from (50) in §2.12 that

$$(5) \qquad \sum_{j=1}^{\infty} (-1)^{j+1} \frac{x^{2j-1}}{2j-1} = x - \frac{x^3}{3} + \frac{x^5}{5} - \cdots = \arctan x \text{ for } |x| \le 1.$$

Setting $x = 1$ here, we get

(6) $$1 - \frac{1}{3} + \frac{1}{5} - \frac{1}{7} + \frac{1}{9} - \cdots = \frac{\pi}{4}.$$

The beautiful series (4) and (6) were known to Leibniz.

Equations (52), (53), and (54) in §2.12 show that for all x we have

(7) $$\sum_{n=0}^{\infty} \frac{x^n}{n!} = 1 + \frac{x}{1!} + \frac{x^2}{2!} + \frac{x^3}{3!} + \cdots = e^x.$$

(8) $$\sum_{n=0}^{\infty} (-1)^n \frac{x^{2n}}{(2n)!} = 1 - \frac{x^2}{2!} + \frac{x^4}{4!} - \frac{x^6}{6!} + \cdots = \cos x,$$

(9) $$\sum_{n=1}^{\infty} (-1)^n \frac{x^{2n+1}}{(2n+1)!} = x - \frac{x^3}{3!} + \frac{x^5}{5!} - \frac{x^7}{7!} + \cdots = \sin x.$$

Indeed, the jth partial sum $1 + \dfrac{x}{1!} + \dfrac{x^2}{2!} + \cdots + \dfrac{x^j}{j!}$ differs from e^x by

at most $\dfrac{e^{|x|}|x|^{j+1}}{(j+1)!}$ and $\lim \dfrac{e^{|x|}|x|^{j+1}}{(j+1)!} = 0$ by Example 3 in §3.5. We verify
(8) and (9) using Example 4 in §3.5.

EXERCISES

1. Evaluate the first 6 partial sums of the infinite series $\displaystyle\sum_{j=0}^{\infty} \frac{j}{j+1}$.

2. Evaluate the first 6 partial sums of the infinite series $\Sigma_{k=0}^{\infty} 2^k$.

3. Evaluate the first 5 partial sums of the infinite series $\Sigma_{\lambda=0}^{\infty} (1)^\lambda \lambda!$.

4. Evaluate the first 4 partial sums of the infinite series $\displaystyle\sum_{n=1}^{\infty} \frac{1 \cdot 3 \cdot 5 \cdots (2n-1)}{2 \cdot 4 \cdot 6 \cdots (2n)}$.

5. Evaluate the first 4 partial sums of the infinite series $\displaystyle\sum_{n=1}^{\infty} \frac{1 \cdot 4 \cdot 7 \cdots (3n-2)}{3^n}$.

► 6. Find an infinite series whose nth partial sum is given by $\dfrac{n}{n+1}$ for
   $n = 1, 2, \cdots$.

7. Find an infinite series whose nth partial sum is given by $\dfrac{2^n}{n!}$ for $n = 1, 2, \cdots$.

In Exercises 8 to 12, find the nth Taylor polynomial, $P_n(x)$, of the given function $f(x)$ around x_0. Then find an infinite series whose sequence of partial sums is given by $P_0(x), P_1(x), P_2(x), \cdots$.

8. $f(x) = e^{x^2}$ at $x_0 = 0$. Does this series converge for all x? See Exercise 65, §2.10.

9. $f(x) = \cos x^3$ at $x_0 = 0$. Does this series converge for all x? See Exercise 67, §2.10.

10. $f(x) = \dfrac{1}{\sqrt{x}}$ at $x_0 = 1$.

11. $f(x) = x^{-1/3}$ at $x_0 = 1$.

12. $f(x) = \dfrac{1}{\sqrt{2-x}}$ at $x_0 = 1$.

4.3 A necessary condition for convergence

There is a simple necessary condition for the convergence of an infinite series.

Theorem 2   IF $\Sigma_{j=0}^{\infty} a_j$ CONVERGES, THEN $\lim_{j \to \infty} a_j = 0$.

  Proof.   Let $S = \Sigma_{j=0}^{\infty} a_j$. Then $\lim_{j \to \infty} (a_0 + a_1 + \cdots + a_j) = S$ and (see §3.5) also $\lim_{j \to \infty} (a_0 + a_1 + \cdots + a_{j-1}) = S$. Hence, by Theorem 1,

$$\lim a_j =$$

$$\lim \left[(a_0 + a_1 + \cdots + a_j) - (a_0 + a_1 + \cdots + a_{j-1}) \right] = S - S = 0.$$

One should *not* think, however, that every series with $\lim a_j = 0$ converges. For instance, the so-called **harmonic series**

$$1 + \frac{1}{2} + \frac{1}{3} + \frac{1}{4} + \frac{1}{5} + \frac{1}{6} + \cdots$$

does not converge. For according to §3.9, the nth partial sum of this series,

$$1 + \frac{1}{2} + \cdots + \frac{1}{n},$$

is, for large n, very close to $\gamma + \log n$, γ being Euler's constant.

4.4 Divergent series

A series that does not converge is called **divergent.** One should remember that, if an infinite series does not converge, the symbol $a_1 + a_2 + \cdots = \Sigma_{j=0}^{\infty} a_j$ is *not* the name of a number.

  If $\lim_{n \to \infty} \Sigma_{j=0}^{n} a_j = +\infty$, we say that the series $\Sigma_{j=0}^{\infty} a_j$ *diverges to* $+\infty$ and we write

$$\sum_{j=0}^{\infty} a_j = +\infty.$$

The meaning of the symbol $\Sigma_{j=0}^{\infty} a_j = -\infty$ needs no separate explanation.

For instance, $1 + 1 + 1 + \cdots = +\infty$, $1 + 2 + 3 + 4 + \cdots = +\infty$. Also

$$1 + \frac{1}{2} + \frac{1}{3} + \frac{1}{4} + \frac{1}{5} + \cdots = +\infty.$$

This follows from the result on Euler's constant, but it can also be seen easily by noting that, for every integer $k > 0$, we have

$$\frac{1}{k+1} + \frac{1}{k+2} + \cdots + \frac{1}{2k} > k\frac{2}{k} = \frac{1}{2}.$$

Hence, if we take n large enough, the partial sum $1 + \cdots + 1/n$ will contain as many segments that add up to more than $\frac{1}{2}$ as we wish.

The series

$$\sum_{j=0}^{\infty} (-1)^j = 1 - 1 + 1 - 1 + 1 - 1 + \cdots$$

diverges, since $a_0 + a_1 + \cdots + a_n = 0$ or 1, depending on whether n is even or odd. But the series does not diverge to either $+\infty$ or $-\infty$.

EXERCISES

Determine which of the following series converge and which diverge.

13. $\displaystyle\sum_{n=1}^{\infty} \frac{n+1}{2n-3}.$

14. $\displaystyle\sum_{n=1}^{\infty} \left(\frac{1}{n} - \frac{1}{n+1}\right).$

15. $\displaystyle\sum_{\lambda=1}^{\infty} \left(\sin\frac{1}{\lambda} - \sin\frac{1}{\lambda+1}\right).$

16. $\displaystyle\sum_{n=1}^{\infty} \left(\sum_{j=1}^{n} \frac{i}{j}\right).$

17. $\displaystyle\sum_{k=1}^{\infty} \frac{k}{\sin k}.$

18. $\displaystyle\sum_{j=0}^{\infty} \binom{1000}{j}.$

19. $\displaystyle\sum_{\beta=1}^{\infty} \cos\frac{1}{\beta}.$

20. $\displaystyle\sum_{n=1}^{\infty} \cos\frac{n\pi}{2}.$

21. $\displaystyle\sum_{n=1}^{\infty} (-1)^n \log\left(\frac{n}{n+2}\right).$

$\left[\text{Hint: Write } \log\left(\dfrac{n}{n+2}\right) \text{ as } \log\left(\dfrac{n}{n+1}\right) + \log\left(\dfrac{n+1}{n+2}\right).\right]$

22. $\displaystyle\sum_{n=1}^{\infty} \frac{1}{\sqrt{n+1} + \sqrt{n}}.$ (Hint: Rationalize the denominator of each term.)

4.5 **The geometric series**

The following result is very simple but also very important.

Theorem 3   THE GEOMETRIC SERIES

$$1 + q + q^2 + \cdots = \sum_{j=0}^{\infty} q^j$$

DIVERGES FOR $|q| \geq 1$, CONVERGES FOR $|q| < 1$, AND

$$(10) \qquad 1 + q + q^2 + \cdots = \sum_{j=0}^{\infty} q^j = \frac{1}{1-q} \qquad \text{if } |q| < 1.$$

Proof. If $|q| \geq 1$, then it is not true that $\lim_{j \to \infty} |q^j| = 0$. Hence the series cannot converge (see Theorem 2). If $|q| < 1$, we note that

$$(11) \qquad \sum_{j=0}^{n} q^j = 1 + q + \cdots + q^n = \frac{1 - q^{n+1}}{1-q} = \frac{1}{1-q} - \frac{q^{n+1}}{1-q}.$$

Since $\lim_{n \to \infty} q^{n+1} = 0$ (see Example 2 in §3.5), we have $\lim_{n \to \infty} \dfrac{q^{n+1}}{1-q}$ $= 0$ and (10) follows.

For $q = \frac{1}{2}$, we obtain the familiar relation $1 + \dfrac{1}{2} + \dfrac{1}{3} + \cdots = 2$.

▶*Examples*   *1.* $q^k + q^{k+1} + q^{k+2} + \cdots = \dfrac{q^k}{1-q}$ $(|q| < 1)$.

Proof. Multiply both sides of (11) by q^k, then take the limits of both sides as $n \to \infty$. One can also subtract $1 + q + \cdots + q^k$ from both sides of (10). It seems simpler to multiply both sides of (10) by q^k. We shall see below (§4.6) that this is legitimate.

2. Find the sum $\sum_{j=1}^{\infty} jq^{j-1} = 1 + 2q + 3q^2 + \cdots$, for $|q| < 1$.

Solution. Differentiating both sides of (11) with respect to q, we see that

$$1 + 2q + \cdots + nq^{n-1} = \frac{-(n+1)q^n(1-q) + 1 - q^{n+1}}{(1-q)^2}$$

$$= \frac{1 - (n+1)q^n + nq^{n+1}}{(1-q)^2}$$

and therefore, taking the limit for $n \to \infty$,

$$1 + 2q + 3q^2 + \cdots = \frac{1}{(1-q)^2}.$$

3. For $|x| < 1$,

$$\frac{1}{1+x^2} = \sum_{j=0}^{\infty} (-1)x^{2j} = 1 - x^2 + x^4 - x^6 + \cdots.$$

This is seen by setting $q = -x^2$ in (10).◄

4.6 Operations on series

The following is an immediate consequence of our definition and of Theorem 1 in §3.6.

Theorem 4   IF

$$\sum_{j=0}^{\infty} a_j = A,$$

THEN

$$\sum_{j=0}^{\infty} ca_j = cA.$$

IF ALSO

$$\sum_{j=0}^{\infty} b_j = B,$$

THEN

$$\sum_{j=0}^{\infty} (a_j + b_j) = A + B.$$

In words: a convergent infinite series can be multiplied term by term by a number, and two convergent infinite series can be added term by term.

In applying this theorem, it is well to remember that every infinite series in which nearly all terms are 0 is convergent, and that every finite sum can be considered as an infinite series with nearly all terms 0. Also, we can insert as many zeros as we want between the terms of a convergent series; this does not change its sum. We can now verify that the manipulations by which we arrived at equation (4) in §3.2 are legitimate.

EXERCISES

Determine which of the following series converge. If the series converges, find the sum.

23. $\sum_{n=1}^{\infty} (10^{-n} + 9^{-n})$.

24. $\sum_{n=1}^{\infty} (10^{-n} + 2^n)$.

25. $\sum_{n=0}^{\infty} x^{n/3}$ for $|x| < 1$.

► 26. $\sum_{k=0}^{\infty} (-1)^k t^{k/2}$ for $0 \leq t < 1$.

27. $\sum_{n=0}^{\infty} nz^n$ for $|z| < 1$.   28. $\sum_{n=0}^{\infty} u^n(1 + u^n)$ for $|u| < 1$.

29. $\sum_{j=0}^{\infty} (2^j + a_j)$, where $\sum_{j=0}^{\infty} a_j$ is any convergent series.

30. $\displaystyle\sum_{n=0}^{\infty} x^n \cos \frac{n\pi}{2}$ for $|x| < 1$.

31. $\displaystyle\sum_{n=0}^{\infty} \frac{(2x)^n + x^n}{2^n}$ for $|x| < \frac{1}{2}$.

32. $\displaystyle\sum_{n=0}^{\infty} \sin^{2n} x$ for $|x| < \frac{\pi}{2}$.

33. $\sum_{n=1}^{\infty} nx^{2n-1}$ for $|x| < 1$.

34. $\sum_{j=2}^{\infty} j(j - 1)q^{j-2}$ for $|q| < 1$.

4.7 Series with positive terms

Consider now an infinite series $a_0 + a_1 + \cdots$ with only positive terms. Then the sequence of partial sums is increasing, for the $(k + 1)$st partial is obtained from the kth by adding to it the positive number a_{k+1}. Hence the basic Theorem 3 of §3.8 is applicable to the sequence of partial sums. This yields the following result.

Theorem 5   IF ALL TERMS IN $\sum_{i=0}^{\infty} a_i$ ARE POSITIVE, THEN EITHER $\sum_{i=0}^{\infty} a_i = +\infty$, OR THE SEQUENCE OF PARTIAL SUMS IS BOUNDED AND THE SERIES CONVERGES. IN THIS CASE THE SUM $\sum_{j=0}^{\infty} a_j$ IS THE SMALLEST NUMBER M SUCH THAT $\sum_{i=0}^{n} a_i \leq M$ FOR ALL n.

This theorem has many important applications. It is clear that it holds also for series with nonnegative terms.

4.8 The comparison test

Theorem 6 (On Dominated Series)   ASSUME THAT

$$(12) \qquad 0 \leq a_i \leq b_i \text{ for } i = 0, 1, 2, \cdots.$$

IF THE SERIES

$$(13) \qquad b_0 + b_1 + b_2 + \cdots$$

CONVERGES, SO DOES THE SERIES

$$(14) \qquad a_0 + a_1 + a_2 + \cdots,$$

AND

$$a_0 + a_1 + a_2 + \cdots \leq b_0 + b_1 + b_2 + \cdots.$$

IF THE SERIES (14) DIVERGES, THEN THE SERIES (13) ALSO DIVERGES.

Remark   If (12) holds, one says that the series (14) is **dominated** by the series (13).

Proof of Theorem 6.   Assume that $b_0 + b_1 + b_2 + \cdots$ converges and set $B = \Sigma_{j=0}^{\infty} b_j$. Then, by Theorem 5, we have $\Sigma_{j=0}^{n} b_j \le B$; hence, also $\Sigma_{j=0}^{n} a_j \le B$ for all n, in view of (12). By Theorem 5, the series $a_0 + a_1 + a_2 + \cdots$ converges and $\Sigma_{j=0}^{\infty} a_j \le B$. The second statement is now obvious.

4.9 Decimals as series

The concept of convergent infinite series is already present, implicitly, in the use of infinite decimals. Indeed, consider the decimal fraction

$$\alpha_0 \cdot \alpha_1 \alpha_2 \alpha_3 \cdots$$

(α_0 being a positive integer, and each other α_j being one of the digits $0, 1, 2, \cdots, 9$). This fraction can be written as an infinite series with nonnegative terms

$$(15) \qquad\qquad \alpha_0 + \frac{\alpha_1}{10} + \frac{\alpha_2}{10^2} + \frac{\alpha_3}{10^3} + \cdots$$

The terms of this series are not greater than those of the convergent series

$$\alpha_0 + \frac{9}{10} + \frac{9}{10^2} + \frac{9}{10^3} + \cdots = \alpha_0 + \frac{9}{10}\left(1 + \frac{1}{10} + \frac{1}{10^2} + \cdots\right).$$

Hence (15) converges. The sum of the series is, of course, the number $a_0 \cdot \alpha_1 \alpha_2 \alpha_3 \cdots$.

If $\alpha_0 \cdot \alpha_1 \alpha_2 \alpha_3 \cdots$ is a repeating decimal, the formula for summing a geometric series shows that this number is rational. For instance,

$$.\overline{32} = .323232 \cdots = \frac{32}{100} + \frac{32}{100^2} + \frac{32}{100^3} + \cdots$$

$$= \frac{32}{100}\left(1 + \frac{1}{100} + \frac{1}{100^2} + \cdots\right) = \frac{32}{100}\frac{1}{1 - (1/100)} = \frac{32}{99}.$$

EXERCISES

Express the following repeating decimals as the quotient of two integers.

35. $.\overline{18}.$

36. $1.2\overline{9}.$

37. $.1\overline{180}.$

38. $.31\overline{17}.$

Express the sum of the following infinite series as the quotient of two integers. These series would correspond to a repeating "decimal" in a base other than 10.

39. $\dfrac{1}{3} + \dfrac{2}{9} + \dfrac{1}{27} + \dfrac{2}{81} + \cdots + \dfrac{1}{3^{2n-1}} + \dfrac{2}{3^{2n}} + \cdots$.

40. $\dfrac{2}{5} + \dfrac{4}{25} + \dfrac{2}{125} + \dfrac{4}{625} + \cdots + \dfrac{2}{5^{2n-1}} + \dfrac{4}{5^{2n}} + \cdots$.

41. $\dfrac{1}{2} + \dfrac{1}{4} + \dfrac{1}{16} + \dfrac{1}{32} + \cdots + \dfrac{1}{2^{3n-2}} + \dfrac{1}{2^{3n-1}} + \dfrac{0}{2^{3n}} + \cdots$.

42. $\dfrac{3}{5} + \dfrac{3}{125} + \dfrac{3}{625} + \dfrac{3}{15{,}625} + \cdots + \dfrac{3}{5^{3n-2}} + \dfrac{0}{5^{3n-1}} + \dfrac{3}{5^{3n}} + \cdots$.

Determine which of the following series converge and which diverge by using the **comparison test**, that is, Theorem 6 on dominated series.

43. $\displaystyle\sum_{n=0}^{\infty} \dfrac{\cos^2 n}{2^n}$.

44. $\displaystyle\sum_{k=1}^{\infty} \dfrac{2 + \cos k}{k}$.

45. $\displaystyle\sum_{n=0}^{\infty} \dfrac{1}{2^n + 1}$.

46. $\displaystyle\sum_{n=1}^{\infty} \dfrac{1}{\sqrt{n}}$.

47. $\displaystyle\sum_{n=1}^{\infty} \dfrac{1}{2n - 1}$.

▶ 48. $\displaystyle\sum_{k=1}^{\infty} \dfrac{1}{3^k - \sin k}$.

49. $\displaystyle\sum_{r=1}^{\infty} \dfrac{\log r}{2^{r-1}}$.

50. $\displaystyle\sum_{n=3}^{\infty} \dfrac{1}{\sqrt{n^2 - 4}}$.

51. $\displaystyle\sum_{n=2}^{\infty} \dfrac{1}{n! - n}$.

52. Assume that $c_i \le a_i \le b_i$ for $i = 0, 1, 2, \cdots$, and $\Sigma_{i=0}^{\infty} c_i$ and $\Sigma_{i=0}^{\infty} b_i$ converge. Show that $\Sigma_{i=0}^{\infty} a_i$ converges.

4.10 The integral test

Here is a useful application of the concept of dominated series. It involves the concept of improper integrals (see Chapter 5, §§4.9 and 6.2).

Theorem 7 (The Integral Test) LET $t \mapsto f(t)$ BE A CONTINUOUS NON-NEGATIVE DECREASING FUNCTION DEFINED FOR $t \ge 1$. IF THE IMPROPER INTEGRAL

$$(16) \qquad\qquad \int_{1}^{+\infty} f(t)\, dt$$

EXISTS, THE SERIES

$$(17) \qquad\qquad \sum_{n=1}^{\infty} f(n)$$

CONVERGES. IF THE INTEGRAL DOES NOT EXIST, THE SERIES DIVERGES.

Before proving the theorem, we note a consequence.

Corollary.   The series

$$1 + \frac{1}{2^s} + \frac{1}{3^s} + \frac{1}{4^s} + \frac{1}{5^s} + \cdots = \sum_{n=1}^{\infty} n^{-s}$$

converges for $s > 1$, diverges for $s \leq 1$.

Indeed, if $s > 0$, then $t \mapsto t^{-s}$ is a decreasing function of t and we can apply the integral test. For $s \neq 1$, we have

$$\int_1^{+\infty} t^{-s}\, dt = \lim_{A \to +\infty} \int_1^A t^{-s}\, dt = \lim_{A \to +\infty} \frac{t^{1-s}}{1-s}\Big|_1^A$$

$$= \lim_{A \to +\infty} \frac{1}{s-1}(1 - A^{1-s}) = \begin{cases} \dfrac{1}{s-1} & \text{if } s > 1, \\[2mm] +\infty & \text{if } s < 1. \end{cases}$$

For $s = 1$, we get the harmonic series, which we know diverges. The integral test also works; the improper integral $\int_1^{+\infty} t^{-1}\, dt$ does not exist. For $s \leq 0$, we have $\lim_{n \to \infty} n^{-s} = +\infty$; the series certainly diverges. For $s > 1$, we can define the function

$$\zeta(s) = 1 + \frac{1}{2^s} + \frac{1}{3^s} + \cdots;$$

this is the so-called **Riemann zeta function**. It plays a fundamental part in number theory.

Proof of Theorem 7.   We note that the graph of $t \mapsto f(t)$ descends. Therefore

$$f(2) + f(3) + \cdots + f(n) \leq \int_1^n f(t)\, dt \leq f(1) + f(2) + \cdots + f(n-1).$$

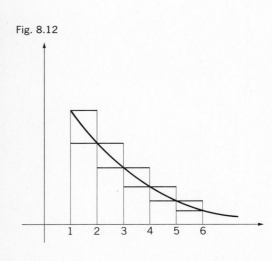

Fig. 8.12

Indeed (see Fig. 8.12, where $n = 6$), the left sum is the area under a step function below the curve $t \mapsto f(t)$ from 1 to n, whereas the right sum is the area from 1 to n under a step function above the curve.

If the integral $\int_1^{+\infty} f(t)\, dt$ exists, let the number I be its value. We have

$$\sum_{j=1}^n f(j) \leq I$$

and the convergence of $\sum_{j=0}^{\infty} f(j)$ follows from Theorem 5 in §4.7.

If the series converges, let S be its sum. We have

$$\int_1^n f(t)\, dt \leq S$$

and the existence of the improper integral follows from Theorem 1 in Chapter 5, §6.4.

EXERCISES

Determine which of the following series converge and which diverge by using the integral test.

53. $\sum_{n=2}^{\infty} \dfrac{1}{n \log n}$.

58. $\sum_{n=0}^{\infty} \dfrac{\text{arc tan } n}{1 + n^2}$.

▶ 54. $\sum_{n=2}^{\infty} \dfrac{1}{n \log^2 n}$.

59. $\sum_{n=0}^{\infty} \dfrac{e^{\text{arc tan } n}}{1 + n^2}$.

55. $\sum_{n=2}^{\infty} \dfrac{1}{n \log^{1/2} n}$.

60. $\sum_{n=1}^{\infty} \dfrac{e^{1/n}}{n^2}$.

56. $\sum_{n=0}^{\infty} \dfrac{n}{(n^2 + 1)^{2/3}}$.

61. $\sum_{n=1}^{\infty} \log \left(\dfrac{n + 5}{n} \right)$.

57. $\sum_{n=0}^{\infty} \dfrac{n^2 + 2n}{n^3 + 3n^2 + 1}$.

62. $\sum_{n=1}^{\infty} \dfrac{1}{\cosh^2 n}$.

63. $\sum_{n=2}^{\infty} \dfrac{1}{n \log^\lambda n}$ for any real number λ. The answer depends on λ; see Exercises 53 to 55.

Show that the following series converge by using Theorem 6 and the series expansion of $\zeta(s)$ for an appropriate value of s.

64. $\sum_{n=2}^{\infty} \dfrac{1}{n^3 - 1}$.

66. $\sum_{n=2}^{\infty} \dfrac{1}{n^2 - n}$.

65. $\sum_{n=2}^{\infty} \dfrac{n}{n^3 - 1}$.

67. $\sum_{n=1}^{\infty} \dfrac{n}{n^{5/2} + 1}$.

4.11 The ratio test

There is another useful method for deciding whether a series of positive terms converges.

Theorem 8 (The Ratio Test) LET THE TERMS IN THE SERIES $\sum_{n=0}^{\infty} a_n$ ALL BE POSITIVE. IF

$$\lim \frac{a_{n+1}}{a_n} < 1,$$

THE SERIES CONVERGES. IF

$$\lim \frac{a_{n+1}}{a_n} > 1,$$

THE SERIES DIVERGES.

Before proving this, we try to apply it to several cases.

▶*Examples*   1. The series $1 + 1 + \dfrac{1}{\sqrt{2!}} + \dfrac{1}{\sqrt{3!}} + \dfrac{1}{\sqrt{4!}} + \cdots =$

$\displaystyle\sum_{n=0}^{\infty} \dfrac{1}{\sqrt{n!}}$ converges.

Proof.   We have $\lim \dfrac{a_{n+1}}{a_n} = \lim \dfrac{\sqrt{n!}}{\sqrt{(n+1)!}} = \lim \dfrac{1}{\sqrt{n+1}} = 0 < 1.$

2. The series $3 + \dfrac{1}{2} + \dfrac{3}{2} + \dfrac{1}{4} + \dfrac{3}{4} + \dfrac{1}{8} + \dfrac{3}{8} + \dfrac{1}{16} + \cdots$ converges (to 7) as the reader is asked to verify. The ratio test is not applicable, however, since $\lim \dfrac{a_{n+1}}{a_n}$ fails to exist. Indeed, the sequence $\dfrac{a_1}{a_0}, \dfrac{a_2}{a_1}, \dfrac{a_3}{a_2},$ $\cdots$ is the sequence $\dfrac{1}{6}, 3, \dfrac{1}{6}, 3, \dfrac{1}{6}, 3, \cdots.$

3. We know that the series $\Sigma_{n=1}^{\infty} n^{-s}$ diverges for $s \le 1$, converges for $s > 1$. But $\lim \dfrac{a_{n+1}}{a_n} = \lim \left(\dfrac{n}{n+1} \right)^s = 1$ for all s. The ratio test is not applicable. ◀

Proof of Theorem 8.   Assume first that $\lim_{n\to\infty} \dfrac{a_{n+1}}{a_n} = L < 1$. Let θ be a number such that $L < \theta < 1$. Then $0 < \dfrac{a_{n+1}}{a_n} < \theta$ for nearly all n. Since convergence or divergence of a series is not affected by changing or dropping finitely many terms, we assume that

$$\frac{a_{n+1}}{a_n} < \theta \qquad \text{for all } n.$$

Then $\dfrac{a_1}{a_0} < \theta$, hence $a_1 < \theta a_0$. Also $\dfrac{a_2}{a_1} < \theta$; hence, $a_2 < \theta a_1 < \theta^2 a_0$. Also $\dfrac{a_3}{a_2} < \theta$; hence $a_3 < \theta a_2 < \theta^2 a_1 < \theta^3 a_0$. In general,

$$0 < a_n < a_0 \theta^n,$$

and since the series $a_0 + a_0\theta + a_0\theta^2 + a_0\theta^3 + \cdots$ converges (we have $0 < \theta < 1$), the series Σa_n converges, by Theorem 6.

Assume next that $\lim \dfrac{a_{n+1}}{a_n} = L > 1$. Let θ be a number such that $1 < \theta < L$.

Then $\dfrac{a_{n+1}}{a_n} > \theta$ for nearly all n. Since convergence or divergence of a series is not affected by dropping finitely many terms, we may assume that

$$\frac{a_{n+1}}{a_n} > \theta \qquad \text{for all } n.$$

Then $a_1 > a_0\theta$, $a_2 > a_1\theta > a_0\theta^2$, and so forth. In general $a_n > a_0\theta^n$. Hence it is not true that $\lim_{n \to \infty} a_n = 0$, and Σa_n diverges.

EXERCISES

Wherever applicable, use the ratio test to determine the convergence or divergence of the following series.

68. $\displaystyle\sum_{n=1}^{\infty} \frac{1 \cdot 3 \cdot 5 \cdots (2n-1)}{2 \cdot 4 \cdot 6 \cdots (2n)}$.

74. $\displaystyle\sum_{n=1}^{\infty} \frac{(2n)!}{n!n!}$.

69. $\displaystyle\sum_{n=1}^{\infty} \frac{1 \cdot 3 \cdot 5 \cdots (2n-1)}{1 \cdot 4 \cdot 7 \cdots (3n-2)}$.

75. $\displaystyle\sum_{n=1}^{\infty} \frac{e^{n^2}}{n!}$.

▶ 70. $\displaystyle\sum_{n=1}^{\infty} \frac{1 \cdot 6 \cdot 11 \cdots (5n-4)}{2 \cdot 6 \cdot 10 \cdots (4n-2)}$.

76. $\displaystyle\sum_{n=1}^{\infty} \frac{(n+1)! - n!}{2^n}$.

71. $\displaystyle\sum_{n=0}^{\infty} \frac{e^n}{n!}$.

77. $\displaystyle\sum_{n=1}^{\infty} \frac{(n!)^2}{(n^2)!}$.

72. $\displaystyle\sum_{n=1}^{\infty} \frac{2^n}{(n!)^\lambda}$ for $\lambda > 0$.

78. $\displaystyle\sum_{n=1}^{\infty} \frac{1}{n} \arctan n$.

73. $\displaystyle\sum_{n=1}^{\infty} \frac{n^3}{e^n}$.

79. $\displaystyle\sum_{n=1}^{\infty} \frac{n!}{n^n}$.

4.12 Alternating series

An **alternating** series is one in which the even terms a_{2j} are positive and the odd terms a_{2j+1} are negative. Examples are the famous series

$$1 - \frac{1}{2} + \frac{1}{3} - \frac{1}{4} + \frac{1}{5} - \frac{1}{6} + \cdots = \log 2,$$

$$1 - \frac{1}{3} + \frac{1}{5} - \frac{1}{7} + \frac{1}{9} - \cdots = \frac{\pi}{4}.$$

The following theorem is known as Leibniz's convergence criterion.

Theorem 9 (On Alternating Series) IF $p_0 > p_1 > p_2 > p_3 > \cdots > 0$ AND $\lim_{n \to \infty} p_n = 0$, THEN THE SERIES

$$p_0 - p_1 + p_2 - p_3 + p_4 - p_5 + \cdots$$

CONVERGES.

Proof. The odd partial sums form an increasing sequence, since the $(2n + 1)$st sum can be written as

$$S_{2n+1} = (p_0 - p_1) + (p_2 - p_3) + (p_4 - p_5) + \cdots + (p_{2n} - p_{2n+1})$$

and all terms $(p_0 - p_1), (p_2 - p_3), \cdots$ are positive. Since also

$$S_{2n+1} = p_0 - (p_1 - p_2) - (p_3 - p_4) - \cdots - (p_{2n-1} - p_{2n}) - p_{2n+1}$$

and the terms $(p_1 - p_2), (p_3 - p_4), \cdots$ are positive, we have $S_{2n+1} < p_0$. Hence $A = \lim_{n \to \infty} S_{2n+1}$ exists, by Theorem 3 in §3.8. We see similarly that $B = \lim_{n \to \infty} S_{2n}$ exists. (The reader should supply the details.) Since

$$\lim_{n \to \infty} (S_{2n+1} - S_{2n}) = \lim_{n \to \infty} p_{2n+1} = 0,$$

we have that $A = B$. Clearly A is the sum of our series.

4.13 Absolute convergence

Given a series $a_0 + a_1 + a_2 + \cdots$, it is of interest to consider also the series of absolute values $|a_0| + |a_1| + |a_2| + \cdots$.

Theorem 10 IF THE SERIES $\Sigma_{j=0}^{\infty} |a_j|$ CONVERGES, THEN THE SERIES $\Sigma_{j=0}^{\infty} a_j$ ALSO CONVERGES, AND WE HAVE

$$(18) \qquad \sum_{j=0}^{\infty} a_j \leq \sum_{j=0}^{\infty} |a_j|.$$

If $\Sigma |a_j|$ converges, we say that the series Σa_j converges **absolutely**.

Proof of Theorem 10. We assume that $\Sigma |a_j|$ converges and set

$$A = \sum_{j=0}^{\infty} |a_j|,$$

$$\alpha_j = \begin{cases} a_j \text{ if } a_j \geq 0, \\ 0 \text{ if } a_j \leq 0; \end{cases} \qquad \beta_j = \begin{cases} -a_j \text{ if } a_j \leq 0, \\ 0 \text{ if } a_j \geq 0. \end{cases}$$

We have

$$0 \leq \alpha_j \leq |a_j|, \qquad 0 \leq \beta_j \leq |\beta_j|.$$

Hence

$$\sum_{j=0}^{n} a_j \leq \sum_{j=0}^{n} |a_j| \leq A, \qquad \sum_{j=0}^{n} \beta_j \leq \sum_{j=0}^{n} |\beta_j| \leq A.$$

Applying Theorem 6 to the series $\Sigma_{j=0}^{\infty} \alpha_j$, $\Sigma_{j=0}^{\infty} \beta_j$, we conclude that these series converge. Set

$$P = \sum_{j=0}^{\infty} \alpha_j, \qquad N = \sum_{j=0}^{\infty} \beta_j.$$

We have $|a_j| = \alpha_j + \beta_j$. Hence, by Theorem 4,

$$P + N = A.$$

Also, $a_j = \alpha_j - \beta_j$. Hence, again by Theorem 4, $\Sigma_{j=0}^{\infty} a_j$ converges and

$$\sum_{j=0}^{\infty} a_j = P - N.$$

Since $P \geq 0, N \geq 0$, we have $|P - N| \leq P + N = A$, which proves (18).

4.14 Rearranging terms

There are convergent series that do not converge absolutely. An example is the Leibniz series $1 - \dfrac{1}{2} + \dfrac{1}{3} - \dfrac{1}{4} + \dfrac{1}{5} - \cdots$. It turns out that the "paradoxical" behavior encountered in §3.2 *cannot* happen in the case of absolute convergence, and *must* happen in the case of a convergent series for which the series of absolute values diverges.

Theorem 11   ASSUME THAT $\Sigma_{j=0}^{\infty} a_j$ CONVERGES. IF $\Sigma_{j=0}^{\infty} |a_j|$ CONVERGES, THEN EVERY SERIES OBTAINED FROM $\Sigma_{j=0}^{\infty} a_j$ BY REARRANGING TERMS CONVERGES, AND TO THE SAME SUM. IF $\Sigma_{j=0}^{\infty} |a_j|$ DIVERGES, THEN THE SERIES $\Sigma_{j=0}^{\infty} a_j$ CAN BE REARRANGED SO AS TO DIVERGE, AND ALSO SO AS TO CONVERGE TO ANY GIVEN SUM.

The proof of this theorem will be found in an appendix to this chapter (see §7.2).

EXERCISES

A convergent series which does not converge absolutely is called conditionally convergent. Determine if the following series converge absolutely, converge conditionally, or diverge.

80. $\displaystyle\sum_{n=0}^{\infty} p^n$ for $|p| < 1$.

81. $\displaystyle\sum_{n=1}^{\infty} \frac{(-1)^n}{\sqrt{n}}$.

82. $\displaystyle\sum_{n=1}^{\infty} \frac{(-1)^{n+1}}{n^{3/2}}$.

83. $\displaystyle\sum_{n=1}^{\infty} \frac{(-1)^n}{2n - 1}$.

84. $\displaystyle\sum_{n=1}^{\infty} \frac{\sin n}{n^2}$.

85. $\displaystyle\sum_{n=0}^{\infty} \frac{\cos n}{2^n}$.

86. $\displaystyle\sum_{n=1}^{\infty} \frac{(-1)^{n-1}n}{2n - 1}$.

87. $\displaystyle\sum_{n=2}^{\infty} \frac{(-1)^{n+1}}{n^2 - n}$.

▶88. $\displaystyle\sum_{n=1}^{\infty} (-1)^n \frac{2n^2 + 1}{n^3 + 3}$.

90. $\displaystyle\sum_{n=2}^{\infty} \frac{(-1)^n}{n \log^2 n}$.

89. $\displaystyle\sum_{n=2}^{\infty} \frac{(-1)^n}{n \log n}$.

91. $\displaystyle\sum_{n=1}^{\infty} (-1)^n \frac{\arctan n}{n}$.

4.15 Cauchy multiplication

Theorem 11 shows that in some sense the commutative law of addition holds for sums of infinitely many terms, provided we deal with absolutely convergent series. Our next result, which will also be proved in §7, asserts that the same is true for the distributive law.

Theorem 12   IF

$$a_0 + a_1 + a_2 + \cdots = \sum_{j=0}^{\infty} a_j = A, \; b_0 + b_1 + b_2 + \cdots = \sum_{k=0}^{\infty} b_k = B,$$

THE TWO SERIES BEING ABSOLUTELY CONVERGENT, THEN

$$AB = \sum_{n=0}^{\infty} \left(\sum_{j+k=n} a_j b_k \right)$$

$$= a_0 b_0 + (a_0 b_1 + a_1 b_0) + (a_0 b_2 + a_1 b_1 + a_2 b_0)$$
$$+ (a_0 b_3 + a_1 b_2 + a_2 b_1 + a_3 b_0) + \cdots$$

AND THIS SERIES, TOO, CONVERGES ABSOLUTELY.

This way of forming the product of two infinite series is called **Cauchy multiplication,** in honor of Cauchy, one of the architects of the rigorous theory of infinite series.

As an application of the Cauchy multiplication, we shall prove the functional equation of the exponential function starting with the formula

$$e^x = 1 + \frac{x}{1!} + \frac{x^2}{2!} + \cdots.$$

We must show that $e^a e^b = e^{a+b}$. Now

$$e^a e^b = \sum_{j=0}^{\infty} \frac{a^j}{j!} \sum_{k=0}^{\infty} \frac{b^k}{k!}.$$

(It is imperative to use different dummy indices of summation in the two series.) Using Cauchy multiplication, the definition of binomial coefficients, and the binomial theorem, we see that the above product equals

$$\sum_{n=0}^{\infty} \left(\sum_{j+k=n} \right) \frac{a^j b^k}{j! k!} = \sum_{n=0}^{\infty} \sum_{j=0}^{n} \frac{a^j b^{n-j}}{j!(n-j)!} = \sum_{n=0}^{\infty} \frac{1}{n!} \sum_{j=0}^{n} \frac{n!}{j!(n-j)!} a^j b^{n-j}$$

$$= \sum_{n=0}^{\infty} \frac{1}{n!} \sum_{j=0}^{n} \binom{n}{j} a^j b^{n-j} = \sum_{n=0}^{\infty} \frac{(a+b)^n}{n!} = e^{a+b},$$

as asserted.

EXERCISES

Prove the following relations using Cauchy multiplication and starting from the formulas $\cos x = \sum_{m=0}^{\infty} (-1)^m \frac{x^{2m}}{(2m)!}$ and $\sin x = \sum_{m=0}^{\infty} (-1)^m \frac{x^{2m+1}}{(2m+1)!}$.

92. (a) $\cos^2 x = \displaystyle\sum_{m=0}^{\infty} \frac{(-1)^m}{(2m)!} \left(\sum_{j=0}^{m} \binom{2m}{2j} \right) x^{2m}$.

(b) $\cos^2 x = 1 + \displaystyle\sum_{m=1}^{\infty} \frac{(-1)^m 2^{2m-1}}{(2m)!} x^{2m}$. [Hint: Use Part (a) and the results of Exercise 89 in §2.11 with $n = 2m$.]

93. (a) $\sin^2 x = \displaystyle\sum_{m=0}^{\infty} \frac{(-1)^m}{(2m+2)!} \left(\sum_{j=0}^{m} \binom{2m+2}{2j+1} \right) x^{2m+2}$.

(b) $\sin^2 x = \displaystyle\sum_{m=0}^{\infty} \frac{(-1)^m 2^{2m+1}}{(2m+2)!} x^{2m+2}$. [Hint: Use Part (a) and the results of Exercise 89 in §2.11 with $n = 2m + 2$.]

94. $\cos^2 x + \sin^2 x = 1$. Use the results of Exercises 92(b) and 93(b).

95. $\cos 2x = 2 \cos^2 x - 1$. Use the result of Exercise 92(b).

5.1 Convergent and divergent power series

A power series is an infinite series of the form

(1)
$$a_0 + a_1(x - x_0) + a_2(x - x_0)^2 + a_3(x - x_0)^3 + \cdots$$
$$= \sum_{n=0}^{\infty} a_n(x - x_0)^n.$$

Here we think of $x_0, a_0, a_1, a_2, \cdots$ as fixed numbers, whereas x will be ranging over an interval. More precisely, we shall call (1) a power series with center at x_0, or a power series **about** x_0 or a power series **in** $(x - x_0)$. The numbers $a_0, a_1, a_2, \cdots$ are called the **coefficients** of the power series. If nearly all coefficients are zero, the power series is a polynomial. We can think of a power series as a polynomial of infinite degree.

If $x_0 = 0$, the power series reduces to

(1')
$$a_0 + a_1 x + a_2 x^2 + a_3 x^3 + \cdots = \sum_{n=0}^{\infty} a_n x^n.$$

In order to simplify writing, we often consider such a series; this is, of course, not very significant.

If we choose a specific number for x, the series (1) is an infinite series of numbers, and it either converges or diverges. For $x = x_0$, we obtain $a_0 + 0 + 0 + \cdots$; hence the series always converges at least at this one point.

5.2 Radius of convergence

We state now the basic theorem on convergence of power series.

Theorem 1   GIVEN A POWER SERIES $\Sigma_{n=0}^{\infty} a_n(x - x_0)^n$, THERE IS A NUMBER $R \geq 0$, WHICH MAY TAKE ON THE "VALUE" $R = +\infty$, SUCH THAT THE SERIES CONVERGES ABSOLUTELY FOR $|x - x_0| < R$, AND DIVERGES FOR $|x - x_0| > R$. (AT $x = x_0 + R$, AND AT $x = x_0 - R$, THE SERIES MAY EITHER CONVERGE OR DIVERGE.)

We shall prove this theorem in an appendix (see §8). The examples below illustrate the validity of the result. We call R the **radius of convergence** of the series considered, and the interval $|x - x_0| < R$ the **interval of convergence,** provided of course that $R > 0$.

▶ *Examples*   **1.** The power series $\Sigma_{n=0}^{\infty} n!\, x^n$ diverges for $x \neq 0$. For we have $\lim |n!\, x^n| = +\infty$ if $x \neq 0$. This is so, since $\lim |x|^{-n}/n! = 0$, by Example 3 in §3.5. Here $R = 0$.

2. The power series $\Sigma_{n=0}^{\infty} x^n$ converges for $|x| < 1$, diverges for $|x| \geq 1$. Here $R = 1$.

3. The power series $\Sigma_{n=0}^{\infty} x^n/n!$ converges for all x (to the sum e^x). We have $R = +\infty$.

4. The power series $\Sigma_{n=1}^{\infty} (-1)^n\, x^n/n$ converges to the sum $\log(1 + x)$, as long as $-1 < x \leq 1$, as we saw above. The series diverges for $x = -1$; it is then the harmonic series $1 + 2^{-1} + 3^{-1} + \cdots$. The series also diverges for $|x| > 1$, since for $|x| > 1$ we have $\lim_{n \to \infty} |x|^n/n = +\infty$. Here $R = 1$.

5. The power series $\Sigma_{n=0}^{\infty} (-1)^n\, x^{2n+1}/(2n + 1)$ converges (to the sum arc tan x) as long as $|x| \leq 1$. It diverges for $|x| > 1$, for the same reasons as above. Again $R = 1$.◀

5.3 Differentiation and integration

We consider now a power series with positive radius of convergence R, and its sum $f(x) = \Sigma_{n=0}^{\infty} a_n(x - x_0)^n$. The function $x \mapsto f(x)$ is defined by this equation for $|x - x_0| < R$.

Theorem 2 IF $f(x) = \sum\limits_{n=0}^{\infty} a_n(x - x_0)^n$ FOR $|x - x_0| < R$,

$R > 0$ BEING THE RADIUS OF CONVERGENCE OF THE POWER SERIES, THEN THE FUNCTION $x \mapsto f(x)$ IS CONTINUOUS AND DIFFERENTIABLE IN THE INTERVAL OF CONVERGENCE, AND

$$f'(x) = \sum_{n=1}^{\infty} na_n(x - x_0)^{n-1},$$

$$\int_{x_0}^{x} f(t)\, dt = \sum_{n=0}^{\infty} \frac{a_n}{n + 1}(x - x_0)^{n+1}.$$

FURTHERMORE, THE SERIES IN THE ABOVE FORMULAS HAVE RADIUS OF CONVERGENCE R.

The theorem, which will be proved in §8, asserts that it is legitimate to differentiate or integrate a convergent power series term by term, as if it were an ordinary polynomial.

▶*Example* Suppose we know nothing of the exponential function but want to find a function $x \mapsto f(x)$ defined near $x = 0$, satisfying the differential equation $f'(x) = f(x)$, and satisfying the condition $f(0) = 1$. We *try* to represent this function as a power series $f(x) = \Sigma_{n=0}^{\infty} a_n x^n$. Since $f(0) = a_0$, we must have $a_0 = 1$. Thus we want to write f as

$$f(x) = 1 + a_1 x + a_2 x^2 + a_3 x^3 + a_4 x^4 + \cdots.$$

If the series converges we must have, by Theorem 2,

$$f'(x) = a_1 + 2a_2 x + 3a_3 x^2 + 4a_4 x^3 + \cdots.$$

We shall have $f'(x) = f(x)$ if we set $a_1 = 1$, $2a_2 = a_1$, $3a_3 = a_2$, $4a_4 = a_3$, and so forth, that is, $a_1 = 1$, $a_2 = \dfrac{1}{2} a_1 = \dfrac{1}{2}$, $a_3 = \dfrac{1}{3} a_2 = \dfrac{1}{2 \cdot 3}$, $a_4 = \dfrac{1}{4} a_3 = \dfrac{1}{2 \cdot 3 \cdot 4}$, and so forth, that is, $a_j = \dfrac{1}{j!}$. Hence we try

$$f(x) = 1 + \frac{x}{1!} + \frac{x^2}{2!} + \frac{x^3}{3!} + \cdots.$$

Does the series converge? We apply the ratio test (Theorem 8 in §4.11) to the series of absolute values and obtain

$$\lim_{n \to \infty} \frac{\left| \dfrac{x^{n+1}}{(n + 1)!} \right|}{\left| \dfrac{x^n}{n!} \right|} = \lim_{n \to \infty} \frac{|x|}{n + 1} = 0.$$

The series converges for all x.

If we now recall that the condition we imposed on $f(x)$ characterizes the exponential function e^x (see Chapter 6, §5.6), we obtain a new proof of the expansion $e^x = 1 + \dfrac{x}{1!} + \dfrac{x^2}{2!} + \cdots$, independent of Taylor's Theorem. ◄

EXERCISES

Find the radius of convergence of the following power series. Usually this is best accomplished by using the ratio test to determine absolute convergence, that is, by applying the ratio test to the associated series of absolute values.

1. $\displaystyle\sum_{n=0}^{\infty} nx^n.$

2. $\displaystyle\sum_{n=0}^{\infty} \frac{2n-1}{3n+1} x^n.$

3. $\displaystyle\sum_{n=0}^{\infty} \frac{n!}{2^n} x^n.$

4. $\displaystyle\sum_{n=1}^{\infty} \frac{1 \cdot 3 \cdot 5 \cdots (2n-1)}{2 \cdot 5 \cdot 8 \cdots (3n-1)} (x-1)^n.$

5. $\displaystyle\sum_{n=1}^{\infty} \frac{1 \cdot 5 \cdot 9 \cdots (4n-3)}{2 \cdot 4 \cdot 6 \cdots (2n)} (x+2)^n.$

6. $\displaystyle\sum_{n=1}^{\infty} \frac{n^n}{n!} \left(x + \frac{1}{2}\right)^n.$

7. Suppose that $\lim_{n\to\infty} \left|\dfrac{a_{n+1}}{a_n}\right| = L$. Show that the power series $\sum_{n=0}^{\infty} a_n(x - x_0)^n$ has radius of convergence $1/L$ if $L \neq 0$, ∞ if $L = 0$.

8. If the power series $\sum_{n=0}^{\infty} a_n x^n$ has radius of convergence R, what is the radius of convergence of $\sum_{n=0}^{\infty} a_n p^n x^n$, where p is any constant?

9. If the power series $\sum_{n=0}^{\infty} a_n x^n$ has radius of convergence R, what is the radius of convergence of $\sum_{n=0}^{\infty} a_n x^{nk}$, where k is a positive integer?

10. If the power series $\sum_{n=0}^{\infty} a_n(x - x_0)^n$ has a radius of convergence R, what is the radius of convergence of $\sum_{n=0}^{\infty} a_n(x - x_0)^{n+k}$, where k is a positive integer?

In the following exercises, use only the series expansions for $\cos x$, $\sin x$, $\arctan x$, and $(1 + x)^{-1}$, and the theorems about manipulating series proved so far.

11. Show that $d \sin x/dx = \cos x$ for all x.
12. Show that $d \cos x/dx = -\sin x$ for all x.
13. Show that $d \arctan x/dx = (1 + x^2)^{-2}$ for $|x| < 1$.
14. Represent $\int_0^x \arctan t \, dt$ as a power series about 0 for $|x| < 1$.

5.4 Taylor series

Let $x \mapsto f(x)$ be a function represented by a convergent power series

$$f(x) = \sum_{n=0}^{\infty} a_n(x - x_0)^n.$$

Applying Theorem 2 several times, we conclude that the function $f(x)$ has derivatives of all orders. Also,

$$f'(x) = \sum_{n=1}^{\infty} na_n(x - x_0)^{n-1}, \qquad f''(x) = \sum_{n=2}^{\infty} n(n-1)a_n(x - x_0)^{n-2},$$

$$f'''(x) = \sum_{n=3}^{\infty} n(n-1)(n-2)a_n(x-x_0)^{n-3},$$

and so forth, and, in general,

$$f^{(j)}(x) = \sum_{n=j}^{\infty} n(n-1)(n-2) \cdots (n-j+1)a_n(x-x_0)^{n-j}.$$

Substituting into the series for $f(x)$, $f'(x)$, $f''(x)$, $f'''(x)$, and so forth, $x = x_0$, we obtain $f(x_0) = a_0$, $f'(x_0) = a_1$, $f''(x_0) = 2a_2$, $f'''(x_0) = 3 \cdot 2 \cdot a_3$, and, in general, $f^{(j)}(x_0) = j!\, a_j$. Thus

$$(2) \qquad\qquad a_j = \frac{f^{(j)}(x_0)}{j!}, \qquad j = 0, 1, 2, \cdots.$$

These formulas show that a given function can be represented by a power series about x_0 *in only one way* (if at all).

Now if $f(x)$ is any function defined near x_0 and having derivatives of all orders, we can compute coefficients a_j by the formulas (2) and then form the power series.

$$\sum a_j(x-x_0)^j = \sum \frac{f^{(j)}(x_0)}{j!}(x-x_0)^j.$$

This is called the **Taylor series** of f at x_0. The partial sums of the Taylor series are the Taylor polynomials of f at x_0. (If $x_0 = 0$, one sometimes calls the Taylor series the **Maclaurin series**.) The formulas (2) have been proved under the assumption that $f(x) = \Sigma a_j(x-x_0)^j$. Thus we have the following theorem.

Theorem 3 A CONVERGENT POWER SERIES (ABOUT x_0) IS THE TAYLOR SERIES OF ITS SUM (AT x_0).

5.5 Analytic functions

A function $x \mapsto f(x)$ defined near x_0 is called **analytic** at x_0 if it can be represented as the sum of a convergent power series about x_0. It follows from Theorem 2 that an analytic function has derivatives of all orders.

A function $x \mapsto f(x)$, defined near x_0 and having derivatives of all orders, *need not* be analytic near x_0. For it may happen that the Taylor series of $f(x)$ has radius of convergence 0. Or, it may happen that the Taylor series has positive radius of convergence, but its sum is not equal to $f(x)$.

We mention an important fact: the sum of a convergent power series is analytic at every point in the interval of convergence. The proof of this result belongs to the general theory of analytic functions and will not be given here.

Analytic functions are in many ways the most "well-behaved" functions imaginable. Fortunately, most functions encountered in applications of calculus are analytic, except at certain points.

▶ *Examples*   *1.* The function $x \mapsto f(x) = 1/x$ is analytic near every point $x_0 \neq 0$, for we can write

$$\frac{1}{x} = \frac{1}{(x - x_0) + x_0} = \frac{1}{x_0} \frac{1}{1 + ((x - x_0)/x_0)}$$

$$= \frac{1}{x_0} \left(1 - \frac{x - x_0}{x_0} + \left(\frac{x - x_0}{x_0} \right)^2 - \cdots \right) = \sum_{j=0}^{\infty} \frac{(-1)^j}{x_0^{2j+1}} (x - x_0)^j,$$

the series converging for $\left| \dfrac{x - x_0}{x_0} \right| < 1$, that is, for $|x - x_0| < x_0$.

The function $x \mapsto f(x) = 1/x$ is not analytic at $x_0 = 0$, for no matter how we define $f(0)$, the function will not be continuous at this point.

2. The exponential function $x \mapsto e^x$ can be expanded in a power series about every point x_0; all these power series have radius of convergence ∞. We may write

$$e^x = e^{x_0} e^{x - x_0} = e^{x_0} \left(1 + \frac{x - x_0}{1!} + \frac{(x - x_0)^2}{2!} + \cdots \right)$$

$$= \sum_{j=0}^{\infty} \frac{e^{x_0}}{j!} (x - x_0)^j.$$

3. The function $x \mapsto f(x)$ defined by $f(x) = e^{-1/x^2}$ for $x \neq 0$, $f(0) = 0$, has derivatives of all orders (see Chapter 6, §8.5) but is not analytic near $x = 0$. Since we have that $f(0) = f'(0) = f''(0) = \cdots = 0$, all Taylor coefficients at $x_0 = 0$ are 0, and the Taylor series converges to 0, but $f(x) > 0$ for $x \neq 0$.

The same function is analytic near every point $x_0 \neq 0$. ◀

EXERCISES

Find the Taylor series of the following functions $f(x)$ around the given point x_0. Then find the radius of convergence of the Taylor series. Note that this does not show where, if at all, the Taylor series of $f(x)$ converges to $f(x)$—except, of course, that this is always true for $x = x_0$.

[Hint: In some cases it will be easier to derive the Taylor series of $f(x)$ by manipulations starting from the Taylor series of another function; in others, the Taylor series is best computed directly from the definition. In all cases, find the general term of the series and write your answer in Σ notation (you may have to write out the first few terms separately).]

15. $f(x) = 1/\sqrt{x}$ at $x_0 = 1$.
16. $f(x) = x^{2/3}$ at $x_0 = 1$.
17. $f(x) = x^{-1/4}$ at $x_0 = 16$.
18. $f(x) = 1/(2 - x)^2$ at $x_0 = 0$.
19. $f(x) = \sqrt{1 - x}$ at $x_0 = 0$.

20. $f(x) = \sin x$ at $x_0 = \pi/6$.
21. $f(x) = 2^x$ at $x_0 = 1$.
22. $f(x) = 1/(1 - x)^2$ at $x_0 = 0$.
23. $f(x) = x^4/(1 - x)$ at $x_0 = 0$.
24. $f(x) = e^{x^2}$ at $x_0 = 0$.

25. Suppose $f(x) = \Sigma_{n=0}^{\infty} a_n x^n$, where the power series is assumed to have positive radius of convergence. If $f(0) = 1$, $f(x) = f'(x/\lambda)$ for constant $\lambda < 1$, find the coefficients a_n. Now find the radius of convergence. This shows that there is one and only one function f, analytic at 0, which satisfies the conditions. If $\lambda = 1$, we have $f(x) = e^x$. What happens if $\lambda > 1$?

5.6 Examples of Taylor series

Given a function $f(x)$ which has a Taylor series about x_0, we can use various methods to determine whether the sum of the series equals $f(x)$ near x_0. The most direct approach is via Taylor's Theorem: the expansion

$$f(x) = \sum_{j=0}^{\infty} \frac{f^{(j)}(x_0)}{j!} (x - x_0)^j$$

is valid for those x for which the nth remainder has the limit 0 for $n \to \infty$. In this way, we obtained the expansions earlier

$$(3) \qquad e^x = 1 + \frac{x}{1!} + \frac{x^2}{2!} + \frac{x^3}{3!} + \cdots \qquad R = \infty,$$

$$(4) \qquad \cos x = 1 - \frac{x^2}{2!} + \frac{x^4}{4!} - \cdots \qquad R = \infty,$$

$$(5) \qquad \sin x = x - \frac{x^3}{3!} + \frac{x^5}{5!} - \cdots \qquad R = \infty,$$

$$(6) \qquad \log (1 + x) = x - \frac{x^2}{2} + \frac{x^3}{3} - \frac{x^4}{4} + \cdots \qquad R = 1,$$

$$(7) \qquad \arctan x = x - \frac{x^3}{3} + \frac{x^5}{5} - \cdots \qquad R = 1.$$

These expansions are valid within the whole interval of convergence of the power series; we indicate the radius of convergence at the right.

Other expansions can be obtained from these by simple algebraic manipulations (using if necessary Theorem 4 in §4.6). For instance, replacing x by $-x$ in (3) yields

$$(8) \qquad e^{-x} = 1 - \frac{x}{1!} + \frac{x^2}{2!} - \frac{x^3}{3!} + \cdots \qquad R = \infty,$$

Adding and subtracting (3) and (8), and dividing by 2, we obtain

$$(9) \qquad \cosh x = 1 + \frac{x^2}{2!} + \frac{x^4}{4!} + \cdots \qquad R = \infty,$$

$$(10) \qquad \sinh x = x + \frac{x^3}{3!} + \frac{x^5}{5!} + \cdots \qquad R = \infty.$$

Dividing (5) by x, we obtain

$$(11) \qquad \frac{\sin x}{x} = 1 - \frac{x^2}{3!} + \frac{x^4}{5!} + \frac{x^6}{7!} - \cdots \qquad R = \infty.$$

Expansion (6) yields

$$(12) \qquad \log (1 - x) = - x - \frac{x^2}{2} - \frac{x^3}{3} - \cdots \qquad R = 1,$$

so that, subtracting (12) from (6), we get

$$(13) \qquad \frac{1}{2} \log \frac{1 + x}{1 - x} = x + \frac{x^3}{3} + \frac{x^5}{5} + \frac{x^7}{7} + \cdots \qquad R = 1.$$

We can obtain Taylor expansions by differentiating and integrating other Taylor series term by term. This is legitimate in view of Theorem 2. Thus integration of the geometric series

$$\frac{1}{1 + x} = 1 - x + x^2 - x^3 + \cdots$$

yields the expansion (6), integration of

$$\frac{1}{1 + x^2} = 1 - x^2 + x^4 - x^6 + \cdots$$

yields the arc tangent series (7).

Differentiating (11), we see that

$$(14) \qquad \frac{x \cos x - \sin x}{x^2} = - \frac{2x}{3!} + \frac{4x^3}{5!} - \frac{6x^5}{7!} + \cdots$$

(which can also be obtained in other ways). Integration of (11) yields

(15) $$\int_0^x \frac{\sin t}{t}\, dt = x - \frac{x^3}{2!\, 3^2} + \frac{x^5}{4!\, 5^2} - \frac{x^7}{6!\, 7^2} + \cdots,$$

as the reader should check. This is interesting, since we know of no way to write $\int_0^x [(\sin t)/t]\, dt$ in terms of elementary functions, without using the integral sign.

A similar situation arises if we replace x by x^2 in (8) and then integrate the resulting series term by term. We obtain

(16) $$e^{-x^2} = 1 - \frac{x^2}{1!} + \frac{x^4}{2!} - \frac{x^6}{3!} + \cdots$$

and

(17) $$\int_0^x e^{-t^2}\, dt = x - \frac{x^3}{1!\, 3} + \frac{x^5}{2!\, 5} - \frac{x^7}{3!\, 7} + \cdots.$$

For small values of x, we can use this expansion to compute the "error integral" $\int^x e^{-t^2}\, dt$ (see §8.3 in Chapter 6).

EXERCISES

Show that the Taylor series of $f(x)$ at x_0 represents $f(x)$ for the given values of x. Compare your result with the radius of convergence of the Taylor series.

► 26. $f(x) = 1/\sqrt{x}$ at $x_0 = 1$ for $1 \le x < 2$. See Exercise 15. The Taylor series actually represents $f(x)$ for $0 < x < 2$, but this is harder to show. See §5.7.

27. $f(x) = x^{-1/4}$ at $x_0 = 16$ for $16 \le x < 32$. See Exercise 17. The Taylor series actually represents $f(x)$ for $0 < x < 32$, but this is harder to show.

28. $f(x) = \sin x$ at $x_0 = \pi/6$ for all x. See Exercise 20.

29. $f(x) = 2^x$ at $x_0 = 1$ for all x. See Exercise 21.

30. $f(x) = x/(1 - x)^2$ at $x_0 = 0$ for $|x| < 1$. See Exercise 22.

31. $f(x) = x^4/(1 - x)$ at $x_0 = 0$ for $|x| < 1$. See Exercise 23.

32. $f(x) = e^{x^2}$ at $x_0 = 0$ for all x. See Exercise 24.

33. $f(x) = \int_0^x u \log (1 + u)\, du$ at $x_0 = 0$ for $|x| < 1$. First find the Taylor series.

34. $f(x) = \int_0^x \cos u^2\, du$ at $x_0 = 0$ for all x. First find the Taylor series.

35. Find a power series representation about $x_0 = 0$ of $f(x) = \cos \sqrt{x}$, valid for $x \ge 0$, that is, valid for the entire domain of $f(x) = \cos \sqrt{x}$. Find the radius of convergence of this power series. This defines a function whose domain includes that of the original function $f(x)$ and agrees with $f(x)$ wherever $f(x)$ is defined.

5.7 Binomial series

Another way of verifying that a function is represented by its Taylor series is by demonstrating that the series converges and the sum of the series has properties characterizing the function. In §5.3 we showed in this way that $e^x = 1 + \dfrac{x}{1!} + \dfrac{x^2}{2!} + \cdots$.

We shall use the same method to establish the **binomial expansion**

$$(1 + x)^\alpha = 1 + \alpha x + \frac{\alpha(\alpha - 1)}{2!} x^2 + \frac{\alpha(\alpha - 1)(\alpha - 2)}{3!} x^3 + \cdots$$

(18)
$$= 1 + \sum_{n=1}^{\infty} \frac{\alpha(\alpha - 1) \cdots (\alpha - n + 1)}{n!} x^n = \sum_{n=0}^{\infty} \binom{\alpha}{n} x^n, \; |x| < 1.$$

(This result is due to Newton himself!) First we apply the ratio test to the series of absolute values $\sum \left| \binom{\alpha}{n} x^n \right|$. We obtain

$$\lim_{n \to \infty} \frac{\left| \binom{\alpha}{n+1} x^{n+1} \right|}{\left| \binom{\alpha}{n} x^n \right|} = \lim_{n \to \infty} \left| \frac{\alpha(\alpha - 1) \cdots (\alpha - n + 1)(\alpha - n)n! \, x^{n+1}}{\alpha(\alpha - 1) \cdots (\alpha - n + 1)(n + 1)! \, x^n} \right|$$

$$= \lim_{n \to \infty} \left| \frac{(\alpha - n)x}{n + 1} \right| = \lim_{n \to \infty} \left| \frac{\frac{\alpha}{n} - 1}{1 + \frac{1}{n}} x \right| = |x|$$

so that the series converges for $|x| < 1$, diverges for $|x| > 1$. For $|x| < 1$, set

$$f(x) = \sum_{n=0}^{\infty} \binom{\alpha}{n} x^n = 1 + \binom{\alpha}{1} x + \binom{\alpha}{2} x^2 + \binom{\alpha}{3} x^3 + \cdots.$$

Then
$$f'(x) = \sum_{n=1}^{\infty} \binom{\alpha}{n} n x^{n-1} = \sum_{n=0}^{\infty} \binom{\alpha}{n+1}(n + 1)x^n$$

(by changing dummy variables), and

$$x f'(x) = \sum_{n=0}^{\infty} \binom{\alpha}{n} n x^n,$$

so that
$$(1 + x)f'(x) = \sum_{n=0}^{\infty} \left\{ \binom{\alpha}{n+1}(n + 1) + \binom{\alpha}{n} n \right\} x^n.$$

But
$$\binom{\alpha}{n+1}(n + 1) + \binom{\alpha}{n} n$$

$$= \frac{\alpha(\alpha - 1)(\alpha - 2) \cdots (\alpha - n)}{(n + 1)!}(n + 1) + \frac{\alpha(\alpha - 1) \cdots (\alpha - n + 1)}{n!} n$$

$$= \frac{\alpha(\alpha - 1) \cdots (\alpha - n + 1)}{n!}(\alpha - n + n) = \alpha \binom{\alpha}{n}.$$

Therefore
$$(1 + x)f'(x) = \alpha f(x)$$

or
$$\frac{f'(x)}{f(x)} = \frac{\alpha}{1 + x},$$

which is the same as

$$\frac{d \log f(x)}{dx} = \alpha \frac{d \log (1 + x)}{dx} = \frac{d(\alpha \log (1 + x))}{dx} = \frac{d \log (1 + x)^{\alpha}}{dx}.$$

Therefore $\log f(x) - \log (1 + x)^{\alpha}$ is a constant; call it c. Thus $\log f(x)$ $= \log (1 + x)^{\alpha} + c$ or $f(x) = e^{c}(1 + x)^{\alpha}$. But $f(0) = 1$ and $(1 + 0)^{\alpha} = 1$. Therefore $e^{c} = 1$ and $f(x) = (1 + x)^{\alpha}$, as asserted.

▶*Examples*   1. For $\alpha = -\frac{1}{2}$, we obtain

$$\frac{1}{\sqrt{1 + x}} = 1 + \frac{-\frac{1}{2}}{1} x + \frac{(-\frac{1}{2})(-\frac{3}{2})}{1 \cdot 2} x^2 + \frac{(-\frac{1}{2})(-\frac{3}{2})(-\frac{5}{2})}{1 \cdot 2 \cdot 3} x^3$$

$$+ \frac{(-\frac{1}{2})(-\frac{3}{2})(-\frac{5}{2})(-\frac{7}{2})}{1 \cdot 2 \cdot 3 \cdot 4} x^4 + \cdots$$

$$= 1 - \frac{x}{2} + \frac{1 \cdot 3}{2!} \left(\frac{x}{2}\right)^2 - \frac{1 \cdot 3 \cdot 5}{3!} \left(\frac{x}{2}\right)^3 + \frac{1 \cdot 3 \cdot 5 \cdot 7}{4!} \left(\frac{x}{2}\right)^4 - \cdots.$$

2. Replacing x by $-x^2$ in Example 1, we get

$$\frac{1}{\sqrt{1 - x^2}} = 1 + \frac{x^2}{2} + \frac{1 \cdot 3 x^4}{2! \, 2^2} + \frac{1 \cdot 3 \cdot 5 x^6}{3! \, 2^3} + \frac{1 \cdot 3 \cdot 5 \cdot 7 x^8}{4! \, 2^4} + \cdots.$$

Integration yields

$$\arcsin x = \int_0^x \frac{dt}{\sqrt{1 - t^2}}$$

$$= x + \frac{x^3}{2 \cdot 3} + \frac{1 \cdot 3 x^5}{2! \, 2^2 5} + \frac{1 \cdot 3 \cdot 5 x^7}{3! \, 2^3 7} + \frac{1 \cdot 3 \cdot 5 \cdot 7 x^9}{4! \, 2^4 9} + \cdots.$$

The expansions are valid for $|x| < 1$. ◀

EXERCISES

Use the binomial expansion to find the Taylor series of $f(x)$ at x_0 and determine where the Taylor series represents $f(x)$.

36. $f(x) = (x + 1)^{-1/3}$ at $x_0 = 0$.

37. $f(x) = \dfrac{1}{\sqrt{1 + x^2}}$ at $x_0 = 0$.

38. $f(x) = (x^2 + 1)^{2/5}$ at $x_0 = 0$.

39. $f(x) = \int_0^x \sqrt{1 + u^2}\, du$ at $x_0 = 0$.

40. $f(x) = \int_0^x (1 + u)^{5/2}\, du$ at $x_0 = 0$.

41. $f(x) = \dfrac{1}{x} \arcsin x$ for $x \neq 0$, 1 for $x = 0$.

5.8 Multiplying power series

Recall that power series converge absolutely inside the intervals of convergence. The theorem on Cauchy multiplication of absolutely convergent series (Theorem 12 in §4.15), when applied to power series, yields the following

Theorem 4   IF

$$f(x) = a_0 + a_1 x + a_2 x^2 + \cdots = \sum_{j=0}^{\infty} a_j x^j \qquad \text{FOR } |x| < \alpha,$$

$$g(x) = b_0 + b_1 x + b_2 x^2 + \cdots = \sum_{k=0}^{\infty} b_k x^k \qquad \text{FOR } |x| < \beta,$$

THEN

$$
\begin{aligned}
f(x)g(x) &= a_0 b_0 + (a_0 b_1 + a_1 b_0)x + (a_0 b_2 + a_1 b_1 + a_2 b_0)x^2 \\
&\quad + (a_0 b_3 + a_1 b_2 + a_2 b_1 + a_3 b_0)x^3 \\
&\quad + (a_0 b_4 + a_3 b_1 + a_2 b_2 + a_3 b_1 + a_4 b_0)x^4 + \cdots \\
&= \sum_{n=0}^{\infty} \Big(\sum_{j+k=n} a_j b_k \Big) x^n = \sum_{n=0}^{\infty} \Big(\sum_{j=0}^{n} a_j b_{n-j} \Big) x^n \qquad \text{for } |x| < \gamma,
\end{aligned}
$$

WHERE γ IS THE SMALLER OF THE TWO NUMBERS α AND β.

As an example, we multiply the two series

$$\log \frac{1}{1 - x} = x + \frac{x^2}{2} + \frac{x^3}{3} + \frac{x^4}{4} + \cdots,$$

$$\frac{1}{1 - x} = 1 + x + x^2 + x^3 + \cdots$$

and obtain

$$
\begin{aligned}
\frac{1}{1 - x} &\log \frac{1}{1 - x} \\
&= \Big(x + \frac{x^2}{2} + \frac{x^3}{3} + \frac{x^4}{4} + \cdots \Big)(1 + x + x^2 + x^3 + \cdots) \\
&= (x \cdot 1) + \Big(x \cdot x + \frac{x^2}{2} \cdot 1 \Big) + \Big(x \cdot x^2 + \frac{x^2}{2} \cdot x + \frac{x^3}{3} \cdot 1 \Big)
\end{aligned}
$$

$$+ \left(x \cdot x^3 + \frac{x^2}{2} \cdot x^2 + \frac{x^3}{3} \cdot x + \frac{x^4}{4} \cdot 1 \right) + \cdots$$

$$= x + \left(1 + \frac{1}{2} \right) x^2 + \left(1 + \frac{1}{2} + \frac{1}{3} \right) x^3 + \left(1 + \frac{1}{2} + \frac{1}{3} + \frac{1}{4} \right) x^4 + \cdots$$

$$= \sum_{n=1}^{\infty} \left(\sum_{j=1}^{n} \frac{1}{j} \right) x^n.$$

This is valid for $|x| < 1$.

5.9 Dividing power series

We stated Theorem 4 for power series in x, rather than for power series in $x - x_0$, merely for the sake of brevity. The theorem implies that the product of two functions analytic near a point is also analytic near this point. It is also true, though we shall not prove it, that the quotient of two analytic functions is analytic, provided, of course, that the denominator is not 0.

For instance, $x \mapsto \dfrac{1}{\cos x} = \sec x$ is analytic near $x_0 = 0$. The Taylor series of $\dfrac{1}{\cos x}$ can be obtained, using Theorem 4, as follows. We write $\sec x = a_0 + a_1 x + a_2 x^2 + \cdots$ and recall that

$$\cos x = 1 - \frac{x^2}{2!} + \frac{x^4}{4!} - \cdots.$$

Now

$$1 = (\sec x)(\cos x) = (a_0 + a_1 x + a_2 x^2 + \cdots) \left(1 - \frac{x^2}{2!} + \frac{x^4}{4!} - \cdots \right)$$

$$= a_0 \cdot 1 + (a_1 \cdot 1)x + \left(a_0 \frac{-1}{2!} + a_2 \cdot 1 \right) x^2 + \left(a_1 \frac{-1}{2!} + a_3 \cdot 1 \right) x^3$$

$$+ \left(a_0 \cdot \frac{1}{4!} + a_2 \frac{-1}{2!} + a_4 \cdot 1 \right) x^4$$

$$+ \left(a_1 \frac{1}{4!} + a_3 \frac{-1}{2!} + a_5 \cdot 1 \right) x^5 + \cdots.$$

Thus we must have

$$a_0 \cdot 1 = 1 \qquad \text{hence } a_0 = 1$$
$$a_1 \cdot 1 = 0 \qquad \text{hence } a_1 = 0$$
$$a_0 \frac{-1}{2!} + a_2 \cdot 1 = 0 \qquad \text{hence } a_2 = \frac{1}{2!}$$

$$a_1 \frac{-1}{2!} + a_3 \cdot 1 = 0 \qquad \text{hence } a_3 = 0$$

$$a_0 \frac{1}{4!} + a_2 \frac{-1}{2!} + a_4 = 0 \qquad \text{hence } a_4 = -\frac{1}{4!} + \frac{1}{(2!)^2} = \frac{5}{24}$$

$$a_1 \frac{1}{4!} + a_3 \frac{-1}{2!} + a_5 \cdot 1 = 0 \qquad \text{hence } a_5 = 0,$$

and so on. This is an effective method to determine the coefficients a_j. We obtain

$$\sec x = 1 - \frac{1}{4} x^2 - \frac{5}{24} x^4 + \cdots .$$

It is clear that the method illustrated by this example is quite general.

EXERCISES

Use the technique of multiplying and dividing series to compute the first 5 terms of the Taylor series of the given function $f(x)$ around $x_0 = 0$.

42. $f(x) = e^x \sin x$. For which x does the series surely represent $f(x)$?

43. $f(x) = e^x \arctan x$. For which x does the series surely represent $f(x)$?

44. $f(x) = \dfrac{e^x}{1 - x}$. For which x does the series surely represent $f(x)$?

45. $f(x) = \tan x = \sin x \left(\dfrac{1}{\cos x} \right)$.

46. $f(x) = \dfrac{\cos x}{1 - 2x}$. For which x does the series surely represent $f(x)$?

47. $f(x) = \dfrac{1}{1 + \sin x}$.

48. $f(x) = \dfrac{x}{1 + \log (1 + x)}$.

49. $f(x) = \dfrac{x}{1 + e^x}$.

50. $f(x) = \dfrac{\cos x}{1 + \sin x}$.

51. (a) $f(x) = \cos x \sin x$. (b) Now show that the Taylor series can be written as $\displaystyle \sum_{m=0}^{\infty} \frac{(-1)^m 2^{2m}}{(2m + 1)!} x^{2m+1}$. Use the results of Exercise 89 in §2.11 and review Cauchy multiplication (§4.15).

5.10 Odd and even functions

The following simple remark may be useful in calculating power series. If $f(x)$ is even, then $f'(x)$ is odd; hence $f'(0) = -f'(-0) = 0$. If $f(x)$ is odd, then $f(0) = 0$ and $f'(x)$ is even, so that $f''(x)$ is odd, hence $f''(0) = 0$, and so forth. We conclude that the Maclaurin series (Taylor series at $x = 0$) of an even function contains only even powers of x, whereas the Maclaurin series of an odd function contains only odd powers.

▶ *Example* The function

$$\frac{e^x - 1}{x} = \frac{1}{1!} + \frac{x}{2!} + \frac{x^2}{3!} + \frac{x^3}{4!} + \cdots$$

is analyltic at $x = 0$. So is its reciprocal, which we define to be 1 at $x = 0$, by the result stated above without proof. Also, the function

$$x \mapsto \frac{x}{1 - e^{-x}} - \frac{x}{2}$$

is even, as the reader should verify. Therefore

$$\frac{x}{e^x - 1} + \frac{x}{2} = \sum_{j=0}^{\infty} b_j x^{2j}.$$

One calls the numbers $B_j = j! \, b_j$ the **Bernoulli numbers.** They appear in many questions in analysis.◀

EXERCISES

Compute the following Bernoulli numbers.

52. B_0. 53. B_1. 54. B_2. 55. B_3.

56. From the expansion $\dfrac{x}{\sin x} - 1 = \sum_{j=1}^{\infty} \alpha_j x^{2j}$ (for small $|x|$) determine $\alpha_1, \alpha_2, \alpha_3$.

57. Find the first 3 terms of the Taylor expansion of $\dfrac{\sin x - x \cos x}{\sin^2 x}$ at $x_0 = 0$.

58. If $\dfrac{\sin x}{x} = (\sum_{j=0}^{\infty} a_j x^j)^2$ for small $|x|$, and $a_0 = 1$, find a_2, a_3, and a_4.

59. If $(\sum_{j=0}^{\infty} b_j x^j)(\sum_{j=0}^{\infty} a_j x^j) = 1$ for small $|x|$, where the a_j have the same meaning as in the preceding problem, find b_0, b_1, b_2, b_3.

5.11 Substitution

If, for small $|x|$, $f(x) = \sum_{n=0}^{\infty} a_n x^n$ and $g(x) = \sum_{j=1}^{\infty} b_j x^j$, so that $g(0) = 0$, then the function $f(g(x))$ is analytic at $x = 0$. We do not prove this either, but show how the Taylor series for the composite function $x \mapsto f(g(x))$ can be computed. For the sake of simplicity, we consider an example, $f(x) = e^x$, $g(x) = \sin x$. We have

$$e^{\sin x} = 1 + \frac{\sin x}{1!} + \frac{\sin^2 x}{2!} + \frac{\sin^3 x}{3!} + \cdots.$$

By repeated application of Cauchy multiplication, we compute the Taylor expansions of $\sin^2 x$, $\sin^3 x$, $\sin^4 x$, $\cdots$, substitute these in the series, and then "collect terms." We deal, of course, with infinite sums, but the calculation of each coefficient in the Maclaurin series of $e^{\sin x}$ involves only finitely many steps. Indeed,

$$\sin x = x - \frac{x^3}{3!} + \frac{x^5}{5!} - \frac{x^7}{7!} + \cdots = x - \frac{x^3}{6} + \frac{x^5}{120} - \cdots,$$

$$\sin^2 x = \left(x - \frac{x^3}{3!} + \frac{x^5}{5!} + \frac{x^7}{7!} + \cdots\right)\left(x - \frac{x^3}{3!} + \frac{x^5}{5!} + \cdots\right)$$

$$= x^2 - \frac{2}{3!}x^4 + \left(\frac{1}{(3!)^2} + \frac{2}{5!}\right)x^6 + \cdots = x^2 - \frac{x^4}{3} + \frac{2x^6}{45} + \cdots,$$

$$\sin^3 x = \left(x - \frac{x^3}{6} + \frac{x^5}{120} - \cdots\right)\left(x^2 - \frac{x^4}{3} + \frac{2x^6}{45} + \cdots\right)$$

$$= x^3 + \left(-\frac{1}{6} - \frac{1}{3}\right)x^5 + \left(+\frac{2}{45} + \frac{1}{3}\cdot\frac{1}{6} + \frac{1}{120}\right)x^7 + \cdots$$

$$= x^3 - \frac{x^5}{2} + \frac{13x^7}{120} + \cdots,$$

and so forth. Similarly, the expansion of $\sin^k x$ begins with the term x^k. Hence, in order to obtain the term with x^k in the expansion of $e^{\sin x}$, we need consider only the terms up to $\frac{\sin^k x}{k!}$. Thus we obtain

$$e^{\sin x} = 1 + x - \frac{x^3}{6} + \frac{x^5}{120} - \cdots + \frac{1}{2}x^2 - \frac{x^4}{6} + \cdots$$

$$+ \frac{1}{6}x^3 - \frac{x^5}{12} + \cdots + \frac{1}{24}(x^4 + \cdots) + \frac{1}{120}(x^5 + \cdots) + \cdots$$

$$= 1 + x + \frac{x^2}{2} - \frac{x^4}{8} - \frac{x^5}{15} + \cdots.$$

The same expansion can, of course, be obtained directly from Taylor's Theorem. The reader may want to check.

EXERCISES

Compute the first 5 terms of the Taylor series of the given function $f(x)$ around $x_0 = 0$ by the technique of this section.

▶ 60. $f(x) = \sin\left(\frac{1}{1 - x^2} - 1\right).$ 61. $f(x) = \cos(e^{x^2} - 1).$

62. $f(x) = \cos(\sin x)$.

63. $f(x) = \arctan \arctan x$.

64. $f(x) = e^{\arctan x}$.

65. $f(x) = \arctan\left(\dfrac{1}{1-x^2} - 1\right)$.

66. $f(x) = \cos(\log(1+x))$.

67. $f(x) = \log(1 + \sin x)$.

5.12 Remark

Before concluding this section, we note the curious similarities between the Maclaurin series of seemingly different functions:

$$\sin x = x - \frac{x^3}{3!} + \frac{x^5}{5!} - \cdots, \qquad \sinh x = x + \frac{x^3}{3!} + \frac{x^5}{5!} + \cdots,$$

$$\cos x = 1 - \frac{x^2}{2!} + \frac{x^4}{4!} - \cdots, \qquad \cosh x = 1 + \frac{x^2}{2!} + \frac{x^4}{4!} + \cdots,$$

$$\arctan x = x - \frac{x^3}{3} + \frac{x^5}{5} - \cdots, \qquad \frac{1}{2}\log\frac{1+x}{1-x} = x + \frac{x^3}{3} + \frac{x^5}{5} + \cdots.$$

The underlying reason for these similarities becomes clear only when one introduces complex numbers and studies power series with complex coefficients.

——————————————————————— Appendixes to **Chapter 8**

§6 L'Hospital's rule

There is a method for evaluating certain limits which is known as l'Hospital's rule. It is based on the corollary to the Generalized Mean Value Theorem stated in §1.4 as Theorem 2, and on its corollary.

6.1 Indeterminate Forms $0/0$

The basic prescription is the l'Hospital rule for evaluating the limit of a fraction $F(x)/G(x)$ at a point a when $F(a) = G(a) = 0$. It is applicable if F and G are continuous at a, if F and G have continuous derivatives, and $G' \neq 0$, near a, except perhaps for $x = a$. The rule reads:

IF $\lim\limits_{x\to a} F(x) = \lim\limits_{x\to a} G(x) = 0$ AND $\lim\limits_{x\to a} \dfrac{F'(x)}{G'(x)}$

EXISTS, THEN

(1) $$\lim_{x\to a}\frac{F(x)}{G(x)} = \lim_{x\to a}\frac{F'(x)}{G'(x)}.$$

GIULAUME FRANCOIS MARQUIS DE L'HOSPITAL (1661–1704), a member of the high French nobility, was an amateur mathematician and a friend and patron of mathematicians. He studied under Johann Bernoulli; this resulted in a textbook of calculus which contains the l'Hospital rule.

The proof is easy: by the corollary to Theorem 2 in §1.4 there is, for every $x \neq a$, a ξ between a and x such that $F(x)/G(x) = F'(\xi)/G'(\xi)$. If x is close to a, then ξ is close to a, and $F'(\xi)/G'(\xi)$ is close to its limit.

L'Hospital's rule is called the rule for evaluating "**indeterminate forms** 0/0." It may happen, of course, that $F'(a) = G'(a) = 0$. In this case, one can try applying l'Hospital's rule once more. If $\lim_{x \to a} [F''(x)/G''(x)]$ exists, we have

$$\lim_{x \to a} \frac{F(x)}{G(x)} = \lim_{x \to a} \frac{F''(x)}{G''(x)}.$$

One may have to do this several times. There is, of course, no guarantee of success. There may be no limit of F/G at a, or there may be a limit, but l'Hospital's rule may not be the right way for computing it.

L'Hospital's rule impressed the early practitioners of calculus. The importance of this method in contemporary mathematics is not overwhelming.

▶*Examples* *1.* Find $\lim_{x \to 0} \dfrac{\sin x - e^x + 1}{x^2}$

Answer. Applying l'Hospital's rule twice, we have

$$\lim_{x \to 0} \frac{\sin x - e^x + 1}{x^2} = \lim_{x \to 0} \frac{\cos x - e^x}{2x} = \lim_{x \to 0} \frac{-\sin x - e^x}{2} = -\frac{1}{2}.$$

2. What is wrong with the following argument, based on l'Hospital's rule?

$$\lim_{x \to 0} \frac{1 - 2x}{2 + 4x} = \lim_{x \to 0} \frac{\dfrac{d(1 - 2x)}{dx}}{\dfrac{d(2 + 4x)}{dx}} = \lim_{x \to 0} \frac{-2}{4} = -\frac{1}{2}.$$

Answer. Since $1 - 2x$ and $2 + 4x$ are equal to 1 and 2 at $x = 0$ (and not to 0 and 0), l'Hospital's rule is not applicable. ◀

EXERCISES

Evaluate the following limits using l'Hospital's rule:

1. $\lim\limits_{x \to 0} \dfrac{4x^3 - 3x^2 + 4x - 1}{3x^2 - 4x + 3}$.

2. $\lim\limits_{x \to 1} \dfrac{\log x}{x - \sqrt{x}}$.

3. $\lim\limits_{x \to -8} \dfrac{x + x^{2/3} + 4}{2x^{2/3} + 2x^{1/3} - 4}$.

4. $\lim\limits_{x \to 2} \dfrac{x^3 - 2x^2 - 4x + 8}{x^3 - 4x^2 + 4x}$.

5. $\lim\limits_{x \to 0} \dfrac{e^x - e^{-x}}{x^2}$.

6. $\lim\limits_{x \to 1} \dfrac{\sin (x - 1)}{\log (2x - 1)}$.

7. $\lim\limits_{x \to 0} \dfrac{1 - \cos^2 x}{\sin (x^2)}$.

8. $\lim\limits_{x \to 1} \dfrac{(x - 1)^2}{(x^2 - 1)^{4/3}}$.

9. $\displaystyle\lim_{x\to 2}\frac{x-\sqrt{2x}}{x^3-8}$.

10. $\displaystyle\lim_{x\to 0}\frac{\log\cos x}{x^2}$.

11. $\displaystyle\lim_{x\to 0}\frac{\arctan x^2}{x\sin x}$.

► 12. $\displaystyle\lim_{x\to 0}\frac{(\arctan x)^2}{\log(x^2+1)}$.

13. $\displaystyle\lim_{x\to 0}\frac{x^2}{1-\cos 2x}$.

14. $\displaystyle\lim_{x\to\pi/4}\frac{\cos x-\sin x}{\log\tan x}$.

15. $\displaystyle\lim_{x\to 0}\frac{\sin(2x^3)}{\sin^3(2x)}$.

6.2 Indeterminate forms ∞/∞

There is another case of l'Hospital's rule. It refers to "indeterminate forms ∞/∞" and reads:

IF
$$\lim_{x\to a}F(x)=+\infty,\qquad \lim_{x\to a}G(x)=+\infty\qquad\text{AND}\qquad \lim_{x\to a}\frac{F'(x)}{G'(x)}$$

EXISTS, THEN

$$(2)\qquad\qquad \lim_{x\to a}\frac{F(x)}{G(x)}=\lim_{x\to a}\frac{F'(x)}{G'(x)}.$$

The proof is more complicated; we shall only sketch it. Set

$$\lim_{x\to a}\frac{F'(x)}{G'(x)}=L.$$

We pick a number $x_1\neq a$ and consider a number x between a and x_1. By the Generalized Mean Value Theorem (Theorem 2 in §1.4), there is a number x_0 between x and x_1 such that

$$(3)\qquad\qquad \frac{F(x)-F(x_1)}{G(x)-G(x_1)}=\frac{F'(x_0)}{G'(x_0)}.$$

On the other hand,

$$\frac{F(x)-F(x_1)}{G(x)-G(x_1)}=\frac{F(x)}{G(x)}\frac{1-(F(x_1)/F(x))}{1-(G(x_1)/G(x))},$$

so that we can write (3) as

$$\frac{F(x)}{G(x)}=\frac{1-(G(x_1)/G(x))}{1-(F(x_1)/F(x))}\frac{F'(x_0)}{G'(x_0)}.$$

Now x_0 lies between x_1 and a; hence x_0 is closer to a than x_1 is. Therefore, if x_1 is sufficiently close to a, the fraction $F'(x_0)/G'(x_0)$ will be as close to L as we want. On the other hand, if we keep x_1 fixed and then choose x sufficiently close to a, then $|F(x)|$ and $|G(x)|$ will be as large as we want, the fractions $G(x_1)/G(x)$ and $F(x_1)/F(x)$ will be as close to 0 as we want, and the factor

$$\frac{1 - (G(x_1)/G(x))}{1 - (F(x_1)/F(x))}$$

will be as close to 1 as we want. Hence $F(x)/G(x)$ will be as close to L as we want, provided $x \neq a$ is sufficiently close to a. That is what we set out to prove.

It is clear that rule (2) holds also with $+\infty$ replaced by $-\infty$.

▶**Example**   Find $\lim\limits_{x \to 0} \dfrac{\log \sin^2 x}{\cot x}$.

Answer.   L'Hospital's rule gives

$$\lim_{x \to 0} \frac{\log \sin^2 x}{\cot x} = \lim_{x \to 0} \frac{(1/\sin^2 x)2 \sin x \cos x}{-1/\sin^2 x} = \lim_{x \to 0}(-2 \sin x \cos x) = 0. \blacktriangleleft$$

6.3 Extensions

L'Hospital's rule holds also for one-sided limits; no new proof is needed. The rule holds also if $\lim F'/G'$ is $+\infty$ or $-\infty$; no new proof is needed. The rule applies also if $a = +\infty$ or if $a = -\infty$. Indeed, if $\lim_{x \to +\infty} F(x) = 0$, $\lim_{x \to +\infty} G(x) = 0$, we have by the definition of limits at infinity:

$$\lim_{x \to +\infty} \frac{F(x)}{G(x)} = \lim_{x \to 0^+} \frac{F(1/x)}{G(1/x)},$$

and by l'Hospital's rule:

$$\lim_{x \to +\infty} \frac{F(x)}{G(x)} = \lim_{x \to 0^+} \frac{dF(1/x)/dx}{dG(1/x)/dx}$$

$$= \lim_{x \to 0^+} \frac{F'(1/x)(-1/x^2)}{G'(1/x)(-1/x^2)} = \lim_{x \to 0^+} \frac{F'(1/x)}{G'(1/x)} = \lim_{x \to +\infty} \frac{F'(x)}{G'(x)},$$

provided that the last limit exists.

EXERCISES

Evaluate the following limits using l'Hospital's rule:

16. $\lim\limits_{x \to 0} \dfrac{\log x^2}{1/x}$.

17. $\lim\limits_{x \to \pi/2^-} \dfrac{\log \tan x}{\tan x}$.

18. $\lim\limits_{x \to 1^+} \dfrac{\log (x - 1)}{\log (x^2 - 1)}$.

19. $\lim\limits_{x \to 1^+} \dfrac{\log \log x}{\log (x - 1)}$.

20. $\lim\limits_{x \to 1^+/2} \dfrac{\sqrt{2x - 1}}{\log 2x}$.

21. $\lim\limits_{x \to 1^+} \dfrac{\log (x^2 + x - 2)}{\log (2x^2 - x - 1)}$.

22. $\lim\limits_{x \to 0^+} \dfrac{\log (e^x - 1)}{1/x}$.

23. $\lim\limits_{x \to 0^+} \dfrac{\log (e^x - 1)}{\log 3x}$.

24. $\lim\limits_{x\to 1^+} \dfrac{\log\,(2x^3 - x - 1)}{\log\,(x^2 - 1)}$.

28. $\lim\limits_{x\to +\infty} \dfrac{\log\,(x + e^x)}{x}$.

25. $\lim\limits_{x\to +\infty} \dfrac{x \log x}{(x + 1)^2}$.

29. $\lim\limits_{x\to +\infty} \dfrac{e^x + x}{e^x + \log x}$.

► 26. $\lim\limits_{x\to +\infty} \dfrac{\arctan 2/x}{1/x}$.

30. $\lim\limits_{x\to 0^+} \dfrac{e^{1/x}}{\log x}$.

27. $\lim\limits_{x\to +\infty} \dfrac{e^{e^x}}{e^x}$.

6.4 Other indeterminate forms

Other "indeterminate forms," that is, other limits, can also sometimes be determined by l'Hospital's rule. In each case, we must first reduce the expression to be evaluated to the form F/G. The procedure is best explained by the use of examples.

► **Examples** **1.** A case of $\dfrac{\infty}{\infty}$. Find $\lim_{x\to +\infty} \dfrac{x^3 + 3}{2x^3 + x}$.

One can easily do this directly. We show that l'Hospital's rule works, too:

$$\lim_{x\to +\infty} \frac{x^3 + 3}{2x^3 + x} = \lim_{x\to +\infty} \frac{3x^2}{6x^2 + 1} = \lim_{x\to +\infty} \frac{6x}{12x} = \frac{1}{2}.$$

2. A case of $0 \cdot \infty$. Find $\lim_{x\to 0^+} x \log x$.

We reduce this to the case $\dfrac{\infty}{\infty}$, namely:

$$\lim_{x\to 0^+} x \log x = \lim_{x\to 0^+} \frac{\log x}{\dfrac{1}{x}} = \lim_{x\to 0^+} \frac{\dfrac{1}{x}}{-\dfrac{1}{x^2}} = \lim_{x\to 0^+} (-x) = 0.$$

(The reader may want to compare this with Chapter 6, §5.7.)

3. A case of $\infty - \infty$. Find $\lim_{x\to 0} \left(\cot x - \dfrac{1}{x}\right)$.

We reduce this to $\dfrac{0}{0}$. Since $\cot x - \dfrac{1}{x} = \dfrac{\cot x}{x}\left(x - \dfrac{1}{\cot x}\right) = \dfrac{x - \tan x}{x \tan x}$, we have

$$\lim_{x\to 0}\left(\cot x - \frac{1}{x}\right) = \lim_{x\to 0} \frac{x - \tan x}{x \tan x} = \lim_{x\to 0} \frac{1 - \dfrac{1}{\cos^2 x}}{\tan x + \dfrac{x}{\cos^2 x}} = \lim_{x\to 0} \frac{\cos^2 x - 1}{\sin x \cos x + x}$$

$$= \lim_{x\to 0} \frac{-\sin^2 x}{\frac{1}{2}\sin 2x + x} = \lim_{x\to 0} \frac{-2\sin x \cos x}{\cos 2x + 1} = \frac{0}{2} = 0.$$

4. A case of 1^∞. Find $\lim_{x\to +\infty} \left(1 + \dfrac{1}{x}\right)^x$.

This is treated by taking logarithms. Set $\phi(x) = \log\left\{\left(1 + \frac{1}{x}\right)^x\right\}$. Then

$$\lim_{x \to +\infty} \phi(x) = \lim_{x \to +\infty} \log\left\{\left(1 + \frac{1}{x}\right)^x\right\} = \lim_{x \to +\infty} x \log\left(1 + \frac{1}{x}\right) = \lim_{x \to +\infty} \frac{\log\left(1 + \frac{1}{x}\right)}{\frac{1}{x}}$$

$$= \lim_{x \to +\infty} \frac{\frac{1}{1 + \frac{1}{x}}\left(-\frac{1}{x^2}\right)}{-\frac{1}{x^2}} = \lim_{x \to +\infty} \frac{1}{1 + \frac{1}{x}} = 1$$

and, since the logarithm and the exponential are continuous functions,

$$\lim_{x \to +\infty} \left(1 + \frac{1}{x}\right)^x = e^{\log \lim \phi} = e^{\lim \log \phi} = e^1 = e.$$

(The reader may want to compare this with the proof of Theorem 8 in Chapter 6, §5.9.)

5. A case of ∞^0. Find $\lim_{x \to +\infty} x^{e^{-x}}$. We proceed as in Case 4:

$$\lim_{x \to +\infty} \log x^{e^{-x}} = \lim_{x \to +\infty} e^{-x} \log x = \lim_{x \to +\infty} \frac{\log x}{e^x} = \lim_{x \to +\infty} \frac{\frac{1}{x}}{e^x} = 0.$$

Hence $\lim_{x \to +\infty} x^{e^{-x}} = e^0 = 1.$ ◄

EXERCISES

Evaluate the following limits using l'Hospital's rule:

31. $\lim\limits_{x \to 0^+} \dfrac{\log \sin x}{\log (e^x - 1)}$.

32. $\lim\limits_{x \to 0}\left(\dfrac{1}{x} - \dfrac{1}{\sin x}\right)$.

33. $\lim_{x \to 0^+} x^{1/3} \log x^3$.

34. $\lim_{x \to 0^+} x \log^2 x$.

35. $\lim_{x \to 0^+} (\log x)(\log (x + 1))$.

36. $\lim_{x \to \pi/2} (\sec x - \tan x)$.

37. $\lim_{x \to 0^+} (\sin x)(\log x)$.

38. $\lim\limits_{x \to 0}\left(1 + \dfrac{1}{x^2}\right)^{x^2}$.

39. $\lim_{x \to 0} (1 + 2x^3)^{3/x^3}$.

40. $\lim\limits_{x \to +\infty}\left(1 + \dfrac{1}{x}\right)^{x^2}$.

41. $\lim_{x \to 0} (\log (x + 1))^x$.

▶ 42. $\lim_{x \to 0} (\sin x)^x$.

43. $\lim_{x \to 1} x^{1/1-x}$.

44. $\lim_{x \to 0^+} x^2 e^{1/x}$.

45. $\lim_{x \to 0^+} (\arctan x)^{1/\log x}$.

46. $\lim_{x \to +\infty} \left(\dfrac{2x}{x+4}\right)^x$.

§7 Convergence Proofs

In this section, we prove some statements about sequences and about series made in the text without proofs.

7.1 Monotone sequences

We begin by establishing the basic theorem in §3.8, Theorem 3. It states that a nondecreasing sequence

$$a_1 \le a_2 \le a_3 \cdots$$

has a limit if it is bounded, and that this limit is the smallest number that is $\ge$ all terms a_j.

The proof is quite simple. We assume the sequence to be bounded. Then the set of all numbers a_j is a bounded set and has, by the least upper bound principle, a least upper bound M. We must show that

$$\lim_{j \to \infty} a_j = M.$$

To obtain the verification of Condition I in the definition of limits (see §3.4), assume that $M < A$. Since M is an upper bound, $a_j \le M$ for all j. Hence $a_j < A$ for all j.

The verification of Condition II begins with the assumption that $B < M$. Since M is the least upper bound, B is not an upper bound. There is a term in the sequence, call it a_n, with $a_n > B$. Since the sequence is not decreasing, $a_j > a_n > B$ if $j > n$. Hence nearly all of the sequence terms are $> B$.

EXERCISE

1. Prove directly that a nonincreasing sequence $b_1 \ge b_2 \ge b_3 \ge \cdots$ has a limit if it is bounded. Modify the proof given for nondecreasing sequences.

7.2 Rearranging terms in a series

We consider next a convergent infinite series

$$\sum_{j=0}^{\infty} a_j$$

and investigate the effect of rearranging the terms. This will lead to a proof of

Theorem 11 in §4.14. We may assume that no a_j is 0, since such terms could be dropped. Since the series converges, we have (see Theorem 2 in §4.3)

$$\lim_{j \to \infty} a_j = 0.$$

It follows that, for every number $T > 0$, nearly all a_j satisfy $|a_j| < T$. Therefore we can collect all positive terms among the a_j and arrange them in decreasing order. We call these terms $\alpha_0, \alpha_1, \alpha_2, \cdots$. Thus

$$\alpha_0 \geq \alpha_1 \geq \alpha_2 \geq \cdots, \alpha_j > 0, \lim_{j \to \infty} \alpha_j = 0.$$

Similarly, we denote the negative numbers among the a_j, arranged in descending order of absolute values, by $-\beta_0, -\beta_1, -\beta_2, \cdots$. We have

$$\beta_0 \geq \beta_1 \geq \beta_2 \geq \cdots, \beta_j > 0, \lim_{j \to \infty} \beta_j = 0.$$

Set
$$P = \sum_{j=0}^{\infty} \alpha_j, \qquad N = \sum_{j=0}^{\infty} \beta_j$$

(if there are no α's, set $P = 0$; if there are no β's, set $N = 0$). Of course, we must admit $+\infty$, as a possible value for P and N. However, either $P = +\infty$ and $N = +\infty$, or $P < +\infty$ and $N < +\infty$. (In other words, the series of positive terms and the series of negative terms are either both divergent or both convergent.) Indeed, assume that $P = +\infty$ and $N < +\infty$. For every m, we have that $\Sigma_{j=0}^{m} a_j = S_m - T_m$, where S_m is the sum of all positive numbers among $a_0, a_1, \cdots, a_m$ and $-T_m$ is the sum of all negative numbers among $a_0, a_1, \cdots, a_m$. Clearly $0 \leq T_m \leq N$, so that $\Sigma_{j=0}^{m} a_j \geq S_m - N$. But if m is large enough, the finite sequence $a_0, a_1, \cdots, a_m$ will contain as many terms $\alpha_0, \alpha_1, \alpha_2, \cdots$ as we like and S_m will be as large as we like. Hence so will $\Sigma_{j=0}^{m} a_j$. This contradicts the assumption that $\Sigma_{j=0}^{\infty} a_j$ converges.

We showed that the case $P = +\infty, N < +\infty$ is impossible. The case $P < +\infty$ is also impossible, for the same reason.

We show next that, if $P < +\infty, N < +\infty$, then $\Sigma_{j=0}^{\infty} a_j = P - N$. Indeed, if we use the same notations as before, we observe that for large m the sum S_m will be as close to P as we like, and the sum T_m will be as close to N as we like.

Finally, we note that $\Sigma_{j=0}^{\infty} |a_j| = P + N$. (In other words, the series $\Sigma_{j=0}^{\infty} a_j$ converges absolutely if and only if $P < +\infty$ and $N < +\infty$.) The proof is left to the reader.

The preceding remarks constitute a proof of the first part of Theorem 11 in §4.14: an absolutely convergent series remains convergent if its terms are rearranged; the rearrangement does not affect the sum of the series.

We assume now that

$$P = +\infty, \qquad N = +\infty$$

and show how to arrange the terms α_k and $-\beta_j$ into a divergent series.

Since $\Sigma_{k=0}^{\infty} \alpha_k = +\infty$, we can divide this series into blocks as follows:

$$\underbrace{\alpha_0 + \alpha_1 + \alpha_2 + \cdots + \alpha_{n_1}}_{>1} + \underbrace{\alpha_{n_1+1} + \alpha_{n_1+2} + \cdots + \alpha_{n_2}}_{>\beta_1 + 1}$$

$$+ \underbrace{\alpha_{n_2+1} + \cdots + \alpha_{n_3}}_{>\beta_2 + 1} + \underbrace{\alpha_{n_3+1} + \cdots + \alpha_{n_4}}_{>\beta_3 + 1} + \cdots.$$

The series

$$\underbrace{\alpha_0 + \alpha_1 + \cdots + \alpha_{n_1}}_{} + \underbrace{(-\beta_1)}_{} + \alpha_{n+1} + \cdots + \alpha_{n_2} +$$

$$\underbrace{(-\beta_2)}_{} + \alpha_{n_2+1} + \cdots + \alpha_{n_3} + \underbrace{(-\beta_3)}_{} + \alpha_{n_3+1} + \cdots + \alpha_{n_4} + \cdots$$

is a rearrangement of the original series $\alpha_0 + \alpha_1 + \cdots$, and diverges.

Now let s be any given number. The terms $a_0, \alpha_1, \cdots, -\beta_0, -\beta_1, \cdots$ can be arranged into a series that converges to s:

$$\underbrace{\alpha_0 + \alpha_1 + \cdots + \alpha_{n_1}}_{\sigma_1} \underbrace{-\beta_0 - \beta_1 - \cdots - \beta_{m_1}}_{\tau_1} + \underbrace{\alpha_{n_1+1} + \cdots + \alpha_{n_2}}_{\sigma_2}$$

$$\underbrace{-\beta_{m_1+1} - \cdots - \beta_{m_2}}_{\tau_2} + \underbrace{\alpha_{n_2+1} + \cdots + \alpha_{n_3}}_{\sigma_3} \underbrace{-\beta_{m_2+1} - \cdots - \beta_{m_3}}_{\tau_3} + \alpha_{n_3+1} + \cdots.$$

Here the numbers $n_1, m_1, n_2, m_2, \cdots$ are chosen as the smallest numbers satisfying the conditions: $\sigma_1 > s$, $\sigma_1 + \tau_1 < s$, $\sigma_1 + \tau_1 + \sigma_2 > s$, $\sigma_1 + \tau_1 + \sigma_2 + \tau_2 < s$, $\sigma_1 + \tau_1 + \sigma_2 + \tau_2 + \sigma_3 > s$, $\cdots$. One can find σ_1, since $\Sigma_{j=0}^{\infty} \alpha_j = +\infty$; one can find τ_1, since $\Sigma_{j=0}^{\infty} \beta_j = +\infty$; one can find σ_2, since $\Sigma_{j=n_1+1}^{\infty} \alpha_j = +\infty$; one can find τ_2 since $\Sigma_{j=m_1+1}^{\infty} \beta_j = +\infty$; and so forth. The partial sums of the series just constructed oscillate around s.

Also, since n_1 is the smallest number such that $\sigma_1 > s$, we have $|\sigma_1 - s| \leq \alpha_{n_1}$. Similarly, since m_1 is the smallest number such that $\sigma_1 + \tau_1 < s$, we must have $|(\sigma_1 + \tau_1) - s| \leq \beta_{m_1}$. For the same reason $|(\sigma_1 + \tau_1 + \sigma_2) - s| \leq \alpha_{n_2}$, and $|(\sigma_1 + \tau_1 + \sigma_2 + \tau_2) - s| \leq \beta_{m_2}$, and so forth. Since $\lim \alpha_j = \lim \beta_j = 0$, we conclude that the series constructed really converges to s.

We have now established the second part of Theorem 11: a series that converges, but not absolutely; can be rearranged so as to diverge; or so as to converge to any given number.

EXERCISES

2. Show that $P < +\infty$ and $N < +\infty$ if $\Sigma_{j=0}^{\infty} |a_j|$ converges.
3. Show that $\Sigma_{j=0}^{\infty} |a_j|$ converges if $P < +\infty$ and $N < +\infty$.
4. Rearrange the terms of the series $1 - \frac{1}{2} + \frac{1}{3} - \frac{1}{4} + \cdots$ so as to form a divergent series.
5. Rearrange the terms of the series $1 - \frac{1}{2} + \frac{1}{3} - \frac{1}{4} + \cdots$ so as to form a series converging to 1.

7.3 Cauchy products

Our next task is to establish the theorem on multiplying absolutely convergent series (Theorem 12 in §4.15).

Assume that $\Sigma a_j = A$, $\Sigma |a_j| = R < +\infty$, $\Sigma b_j = B$, $\Sigma |b_j| = S < +\infty$ and form first the series

$$(1) \qquad |a_0||b_0| + (|a_0||b_1| + |a_1||b_0|) + (|a_0||b_2| + |a_1||b_1| + |a_2||b_0|) + \cdots$$

$$\sum_{n=0}^{\infty} \left(\sum_{j=0}^{n} |a_j||b_{n-j}| \right).$$

A partial sum of this series $\Sigma_{n=0}^{N} (\Sigma_{j=0}^{n} |a_j||b_{n-j}|)$ is a sum of positive numbers. These are *some* of the terms occurring in the product $\Sigma_{j=0}^{N} |a_j| \Sigma_{k=0}^{N} |b_k|$. Hence this partial sum is at most RS. It follows from Theorem 5 in §4.7 that the series (1) converges. Thus the series (Cauchy product)

$$(2) \qquad \sum_{n=0}^{\infty} \left(\sum_{j=0}^{n} a_j b_{n-j} \right)$$

converges absolutely.

Let C denote the sum of the series (2). To complete the proof of Theorem 12, we must show that $C = AB$. Let A_N and B_N denote the Nth partial sums of the series Σa_j, Σb_j and let C_N denote the Nth partial sum of the series (2). Now, for every positive integer N,

$$|A_{2N}B_{2N} - C_{2N}| = \left| \left(\sum_{j=0}^{2N} a_j \right) \left(\sum_{k=0}^{2N} b_k \right) - \sum_{n=0}^{2N} \left(\sum_{j+k=n} a_j b_k \right) \right|$$

$= |\text{sum of all terms } a_j b_k \text{ with } j \le 2N, k \le 2N, j + k > 2N|$
$\le$ sum of all terms $|a_j||b_k|$ with $j \le 2N, k \le 2N, j + k > 2N$
$\le$ sum of all terms $|a_j||b_k|$ with $j \le 2N, k \le 2N$ and either $j > N$ or $k > N$

$$= \left(\sum_{j=0}^{2N} |a_j| \right) \left(\sum_{h=0}^{2N} |b_k| \right) - \left(\sum_{j=0}^{N} |a_j| \right) \left(\sum_{k=0}^{N} |b_k| \right).$$

Since

$$\lim_{N \to \infty} \left\{ \left(\sum_{j=0}^{2N} |a_j| \right) \left(\sum_{k=0}^{2N} |b_k| \right) - \left(\sum_{j=0}^{N} |a_j| \right) \left(\sum_{h=0}^{N} |b_h| \right) \right\} = RS - RS = 0$$

we conclude that $AB - C = \lim_{N \to \infty} (A_{2N}B_{2N} - C_{2N}) = 0$ as expected.

§8 Radius of Convergence

We proceed to prove the assertions about power series made without proof in §5.2 (Theorems 1 and 2). To simplify writing, we consider only power series about $x_0 = 0$; this is clearly sufficient.

8.1 Abel's lemma

Lemma 1 (Abel's lemma)   IF THE POWER SERIES

$$(1) \qquad\qquad \sum_{n=0}^{\infty} a_n x^n$$

CONVERGES FOR SOME $x = \rho \neq 0$, THEN THE SERIES CONVERGES ABSOLUTELY FOR $|x| < |\rho|$ AND THERE IS A CONSTANT M SUCH THAT

$$(2) \qquad\qquad |a_n| \leq \frac{M}{|\rho|^n} \qquad \text{for } n = 0, 1, 2, \cdots.$$

This is the basic result in the theory of power series. The proof is quite simple. If $\Sigma_{n=0}^{\infty} a_n \rho^n$ converges, then, by Theorem 2 in §4.3, we have $\lim a_n \rho^n = 0$. Therefore, by the definition of limit, $|a_n \rho^n| < 1$ for nearly all n, say for $n > N$. Let M be the largest of the numbers $|a_0|$, $|a_1 \rho|$, $|a_2 \rho^2|$, $\cdots$, $|a_N \rho^N|$, and 1. Then $|a_n \rho^n| \leq M$ or $|a_n||\rho^n| \leq M$. Since $\rho \neq 0$, assertion (2) follows.

Now let x be a number such that $|x| < |\rho|$. Then

$$\left| \frac{x}{\rho} \right| < 1$$

and, by (2),

$$|a_n x^n| \leq M \left| \frac{x}{\rho} \right|^n.$$

Thus the terms of (1) are, in absolute value, not greater than those of the convergent series

$$M + M \left| \frac{x}{\rho} \right| + M \left| \frac{x}{\rho} \right|^2 + \cdots.$$

Hence the series $\Sigma |a_n x^n|$ converges (by Theorem 6 in §4.8) and (1) converges absolutely (compare Theorem 10 in §4.13).

8.2 Radius of convergence

Lemma 2   IF THE POWER SERIES (1) CONVERGES FOR SOME BUT NOT ALL $x \neq 0$, THEN THERE IS A NUMBER $R > 0$ (THE RADIUS OF CONVERGENCE) SUCH THAT THE SERIES (1) CONVERGES ABSOLUTELY FOR $|x| < R$ AND DIVERGES FOR $|x| > R$.

Proof.   Assume that (1) converges for $x = \rho_1 \neq 0$ and diverges for $x = \rho_2$. Then (1) converges for all x such that $|x| < |\rho_1|$ and diverges for all $|x| > |\rho_2|$. This follows from Abel's lemma. Let S be the set of all positive numbers ρ such that (1) converges for $|x| < \rho$. This set is not empty; it contains $|\rho_1|$. It is bounded; ρ_2 is an upper bound. Let R be the least upper bound of S. If $|x_0| < R$, then $\Sigma_{n=0}^{\infty} a_n x_0^n$ converges, for otherwise $|x_0|$ would be an upper bound for S.

By the same token, for every x_1, $|x_0| < |x_1| < R$, the series $\Sigma_{n=0}^{\infty} a_n x_1^n$ converges, hence $\Sigma_{n=0}^{\infty} a_n x_0^n$ converges absolutely, by Abel's lemma. Assume next that $|x_2| > R$. Then $\Sigma_{n=0}^{\infty} a_n x_2^n$ diverges. For, if this series would converge, $|x_2|$ would belong to S, by Abel's lemma.

We see that R is the desired radius of convergence.

If the series converges for all values of x, one sets $R = +\infty$; if it diverges for all $x \neq 0$, one sets $R = 0$.

8.3 Differentiated and integrated series

Lemma 3   THE THREE POWER SERIES

(3) $$\sum a_n x^n, \qquad \sum n a_n x^{n-1}, \qquad \sum \frac{a_n}{n+1} x^{n+1}$$

HAVE EQUAL RADII OF CONVERGENCE.

Proof.   Denote the three radii of convergence by R, R_*, and R^*, respectively. (These symbols may be nonnegative numbers or $+\infty$.)

If $R > 0$, let ρ be a number such that $0 < \rho < R$. Then, by Abel's lemma, we have inequality (2) and, for $|x| < \rho$,

$$|na_n x^{n-1}| \leq n\frac{M}{\rho}\left(\frac{|x|}{\rho}\right)^{n-1}, \qquad \left|\frac{a_n x^{n+1}}{n+1}\right| \leq \frac{M|x|}{(n+1)}\left(\frac{|x|}{\rho}\right)^n \leq M|x|\left(\frac{x}{\rho}\right)^n$$

so that the terms of the second and third series in (3) are in absolute value not greater than the corresponding terms of the convergent series

$$\frac{M}{\rho} + 2\frac{M}{\rho}\frac{|x|}{\rho} + 3\frac{M}{\rho}\frac{|x|^2}{\rho^2} + \cdots = \frac{M}{\rho}\left(1 - \frac{|x|}{\rho}\right)^{-2}$$

and $$M|x| + M|x|\frac{|x|}{\rho} + M|x|\left(\frac{|x|}{\rho}\right)^2 + \cdots = M|x|\left(1 - \frac{|x|}{\rho}\right)^{-1},$$

respectively. We apply Theorem 6 in §4.8 and conclude that the second and third series (3) converge for $|x| < \rho < R$, hence for *all* $|x| < R$. Therefore $R^* \geq R$ and $R_* \geq R$.

But the first series in (3) is obtained from the third in the same way as the second is obtained from the first, namely, by term-by-term differentiation. Therefore we also have $R \geq R_*$ and $R \geq R^*$. Thus $R = R_* = R^*$.

8.4 Continuity proof

Lemma 4   ASSUME THAT $\Sigma_{n=0}^{\infty} a_n x^n$ CONVERGES FOR $|x| < A$ (FOR SOME $A > 0$) AND SET

(4) $$f(x) = \sum a_n x^n.$$

THEN $f(x)$ IS CONTINUOUS FOR $|x| < A$.

Proof. We shall establish the continuity of $f(x)$ at some fixed but arbitrary point ξ, $|\xi| < A$. More precisely, we shall show that for every given $\epsilon > 0$, we have $|f(x) - f(\xi)| < \epsilon$ for x near ξ.

We choose numbers σ and ρ such that

(5)
$$|\xi| < \sigma < \rho < A.$$

In view of Abel's lemma, inequality (2) holds. Set

(6)
$$R_m(x) = \sum_{n=m+1}^{\infty} a_n x^n.$$

For $|x| \leq \sigma$, we have (using Theorem 10 in §4)

(7)
$$|R_m(x)| \leq \sum_{n=m+1}^{\infty} |a_n x^n| \leq \sum_{n=m+1}^{\infty} \frac{M}{\rho^n} \sigma^n = M\left(\frac{\sigma}{\rho}\right)^{m+1} \sum_{0}^{\infty} \left(\frac{\sigma}{\rho}\right)^j$$
$$= \frac{M\left(\frac{\sigma}{\rho}\right)^{m+1}}{1 - \frac{\sigma}{\rho}}.$$

Let $\epsilon > 0$ be a given number. We choose a fixed m such that

(8)
$$|R_m(x)| < \frac{\epsilon}{2} \qquad \text{for } |x| < \sigma.$$

This is possible in view of (7), since $\lim_{m \to \infty} \left(\frac{\sigma}{\rho}\right)^{m+1} = 0$. Now set

(9)
$$S_m(x) = \sum_{n=0}^{m} a_n x^n.$$

Then
(10)
$$f(x) = S_m(x) + R_m(x).$$

In particular, $f(\xi) = S_m(\xi) + R_m(\xi)$ and, by (8), we have that

$$f(\xi) - \frac{\epsilon}{2} < S_m(\xi) < f(\xi) + \frac{\epsilon}{2}.$$

Since $S_m(x)$ is a polynomial, hence continuous, we have

(11)
$$f(\xi) - \frac{\epsilon}{2} < S_m(x) < f(\xi) + \frac{\epsilon}{2} \text{ for } x \text{ near } \xi.$$

Hence, by (8), (10), and (11),

$$f(\xi) - \epsilon < f(x) < f(\xi) + \epsilon \text{ for } x \text{ near } \xi.$$

This completes the proof.

8.5 Derivatives and integrals

Lemma 5 UNDER THE HYPOTHESIS OF LEMMA 4,

(12)
$$\int_0^x f(t)\,dt = \sum_{n=0}^{\infty} \frac{a_n x^{n+1}}{n+1}$$

FOR $|x| < A$.

Proof. It suffices to establish (12) under the assumption that $|x| < \sigma < \rho < A$. We use the notations of the previous proof and have

$$\int_0^x f(t)\,dt = \int_0^x S_m(t)\,dt + \int_0^x R_m(t)\,dt.$$

Since (7) holds,

$$\left| \int_0^x R_m(t)\,dt \right| \le \frac{M\left(\frac{\sigma}{\rho}\right)^{m+1}}{1 - \frac{\sigma}{\rho}} |x| \text{ for } |x| \le \sigma.$$

Therefore $\lim_{m\to\infty} |\int_0^x R_m(t)\,dt| = 0$ and

$$\int_0^x f(t)\,dt = \lim_{m\to\infty} \int_0^x S_m(t)\,dt = \lim_{m\to\infty} \int_0^x \sum_{n=0}^{m} a_n t^n\,dt$$

$$= \lim_{m\to\infty} \sum_{n=0}^{m} \frac{a_n x^{n+1}}{n+1} = \sum_{n=0}^{\infty} \frac{a_n x^{n+1}}{n+1},$$

as asserted.

Lemma 6 UNDER THE HYPOTHESIS OF LEMMA 4,

$$f'(x) = \sum_{n=1}^{\infty} n a_n x^{n-1}.$$

Proof. Set, for $|x| < A$, $g(x) = \sum_{n=1}^{\infty} n a_n x^{n-1}$, the series being convergent, by Lemma 3, and $g(x)$ being continuous, by Lemma 4. By Lemma 5, we have

$$\int_0^x g(t)\,dt = \sum_{n=1}^{\infty} a_n x^n = f(x) - a_0.$$

Hence, $g(x) = f'(x)$ by the fundamental theorem of calculus.

Now Lemmas 1 to 6 contain the proof of Theorems 1 and 2 of §5.2 for the case $x_0 = 0$; but, if these theorems hold for $x_0 = 0$, they also hold for every x_0.

EXERCISES

1. Let k be a fixed positive integer. Prove that the series

$$\sum_{n=0}^{\infty} a_n x^n \quad \text{and} \quad \sum_{n=k+1}^{\infty} n(n-1)(n-2) \cdots (n-k) x^{n-k-1}$$

have the same radius of convergence.

2. State and prove the analogue of Abel's lemma for a power series $\Sigma_{n=0}^{\infty} a_n(x - x_0)^n$.
3. State and prove the analogues of Lemmas 2 to 6 for the power series about x_0.

§9 The Truncation Error in Numerical Integration

In this section, we prove the estimates for the truncation error in numerical integration stated in §1 of Chapter 7. We emphasize that these estimates do not take into account the round-off errors that are practically always present in numerical calculations.

9.1 Estimate for the trapezoidal rule

Let $f(x)$ be a continuous function defined in the finite interval $[a, b]$, and let $\phi(x)$ be a linear function that coincides with $f(x)$ at the endpoints, that is, satisfies the conditions $\phi(a) = f(a)$, $\phi(b) = f(b)$. We set

$$E^* = \int_a^b f(t) \, dt - \int_a^b \phi(t) \, dt$$

and want to estimate E^*, assuming that $f(x)$ has a continuous second derivative $f''(x)$ in $[a, b]$, and

$$|f''(x)| \le M_2. \tag{1}$$

To achieve this, we set

$$g(x) = \int_a^x f(t) \, dt - \int_a^x \phi(t) \, dt + \left\{ 2\left(\frac{x - a}{b - a}\right)^3 - 3\left(\frac{x - a}{b - a}\right)^2 \right\} E^*.$$

We have $g(a) = 0$, $g(b) = 0$, so that by Rolle's Theorem there is a point c with $a < c < b$ and $g'(c) = 0$. We compute easily that $g'(a) = g'(b) = 0$. Hence, again by Rolle's Theorem, there are points c_1 and c_2 such that $a < c_1 < c$, $c < c_2 < b$ and $g''(c_1) = 0$, $g''(c_2) = 0$. We conclude, once more by Rolle, that there is a point y between c_1 and c_2 with $g'''(y) = 0$. Thus, since $\phi''(x) \equiv 0$, we have

$$0 = g'''(y) = f''(y) + \frac{12}{(b - a)^3} E^*.$$

Noting (1), we conclude that

$$|E^*| \le \frac{(b - a)^3}{12} M_2. \tag{2}$$

Suppose next that we divide the interval $[a, b]$ into n equal parts, each of length h, and use the trapezoidal rule to obtain an approximate value for the integral $\int_a^b f(t) \, dt$. This means that, in each subinterval, we replace f by a linear function that coincides with f at the endpoints, compute the integral of this linear function over the subinterval, and add the n numbers so obtained. According to the result established above, the error committed in computing the integral over a subinterval is, in absolute value, at most $h^3 M_2/12$. The total absolute error, E, is therefore not greater than

$$\frac{nh^3 M_2}{12} = (b - a)h^2 \frac{M_2}{12}.$$

This is the estimate stated in §1.5 of Chapter 7.

EXERCISES

Find a number y satisfying $0 = g'''(y) = f''(y) + 12E^*/(b - a)^3$ for the given function f and points a and b. [Hint: Compute $\phi(t)$ and then E^* exactly.]

1. $f(x) = x^3 - x + 1$, $a = 0$, $b = 1$.
2. $f(x) = \sqrt{x}$, $a = 1$, $b = 4$.
3. $f(x) = \log x$, $a = 1$, $b = e$.

9.2 Estimate for Simpson's rule

Let f be as before, except that now we assume that f has in $[a, b]$ a continuous fourth derivative $f^{(4)}$ and

(3) $$|f^{(4)}(x)| \leq M_4.$$

We denote by $\phi(x)$ a quadratic polynomial which coincides with $f(x)$ at the endpoints and at the midpoint of the interval considered, that is, satisfies the conditions

$$\phi(a) = f(a), \qquad \phi\left(\frac{a + b}{2}\right) = f\left(\frac{a + b}{2}\right), \qquad \phi(b) = f(b).$$

We set $$E^* = \int_a^b f(t)\, dt - \int_a^b \phi(t)\, dt$$

and proceed to estimate E^*.

We consider first the function

$$q(x) = (x - a)\left(x - \frac{a + b}{2}\right)(x - b)$$

and compute that

$$q(a) = q\left(\frac{a + b}{2}\right) = q(b) = 0, \quad \int_a^b q(t)\, dt = 0, \quad \int_a^{(a+b)/2} q(t)\, dt > 0.$$

It follows that, for every choice of the number k, the function $\phi_0(x) = \phi(x) + kq(x)$ satisfies the conditions

$$\phi_0(a) - f(a) = \phi_0\left(\frac{a + b}{2}\right) - f\left(\frac{a + b}{2}\right) = \phi_0(b) - f(b) = 0$$

and $$E^* = \int_a^b f(t)\, dt - \int_a^b \phi_0(t)\, dt.$$

Also, k can be chosen so that

$$\int_a^{(a+b)/2} f(t)\, dt = \int_a^{(a+b)/2} \phi_0(t)\, dt.$$

Next, we form the fifth degree polynomial

$$p(x) = \frac{4(x-a)^2\left(x - \dfrac{a+b}{2}\right)^2 (7b - a - 6x)}{(b-a)^5}$$

and verify that it satisfies the conditions

$$p(a) = p'(a) = p\left(\frac{a+b}{2}\right) = p'\left(\frac{a+b}{2}\right) = p'(b) = 0,\ p(b) = 1.$$

Finally, we consider the function

$$g(x) = \int_a^x f(t)\, dt - \int_a^x \phi_0(t)\, dt - E^* p(x).$$

One verifies easily that

$$g(a) = g\left(\frac{a+b}{2}\right) = g(b) = 0$$

and also

(4)
$$g'(a) = g'\left(\frac{a+b}{2}\right) = g'(b) = 0.$$

By repeated applications of Rolle's Theorem, we conclude that g' is 0 at a point between a and $(a+b)/2$ and at a point between $(a+b)/2$ and b, so that, in view of (4), the function g'' is 0 at least four points in our interval, the function g''' is 0 at at least three points, the function $g^{(4)}$ is 0 at at least two points, and, finally, the function $g^{(5)}$ is 0 at at least one point. Call this point y. Since ϕ_0 is a cubic polynomial, $\phi_0^{(4)}(x) \equiv 0$. Also, $p^{(5)}(x) \equiv -2880(b-a)^{-5}$. Thus

$$0 = g^{(5)}(y) = f^{(4)}(y) - \frac{2880 E^*}{(b-a)^5},$$

and by (3) we obtain the estimate

$$|E^*| \le \frac{(b-a)^5 M_4}{2880}.$$

Now if we divide $[a, b]$ into $2m$ equal intervals (each of length h) and compute the integral $\int_a^b f(t)\, dt$ by Simpson's rule, we are using in each of the m pairs of successive intervals the construction described above. The absolute error over each pair of subintervals is, therefore, at most $(2h)^5 M_4/2880$ and the total absolute error does not exceed

$$\frac{m(2h)^5 M_4}{2880} = (b-a)\frac{h^4 M_4}{180}.$$

This is the estimate stated in §1.7 of Chapter 7.

EXERCISES

Find a number y satisfying $0 = g^{(5)}(y) = f^{(4)}(y) - 2880E^*/(b-a)^5$ for the given function f and points a and b.

4. $f(x) = x^3 - x + 1$, $a = 0$, $b = 1$.
5. $f(x) = x^{3/2}$, $a = 1$, $b = 4$.
6. $f(x) = x^5$, $a = 0$, $b = 1$.

Problems

1. Give a geometric interpretation for the function $\phi(x)$ defined in §1.3, equation (4), and used in deriving the Mean Value Theorem from Rolle's Theorem.
2. Give a geometric interpretation of the Generalized Mean Value Theorem (Theorem 2 in §1.4). [Hint: Consider the curve in the u, v plane defined by $u = G(x)$, $v = F(x)$. You might want to postpone doing this problem until you read Chapter 11.]
3. Prove the so-called **Mean Value Theorem for Integrals:** If $f(x)$ is continuous and bounded in (a, b), then there is a number ξ such that $a < \xi < b$ and

$$\frac{1}{b-a} \int_a^b f(x)\, dx = f(\xi).$$

   Prove this without using the Mean Value Theorem for Derivatives.
4. Show that the Mean Value Theorem for Integrals stated in Problem 3 is a consequence of the fundamental theorem of calculus and of the Mean Value Theorem for Derivatives. [Hint: Consider the function $F(x) = \int_a^x f(t)\, dt$.]
5. Show that the Mean Value Theorem for Integrals may be false for a function that fails to be continuous at a single point.
6. Prove the **Generalized Mean Value Theorem for Integrals:** let $f(x)$ be as in Problem 3, and let the function $p(x)$ be continuous, bounded, and positive in (a, b). Then there is a number ξ, $a < \xi < b$, such that

$$\int_a^b p(x) f(x)\, dx = f(\xi) \int_a^b p(x)\, dx.$$

7. Does the theorem in Problem 6 remain valid for a negative $p(x)$? Does the theorem remain valid if $p(x)$ is nonnegative or nonpositive, but is permitted to be 0 at some points of the interval?
8. Prove the **Second Mean Value Theorem for Integrals:** for a bounded continuous function $f(x)$ and for a monotone continuously differentiable function $\phi(x)$, both defined in $[a, b]$, there is a point ξ between a and b such that

$$\int_a^b f(x)\phi(x)\, dx = \phi(a) \int_a^\xi f(x)\, dx + \phi(b) \int_\xi^b f(x)\, dx.$$

9. The remainder R in equation (11) in §1.5 depends on the point x_0. We make this dependence explicit and write

$$f(x) = f(x_0) + f'(x_0)(x - x_0) + R(x, x_0).$$

For a fixed x, this defines R as a function of x_0. Compute the derivative of R with respect to x_0. Compute the value of R for $x_0 = x$. Now apply the mean value theorem to R considered as a function of x_0 and obtain for R the representation

$$R = f''(\xi)(\xi - x)(x_0 - x), \qquad \xi \text{ a number between } x \text{ and } x_0.$$

Is the number ξ in this formula the same as the number ξ in equation (9) in §1.5? Give reasons for your answer.

10. (Continuation of 9). Derive the integral form of the remainder:

$$R = \int_{x_0}^{x} f''(t)(x - t)\, dt.$$

11. Show how one can obtain the formula for R given in the text [equation (9) in §1.5] and the formula for R derived in Problem 9 from the integral form of the remainder (compare Problem 10). [Hint: Use the Generalized Mean Value Theorem for Integrals, once with $p(t) = t - x$ and then with $p(t) = 1$.]

12. Use Rolle's Theorem to show that a polynomial of degree n can have at most n distinct real roots. (Hint: If a polynomial has k distinct real roots, what is the smallest number of real roots of its derivative?)

13. The remainder R in the Taylor formula

$$f(x) = \sum_{j=0}^{j=n} \frac{f^{(j)}(x_0)}{j!}(x - x_0)^j + R$$

is, for a fixed x, a function of x_0. Find the derivative of R with respect to x_0, and the value of R for $x_0 = x$. Using this and the Mean Value Theorem, obtain the so-called **Cauchy form** of the remainder:

$$R = \frac{f^{(n+1)}(\xi)}{n!}(x - \xi)^n(x - x_0),$$

where ξ is a number between x and x_0.

14. (Continuation of 13). Derive the **integral form** of the remainder:

$$R = \frac{1}{n!} \int_{x_0}^{x} f^{(n+1)}(t)(x - t)^n\, dt.$$

15. Show that one can obtain the Cauchy form of the remainder (compare Problem 13) from the integral form. Show also how one can obtain from the integral form the **Lagrange form** of the remainder, the form given in the text:

$$R = \frac{f^{(n+1)}(\xi)}{(n + 1)!}(x - x_0)^{n+1},$$

where ξ is a number between x and x_0, in general, not the same number as in the Cauchy form. (Hint: Compare Problem 11.)

16. Show that, if $x \mapsto f(x)$ is a function defined for x near x_0 (except perhaps for $x = x_0$), and if for every sequence of numbers $a_1, a_2, a_3, \cdots$ such that $a_j \neq x_0$ for all j, $\lim a_j = x_0$ and $f(a_j)$ is defined, we have $\lim f(a_j) = A$, then $\lim_{x \to x_0} f(x) = A$.

17. Formulate and prove analogues of the theorem in Problem 16 for one-sided limits, limits at infinity, and infinite limits.

18. Assuming Theorem 3 in §3.8, prove Theorem A in Chapter 5, §9.1, on monotone, bounded functions.

19. In Chapter 5, we defined improper integrals for infinite intervals of integration, but only for nonnegative integrands. Now let $f(x)$ be a piecewise continuous function defined for $x \geq 0$, and assume that it is bounded in every finite interval $[0, A]$. We say that the improper integral $\int_0^{+\infty} f(x)\, dx$ converges (or exists) if the finite limit

$$\lim_{A \to +\infty} \int_0^A f(x)\, dx$$

exists. We say that the above integral converges absolutely if the integral $\int_0^{+\infty} |f(x)|\, dx$ exists.

Prove that, if the integral $\int_0^{+\infty} f(x)\, dx$ converges absolutely, it converges, but the converse statement need not be true.

20. Construct a continuous function $f(x)$ defined for $x \geq 0$ such that the integral $\int_0^{+\infty} f(x)\, dx$ exists, and even converges absolutely, but it is not true that $\lim_{x \to +\infty} f(x) = 0$. (Compare Theorem 2 in §4.3.)

21. Does the integral $\int_0^{+\infty} [(\sin x)/x]\, dx$ converge? Does it converge absolutely?

22. Find a divergent series $a_1 + a_2 + a_3 + \cdots$ such that the series $(a_1 + a_2) + (a_3 + a_4) + (a_5 + a_6) + \cdots$ converges, but not absolutely.

23. Let $f(x)$ be an analytic function defined near $x = 0$. Assume that $f(0) = 0$ and $f'(0) \neq 0$. Then $f(x)$ has a constant sign near 0 (why?) so that there is an inverse function, call it $g(y)$ defined near $y = 0$, such that $f(g(y)) = y$, $g(f(x)) = x$. This function is known to be analytic. Assuming this result, show how the coefficients of the Taylor series for $g(y)$ about $y = 0$ can be computed from the Taylor coefficients of $f(x)$ at $x = 0$.

24. State a result similar to the one in Problem 23 for the case in which f is defined near $x = a$, and $f(a) = b$.

25. Use the method of Problem 23 to obtain the first 3 coefficients of the function $y \mapsto \tan y$ from the expansion of the arc tan function. Verify the result by Taylor's Theorem.

26. Use the method of Problem 23 to obtain the first few terms of the Taylor expansion of the function $g(y)$ defined near $y = 0$ and inverse to the function

$$x \mapsto \int_0^x \left(\frac{\sin t}{t} \right) dt.$$

9/Vectors

This chapter introduces the language of vectors and applies this language to the study of analytic geometry, in particular, to the elements of solid analytic geometry. We also discuss polar, cylindrical, and spherical coordinates. The chapter can be read immediately following Chapter 2.

The material on solid geometry will be used primarily in Chapters 12 and 13 on multidimensional calculus.

The appendixes contain brief introductions to inner products of vectors and to vectors in n-space.

Before continuing the development of calculus, we devote several sections to the elements of vector analysis, a subject that is interesting for its own sake and also useful in applications of calculus. We combine the discussion of vectors with the study of analytic geometry begun in Chapter 2.

Vector analysis was invented in the nineteenth century. Its foundations were laid by Hamilton; the present form is due largely to the American mathematician Gibbs. At a first glance, vectors seem merely a special way of writing the formulas of analytic geometry. Yet the development of vector analysis had a far-reaching effect on science and mathematics.

1.1 Directed segments

A **directed segment** is an ordered pair of points, P and Q. We denote such a directed segment by the symbol $\overrightarrow{PQ}$, and we usually draw it as an arrow (Fig. 9.1). The first point is called the **initial point,** the second the **terminal point.** The distance between the points P and Q is called the **length** or **magnitude** of $\overrightarrow{PQ}$; we shall denote it by $|\overrightarrow{PQ}|$. The length may be 0, since the initial and terminal points may coincide. Directed segments are also called **bound vectors.**

The meaning of the statement "two directed segments have the **same magnitude and direction,** though perhaps different initial points" is geometrically obvious (see Fig. 9.2, where five segments have the same magnitude and direction). A precise definition follows.

§1 Algebra of Vectors

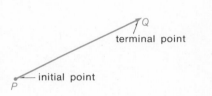

Fig. 9.1

terminal point

initial point

WILLIAM ROWAN HAMILTON (1805–1865). His life began with an amazingly rapid rise to intellectual eminence and academic success. As a boy of five, he read Hebrew, Greek, and Latin; at 17 he wrote his first important paper. Hamilton was 22 when he became professor at Trinity College, Dublin, 30 when he was knighted. But the second half of his life was marred by domestic and personal difficulties.

Hamilton invented quaternions, a generalization of complex numbers; vector calculus is an outgrowth of this theory. His most important achievements, however, are his earlier investigations in optics and mechanics.

JOSIAH WILLARD GIBBS (1839–1903) was a towering and somewhat lonely figure in American nineteenth-century science. Gibbs was educated at Yale, studied for several years in Europe, and then returned to Yale where he spent the rest of his life.

Gibbs' main work was in theoretical thermodynamics and its applications to chemistry, but he also made several contributions to mathematics.

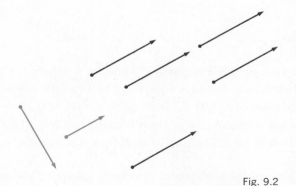

Fig. 9.2

Fig. 9.3

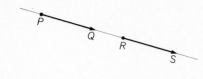

Fig. 9.3

Fig. 9.4

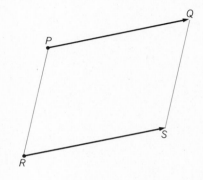

Fig. 9.5

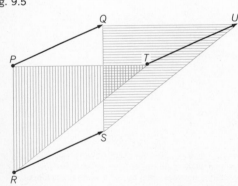

Fig. 9.6

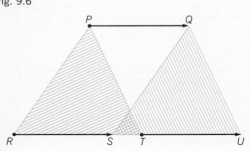

A directed segment $\overrightarrow{PQ}$ is said to have the same magnitude and direction as the directed segment $\overrightarrow{RS}$ if either:

1. $P = Q$ and $R = S$, or
2. $P \neq Q$, R, and S lie on the line through P and Q, $|\overrightarrow{RS}| = |\overrightarrow{PQ}|$, and S lies "to the left" or "to the right" of R according to whether Q lies to the left or right of P (see Fig. 9.3), or
3. P, Q, R, and S are four distinct points, no three of which lie on one line; the line through P, Q is parallel to that through R, S; the line through P, R is parallel to the line through Q, S (see Fig. 9.4).

If $\overrightarrow{PQ}$ has the same magnitude and direction as $\overrightarrow{RS}$, we write $\overrightarrow{PQ} \sim \overrightarrow{RS}$. We note the following properties of the relation $\sim$.

1. $\overrightarrow{PQ} \sim \overrightarrow{PQ}$ ($\sim$ is **reflexive**).
2. If $\overrightarrow{PQ} \sim \overrightarrow{RS}$, then $\overrightarrow{RS} \sim \overrightarrow{PQ}$ ($\sim$ is **symmetric**).
3. If $\overrightarrow{PQ} \sim \overrightarrow{RS}$ and $\overrightarrow{RS} \sim \overrightarrow{TU}$, then $\overrightarrow{PQ} \sim \overrightarrow{TU}$ ($\sim$ is **transitive**).

These three properties are abbreviated by saying that $\sim$ is an **equivalence relation.**

4. If $\overrightarrow{PQ} \sim \overrightarrow{RS}$, then $|\overrightarrow{PQ}| = |\overrightarrow{RS}|$.
5. If P, Q, and R are given, there is exactly one point S such that $\overrightarrow{PQ} \sim \overrightarrow{RS}$.

Only the third property needs a proof, and this proof would involve only very simple geometry and a somewhat tedious examination of cases. One case is represented in Fig. 9.5. We must prove that the lines through P, T, and through Q, U are parallel; this is so, because the shaded triangles are congruent. Another case is represented in Fig. 9.6; again the proof amounts to showing that the shaded triangles are congruent.

If $\overrightarrow{PQ} \sim \overrightarrow{RS}$, we shall also say, for the sake of brevity, that the two directed segments are **equivalent.**

1.2 **Vectors**

Consider now a directed segment $\overrightarrow{PQ}$ and let **a** denote the set (collection) of all directed segments equivalent to $\overrightarrow{PQ}$. One often calls **a** the **equivalence class** of $\overrightarrow{PQ}$. If $\overrightarrow{RS}$ belongs to **a**, that is, if $\overrightarrow{RS} \sim \overrightarrow{PQ}$, then every directed segment in **a** is equivalent to $\overrightarrow{RS}$, and all directed segments equivalent to $\overrightarrow{RS}$ belong to **a.** Hence **a** is also the equivalence class of $\overrightarrow{RS}$.

We call an equivalence class of directed segments a **vector** (or a **free**

vector, to distinguish it from a bound vector, which is just a directed segment). It is, of course, legitimate and standard to visualize free vectors as directed segments or arrows, keeping in mind that the initial point is not important, and that any two arrows with the same magnitude and direction are "equivalent."

Vectors will be denoted by boldface letters.

If **a** is a vector and $\overrightarrow{PQ}$ any directed segment that belongs to **a**, we say that $\overrightarrow{PQ}$ **represents** or **determines a.** Given a (free) vector **a** and a point P, by Property 5 above, there is exactly one point Q such that the bound vector $\overrightarrow{PQ}$ represents **a.** By "abuse of language," we sometimes write $\mathbf{a} = \overrightarrow{PQ}$, meaning that $\overrightarrow{PQ}$ represents **a.**

The **length** of a vector **a,** denoted by $|\mathbf{a}|$, is the length of any representative of **a.** This definition is unambiguous in view of Property 4 above.

EXERCISES

In Exercises 1 to 8, assume that a fixed Cartesian coordinate system is given on the plane. For each exercise, make a sketch illustrating the situation.

1. If P has coordinates $(0, 1)$ and Q has coordinates $(2, 1)$, find R such that $\overrightarrow{PQ} \sim \overrightarrow{QR}$.

2. If P has coordinates $(-1, 2)$ and Q has coordinates $(-1, -4)$, find R such that $\overrightarrow{PQ} \sim \overrightarrow{RP}$.

3. If P has coordinates $(2, 1)$, Q has coordinates $(2, -3)$, and R has coordinates $(-1, 2)$, find the coordinates of the point S such that $\overrightarrow{PQ} \sim \overrightarrow{RS}$.

▶ 4. If P has coordinates $(1, 1)$ and Q has coordinates $(4, 5)$, find the coordinates of the point R such that $\overrightarrow{PQ} \sim \overrightarrow{QR}$.

5. If P has coordinates $(2, -6)$ and Q has coordinates $(1, -5)$, and if $\overrightarrow{PQ} \sim \overrightarrow{RP}$, find the coordinates of R.

6. If P has coordinates $(\sqrt{2}, \sqrt{3})$, Q has coordinates $(\sqrt{2} + 4, \sqrt{3} - 2)$, and R has coordinates $(3, \frac{1}{2})$, find the coordinates of the point S such that $\overrightarrow{PQ} \sim \overrightarrow{RS}$.

7. If P lies on the line $y = ax$ and the vector represented by $\overrightarrow{OP}$ has length 20, find all possible coordinates of P. (Here, as usual, O is the origin.)

8. If P has coordinates $(2 \cos \frac{1}{9}\pi)$, $(2 \sin \frac{1}{9}\pi)$, and $\overrightarrow{OP} \sim \overrightarrow{PQ}$, find the coordinates of Q. (Again, O is the origin.)

1.3 Addition of vectors

We proceed to define the **sum** of two vectors **a** and **b.** The sum, denoted by $\mathbf{a} + \mathbf{b}$, will again be a vector.

Let **a** and **b** be given. We choose a point P and another point Q such that $\overrightarrow{PQ}$ represents **a.** Then we choose a point R such that $\overrightarrow{QR}$ repre-

Fig. 9.7

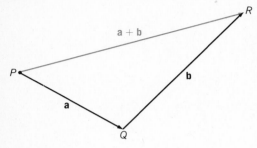

sents **b**. The directed segment $\overrightarrow{PR}$ represents a vector; this vector is called **a** + **b** (see Fig. 9.7).

We shall prove that vector addition satisfies the same formal rules as the addition of numbers, that is, the first five postulates of a field stated in Chapter 1, §2.2. In what follows, boldface letters represent vectors; ordinary letters represent numbers.

Theorem 1   VECTOR ADDITION OBEYS THE FOLLOWING POSTULATES.

1. FOR ANY TWO VECTORS **a** AND **b**, THERE IS A UNIQUE VECTOR CALLED THE **sum** OF **a** AND **b**, DENOTED BY **a** + **b**.
2. **a** + **b** = **b** + **a** FOR ALL **a** AND **b**.
3. (**a** + **b**) + **c** = **a** + (**b** + **c**) FOR ALL **a**, **b** AND **c**.
4. THERE IS A UNIQUE VECTOR CALLED THE **null vector**, DENOTED BY **0**, SUCH THAT **0** + **a** = **a** FOR ALL **a**.
5. FOR EVERY VECTOR **a**, THERE IS A UNIQUE VECTOR CALLED ITS **negative**, AND DENOTED BY (−**a**) SUCH THAT **a** + (−**a**) = **0**.

Let us first prove Postulate 1. The definition of addition was given above. It involved the choice of a point P. We must show that this choice does not affect the final result. In other words, we must verify that, if $\overrightarrow{PQ} \sim \overrightarrow{P_1Q_1}$ and $\overrightarrow{QR} \sim \overrightarrow{Q_1R_1}$, then $\overrightarrow{PR} \sim \overrightarrow{P_1R_1}$.

In the case shown in Fig. 9.8, when the points P, Q, and R do not lie on the same line, the proof is accomplished by noting that PQQ_1P_1 and QRR_1Q_1 are parallelograms, and therefore so is PRR_1P_1. In other cases the proof is similar, but simpler.

Fig. 9.8

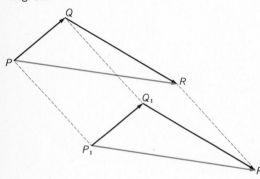

1.4 **Proofs of the addition rules**

In order to prove the **commutative law 2**, we shall give an alternative but equivalent definition of vector addition. The two vectors that have been added appear in this alternative definition in a symmetric way. This shows that **a** + **b** = **b** + **a**.

We represent **a** and **b** by directed segments $\overrightarrow{OP}$ and $\overrightarrow{OQ}$ with the same initial point and define **a** + **b** to be the vector represented by $\overrightarrow{OR}$, where:

(α) $R = Q$ if $P = O$ and $R = P$ if $Q = O$.

(β) If O, P, and Q are three distinct points on one line, then R also lies on this line. If P and Q lie on the same side of O, then R also lies on this side and $|\overrightarrow{OR}| = |\overrightarrow{OP}| + |\overrightarrow{OQ}|$ (see Fig. 9.9). If P and Q lie on opposite sides of O, then $|\overrightarrow{OR}| = ||\overrightarrow{OP}| - |\overrightarrow{OQ}||$ and R lies on the same

Fig. 9.9

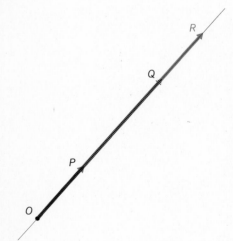

side of O as P if $|\overrightarrow{OP}| > |\overrightarrow{OQ}|$, on the same side as Q if $|\overrightarrow{OQ}| > |\overrightarrow{OP}|$, and $R = O$ if $|\overrightarrow{OP}| = |\overrightarrow{OQ}|$ (see Fig. 9.10).

(γ) If O, P, and Q do not lie on one line, then O, P, Q, and R are the vertices of a **parallelogram** (see Fig. 9.11).

The interesting case is case (γ). The addition rule just stated is sometimes called the **Parallelogram Law.**

It is obvious, and easily checked, that the new definition of addition is equivalent to the old one.

The proof of the **associative law 3** is immediate (see Fig. 9.12). Let $\mathbf{a}$, $\mathbf{b}$, and $\mathbf{c}$ be represented by $\overrightarrow{PQ}$, $\overrightarrow{QR}$, and $\overrightarrow{RS}$, respectively. Then $\overrightarrow{PR}$ represents $\mathbf{a} + \mathbf{b}$ and $\overrightarrow{PS}$ represents $(\mathbf{a} + \mathbf{b}) + \mathbf{c}$. But $\overrightarrow{QS}$ represents $\mathbf{b} + \mathbf{c}$ and therefore $\overrightarrow{PS}$ also represents $\mathbf{a} + (\mathbf{b} + \mathbf{c})$. (Note that a proof of the associative law **3** based on the Parallelogram Law would be more cumbersome.)

In view of **3**, expressions like $\mathbf{a} + \mathbf{b} + \mathbf{c}$ make unambiguous sense; similarly for sums of more vectors.

Let $\mathbf{0}$ denote the vector represented by $\overrightarrow{PP}$. If $\mathbf{a}$ is represented by $\overrightarrow{PQ}$, then $\mathbf{0} + \mathbf{a}$ is also represented by $\overrightarrow{PQ}$, hence equal to $\mathbf{a}$. If $P_1 \neq P$, and $\overrightarrow{P_1Q_1}$ represents $\mathbf{a}$, then the vector represented by $\overrightarrow{PQ_1}$ is not $\mathbf{a}$. This proves **4**. We call $\mathbf{0}$ the **null vector**; its length is 0.

If $\mathbf{a}$ is represented by $\overrightarrow{PQ}$, let $-\mathbf{a}$ be the vector represented by $\overrightarrow{QP}$. Then $\mathbf{a} + (-\mathbf{a})$ is represented by $\overrightarrow{PP}$, and hence equals $\mathbf{0}$. Also, if $\mathbf{x}$ is represented by $\overrightarrow{QX}$, $X \neq P$, then $\mathbf{a} + \mathbf{x}$ is represented by $\overrightarrow{PX}$ and is not $\mathbf{0}$. This shows that the rule **5** on negatives holds.

We write $\mathbf{a} - \mathbf{b}$ for $\mathbf{a} + (-\mathbf{b})$, as with numbers.

There is an interesting interpretation of $\mathbf{a} - \mathbf{b}$ by a parallelogram. Represent $\mathbf{a}$ and $\mathbf{b}$ by directed segments $\overrightarrow{OP}$ and $\overrightarrow{OQ}$ with the same initial point, and assume that the three points O, P, and Q do not lie on the same line. Let R be the fourth vertex of a parallelogram O, P, R,

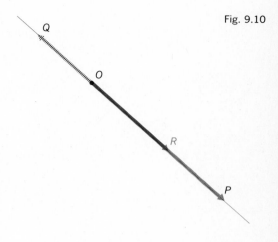

Fig. 9.10

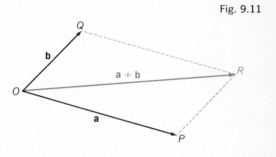

Fig. 9.11

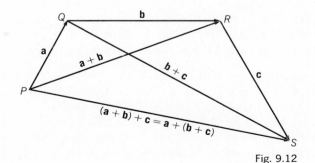

Fig. 9.12

Fig. 9.13

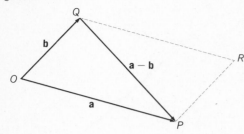

and Q. Then, as we know, the segment $\overrightarrow{OR}$ represents $\mathbf{a} + \mathbf{b}$. We claim that the segment $\overrightarrow{QP}$ represents $\mathbf{a} - \mathbf{b}$. Indeed, the segment $\overrightarrow{OP}$, which represents $\mathbf{a}$, also represents $\mathbf{b} + $ (the vector represented by $\overrightarrow{QP}$). (See Fig. 9.13.) Thus one diagonal of the parallelogram "spanned by two vectors" represents their sum, and the other diagonal represents their difference.

1.5 Product of a number and vector

We define next the **product** of a number α and a vector $\mathbf{a}$. The product, denoted by $\alpha\mathbf{a}$, is again a vector.

If $\mathbf{a} \neq \mathbf{0}$ and $\alpha > 0$, we choose a point P, a directed segment $\overrightarrow{PQ}$ representing $\mathbf{a}$, and a point R such that (a) R lies on the line through P and Q, (b) R lies on the same side of P as Q, and (c) $|\overrightarrow{PR}| = \alpha|\overrightarrow{PQ}|$. Then $\alpha\mathbf{a}$ is the vector represented by $\overrightarrow{PR}$. (See Fig. 9.14.) If $\mathbf{a} \neq \mathbf{0}$ and $\alpha < 0$, we define $\alpha\mathbf{a} = -((-\alpha)\mathbf{a})$. Finally, we set

Fig. 9.14

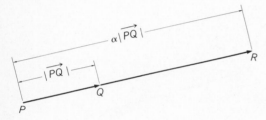

(1) $0\mathbf{a} = \mathbf{0}$ for all vectors $\mathbf{a}$, $\alpha\mathbf{0} = \mathbf{0}$ for all numbers α.

In other words, $\alpha\mathbf{a}$ is obtained from $\mathbf{a}$ by preserving the direction of the vector if $\alpha > 0$ (and reversing it if $\alpha < 0$) and by multiplying the length of the vector by $|\alpha|$. The definition implies that

(2)   if $\alpha\mathbf{a} = \mathbf{0}$, then either $\alpha = 0$ or $\mathbf{a} = \mathbf{0}$.

Theorem 2   THE FOLLOWING POSTULATES HOLD FOR THE MULTIPLICA-TION OF VECTORS BY NUMBERS

> **6.** FOR ANY NUMBER α AND ANY VECTOR $\mathbf{a}$, THERE IS A UNIQUE VECTOR CALLED THE **product** OF α AND $\mathbf{a}$. IT IS DENOTED BY $\alpha\mathbf{a}$.
> **7.** $1\mathbf{a} = \mathbf{a}$ FOR ALL $\mathbf{a}$.
> **8.** $\alpha(\beta\mathbf{a}) = (\alpha\beta)\mathbf{a}$ FOR ALL α, β AND $\mathbf{a}$.
> **9.** $(\alpha + \beta)\mathbf{a} = \alpha\mathbf{a} + \beta\mathbf{a}$ FOR ALL α, β AND $\mathbf{a}$.
> **10.** $\alpha(\mathbf{a} + \mathbf{b}) = \alpha\mathbf{a} + \alpha\mathbf{b}$ FOR ALL α, $\mathbf{a}$ AND $\mathbf{b}$.

Before proving these rules, we make two remarks. It makes no sense to ask whether $\mathbf{a}\alpha = \alpha\mathbf{a}$, since $\mathbf{a}\alpha$ has not been defined. Also the two distributive laws, **9** and **10**, express different properties.

The rules **1** to **10** formulated above are called postulates of a **vector space**. In talking about vectors, numbers are often called **scalars**. The vector $\alpha\mathbf{a}$ is called a **scalar multiple** of the vector $\mathbf{a}$.

1.6 **Proofs of the multiplication rules**

We proceed to prove the rules **6** to **10**. The product $\alpha\mathbf{a}$ has been defined above. To show that the definition is legitimate and that **6** holds, we must show that the choice of a point P used in the definition does not affect the result. It suffices to show the following. Let P, $Q \neq P$, and R be three points on a line, Q and R lying on the same side of P; and let P_1, $Q_1 \neq P_1$, and R_1 be three points on a line, Q_1 and R_1 lying on the same side of P. If $\overrightarrow{PQ} \sim \overrightarrow{P_1Q_1}$ and $|\overrightarrow{PR}|/|\overrightarrow{PQ}| = |\overrightarrow{P_1R_1}|/|\overrightarrow{P_1Q_1}|$, then $\overrightarrow{PR} \sim \overrightarrow{P_1R_1}$.

We do not carry out the tedious discussion of all possible cases. A typical case is shown in Fig. 9.15. We know that segment $\overrightarrow{PQ}$ is parallel to $\overrightarrow{P_1Q_1}$ and segment $\overrightarrow{PP_1}$ is parallel to $\overrightarrow{QQ_1}$. We must show that $\overrightarrow{PP_1}$ is parallel to $\overrightarrow{RR_1}$. This is so, since the distance from P to R equals the distance from P_1 to R_1, by hypothesis.

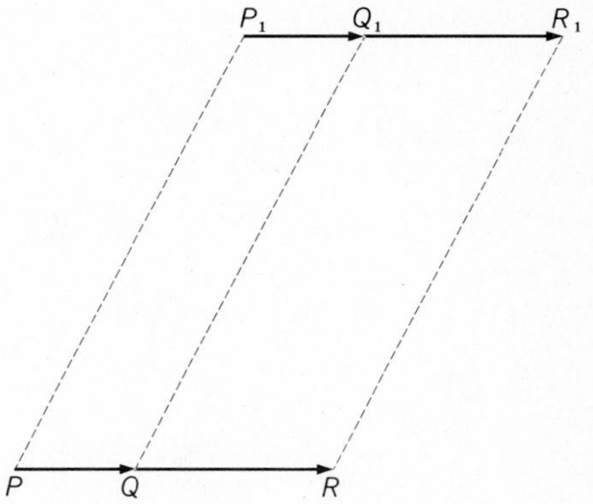

The validity of **7**, **8**, and **9** is obvious if $\mathbf{a} = \mathbf{0}$. If $\mathbf{a} \neq \mathbf{0}$, let $\overrightarrow{OP}$ be a segment representing $\mathbf{a}$. We make the line through O and P into a number line by making O the origin and P the point 1. For every number λ, let Q_λ be the point on the line representing the number λ. Then $\overrightarrow{OQ_\lambda}$ represents the vector $\lambda\mathbf{a}$, as is seen from our definition. The validity of rules **7**, **8**, and **9** is now obvious.

Note that, in view of the **associative law 8**, an expression like $\lambda\mu\,\mathbf{a}$ makes unambiguous sense.

We shall prove the **distributive law 10** for the case when $\alpha > 0$, $\mathbf{a} \neq \mathbf{0}$, and $\mathbf{b} \neq \mathbf{0}$, $\mathbf{a}$ and $\mathbf{b}$ not scalar multiples of each other. (The remaining cases are easier and are left to the reader.)

Let $\overrightarrow{OP}$ and $\overrightarrow{PQ}$ represent $\mathbf{a}$ and $\mathbf{b}$, respectively, so that $\overrightarrow{OQ}$ represents $\mathbf{a} + \mathbf{b}$ (see Fig. 9.16; in drawing it we assumed that $\alpha > 1$). Choose the points R and S so that $\overrightarrow{OR}$ represents $\alpha\mathbf{a}$ and $\overrightarrow{OS}$ represents $\alpha(\mathbf{a} + \mathbf{b})$. Then $|\overrightarrow{OR}|/|\overrightarrow{OP}| = |\overrightarrow{OS}|/|\overrightarrow{OQ}| = \alpha$ so that, by a familiar geometric theorem, $\overrightarrow{RS}$ is parallel to $\overrightarrow{PQ}$. Since the two triangles OPQ and

Fig. 9.15

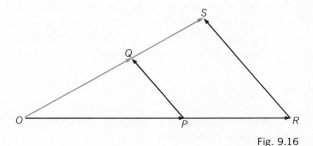

Fig. 9.16

ORS are similar, we conclude that $|\overrightarrow{RS}| = \alpha|\overrightarrow{PQ}|$. Therefore $\overrightarrow{RS}$ represents $\alpha\mathbf{b}$ and $\overrightarrow{OS}$ represents $\alpha\mathbf{a} + \alpha\mathbf{b}$. But $\overrightarrow{OS}$ represents $\alpha(\mathbf{a} + \mathbf{b})$. We see that **10** holds.

EXERCISES

In Exercises 9 to 15, assume that a fixed Cartesian coordinate system is given on the plane. For each exercise, make a sketch illustrating the situation.

9. Suppose $A = (1, 2)$, $B = (2, 3)$, $C = (-1, 2)$, and $D = (0, -1)$. Find P if the vector represented by $\overrightarrow{AP}$ equals the vector represented by $\overrightarrow{AB}$ + the vector represented by $\overrightarrow{CD}$.

10. Suppose $A = (0, 1)$, $B = (1, 3)$, $C = (1, 0)$, and $D = (0, -2)$. Find the coordinates of P if the vector represented by $\overrightarrow{DP}$ equals the vector represented by $\overrightarrow{AB}$ + the vector represented by $\overrightarrow{CD}$.

11. Suppose $A = (1, -1)$, $B = (2, -1)$, $C = (-1, 1)$, and $D = (0, 4)$. Find the coordinates of P if the vector represented by $\overrightarrow{BP}$ equals the vector represented by $\overrightarrow{AB}$ + the vector represented by $\overrightarrow{CD}$.

12. Suppose $A = (-2, 1)$, $B = (1, -3)$, and $C = (1, 0)$. Find the coordinates of P if the vector represented by $\overrightarrow{CP}$ equals $\frac{1}{6}$ times the vector represented by $\overrightarrow{AB}$.

13. Suppose $A = (\sqrt{2}, \sqrt{3})$, $B = (2 + \sqrt{2}, -1 + \sqrt{3})$, and $C = (\sqrt{7}, \sqrt{11})$. Find the coordinates of P if the vector represented by $\overrightarrow{CP}$ equals -3 times the vector represented by $\overrightarrow{AB}$.

14. Suppose $A = (2, -4)$, $B = (1, 1)$, $C = (-1, 2)$, and $D = (0, -2)$. Find the coordinates of P if the vector represented by $\overrightarrow{AP}$ equals 2 times the vector represented by $\overrightarrow{AB}$ + 3 times the vector represented by $\overrightarrow{CD}$.

15. Suppose $A = (0, 1)$, $B = (4, -1)$, $C = (3, 2)$, and $D = (2, 3)$. Find $|\mathbf{a}|$ if $\mathbf{a}$ equals 2 times the vector represented by $\overrightarrow{AB}$ − two times the vector represented by $\overrightarrow{CD}$.

1.7 Vectors in the plane and vectors in space

We have tacitly assumed that all points, and thus all segments, considered lie in the same plane. If this is so, one deals with **vectors in the plane.**

But everything we have said thus far applies equally well to **vectors in space,** that is, to equivalence classes of directed segments that need not lie in one plane. In particular, the definitions of addition and multiplication, and Theorems 1 and 2, remain valid.

1.8 Bases and frames in the plane

In order to compute with vectors efficiently, we must express vectors by numbers. Here is the basic proposition.

Theorem 3   THE FOLLOWING HOLDS FOR VECTORS IN THE PLANE.

11_2. ONE CAN FIND TWO VECTORS, $\mathbf{e}_1$ AND $\mathbf{e}_2$, SUCH THAT EVERY VECTOR $\mathbf{a}$ CAN BE WRITTEN, UNIQUELY, IN THE FORM

(3)
$$\mathbf{a} = \alpha_1\mathbf{e}_1 + \alpha_2\mathbf{e}_2,$$

WHERE α_1 AND α_2 ARE NUMBERS.

This is the first time we use the hypothesis that all points considered lie in one plane; Postulate 11_2, added to 1 to 10, characterizes plane vectors. We call 1 to 11_2 the postulates of a **two-dimensional vector space**.

Proof.   We introduce a Cartesian coordinate system in the plane. Let O be the origin, E_1 the point with coordinates $(1, 0)$, and E_2 the point with coordinates $(0, 1)$. Let $\mathbf{e}_1$ and $\mathbf{e}_2$ be the vectors represented by $\overrightarrow{OE_1}$ and $\overrightarrow{OE_2}$, respectively.

Now let $\mathbf{a}$ be any vector. Represent it by the directed segment $\overrightarrow{OP}$, and let the point P have coordinates (α_1, α_2). Also, let P_1 be the point with the coordinates $(\alpha_1, 0)$ and P_2 the point with the coordinates $(0, \alpha_2)$. Let $\mathbf{a}_1$ be the vector represented by $\overrightarrow{OP_1}$ and $\mathbf{a}_2$ the vector represented by $\overrightarrow{P_1P}$. Then

$$\mathbf{a} = \mathbf{a}_1 + \mathbf{a}_2$$

by the definition of vector addition. Recalling the definition of vector multiplication, we see that

$$\mathbf{a}_1 = \alpha_1\mathbf{e}_1.$$

The vector $\mathbf{a}_2$ is represented by the segment $\overrightarrow{P_1P}$ and hence also by the equivalent vector $\overrightarrow{OP_2}$ (see Fig. 9.17). We conclude now that

$$\mathbf{a}_2 = \alpha_2\mathbf{e}_2.$$

Combining the three displayed equations, we see that (3) holds.

Reversing the argument we see that if (3) holds, then the vector $\mathbf{a}$ can be represented by the segment $\overrightarrow{OP}$, where P has coordinates (α_1, α_2). This shows that the numbers α_1 and α_2 are determined uniquely. Now Postulate 11_2 is completely proved.

An ordered pair of vectors $\{\mathbf{e}_1, \mathbf{e}_2\}$ is called a **basis** if every vector can be represented in the form (3), in exactly one way. It follows from the proof given above that if two vectors, $\mathbf{e}_1$ and $\mathbf{e}_2$, are represented by

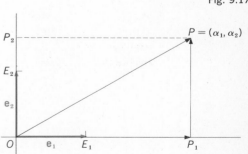

Fig. 9.17

mutually perpendicular segments of length 1, then $\{\mathbf{e}_1, \mathbf{e}_2\}$ is a basis. Such a basis is called orthonormal, or a **frame**. We use such bases almost exclusively in what follows. It is worthwhile to note, however, that there are also other bases. (We shall give examples later.)

Remark   In many books, two plane vectors that form a frame are denoted by $\mathbf{i}$ and $\mathbf{j}$.

1.9   Components

Choose a basis $\{\mathbf{e}_1, \mathbf{e}_2\}$; then every vector $\mathbf{a}$ can be written uniquely as $\mathbf{a} = a_1\mathbf{e}_1 + a_2\mathbf{e}_2$. The numbers a_1 and a_2 are called the **components** of $\mathbf{a}$, with respect to $\{\mathbf{e}_1, \mathbf{e}_2\}$. Thus we have a one-to-one correspondence between plane vectors and ordered pairs of numbers. We sometimes use an abbreviated notation and write

$$\mathbf{a} = (a_1, a_2), \qquad \text{instead of} \qquad \mathbf{a} = a_1\mathbf{e}_1 + a_2\mathbf{e}_2.$$

This can be done *only if the basis is kept fixed, since the components of* $\mathbf{a}$ *depend on the basis.* An exception is the null vector $\mathbf{0}$; its components are always $(0, 0)$.

Theorem 4   GIVEN A BASIS $\{\mathbf{e}_1, \mathbf{e}_2\}$, WE HAVE

$$(4) \qquad (a_1\mathbf{e}_1 + a_2\mathbf{e}_2) + (b_1\mathbf{e}_1 + b_2\mathbf{e}_2) = (a_1 + b_1)\mathbf{e} + (a_2 + b_2)\mathbf{e}_2$$

AND

$$(5) \qquad \lambda(a_1\mathbf{e}_1 + a_2\mathbf{e}_2) = (\lambda a_1)\mathbf{e}_1 + (\lambda a_2)\mathbf{e}_2.$$

The proof of Theorem 1 is self-evident, given rules 1 to 10. The left-hand side of (4) equals $(a_1\mathbf{e}_1 + b_1\mathbf{e}_1) + (a_2\mathbf{e}_2 + b_2\mathbf{e}_2)$, by the commutative law **2** and the associative law **3**, and equals $(a_1 + b_1)\mathbf{e}_1 + (a_2 + b_2)\mathbf{e}_2$ by the distributive law **9**. Statement (5) is essentially the distributive law **10**, together with the associative law **7** which permits us to write $(\lambda a_1)\mathbf{e}_1 + (\lambda a_2)\mathbf{e}_2$ instead of $\lambda(a_1\mathbf{e}_1) + \lambda(a_2\mathbf{e}_2)$. In the abbreviated notation, neglecting the mention of the basis, Theorem 2 reads

$$(4') \qquad (a_1, a_2) + (b_1, b_2) = (a_1 + a_2, b_1 + b_2),$$
$$(5') \qquad \lambda(a_1, a_2) = (\lambda a_1, \lambda a_2).$$

In words, *we add vectors by adding their components; we multiply a vector by a number by multiplying each component by the number.*

EXERCISES

In Exercises 16 to 27, assume that a fixed basis $\{e_1, e_2\}$ has been chosen for vectors in the plane, and the components of all vectors are given with respect to this basis.

16. If **a** has components $(2, -4)$ and **b** has components $(4, -3)$, what are the components of $\mathbf{a} + \mathbf{b}$?   Of $\mathbf{a} - \mathbf{b}$?   Of $2\mathbf{a}$?   Of $(-3)\mathbf{b}$?

17. If **a** has components $(3, \sqrt{2})$ and **b** has components $(1, 3\sqrt{2})$, what are the components of $-2\mathbf{a} + 3\mathbf{b}$?

18. If **a** has components $(2, 3)$ and $\mathbf{b} = \mathbf{e}_1 - \mathbf{e}_2 + \mathbf{a}$, what are the components of $3\mathbf{a} - 2\mathbf{b}$?

19. Compute the components of $(2, 3) + 4(6, 1) - 3(0, 2)$.

20. Compute the components of $2(0, 1) - 4(3, 2) + 3(3, 1)$.

21. Compute the components of $4(2, \frac{1}{2}) - (0, 2) + 2(-1, -\frac{1}{2})$.

▶ 22. Find α_1 and α_2 if $3(\alpha_1, \alpha_2) + 2(3, 0) = (9, 12)$.

23. Find α_1 and α_2 if $3(2, 0) - 2(\alpha_1, \alpha_2) = (8, 4)$.

24. Find α_1 and α_2 if $5(0, -1) + 2(\alpha_1, \alpha_2) - 3(1, -1) = (7, 3)$.

25. Suppose $\mathbf{a} = 3\mathbf{e}_1 - 2\mathbf{e}_2$, $\mathbf{b} = -\mathbf{e}_1 + 4\mathbf{e}_2$, $\mathbf{c} = \frac{1}{2}\mathbf{e}_1$, and $2\mathbf{a} - \mathbf{b} + 3\mathbf{v} = 2\mathbf{c}$. Find the components of **v**.

26. Suppose $\mathbf{a} + \mathbf{b} = \mathbf{e}_1$ and $\mathbf{a} - \mathbf{b} = \mathbf{e}_2$. Find the components of **a** and **b**.

27. Suppose $\mathbf{e}_1 = 2\mathbf{a} - 3\mathbf{b}$ and $\mathbf{e}_2 = \mathbf{a} + \mathbf{b}$. If $\mathbf{c} = 2\mathbf{e}_1 - 3\mathbf{e}_2$, can you find numbers α and β so that $\mathbf{c} = \alpha\mathbf{a} + \beta\mathbf{b}$?

1.10 Ordered pairs of numbers as a vector space

We may forget for a while about the geometric reasoning that led to relations (4′) and (5′), and instead look at them as definitions.

Theorem 5 ORDERED PAIRS OF NUMBERS, WITH ADDITION AND MULTIPLICATION DEFINED BY (4′) AND (5′), FORM A TWO-DIMENSIONAL VECTOR SPACE.

The theorem means that Postulates 1 to 11_2 hold if we agree that boldface letters represent ordered pairs of numbers.

First Proof.   Recall how we arrived at relations (4′) and (5′).

Second Proof.   Verify **1** to **10** from the properties of numbers, without reference to geometry. We carry this out for **4** and **8** only; the reader is asked to prove the remaining statements.

Statement **4** says: there are two numbers, x_1 and x_2, such that $(x_1, x_2) + (a_1, a_2) = (a_1, a_2)$ for any two numbers a_1, a_2. But, by our definition, $(x_1, x_2) + (a_1, a_2) = (x_1 + a_1, x_2 + a_2)$. For this to equal (a_1, a_2), we set $x_1 = 0$, $x_2 = 0$. No other pair (x_1, x_2) will do.

Statement **8** says that

$$\alpha(\beta(a_1, a_2)) = (\alpha\beta)(a_1, a_2).$$

By our definition, the left-hand side equals $\alpha(\beta a_1, \beta a_2) = (\alpha\beta a_1, \alpha\beta a_2)$. The right-hand side equals $(\alpha\beta a_1, \alpha\beta a_2)$. Thus the statement is true.

It remains to prove $\mathbf{11}_2$ for ordered pairs of numbers. We note that $(a_1, a_2) = (a_1, 0) + (0, a_2)$ and therefore

$$(a_1, a_2) = a_1(1, 0) + a_2(0, 1).$$

Set
$$\mathbf{e}_1 = (1, 0), \qquad \mathbf{e}_2 = (0, 1).$$

The vectors $\mathbf{e}_1$ and $\mathbf{e}_2$ have the properties required by $\mathbf{11}_2$.

EXERCISES

28. Consider the collection of ordered pairs of numbers with addition and multiplication defined by (4′) and (5′). Verify Statement **3** directly from the properties of numbers.
29. Repeat Exercise 28 for Statement **5**.
30. Repeat Exercise 28 for Statement **9**.
31. Repeat Exercise 28 for Statement **10**.
32. Suppose that we again consider the collection of ordered pairs of numbers with addition defined by (4′), but now we define multiplication by $\lambda(a_1, a_2) = (\lambda a_1, a_2)$. Does this system form a vector space? (Hint: It is unnecessary to reconsider the validity of Postulates **1** to **5**, since these postulates concern only the addition.)

1.11 General bases. Independence

We now give an example of a nonorthonormal basis. Let $\{\mathbf{e}_1, \mathbf{e}_2\}$ be a frame, and set $\mathbf{f}_1 = 2\mathbf{e}_1$, $\mathbf{f}_2 = \mathbf{e}_1 + \mathbf{e}_2$, or, in abbreviated notation,

$$\mathbf{f}_1 = (2, 0), \qquad \mathbf{f}_2 = (1, 1)$$

(see Fig. 9.18). Every ordered pair of numbers can be written as a number times $\mathbf{f}_1$ plus a number times $\mathbf{f}_2$, and in one way only. Indeed, let the numbers (x_1, x_2) be given. We want to write

$$(x_1, x_2) = \alpha_1(2, 0) + \alpha_2(1, 1).$$

This is the same as

$$(x_1, x_2) = (2\alpha_1 + \alpha_2, \alpha_2)$$

or
$$x_1 = 2\alpha_1 + \alpha_2, \qquad x_2 = \alpha_2.$$

Fig. 9.18

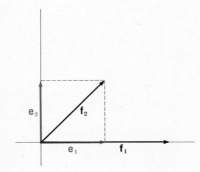

These conditions can be satisfied by setting $\alpha_2 = x_2$, $\alpha_1 = (x_1 - x_2)/2$, and in no other way.

We conclude that $\{\mathbf{f}_1, \mathbf{f}_2\}$ is a basis. A glance at Fig. 9.18 shows that it is not a frame. This example illustrates a general proposition.

Two vectors are called **independent** if they can be represented by directed segments $\overrightarrow{OP}$ and $\overrightarrow{OQ}$, such that the points O, P, and Q are not collinear. *Two vectors form a basis for plane vectors if and only if they are independent.* We shall make no use of this theorem; therefore we omit the proof.

EXERCISES

In Exercises 33 to 35, it is assumed that a fixed frame $\{\mathbf{e}_1, \mathbf{e}_2\}$ has been chosen. All components are with respect to this frame.

33. Verify that the vectors $(3, 4)$ and $(2, 6)$ are independent and form a basis.
34. Verify that the vectors $(3, 4)$ and $(-6, -8)$ are not independent and do not form a basis.
35. Find all numbers u such that the vectors $(-1, 3)$ and $(2, u)$ are dependent (that is, not independent).
36. Given a point P in the plane, can you ever find a point Q so that the vector determined by $\overrightarrow{PP}$ and the vector determined by $\overrightarrow{PQ}$ are independent?
37. Suppose $\mathbf{a}$ and $\mathbf{b}$ are independent vectors and α is a nonzero number. Are $\mathbf{a}$ and $\alpha\mathbf{b}$ always independent, or sometimes independent, or never independent? Justify your answer.
38. Suppose $\mathbf{a}$ and $\mathbf{b}$ are independent vectors. Are $\mathbf{a}$ and $\mathbf{a} + \mathbf{b}$ always independent, sometimes independent, or never independent? Justify your answer.

1.12 Bases and frames in space

The preceding considerations can be repeated, with obvious changes, for vectors in space.

Theorem 6 THE FOLLOWING HOLDS FOR VECTORS IN SPACE.

$\mathbf{11}_3$. ONE CAN FIND THREE VECTORS, $\mathbf{e}_1$, $\mathbf{e}_2$, AND $\mathbf{e}_3$, SUCH THAT EVERY VECTOR $\mathbf{a}$ CAN BE WRITTEN, UNIQUELY, IN THE FORM

$$(6) \qquad \mathbf{a} = \alpha_1\mathbf{e}_1 + \alpha_2\mathbf{e}_2 + \alpha_3\mathbf{e}_3,$$

WHERE α_1, α_2, AND α_3 ARE NUMBERS.

An ordered triple $\{\mathbf{e}_1, \mathbf{e}_2, \mathbf{e}_3\}$ of vectors is called a **basis** for vectors in space if it has the property described in Property $\mathbf{11}_3$.

Proof of $\mathbf{11}_3$. Introduce a Cartesian coordinate system in space. Let O be the origin, and let E_1, E_2, and E_3 be the points with coordinates

Fig. 9.19

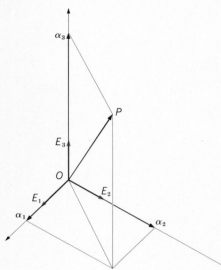

$(1, 0, 0)$, $(0, 1, 0)$, and $(0, 0, 1)$, respectively. Let $\mathbf{e}_1$, $\mathbf{e}_2$, and $\mathbf{e}_3$ be the vectors represented by $\overrightarrow{OE_1}$, $\overrightarrow{OE_2}$, and $\overrightarrow{OE_3}$. See Fig. 9.19. One sees, just as in the proof of Theorem 3 that $\{\mathbf{e}_1, \mathbf{e}_2, \mathbf{e}_3\}$ is a basis. More precisely, if $\mathbf{a}$ is a vector represented by $\overrightarrow{OP}$ and P has coordinates $(\alpha_1, \alpha_2, \alpha_3)$, then $\mathbf{a} = \alpha_1\mathbf{e}_1 + \alpha_2\mathbf{e}_2 + \alpha_3\mathbf{e}_3$, and there is no other way of writing $\mathbf{a}$ as a sum of products of numbers by the vectors $\mathbf{e}_1$, $\mathbf{e}_2$, and $\mathbf{e}_3$.

A basis of vectors obtained in this way from a Cartesian coordinate system is called a **frame in space.**

Three vectors are called **independent** if they can be represented by three directed segments with a common initial point that do not lie in one plane. *Three vectors $\{\mathbf{e}_1, \mathbf{e}_2, \mathbf{e}_3\}$ form a basis if and only if they are independent.* We shall not use this theorem; therefore we omit the proof.

If we choose a fixed basis $\{\mathbf{e}_1, \mathbf{e}_2, \mathbf{e}_3\}$, every vector $\mathbf{a}$ in space is represented uniquely by an ordered triple of numbers $(\alpha_1, \alpha_2, \alpha_3)$, the **components of a,** such that

$$\mathbf{a} = \alpha_1\mathbf{e}_1 + \alpha_2\mathbf{e}_2 + \alpha_3\mathbf{e}_3.$$

To add two vectors, we add the components by the rule

$$(\alpha_1, \alpha_2, \alpha_3) + (\beta_1, \beta_2, \beta_3) = (\alpha_1 + \beta_1, \alpha_2 + \beta_2, \alpha_3 + \beta_3).$$

To multiply a vector by a number, we multiply the components by the rule

$$\lambda(\alpha_1, \alpha_2, \alpha_3) = (\lambda\alpha_1, \lambda\alpha_2, \lambda\alpha_3).$$

The proof is the same as for plane vectors.

Postulates **1** to **10** of §§1.3 and 1.5, together with **11₃**, are the postulates of a **three-dimensional vector space.**

Remark   In many books, the vectors forming a frame are denoted not by $\mathbf{e}_1$, $\mathbf{e}_2$, $\mathbf{e}_3$ but by $\mathbf{i}$, $\mathbf{j}$, $\mathbf{k}$.

EXERCISES

39. Suppose $\mathbf{a} = \mathbf{b} + \mathbf{c}$. Are the vectors $\mathbf{a}$, $\mathbf{b}$, and $\mathbf{c}$ ever independent?
40. Suppose $\mathbf{a} + \mathbf{b} + \mathbf{c} = 0$. Are the vectors $\mathbf{a}$, $\mathbf{b}$, and $\mathbf{c}$ ever independent?
41. Suppose $\mathbf{a} = 2\mathbf{b}$. Let $\mathbf{c}$ be any vector. Can the vectors $\mathbf{a}$, $\mathbf{b}$, and $\mathbf{c}$ be independent?

In Exercises 42 to 52, assume that a fixed basis $\{\mathbf{e}_1, \mathbf{e}_2, \mathbf{e}_3\}$ has been chosen for vectors in space and the components of all vectors are given with respect to this basis.

42. If **a** has components $(-1, 2, 3)$ and **b** has components $(3, -3, 0)$, what are the components of $2(\mathbf{a} + \mathbf{b}) - 3(2\mathbf{a} - \mathbf{b})$?

43. If **a** has components $(2, 0, 1)$ and **b** has components $(-1, 4, -1)$, what are the components of $2\mathbf{a} - 3(\mathbf{b} + \mathbf{e}_1) + 2(\mathbf{a} - \mathbf{e}_3)$?

44. Compute the components of $2(1, -1, 3) + 3(2, 1, 4) - 3(1, 0, -1) + 2(0, 0, -3)$.

45. Compute the components of $\frac{1}{2}(\sqrt{3}, 0, -1) - 2(1, -1, 2) + \sqrt{3}(\frac{1}{2}, 0, 0)$.

46. Find α_1, α_2, and α_3 if $2(\alpha_1, \alpha_2, \alpha_3) + 3(\alpha_1, 0, \alpha_3) = (10, -6, -3)$.

47. Suppose $\mathbf{a} + \mathbf{b} - 2\mathbf{c} = \mathbf{e}_1$, $2\mathbf{a} - \mathbf{b} + \mathbf{c} = \mathbf{e}_2$, and $2\mathbf{a} - \mathbf{b} + 3\mathbf{c} = \mathbf{e}_3$. Find the components of **a**, **b**, and **c**.

48. Suppose $\mathbf{a} + \mathbf{b} + \mathbf{c} = \mathbf{e}_1$, $\mathbf{a} - \mathbf{b} + \mathbf{c} = \mathbf{e}_2$, and $3\mathbf{a} - \mathbf{b} - 2\mathbf{c} = \mathbf{e}_3$. Find the components of **a**, **b**, and **c**.

49. Verify that the vectors $(1, 1, 1)$, $(1, 1, 0)$, and $(1, 0, 0)$ are independent and form a basis.

50. Verify that the vectors $(1, 1, 1)$, $(1, 1, 0)$, and $(2, 2, 1)$ are not independent, and do not form a basis.

51. Suppose $\mathbf{v} = \mathbf{e}_1 + \mathbf{e}_2 + \alpha\mathbf{e}_3$. For what values of α is $\{\mathbf{e}_1, \mathbf{e}_2, \mathbf{v}\}$ a basis?

► 52. Suppose $\mathbf{v} = \alpha\mathbf{e}_1 + (1 - \alpha)\mathbf{e}_3$. For what values of α is $\{2\mathbf{e}_1, \mathbf{e}_2, \mathbf{v}\}$ a basis?

2.1 Position vector of a point

Let O be a fixed point; we usually think of it as the origin of a Cartesian coordinate system. Let P be any other point. The (free) vector represented by the directed segment $\overrightarrow{OP}$ is called the **position vector** of P (with respect to O).

If O is kept fixed, we denote the position vector of a point P simply by **P**. Clearly, every vector is, for a fixed O, the position vector of some point, and distinct points have distinct position vectors. The null vector **0** is the position vector of O. In applying vectors to geometry, we always assume that an origin O has been chosen.

Suppose that a point O and a frame $\{\mathbf{e}_1, \mathbf{e}_2\}$ have been chosen. This determines a Cartesian coordinate system in the plane, namely, the system in which the point with position vector $\mathbf{e}_1$ has coordinates $(1, 0)$ and the point with the position vector $\mathbf{e}_2$ has coordinates $(0, 1)$. If P is any point, its position vector can be written as

$$\mathbf{P} = x_1\mathbf{e}_1 + x_2\mathbf{e}_2.$$

The numbers (x_1, x_2) are the coordinates of P in the coordinate system $(O, \mathbf{e}_1, \mathbf{e}_2)$.

We obtain a Cartesian coordinate system if we start with a frame. If we start with a basis $\{\mathbf{e}_1, \mathbf{e}_2\}$ that is not a frame, we arrive at a more general coordinate system. We do not use such systems in this book.

Fig. 9.20

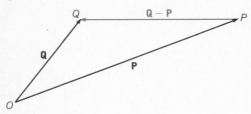

▶ **Examples**   *1.* Let P and Q be two points; the vector determined by $\overrightarrow{PQ}$ is $\mathbf{Q} - \mathbf{P}$ (so that $|\overrightarrow{PQ}| = |\mathbf{Q} - \mathbf{P}|$). The proof is illustrated in Fig. 9.20.

2. If P and Q have coordinates (p_1, p_2) and (q_1, q_2), respectively (in some coordinate system $\{O, \mathbf{e}_1, \mathbf{e}_2\}$), then the vector $\mathbf{a}$ represented by $\overrightarrow{PQ}$ has components $q_1 - p_1$ and $q_2 - p_2$ (with respect to the basis $\{\mathbf{e}_1, \mathbf{e}_2\}$).

Proof.   The components of the position vectors $\mathbf{P}$ and $\mathbf{Q}$ are (p_1, p_2) and (q_1, q_2), by definition. Since $\mathbf{a} = \mathbf{Q} - \mathbf{P}$, by Example 1, the assertion follows.

3. Let the points P, Q, R, and S have coordinates (p_1, p_2), (q_1, q_2), (r_1, r_2), and (s_1, s_2), respectively. Then $\overrightarrow{PQ} \sim \overrightarrow{RS}$ if and only if $q_1 - p_1 = s_1 - r_1$ and $q_2 - p_2 = s_2 - r_2$.

Fig. 9.21

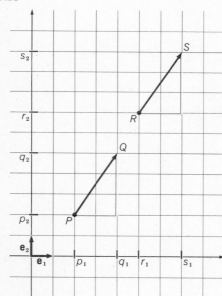

Proof.   By Example 1, we know that $\overrightarrow{PQ}$ represents the vector $\mathbf{Q} - \mathbf{P}$ and $\overrightarrow{RS}$ represents the vector $\mathbf{S} - \mathbf{R}$. Two directed segments are equivalent if and only if they represent the same vector (this is simply a restatement of the definition of vectors; see §1.2). Hence $\overrightarrow{PQ} \sim \overrightarrow{RS}$ means that $\mathbf{Q} - \mathbf{P} = \mathbf{S} - \mathbf{R}$. By Example 2, the components of $\mathbf{Q} - \mathbf{P}$ are $(q_1 - p_1, q_2 - p_2)$, those of $\mathbf{S} - \mathbf{R}$ are $(s_1 - r_1, s_2 - r_2)$. Therefore $\mathbf{Q} - \mathbf{P} = \mathbf{S} - \mathbf{R}$ if and only if $q_1 - p_1 = s_1 - r_1$ and $q_2 - p_2 = s_2 - r_2$.

The situation is illustrated in Fig. 9.21. (We could have carried out the proof earlier, using only the definition of $\sim$, but this would have taken longer.)◀

2.2 Dividing a segment in a given ratio

Many applications of vectors to geometry make use of the following result.

Theorem 1   LET P AND Q BE TWO DISTINCT POINTS. A POINT X LIES ON THE LINE THROUGH P AND Q IF AND ONLY IF THERE ARE NUMBERS α, β SUCH THAT THE POSITION VECTORS $\mathbf{P}$, $\mathbf{Q}$, $\mathbf{X}$ SATISFY

$$(1) \qquad \mathbf{X} = \alpha\mathbf{P} + \beta\mathbf{Q} \qquad \alpha + \beta = 1.$$

IF SO, THE POINT X DIVIDES THE SEGMENT $\overrightarrow{PQ}$ IN THE RATIO β/α.

Fig. 9.22

(Let us clarify the meaning of the last statement. The points P and Q divide the line through them in three intervals shown in Fig. 9.22. The point X on the line is said to divide $\overrightarrow{PQ}$ in the ratio β/α if (i) X lies in I,

II, or III according to whether $\beta < 0$, $0 < \beta < 1$, or $\beta > 1$ and (ii) we have $|\overrightarrow{PX}|/|\overrightarrow{QX}| = |\beta/\alpha|$. Here α and β are such that $\alpha + \beta = 1$.)

Proof. It follows from the definition of multiplication (see §1.5) that a point X lies on the line through P and Q if and only if the vector represented by $\overrightarrow{PX}$ is a scalar multiple of the vector represented by $\overrightarrow{PQ}$. In view of Example 1 in §2.1, this condition reads: there is a number β such that

$$\mathbf{X} - \mathbf{P} = \beta(\mathbf{Q} - \mathbf{P})$$

or (using the rules for computing implied by **1** to **10**)

$$\mathbf{X} = \mathbf{P} + \beta(\mathbf{Q} - \mathbf{P}) = \mathbf{P} + \beta\mathbf{Q} - \beta\mathbf{P} = \mathbf{P} - \beta\mathbf{P} + \beta\mathbf{Q}$$
$$= (1 - \beta)\mathbf{P} + \beta\mathbf{Q} = \alpha\mathbf{P} + \beta\mathbf{Q},$$

where we set $\alpha = 1 - \beta$, so that $\alpha + \beta = 1$. Thus (1) holds.

Assume now that (1) holds. We choose P as the origin of a coordinate system, and we choose the length $|\overrightarrow{PQ}|$ as a unit of length. Then we may choose the vector $\mathbf{Q}$ as the first vector of a frame. Thus we have $\mathbf{P} = \mathbf{0}$, $\mathbf{Q} = \mathbf{e}_1$ and $\mathbf{X} = \beta\mathbf{e}_1$. In other words, P, Q, X lie on the x axis and have x coordinates 0, 1, and β, respectively. Hence $|\overrightarrow{PX}| = |\beta|$, $|\overrightarrow{QX}| = |1 - \beta| = |\alpha|$. It is now clear that X divides the segment $\overrightarrow{PQ}$ in the ratio β/α.

We note some consequences of this theorem.

Corollary 1. *Let m and n be two numbers, not both zero. The point that divides the segment PQ in the ratio n/m has the position vector*

$$\frac{m}{m + n}\mathbf{P} + \frac{n}{m + n}\mathbf{Q}.$$

This follows from the theorem by setting $\alpha = m/(m + n)$ and $\beta = n/(m + n)$. Note that $n/m = \beta/\alpha$ unless $\alpha = m = 0$.

Corollary 2. *The midpoint of the segment PQ has the position vector*

$$\frac{1}{2}\mathbf{P} + \frac{1}{2}\mathbf{Q}.$$

Proof. Set $m = n = 1$ in the preceding formula.

We remark that if P has coordinates (p_1, p_2) and Q has coordinates

(q_1, q_2), then the point that divides the segment PQ in the ratio n/m has the coordinates

$$\frac{mp_1 + nq_1}{n + m}, \qquad \frac{mp_2 + nq_2}{n + m}.$$

The midpoint of the segment has coordinates

$$\frac{p_1 + q_1}{2}, \qquad \frac{p_2 + q_2}{2}.$$

EXERCISES

In Exercises 1 to 8, assume that a fixed Cartesian coordinate system with origin O is given on the plane. For each point P, **P** denotes the vector represented by $\overrightarrow{OP}$.

1. Find X, if X is the midpoint of the line segment joining $(2, 1)$ to $(-3, 5)$.
2. Find the midpoint of the line segment joining the point $(\frac{1}{2}, 3)$ to the point $(\frac{3}{2}, -1)$.

In Exercises 3 to 7, find the coordinates of X and express **X** in terms of **P** and **Q** as in §2.2.

3. P has coordinates $(2, 3)$, Q has coordinates $(-1, 9)$, X lies on the line through P and Q between P and Q, and $|\overrightarrow{PX}| = \frac{2}{3}|\overrightarrow{PQ}|$.
▶ 4. P has coordinates $(2, 0)$, Q has coordinates $(3, 2)$, X lies on the line through P and Q, P lies between Q and X, and $|\overrightarrow{PX}| = \frac{1}{4}|\overrightarrow{PQ}|$.
5. P has coordinates $(-1, 2)$, Q has coordinates $(2, 4)$, X lies on the line through P and Q, Q and X lie on the same side of P, and $|\overrightarrow{PX}| = 3|\overrightarrow{PQ}|$.
6. P has coordinates $(-2, 1)$, Q has coordinates $(2, 5)$, X lies on the line through P and Q, P and X lie on the same side of Q, and $|\overrightarrow{QX}| = \frac{1}{3}|\overrightarrow{QP}|$.
7. P has coordinates $(1, 0)$, Q has coordinates $(0, 4)$, X lies on the line through P and Q, P and X lie on opposite sides of Q, and $|\overrightarrow{PQ}| = 2|\overrightarrow{QX}|$.
8. If X divides $\overrightarrow{PQ}$ in the ratio $-2/5$, in what ratio does P divide $\overrightarrow{QX}$?

2.3 Proving theorems by vectors

We explain the use of Theorem 1 in proving geometric theorems by giving several samples.

▶ **Examples** **1.** *In a triangle, the three medians (segments joining a vertex to the midpoint of the opposite side) intersect at one point. This point divides each median in the ratio 2/1. (See Fig. 9.23.)*

Proof. Let the vertices of the triangle be A, B, and C. The midpoint A^* of the side opposite the vertex A has, by Corollary 2 to Theorem 1,

Fig. 9.23

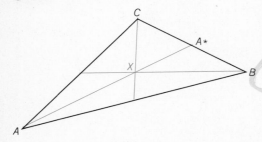

the position vector $\frac{1}{2}\mathbf{B} + \frac{1}{2}\mathbf{C}$. Let X be the point dividing the segment $\overrightarrow{AA^*}$ in the ratio $2/1$. Then, by Corollary 1 of Theorem 1, X has the position vector

$$\frac{1}{3}\mathbf{A} + \frac{2}{3}\mathbf{A^*} = \frac{1}{3}\mathbf{A} + \frac{2}{3}\left(\frac{1}{2}\mathbf{B} + \frac{1}{2}\mathbf{C}\right) = \frac{1}{3}(\mathbf{A} + \mathbf{B} + \mathbf{C}).$$

Since this expression is symmetric in A, B, C, we would have obtained the same result had we started with the median from vertex B or C. Hence the theorem is true.

One does not need vectors to give an analytic proof of this theorem, but without vectors the proof is longer. The reader is urged to try a proof without vectors.

2. *Four points A, B, C, D are the vertices of a parallelogram, in that order, if and only if the two diagonals $\overrightarrow{AC}$ and $\overrightarrow{BD}$ intersect at their midpoints.* (See Fig. 9.24.)

Proof. The midpoints of $\overrightarrow{AC}$ and $\overrightarrow{BD}$ have position vectors $\frac{1}{2}(\mathbf{A} + \mathbf{C})$ and $\frac{1}{2}(\mathbf{B} + \mathbf{D})$. The two midpoints coincide if and only if

$$\frac{1}{2}\mathbf{A} + \frac{1}{2}\mathbf{C} = \frac{1}{2}\mathbf{B} + \frac{1}{2}\mathbf{D},$$

that is, if and only if

$$\mathbf{A} - \mathbf{B} = \mathbf{D} - \mathbf{C}$$

But $\mathbf{A} - \mathbf{B}$ is the vector represented by $\overrightarrow{BA}$ and $\mathbf{D} - \mathbf{C}$ is the vector represented by $\overrightarrow{CD}$. The two free vectors are equal if and only if the directed segments are equivalent, that is, have the same magnitude and direction, and this is true if and only if $ABCD$ is a parallelogram. This proves the theorem.

Note that our formulation also covers certain *degenerate cases;* we did not assume that the four points were distinct or that they did not lie on the same line. The same applies to other geometric theorems considered here and later.

3. *Let A, B, C, and D be four points. The midpoints of the segments $\overrightarrow{AB}$, $\overrightarrow{BC}$, $\overrightarrow{CD}$, and $\overrightarrow{DA}$ are the vertices of a parallelogram, in that order.* (See Fig. 9.25.)

Proof. The position vectors of the midpoints in question are

$$\frac{1}{2}(\mathbf{A} + \mathbf{B}), \frac{1}{2}(\mathbf{B} + \mathbf{C}), \frac{1}{2}(\mathbf{C} + \mathbf{D}), \frac{1}{2}(\mathbf{A} + \mathbf{D}).$$

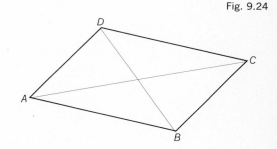

Fig. 9.24

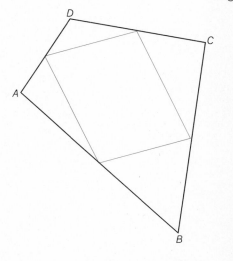

Fig. 9.25

The midpoints of the two diagonals of the figure formed by the four points have the same position vector, namely,

$$\frac{1}{2}\left(\frac{1}{2}(\mathbf{A} + \mathbf{B}) + \frac{1}{2}(\mathbf{C} + \mathbf{D})\right) = \frac{1}{2}\left(\frac{1}{2}(\mathbf{B} + \mathbf{C}) + \frac{1}{2}(\mathbf{A} + \mathbf{D})\right)$$

$$= \frac{1}{4}(\mathbf{A} + \mathbf{B} + \mathbf{C} + \mathbf{D})$$

Now apply the theorem in Example 2. ◄

EXERCISES

Use vector algebra to do the following exercises.

9. Let $A = (1, 4)$, $B = (-1, 2)$, and $C = (3, 3)$. Find the point where the three segments joining each vertex of the triangle ABC to the midpoint of the opposite side intersect.

10. Let A, B, C, D be the vertices of a parallelogram, in that order. Let P be the midpoint of $\overrightarrow{CD}$ and suppose that the line through A and P intersects the diagonal through B and D at X. Show that X divides $\overrightarrow{AP}$ in the ratio 2/1. In what ratio does X divide $\overrightarrow{DB}$? (Hint: Remember that $\mathbf{A} + \mathbf{C} = \mathbf{B} + \mathbf{D}$.)

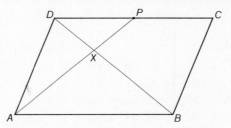

11. Let A, B, C, D be the vertices of a parallelogram, in that order. Let P be the midpoint of $\overrightarrow{CD}$ and Q the midpoint of $\overrightarrow{BC}$. Suppose the line through A and P intersects the line through D and Q at X. Show that X divides $\overrightarrow{AP}$ in the ratio $\frac{4}{5}/\frac{1}{5}$. In what ratio does X divide $\overrightarrow{DQ}$? (Hint: Remember that $\mathbf{A} + \mathbf{C} = \mathbf{B} + \mathbf{D}$.)

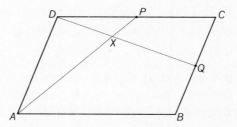

12. Let A, B, C, D be the vertices of a parallelogram, in that order. Let P be the midpoint of $\overrightarrow{CD}$ and Q the midpoint of $\overrightarrow{AD}$. Suppose the line through P and Q intersects the diagonal through B and D at X. Show that X divides $\overrightarrow{DB}$ in the ratio $\frac{1}{4}/\frac{3}{4}$. In what ratio does X divide $\overrightarrow{PQ}$? (Hint: Remember that $\mathbf{A} + \mathbf{C} = \mathbf{B} + \mathbf{D}$.)

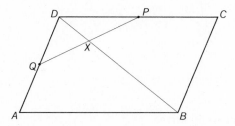

13. Let A, B, C be the vertices of a triangle. Let P divide $\overrightarrow{BC}$ in the ratio $\frac{2}{3}/\frac{1}{3}$ and let Q divide $\overrightarrow{AC}$ in the same ratio. Suppose the line through A and P intersects the line through B and Q at X. Show that X divides $\overrightarrow{AP}$ and $\overrightarrow{BQ}$ in the ratio $\frac{3}{4}/\frac{1}{4}$.

14. Let A, B, C be the vertices of a triangle. Let P divide $\overrightarrow{BC}$ in the ratio $(1 - \alpha) \div \alpha$, where $0 < \alpha < 1$ and let Q divide $\overrightarrow{AC}$ in the same ratio. Suppose the line through A and P intersects the line through B and Q at X. Show that X divides $\overrightarrow{AP}$ and $\overrightarrow{BQ}$ in the ratio $1/\alpha$.

15. Let A, B, C, D be the vertices of a parallelogram, in that order. Let P divide $\overrightarrow{AB}$ in the ratio 2/1, let Q divide $\overrightarrow{BC}$ in the same ratio, and similarly, let R divide $\overrightarrow{CD}$ and S divide $\overrightarrow{DA}$. Show that P, Q, R, S, in that order, are the vertices of a parallelogram.

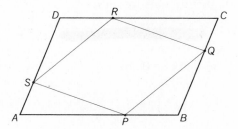

16. Let A, B, C, D, in that order, be the vertices of a quadrilateral and let P, Q, R, S be determined as in Exercise 15. Suppose P, Q, R, S, in that order, are the vertices of a parallelogram. Show that the given quadrilateral must also be a parallelogram.

17. Let A, B, C, D, in that order, be the vertices of a trapezoid with base AB. Since the line through A and B is parallel to the line through C and D, we can

write $\overrightarrow{DC} = \alpha\overrightarrow{AB}$, where $\alpha > 0$. Suppose the diagonals intersect at X. Show that X divides the diagonal $\overrightarrow{AC}$ in the ratio $1/\alpha$ and similarly for the diagonal $\overrightarrow{BD}$.

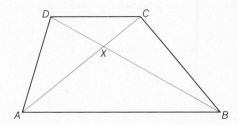

2.4 Position vectors in space

The concept of position vector is, of course, valid in space, too. If we choose a point O and a frame $\{\mathbf{e}_1, \mathbf{e}_2, \mathbf{e}_3\}$, this determines a Cartesian coordinate system in space. The coordinates of a point P, in this coordinate system, are the numbers (x_1, x_2, x_3) such that the position vector $\mathbf{P}$ of P (with respect to O) is given by $\mathbf{P} = x_1\mathbf{e}_1 + x_2\mathbf{e}_2 + x_3\mathbf{e}_3$.

We shall discuss applications of vectors to analytic geometry in space in §4.

§3 Polar Coordinates

3.1 Polar angle of a vector

Let $\{\mathbf{e}_1, \mathbf{e}_2\}$ be a frame and $\mathbf{a} = a_1\mathbf{e}_1 + a_2\mathbf{e}_2$ a vector. If it is represented by a segment $\overrightarrow{PQ}$ with $P = (p_1, p_2)$, $Q = (q_1, q_2)$, then $a_1 = q_1 - p_1$, $a_2 = q_2 - p_2$ (see Example 2 in §2.1 and Fig. 9.26). We compute the length of $|\mathbf{a}|$ of $\mathbf{a}$ by the distance formula:

$$(1) \qquad r = |\mathbf{a}| = \sqrt{a_1{}^2 + a_2{}^2}.$$

Assume that $\mathbf{a} \neq \mathbf{0}$, so that $r > 0$. Since

$$\left(\frac{a_1}{r}\right)^2 + \left(\frac{a_2}{r}\right)^2 = 1,$$

there is a number θ such that

$$(2) \qquad \cos\theta = \frac{a_1}{r}, \qquad \sin\theta = \frac{a_2}{r}.$$

(Compare Theorem 1 in Chapter 6, §1.8.) The number θ is not determined uniquely, but we can make it unique if we insist that $0 \leq \theta < 2\pi$;

Fig. 9.26

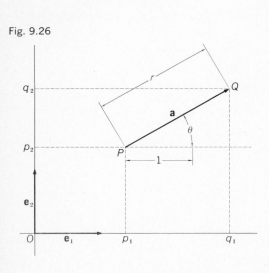

it is then the length of the circular arc shown in Fig. 9.26. One calls θ the **argument** (abbreviated arg) or **polar angle** of **a**, with respect to $\{e_1, e_2\}$, and writes

$$\theta = \arg \mathbf{a}.$$

If $r = 0$, θ is not defined.

The vector **a** is said to have magnitude r and direction θ. In terms of r and θ, we obtain the components a_1, a_2 of **a** as

(3) $$a_1 = r \cos \theta, \qquad a_2 = r \sin \theta.$$

The changing of θ by adding a multiple of 2π does not affect **a**. We note that, if $r = |\mathbf{a}|$ and $\theta = \arg \mathbf{a}$, then

$$\mathbf{a} = r \cos \theta \, \mathbf{e}_1 + r \sin \theta \, \mathbf{e}_2.$$

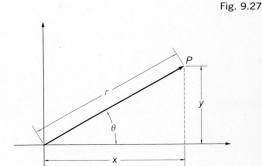

Fig. 9.27

3.2 Polar coordinates

The length r and the polar angle θ of the position vector of a point P are called the **polar coordinates** of P. The connection between polar coordinates (r, θ) and Cartesian coordinates (x, y) is given by the formulas

(4)
$$x = r \cos \theta, \qquad y = r \sin \theta,$$
$$r = \sqrt{x^2 + y^2}, \qquad \cos \theta = \frac{x}{r}, \qquad \sin \theta = \frac{y}{r}.$$

These can be read off Fig. 9.27.

It is convenient to consider *every* pair (r, θ) of numbers as a set of polar coordinates of the point with Cartesian coordinates (x, y) if equation (6) holds. This **convention**, which we shall follow, means that

1. *if (r, θ) are polar coordinates of P, so are $(r, \theta + 2\pi)$;*
2. *if (r, θ) are polar coordinates of P, so are $(-r, \theta + \pi)$;*
3. *for every θ, the numbers $(0, \theta)$ are polar coordinates of the origin.*

The grid formed by curves with polar coordinates $r = $ const (circles with center at 0) and the curves with polar coordinates $\theta = $ const (rays emanating from 0) are shown in Fig. 9.28.

▶*Examples 1.* The **distance formula** in polar coordinates: if two points have polar coordinates (r_1, θ_1) and (r_2, θ_2), respectively, then the distance d between them is

(5) $$d = \sqrt{r_1^2 - 2r_1 r_2 \cos(\theta_2 - \theta_1) + r_2^2}.$$

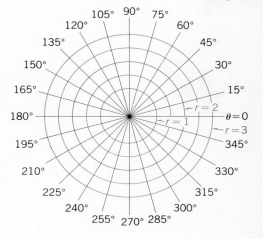

Fig. 9.28

Proof. The Cartesian coordinates of the two points are

$$(r_1 \cos \theta_1, r_1 \sin \theta_1) \qquad \text{and} \qquad (r_2 \cos \theta_2, r_2 \sin \theta_2).$$

The distance d is found using the Cartesian distance formula:

$$
\begin{aligned}
d^2 &= (r_1 \cos \theta_1 - r_2 \cos \theta_2)^2 + (r_1 \sin \theta_1 - r_2 \sin \theta_2)^2 \\
&= r_1{}^2 \cos^2 \theta_1 - 2r_1r_2 \cos \theta_1 \cos \theta_2 + r_2{}^2 \cos^2 \theta_2 + r_1{}^2 \sin^2 \theta_1 \\
&\qquad\qquad\qquad\qquad\qquad - 2r_1r_2 \sin \theta_1 \sin \theta_2 + r_2{}^2 \sin^2 \theta_2 \\
&= r_1{}^2 - 2r_1r_2 (\cos \theta_1 \cos \theta_2 + \sin \theta_1 \sin \theta_2) + r_2{}^2 \\
&= r_1{}^2 - 2r_1r_2 \cos (\theta_2 - \theta_1) + r_2{}^2,
\end{aligned}
$$

as claimed.

2. The **cosine law** in trigonometry. In a triangle with vertices A, B, C, we have

$$c^2 = a^2 + b^2 - 2ab \cos C.$$

(Here, as usual, C denotes the measure of the angle ACB, c denotes the length $|\overrightarrow{AB}|$, and so forth; see Fig. 9.29.)

Proof. Choose a Cartesian coordinate system with C as the origin, A on the positive x axis and B in the upper half-plane. Then A has polar coordinates $(b, 0)$ and B has polar coordinates (a, C). Now apply Example 1 with $r_1 = b$, $\theta_1 = 0$, $r_2 = a$, $\theta_2 = C$, $d = c$. (Or, apply directly the Cartesian distance formula, noting that A has Cartesian coordinates $(b, 0)$ and B has Cartesian coordinates $(a \cos C, a \sin C)$.)◄

Fig. 9.29

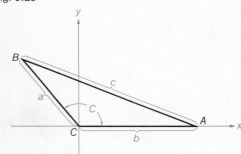

EXERCISES

In each of Exercises 1 to 6, the coordinates of a point in a Cartesian coordinate system are given. Find the corresponding polar coordinates.

1. $(0, \tfrac{1}{2})$.   4. $(-3, 3)$.
2. $(1, 1)$.   5. $(1, -\sqrt{3})$.
3. $(-\sqrt{3}, 1)$.   6. $(-2 \sin \tfrac{1}{9}\pi, 2 \cos \tfrac{1}{9}\pi)$.

In each of Exercises 7 to 12, the polar coordinates of a point are given. Find the coordinates of the point in the corresponding Cartesian coordinate system.

7. $(2, \pi/2)$.   9. $(3, 3\pi/2)$.   11. $(\sqrt{3}, 5\pi/6)$.
8. $(2, 3\pi/4)$.   10. $(\tfrac{1}{2}, \pi/4)$.   12. $(\sqrt{2}, 4\pi/3)$.

13. If P has polar coordinates $(1, \pi/2)$ and Q has polar coordinates $(2, \pi/6)$, find the distance between P and Q.
14. If P has polar coordinates $(3, \pi/3)$ and Q has polar coordinates $(1, \pi/12)$, find the distance between P and Q.

15. If P has polar coordinates $(\sqrt{3}, 7\pi/6)$ and Q has polar coordinates $(2, \pi/3)$, find the distance between P and Q.

16. For a triangle, like the one in Fig. 9.29, write a formula expressing a^2 in terms of b, c, and $\cos A$.

17. In a triangle, $a = 2$, $b = 3$, $c = 6$, find the angle C. (Use tables of the function cos.)

18. In a triangle, $a = 2.3$, $b = 4$, and $C = \pi/3$, find c.

19. In a triangle, $b = 1$, $c = 2$, and $A = 45°$, find a.

3.3 Equations in polar coordinates

Polar coordinates, as well as Cartesian coordinates, can be used to represent equations and functions by geometric figures as we shall show by examples. Since the equations considered may involve negative values of r, the convention stated above should be remembered.

▶ *Examples* *1.* Find an equation, in polar coordinates, of a circle with center at the point P (polar coordinates: r_0, θ_0) and radius R.

First Solution. We must write down the condition: the distance from a point with polar coordinates (r, θ) to P is R. By equation (5), this reads

$$r^2 - 2r_0 r \cos(\theta - \theta_0) + r_0{}^2 - R^2 = 0.$$

Second Solution. We use our knowledge of the equation of a circle in Cartesian coordinates (see Chapter 2, §3.1). The Cartesian coordinates of the center are $r_0 \cos \theta_0$ and $r_0 \sin \theta_0$ [by equation (4)]. Therefore the equation of our circle in Cartesian coordinates reads

$$(x - r_0 \cos \theta_0)^2 + (y - r_0 \sin \theta_0)^2 = R^2.$$

Set $x = r \cos \theta$, $y = r \sin \theta$ and simplify. This yields the same equation as before.

2. What is the equation, in polar coordinates, of the line through the origin of slope m?

Solution. The equation of this line in Cartesian coordinates reads $y = mx$. Using (4), we obtain the equation

$$r \cos \theta = mr \sin \theta.$$

If $r \neq 0$, we obtain $\cos \theta = m \sin \theta$ or $\tan \theta = m$. Thus we have the answer: $\tan \theta = m$ or $r = 0$. [This means: all points with $\tan \theta = m$ and the origin: $r = 0$.]

3. What points satisfy the relation: $\cos \theta = 0$ or $r = 0$ (r, θ polar coordinates)?

Answer. The points on the vertical line through the origin. The reader is asked to verify this.

4. What points satisfy the equation (in polar coordinates r, θ) $r = 2 \csc \theta$?

Solution. The equation reads $r = 2/\sin \theta$ or $r \sin \theta = 2$. Setting $x = r \cos \theta$, $y = r \sin \theta$, we rewrite the equation as $y = 2$. This represents a horizontal line.

5. The equation of a curve, in Cartesian coordinates, reads $y^2 = 4x$. Find the equation of this curve in polar coordinates.

Answer. Using (4), we obtain the equation $r^2 \cos^2 \theta = 4r \sin \theta$ or, dividing by $r \sin \theta$, $r \cos \theta \tan \theta = 4$; we may want to add: and the point $r = 0$. The curve is, of course, a parabola. ◄

I skipped

3.4 Sketching curves

In plotting a curve $r = f(\theta)$ in polar coordinates, it is convenient to make a table of $f(\theta)$ for various values of θ. It is also convenient to obtain or prepare a "polar graph paper" with a grid like the one in Fig. 9.28.

One should remember the following points.

If $f(\theta)$ is an even function, the curve is symmetric with respect to the x axis. If $f(\theta)$ is an odd function, the curve is symmetric with respect to the y axis. If $f(\theta)$ satisfies the relation $f(\theta + \pi) = f(\theta)$ for all θ, then the curve is symmetric with respect to the origin.

We prove only the first statement. The other two are proved similarly. Suppose that f is an even function. Let (x_0, y_0) be a point on the curve, x_0 and y_0 being Cartesian coordinates. Then there is a number θ_0 such that $x_0 = f(\theta_0) \cos \theta_0$, $y_0 = f(\theta_0) \sin \theta_0$. Since $f(-\theta_0) \cos(-\theta_0) = x_0$ and $f(-\theta_0) \sin(-\theta_0) = -y_0$, the point $(x_0, -y_0)$ also lies on our curve. But the points (x_0, y_0) and $(x_0, -y_0)$ are symmetrical with respect to the x axis.

If there is a number θ_0 such that $f(\theta_0) = 0$, then the curve passes through the origin. *In this case, it has at the origin a tangent with slope $\tan \theta_0$ (a vertical tangent if θ_0 is an odd multiple of $\pi/2$).* We shall prove this in Chapter 11, §3.5, see equation (9). (It is assumed here that the function $f(\theta)$ has a continuous derivative.)

►*Examples   1.* $r = \cos \theta$.

The graph is a circle of radius $\frac{1}{2}$ with center at $x = \frac{1}{2}$, $y = 0$. This is seen, for instance, by going over to Cartesian coordinates: if $r = \cos \theta$, then $r^2 = r \cos \theta$ or $x^2 + y^2 = y$. The latter equation can be written (see Chapter 2, §3.1) as $(x - \frac{1}{2})^2 + y^2 = \frac{1}{4}$.

(We also note that the equation in Example 1, §3.3 becomes $r = \cos \theta$ if we set $\theta_0 = 0$, $R = r_0 = \frac{1}{2}$.)

The graph is shown in Fig. 9.30. Note that, for $\theta = 90°$, we have $r = 0$. The tangent at this point is vertical. Note also that every point of the circle corresponds to two values of θ such that $-180° < \theta \le 180°$. For instance, if $\theta = -45°$, then $r = 1/\sqrt{2}$, and if $\theta = 135°$ then $r = -1/\sqrt{2}$. These two values of θ yield the same point, the point with Cartesian coordinates $x = (1/\sqrt{2}) \cos(-45°) = (-1/\sqrt{2}) \cos 135° = \frac{1}{2}$, $y = (1/\sqrt{2}) \sin(-45°) = (-1/\sqrt{2}) \sin 135° = \frac{1}{2}$.

2. $r = \cos 2\theta$.

The graph, shown in Fig. 9.31 is called a **four-leaved rose.** Since $\cos 2\theta$ is even, the graph is symmetric about the x axis. Since $\cos 2(\theta + \pi) = \cos(2\theta + 2\pi) = \cos 2\theta$, the graph is symmetric about the origin. Hence the graph is also symmetric about the y axis. (The function $\cos 2\theta$ is not odd, but there is no contradiction. We said above that, if $f(\theta)$ is odd, then the graph $r = f(\theta)$ is symmetric about the y axis; we did *not* say that, if the graph is symmetric, then $f(\theta)$ is odd.)

3. $r = \cos 3\theta$.

The graph is called a **three-leaved rose;** see Fig. 9.32.

4. $r = Ae^{\alpha\phi}$.

The graph is shown in Fig. 9.33. The curve is called a **logarithmic spiral,** since its equation may be written in the form

$$\theta = \frac{1}{\alpha}\left(\log \frac{r}{A}\right)$$

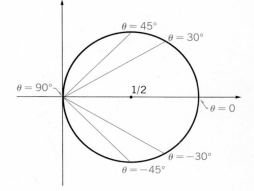

Fig. 9.30

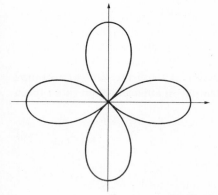

Fig. 9.31

Fig. 9.32

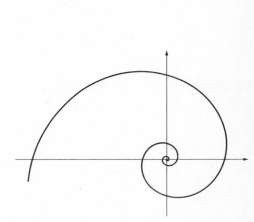

Fig. 9.33

Fig. 9.34

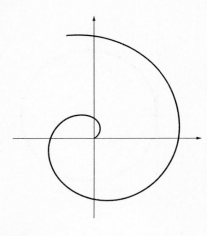

and since it spirals infinitely often both in approaching the origin and in receding from it.

5. $r = k\theta$.

This curve (shown in Fig. 9.34 for a positive value of k) is called the **spiral of Archimedes.** It enters the origin but spirals infinitely often as it recedes from the origin. ◄

EXERCISES

In each of Exercises 20 to 25, the equation of a curve is given with respect to Cartesian coordinates (x, y). Find an equation for the curve with respect to the corresponding polar coordinates (r, θ).

20. $3y - 2x = 8$.   23. $x^2 y^2 = x^2 + y^2$.

21. $(x - 1)^2 + (y - 2)^2 = 4$.   24. arc cos $(x/\sqrt{x^2 + y^2}) = \sqrt{x^2 + y^2}$.

► 22. $(x^2 + y^2)^{3/2} = x + y$.   25. $x^2 + y^2 - 2\sqrt{x^2 + y^2} - 3 = 0$.

In each of Exercises 26 to 31, the equation of a curve is given with respect to polar coordinates (r, θ). Find an equation for the curve with respect to the corresponding Cartesian coordinates (x, y).

26. $r = 1/(2 \cos \theta + 3 \sin \theta)$.   29. $r^2 \cos \theta \sin \theta = 1$.

27. $r = 4/(6 \cos \theta - 5 \sin \theta)$.   30. $r^2 = \sin \theta$.

28. $r^2 - 2r \cos \theta - 3 = 0$.   31. $r = \cos \theta + \sin \theta$.

In each of Exercises 32 to 46, sketch the curve whose equation is given with respect to the polar coordinates (r, θ). Make a table listing easily computable values of r and θ to aid in your sketch. Try to exploit any symmetry in the equation.

32. $r = \sin 2\theta$.   40. $r^2 = \sin \theta$.

33. $r = \cos 4\theta$.   41. $r^2 = \cos (\theta/2)$.

34. $r = 1 + \cos \theta$.   42. $r = 2/\cos \theta$.

35. $r = \cos (\theta/2)$.   43. $r = 1/(\sin \theta - \cos \theta)$.

36. $r = 1 - \sin \theta$.   44. $r = \cos \theta \cos 2\theta$.

37. $r = 2 - \sin \theta$.   45. $r = \cos (2\theta/3)$.

38. $r = 1 - 2 \sin \theta$.   46. $r = 3 + 2 \cos \theta$.

39. $r^2 = 4 \cos 2\theta$.

§4   Lines and Planes in Space

In this section, we use the language of the vectors to introduce the basic concepts of solid analytic geometry. This discussion is continued in the following section. Questions relating to angles are relegated to an appendix (§6).

Throughout this section, we assume that a fixed Cartesian coordinate system in space defined by an origin O and a frame $\{e_1, e_2, e_3\}$ has been chosen. Every point P has a position vector $\mathbf{P}$; the components of $\mathbf{P}$ are the coordinates of the point P. The coordinates of a point will be denoted sometimes by (x_1, x_2, x_3) and sometimes by (x, y, z).

4.1 Division formula

Let P and Q be two distinct points in space. Then every point X on the line through P and Q has a position vector of the form

$$(1) \qquad \mathbf{X} = \alpha\mathbf{P} + \beta\mathbf{Q}, \qquad \alpha + \beta = 1,$$

and X divides the segment $\overrightarrow{PQ}$ in the ratio β/α. The statement just made is Theorem 1 in §2.2. There we talked about plane vectors. But the proof goes through in space, without any changes. The corollaries of Theorem 1 in §2.2 are also valid in space. Thus if $m^2 + n^2 \neq 0$, then $[m/(m + n)]\mathbf{P} + [n/(m + n)]\mathbf{Q}$ is the position vector of the point that divides $\overrightarrow{PQ}$ in the ratio n/m; in particular $\frac{1}{2}\mathbf{P} + \frac{1}{2}\mathbf{Q}$ is the position vector of the midpoint of $\overrightarrow{PQ}$.

Suppose now that P has coordinates (p_1, p_2, p_3); then $\mathbf{P} = p_1\mathbf{e}_1 + p_2\mathbf{e}_2 + p_3\mathbf{e}_3$ (see §2.4). If Q has coordinates (q_1, q_2, q_3) and X the coordinates (x_1, x_2, x_3), then $\mathbf{Q} = q_1\mathbf{e}_1 + q_2\mathbf{e}_2 + q_3\mathbf{e}_3$ and $\mathbf{X} = x_1\mathbf{e}_1 + x_2\mathbf{e}_2 + x_3\mathbf{e}_3$. Hence (1) may be rewritten as

$$(2) \quad x_1 = \alpha p_1 + \beta q_1,\ x_2 = \alpha p_2 + \beta q_2,\ x_3 = \alpha p_3 + \beta q_3 \quad (\alpha + \beta = 1).$$

▶**Example** Do the three points $(1, 2, 3)$, $(4, -10, 8)$, and $(-7, 22, 9)$ lie on one line?

Solution. Applying (2) with $P = (1, 2, 3)$, $Q = (4, -10, 8)$ and $X = (-7, 22, 9)$ we see that the three points lie on one line if and only if there are numbers α and β such that

$$(3) \quad -7 = \alpha + 4\beta, \quad 22 = 2\alpha - 10\beta, \quad 9 = 3\alpha + 8\beta, \quad \alpha + \beta = 1.$$

The first and the last equations show that we must have $\alpha = 1$, $\beta = -2$. With these values, the second equation is true but the third is false. Hence the answer is no. ◀

EXERCISES

1. Let the points P and Q have coordinates $(1, -3, 0)$ and $(2, 5, -7)$. Find the midpoint of the segment $\overrightarrow{PQ}$.

▶ 2. Let P and Q be as in Exercise 1. Find the point that divides the segment $\overrightarrow{PQ}$ in the ratio $4/5$.

3. Given four points: $P = (1, 2, 3)$, $Q = (3, 2, 1)$, $R = (2, 2, 2)$ and $S = (2, 2, 3)$, which three lie on one line?

4. Find a number u such that $(1, 2, u)$ lies on the straight line passing through $(2, 0, 5)$ and $(0, 4, 7)$.

5. Find numbers x and y such that $(x, y, 0)$ lies on the straight line passing through $(4, 3, -1)$ and $(2, 8, 2)$.
6. Consider the line through the points $(1, 5, 4)$ and $(-2, 7, 9)$. Find the points at which this line intersects the coordinate planes $(x_1 = 0, x_2 = 0,$ and $x_3 = 0)$.
7. Find a point P on the coordinate plane $x_2 = 0$ and a point Q on the coordinate plane $x_3 = 0$, such that the point $(1, 2, 3)$ is the midpoint of the segment $\overrightarrow{PQ}$.

4.2 Parametric representation of a line

In the plane, a line is represented by a single linear equation satisfied by the Cartesian coordinates of its points. In space, we must represent a line by equations in a different way.

Theorem 1   LET P AND Q BE TWO DISTINCT POINTS. EVERY POINT X ON THE LINE THROUGH P AND Q HAS A POSITION VECTOR OF THE FORM

$$(4) \qquad \mathbf{X} = \mathbf{P} + t(\mathbf{Q} - \mathbf{P})$$

WHERE t IS A NUMBER. CONVERSELY, FOR EVERY t THE VECTOR $\mathbf{X}$ IS THE POSITION VECTOR OF A POINT ON THE LINE, AND DISTINCT VALUES OF t CORRESPOND TO DISTINCT POINTS.

This follows simply from relation (1) by setting $\beta = t$, $\alpha = 1 - t$. We have then $\alpha\mathbf{P} + \beta\mathbf{Q} = (1 - t)\mathbf{P} + t\mathbf{Q} = \mathbf{P} + t(\mathbf{Q} - P)$. The variable t occurring in (4) is called a **parameter,** and (4) is called a parametric representation of a line.

Let us rewrite (4) in terms of the coordinates (x_1, x_2, x_3) of X and the coordinates (p_1, p_2, p_3) of P and (q_1, q_2, q_3) of Q. We get

$$(5) \quad x_1 = p_1 + (q_1 - p_1)t, \; x_2 = p_2 + (q_2 - p_2)t, \; x_3 = p_3 + (q_3 - p_3)t.$$

One often denotes the coordinates of a point X not by (x_1, x_2, x_3) but by (x, y, z). In order to conform to this convention, let us denote the coordinates of P and Q by (x_1, y_1, z_1) and (x_2, y_2, z_2), respectively. Then relation (5) may be rewritten as

$$(5') \quad x = x_1 + (x_2 - x_1)t, \;\; y = y_1 + (y_2 - y_1)t, \;\; z = z_1 + (z_2 - z_1)t.$$

Theorem 2   LET α, β, γ, a, b, AND c BE 6 NUMBERS SUCH THAT $a^2 + b^2 + c^2 \neq 0$ (THIS MEANS THAT a, b, c ARE NOT ALL 0). THEN ALL POINTS (x, y, z) WITH

$$(6) \qquad x = \alpha + at, \qquad y = \beta + bt, \qquad z = \gamma + ct$$

LIE ON ONE LINE, AND ALL POINTS ON THIS LINE CAN BE SO REPRESENTED. EVERY LINE HAS A REPRESENTATION OF THE FORM (6).

Proof. Set $x_1 = \alpha$, $y_1 = \beta$, $z_1 = \gamma$, $x_2 = \alpha + a$, $y_2 = \beta + b$, $z_2 = \gamma + c$. Then (6) becomes identical with (5'). The condition $a^2 + b^2 + c^2 \neq 0$ insures that (x_1, y_1, z_1) and (x_2, y_2, z_2) are distinct points.

►*Examples* **1.** Find a parametric representation of the line through the points $(2, -1, 4)$ and $(3, 2, 6)$.

Solution. Using (5') with $(x_1, y_1, z_1) = (2, -1, 4)$ and $(x_2, y_2, z_2) = (3, 2, 6)$, we obtain the representation

$$(7) \qquad x = 2 + t, \qquad y = -1 + 3t, \qquad z = 4 + 2t.$$

2. At which points does the line in Example 1 intersect the coordinate planes $x = 0$, $y = 0$, and $z = 0$?

Answer. We see from (7) that $x = 0$ for $t = -2$. But for $t = -2$ we have, again by (7), $y = -7$, $z = 0$. Hence our line intersects the plane $x = 0$ at the point $(0, 7, 0)$. We note that this is also the intersection of the line (7) with the plane $z = 0$. In the same way, we see that $y = 0$ for $t = \frac{1}{3}$. For this value of t, relations (7) yield: $x = \frac{7}{3}$, $z = \frac{14}{3}$. Hence our line intersects the x, z plane (the plane $y = 0$) at $(\frac{7}{3}, 0, \frac{14}{3})$.

3. Does the line given by

$$(8) \qquad x = 1 + 2t, \qquad y = -4 + 6t, \qquad z = 2 + 4t$$

coincide with the line in Example 1?

Answer. First of all, in order to prevent misunderstandings, it is much better to denote the parameters in two representations by the different letters. Therefore we rewrite (8) in the form

$$(8') \qquad x = 1 + 2s, \qquad y = -4 + 6s, \qquad z = 2 + 4s.$$

In order to obtain two points of (8'), we choose two values of the parameter s, for instance, $s = 0$ and $s = 1$. This gives us the points $(1, -4, 2)$ and $(3, 2, 6)$. We ask whether these points lie on the first line, that is, whether their coordinates can be represented in the form (7).

Does there exist a t such that

$$1 = 2 + t, \qquad -4 = -1 + 3t, \qquad 2 = 4 + 2t?$$

The first equation gives $t = -1$; for this value of t the other two equations also hold. Hence $(1, -4, 2)$ lies on (7). In the same way, we see that $(3, 2, 6)$ is a point of (7), corresponding to the parameter value $t = 1$. Since the two lines have two points in common, they coincide. ◄

Remark   One can obtain, in a similar way, parametric representations for lines in the plane.

EXERCISES

In Exercises 8 to 13, find a parametric representation for the line passing through the indicated points.

8. $(3, 4, 5)$ and $(5, 4, 3)$.
9. $(-1, 2, 3)$ and $(0, 4, 5)$.
10. $(0, 1, 0)$ and $(1, 0, 0)$.

11. $(0, 12, 7)$ and $(-5, 0, 3)$.
12. $(-3, 2, 0)$ and $(12, -4, 0)$.
13. $(-1, 6, 8)$ and $(-1, -2, -2)$.

► 14. At which points does the line in Exercise 8 meet the coordinate planes?
15. At which points does the line in Exercise 11 meet the coordinate planes?
16. Do the lines in Exercises 10 and 11 coincide?
17. Do the lines in Exercises 9 and 14 coincide?
18. At which point does the line $x = 1 - t$, $y = 1 + t$, $z = 2 - 2t$ meet the plane $x = 2$?
19. Write down a parametric representation of the z axis.
20. Does the line through $(1, 3, 5)$ and $(1, 3, 6)$ intersect the z axis?
21. Find a number u such that the line $x = 1 - ut, y = 1 + ut, z = 2 - t$ coincides with the line in Exercise 18.

4.3 Equations of planes

The next theorem is the analogue of Theorem 1 in Chapter 2, §2.1.

Theorem 3   IF A, B, C, D ARE NUMBERS SUCH THAT $A^2 + B^2 + C^2 \neq 0$, THEN THE SET OF POINTS WHOSE COORDINATES SATISFY THE EQUATION

$$(9) \qquad\qquad Ax + By + Cz + D = 0$$

IS A PLANE. EVERY PLANE CAN BE SO REPRESENTED.

Remark   An equation of the form (9) is called a **linear equation** in x, y, z.

Proof.   Let equation (9) be given, and let σ denote the set of all points whose coordinates satisfy (9). We show first that (i) if σ contains two distinct points of a line l, it contains all points of l. Indeed, let (x_1, y_1, z_1) and (x_2, y_2, z_2) be distinct points of σ; then

$$Ax_1 + By_1 + Cz_1 + D = 0, \qquad Ax_2 + By_2 + Cz_2 + D = 0.$$

Let (x_3, y_3, z_3) be a point on l. Then (see §6.1) there are numbers α and β with $\alpha + \beta = 1$ and $x_3 = ax_1 + \beta x_2$, $y_3 = \alpha y_1 + \beta y_2$, $z_3 = \alpha z_1 + \beta z_2$. Thus

$$Ax_3 + By_3 + Cz_3 + D$$
$$= A(\alpha x_1 + \beta x_2) + B(\alpha y_1 + \beta y_2) + C(\alpha z_1 + \beta z_2) + (\alpha + \beta)D$$
$$= \alpha(Ax_1 + By_1 + Cz_1 + D) + \beta(Ax_2 + By_2 + Cz_2 + D) = 0$$

so that (x_3, y_3, z_3) belongs to σ.

Next we show that (ii) there are 3 points, not in one line, which belong to σ, and there are points that do not belong to σ. Indeed, one of the numbers A, B, C is not 0, by hypothesis. Assume, for the sake of definiteness, that $C \neq 0$; dividing both sides of (9) by C, we obtain the equivalent equation

$$(10) \qquad\qquad A'x + B'y + z + D' = 0$$

(with $A' = A/C$, $B' = B/C$, $D' = D/C$). The points $(0, 0, -D')$, $(1, 0, -A' - D')$ and $(0, 1, -B' - D')$ satisfy (10); these points do not lie on one line. On the other hand, the coordinates of $(0, 0, 1 - D')$ do not satisfy (10).

Properties (i) and (ii) identify σ as a plane.

We show now that any given plane σ can be represented by equation (9) for an appropriate choice of A, B, and C. We assume first that σ is neither the x, y plane nor parallel to this plane. Then σ intersects the x, y plane in a line l (see Fig. 9.35). Let the equation of l be $ax + by + c = 0$, with $a^2 + b^2 \neq 0$.

Also, let (x_1, y_1, z_1) be the coordinates of a point P on σ, which does not lie on l. Then $z_1 \neq 0$, for otherwise σ would contain l and a point P in the x, y plane not on l, and so σ would be the x, y plane, contrary to assumption. We consider now the equation

$$(11) \qquad\qquad ax + by + c - \frac{ax_1 + by_1 + c}{z_1}z = 0.$$

It is of the form (9) with $A = a$, $B = b$, $C = -(ax_1 + by_1 + c)/z_1$ and $D = c$. Hence (11) defines a plane, call it τ. Every point of l has coordinates $(x, y, 0)$ with $ax + by + c = 0$; these coordinates satisfy (11). Also, the coordinates (x_1, y_1, z_1) of P satisfy (11). Thus τ is a plane through the line l and a point P not on l. This determines τ uniquely. Hence τ is the plane σ with which we started.

If the given plane σ is parallel to the x, y plane (see Fig. 9.36) or coincides with the x, y plane, then all points P of σ have the same z coor-

Fig. 9.35

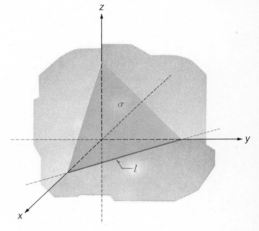

Fig. 9.36

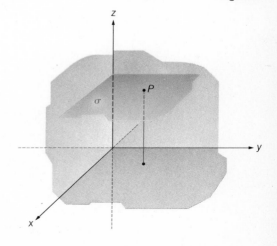

dinate, call it h. Hence σ is defined by the equation

$$z = h$$

which is of the form (9), with $A = B = 0$, $C = 1$, and $D = -h$. This completes the proof.

▶ *Examples* **1.** Find the equation of a plane that passes through the points $(5, 4, 1)$, $(4, -2, -3)$ and $(0, 6, 5)$.

Solution. The three points do not lie on one line, so that the plane will be uniquely determined. Its equation must be of the form (9) and, since the three points must lie on the plane, we should have

$$5A + 4B + C + D = 0,$$
$$4A - 2B - 3C + D = 0,$$
$$6B + 5C + D = 0.$$

There are three linear equations for four unknowns; A, B, C, and D. (This is a reflection of the fact that we can multiply all coefficients in (9) by a number $\alpha \neq 0$ and obtain an equivalent equation.) We try to express three unknowns in terms of the fourth. The last equation shows that

$$D = -6B - 5C.$$

Substituting this expression in the first two equations, we obtain

$$5A + 4B + C - 6B - 5C = 0, \qquad 4A - 2B - 3C - 6B - 5C = 0$$

or $\qquad\qquad 5A - 2B - 4C = 0, \qquad 4A - 8B - 8C = 0.$

The second equation here says that

$$A = 2B + 2C;$$

we substitute this in the first equation above and obtain

$$5(2B + 2C) - 2B - 4C = 0$$

or $\qquad\qquad\qquad B = -\dfrac{3}{4}C.$

Hence $\qquad\qquad A = 2\left(-\dfrac{3}{4}C\right) + 2C = \dfrac{1}{2}C$

and $\qquad\qquad D = -6\left(-\dfrac{3}{4}C\right) - 5C = -\dfrac{1}{2}C.$

Now we choose for C the value 4 and obtain $A = 2$, $B = -3$, and $D = -2$. The desired equation is

$$2x - 3y + 4z - 2 = 0.$$

One checks easily that the given points actually satisfy this equation.

2. If $a \neq 0$, $b \neq 0$, and $c \neq 0$, then the plane through the points $(a, 0, 0)$, $(0, b, 0)$, $(0, 0, c)$ has the equation

$$\frac{x}{a} + \frac{y}{b} + \frac{z}{c} = 1.$$

Proof. This is an equation of the form (9) satisfied by the coordinates of the points considered. The result should be compared with the intercept form of the equation of a line in the x, y plane (Chapter 2, §2.4). ◄

EXERCISES

In Exercises 22 to 26, find the equation of the plane that passes through the given points

22. $(2, -1, 3)$, $(4, 0, 5)$, $(2, 1, 7)$.   25. $(-1, 0, 0)$, $(0, 3, 0)$, $(0, 0, 2)$.
23. $(1, -4, 5)$, $(6, 7, 0)$, $(3, -2, 5)$.   26. $(0, 0, 0)$, $(2, 2, 2)$, $(3, 3, 1)$.
24. $(6, 1, 5)$, $(-6, 2, 7)$, $(1, 2, 3)$.

In Exercises 27 to 30, find the equation of the plane that passes through the line l and through the point P.

27. $P = (0, 0, 0)$, l: $x = 2 - t$, $y = 3 - t$, $z = 2t$.
► 28. $P = (4, -1, 0)$, l: $x = t$, $y = 2t$, $z = 3t$.
29. $P = (1, 2, 3)$, l: $x = 2 + t$, $y = 2 + 2t$, $z = -3t$.
30. Find a number u such that the plane $ux + y - z - 1 = 0$ passes through the point $(100, u, u)$.

4.4 Parallel planes

Theorem 4   LET

(12) $$Ax + By + Cz + D = 0$$

(13) $$A_1 x + B_1 y + C_1 z + D_1 = 0$$

BE THE EQUATIONS OF TWO DISTINCT PLANES. THE TWO PLANES ARE PARALLEL IF AND ONLY IF THERE IS A NUMBER $\alpha \neq 0$ SUCH THAT

(14) $$A_1 = \alpha A, \qquad B_1 = \alpha B, \qquad C_1 = \alpha C.$$

Proof. Assume that there is an $\alpha \neq 0$ satisfying (14). Dividing both sides of (13) by α, we write (13) in the equivalent form

(13′)
$$Ax + By + Cz + \frac{D_2}{\alpha} = 0.$$

Now $D_2/\alpha \neq D$, for otherwise (12) and (13′) would define the same plane. Therefore no point (x, y, z) can satisfy both (12) and (13′). The two planes have no point in common, that is, they are parallel.

Now suppose that there is no $\alpha \neq 0$ satisfying (14). We shall show that the two planes have points in common and are thus not parallel. If there is no $\alpha \neq 0$ with the property (14), then at least one of the numbers

(15)
$$AB_1 - A_1B, \qquad BC_1 - B_1C, \qquad AC_1 - A_1C$$

is different from 0. (Indeed, if all numbers (15) are 0, and, say $A \neq 0$, we set $\alpha = A_1/A$ and verify that $B_1 = \alpha B$, $C_1 = \alpha C$. Then $\alpha \neq 0$, since one of the numbers A_1, B_1, C_1 is not 0.) Assume for the sake of definiteness that

(16)
$$AB_1 - A_1B \neq 0.$$

We claim that, for every given number z, we can find numbers x and y such that the point (x, y, z) lies on both planes (12) and (13). Indeed, if we choose a value for z and consider (12) and (13) as a system of simultaneous equations for x and y, we can solve these equations. We multiply (12) by B_1 and (13) by B. We subtract the resulting equations and obtain

$$(AB_1 - A_1B)x + (CB_1 - C_1B)z + DB_1 - D_1B = 0.$$

In view of (16), this determines x, and y can be found by substituting this value of x into (12) and solving for y.

▶ *Example* Find a plane through the point $(2, -1, 4)$ that is parallel to the plane

$$5x - 7y + 8z - 6 = 0.$$

Solution. The coefficients of x, y, z in the equation of the desired plane must be of the form 5α, -7α, 8α, in view of Theorem 4. We may choose $\alpha = 1$ and write the desired equation as

$$5x - 7y + 8z + D = 0$$

where D is to be determined. Substituting $x = 2$, $y = -1$, and $z = 4$, we see that $(5 \cdot 2) + (-7)(-1) + (8 \cdot 4) + D = 0$. Hence $D = -49$ and the desired plane has the equation

$$5x - 7y + 8z - 49 = 0. \blacktriangleleft$$

4.5 Equations of lines

Two nonparallel planes, say

(17) $\quad 2x - 3y + 4z - 5 = 0 \quad$ and $\quad 6x - 9y + 5z - 1 = 0$

intersect in a line l. We can therefore describe the line l by the two equations (17): the line consists of all points (x, y, z) satisfying *both* equations.

In order to get a parametric representation of this line, we consider one of the three variables x, y, z as given and solve for the other two. We must decide which variable may be prescribed; this depends upon the numbers (15) that we used in the proof of Theorem 4. (Fig. 9.37 may help in remembering these expressions.) In our case, $A = 2$, $B = -3$, $C = 4$, $A_1 = 6$, $B_1 = -9$, $C_1 = 5$, and the numbers (15) are

$$2(-9) - (-3)6 = 0, \quad (-3)5 - 4(-9) \neq 0, \quad 2 \cdot 5 - 4 \cdot 6 \neq 0.$$

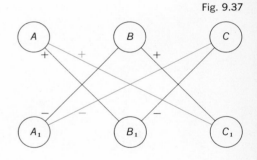

Fig. 9.37

This means that we can solve equations (17) for y and z in terms of x, and for x and z in terms of y. But we *cannot* solve for x and y in terms of z (as the reader is asked to verify). We give to x an arbitrary value, call it t and consider (17) as a system of two equations for y and z:

(18)
$$-3y + 4z = 5 - 2t$$
$$-9y + 5z = 1 - 6t.$$

Multiply the first equation by 5, the second by 4, and subtract. This yields

$$[(-3)5 - 4(-9)]y = 5(5 - 2t) - 4(1 - 6t)$$

so that $21y = 21 + 14t$ or

$$y = 1 + \frac{2}{3}t.$$

Substituting this into the first equation (18) and solving for z, we obtain

$$-3\left(1 + \frac{2}{3}t\right) + 4z = 5 - 2t$$

or
$$z = 2.$$

Thus the intersection of the planes (17) has a parametric representation

$$x = t, \qquad y = 1 + \frac{2}{3}t, \qquad z = 2.$$

If a line l is given by a parametric representation, it is possible to represent it by two linear equations, that is, as the intersection of two planes. This can be done in infinitely many ways, since there are infinitely many planes through a given line.

For instance, if a line l is given by

$$(19) \qquad x = 2 - 2t, \qquad y = -3 + 4t, \qquad z = 6 - t,$$

we can obtain linear equations satisfied by all points of l as follows. We "solve" each equation (19) for t, that is, we rewrite (19) in the form

$$t = \frac{2 - x}{2}, \qquad t = \frac{3 + y}{4}, \qquad t = 6 - z.$$

Now we "eliminate" t: if (x, y, z) is a point on the line l, we must have

$$\frac{2 - x}{2} = \frac{3 + y}{4} = 6 - z$$

or, multiplying all terms by 4,

$$(20) \qquad 4 - 2x = 3 + y = 24 - 4z.$$

These are called equations of the line (19) in *symmetric form*. We can now consider our line to be the intersection of the two planes

$$4 - 2x = 3 + y \qquad \text{and} \qquad 3 + y = 24 - 4z,$$

that is,

$$2x + y - 1 = 0 \qquad \text{and} \qquad y + 4z - 21 = 0.$$

We can also consider our line to be the intersection of the planes

$$4 - 2x = 24 - 4z \qquad \text{and} \qquad 3 + y = 24 - 4z,$$

that is,

$$x - 2z + 10 = 0 \quad \text{and} \quad y + 4z - 21 = 0,$$

and so forth.

▶ *Example* Represent the line $x = 2 + t$, $y = 2 - t$, $z = 2$ as the intersection of two planes.

Answer. The parameter t does not occur in the expression for z. We eliminate t from the expressions for x and for y and conclude that the line considered is the intersection of the planes $x + y = 4$ and $z = 2$. ◀

EXERCISES

In Exercises 31 to 33, find the plane passing through the given point and parallel to the given plane.

31. $(7, 4, 5)$, $2x - 3y - 6 = 0$.
32. $(4, 0, 6)$, $x + y + z = 0$.
33. $(1, 1, 1)$, $x - 20y + 7z - 5 = 0$.
34. Find a number a such that the plane $ax + 2ay + 10z - 2 = 0$ be parallel to the plane $x + 2y + 5z - 7 = 0$.
35. The planes $x - y = 2$ and $2x - ay + 7 = 0$ are parallel. Find a.
36. Complete the sentence: "The plane $Ax + By + Cz + D = 0$ is parallel to the y, z plane if and only if . . . ."

In Exercises 37 to 43, determine whether the given planes are parallel. If they are not, find a parametric representation for the intersection line.

37. $x = 0$, $y - z = 0$.
38. $x - y = 1$, $y - z = 1$.
39. $2x - 3y - 5 = 0$, $-4x + 6y - 4 = 0$.
40. $7x - 5y + 6z = 0$, $21x - 15y + 18z - 62 = 0$.
41. $x + y + 2z - 3 = 0$, $x + y - 2z + 3 = 0$.
▶ 42. $2x - 6y + 7z - 5 = 0$, $6x + y + 4z - 1 = 0$.
43. $x - y - z = 5$, $x - y - z = 5.001$.

In Exercises 44 to 50, find two linear equations representing the given line.

44. $x = t$, $y = -t$, $z = 0$.
45. $x = 2 - t$, $y = 2 - 2t$, $z = 6t$.
46. The line through $(0, 0, 0)$ and $(7, 6, 3)$.
47. $x = -1 + 2t$, $y = -2 + 3t$, $z = 6 - 7t$.
48. The z axis.
49. The line through $(2, -1, 4)$ and $(6, 2, -3)$.
50. $x = 7 - 6t$, $y = 5t$, $z = 6 - t$.

In Exercises 51 to 57, the equations of a line are given in symmetric form. Find a parametric representation of the line.

51. $x = y = 0$.
52. $x = y = 1$.
53. $x = y = z$.
54. $x - y = x = y$.
55. $2x - y = x + z = z - 4y$.
▶ 56. $x - y + z = 2x - 3y + z = x - 10y + 5z$.
57. $2x + y - 4 = x - 3y + z + 5 = x + 2y - z - 10$.
58. What points (x, y, z) satisfy the equation $(x - y + 1)^2 + (x + y - z + 4)^2 = 0$?
59. For which values of q does the equation

$$(2x + 3y + 4z - 1)^2 + (qx + 3y + 4z - 1)^2 = 0$$

represent a line? What does it represent if it does not represent a line?

4.6 Mutual position of lines and planes

If l is a line and σ a plane, then l either lies in σ, or meets σ in one point, or is parallel to σ; see Fig. 9.38.

If l and l_1 are two lines, then l_1 either coincides with l, or intersects l, or is parallel to l; or l_1 and l do not lie in one plane (in the latter case l_1 and l are called **skew**); see Fig. 9.39. We shall show by examples how the mutual position of lines and planes can be determined by calculations.

▶ *Examples* **1.** Determine the mutual position of the line

$$(21) \qquad x = 2 - 3t, \qquad y = 6 + t, \qquad z = -1 - 2t$$

and the plane

$$(22) \qquad\qquad 3x - y + 2z - 5 = 0.$$

Solution. We ask which points (21) lie on (22). To find out, replace x, y, z in (22) by their expression in (21). We obtain

$$3(2 - 3t) - (6 + t) + 2(-1 - 2t) - 5 = 0$$

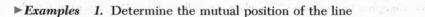

or

$$t + \frac{1}{2} = 0.$$

This is true if $t = -\frac{1}{2}$, and for no other value of t. Thus the line intersects the plane in one point, the point with the coordinates $x = 2 - 3(-\frac{1}{2})$, $y = 6 + (-\frac{1}{2})$, $z = -1 - 2(-\frac{1}{2})$, that is, the point $(\frac{7}{2}, \frac{11}{2}, 0)$.

2. Determine the mutual position of the line (21) and the planes

Fig. 9.38

Fig. 9.39

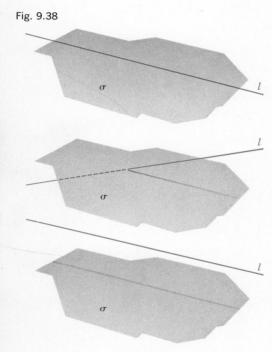

(23) $$3x + 13y + 2z - 5 = 0$$
(24) $$3x + 13y + 2z - 82 = 0.$$

Solution. We substitute (21) into (23) and obtain the relation $77 = 0$. This is never true; the line (21) and the plane (23) never meet—they are parallel.

Substituting (21) into (24), we obtain the relation $0 = 0$, which is true (and does not involve t). Hence all points of the line (21) lie on the plane (24).

3. Determine the mutual position of the line (21) and the line

(25) $$x = -1 + 6s, \qquad y = -2s, \qquad z = 4 + 4s.$$

(Note that we use a new letter for the parameter, as we should!)

Solution. We determine first whether the two lines have a point in common. If there is such a point, then it corresponds to a value of t and a value of s such that both give the same values to x, y, z. Thus we must have

$$2 - 3t = -1 + 6s$$
$$6 + t = -2s$$
$$-1 - 2t = 4 + 4s.$$

These equations are not satisfied for any values of s and t. (For instance, the first equation says that $3t + 6s = 3$, that is, $t + 2s = 1$. But the second equation asserts that $t + 2s = -6$.) Hence the two lines do not meet. They are either parallel (if they lie in the same plane) or skew.

To determine what takes place, we choose two points on (21), say $(2, 6, -1)$ and $(-1, 7, -3)$, and a point on (25), say $(-1, 0, 4)$. The plane through these points contains the line (21). The equation of this plane is found to be

$$x - 3y - 3z + 13 = 0.$$

We choose another point on (25), say $(-7, 2, 0)$, and check whether it too lies on this plane. It does. Hence the two lines are parallel. ◀

EXERCISES

In Exercises 60 to 68, determine whether the line l lies in the plane σ, is parallel to it, or intersects it at one point. In the latter case, find the point of intersection.

60. l: $x = 2y = 3z$; σ: $8x - 6y + 7z = 330$.
61. l: $x = t$, $y = 2 + 2t$, $z = 4t$; σ: $2x - 3y - 5z + 6 = 0$.

62. $l: x = 2 - 2t, y = 3 + 2t, z = 4 + t$; $\sigma: x + 2y - 4z + 8 = 0$.
63. $l: x - 2y = 3$ and $x + z = 5$; $\sigma: x + y + z = 1$.
64. $l: x - 2y + z = 7$ and $x - y - 2z = 0$; $\sigma: 2x - 4y = 0$.
65. $l: x - 2 = y - 3 = 2z - 5$; $\sigma: 2x + 3y - 4z + 5 = 0$.
66. $l: x - 2077y + 103z - 4 = 0$ and $x + 2y = 3$; $\sigma: x - 2077y + 103z - 4 = 0$.
67. l passes through $(0, 2, 1)$ and $(4, -1, 0)$; σ passes through $(0, 0, 0)$, $(2, -1, 6)$, and $(-2, 1, 7)$.
68. l as in Exercise 67, $\sigma: x + y - z = 2$.

In Exercises 69 to 78, determine whether the lines l_1 and l_2 are skew, or parallel, or coincide, or meet at precisely one point. In the last case, determine the point.

69. $l_1: x = 1 - t, y = 2 + t, z = 2t$; $l_2: x = 1 - s, y = 1 + 2s, z = 4 - s$.
70. $l_1: x = 1 - t, y = 2 + t, z = 2t$; $l_2: x = 3 - 2s, y = 4 + 2s, z = 6 + 4s$.
71. $l_1: x - 1 = y + 2 = 2z - 3$; $l_2: x = s, y = s - 3, z = \frac{1}{3}s - 1$.
72. $l_1: 2x - 3y + 4z - 1 = x - y + 2z = 0$; $l_2: 3x = 3y = z - 2x$.
73. $l_1: x = 2 + 3t, y = -2 + 3t, z = 4$; $l_2: x + y - 2 = z - 4 = 0$.
74. $l_1: x - 2y = z - 4 = 3x + z - 5$, $l_2: x + y - z = x + y - 2z = 1$.
75. l_1 passes through $(2, 3, 4)$ and $(1, 6, 7)$; $l_2: x = 1 + t, y = z = t$.
▶ 76. l_1 passes through $(1, -1, 0)$ and $(0, 1, 0)$; $l_2: x + y + z - 2 = x - y + z - 2 = x + y - z - 2$.
77. l_1 is the x axis; $l_2: 1000x = 10^{10}y = 100^{100}z$.
78. $l_1: x - y = x - 2z = y + z$; $l_2: 2x - 3y = z + 4y - 5 = x - y + 1$.

4.7 Spheres

In space, the set of all points that have the same distance $r > 0$ from a fixed point (center) is called a **sphere**; the number r is called the **radius**. The sphere is the set of all points (x, y, z) such that

$$(x - a)^2 + (y - b)^2 + (z - c)^2 = r^2,$$

where (a, b, c) is the center. The solution sets of the inequalities

$$(x - a)^2 + (y - b)^2 + (z - c)^2 < r^2,$$
$$(x - a)^2 + (y - b)^2 + (z - c)^2 > r^2$$

consists of all points interior or exterior to the sphere, respectively.
    The solution set of any equation of the form

$$x^2 + y^2 + z^2 + ax + by + cz + d = 0$$

is a sphere, or a point, or the empty set. This can be proved by "completing the square," just as we proved the corresponding statement in the plane (Theorem 1 in Chapter 2, §3.1).

▶*Examples* **1.** What is the solution set of

$$x^2 + 2x + y^2 - 2y + z^2 - 4z + 6 = 0?$$

Answer. Completing the squares, we rewrite the equation in the form

$$(x + 1)^2 - 1 + (y - 1)^2 - 1 + (z - 2)^2 - 4 + 6 = 0$$

or $$(x + 1)^2 + (y - 1)^2 + (z - 2)^2 = 0. \quad \text{Sphere with } r = 0.$$

The solution set consists of one point, the point $(-1, 1, 2)$.

2. Find the intersection of the plane $z = 10$ and the sphere

$$(x - 2)^2 + (y - 3)^2 + (z - 4)^2 = 100.$$

Solution. We set $z = 13$ in the equation of the sphere and obtain the equation

$$(x - 2)^2 + (y - 3)^2 = 19.$$

In the plane $z = 13$, one can use x and y as Cartesian coordinates (see Fig. 9.40); hence the intersection is a circle. Its center is $(2, 3, 13)$; its radius is $\sqrt{19}$.

One can use the method of this example for an analytic proof of the theorem: *If a plane intersects a sphere, the intersection is a point or a circle.* (See Fig. 9.41.)◀

Fig. 9.40

EXERCISES

79. The plane $z = 1$ intersects the sphere $x^2 + y^2 + z^2 = 10$ in a circle. Find the radius and center of this circle.
80. The plane $y = 3$ intersects the sphere $x^2 + y^2 + z^2 - 2x + 2z = 16$ in a circle. Find the radius and center of this circle.
81. Find the equation of the sphere whose center is at $(1, -2, 2)$ and which passes through the point $(2, 0, 3)$.
82. Find the center and radius of the sphere whose equation is $x^2 + y^2 + z^2 - 4x + 6y - z + \frac{49}{4} = 1$.
83. Find the center and radius of the sphere whose equation is $4x^2 + 4y^2 + 4z^2 + 40x - 32y - 4z = 31$.
84. The spheres $x^2 + y^2 + z^2 = 19$ and $x^2 + y^2 + z^2 - 10z = -21$ intersect in a circle. Find the radius and center of this circle.
85. Find the equation of a sphere of radius $\sqrt{114}$ which passes through the points $(2, 6, -3)$ and $(1, -1, 5)$.
86. Find the intersection of the line $x = 2 + t, y = 3 - t, z = t$ with the sphere of radius 3 with center at the origin.
87. Find the intersection of the line $x = y = 2z$ with the sphere $(x - 1)^2 + y^2 + z^2 = 100$.

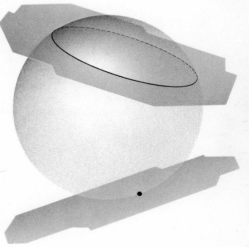

Fig. 9.41

88. Find all planes σ such that σ is parallel to the x, y plane, and σ intersects the sphere $x^2 + y^2 + z^2 = 100$ in a circle of radius 4.

§5 Cylindrical and Spherical Coordinates

In this section, we describe two coordinate systems in space that are analogous to polar coordinates in the plane. We shall use these coordinates only in Chapter 10, §4 where we discuss quadric surfaces, and in Chapter 13 where we compute multiple integrals.

5.1 Cylindrical coordinates

If a point P has Cartesian coordinates (x, y, z), its **cylindrical coordinates** are, by definition, the three numbers (r, θ, z), where (r, θ) are the polar coordinates of the point (x, y) in the x, y plane. Note that $r = \sqrt{x^2 + y^2}$ is the distance from the point $P = (x, y, z)$ to the point $(0, 0, z)$, that is, the distance from P to the z axis (see Fig. 9.42).

The surfaces corresponding to fixed values of r are circular cylinders (see Fig. 9.43), hence the name of this coordinate system. The surfaces corresponding to fixed values of θ are planes through the z axis (see Fig. 9.44); the surfaces corresponding to fixed values of z are, of course, planes parallel to the x, y planes.

Fig. 9.42

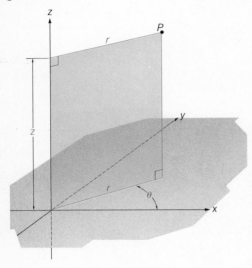

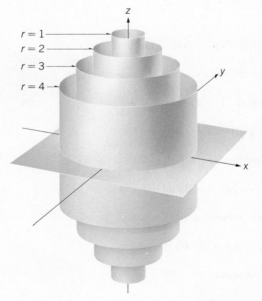

Fig. 9.43

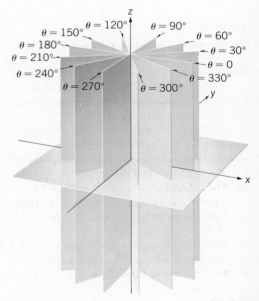

Fig. 9.44

The connection between the cylindrical coordinates (r, θ, z) and Cartesian coordinates (x, y, z) of a point is given by the formulas

(1)
$$r^2 = x^2 + y^2,$$
$$\cos \theta = \frac{x}{r}, \quad \sin \theta = \frac{y}{r},$$
$$z = z,$$

and

(2)
$$x = r \cos \theta,$$
$$y = r \sin \theta,$$
$$z = z.$$

As in the case of polar coordinates, it is also convenient to permit negative values of r.

▶*Example* In cylindrical coordinates, the equation of a plane reads

$$Rr \cos (\theta - \alpha) + Cz + D = 0$$

where R, α, C, D are numbers with $R^2 + C^2 \neq 0$.
    To see this, start with a linear equation

$$Ax + By + Cz + D = 0$$

where $A^2 + B^2 + C^2 \neq 0$. Let (R, α) be the polar coordinates of the point (A, B). Now replace x, y, z by the expressions (2), and replace A and B by $R \cos \alpha$ and by $R \sin \alpha$.◀

5.2 Surfaces of revolution

Assume that we are given a curve C in the x, z plane, that is, in the plane $y = 0$, located in the half-plane $x \geq 0$. Suppose we rotate it about the z axis to obtain a **surface of revolution** S. Then consider a point P on S, with the distance r from the z axis. Let P_0 be the point on C with the same z coordinate as P (see Fig. 9.45). Then $r =$ the distance x from P_0 to the z axis. We conclude that:

The equation of S in cylindrical coordinates is obtained from that of C by replacing x by r.

Once one has the equation of S in cylindrical coordinates, it is, of

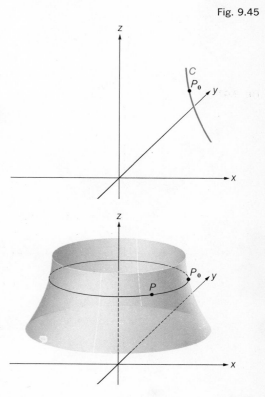

Fig. 9.45

Fig. 9.46

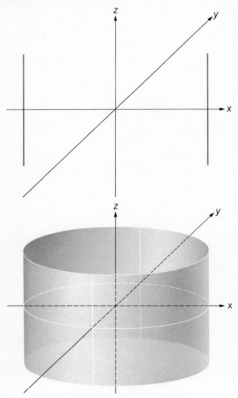

course, easy to replace r^2 by $x^2 + y^2$ and obtain the equation of S in Cartesian coordinates.

It is clear that, instead of assuming the curve C to lie in the half-plane $x \geq 0$, we may assume it to be symmetric about the z axis.

For instance, if, in the x, z plane, C is the line $x = p$, p a positive number, then the surface S obtained by rotating C about the z axis is the same surface we would obtain by rotating the pair consisting of the lines $x = p$ and $x = -p$ (see Fig. 9.46). The equation of this pair is $x^2 = p^2$. The equation of S is therefore $r^2 - p^2 = 0$ or

$$x^2 + y^2 = p^2.$$

The surface S is, of course, a circular cylinder.

▶ **Example** Let m be a positive number. We rotate about the z axis the line $z = mx$ or (which is the same) the pair of lines $z = mx$ and $z = -mx$ (see Fig. 9.47). The equation of the pair is

$$(3) \qquad (z + mx)(z - mx) = 0 \qquad \text{or} \qquad z^2 - m^2 x^2 = 0.$$

The equation of the surface of revolution is therefore $z^2 - m^2 r^2 = 0$, or, setting $a = 1/m$,

$$(4) \qquad \frac{x^2}{a^2} + \frac{y^2}{a^2} - z^2 = 0.$$

The surface is called a circular cone, with the z axis as axis. ◀

Fig. 9.47

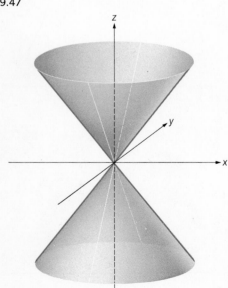

EXERCISES

In Exercises 1 to 8, the Cartesian coordinates of a point in space are given. Find the corresponding cylindrical coordinates of the point.

1. $(2, -2, -3)$.
2. $(-\sqrt{2}, \sqrt{2}, 1)$.
3. $(2, -2\sqrt{3}, 2)$.
4. $(-\sqrt{3}/2, \frac{1}{2}, \frac{1}{4})$.
5. $(4\cos 15°, -4\sin 15°, 1)$.
6. $(1, 2, 3)$.
7. $(\frac{1}{2}\sin \pi/8, \frac{1}{2}\cos \pi/8, \sqrt{3}/2)$.
8. $(3, 4, -\frac{1}{2})$.

In Exercises 9 to 12, find the equation of the surface S in cylindrical coordinates.

9. S is the y, z plane.
10. S: $x^2 + y^2 + 2z^2 + 2z - 5 = 0$.
11. S is the sphere of radius 1 about the origin.
12. S is the plane $x - 2y + 3z - 5 = 0$.

In Exercises 13 to 17, a surface S is obtained by rotating a curve C in the x, z plane about the z axis. Find the equation of S in Cartesian coordinates. Sketch S.

13. $C: z = e^{-x^2}$.   16. $C: z = x$.
► 14. $C: z = |x^3|$.   17. $x^2 + 4z^2 = 100$.
15. $C: (x - 1)^2 + z^2 = 1$.

Fig. 9.48

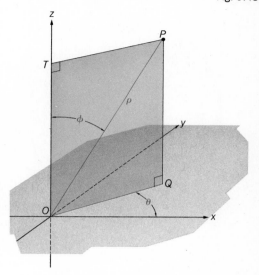

5.3 Spherical coordinates

The spherical coordinates of a point P with Cartesian coordinates (x, y, z) are three numbers, ρ, θ, and ϕ, defined as follows (see Fig. 9.48): ρ is the distance from the origin O to P, θ has the same significance as in cylindrical coordinates, ϕ is the angle between the positive z axis and the ray from O to P; θ is not defined if $x^2 + y^2 = 0$; ϕ is not defined if $\rho = 0$.

Denote by Q the point $(x, y, 0)$ and by T the point $(0, 0, z)$. We see easily from Fig. 9.48 that $z = \rho \cos \phi$. On the other hand, $x^2 + y^2 = |OQ|^2 = |\overrightarrow{TP}|^2 = \rho^2 \sin^2 \phi$. Thus the polar coordinates of Q are $\rho \sin \phi$ and θ, so that $x = \rho \sin \phi \cos \theta$ and $y = \rho \sin \phi \sin \theta$. We summarize the formulas connecting the spherical coordinates of a point with the Cartesian coordinates:

$$\rho^2 = x^2 + y^2 + z^2$$

(5)
$$\cos \theta = \frac{x}{\sqrt{x^2 + y^2}}, \qquad \sin \theta = \frac{y}{\sqrt{x^2 + y^2}}$$

$$\cos \phi = \frac{z}{\rho}, \qquad \sin \phi = \frac{\sqrt{x^2 + y^2}}{\rho}$$

and

(6)
$$x = \rho \sin \phi \cos \theta$$
$$y = \rho \sin \phi \sin \theta$$
$$z = \rho \cos \phi.$$

Fig. 9.49

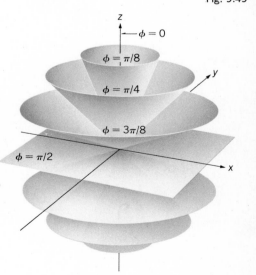

By construction, we have $\rho \geq 0$ and $0 \leq \phi \leq \pi$. But it is convenient to call any set of numbers (ρ, θ, ϕ) for which (5) holds spherical coordinates of (x, y, z).

All points with $\phi = 0$ or $\phi = 180°$ lie on the z axis. The surfaces corresponding to a fixed value of ϕ between 0 and $\pi/2$, or between $\pi/2$ and π are circular cones, with the z axis as axis (see Fig. 9.49). This is geometrically obvious and can be seen analytically as follows.

Let a be a given positive number; by (6), we have

$$\frac{x^2}{a^2} + \frac{y^2}{a^2} - z^2 = z^2 \left(\frac{\tan^2 \phi}{a^2} - 1 \right).$$

Thus the condition $|\tan \phi| = a$ is equivalent to the equation (4) of a cone (see §5.2). For $\phi = 90°$, we obtain the x, y plane.

The surfaces corresponding to a fixed value of θ are the same as in the case of cylindrical coordinates. The surfaces corresponding to fixed

Fig. 9.50

values of r are, of course, spheres about the origin, hence the name of this coordinate system.

Consider now a point P on a fixed sphere about the origin. One can specify the position of P by specifying the numbers ϕ and θ. We restrict ϕ and θ by the inequalities $0 \leq \phi \leq 180°$, $-180° < \theta \leq 180°$. In this case, one calls θ the *longitude* of P and $90° - \phi$ the *latitude* of P. This corresponds to the use of these terms in geography where $\phi = 0$, and $\phi = 90°$ and $\phi = 180°$ correspond to the North Pole, the equator, and the South Pole, respectively. The curves on the sphere corresponding to fixed values of ϕ are called *parallels*, those corresponding to fixed values of θ *meridians*; see Fig. 9.50.

EXERCISES

In Exercises 18 to 24 find spherical coordinates (ρ, θ, ϕ) for a point whose Cartesian coordinates (x, y, z) are given.

18. $(1, 1, 1)$.   22. $(1, 2, 3)$.
19. $(7, -7, 5)$.   23. $(\cos 77, \sin 77, 0)$.
20. $(5/\sqrt{2}, -5/\sqrt{2}, 5\sqrt{3}/\sqrt{2})$.   24. $(0, 1, 0)$.
21. $(0, 0, -\pi)$.

In Exercises 25 to 29, find the equation of the surface S in spherical coordinates.

25. S is the plane $y = 0$.
26. S is the plane $z = 1$.
27. S is the sphere of radius 1; the center has the Cartesian coordinates $(0, 1, 0)$.
28. S is obtained by rotating the line $x = 3$ (in the x, z plane) about the z axis.
29. S is obtained by rotating the line $x = z$ (in the x, z plane) about the z axis.

Appendixes to Chapter 9

§6 Inner Product

There are various ways of multiplying vectors. Of these, the so-called inner product of two vectors is particularly important for calculus of several variables, and for various applications to algebra, geometry, and physics. In the present volume, however, inner products will be used only in a few places.

6.1 Definition of inner product

Let $\mathbf{a}$ and $\mathbf{b}$ be two vectors, and assume that $\mathbf{a} \neq \mathbf{0}$, $\mathbf{b} \neq \mathbf{0}$. If O is any point, we can find points P and Q such that $\overrightarrow{OP}$ represents $\mathbf{a}$ and $\overrightarrow{OQ}$ represents $\mathbf{b}$; (see Fig.

9.51). The angle $\angle POQ$ does not depend upon the choice of O, but only on the vectors **a** and **b**. This angle is, by definition, the angle between **a** and **b**; it will be denoted by $\angle (\mathbf{a}, \mathbf{b})$.

One associates with any two vectors **a** and **b** a number called their **inner product** and denoted by $(\mathbf{a}, \mathbf{b})$. The definition reads:

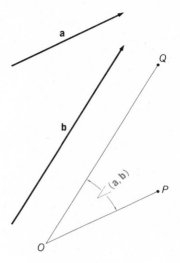

Fig. 9.51

(1)
$$(\mathbf{a}, \mathbf{b}) = |\mathbf{a}||\mathbf{b}| \cos \theta, \ \theta = \text{angle between } \mathbf{a} \text{ and } \mathbf{b},$$
$$(\mathbf{a}, \mathbf{b}) = 0 \text{ if either } \mathbf{a} = \mathbf{0} \text{ or } \mathbf{b} = \mathbf{0}.$$

The inner product is also called the "scalar product" and the "dot product;" it is often denoted by $\mathbf{a} \cdot \mathbf{b}$. The definition just given applies to plane vectors and to vectors in space.

It follows from (1) that

(2)
$$\text{if } (\mathbf{a}, \mathbf{b}) = 0, \text{ then either } |\mathbf{a}| = 0, \text{ or } |\mathbf{b}| = 0, \text{ or } \theta = 90°.$$

Vectors **a**, **b** with $(\mathbf{a}, \mathbf{b}) = 0$ are called **orthogonal** or perpendicular.

For $\mathbf{a} = \mathbf{b}$, we have $\theta = 0$, $\cos \theta = 1$. Therefore

$$(\mathbf{a}, \mathbf{a}) = |\mathbf{a}|^2$$

or

(3)
$$|\mathbf{a}| = \sqrt{(\mathbf{a}, \mathbf{a})}.$$

6.2 Computing the inner product

Theorem 1   LET $\{\mathbf{e}_1, \mathbf{e}_2, \mathbf{e}_3\}$ BE A FRAME AND LET THE VECTORS **a** AND **b** HAVE THE COMPONENTS a_1, a_2, a_3 AND b_1, b_2, b_3, RESPECTIVELY, WITH RESPECT TO THIS FRAME. THEN

(4)
$$(\mathbf{a}, \mathbf{b}) = a_1 b_1 + a_2 b_2 + a_3 b_3.$$

Proof.   Relation (4) is certainly true if either $\mathbf{a} = \mathbf{0}$ or $\mathbf{b} = \mathbf{0}$, for then both sides of (4) are 0. Assume that $\mathbf{a} \neq \mathbf{0}$, $\mathbf{b} \neq \mathbf{0}$, and consider the Cartesian coordinate system $\{O, \mathbf{e}_1, \mathbf{e}_2, \mathbf{e}_3\}$. In this system, **a** is the position vector of a point P with coordinates a_1, a_2, a_3 and **b** is the position vector of a point Q with coordinates b_1, b_2, b_3; see Fig. 9.52. By the distance formula,

$$|\mathbf{a}|^2 = |\overrightarrow{OP}|^2 = a_1{}^2 + a_2{}^2 + a_3{}^2, \ |\mathbf{b}|^2 = |\overrightarrow{OQ}|^2 = b_1{}^2 + b_2{}^2 + b_3{}^2$$
$$|\mathbf{b} - \mathbf{a}|^2 = |\overrightarrow{QP}|^2 = (b_1 - a_1)^2 + (b_2 - a_2)^2 + (b_3 - a_3)^2$$
$$= b_1{}^2 + b_2{}^2 + b_3{}^2 + a_1{}^2 + a_2{}^2 + a_3{}^2 - 2(a_1 b_1 + a_2 b_2 + a_3 b_3).$$

Since $\theta = \angle (\mathbf{a}, \mathbf{b}) = \angle (POQ)$, we can write the cosine law (see §3.2, Example 2) in the form

$$|\mathbf{b} - \mathbf{a}|^2 = |\mathbf{a}|^2 + |\mathbf{b}|^2 - 2|\mathbf{a}||\mathbf{b}| \cos \theta = |\mathbf{a}|^2 + |\mathbf{b}|^2 - 2(\mathbf{a}, \mathbf{b}).$$

Thus
$$(\mathbf{a}, \mathbf{b}) = \frac{1}{2}(|\mathbf{a}|^2 + |\mathbf{b}|^2 - |\mathbf{b} - \mathbf{a}|^2).$$

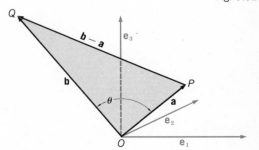

Fig. 9.52

Substituting for $|\mathbf{a}|^2$, $|\mathbf{b}|^2$ and $|\mathbf{b} - \mathbf{a}|^2$ their expressions in terms of components, we obtain (4).

Corollary 1. If $\mathbf{a}$ *and* $\mathbf{b}$ *are plane vectors with components* a_1, a_2 *and* b_1, b_2 *with respect to a frame* $\{\mathbf{e}_1, \mathbf{e}_2\}$, *then*

$$(5) \qquad\qquad (\mathbf{a}, \mathbf{b}) = a_1 b_1 + a_2 b_2.$$

Indeed, if $\mathbf{e}_3$ is a vector such that $\{\mathbf{e}_1, \mathbf{e}_2, \mathbf{e}_3\}$ is a frame in space, then $\mathbf{a}$ and $\mathbf{b}$ may be considered as space vectors with components a_1, a_2, 0 and b_1, b_2, 0 with respect to this frame. Hence (5) follows from (4).

Theorem 1 permits us to calculate the angle θ between two vectors $\mathbf{a}$ and $\mathbf{b}$ (of positive length) or between the directed segments representing these vectors. Indeed, by (1) and (3), we have

$$(6) \qquad \theta = \angle(\mathbf{a}, \mathbf{b}) = \text{arc cos}\ \frac{(\mathbf{a}, \mathbf{b})}{|\mathbf{a}||\mathbf{b}|} = \text{arc cos}\ \frac{(\mathbf{a}, \mathbf{b})}{\sqrt{(\mathbf{a}, \mathbf{a})}\sqrt{(\mathbf{b}, \mathbf{b})}},$$

and by (4) we are able to compute the numbers $(\mathbf{a}, \mathbf{b})$, $(\mathbf{a}, \mathbf{a})$ and $(\mathbf{b}, \mathbf{b})$.

▶ *Examples* *1.* If $\{\mathbf{e}_1, \mathbf{e}_2, \mathbf{e}_3\}$ is a frame, then $(\mathbf{e}_i, \mathbf{e}_j) = 1$ if $i = j$, 0 if $i \neq j$.
Indeed, $|\mathbf{e}_i| = 1$ for $i = 1, 2, 3$ and $\angle(\mathbf{e}_i, \mathbf{e}_j) = 90°$ if $i \neq j$.

2. Let points X, Y, Z have the Cartesian coordinates $(1, 2, 3)$, $(4, -1, 7)$ and $(6, -1, 4)$. Find the angle $\theta = \angle XYZ$.

Solution. We have $\theta = \angle(\mathbf{X} - \mathbf{Y}, \mathbf{Z} - \mathbf{Y})$; see Fig. 9.53. Since $\mathbf{X} - \mathbf{Y}$ has components $(1 - 4, 2 + 1, 3 - 7) = (-3, 3, -4)$ and $\mathbf{Z} - \mathbf{Y}$ has components $(2, 0, -3)$, we have

$$|\mathbf{X} - \mathbf{Y}|^2 = (-3)^2 + 3^2 + 4^2 = 34, |\mathbf{Z} - \mathbf{Y}|^2 = 2^2 + 0^2 + (-3)^2 = 13$$
$$(\mathbf{X} - \mathbf{Y}, \mathbf{Z} - \mathbf{Y}) = (-3)2 + 3 \cdot 0 + (-4)(-3) = 6$$

so that $\cos \theta = 6/(\sqrt{34}\sqrt{13}) \approx .28$ and $\theta \approx 74°$. ◀

Fig. 9.53

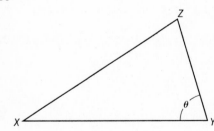

EXERCISES

In Exercises 1 to 4, find the number $(\mathbf{a}, \mathbf{b})$ if $\mathbf{a}$ and $\mathbf{b}$ have the indicated components, with respect to some frame.

1. $(2, 0, 7)$, $(-2, 0, -7)$.   3. $(2, 4)$, $(6, -1)$.
2. $(3, -5, 6)$, $(.01, .02, .3)$.   4. $(51, 6, -3)$, $(0, 7, 4)$.

In the following exercises $\{\mathbf{e}_1, \mathbf{e}_2, \mathbf{e}_3\}$ is a frame.

5. Find a number u such that $\cos \angle(\mathbf{e}_1 + 2\mathbf{e}_2 + u\mathbf{e}_3, 3\mathbf{e}_1 + \mathbf{e}_2) = \frac{5}{12}$.
6. Find all numbers u such that the vectors $\mathbf{e}_1 + 5\mathbf{e}_2 - 6\mathbf{e}_3$ and $2\mathbf{e}_1 - \mathbf{e}_2 + u\mathbf{e}_3$ are orthogonal.
7. Suppose P, Q, R are points with Cartesian coordinates $(0, 1, 3)$, $(2, -1, 6)$ and $(8, -2, 4)$, respectively. Find the three angles in the triangle PQR. (Since all

such calculations must be approximate, the sum of the computed angles may not be precisely 180°.)

In Exercises 8 to 12, find the angle between the two given vectors.

8. $\mathbf{a} = 2\mathbf{e}_1 + (\cos 10°)\mathbf{e}_2 - (\sin 10°)\mathbf{e}_3$, $\mathbf{b} = (1/\sqrt{2})\mathbf{e}_1 + (\sin 10°)\mathbf{e}_2 + (\cos 10°)\mathbf{e}_3$.
9. $\mathbf{a} = 2\mathbf{b}$, $\mathbf{b} = \mathbf{e}_1 + \mathbf{e}_2 + \mathbf{e}_3$.
10. $\mathbf{a} = \mathbf{e}_1 - \mathbf{e}_2$, $\mathbf{b} = \mathbf{e}_2 - \mathbf{e}_3$.
11. $\mathbf{a} = 4\mathbf{e}_1 - .5\mathbf{e}_2 + 6\mathbf{e}_3$, $\mathbf{b} = 2(3\mathbf{e}_1 - 4\mathbf{e}_2 + .5\mathbf{e}_3)$.
12. $\mathbf{a} = (\cos 3°)\mathbf{e}_1 + (\cos 20°)\mathbf{e}_2$, $\mathbf{b} = -\mathbf{e}_1 + 6\mathbf{e}_3$.

6.3 Rules for inner products

We note the formal properties of inner products.

Theorem 2 INNER PRODUCTS OBEY THE FOLLOWING RULES:

    I. FOR ANY TWO VECTORS $\mathbf{a}$ AND $\mathbf{b}$, THERE IS A UNIQUE NUMBER CALLED THEIR **inner product** AND DENOTED BY $(\mathbf{a}, \mathbf{b})$.

    II. $(\mathbf{a}, \mathbf{b}) = (\mathbf{b}, \mathbf{a})$ FOR ALL $\mathbf{a}$ AND $\mathbf{b}$.

    III. $(\lambda\mathbf{a}, \mathbf{b}) = \lambda(\mathbf{a}, \mathbf{b})$ FOR ALL $\mathbf{a}$, $\mathbf{b}$ AND ALL NUMBERS λ.

    IV. $(\mathbf{a}, \mathbf{b} + \mathbf{c}) = (\mathbf{a}, \mathbf{b}) + (\mathbf{a}, \mathbf{c})$ FOR ALL $\mathbf{a}$, $\mathbf{b}$, $\mathbf{c}$.

    V. $(\mathbf{a}, \mathbf{a}) > 0$ UNLESS $\mathbf{a} = \mathbf{0}$.

Proof. Rule **I** needs no proof. The other rules can be proved either geometrically, from the definition (1), or analytically, using (5) and the properties of numbers. We carry out the second approach.

Let $\mathbf{a}$, $\mathbf{b}$, and $\mathbf{c}$ be vectors, and let their components, in some frame, be (a_1, a_2, a_3), (b_1, b_2, b_3) and (c_1, c_2, c_3), respectively. The commutative law **II** follows from (5) by noting that each term on the right-hand side of (5) is unchanged if we interchange the a's and b's. The associative law **III** is proved by noting that $\lambda\mathbf{a}$ has components $(\lambda a_1, \lambda a_2, \lambda a_3)$. Hence **III** says that

$$(\lambda a_1)b_1 + (\lambda a_2)b_2 + (\lambda a_3)b_3 = \lambda(a_1 b_1 + a_2 b_2 + a_3 b_3).$$

Since $\mathbf{b} + \mathbf{c}$ has the components $(b_1 + c_1, b_2 + c_2, b_3 + c_3)$, the distributive law **IV** asserts that

$$a_1(b_1 + c_1) + a_2(b_2 + c_2) + a_3(b_3 + c_3)$$
$$= (a_1 b_1 + a_2 b_2 + a_3 b_3) + (a_1 c_1 + a_2 c_2 + a_3 c_3).$$

This is true. Finally, $a_1{}^2 + a_2{}^2 + a_3{}^2$ is 0 only if $a_1 = a_2 = a_3 = 0$. This proves **V**.

We note that

(7) $$(\mathbf{a}, \mathbf{0}) = 0 \qquad \text{for all } \mathbf{a},$$

(8) $$(\mathbf{b} + \mathbf{c}, \mathbf{a}) = (\mathbf{b}, \mathbf{a}) + (\mathbf{c}, \mathbf{a}).$$

The trivial proofs are left to the reader.

A vector space in which postulates **I** to **V** hold is called an **inner product space**. We shall give examples of such spaces in the next section.

6.4 Applications of the inner product rules

We shall derive some consequences of Theorem 2. [In doing this, we shall use only the Postulates 1 to 10 of a vector space (see §1.5) and the rules I to V of Theorem 2. Hence our results will be valid in every inner product space.]

Theorem 3 (*Schwarz's inequality*) FOR ANY TWO VECTORS $\mathbf{a}$ AND $\mathbf{b}$,

$$(9) \qquad\qquad |(\mathbf{a}, \mathbf{b})| \leq |\mathbf{a}||\mathbf{b}|.$$

This is clearly true, by (1), since $-1 \leq \cos \theta \leq 1$. But we want to prove (9) using only the rules I to V. To do this, we note that for every number λ:

$$0 \leq (\mathbf{a} + \lambda\mathbf{b}, \mathbf{a} + \lambda\mathbf{b}) \qquad\qquad \text{by V}$$
$$= (\mathbf{a}, \mathbf{a}) + 2(\mathbf{a}, \mathbf{b})\lambda + (\mathbf{b}, \mathbf{b})\lambda^2 \qquad \text{by repeated application of II, III, IV}$$
$$= |\mathbf{a}|^2 + 2(\mathbf{a}, \mathbf{b})\lambda + |\mathbf{b}|^2\lambda^2.$$

Since this quadratic polynomial in λ has no (real) roots, $|\mathbf{a}|^2|\mathbf{b}|^2 - (\mathbf{a}, \mathbf{b})^2 \geq 0$, which is the same as (9).

Theorem 4   THE **length** OF A VECTOR

$$(10) \qquad\qquad |\mathbf{a}| = \sqrt{(\mathbf{a}, \mathbf{a})},$$

HAS THE FOLLOWING FORMAL PROPERTIES.

(i) FOR EVERY VECTOR $\mathbf{a}$, THE LENGTH $|\mathbf{a}|$ IS A NONNEGATIVE NUMBER.

(ii) $|\mathbf{a}| = 0$ IF AND ONLY IF $\mathbf{a} = \mathbf{0}$,

(iii) $|\lambda\mathbf{a}| = |\lambda||\mathbf{a}|$ FOR ALL $\mathbf{a}$ AND ALL NUMBERS λ,

(iv) $|\mathbf{a} + \mathbf{b}| \leq |\mathbf{a}| + |\mathbf{b}|$ FOR ALL $\mathbf{a}, \mathbf{b}$ AND $\mathbf{c}$. (TRIANGLE INEQUALITY)

All statements are geometrically obvious. But they can also be derived, algebraically, from (10) and the properties of scalar products. Thus (ii) follows from V and (iii) follows from III. The triangle inequality (iv) is equivalent to

$$|\mathbf{a} + \mathbf{b}|^2 \leq |\mathbf{a}|^2 + 2|\mathbf{a}||\mathbf{b}| + |\mathbf{b}|^2$$

or

$$(\mathbf{a} + \mathbf{b}, \mathbf{a} + \mathbf{b}) \leq |\mathbf{a}|^2 + 2|\mathbf{a}||\mathbf{b}| + |\mathbf{b}|^2.$$

Applying the rules III and IV to $(\mathbf{a} + \mathbf{b}, \mathbf{a} + \mathbf{b})$, we may write this as

$$(\mathbf{a}, \mathbf{a}) + 2(\mathbf{a}, \mathbf{b}) + (\mathbf{b}, \mathbf{b}) \leq |\mathbf{a}|^2 + 2|\mathbf{a}||\mathbf{b}| + |\mathbf{b}|^2,$$

which is the same as the Schwarz inequality.

We now express the components of a vector in terms of inner products. (Here we assume 11_3.)

Theorem 5   IF $\{\mathbf{e}_1, \mathbf{e}_2, \mathbf{e}_3\}$ IS A FRAME AND $\mathbf{a}$ A VECTOR WITH COMPONENTS (a_1, a_2, a_3), THEN

(11) $$a_1 = (\mathbf{a}, \mathbf{e}_1),\; a_2 = (\mathbf{a}, \mathbf{e}_2),\; a_3 = (\mathbf{a}, \mathbf{e}_3).$$

Proof.   By the definition of components (see §1.12)

$$\mathbf{a} = a_1\mathbf{e}_1 + a_2\mathbf{e}_2 + a_3\mathbf{e}_3.$$

Therefore, making use of the rules **III**, **IV**, and of the relation $(\mathbf{e}_1, \mathbf{e}_1) = 1$, $(\mathbf{e}_2, \mathbf{e}_1) = 0$, $(\mathbf{e}_3, \mathbf{e}_1) = 0$ (see Example 1 in §6.2), we have

$$(\mathbf{a}, \mathbf{e}_1) = (a_1\mathbf{e}_1 + a_2\mathbf{e}_2 + a_3\mathbf{e}_3, \mathbf{e}_1) = (a_1\mathbf{e}_1, \mathbf{e}_1) + (a_2\mathbf{e}_2, \mathbf{e}_1) + (a_3\mathbf{e}_3, \mathbf{e}_1)$$
$$= a_1(\mathbf{e}_1, \mathbf{e}_1) + a_2(\mathbf{e}_2, \mathbf{e}_1) + a_3(\mathbf{e}_3, \mathbf{e}_1) = a_1.$$

The other relations (11) are proved similarly.

▶ *Examples*   *1.* What is the angle between the diagonal of a cube and an edge?

Answer.   Let $\{\mathbf{e}_1, \mathbf{e}_2, \mathbf{e}_3\}$ be a frame. We must find the angle α between $\mathbf{e}_1 + \mathbf{e}_2 + \mathbf{e}_3$ and $\mathbf{e}_1$ (this is seen from Fig. 9.54). We have

Fig. 9.54

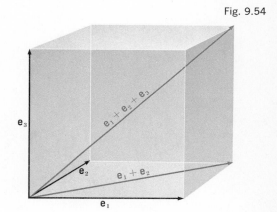

$$\cos \alpha = \frac{(\mathbf{e}_1 + \mathbf{e}_2 + \mathbf{e}_3, \mathbf{e}_1)}{|\mathbf{e}_1 + \mathbf{e}_2 + \mathbf{e}_3||\mathbf{e}_1|} = \frac{1}{\sqrt{3}}$$

so that $\alpha = 54°41'$ approximately.

2. What is the angle between the diagonal of a cube and the diagonal of a face?

Now we need the angle β between $\mathbf{e}_1 + \mathbf{e}_2 + \mathbf{e}_3$, and $\mathbf{e}_1 + \mathbf{e}_2$. A similar calculation gives $\cos \beta = \sqrt{\frac{2}{3}}$, $\beta = 35°25'$, approximately.

3. **In a parallelogram, the two diagonals are equal (have the same length) if and only if the parallelogram is a rectangle.**

Proof.   Let the two sides issuing from a vertex, P, of the parallelogram represent the vectors $\mathbf{a}$ and $\mathbf{b}$ (see Fig. 9.55). The two diagonals represent the vectors $\mathbf{a} + \mathbf{b}$ and $\mathbf{a} - \mathbf{b}$. The equality of the diagonals means that $|\mathbf{a} + \mathbf{b}|^2 = |\mathbf{a} - \mathbf{b}|^2$ or

Fig. 9.55

$$0 = (\mathbf{a} + \mathbf{b}, \mathbf{a} + \mathbf{b})^2 - (\mathbf{a} - \mathbf{b}, \mathbf{a} - \mathbf{b})$$
$$= (\mathbf{a}, \mathbf{a}) + 2(\mathbf{a}, \mathbf{b}) + (\mathbf{b}, \mathbf{b}) - [(\mathbf{a}, \mathbf{a}) - 2(\mathbf{a}, \mathbf{b}) + (\mathbf{b}, \mathbf{b})] = 4(\mathbf{a}, \mathbf{b}).$$

But $(\mathbf{a}, \mathbf{b}) = 0$ means that $\mathbf{a}$ is perpendicular to $\mathbf{b}$, that is, that the parallelogram is a rectangle.

4. *In a parallelogram the two diagonals are perpendicular if and only if the parallelogram is a rhombus.*

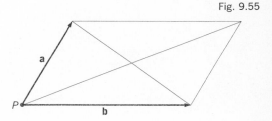

Proof. We use the notation of **3**. Perpendicularity of the diagonals means that

$$0 = (\mathbf{a} + \mathbf{b}, \mathbf{a} - \mathbf{b}) = (\mathbf{a}, \mathbf{a}) - (\mathbf{b}, \mathbf{b}) = |\mathbf{a}|^2 - |\mathbf{b}|^2,$$

while $|\mathbf{a}| = |\mathbf{b}|$ means that the parallelogram is a rhombus. ◀

EXERCISES

In Exercises 13 to 21, we assume a fixed frame $\{\mathbf{e}_1, \mathbf{e}_2, \mathbf{e}_3\}$.

13. Suppose that $(\mathbf{a}, \mathbf{e}_1) = 1$, $(\mathbf{a}, \mathbf{e}_2) = 2$, $(\mathbf{a}, \mathbf{e}_3) = 3$, $(\mathbf{b}, \mathbf{e}_1) = 3$, $(\mathbf{b}, \mathbf{e}_2) = 2$, $(\mathbf{b}, \mathbf{e}_3) = 1$. Find $\angle(\mathbf{a}, \mathbf{b})$.

14. Suppose that $(\mathbf{a}, \mathbf{e}_1) = (\mathbf{a}, \mathbf{e}_2) = 2$ and $|\mathbf{a}| = \sqrt{17}$. Find $(\mathbf{a}, \mathbf{e}_3)$.

15. $(\mathbf{a} + \mathbf{b} + \mathbf{c}, \mathbf{a} + \mathbf{b} + \mathbf{c}) = |\mathbf{a}|^2 + |\mathbf{b}|^2 + |\mathbf{c}|^2 + 2(\mathbf{a}, \mathbf{b}) + 2(\mathbf{a}, \mathbf{c}) + 2(\mathbf{b}, \mathbf{c})$. Prove this identity.

16. In a cube, find the angle between two diagonals of faces emanating from the same vertex. (Hint: Compare Examples 1 and 2.)

17. In a cube, find the angle between an edge and a diagonal of a face.

▶ 18. Find a vector $\mathbf{a}$ such that $(\mathbf{a}, \mathbf{e}_1 + \mathbf{e}_2) = 2$, $(\mathbf{a}, \mathbf{e}_1 - \mathbf{e}_2) = 3$ and $(\mathbf{a}, \mathbf{e}_3) = 0$.

19. Find a vector $\mathbf{a}$ with $(\mathbf{a}, \mathbf{e}_1) = (\mathbf{a}, \mathbf{e}_2) = (\mathbf{a}, \mathbf{e}_3)$ and $|\mathbf{a}| = 100$.

20. If $(\mathbf{a}, \mathbf{e}_1) = (\mathbf{a}, \mathbf{e}_2) = 5$, can one have $|\mathbf{a}| \leq 7$?

21. Show that $(\mathbf{a}, \mathbf{e}_1)^2 \leq (\mathbf{a}, \mathbf{a})$ for all $\mathbf{a}$. When is $(\mathbf{a}, \mathbf{e}_1)^2 = (\mathbf{a}, \mathbf{a})$?

6.5 Direction cosines. Unit vectors

In this and the next subsections, we use a fixed frame $\{\mathbf{e}_1, \mathbf{e}_2, \mathbf{e}_3\}$ and a fixed origin O.

Let $\mathbf{a} \neq \mathbf{0}$ be a vector and let $\alpha_1, \alpha_2, \alpha_3$ be the angles between $\mathbf{a}$ and $\mathbf{e}_1, \mathbf{e}_2, \mathbf{e}_3$; (see Fig. 9.56). These angles are called the **direction angles** of $\mathbf{a}$ and the numbers $\cos \alpha_1, \cos \alpha_2, \cos \alpha_3$ are called the **direction cosines** of $\mathbf{a}$.

For any vector, the direction cosines have properties similar to those of the cosine and sine of the polar angle of a vector in the plane.

If $\mathbf{a} = a_1\mathbf{e}_1 + a_2\mathbf{e}_2 + a_3\mathbf{e}_3$, then the direction cosines of $\mathbf{a}$ are

(10)
$$\cos \alpha_1 = \frac{a_1}{|\mathbf{a}|}, \qquad \cos \alpha_2 = \frac{a_2}{|\mathbf{a}|}, \qquad \cos \alpha_3 = \frac{a_3}{|\mathbf{a}|}.$$

For we have (see Theorem 3 in §6.3)

$$\cos \alpha_1 = \cos \angle(\mathbf{a}, \mathbf{e}_1) = \frac{(\mathbf{a}, \mathbf{e}_1)}{|\mathbf{a}||\mathbf{e}_1|} = \frac{a_1}{|\mathbf{a}|},$$

since $|\mathbf{e}_1| = 1$. The other two relations (10) are proved in the same way. Knowing the length $|\mathbf{a}|$ of a vector $\mathbf{a}$ and its direction angles $\alpha_1, \alpha_2, \alpha_3$, we can write the vector in the form

(12)
$$\mathbf{a} = |\mathbf{a}|(\cos \alpha_1)\mathbf{e}_1 + |\mathbf{a}|(\cos \alpha_2)\mathbf{e}_2 + |\mathbf{a}|(\cos \alpha_3)\mathbf{e}_3.$$

The formulas (10) and (11) should be compared with the corresponding formulas for plane vectors (see §3.1):

Fig. 9.56

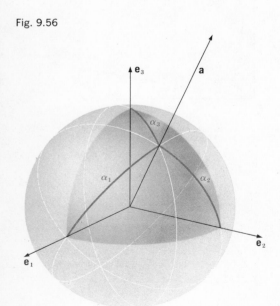

(13) $\qquad \cos \theta = \dfrac{a_1}{|\mathbf{a}|}, \qquad \sin \theta = \dfrac{a_2}{|\mathbf{a}|}, \qquad a = |\mathbf{a}|(\cos \theta)\mathbf{e}_1 + |\mathbf{a}|(\sin \theta)\mathbf{e}_2.$

The relation $\cos^2 \phi + \sin^2 \phi$ also has an analogue. Since $|\mathbf{a}|^2 = a_1{}^2 + a_2{}^2 + a_3{}^2$ we conclude from (10) that

(14) $\qquad\qquad\qquad \cos^2 \alpha_1 + \cos^2 \alpha_2 + \cos^2 \alpha_3 = 1.$

Thus the three direction cosines are not independent.

On the other hand, if α_1, α_2, α_3 are any numbers between 0 and π, inclusive, such that (14) holds, then they are the direction angles of a unique **unit vector** (that is, vector of length one). This vector is, of course,

$$\mathbf{a_0} = (\cos \alpha_1)\mathbf{e_1} + (\cos \alpha_2)\mathbf{e_2} + (\cos \alpha_3)\mathbf{e_3}.$$

Every other vector **a** with the direction angles α_1, α_2, α_3 is of the form $\mathbf{a} = \lambda\mathbf{a_0}$, where λ is some positive number.

Remark Let **a** be a vector orthogonal to $\mathbf{e_3}$, so that $\alpha_3 = 90°$, $\cos \alpha_3 = 0$. In this case, we may think of **a** as a vector in the plane. Let θ be the polar angle of **a** in the coordinate system $\{O, \mathbf{e_1}, \mathbf{e_2}\}$. Then $\cos \alpha_1 = \cos \theta$ and $\cos \alpha_2 = \sin \theta$, as is seen by comparing (12) with (13). Note, however, that direction angles must satisfy the inequality $0 \le \alpha \le \pi$, while the polar angle can take any value. In the case shown in Fig. 9.57, we have $\alpha_1 = \theta$, $\alpha_2 = 90° - \theta$, whereas in the case shown in Fig. 9.58, we have $\alpha_1 = 360° - \theta$, $\alpha_2 = 90° + \alpha_1$.

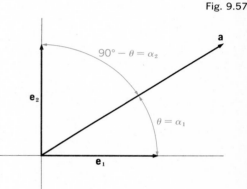

Fig. 9.57

▶*Example* Let θ be the angle between the vectors **a** and **b** with direction angles α_1, α_2, α_3 and β_1, β_2, β_3, respectively. Then

(15) $\qquad\qquad \cos \theta = \cos \alpha_1 \cos \beta_1 + \cos \alpha_2 \cos \beta_2 + \cos \alpha_3 \cos \beta_3.$

Proof. We note equation (12) and the analogous equation for **b**,

$$\mathbf{b} = |b|(\cos \beta_1)\mathbf{e_1} + |b|(\cos \beta_2)\mathbf{e_2} + |b|(\cos \beta_3)\mathbf{e_3}.$$

Substituting (12) and the equation just written into (6), we obtain (15). ◀

Fig. 9.58

6.6 Angle between two lines

A vector **a** is said to be parallel to a line l if it can be represented by a directed segment of l; if so, then $(-\mathbf{a})$ is also a vector parallel to l. If $\mathbf{a} \ne 0$, then the direction cosines of **a** are called direction cosines of l.

Let l and k be two lines, let **a** and **b** be nonnull vectors parallel to these lines, and set $\theta = \angle(\mathbf{a}, \mathbf{b})$. If $0 < \theta \le 90°$, we call $\phi = \theta$ the acute angle between l and k. If $90° < \theta \le 180°$, then we set $\phi = 180° - \theta$ and call ϕ the acute angle between the two lines. Thus $\cos \phi = |\cos \theta|$ in all cases; hence

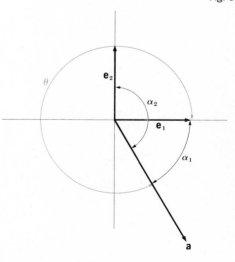

(16) $\qquad\qquad\qquad \cos \phi = \dfrac{|(\mathbf{a}, \mathbf{b})|}{|\mathbf{a}||\mathbf{b}|}.$

Fig. 9.59

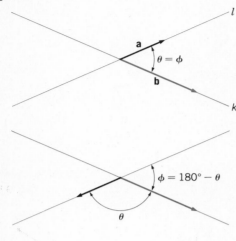

Fig. 9.59 shows that, if l and k intersect, our definition makes geometric sense. If l and k are parallel, then $\phi = 0$.

EXERCISES

22. Find the direction cosines of $\mathbf{a} = 2\mathbf{e}_1 - 3\mathbf{e}_2 + 4\mathbf{e}_3$.

23. Find the direction cosines of $\mathbf{a} = -\mathbf{e}_1 + 2\mathbf{e}_2 - 10\mathbf{e}_3$.

24. $\mathbf{a}$ has direction angles satisfying $\alpha_1 = \alpha_2 = \alpha_3$, $0 < \alpha_1 < 90°$. Find α_1.

25. $\mathbf{a}$ has direction angles satisfying $\alpha_1 = \alpha_2 = \alpha_3$ and $90° < \alpha_1 < 180°$. Find α_1.

26. P has coordinates $(7, -11, 6)$. Find the direction cosines of the vector represented by $\overrightarrow{OP}$.

27. P has coordinates $(3\cos 13°, -3\sin 13°, 0)$. Find the direction cosines and direction angles of the vector represented by $\overrightarrow{OP}$.

28. Find the direction cosines and direction angles of the vector $\mathbf{a}$ with components $(0, 20\cos 20°, 20\sin 20°)$.

29. A point P has spherical coordinates (ρ, θ, ϕ) with $\rho > 0$. Find the direction cosines of the vector represented by $\overrightarrow{OP}$.

30. A point P has cylindrical coordinates (r, θ, z) with $r > 0$. Find the direction cosines of the vector represented by $\overrightarrow{OP}$.

31. Suppose P and Q have spherical coordinates $(1, \theta_1, \phi_1)$ and $(1, \theta_2, \phi_2)$, respectively. Find a formula for $\cos \angle(POQ)$. (Hint: Note Exercise 29 and the Example in §6.4.)

32. Find direction cosines of the line represented by $x_1 = 2 - 2t$, $x_2 = 3 + t$, $x_3 = 4 - 5t$.

33. Find the acute angle between the line in Exercise 32 and the line passing through the points $(4, -1, 6)$ and $(5, 2, -1)$.

6.7 Normals to planes. Angles between planes

Let σ be a plane, l the line through the origin O perpendicular to σ, and P the intersection point of σ and l; see Fig. 9.60. There are two points on l, Q, and T, whose distance from P is 1; the segments $\overrightarrow{PQ}$ and $\overrightarrow{PT}$ define vectors which we denote by $\mathbf{n}$ and $-\mathbf{n}$, respectively. These vectors are called the **unit normals** to σ. If σ does not pass through O, $P \neq O$, we choose Q so that O and Q lie on different sides of σ. In this case, we call $\mathbf{n}$ the unit normal to σ *pointing from the origin*. If σ passes through O, then $P = 0$ and the choice of Q is arbitrary.

Fig. 9.60

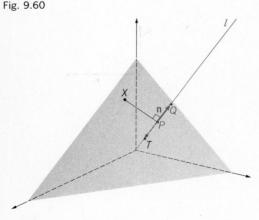

Let $\cos \alpha_1$, $\cos \alpha_2$, $\cos \alpha_3$ be the direction cosines of $\mathbf{n}$. The position vector of P is $p\mathbf{n}$ where $p = |OP|$ is the distance from O to σ. Let X be a point with position vector $\mathbf{X}$ and coordinates x_1, x_2, x_3. The point X lies on σ if and only if the segment $\overrightarrow{PX}$ is orthogonal to l. Since $\overrightarrow{PX}$ defines the vector $\mathbf{X} - p\mathbf{n}$, this condition can be written as

$$(\mathbf{X} - p\mathbf{n}, \mathbf{n}) = 0 \qquad \text{or} \qquad (\mathbf{X}, \mathbf{n}) - p(\mathbf{n}, \mathbf{n}) = 0$$

and, since $(\mathbf{n}, \mathbf{n}) = 1$, this means that

$$(17) \qquad\qquad (\mathbf{X}, \mathbf{n}) - p = 0$$

or, in components,

$$(17') \qquad x_1 \cos \alpha_1 + x_2 \cos \alpha_2 + x_3 \cos \alpha_3 - p = 0 \qquad (p \geq 0).$$

This linear equation in x_1, x_2, x_3 is an equation of the plane σ.

Conversely, every linear equation

$$(18) \qquad A_1 x_1 + A_2 x_2 + A_3 x_3 + D = 0$$

(with A_1, A_2, A_3 not all equal to 0) can be brought into the **normal form** $(17')$. We need only multiply both sides by ρ where

$$(19) \quad \rho = \frac{1}{\sqrt{A_1{}^2 + A_2{}^2 + A_3{}^2}} \text{ if } D \leq 0, \; \rho = \frac{-1}{\sqrt{A_1{}^2 + A_2{}^2 + A_3{}^2}} \text{ if } D > 0;$$

(if $D = 0$ either sign may be used). We obtain an equation of the form $(17')$, which is equivalent to (18), with

$$(20) \qquad \cos \alpha_1 = \rho A_1, \cos \alpha_2 = \rho A_2, \cos \alpha_3 = \rho A_3, \qquad p = -\rho D \geq 0.$$

Since $\cos^2 \alpha_1 + \cos^2 \alpha_2 + \cos^2 \alpha_3 = 1$, there is a vector with direction angles α_1, α_2, α_3. We conclude that the direction angles α_1, α_2, α_3 of the unit normal to the plane (18), and the distance p of this plane from the origin are given by formulas (19) and (20). If the plane does not pass through the origin, the formulas give the normal pointing from the origin.

The above statement implies, in particular, that the vector with the components (A_1, A_2, A_3), that is the vector $(1/p)\mathbf{n}$, is parallel to the unit normal to the plane σ. Using this remark, we can compute the acute angle between two planes. This is defined as the acute angle between the lines normal to the planes. The acute angle ϕ between the planes

$$A_1 x_1 + A_2 x_2 + A_3 x_3 + D = 0 \qquad \text{and} \qquad \hat{A}_1 x_1 + \hat{A}_2 x_2 + \hat{A}_3 x_3 + \hat{D} = 0$$

is therefore given by the equation

$$(21) \qquad \cos \phi = \frac{|A_1 \hat{A}_1 + A_2 \hat{A}_2 + A_3 \hat{A}_3|}{\sqrt{(A_1{}^2 + A_2{}^2 + A_3{}^2)(\hat{A}_1{}^2 + \hat{A}_2{}^2 + \hat{A}_3{}^2)}}.$$

This follows from (16).

6.8 Distance from a point to a plane

We conclude this section by deriving a beautiful formula.

Theorem 6 THE DISTANCE δ FROM A POINT P WITH COORDINATES (a_1, a_2, a_3) TO THE PLANE σ: $A_1 x_1 + A_2 x_2 + A_3 x_3 + D = 0$ IS

$$(22) \qquad \delta = \frac{|A_1 a_1 + A_2 a_2 + A_3 a_3 + D|}{\sqrt{A_1{}^2 + A_2{}^2 + A_3{}^2}}.$$

Fig. 9.61

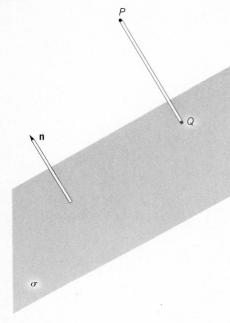

Proof. We define the numbers $\cos \alpha_1$, $\cos \alpha_2$, $\cos \alpha_3$ and p by (19) and (20) and denote the vector with the components $\cos \alpha_1$, $\cos \alpha_2$, $\cos \alpha_3$, that is, the unit normal to σ, by $\mathbf{n}$. Then we can write the equation of σ in the form (17), and we can write the relation (22) to be established as

(22′) $$\delta = |(\mathbf{n}, \mathbf{P}) - p|$$

where $\mathbf{P}$ is the position vector of P.

Let k be the line through P orthogonal to σ and let Q be the intersection of k and σ (see Fig. 9.61). Then $\delta = \overrightarrow{PQ} = |\mathbf{P} - \mathbf{Q}|$ where $\mathbf{Q}$ is the position vector of Q. Since $\mathbf{P} - \mathbf{Q}$ is parallel to $\mathbf{n}$, we have $\mathbf{P} - \mathbf{Q} = t\mathbf{n}$ where t is a number. Since Q lies on σ, we have, by (17), $(\mathbf{Q}, \mathbf{n}) = p$, that is, $(\mathbf{P} - t\mathbf{n}, \mathbf{n}) = p$ or $(\mathbf{P}, \mathbf{n}) - t(\mathbf{n}, \mathbf{n}) = p$, and since $(\mathbf{n}, \mathbf{n}) = 1$, $t = (\mathbf{n}, \mathbf{P}) - p$. But $\delta = |\mathbf{P} - \mathbf{Q}| = |t\mathbf{n}| = |t||\mathbf{n}| = |t|$, and (22′) follows.

6.9 Distance from a point to a line

Equation (22) has an analogue in plane analytic geometry. If a line l in the x, y plane is given by the equation $Ax + By + C = 0$, then the distance δ between a point (a, b) in the plane and the line l is

(23) $$\delta = \frac{|Aa + Bb + C|}{\sqrt{A^2 + B^2}}$$

One can prove this by imitating the proof of Theorem 6 above. It is simpler to observe that the desired distance δ is also the distance from the point $(a, b, 0)$ in space to the plane σ with the equation $Ax_1 + Bx_2 + C = 0$ (see Fig. 9.62). Hence (23) follows from (22) by substituting $a_1 = a$, $a_2 = b$, $a_3 = 0$, $A_1 = A$, $A_2 = B$, $A_3 = 0$, and $C = D$.

Fig. 9.62

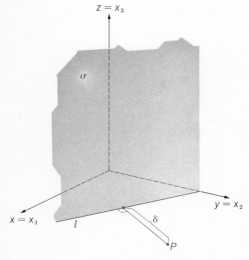

EXERCISES

The following exercises refer to the x_1, x_2, x_3 space.

34. Find the components of the unit normal to the plane $3x_1 - 2x_2 + 5x_3 - 6 = 0$.
35. Find the components of the unit normal to the plane which passes through the points $(1, 0, 2)$, $(4, 0, 0)$ and $(1, 1, 1)$.
36. Find the acute angles between the plane $x_1 + x_2 + x_3 = 0$ and the coordinate planes.
37. Find the acute angle between the planes in Exercises 34 and 35.
38. Find the distance from the point $(1, 2, 3)$ to the plane $2x_1 - 3x_2 + 4x_3 - 7 = 0$.
39. What is the distance from the origin to the plane through the points $(6, 1, 0)$, $(0, 0, 7)$ and $(2, 0, 3)$?
40. What is the distance from the point $(10, 10, -100)$ to the plane in Exercise 39?
41. Find a positive number u such that the point $(2, u, 3)$ has distance 100 from the plane $x_1 + x_2 + x_3 = 8$.
► 42. Find a number $v > 0$ such that the point $(v, 2v, v^2)$ has distance 1000 from the plane $2x_1 - x_2 + 2x_3 = 6$.

The following problems refer to the x, y plane.

43. Find the distance from the point $(2, 3)$ to the line $y - 7x = 5$.
44. On the parabola $y = x^2$, find a point (or points) whose distance from the line $3y = -5 - 4x$ is $\frac{4}{5}$.
45. Find a number t such that the point (t, t^2) has distance 6 from the line $3x - 4y = 0$. How many such numbers are there?
46. Find all points on the line $y - 2x + 5 = 0$ with distance 7 from the line $y + x - 6 = 0$.

§7 Spaces of n Dimensions

7.1 Space of n-tuples

We mentioned in Chapter 2, §6.4 that one calls an ordered sequence of n real numbers, say $(\alpha_1, \alpha_2, \cdots, \alpha_n)$, a point in an n-dimensional space (or n-space). Here n may be any positive integer. At this stage, it pays to identify a point with its position vector. A vector is therefore simply an ordered n-tuple of numbers

$$\mathbf{a} = (\alpha_1, \alpha_2, \cdots, \alpha_n).$$

Addition, and multiplication by numbers, are *defined* as follows:

(1) $\quad (\alpha_1, \alpha_2, \cdots, \alpha_n) + (\beta_1, \beta_2, \cdots, \beta_n) = (\alpha_1 + \beta_1, \alpha_2 + \beta_2, \cdots, \alpha_n + \beta_n);$

(2) $\quad\quad\quad\quad\quad\quad \lambda(\alpha_1, \alpha_2, \cdots, \alpha_n) = (\lambda\alpha_1, \lambda\alpha_2, \cdots, \lambda\alpha_n).$

The null vector $\mathbf{0}$ is, of course, the set of n zeros $(0, 0, \cdots, 0)$. The postulates $\mathbf{1}$ to $\mathbf{10}$ still hold. (This is proved exactly as for $n = 2$; see §1.10.) We also have an additional postulate:

$\mathbf{11}_n$. *One can find n vectors $\mathbf{e}_1, \cdots, \mathbf{e}_n$ such that every vector $\mathbf{a}$ can be written, uniquely, as*

$$\mathbf{a} = \alpha_1\mathbf{e}_1 + \alpha_2\mathbf{e}_2 + \cdots + \alpha_n\mathbf{e}_n.$$

For instance, we may choose:

$$\mathbf{e}_1 = (1, 0, 0, \cdots, 0)$$
$$\mathbf{e}_2 = (0, 1, 0, \cdots, 0)$$
(3) $$\mathbf{e}_3 = (0, 0, 1, \cdots, 0)$$
$$\cdots$$
$$\mathbf{e}_n = (0, 0, 0, \cdots, 1).$$

Postulates $\mathbf{1}$ to $\mathbf{10}$ and $\mathbf{11}_n$ are called the postulates of a **vector space of n dimensions**. A sequence of vectors $\mathbf{e}_1, \cdots, \mathbf{e}_n$ having the property $\mathbf{11}_n$ is called a **basis**.

One cannot, of course, draw pictures of an n-dimensional vector space, for $n > 3$. But one can compute with n-dimensional vectors, and it turns out that n-dimensional vector spaces play an ever-increasing part in mathematics and its applications. (So do vector spaces of infinitely many dimensions, which we do not describe here.)

1. If $\mathbf{a} = (2, 7, 0, 6)$, $\mathbf{b} = (0, 0, 2, -3)$, $\mathbf{c} = (1, 1, -1, 5)$, compute $\mathbf{a} + \mathbf{b}$, $\mathbf{a} - \mathbf{c} = \mathbf{a} + (-1)\mathbf{c}$, $2\mathbf{a} - 3\mathbf{b}$ and $7\mathbf{c} - 4\mathbf{b}$.

2. Compute $33(0, 6, -4, 5) - 32(6, -7, 4, 6) + (2, 2, 100, 0)$.

3. Do the vectors $(0, 1, 0, 0)$, $(1, 0, 0, 0)$, $(0, 0, 1, 1)$, and $(0, 0, 0, 1)$ form a basis for the vector space of 4-tuples of numbers?

4. Do the vectors $(1, 0, 1, 0)$, $(1, 0, 0, 0)$, $(0, 0, 1, 0)$, and $(1, 0, 0, 1)$ form a basis for the vector space of 4-tuples of numbers?

5. Do the vectors $(1, 1, 0, 0)$, $(1, -1, 0, 0)$, $(0, 0, 1, 1)$, and $(0, 0, 1, -1)$ form a basis for the vector space of 4-tuples of numbers?

► 6. Do the vectors $(1, 1, 1, 0)$, $(0, 1, 0, 0)$, $(1, 0, 1, 0)$, and $(0, 0, 0, 1)$ form a basis for the vector space of 4-tuples of numbers?

7. Do the vectors $(1, 0, 0, 0, 1)$, $(1, 0, 0, 0, -1)$, $(0, 1, 1, 0, 0)$, $(0, 1, 0, 0, 0)$, and $(0, 0, 0, 1, 0)$ form a basis for the vector space of 5-tuples of numbers?

8. Do the vectors $(1, 0, 2, 0, 1)$, $(1, 0, 1, 0, 0)$, $(0, 0, 1, 0, 1)$, $(0, 1, 0, 0, 1)$, and $(0, 0, 0, 0, 1)$ form a basis for the vector space of 5-tuples of numbers?

7.2 Lines and hyperplanes

A **line** in n-space is, *by definition*, the set of all points (vectors) $\mathbf{x} = (x_1, x_2, \cdots, x_n)$ that can be written in the form

$$(4) \qquad\qquad \mathbf{x} = \mathbf{a} + t\mathbf{b}$$

where $\mathbf{a}$ and $\mathbf{b} \neq 0$ are fixed vectors and t is any real number. The vector equation (3) is equivalent to n equations for numbers:

$$(4') \qquad\qquad x_j = a_j + tb_j, \qquad j = 1, 2, \cdots, n.$$

A **hyperplane** in n-space is, *by definition*, the set of all points (vectors) $\mathbf{x}$ which satisfy a linear equation, that is, an equation of the form

$$(5) \qquad\qquad A_1 x_1 + A_2 x_2 + \cdots + A_n x_n + D = 0$$

where $A_1, A_2, \cdots, A_n$ and D are numbers, and $A_1^2 + A_2^2 + \cdots + A_n^2 \neq 0$. (If $n = 1$, a hyperplane is a point; if $n = 2$, a hyperplane is a line; if $n = 3$, a hyperplane is a plane.)

Familiar geometric theorems still hold. For instance, there is exactly one line through two distinct points; if two distinct points on a line lie in a hyperplane, so does the whole line.

9. If α and β are numbers such that $\alpha + \beta = 1$, the point $\mathbf{x} = \alpha\mathbf{a} + \beta\mathbf{b}$ lies on the line through $\mathbf{a}$ and $\mathbf{b}$. Prove this statement.

10. Find numbers u, v, w such that the point $(2, 4, u, v, w)$ lies on the line through the points $(0, 2, 3, 4, 5)$ and $(2, 2, -1, 6, 3)$.

11. Find numbers u, v such that the points $(3, 3, u, v)$, $(4, 0, 5, 3)$, $(2, 6, 3, 5)$ lie on one line.

12. Find an equation for the hyperplane in 4-space which contains the points $(2, 0, 0, 0)$, $(0, 4, 0, 0)$, $(0, 0, 1, 0)$ and $(0, 0, 0, 5)$.

13. Find the intersection of the line $x_1 = x_2 = x_3 = t$, $x_4 = 2 - 2t$ with the hyperplane $2x_1 - x_2 - x_3 - x_4 - 6 = 0$.

14. Find the point at which the line $x_1 = 2t$, $x_2 = x_3 = 1 - t$, $x_4 = 1 + t$ intersect the hyperplane of Exercise 12.

15. Do the hyperplanes in 5-space, $x_1 = 0$ and $x_2 + x_3 = 0$, intersect? Try to describe all points that lie on both hyperplanes.

16. Find a hyperplane in 4-space that passes through the origin $(0, 0, 0, 0)$ and does not intersect the hyperplane $2x_3 - x_4 - 7 = 0$.

7.3 Euclidean n-space

The space of ordered n-tuples of numbers $\mathbf{a} = (\alpha_1, \alpha_2, \cdots, \alpha_n)$, $\mathbf{b} = (\beta_1, \beta_2, \cdots, \beta_n)$, and so forth, can be made into an inner product space (compare §6.3) by *defining* the inner product $(\mathbf{a}, \mathbf{b})$ to be

$$(6) \qquad (\mathbf{a}, \mathbf{b}) = \alpha_1\beta_1 + \alpha_2\beta_2 + \cdots + \alpha_n\beta_n.$$

Note that what was a theorem before is now a definition! The rules **I** to **V** of §6.3 hold. The easy proof is left to the reader; it involves the same arguments as the ones used in §6.3 for $n = 3$.

We define the length of a vector $\mathbf{a} = (a_1, a_2, \cdots, a_n)$ by the formula

$$|\mathbf{a}| = \sqrt{(\mathbf{a}, \mathbf{a})} = \sqrt{\alpha_1^2 + \alpha_2^2 + \cdots + \alpha_n^2}.$$

The Schwarz inequality (compare Theorem 3 in §6.4)

$$(7) \qquad |(\mathbf{a}, \mathbf{b})| \le |\mathbf{a}||\mathbf{b}|$$

is valid, so are the rules (**i**) to (**iv**) for length from §6.4, in particular the triangle inequality

$$(8) \qquad |\mathbf{a} + \mathbf{b}| \le |\mathbf{a}| + |\mathbf{b}|.$$

Indeed, the proofs in §6.4 used only the postulates of an inner product space.

The distance d between two points (vectors) $\mathbf{a}$ and $\mathbf{b}$ is now defined as

$$(9) \qquad d = |\mathbf{a} - \mathbf{b}| = \sqrt{(\alpha_1 - \beta_1)^2 + \cdots + (\alpha_n - \beta_n)^2}.$$

The space of n-tuples with the distance definition (9) is called the *n-dimensional Euclidean space*. The triangle inequality (8) has a simple "geometric" meaning: the distance between two points ($\mathbf{a}$ and $-\mathbf{b}$) is not greater than the sum of their distances from a third point (the origin $\mathbf{0}$).

The set of points $\mathbf{x}$ in Euclidean n-space having a fixed distance $r > 0$ from a point $\mathbf{a}$ is called a sphere of radius r and center a. (For $n = 2$, this is a circle.)

Of course, Schwarz's inequality (7) and the triangle inequality (8) are theorems about real numbers. We restate them as such:

For any $2n$ numbers $\alpha_1, \alpha_2, \cdots, \alpha_n, \beta_1, \beta_2, \cdots, \beta_n$

$$(\alpha_1\beta_1 + \cdots + \alpha_n\beta_n)^2 \leq (\alpha_1{}^2 + \cdots + \alpha_n{}^2)(\beta_1{}^2 + \cdots + \beta_n{}^2).$$

and

$$\sqrt{(\alpha_1 - \beta_1)^2 + \cdots + (\alpha_n - \beta_n)^2} \leq \sqrt{\alpha_1{}^2 + \cdots + \alpha_n{}^2} + \sqrt{\beta_1{}^2 + \cdots + \beta_n{}^2}.$$

7.4 Angles

We can now define the angle θ between two vectors **a** and **b** in n-space by the formula

(10)
$$\theta = \text{arc cos} \frac{(\mathbf{a}, \mathbf{b})}{|\mathbf{a}||\mathbf{b}|}$$

(provided $\mathbf{a} \neq \mathbf{0}$, $\mathbf{b} \neq \mathbf{0}$) and pursue analytic geometry in n dimensions. Note that Schwarz's inequality (7) insures that the right-hand side of (10) is always a number between -1 and $+1$, inclusive. Note also that the basis (3) in §7.1 has been made into a frame. The $\mathbf{e}_j$ are mutually orthogonal unit vectors.

▶ *Example*   What is the angle θ between the vectors $\mathbf{a} = (1, 3, -7, 1)$ and $\mathbf{b} = (6, 0, -2, -20)$?

    Answer.   Since $(\mathbf{a}, \mathbf{b}) = 1 \cdot 6 + 3 \cdot 0 + (-7)(-2) + 1(-20) = 0$, $\theta = 90°$.◀

EXERCISES

17. Find a number u such that the vectors $(2, 7, -1, u)$ and $(4, 6, 2, u^2)$ become orthogonal.
18. In 4-space, find the angle between the vector $(1, 1, 1, 1)$ and the frame vector $\mathbf{e}_1 = (1, 0, 0, 0)$.
19. Define the direction cosines of a vector in n-space following the definition in §6.5. Prove that the sum of the squares of the n direction cosines is 1.
20. The set of all points **x** in n-space with $(\mathbf{a}, \mathbf{x}) = 0$, **a** a fixed vector; $\mathbf{a} \neq \mathbf{0}$, is a hyperplane through the origin. Prove this.
21. Write the equation of the hyperplane $2x_1 + 3x_2 - x_3 - x_4 + x_5 = 100$ in the form $(\mathbf{n}, \mathbf{x}) - p = 0$, where n is a unit vector and $p > 0$.
22. What figure in n-space is described by the equation $(\mathbf{x} - \mathbf{a}, \mathbf{x} - \mathbf{a}) = \alpha$ where a is a fixed number and **a** a fixed vector. (Consider the cases $\alpha > 0$, $\alpha = 0$, and $\alpha < 0$.)

Problems

Problems 1 to 4 refer to vectors in the *plane*.

1. In the text, we defined the product of a scalar and a vector and then verified that the rules **6** to **10** hold (see §1). Show that, if we assume the rules **1** to **10**, then one can prove that the product of a rational number and a vector must

be what we defined it to be. (Hint: Consider first the case in which the rational number is an integer.)

2. Show that two vectors are independent if and only if one is not a scalar multiple of another.

3. Let $\{e_1, e_2\}$ be a frame, and $a = \alpha_1 e_1 + \alpha_2 e_2$ and $b = \beta_1 e_1 + \beta_2 e_2$ two vectors. Show that the two vectors are independent if and only if $\alpha_1 \beta_2 - \alpha_2 \beta_1 \neq 0$.

4. Prove the theorem stated in the text without a proof: two vectors form a basis of vectors in the plane if and only if they are independent.

5. Given a sequence of vectors, $a_1, a_2, a_3, \cdots$ (in the plane or in space) and a vector a, the statement $a = \lim a_j$ is defined to mean that $\lim |a_j - a| = 0$. Which of the theorems in Chapter 8, §3 have valid generalizations for vectors?

6. Define convergence and divergence of an infinite series of vectors. State which of the theorems in Chapter 8, §4 have valid generalizations for vectors. In particular, prove that, if the series of numbers $\Sigma_{j=1}^{j=\infty} |a_j|$ converges, then the series of vectors $\Sigma_{j=1}^{j=\infty} a_j$ converges. What else can you say about the latter series?

Problems 7 and 8 refer to a **general linear coordinate system** in the plane. Such a system is defined by a point O, called the origin, and by a basis of vectors $\{e_1, e_2\}$. Let P be a point and let P be its position vector with respect to O, that is, the vector defined by the segment $\overrightarrow{OP}$. Then there are numbers x_1 and x_2 such that $P = x_1 e_1 + x_2 e_2$. The numbers (x_1, x_2) are called the coordinates of P in the system (O, e_1, e_2).

Let E_1 and E_2 be the points with coordinates $(1, 0)$ and $(0, 1)$, in the system considered. The coordinate axes are now the lines through O and E_1 (the x_1 axis) and through O and E_2 (the x_2 axis). They are not necessarily perpendicular. Neither do we have one unit of length: along the x_1 axis, we compare lengths with that of e_1, along the x_2 axis with that of e_2.

7. Let E_1 and E_2 have Cartesian coordinates $(2, 0)$ and $(1, 1)$, respectively, and let O be the origin in the Cartesian system. Draw the x_1 axis and the x_2 axis. Draw the points with coordinates, in the (O, e_1, e_2) system, equal to $(1, 1)$, $(2, -3)$, $(-1, 0)$, $(-2, -1)$. Find the Cartesian coordinates of these points.

8. Repeat Problem 7, assuming this time that the Cartesian coordinates of E_1 and E_2 are $(-1, 1)$ and $(1, 1)$, respectively.

Problems 9 to 11 refer to vectors in *ordinary space*.

9. Let a_1, a_2, a_3 be three vectors. Show that the following conditions are equivalent: (1) the three vectors are not independent, (2) one of the three vectors can be written as a linear combination of the other two, (3) there are three numbers $\alpha_1, \alpha_2, \alpha_3$, not all of them 0, such that $\alpha_1 a_1 + \alpha_2 a_2 + \alpha_3 a_3 = 0$. (We say that a vector x is a **linear combination** of vectors $y_1, y_2, \cdots, y_k$ if $x = \xi_1 y_1 + \xi_2 y_2 + \cdots + \xi_k y_k$, where the ξ_j are numbers.)

10. Prove the theorem stated in §1 without proof: three vectors in space form a basis if and only if they are independent.

11. Define a general linear coordinate system in space, by generalizing the definition given before Problem 7.

Problems 12, 13 refer to a *vector space*, that is, to any collection of elements, called vectors, for which addition and multiplication by numbers (scalars) is defined in such a way that the postulates **1** to **10** in §1 hold. A vector space is called n-dimensional if Postulate 11_n in §7.1 is valid.

12. Given m vectors $\mathbf{a}_1, \cdots, \mathbf{a}_m$, show that the following two statements are equivalent: (1) there are numbers $\alpha_1, \cdots, \alpha_m$, not all of them 0, such that $\alpha_1 \mathbf{a}_1 + \alpha_2 \mathbf{a}_2 + \cdots + \alpha_m \mathbf{a}_m = 0$, (2) one of the m vectors is a linear combination of the other $m - 1$. (Hint: Compare Problem 9.) Vectors $\mathbf{a}_1, \cdots, \mathbf{a}_m$ having these properties are called **dependent**.

13. Show that, if m vectors are independent, so are any $m - 1$ of them.

14. Show that, in the space of n-tuples of real numbers, any $n + 1$ vectors are dependent. (If you never saw the proof, you might find this a difficult problem.)

15. Show that n vectors in the space of n-tuples form a basis if and only if they are independent.

16. Consider the set of all polynomials $x \mapsto p(x)$. Polynomials can be added, and a polynomial can be multiplied by a number. Show that the polynomials form a vector space, that is, that the postulates **1** to **10** of §1 are valid. Show that the same is true for polynomials of degree not greater than some integer k.

17. Show that the space of polynomials of degree not greater than k is $(k + 1)$-dimensional, and conclude that the space of all polynomials is not n-dimensional for any n. (Such a space is called **infinitely dimensional**.)

18. Suppose that, in the space of 4-tuples of real numbers, we define the scalar product of two vectors $\mathbf{a} = (\alpha_1, \alpha_2, \alpha_3, \alpha_4)$ and $\mathbf{b} = (\beta_1, \beta_2, \beta_3, \beta_4)$ not as in §7.3 but by the formula

$$(a, b) = \alpha_1 \beta_1 + 2\alpha_2 \beta_2 + 3\alpha_3 \beta_3 + 100\alpha_4 \beta_4.$$

Does this make the space into an inner product space, that is, do the rules **I** to **V** of §6.3 hold? (If not, which postulate fails to be true?)

19. Answer the question in Problem 18 for the following definitions:

$$(a, b) = \alpha_1 \beta_1 + \alpha_2 \beta_2 - \alpha_3 \beta_3 - \alpha_4 \beta_4,$$
$$(a, b) = \alpha_1 \beta_1 + 100\alpha_4 \beta_4.$$

Problems 20 to 23 depend on calculus.

20. Show that all bounded continuous functions $x \mapsto f(x)$ defined for $0 \leq x \leq 1$ form an infinitely dimensional vector space. (Hint: Remember that polynomials are continuous functions.)

21. In the vector space of Problem 20, define

$$(f, g) = \int_0^1 f(x)g(x)\, dx.$$

Show that the postulates of an inner product space hold.

22. Write out the Schwarz inequality and the triangle inequality for the inner product just defined. (Compare Problems 12 and 13 in Chapter 5.)

23. Compute the angle between the "vectors" $x \mapsto x^2$ and $x \mapsto 2x - 1$ in the vector space of Problems 20 and 21.

10/Quadrics

This chapter continues the development of plane analytic geometry. It makes practically no use of calculus and may be read immediately after the first three sections of Chapter 9. Students familiar with analytic geometry may omit most of the sections.

The chapter presents the theory of quadric curves (conic sections) together with a discussion of rotating a coordinate system. Only the material in §1 will be used in subsequent chapters.

The appendix continues the development of solid analytic geometry. It contains a brief introduction to quadric surfaces.

The parabola, discussed in Chapter 2, §4.1, can be defined as a curve that, in a suitably chosen Cartesian coordinate system, satisfies the equation

$$y^2 = 4ax \qquad a > 0 \text{ a constant.}$$

(We interchange x and y, and write a instead of p, for convenience.) The parabola is one of the beautiful curves discovered by the Greeks and named by them conic sections or **conics.** The others are ellipses and hyperbolas. An **ellipse** is a curve that, in an appropriate Cartesian coordinate system, is defined by an equation

$$\frac{x^2}{a^2} + \frac{y^2}{b^2} = 1 \qquad a \geq b > 0 \text{ constants.}$$

A **hyperbola** is a curve that, in an appropriate Cartesian coordinate system, is defined by an equation

$$\frac{x^2}{a^2} - \frac{y^2}{b^2} = 1 \qquad a > 0,\, b > 0 \text{ constants.}$$

These two curves, for $a = 4$ and $b = 3$, are shown in Fig. 10.1. An ellipse with $a = b$ is a **circle.** A hyperbola with $a = b$ is an **equilateral hyperbola.**

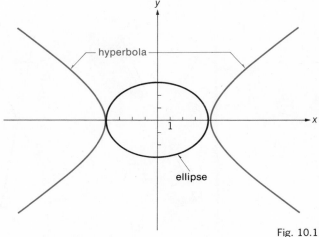

Fig. 10.1

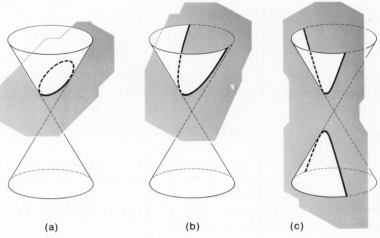

(a)   (b)   (c)

Fig. 10.2

The Greek mathematicians did not, of course, use the algebraic definition. They started instead, with a circular cone (a surface obtained by rotating a line about a line intersecting it) and asked what curves are obtained by intersecting the cone by a plane. It turns out that if the plane does not pass through the vertex ("tip") of the cone, then the intersection is either (a) an ellipse or (b) a parabola or (c) a hyperbola, see Fig. 10.2. In cases (a) and (b), the plane meets only one "nappe" of the cones; in case (b), it meets both nappes. If the plane passes through the vertex of the cone, the intersection is either (a) a point, or (b) a line, or (c) a pair of lines, see Fig. 10.3. We shall prove the statement just made in the appendix to this chapter (§4.9).

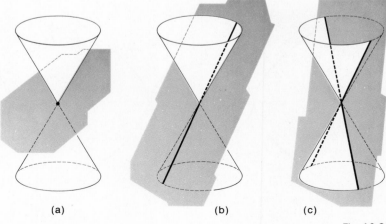

(a)   (b)   (c)

Fig. 10.3

1.2 Standard equations

We develop the theory of conics using the three **standard equations**

(P) $$y^2 = 4ax \qquad a > 0,$$

(E) $$\frac{x^2}{a^2} + \frac{y^2}{b^2} = 1 \qquad a \geq b > 0,$$

(H) $$\frac{x^2}{a^2} - \frac{y^2}{b^2} = 1 \qquad a > 0, b > 0.$$

The Greeks developed this theory by purely geometrical methods. Their results are summarized in a famous treatise by Apollonius, one of the most striking achievements of Greek mathematics. The invention of analytic geometry transformed a subject accessible only to the best minds of antiquity into a source of exercises for beginning students.

All information on conics is contained in the three simple equations (P), (E), and (H), and can be extracted from these equations by the rules of algebra. In mathematics, each succeeding generation builds on the work of its predecessors, preserving in a compact and remarkably efficient way all the knowledge that has been gathered in the past. The work of Apollonius is alive in analytic geometry, and is in some sense better understood by us than it was by him.

The curves described by the equations (P), (E), and (H) will be called **conics in standard position.**

It is often convenient to discuss the ellipse (E) and the hyperbola (H) together. The numbers a and b are called lengths of the **semiaxes.** The x axis is called the **axis** of the hyperbola, the **major axis** of the ellipse. The y axis is called the **conjugate axis** of the hyperbola, the **minor axis** of the ellipse. The ellipse and hyperbola are both symmetric about both axes, since replacing x by $-x$ or y by $-y$ does not affect equations (E) and (H). The origin is called the **center** of both curves.

We recall that the parabola (P) has the x axis as axis; it has no center. The equation shows that the parabola is symmetric about its axis.

Equation (E) implies that $|x| \leq a$, $|y| \leq b$. Thus the ellipse lies completely inside the box formed by the lines $x = \pm a$, $y = \pm b$. The four points at which the ellipse intersects the axes (and touches the box) are called the **vertices** of the ellipse (see Fig. 10.4).

Equation (H) shows that $|x| \geq a$. Thus the hyperbola consists of two halves, called its **branches.** The hyperbola intersects the axis at two points called **vertices** (see Fig. 10.5).

We recall that the parabola (P) has one vertex, the origin. The equation shows that the curve lies entirely to the right of the line $x = 0$.

APOLLONIUS of Perga (ca. 260–170 B.C.) taught in Alexandria and at Pergammon. Of the eight books (that is, parts) of *Conics*, seven survived, some only in Arabic. Other works of Apollonius also survived in parts.

Fig. 10.4

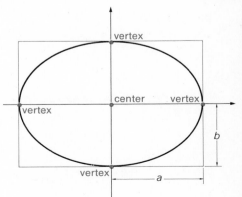

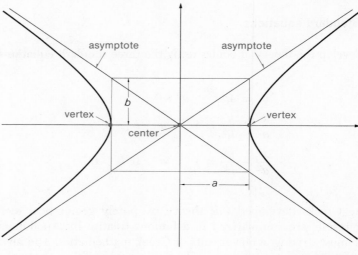

Fig. 10.5

1.3 Asymptotes

The diagonals of the box $|x| < a$, $|y| < b$ are the lines $y = (b/a)x$ and $y = -(b/a)x$. These are called **asymptotes** of the hyperbola (H). It is convenient to write the equation of the asymptotes in the form

(1)
$$\frac{x^2}{a^2} - \frac{y^2}{b^2} = \left(\frac{x}{a} + \frac{y}{b}\right)\left(\frac{x}{a} - \frac{y}{b}\right) = 0.$$

Let (x, y) be a point on the upper half of the right branch of the hyperbola (everything we shall say can be repeated, with obvious modifications, for the lower half and for the left branch). For the distance δ shown in Fig. 10.6, we have

$$\delta = \frac{b}{a}x - y = \frac{b}{a}x - b\sqrt{\frac{x^2}{a^2} - 1} = \frac{b}{a}\left(x - a\sqrt{\frac{x^2}{a^2} - 1}\right)$$
$$= \frac{b}{a}(x - \sqrt{x^2 - a^2}).$$

This shows that the curve lies under the asymptote $x \mapsto (b/a)x$, since $x - \sqrt{x^2 - a^2} > 0$. Also,

$$x - \sqrt{x^2 - a^2} = \frac{a^2}{x + \sqrt{x^2 - a^2}},$$

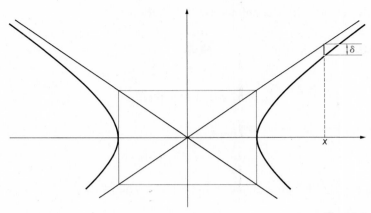

Fig. 10.6

which shows that

$$\lim_{x \to +\infty} (x - \sqrt{x^2 - a^2}) = 0,$$

so that

$$\lim_{x \to +\infty} \delta = 0.$$

Therefore the vertical distance δ between the curve and the asymptote is as small as we like if we go far enough to the right.

1.4 Foci and directrices. Eccentricity

All interesting geometric properties of the conics are connected with certain points called **foci**. The ellipse (E) and the hyperbola (H) have two foci each, a left focus F_-, and a right focus F_+:

$$F_- = (-c, 0), \qquad F_+ = (c, 0),$$

where the **focal distance** c is

(2) $\qquad c = \sqrt{a^2 - b^2} \qquad$ for the ellipse,

(3) $\qquad c = \sqrt{a^2 + b^2} \qquad$ for the hyperbola.

The geometric construction of the foci is shown in Fig. 10.7. Note that the two foci of the ellipse coincide with the center if $a = b$, that is, if the ellipse is a circle. We recall (see Chapter 2, §4.1) that the parabola (P) has one focus,

(4) $\qquad F = (a, 0);$

it will be convenient to call it a left focus.

Fig. 10.7

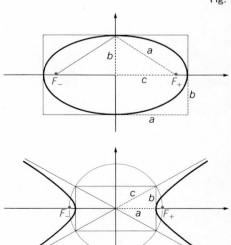

The **eccentricity** ϵ is defined by the relation

(5) $$c = \epsilon a,$$

that is, $$\epsilon = \frac{c}{a}.$$

Thus, $$0 \le \epsilon < 1 \text{ for an ellipse}$$

(and $\epsilon = 0$ for a circle) and

$$1 < \epsilon \text{ for a hyperbola}$$

(with $\epsilon = \sqrt{2}$ for an equilateral hyperbola). The relation

$$\epsilon = 1 \text{ for a parabola}$$

holds, by definition.

The lines

(6) $$l_-: x = -\frac{a}{\epsilon}, \qquad l_+: x = \frac{a}{\epsilon}$$

are called the left and right **directrices** of the ellipse (E) or hyperbola (H). The circle has no directrices. We recall that the line

(7) $$l: x = -a$$

is called the directrix of the parabola (P). It will be convenient to call it the left directrix. No point on a conic lies on its directrix (see Fig. 10.8).

Lemma   IF $P = (x, y)$ IS A POINT ON A CONIC IN STANDARD POSITION, THEN

(8) $$|\overrightarrow{F_-P}| = |a + \epsilon x|,$$
(9) $$|\overrightarrow{F_+P}| = |a - \epsilon x|,$$

(10)   DISTANCE FROM P TO $l_- = \left| \dfrac{a}{\epsilon} + x \right|,$

(11)   DISTANCE FROM P TO $l_+ = \left| \dfrac{a}{\epsilon} - x \right|.$

The relations (10) and (11) are seen in Fig. 10.8. We prove (8) only for an ellipse. The proof for a hyperbola is similar, that for a parabola even simpler. These proofs, as well as the analogous proof of (9), are left to the reader.

Fig. 10.8

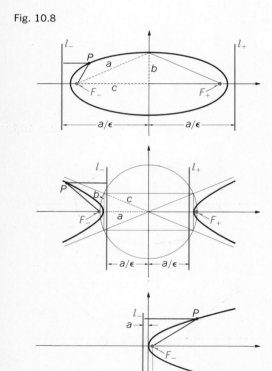

The proof of (8) for an ellipse proceeds as follows. If $P = (x, y)$ satisfies (E), then $y^2 = b^2 - (b^2/a^2)x^2$, and therefore

$$\begin{aligned}
|\overrightarrow{F_-P}|^2 &= (x + c)^2 + y^2 = x^2 + 2cx + c^2 + b^2 - \frac{b^2}{a^2}x^2 \\
&= x^2\left(1 - \frac{b^2}{a^2}\right) + 2cx + a^2 \qquad \text{since } c^2 + b^2 = a^2 \\
&= x^2\frac{a^2 - b^2}{a^2} + 2cx + a^2 = a^2 + 2cx + \frac{c^2}{a^2}x^2 \\
&= \left|a + \frac{c}{a}x\right|^2 = |a + \epsilon x|^2 \qquad \text{since } c = a\epsilon.
\end{aligned}$$

This is the assertion to be established.

EXERCISES

In Exercises 1 to 8, sketch the conic whose equation is given. Find whichever of the following are relevant and clearly indicate these on your sketch: center, focus, vertex, asymptote, axis, and directrix.

1. $x^2/4 - y^2/16 = 1$.
2. $x^2 - 4y^2 = 36$.
3. $x^2 + 4y^2 = 36$.
4. $y^2 = 36x$.
5. $16x^2 - y^2 = 4$.
6. $x^2/25 + y^2/4 = 1$.
7. $x^2 - 4y^2 = 100$.
8. $3x^2 + 4y^2 = 12$.

1.5 Geometric properties of conics

We can now state the main geometric theorem on conics.

Theorem 1 CONSIDER A CONIC OF ECCENTRICITY ϵ.

(1) IF $0 \leq \epsilon < 1$ (ELLIPSE), THEN FOR ALL POINTS ON THE CONIC THE SUM OF ITS DISTANCES FROM THE TWO FOCI HAS THE SAME VALUE.

(2) IF $\epsilon > 1$ (HYPERBOLA), THEN FOR ALL POINTS ON THE CONIC THE DIFFERENCE OF THE DISTANCES FROM THE TWO FOCI HAS THE SAME ABSOLUTE VALUE.

(3) IF $\epsilon > 0$ (NOT A CIRCLE), THEN FOR EVERY POINT ON THE CONIC THE DISTANCE FROM A FOCUS IS ϵ TIMES THE DISTANCE FROM A DIRECTRIX (RIGHT FOCUS AND RIGHT DIRECTRIX OR LEFT FOCUS AND LEFT DIRECTRIX).

EACH OF THESE PROPERTIES CHARACTERIZES THE CONIC.

The meaning of the theorem is shown in Figs. 10.9, 10.10, and 10.11.

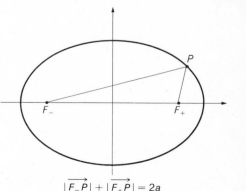

Fig. 10.9

$$|\overrightarrow{F_-P}| + |\overrightarrow{F_+P}| = 2a$$

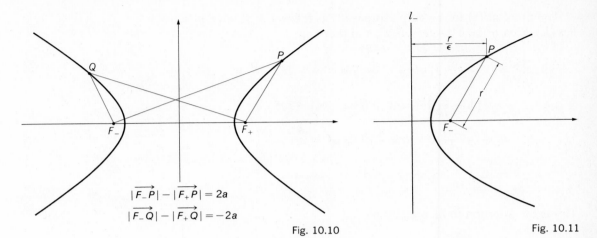

$$|\overrightarrow{F_-P}| - |\overrightarrow{F_+P}| = 2a$$
$$|\overrightarrow{F_-Q}| - |\overrightarrow{F_+Q}| = -2a$$

Fig. 10.10   Fig. 10.11

Proof. It will suffice to verify (1), (2), and (3) for conics in standard position.

(1) If $P = (x, y)$ satisfies (E), then $|x| \leq a$, and since $0 \leq \epsilon < 1$ we have $a + \epsilon x > 0$, $a - \epsilon x > 0$. Therefore, by (8) and (9), $|\overrightarrow{F_-P}| = a + \epsilon x$, $|\overrightarrow{F_+P}| = a - \epsilon x$, and $|\overrightarrow{F_-P}| + |\overrightarrow{F_+P}| = 2a$, the same value for all P.

(2) If $P = (x, y)$ satisfies (H) and lies on the right branch of the hyperbola, then $a < x$. Since $\epsilon > 1$, we obtain from (8) and (9) that $|\overrightarrow{F_-P}| = a + \epsilon x$, $|\overrightarrow{F_+P}| = -a + \epsilon x$, and $|\overrightarrow{F_-P}| - |\overrightarrow{F_+P}| = 2a$. In the same way one shows that, on the left branch, $|\overrightarrow{F_-P}| - |\overrightarrow{F_+P}| = -2a$.

(3) The statement follows at once from equations (8) to (11). Recall that the focus and directrix of the parabola are left ones, by definition.

We must next show that each of the geometric conditions (1), (2), and (3) leads to the corresponding algebraic equation. We shall consider only two cases, Condition (1) and Condition (3) for $\epsilon > 1$. The reader is urged to treat the other cases.

(1) We are given two points F_1 and F_2 and a number a such that $2a > |\overrightarrow{F_1F_2}|$. We must show that the set of all points P such that

$$|\overrightarrow{PF_1}| + |\overrightarrow{PF_2}| = 2a$$

is an ellipse with foci F_1 and F_2. The case $F_1 = F_2$ is trivial; we obtain a circle of radius a. Assume that $F_1 \neq F_2$ and set $|\overrightarrow{F_1F_2}| = 2c$. We choose a Cartesian coordinate system such that the midpoint of $\overrightarrow{F_1F_2}$ is the origin and F_2 lies on the positive ray of the x axis. Then $F_1 = (-c, 0)$ and $F_2 = (c, 0)$ and, if $P = (x, y)$, we have

$$|\overrightarrow{PF_1}|^2 = (x + c)^2 + y^2, \qquad |\overrightarrow{PF_2}|^2 = (x - c)^2 + y^2.$$

The condition $|\overrightarrow{F_1P}| + |\overrightarrow{F_2P}| = 2a$ reads

$$\sqrt{x^2 + 2cx + c^2 + y^2} + \sqrt{x^2 - 2cx + c^2 + y^2} = 2a.$$

If this holds, then also $(\sqrt{\cdots} + \sqrt{\cdots})^2 = 4a^2$, or

$$x^2 + 2cx + c^2 + y^2 + x^2 - 2cx + c^2 + y^2$$
$$+ 2\sqrt{(x^2 + c^2 + y^2)^2 - 4c^2x^2} = 4a^2$$

or (collecting terms, transposing, and dividing by 2)

$$\sqrt{(x^2 + y^2 + c^2)^2 - 4c^2x^2} = 2a^2 - (x^2 + y^2 + c^2).$$

If so, then (squaring both sides)

$$(x^2 + y^2 + c^2)^2 - 4c^2x^2 = 4a^4 + (x^2 + y^2 + c^2)^2 - 4a^2(x^2 + y^2 + c^2)$$

or (after obvious manipulations)

$$a^4 - a^2x^2 - a^2y^2 - a^2c^2 + c^2x^2 = 0,$$

which is the same as

$$a^2(a^2 - c^2) = (a^2 - c^2)x^2 + a^2y^2.$$

Set $b = \sqrt{a^2 - c^2}$. Our equation becomes $b^2x^2 + a^2y^2 = a^2b^2$ or

$$\frac{x^2}{a^2} + \frac{y^2}{b^2} = 1.$$

This is the standard equation (E) of an ellipse; the foci are $(-c, 0)$ and $(c, 0)$. We already know that all points (x, y) satisfying (E) have the required property.

(3) We are given a line l, a point F not on the line, and a number $\epsilon > 1$. We want to find all points P such that

(12) $\qquad |\overrightarrow{PF}| = \epsilon$ times the distance from P to l.

We choose a coordinate system such that F is the origin and l is vertical and intersects the positive ray of the x axis at a point $x = -\alpha$, $\alpha > 0$ (see Fig. 10.12). If $P = (x, y)$, then the distance from P to l is $|x + \alpha|$ and $|\overrightarrow{FP}| = \sqrt{x^2 + y^2}$. If (12) holds, then

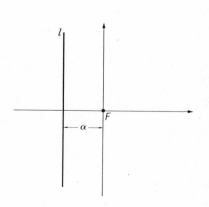

Fig. 10.12

$$x^2 + y^2 = \epsilon^2(x + \alpha)^2 = \epsilon^2 x^2 + 2\alpha\epsilon^2 x + \epsilon^2\alpha^2,$$

or
$$(\epsilon^2 - 1)x^2 + 2\alpha\epsilon^2 x - y^2 + \epsilon^2\alpha^2 = 0,$$

that is,

$$(\epsilon^2 - 1)\left(x^2 + \frac{2\alpha\epsilon^2}{\epsilon^2 - 1}x + \frac{\alpha^2\epsilon^4}{(\epsilon^2 - 1)^2}\right) - y^2 - \frac{\alpha^2\epsilon^4}{\epsilon^2 - 1} + \epsilon^2\alpha^2 = 0,$$

or
$$(\epsilon^2 - 1)\left(x + \frac{\alpha\epsilon^2}{\epsilon^2 - 1}\right)^2 - y^2 = \frac{\alpha^2\epsilon^2}{\epsilon^2 - 1}.$$

Set

(13)
$$\xi = x + \frac{\alpha\epsilon^2}{\epsilon^2 - 1}, \qquad \eta = y.$$

Our equation becomes (in the new "translated" coordinates ξ, η)

$$(\epsilon^2 - 1)^2\xi^2 - \eta^2 = \frac{\alpha^2\epsilon^2}{\epsilon^2 - 1}$$

or

(14)
$$\frac{(\epsilon^2 - 1)}{\alpha^2\epsilon^2}\xi^2 - \frac{\epsilon^2 - 1}{\alpha^2\epsilon^2}\eta^2 = 1.$$

This is a standard equation of a hyperbola. To show that all points of this hyperbola have the desired property, we must verify that the eccentricity is ϵ and that F and l are the focus and directrix, respectively.

Now, from (14), we have

$$a = \frac{\alpha\epsilon}{\epsilon^2 - 1}, \qquad b = \frac{\alpha\epsilon}{\sqrt{\epsilon^2 - 1}}.$$

Therefore

$$c^2 = a^2 + b^2 = \alpha^2\epsilon^2\left(\frac{1}{(\epsilon^2 - 1)^2} + \frac{1}{\epsilon^2 - 1}\right) = \frac{\alpha^2\epsilon^4}{(\epsilon^2 - 1)^2}.$$

Hence $c = \dfrac{\alpha\epsilon^2}{\epsilon^2 - 1} = a\epsilon$, so that ϵ is indeed the eccentricity. The foci lie on the ξ axis, which is also the x axis. They have the ξ coordinates $\xi = \pm c = \pm\dfrac{\alpha\epsilon^2}{\epsilon^2 - 1}$. Hence the x coordinates of the foci are [compare (13)] $x = \pm\dfrac{\alpha\epsilon^2}{\epsilon^2 - 1} - \dfrac{\alpha\epsilon^2}{\epsilon^2 - 1}$. Hence $F = (0, 0)$ is the right focus. Finally, the right directrix of (14) is the line

$$\xi = \frac{a}{\epsilon} = \frac{\alpha}{\epsilon^2 - 1},$$

that is,
$$x = \frac{\alpha}{\epsilon^2 - 1} - \frac{\alpha\epsilon^2}{\epsilon^2 - 1} = -\alpha.$$

But this is the line l with which we started. This completes the argument.

▶ *Example*   Find the equation of the ellipse with foci at $(0, 0)$ and $(0, 8)$ and major semiaxis $a = 5$.

First Solution.   We use the property (1) of Theorem 1. The distances from a point (x, y) to the foci are $\sqrt{x^2 + y^2}$ and $\sqrt{x^2 + (y - 8)^2}$. Hence condition (1) reads

$$\sqrt{x^2 + y^2} + \sqrt{x^2 + (y - 8)^2} = 10.$$

Simplifying, we get

$$\sqrt{x^2 + y^2 - 16y + 64} = 10 - \sqrt{x^2 + y^2}$$

or $\qquad x^2 + y^2 - 16y + 64 = 100 + x^2 + y^2 - 20\sqrt{x^2 + y^2}$

or $\qquad\qquad 16y + 36 = 20\sqrt{x^2 + y^2}$

or $\qquad\qquad\quad 4y + 9 = 5\sqrt{x^2 + y^2}$

or $\qquad\qquad 16y^2 + 72y + 81 = 25x^2 + 25y^2$

or $\qquad\qquad 25x^2 + 9y^2 - 72y - 81 = 0.$

Second Solution.   We shall use the standard equation. To make the major axis horizontal, we introduce new coordinates X, Y by rotating the coordinate system by $90°$:

(15) $$X = y, \qquad Y = -x.$$

The foci are $X = 0$, $Y = 0$ and $X = 8$, $Y = 0$. In order to make the foci symmetric with respect to the origin, we introduce new coordinates by a translation (see Fig. 10.13).

$$\xi = X - 4, \qquad \eta = Y.$$

The foci are $\xi = \pm 4$, $\eta = 0$. In the ξ, η coordinates the desired ellipse is in standard position. We have $a = 5$, $c = 4$. Therefore $b = \sqrt{a^2 - c^2} = 3$. The desired equation reads

$$\frac{\xi^2}{25} + \frac{\eta^2}{9} = 1,$$

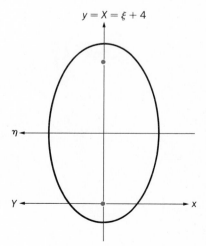

Fig. 10.13

$y = X = \xi + 4$

that is,
$$\frac{(X-4)^2}{25} + \frac{Y^2}{9} = 1,$$

that is,
$$\frac{X^2 - 8X + 16}{25} + \frac{Y^2}{9} = 1$$

or, using (15) and simplifying,

$$25x^2 + 9y^2 - 72y - 81 = 0,$$

as before. ◄

EXERCISES

9. Find the equation of the ellipse with foci at $(0, 0)$ and $(4, 0)$ and major semi-axis of length $a = \sqrt{5}$.
10. Find the equation of the ellipse with foci at $(-1, 0)$ and $(5, 0)$ and major semiaxis of length $a = 2\sqrt{3}$.
11. Find the equation of the ellipse with foci at $(-2, 1)$ and $(6, 0)$ and eccentricity $\epsilon = \frac{4}{5}$.
12. Find the equation of the ellipse with foci at $(0, 0)$ and $(0, 10)$ and major semi-axis of length $a = 5\sqrt{5}$.
13. Find the equation of the ellipse with center at $(1, -1)$, major semiaxis of length $a = 6$ and eccentricity $\epsilon = \frac{2}{3}$ if the major axis is horizontal.
14. Find the equation of each ellipse with one vertex at $(1, 1)$ and one vertex at $(3, 5)$ if the major axis is vertical.
15. Find the foci, vertices, eccentricity, and major and minor axes of the ellipse $9x^2 + 25y^2 - 225 = 0$.
► 16. Find the equation of the hyperbola with foci at $(0, 0)$ and $(6, 0)$ and eccentricity $\epsilon = \frac{3}{2}$.
17. Find the equation of the hyperbola with foci at $(-2, -1)$ and $(6, -1)$ and one vertex at $(3, -1)$.
18. Find the equation of the hyperbola with foci at $(0, -3)$ and $(0, 5)$ and one vertex at $(0, 1 + \sqrt{10})$.
19. Find the equation of the hyperbola with vertices at $(1, 2)$ and $(1, 12)$ and eccentricity $\epsilon = 2$.
20. Find the equation of the hyperbola with vertices at $(-2, 0)$ and $(-2, 16)$ and one focus at $(-2, 18)$.
21. Find the equation of the hyperbola whose asymptotes are given by $y = \pm\frac{3}{2}(x - 1) + 2$ and which passes through the point $(4\sqrt{3}, 6\sqrt{2})$.
22. Find the foci, vertices, eccentricity, and asymptotes of the hyperbola $25x^2 - 9y^2 = 225$.

1.6 Tangents

The theory of tangents to conics can be developed either by "elementary" methods, or by calculus. (We already know that this is so for circles and parabolas.) We shall describe both approaches, and we start with the approach via calculus.

(Readers who did not yet read Chapter 4 may omit the following lines, for the time being.)

The upper and lower halves of the parabola (P) are graphs of the functions

$$x \mapsto y = 2\sqrt{ax} \qquad \text{and} \qquad x \mapsto y = -2\sqrt{ax}.$$

For $x \neq 0$, these functions have the derivatives

$$\frac{dy}{dx} = \sqrt{\frac{a}{x}}, \qquad \frac{dy}{dx} = -\sqrt{\frac{a}{x}}.$$

Note that both relations may be written as

$$(16) \qquad \frac{dy}{dx} = \frac{1}{2}\frac{y}{x} = \frac{2a}{y} \qquad \text{on the parabola.}$$

The upper and lower half of the ellipse (E) and hyperbola (H) are graphs of the functions

$$x \mapsto y = b\sqrt{1 - \frac{x^2}{a^2}}, \qquad x \mapsto y = -b\sqrt{1 - \frac{x^2}{a^2}}$$

and

$$x \mapsto y = b\sqrt{\frac{x^2}{a^2} - 1}, \qquad x \mapsto y = -b\sqrt{\frac{x^2}{a^2} - 1}.$$

Differentiation yields, for $x^2 \neq a^2$,

$$\frac{dy}{dx} = -\frac{b}{a^2}\frac{x}{\sqrt{1 - x^2/a^2}}, \qquad \frac{dy}{dx} = \frac{b}{a^2}\frac{x}{\sqrt{1 - x^2/a^2}}$$

and

$$\frac{dy}{dx} = \frac{b}{a^2}\frac{x}{\sqrt{x^2/a^2 - 1}}, \qquad \frac{dy}{dx} = -\frac{b}{a^2}\frac{x}{\sqrt{x^2/a^2 - 1}}.$$

The formulas can be combined into

$$(17) \qquad \frac{dy}{dx} = -\frac{b^2}{a^2}\frac{x}{y} \qquad \text{on the ellipse,}$$

$$(18) \qquad \frac{dy}{dx} = \frac{b^2}{a^2}\frac{x}{y} \qquad \text{on the hyperbola.}$$

Theorem 2   LET (x_0, y_0) BE A POINT ON A CONIC IN STANDARD POSITION. THE EQUATION OF THE TANGENT TO THE CONIC PASSING THROUGH THIS POINT IS

$$(19) \qquad yy_0 = 2a(x + x_0) \qquad (\text{PARABOLA}),$$

$$(20) \qquad \frac{xx_0}{a^2} + \frac{yy_0}{b^2} = 1 \qquad (\text{ELLIPSE}),$$

$$(21) \qquad \frac{xx_0}{a^2} - \frac{yy_0}{b^2} = 1 \qquad (\text{HYPERBOLA}).$$

We prove this for the hyperbola. The reader should verify the other two cases.

If x_0, y_0 is a vertex of the hyperbola, then $y_0 = 0$, $x_0 = a$, or $-a$. Equation (21) is then the equation of the vertical line $x = a$ or $x = -a$, as required.

We assume now that (x_0, y_0) lies on the upper half of the right branch of the hyperbola (other points are treated similarly). The slope of the tangent at $x = x_0$ is, by (18),

$$m = \left(\frac{dy}{dx}\right)_{x=x_0} = \frac{b^2}{a^2}\frac{x_0}{y_0}.$$

The equation of the tangent is

$$y - y_0 = m(x - x_0),$$

or

$$y - y_0 = \frac{b^2}{a^2}\frac{x_0}{y_0}(x - x_0),$$

that is,

$$a^2 yy_0 - a^2 y_0{}^2 = b^2 x_0 x - b^2 x_0{}^2,$$

or, dividing by $a^2 b^2$ and transposing,

$$\frac{xx_0}{a^2} - \frac{yy_0}{b^2} = \frac{x_0{}^2}{a^2} - \frac{y_0{}^2}{b^2}.$$

Since (x_0, y_0) lies on the hyperbola, the right-hand side equals 1, and equation (21) results.

The next theorem gives an elementary description of tangents. We forget calculus, *define* the tangents by equations (19), (20), or (21), and show that these lines have a remarkable property.

Theorem 3   LET P BE A POINT ON A CONIC. THE TANGENT TO THE CONIC AT P IS THE ONLY LINE THROUGH P THAT DOES NOT MEET THE CONIC AT ONE MORE POINT (AND, IN THE CASE OF A PARABOLA, IS NOT PARALLEL TO THE AXIS).

We already know this to be true for a circle and for a parabola (see Chapter 2, §§3.2 and 4.4). We shall give the proof for the hyperbola $x^2 - y^2 = 1$. The proofs for other hyperbolas and for ellipses are similar, but the calculations look more formidable (see also §4.9).

It is evident that a vertical line either meets $x^2 - y^2 = 1$ at two points, or not at all, or is a tangent to the hyperbola at one of the vertices. We ask now when a nonvertical line $y = mx + p$ meets $x^2 - y^2 = 1$ in exactly one point (x_0, y_0). We substitute $y = mx + p$ into the equation of the conic and obtain the quadratic equation

$$x^2 - (mx + p)^2 = 0$$

or
$$(1 - m^2)x^2 - 2mpx - (1 + p^2) = 1.$$

There will be precisely one root (see Chapter 3, §2.2) if

$$m^2p^2 + (1 - m^2)(1 + p^2) = 0$$

or
$$1 - m^2 + p^2 = 0.$$

The roots are the x coordinates of the intersection of the line and the conic. If so, the single root x_0 is $x_0 = -m/p$, the corresponding value of y_0 is $y_0 = mx_0 + p = -1/p$, and $m = x_0/y_0$, $p = -1/y_0$ as the reader is asked to verify. The equation of our line reads

$$y = \frac{x_0}{y_0} x - \frac{1}{y_0}$$

or
$$x_0 x - y_0 y = 1,$$

which is precisely (21) for $a = b = 1$.

▶ **Example**   Find the tangent to the ellipse $x^2 + 4y^2 = 32$ which has slope $\frac{1}{2}$.

First Solution.   By Theorem 2, every tangent has the equation

$$xx_0 + 4yy_0 = 32,$$

where (x_0, y_0) is the point of tangency. The slope of this line must be $\frac{1}{2}$; thus we need $-x_0/4y_0 = \frac{1}{2}$ or $x_0 = -2y_0$. Since (x_0, y_0) lies on our ellipse, we must also have $x_0^2 + 4y_0^2 = 32$ or $8y_0^2 = 32$. Hence either $y_0 = 2$ and $x_0 = -4$ or $y_0 = -2$ and $x_0 = 4$. We get two tangents with the desired slope, $-4x + 8y = 32$ and $4x - 8y = 32$, that is,

$$x - 2y = -8, \qquad x - 2y = 8;$$

they are shown in Fig. 10.14.

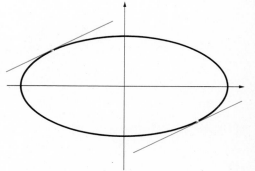

Fig. 10.14

Second Solution. We make use of Theorem 3. The equation of a line with slope $\frac{1}{2}$ reads $y = \frac{1}{2}x + \text{const}$, or $x - 2y = p$ where p is some number. We shall determine p so that the line meets our ellipse in exactly one point. Substituting $x = 2y + p$ into $x^2 + 4y^2 = 32$, we obtain the quadratic equation

$$(2y + p)^2 + 4y^2 = 32 \qquad \text{or} \qquad 8y^2 + 4py + p^2 - 32 = 0.$$

There will be exactly one root if $(2p)^2 - 8(p^2 - 32) = 0$, that is, for $p = 8$ and $p = -8$. We obtain the same solutions as above. ◀

EXERCISES

In Exercises 23 to 30, find the equation of the tangent line l to the given conic at the given point, by using Theorem 2 and by using a direct calculus approach. Verify that the answers obtained by these two methods are equivalent.

23. $9y^2 = x$, at $(1, \frac{1}{3})$.
24. $x^2 + 25y^2 = 100$, at $(6, \frac{8}{5})$.
25. $25x^2 - y^2 = 100$, at $(-3, 5\sqrt{5})$.
26. $y^2 = 25x$, at $(9, -15)$.
27. $4x^2 + 25y^2 = 1$, at $(\frac{1}{4}, \sqrt{3}/10)$.
28. $4y^2 - x^2 = 1$, at $(-1, -\sqrt{2}/2)$.
29. $xy = 1$, at $(4, \frac{1}{4})$.
30. $x^2/4 + y^2/49 = 1$, at $(\sqrt{3}, \frac{7}{2})$.
31. Find the angle between the tangent to the ellipse $x^2 + 4y^2 = 5$ at $(-1, 1)$ and the tangent to the hyperbola $x^2 - 10y^2 = 90$ at $(10, 1)$.
32. Find all tangents to the ellipse $9x^2 + 25y^2 = 225$ that have slope $\frac{9}{25}$.
33. Find all tangents to the hyperbola $x^2 - y^2 = 1$ that have slope 2.
▶ 34. Find a tangent to the parabola $y^2 = 10x$ that has slope 2.
35. Find the tangents to the parabola $y^2 = 4x$ that pass through the point $(-2, 0)$.
36. There are two tangents to the ellipse $x^2 + 4y^2 = 32$ that pass through the point $(10, 1)$. Find them.
37. Find the tangents to the hyperbola $x^2 - y^2 = 16$ that pass through the point $(3, -\frac{1}{3})$.

1.7 Conics in polar coordinates

The theory of conics as developed by the Greeks, although probably motivated by observing sun dials, was a branch of pure mathematics. Two thousand years later, mathematicians discovered that conics had many applications in natural sciences and engineering. The most spectacular of these is to planetary motion, which we shall consider in Chapter 11. As a preparation, we derive here the equations of the conics in polar coordinates.

Consider first a conic of eccentricity $\epsilon \neq 0$, that is, not a circle. Let O be a focus of the conic and l the corresponding directrix. We chose a

Cartesian coordinate system with O as the origin, with the axis of the conic (the major axis in the case of an ellipse) as the x axis, and with the positive x direction chosen so that the equation of l reads $x = -p$, where p is the distance from O to l. See Fig. 10.15. A point P with Cartesian coordinates (x, y) has distance

$$\delta = |x - (-p)| = |x - p|$$

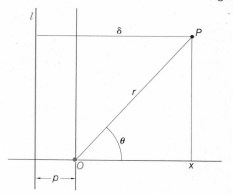

Fig. 10.15

from the directrix l. If P has polar coordinates (r, θ), then $x = r \cos \theta$ and

$$\delta = |r \cos \theta = p|.$$

The distance from P to O is r. By Theorem 1, P lies on our conic if and only if $r = \epsilon\delta$. If $r \cos \theta + p > 0$, then $\delta = r \cos \theta + p$ and the condition $r = \epsilon\delta$ becomes

$$(22) \qquad\qquad r = \epsilon(r \cos \theta + p).$$

If $r \cos \theta + p < 0$, then $\delta = -r \cos \theta - p$ and the condition $r = \epsilon\delta$ becomes

$$(23) \qquad\qquad r = -\epsilon(r \cos \theta + p).$$

Solving (22) and (23) for r, we obtain

$$(24) \qquad\qquad r = \frac{\epsilon p}{1 - \epsilon \cos \theta}$$

or

$$(25) \qquad\qquad r = \frac{-\epsilon p}{1 + \epsilon \cos \theta}.$$

Now we recall the convention made in Chapter 9, §3.2. According to this convention, we can replace, in (25), r by $(-r)$ and θ by $(\theta + \pi)$. This amounts to replacing the numerator $(-\epsilon p)$ by (ϵp) and the denominator $1 + \epsilon \cos \theta$ by $1 - \epsilon \cos \theta$. Then (25) becomes identical with (24).

Theorem 4   THE EQUATION, IN SUITABLE POLAR COORDINATES, OF A CONIC OF ECCENTRICITY ϵ, WITH A FOCUS AT O, READS

$$(26) \qquad\qquad r = \frac{A}{1 - \epsilon \cos \theta}$$

WITH A POSITIVE CONSTANT A. CONVERSELY, (26) IS ALWAYS THE EQUATION OF A CONIC.

We have already proved this for a noncircle [set $A = \epsilon p$ in (24)]. For $\epsilon = 0$, equation (26) reads $r = A$. This is, of course, the equation of a circle.

Remark 1   For an ellipse $(0 \le \epsilon < 1)$, the denominator in (26) is never 0.

For a parabola $(\epsilon = 1)$, the denominator is never negative. It is 0, as it should be, for $\cos \theta = 1$. For $\theta = 0$ is the polar angle of points on the axis, far away from the focus, and there are no points on the parabola with this polar angle.

For the hyperbola $(\epsilon > 1)$, the denominator is sometimes negative, sometimes positive; this corresponds to the two branches of the hyperbola. The denominator is 0 for $\cos \theta = 1/\epsilon$, that is, for the two values of θ corresponding to the asymptotes.

Remark 2   By changing the unit of length, we can replace the numerator A in (26) by 1. We conclude that

the shape of a conic depends only on the eccentricity ϵ.

EXERCISES

In Exercises 38 to 44, find the polar equation of the given conic.

38. Parabola with focus at 0 and directrix $x = -5$.
39. Parabola with focus at 0 and vertex at $(-3, 0)$.
40. Ellipse of eccentricity $\frac{1}{4}$ and with foci 0 and $(6, 0)$.
41. Ellipse with one vertex at $(-1, 0)$ and foci 0 and $(4, 0)$.
42. Hyperbola of eccentricity 2 with one focus at 0 and the corresponding directrix $x = -1$.
43. Equilateral hyperbola with foci at 0 and at $(-5, 0)$.
44. Hyperbola of eccentricity 10 with focus at 0 and with corresponding vertex at $(-1, 0)$.

In Exercises 45 to 50, identify the conic whose polar equation is given as an ellipse, hyperbola, or parabola, and find the Cartesian coordinates of the foci and vertices. Sketch the conic, showing these points, the axis, and the asymptotes, if relevant.

45. $r = 10/(1 - \cos \theta)$.
46. $r = 2/(1 - \cos \theta)$.
47. $r = 4/(1 - 20 \cos \theta)$.
48. $r = 1/(1 - \frac{1}{2} \cos \theta)$.
49. $r = 1/(2 - 3 \cos \theta)$.
▶ 50. $r = 1/(20 - 25 \cos \theta)$.

1.8 Parametric equations

We conclude this section by giving a so-called **parametric representation** of ellipses and hyperbolas. We know from the results of Chapter 6, §1.8 that every point $P = (x, y)$ on the unit circle $x^2 + y^2 = 1$ can be written as

(27) $$x = \cos t, \qquad y = \sin t.$$

If we require that $0 \le t < 2\pi$, then every point P of the circle corresponds to a unique value of t. Also, whenever (27) holds, (x, y) lies on the unit circle. We also know (see Chapter 6, §7.4) that every point P on the right branch of the equilateral hyperbola $x^2 - y^2 = 1$ can be written, uniquely, as

(28) $$x = \cosh t, \qquad y = \sinh t,$$

and all points with coordinates (x, y) given by (28) lie on this branch. The points (x, y) on the left branch of $x^2 - y^2 = 1$ may be represented as

$$x = -\cosh t, \qquad y = \sinh t.$$

Let $a > 0$ be a number. It is clear that the relations

(29) $$x = a \cos t, \qquad y = a \sin t \qquad 0 \le t < 2\pi$$

represent the points on the circle

(30) $$x^2 + y^2 = a^2,$$

while the relations

(31) $$x = \pm a \cosh t, \qquad y = a \sinh t \qquad -\infty < t < +\infty$$

represent the points on the equilateral hyperbola

(32) $$x^2 - y^2 = a^2.$$

Now the standard equation of an ellipse

(E) $$\frac{x^2}{a^2} + \frac{y^2}{b^2} = 1$$

may be rewritten as $x^2 + (a/b)^2 y^2 = a^2$. Thus (x, y) lies on (E) if and only if the point $(x, (a/b)y)$ lies on the circle (29), that is, if and only if $x = a \cos t$, $(a/b)y = a \sin t$ for some t. We conclude that the relations

(33) $$x = a \cos t, \qquad y = b \sin t \qquad 0 \le t < 2\pi$$

represent the points on the ellipse (E).

One sees in the same way that the points on the hyperbola

(H) $$\frac{x^2}{a^2} - \frac{y^2}{b^2} = 1$$

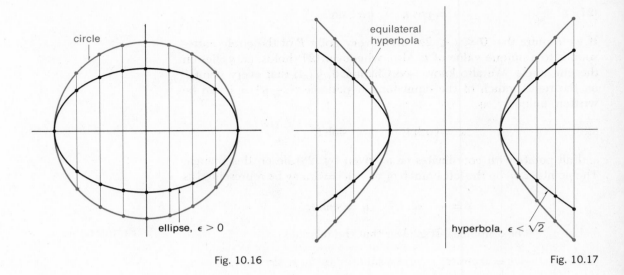

Fig. 10.16   Fig. 10.17

can be represented as

$$(34) \qquad x = \pm a \cosh t, \qquad y = b \sinh t \qquad -\infty < t < +\infty.$$

In Figs. 10.16 and 10.17, we indicate points of (30) and (E) and of (32) and (H) corresponding to the same values of t. Comparing (33) with (29), we see that *the ellipse* E *is obtained from the circle by "contracting in the y direction in the ratio* b/a." If we compare (34) and (31), we see that *the hyperbola* (H) *is obtained from the equilateral hyperbola by "contracting (or expanding) in the y direction in the ratio* b/a."

(This remark shows, by the way, that in proving Theorem 3 in §1.6, it suffices to consider a circle, rather than an arbitrary ellipse, and an equilateral hyperbola, rather than an arbitrary hyperbola.)

▶ *Example* Show that *the area of an ellipse* (that is, of the region bounded by an ellipse) *with semiaxes a and b, is* πab.

First Proof (*without calculus*).   We assume as known the area of a circle of radius a, namely πa^2. Our ellipse is obtained from this circle by a contraction in the vertical direction in the ratio b/a. We assume as intuitively obvious the fact that in such a contraction all areas get multiplied by b/a. The area of the ellipse is, therefore, $(b/a)\pi a^2 = \pi ab$.

Second Proof (*using calculus*).   We use the standard equation of the ellipse, and the area formula in Chapter 5, §4.1. According to this formula, the area of the ellipse is

$$2 \int_{\underline{a}}^{a} b \sqrt{1 - \frac{x^2}{a^2}} \, dx = 2 \int_{\underline{a}}^{a} ab \sqrt{1 - \left(\frac{x}{a}\right)^2} \frac{dx}{a}$$
$$= 2ab \int_{-1}^{1} \sqrt{1 - u^2} \, du = \pi ab.$$

(Here we used the definition of π from Chapter 5, §4.3.)◄

EXERCISES

In Exercises 51 to 56, find the Cartesian coordinates of the foci and vertices of the conic whose parametric equation is given.

51. $x = 7 \cos t, y = 8 \sin t.$ 53. $x = \pm 2 \cosh t, y = 3 \sinh t.$
52. $x = \frac{3}{2} \cos t, y = \frac{1}{2} \sin t.$ 54. $x = \pm \frac{2}{3} \cosh t, y = \frac{1}{6} \sinh t.$
55. $x = \sqrt{2} \cos t, y = \sqrt{3} \sin t.$
56. $x = (\cos t)/100, y = (\sin t)/10{,}000.$

In §1, we used Cartesian coordinate systems chosen so as to obtain particularly simple equations for the curves considered. It is desirable to be able to discuss conics in arbitrary coordinate systems. As a preparation for this, we consider now the important topic of changing coordinates.

§2 Rotations

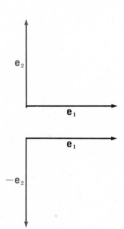

Fig. 10.18

2.1 Reflections and rotations. Orientation

By means of polar coordinates, it is easy to describe all frames in the plane. We start with a fixed frame $\{e_1, e_2\}$ and note that $\{e_1, -e_2\}$ is also a frame (see Fig. 10.18). We say that $\{e_1, -e_2\}$ is obtained from $\{e_1, e_2\}$ by **reflection** about the horizontal axis.

Let $a \neq 0$ be any vector. If its polar angle with respect to $\{e_1, e_2\}$ is ϕ, then its polar angle with respect to $\{e_1, -e_2\}$ is $-\phi$. This statement is clear from Fig. 10.19. We also give an analytic proof. Let $r = |a|$. The statement "the polar angle of a with respect to $\{e_1, e_2\}$ is ϕ" means that

$$a = (r \cos \phi) e_1 + (r \sin \phi) e_2.$$

But in this case,

$$a = (r \cos \phi) e_1 + (-r \sin \phi)(-e_2)$$
or
$$a = r \cos (-\phi) e_1 + r \sin(-\phi)(-e_2).$$

Fig. 10.19

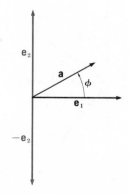

This means that the polar angle of a with respect to $\{e_1, -e_2\}$ is $-\phi$.

Suppose that $\{f_1, f_2\}$ is another frame. Then $|f_1| = 1$ and $|f_2| = 1$. Let θ be the polar angle of f_1 with respect to the frame $\{e_1, e_2\}$. Then the polar angle of f_2 is either $\theta + 90°$ or $\theta - 90°$. This is seen from Fig. 10.20. (An analytic proof could be given, but we omit it.)

Fig. 10.20

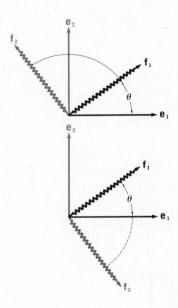

If the polar angle of $\mathbf{f}_2$ is $\theta + 90°$, we have

$$\mathbf{f}_1 = \cos\theta\,\mathbf{e}_1 + \sin\theta\,\mathbf{e}_2,$$
$$\mathbf{f}_2 = \cos(\theta + 90°)\,\mathbf{e}_1 + \sin(\theta + 90°)\,\mathbf{e}_2;$$

or, since $\cos(\theta + 90°) = -\sin\theta$, $\sin(\theta + 90°) = \cos\theta$,

$$(1) \qquad \begin{cases} \mathbf{f}_1 = \cos\theta\,\mathbf{e}_1 + \sin\theta\,\mathbf{e}_2, \\ \mathbf{f}_2 = -\sin\theta\,\mathbf{e}_1 + \cos\theta\,\mathbf{e}_2. \end{cases}$$

In this case we say that the frame $\{\mathbf{f}_1, \mathbf{f}_2\}$ is obtained from $\{\mathbf{e}_1, \mathbf{e}_2\}$ by a **rotation by** θ. Figure 10.21 shows that the terminology is appropriate; we show there, from right to left, the frame $\{\mathbf{e}_1, \mathbf{e}_2\}$ rotated by various angles θ.

The frame $\{\mathbf{e}_1, -\mathbf{e}_2\}$ *cannot* be obtained from $\{\mathbf{e}_1, \mathbf{e}_2\}$ by a rotation. For if there were a number θ such that $\mathbf{e}_1 = \cos\theta\,\mathbf{e}_1 + \sin\theta\,\mathbf{e}_2$ and $-\mathbf{e}_2 = -\sin\theta\,\mathbf{e}_1 + \cos\theta\,\mathbf{e}_2$, then the first equation would require that $\cos\theta = 1$ and the second that $\cos\theta = -1$, which is absurd.

Consider next a frame $\{\mathbf{f}_1, \mathbf{f}_2\}$ such that the polar angle of $\mathbf{f}_1$, with respect to $\{\mathbf{e}_1, \mathbf{e}_2\}$, is θ and the polar angle of $\mathbf{f}_2$ is $\theta - 90°$. Such a

| $\theta = 0$ | $\theta = 45°$ | $\theta = 90°$ | $\theta = 135°$ | $\theta = 180°$ | $\theta = 225°$ | $\theta = 270°$ | $\theta = 315°$ |

Fig. 10.21

frame is shown in Fig. 10.22. One sees that $\{\mathbf{f}_1, \mathbf{f}_2\}$ is obtained from $\{\mathbf{e}_1, -\mathbf{e}_2\}$ by a rotation by θ (this can also be proved analytically).

The following statements are geometrically obvious (and easy to prove analytically): a rotation of a frame by $0°$ or by $360°$ leaves the frame unchanged; a rotation of a frame by α followed by a rotation by β is equivalent to a rotation by $\alpha + \beta$.

Fig. 10.22

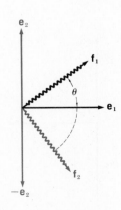

Theorem 1   CHOOSE ONCE AND FOR ALL A FIXED FRAME $\{\mathbf{e}_1, \mathbf{e}_2\}$. CALL EVERY FRAME OBTAINED FROM $\{\mathbf{e}_1, \mathbf{e}_2\}$ BY A ROTATION **RIGHT-HANDED**. CALL EVERY FRAME OBTAINED FROM $\{\mathbf{e}_1, -\mathbf{e}_2\}$ BY A ROTATION **LEFT-HANDED**. THEN (a) ANY ROTATION OF A RIGHT-HANDED FRAME IS RIGHT-HANDED, (b) ANY ROTATION OF A LEFT-HANDED FRAME IS LEFT-HANDED, AND (c) EVERY FRAME IS EITHER RIGHT-HANDED OR LEFT-HANDED, BUT NOT BOTH.

This is merely a restatement of the preceding discussion.

While the difference between right-handed and left-handed frames can be defined mathematically, one cannot define a right-handed frame; one can only exhibit one. Sometimes one says that in a right-handed frame the swing by 90° which brings the vector $\mathbf{e}_1$ into the position $\mathbf{e}_2$ is "counterclockwise." But not all clocks look alike. (A famous medieval clock on the so-called Jewish City Hall in Prague is marked by Hebrew letters rather than numerals. Its hands move "counterclockwise." The author's teacher, C. Loewner, liked to ask whether a man who knew only this clock and read only mathematics books without pictures would ever discover that other clocks move differently?)

When we choose a frame, which will be called right-handed, we **orient** the plane. In an **oriented plane,** one can talk about a "counterclockwise" direction; it is the direction of the 90° swing that brings the first vector of a right-handed frame into the position of the second vector. We assume from now on that the plane has been oriented.

CHARLES LOEWNER (1895–1968) came to the United States after his native Czechoslovakia was overrun by the Nazis. He made important discoveries in several fields of mathematics.

EXERCISES

In Exercises 1 to 5 assume that $\{\mathbf{e}_1, \mathbf{e}_2\}$ is a fixed right-handed frame for vectors in this plane. Determine if the given frame $\{\mathbf{f}_1, \mathbf{f}_2\}$ is right- or left-handed.

1. $\mathbf{f}_1 = \mathbf{e}_2, \mathbf{f}_2 = -\mathbf{e}_1$.

2. $\mathbf{f}_1 = \dfrac{\sqrt{2}}{2}\mathbf{e}_1 + \dfrac{\sqrt{2}}{2}\mathbf{e}_2, \mathbf{f}_2 = -\dfrac{\sqrt{2}}{2}\mathbf{e}_1 + \dfrac{\sqrt{2}}{2}\mathbf{e}_2$.

3. $\mathbf{f}_1 = \dfrac{\sqrt{3}}{2}\mathbf{e}_1 + \dfrac{1}{2}\mathbf{e}_2, \mathbf{f}_2 = \dfrac{1}{2}\mathbf{e}_1 - \dfrac{\sqrt{3}}{2}\mathbf{e}_2$.

▶ 4. $\mathbf{f}_1 = -\dfrac{\sqrt{3}}{2}\mathbf{e}_1 + \dfrac{1}{2}\mathbf{e}_2, \mathbf{f}_2 = -\dfrac{1}{2}\mathbf{e}_1 - \dfrac{\sqrt{3}}{2}\mathbf{e}_2$.

5. $\mathbf{f}_1 = -\dfrac{1}{2}\mathbf{e}_1 - \dfrac{\sqrt{3}}{2}\mathbf{e}_2, \mathbf{f}_2 = -\dfrac{\sqrt{3}}{2}\mathbf{e}_1 + \dfrac{1}{2}\mathbf{e}_2$.

2.2 Changing coordinates

We want to investigate how the coordinates of a point change when one changes the Cartesian coordinate system. In Chapter 2, §1.2, we described what happens when one translates the axis, that is, when one goes over from a system $(O, \mathbf{e}_1, \mathbf{e}_2)$, where $\{\mathbf{e}_1, \mathbf{e}_2\}$ is a frame, to a system $(O', \mathbf{e}_1, \mathbf{e}_2)$ with a different origin and same direction of the axis. We recall that the old coordinates (x, y) of a point P and the new coordinates (X, Y) of the same point are connected by the formulas

$$(2) \qquad\qquad x = a + X, \qquad y = b + Y$$

where (a, b) are the components of O' in the system $(O, \mathbf{e}_1, \mathbf{e}_2)$. Let us derive this result again, using vectors. Let $\mathbf{T}$ be the vector determined by $\overrightarrow{OO'}$; then $\mathbf{T} = a\mathbf{e}_1 + b\mathbf{e}_2$. Let $\mathbf{P}$ be the position vector of P with respect to O and $\mathbf{P}'$ the position vector of P with respect to O'. Then $\mathbf{P} = \mathbf{T} + \mathbf{P}'$. But we have $\mathbf{P} = x\mathbf{e}_1 + y\mathbf{e}_2$ and $\mathbf{P}' = X\mathbf{e}_1 + Y\mathbf{e}_2$. Therefore

$$x\mathbf{e}_1 + y\mathbf{e}_2 = a\mathbf{e}_1 + b\mathbf{e}_2 + X\mathbf{e}_1 + Y\mathbf{e}_2 = (a + X)\mathbf{e}_1 + (b + Y)\mathbf{e}_2,$$

and (2) follows.

Consider next two frames $\{\mathbf{e}_1, \mathbf{e}_2\}$ and $\{\mathbf{f}_1, \mathbf{f}_2\}$ and two Cartesian coordinate systems with the same origin: the "old" system $\{O, \mathbf{e}_1, \mathbf{e}_2\}$ and the "new" system $\{O, \mathbf{f}_1, \mathbf{f}_2\}$. One says that the new system is obtained from the old by **changing axes.**

A change of axis is either a rotation, or a reflection, or a reflection followed by a rotation. Consider first a reflection: $\mathbf{f}_1 = \mathbf{e}_1, \mathbf{f}_2 = -\mathbf{e}_2$. Let the old coordinates of a point P be (x, y) and let the new coordinates of the same point be (X, Y). Then

$$\mathbf{P} = x\mathbf{e}_1 + y\mathbf{e}_2 = x\mathbf{f}_1 - y\mathbf{f}_2 = X\mathbf{f}_1 + Y\mathbf{f}_2.$$

Thus

$$(3) \qquad\qquad x = X, \qquad y = -Y.$$

It remains to compute the effect of rotating a frame. This will be done below. Then we shall have exhausted all cases. Indeed, given any two coordinate systems $\{O, \mathbf{e}_1, \mathbf{e}_2\}$ and $\{O', \mathbf{f}_1, \mathbf{f}_2\}$ we can go from one to the other by the translation followed by a change of axes, or also by a change of axes followed by a translation.

2.3 Rotating coordinate axes

Theorem 2 LET THE FRAME $\{\mathbf{f}_1, \mathbf{f}_2\}$ BE OBTAINED FROM THE FRAME $\{\mathbf{e}_1, \mathbf{e}_2\}$ BY A ROTATION BY θ. LET A POINT P HAVE CARTESIAN COORDINATES (x, y) AND POLAR COORDINATES (r, ϕ) IN THE SYSTEM $\{O, \mathbf{e}_1, \mathbf{e}_2\}$, AND LET ITS CARTESIAN AND POLAR COORDINATES IN THE SYSTEM $\{O, \mathbf{f}_1, \mathbf{f}_2\}$ BE (X, Y) AND (R, Φ), RESPECTIVELY. THEN

$$(4) \qquad\qquad r = R \qquad \phi = \Phi + \theta$$

AND

$$(5) \qquad\qquad \begin{cases} x = X \cos\theta - Y \sin\theta, \\ y = X \sin\theta + Y \cos\theta. \end{cases}$$

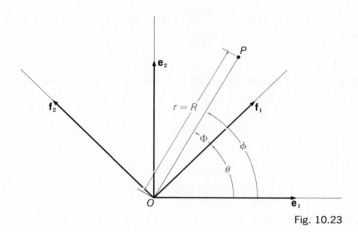

Fig. 10.23

We give two proofs. The first proof begins by noting that the relations (4) are geometrically obvious; they are shown in Fig. 10.23. Now we prove (5). Since $X = R \cos \Phi$, $Y = R \sin \Phi$, we have

$$
\begin{aligned}
x &= r \cos \phi = R \cos (\Phi + \theta) = R \cos \Phi \cos \theta - R \sin \Phi \sin \theta \\
&= X \cos \theta - Y \sin \theta, \\
y &= r \sin \phi = R \sin (\Phi + \theta) = R \sin \Phi \cos \theta + R \cos \Phi \sin \theta \\
&= X \sin \theta + Y \cos \theta.
\end{aligned}
$$

The second proof is purely analytic. The position vector of P is $x\mathbf{e}_1 + y\mathbf{e}_2 = X\mathbf{f}_1 + Y\mathbf{f}_2$. Also, we have the relations (1). Therefore,

$$
\begin{aligned}
x\mathbf{e}_1 + y\mathbf{e}_2 &= X(\cos \theta \, \mathbf{e}_1 + \sin \theta \, \mathbf{e}_2) + Y(-\sin \theta \, \mathbf{e}_1 + \cos \theta \, \mathbf{e}_2) \\
&= (X \cos \theta - Y \sin \theta) \, \mathbf{e}_1 + (X \sin \theta + Y \cos \theta) \, \mathbf{e}_2.
\end{aligned}
$$

This implies (5). Now, using Chapter 9, §3.2, we have

$$
\begin{aligned}
r^2 = x^2 + y^2 &= (X \cos \theta - Y \sin \theta)^2 + (X \sin \theta + Y \cos \theta)^2 \\
&= X^2(\cos^2 \theta + \sin^2 \theta) + 2XY(-\cos \theta \sin \theta + \sin \theta \cos \theta) \\
&\qquad\qquad\qquad + Y^2(\sin^2 \theta + \cos^2 \theta) = X^2 + Y^2 = R^2,
\end{aligned}
$$

so that $r = R$. Also

$$
\begin{aligned}
\cos \phi = \frac{x}{r} &= \frac{X \cos \theta - Y \sin \theta}{R} \\
&= \frac{R \cos \Phi \cos \theta - R \sin \Phi \sin \theta}{R} = \cos (\Phi + \theta).
\end{aligned}
$$

One shows similarly that $\sin \phi = \sin (\Phi + \theta)$. Hence $\phi = \Phi + \theta$. Now (4) is proved.

▶ *Examples*   *1.* Let the frame $\{f_1, f_2\}$ be the frame $\{e_1, e_2\}$ rotated by θ, so that (1) holds. Express e_1 and e_2 in terms of f_1 and f_2.

Solution.   Since $\{e_1, e_2\}$ is obtained by rotating $\{f_1, f_2\}$ by $(-\theta)$, we have

$$e_1 = \cos \theta\, f_1 - \sin \theta\, f_2$$
$$e_2 = \sin \theta\, f_1 + \cos \theta\, f_2.$$

This can also be obtained by solving (1) for e_1 and e_2 as follows. Multiply the first equation by $\cos \theta$, the second by $-\sin \theta$, and add. Then multiply the first equation by $\sin \theta$, the second by $\cos \theta$, and add.

2. If $\{e_1, e_2\}$ is a right-handed frame, are the frames $\{-e_1, e_2\}$ and $\{-e_1, -e_2\}$ right-handed or left-handed?

Answer.   The frame $\{-e_1, -e_2\}$ is $\{e_1, e_2\}$ rotated by $180°$, and hence right-handed. The frame $\{-e_1, e_2\}$ is left-handed; in fact, it is $\{e_1, -e_2\}$ rotated by $180°$.

3. If P has coordinates (x, y) in the $\{O, e_1, e_2\}$ system and (X, Y) in the system obtained by rotating $\{O, e_1, e_2\}$ by $-45°$, what is the connection between (x, y) and (X, Y)?

Answer.   Using the formulas (5) with $\theta = -45°$, we obtain $x = (X + Y)/\sqrt{2}$, $y = (X - Y)/\sqrt{2}$. Therefore

$$X = \frac{x + y}{\sqrt{2}}, \qquad Y = \frac{x - y}{\sqrt{2}}.$$

4. Show that the set of points (x, y) satisfying $x^2 - y^2 = 1$ is an equilateral hyperbola.

Solution.   Rotate the coordinate system by $-45°$; denote the new coordinates by (X, Y). By Example 3, $x^2 - y^2 = (x - y)(x + y) = 2XY$. In the new coordinates, the equation of our curve reads $XY = \frac{1}{2}$ or $Y = 1/2X$. This is, by definition, an equilateral hyperbola. ◀

EXERCISES

In Exercises 6 to 13, assume that the frame $\{f_1, f_2\}$ is obtained from the frame $\{e_1, e_2\}$ by a rotation by θ.

6. If $\theta = 5\pi/6$ and P has Cartesian coordinates $(2, -1)$ in the system $\{O, \mathbf{e}_1, \mathbf{e}_2\}$, what are the Cartesian coordinates of P in the system $\{O, \mathbf{f}_1, \mathbf{f}_2\}$?

7. If $\theta = 3\pi/4$ and P has polar coordinates $(2, \pi/3)$ in the system $\{O, \mathbf{e}_1, \mathbf{e}_2\}$, what are the Cartesian coordinates of P in the system $\{O, \mathbf{f}_1, \mathbf{f}_2\}$?

8. If $\theta = \pi/3$ and P has coordinates $(-2, 3)$ in the system $\{O, \mathbf{e}_1, \mathbf{e}_2\}$, what are the Cartesian coordinates of P in the system $\{O, \mathbf{f}_1, \mathbf{f}_2\}$?

9. If $\theta = 4\pi/3$ and P has Cartesian coordinates $(\tfrac{1}{2}, 3)$ in the system $\{O, \mathbf{e}_1, \mathbf{e}_2\}$, what are the Cartesian coordinates of P in the system $\{O, \mathbf{f}_1, \mathbf{f}_2\}$?

10. If $\theta = 7\pi/6$ and P has polar coordinates $(\sqrt{2}, \pi/4)$ in the system $\{O, \mathbf{e}_1, \mathbf{e}_2\}$, what are the Cartesian coordinates of P in the system $\{O, \mathbf{f}_1, \mathbf{f}_2\}$?

11. If $\theta = -\pi/4$ and P has Cartesian coordinates $(3, -8)$ in the system $\{O, \mathbf{e}_1, \mathbf{e}_2\}$, what are the Cartesian coordinates of P in the system $\{O, \mathbf{f}_1, \mathbf{f}_2\}$?

12. If $\theta = \pi/9$ and P has Cartesian coordinates $(2 \cos \tfrac{1}{9}\pi, 2 \sin \tfrac{1}{9}\pi)$ in the system $\{O, \mathbf{e}_1, \mathbf{e}_2\}$, what are the polar coordinates of P in the system $\{O, \mathbf{f}_1, \mathbf{f}_2\}$?

13. If $\theta = \pi/10$ and P has polar coordinates $(\tfrac{1}{2}, \pi/4)$ in the system $\{O, \mathbf{e}_1, -\mathbf{e}_2\}$, what are the polar coordinates of P in the system $\{O, \mathbf{f}_1, \mathbf{f}_2\}$?

2.4 Directed angle between two lines

As another application of polar coordinates, we shall define the **directed angle between two lines** l_1 and l_2. The definition reads:

Let l_1 and l_2 be two lines; choose a right-handed coordinate system in which l_1 is the x axis; the angle α between l_1 and l_2 [in symbols $\angle\,(l_1, l_2)$] is any number α such that $\tan \alpha$ is the slope of l_2. If l_2 is vertical, the $\angle\,(l_1, l_2) = 90°$ or $-90°$. (Since the function $\tan$ has period $\pi = 180°$, α is determined up to a multiple of π.)

Figure 10.24 shows that this definition corresponds to our previous observations. Note that $\angle\,(l_2, l_1) = -\angle\,(l_1, l_2)$. Also, $\angle\,(l_1, l_2) = 0$ or $180°$ if l_1 and l_2 are parallel. One may always choose for $\angle\,(l_1, l_2)$ a number such that $-90° \leq \angle\,(l_1, l_2) \leq 90°$. If so, then the nonnegative number $|\angle\,(l_1, l_2)|$ is called the acute angle between l_1 and l_2.

Theorem 3 LET l_2 AND l_2 BE TWO LINES WITH SLOPES m_1 AND m_2, RESPECTIVELY. THEN

$$(6) \qquad \tan \angle\,(l_1, l_2) = \frac{m_2 - m_1}{1 + m_1 m_2}.$$

Proof. Let l_0 denote the x axis and set $\alpha_1 = \angle\,(l_0, l_1)$, $\alpha_2 = \angle\,(l_0, l_2)$. One sees from Fig. 10.25 that

$$\angle\,(l_1, l_2) = \alpha_2 - \alpha_1,$$

so that, by the addition theorem for $\tan$ (see Chapter 6, §2.2),

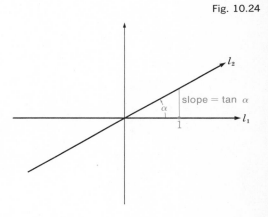

Fig. 10.24

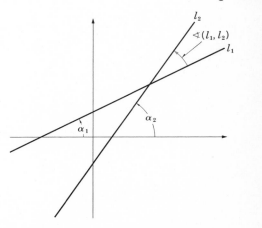

Fig. 10.25

$$\tan \angle(l_1, l_2) = \frac{\tan \alpha_2 - \tan \alpha_1}{1 + \tan \alpha_2 \tan \alpha_1}.$$

Since $\tan \alpha_1 = m_1$, $\tan \alpha_2 = m_2$, by definition, the desired relation (6) follows. If $m_1 m_2 = -1$, the lines l_1 and l_2 are perpendicular. In this case, $\tan \angle(l_1, l_2)$ is not defined.

Remark When we say "the (directed) angle between l_1 and l_2," we mean $\angle(l_1, l_2)$ and not $\angle(l_2, l_1)$.

▶ *Examples* *1.* Find the angle between the tangents to the parabola $y = \frac{1}{2}x^2$ at the points $(1, \frac{1}{2})$ and $(2, 2)$.

Solution. The slopes of the tangents are

$$m_1 = \left(\frac{dy}{dx}\right)_{x=1} = 1, \qquad m_2 = \left(\frac{dy}{dx}\right)_{x=2} = 2.$$

Let the desired angle be ϕ. Then, by (6),

$$\tan \phi = \frac{2 - 1}{1 + 2} = \frac{1}{3},$$

so that $\phi = \arctan .33 \cdots = 18°16'$, approximately.

2. Find the angle between the parabola $y = x^2$ and the curve $y = 2x^3 - x$ at their points of intersection. (The angle between two curves is, by definition, the angle between their tangents.)

Solution. To find the intersections, we solve the equation $x^2 = 2x^3 - x$ or

$$2x^3 - x^2 - x = 0,$$

that is,

$$x(x - 1)(2x + 1) = 0.$$

There are intersection points for $x = -\frac{1}{2}$, $x = 0$, $x = 1$. Call the angles at these points α, β, and γ. Since

$$\frac{dx^2}{dx} = -1, 0, 2 \qquad \text{at } x = -\frac{1}{2}, 0, 1,$$

and

$$\frac{d(2x^3 - x)}{dx} = \frac{1}{2}, -1, 5 \qquad \text{at } x = -\frac{1}{2}, 0, 1,$$

we have

$$\tan \alpha = \frac{\frac{1}{2} + 1}{1 - \frac{1}{2}} = 3, \qquad \tan \beta = \frac{-1 + 0}{1} = -1,$$

$$\tan \gamma = \frac{5 - 2}{1 + 10} = \frac{3}{11},$$

so that $\alpha = 72°$ approximately, $\beta = 135°$, and $\gamma = 15°$, approximately. ◄

EXERCISES

In Exercises 14 to 25, assume that a fixed Cartesian coordinate system is given for the plane and the coordinates of a general point are written (x, y). Use the tables in the back of the book to get numerical answers.

14. Find the angle between the y axis and the line $2x - 3y - 5 = 0$.
15. If l_1 is defined by $x - 2y = 0$ and l_2 by $2x - 3y - 6 = 0$, find $\angle(l_1, l_2)$.
16. Find the angle between the line $3y - 6x = 2$ and the line $2y + 3x = -1$.
17. Let the line l_1 have the equation $\frac{1}{2}y - \frac{1}{3}x = 1$ while l_2 has the equation $\frac{3}{2}y + 6x = \frac{1}{3}$. Find $\angle(l_1, l_2)$.
18. Let l_1 be the line tangent to $y = x^3$ at $(1, 1)$ and let l_2 be the line tangent to $y = x^3$ at $(2, 8)$. Find $\angle(l_1, l_2)$.
19. Find the angle between the lines tangent to the circle $x^2 + y^2 = 1$ at $x = \sqrt{2}/2$ and $x = \frac{4}{5}$, in that order.
20. Find the angle between the lines tangent to $y = \frac{1}{4}x^4$ at $x = 1$ and $x = 2$, in that order.
21. Find the angle between the curves $y = x^2$ and $y = \sqrt{x}$ at their points of intersection.
22. Find the angle between the curves $y = x^3 - 4x^2 + 3x$ and $y = x$ at their points of intersection.
23. Find the angle between the curves $y = x^{1/3}$ and $y = x^{2/3}$ at their points of intersection.
24. If l_1 has the equation $-y + \frac{4}{3}x = -1$ and $\angle(l_1, l_2) = 25°$, find the slope of l_2.
25. If l_1 has the equation $6y - 5x = 4$ and $\angle(l_1, l_2) = 100°$, find the slope of l_2.

2.5 Angle bisectors

As an application of Theorem 3, we proceed to find the slope n of the line l bisecting the angle between two lines l_1 and l_2 of slope m_1 and m_2, respectively.

We must have $\angle(l_1, l) = \angle(l, l_2)$, hence $\tan \angle(l_1, l) = \tan \angle(l, l_2)$ or, by Theorem 3,

$$\frac{m_2 - n}{1 + m_2 n} = \frac{n - m_1}{1 + m_1 n}.$$

Obvious manipulations transform this into

Fig. 10.26

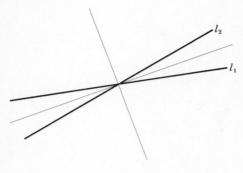

(7) $(m_1 + m_2)n^2 + 2(1 - m_1m_2)n - (m_1 + m_2) = 0.$

This is a quadratic equation for n; it will have in general two distinct roots. This is as it should be, because, given l_1 and l_2, there are two bisectors and not one (see Fig. 10.26). Here is an example of the pen's being smarter than we (Euler's comment). We forgot about the second bisector; the formulas did not.

If $m_1 = -m_2$, the lines l_1 and l_2 are symmetric about both the x and y axes. In this case, equation (7) reads $(1 + m_1{}^2)n = 0$, that is, $n = 0$. One bisector is horizontal, the other vertical (and has no slope).

If $m_1 + m_2 \neq 0$, we can rewrite (7) as

(8) $$n^2 + 2\frac{1 - m_1m_2}{m_1 + m_2}n - 1 = 0.$$

Fig. 10.27

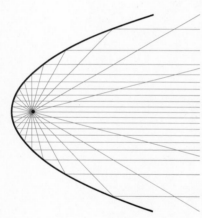

There will be two real roots n_1 and n_2 with $n_1n_2 = -1$. This shows that the two bisectors are perpendicular.

EXERCISES

In Exercises 26 to 31, find the Cartesian equations of the lines bisecting the angles between the pair of given lines, l_1 and l_2.

▶ 26. l_1: $y = x + 1$, l_2: $y = 7x - 2$.
  27. l_1: $2y + 4x = 5$, l_2: $y = \frac{1}{2}x$.
  28. l_1: $y = x$, l_2: $y = \sqrt{7}x$.
  29. l_1: $y = 2x$, l_2: $y = \sqrt{19}x$.
  30. l_1: $y = \sqrt{2}x + 1$, l_2: $2y - \sqrt{2}x = 1$.
  31. l_1 passes through $(2, 3)$ and $(4, 5)$, l_2 passes through $(4, 1)$ and $(3, -2)$.

2.6 Reflecting property

Fig. 10.28

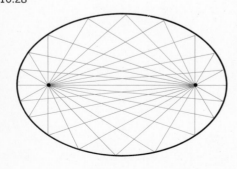

We use the results established above to prove the so-called **reflecting property** of conics. This is used in parabolic mirrors, which have the shape of a paraboloid of revolution, obtained by rotating a parabola about its axis. Such a mirror either transforms a parallel beam of light into a beam concentrated at the focus (in a telescope) or transforms light emanating from a source at the focus into a parallel beam (in a searchlight); see Fig. 10.27. A less serious application is a whispering gallery, a room with ellipsoidal walls and ceiling. Here a whisper at one focus is clearly audible to a man standing at the other focus, since the sound waves emanating from one focus are reflected by the wall into the other focus (see Fig. 10.28).

Both phenomena depend upon a physical law governing wave motion ("the angle of reflection equals the angle of incidence") and upon the following mathematical theorem.

Theorem 4 LET P BE A POINT ON A CONIC AND l THE TANGENT TO THE CONIC AT P. IF THE CONIC IS AN ELLIPSE OR HYPERBOLA, LET l_1 AND l_2 BE THE LINES JOINING P TO THE FOCI (SEE FIG. 10.29). IF THE CONIC IS A PARABOLA, LET l_1 JOIN P TO THE FOCUS AND LET l_2 BE THE LINE THROUGH P PARALLEL TO THE AXIS (SEE FIG. 10.30). THEN l BISECTS THE ANGLE BETWEEN l_1 AND l_2.

Fig. 10.29

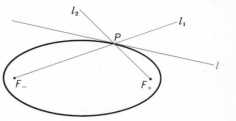

Proof. We assume the conic to be in standard position (see §1.1). Let m, m_1, m_2 be the slopes of l, l_1, l_2, respectively. We must show (see §2.5) that

(9) $$(m_1 + m_2)m^2 + 2(1 - m_1 m_2)m - (m_1 + m_2) = 0.$$

Fig. 10.30

Now let P have coordinates (x_0, y_0). We recall (see §1.4) that for an ellipse or for a hyperbola the two foci are $(-c, 0)$ and $(c, 0)$, whereas the parabola has its focus at $(a, 0)$. Therefore we have

$$m_1 = \frac{y_0}{x_0 + c}, \qquad m_2 = \frac{y_0}{x_0 - c} \qquad \text{if } \epsilon \neq 1,$$

$$m_1 = \frac{y_0}{x_0 - a}, \qquad m_2 = 0 \qquad \text{if } \epsilon = 1.$$

In §1.6, we found for the tangent l through (x_0, y_0) the equations (19), (20), and (21). From these equations, we obtain that

$$m = -\frac{b^2}{a^2}\frac{x_0}{y_0} \qquad \text{if } 0 \leq \epsilon < 1,$$

$$m = \frac{2a}{y_0} \qquad \text{if } \epsilon = 1,$$

$$m = \frac{b^2}{a^2}\frac{x_0}{y_0} \qquad \text{if } \epsilon > 1.$$

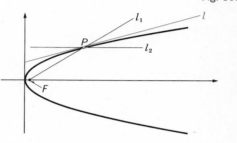

Substitute into (9) and verify.

We give an application of the reflecting property. Two conics are called **confocal** if they have the same foci.

Fig. 10.31

Theorem 5 A HYPERBOLA INTERSECTS A CONFOCAL ELLIPSE AT RIGHT ANGLES.

Proof. Let P be an intersection point, say, one in the first quadrant (see Fig. 10.31). Let l_1 and l_2 be the lines joining the two foci to P. The tangents to the ellipse and to the hyperbola at P are (by Theorem 4) the

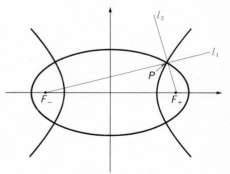

two bisectors of the angles formed by the lines l_1 and l_2; hence these tangents (see §2.5) are perpendicular. [The two tangents are distinct, since one has a negative and the other a positive slope, as the reader should verify using the relations (20) and (21) in §1.6.]

A purely analytic proof of this theorem would be longer and less elegant.

EXERCISES

32. Verify the statement of Theorem 4 for the parabola $y^2 = 4x$ and the ray leading from the focus to $(4, 4)$.
33. Verify the statement of Theorem 4 for the ellipse $x^2 + 4y^2 = 5$ and the point $(1, 1)$ on the ellipse.
34. Verify the statement of Theorem 5 for the ellipse $(x^2/25) + (y^2/16) = 1$ and the hyperbola $x^2 - (y^2/8) = 1$.

2.7 Orientation in space

Space is **oriented** by choosing one frame and calling it **right-handed.** Then every other frame is either right-handed or left-handed, but not both. In Fig. 10.32, the frame (a) is chosen to be right-handed; then the frames (b), (c), (d), and (e) are also right-handed, while (f) and (g) are left-handed.

Intuitively, if we imagine the frames to be made of a rigid material, then any right-handed frame can be moved into the position of any other right-handed frame, and similarly for left-handed frames. But it is impossible to turn a right-handed frame into a left-handed one, just as it is impossible to fit a right glove over a left hand.

The statements just made must be reinforced by precise definitions and by an analytic proof. We do not do it here, since we shall make no

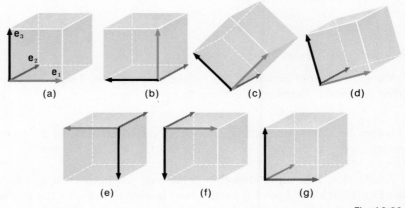

(a)   (b)   (c)   (d)

(e)   (f)   (g)

Fig. 10.32

use of orientation in space, and since the proof would involve us in the theory of determinants. (See, however, the appendix to this chapter, §4.10.)

3.1 Algebraic curves

Parabolas, hyperbolas, and ellipses are all special cases of algebraic curves. Let us define this concept.

A **monomial** in two variables, x and y, is an expression of the form $\alpha x^p y^q$, where α is a number and p and q are nonnegative integers. The sum $p + q$ is called the **degree** of the monomial, and α its **coefficient**. A **polynomial in two variables** is a sum of monomials. A polynomial is of degree n if none of the monomials with coefficient $\neq 0$ has degree greater than n and at least one has degree n (for example: $3x^3 + 4x^2y^2 - x - y$ is a polynomial in x, y of degree 4).

A **plane algebraic curve** of degree n is the set of points P whose coordinates (x, y) in some Cartesian coordinate system satisfy an equation of the form

$$\text{(a polynomial of degree } n \text{ in } x, y) = 0.$$

A straight line, for instance, is an algebraic curve of degree 1, while circles are algebraic curves of degree 2.

Theorem 1   THE DEGREE OF AN ALGEBRAIC CURVE DOES NOT DEPEND ON THE CARTESIAN COORDINATE SYSTEM USED.

Proof.   Consider two coordinate systems. Then one is obtained from the other by a translation and change of axes. Hence the coordinates of a point P in the first system (call them x and y) and in the second system (call them X and Y) are connected by relations of the form

$$x = \alpha X + \beta Y + a, \qquad y = \gamma X + \delta Y + b$$

where the numbers α, β, γ, δ, a, and b depend on the two coordinate systems, but not on the point considered (see §§2.2, 2.3). Suppose we are given a curve of degree n in the first system. We write down its equation, say,

$$21x^{50}y^{50} - 2x^{17}y^{83} - x^3 + 3 = 0$$

(in this example $n = 100$). In order to obtain the equation of our curve in the new system, we must simply substitute for x and y their expressions in terms of X and Y. Now, if we take an expression of the form

$$(\alpha X + \beta Y + a)^p(\gamma X + \delta Y + b)^q,$$

carry out the indicated multiplications, and collect terms, we get a polynomial in X, Y of degree at most $p + q$. Hence we obtain as the equation of our curve in the new coordinates an equation of degree at most n. It may at first seem conceivable that the new equation has a degree less than n (for instance, that in our example all terms in X and Y of degree 100 cancel each other). But this cannot be, for it is up to us which coordinates we call "old" and which "new"; and if the equation in X, Y had degree less than n, we could transform it into an equation in x, y of degree less than n.

Our assertion is proved.

The theory of algebraic curves is a fascinating chapter in so-called "algebraic geometry" that has occupied mathematicians for more than 300 years. But in this book we are concerned only with a simple (and much older) special case, that of curves of second degree.

EXERCISES

In Exercises 1 to 6, find the degree of the given polynomial in x and y.

1. $50x^3y^3$.
2. $10x + 8y + 4xy$.
3. $x^2y + x^4 - y^2 - 100$.
4. $x^3 - y^3 + x^2 - xy + 4x^3y$.
5. $2x^2y - xy^2 + 10xy + 1$.
6. $x^2y^3 - 2x^3y + 4xy^5$.

7. For what real numbers a is $x^2 - y^2 - 5ax^6 - 7$ a polynomial of degree 6? Of degree 2?
8. For what real numbers t is $x^2y^3 - tx + (t - 1)y^6 - 4t$ a polynomial of degree 6? Of degree 5?
9. For what real numbers w does the algebraic curve whose equation is given by $w^2x^5 + wx^2y + w^2xy = wx^5 + x^2y + 1$ have degree 5? Degree 3? Degree 2?
10. (a) For any fixed real number $w \neq 0$, what is the degree of the algebraic curve (in the variables x, y) whose equation is given by $w^2x^3y - w^4x^2y + 1 = 0$?
    (b) For any fixed real number $x \neq 0$, what is the degree of the algebraic curve (in the variables w, y) whose equation is given by $w^2x^3y - w^4x^2y + 1 = 0$?
    (c) For any fixed real number $y \neq 0$, what is the degree of the algebraic curve (in the variables w, x) whose equation is given by $w^2x^3y - w^4x^2y + 1 = 0$?

3.2 Curves of second degree

A **quadric curve** is a plane algebraic curve of degree 2, that is, the set of points with Cartesian coordinates (x, y) satisfying an equation of the second degree:

(1) $$Ax^2 + 2Bxy + Cy^2 + Dx + Ey + F = 0,$$

where A, B, $\cdots$, F are some numbers. In such an equation, we call the terms Ax^2, $2Bxy$ (often called the mixed term), and Cy^2 the quadratic terms, the terms Dx and Ey the linear terms, and F the constant term. The coefficient of the mixed term is called $2B$ for the sake of convenience. We assume, of course, that A, B, and C are not all three equal to 0.

Conic sections are quadric curves, since the standard equations, (P), (E), and (H) in §1.2 are special cases of (1).

A quadric curve may also be empty (for instance, the equation $x^2 + 1 = 0$ defines the empty set), or it may consist of a single point (for instance, $x^2 + y^2 = 0$), of a single line (for instance, $x^2 = 0$), of two parallel lines (for instance, $x^2 - 1 = 0$), or of two intersecting lines (for instance, $xy = 0$). We call such curves **degenerate.**

In this section, we prove the remarkable fact that a nondegenerate second order curve is always a conic (see Theorem 3 below).

3.3 Discriminant and trace

The numbers

$$\delta = AC - B^2 \qquad \text{and} \qquad t = A + C$$

formed from the coefficients of the quadratic terms in (1) are called the **discriminant** and **trace** of the equation, respectively. We may assume that discriminant and trace are not both 0. For, if $t = 0$, then $C = -A$, and if also $\delta = 0$, then $-C^2 - B^2 = 0$, so that $A = B = C = 0$. In this case, our equation is linear. It will turn out that the discriminant and trace determine the shape of the curve described by (1).

If equation (1) is multiplied by a number $\alpha \neq 0$, then the discriminant is multiplied by α^2 and the trace by α, since

$$(\alpha A)(\alpha C) - (\alpha B)^2 = \alpha^2(AC - B^2) = \alpha^2\delta, \quad \alpha A + \alpha C = \alpha(A + C) = \alpha t.$$

Theorem 2   THE DISCRIMINANT AND TRACE (OF A QUADRIC CURVE) ARE UNCHANGED BY A ROTATION OR TRANSLATION OF THE COORDINATE SYSTEM.

We prove this first for translations. Suppose the coordinate system is translated to a new position. If the coordinates of (x, y) in the new system are (X, Y), then (see §2.2)

$$x = X + a, \qquad y = Y + b$$

where (a, b) is the origin of the new system. Therefore

$$x^2 = X^2 + 2aX + a^2,$$
$$xy = XY + bX + aY + ab,$$
$$y^2 = Y^2 + 2bY + b^2.$$

Substituting into (1), we obtain the transformed equation

$$AX^2 + 2BXY + CY^2 + (2Aa + 2Bb + D)X + (2Ba + 2Cb + E)Y$$
$$+ Aa^2 + 2Bab + Cb^2 + Da + Eb + F = 0.$$

Since this equation has the same second order terms as (1), its discriminant is again $AC - B^2$, and its trace is again $A + C$.

Consider next a rotation of the coordinate system by θ. Then (see Theorem 2 in §2.3) the new coordinates X, Y are connected with the old ones by the relations

$$x = X \cos \theta - Y \sin \theta, \qquad y = X \sin \theta + Y \cos \theta.$$

Therefore

$$x^2 = X^2 \cos^2 \theta - 2XY \cos \theta \sin \theta + Y^2 \sin^2 \theta,$$
$$xy = X^2 \cos \theta \sin \theta + XY(\cos^2 \theta - \sin^2 \theta) - Y^2 \sin \theta \cos \theta,$$
$$y^2 = X^2 \sin^2 \theta + 2XY \sin \theta \cos \theta + Y^2 \cos^2 \theta.$$

Substituting into (1), we obtain the transformed equation in the form

$$(2) \qquad \hat{A}X^2 + 2\hat{B}XY + \hat{C}Y^2 + \hat{D}X + \hat{E}Y + \hat{F},$$

where

$$(3) \qquad \begin{cases} \hat{A} = A \cos^2 \theta + 2B \sin \theta \cos \theta + C \sin^2 \theta, \\ \hat{B} = (C - A) \sin \theta \cos \theta + B(\cos^2 \theta - \sin^2 \theta), \\ \hat{C} = A \sin^2 \theta - 2B \sin \theta \cos \theta + C \cos^2 \theta, \\ \hat{D} = D \cos \theta + E \sin \theta, \hat{E} = -D \sin \theta + E \cos \theta, \hat{F} = F. \end{cases}$$

The new trace is

$$\hat{A} + \hat{C} = A(\cos^2 \theta + \sin^2 \theta) + C(\cos^2 \theta + \sin^2 \theta) = A + C.$$

The new discriminant is

$$\hat{A}\hat{C} - \hat{B}^2 = A^2 \sin^2 \theta \cos^2 \theta - 2AB \sin \theta \cos^3 \theta + AC \cos^4 \theta$$
$$+ 2AB \sin^3 \theta \cos \theta - 4B^2 \sin^2 \theta \cos^2 \theta + 2BC \sin \theta \cos^3 \theta$$
$$+ AC \sin^4 \theta - 2BC \sin^3 \theta \cos \theta + C^2 \sin^2 \theta \cos^2 \theta$$
$$- (C^2 - 2AC + A^2) \sin^2 \theta \cos^2 \theta$$
$$- 2(BC - AB)(\sin \theta \cos^3 \theta - \sin^3 \theta \cos \theta)$$
$$- B^2(\cos^4 \theta - 2 \sin^2 \theta \cos^2 \theta + \sin^4 \theta)$$
$$= AC(\cos^4 \theta + \sin^4 \theta + 2 \sin^2 \theta \cos^2 \theta)$$
$$- B^2(4 \sin^2 \theta \cos^2 \theta + \cos^4 \theta - 2 \sin^2 \theta \cos^2 \theta + \sin^4 \theta)$$
$$= (AC - B^2)(\cos^2 \theta + \sin^2 \theta)^2 = AC - B^2.$$

This completes the proof.

▶ **Examples   1.** We showed in §2.3, Example 4, that the equation $x^2 - y^2 = 1$ is transformed, by a rotation of the coordinate system, into $XY = \frac{1}{2}$. Both equations have discriminant -1.

2. If $t^2 = 4\delta$, equation (1) represents a circle, a point, or the empty set.

Proof.   If $(A + C)^2 = A^2 + 2AC + C^2 = 4AC - 4B^2$, then $(A - C)^2 = -4B^2$; hence $A = C$ and $B = 0$. Now apply the result in Chapter 2, §3.1. ◀

3.4 Main theorem. Removing the mixed term

Now we state the main result.

Theorem 3   THE SET OF POINTS (x, y) SATISFYING A SECOND ORDER EQUATION (1) IS EITHER EMPTY, OR A POINT, OR CONSISTS OF ONE OR TWO LINES; OR IS A PARABOLA, AN ELLIPSE, OR A HYPERBOLA.

We shall prove this by showing how to transform the general second order equation $Ax^2 + 2Bxy + Cy^2 + Dx + Ey + F = 0$ into one of the standard equations (P), (E), (H). This will involve *rotating* the coordinate system, *translating* the coordinate system, and *multiplying* both sides of the equation by a number $\alpha \neq 0$. The proof consists of several steps; the procedure described is called "reducing a second order equation to standard form." The first step is the most important.

Proof of Theorem 3. Step I.   *By a rotation of the coordinate system, a second order equation can be brought into a form with no mixed term* (no term with xy).

Indeed, a rotation by θ transforms (1) into (2), the new coefficients being given by (3). We rewrite the formula for $\hat{B}$ (pronounced "B hat") as

$$2\hat{B} = (C - A)\sin 2\theta + 2B\cos 2\theta.$$

We want to choose θ so as to have $\hat{B} = 0$. If $A \neq C$, then we must have

(4) $$\tan 2\theta = \frac{2B}{A - C}.$$

In this case, there will be four values θ with $\hat{B} = 0$, namely,

$$\theta_0 = \frac{1}{2}\arctan\frac{2B}{A - C}, \quad \theta_0 + 90°, \quad \theta_0 + 180°, \quad \text{and } \theta° + 270°.$$

If $A = C$ and $B \neq 0$, then $\hat{B} = 0$ if $\cos 2\theta = 0$, that is, if θ has one of the values $45°$, $135°$, $225°$, or $315°$. Finally, if $A = C$ and $B = 0$, then $\hat{B} = 0$ for all θ. Of course, if $B = 0$, there is no need to introduce new coordinates.

▶ *Example*   Transform the equation

$$3x^2 + 2xy + 3y^2 - \sqrt{2}x = 0$$

into an equation without a mixed term, by rotating the coordinate system.

Solution.   Since $A = 3$, $2B = 2$, and $C = 3$, an appropriate rotation angle θ is $45°$. Since $\cos 45° = \sin 45° = 1/\sqrt{2}$, the "new" coordinates X, Y are given by

$$x = \frac{1}{\sqrt{2}}(X - Y), \qquad y = \frac{1}{\sqrt{2}}(X + Y).$$

Substituting into our equation [*this is preferable to remembering and using formulas* (3)], we get

$$\frac{3}{2}(X^2 - 2XY + Y^2) + (X^2 - Y^2) + \frac{3}{2}(X^2 + 2XY + Y^2) - (X - Y) = 0$$

or, collecting terms,

$$4X^2 + 2Y^2 - X + Y = 0. ◀$$

EXERCISES

In each of Exercises 11 to 22, find a rotation of the given (x, y) Cartesian coordinate system that will eliminate the mixed terms of the given second degree equation in x and y, and write down the equivalent equation in the new (X, Y) coordinate system. Sketch the old and new coordinate axes. In some cases, it will be easiest to determine $\cos 2\theta$ and $\sin 2\theta$ from $\tan 2\theta$ and then use the half-angle formulas for finding $\sin \theta$ and $\cos \theta$. This avoids the necessity of explicitly finding θ.

11. $2x^2 + 3xy + 2y^2 - x + 1 = 0$.

12. $xy + x + y = 0$.

13. $x^2 + \sqrt{3}xy - 1 = 0$.

14. $x^2 + xy + y^2 - 3 = 0$.

15. $3x^2 + 2\sqrt{3}xy + y^2 - 4 = 0$.

16. $3x^2 - 4\sqrt{3}xy - y^2 + 20y - 25 = 0$.

17. $3x^2 - 2\sqrt{3}xy - y^2 - 3\sqrt{3}x + 7y = 0$.

▶18. $4x^2 + 4xy + y^2 - x = 0$.

19. $5x^2 - 3xy + y^2 - \dfrac{33}{\sqrt{10}}x + \dfrac{11}{\sqrt{10}}y = 0$.

20. $4xy + 3y^2 + 2\sqrt{5}x + 4\sqrt{5}y = 0$.

21. $4x^2 + 12xy + 9y^2 + 2\sqrt{13}x + 2\sqrt{13}y = 0$.

22. $14x^2 + 5xy + 2y^2 - 2 = 0$.

3.5 Removing linear terms

Proof of Theorem 3. Step II.   We consider a second order equation without a mixed term

$$(5) \qquad\qquad Ax^2 + Cy^2 + Dx + Ey + F = 0.$$

If the discriminant is not 0, then the linear terms (terms with x or y) *can be removed by a translation of the coordinate system.*

Indeed, since the discriminant $AC \neq 0$, we have $A \neq 0$, $C \neq 0$. "Completing the square," we obtain

$$Ax^2 + Dx = A\left(x^2 + \frac{D}{A}x + \left(\frac{D}{2A}\right)^2\right) - \frac{D^2}{4A} = A\left(x + \frac{D}{2A}\right)^2 - \frac{D^2}{4A}.$$

Similarly,

$$Cy^2 + Ey = C\left(y^2 + \frac{E}{C}y + \left(\frac{E}{2C}\right)^2\right) - \frac{E^2}{4C} = C\left(y + \frac{E}{2C}\right)^2 - \frac{E^2}{4C},$$

so that our equation reads

$$A\left(x + \frac{D}{2A}\right)^2 + C\left(y + \frac{E}{2C}\right)^2 + F_1 = 0,$$

where $F_1 = F - (D^2/4A) - (E^2/4C)$. Now translate the coordinate system so that the new origin is at $(-D/2A, -E/2C)$. Then the new coordinates X, Y are given by

$$x = X - \frac{D}{2A}, \qquad y = Y - \frac{E}{2C}$$

and the equation becomes

$$AX^2 + CY^2 + F_1 = 0.$$

There are no linear terms.

▶ **Example** (Continuation)   Transform the equation

$$3x^2 + 2xy + 3y^2 - \sqrt{2}x = 0$$

into one without mixed or linear terms.

Solution.   The discriminant is $9 - 1 > 0$. In the previous example, we had already removed the mixed term by a rotation and obtained the equation

$$4X^2 + 2Y^2 - X + Y = 0.$$

We rewrite this as

$$4\left(X^2 - \frac{1}{4}X + \frac{1}{64}\right) - \frac{1}{16} + 2\left(Y^2 + \frac{1}{2}Y + \frac{1}{16}\right) - \frac{1}{8} = 0$$

or
$$4\left(X - \frac{1}{8}\right)^2 + 2\left(Y + \frac{1}{4}\right)^2 = \frac{3}{16}.$$

Set
$$\xi = X - \frac{1}{8}, \qquad \eta = Y + \frac{1}{4}$$

(this amounts to a translation of the coordinate system). The equation becomes

$$4\xi^2 + 2\eta^2 = \frac{3}{16}. \blacktriangleleft$$

EXERCISES

In each of Exercises 23 to 34, determine the discriminant of the given second degree equation in X and Y without mixed terms. Then, if possible, find a translation of the (X, Y) Cartesian coordinate system that will eliminate the linear terms and write down the equivalent equation in the new (ξ, η) coordinate system. Sketch the old and new coordinate axes.

23. Equation obtained in Exercise 11.
24. Equation obtained in Exercise 12.
25. Equation obtained in Exercise 13.
26. Equation obtained in Exercise 14.
27. Equation obtained in Exercise 15.
28. Equation obtained in Exercise 16.

29. Equation obtained in Exercise 17.
30. Equation obtained in Exercise 18.
31. Equation obtained in Exercise 19.
32. Equation obtained in Exercise 20.
33. Equation obtained in Exercise 21.
34. Equation obtained in Exercise 22.

3.6 Positive discriminant

Proof of Theorem 3. Step III. *If the discriminant of a second degree equation is positive, then the equation represents the empty set, or a point, or an ellipse* (which may be a circle).

Indeed, by Steps I and II, we can assume that the equation is already in the form

$$(6) \qquad Ax^2 + Cy^2 = F, \qquad AC > 0.$$

We assume that $A > 0$ and $C > 0$. This can be achieved by multiplying, if needed, both sides of our equation by (-1).

If $F < 0$, equation (6) is never true; it represents the empty set. If $F = 0$, equation (6) is equivalent to $x = y = 0$; it represents a point.

Assume finally that $F > 0$. We also assume that $A \leq C$. (For if $C < A$, we rotate the coordinate system by $90°$. The new coordinates are $X = y$, $Y = -x$. Equation (6) becomes $CX^2 + AY^2 = F$; now the coefficient of X^2 is smaller than that of Y^2.) Set $a = \sqrt{F/A}$, $b = \sqrt{F/C}$. Then $a \geq b \geq 0$. Multiplying both sides of equation (6) by $1/F$, we obtain

$$\frac{x^2}{a^2} + \frac{y^2}{b^2} = 1,$$

a standard equation of an ellipse.

►*Example* (Continuation) Reduce the equation

$$3x^2 + 2xy + 3y^2 - \sqrt{2}x = 0$$

to standard form.

Solution. In the previous examples we obtained, after rotating and translating the coordinate system, the equation

$$4\xi^2 + 2\eta^2 = \frac{3}{16}$$

or, multiplying by $\frac{16}{3}$,

$$\frac{64}{3}\xi^2 + \frac{32}{3}\eta^2 = 1.$$

Since the coefficient of ξ^2 is larger than that of η^2, we rotate the coordinate system by $90°$ and obtain the new coordinates

$$u = \eta, \qquad v = -\xi.$$

Our equation becomes

$$\frac{32}{3}u^2 + \frac{64}{3}v^2 = 1.$$

This is a standard equation of an ellipse, with $a = \sqrt{\frac{3}{32}}, b = \frac{1}{8}\sqrt{3}.$ ◄

3.7 Negative discriminant

Proof of Theorem 3. Step IV.   *If the discriminant of a second degree equation is negative, then the equation represents either two intersecting lines or a hyperbola.*

Indeed, by Steps I and II, we may assume that our equation has the form

$$(7) \qquad\qquad Ax^2 + Cy^2 = F, \qquad AC < 0.$$

Assume first that $F=0$. If $A>0$ and $C<0$, the equation $Ax^2 + Cy^2 = 0$ is equivalent to $(\sqrt{A}x + \sqrt{-Cy})(\sqrt{A}x - \sqrt{-Cy}) = 0$. If $A < 0$ and $C > 0$, it is equivalent to $(\sqrt{-A}x + \sqrt{Cy})(\sqrt{-A}x - \sqrt{Cy})$. In both cases, the equation represents two lines passing through the origin, with slopes $\sqrt{-A/C}$ and $-\sqrt{-A/C}$.

If $F \neq 0$, we may assume that $F = 1$. (If not, multiply the equation by $1/F$.) In the equation

$$Ax^2 + Cy^2 = 1, \qquad AC < 0$$

A and C have opposite signs. We may assume that $A > 0$. (If not, rotate the coordinate system by $90°$.) Set $a = \sqrt{1/A}, b = \sqrt{-1/C}$. Then our equation becomes

$$\frac{x^2}{a^2} - \frac{y^2}{b^2} = 1,$$

a standard equation of a hyperbola.

3.8 Discriminant zero

Proof of Theorem 3. Step V.   *If the discriminant of a second order equation is zero, the equation represents the empty set, or a line, or two parallel lines, or a parabola.*

Indeed, in view of Step I we can assume that the equation has no mixed terms. Then $\delta = AC$; since $\delta = 0$, we have $A = 0$ or $C = 0$. We may assume that A and C are not both zero; otherwise the equation is linear. We may assume that $A = 0$ and $C \neq 0$ (for, if $A \neq 0$ and $C = 0$, we may rotate the coordinate system by $90°$; this interchanges x^2 and y^2). We may even assume that $A = 0$ and $C = 1$; this can be achieved by multiplying the equation by $1/C$.

Thus we have to consider the equation

(8) $$y^2 + Dx + Ey + F = 0.$$

Assume first that $D = 0$. The equation is simply a quadratic equation in y alone. If it has no (real) roots, it represents the empty set. If it has exactly one real root α, the equation represents the line $y = \alpha$. If the equation has two real roots, β and γ, the equation represents two parallel lines $y = \beta$ and $y = \gamma$.

Assume next that $D \neq 0$. We "complete the square" and write equation (8) as

$$\left(y + \frac{E}{2}\right)^2 + Dx - \frac{E^2}{4} + F = 0$$

or $$\left(y + \frac{E}{2}\right)^2 + D\left(x - \frac{E^2}{4D} + \frac{F}{D}\right) = 0.$$

A translation of the coordinate system, which leads to the new coordinates, $X = x - (E^2 - 4F)/4D$, $Y = y + (E/2)$ transforms the equation into

$$Y^2 + DX = 0.$$

We may assume that $D < 0$; otherwise rotate the coordinate system by $180°$, which amounts to replacing X by $-X$ and Y by $-Y$. Set $a = -D/4$; then $a > 0$ and we obtain the equation

$$Y^2 = 4aX,$$

a standard equation of the parabola.

The proof of our theorem is now complete.

▶**Examples**   **1.** Find the vertices of the ellipse

$$3x^2 + 2xy + 3y^2 - \sqrt{2}x = 0.$$

Solution.   In the previous examples, we transformed this equation to the standard form

$$\frac{u^2}{(\sqrt{\frac{3}{32}})^2} + \frac{v^2}{(\frac{1}{8}\sqrt{3})^2} = 1.$$

This involved the transition from (x, y) to coordinates X, Y given by

$$x = \frac{1}{\sqrt{2}}(X - Y), \qquad y = \frac{1}{\sqrt{2}}(X + Y),$$

to coordinates ξ, η given by

$$\xi = X - \frac{1}{8}, \qquad \eta = Y + \frac{1}{4},$$

and to coordinates u, v defined by

$$u = \eta, \qquad v = -\xi.$$

In the u, v system, the four vertices are

$$\left(\sqrt{\frac{3}{32}}, 0\right), \left(-\sqrt{\frac{3}{32}}, 0\right), \left(0, \frac{\sqrt{3}}{8}\right), \left(0, -\frac{\sqrt{3}}{8}\right).$$

Now, if $u = \sqrt{\frac{3}{32}}, v = 0,$ then $\xi = 0, \eta = \sqrt{\frac{3}{32}}$

and therefore $X = \frac{1}{8}, \qquad Y = \sqrt{\frac{3}{32}} - \frac{1}{4}$

and
$$x = \frac{1}{\sqrt{2}}\left(\frac{1}{8} - \sqrt{\frac{3}{32}} + \frac{1}{4}\right) = -\frac{1}{\sqrt{2}}\left(\sqrt{\frac{3}{32}} - \frac{3}{8}\right)$$

$$= -\sqrt{\frac{3}{64}} + \frac{3}{8\sqrt{2}} = -\frac{\sqrt{3}}{8} + \frac{3}{8\sqrt{2}} = -\frac{1}{8}\left(\sqrt{3} - \frac{3}{\sqrt{2}}\right)$$

$$y = \frac{1}{\sqrt{2}}\left(\frac{1}{8} + \sqrt{\frac{3}{32}} - \frac{1}{4}\right) = \frac{1}{\sqrt{2}}\left(\sqrt{\frac{3}{32}} - \frac{1}{8}\right)$$

$$= \sqrt{\frac{3}{64}} - \frac{1}{8\sqrt{2}} = \frac{\sqrt{3}}{8} - \frac{1}{8\sqrt{2}} = \frac{1}{8}\left(\sqrt{3} - \frac{1}{\sqrt{2}}\right).$$

Hence one vertex has the (x, y) coordinates

$$x = -\frac{\sqrt{6} - 3}{8\sqrt{2}}, \qquad y = \frac{\sqrt{6} - 1}{8\sqrt{2}}.$$

The reader can easily find the (x, y) coordinates of the three others.

2. Find the asymptotes of the hyperbola

$$\sqrt{3}xy - y^2 = 1.$$

Solution. First reduce the equation to standard form. Formula (4) with $A = 0$, $B = \sqrt{3}/2$, and $C = -1$ gives the rotation angle $\theta = \frac{1}{2}$ arc tan $\sqrt{3} = 30°$. Since $\sin 30° = \frac{1}{2}$, $\cos 30° = \frac{1}{2}\sqrt{3}$, a rotation of the coordinate system by $30°$ gives new coordinates X, Y such that

$$x = \frac{\sqrt{3}X - Y}{2}, \qquad y = \frac{X + \sqrt{3}Y}{2}.$$

Substituting into the equation, we obtain

$$\sqrt{3}\,\frac{\sqrt{3}X^2 - \sqrt{3}Y^2 + 2XY}{4} - \frac{X^2 + 2\sqrt{3}XY + 3Y^2}{4} = 1$$

or
$$\frac{1}{2}X^2 - \frac{3}{2}Y^2 = 1,$$

that is, the standard equation of a hyperbola (with $a = \sqrt{2}$ and $b = \sqrt{\frac{2}{3}}$). The equation of the asymptotes is

$$\frac{1}{2}X^2 - \frac{3}{2}Y^2 = 0 \qquad \text{or} \qquad \sqrt{3}xy - y^2 = 0.$$

The asymptotes are: the x axis and the line $y = \sqrt{3}x$. ◀

EXERCISES

In Exercises 35 to 46, complete the work started in Exercises 11 to 34 as follows: identify the given quadric curve as an ellipse, a hyperbola, a parabola, a point, one line, two lines, or the empty set. In the nondegenerate cases, find the vertices, foci, and asymptotes, if relevant, in the original (x, y) coordinate system. In all cases, sketch the quadric curve in the original (x, y) coordinate system.

35. Equation in 11 (see 23).
36. Equation in 12 (see 24).
37. Equation in 13 (see 25).
▶ 38. Equation in 14 (see 26).
39. Equation in 15 (see 27).
40. Equation in 16 (see 28).
41. Equation in 17 (see 29).
42. Equation in 18 (see 30).
43. Equation in 19 (see 31).
44. Equation in 20 (see 32).
45. Equation in 21 (see 33).
46. Equation in 22 (see 34).

In Exercises 47 to 54, find a second degree equation in x and y defining the quadric curve described and compute the discriminant in each case. Determine geometrically the angle of rotation required to eliminate mixed terms, if present.

47. The two lines of slope 1 through the points $(0, 0)$ and $(0, 1)$.
48. The two lines of slope 1 and -1 through the point $(1, 1)$.
49. The point $(2, -1)$.
50. The line of slope $\frac{1}{2}$ through the point $(2, 3)$.
51. The two lines of slope $\sqrt{3}/3$ and $\sqrt{3}$ through the point $(1, 0)$.
52. The ellipse with vertices at $(\pm 3, \mp 4)$ and $(\pm \frac{4}{5}, \pm 1)$.
53. The hyperbola with vertices at $(\pm 12, \pm 5)$ and foci at $(\pm 24, \pm 10)$.
54. The parabola with vertex at $(1, 2)$ and focus at $(2, 3)$.

Appendix to Chapter 10

§4 Quadric Surfaces

4.1 Surfaces of second degree

The definition of a polynomial of degree n in three variables is an obvious generalization of the definition given in §3.1 for two variables. An **algebraic surface of degree** n is the set of points in space whose coordinates x, y, z satisfy an equation of the form (a polynomial of degree n) $= 0$. For example, a plane is an algebraic surface of degree 1, a sphere one of degree 2.

A **quadric surface** is an algebraic surface of degree 2, that is, the set of points with Cartesian coordinates (x, y, z) satisfying a second degree equation, that is, an equation of the form

(1) $Ax^2 + By^2 + Cz^2 + 2Dxy + 2Eyz + 2Fxz + Gx + Hy + Iz + K = 0.$

We begin by considering some examples.

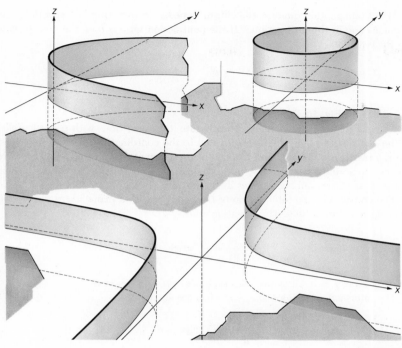

Fig. 10.33

4.2 Degenerate cases. Cylinders

First we look at some degenerate cases. Equation (1) may represent the empty set ($x^2 = -1$), a point ($x^2 + y^2 + z^2 = 0$), a line ($x^2 + y^2 = 0$), a plane ($z^2 = 0$), two parallel planes ($z^2 = 1$), or two intersecting planes ($xy = 0$).

If no term with z appears in (1), the surface is called a second degree **cylinder.** To construct such a cylinder, consider some second order curve γ in the x, y plane and "draw" all points (x, y, z) in space with (x, y) on γ. Some examples (**parabolic cylinder, elliptic cylinder,** and **hyperbolic cylinder**) are shown in Fig. 10.33.

More generally, a cylinder is a surface S such that, for a suitable coordinate system, S consists of all lines, orthogonal to the plane $z = 0$, which pass through some curve γ in that plane.

4.3 Cones

In Chapter 9, §5.2, we noted that, by rotating the line $z = mx$ in the x, z plane about the z axis, we obtain the surface (see Fig. 10.34)

$$(2) \qquad \frac{x^2}{a^2} + \frac{y^2}{a^2} - z^2 = 0.$$

The quadric surface is called a **circular cone.**

Fig. 10.34

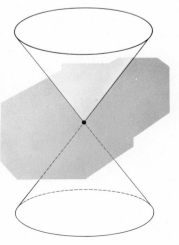

Fig. 10.35

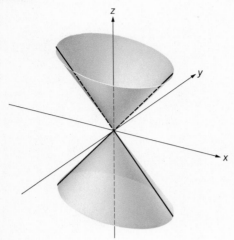

We obtain another quadric, the **elliptic cone** by "contracting (or expanding) the circular cone in the y direction." If the contraction ratio is b/a, the new surface has the equation

(3)
$$\frac{x^2}{a^2} + \frac{y^2}{b^2} - z^2 = 0;$$

it is shown in Fig. 10.35.

In order to study this surface, we compute its intersections with planes parallel to the coordinate planes. A plane parallel to the coordinate plane $z = 0$ (that is, the x, y plane) has the equation $z = \gamma$ (γ some constant). In this plane, x and y are Cartesian coordinates. Similarly, (y, z) are Cartesian coordinates in a plane $x = \alpha$, and (x, z) are Cartesian coordinates in a plane $y = \beta$.

The equation of the intersection of the cone (3) with a plane $z = \gamma$ is obtained by replacing z in (3) by γ. This equation reads

$$\frac{x^2}{a^2} + \frac{y^2}{b^2} = \gamma^2.$$

It is a point if $\gamma = 0$, an ellipse with semiaxes $|\gamma| a$, $|\gamma| b$ if $\gamma \neq 0$.

The intersection with a plane $x = \alpha$ has the equation

$$\frac{\alpha^2}{a^2} + \frac{y^2}{b^2} - z^2 = 0$$

or
$$z^2 - \frac{y^2}{b^2} = \frac{\alpha^2}{a^2}.$$

This is, for $\alpha \neq 0$, a hyperbola with semiaxes α/a and $b\alpha/a$. For $\alpha = 0$, we obtain a pair of lines. The intersection with a plane $y = \beta$ is also either a hyperbola or a pair of lines.

4.4 Ellipsoids

Rotating a conic about an axis, we also obtain a quadric surface. The rotating of an ellipse

$$\frac{x^2}{a^2} + \frac{z^2}{c^2} = 1$$

about the z axis yields an ellipsoid of revolution. The equation of this surface in cylindrical coordinates (see Chapter 9, §5.2) is

$$\frac{r^2}{a^2} + \frac{z^2}{c^2} = 1.$$

The equation in Cartesian coordinates is therefore (since $r^2 = x^2 + y^2$)

(4)
$$\frac{x^2}{a^2} + \frac{y^2}{a^2} + \frac{z^2}{c^2} = 1.$$

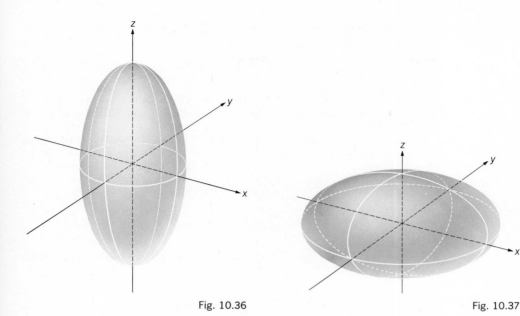

Fig. 10.36

Fig. 10.37

The surface is called an **ellipsoid of revolution,** an **oblong** ellipsoid if $c > a$ (Fig. 10.36), an **oblate** ellipsoid if $c < a$ (Fig. 10.37), and, of course, a sphere if $a = c$. In other words, an oblong ellipsoid is obtained by rotating an ellipse about its major axis; an oblate one by rotating an ellipse about its minor axis. The surface of the earth is, approximately, an oblate ellipsoid.

A general ellipsoid is obtained from (4) by a contraction in the y direction; its equation reads

(5)
$$\frac{x^2}{a^2} + \frac{y^2}{b^2} + \frac{z^2}{c^2} = 1$$

(see Fig. 10.38, where $a > b > c$). The intersection of a plane $x = \alpha$, $y = \beta$, or $z = \gamma$ with (5) is either empty, or a point, or an ellipse, as the reader should verify.

Fig. 10.38

4.5 Hyperboloids

Rotating a hyperbola *around its axis* gives a **hyperboloid of revolution of two sheets.** If the equation of the hyperbola is

$$\frac{z^2}{c^2} - \frac{x^2}{a^2} = 1,$$

the equation of the hyperboloid reads

$$-\frac{x^2}{a^2} - \frac{y^2}{a^2} + \frac{z^2}{c^2} = 1.$$

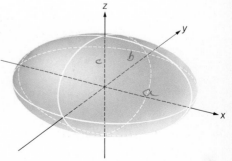

A contraction in the y direction gives the **general hyperboloid of two sheets:**

(6)
$$-\frac{x^2}{a^2} - \frac{y^2}{b^2} + \frac{z^2}{c^2} = 1$$

(Fig. 10.39). It is easy to see that the intersection of this surface with a plane $z = \gamma$ is empty, a point, or an ellipse, while the intersection with either $x = \alpha$ or $y = \beta$ is always a hyperbola.

A **hyperboloid of revolution of one sheet** is obtained by rotating a hyperbola *about its conjugate axis*. Starting with

$$\frac{x^2}{a^2} - \frac{z^2}{c^2} = 1,$$

we obtain
$$\frac{x^2}{a^2} + \frac{y^2}{a^2} - \frac{z^2}{c^2} = 1.$$

A contraction in the y direction gives us the **general hyperboloid of one sheet** (see Fig. 10.40).

(7)
$$\frac{x^2}{a^2} + \frac{y^2}{b^2} - \frac{z^2}{c^2} = 1.$$

The intersections with planes $z = \gamma$ are ellipses, those with planes $x = \alpha$ or $y = \beta$ are hyperbolas which for $\alpha = a$ or $\beta = b$ degenerate into pairs of straight lines.

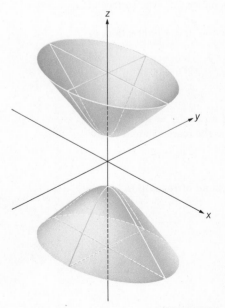

Fig. 10.39

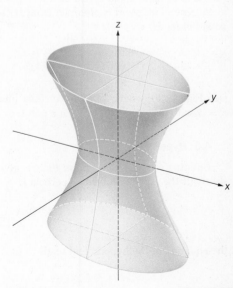

Fig. 10.40

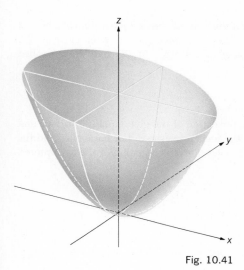

Fig. 10.41

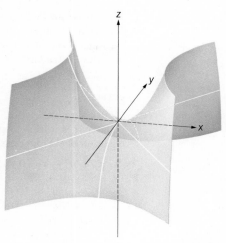

Fig. 10.42

4.6 Paraboloids

There are two kinds of quadric surfaces called paraboloids. An **elliptic paraboloid** has the equation

$$(8) \qquad \frac{x^2}{a^2} + \frac{y^2}{b^2} = z.$$

The surface lies in the half-space $z \geq 0$ (see Fig. 10.41). For $\gamma > 0$, the plane $z = \gamma$ intersects the surface in an ellipse. The intersections with the planes $x = \alpha$ and $y = \beta$ are parabolas. We may think of (8) as having been obtained by a contraction in the y direction of a paraboloid of revolution

$$x^2 + y^2 = a^2 z,$$

which is the result of rotating the parabola

$$x^2 - a^2 z = 0$$

about its axis.

The **hyperbolic paraboloid** (Fig. 10.42) is not the result of contracting a surface of revolution. Its equation is of the form

$$(9) \qquad \frac{x^2}{a^2} - \frac{y^2}{b^2} = z.$$

The intersections with the planes $x = \alpha$ or $y = \beta$ are parabolas, those with planes $z = \gamma$ are hyperbolas, except that for $z = 0$ we get two lines.

It turns out that the examples given thus far *exhaust all possibilities*. This is a central result in the theory of quadric surfaces, analogous to Theorem 2 in §3. We shall prove it in §4.11. The surfaces (3), (4), (5), (6), (7), and (8) are called quadrics in standard position.

EXERCISES

In Exercises 1 to 14, identify the intersection of the given quadric surface and plane and, if it is nondegenerate, find the foci and vertices. For parabolas, find the equation of the directrix, and for hyperbolas, find the equation of the asymptotes.

1. $x^2 + y^2 - z = 0$ and $x = 2$.
▶ 2. $4x^2 + y^2 + z^2/9 = 1$ and $z = 2$.
3. $4x^2 + 25y^2 - 100z^2 = 0$ and $x = 1$.
4. $4x^2 + 25y^2 - 100z^2 = 0$ and $z = -1$.
5. $x^2/36 + 4y^2 + z^2/9 = 1$ and $4y + 1 = 0$.
6. $z^2 - y^2/4 - x^2/36 = 1$ and $x = 8$.
7. $z^2 - y^2/4 - x^2/36 = 1$ and $z = -\sqrt{5}$.
8. $4x^2 + y^2 - z^2/16 = 1$ and $z = 3$.
9. $4x^2 + y^2 - z^2/16 = 1$ and $y = \sqrt{3}/2$.
10. $4x^2 + y^2 - z^2/16 = 1$ and $y = \sqrt{5}/2$.
11. $z = x^2/4 + y^2/9$ and $x = -2$.
12. $4x^2 - 16y^2 = z$ and $z = 1$.
13. $4x^2 - 16y^2 = z$ and $z = -1$.
14. $4x^2 - 16y^2 = z$ and $2x - 1 = 0$.

4.7 Ruled surfaces

A surface S is called **ruled** if for every point P on S there is a line l that passes through P and lies on S (that is, such that every point on the line is a point on the surface). It is obvious that all planes and cylinders are ruled surfaces, and so are all cones. But there are also other ruled quadric surfaces.

Theorem 1 HYPERBOLOIDS OF ONE SHEET AND HYPERBOLIC PARABOLOIDS ARE RULED SURFACES.

Proof. Let α be a fixed number. The line given parametrically by

$$(10) \qquad x = a \cos \alpha + (a \sin \alpha)t, \qquad y = b \sin \alpha - (b \cos \alpha)t, \qquad z = t$$

lies on the surface (7). This is verified by substituting the values of x, y, z given by (10) into the equation (7); we obtain a correct statement, for all values of the parameter t, as the reader should check.

Assume now that the point $P = (x_0, y_0, z_0)$ lies on the hyperboloid (7). We want to find a number α such that, for this value of α, the line (10) passes through P. The value of t corresponding to P must be z_0. Hence we must have

$$x_0 = a \cos \alpha + (a \sin \alpha)z_0, \qquad y_0 = b \sin \alpha - (b \cos \alpha)z_0.$$

We treat this as a system of two equations for the "unknowns" $\cos \alpha$ and $\sin \alpha$. Solving the equations, we obtain

$$\cos \alpha = \frac{(x_0/a) - (y_0/b)z_0}{1 + z_0^2}, \qquad \sin \alpha = \frac{(x_0/a)z_0 + (y_0/b)}{1 + z_0^2}.$$

In order that there be an α satisfying these relations, we must have

$$\left[\frac{(x_0/a) - (y_0/b)z_0}{1 + z_0^2}\right]^2 + \left[\frac{(x_0/a)z_0 + (y_0/b)}{1 + z_0^2}\right]^2 = 1.$$

This is equivalent to the condition that (x_0, y_0, z_0) satisfies (7).

There is another family of straight lines covering the hyperboloid (7), namely the lines

$$(11) \qquad x = a \cos \beta - (a \sin \beta)s, \qquad y = b \sin \beta + (b \cos \beta)s, \qquad z = s;$$

here β is a fixed number and s the parameter.

Our statement concerning the hyperboloid of one sheet is proved; it follows that one can make a model of this surface out of straight rods, such a model is shown in Fig. 10.43.

Next we verify that, for every fixed number A, the line

$$(12) \qquad x = Aa + at, \qquad y = Ab - bt, \qquad z = 4At$$

lies on the hyperbolic paraboloid (9). If $P = (x_0, y_0, z_0)$ is a point on (9), the line (12) through this point is found by setting

$$A = \frac{(x_0/a) + (y_0/b)}{2}.$$

There is also another family of lines covering (9), namely, the lines

$$(13) \qquad x = Ba + as, \qquad y = -Bb + bs, \qquad z = 4Bs,$$

with fixed B and parameter s. The straight rod model of the paraboloid is shown in Fig. 10.44.

Fig. 10.43

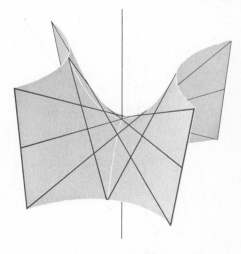

Fig. 10.44

EXERCISES

15. Find two lines through the point $(\sqrt{\frac{3}{2}}, \sqrt{\frac{1}{2}}, 0)$ that lie on the hyperboloid $x^2 + y^2 - 100z^2 = 1$.

16. Find two lines through the point $(2, 2, \sqrt{3})$ that lie on the hyperboloid $x^2 + y^2 - z^2 = 1$.

17. Find two lines through the point $(0, 0, 0)$ that lie on the paraboloid $4x^2 - 5y^2 = z$.

18. Find two lines through the point $(4, 5, -9)$ that lie on the paraboloid $x^2 - y^2 = z$.

19. Show that two lines of the form (10), corresponding to 2 different values of α from the range $0 \leq \alpha < 2\pi$, are skew to each other.
20. Show that no line in the family (10) coincides with a line in the family (11).
21. Show that two lines in the family (12), corresponding to different values of A, are skew to each other.
22. Show that no line in the family (12) coincides with a line in the family (13).

4.8 Changing coordinates in space

Thus far we have considered quadric surfaces in standard position, that is, we have assumed that the coordinate system is chosen so as to make the equations of our surfaces as simple as possible. In order to consider all quadric surfaces, we first make some remarks about the change in coordinates resulting from changing the origin and the axes in a system (compare the discussion about coordinates in the plane in §§2.2, 2.3).

Suppose first that we translate the coordinate axes, that is, we pass from the system $(O, \mathbf{e}_1, \mathbf{e}_2, \mathbf{e}_3)$ to the system $(O', \mathbf{e}_1, \mathbf{e}_2, \mathbf{e}_3)$ where $\{\mathbf{e}_1, \mathbf{e}_2, \mathbf{e}_3\}$ is a frame. Let P be a point, and let $\mathbf{P}$ and $\mathbf{P}'$ be its position vectors with respect to O and O', respectively. Then $\mathbf{P} = \mathbf{T} + \mathbf{P}'$ where $\mathbf{T}$ is determined by $\overrightarrow{OO'}$; see Fig. 10.45. Let P have coordinates x, y, z in the "old" system $(O, \mathbf{e}_1, \mathbf{e}_2, \mathbf{e}_3)$ and coordinates (X, Y, Z) in the "new" system $(O', \mathbf{e}_1, \mathbf{e}_2, \mathbf{e}_3)$ and let O' have "old" coordinates (a, b, c). Then $\mathbf{P} = x\mathbf{e}_1 + y\mathbf{e}_2 + z\mathbf{e}_3$. $\mathbf{P}' = X\mathbf{e}_1 + Y\mathbf{e}_2 + Z\mathbf{e}_3$ and $\mathbf{T} = a\mathbf{e}_1 + b\mathbf{e}_2 + c\mathbf{e}_3$, so that

$$(14) \qquad x = a + X, \qquad y = b + Y, \qquad z = c + Z.$$

Consider next the case in which the new coordinate system $(O, \mathbf{f}_1, \mathbf{f}_2, \mathbf{f}_3)$ is obtained from the old by changing axes, that is, by going over to a new frame $(\mathbf{f}_1, \mathbf{f}_2, \mathbf{f}_3)$. Let (X, Y, Z) be the coordinates of P in the system $(O, \mathbf{f}_1, \mathbf{f}_2, \mathbf{f}_3)$; see Fig. 10.46. Then

$$(15) \qquad x\mathbf{e}_1 + y\mathbf{e}_2 + z\mathbf{e}_3 = X\mathbf{f}_1 + Y\mathbf{f}_2 + Z\mathbf{f}_3.$$

Now we have

$$(16) \qquad \begin{aligned} \mathbf{f}_1 &= \alpha_1\mathbf{e}_1 + \alpha_2\mathbf{e}_2 + \alpha_3\mathbf{e}_3, \qquad \mathbf{f}_2 = \beta_1\mathbf{e}_1 + \beta_2\mathbf{e}_2 + \beta_3\mathbf{e}_3, \\ \mathbf{f}_3 &= \gamma_1\mathbf{e}_1 + \gamma_2\mathbf{e}_2 + \gamma_3\mathbf{e}_3 \end{aligned}$$

where the α's, β's, and γ's are numbers that depend on the two frames $\{\mathbf{e}_1, \mathbf{e}_2, \mathbf{e}_3\}$ and $\{\mathbf{f}_1, \mathbf{f}_2, \mathbf{f}_3\}$ but not on the point P. If we substitute (16) into the right-hand side of (15), rearrange the terms, and observe that the coefficients of $\mathbf{e}_1, \mathbf{e}_2, \mathbf{e}_3$ on both sides must be the same, we obtain the equation

$$(17) \qquad \begin{aligned} x &= \alpha_1 X + \beta_1 Y + \gamma_1 Z, \qquad y = \alpha_2 X + \beta_2 Y + \gamma_2 Z, \\ z &= \alpha_3 X + \beta_3 Y + \gamma_3 Z. \end{aligned}$$

The formulas (14) and (17) should be compared with the formulas in §3.1. Arguing in exactly as in §3.1, we can establish the following result.

Fig. 10.45

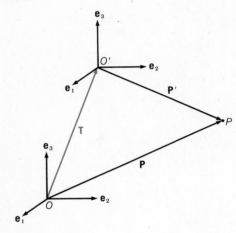

Fig. 10.46

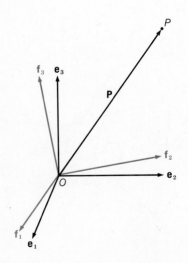

Theorem 2 THE DEGREE OF AN ALGEBRAIC SURFACE DOES NOT DEPEND ON THE COORDINATE SYSTEM USED.

4.9 Plane sections of quadrics

We are now in a position to prove the following result.

Theorem 3 THE INTERSECTION OF A QUADRIC SURFACE WITH A PLANE (NOT CONTAINED IN IT) IS A QUADRIC CURVE.

The proof is quite simple. By Theorem 2 a quadric surface remains a quadric surface if we change the coordinate system. Therefore we may choose the coordinate system so that the plane in question is the plane $z = 0$. The equation of our quadric surface is of the form (1). To obtain the equation of the intersection, we set $z = 0$ in (1) and obtain

$$(18) \qquad Ax^2 + By^2 + 2Dxy + Gx + Hy + K = 0.$$

Not all numbers A, B, $\cdots$, K are 0; otherwise the plane $z = 0$ would be part of the surface. The set of points satisfying (18) is a quadric curve. Theorem 3 is proved.

In particular, since a circular cone is a quadric surface, the intersection of a plane and a circular cone is a parabola, an ellipse, a hyperbola, a pair of lines, a line, or a point. Thus we have now obtained, as a special case of a very general result, the property with which the Greek mathematicians started their investigations.

4.10 Rotations and reflections in space

Let $\{\mathbf{e}_1, \mathbf{e}_2, \mathbf{e}_3\}$ be a frame that we choose to call right-handed. Let θ be a number and set

$$(19) \qquad \mathbf{f}_1 = \cos\theta\,\mathbf{e}_1 + \sin\theta\,\mathbf{e}_2, \qquad \mathbf{f}_2 = -\sin\theta\,\mathbf{e}_1 + \cos\theta\,\mathbf{e}_2, \qquad \mathbf{f}_3 = \mathbf{e}_3.$$

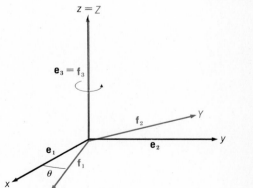

Fig. 10.47

One checks easily that $\{\mathbf{f}_1, \mathbf{f}_2, \mathbf{f}_3\}$ is again a frame; we call it the frame obtained from $\{\mathbf{e}_1, \mathbf{e}_2, \mathbf{e}_3\}$ by a **rotation about the z axis by the angle** θ (see §2.1 and Fig. 10.47). If a point P has coordinates x, y, z in the system $(O, \mathbf{e}_1, \mathbf{e}_2, \mathbf{e}_3)$ and coordinates X, Y, Z in the system $(O, \mathbf{f}_1, \mathbf{f}_2, \mathbf{f}_3)$, then

$$(19') \qquad x = X\cos\theta - Y\sin\theta, \qquad y = X\sin\theta + Y\cos\theta, \qquad z = Z.$$

This is proved exactly as the relations (5) in §2.3. Indeed, we simply have a rotation in the x, y plane, whereas nothing happens in the z direction.

Similarly, we may say that $\{\mathbf{f}_1, \mathbf{f}_2, \mathbf{f}_3\}$ is obtained from $\{\mathbf{e}_1, \mathbf{e}_2, \mathbf{e}_3\}$ by a rotation by θ about the x axis, or about the y axis, if

$$(20) \qquad \mathbf{f}_1 = \mathbf{e}_1, \qquad \mathbf{f}_2 = \cos\theta\,\mathbf{e}_2 + \sin\theta\,\mathbf{e}_3, \qquad \mathbf{f}_3 = -\sin\theta\,\mathbf{e}_2 + \cos\theta\,\mathbf{e}_3,$$

or

(21) $\mathbf{f}_3 = \cos\theta\,\mathbf{e}_3 + \sin\theta\,\mathbf{e}_1, \qquad \mathbf{f}_2 = \mathbf{e}_2, \qquad \mathbf{f}_1 = -\sin\theta\,\mathbf{e}_3 + \cos\theta\,\mathbf{e}_1.$

In this case, the connection between the old and the new coordinates of a point is given by the formulas

(20') $x = X, \qquad y = Y\cos\theta - Z\sin\theta, \qquad z = Y\sin\theta + Z\cos\theta$

or

(21') $x = X\cos\theta - Z\sin\theta, \qquad y = Y, \qquad z = X\sin\theta + Z\cos\theta.$

Fig. 10.48

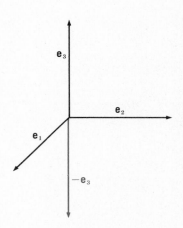

A **rotation in space** may be described as a rotation about the z axis, followed by a rotation about the x axis, followed by a rotation about the y axis; the rotation angles need not be the same. A **reflection** (about the x, y plane) is the transition from the frame $\{\mathbf{e}_1, \mathbf{e}_2, \mathbf{e}_3\}$ to the frame $\{\mathbf{e}_1, \mathbf{e}_2, -\mathbf{e}_3\}$; see Fig. 10.48. It can be shown that *every frame in space is obtained from* $\{\mathbf{e}_1, \mathbf{e}_2, \mathbf{e}_3\}$ *either by a rotation or by a reflection followed by a rotation.* (In the first case, the new frame is called right-handed, in the second case, left-handed; see §2.7.) We shall not prove this here.

EXERCISES

In Exercises 23 to 28, identify the intersection of the given cone and plane and, if it is nondegenerate, find the foci and vertices. In Exercises 25 to 28, begin by making a translation and a rotation about a coordinate axis which takes the given plane into a coordinate plane.

23. $z^2 = x^2 + y^2$ and $x = 2$.
24. $9z^2 = x^2 + y^2$ and $z = -2$.
25. $z^2 = x^2 + y^2$ and $2z = x + 1$.
26. $z^2 = 9x^2 + 9y^2$ and $z = x + 1$.
27. $z^2 = 4x^2 + 9y^2$ and $z = 2x$.
28. $4z^2 = 36x^2 + y^2$ and $z = 3x + 1$.
29. Show that there is a rotation in space that takes $\{\mathbf{e}_1, \mathbf{e}_2, \mathbf{e}_3\}$ into $\{\mathbf{e}_1, -\mathbf{e}_2, -\mathbf{e}_3\}$.
30. Show that there are rotations in space that take $\{\mathbf{e}_1, \mathbf{e}_2, \mathbf{e}_3\}$ into $\{\mathbf{e}_2, \mathbf{e}_3, \mathbf{e}_1\}$, and into $\{\mathbf{e}_3, \mathbf{e}_1, \mathbf{e}_2\}$.

4.11 Reducing a quadric surface to standard form

We shall establish the main theorem on quadric surfaces, an analogue of Theorem 3 in §3.4.

Theorem 4   THE SET OF POINTS (x, y, z) SATISFYING A SECOND ORDER EQUATION

(1) $Ax^2 + By^2 + Cz^2 + 2Dxy + 2Eyz + 2Fxz + Gx + Hy + Iz + K = 0$

IS EITHER EMPTY, OR A POINT, OR A LINE, OR A SECOND DEGREE CYLINDER, OR AN ELLIPTIC CONE, OR AN ELLIPSOID, HYPERBOLOID, OR PARABOLOID.

Proof. We shall show how equation (1) can be reduced to an equation of a quadric in standard form, by rotating and translating the coordinate system, and by multiplying the equation by a number.

For a fixed value of z, we may consider (1) as a second order equation in x and y. We learned in §3.4 how to rotate the x, y coordinate system so as to "eliminate" the "mixed term" $2Dxy$; the angle of rotation depended only on A, B, and D. It follows that by a rotation about the z axis [see equations (19′) above] we can transform (1) into an equivalent equation without a term with xy. In the same way, we can then get rid of the term with yz, by a rotation about the x axis and we can follow this by a rotation about the y axis that will eliminate the term with xz.

Now we have an equation of the form

$$(22) \qquad Ax^2 + By^2 + Cz^2 + Gx + Hy + Iz + K = 0.$$

If $C^2 + I^2 = 0$, (22) looks like a second degree equation in x and y. It may define in the x, y plane a nondegenerate conic, or a pair of lines, or a line. In these cases, it defines in the x, y, z space a cylinder (which may be a plane or a pair of planes). Or (22), with $C^2 + I^2 = 0$, may represent in the x, y plane a point or the empty set. In this case, it represents in the x, y, z space either a line or the empty set. A similar result holds if (22) contains no terms with y or no terms with z. We assume from now on that neither of the numbers $A^2 + G^2$, $B^2 + H^2$, $C^2 + I^2$ is 0.

Not all three numbers A, B, C are 0, for otherwise the degree of (22) would be 1 and not 2. It will suffice to consider the following cases: (i) $A \neq 0$, $B = C = 0$, (ii) $A \neq 0$, $B \neq 0$, $C = 0$, and (iii) $A \neq 0$, $B \neq 0$, $C \neq 0$. [For if, for example, $A = 0$, $B \neq 0$, $C = 0$, we can use a rotation which replaces x, y, z by z, x, y, respectively, and arrive at case (i).]

In case (i), we can get rid of the term Gx by completing the square and translating the coordinate system. More precisely, we write

$$Ax^2 + Gx = A\left(x + \frac{G}{2A}\right)^2 - \frac{G^2}{4A},$$

introduce the new variable $X = x + G/2A$ (this amounts to a translation) and absorb $-G^2/4A$ in the constant term. We arrive at an equation of the form

$$Ax^2 + Hy + Iz + K = 0 \qquad (A \neq 0, H \neq 0, I \neq 0).$$

We may assume that $H^2 + I^2 = 1$, this can be achieved by dividing the equation by $\sqrt{H^2 + I^2}$. Then there is a number α such that $H = \cos\alpha$ and $I = \sin\alpha$. We can introduce new variables X, Y, Z such that $X = X$, $Y = y\cos\alpha + z\sin\alpha + K$ (this amounts to a rotation about the x axis followed by a translation). The equation becomes

$$AX^2 + Y = 0, \qquad A \neq 0.$$

This defines a parabolic cylinder.

In case (ii), we eliminate the terms with x and y by translations, as above; this gives an equation of the form

$$AX^2 + BY^2 + Iz + K = 0 \qquad (A \neq 0, B \neq 0, I \neq 0).$$

We may assume that $I = 1$ (otherwise divide the equation by I). A new translation $(Z = z + K)$ yields

$$AX^2 + BY^2 + Z = 0 \qquad (AB \neq 0).$$

This defines an elliptic paraboloid if $AB > 0$, a hyperbolic paraboloid if $AB < 0$; compare §4.6.

In case (ii), we can eliminate all linear terms by a translation; this yields an equation of the form

$$AX^2 + BY^2 + CZ^2 + K = 0 \qquad (ABC \neq 0).$$

If $K = 0$ and all three numbers A, B, C have the same sign, the equation defines a single point. If $K = 0$ and not all three numbers A, B, C have the same sign, the equation defines an elliptic cone; compare §4.3. If $K \neq 0$, we may assume that $K = -1$, otherwise divide the equation by $(-K)$. Let $K = -1$, and assume that precisely j of the numbers A, B, C are positive, so that $3 - j$ are negative. Our equation represents the empty set if $j = 0$, a hyperboloid of two sheets if $j = 1$, a hyperboloid of one sheet if $j = 2$, and an ellipsoid if $j = 3$; compare §§4.4, 4.5. The theorem is proved.

Theorem 4 has a generalization for an n-dimensional space. This generalization has no "visible" geometric content, but is very important for mathematics, physics, and engineering.

EXERCISES

Identify the following quadrics. (Hint: Use the method of §4.11 in order to simplify the equation, if necessary, but also use ingenuity.)

31. $(x - y)^2 - (x + y)^2 - z = 0$.
32. $100(x + y)^2 + 100(x - y)^2 + (z - 100)^2 = 1$.
33. $xy + yz + zx = 1$.
34. $xy + yz + zx = 0$.
35. $x - yz = 5$.
36. $10x^2 - 2xy + y^2 + z^2 + 2z = 99$.
37. $x^2 + y^2 + 5z^2 - 10x - y - z = 0$.
38. $x^2 - y^2 - 2z^2 + 3x - z + 5 = 0$.
39. $2xy - z^2 + x - y = 100$.
40. $(x - y)^2 + yz - 5x = 1$.

Problems

Problems 1 to 11 refer to general rectilinear coordinate systems in the plane and in space. These systems have been discussed in a remark preceding Problem 5 at the end of Chapter 9, and in Problem 9 at the end of that chapter. We emphasize that in the problems below $\{e_1, e_2\}$ and $\{e_1, e_2, e_3\}$ are bases but not necessarily frames.

1. Let a point P have coordinates (x_1, x_2) in the system $(O, \mathbf{e}_1, \mathbf{e}_2)$ and coordinates (X_1, X_2) in the system $(O', \mathbf{e}_1, \mathbf{e}_2)$. (This second system is said to have been obtained from the first by a *translation*.) Let the point O' have the coordinates (a_1, a_2) in the system $(O, \mathbf{e}_1, \mathbf{e}_2)$. Find the relation between the old and the new coordinates of P. Compare with the result in §2.2.

2. Let a point P have coordinates (x_1, x_2) in the system $(O, \mathbf{e}_1, \mathbf{e}_2)$ and coordinates (X_1, X_2) in the system $(O, \mathbf{f}_1, \mathbf{f}_2)$. (The second system is said to have been obtained from the first by a *change of axes*.) Let the vectors $\mathbf{e}_1$ and $\mathbf{e}_2$ have the components (α, β) and (γ, δ) with respect to the basis $(\mathbf{f}_1, \mathbf{f}_2)$. Find the relation between the old and the new coordinates of P. Compare with the results in §2.3.

3. In the text, we proved (see §2.1) that the degree of an algebraic curve does not depend on the Cartesian coordinates used. Show that the same result holds if one also uses general linear coordinate systems. (Hint: Use the results of Problems 1 and 2.)

4. Prove that a straight line is the set of points whose coordinates, in some linear coordinate system, satisfy a linear equation. Give two proofs, one a direct proof, and the other a proof using the result in Problem 3.

5. Suppose x_1 and x_2 are general linear coordinates. Show that the curve $x_1^2 + x_2^2 = 1$ is an ellipse. Can every ellipse be so represented by choosing a suitable general rectilinear coordinate system?

6. State and prove statements analogous to Problem 5 for hyperbolas and for parabolas.

7. Describe how the coordinates of a point change if we translate the coordinate system in space, that is, if we pass from a system $(O, \mathbf{e}_1, \mathbf{e}_2, \mathbf{e}_3)$ to a system $(O', \mathbf{e}_1, \mathbf{e}_2, \mathbf{e}_3)$. Compare with the result in §4.8.

8. Describe how the coordinates of a point change when we leave the origin fixed and change the coordinate axes, that is, when we pass from the system $(O, \mathbf{e}_1, \mathbf{e}_2, \mathbf{e}_3)$ to a system $(O, \mathbf{f}_1, \mathbf{f}_2, \mathbf{f}_3)$. (Hint: Write the base vectors $\mathbf{f}_1, \mathbf{f}_2, \mathbf{f}_3$ as linear combinations of the vectors $\mathbf{e}_1, \mathbf{e}_2, \mathbf{e}_3$.) Compare with the result in §4.8.

9. Using the results of the preceding two problems, show that the degree of an algebraic surface is unchanged when we pass from one general linear coordinate system to another. (The result has been stated in the text for Cartesian coordinates.)

10. Show that every plane is the set of points satisfying a linear equation in rectilinear coordinates, and vice versa. (Hint: Use the preceding problem; also note that, given a plane, we can always find a coordinate system in which the plane is a coordinate plane.) This result has been proved in the text for Cartesian systems only.

11. Examine all quadric surfaces discussed in §4. In each case, find the simplest form to which the equation can be reduced by introducing a suitable general linear coordinate system. (For instance, show that every ellipsoid can be written in the form $x_1^2 + x_2^2 + x_3^2 = 1$.)

12. Prove that through every point on the hyperboloid, $x^2 + y^2 - z^2 = 1$, pass *exactly* two lines that lie on the hyperboloid. (In the text, §4.7, we prove that there are at least two such lines.)

13. Prove the same theorem as in Problem 12 for the hyperbolic paraboloid $z = x^2 - y^2$.

11/Vector-valued Functions

In this chapter, we combine the concepts of calculus and vector algebra to study plane curves and the motion of particles. The discussion culminates in one of the most important applications of calculus—the theory of planetary motion.

The appendixes to this chapter contain other applications of calculus to mechanics.

A vector-valued function

$$(1) \qquad\qquad t \mapsto \mathbf{f}(t)$$

is a rule that associates a vector $\mathbf{f}(t)$ with every real number t in some interval or collection of intervals. For the sake of simplicity, we assume throughout this section that "vector" means "plane vector." However, everything we say will apply, with obvious modifications, to vectors in space (and also to vectors from an n-dimensional vector space).

It is often convenient to think of vectors as position vectors of points, with respect to some fixed point O. One represents a vector-valued function (1) geometrically by drawing all points with position vector $\mathbf{f}(t)$. This set of points is said to be **represented parametrically** by the function $\mathbf{f}(t)$.

An example of a vector-valued function is a nonconstant linear function

$$(2) \qquad\qquad \mathbf{f}(t) = t\mathbf{a} + \mathbf{b}, \qquad -\infty < t < +\infty,$$

where $\mathbf{a} \neq \mathbf{0}$ and $\mathbf{b}$ are fixed vectors. It represents a straight line. Indeed, we can write (2) in the form $\mathbf{f}(t) = t(\mathbf{a} + \mathbf{b}) + (1 - t)\mathbf{b}$; this shows that all points with position vector $\mathbf{f}(t)$ lie on the line through the points with position vectors $\mathbf{a} + \mathbf{b}$ and $\mathbf{b}$; see Chapter 9, §2.2.

As another example, let $\{\mathbf{e}_1, \mathbf{e}_2\}$ be a frame (compare Chapter 9, §1.8) and consider the vector-valued function

$$(3) \qquad\qquad \mathbf{f}(t) = (\cos t)\mathbf{e}_1 + (\sin t)\mathbf{e}_2, \qquad 0 \leq t < 2\pi.$$

The set of points represented parametrically by this function is, of course, the unit circle.

1.2 Kinematic interpretation

It is often convenient to think of a vector-valued function $\mathbf{f}(t)$ as describing the motion of a point in the plane; namely, the point with the position vector $\mathbf{f}(t)$. In this case, the independent variable t (also called the **parameter**) is interpreted as time.

Equation (2) describes the motion of a point along a straight line. At $t = 0$, the moving point is at the point P with position vector $\mathbf{b}$, at time $t = 1$ at the point Q with position vector $\mathbf{a} + \mathbf{b}$. More generally, at any time t the point divides the segment $\overrightarrow{PQ}$ in the ratio $t/(1 - t)$.

The function (3) describes a motion of a point along the unit circle, starting at the point $(0, 1)$ at $t = 0$, proceeding in the counterclockwise direction, and traversing the whole circle in 2π time units.

1.3 Components of a vector-valued function

We choose a Cartesian coordinate system with origin O, and base vectors $\{\mathbf{e}_1, \mathbf{e}_2\}$. Given a vector-valued function $t \mapsto \mathbf{f}(t)$, we can write it as

$$(4) \qquad \mathbf{f}(t) = x(t)\mathbf{e}_1 + y(t)\mathbf{e}_2.$$

Thus a vector-valued function is simply a pair of ordinary (number-valued) functions

$$(5) \qquad t \mapsto x(t), \qquad t \mapsto y(t).$$

One calls these functions the **components** of $\mathbf{f}(t)$.

In the case of the linear function (2), for instance, let the components of $\mathbf{a}$ and $\mathbf{b}$ be (a_1, a_2) and (b_1, b_2), respectively. Then

$$\mathbf{a} = a_1\mathbf{e}_1 + a_2\mathbf{e}_2, \qquad \mathbf{b} = b_1\mathbf{e}_1 + b_2\mathbf{e}_2,$$

$$\mathbf{f}(t) = t(a_1\mathbf{e}_1 + a_2\mathbf{e}_2) + (b_1\mathbf{e}_1 + b_2\mathbf{e}_2) = (a_1t + b_1)\mathbf{e}_1 + (a_2t + b_2)\mathbf{e}_2.$$

The components of the function (2) are therefore

$$x(t) = a_1t + b_1, \qquad y(t) = a_2t + b_2.$$

In the case of the function (3), the components are, of course,

$$x(t) = \cos t, \qquad y(t) = \sin t.$$

A vector-valued function $t \mapsto \mathbf{f}(t)$ can also be written in the form

$$t \mapsto r(t), \qquad t \mapsto \theta(t),$$

where $r(t)$ and $\theta(t)$ are the magnitude and polar angle, respectively, of $\mathbf{f}(t)$. For the function (3), we obtain the very simple formulas

$$(6) \qquad r(t) = 1, \qquad \theta(t) = t.$$

EXERCISES

In Exercises 1 to 13, find the vector-valued function describing the given motion. Assume that (x, y) denotes the coordinates with respect to a fixed Cartesian coordinate system, (r, θ) denotes the corresponding polar coordinates, and $\{e_1, e_2\}$ are the corresponding base vectors.

1. A point moves on the parabola $y = x^2$, starting at $(1, 1)$ at time $t = 0$, moving to the right, and reaching $(2, 4)$ at time $t = 3$. Suppose the horizontal displacement of the point from the start of the motion is proportional to the square of the elapsed time from the start of the motion.

2. A point moves on the curve $y = x^3 + 1$, starting at $(-1, 0)$ at time $t = 1$, moving to the right, and reaching $(1, 2)$ at time $t = 5$. Suppose the horizontal displacement of the point from the start of the motion is proportional to the square root of the elapsed time from the start of the motion.

3. A point moves on the line $y = 8x$, starting at the origin at time $t = 1$, moving to the right, and reaching $(1, 8)$ at time $t = 2$. Suppose that the area of the right triangle formed by the origin, the point, and the projection of the point on the x axis is proportional to the cube of the elapsed time from the start of the motion.

▶ 4. A point moves on the curve $y = \sqrt{x^2 + 1}$, starting at $(0, 1)$ at time $t = \frac{1}{4}$, and moving to the right. Suppose the distance of the point from the origin is proportional to t.

5. A point moves on the curve $3y - 2x^{3/2} = 0$, starting at the origin at time $t = 2$, moving to the right, and reaching $x = 3$ at time $t = 16$. Suppose the total distance traveled by the point (measured along the curve) is proportional to the elapsed time from the start of the motion.

6. A point moves on the curve $y = x^3$, starting at the origin at time $t = 2$, moving to the right, and reaching $(2, 8)$ at time $t = 4$. Suppose the slope of the line through the particle tangent to the path of the particle is proportional to the cube of the elapsed time from the start of the motion.

7. A point moves on the parabola $y = x^2 - x$, starting at the origin at time $t = -1$, moving to the left, and reaching $(-6, 42)$ at time $t = 1$. Suppose the sum of the horizontal and vertical displacements of the point from the start of the motion is proportional to the square of the elapsed time from the start of the motion.

8. A point moves on the curve $y = \log x$, starting at $(1, 0)$ at time $t = -3$, moving to the right, and reaching $(10, \log 10)$ at time $t = 0$. Suppose the rate of change of the horizontal displacement of the point is proportional to the elapsed time from the start of the motion.

9. A point moves on the circle $r = 2$, starting at $r = 2$, $\theta = 0$ at time $t = 4$, moving counterclockwise, and reaching $r = 2$, $\theta = \pi/2$ at time $t = 5$. Suppose the polar angle of the point is proportional to the square of the elapsed time from the start of the motion.

10. A point moves on the circle $r = 1 + \cos\theta$, starting at $r = 2$, $\theta = 0$ at time $t = 1$, moving counterclockwise, and reaching $r = 1$, $\theta = \pi/2$ at time $t = 10$. Suppose the polar angle of the point is proportional to the square root of the elapsed time.

11. A point moves on the spiral $r = \theta$, starting at $r = \theta = 2\pi$ at time $t = 1$, spiraling in to the origin, and reaching $r = \theta = \pi$ at time $t = 4$. Suppose the rate of change of the distance of the point from the pole is inversely proportional to t^2.

12. A point moves on the arc of the rose $r = \cos 3\theta$ given by $0 \le \theta \le \pi/6$, starting at $r = \sqrt{3}/2$, $\theta = \pi/18$, moving to the right, and reaching $r = \sqrt{2}/2$, $\theta = \pi/12$ at time $t = 144$. Suppose the rate of change of the polar angle of the point is inversely proportional to $t^{3/2}$.

13. A point moves on the curve $r = 1 - \cos\theta$, starting at the origin at time $t = -1$, moving counterclockwise, and reaching $r = 1$, $\theta = \pi/2$ at time $t = 3$. Suppose the rate of change of the polar angle is inversely proportional to the square root of the elapsed time from the start of the motion.

1.4 Calculus of vector-valued functions

A vector-valued function is equivalent to two numerical functions, its components, $x(t)$ and $y(t)$:

$$t \mapsto \mathbf{f}(t) = x(t)\mathbf{e}_1 + y(t)\mathbf{e}_2.$$

Using this fact, we transfer to vector-valued functions most concepts of calculus. The rule of thumb is: *Calculate in the usual way, treating the base vectors $\mathbf{e}_1$ and $\mathbf{e}_2$ as constants.* Let us make this explicit.

The function $\mathbf{f}(t)$ is said to be **bounded** in an interval if both components are. It is said to be **continuous** at a point if both components $x(t)$ and $y(t)$ are continuous at the point. The relation

$$\lim_{t \to t_0} \mathbf{f}(t) = a\mathbf{e}_1 + b\mathbf{e}_2$$

is defined to mean that

$$\lim_{t \to t_0} x(t) = a, \qquad \lim_{t \to t_0} y(t) = b;$$

similarly for one-sided limits and limits at infinity.

The **derivative** is defined by the relation

(7)
$$\frac{d\mathbf{f}}{dt} = \mathbf{f}'(t) = \frac{dx}{dt}\mathbf{e}_1 + \frac{dy}{dt}\mathbf{e}_2$$

if x' and y' exist at the point considered. For a fixed t, the derivative $\mathbf{f}'(t)$ is a vector; if we consider it for all allowable values of t, we have a vector-valued function. Therefore

$$\frac{d^2\mathbf{f}}{dt^2} = \frac{d^2x}{dt^2}\mathbf{e}_1 + \frac{d^2y}{dt^2}\mathbf{e}_2,$$

and so forth.

We often denote differentiation with respect to the parameter t by a dot:

$$\frac{d\mathbf{f}}{dt} = \dot{\mathbf{f}} = \mathbf{f}, \qquad \frac{dx}{dt} = \dot{x} = x^{\cdot}$$

(this tradition goes back to Newton). The preceding relations can be re-written as

$$\dot{\mathbf{f}} = \dot{x}\mathbf{e}_1 + \dot{y}\mathbf{e}_2, \qquad \ddot{\mathbf{f}} = \ddot{x}\mathbf{e}_1 + \ddot{y}\mathbf{e}_2.$$

It is often not necessary to specify the base vectors. One can also write

$$\mathbf{f} = (x, y), \qquad \dot{\mathbf{f}} = (\dot{x}, \dot{y}), \qquad \ddot{\mathbf{f}} = (\ddot{x}, \ddot{y}),$$

and so forth.

The vector-valued function $t \mapsto \mathbf{F}(t)$ is called a **primitive function** of $\mathbf{f}(t)$ if

$$\mathbf{F}'(t) = \mathbf{f}(t).$$

The **integral** of a vector-valued function is defined by the obvious formula:

$$\int_a^b \mathbf{f}(t)\, dt = \left(\int_a^b x(t)\, dt \right)\mathbf{e}_1 + \left(\int_a^b y(t)\, dt \right)\mathbf{e}_2.$$

The integral is, of course, a vector.

1.5 Rules of calculus

The rules of calculus that we derived in the preceding chapters can be immediately transferred to vector-valued functions. We give only a few examples.

(a) *Differentiation of a product of a number by a vector.* We have

$$\frac{d\,(\alpha(t)\mathbf{f}(t))}{dt} = \frac{d\,\alpha(t)}{dt}\mathbf{f}(t) + \alpha(t)\frac{d\,\mathbf{f}(t)}{dt}$$

or

$$(\alpha\mathbf{f})^{\cdot} = \dot{\alpha}\mathbf{f} + \alpha\dot{\mathbf{f}}.$$

Proof. We have, of course, assumed that $\dot{\mathbf{f}}$ and $\alpha^{\cdot}$ exist. Now

$$\begin{aligned}
(\alpha\mathbf{f})^{\cdot} &= (\alpha(x\mathbf{e}_1 + y\mathbf{e}_2))^{\cdot} = (\alpha x\mathbf{e}_1 + \alpha y\mathbf{e}_2)^{\cdot} = (\alpha x)^{\cdot}\mathbf{e}_1 + (\alpha y)^{\cdot}\mathbf{e}_2 \\
&= (\dot{\alpha}x + \alpha\dot{x})\mathbf{e}_1 + (\dot{\alpha}y + \alpha\dot{y})\mathbf{e}_2 = \dot{\alpha}x\mathbf{e}_1 + \alpha\dot{x}\mathbf{e}_1 + \dot{\alpha}y\mathbf{e}_2 + \alpha\dot{y}\mathbf{e}_2 \\
&= \dot{\alpha}(x\mathbf{e}_1 + y\mathbf{e}_2) + \alpha(\dot{x}\mathbf{e}_1 + \dot{y}\mathbf{e}_2) = \dot{\alpha}\mathbf{f} + \alpha\dot{\mathbf{f}}.
\end{aligned}$$

(b) *If $\dot{\mathbf{f}}(t) \equiv 0$, then $\mathbf{f}(t) = constant$.*

Proof. If $\dot{\mathbf{f}} \equiv 0$, then $\dot{x} \equiv 0$ and $\dot{y} \equiv 0$; therefore there are numbers a_1 and a_2 such that $x(t) = a_1$ and $y(t) = a_2$ for all t. Therefore $\mathbf{f}(t) = a_1\mathbf{e}_1 + a_2\mathbf{e}_2$ for all t.

(c) *Fundamental theorem of calculus, first part.* We have

$$\frac{d}{d\tau}\int_a^\tau \mathbf{f}(t)\,dt = \mathbf{f}(\tau)$$

if $\mathbf{f}(t)$ is continuous at $t = \tau$.

The proof is left to the reader.

(d) *Fundamental theorem of calculus, second part.* If $\mathbf{f}(t)$ is continuous and bounded for $a < t < b$, and $\mathbf{F}(t)$ is a primitive function for $\mathbf{f}(t)$, then

$$\int_a^b \mathbf{f}(t)\,dt = \mathbf{F}(b) - \mathbf{F}(a).$$

The proof is again left to the reader.

(e) *Limit of length is length of limit.* If $\lim_{t \to t_0} \mathbf{f}(t) = \mathbf{a}$, then $\lim_{t \to t_0} |\mathbf{f}(t)| = |\mathbf{a}|$.

Proof. Set $\mathbf{a} = a_1\mathbf{e}_1 + a_2\mathbf{e}_2$. The hypothesis means that, for $t \to t_0$, we have $\lim x(t) = a_1$, $\lim y(t) = a_2$. Thus $\lim x(t)^2 = a_1{}^2$, $\lim y(t)^2 = a_2{}^2$, and $\lim \sqrt{x(t)^2 + y(t)^2} = \sqrt{a_1{}^2 + a_2{}^2} = |\mathbf{a}|$. This is the desired conclusion.

▶ *Example* Compute $\int_0^{2\pi}(t^3\mathbf{a} + \sin t\,\mathbf{b})\,dt$, where $\mathbf{a}$ and $\mathbf{b}$ are fixed (constant) vectors.

Solution. The integral equals $\left.\frac{1}{4}t^4\mathbf{a} - \cos t\,\mathbf{b}\right|_0^{2\pi} = 4\pi^4\mathbf{a}$. ◀

EXERCISES

In Exercises 14 to 33, assume that $\{\mathbf{e}_1, \mathbf{e}_2\}$ is a fixed frame in the plane. Evaluate each of the following derivatives.

14. $\dfrac{d}{dt}[(\cos t)^2\,\mathbf{e}_1 + (\sin t)\,\mathbf{e}_2]$.

15. $\dfrac{d}{dt}[t\mathbf{e}_1 - (\log t)^3\,\mathbf{e}_2]$.

16. $\dfrac{d}{dt}[(t(\sqrt{t}\,\mathbf{e}_1 - (\sec t)\,\mathbf{e}_2)]$.

17. $\dfrac{d}{dt}\left\{\dfrac{1}{t}(t\mathbf{e}_1 - \mathbf{e}_2) + (\cos t)\,(t^2\mathbf{e}_1 + \mathbf{e}_2)\right\}$.

18. The second derivative of $t \mapsto t^{5/2}\mathbf{e}_1 + (t - 1)\mathbf{e}_2$.

19. $\mathbf{f}''(t)$ if $\mathbf{f}(t) = (\cos t)\,\mathbf{e}_1 + (\tan t)\,\mathbf{e}_2$.

20. $\dfrac{d^2}{dt^2}(\sqrt{t}\,\mathbf{e}_1 + e^t\mathbf{e}_2)$.

21. $\dfrac{d}{dt}\left(\dfrac{t\mathbf{e}_1 + \mathbf{e}_2}{|t\mathbf{e}_1 + \mathbf{e}_2|}\right)$.

22. $\mathbf{f}'(0)$ if $\mathbf{f}(t) = (a(t) + t)(t\mathbf{e}_1 - a(t)\mathbf{e}_2)$ and $a(0) = 1$, $a'(0) = -2$.

23. $\mathbf{f}''(0)$ if $\mathbf{f}(t) = a(t)(t^2\mathbf{e}_1 + (\cos t)\mathbf{e}_2)$ and $a(0) = 1$, $a'(0) = -2$, $a''(0) = 4$.

Evaluate each of the following integrals.

24. $\displaystyle\int_0^{1/2}\left(t^2\mathbf{e}_1 - \dfrac{4t}{1 + t^2}\,\mathbf{e}_2\right)dt$.

25. $\displaystyle\int_0^{\pi/3}((\cos 2t)\,\mathbf{e}_1 + (\sin 3t)\,\mathbf{e}_2)\,dt$.

▶ 26. $\displaystyle\int_1^2\left(\dfrac{1}{t}\,\mathbf{e}_1 + \dfrac{1}{t + 1}\,\mathbf{e}_2\right)dt$.

27. $\displaystyle\int_1^3\left(2^t\mathbf{e}_1 - \dfrac{1}{\sqrt{t + 1}}\,\mathbf{e}_2\right)dt$.

28. $\displaystyle\int_0^{\sqrt{\pi}} t(\sqrt{t^2 + 1}\,\mathbf{e}_1 + (\cos t^2)\,\mathbf{e}_2)\,dt$.

29. $\displaystyle\int_0^{\sqrt{3}} \dfrac{t\mathbf{e}_1 + 4\mathbf{e}_2}{t^2 + 1}\,dt$.

30. $\displaystyle\int_0^{\log 2} e^t(\mathbf{e}_1 + t\mathbf{e}_2)\,dt$.

31. Find a vector-valued function $\mathbf{f}(t)$ such that $\mathbf{f}'(t) = t^{1/3}\mathbf{e}_1 - t^2\mathbf{e}_2$ and $\mathbf{f}(0) = 2\mathbf{e}_1 + \mathbf{e}_2$.

32. Find a vector-valued function $\mathbf{f}(t)$ such that $\mathbf{f}''(t) = t^2\mathbf{e}_1$ and $\mathbf{f}(0) = \mathbf{e}_1 + \mathbf{e}_2$, $\mathbf{f}(1) = 2\mathbf{e}_1 - \mathbf{e}_2$.

33. Find a vector-valued function $\mathbf{f}(t)$ such that $\mathbf{f}''(t) = (\cos t)\,\mathbf{e}_1 + \mathbf{e}_2$ and $\mathbf{f}(0) = 3\mathbf{e}_1 - \mathbf{e}_2$, $\mathbf{f}'(0) = \frac{3}{2}\,\mathbf{e}_1 - \mathbf{e}_2$.

34. Prove that $\dfrac{d}{dt}\displaystyle\int_a^\tau \mathbf{f}(t)\,dt = \mathbf{f}(\tau)$ if $\mathbf{f}(t)$ is continuous at $t = \tau$.

35. Prove that $\int_a^b \mathbf{f}(t)\,dt = \mathbf{F}(b) - \mathbf{F}(a)$ if $\mathbf{f}(t)$ is continuous and bounded for $a < t < b$ and $\mathbf{F}(t)$ is a primitive function for $\mathbf{f}(t)$.

36. Prove that $\lim_{t\to t_0} \mathbf{f}(\phi(t)) = \mathbf{a}$ if $\lim_{t\to t_0} \phi(t) = u_0$ and $\lim_{u\to u_0} \mathbf{f}(u) = \mathbf{a}$.

37. Prove that $(\mathbf{f} \circ \phi)'(t_0) = \mathbf{f}'(u_0)\phi'(t_0)$ if $\phi(t_0) = u_0$ and ϕ is differentiable at t_0, while $\mathbf{f}$ is differentiable at u_0.

38. Let $\mathbf{f} = \alpha\mathbf{e}_1 + \beta\mathbf{e}_2$.
    (a) Prove that $|\alpha| \le M$ and $|\beta| \le M$ if $|\mathbf{f}| \le M$.
    (b) Prove that $|\mathbf{f}| \le M\sqrt{2}$ if $|\alpha| \le M$ and $|\beta| \le M$.

Explain how these results can be visualized in terms of a circle inscribed in a square inscribed in a circle.

39. Suppose that $|\mathbf{f}(t)| \leq M$ for $a \leq t \leq b$. Show that $|\int_a^b \mathbf{f}(t) \, dt| \leq (b - a)M\sqrt{2}$. See Exercise 38.

40. Suppose that $\mathbf{f}(t)$ and $\mathbf{f}'(t)$ are continuous for $0 \leq t \leq 1$, and that $|\mathbf{f}'(t)| > 0$ for all t. Can it happen that $\mathbf{f}(0) = \mathbf{f}(1)$?

1.6 Independence of bases

The components of a vector depend upon the choice of base vectors. It is essential to verify that the concepts introduced above (boundedness, continuity, limit, derivative, and so forth) *do not* depend upon this choice.

We accomplish this by giving new descriptions of the concepts, equivalent to the preceding ones, and not involving base vectors or components. We shall make use of the fact that the length of a vector does not depend on the frame used.

1. *A vector-valued function $t \mapsto \mathbf{f}(t)$ is bounded in an interval if and only if the number-valued function $t \mapsto |\mathbf{f}(t)|$ is bounded.*

Proof.   Here and hereafter we set

$$\mathbf{f}(t) = x(t)\mathbf{e}_1 + y(t)\mathbf{e}_2.$$

If $\mathbf{f}$ is bounded, then $x(t)$ and $y(t)$ are bounded, by the definition given above. Hence there are numbers M_1 and M_2 such that in the interval considered $|x(t)| \leq M_1$, $|y(t)| \leq M_2$. Then, since

$$(8) \qquad\qquad |\mathbf{f}(t)|^2 = x(t)^2 + y(t)^2,$$

we have $|\mathbf{f}(t)|^2 \leq M_1{}^2 + M_2{}^2$ or $|\mathbf{f}(t)| \leq \sqrt{M_1{}^2 + M_2{}^2}$, so that $|\mathbf{f}(t)|$ is bounded. Conversely, if $|\mathbf{f}(t)|$ is bounded, then there is a number M with $|\mathbf{f}(t)| \leq M$ and $|\mathbf{f}(t)|^2 \leq M^2$. By (8) we have $|x(t)| \leq M$ and $|y(t)| \leq M$. Hence $x(t)$ and $y(t)$ are bounded and, by our previous definition, so is $\mathbf{f}(t)$.

2. *The function $\mathbf{f}(t)$ defined near t_0 is continuous at t_0 if and only if*

$$(9) \qquad\qquad \lim_{t \to t_0} |\mathbf{f}(t) - \mathbf{f}(t_0)| = 0.$$

Proof.   Suppose $\mathbf{f}(t)$ is continuous at t_0. By the definition given above, this means that the number-valued functions $x(t)$ and $y(t)$ are continuous at t_0, and this means that

$$(10) \qquad\qquad \lim_{t \to t_0} x(t) = x(t_0), \qquad \lim_{t \to t_0} y(t) = y(t_0),$$

hence

(11)
$$\lim_{t \to t_0} (x(t) - x(t_0)) = 0, \qquad \lim_{t \to t_0} (y(t) - y(t_0)) = 0,$$

hence

(12)
$$\lim_{t \to t_0} \{(x(t) - x(t_0))^2 + (y(t) - y(t_0))^2\} = 0,$$

and hence

(13)
$$\lim_{t \to t_0} \sqrt{(x(t) - x(t_0))^2 + (y(t) - y(t_0))^2} = 0,$$

which is precisely (9).

Conversely, if (9) holds, so do (13), (12), (11), and (10), so that $x(t)$ and $y(t)$ are continuous at t_0. Then, by our original definition, $\mathbf{f}(t)$ is continuous at t_0.

3. *Let $\mathbf{f}(t)$ be defined near t_0. Then*

(14)
$$\lim_{t \to t_0} \mathbf{f}(t) = \mathbf{a}$$

if and only if

(15)
$$\lim_{t \to t_0} |\mathbf{f}(t) - \mathbf{a}| = 0.$$

Proof. Set $\mathbf{a} = a_1 \mathbf{e}_1 + a_2 \mathbf{e}_2$. Statement (14) means, according to our definition, that

$$\lim_{t \to t_0} x(t) = a_1 \qquad \text{and} \qquad \lim_{t \to t_0} y(t) = a_2,$$

which is the same as

$$\lim_{t \to t_0} |x(t) - a_1|^2 = 0 \qquad \text{and} \qquad \lim_{t \to t_0} |y(t) - a_2|^2 = 0,$$

or
$$\lim_{t \to t_0} \sqrt{|x(t) - a_1|^2 + |y(t) - a_2|^2} = 0,$$

and this is another way of writing (15).

4. *Let $\mathbf{f}(t)$ be defined near t_0. Then*

(16)
$$\mathbf{f}'(t_0) = \mathbf{a}$$

if and only if

(17)
$$\lim_{h \to 0} \frac{1}{h} \{\mathbf{f}(t_0 + h) - \mathbf{f}(t_0)\} = \mathbf{a}.$$

Note that, since we already know by Statement 3 that the limit concept is independent of the basis used, Statement 4 establishes the same for the derivative and also, in view of the fundamental theorem of calculus [see (b), (c), and (d) in §1.5], for the integral.

To prove Statement 4, set $a = a_1\mathbf{e}_1 + a_2\mathbf{e}_2$. Equation (16) is, by our definition, equivalent to

$$x'(t_0) = a_1 \qquad \text{and} \qquad y'(t_0) = a_2,$$

which is the same as

$$\lim_{h \to 0} \frac{x(t_0 + h) - x(t_0)}{h} = a_1 \qquad \text{and} \qquad \lim_{h \to 0} \frac{y(t_0 + h) - y(t_0)}{h} = a_2.$$

But this is equivalent to

$$\lim_{h \to 0} \left[\frac{x(t_0 + h) - x(t_0)}{h} \mathbf{e}_1 + \frac{y(t_0 + h) - y(t_0)}{h} \mathbf{e}_2 \right] = \mathbf{a},$$

which is the same as (17).

(One-sided derivatives and limits, and limits at infinity, are treated similarly.)

EXERCISES

In Exercises 41 to 46, use Statements 1, 2, 3, and 4 from the text as *definitions*. Try to prove the results in these exercises directly from these definitions without using bases.

41. Suppose $\mathbf{f}(t)$ and $\mathbf{g}(t)$ are bounded in an interval. Prove that $\mathbf{f}(t) + \mathbf{g}(t)$ is bounded in the same interval.
42. Suppose $\alpha(t)$ and $\mathbf{f}(t)$ are bounded in an interval. Prove that $\alpha(t)\mathbf{f}(t)$ is bounded in the same interval.
43. Prove that $\lim_{t \to t_0} (\mathbf{f}(t) + \mathbf{g}(t)) = \mathbf{a} + \mathbf{b}$ if $\lim_{t \to t_0} \mathbf{f}(t) = \mathbf{a}$ and $\lim_{t \to t_0} \mathbf{g}(t) = \mathbf{b}$.
► 44. Prove that $(\mathbf{f} + \mathbf{g})'(t_0) = \mathbf{a} + \mathbf{b}$ if $\mathbf{f}'(t_0) = \mathbf{a}$ and $\mathbf{g}'(t_0) = \mathbf{b}$.
45. Prove that $(\alpha\mathbf{f})'(t_0) = \alpha_0\mathbf{f}(t_0) + \alpha(t_0)\mathbf{a}$ if $\alpha'(t_0) = \alpha_0$ and $\mathbf{f}'(t_0) = \mathbf{a}$.
46. Prove that the function $t \mapsto |\mathbf{f}(t)|$ is continuous at t_0 if $\mathbf{f}(t)$ is continuous at t_0.

§2 Oriented Curves

In this and the following section, we apply vector-valued functions to the study of **plane curves**. It is convenient to choose once and for all an origin O and a frame $\{\mathbf{e}_1, \mathbf{e}_2\}$.

2.1 Parametric representations

The graph of a number-valued function $x \mapsto y = f(x)$ contains *all* the information about the function; knowing the graph, we can reconstruct the function. The set of points represented parametrically by a vector-valued function $t \mapsto \mathbf{f}(t)$ contains some, but *not all*, information about the function. Two distinct vector-valued functions may represent the same point set. This is so, for a point may describe the same trajectory in different ways.

For instance, each of the two functions

$$t \mapsto t\mathbf{a}, \qquad 0 \le t \le 1,$$

and

$$t \mapsto t^3\mathbf{a}, \qquad 0 \le t \le 1,$$

describes a straight segment joining the point O with the point with position vector $\mathbf{a}$. Both functions describe the motions of a point from O to $\mathbf{a}$, but the motions are different.

The functions

$$t \mapsto (\cos t)\,\mathbf{e}_1 + (\sin t)\,\mathbf{e}_2, \qquad 0 \le t \le 2\pi,$$

and

$$t \mapsto (\cos 2t)\,\mathbf{e}_1 + (\sin 2t)\,\mathbf{e}_2, \qquad 0 \le t \le 2\pi,$$

both describe the motion of a point on the unit circle, starting and ending at the point $(1, 0)$. In both cases the point moves counterclockwise. But in one case the circle is traversed once, in the second twice. The function

$$t \mapsto (\cos t)\,\mathbf{e}_1 - (\sin t)\,\mathbf{e}_2, \qquad 0 \le t \le 2\pi,$$

describes a point moving again along the unit circle, but in the clockwise direction.

2.2 Definition of an oriented curve

These examples suggest a definition: two *continuous* vector-valued functions

(1)
$$t \mapsto \mathbf{f}(t), \qquad a \le t \le b$$

and

(2)
$$\tau \mapsto \mathbf{g}(\tau), \qquad \alpha \le \tau \le \beta$$

define the *same oriented curve* if there exists an increasing continuous function

(3)
$$\tau \mapsto \phi(\tau), \qquad \alpha \le \tau \le \beta$$

with

(4)
$$\phi(\alpha) = a, \qquad \phi(\beta) = b,$$

such that

(5)
$$\mathbf{f}(\phi(\tau)) = \mathbf{g}(\tau), \qquad \alpha \le \tau \le \beta.$$

We also say: $\mathbf{f}(t)$ is obtained from $\mathbf{g}(\tau)$ by **parameter change** $\tau \mapsto t = \phi(\tau)$.

By an **oriented curve** C, we mean a continuous vector-valued function $t \mapsto \mathbf{f}(t)$ defined over some closed finite interval $[a, b]$, *together* with all vector-valued functions obtained from this one by parameter changes.

Note that *only increasing functions are allowed as parameter changes.* If

$$\xi \mapsto \psi(\xi), \qquad \alpha \leq \xi \leq \beta,$$

is a continuous decreasing function with

$$\psi(\alpha) = b, \qquad \psi(\beta) = a,$$

and we set
$$\mathbf{h}(\xi) = \mathbf{f}(\psi(\xi)),$$

then the oriented curve defined by

$$\xi \mapsto \mathbf{h}(\xi), \qquad \alpha \leq \xi \leq \beta,$$

is said to be the **opposite** of that oriented curve defined by $t \mapsto \mathbf{f}(t)$, $a \leq t \leq b$. The endpoint of the curve defined by $\mathbf{h}$ is the initial point of the curve defined by $\mathbf{f}$, and vice versa.

The set of points defined by $t \mapsto \mathbf{f}(t)$, $a \leq t \leq b$, will be called the **trajectory of the oriented curve** defined by the function $\mathbf{f}$ over the interval $[a, b]$. We note that an oriented curve and its opposite have the same trajectory.

The word "curve" without an adjective is often used to denote *either* an oriented curve, *or* the trajectory of an oriented curve. We shall adhere to this custom. Confusion seldom arises, since the context usually makes the meaning clear.

Remark   Occasionally we permit functions defining oriented curves to be defined also in open intervals, that is, not including the endpoints, and in infinite intervals. In this case, relations (4) are to be interpreted as limits.

2.3 Subarcs

Suppose that A is the oriented curve defined by

$$A: \ t \mapsto \mathbf{f}(t), \qquad t_0 \leq t \leq t_1.$$

Let λ be a number such that $t_0 < \lambda < t_1$. It is clear that the function $\mathbf{f}(t)$ considered only for $t_0 \leq t \leq \lambda$ also defines a curve

$$B: \quad t \mapsto \mathbf{f}(t), \qquad t_0 \le t \le \lambda.$$

One says that B is defined by the **restriction** of $\mathbf{f}$ to the interval $[t_0, \lambda]$.
Similarly, we have a curve

$$C: \quad t \mapsto \mathbf{f}(t), \qquad \lambda \le t \le t_1,$$

obtained by restricting $\mathbf{f}$ to the interval $[\lambda, t_1]$. Under these conditions, we say that B and C are **subarcs** of A, and that

$$A = B + C.$$

[Read: "A is B followed by C." Note that the endpoint of B is the initial point of C (see Fig. 11.1).]
More generally, if we subdivide the interval $[t_0, t_1]$ into k subintervals,

$$t_0 = \lambda_0 < \lambda_1 < \lambda_2 < \cdots < \lambda_k = t_1,$$

then the restriction of $\mathbf{f}$ to each subinterval $[\lambda_{j-1}, \lambda_j]$ defines a curve A_j,

$$A_j: \quad t \mapsto \mathbf{f}(t), \qquad \lambda_{j-1} \le t \le \lambda_j.$$

One says that each A_j is a subarc of A, and that

$$A = A_1 + A_2 + \cdots + A_k.$$

[Read "A is A_1 followed by A_2, followed by A_3, $\cdots$, followed by A_k."]
The geometric meaning of such a decomposition is shown in Fig. 11.2.

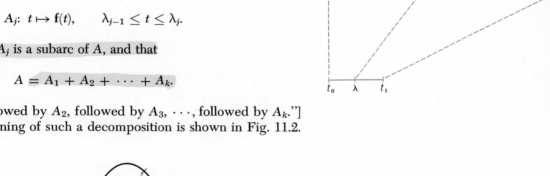

Fig. 11.1

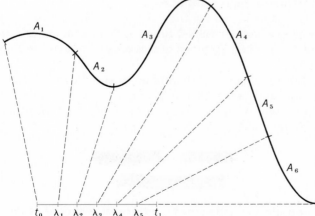

Fig. 11.2

EXERCISES

In each of Exercises 1 to 10, decide whether the two given vector-valued functions define (a) the same oriented curve or (b) opposite curves; whether (c) one defines a subarc or the opposite of a subarc of the oriented curve defined by the other or (d) one defines an oriented curve that can be written as a sum of subarcs or the opposite of subarcs of the other; or whether (e) none of these relations hold.

1. $t \mapsto t^2\mathbf{e}_1 + t^4\mathbf{e}_2$ for $0 \leq t \leq 1$; $u \mapsto (\sin u)\mathbf{e}_1 + (\sin u)^2\mathbf{e}_2$ for $0 \leq u \leq \frac{\pi}{2}$.

▶ 2. $t \mapsto \dfrac{t}{t+1}\mathbf{e}_1 + \dfrac{t+1}{t-1}\mathbf{e}_2$ for $2 \leq t \leq 4$, $u \mapsto \dfrac{1}{u+1}\mathbf{e}_1 + \dfrac{1+u}{1-u}\mathbf{e}_2$ for $\frac{1}{4} \leq u \leq \frac{1}{2}$.

3. $t \mapsto (\log(t-1))\mathbf{e}_1 + \dfrac{1}{t-1}\mathbf{e}_2$ for $2 \leq t \leq 8$, $u \mapsto 2(\log u)\mathbf{e}_1 + u^{-2}\mathbf{e}_2$ for $1 \leq u \leq 2$.

4. $t \mapsto \dfrac{t}{t^2+1}\mathbf{e}_1 + t\mathbf{e}_2$ for $0 \leq t \leq 1$, $u \mapsto u^2\mathbf{e}_1 + \dfrac{2u}{u+1}\mathbf{e}_2$ for $0 \leq u \leq 1$.

5. $t \mapsto (t^2+1)\mathbf{e}_1 + t^2\mathbf{e}_2$ for $0 \leq t \leq 2$, $u \mapsto (\sec^2 u)\mathbf{e}_1 + (\sec^2 u - 1)\mathbf{e}_2$ for $0 \leq u \leq \frac{\pi}{4}$.

6. $t \mapsto e^t\mathbf{e}_1 + t\mathbf{e}_2$ for $-1 \leq t \leq 0$, $u \mapsto \dfrac{1}{u}\mathbf{e}_1 - (\log u)\mathbf{e}_2$ for $1 \leq u \leq e$.

7. $t \mapsto (t^4 - 2t^2)\mathbf{e}_1 + t^2\mathbf{e}_2$ for $0 \leq t \leq 1$, $u \mapsto (u^4 - 1)\mathbf{e}_1 + (u^2 - 1)\mathbf{e}_2$ for $1 \leq u \leq \sqrt{2}$.

8. $t \mapsto \sqrt{t+1}\mathbf{e}_1 + (t+1)\mathbf{e}_2$ for $-1 \leq t \leq 3$, $u \mapsto u\mathbf{e}_1 + u^2\mathbf{e}_2$ for $-2 \leq u \leq 2$.

9. $t \mapsto (\cos \pi t^2)\mathbf{e}_1 + (\sin \pi t^2)\mathbf{e}_2$ for $-1 \leq t \leq 1$, $u \mapsto (\cos \pi u^3)\mathbf{e}_1 + (\sin \pi u^3)\mathbf{e}_2$ for $-1 \leq u \leq 1$.

10. $t \mapsto t^2\mathbf{e}_1 - \sqrt{t}\mathbf{e}_2$ for $0 \leq t \leq 1$, $u \mapsto e^{-2u}\mathbf{e}_1 - e^{-u/2}\mathbf{e}_2$ for $0 \leq u \leq 1$.

2.4 Nonparametric representations

Every continuous number-valued function defined in an interval defines an oriented curve: the trajectory of this curve is given by the graph of the function.

A nonparametric representation of an oriented curve is a representation of an oriented curve as the graph of a number-valued function $x \mapsto y = \phi(x)$. Such a representation can *always* be transformed into a parametric representation. We simply write

$$x(t) = t, \qquad y(t) = \phi(t),$$

or

$$t \mapsto t\mathbf{e}_1 + \phi(t)\mathbf{e}_2.$$

(Note that we did not indicate the interval in which ϕ is defined. We shall usually not mention the interval of definition, when this interval is not important.)

A parametric representation can *sometimes* be transformed into a nonparametric representation. For instance, consider the straight line represented parametrically by

$$t \mapsto t\mathbf{a} + \mathbf{b},$$

where $\mathbf{a}$ and $\mathbf{b}$ are two fixed vectors and $\mathbf{a} \neq 0$. Let $\mathbf{a} = (a_1, a_2)$ and $\mathbf{b} = (b_1, b_2)$. Our parametric representation reads

$$x = a_1 t + b_1, \qquad y = a_2 t + b_2.$$

If a_1 and a_2 are both not 0, we can "eliminate t" as follows. For any given t, the values of $x(t)$ and $y(t)$ are such that

$$t = \frac{x - b_1}{a_1} \qquad \text{and} \qquad t = \frac{y - b_2}{a_2}.$$

Hence
$$\frac{x - b_1}{a_1} = \frac{y - b_2}{a_2}$$

or
$$y = \frac{a_2}{a_1} x + \frac{a_1 b_2 - a_2 b_1}{a_1}.$$

This is the desired nonparametric representation. The line appears as the graph of a number-valued function $x \mapsto y$.

If $a_2 = 0$, the line admits the nonparametric representation $y = b_2$. If $a_1 = 0$, the equation of our line reads $x = b_1$. This is not a function $x \mapsto y$, but the line may be thought of as the graph of a constant function $y \mapsto x = g(y) = b_1$.

On the other hand, the unit circle $x^2 + y^2 = 1$ admits a parametric representation, say,

$$x = \cos t, \qquad y = \sin t,$$

but a nonparametric representation of the whole circle is, as we know, impossible, no matter how we choose the coordinate system.

We also note that a nonparametric representation, in polar coordinates (r, θ),

$$r = g(\theta)$$

may be rewritten as a parametric representation by setting

$$\theta = t, \qquad r = g(t).$$

▶*Examples*   *1.* What point set is represented parametrically by the vector-valued function

$$\mathbf{f}(t) = (\cos t)\mathbf{a} + (\sin t)\mathbf{b},$$

where **a** and **b** are two fixed vectors?

Answer.   Let $\mathbf{a} = a_1\mathbf{e}_1 + a_2\mathbf{e}_2$, $\mathbf{b} = b_1\mathbf{e}_1 + b_2\mathbf{e}_2$. Then

$$\mathbf{f}(t) = (a_1 \cos t + b_1 \sin t)\mathbf{e}_1 + (a_2 \cos t + b_2 \sin t)\mathbf{e}_2$$

or, going over to components,

(6) $x(t) = a_1 \cos t + b_1 \sin t, \qquad y(t) = a_2 \cos t + b_2 \sin t.$

From now on, we must consider two cases:

$$a_1 b_2 - a_2 b_1 = 0 \qquad \text{or} \qquad a_1 b_2 - a_2 b_1 \neq 0.$$

In the first case, it may happen that $a_1 = a_2 = b_1 = b_2 = 0$. In this uninteresting case $\mathbf{f}(t) = \mathbf{0}$ for all t, the set is a point. Otherwise, multiply the first equation (6) by b_2, the second by b_1, and subtract, or multiply the first by a_2 the second by a_1, and subtract. We obtain

$$b_2 x(t) - b_1 y(t) = 0 \qquad \text{or} \qquad a_2 x(t) - a_1 y(t) = 0.$$

Since $a_1 b_2 - a_2 b_1 = 0$, both equations mean the same thing: the point $(x(t), y(t))$ lies on the line $a_2 x - a_1 y = 0$. Our function represents parametrically a segment of this line (never the whole line, since $\sin t$ and $\cos t$ remain between -1 and 1, for all values of t).

In the second case, $a_1 b_2 - a_2 b_1 \neq 0$, we can "solve" equations (6) for $\cos t$ and $\sin t$; this yields

$$\cos t = \frac{b_2 x(t) - b_1 y(t)}{a_1 b_2 - a_2 b_1}, \qquad \sin t = \frac{-a_2 x(t) + a_1 y(t)}{a_1 b_2 - a_2 b_1}.$$

Since $\cos^2 t + \sin^2 t = 1$, we obtain for $x(t)$, $y(t)$ the second degree equation

$$(b_2{}^2 + a_2{}^2)x(t)^2 - 2(b_1 b_2 + a_1 a_2)x(t)y(t) + (b_1{}^2 + a_1{}^2)y(t)^2$$
$$= (a_1 b_2 - a_2 b_1)^2.$$

The discriminant is

$(a_1{}^2 + b_1{}^2)(a_2{}^2 + b_2{}^2) - (a_1a_2 + b_1b_2)^2$

$= a_1{}^2a_2{}^2 + b_1{}^2b_2{}^2 + a_1{}^2b_2{}^2 + a_2{}^2b_1{}^2 - a_1{}^2a_2{}^2 - b_1{}^2b_2{}^2 - 2a_1a_2b_1b_2$

$= (a_1b_2 - a_2b_1)^2 > 0.$

We conclude that the equation is that of an ellipse. This ellipse is represented parametrically by our function.

2. What point set is represented parametrically by the function

$$\mathbf{f}(t) = \frac{a(1 - t^2)}{1 + t^2}\,\mathbf{e}_1 + \frac{2bt}{1 + t^2}\,\mathbf{e}_2, \qquad -\infty < t < +\infty,$$

where $a > 0$ and $b > 0$ are some numbers?

Solution. The components of $\mathbf{f}$ are

$$x(t) = a\,\frac{1 - t^2}{1 + t^2}, \qquad y(t) = b\,\frac{2t}{1 + t^2}.$$

One verifies that $\qquad \dfrac{x(t)^2}{a^2} + \dfrac{y(t)^2}{b^2} = 1;$

this is the equation of an ellipse.

It is more instructive to introduce a new parameter ϕ such that $t = \tan \frac{1}{2}\phi$ (that is, $\phi = 2 \text{ arc tan } t$). As t takes on all values between $-\infty$ and $+\infty$, the parameter ϕ takes on all values between $-\pi$ and π, exclusively. Now we have

$$x(t) = a \cos \phi, \qquad y(t) = b \sin \phi, \qquad -\pi < \phi < \pi,$$

which is a parametric representation of an ellipse (see Chapter 10, §1.8).

Note that the point $x = -a$, $y = b$ of the ellipse is not represented for any value of ϕ. ◀

EXERCISES

In Exercises 11 to 18, represent the oriented curve determined by the given vector-valued function as the graph of a number-valued function, or as the opposite of such a graph, or as a sum of subarcs, each represented by such a graph or its opposite.

11. $t \mapsto t^2\mathbf{e}_1 + t^3\mathbf{e}_2$ for $0 \leq t \leq +\infty$.
12. $t \mapsto \sqrt{t^2 + 1}\,\mathbf{e}_1 + t^3\mathbf{e}_2$ for $0 \leq t \leq +\infty$.
13. $t \mapsto (1/t)\mathbf{e}_1 + (t^2 - 2)\mathbf{e}_2$ for $0 < t < +\infty$.
14. $t \mapsto (\cos t)\,\mathbf{e}_1 + (\sin 2t)\,\mathbf{e}_2$ for $0 \leq t \leq \pi/2$.

15. $t \mapsto (\cos 2t)\, \mathbf{e}_1 + (\cos t)\, \mathbf{e}_2$ for $-\pi/2 \le t \le \pi/2$.
16. $t \mapsto t^{2/3}\mathbf{e}_1 + (t+1)^3\mathbf{e}_2$ for $-\infty < t < +\infty$.
17. $t \mapsto t^2\mathbf{e}_1 + \sqrt{t^2+1}\,\mathbf{e}_2$ for $-\infty < t < \infty$.
18. $t \mapsto (e^{\cos^2 t})\mathbf{e}_1 + (e^{\sin^2 t})\mathbf{e}_2$ for $0 \le t \le 2\pi$.

In Exercises 19 to 24, sketch the trajectory determined by the given vector-valued function. If possible, determine if your sketch is all or part of a line, a circle, an ellipse, a hyperbola, or a parabola.

19. $t \mapsto e^t\mathbf{e}_1 + e^{t/2}\mathbf{e}_2$ for $-\infty < t < +\infty$.
20. $t \mapsto (t^2 - \frac{1}{2})\mathbf{e}_1 + (t^4 - \frac{1}{4})\mathbf{e}_2$ for $-\infty < t < +\infty$.
21. $t \mapsto \sqrt{\log t}\,\mathbf{e}_1 + \sqrt{\log (e/t)}\,\mathbf{e}_2$ for $1 \le t \le e$.
22. $t \mapsto \sqrt{\log et}\,\mathbf{e}_1 + \sqrt{\log t}\,\mathbf{e}_2$ for $1 \le t \le +\infty$.
23. $t \mapsto (\sin 2t)\, \mathbf{e}_1 + (3 - \sin 2t)\, \mathbf{e}_2$ for $-\infty < t < +\infty$.
24. $t \mapsto \sqrt{1 - t^2}\,\mathbf{e}_1 + \sqrt{1 + t^2}\,\mathbf{e}_2$ for $-1 \le t \le 1$.

2.5 The cycloid

Fig. 11.3a

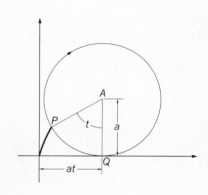

We shall define an interesting curve, the **cycloid,** for which we can easily obtain a parametric representation. (Finding a nonparametric representation would be difficult.)

Consider a circle of radius a which rolls along a straight line without slipping or sliding. The cycloid is the trajectory described by a fixed point P on this circle.

To derive the parametric representation of the cycloid, we assume that the circle rolls along the x axis, and that the motion begins with P at the origin (see Fig. 11.3a). Assume that the circle traveled a certain distance at to the right, $0 \le t \le 2\pi$, and let Q be the point on the circle which is now on the "ground," that is, on the x axis. Then the length of the circular arc between P and Q is precisely at; this is what is meant by saying that there is no slipping or sliding. (If $t > 2\pi$, the length of the arc is at minus a multiple of $2\pi a$.) Let A be the present position of the center. Clearly A has the coordinates (at, a) and $t = \angle QAP$. Thus the coordinates (x, y) of P are $x = at - a \sin t$ and $y = a - a \cos t$. Then we have the representation

(7) $$x = a(t - \sin t), \qquad y = a(1 - \cos t).$$

The curve is shown in Fig. 11.3b. The points at which the cycloid reaches the x axis are called *cusps*. They correspond to the parameter values $t = 0, \pm 2\pi, \pm 4\pi, \cdots$.

The cycloid was a favorite subject of mathematicians during the period when calculus was being created. We shall mention some of the remarkable properties of this curve in the appendix (see §5.4).

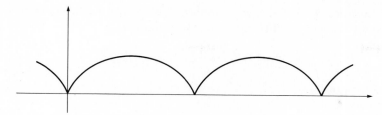

Fig. 11.3b

EXERCISES

25. For every point Q on the upper unit semicircle (centered at the origin), a line is drawn from the origin, through Q, and meeting the line l whose equation is $y = 2$ at the point P. Let R be the midpoint of the line segment PQ. Find a parametric representation for the curve described by the point R.

▶ 26. For every point Q on the upper unit semicircle (centered at the origin), a line is drawn from the point P with Cartesian coordinates $(-1, 0)$, through Q, and ending in the point R whose distance from Q is 1. Find a parametric representation for the curve described by the point R.

27. Consider a disk of radius a which rolls along a straight line without slipping or sliding. Suppose that this disk is glued face to face and concentrically to a second disk of radius b. Fix a point R on the circumference of the second disk. Find a parametric representation for the curve described by the point R. This curve is called a **trochoid**.

28. Suppose a string of length $2\pi a$ is unwound from around the circumference of a disk of radius a whose center is kept fixed. Find a parametric equation for the curve described by the free end of the string. This curve is called the **involute** of the circle.

29. Suppose a disk of radius b rolls along the outside of a fixed circle of radius a without slipping or sliding. Find the parametric representation for the curve described by a fixed point R on the circumference of the second disk. This curve is called an **epicycloid** and the special case $a = b$ is called a **cardioid**.

30. In Exercise 29, suppose $b < a$ and the moving disk rolls along the inside of the circle. This curve is called a **hypocycloid**.

2.6 Area formula for polar coordinates

We conclude this section by deriving a special but important result, the formula for computing **areas in polar coordinates**.

Consider a curve defined by an equation in polar coordinates

$$(8) \qquad r = f(\theta), \qquad 0 \le \theta \le 2\pi, \qquad f(\theta) > 0$$

with

$$(9) \qquad f(2\pi) = f(0).$$

We assume, of course, that f is continuous. Using θ as a parameter, we have for the curve the parametric representation

$$\theta \mapsto f(\theta) \cos \theta \, \mathbf{e}_1 + f(\theta) \sin \theta \, \mathbf{e}_2$$

or, in components,

$$x = f(\theta) \cos \theta, \qquad y = f(\theta) \sin \theta.$$

In view of condition (9), the endpoint and initial point of our curve coincide; such a curve is called **closed.**

It is clear when a point P in the plane is interior to our curve. The set of all interior points is called the region bounded by the curve; it consists of all points with polar coordinates (r, θ) satisfying $0 \leq r < f(\theta)$.

We want to compute the area A of this region. We shall first assume, as usual, that we know what area is and then use our intuitive notions about area to derive a formula for it. Next we should have to verify that this formula leads to the same number as the area formula in Chapter 5, §5. This, however, will not be done here but only in Chapter 13, §1.11.

Fig. 11.4

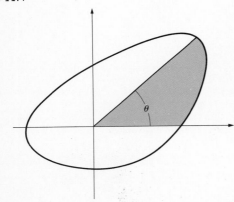

Let $A(\theta)$ be the area of the region bounded by the rays from O corresponding to the polar angles 0 and θ and the arc of the curve corresponding to the interval $[0, \theta]$; this is the shaded region in Fig. 11.4. We set $A(0) = 0$. Our intuitive notions about area tell us that the function $\theta \mapsto A(\theta)$ is continuous. Since $A(2\pi) = A$, we have

$$A = \int_0^{2\pi} A'(\theta) \, d\theta,$$

assuming that $A'(\theta)$ exists and is piecewise continuous. We proceed to find or rather guess the value of the derivative $A'(\theta)$.

Fig. 11.5

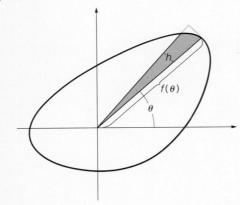

Let h be a small positive number. The number $A(\theta + h) - A(\theta)$ is the area of the narrow sector shown in Fig. 11.5. This area differs very little from the area of the circular sector shown in the same figure; the latter area is $\frac{1}{2}$ of the radius squared times the central angle, that is, $\frac{1}{2} f(\theta)^2 h$. Hence

$$\frac{A(\theta + h) - A(\theta)}{h} = \frac{1}{2} f(\theta)^2 \qquad \text{approximately.}$$

This suggests that

(10) $$A'(\theta) = \frac{1}{2} f(\theta)^2$$

and

(11) $$A = \frac{1}{2} \int_0^{2\pi} f(\theta)^2 \, d\theta, \qquad r = f(\theta).$$

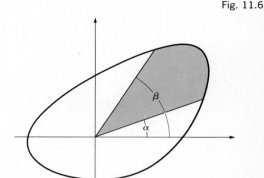

Fig. 11.6

This is the desired formula.

Clearly,

(12) $$A = \frac{1}{2} \int_\alpha^\beta f(\theta)^2 \, d\theta$$

is the area of the "sector" bounded by an arc of the curve and by the two radii with polar angles α and $\beta > \alpha$; see Fig. 11.6. This formula is obtained in exactly the same way as (11). It can be applied also in cases when the origin O lies on the curve.

▶**Examples   1.** Find the area of the region bounded by the curve $r = 2 + \sin \theta$, $0 \le \theta \le 2\pi$.

Fig. 11.7

Solution.   We have

$$A = \frac{1}{2} \int_0^{2\pi} (2 + \sin \theta)^2 \, d\theta = \frac{1}{2} \int_0^{2\pi} (4 + 4 \sin \theta + \sin^2 \theta) \, d\theta$$

$$= \int_0^{2\pi} \left(2 + 2 \sin \theta + \frac{1 - \cos 2\theta}{4} \right) d\theta$$

$$= \int_0^{2\pi} \left(\frac{9}{4} + 2 \sin \theta - \frac{1}{4} \cos 2\theta \right) d\theta$$

$$= \left(\frac{9\theta}{4} - 2 \cos \theta - \frac{\sin 2\theta}{8} \right) \Bigg|_0^{2\pi} = \frac{9\pi}{2}.$$

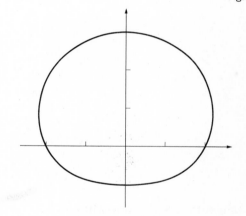

Here we used the identity $2 \sin^2 x = 1 - \cos 2x$. The curve is shown in Fig. 11.7.

2. Find the area of a right triangle. Let the triangle be bounded by the x axis, the line $x = 1$ and the line $y = mx$, with $m > 0$; see Fig. 11.8. The equation of the line $x = 1$ in polar coordinates is $r \cos \theta = 1$, and our triangle is obtained for $0 \le \theta \le \text{arc tan } m$. Using (11), we obtain the desired area A:

Fig. 11.8

$$\frac{1}{2} \int_0^{\text{arc tan } m} \frac{d\theta}{\cos^2 \theta} = \frac{1}{2} \tan \theta \Bigg|_0^{\text{arc tan } m} = \frac{1}{2} m,$$

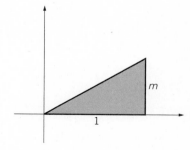

as expected.

Fig. 11.9

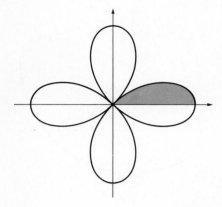

3. Find the area enclosed by the four-leaved rose, $r = 2 \cos 2\theta$.

Solution. We plot the curve (see Chapter 9, §3.4, Example 2) and note that, by symmetry, the desired area is 8 times the area of the shaded half-petal in Fig. 11.9. The curved boundary of this half-petal is described as the polar angle goes from 0 to $\pi/4$. The desired area is therefore

$$8 \cdot \frac{1}{2} \int_0^{\pi/4} (2 \cos 2\theta)^2 \, d\phi = 16 \int_0^{\pi/4} \cos^2 2\theta \, d\theta = 8 \int_0^{\pi/4} (1 + \cos 4\theta) \, d\theta$$

$$= 8 \left(\theta + \frac{1}{4} \sin 4\theta \right) \Big|_0^{\pi/4} = 2\pi.$$

Here we used the identity $2 \cos^2 x = 1 + \cos 2x$.

Note that the area of the rose is half of the area of the circle circumscribed around it. ◄

EXERCISES

Find the area of each of the regions bounded by the curve or curves described below. The coordinates in each case are polar coordinates. It is advisable to sketch each curve before computing the area.

31. $r = 1 + \cos \theta$, $0 \le \theta \le 2\pi$.   33. $r = \cos 3\theta$, $0 \le \theta \le \pi$.
32. $r = 2 \cos \theta$, $0 \le \theta \le \pi$.   34. $r = \sqrt{\cos 2\theta}$, $-\pi/4 \le \theta \le \pi/4$.
35. $r = \theta$, $0 \le \theta \le \pi/2$ and the line $\theta = \pi/2$.
36. $r = \tan \theta$ and the horizontal lines whose Cartesian equations are $y = \sqrt{2}$ and $y = -\sqrt{2}$.
37. The smaller (inner) loop of the curve $r = 1 + 2 \cos \theta$.
38. $r = \theta$, $2\pi \le \theta \le 6\pi$ and the line segment $\theta = 0$, $2\pi \le r \le 4\pi$.
39. One of the smaller (inner) loops of the curve $r = \cos \frac{1}{2}\theta$.
40. One of the smaller (left-hand) loops of the curve $r = \cos \theta \cos 2\theta$.

§3 Tangents, Length, Curvature

3.1 Smooth curves

The continuous vector-valued functions considered thus far are too general for many applications. The curves defined by them include curves without tangents, curves no arc of which has a finite length, and even weird curves that pass through every point of a region (such curves are called Peano curves). We obtain a more manageable class of curves if we insist that all functions used for definitions and in changing parameters have continuous derivatives.

An oriented curve will be called **smooth** if it is defined by a vector-valued function

$$t \mapsto \mathbf{f}(t), \qquad a \le t \le b,$$

such that the derivative $\mathbf{f}'(t)$ exists, is continuous, and

GIUSEPPE PEANO (1858–1932) was a brilliant mathematician with many interests (including the invention of an artificial world language based on Latin). He formulated the celebrated *Peano Axioms* which describe positive integers.

(1) $$\mathbf{f}'(t) \neq \mathbf{0}$$

for all values of t considered. A parameter change $\tau \mapsto t = \phi(\tau)$ is called
allowable if $\phi'(\tau)$ exists and is continuous and positive. Such a parameter
change transforms $\mathbf{f}(t)$ into another representation, $\mathbf{g}(\tau)$: we have

$$\mathbf{g}(\tau) = \mathbf{f}(\phi(\tau)),$$

and since $\mathbf{g}'(\tau) = \mathbf{f}'(\phi(\tau))\phi'(\tau)$, we have

$$\mathbf{g}'(\tau) \neq \mathbf{0}.$$

The vector-valued function $\mathbf{f}(t)$ can be written as

$$\mathbf{f}(t) = x(t)\mathbf{e}_1 + y(t)\mathbf{e}_2.$$

Whenever convenient, we shall denote it by $(x(t), y(t))$. The derivative
of the function considered is $(x'(t), y'(t))$ and the condition $\mathbf{f}'(t) \neq \mathbf{0}$ is
equivalent to

(2) $$x'(t)^2 + y'(t)^2 \neq 0.$$

3.2 Nonparametric representation

Let $x \mapsto f(x)$ be a number-valued function defined for $a \leq x \leq b$ and
having there a continuous derivative $f'(x)$. The graph of this function is
a smooth oriented curve, the vector-valued function defining it is

(3) $$t \mapsto t\mathbf{e}_1 + f(t)\mathbf{e}_2, \qquad a \leq t \leq b.$$

Note that the derivative of this vector-valued function is $\mathbf{e}_1 + f'(t)\mathbf{e}_2$,
and hence never the null vector.

3.3 The unit tangent vector

For a smooth curve defined by $t \mapsto \mathbf{f}(t)$, we consider the vector-valued
function

(4) $$\mathbf{T}(t) = \frac{1}{|\mathbf{f}'(t)|}\mathbf{f}'(t);$$

it is well defined, since $|\mathbf{f}'(t)| \neq 0$ by hypothesis. We call this vector $\mathbf{T}(t)$
the **unit tangent vector** to the curve at the point P with position vector
$\mathbf{f}(t)$. This name is justified by the following observations.

1. **T** is *not altered* by an allowable parameter change.

Indeed, consider the parameter change $\tau \mapsto \phi(\tau) = t$, which transforms the function $\mathbf{f}(t)$ into the function

$$\tau \mapsto \mathbf{g}(\tau) = \mathbf{f}(\phi(\tau)).$$

This function also represents our curve. Let P be a point on the curve with position vector $\mathbf{f}(t_0) = \mathbf{g}(\tau_0)$, so that $t_0 = \phi(\tau_0)$. By the chain rule,

$$\mathbf{g}'(\tau) = \phi'(\tau)\mathbf{f}'(\phi(\tau)).$$

By assumption $\phi'(\tau) > 0$, since we consider only allowable parameter changes. Therefore

$$|\mathbf{g}'(\tau)| = \phi'(\tau)|\mathbf{f}'(\phi(\tau))|.$$

Thus we have, for $\tau = \tau_0$,

$$\frac{1}{|\mathbf{g}'(\tau_0)|}\mathbf{g}'(\tau_0) = \frac{1}{|\mathbf{f}'(\phi(\tau_0))|}|\mathbf{f}'(\phi(\tau_0))| = \frac{1}{|\mathbf{f}'(t_0)|}\mathbf{f}'(t_0).$$

But the left-hand side of the equation represents the vector **T** at P, computed by using the function **g**, and the right-hand side is **T** at P computed by using **f**. Our assertion is proved.

2. **T** is a *unit vector*, that is, $|\mathbf{T}| = 1$.

This is so, since

$$|\mathbf{T}| = \left|\frac{1}{|\mathbf{f}'|}\mathbf{f}'\right| = \frac{1}{|\mathbf{f}'|}|\mathbf{f}'| = 1.$$

3. **T** is a *tangent vector*.

We mean by this that, if we choose a coordinate system $(O, \mathbf{e}_1, \mathbf{e}_2)$ such that a subarc which contains P is the graph of a number-valued function $x \mapsto F(x)$, and if we represent **T** by a directed segment $\overrightarrow{PQ}$, the straight line through P and Q is tangent to the curve. To verify the statement, note that the subarc has the parametric representation

$$t \mapsto \mathbf{f}(t) = t\mathbf{e}_1 + F(t)\mathbf{e}_2.$$

Let P be the point $x_0 = t_0$, $y_0 = F(t_0)$. Then the slope of the tangent to the curve at P is $m = F'(t_0)$. We have $\mathbf{f}'(t) = \mathbf{e}_1 + F'(t)\mathbf{e}_2$ so that $\mathbf{f}'(t_0) = \mathbf{e}_1 + m\mathbf{e}_2$, and the vector $\mathbf{f}'(t_0)$ has the direction of the tangent line at P. So does the vector $\mathbf{T}(t_0)$, which differs from $\mathbf{f}'(t_0)$ only in length.

The geometric meaning of **T** could also have been seen as follows. Let h be a small positive number. Then the vector

$$\frac{1}{h}(\mathbf{f}(t_0 + h) - \mathbf{f}(t_0))$$

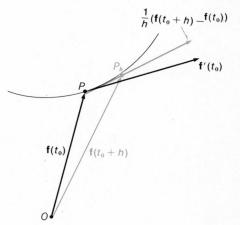

Fig. 11.10

is close to the vector $\mathbf{f}'(t_0)$. But (see Fig. 11.10) the vector $\mathbf{f}(t_0 + h) - \mathbf{f}(t)$ is represented by the directed segment $\overrightarrow{PP_h}$, where P_h has the position vector $\mathbf{f}(t_0 + h)$. Thus P_h is a point close to P and located in the direction of increasing t. We conclude that $\mathbf{f}'(t_0)$ has the direction of the tangent line and points in the direction of increasing parameter values, or, as one says, in the direction of the curve. $\mathbf{T}(t_0)$ has the same direction as $\mathbf{f}'(t_0)$.

Remark Given a point P on a smooth oriented curve, there *is* a sufficiently small subarc containing P, which is the graph of a continuously differentiable number-valued function, in a suitably chosen coordinate system. This is obvious geometrically; see Fig. 11.11. Given P, we may choose P as the origin of a coordinate system, and we may pick a frame $\{\mathbf{e}_1, \mathbf{e}_2\}$ such that $\mathbf{e}_1$ has the direction of the tangent vector at P. (An analytic proof is not difficult, but will be omitted.)

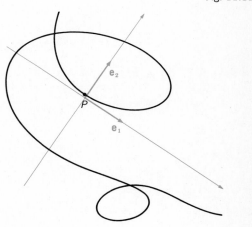

Fig. 11.11

3.4 The polar angle of the tangent vector

We note the useful formulas: if a smooth curve is represented by $x = x(t)$, $y = y(t)$, then $\mathbf{T}(t)$ has the components

$$(5) \qquad \frac{x'(t)}{\sqrt{x'(t)^2 + y'(t)^2}}, \qquad \frac{y'(t)}{\sqrt{x'(t)^2 + y'(t)^2}}.$$

Indeed, if $\mathbf{f} = x\mathbf{e}_1 + y\mathbf{e}_2$, then $\mathbf{f}' = x'\mathbf{e}_1 + y'\mathbf{e}_2$ and $|\mathbf{f}'|^2 = (x')^2 + (y')^2$. Let $\Theta(t)$ be the **polar angle** of $\mathbf{T}(t)$. Then

$$(6) \qquad \cos \Theta = \frac{x'}{\sqrt{(x')^2 + (y')^2}}, \qquad \sin \Theta = \frac{y'}{\sqrt{(x')^2 + (y')^2}}$$

and therefore

$$(7) \qquad \tan \Theta = \frac{y'(t)}{x'(t)}.$$

If the same curve is represented nonparametrically as a graph of the function $x \mapsto y = F(x)$, then we have $\tan \Theta = F'(x) = dy/dx$, since $\tan \Theta$ is the slope of the tangent to the curve. Thus we may rewrite (7) as

$$(8) \qquad \frac{dy}{dx} = \frac{y'(t)}{x'(t)} = \frac{dy/dt}{dx/dt}.$$

This relation shows once more the usefulness of the Leibniz notation.

▶ *Example* An ellipse in standard position has a parametric representation

$$x = a \cos t, \qquad y = b \sin t.$$

Hence $$\frac{dx}{dt} = -a \sin t = -\frac{a}{b} y, \qquad \frac{dy}{dt} = b \cos t = \frac{b}{a} x$$

and by (8) $$\frac{dy}{dx} = -\frac{b^2}{a^2} \frac{x}{y}.$$

Before, we obtained this in a different way (see Chapter 10, §1.6).◀

EXERCISES

1. Find the unit tangent vector to the curve represented by $\mathbf{f}(t) = (t^3 + t)\mathbf{e}_1 + t^2\mathbf{e}_2$ at the point P with Cartesian coordinates $(-2, 1)$.
2. Find the unit tangent vector to the curve represented by $\mathbf{f}(t) = e^{2t}\mathbf{e}_1 - (t + 8)^{4/3}\mathbf{e}_2$ at the point P with Cartesian coordinates $(1, -16)$.
3. Find the unit tangent vector to the curve represented by $\mathbf{f}(t) = (\cos 2t)\mathbf{e}_1 + (\sin 3t)\mathbf{e}_2$ (for $-\pi/6 < t < \pi/6$) at the point P with Cartesian coordinates $(\sqrt{3}/2, \sqrt{2}/2)$.
4. Find the unit tangent vector to the graph of $y = \log(x^2 + 1)$ at the point P with Cartesian coordinates $(-2, \log 5)$.
5. Find $\mathbf{T}(t)$ for the curve represented by $\mathbf{f}(t) = (\cos t)\mathbf{e}_1 + t\mathbf{e}_2$.
6. Find $\mathbf{T}(t)$ for the curve represented by $\mathbf{f}(t) = \sqrt{t^2 + 1}\,\mathbf{e}_1 + t\mathbf{e}_2$.
7. Find $\mathbf{T}(x)$ for the graph $y = x^3 + 2$.
▶ 8. Find $\mathbf{T}(y)$ for the graph $y = x^{5/3} - 2$.
9. Consider the curve represented by $\mathbf{f}(t) = e^t\mathbf{e}_1 - t^3\mathbf{e}_2$ and let $\mathbf{g}(u)$ be the parameterization of this curve obtained from $\mathbf{f}(t)$ by the allowable parameter change $t = \phi(u) = u^2 + 1$. Find $\mathbf{T}(u)$.
10. Find $\mathbf{T}(t)$ for the curve represented by $\mathbf{f}(t) = (t^3 - 1)\mathbf{e}_1 - t^2\mathbf{e}_2$.

3.5 Formulas in polar coordinates

Let us compute $\mathbf{T}$ in polar coordinates (r, θ). If the curve is defined by

$$r = f(\theta),$$

it admits the parametric representation

$$x = f(\theta) \cos \theta, \qquad y = f(\theta) \sin \theta,$$

the polar angle θ being the parameter. We have

$$\frac{dx}{d\theta} = f'(\theta) \cos\theta - f(\theta) \sin\theta, \qquad \frac{dy}{d\theta} = f'(\theta) \sin\theta + f(\theta) \cos\theta.$$

Hence $\qquad \left(\frac{dx}{d\theta}\right)^2 + \left(\frac{dy}{d\theta}\right)^2 = f(\theta)^2 + f'(\theta)^2 = r^2 + \left(\frac{dr}{d\theta}\right)^2,$

as the reader will verify. The components of $\mathbf{T}$ are

$$\frac{f'\cos\theta - f\sin\theta}{\sqrt{f^2 + (f')^2}}, \qquad \frac{f'\sin\theta + f\cos\theta}{\sqrt{f^2 + (f')^2}}$$

and the slope of the tangent is

(9) $$\tan\Theta = \frac{dy}{dx} = \frac{f'(\theta)\sin\theta + f(\theta)\cos\theta}{f'(\theta)\cos\theta - f(\theta)\sin\theta}.$$

Let P be a point on the curve and let us compute the angle Ψ between the line $\overrightarrow{OP}$ and the tangent to the curve through P (see Fig. 11.12). Since $\Psi = \Theta - \theta$, and since the slope of the tangent is dy/dx and that of $\overrightarrow{OP}$ is $\tan\theta$, we have (compare Chapter 10, §2.4)

Fig. 11.12

$$\tan\Psi = \frac{\tan\Theta - \tan\theta}{1 + \tan\Theta \tan\theta}$$
$$= \frac{\dfrac{f'\sin\theta + f\cos\theta}{f'\cos\theta - f\sin\theta} - \dfrac{\sin\theta}{\cos\theta}}{1 + \left(\dfrac{f'\sin\theta + f\cos\theta}{f'\cos\theta - f\sin\theta}\right)\left(\dfrac{\sin\theta}{\cos\theta}\right)} = \frac{f}{f'}$$

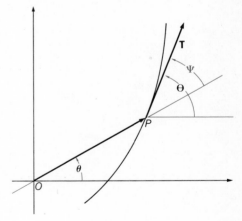

which may be written as

(10) $$\tan\Psi = \frac{r}{dr/d\theta} \qquad \text{or} \qquad \cot\Psi = \frac{1}{r}\frac{dr}{d\theta}.$$

EXERCISES

Compute $\mathbf{T}(\theta)$ and $\tan\Psi$ for each of the following curves given in polar coordinates.

11. $r = 2 + \cos\theta$ for $-\pi \le \theta \le \pi$.
▶ 12. $r = 1 + \cos\theta$ for $-\pi < \theta < \pi$.
13. $r = \cos^2\theta$ for $-\pi/2 < \theta < \pi/2$.
14. $r = \theta$ for $0 < \theta < +\infty$.
15. $r = e^\theta$ for $-\infty < \theta < +\infty$.
16. $r = 1/(\cos\theta + \sin\theta)$ for $-\pi/4 < \theta < 3\pi/4$.
17. $r = \log\theta$ for $0 < \theta < 1$.

Fig. 11.13

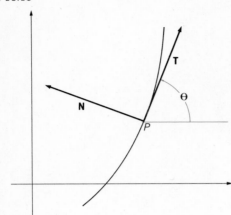

3.6 The unit normal vector

Let $\mathbf{T}$ be the unit tangent vector to an oriented curve at the point P. Its polar angle Θ is given by (6). The vector $\mathbf{N}$ of length 1 and polar angle $\Theta + 90°$ is called the **unit normal vector** at P. We can think of $\mathbf{N}$ as obtained from $\mathbf{T}$ by a counterclockwise rotation by $90°$ (see Fig. 11.13). Since $\cos(\Theta + 90°) = -\sin\Theta$, $\sin(\Theta + 90°) = \cos\Theta$, the components of $\mathbf{N}(t)$ are

$$(11) \qquad -\frac{y'}{\sqrt{(x')^2 + (y')^2}}, \qquad \frac{x'}{\sqrt{(x')^2 + (y')^2}}.$$

Note that the definition of $\mathbf{N}$ depends on the orientation of the plane.

EXERCISES

18. Find the unit normal vector to the curve represented by $\mathbf{f}(t) = (t^3 + t)\mathbf{e}_1 + t^2\mathbf{e}_2$ at the point P with Cartesian coordinates $(-2, 1)$.
19. Find the unit normal vector to the curve represented by $\mathbf{f}(t) = e^{2t}\mathbf{e}_1 - (t + 8)^{4/3}\mathbf{e}_2$ at the point P with Cartesian coordinates $(1, -16)$.
20. Find the unit normal vector to the curve represented by $\mathbf{f}(t) = (\cos 2t)\mathbf{e}_1 + (\sin 3t)\mathbf{e}_2$ (for $-\pi/6 < t < \pi/6$) at the point P with Cartesian coordinates $(\sqrt{3}/2, \sqrt{2}/2)$.
21. Find the unit normal vector to the graph of $y = \log(x^2 + 1)$ at the point P with Cartesian coordinates $(-2, \log 5)$.
22. Find $\mathbf{N}(t)$ for the curve represented by $\mathbf{f}(t) = (\cos t)\mathbf{e}_1 + t\mathbf{e}_2$.
23. Find $\mathbf{N}(t)$ for the curve represented by $\mathbf{f}(t) = \sqrt{t^2 + 1}\,\mathbf{e}_1 + t\mathbf{e}_2$.
24. Find $\mathbf{N}(x)$ for the graph $y = x^3 + 2$.
25. Find $\mathbf{N}(y)$ for the graph $y = x^{5/3} - 2$.

3.7 Length of a curve

In Chapter 5, §4.8, we obtained a formula for the length of a graph of a function. Now we consider the more general case of a smooth curve defined parametrically. First we assume that we know what length is, and, using our intuitive idea of length, we ask only how to compute it. Then we turn the method of computing into a definition. Finally, we check that our new definition of length coincides with the one given in §4.8 of Chapter 5.

Let A be a smooth curve defined by $t \mapsto \mathbf{f}(t)$, $a \leq t \leq b$, and let L denote the length of A. Also, let A_ξ denote the subarc defined by the restriction $\mathbf{f}$ to $[a, \xi]$, and let $L(\xi)$ denote the length of A_ξ. In other words: $L(\xi)$ is the length of the arc defined by $t \mapsto \mathbf{f}(t)$, $a \leq t \leq \xi$ and $L = L(b)$. We set $L(a) = 0$. Our intuitive concept of length tells us that $L(\xi)$ is a continuous function. Thus

(12)
$$L = \int_a^b L'(t)\, dt,$$

assuming $L'(t)$ exists and is continuous. We proceed to "calculate" $L'(t)$, or rather to guess it.

Let t_0 be a fixed number between a and b, h a small positive number (or a negative number of small absolute value). Our intuitive length concept tells us that

$$L(t_0 + h) - L(t_0) = \text{length of the curve defined by } t \mapsto \mathbf{f}(t),$$
$$t_0 \le t \le t_0 + h.$$

The length of this short subarc will, we expect, differ little from the distance between its endpoints (see Fig. 11.14). This distance is $|\mathbf{f}(t_0 + h) - \mathbf{f}(t_0)|$, for the distance between two points is the length of the difference of their position vectors. Therefore we expect that

$$\frac{L(t_0 + h) - L(t_0)}{h} = \frac{1}{h} |\mathbf{f}(t_0 + h) - \mathbf{f}(t_0)| + \;\begin{array}{l}\text{an error that gets as small as we}\\ \text{like as } |h| \text{ gets sufficiently small,}\end{array}$$

or
$$L'(t_0) = \lim_{h \to 0} \frac{L(t_0 + h) - L(t_0)}{h} = \lim_{h \to 0} \left(\frac{1}{h} |\mathbf{f}(t_0 + h) - \mathbf{f}(t_0)| \right)$$

$$= \lim_{h \to 0} \left| \frac{1}{h} (\mathbf{f}(t_0 + h) - \mathbf{f}(t_0)) \right|.$$

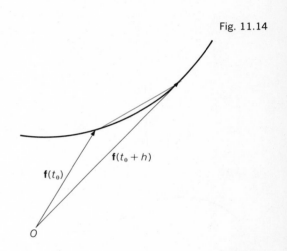

Fig. 11.14

Using rule (e) from §1.5, we conclude that

$$L'(t_0) = \left| \lim_{h \to 0} \frac{1}{h} (\mathbf{f}(t_0 + h) - \mathbf{f}(t_0)) \right| = |\mathbf{f}'(t_0)|.$$

Substitution in (12) yields

(13)
$$L = \int_a^b |\mathbf{f}'(t)|\, dt$$

or, in components,

(14)
$$L = \int_a^b \sqrt{x'(t)^2 + y'(t)^2}\, dt.$$

We accept this formula as the *definition* of length.

3.8 Independence of parameter

In order to justify our definition, we show that the value of L is not changed by an allowable parameter change. This must be so in order that the length be a number determined by the curve and not by a particular way of representing it.

Indeed, let $\tau \mapsto \phi(\tau)$, $\alpha \leq \tau \leq \beta$, be a function such that $\phi'(\tau)$ exists and is positive and continuous, and $\phi(\alpha) = a$, $\phi(\beta) = b$. Set $\mathbf{g}(\tau) = \mathbf{f}(\phi(\tau))$ so that $\tau \mapsto \mathbf{g}(\tau)$, $\alpha \leq \tau \leq \beta$, again defines the curve A. We have now, by the chain rule and by the substitution rule for integrals,

$$\int_\alpha^\beta |\mathbf{g}'(\tau)|\, d\tau = \int_\alpha^\beta |\mathbf{f}'(\phi(\tau))\phi'(\tau)|\, d\tau = \int_\alpha^\beta |\mathbf{f}'(\phi(\tau))|\, \phi'(\tau)\, d\tau = \int_a^b |\mathbf{f}'(t)|\, dt,$$

as asserted.

Using the shorthand notations,

$$d\mathbf{f}(t) = \mathbf{f}'(t)\, dt, \ dx(t) = x'(t)\, dt, \ dy(t) = y'(t)\, dt$$

we sometimes write (13) and (14) as

$$L = \int_A |d\mathbf{f}| \qquad \text{and} \qquad L = \int_A \sqrt{dx^2 + dy^2}.$$

This notation indicates that L is independent of the choice of parameter.

3.9 Additivity

If $A = B + C$ (that is, if A is the curve B followed by the curve C; see Fig. 11.15) we would expect that the length of A equals the length of B plus that of C. This is indeed so, for if A is defined by $t \mapsto \mathbf{f}(t)$, $t_0 \leq t \leq t_1$, B by the restriction of $\mathbf{f}$ to $[t_0, t_2]$, and C by the restriction of $\mathbf{f}$ to $[t_2, t_1]$, we have

$$\begin{aligned}
\text{length of } A &= \int_{t_0}^{t_1} |\mathbf{f}'(t)|\, dt && \text{by definition} \\
&= \int_{t_0}^{t_2} |\mathbf{f}'(t)|\, dt + \int_{t_2}^{t_1} |\mathbf{f}'(t)|\, dt && \left\{ \begin{array}{l} \text{by the additivity} \\ \text{property of integrals} \end{array} \right. \\
&= \text{length of } B + \text{length of } C && \text{by definition.}
\end{aligned}$$

One can show similarly that A and the curve opposite to A (see §2.2), that is, the curve A traversed in the opposite direction, have the same length.

3.10 Comparison with the previous length formula

Let us verify that our new definition of length conforms to what we agreed upon in Chapter 5, §4.8. We consider a continuously differentiable function $x \mapsto y = f(x)$, $a \leq x \leq b$. Its graph is represented parametrically by

$$t \mapsto \mathbf{f}(t) = t\mathbf{e}_1 + f(t)\mathbf{e}_2, \qquad a \leq t \leq b;$$

Fig. 11.15

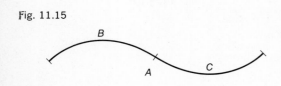

the length of the graph is, by the present definition,

$$L = \int_a^b |\mathbf{f}'(t)| \, dt = \int_a^b |\mathbf{e}_1 + f'(t)\mathbf{e}_2| \, dt = \int_a^b \sqrt{1 + f'(t)^2} \, dt,$$

which is precisely the formula arrived at in Chapter 5, §4.8.

We note that the definition of length by the formula (13) is given in terms of vectors. It presupposes no special choice of a coordinate system. Thus the length of a curve does not depend on the position of the Cartesian coordinates used to calculate it. This is, of course, as it should be. We have now settled a point left open in Chapter 5, §4.8.

▶*Examples* 1. Length of a *circular arc*. The arc is defined by

$$x = r \cos \theta, \qquad y = r \sin \theta, \qquad 0 \le \theta \le \alpha,$$

where a is the central angle (see Fig. 11.16). Hence

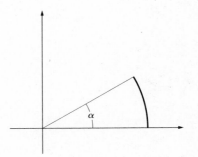

Fig. 11.16

$$L = \int_0^\alpha \sqrt{\left(\frac{dx}{d\theta}\right)^2 + \left(\frac{dy}{d\theta}\right)^2} \, d\theta = \int_0^\alpha r \, d\theta = \alpha r.$$

For the full circle ($\alpha = 2\pi$), one gets $L = 2\pi r$.

2. Length of an *ellipse*. Let a and b be the semiaxes. The parametric representation of the ellipse

$$x = a \cos \theta, \qquad y = b \sin \theta, \qquad 0 \le \theta \le 2\pi$$

leads to the formula

$$L = \int_0^{2\pi} \sqrt{\left(\frac{dx}{d\theta}\right)^2 + \left(\frac{dy}{d\theta}\right)^2} \, d\theta = \int_0^{2\pi} \sqrt{a^2 \sin^2 \theta + b^2 \cos^2 \theta} \, d\theta.$$

Writing $b^2 = a^2 - c^2$ and noting that $c = a\epsilon$, where ϵ is the eccentricity, we obtain

$$L = a \int_0^{2\pi} \sqrt{1 - \epsilon^2 \cos^2 \theta} \, d\theta = 2a \int_0^{\pi} \sqrt{1 - \epsilon^2 \cos^2 \theta} \, d\theta,$$

the latter since $\theta \mapsto \cos^2 \theta$ has period π. This is a so-called elliptic integral. It cannot be expressed in terms of elementary functions. ◀

EXERCISES

In each of Exercises 26 to 35, find the length of the curve determined by the given vector-valued function between the indicated limits.

26. $\mathbf{f}(t) = (t^2 - 1)\mathbf{e}_1 + t^3\mathbf{e}_2$ for $\frac{1}{3} \le t \le \frac{2}{3}$.

27. $\mathbf{f}(t) = \frac{1}{2}t^2\mathbf{e}_1 + \frac{1}{3}(2t + 1)^{3/2}\mathbf{e}_2$ for $0 \le t \le 1$.

28. $\mathbf{f}(t) = \frac{1}{6}t^6\mathbf{e}_1 + \frac{1}{4}t^4\mathbf{e}_2$ for $1 \leq t \leq \sqrt[4]{5}$.
29. $\mathbf{f}(t) = 2e^{t/2}\mathbf{e}_1 + 4e^{3t/4}\mathbf{e}_2$ for $0 \leq t \leq 2$.
30. $\mathbf{f}(t) = \frac{1}{3}(\cos^3 t)3\mathbf{e}_1 + (\sin t - \frac{1}{3}(\sin^3 t))\dot{\mathbf{e}}_2$ for $\pi/6 \leq t \leq \pi/3$.
31. $\mathbf{f}(t) = (\sin t)\mathbf{e}_1 + (\sin^{3/2} t)\mathbf{e}_2$ for $0 \leq t \leq \pi/4$.
32. $\mathbf{f}(t) = t^2\mathbf{e}_1 + (t^3 + t)\mathbf{e}_2$ for $0 \leq t \leq 1$. Leave the answer in integral form.
33. $\mathbf{f}(t) = \sqrt{t^2 + 1}\mathbf{e}_1 + \sqrt{t}\mathbf{e}_2$ for $1 \leq t \leq 4$. Leave the answer in integral form.
34. $\mathbf{f}(t) = (\cos t^2)\mathbf{e}_1 + (\sin t^3)\mathbf{e}_2$ for $\sqrt{\pi}/4 \leq t \leq \sqrt{\pi}/2$. Leave the answer in integral form.
35. $\mathbf{f}(t) = (\cos t)\mathbf{e}_1 + (\log t)\mathbf{e}_2$ for $\pi/4 \leq t \leq e/3$. Leave the answer in integral form.

3.11 Length formula in polar coordinates

Assume that a curve is defined by its equation in polar coordinates (r, θ)

$$r = f(\theta), \qquad a \leq \theta \leq \beta.$$

The same curve is defined parametrically by

$$x(\theta) = f(\theta) \cos \theta, \qquad y(\theta) = f(\theta) \sin \theta.$$

Noting (13), we obtain easily the formula for length in terms of polar coordinates

(15) $$L = \int_\alpha^\beta \sqrt{f'(\theta)^2 + f(\theta)^2}\, d\theta.$$

▶ **Example**   The *equiangular spiral* is defined in polar coordinates by the equation $r = f(\theta) = a\mathbf{e}^\theta$. We obtain the length of an arc of the spiral as

$$L = \int_\alpha^\beta \sqrt{2a^2 e^{2\theta}}\, d\theta = \sqrt{2}a(e^\beta - e^\alpha).◀$$

EXERCISES

In each of Exercises 36 to 41, find the length of the curve given in polar coordinates (r, θ) between the indicated limits.

36. $r = \theta^2$ for $\sqrt{5} \leq \theta \leq 2\sqrt{3}$.
37. $r = e^{a\theta}$ for $0 \leq \theta \leq 1$, a a fixed number.
38. $r = \sec \theta$ for $0 \leq \theta \leq \pi/4$.
39. $r = 1 + \cos \theta$ for $0 \leq \theta \leq \pi/2$. Leave the answer in integral form.
40. $r = \sqrt{\cos \theta}$ for $\pi/6 \leq \theta \leq \pi/3$. Leave the answer in integral form.
41. $r = \cos \theta + \sin \theta$ for $0 \leq \theta \leq \pi/2$.

3.12 Arc length as parameter

A given oriented smooth curve has many parametric representations. But there is one representation that is in a certain sense particularly simple; in this representation the parameter is the arc length on the curve measured from some point on the curve. We shall show how one can obtain this representation.

Consider a smooth curve defined by $t \mapsto \mathbf{f}(t)$, $a \leq t \leq b$; let L be its length, and let $s(\xi)$ be the length of the arc $t \mapsto \mathbf{f}(t)$, $a \leq t \leq \xi$. Then

$$(16) \qquad\qquad s(\xi) = \int_a^\xi |\mathbf{f}'(t)|\, dt.$$

Thus $\xi \mapsto s(\xi)$ is a continuous increasing function with $s(a) = 0$, $s(b) = L$, and its derivative

$$s'(\xi) = |\mathbf{f}'(\xi)|$$

is continuous and positive. Hence there exists an inverse function $t = \phi(s)$, with $\phi(0) = a$, $\phi(L) = b$, and $\phi'(s(t))s'(t) = 1$, that is, $\phi'(s(t)) = 1/|\mathbf{f}'(t)|$. This function describes an allowable parameter change. Set $\mathbf{g}(s) = \mathbf{f}(\phi(s))$. Then

$$s \mapsto \mathbf{g}(s), \qquad 0 \leq s \leq L$$

defines our curve and

$$|\mathbf{g}'(s)| = 1,$$

since
$$\mathbf{g}'(s) = \phi'(s)\mathbf{f}'(\phi(s)) = \frac{1}{|\mathbf{f}'(t)|}\, \mathbf{f}'(t).$$

This shows that the length of the curve between the points $\mathbf{g}(0)$ and $\mathbf{g}(s)$ is s (or $|s|$ if s is negative).

Using the arc length s as parameter simplifies certain formulas. If

$$\mathbf{g}(s) = x(s)\mathbf{e}_1 + y(s)\mathbf{e}_2$$

and $|\mathbf{g}'(s)| = 1$, then $x'(s)^2 + y'(s)^2 = 1$ and the formulas (5), (6), (7), and (11) simplify to

$$(17) \qquad \mathbf{T}(s) = x'(s)\mathbf{e}_1 + y'(s)\mathbf{e}_2, \qquad \mathbf{N}(s) = -y'(s)\mathbf{e}_1 + x'(s)\mathbf{e}_2,$$

$$(18) \qquad\qquad \cos\Theta = x'(s), \qquad \sin\Theta = y'(s).$$

EXERCISES

In each of Exercises 42 to 46, reparametize the curve represented by a vector-valued function or as a Cartesian graph using the arc length s as a parameter.

▶ 42. $2y = 3x - 1$ for $0 \leq x$.
43. $y = 2^{3/2}3^{-3/2}x^{3/2}$ for $x \geq 0$.
44. $\mathbf{f}(t) = \frac{1}{2}t\mathbf{e}_1 + \frac{1}{3}(2t + 1)^{3/2}\mathbf{e}_2$ for $0 \leq t$.
45. $y = \int_1^x \sqrt{u^2 - 1}\, du$ for $1 \leq x$.
46. $y = \sqrt{1 - x^2}$ for $0 \leq x < 1$.

3.13 Curvature

We consider now an oriented curve defined by

$$t \mapsto \mathbf{f}(t), \qquad a \leq t \leq b,$$

and we assume that the function $\mathbf{f}(t)$ has a continuous *second* derivative $\mathbf{f}''(t)$. Two such curves are shown in Fig. 11.17. It is evident that the first is "more curved" than the second, and that the first curve is "more curved" at the point P than at the point Q. We want to find a way to measure the property of "being curved" numerically.

To this end, we recall that we defined above the polar angle $\Theta(t)$ of the unit tangent vector $\mathbf{T}(t)$ to the curve. If the curve is a straight segment, the angle Θ is constant. It is natural to take the rate of change of Θ as we move along the curve as the measure of how much the curve is curved. But the rate of change with respect to what? There is only one "natural parameter" on a curve; this is the arc length. We therefore define the **curvature** κ of the curve at a point P as the value, at this point, of the derivative of the polar angle of the unit tangent vector with respect to arc length.

In order to write this in a formula, let $t \mapsto s(t)$ be the function defining the arc length as a function of the parameter t. We may consider the polar angle Θ and also t as functions of s. By the chain rule

(19)
$$\kappa = \frac{d\Theta}{dt}\frac{dt}{ds},$$

and, since $s(t)$ is defined by (16), we have

$$\frac{dt}{ds} = \frac{1}{ds/dt} = \frac{1}{|\mathbf{f}'(t)|},$$

and therefore

(20)
$$\kappa(t) = \frac{d\Theta}{ds} = \frac{\Theta'(t)}{|\mathbf{f}'(t)|}.$$

Fig. 11.17

Since we have [see (7)] $\tan \Theta(t) = y'(t)/x'(t)$, we also have

$$(21) \qquad \Theta(t) = \text{arc} \tan \frac{y'(t)}{x'(t)}$$

and therefore

$$\Theta'(t) = \frac{1}{1 + y'(t)^2/x'(t)^2} \frac{x'(t)y''(t) - x''(t)y'(t)}{x'(t)^2} = \frac{x'(t)y''(t) - x''(t)y'(t)}{x'(t)^2 + y'(t)^2}.$$

At points at which $x'(t) = 0$, we arrive at the same result starting with the formula $\cot \Theta(t) = x'(t)/y'(t)$; recall that we assumed that $x'(t)$ and $y'(t)$ cannot both be zero.

Substituting into (20) and noting that $|\mathbf{f}'| = \sqrt{(x')^2 + (y')^2}$, we obtain the final formula:

$$(22) \qquad \kappa(t) = \frac{x'(t)y''(t) - x''(t)y'(t)}{(x'(t)^2 + y'(t)^2)^{3/2}}.$$

3.14 Sign of the curvature

It may appear that the value of the curvature depends on the position of the coordinate axes, since the polar angle of a vector depends on this position, but this is not so. For if we rotate the coordinates we must add a constant to $\Theta(t)$, and this does not affect the derivative of Θ. On the other hand, the sign of the curvature depends on the orientation of the plane. For this reason, some authors call the absolute value $|\kappa(t)|$ the curvature.

If $\kappa = d\Theta/ds > 0$, then the angle Θ increases as we proceed along the curve in the direction of increasing parameter values; that means the unit tangent vector is rotated counterclockwise (see Fig. 11.18). If $\kappa < 0$, then Θ decreases, that is, $\mathbf{T}$ is rotated in the clockwise direction. A point at which the curvature changes sign is called an **inflection point** (see Fig. 11.19).

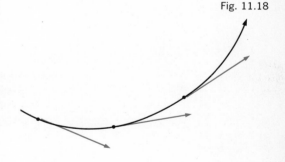

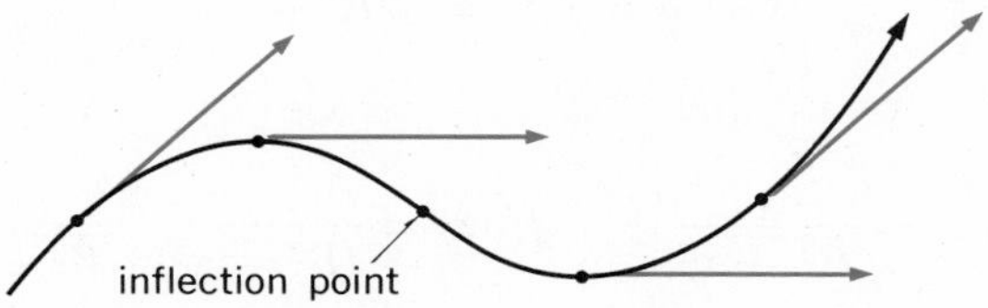

Fig. 11.18

inflection point

Fig. 11.19

In Chapter 4, §4.9, we used the word inflection point to designate a point (x_0, y_0) on a graph $y = f(x)$, with $f''(x_0) = 0$ and $f''(x)$ not changing sign at x_0. These inflection points are special cases of the points defined above.

3.15 Curvature of graphs

The graph of a number-valued function $x \mapsto f(x)$, $a \leq x \leq b$ has the parametric representation

$$x(t) = t, \qquad y(t) = f(t),$$

so that $\quad x'(t) = 1, \qquad x''(t) = 0, \qquad y'(t) = f'(t), \qquad y''(t) = f''(t),$

and we obtain from (22) the following formula for the curvature of the graph of a function f at point $(x, f(x))$

$$(23) \qquad\qquad \kappa(x) = \frac{f''(x)}{[1 + f'(x)^2]^{3/2}}.$$

▶ *Examples* **1. Circle** of radius R. A parametric representation is

$$(24) \qquad\qquad x = R \cos t, \qquad y = R \sin t, \qquad 0 \leq t \leq 2\pi;$$

hence $\ x' = -R \sin t, \ x'' = -R \cos t, \ y' = R \cos t, \ y'' = -R \sin t$, and by (22)

$$\kappa = \frac{1}{R}.$$

The curvature at all points of the circle is $\dfrac{1}{R}$. Of course, we could have obtained this directly by noting that $\Theta(t) = t + \dfrac{\pi}{2}$ so that $\dfrac{d\Theta}{dt} = 1$, and $s(t) = Rt$ so that $\dfrac{dt}{ds} = \dfrac{1}{R}$, and hence $\dfrac{d\Theta}{ds} = \dfrac{1}{R}$.

The upper half of our circle is the graph of the function

$$(25) \qquad\qquad x \mapsto y = f(x) = \sqrt{R^2 - x^2}.$$

We have

$$f'(x) = \frac{-x}{\sqrt{R^2 - x^2}}, \qquad f''(x) = \frac{-R^2}{(R^2 - x^2)\sqrt{R^2 - x^2}},$$

$$[1 + f'(x)^2]^{3/2} = \frac{R^3}{(R^2 - x^2)\sqrt{R^2 - x^2}},$$

and, by (23), $\kappa = -1/R$. The apparent discrepancy is easily explained. In the parametric representation (24), the parameter t increases as we proceed along the circle in the counterclockwise direction. In the representation (25), the parameter is x; here x increases as we proceed along the circle in the clockwise direction.

2. The **parabola** $y = x^2$. Here

$$\kappa(x) = \frac{2}{(1 + 4x^2)^{3/2}}.$$

There are no inflection points.

3. The **cubic parabola.** This is the graph of $x \mapsto x^3 = y$. We have

$$\kappa(x) = \frac{6x}{(1 + 9x^2)^{3/2}}.$$

There is an inflection point at $x = y = 0.$ ◄

3.16 Radius of curvature. Circle of curvature

If the curvature κ at a point P on a curve A is not 0, we call

(26) $$\rho = \frac{1}{|\kappa|}$$

the **radius of curvature** of A at P. The name is chosen because, if A is a circle, then ρ is indeed the radius (see Example 1 above).

The **circle of curvature** C of A at P is defined as the circle of radius ρ which passes through P, is tangent to A at P, and such that C and A are convex in the same direction. The center of C is called the **center of curvature** (of A at P). It is evident that, if $\kappa > 0$, then the center of curvature is reached from P by proceeding the distance ρ in the direction of the unit normal vector $\mathbf{N}$; if $\kappa < 0$, we must go the distance ρ in the direction opposite to $\mathbf{N}$ (see Fig. 11.20). Hence the position vector of the center of curvature is the position vector of P plus $(1/\kappa)\mathbf{N}$, that is,

(27) $$\mathbf{f}(t) + \frac{1}{\kappa(t)}\mathbf{N}(t).$$

EXERCISES

In Exercises 47 to 54, find the curvature and the radius of curvature of the given curve. Find any inflection points on the curve.

47. $y = e^x$ for $-\infty < x < +\infty$.
48. $y = 1/x$ for $0 < x < +\infty$.

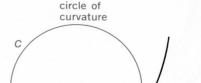

Fig. 11.20

circle of curvature

49. $y = \log x$ for $0 < x < +\infty$. What is the limit of $\kappa(x)$ as x approaches 0 from the right? As x approaches $+\infty$?

50. $y = -\log \cos x$ for $-\pi/2 < x < \pi/2$. Find any points on the curve where the curvature attains a local maximum or minimum.

51. $\mathbf{f}(t) = (2 \cos t)\mathbf{e}_1 + (3 \sin t)\mathbf{e}_2$ for $0 \le t \le 2\pi$. Find any points on the curve where the curvature attains a local maximum or minimum.

52. $\mathbf{f}(t) = e^t\mathbf{e}_1 + \cos t\,\mathbf{e}_2$ for $-\infty < t < +\infty$.

53. $\mathbf{f}(t) = (t^3 + t)\mathbf{e}_1 + (\frac{1}{3}t^3 - \frac{1}{2}t^2)\mathbf{e}_2$ for $-\infty < t < +\infty$.

54. $\mathbf{f}(t) = (t^3 - t)\mathbf{e}_1 + t^4\mathbf{e}_2$ for $-\infty < t < +\infty$.

55. Find the circle of curvature of the graph of $y = e^{-x^2}$ at $(0, 1)$. Sketch the curve and the circle.

56. Find the centers of curvature of the ellipse $\mathbf{f}(t) = (a \cos t)\mathbf{e}_1 + (b \sin t)\mathbf{e}_2$ at $(a, 0)$ and $(0, b)$. Here a and b are arbitrary positive constants.

57. Find the circle of curvature of the curve determined by $y = 1/x$ at $(1, 1)$. Sketch the curve and the circle.

58. Find the center of curvature of the curve determined by $\mathbf{f}(t) = e^t\mathbf{e}_1 + (\sin 3t)\mathbf{e}_2$ at $(1, 0)$.

3.17 Piecewise smooth curves. Singular points

The concept of a smooth curve is somewhat too restrictive for many purposes. A triangle, for instance, is not a smooth curve. We say that a continuous vector-valued function $t \mapsto \mathbf{f}(t)$, $a \le t \le b$ defines a **piecewise smooth** curve if the derivative $\mathbf{f}'(t)$ exists, is continuous, and is different from 0, except perhaps at finitely many points in the interval $[a, b]$. The points on the curve A corresponding to such "bad" parameter values are called **singular points** of A.

An example is the cycloid discussed in §2.5. This curve is defined by

$$\mathbf{f}(t) = (t - \sin t)\mathbf{e}_1 + (1 - \cos t)\mathbf{e}_2.$$

Hence $\mathbf{f}'(t) = (1 - \cos t)\mathbf{e}_1 + \sin t\,\mathbf{e}_2, \ |\mathbf{f}'(t)|^2 = 2(1 - \cos t).$

The cycloid has singularities for $t = 0, \pm 2\pi, \pm 4\pi, \cdots$; these are the points at which the curve meets the x axis (see Fig. 11.3b). These singular points are called **cusps** (compare Chapter 4, §2.9).

The unit tangent vector $\mathbf{T}$ and the unit normal vector $\mathbf{N}$ can of course be defined at all nonsingular points of a piecewise smooth curve. The arc length parameter can be used on any subarc free of singular points, and the length formula (13) may be applied to the whole curve provided the integral makes sense, perhaps as an improper integral.

3.18 Space curves

Everything in this section, except the considerations involving nonparametric curves, polar angles, the unit normal, and curvature, extends, with only obvious changes, to curves in space.

Let $\{O, \mathbf{e}_1, \mathbf{e}_2, \mathbf{e}_3\}$ be a Cartesian coordinate system in space. An oriented **space curve** is defined by a continuous vector-valued function

$$t \mapsto \mathbf{f}(t) = x(t)\mathbf{e}_1 + y(t)\mathbf{e}_2 + z(t)\mathbf{e}_3, \qquad a \le t \le b.$$

If the function $\mathbf{f}(t)$ has a continuous derivative $\mathbf{f}'(t)$ which is never $\mathbf{0}$, the curve is smooth. In this case, we can define the unit tangent vector

$$\mathbf{T}(t) = \frac{1}{|\mathbf{f}'(t)|}\,\mathbf{f}'(t) = \frac{1}{\sqrt{x'(t)^2 + y'(t)^2 + z'(t)^2}}(x'(t)\mathbf{e}_1 + y'(t)\mathbf{e}_2 + z'(t)\mathbf{e}_3),$$

and the arc length parameter

$$s(t) = \int_a^t |\mathbf{f}'(\xi)|\, d\xi.$$

The length of A is

$$\int_A |\mathbf{f}'(t)|\, dt = \int_A ds = \int_a^b \sqrt{x'(t)^2 + y'(t)^2 + z'(t)^2}\, dt.$$

The extension of these concepts to an n-dimensional vector space presents no difficulties whatsoever.

EXERCISES

59. Find the unit tangent vector to the space curve determined by $\mathbf{f}(t) = (t^2 + 1)\mathbf{e}_1 + (\cos t)\mathbf{e}_2 + e^t\mathbf{e}_3$ at the point with Cartesian coordinates $(1, 1, 1)$.
60. Find the unit tangent vector to the space curve determined by $\mathbf{f}(t) = (t + 1)^2\mathbf{e}_1 + t^3\mathbf{e}_2 + \sqrt{t^2 + 1}\,\mathbf{e}_3$ at the point with Cartesian coordinates $(1, -8, \sqrt{5})$.
61. Find $\mathbf{T}(t)$ for the space curve given by $\mathbf{f}(t) = (\cos^2 t)\mathbf{e}_1 + (\cos t \sin t)\mathbf{e}_2 + (\sin t)\mathbf{e}_3$.
62. Find $\mathbf{T}(t)$ for the space curve given by $\mathbf{f}(t) = (t \cos t)\mathbf{e}_1 + (t \sin t)\mathbf{e}_2 + t\mathbf{e}_3$.
63. Find $\mathbf{T}(t)$ for the space curve given by $\mathbf{f}(t) = (t + \cos t)\mathbf{e}_1 + (\sin t)\mathbf{e}_2 + t\mathbf{e}_3$.

In Exercises 64 to 68, find the length of the space curve determined by the given vector-valued function between the indicated limits.

64. $\mathbf{f}(t) = (t^3/3)\mathbf{e}_1 + (t^2/\sqrt{2})\mathbf{e}_2 + t\mathbf{e}_3$ for $0 \le t \le 1$.
65. $\mathbf{f}(t) = (\sin t)\mathbf{e}_1 + (\cos t)\mathbf{e}_2 + t^{3/2}\mathbf{e}_3$ for $0 \le t \le 4$.
66. $\mathbf{f}(t) = e^t\mathbf{e}_1 + 2\sqrt{2}\,e^{t/2}\mathbf{e}_2 + t\mathbf{e}_3$ for $0 \le t \le 1$.
67. $\mathbf{f}(t) = (t^2/2)\mathbf{e}_1 + (1/3)(t^2 - 1)^{3/2}\mathbf{e}_2 + (t^2/2)\mathbf{e}_3$ for $1 \le t \le \sqrt{3}$.
68. $\mathbf{f}(t) = (\cos^2 t)\mathbf{e}_1 + (\cos t \sin t)\mathbf{e}_2 + (\sin t)\mathbf{e}_3$ for $0 \le t \le \pi/2$.

4.1 Curvilinear motion   §4 Motion

In Chapter 4, we treated the motion of a particle (point with a mass) along a straight line. In this section, we consider the more complicated case of **curvilinear motion.** Such a motion is described by a vector-valued function

$$t \mapsto \mathbf{r}(t) = x(t)\mathbf{e}_1 + y(t)\mathbf{e}_2 + z(t)\mathbf{e}_3,$$

which indicates the position vector $\mathbf{r}$ of the particle at the time t. (We assume that this function has a continuous derivative of the second order. The position vector is denoted by the letter $\mathbf{r}$ in order to reserve $\mathbf{f}$ for force.) A special case is that of **plane motion;** during such motion, the particle remains in one plane. We may then choose the coordinate system so that this plane is the x, y plane ($z = 0$). Thus a plane motion is described by a vector-valued function

$$t \mapsto \mathbf{r}(t) = x(t)\mathbf{e}_1 + y(t)\mathbf{e}_2.$$

4.2 Velocity and acceleration

The derivative

$$\mathbf{v}(t) = \frac{d\mathbf{r}(t)}{dt}$$

is called **velocity.** It is again a vector-valued function. Throughout this and the following section, we shall denote differentiation with respect to time by a dot. Thus we have

$$\mathbf{v} = \dot{\mathbf{r}}$$

or, in components,

$$\mathbf{v} = \dot{x}\mathbf{e}_1 + \dot{y}\mathbf{e}_2 + \dot{z}\mathbf{e}_3.$$

In plane motion $z \equiv 0$, $\dot{z} \equiv 0$, and

$$\mathbf{v} = \dot{x}\mathbf{e}_1 + \dot{y}\mathbf{e}_2.$$

The length of the velocity vector,

$$|\mathbf{v}| = |\dot{\mathbf{r}}| = \sqrt{\dot{x}^2 + \dot{y}^2 + \dot{z}^2}$$

or, for plane motion,

$$|\mathbf{v}| = \sqrt{\dot{x}^2 + \dot{y}^2},$$

is called **speed.** Finally, the time derivative of the velocity vector

$$\mathbf{a}(t) = \frac{d\mathbf{v}(t)}{dt} = \frac{d^2\mathbf{r}(t)}{dt^2},$$

that is,

$$\mathbf{a} = \dot{\mathbf{v}} = \ddot{\mathbf{r}} = \ddot{x}\mathbf{e}_1 + \ddot{y}\mathbf{e}_2 + \ddot{z}\mathbf{e}_3$$

or, for plane motion,

$$\mathbf{a} = \ddot{x}\mathbf{e}_1 + \ddot{y}\mathbf{e}_2,$$

is called **acceleration.**

The function $t \mapsto \mathbf{r}(t)$ describing the motion of our particle defines an oriented curve, the **trajectory** of the motion. Now, however, the parameter t has a definite meaning: it is the time at which the particle is at the point $\mathbf{r}(t)$.

The scientific use of the terms velocity, speed, and acceleration corresponds rather closely to the everyday meaning of these words. Indeed, consider a moving particle and assume that it is at the point P with position vector $\mathbf{r}(t_0)$ at the time t_0, and at the point Q with position vector $\mathbf{r}(t_0 + h)$ after a short time interval h (see Fig. 11.21). The vector

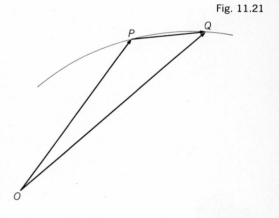

Fig. 11.21

$$(1) \qquad \frac{1}{h}[\mathbf{r}(t_0 + h) - \mathbf{r}(t_0)]$$

is nearly the velocity vector $\mathbf{v}(t_0)$. Its direction is that from P to Q, which is nearly the direction of motion. The magnitude of (1) is $|\overrightarrow{PQ}|/h$ which is nearly the distance traveled, divided by the time elapsed. Thus the velocity vector $\mathbf{v}(t_0)$ may be said to describe the instantaneous change of position. One sees similarly that the acceleration vector describes the instantaneous change in velocity.

Consider now a motion described by the function $t \mapsto \mathbf{r}(t)$ and assume that at the time t_0 the velocity vector $\mathbf{v}(t_0)$ is not the null vector. Then the velocity vector $\mathbf{v}(t_0)$ is tangent to the trajectory at the point $\mathbf{r}(t_0)$. Indeed, as we saw in §3.3, the unit tangent vector $\mathbf{T}(t_0)$ is

$$\mathbf{T}(t_0) = \frac{1}{|\dot{\mathbf{r}}(t_0)|}\,\dot{\mathbf{r}}(t_0) = \frac{1}{|\mathbf{v}(t_0)|}\,\mathbf{v}(t_0),$$

so that $\mathbf{T}$ and $\mathbf{v}$ have the same direction.

Let $s(t)$ denote the length on the trajectory, measured from some fixed time t_0, in the direction of the motion. Then, as we saw in §3.7,

$$s(t) = \int_{t_0}^{t} |\dot{\mathbf{r}}(\tau)|\, d\tau = \int_{t_0}^{t} |\mathbf{v}(\tau)|\, d\tau$$

or
$$\frac{ds}{dt} = |\mathbf{v}(t)|;$$

thus the speed is the time rate of change of the distance traveled.

4.3 Circular motion

An important example is a plane motion given by

$$x = R\cos 2\pi\nu t, \qquad y = R\sin 2\pi\nu t,$$

where R and ν and positive constants. The trajectory is a circle of radius R. The velocity has the components

$$\dot{x} = -2\pi\nu R\sin 2\pi\nu t, \qquad \dot{y} = 2\pi\nu R\cos 2\pi\nu t;$$

the speed
$$|\mathbf{v}(t)| = \sqrt{\dot{x}^2 + \dot{y}^2} = 2\pi\nu R$$

is constant; per unit time the moving particle covers the distance $2\pi\nu R$, that is, ν times the length of the full circle. For this reason, ν is called the **frequency**. The **period** of the motion, that is, the time required for covering the circle once, is

$$T = \frac{1}{\nu}.$$

We note the interesting connection between a circular motion and harmonic motion (see Chapter 6, §1.14). When a particle moves along a circle with constant speed, each of its components (in a Cartesian system) executes a harmonic motion.

The acceleration of our motion has the components

$$\ddot{x} = -4\pi^2\nu^2 R\cos 2\pi\nu t, \qquad \ddot{y} = -4\pi^2\nu^2 R\sin 2\pi\nu t$$

or
$$\ddot{x} = -4\pi^2\nu^2 x, \qquad \ddot{y} = -4\pi^2\nu^2 y.$$

This shows that acceleration vector $\mathbf{a}$ has the direction opposite to that of the position vector $\mathbf{r}$. The acceleration is not null, although the speed is constant. There is no paradox: the direction of the velocity vector changes in time. The magnitude of the acceleration, however, is constant; we have

$$(2) \qquad |\mathbf{a}| = \sqrt{\ddot{x}^2 + \ddot{y}^2} = 4\pi^2\nu^2 R = \frac{4\pi^2 R}{T^2}.$$

▶ *Example* The motion of an **artificial satellite** may be considered, in first approximation, as a circular motion with radius R equal to that of the earth. The satellite experiences, therefore, an acceleration directed toward the center of the earth, of magnitude $4\pi^2 R/T^2$. But the satellite in orbit is subject only to gravity. Thus $4\pi^2 R/T^2 = g = 980$ cm/sec². Hence the period $T = 2\pi\sqrt{R/2}$ can be computed if R is known. This explains why most artificial satellites have periods of the same order of magnitude.◀

4.4 Tangential and normal accelerations

Consider now a plane motion

$$t \mapsto \mathbf{r}(t) = x(t)\mathbf{e}_1 + y(t)\mathbf{e}_2.$$

At any instant t, the unit tangent vector $\mathbf{T}(t)$ and the unit normal vector $\mathbf{N}(t)$ form a frame. We noticed above that the velocity vector has the direction of $\mathbf{T}(t)$. The acceleration vector $\mathbf{a}(t)$ will have, in general, components in both the tangential and the normal directions; that is,

$$\mathbf{a} = \alpha\mathbf{T} + \beta\mathbf{N}.$$

We call $\alpha\mathbf{T}$ and $\beta\mathbf{N}$ the **tangential** and **normal accelerations**, respectively,

$$\alpha\mathbf{T} = \mathbf{a}_T, \qquad \beta\mathbf{N} = \mathbf{a}_N,$$

and note that

$$\mathbf{a} = \mathbf{a}_T + \mathbf{a}_N.$$

In order to find $\mathbf{a}_T$ and $\mathbf{a}_N$, we introduce a frame $\{\mathbf{e}_1, \mathbf{e}_2\}$ such that at the time instant t_0 considered, $\mathbf{e}_1 = \mathbf{T}$ and $\mathbf{e}_2 = \mathbf{N}$. The motion is given, in components, by

$$x = x(t), \qquad y = y(t).$$

Since $\mathbf{v}(t_0) = |\mathbf{v}(t_0)|\mathbf{T}(t_0) = |\mathbf{v}(t_0)|\mathbf{e}_1$, we have

(3) $$\dot{x}(t_0) > 0, \qquad \dot{y}(t_0) = 0.$$

The acceleration vector at t_0 has the components $\ddot{x}(t_0)$, $\ddot{y}(t_0)$, so that

$$\mathbf{a}_T = \ddot{x}(t_0)\mathbf{e}_1, \qquad \mathbf{a}_N = \ddot{y}(t_0)\mathbf{e}_2.$$

Since

$$\frac{ds}{dt} = \sqrt{\dot{x}^2 + \dot{y}^2},$$

we have
$$\frac{d^2s}{dt^2} = \frac{\dot{x}\ddot{x} + \dot{y}\ddot{y}}{\sqrt{\dot{x}^2 + \dot{y}^2}}$$

and, at $t = t_0$, we have in view of (3)

$$\frac{ds}{dt} = \dot{x}, \qquad \frac{d^2s}{dt^2} = \ddot{x}.$$

The curvature of the trajectory is, according to equation (22) in §3.13,

$$\kappa = \frac{\dot{x}\ddot{y} - \ddot{x}\dot{y}}{(\dot{x}^2 + \dot{y}^2)^{3/2}},$$

so that, for $t = t_0$, we have

$$\kappa = \frac{\ddot{y}}{\dot{x}^2}, \qquad \ddot{y} = \kappa\dot{x}^2.$$

We conclude that, for $t = t_0$,

(4)
$$\mathbf{a}_T = \frac{d^2s}{dt^2}\mathbf{T}, \qquad \mathbf{a}_N = \kappa\left(\frac{ds}{dt}\right)^2\mathbf{N}.$$

If $\kappa > 0$, the latter equation may also be written as

(5)
$$\mathbf{a}_N = \frac{1}{R}\left(\frac{ds}{dt}\right)^2\mathbf{N},$$

where R is the radius of curvature of the trajectory. Since the relations (4) are written in terms of quantities that do not depend on the choice of a coordinate system, they hold in general.

Thus the acceleration in plane motion is a sum of two vectors: the **tangential acceleration $\mathbf{a}_T$,** which is equal to the time derivative of the speed, times the unit tangent vector, and the **normal acceleration $\mathbf{a}_N$,** which is equal to the product of the curvature of the trajectory and the square of the speed, times the unit normal vector. (Note that the vector $\mathbf{a}_N$ points toward the center of curvature of the path if $\kappa > 0$ and $ds/dt \neq 0$; compare Fig. 11.22.)

▶*Example*   The motion of a particle is described by

$$x = \phi(t), \qquad y = -\phi(t)^3,$$

Fig. 11.22

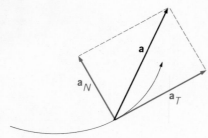

where ϕ is an increasing function. Find the trajectory, the speed, and the normal and tangential accelerations.

Solution. The trajectory is clearly the cubic parabola $y = -x^3$. The curvature at a point (x, y) is (compare equation (23) in §3.15):

$$\kappa(x) = \frac{d^2y/dx^2}{[1 + (dy/dx)^2]^{3/2}} = -\frac{6x}{\sqrt{(1 + 9x^4)^3}}.$$

The speed equals

$$\frac{ds}{dt} = \sqrt{\dot{x}^2 + \dot{y}^2} = \sqrt{1 + 9x^4}\,\phi'(t) \qquad \text{with } x = \phi(t).$$

The unit tangent vector and unit normal vector have the components

$$\left(\frac{\dot{x}}{\sqrt{\dot{x}^2 + \dot{y}^2}}, \frac{\dot{y}}{\sqrt{\dot{x}^2 + \dot{y}^2}}\right) \text{ and } \left(-\frac{\dot{y}}{\sqrt{\dot{x}^2 + \dot{y}^2}}, \frac{\dot{x}}{\sqrt{\dot{x}^2 + \dot{y}^2}}\right),$$

respectively. Therefore

$$\mathbf{T} = \left(\frac{1}{\sqrt{1 + 9x^4}}, \frac{-3x^2}{\sqrt{1 + 9x^4}}\right), \mathbf{N} = \left(\frac{3x^2}{\sqrt{1 + 9x^4}}, \frac{1}{\sqrt{1 + 9x^4}}\right).$$

Next,

$$\frac{d^2s}{dt^2} = \frac{d}{dt}[\sqrt{1 + 9x^4}\phi'(t)] = \frac{18x^3}{\sqrt{1 + 9x^4}}\,\phi'(t)^2 + \sqrt{1 + 9x^4}\phi''(t).$$

We conclude by (4) that

$$\mathbf{a}_T = \left(\frac{18x^3\phi'(t)^2}{1 + 9x^4} + \phi''(t)\right)(\mathbf{e}_1 - 3x^2\mathbf{e}_2), \mathbf{a}_N = -\frac{6x\phi'(t)^2}{1 + 9x^4}(3x^2\mathbf{e}_1 + \mathbf{e}_2).$$

In these formulas $x = \phi(t)$.◄

EXERCISES

For each of the following motions, find the velocity, speed, acceleration, and normal and tangential components of acceleration. Sketch the trajectory.

1. $\mathbf{r}(t) = t\mathbf{e} + \frac{1}{2}t^2\mathbf{e}_2$.
2. $\mathbf{r}(t) = (\cos^2 t)\mathbf{e}_1 + (\sin^2 t)\mathbf{e}_2$ for $0 \le t \le \pi/2$.
3. $\mathbf{r}(t) = (\cos t)\mathbf{e}_1 + (\cos 2t)\mathbf{e}_2$.
► 4. $\mathbf{r}(t) = (a \cos t)\mathbf{e}_1 + (b \sin t)\mathbf{e}_2$, where a and b are arbitrary positive constants.
5. $\mathbf{r}(t) = \frac{1}{2}t^2\mathbf{e}_1 + \frac{1}{3}(2t + 1)^{3/2}\mathbf{e}_2$ for $-\frac{1}{2} \le t$.

6. $\mathbf{r}(t) = e^t\mathbf{e}_1 + e^{-t}\mathbf{e}_2$.

7. $\mathbf{r}(t) = (\cos^3 t)\mathbf{e}_1 + (\cos t)\mathbf{e}_2$ for $0 \le t \le \pi$.

8. $\mathbf{r}(t) = t\mathbf{e}_1 + (\log t)\mathbf{e}_2$ for $0 < t$.

9. $\mathbf{r}(t) = (t\cos t)\mathbf{e}_1 + (t\sin t)\mathbf{e}_2$. (Hint: To sketch the trajectory, use polar coordinates.)

10. $\mathbf{r}(t) = \cos(t + 1)\mathbf{e}_1 + (t + 1)\mathbf{e}_2$.

11. $\mathbf{r}(t) = t\mathbf{e}_1 + \sqrt{1 + t^2}\,\mathbf{e}_2$.

4.5 Law of motion

We formulate now **Newton's Law of Motion:** at any moment the acceleration of a particle of mass m equals $1/m$ times the force acting on the particle. Here m is a number; since the acceleration $\mathbf{a}$ is a vector, the force must be also a vector. If we denote it by $\mathbf{f}$, the mathematical expression of Newton's Law reads

$$(6) \qquad\qquad\qquad m\mathbf{a} = \mathbf{f}.$$

Of course, this becomes a law of nature that can be confirmed or disproved by experiments only if something is said about finding $\mathbf{f}$.

One statement about forces is implicit in the usual formulation of Newton's Law, but should be made explicit: if two forces $\mathbf{F}_1$ and $\mathbf{F}_2$ act on a particle, the particle moves as if it were acted upon by the sum $\mathbf{f} = \mathbf{F}_1 + \mathbf{F}_2$.

The reader will observe that the "parallelogram law" for vector addition (see Chapter 9, §1.4) is a mathematical definition, but the statement that the effect of two forces can be computed by forming their sum according to this definition is a physical law.

If no force acts on the particle, Newton's Law gives $\mathbf{a} \equiv 0$, that is, $\dot{\mathbf{v}}(t) \equiv 0$ and $\mathbf{v} = $ const, and therefore $\mathbf{r}(t) = t\mathbf{v} + $ const. A particle subject to no force moves along a straight line and with constant speed. This statement is the **Law of Inertia.** It is also called Newton's **First Law,** whereas relation (6) is called the **Second Law.** There is also a third law, which refers to more than one particle. It need not be considered here (however, see §6.3).

4.6 Motion of a projectile

As the first application of Newton's Law, consider plane motion under the influence of gravity. We choose the x, y plane as the plane of motion with the y axis pointing "up," so that the weight of the particle is $-mg\mathbf{e}_2$, where g is the acceleration of gravity (compare Chapter 4, where we considered only vertical motion). If weight is the only force present, then the equation of motion reads

$$m\ddot{\mathbf{r}} = m\mathbf{a} = -mg\mathbf{e}_2$$

or, canceling m and using components,

$$\ddot{x} = 0, \qquad \ddot{y} = -g.$$

The two equations are independent of each other and can be solved independently. We have

(7) $$\dot{x}(t) = \alpha, \qquad \dot{y}(t) = -gt + \beta$$

where α and β are constants, and therefore

(8) $$x(t) = \alpha t + \gamma, \qquad y(t) = -\frac{1}{2}gt^2 + \beta t + \delta$$

with some new constants γ, δ. It is seen from (8) that (γ, δ) are the coordinates of our particle at time $t = 0$, whereas (7) shows that (α, β) are the components of the velocity vector at time $t = 0$. The same formula shows that the horizontal velocity $\dot{x}$ is constant during the motion. (It must be, since there is no horizontal force.) The whole motion is determined if we know the **initial position** (γ, δ) and the **initial velocity** (α, β).

Formulas (8) are a parametric representation of the trajectory. To find a nonparametric representation of this curve, we "eliminate t." The first equation yields $t = (x - \gamma)/\alpha$, provided that $\alpha \neq 0$. Substituting this value of t in the second equation, we obtain

$$y = -\frac{1}{2}g\left(\frac{x-\gamma}{\alpha}\right)^2 + \beta\frac{x-\gamma}{\alpha} + \delta$$

or $$y = -\frac{g}{2\alpha^2}x^2 + \frac{g\gamma + \alpha\beta}{\alpha^2}x - \frac{g\gamma^2 + 2\alpha\beta\gamma - 2\alpha^2\delta}{2\alpha^2}.$$

Fig. 11.23

Hence the trajectory is a **parabola,** with a vertical axis, that is convex upward (see Fig. 11.23). If $\alpha = 0$, then x is constant and we have rectilinear vertical motion as considered in Chapter 4.

It should be remembered, of course, that in deriving (8) we neglected air resistance, wind, the influence of the rotation of the earth, and so forth.

▶ *Examples* *1.* A particle moves along a parabola under the influence of gravity alone. Show that the curvature κ of the trajectory at the vertex depends only on the horizontal velocity. (Assume that $\kappa > 0$.)

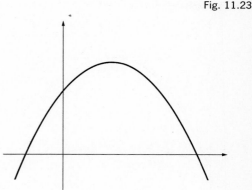

First solution. Let the horizontal velocity be α. At the vertex, the velocity vector is horizontal. At this point, therefore, the speed equals $|\alpha|$. The acceleration is vertical and has magnitude g. At the vertex, the acceleration vector is normal to the curve ($\mathbf{a} = \mathbf{a}_N$), so that by (4) $g = \kappa\alpha^2$ or $\kappa = g/\alpha^2$. (Let us check the units: g is measured in cm/sec^2, α in cm/sec, hence κ in cm^{-1}. This is correct; the reciprocal of κ is a length, the radius of curvature.)

Second solution. Using (8) and equation (22) in §3.15, we compute

$$\kappa = \frac{\dot{x}\ddot{y} - \ddot{x}\dot{y}}{(\dot{x}^2 + \dot{y}^2)^{3/2}};$$

since $\dot{y} = 0$ at the vertex, we obtain

$$\kappa = \frac{\dot{x}\ddot{y}}{\dot{x}^3} = \frac{g}{\alpha^2}.$$

2. The initial speed of a projectile is σ cm/sec. It is shot from ground level at an angle θ, $0 < \theta < 90°$, and hits the ground again at a distance d cm. Find θ. [We assume, of course, that the projectile moves according to the law (8).]

Fig. 11.24

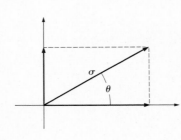

Solution. We have $\delta = 0$ and may assume $\gamma = 0$. The initial velocity has components $\alpha = \sigma \cos \theta$ and $\beta = \sigma \sin \theta$ (see Fig. 11.24). Therefore $y(t) = -\frac{1}{2}gt^2 + \beta t$; this equals 0 for $t = 0$, the initial moment, and for $t = 2\beta/g = 2\sigma \sin \theta/g$. For this value of t, we have

$$x(t) = \alpha t = \frac{2\sigma^2 \sin \theta \cos \theta}{g} = \frac{\sigma^2}{g} \sin 2\theta,$$

Fig. 11.25

and this should be d. Thus θ is found from the relation $\sin 2\theta = gd/\sigma^2$. This gives two possible values for θ, $\theta = \theta_1$, and $\theta = 90° - \theta_1$ (see Fig. 11.25). ◀

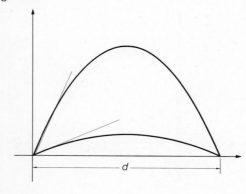

EXERCISES

12. A projectile is shot from ground level with an initial speed σ_0 and strikes the ground again with a terminal speed σ_1. Show that $\sigma_0 = \sigma_1$.

13. A projectile is shot from ground level at time $t = 0$ and strikes the ground again at time $t = t_1$. Show that the projectile attains its maximum height at time $t = t_1/2$.

14. A projectile is shot from point O at ground level with an initial velocity $\alpha\mathbf{e}_1 + \beta\mathbf{e}_2$. Let P be the highest point on the trajectory and let Q be the point where the particle strikes the ground again. Let h be the height of P above ground level, let r ($=$ range) be the distance from O to Q, and let θ ($=$ angle of inclination) be the polar angle of the initial velocity. Also, let ϕ be the angle $\angle QOP$. (a) Compute h and r. (b) Show that $2 \tan \phi = \tan \theta$. (Hint: Sketch the trajectory. Mark the points O, P, and Q.)

15. A projectile is shot from ground level with an initial velocity $30\mathbf{e}_1 + 40\mathbf{e}_2$, where the units are feet and seconds. How far away will the projectile again hit the ground?

16. A projectile is shot from ground level with an initial velocity $25\mathbf{e}_1 + 48\dot{\mathbf{e}}_2$, where the units are feet and seconds. How far above ground level will the projectile strike a sheer vertical cliff which is 50 ft away?

17. A projectile is shot from ground level with an initial velocity $50\mathbf{e}_1 + 32\mathbf{e}_2$, where the units are feet and seconds. The projectile strikes a sheer vertical cliff 12 ft above ground level. How far away is the cliff? Explain why there are two possible answers.

18. A projectile is shot from ground level with an angle of elevation of $\pi/6$ radians (angle of elevation = polar angle of initial velocity vector). The projectile strikes a sheer vertical cliff which is $80\sqrt{3}$ ft away at a height of 64 ft above ground level. Find the initial speed of the projectile. Find the speed with which the projectile strikes the cliff.

19. A projectile shot from ground level attains a maximum height of 100 ft above ground level and a range (distance to the point where the projectile strikes the ground again) of 200 ft. Find the initial velocity.

▶ 20. A projectile shot from a 16-ft-high platform strikes the ground after 2 sec. If the tangent of the angle of inclination is 2, find the horizontal distance traveled by the projectile.

21. A projectile shot from a cannon at ground level strikes a target 240 ft above ground level. The velocity of the projectile at the instant of impact is $48\mathbf{e}_1 - 32\mathbf{e}_2$. Find the initial velocity of the projectile and the horizontal distance from the cannon to the target.

4.7 Motion on a rigid path

We consider next a heavy particle that is constrained to move along a perfectly **smooth rigid path** (see Fig. 11.26). The path is represented by a curve

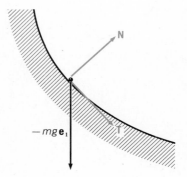

Fig. 11.26

$$x = x(s), \qquad y = y(s),$$

where s is the arc length. The forces acting are the weight $-mg\mathbf{e}_2$ and an unknown force $\mathbf{G}$, the constraint exerted by the path on the particle. We assume that this force is always *perpendicular* to the path; this is what is meant by saying that the path is perfectly smooth.

At every point of the path let $\mathbf{T}(s)$ and $\mathbf{N}(s)$ be the unit tangent and unit normal vectors, respectively. We have

$$\mathbf{T} = x'\mathbf{e}_1 + y'\mathbf{e}_2, \qquad \mathbf{N} = -y'\mathbf{e}_1 + x'\mathbf{e}_2, \qquad (x')^2 + (y')^2 = 1,$$

where the prime denotes differentiation with respect to s [equations (17) in §3.12]. Multiplying the first equation above by y' and the second by x' and adding, we obtain $\mathbf{e}_2 = y'\mathbf{T} + x'\mathbf{N}$. Therefore the weight equals

$$(9) \qquad\qquad -mg\mathbf{e}_2 = -mgy'(s)\mathbf{T}(s) - mgx'(s)\mathbf{N}(s).$$

By our assumption, $\mathbf{G}$ has the direction of $\mathbf{N}$. There is a function $\gamma(s)$ such that

$$(10) \qquad\qquad \mathbf{G} = \gamma(s)\mathbf{N}(s).$$

We saw above [see equation (4)] that the acceleration vector is

$$(11) \qquad\qquad \mathbf{a} = \frac{d^2s}{dt^2}\,\mathbf{T} + \kappa\left(\frac{ds}{dt}\right)^2\mathbf{N},$$

where $\kappa = \kappa(s)$ is the curvature of the path at the point considered. Newton's law applied to our motion gives

$$-mg\mathbf{e}_2 + \mathbf{G} = m\mathbf{a}.$$

Using (9), (10), and (11), we write this as

$$-\left(mgy'(s) + m\frac{d^2s}{dt^2}\right)\mathbf{T}(s) + \left(\gamma(s) - mgx'(s) - m\kappa(s)\left(\frac{ds}{dt}\right)^2\right)\mathbf{N}(s) = 0.$$

Since $\{\mathbf{T}, \mathbf{N}\}$ is a frame, this vector equation is equivalent to two scalar equations:

$$(12) \quad mgy'(s) + m\frac{d^2s}{dt^2} = 0, \qquad \gamma(s) = mgx'(s) + m\kappa(s)\left(\frac{dt}{dt}\right)^2.$$

The first equation can be rewritten as

$$(13) \qquad\qquad \frac{d^2s}{dt^2} = -gy'(s),$$

and is a second order differential equation for the unknown function $t \mapsto s(t)$, which describes the motion of our particle. We shall show later how it can be solved. If we know $s(t)$, we can use the second equation (12) to compute the force $\mathbf{G} = \gamma\mathbf{N}$ exerted by the path on the particle.

4.8 Motion on an inclined plane

The general equation (13) will be solved in an appendix to this chapter (see §5). Here we consider only two simple special cases. Assume first that the path is straight; we may think of a particle sliding down an inclined plane. Then (see Fig. 11.27):

$$x = x_0 + s \cos \alpha, \qquad y = y_0 + s \sin \alpha, \qquad \alpha = \text{arc tan } \mu,$$

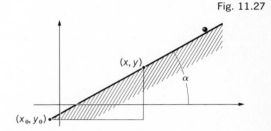

Fig. 11.27

μ being the slope of the path. Equation (13) becomes

$$s''(t) = -g \sin \alpha.$$

This has the general solution

$$s(t) = -g \frac{\sin \alpha}{2} t^2 + at + b.$$

For $t = 0$, we have $s = b$; thus b is the value of s at the "initial moment." Similarly, since $s'(t) = -g (\sin \alpha)t + a$, a is the "initial value" of $s'(t)$, that is, $s'(0)$. If the particle is released at $t = 0$ with initial speed 0, we have

$$s(t) = \frac{g \sin \alpha}{2} t^2.$$

Since $\sin \alpha = \mu/\sqrt{1 + \mu^2}$, this is Galileo's Law:

$$s = -\frac{g\mu}{2\sqrt{1 + \mu^2}} t^2,$$

which we stated without proof in Chapter 4, §1.14.

4.9 Mathematical pendulum

We consider next the "mathematical pendulum." This is a particle of mass m attached to a weightless rod of length l, which hangs on a frictionless pivot (see Fig. 11.28). We assume that the particle remains below the pivot; then the particle is constrained to move on the semicircle

(14) $$x = l \sin \theta, \qquad y = -l \cos \theta,$$

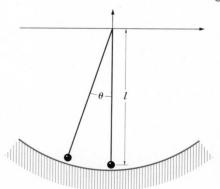

Fig. 11.28

where the angle θ measures the deviation from the vertical position. The particle moves as if it were sliding along a smooth circular arc defined by (14).

The arc length parameter s, chosen so that $s = 0$ for $\theta = 0$, is $s = l\theta$. Hence

$$x = l \sin \frac{s}{l}, \qquad y = -l \cos \frac{s}{l}, \qquad \frac{dy}{ds} = \sin \frac{s}{l}.$$

Now equation (13) becomes

$$(15) \qquad\qquad \frac{d^2s}{dt^2} + g \sin \frac{s}{l} = 0.$$

This is a rather complicated equation, whose solution involves so-called elliptic functions. Let us, however, restrict ourselves to motions during which the pendulum remains *nearly vertical* (as is the case in a clock). Then $\theta = s/l$ is small and we commit a very small error if we replace $\sin (s/l)$ by s/l. The simplified equation reads

$$\frac{d^2s}{dt^2} + \frac{g}{l} s = 0.$$

This equation we know well (see Chapter 6, §1.12). Its general solution is

$$s = A \cos \sqrt{\frac{g}{l}} t + B \sin \sqrt{\frac{g}{l}} t.$$

Thus a pendulum executes, within the error committed when we replaced $\sin \theta$ by θ, a simple harmonic motion with period

$$(16) \qquad\qquad T = 2\pi \sqrt{\frac{l}{g}};$$

the period does *not* depend on the amplitude, but only on the length l and the value of g.

This formula explains how a pendulum stabilizes the movement of a clock, and how a pendulum is used to measure the acceleration of gravity.

4.10 Planetary motion. Kepler's Second Law

We come now to one of the most dramatic and most important applications of calculus—the derivation of the Law of Universal Gravitation from astronomical observations. This application was made by Newton himself, but when he published it he did not mention calculus.

The first step toward this momentous discovery was purely conceptual. This was the recognition, obvious to us, but against all traditions of antiquity, that the laws of physics abstracted from observations on ordinary (terrestrial) objects are applicable also to heavenly (celestial) bodies. For instance, the same force that makes an apple fall to the ground keeps the moon in orbit around the earth.

The second step was mathematical. Kepler analyzed the astronomical observation of Tycho Brahe and extracted from them his three laws of planetary motion. The first two state that

(I) PLANETS MOVE IN PLANES CONTAINING THE SUN, AND THEIR ORBITS ARE ELLIPSES WITH THE SUN IN A FOCUS (see Fig. 11.29).

(II) A SEGMENT JOINING THE SUN AND A PLANET SWEEPS OUT EQUAL AREAS IN EQUAL TIME INTERVALS.

Newton calculated (presumably at the age of 23) that Kepler's laws imply that the acceleration of a planet is directed toward the sun and has a magnitude inversely proportional to the square of the distance from the sun.

The statement that the planet's acceleration is directed along the line joining it to the sun follows from the Second Law alone. To see this, we introduce a coordinate system with the sun at the origin O. Let r and θ be the polar coordinates of the planet P. The motion of the planet is then described by two functions

$$r = r(t), \qquad \theta = \theta(t).$$

The area swept out by the segment $\overrightarrow{OP}$ from the time t_0 to the time t equals [see Fig. 11.30 and equation (12) in §2.6]

$$\frac{1}{2} \int_{t_0}^{t} r(t)^2 \, d\theta(t) = \frac{1}{2} \int_{t_0}^{t} r(t)^2 \dot{\theta}(t) \, dt.$$

Law II says that the time derivative of this area,

$$\frac{d}{dt} \frac{1}{2} \int_{t_0}^{t} r(t)^2 \dot{\theta}(t) \, dt = \frac{1}{2} r(t)^2 \dot{\theta}(t),$$

is a constant, call it α. Thus

(17) $$r^2 \dot{\theta} = 2\alpha.$$

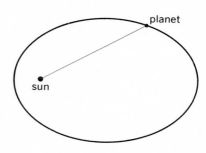

Fig. 11.29

planet

sun

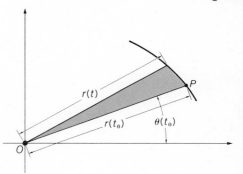

Fig. 11.30

$r(t)$

$r(t_0)$ $\theta(t_0)$

O

JOHANNES KEPLER (1571–1630) devoted years of laborious calculations to uncovering the hidden mathematical harmonies which, he was sure, underlay the motion of the planets. Not all of his guesses were right, but in 1609 he published his first two laws and 10 years later he announced the third. Kepler also made pioneering contributions to optics; his treatise on measuring the volumes of wine barrels anticipates calculus.

Ancient superstition influenced the life of this creator of modern astronomy. His official duties as mathematician to Emperor Rudolph II included the preparation of horoscopes, and his mother was tried for witchcraft (and acquitted).

TYCHO BRAHE (1546–1607), a Danish nobleman, perfected the art of astronomical observations, without telescopes, and built a big observatory on the island of Ven. He also had the insight to entrust the records of his observations to the man who could make the best use of them, Kepler.

Differentiating this, we obtain

$$2r\dot{r}\dot{\theta} + r^2\ddot{\theta} = 0,$$

so that

(18) $$2\dot{r}\dot{\theta} + r\ddot{\theta} = 0.$$

The Cartesian coordinates of P are

$$x = r\cos\theta, \qquad y = r\sin\theta,$$

so that the velocity vector **v** has the components

$$\dot{x} = \dot{r}\cos\theta - r\dot{\theta}\sin\theta, \qquad \dot{y} = \dot{r}\sin\theta + r\dot{\theta}\cos\theta.$$

The components of the acceleration vector **a** are therefore

$$\ddot{x} = \ddot{r}\cos\theta - 2\dot{r}\dot{\theta}\sin\theta - r\dot{\theta}^2\cos\theta - r\ddot{\theta}\sin\theta,$$
$$\ddot{y} = \ddot{r}\sin\theta + 2\dot{r}\dot{\theta}\cos\theta - r\dot{\theta}^2\sin\theta - r\ddot{\theta}\cos\theta.$$

Using (18), we obtain

$$\ddot{x} = [\ddot{r} - r\dot{\theta}^2]\cos\theta, \qquad \ddot{y} = [\ddot{r} - r\dot{\theta}^2]\sin\theta,$$

that is, $$\ddot{x} = \left[\frac{\ddot{r}}{r} - \dot{\theta}^2\right]x, \qquad \ddot{y} = \left[\frac{\ddot{r}}{r} - \dot{\theta}^2\right]y.$$

This shows that the acceleration vector $(\ddot{x}, \ddot{y})$ is a scalar multiple of the position vector (x, y). Thus the acceleration of the planet is directed toward the sun.

The magnitude of the acceleration vector is

$$|a| = |\ddot{r} - r\dot{\theta}^2|.$$

To complete the argument, we must show that $|a|$ is inversely proportional to the square of the distance from the sun. This will follow if we can show that

(19) $$\ddot{r} - r\dot{\theta}^2 = -\frac{\mu}{r^2},$$

where μ is a positive constant. Now we must use the First Law.

4.11 Application of Kepler's First Law

The trajectory of the planet is an ellipse with focus O; let ϵ be its eccentricity. We recall the formulas for ellipses in polar coordinates (see §1.7 in Chapter 10) and conclude that, if the position of the axes is chosen suitably, then

$$r(t) = \frac{A}{1 - \epsilon \cos \theta(t)},$$

where A is a positive constant. Thus

(20) $$r(1 - \epsilon \cos \theta) = A.$$

Differentiating with respect to time, we obtain

$$\dot{r}(1 - \epsilon \cos \theta) + \epsilon r \dot{\theta} \sin \theta = 0.$$

We multiply both sides of this equation by r and obtain

$$\dot{r}r(1 - \epsilon \cos \theta) + \epsilon r^2 \dot{\theta} \sin \theta = 0,$$

which, in view of (17) and (20), may be written as

$$A\dot{r} + 2\alpha\epsilon \sin \theta = 0.$$

Another differentiation gives

(21) $$A\ddot{r} + 2\alpha\epsilon\dot{\theta} \cos \theta = 0.$$

Since, by virtue of (17),

(22) $$\dot{\theta} = \frac{2\alpha}{r^2},$$

we have from (21) that

$$\ddot{r} = -\frac{4\alpha^2}{r^2} \frac{\epsilon \cos \theta}{A}.$$

Since, by (20), $-\epsilon \cos \theta = (A/r) - 1$, we obtain

$$\ddot{r} = \frac{4\alpha^2}{r^2} \left(\frac{1}{r} - \frac{1}{A} \right).$$

Now
$$r\dot\theta^2 = \frac{4\alpha^2}{r^3}$$

by (22), so that

(23)
$$\ddot r - r\dot\theta^2 = -\frac{4\alpha^2}{A}\frac{1}{r^2}.$$

This is the same as relation (19) which we wanted to prove, with

(24)
$$\mu = \frac{4\alpha^2}{A}.$$

4.12 Application of Kepler's Third Law

We have seen that Kepler's first two laws show that a planet P moving around the sun O experiences an acceleration of magnitude

$$\frac{4\alpha^2}{A}\frac{1}{r^2}.$$

Here $r = |\overrightarrow{OP}|$ is the distance from the sun and

$$r = \frac{A}{1 - \epsilon\cos\phi}$$

is the equation of the orbit in polar coordinates. The constant α is given by (17), that is, α is the area swept out by the segment $\overrightarrow{OP}$ per unit time. Let T be the period of the planet. Then αT is the area of the ellipse. If the semiaxes of the ellipse are a and b, its area is πab (see the example in Chapter 10, §1.8). Therefore

(25)
$$\alpha = \frac{\pi ab}{T}.$$

From (20), we conclude that the largest and smallest distances of P from O are $\dfrac{A}{1 - \epsilon}$ and $\dfrac{A}{1 + \epsilon}$. Therefore $\dfrac{A}{1 - \epsilon} + \dfrac{A}{1 + \epsilon} = 2a$ (see Fig. 11.31), or

$$A = a(1 - \epsilon^2).$$

Also,
$$b = a\sqrt{1 - \epsilon^2}$$

Fig. 11.31

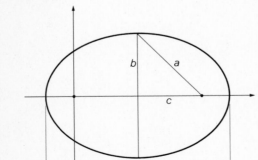

so that, by (25),

$$\alpha = \frac{\pi a^2 \sqrt{1 - \epsilon^2}}{T}$$

and

(26)
$$\mu = \frac{4\alpha^2}{A} = \frac{4\pi^2 a^3}{T^2}.$$

The magnitude of the attraction exerted by the sun on the planet is therefore (being equal to mass times acceleration)

(27)
$$\frac{4\pi^2 a^3 m}{T^2} \frac{1}{r^2}.$$

At this point, one uses Kepler's Third Law:

(III) THE RATIO a^3/T^2 IS THE SAME FOR ALL PLANETS.

Newton concluded from this that the attractive force of the sun per unit mass of a planet is "universal," the same for all planets, and dependent only on the distance from the sun. Furthermore, Kepler's laws hold also for the moons of planets, in particular for the moons of Jupiter. The value of a^3/T^2 for the moons of Jupiter, however, is different from the value of this ratio for planets.

These considerations led Newton to postulate that an attractive force exists between *any* two masses, M and m, and has the magnitude

(28)
$$\frac{\gamma M m}{r^2},$$

r being the distance between the masses and γ a universal constant.

Comparing (27) and (28), we conclude that the mass of the sun is

$$M = \frac{4\pi^2}{\gamma} \frac{a^3}{T^2}.$$

If M_1 is the mass of a planet (or of another sun) and a_1 and T_1 the major semiaxis and period of a moon (or planet) rotating about it, we have

$$M_1 = \frac{4\pi^2}{\gamma} \frac{a_1{}^3}{T_1{}^2}$$

and therefore
$$\frac{M}{M_1} = \frac{a^3 T_1{}^2}{a_1{}^3 T^2}.$$

Thus astronomical observations permit us to compare masses of two heavenly bodies. In order to determine the actual values of these masses, however, one needs a direct measurement of the gravitational constant γ. This was first done accurately by Cavendish in 1798.

4.13 Celestial mechanics

The reader will note that, in deriving Newton's Gravitation Law from Kepler's Law, we never used the fact that the eccentricity of the orbit was $\epsilon < 1$. The same argument would hold if the orbit were a parabola ($\epsilon = 1$) or a hyperbola ($\epsilon > 1$). For a circular orbit ($\epsilon = 0$), the argument would give little information, since then r would be constant.

One can reverse the argument and show that a particle subject to an attractive force directed toward a fixed center O, and varying as the inverse square of the distance from this center, *always* moves on a conic with a focus at O and obeys Kepler's Second Law of Constant "Areal Velocity." We shall not carry out this calculation.

Actual planets, however, satisfy Kepler's laws only approximately. For each planet is also subject to the attraction of all other planets, and so is the sun. Beginning with Newton himself, generations of mathematicians have developed methods, based on calculus, for taking these effects into account and for solving the problems of "celestial mechanics." An important triumph of this science was the discovery of the planet Neptune in 1846. Before anyone saw this planet through a telescope, Adams and Leverrier had calculated its mass and its orbit from observed perturbations in the motion of the planet Uranus. Today space exploration poses new challenges to celestial mechanics, and modern computers permit us to meet these challenges.

There is one planet, however, whose motion defies Newtonian mechanics. A certain peculiarity in the motion of Mercury can be explained only on the basis of Einstein's general theory of relativity.

HENRY CAVENDISH (1731–1810) was a very rich, very eccentric English aristocrat who lived almost as a recluse and performed some of the most brilliant experiments in the history of science.

JOHN COUCH ADAMS (1819–1892) was educated at Cambridge University and spent the rest of his life there. As a student, he became interested in the unexplained irregularities in the motion of Uranus, and within a few years he calculated the motion of a more distant planet (later named Neptune) whose gravitational pull would account for these irregularities.

U. J. J. LEVERRIER (1811–1877) was director of the Paris Observatory when he solved the same problem as Adams. The two men worked without knowledge of each other. A German astronomer directed his telescope at the point of the sky indicated by Leverrier and saw the "new" planet. It seems that Leverrier himself never cared to take a look.

EXERCISES

22. Let $A(t)$ be a function defined for $0 \leq t < +\infty$ and suppose that we have $A(t_1 + \alpha) - A(t_1) = A(t_2 + \alpha) - A(t_2)$ for any t_1, t_2, and α. If we think of t as time, this means that A changes equal amounts in equal times. Let $a = A(1) - A(0)$ and $b = A(0)$.

(a) Show that, for any $t \geq 0$ and any positive integer m, one has $A(mt) = mA(t) + (1 - m)b$. [Hint: Write $A(mt) - A(0) = A(mt) - A((m - 1)t)) + (A((m - 1)t) - A((m - 2)t)) + \cdots + (A(t) - A(0)).$]

(b) Show that, for any positive integer n, one has $A\left(\dfrac{1}{n}\right) = \dfrac{a}{n} + b$. $\Big[$Hint: Write

$$A(1) - A(0) = \left(A(1) - A\left(\frac{n-1}{n}\right)\right) + \left(A\left(\frac{n-1}{n}\right) - A\left(\frac{n-2}{n}\right)\right) +$$

$$\cdots + \left(A\left(\frac{1}{n}\right) - A(0)\right).\Big]$$

(c) Show that, for any positive rational number r, we have $A(r) = ar + b$. If we assume that A is continuous, then we have $A(t) = at + b$ for all t. This result was used in applying Kepler's Second Law in the text.

23. Suppose, in place of Kepler's First Law, we assume that the orbits of planets are spirals that have the form $r = \beta e^{\theta}$ in a polar coordinate system with origin at the sun, and that we also assume the Second Law.
    (a) Show that $r = (4\alpha t + r_0{}^2)^{1/2}$, where $r_0 = r(0)$.
    (b) Show that the acceleration is inversely proportional to the cube of the distance from the sun.

24. In Exercise 23, suppose that the spirals have the form $r = \beta\theta$.
    (a) Show that $r = (6\alpha\beta t + r_0{}^3)^{1/3}$, where $r_0 = r(0)$.
    (b) Show that the acceleration is not proportional to any power (positive or negative) of the distance from the sun.

25. Suppose that, instead of Newton's Law of Gravitation, we have the law: the sun attracts a planet with a force proportional to the distance of the planet to the sun. Find the motion of the planet.

Appendixes to Chapter 11

§5 Motion on a Prescribed Path

In §4.7, we derived the differential equation for the motion of a heavy particle constrained to slide along a frictionless path represented parametrically by $x = x(s)$, $y = y(s)$. Here s is the arc length along the path and the negative y direction is the direction of gravity. The equation reads [see (13) in §4.7]

(1)
$$\frac{d^2s}{dt^2} = -gy'(s),$$

where g is the acceleration of gravity. We discuss this equation in the present section.

5.1 Solution of the equation

Equation (1) can be solved by a trick. We multiply both sides of the equation by ds/dt, which yields

$$\frac{ds}{dt}\frac{d^2s}{dt^2} + gy'(s)\frac{ds}{dt} = 0$$

or

$$\frac{d}{dt}\left(\frac{1}{2}\left(\frac{ds}{dt}\right)^2 + gy(s)\right) = 0.$$

We conclude that our equation is equivalent to the statement: the quantity $\frac{1}{2}s'(t)^2 + gy(s)$ remains constant during the motion. Call this constant ϵ, then

(2)
$$\frac{1}{2}\left(\frac{ds}{dt}\right)^2 + gy(s) = \epsilon.$$

(This is a special case of conservation of energy, but we do not need the concept of energy here.)

Assume now that we follow the motion during a time interval when $s'(t)$ has a fixed sign, say, is positive. Then (2) implies that

(3)
$$\frac{ds}{dt} = \sqrt{2\epsilon - 2gy(s)}.$$

Since $s'(t) > 0$, the function $t \mapsto s(t)$ has a differentiable inverse; we may consider t (= time) as a function of s (= distance measured along the path). By (3),

$$\frac{dt}{ds} = \frac{1}{\sqrt{2[\epsilon - gy(s)]}},$$

and, if $t = 0$ for $s = s_0$, then

(4)
$$t = \int_{s_0}^{s} \frac{d\xi}{\sqrt{2[\epsilon - gy(\xi)]}}.$$

Inverting the function $s \mapsto t$, we obtain the desired solution.

We still do not know the value of ϵ; it can be determined from the value σ of the speed ds/dt at $t = 0$. Indeed, by (2) we have

(5)
$$\epsilon = \frac{1}{2}\sigma^2 + gy(s_0).$$

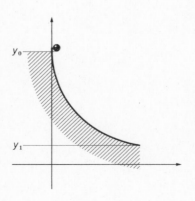

Fig. 11.32

Let us assume that, for $s_0 \le s \le s_1$, the function $y(s)$ is decreasing (see Fig. 11.32) and that the particle is released from the height $y_0 = y(s_0)$ with initial speed 0. We want to compute the time τ it will take for the particle to reach the height $y_1 = y(s_1)$. We must merely choose for ϵ the value $gy(s_0)$ (since the initial speed is 0) and apply relation (4) with $s = s_1$. This yields

(6)
$$\tau = \int_{s_0}^{s_1} \frac{ds}{\sqrt{2g[y(s_0) - y(s)]}}.$$

5.2 Periodic motion

Assume next that the path considered is the graph of an even function $x \mapsto y$, which is first decreasing to a minimum and then increasing, as in Fig. 11.33. Let $s = 0$ and $y = 0$ correspond to the lowest point on the curve. We put the particle at a point P corresponding to the parameter value $-s_0 < 0$, of height y_0, and release it with initial speed 0, at the time $t = 0$. It will slide down to the point O, and this descent will take

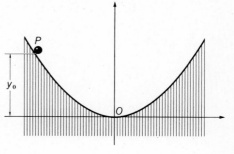

Fig. 11.33

$$(7) \qquad \tau = \int_{-s_0}^{0} \frac{ds}{\sqrt{2g[y_0 - y(s)]}}$$

time units. At the time τ, the particle will have a certain speed, σ, which can be easily computed. Indeed, by (2), the quantity $(ds/dt)^2 + 2gy$ is constant during the motion; it equals $2gy_0$ at $t = 0$. Hence $\sigma = \sqrt{2gy_0}$. Now the particle will continue to rise till the speed becomes 0. This will happen, of course, as soon as the height y_0 is reached. Next, the particle will slide down again and continue to rise until it reaches the initial position P. Then the whole process will repeat itself. This perpetual motion is possible only because we neglect friction.

The first upswing, the second downswing, and the second upswing each last as long as the first downswing, that is, τ time units. This is clear by symmetry, and can also be computed directly. The **period** T of the whole motion is therefore 4τ, that is,

$$T = 4 \int_{-s_0}^{0} \frac{ds}{\sqrt{2g[y_0 - y(s)]}}$$

or, since y is an even function of s,

$$(8) \qquad T = 2 \int_{-s_0}^{s_0} \frac{ds}{\sqrt{2g[y_0 - y(s)]}}.$$

The above formula was derived under the assumption that we use the parametric representation of the path in terms of arc length. It is convenient to rewrite it in terms of an arbitrary parameter θ. (We do not want to use the letter t reserved for time.) If

$$s \mapsto s(\theta), \qquad s(0) = 0, \qquad s(\theta_0) = s_0$$

is an allowable parameter change, we obtain

$$T = 2 \int_{-\theta_0}^{\theta_0} \frac{1}{\sqrt{2g[y_0 - y(s(\theta))]}} \frac{ds}{d\theta} d\theta.$$

Noting that

$$\frac{ds}{d\theta} = \sqrt{\left(\frac{dx}{d\theta}\right)^2 + \left(\frac{dy}{d\theta}\right)^2}$$

and that we may consider x and y as functions of θ, we can write this relation as

$$(9) \qquad T = 2 \int_{-\theta_0}^{\theta_0} \sqrt{\frac{(dx/d\theta)^2 + (dy/d\theta)^2}{2g(y_0 - y)}}\, d\theta$$

or, in an abbreviated notation,

$$T = 2 \int_{-\theta_0}^{\theta_0} \sqrt{\frac{(dx)^2 + (dy)^2}{2g(y_0 - y)}}.$$

In general, T will depend on the path and on the amplitude y_0.

5.3 The circular pendulum

Let us compute T for a circular path, that is, the period of a mathematical pendulum (see §4.9). The curve is defined by $x = l \sin \theta$, $y = -l \cos \theta$. Therefore

$$T = 2 \int_{-\theta_0}^{\theta_0} \sqrt{\frac{l^2 \cos^2 \theta + l^2 \sin^2 \theta}{2g(l \cos \theta - l \cos \theta_0)}}\, d\theta = 2 \sqrt{\frac{l}{2g}} \int_{-\theta_0}^{\theta_0} \frac{d\theta}{\sqrt{\cos \theta - \cos \theta_0}}$$

$$= \sqrt{\frac{l}{g}} \int_{-\theta_0}^{\theta_0} \frac{d\theta}{\sqrt{\sin^2 \tfrac{1}{2}\theta_0 - \sin^2 \tfrac{1}{2}\theta}} = \sqrt{\frac{l}{g}} \int_{-\theta_0}^{\theta_0} \frac{d\theta}{\lambda \sqrt{1 - (1/\lambda^2) \sin^2 \tfrac{1}{2}\theta}}$$

where $\lambda = \sin \tfrac{1}{2}\theta_0$, and we made one of the identity $\cos x = 1 - 2 \sin^2 \tfrac{1}{2}x$. If we now make the substitution

$$u = \frac{1}{\lambda} \sin \frac{\theta}{2},$$

we obtain that

$$T = 2 \sqrt{\frac{l}{g}} \int_{-1}^{1} \frac{du}{\sqrt{(1 - u^2)(1 - \lambda^2 u^2)}}.$$

This is a so-called "elliptic integral"; its value depends on the initial displacement θ_0. But if θ_0 is small, then

$$\lambda^2 = \sin^2 \frac{\theta_0}{2}$$

is *very* small (for instance, $\lambda = .0019$ for $\theta_0 = 5°$). In this case, we may neglect the term $\lambda^2 u^2$ in the above expression. Then we get, approximately,

$$T = 2 \sqrt{\frac{l}{g}} \int_{-1}^{1} \frac{du}{\sqrt{1 - u^2}} = 2\pi \sqrt{\frac{l}{g}},$$

in accordance with what was said in §4.9.

Remark   The integrals (6), (7), (8), and (9) are improper integrals (see Chapter 6, §6.1) since $y = y_0$ for $s = \pm s_0$ or $\theta = \pm \theta_0$. But if $y'(s)$ is continuous and $y'(s_0) \neq 0$, as we assume, the integrals converge. This can be proved using the comparison theorem in Chapter 6, §6.4. We omit the details.

Fig. 11.34

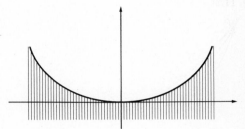

5.4 The cycloidal pendulum

Huygens asked whether there are curves for which T would be actually independent of the amplitude (initial displacement). He discovered that the cycloid has this property, which is easy to verify, and that no other curve will do. The latter statement is difficult to prove, and we will not consider it. The cycloid must, of course, be placed so that the cusps are on top (see Fig. 11.34). A parametric representation of this curve is (see §2.5)

Fig. 11.35

$$(10) \qquad x = a(\theta + \sin \theta), \qquad y = a(1 - \cos \theta).$$

[One obtains (10) from the parametric representation (7) in §2.5 by first replacing x by $x = a\pi$ and y by $2 - y$, and then replacing the parameter θ by $\theta + \pi$.] We have, by equation (9),

$$T = 2 \int_{-\theta_0}^{\theta_0} \sqrt{\frac{a^2(1 + \cos \theta)^2 + a^2 \sin^2 \theta}{2ga(\cos \theta - \cos \theta_0)}}\, d\theta = 2\sqrt{\frac{a}{g}} \int_{-\theta_0}^{\theta_0} \sqrt{\frac{1 + \cos \theta}{\cos \theta - \cos \theta_0}}\, d\theta$$

$$= 2\sqrt{\frac{a}{g}} \int_{-\theta_0}^{\theta_0} \frac{\cos \tfrac{1}{2}\theta\, d\theta}{\sqrt{\sin^2 \tfrac{1}{2}\theta_0 - \sin^2 \tfrac{1}{2}\theta}} = 4\sqrt{\frac{a}{g}} \int_{-1}^{1} \frac{dv}{\sqrt{1 - v^2}} = 4\pi\sqrt{\frac{a}{g}}.$$

where we used the substitution

$$v = \frac{\sin \tfrac{1}{2}\theta}{\sin \tfrac{1}{2}\theta_0}.$$

Thus T does not depend on θ_0, as asserted.

Huygens also indicated how a cycloidal pendulum can be realized: one should let a flexible string, with a mass attached at the end, vibrate between two rigid cycloids (see Fig. 11.35). The mass will describe a so-called involute of the cycloid; this is again a cycloid. (We do not prove this here; compare, however, Problems 21 and 24 at the end of this chapter.) Cycloidal pendulums are not used in actual clocks.

The cycloid (10) has another remarkable property which we mention without proof. If we join a point P on the cycloid to the lowest point Q by *any* curve C other than the cycloid (see Fig. 11.36), then the descent from P to Q along this other curve will take *longer* than the descent along the cycloid, that is, longer than $\pi\sqrt{a/g}$. Several famous mathematicians established this result, independently of each other, among them two Bernoulli brothers (Jacob and Johann) and Pascal. This discovery was a significant step in creating a branch of mathematics called calculus of variations.

CHRISTIAN HUYGENS (1629–1695), a mathematician, astronomer, and physicist, developed the wave theory of light. (This theory was the only one accepted by physicists until the advent of quantum theory, which reconciled Huygens' wave theory with the particle theory of light advanced by Newton.)

A Dutchman of independent means, Huygens lived for many years in Paris. There he met the 26-year-old Leibniz and tutored him in the then-modern mathematics.

BLAISE PASCAL (1623–1662) was perhaps the most amazing prodigy in the history of mathematics. As a child he is supposed to have proved, entirely on his own, that the sum of the angles in a triangle is 180°. At 16 he discovered a fundamental theorem about conics and then wrote a long treatise about these curves, now lost. Pascal's most notable achievements are the creation of the theory of probability (with Fermat) and his work on the pressure of gases and liquids. He also built the first computing machine.

Later, however, mystical and religious interest became predominant in Pascal's life. His *Provincial Letters* (a brilliant theological polemic) and his posthumous *Pensées* are considered masterpieces of French literature.

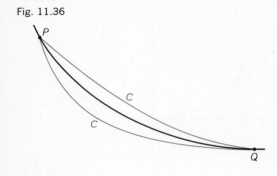

Fig. 11.36

5.5 Nonsymmetric paths and springs

The discussion of the motion of a heavy particle on a symmetric path with one minimum, like the one shown in Fig. 9.33, can be repeated without significant changes for the case when the curve first falls and then rises, without being symmetric.

The new interpretation of (1) shows how one could have obtained Huygens' result about the cycloid. We ask: for which curve $x = x(s)$, $y = y(s)$ is (1) the equation of an *elastic spring* that executes simple harmonic motions with a period independent of the amplitude (see Chapter 6, §1.4)? We must have $y'(s) = bs$, $b > 0$ a constant. Since we want that $x'(s)^2 + y'(s)^2 = 1$, we must have $x'(s) = \sqrt{1 - b^2 s^2}$. If we require that $x = y = 0$ for $s = 0$, simple integrations yield

$$(11) \qquad x = \frac{\text{arc sin } (bs) + bs\sqrt{1 - b^2 s^2}}{2b}, \, y = \frac{bs^2}{2}.$$

An allowable parameter change, $\theta = 2$ arc sin (bs), transforms this into (10), with $a = \frac{1}{4}b$. Thus the desired curve is a cycloid.

Equation (1) can also be interpreted as the equation of motion of a spring with mass m and restoring force $-(g/m)y'(s)$. This is seen by noting the definition of a spring given in Chapter 5, §5.1. In this interpretation, the form (2) of our equation is actually the law of conservation of energy discussed in Chapter 5, §5.4.

EXERCISES

1. Assume that the kinetic energy of a particle sliding down a smooth rigid path is $\frac{1}{2}m(ds/dt)^2$. How must one define potential energy so that equation (2) will be a conservation of energy law?
2. Verify the cycloid's property of being the path of fastest descent by comparing the descent along a cycloid with a descent along a straight line.
3. Verify the cycloid's property of being the path of fastest descent by comparing the cycloid with a circular arc. (Note: This is not very easy.)
4. Apply equation (6) to the case $y(s) = -s$. What does the result mean?
5. Carry out the change of variables $s \mapsto \theta$ suggested at the end of §5.5 and transform (11) into (10).
6. Without assuming any knowledge of trigonometric functions, show how we arrive at the primitive function of $u \mapsto (1 - u^2)^{-1/2}$ if we try to solve the differential equation $s''(t) + s(t) = 0$.

§6 Motion of Systems

In this section, we consider briefly the motion of a system consisting of several particles. This leads to the important concept of centers of mass.

6.1 Two particles subject to mutual attraction

Let us consider first two particles located at points P_1 and P_2, with masses m_1 and m_2. We assume that the particles are subject only to their mutual gravitational attraction. Let $\mathbf{r}_1$ and $\mathbf{r}_2$ be the position vectors of the points P_1 and P_2. The directed segment $\overrightarrow{P_1P_2}$ represents the vector $\mathbf{r}_2 - \mathbf{r}_1$ (see Fig. 11.37) and the distance between P_1 and P_2 is $|\mathbf{r}_2 - \mathbf{r}_1|$. The first particle is subject to the gravitational attraction of the second; this force, which we denote by $\mathbf{f}_{12}$, is given by

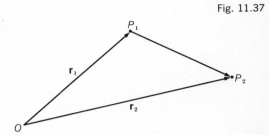

Fig. 11.37

$$\mathbf{f}_{12} = \frac{\gamma m_1 m_2}{|\mathbf{r}_1 - \mathbf{r}_2|^2}\, \mathbf{u},$$

where γ is the gravitational constant and $\mathbf{u}$ the unit vector (vector of length 1) pointing from the second particle to the first. Clearly, $\mathbf{u} = (\mathbf{r}_1 - \mathbf{r}_2)/|\mathbf{r}_1 - \mathbf{r}_2|$. Thus

(1) $$\mathbf{f}_{12} = \gamma m_1 m_2 \frac{\mathbf{r}_1 - \mathbf{r}_2}{|\mathbf{r}_1 - \mathbf{r}_2|^3}.$$

By the same token,

$$\mathbf{f}_{21} = \gamma m_1 m_2 \frac{\mathbf{r}_2 - \mathbf{r}_1}{|\mathbf{r}_1 - \mathbf{r}_2|^3}.$$

We note that

(2) $$\mathbf{f}_{21} + \mathbf{f}_{12} = 0.$$

The point with the position vector

(3) $$\mathbf{R} = \frac{m_1 \mathbf{r}_1 + m_2 \mathbf{r}_2}{m_1 + m_2}$$

is called the **center of mass** (or **centroid**) of the two particles. We recall (see Chapter 9, §2.1) that this point lies on the segment $\overrightarrow{P_1P_2}$ and divides this segment in the ratio m_2/m_1 (see Fig. 11.38, where $m_1 = 4m_2$). We claim that the two particles move so that the mass center is not accelerated.

Indeed, by Newton's Second Law

Fig. 11.38

center of mass

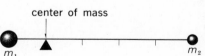

$$m_1 \ddot{\mathbf{r}}_1 = \mathbf{f}_{12}, \qquad m_2 \ddot{\mathbf{r}}_2 = \mathbf{f}_{21}.$$

Adding these two equations and noting (2), we obtain

$$m_1 \ddot{\mathbf{r}}_1 + m_2 \ddot{\mathbf{r}}_2 = \mathbf{f}_{12} + \mathbf{f}_{21} = 0,$$

which is the same as

$$\ddot{\mathbf{R}} = 0.$$

Thus the mass center moves along a straight line with constant velocity.

6.2 The two-body problem

In §4, we discussed the motion of a planet around the sun. In this discussion, we treated the sun as an unmovable body and took no account of the gravitational pull exerted by the planet on the sun. Now we consider the so-called **two-body problem:** the motion of two particles (position vectors $\mathbf{r}_1$ and $\mathbf{r}_2$, masses m_1 and m_2) that attract each other according to the law of universal gravitation.

We proved above that the mass center of the two particles has zero acceleration. We can therefore introduce a coordinate system in which this mass center is at rest. We may apply the laws of physics in this coordinate system—this is the so-called relativity principle of classical mechanics. Hence we assume from now on that the mass center is used as the origin, in other words that

$$(4) \qquad\qquad \mathbf{R} = \mathbf{0}.$$

From (2), we have

$$m_1 \mathbf{r}_1 + m_2 \mathbf{r}_2 = \mathbf{0}$$

or

$$\mathbf{r}_2 = -\frac{m_1}{m_2}\mathbf{r}_1.$$

Therefore

$$\mathbf{r}_2 - \mathbf{r}_1 = -\left(1 + \frac{m_1}{m_2}\right)\mathbf{r}_1$$

and the force $\mathbf{f}_{12}$ exerted on the first particle is [compare (1)]

$$\mathbf{f}_{12} = -\gamma m_1 \frac{m_2}{\left(1 + \dfrac{m_1}{m_2}\right)^2} \frac{\mathbf{r}_1}{|\mathbf{r}_1|^3}.$$

Set

$$(5) \qquad\qquad M_1 = \frac{m_2}{\left(1 + \dfrac{m_1}{m_2}\right)^2}.$$

Then

$$\mathbf{f}_{12} = -\gamma m_1 M_1 \frac{\mathbf{r}_1}{|\mathbf{r}_1|^3}.$$

This means: the first particle moves as if it were attracted by an unmovable body of mass M_1 located at the center of mass of the two particles. Therefore it moves according to Kepler's laws about this mass center; its *trajectory is a conic with the mass center at a focus,* and the segment from the mass center to the particle describes equal areas in equal time intervals.

Assume that m_1 is negligibly small with respect to m_2, as is the case for a planet circling our sun, as long as the planet is not the enormous Jupiter. Then the center of mass of the planet-sun system is nearly the center of the sun, and the apparent mass (5) is nearly the mass of the sun m_2. We commit no serious error if we treat the sun as unmovable. In the case of the sun-Jupiter system, however, and even more important in the case of a double star with two "components" of comparable mass, it is essential to describe the motion as Keplerian motion of each component about the center of mass.

We have just shown that the two-body problem is easily reduced to the one-body problem of motion of a particle about a fixed center of attraction. The situation is much more complicated in the case of three or more mutually attracting bodies (the three or n-body problem). A useful solution of the n-body problem can be obtained only by extensive numerical calculations.

EXERCISES

1. Suppose two particles of equal mass, which attract each other according to Newton's Law of Gravitation, move along a circular orbit with the same constant speed. Assume that there are no other forces acting on the two particles. What can you say about the mutual position of the two particles?
▶ 2. A double star consists of two components. Each moves along an elliptic orbit with the center of mass in a focus. Astronomical observations disclose the periods of both motions and the lengths of the axes of the orbits. Is it possible to calculate the ratio of the two masses? If so, how?
3. We defined the center of mass of two particles making use of position vectors. It is conceivable that the center of mass depends on the origin chosen in order to represent points by vectors. Show that this is not so.

6.3 Newton's Third Law. Linear momentum

Newton's Law of Motion ($\mathbf{f} = m\mathbf{a}$) is not sufficient for describing the motion of a system consisting of more than one particle. It must be supplemented by **Newton's Third Law.** This is a symmetry condition; it states that

FORCES EXERTED BY TWO PARTICLES ON EACH OTHER ARE NEGATIVES OF EACH OTHER.

More precisely, let us consider a system consisting of N particles with masses $m_1, m_2, \cdots, m_N$ and position vector $\mathbf{r}_1, \mathbf{r}_2, \cdots, \mathbf{r}_N$. We assume that each particle m_j is subject to $N - 1$ forces $\mathbf{f}_{j1}, \mathbf{f}_{j2}, \cdots, \mathbf{f}_{jN}$ exerted on it by the masses $m_1, m_2, \cdots, m_N$ and to an "external" force $\mathbf{f}_j$. (There is no force $\mathbf{f}_{jj}$.) The "internal" forces obey the symmetry condition

$$(6) \qquad\qquad \mathbf{f}_{jk} + \mathbf{f}_{kj} = 0.$$

(A critically minded reader might object to the somewhat anthropomorphic form of the Third Law, which mentions particles "exerting" forces. The real content of the law, however, is the possibility of dividing the forces acting on a set of particles into external and internal ones, the internal being subject to the symmetry requirement.)

Newton's Second Law, applied to each of the N particles, yields N vector equations

$$m_1\ddot{\mathbf{r}}_1 = \mathbf{f}_{12} + \mathbf{f}_{13} + \cdots + \mathbf{f}_{1N} + \mathbf{f}_1,$$
$$m_2\ddot{\mathbf{r}}_2 = \mathbf{f}_{21} + \mathbf{f}_{23} + \cdots + \mathbf{f}_{2N} + \mathbf{f}_2,$$
$$\cdots\cdots\cdots\cdots\cdots\cdots\cdots\cdots\cdots\cdots\cdots$$
$$m_N\ddot{\mathbf{r}}_N = \mathbf{f}_{N1} + \mathbf{f}_{N2} + \cdots + \mathbf{f}_{N,N-1} + \mathbf{f}_N.$$

We add all of these equations and note that, because of the Third Law (6), all internal forces cancel each other. Thus we obtain

(7) $$m_1\ddot{\mathbf{r}}_1 + m_2\ddot{\mathbf{r}}_2 + \cdots + m_N\ddot{\mathbf{r}}_N = \mathbf{f}_1 + \mathbf{f}_2 + \cdots + \mathbf{f}_N.$$

The vector $\mathbf{v}_j = \dot{\mathbf{r}}_j$ is the velocity of the jth particle. The quantity $m_j\mathbf{v}_j$ is called the **linear momentum of the jth particle**, and the sum

$$\mathbf{Q} = m_1\mathbf{v}_1 + m_2\mathbf{v}_2 + \cdots + m_N\mathbf{v}_N$$

is called the (total) linear momentum of the system.

Now let

$$\mathbf{F} = \mathbf{f}_1 + \mathbf{f}_2 + \cdots + \mathbf{f}_N$$

denote the sum of all external forces. We can write (8) in the form

(8) $$\dot{\mathbf{Q}} = \mathbf{F};$$

the time derivative of the linear momentum of the system is the sum of all external forces. If there are no external forces, then $\dot{\mathbf{Q}} = \mathbf{0}$ and

$$\mathbf{Q} = m_1\mathbf{v}_1 + m_2\mathbf{v}_2 + \cdots + m_N\mathbf{v}_N = \text{const.}$$

This statement is called the **principle of conservation of linear momentum.** It implies that if, in the absence of external forces, the velocities of some particles are increased in one direction, then the velocities of other particles must be increased in the opposite direction. This explains such phenomena as recoil of guns and jet propulsion.

6.4 Center of mass

The center of mass of our N particles is, by definition, the point with position vector

(9) $$\mathbf{R} = \frac{m_1\mathbf{r}_1 + m_2\mathbf{r}_2 + \cdots + m_N\mathbf{r}_N}{m_1 + m_2 + \cdots + m_N}.$$

The definition is justified by the following theorem:

THE MASS CENTER OF A SYSTEM OF PARTICLES MOVES AS IF IT WERE A PARTICLE OF MASS $m_1 + m_2 + \cdots + m_N = M$, SUBJECT TO THE FORCE $\mathbf{F} = \mathbf{f}_1 + \cdots + \mathbf{f}_N$ (TOTAL EXTERNAL FORCE).

To prove this, we need only to note that the time derivative of the position vector of the mass center is

$$\dot{\mathbf{R}} = \frac{m_1\dot{\mathbf{r}}_1 + \cdots + m_N\dot{\mathbf{r}}_N}{m_1 + \cdots + m_N} = \frac{\mathbf{Q}}{M}.$$

Therefore $M\ddot{\mathbf{R}} = \dot{\mathbf{Q}}$ and, by (8),

$$(10) \qquad\qquad M\ddot{\mathbf{R}} = \mathbf{F}.$$

This is precisely the differential equation for the motion of a particle of mass M with position vector $\mathbf{R}$ subject to a force $\mathbf{F}$.

EXERCISES

4. Show that the location of the center of mass of several particles does not depend on the location of the origin (the point with position vector $\mathbf{0}$).

5. Under normal circumstances, we are aware of our weight. Does Newton's Third Law have something to do with it?

► 6. Using the answer to the preceding question, explain why a man in free fall (for instance, an astronaut in his capsule) is "weightless."

7. Explain why conservation of linear momentum "causes" the recoil of guns. Explain how conservation of linear momentum is used in jet propulsion.

6.5 Rigid bodies

The theorem just proved applies, in particular, to rigid bodies.

A **rigid body** is a system of particles in which the internal forces are such that the distances between the particles are constant. Such concepts as "particle" or "rigid body" are, of course, idealization. No actual body is completely rigid, but many bodies can be considered as rigid under a wide variety of circumstances. The preceding considerations show that the mass center of a rigid body moves as if it were a particle in which the whole mass of the body is concentrated, and as if the sum of all external forces applied to the body were acting on it.

In order to describe the motion of a rigid body or of any other system completely, we must also compute the motion of the particles with respect to the mass center. This is a very interesting problem, but it would lead us too far to consider it here.

6.6 Additivity property

We list now three properties of the center of mass.

1. Suppose we divide all particles of the system into two classes. Suppose that the mass center of the first class is the point P_1, that of the second class the point P_2. Suppose also that the total mass of the first class is M_1, that of the second class M_2. The center of mass of the whole system is also the center of mass of the two particles, one of mass M_1 at P_1, and the other of mass M_2 at P_2.

This property is called the **additivity** of the mass center. It is easy to extend it to a division of all particles into several classes.

Proof of Property 1.   We may assume that the first class consists of the particles $m_1, m_2, \cdots, m_k$, and the second class of the particles $m_{k+1}, m_{k+2}, \cdots, m_N$. Then $M_1 = m_1 + m_2 + \cdots + m_k$ and $M_2 = m_{k+1} + m_{k+2} + \cdots + m_N$. Let $\mathbf{R}_1$ and $\mathbf{R}_2$ be the position vectors of P_1 and P_2, respectively. By the definition of the mass center, we have

$$\mathbf{R}_1 = \frac{m_1\mathbf{r}_1 + \cdots + m_k\mathbf{r}_k}{M_1}, \qquad \mathbf{R}_2 = \frac{m_{k+1}\mathbf{r}_{k+1} + \cdots + m_N\mathbf{r}_N}{M_2}.$$

Now the mass center of the system consisting of a particle with position vector $\mathbf{R}_1$ and mass M_1 and of a particle with position vector $\mathbf{R}_2$ and mass M_2 is

$$\frac{M_1\mathbf{R}_1 + M_2\mathbf{R}_2}{M_1 + M_2} = \frac{m_1\mathbf{r}_1 + \cdots + m_N\mathbf{r}_N}{m_1 + \cdots + m_N}$$

and this is the mass center of $m_1, m_2, \cdots, m_N$.

6.7 Location of the center of mass

2. If all particles of a system lie in one plane, the mass center lies in this plane. If all particles of the system lie on one line, the mass center lies on this line.

In order to prove this, we first write down the coordinates of the center of mass of a system of masses $m_1, m_2, \cdots, m_N$ located at points with coordinates (x_1, y_1, z_1), $(x_2, y_2, z_2), \cdots, (x_N, y_N, z_N)$. We may think of (x_j, y_j, z_j) as the components of the position vector $\mathbf{r}_j$ in an appropriate frame. Let (X, Y, Z) be the components of the position vector $\mathbf{R}$ of the mass center, that is, the set of Cartesian coordinates of the mass center. From (9), we obtain the formulas

$$(11) \qquad \begin{aligned} X &= \frac{m_1x_1 + m_2x_2 + \cdots + m_Nx_N}{m_1 + m_2 + \cdots + m_N}, \\[1em] Y &= \frac{m_1y_1 + m_2y_2 + \cdots + m_Ny_N}{m_1 + m_2 + \cdots + m_N}, \\[1em] Z &= \frac{m_1z_1 + m_2z_2 + \cdots + m_Nz_N}{m_1 + m_2 + \cdots + m_N}. \end{aligned}$$

Proof of Property 2. If all particles lie in one plane, use a coordinate system in which this is the plane $x = 0$. Then $x_j = 0$ for all j, hence $X = 0$, by (11). The center of mass lies on the plane $x = 0$.

If all particles lie on one line, choose two planes intersecting in this line. By what was proved above, the mass center lies on both planes, and hence on the line.

6.8 Symmetry properties

3. If the masses are situated symmetrically with respect to a plane, the mass center lies on this plane. If all masses lie in one plane and are situated symmetrically with respect to a line on this plane, the mass center lies in this line. If all masses lie in one line and are situated symmetrically with respect to a point on this line, this point is the mass center.

Proof of Property 3. To say that the masses are situated symmetrically with respect to a plane σ means that, if the mass m_i is at the point P_i, not on σ, then there is a mass m_j at a point P_j such that $m_j = m_i$ and the segment $\overrightarrow{P_jP_i}$ is perpendicular to σ and meets this plane at its midpoint.

Assume now that the symmetry plane σ is the plane $x = 0$; this can be achieved by choosing an appropriate coordinate system. If the mass m_i has coordinates (x_i, y_i, z_i) with $x_i \neq 0$, then there is a mass m_j with $m_j = m_i$, $x_j = -x_i$, $y_j = y_i$, and $z_j = z_i$. We conclude that, in the expression for X in (11), all terms in the numerator cancel each other. Thus $X = 0$ and the mass center lies on the plane $x = 0$, that is, on the plane of symmetry.

Suppose that all masses lie on the plane τ and are symmetrical about a line l in τ. Then the masses are symmetrical about the plane σ which passes through l and is perpendicular to τ. The mass center lies on σ by what was proved above, and it lies on τ by Property 2. Hence it lies on l.

Suppose all masses lie on the line l and are symmetrical about a point P on the line. This means that the masses are situated symmetrically with respect to the plane σ, which passes through P and is perpendicular to l. The mass center lies on σ by what was proved above, and it lies on l by Property 2. Hence the mass center is at P.

EXERCISES

8. Prove the following generalization of Property 1. If a system of N particles is divided into K sets, and if the centers of mass of those K sets are the points $P_1, P_2, \cdots, P_K$, then the center of mass of the original system is also the center of mass of a system consisting of K particles, located at $P_1, P_2, \cdots, P_K$ with masses $M_1, M_2, \cdots, M_K$, where M_j is the total mass of the jth system.
9. Show that two positive masses can be put at the endpoints of a given segment, so that their center of mass should lie at a prescribed inner point of the segment. Are the two masses determined uniquely? Is their ratio?
10. Let P_1, P_2, P_3 be three vertices of a triangle, and let Q be a given inner point of this triangle. Show that there are three positive numbers, m_1, m_2, m_3 such that $m_1 + m_2 + m_3 = 1$ and the center of mass of masses m_1, m_2, m_3 located at P_1, P_2, P_3, respectively, is Q. Are the three numbers determined uniquely?
11. Show that every inner point of a square is the center of mass of four particles located at the four vertices of the square. Show that the masses of these particles can be chosen in infinitely many different ways, even if one requires that the sum of the masses be 1.

Problems

1. Using the definition of the derivative of a vector-valued function contained in equation (16) of §1.6, find the derivatives of the following vector-valued functions, without using components: $t \mapsto t\mathbf{a} + \mathbf{b}$, $t \mapsto t^2\mathbf{a} + t^3\mathbf{b} + e^t\mathbf{c}$, $t \mapsto |\mathbf{a}| (\sin t)\mathbf{a}$. Here $\mathbf{a}$, $\mathbf{b}$, and $\mathbf{c}$ are fixed vectors.
2. Prove that, if $t \mapsto \mathbf{f}(t)$ is a continuous vector-valued function defined in $[a, b]$, then

$$\left| \int_a^b \mathbf{f}(t) \, dt \right| \leq \int_a^b |\mathbf{f}(t)| \, dt.$$

3. Let $t \mapsto \mathbf{r}(t)$ be a continuously differentiable vector-valued function defined in $[a, b]$. It defines a curve C. Show analytically that the length of the curve is not greater than the distance between its initial point and its endpoint. In other words, prove that a straight segment is the shortest path between its endpoints.

4. Define oriented curves whose images look like: the numeral 8, the letter H, the Greek letter θ, a square together with its two diagonals. Which of these curves may be required to be smooth?

5. Let $x \mapsto f(x)$ be a continuous number-valued function defined for $0 \leq x \leq 1$. Assume that there is a number a, $0 < a < 1$, such that $f(x)$ is increasing on $(0, a)$ and decreasing on $(a, 1)$ and that $f(0) = f(1) = 0$. Compute the area of the region enclosed by the graph of f and by the segment $\overrightarrow{01}$, using the formula (12) in §2.6, and show that this area equals $\int_a^b f(x)\, dx$.

6. Show that the result of the preceding problem remains valid if $f(0)$ and $f(1)$ are positive numbers.

7. Show that a smooth oriented curve with a constant unit tangent vector $\mathbf{T}$ is a straight segment.

8. Find all curves (in the plane) for which the angle between the unit tangent vector at a point P on the curve and the vector represented by $\overrightarrow{OP}$, O a fixed point, is constant. (Hint: Use polar coordinates.)

9. Prove that a curve whose curvature is identically 0 is a straight segment.

10. Prove that a plane curve whose curvature is constant is a circular arc.

11. Let $t \mapsto \mathbf{f}(t)$, $t \mapsto \mathbf{g}(t)$ be two differentiable vector-valued functions defined in the same interval. Prove that $(\mathbf{f}, \mathbf{g})^{\cdot} = (\mathbf{f}^{\cdot}, \mathbf{g}) + (\mathbf{f}, \mathbf{g}^{\cdot})$. (Here the dot denotes the differentiation with respect to t_1 and $(,)$ is the scalar product defined in Chapter 9, §6.)

12. Suppose a particle moves in the plane in such a way that its speed has a constant value. Prove that, at each time instant, the acceleration is perpendicular to the velocity.

13. The force between two electrically charged particles is proportional to their charges, and inversely proportional to the square of the distance between them; it is an attraction or a repulsion according to whether the charges are unlike or like (Coulomb's Law). Consider a fixed positively charged particle, located at 0, and another positively charged particle moving along a hyperbola, with 0 as focus, and obeying Kepler's Second Law of Constant "Areal Velocity." (Such motions occurred in Rutherford's experiments, which initiated atomic physics.) Show that this observation is consistent with Coulomb's Law. (Hint: Remember that a hyperbola has two foci.)

14. The concept of center of mass of several particles can be extended to negative masses. One simply permits the numbers m_j in the formulas (9) and (11) in §6 to be 0 or negative. Show that, if P_1, P_2, and P_3 are three points in the plane, not on one line, then every point Q in the plane is the center of mass of (not necessarily positive) masses m_1, m_2, m_3 located at P_1, P_2, P_3, respectively. Show that these masses are uniquely determined by the point Q if we require that $m_1 + m_2 + m_3 = 1$. In this case, the numbers m_1, m_2, m_3 are called the *baricentric coordinates* of Q, with respect to P_1, P_2, P_3.

15. Let m_1, m_2, m_3 be the baricentric coordinates of a point Q with respect to

CHARLES AUGUSTIN COULOMB (1736–1806), a French military engineer, civil servant, and physicist, made some of the fundamental discoveries about electrical phenomena.

ERNEST RUTHERFORD (1871–1937), a New Zealander who spent most of his adult life in England, discovered the nucleus of the atom. On the basis of his experiments he proposed the first model for the internal structure of atoms.

P_1, P_2, P_3. Translate the following statements into statements about m_1, m_2, m_3: $Q = P_1$; Q lies on the line through P_1 and P_2; Q lies inside or on the boundary of the triangle with vertices P_1, P_2, P_3; Q is an inner point of that triangle.

16. Let P_1, P_2, P_3, and P_4 be four points in space not in one plane. Define baricentric coordinates of a point in space with respect to the points P_j. Answer questions similar to the ones asked in Problem 15 for the case of the plane.

17. Let A be a plane curve defined by $x = x(t)$, $y = y(t)$ where $x(t)$, $y(t)$ have continuous derivatives of the second order. Let $\kappa(t)$ denote the curvature at the point $(x(t), y(t))$, and let $\xi(t)$, $\eta(t)$ be the coordinates of the corresponding center of curvature. Assuming that $\kappa \neq 0$, prove that

$$\xi = x - \frac{(x'^2 + y'^2)y'}{x'y'' - x''y'}, \qquad \eta = y + \frac{(x'^2 + y'^2)x'}{x'y'' - x''y'}.$$

(We are suppressing t for the sake of brevity.) The curve defined by $t \mapsto \xi(t)\mathbf{e}_1 + \eta(t)\mathbf{e}_2$ is called the **evolute** of A. It may have singularities and it may degenerate into a point (if A is a circle).

18. How do the formulas in Problem 17 simplify if (a) t is the arc length parameter on A or if (b) the curve A is defined as the graph of a function $x \mapsto y = f(x)$?

19. Find the nonparametric representation for the evolute of the parabola $x \mapsto y = x^2$. (The parabola and its involute are shown in the figure at the right.)

20. Show that the evolute of the cycloid

$$x = t - \sin t, \qquad y = 1 - \cos t$$

is again a cycloid congruent to the original one, but displaced 2 units down and π units to the right. The two curves are shown in the figure below.

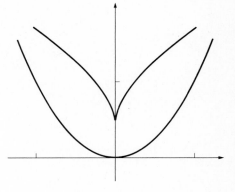

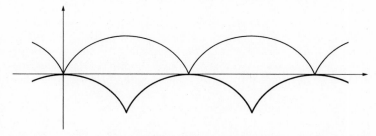

21. Let A be a smooth curve with a nonvanishing curvature. At every point Q on A, draw a line tangent to A and mark on it the point S such that Q is between P and S and the curve consisting of the arc of A between P and Q and the straight segment from Q to S has length α. The curve B consisting of all points S is called the **involute** of A. Note that, if A is thought of as made of a rigid material, the involute B can be constructed "mechanically" by unwinding a string stretched along A (see the figure at the right). The string is thought of as being flexible but of an unchanging length.

Let A have the parametric representation $u = u(t)$, $v = v(t)$ where (u, v)

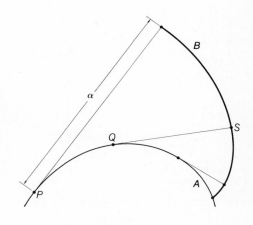

are Cartesian coordinates and t is the arc length on A measured from P. Show that the involute B has the parametric representation

$$x = u(t) + (\alpha - t)u'(t), \qquad y = v(t) + (\alpha - t)v'(t).$$

22. Show that the evolute of the involute of A is A. (Assume that the parametric representation of A in terms of arc length has three continuous derivatives.)
23. Compute and sketch the involute of the circle $u = \cos t$, $v = \sin t$.
24. Show that the involute of the cycloid $\mathbf{f}(t) = (t - \sin t)\mathbf{e}_1 + (1 - \cos t)\mathbf{e}_2$ is again a cycloid. (See Problem 20.) This result explains Huygens' construction of the cycloidal pendulum.

12/Partial Derivatives

Throughout this book thus far, we have considered functions of one variable, that is, rules assigning a number, or a vector, to a number. This chapter and the following one contain an introduction to calculus of several variables. Only the simpler parts of the theory will be considered.

In this chapter, we extend to functions of several variables the concepts of continuity and differentiability, and in §4 we describe one way of integrating such functions, the so-called line integrals.

Applications of line integrals to mechanics are treated in an appendix (§5). Two other appendixes contain proofs of some theorems and Taylor's formula for functions of several variables.

There are significant differences between functions of one and functions of several variables. But in most cases, the transition from two to more than two variables is easy; for this reason, we shall often consider in detail only the case of two variables.

1.1 Functions of two variables

A **function of two variables** is a rule that assigns to every ordered pair of numbers, from some set of such pairs, another number. The set of pairs of numbers to which other numbers are assigned is called the **domain of definition** of the function. Often the domain of definition will be obvious from the context. In working with functions of two variables, and later with functions of more than two variables, we shall use variables, as well as the arrow $\mapsto$, just as we did for functions of one variable.

Also, we shall use the **convention:** If a function is defined by a formula, then, in the absence of specific instructions to the contrary, the domain of definition is assumed to be the largest set on which the formula makes sense.

In dealing with functions of two variables, $(x, y) \mapsto z$, it is usually illuminating to consider the two numbers x and y as the coordinates of a point in the plane, with respect to some Cartesian coordinate system. (We did the same, as the reader will realize, in working with functions of one variable. There we identified a number x with the corresponding point on the number line.) We may think of a function of two variables as a rule by which we assign a number to every point in the plane or to every point in some set of points.

The definitions of sums, differences, products, and quotients of functions, as well as of composite functions, are so similar to the corresponding definitions for functions of one variable, that there is no need to state them explicitly.

▶*Examples* *1.* The rule

$$(x, y) \mapsto x + y$$

assigns to each pair of numbers their sum. If we denote this rule by f, we can write

$$f(x, y) = x + y;$$

for instance, $f(2, 1) = 3$, $f(1, 2) = 3$, $f(x, 0) = x$ for all numbers x.

If we use the variable z to denote the number that is assigned to the pair (x, y), then the function considered can be written as

$$z = x + y.$$

This function is defined for all pairs (x, y) or, as one says, everywhere.

2. The function

$$f(x, y) = \frac{x}{y}$$

is defined for all x and y such that $y \neq 0$. We have $f(4, 2) = 2$, $f(2, 4) = .5$, $f(0, y) = 0$ for all $y \neq 0$, and so forth.

3. The function

$$f(x, y) = \sqrt{x - y}$$

is defined for all (x, y) such that $x \geq y$.

4. Set $f(x, y) = x^2 + y^2$, $g(x, y) = x^2 - y^2$, $\phi(x) = \cos x$, $\psi(x) = \sin x$. Then

$$f(x, y) + g(x, y) = 2x^2, \qquad f(g(x, y), y^2) = x^4 - 2x^2y^2 + 2y^4$$

$$f(\phi(x), \psi(x)) = 1, \quad g(\phi(x), \psi(x)) = \cos 2x$$

$$\phi(g(x, y)) = \cos (x^2 - y^2), \quad g(f(x, y), g(x, y)) = 4x^2y^2,$$

as the reader should check. All these functions are defined everywhere.

5. The function

$$(x, y) \mapsto z = \begin{cases} x^2 + y & \text{if } y \leq 0 \\ \dfrac{x^3 + xy}{y} & \text{if } x > 0, y > 0 \\ 315 & \text{if } x \leq 0, y > 0 \end{cases}$$

is defined everywhere. ◄

EXERCISES

In Exercises 1 to 4, it is assumed that $\phi(x) = \log x$, $f(x, y) = x^2 + y^2 + 1$, $\psi(x) = e^x$, $\lambda(x) = 1/x$. Compute the following functions.

1. $\lambda(\phi[f(3, 1)])$.
2. $\phi(\psi[f(0, 0)])$.
3. $f(\psi(2), \phi(3))$.
4. $\sqrt{[f(x, x)]^2 - 2}$.
5. Let $h(s, t) = |1 - s| \cdot |1 - t| - st - |s| - |t| - 1$. For which s, t is $h(s, t) = 0$?
6. Let $k(a, b) = |a| + |b|$. Sketch the graph in the θ, t-plane of $\theta(t) = k(\sin t, \cos t)$.

7. If $a(\alpha, \beta) = $ arc tan (α/β), $b(\alpha, \beta) = \alpha^2 + \beta^2$, show that $\sqrt{b}\ \sin a = \alpha$ and $\sqrt{b}\ \cos a = \beta$.

▶ 8. If $r(m, n) = e^m \cos n$, $s(m, n) = e^m \sin n$, show that $[r(m, n)]^2 - [s(m, n)]^2 = r(2m, 2n)$ and $2r(m, n)s(m, n) = s(2m, 2n)$.

9. Let $f(x) = x$ for $x \geq 0$, $f(x) = 0$ for $x < 0$; let $g(x, y) = x^2 e^y$. Simplify $h(x, y) = g(f[g(x, y)]g(y, x))$.

Fig. 12.1

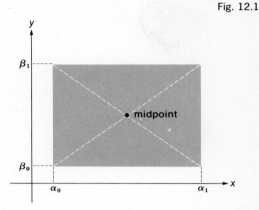

1.2 Intervals

The set of points inside a rectangle, with sides parallel to the coordinate axes is called a finite, open, two-dimensional interval, or simply an **open interval.** Such a set is shown in Fig. 12.1. It can be described as the set of all points (x, y) satisfying the inequalities

$$(1) \qquad \alpha_0 < x < \alpha_1, \qquad \beta_0 < y < \beta_1,$$

where $\alpha_0, \alpha_1, \beta_0, \beta_1$ are numbers such that $\alpha_0 < \alpha_1, \beta_0 < \beta_1$. The point with the coordinates $x = (\alpha_1 + \alpha_0)/2$, $y = (\beta_1 + \beta_0)/2$ is called the **midpoint** of the interval. This is the intersection point of the diagonals of the rectangle. The **boundary** of the interval consists of all points on the rectangle proper.

The set of all points in the open interval and on the boundary is called a **closed interval** (see Fig. 12.2). The closed interval corresponding to (1) is, therefore, the set of all points (x, y) satisfying the inequalities

$$(2) \qquad \alpha_0 \leq x \leq \alpha_1, \qquad \beta_0 \leq x \leq \beta_1.$$

We agree to use the phrase "**near a point** P" (in the plane) to mean "in some interval with P as midpoint." Since every interval with midpoint P contains a circular disk with center P, and conversely (see Fig. 12.3), we conclude that "near P" also means "in some circular disk with center P," or "at all points Q whose distance from P is less than some fixed positive number."

Fig. 12.2

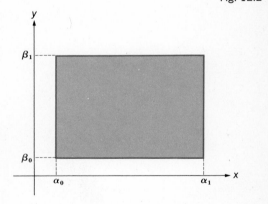

Fig. 12.3

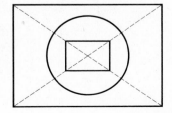

1.3 Continuity

The concept of **continuity** for functions of two variables is an obvious extension of the concept of continuity of functions of one variable (see Chapter 3, §3.2). Roughly speaking, a function $f(x, y)$ is continuous if a small change in x and y results in a small change in $f(x, y)$. A precise definition follows.

A function $f(x, y)$ defined for $x = x_0$, $y = y_0$ is called continuous at the point (x_0, y_0) if it has the two properties:

(I) *if A is any number such that $f(x_0, y_0) < A$, then $f(x, y) < A$ for all (x, y) near (x_0, y_0) for which $f(x, y)$ is defined;*

(II) *if B is any number such that $B < f(x_0, y_0)$, then $B < f(x, y)$ for all (x, y) near (x_0, y_0) for which $f(x, y)$ is defined.*

As in the case of functions of one variable (see Chapter 3, §3.2), this definition is equivalent to the **$\epsilon - \delta$ definition:** *$f(x, y)$ is called continuous at (x_0, y_0) if, given any positive number ϵ, we can find a positive δ such that*

$$|f(x, y) - f(x_0, y_0)| < \epsilon$$

for all (x, y) for which $f(x, y)$ is defined and

$$\sqrt{(x - x_0)^2 + (y - y_0)^2} < \delta.$$

Finally, a function is called **continuous on a set S** if it is defined and continuous at every point of S.

Sums, products, and differences of continuous functions are continuous; so is the quotient of two continuous functions, except where the denominator is 0. Composing continuous functions, we again obtain continuous functions. All this is proved exactly as for functions of one variable (see Chapter 3, §§3 and 6). Also, if $f(x)$ is a continuous function of one variable, the functions of two variables $F(x, y) = f(x)$ and $G(x, y) = f(y)$ are continuous; the proof of this statement is clear. We conclude that every monomial $(x, y) \mapsto cx^n y^m$, c a number, n and m nonnegative integers, is a continuous function. Hence every polynomial in two variables, that is, a finite sum of monomials, is continuous everywhere. A rational function of two variables, that is, a quotient of two polynomials, is continuous where the denominator is not 0. A radical function of two variables (the definition of such functions parallels the one given in Chapter 3, §2.7) is continuous wherever defined.

▶ **Examples** **1.** Where is the function $(x, y) \mapsto e^{\sin(x^2 + 2y)} + \cos(xy)$ continuous?

Answer. It is continuous everywhere, for it is the sum of two functions, each of which is a continuous function of a polynomial.

2. Where is $f(x, y) = \tan(x/y)$ defined? Where is it continuous?

Answer. The quotient x/y is defined for $y \neq 0$. Assuming $y \neq 0$, we see that f is defined if x/y is not an odd multiple of $\pi/2$, that is, if

$$\frac{x}{y} \neq \frac{2n+1}{2}\pi, \qquad n = 0, \pm 1, \pm 2, \cdots.$$

In other words, $f(x, y)$ is defined whenever (x, y) lies neither on the line $y = 0$ nor on one of the lines

$$x = \frac{\pi}{2}y, \ x = \frac{3\pi}{2}y, \ x = -\frac{\pi}{2}y, \ x = -\frac{3\pi}{2}y, \ x = \frac{5\pi}{2}y, \cdots.$$

Also f is continuous wherever it is defined, being a composite of continuous functions.

3. The function

$$g(x, y) = \begin{cases} \dfrac{1}{x^2 + y^2} & \text{if } (x, y) \neq (0, 0) \\ \dfrac{1}{134} & \text{if } x = y = 0 \end{cases}$$

is defined everywhere; it is continuous everywhere except at $x = y = 0$.

4. Consider the function

$$\phi(x, y) = \begin{cases} \dfrac{xy}{(x^2 + y^2)^2} & \text{if } x^2 + y^2 \neq 0 \\ 0 & \text{if } x = y = 0 \end{cases}$$

which is defined everywhere. If we choose a fixed value of x and consider $\phi(x, y)$ as a function of y alone, we always obtain an everywhere continuous function of one variable. Indeed, for a fixed $x = a \neq 0$, we have $\phi(x, y) = ay/(a^2 + y^2)^2$ for all y, and for $x = 0$ we have $\phi(x, y) = 0$ for all y. Similarly, for every fixed value of y, the function $x \mapsto \phi(x, y)$ is an everywhere continuous function of x. But ϕ considered as a function of (x, y) is not continuous at the origin. Indeed, $\phi(0, 0) = 0 < 1$, and in every interval with midpoint $(0, 0)$ there are points (x, y) with $\phi(x, y) > 1$, for instance, the points $x = y = t$ with $0 < t < \frac{1}{2}$.

This example shows that continuity in each variable x and y separately is a weaker condition than continuity in (x, y). ◄

EXERCISES

In Exercises 10 to 23, state for each function where it is defined and where it is continuous.

10. $x^{-1} + y^{-1}$.

11. $(1 + y^2)/\sin x$.

12. $|\sqrt{x} - \sqrt{y}|$.

13. $\sqrt{|x|} - \sqrt{|y|}$.

14. $\sin(x/y)$.   17. $|x|^{1/|y|}$.

15. $\sin(x \tan y)$.   18. $(x^2 + y^2)/(x^2 - y^2)$.

16. $|x|^y$.   19. $\arctan^{-1}(y/x)$.

20. $f(x, y) = xy$ for $x > 0$ and $y > 0$, $= 0$ for $x \leq 0$ or $y \leq 0$.

21. $f(x, y) = xy$ for x, y of the same sign, $= 0$ for all other x and y.

22. $f(x,y) = xy$ for $x > 0$, $= 0$ for $x \leq 0$.

23. $f(x, y) = xy$ for $x + y > 0$, $= 0$ for $x + y \leq 0$.

1.4 Functions of three or more variables

It is hardly necessary to point out that a function of three variables is a rule $(x, y, z) \mapsto u = f(x, y, z)$ which assigns to ordered triples of numbers, from some set of such triples, a number. Similarly, for any positive integer n, a **function of n variables** is a rule

$$(1) \qquad (x_1, x_2, \cdots, x_n) \mapsto u = f(x_1, x_2, \cdots, x_n)$$

assigning numbers u to certain ordered n-tuples of numbers. It is useful to think of the ordered n-tuples of numbers as the coordinates of a point or as the components of the position vector $\mathbf{x}$ of this point (in ordinary, that is, three-dimensional, space for $n = 3$, in n-dimensional space for any $n > 0$). If we do so, we may write instead of (1)

$$\mathbf{x} \mapsto u = f(\mathbf{x}).$$

An **open n-dimensional interval** is the set of all points $(x_1, x_2, \cdots, x_n)$ that satisfy the inequalities

$$\alpha_j < x_j < \beta_j, \qquad j = 1, 2, \cdots, n$$

where α_j and β_j are numbers, $\alpha_j < \beta_j$. The corresponding **closed interval** is the set of all points $(x_1, x_2, \cdots, x_n)$ satisfying

$$\alpha_j \leq x_j \leq \beta_j, \qquad j = 1, 2, \cdots, n.$$

The point with the coordinates

$$x_j = \frac{\alpha_j + \beta_j}{2}, \qquad j = 1, 2, \cdots, n$$

is called the **midpoint** of the interval.

We can now repeat the definition of the term "near" as given in §1.2, and the definition of continuity as given in §1.3. We do it explicitly only for the $\epsilon - \delta$ definition.

A function $\mathbf{x} \mapsto f(\mathbf{x})$ *defined at* $\mathbf{x}_0$ *is called continuous at* $\mathbf{x}_0$ *if, given any positive number* ϵ, *we can find a positive number* δ, *such that* $|f(\mathbf{x}) - f(\mathbf{x}_0)| < \epsilon$ *for all* $\mathbf{x}$ *for which* $f(\mathbf{x})$ *is defined and* $|\mathbf{x} - \mathbf{x}_0| < \delta$.

The theorems on continuity stated in §1.3 are, of course, valid for functions of three or more variables.

Remark A continuous function defined on a set S remains continuous if we consider it on a subset of S. In particular, a continuous function of several variables remains continuous if we keep one or several variables fixed.

EXERCISES

For each function in Exercises 24 to 29, state where it is defined and where it is continuous.

24. $(y/x) + (x/z) + (z/y)$.

25. $(y/xz) + (x/yz) + (z/xy)$.

26. $x \sin (y/z)$.

27. $xy \log z$.

28. $(x^2 + y^2 + z^2)/(x + y + z)$.

29. $(x^z - y^z)/(x - y)$.

1.5 Accumulation points. Limits

Let S be a set of points (on the number line, or in the plane, or in ordinary space, or in an n-dimensional space) and let P be a point. This point may or may not belong to S. We ask: is it possible to find an interval with P as midpoint which contains no points of S different from P? If this is not possible, we say that P is an **accumulation point** of S. In other words: P is an accumulation point of S if every interval with midpoint P contains a point of S distinct from P (and therefore contains infinitely many points of S, since any interval contains infinitely many smaller intervals). For instance: every point on the boundary of an open or closed interval is an accumulation point of the interval. On the other hand, the point $(1, 1)$ is not an accumulation point of the x axis.

Let S be a set, f a function defined on S, and P an accumulation point of S with position vector $\mathbf{x}_0$, at which f may or may not have been defined. We ask: is there a number α such that if we define, or redefine, the value $f(\mathbf{x}_0)$ of f at P to be α, then the function becomes continuous at P? If there is such an α, we call it the **limit** of f at P, and we write

$$\lim_{\mathbf{x} \to \mathbf{x}_0} f(\mathbf{x}) = \alpha.$$

Thus the statement
$$\lim_{\mathbf{x} \to \mathbf{x}_0} f(\mathbf{x}) = f(\mathbf{x}_0)$$

means that $\mathbf{x}_0$ is an accumulation point of the domain of definition of f

and f is defined and continuous at $\mathbf{x}_0$. That f can have at $\mathbf{x}_0$ at most one limit is seen as in the case of functions of one variable (compare Chapter 3, §4.1).

We do not talk about limits of a function f at points that are not accumulation points of the set S on which f is originally defined. For at such points, f may be defined quite arbitrarily, without affecting its continuity. For instance, the function $f(x, y)$ defined to be $x\sqrt{y}$ for $y > 0$, to equal -7 at $(-1, -1)$, and not defined elsewhere, is continuous wherever it is defined. The continuity is not impaired if we set $f(-1, -1) = 5$ or $f(-1, -1) = 100,000$.

The theorems about limits stated in Chapter 3, §4 remain valid, when properly restated, for functions of several variables. The proofs are essentially the same. One can also define infinite limits of functions of several variables. For instance, the statement

$$\lim_{(x,y)\to(a,b)} f(x, y) = +\infty$$

means that $$\lim_{(x,y)\to(a,b)} \frac{1}{f(x, y)} = 0$$

and $f(x, y) \geq 0$ near (a, b). We do not enter into details, since we shall have few occasions to use limits of functions of several variables.

▶*Examples 1.* What is the limit of $f(x, y) = x^2 - 2xy + y^4$ as $(x, y) \to (-7, 2)$?

Answer. Since $f(x, y)$ is a polynomial, it is defined and continuous everywhere, and therefore

$$\lim_{(x,y)\to(-7,2)} f(x, y) = f(-7, 2) = 93.$$

2. What is the limit of $f(x, y) = \sqrt{x^2 y}$ as $(x, y) \to (-1, -1)$?

Answer. The question is meaningless. The function $f(x, y)$ is defined if $x^2 y \geq 0$, that is, if $y \geq 0$. The domain of definition is the right half-plane; the point $(-1, -1)$ is evidently not an accumulation point of this set.

3. What is

$$\lim_{(x,y)\to(0,0)} \frac{27}{x^4 + 2x^2 + y^6}?$$

Answer. $+\infty$, since the function considered is > 0 near $(0, 0)$, and its reciprocal is a polynomial that takes on the value 0 for $x = y = 0$.

4. Does $f(x, y) = \sin \log (x^4 + y^2)$ have a limit as $(x, y) \to (0, 0)$?

Answer. No. For $\log (x^4 + y^2)$ takes on negative values of arbitrarily large absolute value, as $x^4 + y^2$ becomes close to 0, so that $f(x, y)$ takes on all values between -1 and $+1$ in every interval with midpoint at the origin. ◀

1.6 Open sets. Boundaries

In dealing with functions of one variable, we have always assumed that the functions considered were defined in one or in several intervals. Such a restriction would be unreasonable for functions of several variables. For instance, the function $f(x, y) = \sqrt{x^2 + y^2 - 1}$ is defined for $x^2 + y^2 > 1$; its domain of definition is not an interval.

A set S of points (on the line, or in the plane, or in ordinary space, or in n-space) is called **open** if, whenever it contains a point P, it also contains all points (on the line, or in the plane, and so forth) near P, that is, all points of some interval with midpoint P.

For instance, an open 2-dimensional interval is open in the plane (see Fig. 12.4). But the closed interval is not open; the condition is violated, for instance, if P is a corner (see Fig. 12.5). The whole plane is open. The empty set, which contains no points at all, is open by definition.

A **boundary point** of a set S is a point Q such that every interval with midpoint Q contains points in S and points not in S. The set of all boundary points is called the **boundary** of the set S.

For instance, if S is the set of all points (x, y) with $x^2 + y^2 < 1$ (this set is called the open unit disk), then the boundary of S consists of all points with $x^2 + y^2 = 1$ (this is the unit circle). If S is an open or closed 2-dimensional interval determined by a rectangle R, the boundary of S is the rectangle R.

We shall usually deal with functions defined on an open set, or on an open set and on all or part of its boundary. This corresponds to what we did for functions of one variable, since it can be shown that any open subset of the line is a collection of nonoverlapping intervals.

Remark Whether a set S is open, or what its boundary points are, depends on whether we consider S as a subset of a line, or of the plane, and so forth. For instance, the set of all numbers x such that $1 < x < 2$ may be considered a set on the line. It is an open set and its boundary consists of two points: $x = 1$ and $x = 2$ (see Fig. 12.6). But the set of all points $(x, 0)$ with $1 < x < 2$ is not an open set of the plane (see Fig. 12.7), and every point $x, 0$ with $1 \leq x \leq 2$ is a boundary point of this set.

Fig. 12.4

Fig. 12.5

Fig. 12.6

Fig. 12.7

EXERCISES

In Exercises 30 to 33, state whether the set of points in the plane is open. A fixed coordinate system is assumed.

► 30. The set of all points the sum of whose coordinates is negative.

31. The set of all points *at least one* of whose coordinates is negative.

32. The set of all points *both* of whose coordinates are positive.

33. The set of all points *neither* of whose coordinates is negative.

34. The set of points whose x coordinate satisfies $\sin x = 0$.

35. If A and B are open sets, let C be the set of points that belong to either A or B, and let D be the set of points belonging to both A and B. Must C be an open set? Must D be an open set?

► 36. Let $\{A_1, A_2, \cdots\}$ be a collection of infinitely many open sets. Let E be the set of all points belonging to at least one of the A_i, and let F be the set of points belonging to all the A_i. Prove that E is open. Show by a suitable choice of $\{A_i\}$ that F need not be open.

37. Find the boundary of the set given in Exercise 30.

38. Find the boundary of the set given in Exercise 31.

39. Find the boundary of the set given in Exercise 34.

40. If A is the set of all points in the x, y plane such that both x and y are rational numbers, what is the boundary of A?

41. Is the set of points (x, y) with $x^2 + y^2 \neq 1$ open? What is the boundary of this set?

42. There are exactly two sets of points (x, y) which have the property that they possess no boundary points at all. Find them.

1.7 Geometric and physical interpretations

A function of two variables $(x, y) \mapsto f(x, y) = z$ defined on some set S can be represented geometrically by its **graph,** the set of all points (x, y, z) in space such that (x, y) belongs to S and $z = f(x, y)$. The graph of a function of two variables contains all the information about the function, just as in the case of functions of one variable. Unfortunately, it is rather difficult to draw such graphs on paper. If $f(x, y)$ is a continuous function defined on an open set, or on an open set and on all or part of its boundary, the graph of f is called a **surface** or, more precisely, a nonparametric surface. (A set is called a surface if it can be decomposed into several sets each of which is a nonparametric surface, for some position of the coordinate axes.)

►*Examples*   *1.* A function of several variables is called **linear** if it is a polynomial of degree 1. *The graph of a linear function of two variables is a plane.* Indeed, such a function can be written as

$$(x, y) \mapsto z = ax + by + c,$$

where a, b, c are fixed numbers, and we know that the set of points (x, y, z) satisfying $ax + by + c - z = 0$ is a plane (see Chapter 9, §4.3).

2. The graph of

$$(x, y) \mapsto \sqrt{1 - x^2 - y^2}$$

is the upper unit hemisphere. Indeed, if $z = \sqrt{1 - x^2 - y^2}$, then $z \geq 0$ and $x^2 + y^2 + z^2 = 1$.

3. The graph of the function

$$z = x^2$$

considered as a function of two variables, that is, the graph of

$$(x, y) \mapsto x^2$$

is a parabolic cylinder (compare Chapter 10, §4.2).

4. The graph of

$$f(x, y) = x^2 - y^2$$

is a hyperbolic paraboloid. We discussed this surface in Chapter 10, §4.6.◄

For a function of three (or more) variables, one can also define a graph, but this is now a set of points in a space of four (or more) dimensions. It cannot be drawn or modeled or visualized.

It is sometimes useful to think of a function of three variables, $(x, y, z) \mapsto f(x, y, z) = \phi$, as describing the temperature ϕ at a point with coordinates (x, y, z). A function of four variables, $(x, y, z, t) \mapsto \phi$, can be thought of as describing the temperature at the point (x, y, z) at the time t. There are many similar interpretations of functions of three or more variables.

§2 **Derivatives of Functions of Several Variables**

In this section, we apply the first basic process of calculus, differentiation, to functions of several variables.

2.1 Partial derivatives

We consider a function $(x, y) \mapsto f(x, y)$ of two variables and fix the value of the second variable at y_0. In other words, we consider the function of *one* variable $x \mapsto f(x, y_0)$. It may happen that this function has, at $x = x_0$, a derivative; this means that the (finite) limit

$$\lim_{h \to 0} \frac{f(x_0 + h, y_0) - f(x_0, y_0)}{h}$$

exists. If so, this number is called the **partial derivative with respect to** x of $(x, y) \mapsto f(x, y)$ at (x_0, y_0), and is denoted by

$$\left(\frac{\partial f}{\partial x}\right)_{x=x_0,\, y=y_0}$$

or simply by

$$\frac{\partial f}{\partial x},$$

and also by

$$\frac{\partial}{\partial x} f(x, y).$$

(The symbol ∂ is called the "round d.") Thus, writing this time x and y instead of x_0, y_0, we have

$$\frac{\partial f}{\partial x} = \frac{\partial f(x, y)}{\partial x} = \lim_{h \to 0} \frac{f(x + h, y) - f(x, y)}{h}.$$

The partial derivative with respect to y is defined similarly:

$$\frac{\partial f}{\partial y} = \lim_{h \to 0} \frac{f(x, y + h) - f(x, y)}{h}.$$

It is the derivative of the function of one variable $y \mapsto f(x, y)$, where x is a fixed number.

Since partial derivatives are, ultimately, derivatives of functions of one variable, it is easy to compute them. We simply use all differentiation rules learned thus far, taking care to remember in each case which variable is "kept fixed" and is to be treated like a constant, and which is the "variable of differentiation."

▶ *Examples*   *1.* Find the partial derivatives (at $x = 2$, $y = 3$) of

$$f(x, y) = 3x^3 y + 4xy^2 - 2x + 4y - 5.$$

Solution.   We have

$$f(x, 3) = 9x^3 + 36x - 2x + 12 - 5 = 9x^3 + 34x + 7.$$

The derivative of this function of x is $27x^2 + 34$; for $x = 2$, we obtain 142. Thus

$$\left(\frac{\partial f}{\partial x}\right)_{x=2,\, y=3} = 142.$$

Next, $f(2, y) = 24y + 8y^2 - 4 + 4y - 5 = 8y^2 + 28y - 9.$

This function of y has the derivative $16y + 28$. For $y = 3$, we obtain 76; hence

$$\left(\frac{\partial f}{\partial y}\right)_{x=2,\, y=3} = 76.$$

2. Find general formulas for the partial derivatives of the function $f(x, y)$ of Example 1.

Answer.   We treat y as a constant and obtain, by the usual differentiation rules,

$$\frac{\partial f}{\partial x} = \frac{\partial(3x^3y + 4xy^2 - 2x + 4y - 5)}{\partial x} = 9x^2y + 4y^2 - 2.$$

Similarly, treating x as a fixed number, we have

$$\frac{\partial f}{\partial y} = \frac{\partial(3x^3y + 4xy^2 - 2x + 4y - 5)}{\partial y} = 3x^3 + 8xy + 4.$$

Substituting into these formulas $x = 2$, $y = 3$, we have

$$\frac{\partial f}{\partial x} = 142, \frac{\partial f}{\partial y} = 76 \qquad \text{at } x = 2,\, y = 3$$

as before.

3. Find the partial derivatives of $f(x, y) = \sin(x^2 + y)$ at $x = 0$, $y = \pi$.

First solution.   We follow the method of Example 1 and write

$$\left(\frac{\partial \sin(x^2 + y)}{\partial x}\right)_{x=0,\, y=\pi} = \left(\frac{d \sin(x^2 + \pi)}{dx}\right)_{x=0} = (2x)\cos(x^2 + \pi)_{x=0} = 0,$$

$$\left(\frac{\partial \sin(x^2 + y)}{\partial y}\right)_{x=0,\, y=\pi} = \left(\frac{d \sin y}{dy}\right)_{y=\pi} = (\cos y)_{y=\pi} = -1.$$

Second solution. We follow the method of Example 2 and obtain first the general formulas:

$$\frac{\partial \sin (x^2 + y)}{\partial x} = 2x \cos (x^2 + y), \qquad \frac{\partial \sin (x^2 + y)}{\partial y} = \cos (x^2 + y).$$

Setting $x = 0$, $y = \pi$, we obtain for the two partial derivatives the values 0 and -1, respectively.

4. What are the partial derivatives of the function $f(x, y) = |x|$ at $x = y = 0$?

Answer. Since $f(x, 0) = |x|$, and the function $x \mapsto |x|$ has no derivative at $x = 0$, there is no partial derivative with respect to x. Since $f(0, y) = 0$, we have $\partial f/\partial y = 0$ at $x = y = 0$.

5. We exhibit a function of two variables that has partial derivatives at a point, without being continuous at this point. Set

$$f(x, y) = 0 \text{ if } xy = 0, \ 1 \text{ if } xy \neq 0.$$

Clearly $\partial f/\partial x = \partial f/\partial y = 0$ at $(0, 0)$, since $f(x, 0) = f(0, y) = 0$. We observe that $f(0, 0) = 0$ and that, in every interval around $(0, 0)$, there are points (x, y) with $f(x, y) = 1$. Thus f is not continuous at $(0, 0)$. ◄

2.2 Notations for partial derivatives

Many symbols other than ∂ are used to denote partial derivatives. For instance, instead of $\partial f/\partial x$, one writes

$$f_x, \qquad f_{,1} \qquad \text{or} \qquad D_x f.$$

Similarly, $\partial f/\partial y$ is sometimes denoted by

$$f_y, \qquad f_{,2} \qquad \text{or} \qquad D_y f.$$

Thus, if $f(x, y) = 2x^3 y - xy^2$, we have

$$f_x(x, y) = 6x^2 y - y^2, \qquad f_x(1, 2) = 12 - 4 = 8$$
$$D_y f(x, y) = 2x^3 - 2xy, \qquad f_{,2}(1, 2) = 2 - 4 = -2.$$

We shall use only the ∂ notation and the subscript notation, f_x and f_y.

EXERCISES

In Exercises 1 to 13, find the partial derivatives of $\partial f/\partial x$ and $\partial f/\partial y$ of the function $f(x, y)$.

1. $f(x, y) = x + 2y$.

7. $f(x, y) = \sin(x/y)$.

2. $f(x, y) = x^2 y^5$.

8. $f(x, y) = (1/x)e^{x^2 + y^2}$.

3. $f(x, y) = x^2 y^3 - 2xy^2$.

9. $f(x, y) = \arctan(y/x)$.

4. $f(x, y) = \sqrt{1 + x^2 + y^2}$.

► 10. $f(x, y) = \log(x \tan y)$.

5. $f(x, y) = \sin(x^2 + y)$.

11. $f(x, y) = \int_x^y e^{\sin t}\, dt$.

6. $f(x, y) = 5x^{10} - 6xy^7$.

12. $f(x, y) = (x^2 - y^2)/(x^2 + y^2)$.

13. $f(x, y) = (xy)^{-1} - \log y$.

14. For the function $f(x, y)$ defined in Exercise 7, let $g(x) = f(x, 3)$. Compute $g'(2)$ and compare with the value of $\partial f/\partial x$ at $x = 2$, $y = 3$ obtained from Exercise 10.

15. Repeat Exercise 14 for the function in Exercise 11.

16. Repeat Exercise 14 for the function in Exercise 12.

17. Let $f(x, y)$ be the function in Exercise 8. Find $f_x(2, 1)$ and $f_y(2, 1)$.

18. Find $f_x(0, 0)$ and $f_y(0, 0)$ for the function in Exercise 4.

19. If $f(u, v) = \sin(u/v)$ let $g = \partial f/\partial u$, $h = \partial f/\partial v$, and compute $\partial g/\partial v$ and $\partial h/\partial u$.

20. Repeat Exercise 19 for $f = \log(u \tan v)$.

21. Find $\partial e^{2x + \sin y}/\partial x$.

22. Find $\partial \sin(\cos\sqrt{1 + u^2 + v^2})/\partial v$.

23. If $l(\alpha, \beta) = \alpha^2 + e^{3\beta} \cos \alpha$, find $\partial l/\partial \alpha$ and $\partial l/\partial \beta$.

24. If $z = \sqrt{1 + u^2} + \sqrt{1 + v^2}$, find $z_u(0, 0)$ and $z_v(0, 0)$.

25. If $z = \int_x^y \sin t\, dt$, what is $z_x - z_y$?

► 26. If $v = \int_a^{a^2 + b^2} e^x\, dx$, what is v_a at $a = 1$, $b = 2$?

27. If $z = \sqrt{x^2 + y^2}$, show that $xz_x + yz_y = z$.

► 28. If $\phi = \log\sqrt{a^2 + b^2}$, prove that $a\phi_a + b\phi_b = 1$.

2.3 Differentiable functions

In the case of functions of one variable, the existence of a derivative tells us a great deal about the function. If $f'(x_0)$ exists, then the function $x \mapsto f(x)$ is continuous at x_0, and close to x_0 it can be approximated, with only a very small error, by a linear function (compare the Linear Approximation Theorem, Theorem 1 in Chapter 4, §1.8). In the case of functions of two variables, we cannot expect to learn so much about a function $(x, y) \mapsto f(x, y)$ from the mere fact that f has partial derivatives at a point (x_0, y_0). Indeed, in computing the partial derivatives at (x_0, y_0), we make use only of the values of the function on the vertical and the horizontal segments through this point. Thus it is not surprising that the mere existence of partial derivatives does not even imply continuity. (See Example 5 in §2.1.)

Because of this we do not call a function $f(x, y)$ **differentiable** at (x_0, y_0), simply because the partial derivatives exist at this point. Instead we demand that close to (x_0, y_0) we commit only a very small error by considering f to be a linear function. A precise definition will be given presently. It will turn out that all differentiable functions have partial derivatives, and that all functions with *continuous* partial derivatives are differentiable.

A function $f(x, y)$ defined at and near a point (x_0, y_0) is called differentiable at this point if there are numbers z_0, a and b such that

$$(1) \quad f(x, y) = z_0 + a(x - x_0) + b(y - y_0) \\ + \sqrt{(x - x_0)^2 + (y - y_0)^2}\, r(x, y)$$

where

$$(2) \qquad r(x, y) \text{ is continuous at } (x_0, y_0) \text{ and } r(x_0, y_0) = 0$$

or, which is the same,

$$(2') \qquad \lim_{(x,y) \to (x_0, y_0)} r(x, y) = 0.$$

To appreciate the meaning of this definition, note that

$$(3) \qquad z = (x, y) \mapsto z_0 + a(x - x_0) + b(y - y_0)$$

is a linear function. The term

$$\sqrt{(x - x_0)^2 + (y - y_0)^2}\, r(x, y)$$

in (1) is the error committed if, in computing the value of $f(x, y)$, we replace the given function f by the linear function (3). If the distance $\sqrt{(x - x_0)^2 + (y - y_0)^2}$ is small, then this error is *very* small, much smaller than the distance. This is the meaning of relation (2).

Let us note an immediate consequence of the definition: *a function differentiable at a point is continuous at this point.* Indeed, the right-hand side of (1) is a sum of continuous functions, and hence continuous.

We note also that the term z_0 in (1) is simply

$$(4) \qquad z_0 = f(x_0, y_0).$$

One sees this by applying (1) to the case $x = x_0$, $y = y_0$.

2.4 Tangent plane

The graph of the linear function (3) is a plane, called the **tangent plane** to the surface $z = f(x, y)$ at the point $P = (x_0, y_0, z_0)$, with z_0 given by (4). The geometric meaning of the definition of differentiability is shown in Fig. 12.8. The plane and the surface pass through the same point P, and at this point the plane attaches itself to the surface as closely as possible—it "just touches" the surface, like the tangent to a curve "just touches" the curve.

We shall presently learn how to compute the equation of the tangent plane, that is, how to find the numbers a and b in (1), given the function f.

2.5 Partial derivatives of differentiable functions

Suppose that the function $(x, y) \mapsto f(x, y)$ is differentiable at (x_0, y_0), so that relation (1) of §2.3 holds, and the surface $z = f(x, y)$ has a tangent plane at $P = (x_0, y_0, z_0)$, as shown in Fig. 12.8. We intersect the surface and the tangent plane by the plane $y = y_0$. The intersection with the surface is a curve shown in Fig. 12.9; this is the graph of the function $x \mapsto f(x, y_0)$. The intersection with the tangent plane is a straight line. We expect that this line is tangent to the curve, so that the function $x \mapsto f(x, y_0)$ has a derivative at x_0, and $f(x, y)$ has a partial derivative with respect to x, at (x_0, y_0). Let us verify this guess.

Using (1) and (4) and setting $y = y_0$, we have

(5) $$f(x, y_0) = f(x_0, y_0) + a(x - x_0) + |x - x_0| r(x, y_0).$$

Here $r(x, y_0)$ is a continuous function and $r(x_0, y_0) = 0$ by (3). Using Theorem 1 of Chapter 4, we conclude that $x \mapsto f(x, y_0)$ has a derivative a at $x = x_0$. This means that $f(x, y)$ has at (x_0, y_0) the partial derivative a (with respect to x). We could also reason directly: by (5)

$$f(x_0 + h, y_0) - f(x_0, y_0) = ah + |h| r(x_0 + h, y_0)$$

and thus

$$\left(\frac{\partial f}{\partial x}\right)_{x_0, y_0} = \lim_{h \to 0} \frac{f(x_0 + h, y_0) - f(x_0, y_0)}{h}$$

$$= a + \lim_{h \to 0} \frac{|h|}{h} r(x_0 + h, y_0) = a,$$

since $|h|/h$ is either 1 or -1.

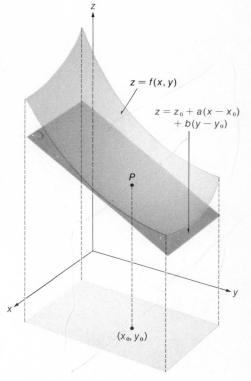

Fig. 12.8

$z = f(x, y)$

$z = z_0 + a(x - x_0) + b(y - y_0)$

P

(x_0, y_0)

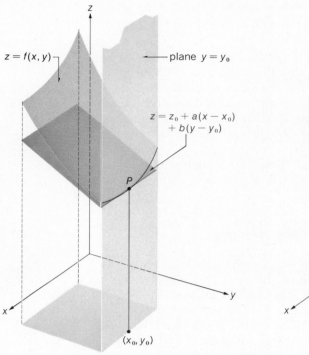

Fig. 12.9   Fig. 12.10

Setting $x = x_0$ in (1), we see that

$$f(x_0, y) = f(x_0, y_0) + b(y - y_0) + |y - y_0| r(x_0, y)$$

and this implies, as before, that $f(x, y)$ has a partial derivative with respect to y at (x_0, y_0); the value of this partial derivative is b. The geometric meaning of this statement is shown in Fig. 12.10. The plane $x = x_0$ intersects the surface $z = f(x, y)$ in a curve, the graph of $y \mapsto f(x_0, y)$, and it intersects the tangent plane in a line. This line is tangent to the curve and has slope b.

We summarize the results of this discussion.

Theorem 1 SUPPOSE $f(x, y)$ IS DIFFERENTIABLE AT (x_0, y_0) AND SATISFIES (1) AND (2). THEN $f(x, y)$ HAS AT (x_0, y_0) THE PARTIAL DERIVATIVES

(6)
$$\frac{\partial f}{\partial x} = a, \qquad \frac{\partial f}{\partial y} = b.$$

This shows, in particular, that the numbers a and b are uniquely determined by the function f; the surface $z = f(x, y)$ can have at x_0, y_0, z_0 at most one tangent plane.

2.6 Continuously differentiable functions

We state next a partial converse of Theorem 1.

Theorem 2   SUPPOSE $f(x, y)$ IS DEFINED AND HAS PARTIAL DERIVATIVES AT (x_0, y_0), AND AT ALL POINTS NEAR (x_0, y_0). SUPPOSE ALSO THAT THESE PARTIAL DERIVATIVES, CONSIDERED AS FUNCTIONS OF (x, y) ARE CONTINUOUS AT (x_0, y_0). THEN f IS DIFFERENTIABLE AT (x_0, y_0).

The proof of this theorem will be found in an appendix to this chapter (see §6.1). The theorem is usually applied to a function that is defined in an open set S, has partial derivatives with respect to both variables at all points of S, and these derivatives are continuous in S. Such functions are called **continuously differentiable.** Fortunately, it is usually easy to recognize, "by inspection," that a function is continuously differentiable.

▶*Examples*   *1.* Is $f(x, y) = \sin (e^x \cos y + 2x^3 y)$ continuously differentiable everywhere?

Answer.   Yes. For f, as well as $\partial f/\partial x$ and $\partial f/\partial y$, which can be computed by the usual rules, are obtained by composition and addition and multiplication from functions of one variable that have derivatives of all orders.

2. The function $f(x, y) = \sqrt{|xy|}$ is continuous everywhere, and $\partial f/\partial x = \partial f/\partial y = 0$ at $(0, 0)$. But the function is not differentiable at $(0, 0)$. Indeed, if it were, we should have

$$\sqrt{|xy|} = \sqrt{x^2 + y^2}\, r(x, y)$$

with $r(x, y)$ continuous at $(0, 0)$ and $r(0, 0) = 0$. For $x = y$, this would give $|x| = \sqrt{2}|x|r(x, x)$ and hence $r(x, x) = 1/\sqrt{2}$ for $x = 0$. This is a contradiction.

3. Find the equation of the tangent plane to the surface $z = f(x, y) = \sin x + e^{xy} + y$ at the point $(0, 2, 3)$.

Solution.   We have

$$\frac{\partial f}{\partial x} = \cos x + ye^{xy}, \qquad \frac{\partial f}{\partial y} = xe^{xy} + 1.$$

These functions are clearly continuous everywhere. Hence Theorems 1 and 2 are applicable. At $x = 0$, $y = 2$, we find

$$f(0, 2) = z_0 = \sin 0 + e^0 + 2 = 3$$

$$\frac{\partial f}{\partial x} = \cos 0 + 2e^0 = 3, \frac{\partial f}{\partial y} = 0 \cdot e^0 + 1 = 1.$$

The equation of the tangent plane (obtained by setting $z_0 = 3$, $x_0 = 0$, $y_0 = 2$, $a = 3$, $b = 1$ in (3)) reads $z = 3 + 3x + (y - 2)$ or

$$z = 3x + y + 1.$$

Note that this plane indeed passes through $(0, 2, 3)$.

4. Find the tangent plane to the sphere $x^2 + y^2 + z^2 = 29$ at the point $(2, 3, 4)$.

Answer. The upper hemisphere is the graph of the function

$$z = \sqrt{29 - x^2 - y^2};$$

the function is continuously differentiable for $x^2 + y^2 < 29$. We obtain the desired equation by substituting into (3) the values $x_0 = 2$, $y_0 = 3$, $z_0 = \sqrt{29 - 2^2 - 3^2} = 4$, and

$$a = \left(\frac{\partial \sqrt{29 - x^2 - y^2}}{\partial x}\right)_{x=2, y=3} = \left(-\frac{x}{\sqrt{29 - x^2 - y^2}}\right)_{x=2, y=3} = -\frac{1}{2},$$

$$b = \left(\frac{\partial \sqrt{29 - x^2 - y^2}}{\partial y}\right)_{x=2, y=3} = \left(-\frac{y}{\sqrt{29 - x^2 - y^2}}\right)_{x=2, y=3} = -\frac{3}{4}.$$

The desired equation reads

$$z = 4 - \frac{1}{2}(x - 2) - \frac{3}{4}(y - 3)$$

or $\qquad\qquad 2x + 3y + 4z - 29 = 0.\blacktriangleleft$

2.7 Surface normal

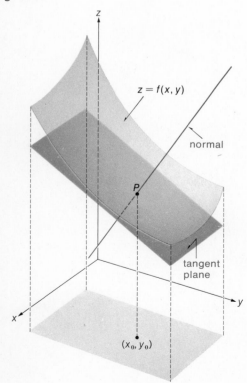

Fig. 12.11

$z = f(x, y)$

normal

P

tangent plane

(x_0, y_0)

Let $P = (x_0, y_0, z_0)$ be a point on the graph of a differentiable function $z = f(x, y)$. The line through P perpendicular to the tangent plane is called the **surface normal** to the graph (see Fig. 12.11).

The surface normal admits the parametric representation

(7) $\qquad\qquad x = x_0 + at, \qquad y = y_0 + bt, \qquad z = z_0 - t$

where

(8) $z_0 = f(x_0, y_0), \qquad a = \left(\dfrac{\partial f}{\partial x}\right)_{x_0, y_0}, \qquad b = \left(\dfrac{\partial f}{\partial y}\right)_{x_0, y_0}.$

This follows by comparing the equation of the tangent plane (see §§2.3 to 2.6) with the results of Chapter 9, §6.7. The result may be stated as follows: *the vector with components*

$$(f_x, f_y, -1)$$

is perpendicular to the surface $z = f(x, y)$, *at the point* (x, y, z).

▶**Examples** **1.** Find the line normal to the surface $z = e^x \cos y$ through the point with the coordinates $x = 1$, $y = \pi$, $z = 1/e$.

Answer. Using the formulas (7) and (8) with $x_0 = 1$, $y_0 = \pi$, $z_0 = 1/e$, and noting that $\partial z/\partial x = \cos y\, e^x \cos y$, $\partial z/\partial y = -x \sin y\, e^x \cos y$, we obtain $a = -1/e$, $b = 0$, so that the normal has a parametric representation

$$x = 1 - \frac{t}{e}, \qquad y = \pi, \qquad z = \frac{1}{e} - t.$$

2. Find the intersection of the normal to the surface $z = x^4 y - 5xy^6$ at the point $(2, 1, 6)$ and the plane $x = 0$.

Solution. We have $\partial z/\partial x = 4x^3 y - 5y^6$, $\partial z/\partial y = x^4 - 30xy^5$. Hence, using the notations (8) with $x_0 = 2$, $y_0 = 1$, $z_0 = 6$, we obtain $a = 27$, $b = -44$. By (7), the normal has a parametric representation

$$x = 2 + 27t, \qquad y = 1 - 44t, \qquad z = 6 - t.$$

If $x = 0$, then $t = -2/27$; hence $y = 115/27$, $z = 164/27$. The desired intersection point is $(0, 115/27, 164/27)$. ◀

EXERCISES

In Exercises 29 to 35, find the equation of the tangent plane and the equations of the normal line to the given surface at the given point.

29. $z = 5x^2 + 2y^2 - 9$; $(3, 2, 4)$.

30. $z = xy$; $(3, -2, -6)$.

31. $z = x^3 + y^3 + 3xy$; $(1, -1, -3)$.

32. $z = 5x - y^2$; $(4, 2, 0)$.

33. $z = e^{x^2 + y^2}$; $(0, 0, 1)$.

34. $z = x^2 + y^2$; $(1, 1, 2)$.

35. $z = \sqrt{1 - x^2 - y^2}$; $(\frac{2}{3}, \frac{2}{3}, \frac{1}{3})$.

36. Show that the surfaces $z = xy/(4x - y)$ and $z = \sqrt{(5x - y)/13}$ intersect at right angles at the point $(1, 2, 1)$. (This means that both surfaces pass through this point and their normals are perpendicular.)

37. Show that the tangent plane to the sphere $x^2 + y^2 + z^2 = 1$, at a point (x_0, y_0, z_0) on this sphere, has the equation $xx_0 + yy_0 + zz_0 = 1$. (Assume that $z_0 > 0$.)

▶ 38. Show that the tangent plane to the ellipsoid $(x/a)^2 + (y/b)^2 + (z/c)^2 = 1$, at a point (x_0, y_0, z_0) on it, has the equation $xx_0 a^{-2} + yy_0 b^{-2} + zz_0 c^{-2} = 1$.

39. Find the equation of the plane tangent to the surface $z = \sqrt{1-(x/10)^2-(y/10)^2}$ at $x = 1$, $y = 2\sqrt{2}$, $z = \frac{1}{10}$ first directly, and then using the result of Exercise 38.

▶ 40. By analogy with Exercise 38, *guess* at the form of the equation of a tangent plane to hyperboloids of the form

$$(x/a)^2 + (y/b)^2 - (z/c)^2 = 1, \qquad (x/a)^2 - (y/b)^2 - (z/c)^2 = 1.$$

Then verify your guess. (Compare this exercise with the result in Chapter 9, §1.6.)

2.8 The chain rule for two variables

The rule for finding partial derivatives of composite functions is similar to, though somewhat more complicated than, the corresponding rule for functions of one variable.

Theorem 3 (Chain rule)   LET $F(x, y)$ BE DIFFERENTIABLE AT (x_0, y_0), LET $\phi(u, v)$ AND $\psi(u, v)$ BE DIFFERENTIABLE AT (u_0, v_0), AND ASSUME THAT $\phi(u_0, v_0) = x_0$, $\psi(u_0, v_0) = y_0$. THEN THE COMPOSED FUNCTION

$$(9) \qquad z = f(u, v) = F(\phi(u, v), \psi(u, v))$$

IS DIFFERENTIABLE AT (u_0, v_0) AND

$$(10) \qquad \frac{\partial f}{\partial u} = \frac{\partial F}{\partial x} \frac{\partial \phi}{\partial u} + \frac{\partial F}{\partial y} \frac{\partial \psi}{\partial u}, \qquad \frac{\partial f}{\partial v} = \frac{\partial F}{\partial x} \frac{\partial \phi}{\partial v} + \frac{\partial F}{\partial y} \frac{\partial \psi}{\partial v}$$

WHERE THE PARTIAL DERIVATIVES WITH RESPECT TO u, v ARE COMPUTED AT (u_0, v_0), THOSE WITH RESPECT TO x, y AT (x_0, y_0).

Another way of expressing this result is as follows: if

$$z = F(x, y), \ x = \phi(u, v), \ y = \psi(u, v)$$

(all functions being differentiable), then

$$\frac{\partial z}{\partial u} = \frac{\partial F}{\partial \phi} \frac{\partial \phi}{\partial u} + \frac{\partial F}{\partial \psi} \frac{\partial \psi}{\partial u},$$

and similarly for $\partial z/\partial v$. If partial derivatives are denoted by subscripts (compare §2.2), the chain rule (10) can be written as follows:

$$f_u(u, v) = F_x(\phi(u, v), \psi(u, v))\phi_u(u, v) + F_y(\phi(u, v), \psi(u, v))\psi_u(u, v),$$

$$f_v(u, v) = F_x(\phi(u, v), \psi(u, v))\phi_v(u, v) + F_y(\phi(u, v), \psi(u, v))\psi_v(u, v).$$

We shall prove Theorem 3, assuming that the functions F, ϕ, and ψ are linear. This should convince us of the truth of the theorem, since differentiable functions may be considered as linear if we consider them very close to some point. Thus we assume that

$$F(x, y) = Ax + By + C, \phi(u, v) = au + bv + c, \psi(u, v) = \alpha u + \beta v + \gamma.$$

Then

$$(11) \qquad A = \frac{\partial F}{\partial x}, B = \frac{\partial F}{\partial y}, a = \frac{\partial \phi}{\partial u}, b = \frac{\partial \phi}{\partial v}, \alpha = \frac{\partial \psi}{\partial u}, \beta = \frac{\partial \psi}{\partial v}.$$

Also

$$\begin{aligned}
z &= F(\phi(u, v), \psi(u, v)) = A\phi(u, v) + B\psi(u, v) + C \\
&= A(au + bv + c) + B(\alpha u + \beta v + \gamma) + C \\
&= (Aa + B\alpha)u + (Ab + B\beta)v + (Ac + B\gamma + C).
\end{aligned}$$

Therefore

$$(12) \qquad \frac{\partial z}{\partial u} = Aa + B\alpha, \qquad \frac{\partial z}{\partial v} = Ab + B\beta.$$

Substituting (11) into (12), we obtain (10). A more formal proof of Theorem 3 will be given in an appendix (§6.2).

▶**Examples** *1.* Suppose that $F(x, y) = 3x^2y$, $\phi(u, v) = u + v$ and $\psi(u, v) = uv$. Set $z = F(\phi(u, v), \psi(u, v))$ and find the partial derivatives $\partial z/\partial u$, $\partial z/\partial v$ at $u = 2$, $v = 3$.

First solution.   We have

$$z(u, v) = 3(u + v)^2uv = 3u^3v + 6u^2v^2 + 3uv^3$$

so that

$$z_u = 9u^2v + 12uv^2 + 3v^3, \qquad z_v = 3u^3 + 12u^2v + 9uv^2$$

and $z_u(2, 3) = 405$, $z_v(2, 3) = 330$.

Second solution.   We have

$$F_x = 6xy, F_y = 3x^2, \phi_u = 1, \phi_v = 1, \psi_u = v, \psi_v = u.$$

For $u = 2$, $v = 3$, we obtain

$$\phi_u = 1, \ \phi_v = 1, \ \psi_u = 3, \ \psi_v = 2$$

and, setting $x = \phi(u, v) = u + v$, $y = \psi(u, v) = uv$,

$$x = 5, \ y = 6, \ F_x = 180, \ F_y = 75.$$

Thus, by the chain rule,

$$z_u = F_x x_u + F_y y_u = F_x \phi_u + F_y \psi_u = 180 \cdot 1 + 75 \cdot 3 = 405$$
$$z_v = F_x x_v + F_y y_v = F_x \phi_v + F_y \psi_v = 180 \cdot 1 + 75 \cdot 2 = 330,$$

as before.

2. Find the partial derivatives with respect to u and v of $z = e^{xy}$ where $x = u^2$ and $y = uv$.

Answer. Direct substitution gives $z = e^{u^3 v}$ so that $z_u = 3u^2 v e^{u^3 v}$ and $z_v = u^3 e^{u^3 v}$. If we apply the chain rule, we must note that a function of u alone may be considered as a function of (u, v) with a partial derivative 0 with respect to v. In our case, we have

$$z_x = y e^{xy}, \ z_y = x e^{xy}, \ x_u = 2u, \ x_v = 0, \ y_u = v, \ y_v = u$$

so that

$$z_u = z_x x_u + z_y y_u = y e^{xy} 2u + x e^{xy} v = 3u^2 v e^{u^3 v}$$
$$z_v = z_x x_v + z_y y_v = y e^{xy} 0 + x e^{xy} u = u^3 e^{u^3 v}$$

as before. ◄

2.9 Directional derivatives

The chain rule is applicable also when we form a function of one variable, $t \mapsto f(t)$, by substituting into a function $F(x, y)$ of two variables, two functions $x = \phi(t)$, $y = \psi(t)$ of one variable. The derivative of the function

(13)
$$z = F(\phi(t), \psi(t))$$

is

(14)
$$\frac{dz}{dt} = \frac{\partial F}{\partial x} \frac{d\phi}{dt} + \frac{\partial F}{\partial y} \frac{d\psi}{dt}$$

or

(14′)
$$z'(t) = F_x(\phi(t), \psi(t))\phi'(t) + F_y(\phi(t), \psi(t))\psi'(t);$$

here we used the usual notations (d instead of ∂ and the prime $'$) for functions of one variable.

We apply formula (14) to a very simple case:

$$x = \phi(t) = x_0 + t \cos \alpha, \qquad y = \psi(t) = y_0 + t \sin \alpha.$$

These equations give a parametric representation of a line through (x_0, y_0); t is the arc length on this line [measured from the point $x_0 = \phi(0)$, $y_0 = \psi(0)$] and α is the polar angle of a vector pointing in the direction of increasing t (see Fig. 12.12). The number (13) is now the value of F at the point reached by going the distance t "in the direction α." The derivative (14) for $t = 0$, that is,

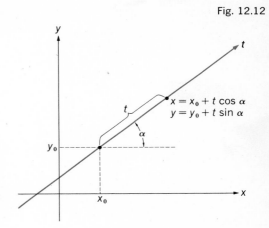

Fig. 12.12

$$(15) \qquad F_x(x_0, y_0) \cos \alpha + F_y(x_0, y_0) \sin \alpha$$

or, simpler,

$$(15') \qquad F_x \cos \alpha + F_y \sin \alpha \qquad \text{(at } x_0, y_0)$$

is called the **directional derivative of F at (x_0, y_0) in the direction α**. It measures the rate of change of F with respect to distance as we proceed from (x_0, y_0) in the direction α. The directional derivative in the x direction is obtained by setting $\alpha = 0$; we obtain the partial derivative F_x. Similarly, F_y is the derivative in the y direction (set $\alpha = 90°$), $-F_x$ is the derivative in the direction opposite to the x direction (set $\alpha = 180°$), and so forth.

A point at which both partial derivatives vanish, $F_x = F_y = 0$, is called a **critical point** of F. At a critical point, all directional derivatives are 0. We shall discuss such points later (see §3.4). At a noncritical point, let us set $a = \sqrt{F_x^2 + F_y^2}$; then there is a unique number θ, $0 \leq \theta < 2\pi$, such that

$$(16) \qquad F_x = a \cos \theta, \; F_y = a \sin \theta, \; (a = \sqrt{F_x^2 + F_y^2} \neq 0).$$

We can write the directional derivative (15) in the form

From trig add formulas.

$$(15'') \qquad a \cos \theta \cos \alpha + a \sin \theta \sin \alpha = a \cos (\theta - \alpha).$$

For which value of α is the directional derivative largest? Clearly for such α that $\cos (\theta - \alpha)$ has as its largest possible value 1, that is, for $\alpha = \theta$; the value of the directional derivative is then a. The directional derivative is smallest and has the value $-a$ if $\cos (\theta - \alpha) = -1$, that is, for $\alpha = \theta + 180°$ [if $\theta + 180°$ does not lie in $(0, 360°)$, add or subtract a multiple of $360°$ from it]. For which α is the directional derivative (15'') equal to 0? For $\theta - \alpha = 90°$ or $\theta - \alpha = -90°$, that is, for $\alpha = \theta \pm 90°$.

Fig. 12.13

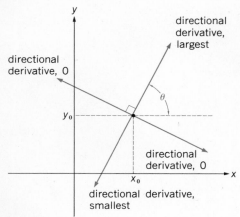

The situation is illustrated in Fig. 12.13. The direction $\alpha = \theta$ is the **direction of steepest increase.**

▶ *Example* For the function $F = x^2y - 2x - y$, find (a) the critical points, (b) the directional derivative at $(3, 1)$ in the direction $\alpha = -45°$, and (c) the direction θ of steepest increase at $(3, 1)$.

  Solution. (a) We have $F_x = 2xy - 2$, $F_y = x^2 - 1$. Hence $F_y = 0$ if $x = 1$ or $x = -1$. If $x = 1$, then $F_x = 0$ for $y = 1$. If $x = -1$, then $F_x = 0$ for $y = -1$. Thus there are two critical points: $(1, 1)$ and $(-1, -1)$.
  (b) At $(3, 1)$ we have $F_x = 4$, $F_y = 8$. Since $\cos(-45°) = \cos 45° = 1/\sqrt{2}$, $\sin(-45°) = -\sin 45° = -1/\sqrt{2}$, the desired directional derivative is, by formula (15′), equal to $(4/\sqrt{2}) - (8/\sqrt{2}) = -4\sqrt{2} \approx 5.66$.
  (c) At $(3, 1)$ we have $a = \sqrt{F_x^2 + F_y^2} = \sqrt{80} = 4\sqrt{5}$. Hence the angle θ of greatest increase, determined by the equations (compare (16) above) $4 = 4\sqrt{5} \cos \theta$, $8 = 4\sqrt{5} \sin \theta$, is $\theta \approx 63°30'$.◀

2.10 Gradient

Let $F(x, y)$ be a continuously differentiable function (defined in some open set). At every point (x, y) at which it is defined, we form the vector with the components (F_x, F_y); this vector is called the **gradient** of F and is denoted by grad F. Thus

$$(17) \qquad\qquad \operatorname{grad} F = \frac{\partial F}{\partial x}\mathbf{e}_1 + \frac{\partial F}{\partial y}\mathbf{e}_2$$

(where $\mathbf{e}_1$ and $\mathbf{e}_2$ are, as usual, unit vectors in the coordinate directions, compare Chapter 9, §1.8). Note that grad F is a rule that assigns a vector to every point in the domain of definition of F; it is an example of a **vector-valued function of two variables.** Recalling relations (16) in §2.9, we see that the polar angle of the gradient is θ. Thus *the gradient points in the direction of steepest increase (or, as one says, steepest ascent). The magnitude of the gradient* $|\operatorname{grad} F| = \sqrt{F_x^2 + F_y^2} = a$ *is the value of the directional derivative in the direction of steepest ascent.* At a critical point the gradient is equal to **0.**

2.11 Level curves

Let us think of a continuously differentiable function F of two variables as describing a mountainous landscape: $F(x, y)$ is the height (elevation) of the point above (x, y); the graph $z = F(x, y)$ is the surface of the mountain. A **level curve** is a smooth arc (compare Chapter 11, §3.1) $x = \phi(t)$, $y = \psi(t)$ such that $F(\phi(t), \psi(t))$ is a constant, say c. In other words, a level curve is the projection on the x, y plane of a trail of constant elevation. We shall assume that *through every point (x_0, y_0) that is not a critical point of F there passes a unique level curve.* (If the partial derivatives F_x and F_y are themselves continuously differentiable, this need not be assumed, but can be proved. But we shall not do so in this book.)

Let (x_0, y_0) be a noncritical point of F. Then we can compute the direction of the level curve through this point. Indeed, the level curve $x = \phi(t)$, $y = \psi(t)$ with $t = t_0$ corresponding to (x_0, y_0) has the property

$$F(\phi(t), \psi(t)) = c$$

(this property makes it into a level curve). We differentiate both sides of the equation with respect to t and obtain [compare (13) and (14) in §2.9], for $t = t_0$,

$$(18) \qquad F_x(x_0, y_0)\phi'(t_0) + F_y(x_0, y_0)\psi'(t_0) = 0.$$

We recall that the vector $\phi'(t_0)\mathbf{e}_1 + \psi'(t_0)\mathbf{e}_2$ is the tangent vector to our level curve at (x_0, y_0). Let b denote the length of this vector and let β be its polar angle. Then $\phi'(t_0) = b \cos \beta$, $\psi'(t_0) = b \sin \beta$. Recalling relations (16) in §2.9, we write (18) as

$$(a \cos \theta)(b \cos \beta) + (a \sin \theta)(b \sin \beta) = ab \cos (\theta - \beta) = 0.$$

Thus $\cos (\theta - \beta) = 0$ so that either $\beta = \theta + 90°$ or $\beta = \theta - 90°$. In other words, the level curve through a (noncritical) point is perpendicular to the gradient at this point (see Fig. 12.14). This is as it should be: the direction of the gradient is that of steepest ascent; the direction perpendicular to it is the level direction; the derivative in this direction is 0.

▶ *Examples*   *1.* The level curves of the function $F(x, y) = x^2 + 4y^2$ are, of course, ellipses $x^2 + 4y^2 = c$. We have $F_x = 2x$, $F_y = 8y$, so that there is one critical point, $x = y = 0$. There is no level curve through this point. Some level curves are shown in Fig. 12.15.

2. The level curves of the function $F(x, y) = x^2 - 4y^2$ are, of course, hyperbolas $x^2 - 4y^2 = c$; the coordinate axes are also level curves. We have $F_x = 2x$, $F_y = -8y$ and $(0, 0)$ is the only critical point. Two level curves intersect at this point. Some level curves are shown in Fig. 12.16. ◀

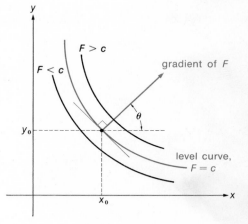

Fig. 12.14

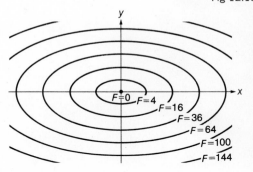

Fig 12.15

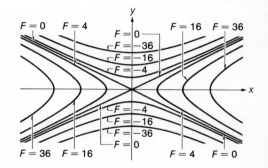

Fig. 12.16

EXERCISES

Each of the Exercises 41 to 45 is to be done twice, once by direct substitution and once by use of the chain rule.

41. Find $\partial u/\partial s$ and $\partial u/\partial t$ if $u = 2x^2 + xy + y^2$, $x = 2s - t$, $y = s - 2t$.
42. Find dz/dx and dz/dy if $z = xy^2 + x^2y$, $y = \log x$.
43. Find dz/dt if $z = \log(x^2 + y^2)$, $x = e^{-t}$, $y = e^t$.
44. Find dz/dt if $z = x^y$, $x = \sin t$, $y = \cos t$.
45. Find dz/dt if $z = x^2 + y^{-2}$, $x = e^t \cos t$, $y = 1 + \log t$.
46. Suppose $f(x, y)$ is differentiable at $(0, 0)$ and $f_x(0, 0) = 2$, $f_y(0, 0) = 3$. Suppose $\phi(u, v)$ is differentiable at $(0, 0)$, $\phi(0, 0) = 0$, $\phi_u(0, 0) = 7$, $\phi_v(0, 0) = 9$. Set $g(u, v) = f[\phi(u, v), u]$. Compute $g_u(0, 0)$ and $g_v(0, 0)$.
47. If $F(\alpha, \beta)$ is differentiable at $\alpha = \beta = 1$ and $F_\alpha = 3$, $F_\beta = 5$ at this point, and if $f(x, y) = F(x^2, y^3)$, what are $f_x(1, 1)$ and $f_y(1, 1)$?
48. If $f(t)$ is a differentiable function, and $F(x, y) = f(x - 2y)$, show that $2F_x(x, y) + F_y(x, y) \equiv 0$.
49. If $f(t)$ is a differentiable function and $f'(t) > 0$, and if $\phi(u, v) = f(u + v)$, what can you say about $\phi_u + \phi_v$?
50. Suppose $f(x, y)$ is differentiable everywhere and $f_x > 0$, $f_y > 0$ everywhere. Set $F(t) = f(t, t^3)$. Is F an increasing function? Give reasons for your answer.

For each of the following five functions F, find the critical points, the gradient at the given point P, and the directional derivative at P in the given direction α.

51. $F = 3x^2 - 6y^2$, $P = (8, 2)$, $\alpha = 120°$.
52. $F = ye^{2x}$, $P = (0, 4)$, $\alpha = 135°$.
53. $F = y + x \cos xy$, $P = (0, 0)$, $\alpha = 2\pi/3$.
54. $F = x^2 + y^2 - 6x + 4y + 25$, $P = (1, -4)$, $\alpha = 45°$.
55. $F = e^x \cos y$, $P = (0, 0)$, $\alpha = 60°$.
56. For the function F and the point P given in Exercise 51, write the equation of the level curve through the point, find the direction of its tangent, and verify that it is perpendicular to the gradient.
57. Repeat Exercise 56 for the functions in Exercise 52.
58. Repeat Exercise 56 for the functions in Exercise 53.
59. Repeat Exercise 56 for the functions in Exercise 54.
60. Repeat Exercise 56 for the functions in Exercise 55.

2.12 Implicit differentiation

Let $F(x, y)$ be as before, and let (x_0, y_0) be a point such that $F_y(x_0, y_0) \neq 0$. This means that the gradient is not horizontal. Therefore the tangent to the level curve at this point, which is perpendicular to the gradient, is not vertical. We conclude that the level curve through (x_0, y_0) can be represented as the graph of a function $y = f(x)$ (compare Chapter 11, §3.2 and Fig. 12.17).

In other words, there is a function $x \mapsto y = f(x)$ defined near x_0 such that $y_0 = f(x_0)$ and

Fig. 12.17

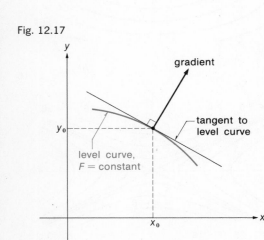

(19) $$F(x, f(x)) = F(x_0, y_0) = \text{const.}$$

(The reader should observe that the statement just made follows from the assumption made in §2.11; we do not give a proof in this book.) This function $f(x)$ is said to be **defined implicitly** by equation (19). The derivative of this function at x_0 can be obtained as follows: differentiate both sides of (19) with respect to x, using the chain rule on the left, and obtain, for $x = x_0$,

(20) $$F_x(x_0, y_0) + F_y(x_0, y_0)f'(x_0) = 0.$$

[We made use of the fact that $f(x_0) = y_0$ and that $dx/dx = 1$. Also, (20) is simply a special case of (18) with $t = x$, $\phi(t) = t$, $\psi(t) = f(t)$.] From (20), we obtain

(21) $$f'(x_0) = -\frac{F_x(x_0, y_0)}{F_y(x_0, y_0)},$$

the denominator being different from 0 by assumption.

In carrying out this process, called **implicit differentiation**, it is advisable not to memorize formula (21) but rather to imitate the method by which we obtained this formula.

Remark If $F_x(x_0, y_0) \neq 0$, the same procedure permits us to find the derivative of the function $x = g(y)$ defined implicitly by the conditions $g(y_0) = x_0$ and $F(g(y), y) = F(x_0, y_0) = \text{const.}$

▶**Examples** **1.** The function $y = f(x)$ satisfies the equation

(22) $$x^2y^3 + 2x^2y + 3x^4 - y = 1.$$

Find the derivative $y'(x)$ at $x = 1$, $y = -1$.

Solution. First we check that $x = 1$, $y = -1$ satisfies (22). Next we differentiate both sides of the given equation with respect to x, remembering that $y = f(x)$ is a function. This yields

$$2xy^3 + 3x^2y^2y' + 4xy + 2x^2y' + 12x^3 - y' = 0$$

or $\quad (3x^2y^2 + 2x^2 - 1)y' + (2xy^3 + 4xy + 12x^3) = 0.$

At $x = 1$, $y = -1$, this becomes $4y' + 6 = 0$. Thus $y'(1) = -\frac{3}{2}$. (Since the given function $F(x, y)$ is composed from elementary functions, we do not have to rely on the chain rule for functions of two variables. This is typical for examples that can be done explicitly.)

2. Find the derivative of the function $y = f(x)$, defined implicitly by the conditions

$$(23) \qquad\qquad e^{-x \cos y} + \sin y = e, \qquad y = \pi \text{ for } x = 1.$$

Solution. We check that $e^{-\cos \pi} + \sin \pi = e$. Differentiating both sides of the first equation (23) with respect to x, we obtain

$$-\cos y \; e^{-x \cos y} + x \sin y \; e^{-x \cos y} \; y' + (\cos y)y' = 0.$$

For $x = 1$, $y = \pi$, this becomes $e - y' = 0$. Thus $y'(1) = e$. ◄

EXERCISES

In Exercises 61 to 68, find the derivative dy/dx at the given point, assuming that the graph of the function $y = y(x)$ passes through the point.

61. $y^2 - 2x - 5 = 0$; $x = 10$, $y = 5$.
62. $2y^4 - 3y^2x - x^5 = 22$; $x = -2$, $y = 1$.
63. $y^{10} + 2yx^{10} - 3 = 0$; $x = 1$, $y = 1$.
64. $1/y + 1/x = 1$; $x = 2$, $y = 2$.
65. $x^3 - xy + y^3 = 1$; $x = 1$, $y = 1$.
66. $x^{2/3} + y^{2/3} = 2$; $x = 1$, $y = 1$.
67. $xy - e^x \sin y = \pi$; $x = 1$, $y = \pi$.
► 68. $\log (x^2 + y^2) = \arctan (y/x)$; $x = 1$, $y = 0$.

2.13 Differentiation of functions of three or more variables

The extension of the concepts of partial derivatives and differentiability to functions of three and more variables is obvious.

For instance, if $(x, y, z) \mapsto f(x, y, z)$ is a function defined in some open set, then the partial derivative with respect to z, at (x_0, y_0, z_0) is the derivative of the function of one variable $z \mapsto f(x_0, y_0, z)$ at $z = z_0$. This partial derivative is denoted by any of the symbols

$$\frac{\partial f}{\partial z}, \qquad \left(\frac{\partial f}{\partial z}\right)_{(x_0, y_0, z_0)}, \qquad \frac{\partial}{\partial z} f(x, y, z), \qquad f_z(x_0, y_0, z_0),$$

and also by $D_z f(x, y, z)$ or $f,_3(x_0, y_0, z_0)$, the 3 denoting differentiation with respect to the third variable. Thus

$$f_z(x_0, y_0, z_0) = \lim_{h \to 0} \frac{f(x_0, y_0, z_0 + h) - f(x_0, y_0, z_0)}{h}.$$

For functions of n variables, the definition is analogous.

The function $f(x, y, z)$ of 3 variables is called **differentiable** at (x_0, y_0, z_0) if, for (x, y, z) near (x_0, y_0, z_0),

$$f(x, y, z) = l(x, y, z) + \sqrt{(x - x_0)^2 + (y - y_0)^2 + (z - z_0)^2}\, r(x, y, z)$$

where l is a linear function and r a function that is continuous at (x_0, y_0, z_0) and equals 0 at this point. The linear function l can be written as

$$l(x, y, z) = ax + by + cz + d$$
$$= a(x - x_0) + b(y - y_0) + c(z - z_0) + (d + ax_0 + by_0 + cz_0).$$

The term $(d + ax_0 + by_0 + cz_0)$ is, clearly, the value of f at x_0, y_0, z_0.

For functions of $n \geq 3$ variables, it is convenient to use the language of vectors. We denote by $\mathbf{x}$ the vector $\mathbf{x} = (x_1, x_2, \cdots, x_n)$. A linear function of $\mathbf{x}$ is of the form

$$l(\mathbf{x}) = a + \alpha_1 x_1 + \alpha_2 x_2 + \cdots + \alpha_n x_n.$$

A function $\mathbf{x} \mapsto f(\mathbf{x})$ is called **differentiable** at $\mathbf{x}_0$ if, near $\mathbf{x}_0$, we have

$$(24) \qquad f(\mathbf{x}) = l(\mathbf{x}) + |\mathbf{x} - \mathbf{x}_0| r(\mathbf{x})$$

where $r(\mathbf{x})$ is continuous at $\mathbf{x}_0$ and $r(\mathbf{x}_0) = 0$. An equivalent definition is: there is a linear function $l(\mathbf{x})$ such that

$$(25) \qquad \lim_{\mathbf{x} \to \mathbf{x}_0} \frac{|f(\mathbf{x}) - l(\mathbf{x})|}{|\mathbf{x} - \mathbf{x}_0|} = 0.$$

This l is called the **linear approximation** to f at $\mathbf{x}_0$. It is clear that a differentiable function is continuous.

Theorem 4   LET $f(\mathbf{x}) = f(x_1, x_2, \cdots, x_n)$ BE DEFINED AT AND NEAR $\mathbf{x}_0$. IF f IS DIFFERENTIABLE AT $\mathbf{x}_0$, THEN IT HAS PARTIAL DERIVATIVES AT THIS POINT; THEIR VALUES ARE THE COEFFICIENTS OF THE COORDINATES IN THE LINEAR APPROXIMATION TO f. IF f HAS PARTIAL DERIVATIVES AT AND NEAR $\mathbf{x}_0$, AND THESE DERIVATIVES ARE CONTINUOUS AT $\mathbf{x}_0$, THEN f IS DIFFERENTIABLE AT $\mathbf{x}_0$.

This generalizes Theorems 1 and 2 stated before; the proof is the same.

We consider mostly continuously differentiable functions, that is, functions that have continuous partial derivatives.

▶*Examples*   *1.* Compute $f_z(2, 6, 0)$ if $f(x, y, z) = \sin(xyz)$.

Answer. Treating x and y as constants and differentiating, we obtain $f_z(x, y, z) = xy \cos(xyz)$. Hence $f_z(2, 6, 0) = 12$.

2. Find the linear approximation to the function $F(x, y, z, t)$: $(x, y, z, t) \mapsto e^{xy} + 2xz^2t - xt^2 - e^2$ at the point $(2, 1, -1, 3)$.

Solution. · The partial derivatives are

$$F_x = ye^{xy} + 2z^2t - t^2, \quad F_y = xe^{xy}, \quad F_z = 4xzt, \quad F_t = 2xz^2 - 2xt.$$

They are clearly continuous everywhere. For $x = 2$, $y = 1$, $z = -1$, $t = 3$, we have

$$F_x = e^2 - 3, \quad F_y = 2e^2, \quad F_z = -24, \quad F_t = -8.$$

We need a linear function with these numbers as coefficients of the variables. At $(2, 1, -1, 3)$, the function should equal $F(2, 1, -1, 3) = -6$. Clearly,

$$-6 + (e^2 - 15)(x - 2) + 2e^2(y - 1) - 24(z + 1) - 8(t - 3)$$

is the desired function. ◄

2.14 The chain rule for any number of variables

The chain rule for more than two variables is the exact analogue of Theorem 3 in §2.6, and is proved in the same way.

Theorem 5 LET THE FUNCTION OF n VARIABLES $F(x_1, x_2, \cdots, x_n)$ BE DIFFERENTIABLE AT THE POINT $P = (\hat{x}_1, \hat{x}_2, \cdots, \hat{x}_n)$. LET THERE BE GIVEN n FUNCTIONS OF m VARIABLES $\phi_1(u_1, \cdots, u_m)$, $\phi_2(u_1, \cdots, u_m)$, $\cdots$, $\phi_n(u_1, \cdots, u_m)$ WHICH ARE DIFFERENTIABLE AT $Q = (\hat{u}_1, \cdots, \hat{u}_m)$ AND SUCH THAT $\phi_j = \hat{x}_j$ AT Q. FORM THE FUNCTION $f(u_1, \cdots, u_m)$ OF m VARIABLES BY SETTING

$$(26) \qquad f(u_1, \cdots, u_m) = F(\phi_1(u_1, \cdots, u_m), \cdots, \phi_n(u_1, \cdots, u_m)).$$

THEN f IS DIFFERENTIABLE AT Q AND, FOR $j = 1, 2, \cdots, m$,

$$\frac{\partial f}{\partial u_j} = \frac{\partial F}{\partial x_1}\frac{\partial \phi_1}{\partial u_j} + \frac{\partial F}{\partial x_2}\frac{\partial \phi_2}{\partial u_j} + \cdots + \frac{\partial F}{\partial x_n}\frac{\partial \phi_n}{\partial u_j}$$

WHERE ALL x DERIVATIVES ARE EVALUATED AT P, AND ALL u DERIVATIVES AT Q.

The same rule applies, of course, if some of the ϕ are functions of only some of the variables u, since a function of less than m variables may be considered as a function of all m variables, with partial derivatives 0 with respect to the "missing" variables.

▶ *Example* Suppose the function $F(x, y, z)$ is differentiable at $x = y = z = 0$ and has the partial derivatives $F_x = 2$, $F_y = F_z = 3$ at this point. Set

$$f(u, v) = F(u - v, u^2 - 1, 3v - 3);$$

find $f_u(1, 1)$ and $f_v(1, 1)$.

Solution. Using Theorem 5, we have

$$\frac{\partial f}{\partial u} = \frac{\partial F}{\partial x}\frac{\partial(u - v)}{\partial u} + \frac{\partial F}{\partial y}\frac{\partial(u^2 - 1)}{\partial u} + \frac{\partial F}{\partial z}\frac{\partial(3v - 3)}{\partial u}$$

$$\frac{\partial f}{\partial v} = \frac{\partial F}{\partial x}\frac{\partial(u - v)}{\partial v} + \frac{\partial F}{\partial y}\frac{\partial(u^2 - 1)}{\partial v} + \frac{\partial F}{\partial z}\frac{\partial(3v - 3)}{\partial v}.$$

For $u = v = 1$, we obtain

$$\left(\frac{\partial f}{\partial u}\right)_{1,1} = 2 \cdot 1 + 3 \cdot 2 + 3 \cdot 0, \quad \left(\frac{\partial f}{\partial v}\right)_{1,1} = 2(-1) + 3 \cdot 0 + 3 \cdot 3.$$

Thus $f_u(1, 1) = 8$, $f_v(1, 1) = 7$. ◀

2.15 Gradients, directional derivatives, and level surfaces

Given a continuously differentiable function $F(x, y, z)$ in an open set of space, the vector whose components are the partial derivatives of F, at some point P is called the **gradient** of F at P, and is denoted by grad F. Thus

$$\operatorname{grad} F = \frac{\partial F}{\partial x}\mathbf{e}_1 + \frac{\partial F}{\partial y}\mathbf{e}_2 + \frac{\partial F}{\partial z}\mathbf{e}_3$$

where $\mathbf{e}_1$, $\mathbf{e}_2$, $\mathbf{e}_3$ are the unit vectors in the direction of the coordinate axes (see Chapter 9, §1.12). The points where grad $F = 0$ are called **critical points** of F.

Now let P have the position vector $\mathbf{P}$ with components (coordinates of P) x_0, y_0, and z_0. Let $\mathbf{n}$ be a unit vector. We recall (see Chapter 9, §6.5) that the components of $\mathbf{n}$ are $\cos \alpha$, $\cos \beta$, $\cos \gamma$, where α, β, and γ are the angles between $\mathbf{n}$ and the coordinate axes:

$$\mathbf{n} = (\cos\alpha)\mathbf{e}_1 + (\cos\beta)\mathbf{e}_2 + (\cos\gamma)\mathbf{e}_3.$$

The point with position vector $\mathbf{P} + t\mathbf{n}$ lies on the line through P in the direction of $\mathbf{n}$, at the distance t from P (see Fig. 12.18); its coordinates are

$$x = x_0 + t\cos\alpha, \; y = y_0 + t\cos\beta, \; z = z_0 + t\cos\gamma.$$

The value of F at this point, that is,

$$F(\mathbf{P} + t\mathbf{n}) = F(x_0 + t\cos\alpha, \; y_0 + t\cos\beta, \; z_0 + t\cos\gamma)$$

is a function of t. The derivative of this function at $t = 0$ is called the **directional derivative** of F at P in the direction $\mathbf{n}$. It measures the rate of change in F as we move from P in the direction $\mathbf{n}$. By Theorem 5, this derivative is equal to

(27) $F_x(x_0, y_0, z_0)\cos\alpha + F_y(x_0, y_0, z_0)\cos\beta + F_z(x_0, y_0, z_0)\cos\gamma.$

Using the concept of inner product introduced in an appendix to Chapter 9 (see §6), we can write the directional derivative as

(28) $(\operatorname{grad} F, \mathbf{n}) = |\operatorname{grad} F|\cos\theta$

where θ is the angle between $\mathbf{n}$ and the gradient.

We conclude from (28) that, at a noncritical point, *the gradient points in the direction of steepest increase of the function, and the length of the gradient is the rate of this increase.* Indeed, the directional derivative (28) is largest if $\cos\theta = 1$, that is, if $\theta = 0$ and $\mathbf{n}$ has the direction of $\operatorname{grad} F$; in this case the directional derivative equals $|\operatorname{grad} F|$.

Suppose now that $z = \phi(x, y)$ is the equation of a surface along which F is constant, that is,

(29) $F(x, y, \phi(x, y)) = c.$

Such a surface is called a **level surface.** Differentiating both sides of (29) with respect to x and to y, we obtain

(30) $F_x + F_z\phi_x = 0, \qquad F_y + F_z\phi_y = 0.$

According to equations (7) of §2.6, the numbers $(\phi_x, \phi_y, -1)$ are the components of a vector perpendicular to the surface $z = \phi(x, y)$ at a point (x, y, z). By (30), we have that $\phi_x = -F_x/F_z$, $\phi_y = -F_y/F_z$, so

Fig. 12.18

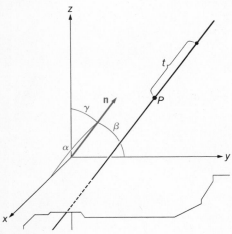

that the vector considered has components

$$\left(-\frac{F_x}{F_z}, \; -\frac{F_y}{F_z}, \; -1\right)$$

and thus is parallel to the gradient (F_x, F_y, F_z). Hence *the gradient at a point is perpendicular to the level surface through this point.*

The same result can be proved if the level surface is represented not by $z = \phi(x, y)$ but by a function $x = \psi(y, z)$ or $y = \chi(z, x)$. One can also prove, though we shall not do this here, that if the derivatives F_x, F_y, F_z are themselves continuously differentiable, then there is a unique level surface through every noncritical point. (If the function represents a temperature distribution, level surfaces are called isothermal surfaces. If the function represents an electric potential, the level surfaces are called equipotentials.)

Everything said in this subsection applies, with obvious changes, to functions of $n > 3$ variables. The details are left to the reader.

EXERCISES

In Exercises 69 to 76 find the partial derivatives f_x, f_y and f_z of the given function.

69. $f(x, y, z) = xyz$.

70. $f(x, y, z) = 2xz^2 + yz^3 - xyz$.

71. $f(x, y, z) = x^2 \sin (yz^2)$.

72. $f(x, y, z) = [e^{x+\sin (zy)}]z^3$.

73. $f(x, y, z) = \arctan (1 + xyz)$.

74. $f(x, y, z) = y \cos [(1 + z^2)x]$.

75. $f(x, y, z) = (x^2 - y^2)/(y^2 + z^2)$.

76. $f(x, y, z) = xyz/(x^2 + 2y^2 + 4z^2)$.

77. If $\phi(r, s, t) = r^{s/t}$, compute $\partial\phi/\partial r$, $\partial\phi/\partial\theta$, $\partial\phi/\partial t$.

78. If $\theta(u, v, w) = e^{uv+w}/w$, compute $\partial\theta/\partial u$, $\partial\theta/\partial v$, $\partial\theta/\partial w$.

79. If $f(\alpha, \beta, \gamma) = \sin (\alpha \cdot \beta^\gamma)$, compute $\partial f/\partial\alpha$, $\partial f/\partial\beta$, $\partial f/\partial\gamma$.

80. If $F(x, y, z) = x^2/a^2 + y^2/b^2 + z^2/c^2$, write the equation of the level surface through the point $(-1, 3, 4)$. Find the normal to this surface at this point, and verify that it is parallel to grad F.

81. Suppose the temperature in x, y, z space is given by $T = x^2y + yz - e^{xy}$. Compute the rate of change of temperature at the point $(1, 1, 1)$ in the direction pointing toward the origin.

▶ 82. Find all critical points of the function $f(x, y, z) = 3x^2y^3z^4$.

83. Compute grad ϕ if $\phi(x, y, z) = \tan yz + e^z \log x$.

84. The density in r, s, t space is given by $\rho = (r^2 + s^2)/(1 + t^2)$. In what direction does the density change most rapidly at the origin? At the point $(-1, 3, 2)$?

85. Compute $\partial F/\partial x$, $\partial F/\partial y$, $\partial F/\partial z$, $\partial F/\partial w$ if $F = xyz/w + xy/zw + x/yzw$.

86. Given that $\phi(u, v, w, t) = \sin (1 + uv^2w^3t)$, find ϕ_u, ϕ_v, ϕ_w, and ϕ_t.

2.16 Functions with vanishing gradients

One of the basic propositions of calculus of one variable states that if $f' = 0$ in an interval, then f is constant in this interval (see Chapter 4, §5.2). An analogous statement holds for functions of several variables.

Theorem 5   IF A FUNCTION f OF SEVERAL VARIABLES HAS PARTIAL DERIVATIVES IN AN OPEN INTERVAL, AND ALL PARTIAL DERIVATIVES ARE 0 THROUGHOUT THE INTERVAL, THEN f IS CONSTANT IN THE INTERVAL.

In other words, if the gradient of a function is **0** in an interval, the function is constant.

Proof.   It suffices to deal with the case of two variables. Let s be a segment, parallel to one of the coordinate axes, in the interval considered. Along this segment s, we may treat f as a function of one variable. The derivative of this function equals either f_x or f_y; hence it is 0, and f is constant along s. Now let P_0 and P_1 be two points in the interval. They can be joined by a path consisting of two such segments, P_0P_2 and P_2P_1 (see Fig. 12.19). Since f is constant along each segment, it takes on the same value at P_0 and at P_2, and the same value at P_2 and at P_1. Thus f takes on the same value at any two points of the interval.

The argument just given applies also to any **connected** open set, that is, to any open set such that any two points of the set can be joined by a path composed of segments parallel to coordinate axes. (Intuitively speaking, a connected set is a set that consists of one piece.)

It is natural to ask whether, or when, one can find a function defined in an interval, with prescribed continuous partial derivatives. We shall answer this question only in the next section (see §§3.2 and 3.10). In preparation, we consider a very useful result.

Fig. 12.19

2.17 Differentiation under the integral sign

We shall state a typical theorem about differentiating a definite integral (with respect to some "variable of integration") when the integrand depends also on another variable, or on several other variables. We call the variable of integration y, and the variable of differentiation x, but it should be clear that the conclusion is independent of the notations used.

It states that under suitable continuity conditions, differentiation with respect to one variable and integration with respect to another can be interchanged.

Theorem 6 LET $f(x, y)$ BE CONTINUOUS, AND HAVE A CONTINUOUS PARTIAL DERIVATIVE $f_x(x, y)$ IN SOME INTERVAL $a \leq x \leq b, c \leq y \leq d$. THEN

$$\frac{d}{dx} \int_c^d f(x, y) \, dy = \int_c^d f_x(x, y) \, dy.$$

The proof of Theorem 6 is not elementary, it will be found in the Supplement (§3).

► *Examples* **1.** Compute $F'(x)$ if

$$F(x) = \int_0^2 (x^2 y + xy^3) \, dy.$$

First solution. We use Theorem 6 with $f(x, y) = x^2 y + xy^3$. We have $f_x(x, y) = 2xy + y^3$, and

$$F'(x) = \frac{d}{dx} \int_0^2 f(x, y) \, dy = \int_0^2 f_x(x, y) \, dy$$

$$= \int_0^2 (2xy + y^3) \, dy = \left[xy^2 + \frac{1}{4} y^4 \right]_{y=0}^{y=2} = 4x + 4.$$

Second solution. We compute directly:

$$F(x) = \int_0^2 (x^2 y + xy^3) \, dy = \left[\frac{1}{2} x^2 y^2 + \frac{1}{4} xy^4 \right]_{y=0}^{y=2} = 2x^2 + 4x,$$

so that $F'(x) = 4x + 4$, as expected.

2. Set

$$u(p, q) = \int_0^1 [e^{x^2} + x \sin (pq^2)] \, dx.$$

Find u_p and u_q.

Answer. Differentiation under the integral sign is legitimate. We have

$$u_p(p, q) = \int_0^1 \frac{\partial}{\partial p} [e^{x^2} + x \sin (pq^2)] \, dx$$

$$= \int_0^1 xq^2 \cos (pq^2) \, dx = q^2 \cos (pq^2) \int_0^1 x \, dx = \frac{1}{2} q^2 \cos (pq^2).$$

In the same way, one sees that $u_q(p, q) = pq \cos (pq^2)$.

3. Find $g'(1)$ if

$$g(x) = \int_0^x e^{xy}\, dy.$$

Answer. The "variable of differentiation" x occurs twice: in the integrand and as the upper limit of integration. It is convenient therefore to consider first the function of two variables

$$G(s, t) = \int_0^s e^{ty}\, dy.$$

We have $g(x) = G(x, x)$, and therefore, by the chain rule

$$g'(x) = G_s(x, x) + G_t(x, x).$$

Now, by the fundamental theorem of calculus,

$$G_s(s, t) = e^{ts}.$$

Also, by Theorem 6,

$$G_t(s, t) = \int_0^s \frac{\partial e^{ty}}{\partial t}\, dy = \int_0^s y e^{ty}\, dy$$

$$= \int_0^s y \left(\frac{d}{dy} \frac{e^{ty}}{t} \right) dy = \left[y \frac{e^{ty}}{t} \right]_{y=0}^{y=s} - \int_0^s \frac{e^{ty}}{t}\, dy$$

$$= \frac{s}{t} e^{ts} - \left[\frac{e^{ty}}{t^2} \right]_{y=0}^{y=s} = \frac{s}{t} e^{ts} - \frac{1}{t^2} e^{ts} + \frac{1}{t^2}.$$

Thus $\quad g'(x) = e^{x^2} + e^{x^2} - \dfrac{1}{x^2} e^{x^2} + \dfrac{1}{x^2} = \left(2 - \dfrac{1}{x^2} \right) e^{x^2} + \dfrac{1}{x^2},$

and finally $g'(1) = e + 1.$◄

EXERCISES

In Exercises 87–92, find $F'(x)$ in two different ways: by integrating first and then differentiating the result; and by differentiating under the integral sign, and then integrating the result.

87. $F(x) = \int_{-1}^2 (t^2/x)\, dt.$

88. $F(x) = \int_a^b (y \cos x + x \cos y)\, dy.$

89. $F(x) = \int_2^3 e^{xy}\, dy.$

90. $F(x) = \int_6^8 \log(xt)\, dt.$

91. $F(x) = \int_0^t (x + y)^n\, dy.$

92. $F(x) = \int_1^2 ds/(s + x).$

93. If $g(x, t) = \int_{-1}^1 \cos(sye^{tv})\, dy$, compute g_s and g_t.

94. If $h(\phi, \psi) = \int_{-1}^1 [(\phi^2 + y)/(\psi + y^2)]\, dy$, compute $\partial h/\partial \phi$ and $\partial h/\partial \psi$.

95. If $f(r) = \int_1^r \cos rt\, dt$, compute $f'(2)$.

96. If $h(\theta) = \int_\theta^{10} \log(\theta + s)\, ds$, compute $h'(5)$.

97. If $H(a) = \int_a^{a^2} e^{ar}\, dr$, compute $H'(2)$.

3.1 Second partial derivatives. Equality of mixed derivatives

If the function $f(x, y)$ has partial derivatives

$$f_x(x, y) = \frac{\partial f(x, y)}{\partial x}, \qquad f_y(x, y) = \frac{\partial f(x, y)}{\partial y},$$

these are themselves functions of two variables and may have partial derivatives. For instance, if $f(x, y) = x^2 y + xy^3$, then

$$f_x = 2xy + y^3, \qquad f_y = x^2 + 3xy^2,$$

and therefore

$$\frac{\partial f_x}{\partial x} = \frac{\partial}{\partial x}\frac{\partial f}{\partial x} = 2y, \qquad \frac{\partial f_x}{\partial y} = \frac{\partial}{\partial y}\frac{\partial f}{\partial x} = 2x + 3y^2$$

$$\frac{\partial f_y}{\partial x} = \frac{\partial}{\partial x}\frac{\partial f}{\partial y} = 2x + 3y^2, \qquad \frac{\partial f_y}{\partial y} = \frac{\partial}{\partial y}\frac{\partial f}{\partial y} = 6xy.$$

Partial derivatives of partial derivatives are called **second partial derivatives,** or partial derivatives of second order.

It may seem that a function of two variables, say x and y, can have four distinct second partials: the x derivative of the x derivative, the y derivative of the x derivative, the x derivative of the y derivative and the y derivative of the y derivative. But in the example just computed ($f = x^2 y + xy^3$), the two "mixed" derivatives:

$$\frac{\partial}{\partial y}\frac{\partial f}{\partial x} \qquad \text{and} \qquad \frac{\partial}{\partial x}\frac{\partial f}{\partial y}$$

turned out to be the same function (namely, $2x + 3y^2$). This was no accident. In fact, the following is true.

Theorem 1 (Equality of mixed derivatives) IF A FUNCTION $f(x, y)$ DEFINED IN SOME OPEN SET HAS ALL SECOND ORDER PARTIAL DERIVATIVES, AND IF THESE DERIVATIVES ARE CONTINUOUS FUNCTIONS, THEN

$$\frac{\partial}{\partial y}\frac{\partial f}{\partial x} = \frac{\partial}{\partial x}\frac{\partial f}{\partial y}.$$

In other words, given a twice continuously differentiable function f, that is, a function whose partial derivatives are themselves continuously differentiable, we obtain the same result if we differentiate f first with respect to x and then with respect to y, or first with respect to y and then with respect to x. One also expresses this by saying: differentiations with respect to two different variables may be interchanged.

It is very easy to verify Theorem 1 for monomials $f(x, y) = Cx^m y^n$ where C is some number and m and n are nonnegative integers. Indeed, in this case, $f_x = mCx^{m-1}y^n$, $(f_x)_y = nmCx^{m-1}y^{n-1}$, $f_y = nCx^m y^{n-1}$, $(f_y)_x = mnCx^{m-1}y^{n-1}$, so that $(f_x)_y = (f_y)_x$ as asserted. It follows that Theorem 1 is true for polynomials, and hence also for rational functions. The proof for the general case will be found in an appendix to this chapter (see §6.3).

The following notations will be used for second partial derivatives:

$$\frac{\partial}{\partial x}\frac{\partial f}{\partial x} = \frac{\partial^2 f}{\partial x^2} = f_{xx}$$

(read: "∂ squared f with respect to x squared" or "second partial of f with respect to x" or "f sub x, x"). Similarly,

$$\frac{\partial}{\partial y}\frac{\partial f}{\partial x} = \frac{\partial^2 f}{\partial y\,\partial x} = f_{xy}, \qquad \frac{\partial}{\partial y}\frac{\partial f}{\partial y} = \frac{\partial^2 f}{\partial y^2} = f_{yy},$$

and so forth. Theorem 1 asserts that $f_{xy} = f_{yx}$.

EXERCISES

1. Verify Theorem 1 for $f(x, y) = \sin(x \cos y)$.
► 2. Prove Theorem 1 for functions of the form $f(x, y) = \Sigma_1^N a_j(x)b_j(y)$, that is, finite sums of products of differentiable functions of one variable.
3. Verify Theorem 1 for $f(x, y) = e^{x^2 y}/(y^2 + x)$.
► 4. Use Theorem 1 to prove that, if $f(x, y)$ has all its third partial derivatives, and these derivatives are all continuous functions, then $(f_{xx})_y = (f_{xy})_x = (f_{yx})_x$.
5. Verify Theorem 1 if $f(x, y) = \int_x^y e^{(tx/y)}\,dt$.

3.2 Functions with prescribed partial derivatives

Let $u(x, y)$ and $v(x, y)$ be continuously differentiable functions defined in some open interval I. Does there exist a function $f(x, y)$ with

$$(1) \qquad\qquad f_x = u, \qquad f_y = v$$

in this interval? Theorem 1 tells us that the answer cannot be yes in all cases, for if (1) holds, then we must also have

$$(2) \qquad\qquad u_y = v_x.$$

One calls this equation the **integrability condition**. If it is satisfied, the answer to our question is affirmative.

Theorem 2   IF $u(x, y)$, $v(x, y)$ ARE CONTINUOUSLY DIFFERENTIABLE FUNC-
TIONS DEFINED IN AN OPEN INTERVAL I, AND $u_y = v_x$ IN I, THEN THERE
EXISTS A FUNCTION $f(x, y)$ IN I SUCH THAT $f_x = u$, $f_y = v$.

Proof.   Note first that, if f satisfies the conditions $f_x = u$, $f_y = v$, and
c is a constant, then the function $f_1 = c + f$ also has the derivatives u
and v. Hence, if our problem (1) has a solution, we may assume that at
some point (x_0, y_0) in I the function f takes on a prescribed value, say A.
   Now let f be a solution with

(3) $$f(x_0, y_0) = A.$$

If (x, y) is some other point of I, then the point (x, y_0) is also in I;
see Fig. 12.20. By (1), (3), and the fundamental theorem of calculus,

$$f(x, y_0) - A = f(x, y_0) - f(x_0, y_0) = \int_{x_0}^{x} f_x(t, y_0)\, dt = \int_{x_0}^{x} u(t, y_0)\, dt,$$

$$f(x, y) - f(x, y_0) = \int_{y_0}^{y} f_y(x, s)\, ds = \int_{y_0}^{y} v(x, s)\, ds$$

and therefore

(4) $$f(x, y) = A + \int_{x_0}^{x} u(t, y_0)\, dt + \int_{y_0}^{y} v(x, s)\, ds.$$

Fig. 12.20

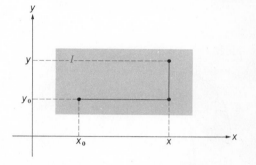

This shows that, given the value $f(x_0, y_0) = A$, our problem has a unique
solution, if any. (That this is so follows also from Theorem 5 in §2.15).
   We show next that, if the integrability condition (2) holds, then the
function f defined by (4) has the required partial derivatives. Indeed, by
the fundamental theorem of calculus and by the theorem about differentiating under the integral sign (Theorem 6 in §2.17) we have, assuming
that $u_y = v_x$,

$$f_x(x, y) = \frac{\partial A}{\partial x} + \frac{\partial}{\partial x} \int_{x_0}^{x} u(t, y_0)\, dt + \frac{\partial}{\partial x} \int_{y_0}^{y} v(x, s)\, ds$$

$$= 0 + u(x, y_0) + \int_{y_0}^{y} v_x(x, s)\, ds = u(x, y_0) + \int_{y_0}^{y} u_y(x, s)\, ds$$

$$= u(x, y_0) + [u(x, y) - u(x, y_0)] = u(x, y).$$

Also,

$$f_y(x, y) = \frac{\partial A}{\partial y} + \frac{\partial}{\partial y} \int_{x_0}^{x} u(t, y_0)\, dt + \frac{\partial}{\partial y} \int_{y_0}^{y} v(x, s)\, ds$$

$$= 0 + 0 + v(x, y) = v(x, y).$$

The theorem is proved.

Fig. 12.21

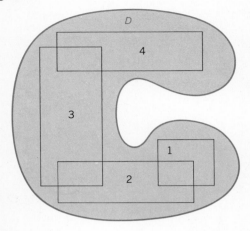

Remark 1   The proof gives a method of computing the function with derivatives u and v, namely, formula (4). It may be easier not to use the formula explicitly, but to proceed as follows. Treat y as a constant, find a "primitive function" ϕ for u treated as a function of x. This gives $f = \phi + c$, the "constant of integration" $c = c(y)$ being a function of y. Now determine $c(y)$ from the conditions: $\phi_y(x, y) + c'(y) = v(x, y)$, $\phi(x_0, y_0) + c(y_0) = A$. One can also interchange the roles of x and y. We illustrate this by Examples 2 and 3 below.

Remark 2   Let D be a connected open set. Given two continuously differentiable functions, u and v, in D satisfying $u_y = v_x$, we may try to find a solution f of $f_x = u$, $f_y = v$ by first defining f in one interval contained in D, then in an overlapping interval, then in a third interval, and so forth (see Fig. 12.21). This may give us a function defined in D, or it may not, for it may happen that going to some point P by two different paths, we arrive at different function values (see Fig. 12.22 and Example 4 below). Such a debacle cannot occur if the set D is what one calls **simply connected,** intuitively speaking: without holes. (In Fig. 12.21, the set D is simply connected; in Fig. 12.22 it is not.) We touch here upon an area in which calculus needs the help of a modern branch of geometry, called topology. We return to this question in §4.

Fig. 12.22

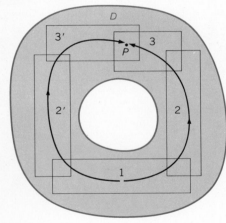

▶ *Examples*   **1.** Is there a function $f(x, y)$ such that $f_x = 5y$, $f_y = 2x$?

Answer.   No. For setting $u = 5y$, $v = 2x$, we have $u_y = 5$, $v_x = 2$; hence $u_y \neq v_x$. Suppose we try to use formula (4) with, say, $x_0 = y_0 = 1$, $A = 3$. This gives

$$f(x, y) = 3 + \int_1^x 5\, dt + \int_1^y 2x\, ds = 3 + 5(x - 1) + 2x(y - 1)$$
$$= 2xy + 3x - 2.$$

Then $f(1, 1) = 3$, $f_y = 2x$, but $f_x = 2y + 3 \neq 5y$.

2. Find a function $f(x, y)$ with

$$f_x = 3x^2y + 2y^2, \qquad f_y = x^3 + 4xy - 1, \qquad f(1, 1) = 4.$$

First solution.   We check the integrability condition:

$$\frac{\partial(3x^2y + 2y^2)}{\partial y} = 3x^2 + 4y, \qquad \frac{\partial(x^3 + 4xy - 1)}{\partial x} = 3x^2 + 4y.$$

Now we apply (4) with $x_0 = y_0 = 1$, $A = 4$, $u(x, y) = 3x^2y + 2y^2$, $v(x, y) = x^3 + 4xy - 1$, and obtain

$$f(x, y) = 4 + \int_1^x (3t^2 + 2)\, dt + \int_1^y (x^3 + 4xs - 1)\, ds$$

$$= 4 + (x^3 - 1) + 2(x - 1) + x^3(y - 1)$$
$$+ (2xy^2 - 2x) - (y - 1)$$

$$= x^3 y + 2xy^2 - y + 2.$$

It is advisable always to check the answer by differentiation:

$$(x^3 y + 2xy^2 - y + 2)_x = 3x^2 y + 2y^2,$$
$$(x^3 y + 2xy^2 - y + 2)_y = x^3 + 4xy - 1,$$

as required.

Second solution. After verifying the integrability condition, we use the equation $f_x = 3x^2 y + 2y^2$ to conclude that

$$f(x, y) = x^3 y + 2xy^2 + c(y).$$

In order to have $f_y = x^3 + 4xy + c'(y) = x^3 + 4xy - 1$,

we need $c(y) = -y + c$, c a constant. Thus

$$f = x^3 + 2xy - y + C,$$

and the condition $f(1, 1) = 4$ gives $C = 2$.

3. Find all functions $\phi(x, y)$ with

$$\phi_x = e^{xy} + xye^{xy} + \cos x + 1, \qquad \phi_y = x^2 e^{xy}.$$

Answer. The integrability condition reads

$$(e^{xy} + xye^{xy} + \cos x + 1)_y = (x^2 e^{xy})_x.$$

It is satisfied; both sides equal $2xe^{xy} + x^2 ye^{xy}$. The second equation ($\phi_y = x^2 e^{xy}$) yields

$$\phi(x, y) = xe^{xy} + c(x)$$

where $c(x)$ is to be determined. In order to satisfy the condition $\phi_x = e^{xy} + xye^{xy} + \cos x + 1$, we need

$$e^{xy} + xye^{xy} + c'(x) = e^{xy} + xye^{xy} + \cos x + 1.$$

Hence $c'(x) = \cos x + 1$, or $c(x) = \sin x + x + C$, C a constant. The final answer

$$\phi(x, y) = xe^{xy} + \sin x + x + C$$

is easily checked by differentiation.

4. The following example illustrates the phenomenon mentioned in Remark 2. Let D denote the set of all points (x, y) with $1 < x^2 + y^2 < 2$. Let D_0 denote the set of all points (x, y) with $1 < x^2 + y^2 < 2$ except the points on the segment from $(1, 0)$ to $(2, 0)$; see Fig. 12.23. Every point (x, y) of D_0 has a unique polar angle θ if we require the inequality $0 < \theta < 2\pi$. Hence $\theta = \theta(x, y)$ is a well-defined function in D_0. This function has the partial derivatives

Fig. 12.23

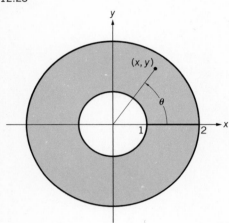

$$\theta_x = -\frac{y}{x^2 + y^2}, \qquad \theta_y = \frac{x}{x^2 + y^2}.$$

This can be verified by expressing $\theta(x, y)$ by inverse trigonometric functions. For instance, for $x > 0$ and $y > 0$, we have $\theta = \arctan(y/x)$; for $y > 0$, we have $\theta = \arccos(x/\sqrt{x^2 + y^2})$, and so forth.

Now set

$$u(x, y) = -\frac{y}{x^2 + y^2}, \qquad v(x, y) = \frac{x}{x^2 + y^2}.$$

These functions have continuous partial derivatives of all orders in D and satisfy the integrability condition $u_y = v_x$. But in D there is no function $f(x, y)$ with $f_x = u$, $f_y = v$. For if there were such a function, then in D_0 it would have to be of the form $\theta + c$, c a constant. Hence at a point like P_1 in Fig. 12.24 it would have to be very close to c, and at a point like P_2 it would be very close to $2\pi + c$. Such an f could not be continuous in D. We remark that the set D_0 is simply connected, while D is not. ◄

Fig. 12.24

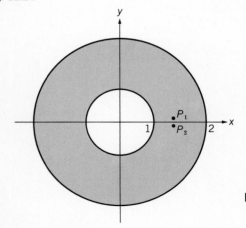

EXERCISES

6. Find a function $f(x, y)$ such that $f_x = 12x^5 + 24x^2y^5$, $f_y = 40x^3y^4 + 80y^9$, $f(0, 0) = 1$.

7. For given constants a, b, c, find $f(x, y)$ such that $f_x \equiv a$, $f_y \equiv b$, $f(x_0, y_0) = c$.

8. Find a function $g(x, y)$ such that $g_x = -e^{x^2}$, $g_y = e^{y^2}$, $g(x, x) \equiv 0$.

9. Suppose $f_x = ax^ny^m$ and $f_y = bx^ry^s$. Prove that $s = m - 1$, $r - 1 = n$, $ma = rb$.

10. Find $f(r, \theta)$ such that $f(1, 0) = 1$, $f_r = r^2 + \sin\theta$, $f_\theta = r\cos\theta + e^\theta$.

11. Find $g(s, t)$ such that $g_s = st + \int_0^s e^{x^2}\, dx$, $g_t = \frac{1}{2}s^2 + \log\sin t$, $g(0, 1) = 2$.

▶ 12. Find a function $f(x, y)$ such that $f_x = \frac{1}{y} e^{x/y}$, $f_y = \frac{-x}{y^2} e^{x/y}$, and $f(1, 1) = 0$.

13. If $f_{xy} = 24x^2y$, $f_{xx} = 24xy^2 + 6x$, $f_y = 8x^3y + 2y$, $f(0, 0) = 0$, find $f(x, y)$.

3.3 Higher derivatives

The process of partial differentiation need not stop with second derivatives. We may be able to find partial derivatives of the second partial derivatives, called third partial derivatives, then fourth partial derivatives, and so forth.

If $f(x, y)$ has all partial derivatives up to, and including those of some order N, and these derivatives are continuous, then f is called an N **times continuously differentiable function.** If so, the "mixed" partial derivatives of f depend only on the number of times we differentiate with respect to each variable, not on the order in which we perform these differentiations. For instance,

$$\frac{\partial}{\partial x}\frac{\partial}{\partial y}\frac{\partial f}{\partial y} = \frac{\partial}{\partial y}\frac{\partial}{\partial x}\frac{\partial f}{\partial y} = \frac{\partial}{\partial y}\frac{\partial}{\partial y}\frac{\partial f}{\partial x},$$

which can be also written as

$$f_{yyx} = f_{yxy} = f_{xyy},$$

and this third order partial derivative is denoted simply by

$$\frac{\partial^3 f}{\partial x\, \partial y^2}$$

(read: "third partial of f, once with respect to x and twice with respect to y"). No new proof is needed; everything follows from Theorem 1 in §3.1. The equality

$$\frac{\partial}{\partial x}\frac{\partial}{\partial y}\frac{\partial f}{\partial y} = \frac{\partial}{\partial y}\frac{\partial}{\partial x}\frac{\partial f}{\partial y}$$

is simply Theorem 1 applied to the function f_y. On the other hand, by Theorem 1 applied to f, we have that

$$\frac{\partial}{\partial x}\frac{\partial f}{\partial y} = \frac{\partial}{\partial y}\frac{\partial f}{\partial x}, \qquad \text{hence} \qquad \frac{\partial}{\partial y}\frac{\partial}{\partial x}\frac{\partial f}{\partial y} = \frac{\partial}{\partial y}\frac{\partial}{\partial y}\frac{\partial f}{\partial x}.$$

A similar reasoning applies to all cases. A function $f(x, y)$ has 4 distinct third partial derivatives (also called "partials"):

$$\frac{\partial^3 f}{\partial x^3}, \qquad \frac{\partial^3 f}{\partial x^2\, \partial y}, \qquad \frac{\partial^3 f}{\partial x\, \partial y^2}, \qquad \frac{\partial^3 f}{\partial y^3},$$

five distinct fourth partials:

$$\frac{\partial^4 f}{\partial x^4}, \qquad \frac{\partial^4 f}{\partial x^3\, \partial y}, \qquad \frac{\partial^4 f}{\partial x^2\, \partial y^2}, \qquad \frac{\partial^4 f}{\partial x\, \partial y^3}, \qquad \frac{\partial^4 f}{\partial y^4},$$

and so forth.

▶ **Example** Find the values of all third partials of $f(x, y) = e^{x^2 y}$ at $x = 0,\ y = 1$.

Answer. We have $f_x = 2xye^{x^2 y}$, $f_y = x^2 e^{x^2 y}$. Hence

$$\begin{aligned} f_{xx} &= (f_x)_x = 2ye^{x^2 y} + 4x^2 y^2 e^{x^2 y}, \\ f_{xy} &= (f_y)_x = 2xe^{x^2 y} + 2x^3 y e^{x^2 y}, \\ f_{yy} &= (f_y)_y = x^4 e^{x^2 y}. \end{aligned}$$

Thus $f_{xx}(0, 1) = 2$, $f_{xy}(0, 1) = f_{yy}(0, 1) = 0$.◀

EXERCISES

14. Find all third partials of $f(x, y) = \sin(xy)$.
15. If $f(x, y) = f(y, x)$ for all x, y, and if we define $g(x, y)$ to be the sum of all third partial derivatives of f, then $g(x, y) = g(y, x)$ for all x, y. Prove this.
16. If $f = xe^{y^2}$, compute f_{xxxx}, f_{xxyy}, and f_{xyxy}.
17. If $f = \log(x^2 + y)$, compute all third derivatives of f.
▶ 18. If $f(x, y) = f(-x, y) = -f(x, -y)$ for all x, y, prove that $f_{xy}(x, y) = -f_{xy}(-x, y) = f_{xy}(x, -y)$ for all x, y and $f_{xxy}(x, y) = f_{xxy}(-x, -y)$ for all x, y.
19. Verify that $f_{xxy} = f_{xyx} = f_{yxx}$ for $f(x, y) = x \cos(xy^2)$.

3.4 Critical points

We recall (see §2.9) that a point at which f_x and f_y are both 0 is called a critical point. For instance, if f has a **local maximum** (or a **local minimum**) at (x_0, y_0), that is, if $f(x, y) \le f(x_0, y_0)$ for all (x, y) near (x_0, y_0) [or if $f(x, y) \ge f(x_0, y_0)$ for all (x, y) near (x_0, y_0)], then (x_0, y_0) is a critical point. Indeed, in this case, the two functions of one variable

$$x \mapsto f(x, y_0), \qquad y \mapsto f(x_0, y)$$

have local maxima (or local minima) at $x = x_0$ and $y = y_0$, and therefore their derivatives are zero at these points. But this means that $f_x(x_0, y_0) = 0$ and $f_y(x_0, y_0) = 0$.

The converse need not be true; at a critical point the function need not have a maximum or a minimum. Whether it does can sometimes be determined by looking at the second partials of f more precisely, at the signs of $f_{xx}f_{yy} - f_{xy}^2$ and $f_{xx} + f_{yy}$ (compare the corresponding statement about functions of one variable in Chapter 4, §4.4).

Before stating the rule, we introduce some terminology. Let $f(x, y)$ be defined near (x_0, y_0). We say that (x_0, y_0) is a **strict minimum point** of $f(x_0, y_0) < f(x, y)$ for all (x, y) near to, and distinct from, (x_0, y_0). If $f(x_0, y_0) > f(x, y)$ for all (x, y) near to, and distinct from, (x_0, y_0), we say that (x_0, y_0) is a **strict maximum point**. Finally, if $f_x = f_y = 0$ at (x_0, y_0) and in *every* interval with midpoint (x_0, y_0) there are points (x_1, y_1) with $f(x_1, y_1) < f(x_0, y_0)$ and also points (x_2, y_2) with $f(x_2, y_2) > f(x_0, y_0)$, then (x_0, y_0) is called a **saddle point.**

For instance, if $f(x, y) = xy$, then the origin is a saddle point. Indeed, $f = f_x = f_y = 0$ at the origin, $f(x, y) > 0$ in the first and third quadrants, and $f(x, y) < 0$ in the second and fourth quadrants.

Theorem 3   IF $f(x, y)$ IS THREE TIMES CONTINUOUSLY DIFFERENTIABLE AT (x_0, y_0), AND IF AT THIS POINT $f_x = f_y = 0$ AND

IF AT THIS POINT	THEN THE POINT IS
$f_{xx}f_{yy} - f_{xy}^2 < 0$	A SADDLE POINT
$f_{xx}f_{yy} - f_{xy}^2 > 0,$ $f_{xx} + f_{yy} > 0$	A STRICT MINIMUM POINT
$f_{xx}f_{yy} - f_{xy}^2 > 0,$ $f_{xx} + f_{yy} < 0$	A STRICT MAXIMUM POINT

The proof will be found in an appendix (§7.3). A more careful analysis shows that the conclusion holds also for twice continuously differentiable functions.

3.5 Maxima and minima

Suppose we are given a "nice" function $(x, y) \mapsto f(x, y)$ defined in a closed interval or in some other "nice" region and on its boundary, and we want to find the points where the function takes on its largest and its smallest values—its **absolute maximum** and **minimum**. (We take the existence of these points for granted, at this stage. A proof will be found in the Supplement, §2.6). The maximum and minimum *may* be achieved on the boundary. If the absolute maximum (or the absolute minimum) of f is attained at a point P not on the boundary of the region considered, it must be a local maximum (or a local minimum) point, and hence a critical point.

This suggests the following procedure for finding the absolute maxima and minima. First, find all critical points inside the region, by solving the equations $f_x(x, y) = 0, f_y(x, y) = 0$. Second, determine which of these critical points are local maxima and local minima; Theorem 3 may be helpful. Third, find the points on the boundary of the region that may be maxima or minima points; this is reduced to finding maxima and minima of functions of one variable. Finally, compare the values of the function at the points considered.

▶ **Example**   What is the largest value (maximum) and smallest value (minimum) of the function

$$f(x, y) = 2x^2 - xy + y^2 + 7x$$

in the closed interval

$$-3 \leq x \leq 3, \qquad -3 \leq y \leq 3,$$

and where is it achieved?

Answer.   We have

$$f_x = 4x - y + 7, \qquad f_y = -x + 2y.$$

Solving the simultaneous equations $f_x = f_y = 0$, we see that the only critical point is at $x = -2$, $y = -1$; it lies in the interval considered. Since $f_{xx} = 4$, $f_{xy} = -1$, $f_{yy} = 2$, we have $f_{xx}f_{yy} - f_{xy}^2 = 7 > 0$, $f_{xx} + f_{yy} = 6 > 0$. Hence (see Theorem 3 in §3.6) the critical point is a local minimum point.

Next we consider the function f for $y = -3$, that is, we look at the function

$$x \mapsto f(x, -3) = 2x^2 + 10x + 9$$

in the interval $-3 \leq x \leq 3$. The derivative is $4x + 10$. This is seen either by differentiation, or by setting $y = -3$ in the expression for f_x. The derivative is 0 at $x = -5/2$. The values of x at which $f(x, -3)$ can achieve extreme values are, therefore, the endpoints of the interval $x = -3, x = +3$, and the root $x = -5/2$ of the derivative.

In a similar way we investigate the other three edges of our interval; on the edge $y = 3$, the function $x \mapsto f(x, 3)$ has a critical point at $x = -1$. On the two vertical edges, the functions $y \mapsto f(-3, y)$ and $y \mapsto f(3, y)$ have no critical points. Thus the only points at which f may take on an extreme value are the critical point inside the interval, the four vertices, and the two points on the horizontal edges at which $f_x = 0$. These seven points are shown in Fig. 12.25. Now we compute the values of f at the seven "interesting" points and obtain the table

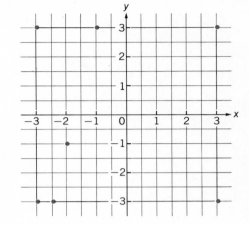

Fig. 12.25

x	-3	-1	3	-2	-3	$-\frac{5}{2}$	3
y	3	3	3	-1	-3	-3	-3
f	15	7	39	-7	-3	$-\frac{7}{2}$	57

Thus the maximum of $f(x, y)$ in the closed interval $|x| \le 3$, $|y| \le 3$ is 57; it is achieved at $x = 3$, $y = -3$. The minimum is -7 and is achieved at $x = -2$, $y = -1$. ◀

EXERCISES

► 20. Show that a rectangular parallelepiped (box) of maximum volume V with pre-scribed surface area S is a cube.

21. Find the volume of the largest rectangular parallelepiped (box) that can be inscribed in the ellipsoid $(x/a)^2 + (y/b)^2 + (z/c)^2 = 1$.

22. Find three numbers x, y, z, such that $x + y + z = 1$ and such that $xy + xz + yz$ is as large as possible.

23. Find the shortest distance from the origin to the plane $ax + by + cz + d = 0$.

24. Find the point on the surface $z = xy - 1$ nearest the origin.

25. a. Find the maximum of $z = 2x + 4y - x^2 - y^2 - 3$ for all x, y.
   b. Find the largest and smallest values of z in the interval $|x| \le 1$, $|y| \le 1$.

► 26. a. Find all critical points of $u = x^3 + y^3 - 3xy$, and classify them as maxima, minima, or saddle points.
   b. Find the largest and smallest values of u in the triangle with vertices $(0, 0)$, $(0, 1)$, $(1, 0)$.

27. If $f(x, y) = e^{x^2 + y^2 + y}$, find the largest and smallest values taken by f in the square with corners at $(1, 0)$, $(0, 1)$, $(-1, 0)$, $(0, -1)$.

28. If $f(x, y) = ax^2 + 2bxy + cy^2$, for which values of a, b, c is there a critical point (x, y) of f with $x^2 + y^2 > 1$?

29. If $f(x, y) = \log(3x^2 + 4y^2 + 2x + 7)$,
   a. Determine the region D in which f is defined.
   b. Find and classify all critical points in D.
   c. Find the maximum and minimum of f in the unit square $0 \le x \le 1$, $0 \le y \le 1$.

3.6 Higher derivatives for more than two variables

The definition of, and notation for, derivatives of higher order for functions of more than two variables are self-evident. If a function of several variables is differentiated several times, the result depends only on how many times we differentiated with respect to each variable, not on the order in which we performed the differentiations. For instance, if $u = f(x, y, z)$, then

$$\frac{\partial}{\partial x}\frac{\partial}{\partial y}\frac{\partial f}{\partial z} = \frac{\partial}{\partial x}\frac{\partial}{\partial z}\frac{\partial f}{\partial y} = \frac{\partial}{\partial y}\frac{\partial}{\partial x}\frac{\partial f}{\partial z} =$$

$$\frac{\partial}{\partial y}\frac{\partial}{\partial z}\frac{\partial f}{\partial x} = \frac{\partial}{\partial z}\frac{\partial}{\partial x}\frac{\partial f}{\partial y} = \frac{\partial}{\partial z}\frac{\partial}{\partial y}\frac{\partial f}{\partial x}$$

and the common value of these six derivatives is denoted by

$$f_{xyz} = \frac{\partial^3 f}{\partial x\,\partial y\,\partial z}.$$

No new proof is needed, the assertion follows by repeating application of Theorem 1 in §3.1. Indeed, if we perform the differentiations with respect to two variables, the others are kept fixed, so that we deal, in effect, with functions of two variables.

▶ **Example**   If $v = xe^{yz}\cos u^2$, find the value $\dfrac{\partial^4 u}{\partial x\,\partial y\,\partial u^2}$ for $x = y = 0$, $z = 1$, $u = \sqrt{\pi}$.

Answer.   We have

$$\frac{\partial v}{\partial x} = e^{yz}\cos u^2, \qquad \frac{\partial^2 v}{\partial x\,\partial y} = \frac{\partial}{\partial y}\frac{\partial v}{\partial x} = ze^{yz}\cos u^2,$$

$$\frac{\partial^3 v}{\partial x\,\partial y\,\partial u} = \frac{\partial}{\partial u}\frac{\partial^2 v}{\partial x\,\partial y} = -ze^{yz}\sin u^2\, 2u,$$

$$\frac{\partial^4 v}{\partial x\,\partial y\,\partial u^2} = \frac{\partial}{\partial u}\frac{\partial^3 v}{\partial x\,\partial y\,\partial u} = -2ze^{yz}\sin u^2 - 4u^2 ze^{yz}\cos u^2.$$

The value of this derivative at $x = y = 0$, $z = 1$, $u = \sqrt{\pi}$ is 4π. ◀

EXERCISES

For the given functions, compute the indicated derivatives at the specified points.

▶ 30. $f = a\cos b\sin t^2$; $\dfrac{\partial^3 f}{\partial a\,\partial b\,\partial t}$, $a = 1$, $b = \dfrac{\pi}{2}$, $t = \sqrt{\pi}$.

31. $g = a^{(bc)}\log(d\sin t)$; g_{abcdt}; $t = \pi/4$, $a = b = c = d = 1$.

32. $H = uvw + uv + uw + vw + u + v + w$; $h_{uv} + h_{uw} + h_{vw}$ at $(1, 1, 1)$.

33. $F = (\cos x_1)(\sin x_1)(\cos x_2)(\sin x_2)\cdots(\cos x_N)(\sin x_N)$. Compute $\dfrac{\partial^N F}{\partial x_1\cdots\partial x_N}$ at

$x_1 = x_2 = \cdots = x_N = \dfrac{\pi}{6}$.

34. $F = \log \{x \log [y(\log z)]\}$. Compute $F_{xyz}(a, a, a)$.

35. $G = (u + v + w + x + y + z)^{100}$. Suppose g is a partial derivative of G of order 100. What is g?

► 36. $H = (3u + 4v + 5w)^t$. Compute all third partials of H.

37. How many distinct third partial derivatives exist in general for a sufficiently differentiable function of r variables?

3.7 Functions with prescribed gradients

If $f(x_1, x_2, \cdots, x_n)$ is a twice continuously differentiable function of n variables, with the partial derivatives

$$(5) \qquad u_i(x_1, \cdots, x_n) = \frac{\partial f(x_1, \cdots, x_n)}{\partial x_i}, \qquad i = 1, 2, \cdots, n,$$

then these derivatives satisfy $\frac{1}{2}n(n - 1)$ **integrability conditions:**

$$(6) \qquad \frac{\partial u_i}{\partial x_j} = \frac{\partial u_j}{\partial x_i}, \, i = 1, \cdots, n; \, j = 1, \cdots, n; \, i \neq j.$$

Indeed, both sides of (6) are equal to

$$\frac{\partial^2 u}{\partial x_i \, \partial x_j}.$$

Conversely, if we are given, in some interval I, n continuously differentiable functions u_i satisfying the integrability conditions, then there *is* a function f in I such that $\partial f/\partial x_i = u_i$ for all i. The solution of (5) is uniquely determined if we prescribe the value of f at some point of I.

All this is proved exactly as for two variables (see §3.2). For instance, in the case $n = 3$, if u, v, w are functions of (x, y, z) defined in I, and we want to find a function f in I with

$$f_x = u, \qquad f_y = v, \qquad f_z = w,$$

then we need the integrability conditions

$$u_y = v_x, \qquad u_z = w_x, \qquad v_z = w_y.$$

If these are satisfied, and (x_0, y_0, z_0) is a point in I, the function

$$f(x, y, z) = A + \int_{x_0}^{x} u(t, y_0, z_0) \, dt + \int_{y_0}^{y} v(x, s, z_0) \, ds + \int_{z_0}^{z} w(x, y, r) \, dr$$

takes on the value A at (x_0, y_0, z_0) and has the required partial derivatives. The reader may want to prove this, imitating the proof of Theorem 2 in §3.2.

We mention without proof that one can find a function with prescribed partial derivatives in a connected open set D provided (1) the prescribed partials satisfy the integrability conditions and (2) the set D is what one calls simply connected. (This means that any closed curve in D can be deformed into a point without leaving D. In the plane, this condition is equivalent to D's having "no holes.") We return to this question in §4.

▶ *Example*   Find a function $f(x, y, z)$ such that

$$f_x = 2x, \qquad f_y = 2y, \qquad f_z = 2z \qquad \text{and } f(0, 0, 0) = 1.$$

Answer.   The integrability conditions are satisfied. Since $f_x = 2x$ we have $f = x^2 + c_1(y, z)$ where the "constant of integration with respect to x" is a function of y and z. The condition $f_y = 2y$ becomes $(c_1)_y = 2y$. Hence, $c_1(y, z) = y^2 + c_2(z)$ and $f = x^2 + y^2 + c_2(z)$. The condition $f_z = 2z$ becomes $c_2'(z) = 2z$. Hence $c_2(z) = z^2 + c_3$; c_3 is a constant. Thus $f = x^2 + y^2 + z^2 + c_3$. Since $f = 1$ at $x = y = z = 0$, we have $c_3 = 1$. So we get

$$f = x^2 + y^2 + z^2 + 1.$$

Of course, this result could have been guessed. ◀

EXERCISES

In Exercises 38 to 43, determine whether there exists a function $f(x, y, z)$ satisfying the prescribed conditions; if one exists, find it.

38. $f_x = e^{y+z^2}, f_y = xe^{z^2} + \cos y, f_z = 2xze^y e^{z^2}, f(0, 0, 0) = 1.$

39. $f_x = \cos(z \log(x + y)), \quad f_y = \sin(z \log(x + y)), \quad f_z = \cos(yz \log(x + y)), f(0, 0, 0) = 1.$

40. $f_x = [\cos(x \cos y)] \cdot \cos y, \qquad f_y = -x[\cos(x \cos y)] \sin y - \sin z \sin(y \sin z), f_z = -y \cos z \sin(y \sin z), f(0, 0, 0) = 1.$

41. $f_x = ye^{xy}, f_y = xe^{xy} + ze^{-yz}, f_z = ye^{-yz}, f(1, 1, 1) = e + 1/e.$

▶ 42. $f_x = (\cos y)e^{x \cos y} - y/z, f_y = -(\sin y)e^{x \cos y} - x/z, f_z = xy/z^2, f(0, 3, 4) = -1.$

43. $f_x = e^{x/y}, f_y = e^{-x^2/y^2}, f_z = 4, f(1, 1, 1) = 0.$

§4 Line Integrals

There are two ways of generalizing integrals of functions of one variable to several variables: multiple integrals, to be discussed in the next chapter, and integration over curves, curved surfaces, and so forth. We discuss here the simplest and most important case: line integrals.

4.1 Definition and notations

Let C be a piecewise smooth curve in three-dimensional space defined, with respect to a Cartesian coordinate system (x, y, z), by three functions

$$(1) \qquad t \mapsto x(t), \qquad t \mapsto y(t), \qquad t \mapsto z(t), \qquad a \leq t \leq b.$$

Also let $P(x, y, z)$, $Q(x, y, z)$, $R(x, y, z)$ be three continuous functions defined in some set containing the trajectory of C. Under these conditions, one defines a number, called "the **line integral** of $P\,dx + Q\,dy + R\,dz$ over C" and denoted by

$$(2) \qquad \int_C P\,dx + Q\,dy + R\,dz.$$

This number is

$$\int_a^b \{ P[x(t), y(t), z(t)]x'(t) + Q[x(t), y(t), z(t)]y'(t) + R[x(t), y(t), z(t)]z'(t) \}\,dt.$$

Note that we could have guessed at the meaning of (2) by trusting the Leibniz notation: what could dx mean besides $x'(t)\,dt$, for instance?

Similarly, if

$$(3) \qquad t \mapsto x(t), \qquad t \mapsto y(t), \qquad a \leq t \leq b,$$

defines a plane smooth curve C, and $P(x, y)$, $Q(x, y)$ are continuous functions, then

$$(4) \qquad \int_C P\,dx + Q\,dy$$

is an abbreviation for the number

$$(4') \qquad \int_a^b \{ P[x(t), y(t)]x'(t) + Q[x(t), y(t)]y'(t) \}\,dt.$$

And if we are given n functions

$$(5) \qquad t \mapsto x_j(t), \qquad j = 1, 2, \cdots, n; \qquad a \leq t \leq b$$

defining a smooth curve C in n-space, and if $P_j(x_1, x_2, \cdots, x_n)$ are n continuous functions, we set

$$
(6) \quad
\begin{aligned}
\int_C P_1\,dx_1 + P_2\,dx_2 + \cdots + P_n\,dx_n &= \int_C \sum_{j=1}^n P_j\,dx_j \\
&= \int_a^b \left\{ \sum_{j=1}^n P_j[x_1(t), x_2(t), \cdots, x_n(t)]x_j'(t) \right\} dt.
\end{aligned}
$$

► *Examples*   *1.* Let C be the curve

$$x = \sin t, \qquad y = \cos t, \qquad z = t^2, \qquad 0 \leq t \leq 1.$$

Compute $\int_C y \, dx - x \, dy - z^2 \, dz.$

Answer.   The line integral equals

$$\int_0^1 \{(\cos t)(\cos t) - (\sin t)(-\sin t) - (t^2)^2 \, 2t\} \, dt = \int_0^1 (1 - 2t^5) \, dt = \frac{2}{3}.$$

2. Let C be the curve

$$x = t, \qquad y = z = 0, \qquad a \leq t \leq b.$$

Compute $\int_C f(x) \, dx + g(y) \, dy + h(z) \, dz$

where f, g, h are continuous functions.

Answer.   Since $dy/dt = dz/dt = 0$, the line integral is

$$\int_a^b f(t) \, dt.$$

Thus the ordinary integral is a special case of the line integral. ◄

EXERCISES

In Exercises 1 to 4, compute $\int_C P \, dx + Q \, dy + R \, dz$ for given P, Q, R, and C.

1. C: $x = t$, $y = t^2$, $z = t^3$, $2 < t < 3$; $P = e^{xy}$, $Q = \sin x$, $R = xy/z$.
2. C: $x = e^t$, $y = e^{-t}$, $z = t^2$, $0 \leq t \leq 1$; $P = xy$, $Q = x^2 z$, $R = xyz$.
3. C: $x = \theta^3$, $y = \theta$, $z = \theta^2$, $2 \leq \theta \leq 3$; $P = yz/x$, $Q = e^y$, $R = \sin z$.
4. C: $x = s^2 + s^3$, $y = \sqrt{s^2 + 1}$, $z = e^{s^2}$, $-1 \leq s \leq 1$; $P = x^2$, $Q = y^3$, $R = \cos z$.
5. Compute $\int_C y^n \, dx + x^n \, dy$, where C is the curve $x = a \sin \theta$, $y = b \cos \theta$, $0 \leq \theta \leq 2\pi$.
► 6. Compute $\int_C x_1 \, dx_1 + x_2 \, dx_2 + \cdots x_n \, dx_n$, where C is the curve $x_1 = a_1 t$, $x_2 = a_2 t \cdots x_n = a_n t$, $t_0 \leq t \leq t_1$.
7. Compute $\int_C \sum_{i=1}^n x_i^2 \, dx_i$, where C is the curve $x_i = t^i$, $i = 1, 2, \cdots, n$, $t_0 \leq t \leq t_1$.
8. Compute $\int_\Gamma \sum_{j=1}^n t_j^2 \, dt_j$, where Γ is the curve $t_j = \theta^j$, $j = 1, 2, \cdots, n$, $t_0 \leq \theta \leq t$.

4.2 Independence of parameter

The value of a line integral over a curve C does *not* depend on the parameter used to represent the curve.

We verify this for $n = 3$ and for a line integral of the form

(7) $\int_C P \, dx,$

that is, for the case when $Q \equiv R \equiv 0$. But the argument is general. Suppose C is represented by $x = x(t)$, $y = y(t)$, $z = z(t)$, $a \leq t \leq b$. Let $t = \phi(\tau)$, $\alpha \leq \tau \leq \beta$, be an allowable parameter change (see Chapter 11, §3.1). Then $\phi(\alpha) = a$, $\phi(\beta) = b$ and $\phi'(\tau) > 0$; the curve C can be represented by the three functions $\tau \mapsto x(\phi(\tau))$, $\tau \mapsto y(\phi(\tau))$, $\tau \mapsto z(\phi(\tau))$, $\alpha \leq \tau \leq \beta$. If we compute the line integral using the t-representation, we obtain

$$(8) \qquad \int_a^b P[x(t), y(t), z(t)] x'(t) \, dt.$$

Computing (7) using τ, we get

$$\int_\alpha^\beta P[x(\phi(\tau)), y(\phi(\tau)), z(\phi(\tau))] x'(\phi(\tau)) \phi'(\tau) \, d\tau;$$

this is the same as (8), by the substitution rule for ordinary integrals (Chapter 5, §3.5).

4.3 Properties of line integrals

We have

$$(9) \qquad \begin{aligned} &\int_C (A_1 + B_1) \, dx_1 + \cdots + (A_n + B_n) \, dx_n \\ &\qquad = \int_C A_1 \, dx_1 + \cdots + A_n \, dx_n + \int_C B_1 \, dx_1 + \cdots + B_n \, dx_n, \end{aligned}$$

$$(10) \qquad \int_C c A_1 \, dx_1 + \cdots + c A_n \, dx_n = c \int_C A_1 \, dx_1 + \cdots + A_n \, dx_n,$$

c a constant,

$$(11) \qquad \int_{-C} A_1 \, dx_1 + \cdots + A_n \, dx_n = - \int_C A_1 \, dx_1 + \cdots + A_n \, dx_n$$

where $-C$ denotes the curve C traversed in the opposite direction (see Chapter 11, §2.2), and

$$(12) \qquad \begin{aligned} &\int_{C_1 + C_2} A_1 \, dx_1 + \cdots + A_n \, dx_n \\ &\qquad = \int_{C_1} A_1 \, dx_1 + \cdots + A_n \, dx_n + \int_{C_2} A_1 \, dx_1 + \cdots + A_n \, dx_n \end{aligned}$$

where $C_1 + C_2$ denotes the curve C_1 followed by C_2 (see Chapter 11, §2.3).

All four relations are proved by direct calculations. We give only the proof of (11); in order to save writing, we consider the integral (7). If C is represented by $t \mapsto x(t)$, $t \mapsto y(t)$, $t \mapsto z(t)$, $a \leq t \leq b$, then $-C$ can be represented by $s \mapsto x(b - s)$, $s \mapsto y(b - s)$, $s \mapsto z(b - s)$, $0 \leq s \leq b - a$. Hence

$$\int_{-C} P\, dx = \int_0^{b-a} P[x(b-s),\, y(b-s),\, z(b-s)][-x'(b-s)]\, ds.$$

Setting $\tau = b - s$, we have $d\tau = -ds$, and $\tau = b$ for $s = 0$ and $\tau = a$ for $s = b - a$. Hence

$$\int_{-C} P\, dx = \int_b^a P[x(\tau),\, y(\tau),\, z(\tau)]x'(\tau)\, d\tau$$

$$= -\int_a^b P[x(\tau),\, y(\tau),\, z(\tau)]x'(\tau)\, d\tau = -\int_C P\, dx,$$

as asserted.

▶ ***Example*** Let C be the curve shown in Fig. 12.26, traversed once starting at $(0, 0)$. Compute

$$\int_C (2 + y)\, dx + x\, dy.$$

Answer. We have $C = C_1 + C_2$ where C_1 is defined by $x = t$, $y = 0$, $0 < t < 1$ and C_2 by $x = \cos\theta$, $y = \sin\theta$, $0 \le \theta \le \pi/2$. On C_1: $dx = dt$, $dy = 0$. On C_2: $dx = -\sin\theta\, d\theta$, $dy = \cos\theta\, d\theta$. Hence, by (9), (10), and (12),

$$\int_C (2 + y)\, dx + x\, dy = 2\int_C dx + \int_C y\, dx + x\, dy$$

$$= 2\int_{C_1} dx + 2\int_{C_2} dx + \int_{C_1} y\, dx + x\, dy + \int_{C_2} y\, dx + x\, dy$$

$$= 2\int_0^1 dt + 2\int_0^{\pi/2}(-\cos\theta)\, d\theta + \int_0^1 0\cdot dt + \int_0^{\pi/2}(-\sin^2\theta + \cos^2\theta)\, d\theta$$

$$= 2 - 2\sin\theta\Big|_0^{\pi/2} + \frac{1}{2}\sin 2\theta\Big|_0^{\pi/2} = 0. \blacktriangleleft$$

EXERCISES

In Exercises 9 to 17, compute $\int_C P\, dx + Q\, dy + R\, dz$ or $\int_C P\, dx + Q\, dy$ for the indicated curves C and integrands P, Q, R or P, Q.

9. C is the triangle with vertices $(0, 0, 1)$, $(0, 1, 0)$, and $(1, 0, 0)$; traversed in this direction, $P = xyz$, $Q = \log(x + y + z)$, $R = 7$.

10. C is the rectangle with vertices $(0, 0, 1)$, $(1, 0, 1)$, $(1, 1, 1)$, $(0, 1, 1)$; traversed in this direction, $P = x^2 + y^2$, $Q = xyz$, $R = x + y + z$.

11. C is the intersection of the unit sphere and the plane $z = 0$; traversed once in the counterclockwise direction, $P = yz$, $Q = xz$, $R = xy$.

12. C is an ellipse with center at the origin, major axis of length $2a$ lying on the x axis, minor axis of length $2b$; traversed once in the counterclockwise direction, $P = x + y$, $Q = x - y$.

13. C is the perimeter of the semicircle with center at $(0, 4)$, radius 2, lying to the right of the y axis; traversed once in the counterclockwise direction, $P = x^2 - y^2$, $Q = x^3 y$.

Fig. 12.26

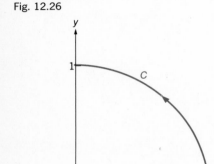

14. C is the unit circle, traversed $1\frac{1}{4}$ times, starting at $(1, 0)$; $P = Ax + By$, $Q = Cx + Dy$ (A, B, C, D are given numbers).

15. C is the perimeter of a regular hexagon with center at the origin and one vertex at $(1, 0)$; it is traversed once in the clockwise direction. $P = 3x^2y - 4y^2x$, $Q = 3y^2x + 4x^2y$.

► 16. C is the circular arc from $(-1, 1)$ through the origin to $(1, 1)$; $P = x + y$, $Q = x - y$.

17. C is the shortest path by which a bug crawling in the x, y plane can get from $(-2, 0)$ to $(1, 0)$ if it is not permitted to enter the unit circle; $P = x^3 + y^2$, $Q = x^2 + y^3$.

► 18. Let C be the curve, $x = \cos\theta$, $y = \sin\theta$, $0 < \theta < 4\pi$, and C_2 the curve $x = \cos t^2$, $y = \sin t^2$, $0 < t < 2\sqrt{\pi}$. For $P = x^3y$, $Q = y^2x$, show that $\int_{C_1} P\,dx + Q\,dy = \int_{C_2} P\,dx + Q\,dy$. Explain why this is so.

19. Let C_1 be the curve $X = e^t$, $y = e^{-t}$, $z = t^2$, $2 \le t \le 4$, and let C_2 be the curve $x = t$, $y = 1/t$, $z = (\log t)^2$, $e^2 \le t \le e^4$. Set $P = e^{xyz}$, $Q = x \sin yz$, $R = y \sin xz$. Prove, without evaluating either integral, that we have $\int_{C_1} P\,dx + Q\,dy + R\,dz = \int_{C_2} P\,dx + Q\,dy + R\,dz$.

20. Let $P(x, y)$ be the larger, and $Q(x, y)$ the smaller, of the two numbers x and y. Let $(x_1(t), y_1(t))$ be the position, at time t, of a bug that crawls around the unit circle at a speed of one radian per hour. Set C_1: $x = x_1(t)$, $y = y_1(t)$, $0 \le t \le 1$, and C_2: $x = x_1(\sqrt{t})$, $y = y_1(\sqrt{t})$, $0 \le t \le 1$. Evaluate $\int_{C_1} P\,dx + Q\,dy$ and $\int_{C_2} P\,dx + Q\,dy$.

21. In Exercises 1 and 2 above (see §4.1), set $t = s^3$ and compute the given integrals using the parameter s.

4.4 The language of differentials

If $t \mapsto \phi(t)$ is a function of one variable, $d\phi$ is an abbreviation for the expression $\phi'(t)\,dt$; we saw in Chapter 5 (see §§3.4 and 3.5) how useful this notation is. We now extend the language to differentials to functions of several variables. If $f(x_1, x_2, \cdots, x_n)$ is such a function, we write formally

$$(13) \qquad df = \frac{\partial f}{\partial x_1}\,dx_1 + \frac{\partial f}{\partial x_2}\,dx_2 + \cdots + \frac{\partial f}{\partial x_n}\,dx_n.$$

"Formally" means: in calculating line integrals, we use the left-hand side of (13) as an abbreviation for the right-hand side.

The chain rule (compare §2.13) is "built into" this language. If the x_i's are functions of m new variables $u_1, u_2, \cdots, u_m$, that is, if $x_i = \phi_i(u_1, \cdots, u_m)$, $i = 1, 2, \cdots, n$, then

$$dx_i = \frac{\partial \phi_i}{\partial u_1}\,du_1 + \frac{\partial \phi_i}{\partial u_2}\,du_2 + \cdots + \frac{\partial \phi_i}{\partial u_m}\,du_m, \qquad i = 1, 2, \cdots, n.$$

If we substitute these expressions into (14) and collect terms, we obtain

$$df = \left(\frac{\partial f}{\partial x_1} \frac{\partial \phi_1}{\partial u_1} + \frac{\partial f}{\partial x_2} \frac{\partial \phi_2}{\partial u_1} + \cdots + \frac{\partial f}{\partial x_n} \frac{\partial \phi_n}{\partial u_1} \right) du_1$$

$$+ \left(\frac{\partial f}{\partial x_1} \frac{\partial \phi_1}{\partial u_2} + \frac{\partial f}{\partial x_2} \frac{\partial \phi_2}{\partial u_2} + \cdots + \frac{\partial f}{\partial x_n} \frac{\partial \phi_n}{\partial u_2} \right) du_2$$

$$+ \cdots\cdots\cdots\cdots\cdots\cdots\cdots\cdots\cdots\cdots\cdots\cdots$$

$$+ \left(\frac{\partial f}{\partial x_1} \frac{\partial \phi_1}{\partial u_m} + \frac{\partial f}{\partial x_2} \frac{\partial \phi_2}{\partial u_m} + \cdots + \frac{\partial f}{\partial x_n} \frac{\partial \phi_n}{\partial u_m} \right) du_m$$

$$= A_1 \, du_1 + A_2 \, du_2 + \cdots + A_m \, du_m$$

where

$$A_j = \frac{\partial f}{\partial x_1} \frac{\partial \phi_1}{\partial u_j} + \frac{\partial f}{\partial x_2} \frac{\partial \phi_2}{\partial u_j} + \cdots + \frac{\partial f}{\partial x_n} \frac{\partial \phi_n}{\partial u_j}.$$

Since A_j is the coefficient of du_j in the expression for df, we have

$$A_j = \frac{\partial f}{\partial u_j},$$

as it must be by the chain rule.

Using the new notation we return to the question considered in §§3.2 and 3.10. Given n functions $A_1, A_2, \cdots, A_n$ of n variables $x_1, x_2, \cdots, x_n$, find, if possible, a function f such that

(14) $$\frac{\partial f}{\partial x_i} = A_i, \; i = 1, 2, \cdots, n.$$

This condition can be written also as

(15) $$A_1 \mathbf{e}_1 + A_2 \mathbf{e}_2 + \cdots + A_n \mathbf{e}_n = \text{grad } F$$

where $\{\mathbf{e}_1, \cdots, \mathbf{e}_n\}$ is a frame, or

(16) $$A_1 \, dx_1 + A_2 \, dx_2 + \cdots + A_n \, dx_n = df.$$

In §3, we solved the problem when all functions considered are defined in an interval. Now we shall consider general regions.

EXERCISES

22. For which x, y is $d(xy) = 0$?
23. If u and v are functions of n variables $(x_1, x_2, \cdots, x_n)$, is it true that $d(u + v) = du + dv$? Is it true that $d(uv) = (du)v + u \, dv$?

In Exercises 24 to 29, compute du for the given function $u = u(x, y, z)$.

24. $u = 8x \cos(zy)$.

25. $u = x^{(\tan yz)}$.

26. $u = \frac{1}{2}\arccos(x + y)$.

27. $u = \log(x^2 + \sin z + ye^x)$.

28. $u = \arctan(xze^y)$.

29. $u = \sinh(xy \operatorname{arc} \tanh z)$.

4.5 Path independence

In this and the next subsection, D denotes some open connected set in n-dimensional space; the most important cases are, of course, $n = 2$ and $n = 3$.

Theorem 1   LET A_1, A_2, $\cdots$, A_n BE CONTINUOUS FUNCTIONS OF $(x_1, x_2, \cdots, x_n)$ DEFINED IN D. IF THERE IS A FUNCTION f DEFINED IN D WITH

$$(17) \qquad df = A_1 \, dx_1 + A_2 \, dx_2 + \cdots + A_n \, dx_n,$$

THEN, FOR EVERY PIECEWISE SMOOTH CURVE IN D LEADING FROM THE POINT P TO THE POINT Q, WE HAVE

$$(18) \qquad \int_C A_1 \, dx_1 + \cdots + A_n \, dx_n = f(Q) - f(P)$$

SO THAT, IN PARTICULAR,

$$(19) \quad \int_C A_1 \, dx_1 + \cdots + A_n \, dx_n = 0 \text{ FOR EVERY CLOSED CURVE } C \text{ IN } D.$$

CONVERSELY, IF (19) HOLDS, THEN THERE IS A FUNCTION f DEFINED IN D AND SATISFYING (17).

The Leibniz notation could have helped us to guess at statement (18), since, if (17) holds, we can rewrite (18) in the form

$$\int_P^Q df = f(Q) - f(P),$$

which looks like the fundamental theorem of calculus. Indeed, the present theorem follows rather simply from the fundamental theorem.

Proof of Theorem 1.   Assume that (17) holds and let C be defined by

$$\mathbf{x} = \mathbf{x}(t), \quad \text{that is,} \quad x_i = x_i(t), \, a \le t \le b, \qquad i = 1, 2, \cdots, n$$

with $\qquad x_i(a) = p_i, \qquad x_i(b) = q_i, \qquad i = 1, 2, \cdots, n$

where $(p_1, p_2, \cdots, p_n)$ and $(q_1, q_2, \cdots, q_n)$ are the coordinates of P and

Q, respectively. By (17), we have that $A_j = \partial f/\partial x_j$, $j = 1, 2, \cdots, n$; this is simply the agreed upon meaning of (17). Set

$$F(t) = f[x_1(t), x_2(t), \cdots, x_n(t)].$$

Then
$$F(a) = f(P), \qquad F(b) = f(Q)$$

and, by the chain rule,

$$F'(t) = \left(\frac{\partial f}{\partial x_1}\right)_{\mathbf{x}(t)} x_1'(t) + \cdots + \left(\frac{\partial f}{\partial x_n}\right)_{\mathbf{x}(t)} x_n'(t)$$
$$= A_1[\mathbf{x}(t)]x_1'(t) + A_2[\mathbf{x}(t)]x_2'(t) + \cdots + A_n[\mathbf{x}(t)]x_n'(t).$$

Now, by the definition of line integrals,

$$\int_C A_1 \, dx_1 + \cdots + A_n \, dx_n = \int_a^b \{A_1[\mathbf{x}(t)]x_1'(t) + \cdots + A_n[\mathbf{x}(t)]x_n'(t)\} \, dt$$
$$= \int_a^b F'(t) \, dt = F(b) - F(a) = f(Q) - f(P).$$

This proves (18).

Assume next that (19) holds. If C_1 and C_2 are two curves in D with the same initial points and the same terminal points, then $C_1 + (-C_2)$ is a closed curve (see Fig. 12.27) so that

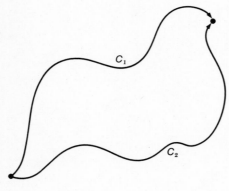

Fig. 12.27

$$0 = \int_{C_1+(-C_2)} A_1 \, dx_1 + \cdots + A_n \, dx_n$$
$$= \int_{C_1} A_1 \, dx_1 + \cdots + A_n \, dx_n + \int_{-C_2} A_1 \, dx_1 + \cdots + A_n \, dx_n$$
$$= \int_{C_1} A_1 \, dx_1 + \cdots + A_n \, dx_n - \int_{-C_2} A_1 \, dx_1 + \cdots + A_n \, dx_n.$$

In other words: condition (19) means that *the line integral*

$$\int_C A_1 \, dx_1 + \cdots + A_n \, dx_n$$

depends only on the endpoints of C; one says in this case that this integral is **path-independent**.

If so, one can define unambiguously a function $f(x_1, x_2, \cdots, x_n)$, by choosing a point $(\hat{x}_1, \hat{x}_2, \cdots, \hat{x}_n)$ and setting

$$f(x_1, \cdots, x_n) = \int_{\hat{\mathbf{x}}}^{\mathbf{x}} A_1 \, dx_1 + \cdots + A_n \, dx_n$$

where the integration is performed along *some* curve in D leading from x̂ to x; since the integral is path-independent, the choice of the curve does not affect the value of $f(x_1, \cdots, x_n)$. It turns out that this f satisfies (17).

We prove this last assertion for the case $n = 2$, though the argument is general. We now write x and y instead of x_1, x_2, and A and B instead of A_1 and A_2. Thus we assume that the integral

$$\int_C A\,dx + B\,dy$$

is path-independent, define a function $f(x, y)$ in D by the relation

$$f(x, y) = \int_{(x_0,y_0)}^{(x,y)} A\,dx + B\,dy,$$

and must show that

(20) $$f_x = A, \qquad f_y = B$$

in D.

Let (x_1, y_1) be some point in D; we show that (20) holds at this point. Let C_0 be a curve leading from (x_0, y_0) to (x_1, y_1), so that

$$f(x_1, y_1) = \int_{C_0} A\,dx + B\,dy.$$

Let C_1 be the horizontal segment joining (x_1, y_1) to $(x_1 + h, y_1)$ where $|h|$ is small. Then (see Fig. 12.28),

$$f(x_1 + h, y_1) = \int_{C_0 + C_1} A\,dx + B\,dy$$

$$= \int_{C_0} A\,dx + B\,dy + \int_{C_1} A\,dx + B\,dy$$

$$= f(x_1, y_1) + \int_{C_1} A\,dx + B\,dy.$$

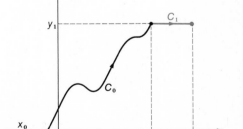

Fig. 12.28

Now C_1 admits the parametric representation $x = x_1 + t$, $y = y_1$, $0 \le t \le h$, if $h \ge 0$ and $x = x_1 - s$, $y = y_1$, $0 \le s \le h$ if $h < 0$. Hence, if $h \ge 0$,

(21) $$f(x_1 + h, y_1) = f(x_1, y_1) + \int_0^h A(x_1 + t, y_1)\,dt.$$

If $h < 0$, $\quad f(x_1 + h, y_1) = f(x_1, y_1) - \int_0^h A(x_1 - s, y_1)\,ds.$

If we set $s = -t$ here, we see that (21) holds in all cases. We differentiate both sides of (21) with respect to h, using the fundamental theorem of calculus, and set $h = 0$. This gives

$$f_x(x_1, y_1) = A(x_1, y_1).$$

The second relation (20) is proved similarly, using for C_1 a vertical segment. The proof of Theorem 1 is complete.

►*Examples*   *1.* Let C be a curve leading from $(1, 2, 3)$ to $(10, 20, 30)$. What is

$$\int_C x\, dx + y\, dy - 2\, dz\,?$$

Answer.   Since, as one can guess almost at once,

$$x\, dx + y\, dy - 2\, dz = d\left(\frac{1}{2}x^2 + \frac{1}{2}y^2 - 2z\right)$$

the integral equals

$$\left(\frac{1}{2}10^2 + \frac{1}{2}20^2 - 2\cdot 30\right) - \left(\frac{1}{2} + \frac{1}{2}2^2 - 2\cdot 3\right) = 193.5.$$

2. We compute once more the example in §4.3. We note that

$$2\, dx = d\, 2x, \quad y\, dx + x\, dy = d\,(xy)$$

so that $\quad\quad (2 + y)\, dx + x\, dy = d\,(2x + xy),$

and that C leads from $(0, 0)$ to $(0, 1)$. At these endpoints $2x + xy = 0$. Therefore the line integral considered is $0 - 0 = 0.$◄

EXERCISES

► 30. Let C be a curve leading from $(-1, 2, 4)$ to $(6, 0, -2)$. Compute the integral $\int_C \cos x\, dx + e^{-y}\, dy + z^2\, dz$?

31. Let C be a curve leading from (x_0, y_0, z_0) to (x_1, y_1, z_1). If A, B, C are given constants, compute $\int_C A\, dx + B\, dy + C\, dz$.

32. Let C be a curve leading from the origin to $(1, 1, 1)$. Compute the integral $\int_C 2\, xy\, dx + (x^2 + 2yz)\, dy + (y^2 + 1)\, dz$.

33. If C is a curve from $(1, 2, 3)$ to $(10, 20, 30)$, compute the integral $\int_C 2xyz^3\, dx + (x^2z^3 + 2y)\, dy + 3x^2\, yz^2\, dz$.

34. Let C be a curve in the x, y plane. Let $R(x, y, z)$ be a given function. Compute $\int_C R(x, y, z)\, dz$.

35. Let C be a curve from (x_0, y_0) to (x_1, y_1). Compute $\int_C e^x\, y^2\, dx + 2ye^x\, dy$.

4.6 Homotopic paths. Simply connected domains

We continue to study the line integral

$$\int_C A_1\, dx_1 + \cdots + A_n\, dx_n,$$

assuming now that the functions A_1, A_2, $\cdots$ are continuously differentiable. We noted in §§3.2 and 3.10 that, if $A_1\,dx_1 + \cdots + A_2\,dx_2 = df$, then the A_i satisfy the $n(n-1)/2$ integrability conditions

(22) $$\frac{\partial A_i}{\partial x_j} = \frac{\partial A_j}{\partial x_i} \qquad \text{for } i \neq j.$$

If the domain we deal with is an interval, the converse is also true, that is, if the integrability conditions hold, then the A_j's are partial derivatives of a function. This is not so for every domain.

Consider, for instance, the two functions

$$A = -\frac{y}{x^2 + y^2}, \qquad B = \frac{x}{x^2 + y^2}$$

which are continuously differentiable in the domain $D =$ the whole plane with the origin removed. Let C be a circle with center at the origin, traversed once in the positive direction. This is a closed curve and

(23) $$\int_C A\,dx + B\,dy = 2\pi.$$

Indeed, C admits the parametric representation $x = R\cos\theta$, $y = R\sin\theta$, $0 \leq \theta \leq 2\pi$. Along C, we have $dx = -R\sin\theta\,d\theta$, $dy = R\cos\theta\,d\theta$, $A = -\sin\theta/R$, $B = \cos\theta/R$. Hence $A\,dx + B\,dy = d\theta$ and (23) follows. But $A\,dx + B\,dy$ satisfies the integrability condition, since

$$\frac{\partial A}{\partial y} = \frac{\partial B}{\partial x} = \frac{y^2 - x^2}{(x^2 + y^2)^2}.$$

The reader will note that the above considerations shed light on Example 4 in §3.2. (The functions A and B are the functions n and v of that example.)

In order to state a positive result valid in any domain, we need a geometric concept for which we give only an intuitive definition. Two curves, C_1 and C_2, in a domain D, with the same endpoints, are called **homotopic** in D if one can be continuously deformed into another, keeping the endpoints fixed, and without leaving the domain D. If we think of the two curves as strings made of a flexible material, the meaning of the term "continuously deformed" becomes clear. The curve C_1 in Fig. 12.29, for instance, can be deformed into the curve C_2 but not into the curve C_3; a deformation of C_1 into C_3 would be obstructed by the hole in D. A precise analytic definition of homotopy may be found in any introductory topology text.

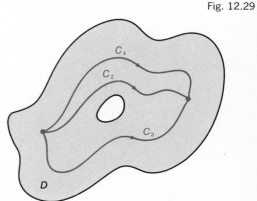

Fig. 12.29

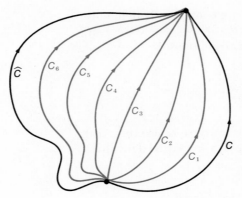

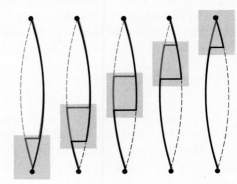

Fig. 12.30

Fig. 12.31

Theorem 2   IF THE FUNCTIONS $A_1, \cdots, A_n$ OF $(x_1, \cdots, x_n)$ ARE CONTINUOUSLY DIFFERENTIABLE IN D, AND SATISFY THE INTEGRABILITY CONDITIONS (22), THEN

$$(24) \qquad \int_C A_1\, dx_1 + \cdots + A_n\, dx_n = \int_{\hat{C}} A_1\, dx_1 + \cdots + A_n\, dx_n$$

FOR ANY TWO CURVES C AND $\hat{C}$ THAT ARE HOMOTOPIC IN D.

We do not give a formal proof but an intuitive argument which shows why Theorem 2 is true. If C can be deformed into $\hat{C}$, then we can find curves C_1, C_2, C_3, $\cdots$, C_N such that not only are all curves C, C_1, C_2, $\cdots$, C_N, $\hat{C}$ homotopic in D, but also C is very close to C_1, C_1 very close to C_2, and so forth (see Fig. 12.30). It suffices to show that the integral of $A_1\, dx_1 + \cdots + A_n\, dx_n$ over C is equal to the integral over C_1, the integral over C_1 is equal to the integral over C_2, and so forth. It will suffice to consider C and C_1.

If two curves are sufficiently close to each other, as we assume C and C_1 to be, one can go from C to C_1 by passing through a succession of several homotopic curves, any two of which run together except inside some small interval that belongs to D (see Fig. 12.31). It suffices to show that the integrals of $A_1\, dx_1 + \cdots + A_n\, dx_n$ over any two such curves are equal.

A typical situation is depicted in Fig. 12.32. We must show that

$$(25) \qquad \int_{K_1+K_2+K_3+K_4} A_1\, dx_1 + \cdots + A_n\, dx_n$$

$$= \int_{K_1+\hat{K}_2+\hat{K}_3+K_4} A_1\, dx_1 + \cdots + A_n\, dx_n.$$

In view of the rules stated in §7.3, this amounts to showing that

$$(26) \qquad \int_{K_2+K_3+(-\hat{K}_2)+(-\hat{K}_3)} A_1\, dx_1 + \cdots + A_n\, dx_n = 0.$$

Fig. 12.32

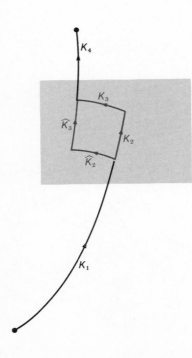

But this is so, for $K_2 + K_3 + (-\hat{K}_2) + (-\hat{K}_3)$ is a closed curve lying in an interval in which the functions $A_1, \cdots, A_n$ are defined and satisfy (22), so that in this interval there is, as we noted before, a function f such that $df = A_1\,dx_1 + \cdots + A_n\,dx_n$. Hence (26) holds, by Theorem 1.

One calls D **simply connected** if any two curves in D with the same endpoints are homotopic. For every n, every n-dimensional interval is simply connected. If $n = 2$, then D is simply connected if and only if D "has no holes."

Theorem 3   IF D IS SIMPLY CONNECTED, AND IF n CONTINUOUSLY DIFFER-ENTIABLE FUNCTIONS $A_1, A_2, \cdots, A_n$ DEFINED IN D SATISFY THE INTEGRA-BILITY CONDITIONS, THEN THERE IS A FUNCTION f DEFINED IN D SUCH THAT $df = A_1\,dx_1 + \cdots + A_n\,dx_n$.

This follows by combining Theorems 1 and 2.

EXERCISES

36. Find a function $f(x, y, z)$ defined for all $x,\ y,\ z$ such that $df = z^2\,dx + 2xz\,dz$.
37. Find a function $f(x, y, z)$ defined for all $x,\ y,\ z$, such that

$$df = (y + e^x y^2)\,dx + (x + 2ye^x - z\sin(yz))\,dy - y\sin(yz)\,dz.$$

► 38. Find a function $f(x, y, z)$ such that

$$df = \left(1 + \frac{1}{x + y}\right)dx + \frac{dy}{x + y} + dz.$$

Specify the domain in which your solution is defined.

39. Find a function $f(x, y)$, defined for all $x,\ y$ except $(0, 0)$, such that

$$df = \frac{x\,dx + y\,dy}{x^2 + y^2}.$$

40. For which constants $A,\ B,\ C,\ D$ can one find a function $f(x, y)$ such that $df = x^A y^B\,dx + x^C y^D\,dy$? For which $A,\ B,\ C,\ D$ is $f(x, y)$ defined everywhere in the $x,\ y$ plane?
41. If $f_1(x, y)$ and $f_2(x, y)$ are such that $df_1 = df_2$ for all $x,\ y$, what can you say about $f_1 - f_2$?

In Exercises 42 to 47, let C_1 be the curve defined by $x = \cos t,\ y = \sin t$, $0 \le t \le \pi$. Let C_2 be the curve defined by $x = \sin t,\ y = \cos t,\ \pi/2 \le t \le 3\pi/2$. Using the intuitive concept of homotopy, decide whether these two curves are homotopic in the given region D.

42. D is the whole $x,\ y$ plane.
43. D is the set $x^2 + y^2 > 0$ in the $x,\ y$ plane.
44. The $x,\ y$ plane is regarded as the plane $z = 0$ in $x,\ y, z$ space, and D is the set $x^2 + y^2 + z^2 > 0$.

45. D is the unit sphere $x^2 + y^2 + z^2 = 1$.
46. D is the unit sphere, with the north pole removed, that is, the set of all (x, y, z) with $x^2 + y^2 + z^2 = 1$, $z < 1$.
47. D is the set of all (x, y, z) with $x^2 + y^2 + z^2 = 1$, $-1 < z < 1$.

In Exercises 48 to 50, use the intuitive meaning of "simply connected."

▶ 48. Suppose D is the connected open set in the plane obtained by removing from the interval $|x| < 2$, $|y| < 2$ all points with $x > 0$, $y = 0$. Is D simply connected? What if we remove all points with $x^2 + y^2 < 1$?
49. Is the set of (x, y, z) with $1 < x^2 + y^2 + z^2 < 7$ simply connected?
50. If a set D in the plane is simply connected and we regard it as a set in 3-space, is it then simply connected? What if it were *not* simply connected when regarded as a set in the plane?

Appendixes to Chapter 12

§5 Energy in Curvilinear Motion

5.1 Vector notation for line integrals

In order to apply line integrals like

$$(1) \qquad \int_C F\, dx + G\, dy + H\, dz$$

to mechanics, it is convenient to use vector language. Let $(\mathbf{e}_1, \mathbf{e}_2, \mathbf{e}_3)$ be a frame belonging to our coordinate system (see Chapter 9, §1.12). We combine the three functions defining the curve C into one vector-valued function

$$(2) \qquad t \mapsto \mathbf{x}(t) = x(t)\mathbf{e}_1 + y(t)\mathbf{e}_2 + z(t)\mathbf{e}_3, \qquad a \le t \le b,$$

so that, formally,

$$(3) \qquad d\mathbf{x} = (\mathbf{x}'(t)\mathbf{e}_1 + y'(t)\mathbf{e}_2 + z'(t)\mathbf{e}_3)\, dt.$$

We combine the functions F, G, H into one vector-valued function of the point $\mathbf{x} = x\mathbf{e}_1 + y\mathbf{e}_2 + z\mathbf{e}_3$:

$$(4) \qquad \mathbf{F}(\mathbf{x}) = \mathbf{F}(x, y, z) = F(x, y, z)\mathbf{e}_1 + G(x, y, z)\mathbf{e}_2 + H(x, y, z)\mathbf{e}_3.$$

Recalling the definition of scalar products (see Chapter 9, §6), we write (1) in the form

$$(1') \qquad \int_C (\mathbf{F}, d\mathbf{x}).$$

This notation suggests a mathematical myth depicted in Fig. 12.33. The curve C consists of infinitely many, infinitely small directed segments; each determines an infinitely short vector $d\mathbf{x}$. At each segment take the corresponding value of $\mathbf{F}$ and form the scalar product

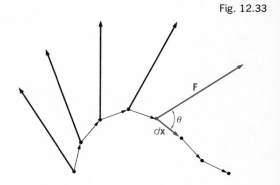

Fig. 12.33

$$(5) \qquad (\mathbf{F}, d\mathbf{x}) = |\mathbf{F}||d\mathbf{x}| \cos \theta$$

where θ is the angle between $\mathbf{F}$ and $d\mathbf{x}$. The integral (2) is the sum of all these infinitely small numbers.

5.2 Work

We recall (see Chapter 5, §5.2) that, if a particle moved the distance s along a line, subject to a constant force $\mathbf{F}$ in the direction of the line, then the work W done against the force is defined to be $-|\mathbf{F}|s$ if the particle moved in the direction of $\mathbf{F}$, and $|\mathbf{F}|s$ if it moved opposite to that direction.

If the particle moves along a directed segment $\overrightarrow{PQ}$, subject to a constant force $\mathbf{F}$, and the angle between $\mathbf{F}$ and $\overrightarrow{PQ}$ is θ (see Fig. 12.34), the work W done against the force is then defined to be $-|\mathbf{F}||\overrightarrow{PQ}| \cos \theta$. If we denote by $\mathbf{s}$ the vector determined by $\overrightarrow{PQ}$, then

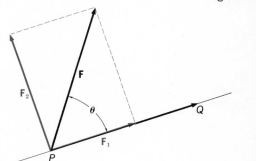

Fig. 12.34

$$(6) \qquad W = -(\mathbf{F}, \mathbf{s}).$$

(One arrives at this definition by setting $\mathbf{F} = \mathbf{F}_1 + \mathbf{F}_2$ where $\mathbf{F}_1$ has the direction of the line PQ and $\mathbf{F}_2$ is perpendicular to this line. The sensible assumption that no work is done against $\mathbf{F}_2$ leads to the conclusion:

$$W = -|\mathbf{F}_1||s| \qquad \text{if } 0 \leq \theta \leq \pi/2, \qquad W = |\mathbf{F}_1||s| \qquad \text{if } \pi/2 \leq \theta \leq \pi.$$

Since $|\mathbf{F}_1| = |\mathbf{F}| \cos \theta$ in the first case and $|\mathbf{F}_1| = -|\mathbf{F}| \cos \theta$ in the second, (6) follows.)

Suppose now that a particle moves in a force field described by the function (4). This means that, when the particle is at the point with position vector $\mathbf{x}$, the force acting on it is $\mathbf{F}(\mathbf{x})$. If our particle is moved along the curve C described by (2), we define the work done against the field on the particle as

$$(7) \qquad W = -\int_C (\mathbf{F}, d\mathbf{x}).$$

(This definition is justified by the mathematical myth told above: W is the sum of numbers (5), each of which represents the work done against the force along an infinitely small segment of C.)

We know (see §4.2) that the integral (7) does not depend on the parametric representation used to define C. This means that the work does not depend on the speed with which the particle moved along C.

▶ *Examples* **1.** Near the surface of the earth, the gravity force acting at the point (x, y, z) is

$$\mathbf{F} = -mg\mathbf{e}_3$$

if we choose $z = 0$ as the surface of the earth, make $\mathbf{e}_3$ point up, and denote by g the acceleration of gravity and by m the mass of the particle. The work done by the gravitational field in moving the particle from the point (x_0, y_0, z_0) to the point (x_1, y_1, z_1) along the curve C is

$$-\int_C -mg \, dz = mg \int_C dz = mg(z_1 - z_0).$$

It depends only on the z coordinates of the endpoints, that is, on the difference in heights.

2. Suppose

$$\mathbf{F}(x, y, z) = \mathbf{e}_1 + x\mathbf{e}_2.$$

Consider two curves with the same endpoints $(0, 0, 0)$ and $(0, 1, 0)$:

$$C_1: \mathbf{x}(t) = t\mathbf{e}_2, \text{ that is, } x = 0, y = t, z = 0, 0 \le t \le 1$$

and

$$C_2: \mathbf{x}(t) = (\sin t)\mathbf{e}_1 + t\mathbf{e}_2, \text{ that is, } x = \sin t, y = t, z = 0, 0 \le t \le \pi.$$

The work done along these two curves is

$$\int_{C_1} (\mathbf{F}, d\mathbf{x}) = \int_{C_1} dx + x \, dy = \int_0^\pi 0 \, dt = 0$$

and

$$\int_{C_2} (\mathbf{F}, d\mathbf{x}) = \int_{C_2} dx + x \, dy = \int_0^\pi \cos t \, dt + \sin t \, dt = 2.$$

Thus the work depends on the path, not only on the endpoints. ◄

EXERCISES

In Exercises 1 to 6, for each given field of force $\mathbf{F}$ and each given curve C, compute the work W needed to move against $\mathbf{F}$ over C.

1. $\mathbf{F} = 3\mathbf{e}_1 + 4\mathbf{e}_2 + 5\mathbf{e}_3$; C is the line segment from (x_0, y_0, z_0) to (x_1, y_1, z_1).
2. $\mathbf{F} = x\mathbf{e}_1 + y\mathbf{e}_2 + z\mathbf{e}_3$; C is the line segment from (x_0, y_0, z_0) to (x_1, y_1, z_1).
3. $\mathbf{F} = 3\mathbf{e}_1 + 4\mathbf{e}_2 + 5\mathbf{e}_3$; C is the broken line that goes from (x_0, y_0, z_0) to (x_1, y_1, z_1) and from there to (x_2, y_2, z_2).
4. $\mathbf{F}$ as in Exercise 2, C as in Exercise 3.
5. $\mathbf{F} = y\mathbf{e}_1 + z\mathbf{e}_2 + y\mathbf{e}_3$; C is the great circle cut out of the unit sphere by the plane $x = y$, traversed once in the direction that is clockwise when seen from the point $(-1/\sqrt{2}, -1/\sqrt{2}, 0)$.
6. $\mathbf{F} = x^2 y \mathbf{e}_1 + x^2 z \mathbf{e}_2 + yz^2 \mathbf{e}_3$; C is the triangle from the origin to $(1, 1, 0)$, to $(0, 0, 1)$ and back to the origin.

5.3 Potential energy

A force field $\mathbf{F}(x, y, z)$ is called **conservative** if no work is done in moving a particle along a closed path:

$$\int_C (\mathbf{F}, d\mathbf{x}) = 0 \qquad \text{if } C \text{ is closed.}$$

For instance the field in Example 1 in §5.2 is conservative but the field in Example 2 is not.

We know (Theorem 1 in §4.5) that a field is conservative if and only if the work done in moving a particle along a path depends only on the endpoints of the path, and if and only if there is a function with gradient $\mathbf{F}$. Then there also is a function whose gradient is $-\mathbf{F}$. This function is

$$(8) \qquad W(x, y, z) = - \int_{(x_0, y_0, z_0)}^{(x, y, z)} (\mathbf{F}, d\mathbf{x})$$

the work done in moving a particle, along any path, from some fixed reference point $x_0 = (x_0, y_0, z_0)$ to $x = (x, y, z)$. That $\mathbf{F}$ is indeed the gradient of $-W$, that is, that the components of the force are the partial derivatives

$$(9) \qquad F = - \frac{\partial W}{\partial x}, \qquad G = - \frac{\partial W}{\partial y}, \qquad H = - \frac{\partial W}{\partial z}$$

follows by the argument used in proving Theorem 1 in §4.5.

One calls $W(x)$ the **potential** of the field, or the **potential energy** of a particle located at x. It is determined by the field only up to an arbitrary additive constant: it is up to us how to choose the point (x_0, y_0, z_0) where $W = 0$, and adding a constant to W does not disturb (9).

If the field has a potential, that is, is conservative, then the integrability conditions are satisfied:

$$F_y = G_x, \qquad F_z = H_x, \qquad G_z = H_y$$

(we always assume that the components of the force field have continuous derivatives). If these conditions hold, and we are in a simply connected region, there is a potential.

The level surfaces of W are called equipotentials. The force is perpendicular to the equipotentials (compare §2.15). At a critical point of W, we have

$$\mathbf{F} = -\operatorname{grad} W = 0$$

and the particle is in equilibrium.

▶ *Examples* *1.* According to Newton's law of universal gravitation (see Chapter 11, §4.12), a mass M located at the origin exerts on a particle with mass m located at (x, y, z) an attractive force $\mathbf{F}$ of magnitude $\gamma mM/r^2$; here m is the mass of the particle, γ the gravitational constant, and $r = \sqrt{x^2 + y^2 + z^2}$, the distance from the origin to (x, y, z). Now, the unit vector pointing from (x, y, z) to the origin is $(-1/|\mathbf{x}|)\mathbf{x}$ and has the components $-x/r, -y/r, -z/r$. Hence F has the components

$$-\gamma mM \frac{x}{r^3}, \qquad -\gamma mM \frac{y}{r^3}, \qquad -\gamma mM \frac{z}{r^3}$$

or

$$-\frac{\gamma mMx}{(x^2 + y^2 + z^2)^{3/2}}, -\frac{\gamma mMy}{(x^2 + y^2 + z^2)^{3/2}}, -\frac{\gamma mMz}{(x^2 + y^2 + z^2)^{3/2}}.$$

One checks easily that these are the negatives of the partial derivatives of

$$W(x, \gamma, z) = \frac{\gamma mM}{r} = \frac{\gamma mM}{\sqrt{x^2 + y^2 + z^2}}.$$

This function is therefore a potential of the gravitational field produced by the mass M.

2. If we restrict ourselves to a small region near the surface of the earth, and choose a coordinate system as in Example 1, §5.2, the gravitational field has the potential

$$W(x, y, z) = mgz, \qquad z \text{ height above earth,}$$

in conformity with Chapter 5, §5.3. Indeed $W_x = W_y = 0$ and $W_z = mg$. ◄

EXERCISES

7. Consider the gravitational force field due to two particles of masses m_1 and m_2, located at $\mathbf{x}_1$ and $\mathbf{x}_2$. Show that this field is conservative by finding a potential.
8. Consider the gravitational force field due to n particles of masses $m_1, m_2, \cdots, m_n$, located at $\mathbf{x}_1, \mathbf{x}_2, \cdots, \mathbf{x}_n$. Show that this field has a potential and is therefore conservative.
9. Is the field given by $\mathbf{F} = x\mathbf{e}_1 + y\mathbf{e}_2 + z^2\mathbf{e}_3$ conservative? If so, what is its potential?

5.4 Conservation of energy

The **kinetic energy** of a moving particle of mass m is defined as

$$K = \frac{1}{2} mv^2, \qquad v = \text{speed,}$$

as in the case of rectilinear motion (see Chapter 5, §5.4). If t represents time, and the motion of our particle is described by

$$t \mapsto \mathbf{x}(t) = x(t)\mathbf{e}_1 + y(t)\mathbf{e}_2 + z(t)\mathbf{e}_3.$$

The velocity vector is $\mathbf{x}'(t)$ and the speed is $|\mathbf{x}'(t)|$, so that

$$K = \frac{1}{2} m |\mathbf{x}'(t)|^2 = \frac{m}{2} (x'(t)^2 + y'(t)^2 + z'(t)^2).$$

We can now state and prove the basic theorem on conservation of energy which contains the result of Chapter 5, §5.4, as a special case.

Theorem 1 IF A PARTICLE MOVES UNDER THE INFLUENCE OF A CONSERVATIVE FORCE FIELD, ITS TOTAL ENERGY

$$E = K + W,$$

(THAT IS, THE SUM OF KINETIC AND POTENTIAL ENERGIES) REMAINS CONSTANT.

Proof. We assume, of course, that the particle obeys Newton's law of motion

$$m\mathbf{x}''(t) = \mathbf{F}(\mathbf{x}(t)).$$

Since $\mathbf{F}$ derives from a potential W, we have $\mathbf{F} = -\operatorname{grad} W$. Hence we also have $m\mathbf{x}'' = -\operatorname{grad} W$ or, in components

$$(10) \quad mx''(t) = -\left(\frac{\partial W}{\partial x}\right)_{\mathbf{x}=\mathbf{x}(t)}, \; my''(t) = -\left(\frac{\partial W}{\partial y}\right)_{\mathbf{x}=\mathbf{x}(t)}, \; mz''(t) = -\left(\frac{\partial W}{\partial z}\right)_{\mathbf{x}=\mathbf{x}(t)}.$$

Now, the energy of our particle at the time t is

$$E(t) = \frac{m}{2} (x'(t)^2 + y'(t)^2 + z'(t)^2) + W(x(t), y(t), z(t)).$$

Using the chain rule, we compute

$$E'(t) = mx'(t)x''(t) + my'(t)y''(t) + mz'(t)z''(t)$$
$$+ \left(\frac{\partial W}{\partial x}\right)_{\mathbf{x}(t)} x'(t) + \left(\frac{\partial W}{\partial y}\right)_{\mathbf{x}(t)} y'(t) + \left(\frac{\partial W}{\partial z}\right)_{\mathbf{x}(t)} z'(t)$$
$$= x'(t)\left\{mx''(t) + \left(\frac{\partial W}{\partial x}\right)_{\mathbf{x}(t)}\right\} + y'(t)\left\{my''(t) + \left(\frac{\partial W}{\partial y}\right)_{\mathbf{x}(t)}\right\}$$
$$+ z'(t)\left\{mz''(t) + \left(\frac{\partial W}{\partial z}\right)_{\mathbf{x}(t)}\right\} = 0$$

by virtue of (10). Hence E is indeed constant.

This theorem is, of course, only a special case of a general principle that dominates all of physics.

EXERCISES

▶ 10. Verify that the motion of a projectile described in Chapter 11, §4.6 obeys the law of conservation of energy.
11. Verify that a planet moving around the sun according to Kepler's laws (see Chapter 11, §§4.10, 4.11, 4.12) obeys the law of conservation of energy.

§6 Proofs of Some Theorems on Partial Derivatives

In this section, we establish the statements made in the preceding sections without proof except for Theorem 6 in §2.16 which is proved in the Supplement, §3.

6.1 Differentiable functions

Suppose that the function $f(x, y)$ has continuous partial derivatives near some point (x_0, y_0). We shall show that f is differentiable at (x_0, y_0) in the sense of §2.3. This will establish Theorem 2 of §2.5. We shall make use of the Mean Value Theorem (Chapter 8, §1.1).

By hypothesis, there is an interval I, with midpoint (x_0, y_0), such that $f(x, y)$ has partial derivatives $f_x(x, y)$ and $f_y(x, y)$ at each point of I. Let (x, y) be any point in I distinct from (x_0, y_0). We have

$$(1) \qquad f(x, y) - f(x_0, y_0) = [f(x, y) - f(x_0, y)] + [f(x_0, y) - f(x_0, y_0)].$$

Keeping y fixed, we consider $f(x, y)$ as a function of x; by the Mean Value Theorem, there is a number u between x and x_0 such that (see Fig. 12.35)

$$(2) \qquad f(x, y) - f(x_0, y) = (x - x_0)f_x(u, y).$$

For the same reason, there is a number v between y and y_0 such that

$$(3) \qquad f(x_0, y) - f(x_0, y_0) = (y - y_0)f_y(x_0, v).$$

Now we set $\qquad f_x(x_0, y_0) = a, \qquad f_y(x_0, y_0) = b, \qquad f(x_0, y_0) = z_0$

and note that

$$(4) \qquad f_x(u, y) = a + [f_x(u, y) - a], \qquad f_y(v, y_0) = b + [f(x_0, v) - b].$$

By (1), (2), (3), and (4), we have

$$
\begin{aligned}
(5) \qquad f(x, y) &= z_0 + a(x - x_0) + b(y - y_0) \\
&\qquad + (x - x_0)[f_x(u, y) - a] + (y - y_0)[f_y(x_0, v) - b] \\
&= z_0 + a(x - x_0) + b(y - y_0) + \sqrt{(x - x_0)^2 + (y - y_0)^2}\, r(x, y)
\end{aligned}
$$

where

$$
r(x, y) = \frac{y - y_0}{\sqrt{(x - x_0)^2 + (y - y_0)^2}}[f_x(u, y) - a]
$$
$$
+ \frac{y - y_0}{\sqrt{(x - x_0)^2 + (y - y_0)^2}}[f_y(x_0, v) - b].
$$

The two fractions to the right have absolute values not exceeding 1. By the triangle inequality, we have

Fig. 12.35

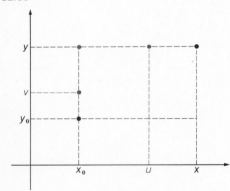

(6) $$|r(x, y)| \leq |f_x(u, y) - a| + |f_y(x_0, v) - b|.$$

The numbers u and v depend on x and y, but u is closer to x_0 than x and v is closer to y_0 than y. Hence, if (x, y) is sufficiently close to (x_0, y_0), the points (u, y) and (x_0, v) will be as close as we want to (x_0, y_0), and the numbers

$$|f_x(u, y) - a| = |f_x(u, y) - f_x(x_0, y_0)|, \qquad |f_y(x_0, v) - b| = |f_y(x_0, v) - f_y(x, y_0)|$$

will be as close to 0 as we want—since u_x and u_y are continuous at (x_0, y_0). We conclude from (6) that

$$\lim_{(x,y) \to (x_0,y_0)} r(x, y) = 0$$

which, together with (5) shows that f is differentiable at (x_0, y_0).

6.2 Chain rule

Now we prove the chain rule (Theorem 3 in §2.7) for functions of two variables. We assume that the functions $F(x, y)$, $\phi(u, v)$, $\psi(u, v)$ are differentiable at (x_0, y_0), (u_0, v_0) and (u_0, v_0), respectively, and that $\phi(u_0, v_0) = x_0$, $\psi(u_0, v_0) = y_0$. We shall show that the composed function $z = f(u, v) = F(\phi(u, v), \psi(u, v))$ is differentiable at (u_0, v_0) and we shall compute its partial derivatives.

To simplify writing, we assume that $x_0 = y_0 = u_0 = v_0 = 0$. The proof in the general case is the same: one must only write $x - x_0$, $y - y_0$, $u - u_0$, $v - v_0$ instead of x, y, u, v. Let

$$F = A, \qquad F_x = B, \qquad F_y = C \qquad \text{at } x = y = 0$$
$$\phi_u = a, \qquad \phi_v = b, \qquad \psi_u = \alpha, \qquad \psi_v = \beta \qquad \text{at } u = v = 0.$$

By hypothesis,

$$F(x, y) = A + Bx + Cy + \sqrt{x^2 + y^2}\, R(x, y)$$
$$\phi(u, v) = au + bv + \sqrt{u^2 - v^2}\, r(u, v)$$
$$\psi(u, v) = \alpha u + \beta v + \sqrt{u^2 + v^2}\, \rho(u, v)$$

where R is continuous and equals 0 at $x = y = 0$, r and ρ are continuous and equal 0 at $u = v = 0$. Therefore

$$
\begin{aligned}
f(u, v) = F(\phi(u, v), \psi(u, v)) &= A + B[au + bv + \sqrt{u^2 + v^2}\, r(u, v)] \\
&\quad + C[\alpha u + \beta v + \sqrt{u^2 + v^2}\, \rho(u, v)] + \sqrt{\phi(u, v)^2 + \psi(u, v)^2}\, R(\phi(u, v), \psi(u, v)) \\
&= A + (Ba + C\alpha)u + (Bb + C\beta)v + B\sqrt{u^2 + v^2}\, r(u, v) \\
&\quad + C\sqrt{u^2 + v^2}\, \rho(u, v) + \sqrt{\phi(u, v)^2 + \psi(u, v)^2}\, R(\phi(u, v), \psi(u, v)) \\
&= A + (Ba + C\alpha)u + (Bb + C\beta)v + \sqrt{u^2 + v^2}\, \hat{R}(u, v)
\end{aligned}
$$

where

$$\hat{R}(u, v) = Br(u, v) + C\rho(u, v) + \left[\frac{\phi(u, v)^2 + \psi(u, v)^2}{u^2 + v^2} \right]^{1/2} R(\phi(u, v), \psi(u, v)).$$

We must show that $\hat{R}(u, v)$ has limit 0 at $u = v = 0$. This will imply that f is differentiable at $u = v = 0$ and has there the partial derivatives $f_x = Ba + C\alpha$, $f_y = Bb + C\beta$ as required by the theorem.

Now $r(u, v)$, $\rho(u, v)$ and $R(\phi(u, v), \psi(u, v))$ are continuous at $u = v = 0$ and are 0 there, the first two functions by hypothesis, the third as the composite of continuous functions. Since B and C are constants, we must only show that the expression $[\cdots]^{1/2}$ remains bounded near $(0, 0)$. But this is so. For if u and v are near 0, then $|r(u, v)| < 1$ and $|\rho(u, v)| < 1$ and therefore

$$|\phi(u, v)| \le (|a| + |b| + 1)\sqrt{u^2 + v^2}, \, |\psi(u, v)| \le (|\alpha| + |\beta| + 1)\sqrt{u^2 + v^2}.$$

Hence if M is the larger of the numbers $(|a| + |b| + 1)$, $(|\alpha| + |\beta| + 1)$, then $[\cdots]^{1/2} \le M$.

6.3 Mixed derivatives

Next we prove Theorem 1 of §3.1 which asserts that, if $f(x, y)$ is defined and has continuous partial derivatives of the first and second order in an open set, then $f_{xy} = f_{yx}$, where

$$f_{xy} = \frac{\partial}{\partial y} \frac{\partial f}{\partial x}, \qquad f_{yx} = \frac{\partial}{\partial x} \frac{\partial f}{\partial y}.$$

We choose a point (x_0, y_0) and consider the expression

(7) $\delta(h) = f(x_0 + h, y_0 + h) - f(x_0 + h, y_0) - f(x_0, y_0 + h) + f(x_0, y_0),$

where $h \ne 0$ and $|h|$ is small. (This is the sum of the values of f at two opposite corners of the square shown in Fig. 12.36 minus the sum of the values at the other corners. In the figure, we have $h > 0$.) We shall show that

(8) $\lim\limits_{h \to 0} \dfrac{\delta(h)}{h} = f_{xy}(x_0, y_0)$

and also

(9) $\lim\limits_{h \to 0} \dfrac{\delta(h)}{h} = f_{yx}(x_0, y_0),$

which proves that the two mixed partials are equal. The proof relies on the Mean Value Theorem (Theorem 1 in Chapter 8, §1.1) for functions of one variable.

Consider the function

(10) $F(x) = f(x, y_0 + h) - f(x, y_0);$

its derivative is

(11) $F'(x) = f_x(x, y_0 + h) - f_x(x, y_0).$

Now

Fig. 12.36

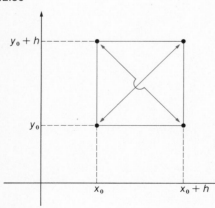

$\delta(h) = F(x_0 + h) - F(x_0)$ [by (7) and (10)]

$\quad = hF'(\xi)$ (where ξ is between 0 and h, by the Mean Value Theorem)

$\quad = h[f_x(\xi, y_0 + h) - f_x(\xi, y_0)]$ [by (11)]

$\quad = h^2 f_{xy}(\xi, \eta)$ (by the Mean Value Theorem; η is between y_0 and $y_0 + h$).

Thus
$$\frac{\delta(h)}{h^2} = f_{xy}(\xi, \eta).$$

The numbers ξ and η depend on h, but if $|h|$ is small enough, the point (ξ, η) is as close as we want to (x_0, y_0). Since f_{xy} is a continuous function, (8) follows.

Next we consider the function

$$G(y) = f(x_0 + h, y) - f(x_0, y)$$

and, using the Mean Value Theorem twice, obtain that

$$\delta(h) = G(y_0 + h) - G(y_0) = hG'(\hat{\eta})$$
$$= hf_y(x_0 + h, \hat{\eta}) - f_y(x_0, \hat{\eta}) = h^2 f_{yx}(\hat{\xi}, \hat{\eta})$$

where $\hat{\xi}$ and $\hat{\eta}$ are numbers, depending on h, which lie between x_0 and $x_0 + h$ and between y_0 and $y_0 + h$, respectively. Hence $\delta(h)/h^2 = f_{yx}(\hat{\xi}, \hat{\eta})$, and (9) follows.

§7 Taylor's Theorem

We shall extend Taylor's Theorem to functions of several variables, and we shall use this theorem to prove Theorem 3 in §3.4 (about critical points).

7.1 Taylor's Theorem for two variables. Special cases

The Mean Value Theorem (see Chapter 8, §1.1) and Taylor's Theorem (see Chapter 8, §2.3) can be extended to functions of two variables by using a very simple device.

Let $f(x, y)$ be defined and several times continuously differentiable near a point (x_0, y_0). We want to use the values of f and of its derivatives up to some order, at (x_0, y_0), in order to represent $f(x, y)$ as a polynomial plus a remainder term which is very small, for (x, y) close to (x_0, y_0). In order to achieve this, we denote the polar coordinates of $(x - x_0, y - y_0)$ by (r, θ), that is, we set

$$x = x_0 + r \cos \theta, \qquad y = y_0 + r \sin \theta$$

(see Fig. 12.37) and consider, for a fixed θ, the value of f at (x, y) as a function of r. In other words, we treat the auxiliary function

$$F(r) = f(x_0 + r \cos \theta, y_0 + r \sin \theta).$$

Using the chain rule we compute the derivatives of F:

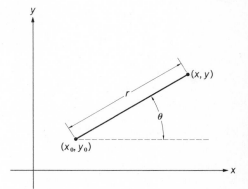

Fig. 12.37

$$F'(r) = f_x(x_0 + r\cos\theta, y_0 + r\sin\theta)\cos\theta + f_y(x_0 + r\cos\theta, y_0 + r\sin\theta)\sin\theta,$$

$$F''(r) = f_{xx}(\cdots)\cos^2\theta + 2f_{xy}(\cdots)\cos\theta\sin\theta + f_{yy}(\cdots)\sin^2\theta,$$

$$F'''(r) = f_{xxx}(\cdots)\cos^3\theta + 3f_{xxy}(\cdots)\cos^2\theta\sin\theta$$
$$+ 3f_{xyy}(\cdots)\cos\theta\sin^2\theta + f_{yyy}(\cdots)\sin^3\theta,$$

$$F^{(4)}(r) = f_{xxxx}(\cdots)\cos^4\theta + 4f_{xxxy}(\cdots)\cos^3\theta\sin\theta + 6f_{xxyy}(\cdots)\cos^2\theta\sin^2\theta$$
$$+ 4f_{xyyy}(\cdots)\cos\theta\sin^3\theta + f_{yyyy}(\cdots)\sin^4\theta,$$

and so forth, as the reader is asked to check. Here the symbol $\cdots$ stands for $x_0 + r\cos\theta, y_0 + r\sin\theta$.

Now we write down the Mean Value Theorem and the various cases of Taylor's Theorem for the function F:

$$F(r) = F(0) + F'(\rho)r,$$

$$F(r) = F(0) + F'(0)r + \frac{1}{2!}F''(\rho)r^2,$$

$$F(r) = F(0) + F'(0)r + \frac{1}{2!}F''(0)r^2 + \frac{1}{3!}F'''(\rho)r^3,$$

$$F(r) = F(0) + F'(0)r + \frac{1}{2!}F''(0)r^2 + \frac{1}{3!}F'''(0)r^3 + \frac{1}{4!}F^{(4)}(\rho)r^4,$$

and so forth; here ρ is a number such that

$$0 < \rho < r,$$

not the same in different formulas. If we substitute the expressions obtained above for F, F', F'', and so forth, note that $r\cos\theta = x - x_0$, $r\sin\theta = y - y_0$, and set

$$\xi = x_0 + \rho\cos\theta, \qquad \eta = y_0 + \rho\sin\theta$$

we obtain the Mean Value Theorem and the various forms of Taylor's Theorem for a function of two variables:

(1_0) $\quad f(x, y) = f(x_0, y_0) + f_x(\xi, \eta)(x - x_0) + f_y(\xi, \eta)(y - y_0),$

(1_1) $\quad f(x, y) = f(x_0, y_0) + f_x(x_0, y_0)(x - x_0) + f_y(x_0, y_0)(y - y_0)$
$$+ \frac{1}{2!}[f_{xx}(\xi, \eta)(x - x_0)^2 + 2f_{xy}(\xi, \eta)(x - x_0)(y - y_0) + f_{yy}(\xi, \eta)(y - y_0)^2],$$

(1_2) $\quad f(x, y) = f(x_0, y_0) + f_x(x_0, y_0)(x - x_0) + f_y(x_0, y_0)(y - y_0)$
$$+ \frac{1}{2!}[f_{xx}(x_0, y_0)(x - x_0)^2 + 2f_{xy}(x_0, y_0)(x - x_0)(y - y_0) + f_{yy}(x_0, y_0)(y - y_0)^2]$$
$$+ \frac{1}{3!}[f_{xxx}(\xi, \eta)(x - x_0)^3 + 3f_{xxy}(\xi, \eta)(x - x_0)^2(y - y_0)$$
$$+ 3f_{xyy}(\xi, \eta)(x - x_0)(y - y_0)^2$$
$$+ f_{yyy}(\xi, \eta)(y - y_0)^3],$$

(1_3) $f(x, y) = f(x_0, y_0) + f_x(x_0, y_0)(x - x_0) + f_y(x_0, y_0)(y - y_0)$

$$+ \frac{1}{2!} [f_{xx}(x_0, y_0)(x - x_0)^2 + 2f_{xy}(x_0, y_0)(x - x_0)(y - y_0) + f_{yy}(x_0, y_0)(y - y_0)^2]$$

$$+ \frac{1}{3!} [f_{xxx}(x_0, y_0)(x - x_0)^3 + 3f_{xxy}(x_0, y_0)(x - x_0)^2(y - y_0)$$

$$+ 3f_{xyy}(x_0, y_0)(x - x_0)(y - y_0)^2 + f_{yyy}(x_0, y_0)(y - y_0)^3]$$

$$+ \frac{1}{4!} [f_{xxxx}(\xi, \eta)(x - x_0)^4 + 4f_{xxxy}(\xi, \eta)(x - x_0)^3(y - y_0)$$

$$+ 6f_{xxyy}(\xi, \eta)(x - x_0)^2(y - y_0)^2 + 4f_{xyyy}(\xi, \eta)(x - x_0)(y - y_0)^3$$

$$+ f_{yyyy}(\xi, \eta)(y - y_0)^4],$$

and so forth. Note that (ξ, η) is some point on the segment joining (x_0, y_0) to (x, y), *not the same* in all formulas (see Fig. 12.38).

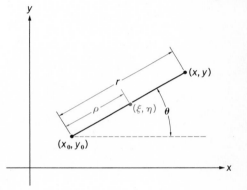

Fig. 12.38

7.2 Taylor's Theorem for two variables. General case

It is now easy to state the general case of Taylor's Theorem. We note that, in the formulas written above, f_x is always multiplied by $(x - x_0)$, f_{xx} by $(x - x_0)^2$, f_{xy} by $(x - x_0)(y - y_0)$, f_{xyy} by $(x - x_0)(y - y_0)^2$, and so forth. Also, the numerical coefficient of a derivative involving p x-differentiations and q y-differentiations is $1/(p!\,q!)$. For instance, f_{xxxx} appears with the coefficient $1/4! = 1/(4!\,0!)$ and f_{xxyy} appears with the coefficient $(1/4!)6 = 1/4 = 1/(2!\,2!)$. Therefore we are led to the following **Taylor formula**:

(2) $$f(x, y) = f(x_0, y_0) + P_n(x, y) + R_{n+1}(x, y)$$

where P_n is the **nth Taylor polynomial** of f at (x_0, y_0), that is, a polynomial of degree n which has at (x_0, y_0) the same value and the same partial derivatives as f, and $R_{n+1}(x, y)$ is a correction term (**remainder**). More precisely, setting

(3) $$a_{pq} = \left(\frac{\partial^{p+q} f}{\partial x^p \, \partial y^q} \right)_{x=x_0, y=y_0}$$

we have

$$P_n = a_{00} + a_{10}(x - x_0) + a_{01}(y - y_0) + \frac{a_{20}}{2!}(x - x_0)^2 + \frac{a_{11}}{1!\,1!}(x - x_0)(y - y_0)$$

$$+ \frac{a_{02}}{2!}(y - y_0)^2 + \frac{a_{30}}{3!}(x - x_0)^3 + \frac{a_{21}}{2!\,1!}(x - x_0)^2(y - y_0)$$

$$+ \frac{a_{12}}{1!\,2!}(x - x_0)(y - y_0)^2 + \frac{a_{03}}{3!}(y - y_0)^3 + \cdots + \frac{a_{n0}}{n!}(x - x_0)^n$$

$$+ \frac{a_{n-1,1}}{(n-1)!\,1!}(x - x_0)^{n-1}(y - y_0) + \frac{a_{n-2,2}}{(n-2)!\,2!}(x - x_0)^{n-2}(y - y_0)^2$$

$$+ \cdots + \frac{a_{0n}}{n!}(y - y_0)^n.$$

This can be written as

$$(4) \qquad P_n = \sum_{j=0}^{n} \sum_{p+q=j} \frac{a_{pq}}{p!\,q!}(x-x_0)^p(y-y_0)^q$$

(remember that $0! = 1$). The remainder term is given by

$$(5) \qquad R_{n+1} = \sum_{p+q=n+1} \frac{\alpha_{pq}}{p!\,q!}(x-x_0)^p(y-y_0)^q$$

where

$$(6) \qquad \alpha_{pq} = \left(\frac{\partial^{p+q}f}{\partial x^p \,\partial y^q}\right)_{x=\xi,\,y=\eta},$$

the point (ξ, η) being some point on the segment joining (x_0, y_0) to (x, y). The formula is valid if $f(x, y)$ is defined and $(n + 1)$ times continuously differentiable in some open set containing this segment. We omit the formal proof involving mathematical induction.

Note that each term in the sum (5) is in absolute value not greater than $[|\alpha_{pq}|/(p!\,q!)]\delta^{n+1}$ where δ is the distance from (x_0, y_0) to (x, y). Hence if we know that the absolute values of all $(n + 1)$st partials of f are not greater than some constant, we can obtain for the remainder an inequality of the form

$$(7) \qquad |R_{n+1}(x, y)| \le \text{const } \delta^{n+1}.$$

This implies that close to (x_0, y_0), that is, for small δ, the remainder is much smaller in absolute value than any term appearing in P_n with a nonzero coefficient.

We mention in passing that it may happen that $\lim_{n \to \infty} R_{n+1}(x, y) = 0$. Then $f(x, y)$ can be represented by a convergent double power series.

We also note that occasionally it is simpler to obtain the Taylor formula for a function of two variables not by using the general formula, but by using the Taylor expansion for some function of one variable, as will be seen from examples.

▶*Examples*   **1.** Find the third Taylor polynomial of the function $f(x, y) = e^{xy}$ at $x = y = 0$.

First solution.   We have $f_x = ye^{xy}$, $f_y = xe^{xy}$, $f_{xx} = y^2e^{xy}$, $f_{xy} = e^{xy} + xye^{xy}$, $f_{yy} = x^2e^{xy}$, $f_{xxx} = y^3e^{xy}$, $f_{xxy} = 2ye^{xy} + xy^2e^{xy}$, $f_{xyy} = 2xe^{xy} + x^2ye^{xy}$, $f_{yyy} = x^3e^{xy}$. Hence, at $x = y = 0$, we have $f = 1$, $f_x = f_y = f_{xx} = f_{yy} = f_{xxx} = f_{xxy} = f_{xyy} = f_{yyy} = 0$, $f_{xy} = 1$. Thus the desired Taylor polynomial is $1 + xy$, so that

$$(8) \qquad e^{xy} = 1 + xy + \text{``terms of order higher than 3.''}$$

Second solution.   Since (by Chapter 8, §5.6)

$$e^{xy} = 1 + xy + \frac{(xy)^2}{2!} + \frac{(xy)^3}{3!} + \cdots,$$

we obtain (8).

2. Find the second Taylor polynomial of the function $f(x, y) = (1 + x + y^2)^{1/2}$ at $x = 1$, $y = 0$.

First solution. We have

$$f_x = (1/2)(1 + x + y^2)^{-1/2}, f_y = (1 + x + y^2)^{-1/2}y,$$
$$f_{xx} = (-1/4)(1 + x + y^2)^{-3/2}, f_{xy} = (-1/2)(1 + x + y^2)^{-3/2}y,$$
$$f_{yy} = (1 + x + y^2)^{-1/2} - (1 + x + y^2)^{-3/2}y^2.$$

At $x = 1$, $y = 0$, we have

$$f = \sqrt{2}, f_x = 1/2\sqrt{2}, f_y = 0, f_{xx} = -1/8\sqrt{2}, f_{xy} = 0, f_{yy} = 1/\sqrt{2}.$$

Hence the desired Taylor polynomial is

$$P_2(x, y) = \sqrt{2} + \frac{x - 1}{2\sqrt{2}} - \frac{(x - 1)^2}{16\sqrt{2}} + \frac{y^2}{2\sqrt{2}}$$

$$= \sqrt{2}\left\{1 + \frac{x - 1}{4} - \frac{(x - 1)^2}{32} + \frac{y^2}{4}\right\}.$$

Second solution. Set $x = 1 + \xi$. By the binomial series (see Chapter 8, §5.7) we have, for small $|\xi|$ and $|y|$,

$$(1 + x + y^2)^{1/2} = (2 + \xi + y^2)^{1/2} = \sqrt{2}\left(1 + \frac{\xi + y^2}{2}\right)^{1/2}$$

$$= \sqrt{2}\left\{1 + \binom{\frac{1}{2}}{1}\left(\frac{\xi + y^2}{2}\right) + \binom{\frac{1}{2}}{2}\left(\frac{\xi + y^2}{2}\right)^2 + \cdots\right\}$$

$$= \sqrt{2}\left\{1 + \frac{1}{4}(\xi + y^2) - \frac{1}{32}(\xi + y^2)^2 + \cdots\right\}$$

$$= \sqrt{2}\left\{1 + \frac{1}{4}\xi + \frac{1}{4}y^2 - \frac{1}{32}\xi^2 + \cdots\right\}$$

$$= \sqrt{2}\left\{1 + \frac{x - 1}{4} + \frac{y^2}{4} - \frac{(x - 1)^2}{32} + \cdots\right\}$$

where $\cdots$ stands for terms of order higher than 2 in $(x - 1)$ and y. Thus

$$\sqrt{2} + \frac{\sqrt{2}}{4}(x - 1) - \frac{\sqrt{2}}{32}(x - 1)^2 + \frac{\sqrt{2}}{4}y^2$$

is the desired Taylor polynomial. ◂

7.3 Classification of critical points

We shall use Taylor's Theorem in order to classify the critical points of a function $f(x, y)$, at which not all second derivatives are 0 and at which $f_{xx}f_{yy} - f_{xy}^2 \neq 0$.

To simplify writing, we assume that $f(x, y)$ has a critical point at $(0, 0)$. We also assume that $f(0, 0) = 0$. If f is three times continuously differentiable, and if we denote the values of f_{xx}, f_{xy}, f_{yy} at $(0, 0)$ by A, B, C, respectively, then we have, near the origin,

$$(9) \qquad f(x, y) = \frac{1}{2}(Ax^2 + 2Bxy + Cy^2) + R(x, y)$$

where the remainder term R satisfies, for some constant M, the inequality

$$(10) \qquad |R(x, y)| \leq M(x^2 + y^2)^{3/2}.$$

This follows from Taylor's Theorem; there are no constant or linear terms in the Taylor polynomial, since $f = f_x = f_y = 0$ at $(0, 0)$. We note that by our hypothesis $A^2 + B^2 + C^2 \neq 0$. Now we introduce a new coordinate system (X, Y), obtained by rotating the (x, y) axes by some angle θ; see Fig. 12.39. Then (see Chapter 10, §2.3)

$$(11) \qquad x = X \cos \theta - Y \sin \theta, \qquad y = X \sin \theta + Y \cos \theta.$$

Consider f as a function of the new variables X and Y, that is, consider the function

$$\hat{f}(X, Y) = f(X \cos \theta - Y \sin \theta, X \sin \theta + Y \cos \theta).$$

Substituting into (12) and noting that $X^2 + Y^2 = x^2 + y^2$, we obtain

$$(12) \qquad \hat{f}(X, Y) = \frac{1}{2}(\hat{A}X^2 + 2\hat{B}XY + \hat{C}Y^2) + \hat{R}(X, Y)$$

where $\hat{A}$, $\hat{B}$, $\hat{C}$ are numbers determined by A, B, C, and θ, and $\hat{R}$ satisfies

$$(13) \qquad |\hat{R}(X, Y)| \leq M(X^2 + Y^2)^{3/2}.$$

There is no need to compute $\hat{A}$, $\hat{B}$, $\hat{C}$; we already did this in Chapter 10, §3.3, and we noted there that

$$\hat{A}\hat{C} - \hat{B}^2 = AC - B^2, \qquad \hat{A} + \hat{C} = A + C.$$

We also showed, in Chapter 10, §3.4, that one can choose θ so that $\hat{B} = 0$. Assume that this has been done; then

$$(14) \qquad \hat{f}(X, Y) = \frac{1}{2}(\hat{A}X^2 + \hat{C}Y^2) + \hat{R}(X, Y).$$

If $AC - B^2 = 0$, then $\hat{A}\hat{C} = 0$. In this case, either $\hat{A}$ or $\hat{C}$ is 0; there is no way to tell from (14) whether $\hat{f}$ is always positive or always negative near the origin. Assume now that $AC - B^2 < 0$. Then $\hat{A}\hat{C} < 0$. If $\hat{A} > 0$, $\hat{C} < 0$, then

$$\hat{f}(X, 0) = \frac{1}{2}\hat{A}X^2 + \hat{R}(X, 0)$$

Fig. 12.39

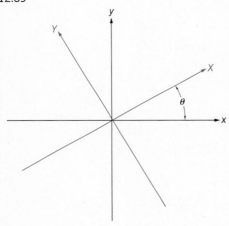

and noting (13) we conclude that, for small positive $|X|$,

$$\hat{f}(X, 0) \geq \frac{1}{2}\hat{A}X^2 - M|X|^3 = \frac{1}{2}\hat{A}|X|^2[1 - (2M/\hat{A})|X|] > 0.$$

One checks similarly that $\hat{f}(0, Y) < 0$ for small positive $|Y|$. Thus f has neither a maximum nor a minimum at $(0, 0)$; one says that f has a saddle point at this point. Near $(0, 0)$ the level curves of f resemble those of $\frac{1}{2}(\hat{A}X^2 - |\hat{C}|Y^2)$, that is, they look like hyperbolas (see §12.39). If $\hat{A} < 0$ and $\hat{C} > 0$, we reach a similar conclusion.

Assume next that $AC - B^2 > 0$, so that $\hat{A}\hat{C} > 0$. If $\hat{A} + \hat{C} > 0$, let α be the smaller of the two positive numbers $\hat{A}$ and $\hat{C}$. We have for small positive $X^2 + Y^2$,

$$\hat{f}(X, Y) \geq \frac{1}{2}(\hat{A}X^2 + \hat{C}Y^2) - M(X^2 + Y^2)^{3/2} \geq \frac{\alpha}{2}(X^2 + Y^2) - M(X^2 + Y^2)^{3/2}$$

$$= \frac{\alpha}{2}(X^2 + Y^2)[1 - \frac{2M}{\alpha}(X^2 + Y^2)^{1/2}] > 0.$$

Thus f has a strict local minimum at the origin: $f(x, y) > f(0, 0)$ for (x, y) near $(0, 0)$ and $(x, y) \neq (0, 0)$. One verifies similarly that if $\hat{A}\hat{C} > 0$ and $\hat{A} + \hat{C} < 0$, then f has at the origin a strict local maximum. Near $(0, 0)$ the level curves of f resemble those of $\frac{1}{2}(\hat{A}X^2 + \hat{C}Y^2)$, that is, they look like ellipses (see §2.11).

The preceding discussion contains the proof of Theorem 3 in §3.4.

7.4 Taylor's Theorem for functions of n variables

One obtains a Taylor formula for a function of n variables from the corresponding result for one variable by repeating the procedure used in §§7.1 and 7.2 for the case $n = 2$. If $f(x_1, x_2, \cdots, x_n) = f(\mathbf{x})$ is defined and $(N + 1)$ times continuously differentiable near a point $\hat{\mathbf{x}} = (\hat{x}_1, \hat{x}_2, \cdots, \hat{x}_n)$, and $\mathbf{x}$ is some point near $\hat{\mathbf{x}}$, we form the unit vector

$$\mathbf{n} = (\cos \alpha_1, \cos \alpha_2, \cdots, \cos \alpha_n)$$

such that the α_j are the direction angles of $\mathbf{x} - \mathbf{x}_0$. Then $\mathbf{x} = \mathbf{x}_0 + r\mathbf{n}$, r a positive number. Now we apply Taylor's Theorem to the function of one variable

$$F(r) = f(\mathbf{x}_0 + r\mathbf{n}).$$

We do not repeat the calculations but state only the final result:

$$f(\mathbf{x}) = P_N(\mathbf{x}) + R_{N+1}(\mathbf{x})$$

where P_N is a polynomial of degree N, the **Taylor polynomial** for f at $\mathbf{x}_0$. This polynomial has at $\mathbf{x}_0$ the same value and the same derivatives, up to the order N, as the function f. It is the sum of all terms of the form

$$\frac{1}{p_1! \, p_2! \, \cdots \, p_n!} \left(\frac{\partial^{p_1 + p_2 + \cdots + p_n}f}{\partial x_1^{p_1} \partial x_2^{p_2} \cdots \partial x_n^{p_n}}\right)_{\mathbf{x}} (x_1 - \hat{x}_1)^{p_1}(x_2 - \hat{x}_2)^{p_2} \cdots (x_n - \hat{x}_n)^{p_n}$$

where

$$p_1 + p_2 + \cdots + p_n \leq N$$

(and, of course, $\partial^0 f / \partial x^0 \cdots \partial x_n{}^0$ is simply f). The **remainder** term R_{N+1} is the sum of all terms of the form

$$\frac{1}{p_1! \, p_2! \, \cdots \, p_n!} \left(\frac{\partial^{N+1} f}{\partial x_1{}^{p_1} \partial x_2{}^{p_2} \cdots \partial p_n{}^{p_n}} \right)_{\xi} (x_1 - \hat{x}_1)^{p_1} (x_2 - \hat{x}_2)^{p_2} \cdots (x_n - \hat{x}_n)^{p_n}$$

where

$$p_1 + p_2 + \cdots + p_n = N + 1$$

and $\boldsymbol{\xi}$ is some point on the segment joining $\hat{\mathbf{x}}$ to $\mathbf{x}$. The remainder satisfies, near $\hat{\mathbf{x}}$, the inequality

$$|R_{N+1}(\mathbf{x})| \leq M|\hat{\mathbf{x}} - \mathbf{x}|^{N+1}$$

with some appropriate constant M.

▶**Example**   Find the second Taylor polynomial of the function $f(x, y, z) = \log (z^2 + xy)$ at $x = y = 0$, $z = 1$.

First answer.   The desired Taylor polynomial is

$$P_2 = f(0, 0, 1) + f_x(0, 0, 1)x + f_y(0, 0, 1)y + f_z(0, 0, 1)(z - 1)$$

$$+ \frac{1}{2!} f_{xx}(0, 0, 1)x^2 + \frac{1}{2!} f_{yy}(0, 0, 1)y^2 + \frac{1}{2!} f_{zz}(0, 0, 1)(z - 1)^2$$

$$+ f_{xy}(0, 0, 1)xy + f_{xz}(0, 0, 1)x(z - 1) + f_{yz}(0, 0, 1)y(z - 1).$$

But

$$f_x = \frac{y}{z^2 + xy}, f_y = \frac{x}{z^2 + xy}, f_z = \frac{2z}{z^2 + xy}$$

$$f_{xx} = -\frac{y^2}{(z^2 + xy)^2}, f_{yy} = -\frac{x^2}{(z^2 + xy)^2}, f_{zz} = \frac{z}{z^2 + xy} - \frac{4z^2}{(z^2 + xy)^2},$$

$$f_{xz} = -\frac{2zy}{(z^2 + xy)^2}, f_{xy} = \frac{1}{z^2 + xy} - \frac{xy}{(z^2 + xy)^2}, f_{yz} = -\frac{2zx}{(z^2 + xy)^2}$$

so that, at $x = y = 0$, $z = 1$

$$f = 0, f_x = f_y = 0, f_z = 2, f_{xx} = f_{yy} = 0,$$
$$f_{zz} = -2, f_{xz} = f_{yz} = 0, f_{xy} = 1$$

and

$$P_2 = 2(z - 1) - (z - 1)^2 + xy.$$

Second answer.   We use the Taylor series for the logarithmic function (see Chapter 8, §5.6). Setting $z = 1 + \zeta$ and neglecting all terms of degree higher than 2, we obtain:

$$\log{(z^2 + xy)} = \log{[(1 + \zeta)^2 + xy]} = \log{[1 + (2\zeta + \zeta^2 + xy)]}$$

$$= (2\zeta + \zeta^2 + xy) - \frac{1}{2}(2\zeta + \zeta^2 + xy)^2 + \cdots$$

$$= 2\zeta + \zeta^2 + xy - \frac{1}{2}(4\zeta^2 + \cdots) = 2\zeta - \zeta^2 + xy + \cdots$$

$$= 2(z - 1) - (z - 1)^2 + xy + \cdots.$$

The terms written above form the desired Taylor polynomial. ◄

EXERCISES

1. Find the third Taylor polynomial (Taylor polynomial of degree 3) of the function $f(x, y) = x \sin{(xy^2)}$ at $x = y = 0$.
2. Find the second Taylor polynomial of the function $f(\alpha, \beta) = \log{(1 + \alpha + \beta)}$ at $\alpha = \beta = 0$.
3. Find the third Taylor polynomial of the function $f(u, v) = (u^2 + v)^{1/3}$ at the point $u = v, v = 1$.
► 4. Find the third Taylor polynomial of the function $f(u, v) = (u - v)^3$ at the point $u = 2, v = -1$.
5. Find the second Taylor polynomial of the function $f(w, z) = e^{w/z}$ at the point $w = v, z = 1$.
6. Find the second Taylor polynomial of the function $f(x, y) = \int_0^x \sin{(yt)}\,dt$ at $x = 0, y = 0$.
7. Find the third Taylor polynomial of the function $f(s, t) = \tan{\pi(s + t^{-1})}$ at the point $s = 1, t = 1$.
► 8. Find the $(2n)$th Taylor polynomial for the function $f(x, y) = (1 + xy)^{-1}$ at $x = 0, y = 0$. [Try to show that the remainder term R_{n+1} goes to zero as n goes to infinity if $|x| < 2, |y| < \frac{1}{2}$.]

In Exercises 9 to 14, find the second Taylor polynomial (that is, the Taylor polynomial of degree 2) of the given function, at the given point.

9. $f(x, y, t) = \tan{(x + \sqrt{y} + z)}$ at $(0, \pi^2/16, 0)$.
10. $g(x, y, z) = 1/(2x + 3y + 4z)$ at $(1, 1, 1)$.
11. $F(\theta, \phi, \psi) = e^{\theta \phi/\psi}$ at $\theta = 0, \phi = 0, \psi = 1$.
12. $f(x, y, z) = \cos{(x^2 + yz)}$ at $(0, 0, 0)$.

13. $g(r, s, t) = \dfrac{r + s - t}{r - s + t}$ at $(1, 1, 1)$.

14. $G(\alpha, \beta, \gamma) = \log{[\alpha\beta - (\gamma/\alpha)]}$ at $\alpha = 1, \beta = 1, \gamma = 0$.

Problems

1. Suppose $f(x, y)$ is defined and continuously differentiable for all (x, y). Show that the two following conditions are equivalent:
   (a) $|f_x(x, y)| \leq 1, |f_y(x, y)| \leq 1$ for all (x, y),
   (b) $|f(x_0, y_0)| - f(x_1, y_1)| \leq \sqrt{(x_0 - x_1)^2 + (y_0 - y_1)^2}$ for all x_0, y_0, x_1, y_1.

2. Suppose $f(x, y)$ is twice continuously differentiable for all (x, y) and $f_{xy}(x, y) \equiv 0$. Show that $f(x, y) = g(x) + h(y)$.

3. Suppose that $f(x, y)$ is twice continuously differentiable for all (x, y) and satisfies the differential equation (called wave equation)

$$c^2 f_{xx} - f_{yy} = 0$$

(where $c > 0$ is a number). Show that

$$f(x, y) = g(x + cy) + h(x - cy).$$

[Hint. Consider the function

$$\phi(\xi, \eta) = f\left(\frac{\xi + \eta}{2}, \frac{\xi - \eta}{2c}\right)$$

and compute $\phi_{\xi\eta}$.]

4. Find all functions $f(x, y, z)$ such that

$$f_{xx} = f_{yy} = f_{zz} = f_{xy} = f_{xz} = f_{yz} = 0.$$

5. Find all functions $f(x, y, z)$ for which all derivatives of order k are 0.

6. A function $f(x_1, x_2, \cdots, x_n)$ defined for $x_1^2 + x_2^2 + \cdots + x_n^2 \neq 0$ is called **homogeneous** of degree n if, for every $t > 0$,

$$f(tx_1, tx_2, \cdots, tx_n) = t^n f(x_1, x_2, \cdots, x_n).$$

Show that, if such a function is continuously differentiable, it satisfies the so-called **Euler equation**

$$\frac{\partial f}{\partial x_1} + \frac{\partial f}{\partial x_2} + \cdots + \frac{\partial f}{\partial x_n} - nf = 0.$$

7. Show that for every continuous function $\xi \mapsto \psi(\xi)$ defined in $[a, b]$, the function

$$f(x, y) = y^{-1/2} \int_a^b e^{-(x-\xi)^2/4y} \psi(\xi)\, d\xi$$

satisfies the differential equation (called **heat equation**)

$$f_{xx} = f_y.$$

[Hint. Note Theorem 6 in §2.17.]

8. Show that the line integral satisfies the inequality:

$$\left| \int_C (\mathbf{F}, d\mathbf{x}) \right| \leq ML$$

if $|\mathbf{F}| \leq M$ and the curve C has length L.

13/Multiple Integrals

This final chapter deals with the definition, properties, and applications of integrals of functions of several variables over intervals and over more general regions. The typical case of double integrals is considered in detail, and then double integrals are used to compute areas of curved surfaces. The extensions to three or more variables are sketched briefly.

There are four appendixes: on centroids; on Green's Theorem, which connects double integrals with line integrals; on improper integrals; and on proofs of some theorems.

The rigorous theory of multiple integrals is more sophisticated than the theory of ordinary integrals; for this reason it is relegated to the Supplement (§§4 and 5).

In this section, we extend the second basic procedure of calculus, integration, to functions of two variables. We proceed as we did in Chapter 5 where we introduced integrals for functions of one variable. First we base the concept of a double integral on geometric intuition; then we give an analytic definition.

1.1 The double integral of a nonnegative function over an interval

Consider a continuous and nonnegative function of two variables, $(x, y) \mapsto z = f(x, y)$, defined in an interval I: $a < x < b$, $c < y < d$ and also on its boundary. Our geometric intuition tells us that the graph of this function, that is, the surface $z = f(x, y)$, together with the five planes, $x = a$, $x = b$, $y = c$, $y = d$, and $z = 0$, bounds a solid which has a definite **volume** V (see Fig. 13.1).

The number V is called the **double integral** of $f(x, y)$ over the interval I; the Leibniz notation for this is

(1) $$V = \iint\limits_{I} f(x, y) \, dx \, dy$$

or

(1') $$V = \iint\limits_{\substack{a < x < b \\ c < y < d}} f(x, y) \, dx \, dy.$$

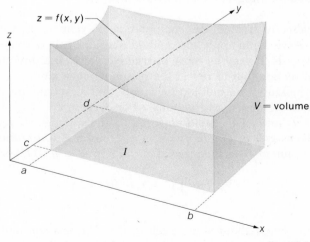

Fig. 13.1

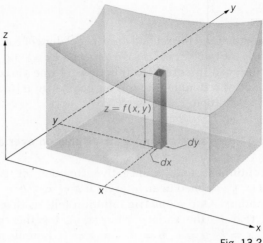

Fig. 13.2

The notation can be justified by a mathematical myth. The volume under the surface is composed of infinitely many infinitely thin three-dimensional boxes (intervals). Each of these is erected over a point (x, y) in I: it has the infinitely small width and breadth, dx and dy, and the height $f(x, y)$. Its volume is therefore $f(x, y)\, dx\, dy$ (see Fig. 13.2). The total volume is the sum of all these infinitely small volumes.

1.2 The double integral as an iterated integral

We have already treated the computation of volumes in Chapter 5, §4.4. Hence we could apply the method developed there to computing double integrals. It may be more instructive not to use directly the result of that section, but to repeat the reasoning once more.

We are looking for a method for computing V, relying on our intuitive grasp of the concept of volumes. For every t, $c < t \le d$, let $V(t)$ denote the volume of the solid enclosed by the surface $z = f(x, y)$ and the five planes $x = a$, $x = b$, $y = c$, $y = t$, and $z = 0$. In other words, $V(t)$ is the volume "over the interval $a < x < b$, $c < y < t$," or,

$$V(t) = \iint\limits_{\substack{a<x<b \\ c<y<t}} f(x, y)\, dx\, dy.$$

Our intuitive ideas about volume tell us that $V(t)$ is a continuous function of t, for $c < t \le d$, and that if we set $V(c) = 0$, then $V(t)$ becomes continuous also at $t = c$. (This means simply that $V(t)$ will be as small as we like provided t is sufficiently close to c.) Hence we have, by the fundamental theorem of calculus,

$$V = V(d) = \int_c^d V'(t)\, dt$$

(provided, of course, that the derivative $V'(t)$ exists and is a bounded piecewise continuous function of t).

Now we must guess at the value of $V'(t)$. Let h be a small positive number. Our intuitive understanding of volumes tells us that

$$V(t + h) - V(t) = \iint\limits_{\substack{a<x<b \\ t<y<t+h}} f(x, y)\, dx\, dy;$$

This is the volume of a thin slice shown in Fig. 13.3. This slice is nearly a cylinder with height h, its volume is nearly $A(t)h$ where $A(t)$ is the "base area," that is, the area of the intersection of our solid with the plane $y = t$. Thus

$$\frac{V(t + h) - V(t)}{h} \approx A(t)$$

(where $\approx$ stands for "approximately equal") and we may expect that

$$V'(t) = \lim_{h \to 0} \frac{V(t + h) - V(t)}{h} = A(t).$$

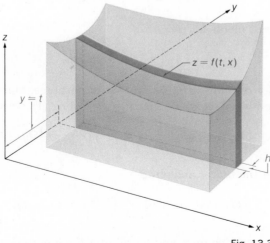

Fig. 13.3

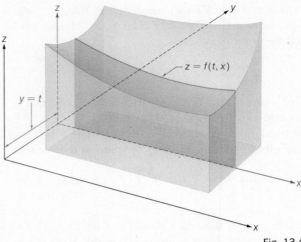

Fig. 13.4

If so, we obtain the formula

$$V = \int_c^d A(t)\, dt.$$

But what is $A(t)$? In the plane $y = t$, one can use x and z as Cartesian coordinates. If $c < t < d$, then the intersection of the plane $y = t$ and our solid is the region under the curve $z = f(x, t)$ from $x = a$ to $x = b$, as is seen from Fig. 13.4. The area $A(t)$ of this intersection is, therefore,

$$A(t) = \int_a^b f(t, x)\, dx,$$

and the desired volume is

$$V = \int_c^d \left\{ \int_a^b f(t, x)\, dx \right\} dt.$$

The name of a dummy variable is of no importance, and we obtain a better looking formula if we replace t by y. Noting (1), we write

(2) $$\iint\limits_I f(x, y)\, dx\, dy = \int_c^d \left\{ \int_a^b f(x, y)\, dx \right\} dy.$$

One may omit the braces and write

(2′) $$\iint\limits_I f(x, y)\, dx\, dy = \int_{y=c}^{y=d} \int_{x=a}^{x=b} f(x, y)\, dx\, dy.$$

Even simpler, agreeing once and for all that *the inner integral sign goes with the inner differential, and the outer integral sign with the outer differential,* one writes

$$(2'') \qquad \iint\limits_I f(x, y)\, dx\, dy = \int_c^d \int_a^b f(x, y)\, dx\, dy.$$

Of course, we could have reversed the parts of x and y. In this case, an entirely analogous argument would have given

$$(3) \qquad \iint\limits_I f(x, y)\, dx\, dy = \int_a^b \left\{ \int_c^d f(x, y)\, dy \right\} dx$$

or

$$(3') \qquad \iint\limits_I f(x, y)\, dx\, dy = \int_a^b \int_c^d f(x, y)\, dy\, dx.$$

Formulas (2) and (3) reduce the calculation of double integrals to two ordinary integrations (one also says: to **iterated integrals,** or repeated integrals). The ordinary integrals can be computed either "formally" (by finding primitives) or numerically.

Remark 1 We state explicitly that formulas (2) and (3) are special cases of the general volume formula in Chapter 5, §4.4. In order to compute a volume by this formula, we must select a directed line L and a point O on it. Once L and O are chosen, denote by $A(t)$ the area of the intersection of the solid under consideration with the plane P perpendicular to L, which intersects L at a point whose distance from O is t. Then the desired volume is given by

$$V = \int_{-\infty}^{+\infty} A(t)\, dt.$$

Consider now the solid whose volume we denoted by (1). We choose for L the y axis, for O the origin of our coordinate system. The intersection of P with our solid is empty if $t < c$ or if $t > d$. Thus $A(t) = 0$ for $t < c$ and for $t > d$, and the integration in the volume formula is to be performed from $t = c$ to $t = d$ only. We compute $A(t)$ as before and obtain relation (2).

If we choose for L the x axis, we obtain, by a similar reasoning, relation (3).

Remark 2 Since we use the Leibniz notation for double integrals, formulas (2) and (3) are rather obvious—one more case in which this marvelously clever notation almost does the thinking for us!

Remark 3 An important special case of formulas (2) and (3) is the statement: *the double integral (over an interval) of a product of functions of one variable is a product of single integrals:*

$$\iiint\limits_{\substack{a<x<b \\ c<y<d}} \phi(x)\psi(y)\ dx\ dy = \int_a^b \phi(x)\ dx \int_c^d \psi(y)\ dy.$$

Indeed, by (2)

$$\iint\limits_{\substack{a<x<b \\ c<y<d}} \phi(x)\psi(y)\ dx\ dy = \int_c^d \left\{ \int_a^b \phi(x)\psi(y)\ dx \right\} dy$$

$$= \int_c^d \psi(y)\left\{ \int_a^b \phi(x)\ dx \right\} dy = \int_a^b \phi(x)\ dx \int_c^d \psi(y)\ dy$$

as asserted.

▶ *Example.*   Find the integral

$$\iint\limits_I (6x^2y + 8xy^3)\ dx\ dy$$

where I is the interval

$$1 < x < 2,\ 3 < y < 4.$$

Answer.   The desired integral equals

$$\int_3^4\!\int_1^2 (6x^2y + 8xy^3)\ dx\ dy = \int_3^4 \left\{ \int_1^2 (6x^2y + 8xy^3)\ dx \right\} dy.$$

But $6x^2y + 8xy^3$, considered as a function of x, with y treated as a constant, has $2x^3y + 4x^2y^3$ as a primitive function. Hence

$$\int_1^2 (6x^2y + 8xy^3)\ dx = 2x^3y + 4x^2y^3 \Big|_{x=1}^{x=2}$$

$$= 16y + 16y^3 - (2y + 4y^3) = 14y + 12y^3$$

and the double integral equals

$$\int_3^4 (14y + 12y^3)\ dy = (7y^2 + 3y^4)\Big|_3^4 = 574.$$

We integrated first with respect to x and then with respect to y, that is, we used equation (2). But we can also integrate first with respect to y, then with respect to x, that is, we can use equation (3). In this case, we obtain

$$\int_1^2\!\int_3^4 (6x^2y + 8xy^3)\ dy\ dx = \int_1^2 \left\{ \int_3^4 (6x^2y + 8xy^3)\ dy \right\} dx$$

$$= \int_1^2 \left\{ 3x^2y^2 + 2xy^4 \Big|_{y=3}^{y=4} \right\} dx$$

$$= \int_1^2 \left\{ (48x^2 + 512x) - (27x^2 + 162x) \right\} dx = \int_1^2 (21x^2 + 350x) \, dx$$

$$= (7x^3 + 175x^2) \Big|_1^2 = 574$$

as expected. ◄

1.3 Functions that take on negative values

If the function $f(x, y)$ also takes on negative values, then the double integral

$$\iint\limits_{\substack{a<x<b \\ c<y<d}} f(x, y) \, dx \, dy$$

is interpreted as the sum of all volumes lying between $z = f(x, y)$ and the plane $z = 0$, and bounded by the planes $x = a$, $x = b$, $y = c$, $y = d$, *the volumes lying about the plane $z = 0$ being counted as positive and those lying below that plane as negative.* In the case shown in Fig. 13.5, for instance, the double integral is the difference of two positive numbers.

We shall show that the reduction of double integrals to iterated integrals remains valid. Let us define two functions,

$$f_+(x, y) = f(x, y) \text{ if } f(x, y) \geq 0, \ = 0 \text{ if } f(x, y) < 0,$$
$$f_-(x, y) = f(x, y) \text{ if } f(x, y) \leq 0, \ = 0 \text{ if } f(x, y) > 0$$

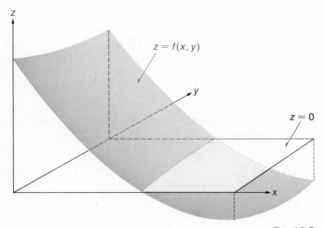

Fig. 13.5

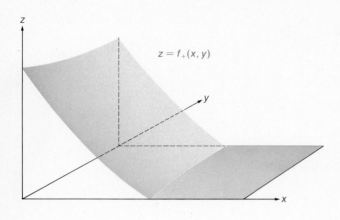

$z = f_+(x, y)$

Fig. 13.6

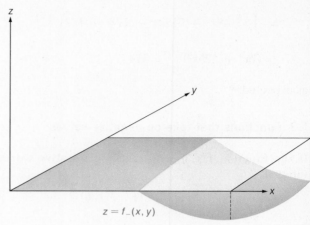

$z = f_-(x, y)$

Fig. 13.7

(for the case of Fig. 13.5, the graphs are shown in Figs. 13.6 and 13.7). Note that

$$f(x, y) = f_+(x, y) + f_-(x, y).$$

We have $f_+ \geq 0$ everywhere; clearly

$$V_+ = \iint_I f_+(x, y) \, dx \, dy$$

is the volume over I, between $z = 0$ and $z = f(x, y)$, lying above the plane $z = 0$. Also, $f_- \leq 0$ everywhere. Clearly, $(-f_-) \geq 0$ and

$$V_- = \iint_I (-f_-(x, y)) \, dx \, dy$$

is the volume under I, between $z = 0$ and $z = f(x, y)$, lying below the plane $z = 0$. Therefore

$$\iint_I f(x, y) \, dx \, dy = V_+ - V_- = \iint_I f_+(x, y) \, dx - \iint_I (-f_-(x, y)) \, dx \, dy.$$

By the results of §1.2 and by familiar properties of ordinary integrals,

$$\iint\limits_{I} f(x, y) \, dx \, dy = \int_c^d \left\{ \int_a^b f_+(x, y) \, dx \right\} dy + \int_c^d \left\{ \int_a^b f_-(x, y) \, dx \, dy \right\}$$

$$= \int_c^d \left\{ \int_a^b f_+(x, y) \, dx + \int_a^b f_-(x, y) \, dx \right\} dy$$

$$= \int_c^d \left\{ \int_a^b [f_+(x, y) + f_-(x, y)] \, dx \right\} dy$$

$$= \int_c^d \left\{ \int_a^b f(x, y) \, dx \right\} dy.$$

This shows that formula (2) remains true for functions that take on negative values. Formula (3) is verified similarly.

EXERCISES

In each of the Exercises 1 to 22, we give an interval and a function t. Integrate f over the given interval. Try to do each problem twice, once by integrating first with respect to one variable and once by integrating first with respect to the other.

1. $0 < x < 1, 2 < y < 3; f(x, y) = xye^x$.

2. $-1 < x < 0, 0.1 < y < 0.2; f(x, y) = x^2 y^3 + \sin x$.

3. $-1 < \theta < 1, -1 < \phi < 1; f(\theta, \phi) = \theta \phi e^{\theta^2}$.

4. $-1 < x < 1, -1 < y < 1, f(x, y) = e^{x+y}$.

5. $\sqrt{2} < l < \pi, .37 < m < .4; f(l, m) = lme^{l+m}$.

6. $10 < s < 11, -2 < t < 0.5; f(s, t) = 2t^2 \sin s. f = st^2 \sin s$.

7. $8 < u < 8.1, 0 < v < 100; f(u, v) = v/(u + v)$.

▶ 8. $.1 < x < .2, .25 < y < .30; f(x, y) = ye^{xy}$.

9. $\pi/2 < x < \pi, \pi/2 < y < \pi, f(x, y) = x \cos xy$.

10. $10 < x < 11, 4 < y < 5, f(x, y) = e^{x+y} + \log(xy)$.

11. $0 < x < 1, 0 < y < 1, f(x, y) = 4x^3\sqrt{y} + 3y^2/\sqrt{x}$.

12. $-5 < z < -4, -3 < w < -2, f(z, w) = 3z - 2w^3$.

13. $-2 < x < 1, -1 < y < 2, f(x, y) = 4y^2\sqrt[3]{x} + \cos y$.

14. $9 < p < 10, -0.1 < q < 0, f(p, q) = \cos(4p + 3q)$.

15. $0 < x < \pi, 0 < y < \pi, f(x, y) = \cos(Ax + By)\sin(Cx + Dy)$.

16. $3 < A < 6.7, 4.9 < B < 5.8, f(A, B) = (A^3 + B^3)^2$.

17. $0 < x < 1, 2 < y < 3, f(x, y) = (e^x + \cos y)^3$.

18. $0 < z < 1, 2 < w < 3, f(z, w) = w \arcsin z$.

19. $1 < x < 2, 2 < y < 3, f(x, y) = (x + y)^3 e^{x+y}$.

20. $A < x < B, C < y < D, f(x, y) = (x + y)^n$.

21. $x_0 < x < x_1, y_0 < y < y_1, f(x, y) = (x + y + 1)^n$.

▶ 22. $x_0 < x < x_1, y_0 < y < y_1, f(x, y) = e^{ax} \sin by$.

1.4 Piecewise continuous functions

Before presenting an analytic definition of double integrals, we define a class of functions which we shall integrate. This class is more extensive than the class of continuous functions, but is sufficiently restricted to avoid unnecessary difficulties.

A function of two variables $(x, y) \to f(x, y)$ will be called **piecewise continuous** in an interval if it is defined and continuous at all points of this interval, except perhaps at points that lie on a finite number of curves each of which is either a vertical segment, or a horizontal segment, or can be represented both as a graph of a continuous function $x \to y = \phi(x)$, $\alpha \le x \le \beta$ and as a graph of a continuous function $y \to x = \psi(y)$, $\gamma \le y \le \delta$.

The definition should be compared with the one given in Chapter 5, §1.5. There we called a function of one variable piecewise continuous if it was continuous at all points of a (finite) interval, except perhaps at finitely many points.

A function $f(x, y)$ of two variables is called **bounded** if there is a number M such that $|f(x, y)| \le M$ wherever $f(x, y)$ is defined. This is the same definition as for functions of one variable.

▶ **Examples**   **1.** Let $x \to f(x)$, $0 < x < 1$, be a piecewise continuous function (say the function $f(x) = 0$ for $0 < x < \frac{1}{2}$, $f(x) = -1$ for $\frac{1}{2} < x < \frac{3}{4}$, $f(x) = 5$ for $\frac{3}{4} < x < 1$). Then $(x, y) \to f(x)$ is a piecewise continuous function for $0 < x < 1$, $0 < y < 1$. Indeed, it is discontinuous only on finitely many vertical segments (the segments $x = \frac{1}{2}$, $x = \frac{3}{4}$, $0 < y < 1$, in our case).

2. Let $f(x, y)$ be continuous for $0 < x < 1$, $0 < y < 1$, except for points lying on the eight-shaped curve shown in Fig. 13.8. Then f is piecewise continuous, since the curve can be decomposed into arcs each of which satisfies the condition of the definition. (In the figure, this decomposition is indicated by colored dots.)

3. The function $f(x, y) = 1$ if $x^2 + y^2 \le 1 = 0$ if $x^2 + y^2 > 0$ is piecewise continuous. The reader should supply the reasons.

4. The function $f(x, y) = 1/(x^2 + y^2)$ is piecewise continuous. The only discontinuity is at the origin, this point can be covered by a horizontal segment.

5. If f and g are piecewise continuous, so are fg, $f + g$, $f - g$. The proof is left to the reader.

Fig. 13.8

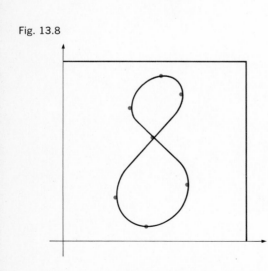

6. The function $f(x, y) = 0$ if x and y are rational, and equals 1 otherwise, is *not* piecewise continuous. ◀

EXERCISES

23. If $f(x, y)$ and $g(x, y)$ are piecewise continuous in an interval I, are the functions $f^2 + g$ and $\sin f$ piecewise continuous? Give reasons for your answers.
24. Under the condition of Exercise 23, under what *additional* condition is f/g certainly piecewise continuous?
25. Prove that $f(x, y) = \dfrac{x^2 + y^4 + 6}{xy}$ if $xy \neq 0$, $f(x, y) = 77$ if $xy = 0$ is piecewise continuous in every interval.
26. Is the function $f(x, y)$ of Exercise 25 bounded?

1.5 Analytic definition of the integral

We shall now define double integrals of bounded piecewise continuous functions. In §§1.1 to 1.3, we used our intuitive concept of volume and we concluded that, for continuous functions, the double integral can be computed as an iterated integral. We want to use this method of computation as a definition. We must be sure, therefore, that the computation can be carried out. The following result tells that this is so.

Theorem 1 LET $f(x, y)$ BE BOUNDED AND PIECEWISE CONTINUOUS IN THE INTERVAL I: $a < x < b$, $c < y < d$. FOR EVERY FIXED y FROM (c, d), WITH PERHAPS FINITELY MANY EXCEPTIONS, $x \to f(x, y)$ IS A PIECEWISE CONTINUOUS FUNCTION IN THE INTERVAL (a, b). SIMILARLY, FOR EVERY FIXED x FROM (a, b), WITH PERHAPS FINITELY MANY EXCEPTIONS, $y \to f(x, y)$ IS A PIECEWISE CONTINUOUS FUNCTION IN THE INTERVAL (c, d). ALSO

$$y \to \int_a^b f(x, y)\, dx$$

IS A BOUNDED PIECEWISE CONTINUOUS FUNCTION IN THE INTERVAL (c, d) AND

$$x \to \int_c^d f(x, y)\, dy$$

IS A BOUNDED PIECEWISE CONTINUOUS FUNCTION IN THE INTERVAL (a, b).

It follows from Theorem 1 that the two iterated integrals

(4) $$\int_c^d \left\{ \int_a^b f(x, y)\, dx \right\} dy, \qquad \int_a^b \left\{ \int_c^d f(x, y)\, dy \right\} dx$$

are well defined.

Theorem 2   UNDER THE HYPOTHESES OF THEOREM 1, THE TWO ITERATED
INTEGRALS (4) ARE EQUAL.

The proofs of Theorem 1 and 2 are not quite elementary. They will
be found in the Supplement (§§4 and 5).

The double integral

$$\iint_I f(x, y) \, dx \, dy$$

is now *defined* as the common value of the two iterated integrals (4):

$$\iint_{\substack{a<x<b \\ c<y<d}} f(x, y) \, dx \, dy = \int_c^d \int_a^b f(x, y) \, dx \, dy = \int_a^b \int_c^d f(x, y) \, dy \, dx.$$

(This definition has the advantage that it tells us, at once, how to compute double integrals. Another definition will be discussed in §1.7 and
in the Supplement, §4.)

1.6   Properties of the integral

Having expressed the double integral by means of integrals of functions
of one variable, we can at once transfer to the double integral the basic
results about single integrals. (These properties are also geometrically
obvious.) Thus, for every interval *I*,

(5) $$\iint_I \alpha \, dx \, dy = \alpha \cdot \text{area of } I \qquad (\alpha \text{ a constant}).$$

If *I* is divided by a vertical or horizontal line into two intervals I_1 and I_2
(see Fig. 13.9), and *f* is bounded and piecewise continuous in *I*, then

(6) $$\iint_I f(x, y) \, dx \, dy = \iint_{I_1} f(x, y) \, dx \, dy + \iint_{I_2} f(x, y) \, dx \, dy.$$

Next,

(7) $$\iint_I f(x, y) \, dx \, dy \le \iint_I g(x, y) \, dx \, dy \qquad \text{if } f(x, y) \le g(x, y) \text{ in } I$$

and *g* is again bounded and piecewise continuous in *I*. Also

(8) $$\iint_I \alpha f(x, y) \, dx \, dy = \alpha \iint_I f(x, y) \, dx \, dy \qquad (\alpha \text{ a constant})$$

and

(9) $$\iint_I [f(x, y) + g(x, y)] \, dx \, dy = \iint_I f(x, y) \, dx \, dy + \iint_I g(x, y) \, dx \, dy.$$

Fig. 13.9

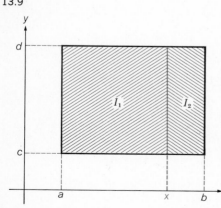

We prove only (6) and leave the other proofs to the reader. Assume that I is the interval $a < x < b$, $c < y < d$ and I is divided into two intervals, I_1 and I_2, by the line $x = x_0$. We have

$$\iint\limits_{I} f(x, y)\, dx\, dy = \int_a^b \left\{ \int_c^d f(x, y)\, dy \right\} dx \qquad \text{by the definition of the double integral}$$

$$= \int_a^{x_0} \left\{ \int_c^d f(x, y)\, dy \right\} dx + \int_{x_0}^b \left\{ \int_c^d f(x, y)\, dy \right\} dx \qquad \text{by the additivity property of the ordinary integral}$$

$$= \iint\limits_{I_1} f(x, y)\, dx\, dy + \iint\limits_{I_2} f(x, y)\, dx\, dy \qquad \text{by the definition of the double integral.}$$

The proofs of statements (5), (7), (8), and (9) follow the same pattern.

1.7 Step functions

In discussing integrals of functions of one variable, we made extensive use of step functions (see Chapter 5, §§1.5, 1.6, and 1.8). We now define step functions of two variables.

If we subdivide the interval $a < x < b$ into several, say p, subintervals and draw vertical lines through each of the dividing lines, and if we subdivide the interval $c < y < d$ into several, say q, subintervals and draw horizontal lines through the division points, then the two-dimensional interval I: $a < x < b$, $c < y < d$ gets subdivided into pq *two-dimensional subintervals.* In Fig. 13.10, for instance, the interval $0 < x < 19$ is subdivided into 4 subintervals, the interval $0 < y < 10$ is subdivided into 3 subintervals, and the two-dimensional interval $0 < x < 19$, $0 < y < 10$ is subdivided into $4 \cdot 3 = 12$ two-dimensional subintervals.

A **step function** $(x, y) \rightarrow \phi(x, y)$ in I is a function which, for a suitable subdivision of I, is constant in each subinterval. For a given subdivision we can define a step function $z = \phi(x, y)$ by giving its values, $z_1, z_2, z_3, \cdots$ in the various subintervals. We did so in Fig. 13.10; the graph of this function is shown in Fig. 13.11. A step function is always piecewise continuous (and bounded). Its values along the division lines are of no interest.

The integral of a step function $z = \phi(x, y)$ which takes on the constant values $z_1, z_2, z_3, \cdots$ in the subintervals $I_1, I_2, I_3, \cdots$ is simply the finite sum

$$(10) \qquad \iint\limits_{I} \phi(x, y)\, dx\, dy = z_1 \cdot \text{area of } I_1 + z_2 \cdot \text{area of } I_2 + \cdots.$$

Indeed, repeated applications of the additivity property (6) yield

$$\iint\limits_{I} \phi(x, y)\, dx\, dy = \iint\limits_{I_1} \phi(x, y)\, dx\, dy + \iint\limits_{I_2} \phi(x, y)\, dx\, dy + \cdots.$$

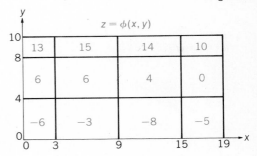

Fig. 13.10

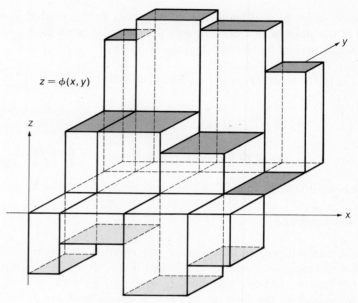

$z = \phi(x, y)$

Fig. 13.11

Since ϕ is constant and equal to z_j in I_j, we have by Property (5) that

$$\iint\limits_{I_j} \phi(x, y)\, dx\, dy = \iint\limits_{I_j} z_j\, dx\, dy = z_j \cdot \text{area of } I_j.$$

Hence (10) follows.

For instance, in the case of the function defined by Fig. 13.10, we have

$$
\begin{aligned}
\iint\limits_{\substack{0<x<19 \\ 0<y<10}} \phi(x, y)\, dx\, dy &= 13 \cdot 6 + 15 \cdot 12 + 14 \cdot 12 + 10 \cdot 8 \\
&\quad + 6 \cdot 12 + 6 \cdot 24 + 4 \cdot 24 + 0 \cdot 16 \\
&\quad + (-6) \cdot 12 + (-3) \cdot 24 + (-8) \cdot 24 + (-5) \cdot 16 \\
&= 818 - 416 = 402.
\end{aligned}
$$

This is, of course, the sum of the volumes of the boxes in Fig. 13.11 above the plane $z = 0$ minus the sum of the volumes of the boxes below that plane.

Let $f(x, y)$ be a bounded piecewise continuous function defined in I, and let $\phi(x, y)$, $\psi(x, y)$ be step functions such that $\phi \le f \le \psi$ in I. By the monotonicity property (7), we then have

$$\iint\limits_{I} \phi(x, y)\, dx\, dy \le \iint\limits_{I} f(x, y)\, dx\, dy \le \iint\limits_{I} \psi(x, y)\, dx\, dy.$$

Since the integrals of the step functions on the left- and right-hand sides are easily computed, this is a way of computing the double integral of f approximately. It turns out that this method gives us the double integral with any desired degree of accuracy; one can enclose f between two step functions whose integrals are as close to each other as we want. This is the content of our next theorem.

Theorem 3   LET $f(x, y)$ BE A BOUNDED PIECEWISE CONTINUOUS FUNCTION DEFINED IN AN INTERVAL I. THEN THERE IS A NUMBER V, AND ONLY ONE SUCH NUMBER, WITH THE PROPERTY: FOR ALL STEP FUNCTIONS $\phi(x, y)$ AND $\psi(x, y)$ SUCH THAT $\phi(x, y) \leq f(x, y) \leq \psi(x, y)$, WE HAVE

(11) $$\iint_I \phi(x, y) \, dx \, dy \leq V \leq \iint_I \psi(x, y) \, dx \, dy.$$

THIS NUMBER IS THE DOUBLE INTEGRAL OF f OVER I. FOR EVERY $\epsilon > 0$, ONE CAN FIND STEP FUNCTIONS ϕ AND ψ WITH $\phi \leq f \leq \psi$ SUCH THAT

$$\iint_I \psi \, dx \, dy - \iint_I \phi \, dx \, dy < \epsilon.$$

This corresponds to Theorem 1 of Chapter 5, §1.8 about single integrals, but the proof for the case of two variables is harder; we give it in the Supplement, §§4 and 5.

Remark   We could use (10) as we did for functions of one variable, as a definition of the double integral of a step function, and then *define* the double integral of a piecewise continuous bounded function as the number having the property stated in Theorem 3. In this case, the equality of a double integral and the corresponding iterated integral would not be a definition but a theorem to be proved. This proof is presented in the Supplement.

EXERCISES

27. If ϕ and ψ are step functions, so are $\phi + \psi$ and $\phi\psi$. Prove this statement.
▶ 28. If $\phi(x, y)$ and $\psi(x, y)$ are step functions, so is $\omega(x, y) = \max [\phi(x, y), \psi(x, y)] =$ the larger of the numbers $\phi(x, y)$, $\psi(x, y)$. Explain why this is so.

1.8 Riemann sums

Let $f(x, y)$ be a (bounded, piecewise continuous) function defined in the open interval I. Suppose we subdivide I, as in §1.7, choose a point (x_j, y_j) in each subinterval I_j or on its boundary, and form the sum

$$S = f(x_1, y_1) \cdot \text{area of } I_1 + f(x_2, y_2) \cdot \text{area of } I_2 + \cdots.$$

The number S is called a **Riemann sum** for the integral

(12) $$V = \iint_I f(x, y) \, dx \, dy.$$

We can define a step function, call it $\omega(x, y)$, which equals $f(x_j, y_j)$ in I_j; S is then the double integral of this function over I. If ϕ and ψ are two step functions such that $\phi \leq f \leq \psi$, then also $\phi \leq \omega \leq \psi$, so that, by the monotonicity property (7) of integrals, we have

(13) $$\iint_I \phi \, dx \, dy \leq S \leq \iint_I \psi \, dx \, dy.$$

By Theorem 3, we can find step functions ϕ and ψ with $\phi \leq f \leq \psi$, such that the difference between the two integrals in (11) and in (13) is as small as we like. Thus, one can approximate a double integral by a Riemann sum as closely as one wants.

▶*Example*   Consider the integral

$$V = \iint_{\substack{0<x<6 \\ 0<y<4}} (x^2 y + xy^2) \, dx \, dy = \int_0^4 \int_0^6 (x^2 y + xy^2) \, dx \, dy$$

$$= \int_0^4 (72y + 18y^2) \, dy = 576 + 384 = 960.$$

We subdivide the interval $0 < x < 6$, $0 < y < 4$ into 6 subintervals and choose one point in each subinterval, as shown in Fig. 13.12. Each subinterval has area 4. The values of $x^2 y + xy^2$ at the points $(0, 0)$, $(3, 1)$, $(5, 1)$, $(1, 3)$, $(3, 3)$, and $(5, 3)$ are 2, 12, 30, 12, 54, and 120. The corresponding Riemann sum is therefore

$$(2 + 12 + 30 + 12 + 54 + 120)4 = 230 \cdot 4 = 920$$

which is about 4 percent less than the integral.◀

Fig. 13.12

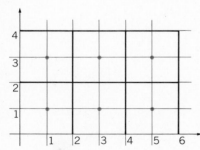

EXERCISES

In Exercises 29 to 32, you are given an integrand $f(x, y)$ and an interval $I = \{x_0 < x < x_0 + n, y_0 < y < y_0 + m\}$. In each case, define nm subintervals by dividing the x and y axes at integer values, and compute a Riemann sum using the step function equal to $f(x_0 + i, y_0 + j)$ in the interval $x_0 + i - 1 < x < x_0 + i$, $y_0 + j - 1 < y < y_0 + j$. Compare this Riemann sum to the exact value of $\int_I f(x, y) \, dx \, dy$.

29. $f = y + x^{-2}$; $100 < x < 102$, $3 < y < 5$.
▶ 30. $f = x^3 y - 4y^2$; $0 < x < 4$, $0 < y < 4$.
31. $f = ax + by$; $7 < x < 9$, $13 < y < 15$.
32. $f = x^2 + xy + y^2$; $10 < x < 13$, $-4 < y < -1$.

1.9 Double integrals over general sets

We have defined the double integral, over an interval, not only for continuous but also for piecewise continuous bounded functions. This has an important advantage: defining and computing integrals over more general sets requires almost no additional work. We explain this first on a simple example.

Let D denote the unit disk, the set of all points (x, y) with $x^2 + y^2 < 1$. We consider the function $z = f(x, y) = 3 + x$, and we want to compute the volume V enclosed between the disk D, the plane $z = 3 + x$, and the cylinder $x^2 + y^2 = 1$ (see Fig. 13.13). It is natural to denote this volume by

$$(14) \qquad V = \iint_D (3 + x)\, dx\, dy.$$

Now define a new function

$$(15) \qquad \hat{f}(x, y) = \begin{cases} 3 + x & \text{if } x^2 + y^2 < 1, \\ 0 & \text{if } x^2 + y^2 \geq 1. \end{cases}$$

If I is any interval containing the disk D, then

$$(16) \qquad \iint_I \hat{f}(x, y)\, dx\, dy = \iint_D (3 + x)\, dx\, dy.$$

This means simply that the volume V may be thought of as the volume enclosed between the interval I in the plane $z = 0$ and the graph of the function $z = \hat{f}(x, y)$; the graph over points outside the disk lies on the plane $z = 0$ and contributes nothing to the volume (compare Fig. 13.14). One can, of course, also verify (16) using the volume formula in Chapter 5, §4.4.

Using (15) and (16), we can compute V by iterated integration. It does not matter how large I is, as long as it encloses D; we may assume that I is the interval $-1 < x < 1$, $-1 < y < 1$; see Fig. 13.15. We have

$$V = \iint_D (3 + x)\, dx\, dy = \int_{-1}^{1}\int_{-1}^{1} \hat{f}(x, y)\, dx\, dy.$$

But, for a fixed y, we have that $\hat{f}(x, y) = 0$ for $x^2 + y^2 > 1$, that is, for $x^2 > 1 - y^2$, that is, for x outside the interval $|x| < |\sqrt{1 - y^2}|$. Hence

$$\int_{-1}^{1} \hat{f}(x, y)\, dx = \int_{-\sqrt{1-y^2}}^{\sqrt{1-y^2}} f(x, y)\, dx = \int_{-\sqrt{1-y^2}}^{\sqrt{1-y^2}} (3 + x)\, dx$$

and

Fig. 13.13

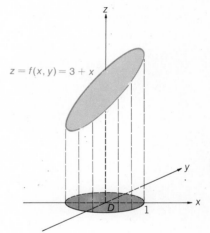

Fig. 13.14

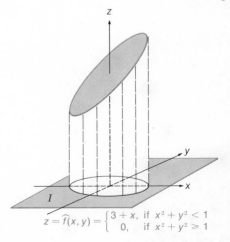

$z = \hat{f}(x, y) = \begin{cases} 3 + x, & \text{if } x^2 + y^2 < 1 \\ 0, & \text{if } x^2 + y^2 \geq 1 \end{cases}$

Fig. 13.15

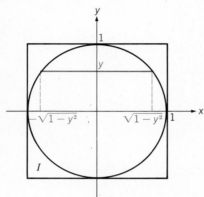

$$V = \int_{-1}^{1} \int_{-\sqrt{1-y^2}}^{\sqrt{1-y^2}} (3 + x) \, dx \, dy = \int_{-1}^{1} 6\sqrt{1 - y^2} \, dy = 3\pi$$

(here we used the result of Chapter 5, §4.3).

The method used above is not limited to a circular disk or to the function $z = 3 + x$. In order to describe this method precisely, it is convenient to use the concept of a characteristic function.

Let D be a bounded set in the plane, that is, a set contained in some interval. The **characteristic function** of D, denoted by $\chi_D(x, y)$, is defined as follows:

$$(17) \qquad \chi_D(x, y) = \begin{cases} 1 & \text{if } (x, y) \text{ is a point in the set } D, \\ 0 & \text{if } (x, y) \text{ is } not \text{ a point in } D. \end{cases}$$

Now let $f(x, y)$ be a function defined on D, and perhaps on a bigger set. The function $\chi_D(x, y) f(x, y)$ is equal to $f(x, y)$ if (x, y) belongs to D and is by definition equal to 0 if (x, y) does not belong to D. For instance, if D is the unit disk $x^2 + y^2 < 1$, then the definition (15) can be written as $\hat{f}(x, y) = \chi_D(x, y) f(x, y)$.

Now if χ_D is piecewise continuous, and if f is such that $\chi_D f$ is a bounded piecewise continuous function, we *define*

$$(18) \qquad \iint\limits_{D} f(x, y) \, dx \, dy = \iint\limits_{I} \chi_D(x, y) f(x, y) \, dx \, dy$$

where I is some interval containing D; it is clear that the choice of I is irrelevant.

In other words, *in order to compute the double integral of f over a bounded set D, set f = 0 outside D and integrate the resulting function over some interval containing D*. In this connection, one often calls D the **region of integration**.

In the following paragraphs we assume that all regions of integration are bounded and have piecewise continuous, characteristic functions and all functions to be integrated are bounded and piecewise continuous.

1.10 Properties of general double integrals

The properties of double integrals over intervals, noted in §1.6, can be extended to double integrals over bounded sets. In particular, a constant can be removed in front of an integral sign, the integral of a sum is the sum of integrals, and the larger function has the larger integral:

$$(19) \qquad \iint\limits_{D} \alpha f \, dx \, dy = \alpha \iint\limits_{D} f \, dx \, dy, \qquad \alpha \text{ a constant,}$$

$$(20) \qquad \iint\limits_{D} (f + g) \, dx \, dy = \iint\limits_{D} f \, dx \, dy + \iint g \, dx \, dy,$$

$$(21) \qquad \iint\limits_{D} f \, dx \, dy \leq \iint\limits_{D} g \, dx \, dy \qquad \text{if } f \leq g \text{ in } D.$$

This follows from the definition (18) and relations (7), (8), and (9) in §1.6. In order to verify (20), for instance we write:

$$\iint\limits_{D} (f + g) \, dx \, dy = \iint (f + g)\chi_D \, dx \, dy$$
$$= \iint f \chi_D \, dx \, dy + \iint g \chi_D \, dx \, dy$$
$$= \iint\limits_{D} f \, dx \, dy + \iint g \, dx \, dy.$$

If χ_D is a piecewise continuous function, the number

$$\iint\limits_{D} dx \, dy = \iint\limits_{D} 1 \, dx \, dy = \iint\limits_{I} \chi_D \, dx \, dy$$

(where I is some interval containing D) is the **area** of D (we discuss this more fully in §2.1). Applying (19) to the function $f \equiv 1$, we see that

$$(22) \qquad \iint\limits_{D} \alpha \, dx \, dy = \alpha \cdot \text{area of } D \qquad (\alpha \text{ a constant}).$$

Now let D_1 and D_2 be two sets; the set consisting of all points belonging to either D_1 or D_2 is called the **union** of D_1 and D_2 and is usually denoted by $D_1 \cup D_2$. If $D = D_1 \cup D_2$ and the sets D_1, D_2 are **disjoint** (have no points in common), then one sees easily that $\chi_D(x, y) = \chi_{D_1}(x, y) + \chi_{D_2}(x, y)$. Using (18) and (20), we conclude that

$$(23) \qquad \iint\limits_{D} f \, dx \, dy = \iint\limits_{D_1} f \, dx \, dy + \iint\limits_{D_2} f \, dx \, dy$$

if D is the union of the two disjoint sets D_1, D_2. This formula generalizes property (6) in §1.6. There is an obvious extension of property (23) to the case where D is decomposed not into two but into several disjoint sets.

Often a double integral over D can be computed, as in the example in §1.9, by carrying out two single integrations. We say that D is **convex** in the x direction if there are numbers γ and δ such that for $y_0 < \gamma$ or $y_0 > \delta$ the horizontal line $y = y_0$ does not intersect D, and for $\gamma < y_0 < \delta$ the intersection of D and the line $y = y_0$ is a single segment, $\alpha(y_0) < x < \beta(y_0)$, see Fig. 13.16. In this case, we have

$$\iint\limits_D f(x, y)\, dx\, dy = \int_\gamma^\delta \left\{ \int_{\alpha(y)}^{\beta(y)} f(x, y)\, dx \right\} dy$$

or, recalling the convention that "the outer integral sign belongs with the outer differential,"

(24) $$\iint\limits_D f(x, y)\, dx\, dy = \int_\gamma^\delta \int_{\alpha(y)}^{\beta(y)} f(x, y)\, dx\, dy.$$

If D is convex in the y direction (in the sense indicated in Fig. 13.17), we can first integrate with respect to y and then with respect to x:

(25) $$\iint\limits_D f(x, y)\, dx\, dy = \int_\alpha^\beta \int_{\gamma(x)}^{\delta(x)} f(x, y)\, dy\, dx;$$

the meaning of α, β, γ, and δ is clear from the figure. In practice, we encounter only regions of integration D that are either nice in one direction, or can be decomposed into several disjoint sets with this property.

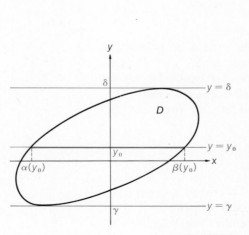

Fig. 13.16

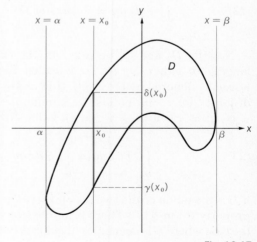

Fig. 13.17

▶ *Examples* **1.** Let D be the triangular region bounded by the lines $2y = x$, $y = 2x$ and $x = \pi$. Compute

$$\iint\limits_{D} \sin y \, dx \, dy.$$

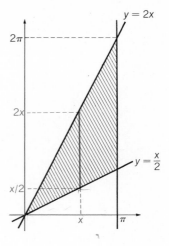

Answer. In doing such problems, it is essential to make a sketch of the region of integration; see Fig. 13.18. The sketch shows that D is convex in both directions, x and y. We shall first integrate with respect to y. For every point (x, y) in D, x lies between 0 and π. If we draw a vertical line at the distance x from the y axis, its intersection with D has endpoints with $y = x/2$ and $y = 2x$. Hence we have

$$\iint\limits_{D} \sin y \, dx \, dy = \int_0^\pi \int_{x/2}^{2x} \sin y \, dy \, dx = \int_0^\pi \left[-\cos y\right]_{y=x/2}^{y=2x} dx$$

$$= \int_0^\pi \left(\cos \frac{x}{2} - \cos 2x\right) dx = 2\sin\frac{x}{2} - \frac{1}{2}\sin 2x \Big|_0^\pi = 2 \sin 90° = 2.$$

Fig. 13.19

If we want to integrate first with respect to x, we note that, for (x, y) in D, we have $0 < y < 2\pi$, and (see Fig. 13.19) that a horizontal line (a line with fixed y) intersects D in a segment with endpoints at

$$x = \frac{y}{2} \text{ and } x = 2y \qquad \text{if } 0 < y < \pi/2$$

$$x = \frac{y}{2} \text{ and } x = \pi \qquad \text{if } \pi/2 < y < 2\pi.$$

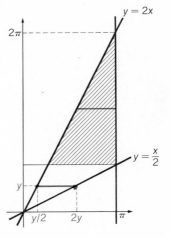

Therefore we think of D as decomposed into two regions, and write

$$\iint\limits_{D} \sin y \, dx \, dy = \int_0^{\pi/2}\int_{y/2}^{2y} \sin y \, dx \, dy + \int_{\pi/2}^{2\pi}\int_{y/2}^{\pi} \sin y \, dx \, dy$$

$$= \int_0^{\pi/2} \left(2y - \frac{y}{2}\right) \sin y \, dy + \int_{\pi/2}^{2\pi} \left(\pi - \frac{y}{2}\right) \sin y \, dy$$

$$= \frac{3}{2}\int_0^{\pi/2} y \sin y \, dy - \frac{1}{2}\int_{\pi/2}^{2\pi} y \sin y \, dy + \pi\int_{\pi/2}^{2\pi} \sin y \, dy.$$

Since $y \mapsto y \sin y$ has a primitive function $y \to \sin y - y \cos y$ [as can be found by integration by parts (compare Chapter 5, §3.3) and can be checked by direct differentiation], the above equals

$$\frac{3}{2}\left[\sin y - y \cos y\right]_0^{\pi/2} - \frac{1}{2}\left[\sin y - y \cos y\right]_{\pi/2}^{2\pi} - \pi\left[\cos y\right]_{\pi/2}^{2\pi}$$

$$= \frac{3}{2} + \frac{1}{2} + \pi - \pi = 2$$

as it must.

Fig. 13.20

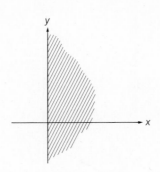

2. Let D be the set of points (x, y) with $x > 0$, $y > x^2$ and $y < 2 - x^2$. Compute

$$\iint_D \sqrt{x}\, y\, dx\, dy.$$

Answer. The set of points (x, y) with $x > 0$ is the right half plane (Fig. 13.20). The set of points (x, y) with $y > x^2$ is the set above the parabola $y = x^2$ (see Fig. 13.21). The set of points (x, y) with $y < 2 - x^2$ is the set below the parabola $y = 2 - x^2$ (see Fig. 13.22). The set D is

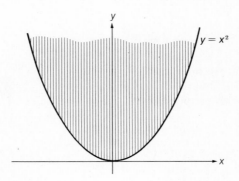

Fig. 13.21

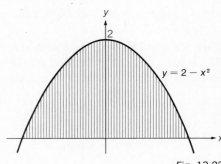

Fig. 13.22

Fig. 13.23

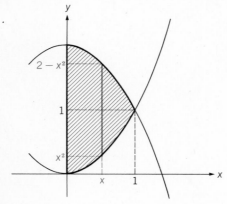

shown in Fig. 13.23; it is the **intersection** of the three sets, that is, it consists of points belonging to all of them. The two parabolas intersect when $2 - x^2 = x^2$, that is, when $2x^2 = 2$; the intersection point in the right half plane has the coordinates $(1, 1)$. In D, the coordinate x varies from $x = 0$ to $x = 1$. A vertical line with a fixed x intersects D in a segment with endpoints $y = x^2$ and $y = 2 - x^2$. Therefore

$$\iint_D \sqrt{x}\, y\, dx\, dy = \int_0^1 \int_{x^2}^{2-x^2} \sqrt{x}\, y\, dy\, dx = \int_0^1 \sqrt{x} \left[\frac{y^2}{2} \right]_{y=x^2}^{y=2-x^2} dx$$

$$= \int_0^1 \sqrt{x}\, \frac{1}{2} [(2 - x^2)^2 - x^4]\, dx = \int_0^1 (1x^{1/2} - 2x^{3/2})\, dx$$

$$= \left[\frac{4}{3} x^{3/2} - \frac{4}{7} x^{7/2} \right]_0^1 = \frac{16}{21}.$$

Our region of integration is also convex in the x direction. The reader should compute the integral, integrating first in the x direction. ◄

EXERCISES

In Exercises 33 to 52, we are given a set D and a function $f(x, y)$. Make a sketch of D and compute $\iint_D f(x, y)\, dx\, dy$.

33. D is the interior of the triangle with vertices at $(0, 0)$, $(0, 1)$, $(1, 0)$; $f = x^2 y$.
34. D is the interior of the triangle with vertices $(0, 0)$, $(0, 1)$, $(1, 1)$; $f = x^2 y$.
35. D is the interior of the square with vertices $(1, 0)$, $(0, 1)$, $(-1, 0)$, $(0, -1)$; $f = y e^x$.
36. D is the region bounded by the curve $y = \sin x$ and the segment $0 \le x \le \pi$; $f = xy$.
37. D is the interior of the triangle with vertices (x_0, y_0), (x_1, y_1), (x_2, y_2); $f(x, y) \equiv 1$. (Give a geometric interpretation of your answer.)
38. D is the interior of the unit circle; $f = 3x + 4y + 5$. (After carrying out the integration, find an argument that would have enabled you to compute the integral without any computation.)
39. D is the bounded region enclosed by the two parabolas $y = x^2$ and $y = -x^2 + 1$; $f(x, y) = x\sqrt{y}$.
40. D is the intersection of the interiors of the two ellipses $(x^2/4) + y^2 = 1$ and $x^2 + (y^2/4) = 1$; $f = x + y + 1$.
41. D is the intersection of the interiors of the two triangles T_1 with vertices $(0, 0)$, $(0, 1)$, $(1, 1)$ and T_2 with vertices $(0, 1)$, $(1, 0)$, $(2, 0)$; $f = y \sin x + x \cos y$. You may leave the answer in the form of iterated integrals.
42. D is the union of the interiors of the two triangles of Exercise 41; $f = y \sin x + x \cos y$. You may leave the answer in the form of iterated integrals.
43. D is the interior of the circle $x^2 + y^2 = R^2$; $f = \sqrt{R^2 - x^2 - y^2}$. (Give a geometric interpretation of your result.)
44. D is the interior of the ellipse $(x/a)^2 + (y/b)^2 = 1$. $f = c\sqrt{1 - (x/a)^2 - (y/b)^2}$. (Give a geometric interpretation of your result.)
45. D is the interior of the circle $x^2 + y^2 = R^2$; $f = R[1 - R^{-1}(x^2 + y^2)^{1/2}]$. (Give a geometric interpretation of your answer.)
► 46. D is the interior of the triangle with vertices $(-7, -6)$, $(5, 3)$, $(0, 0)$; $f = e^{x+y}$.
47. D is the interior of the triangle with vertices $(-1, -1)$, $(-4, -8)$, $(-8, -4)$; $f = (x^2 - 3y)^2$.
► 48. D is the ring between two concentric circles with centers at the origin and radii R_1 and R_2. $f = Ax^2 + Bxy + Cy^2$.
49. D is the region cut off by the x and y axes between the two parallel lines of slope 2 that pass through $(1, 0)$ and $(2, 0)$; $f = \cos(x + y)$.
50. D is the wedge cut out of the unit circle by the rays $\theta = 30°$ and $\theta = 45°$; $f(x, y) = 7x - 9y$.
51. D is the wedge cut out of the unit circle by the rays $\theta = \theta_0$ and $\theta = \theta_1$; $f(x, y) = Ax + By$.
► 52. D is the region bounded by the parabolas $y = x^2$ and $x = y^2$; $f = x^n y^m$ where n and m are positive integers.

1.11 Double integrals in polar coordinates

For certain shapes of regions of integration, it is convenient to compute double integrals using polar coordinates (compare Chapter 9, §3.2).

In order to guess at the correct formula we consider a continuous nonnegative function $(x, y) \mapsto z = f(x, y)$ defined for $x^2 + y^2 \leq R^2$ and try to compute the volume V between the graph of the function and the disk of radius R and center at $(0, 0)$ in the x, y plane. This is the volume of the solid bounded by the plane $z = 0$, the cylinder $x^2 + y^2 = R^2$, and the graph of our function (see Fig. 13.24).

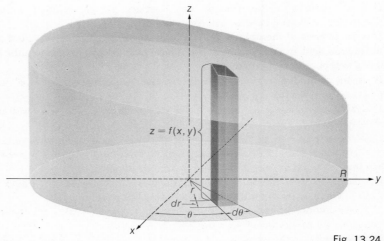

Fig. 13.24

Let (r, θ) denote the polar coordinates of (x, y). We now tell one more mathematical myth: the volume in question consists of infinitely many infinitely thin cylinders. Each cylinder is erected at some point in the disk with coordinates $x = r \cos \theta$, $y = r \sin \theta$. Its base is an infinitely small figure bounded by two segments of infinitely small length dr, lying on rays through the origin, and two circular arcs with center at the origin joining the two rays. The two rays form the infinitely small angle $d\theta$ (see Fig. 13.25). The height of the cylinder is, of course,

$$z = f(x, y) = f(r \cos \theta, r \sin \theta).$$

The base of the cylinder is nearly a rectangle, the lengths of the sides are dr and $r \, d\theta$ (by the formula for the length of a circular arc). Hence the area of the base is $(dr) \cdot (r \, d\theta) = r \, dr \, d\theta$ and the infinitely small volume of the cylinder is height times area of the base, that is,

Fig. 13.25

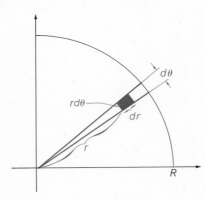

$$f(r \cos \theta, r \sin \theta) r \, dr \, d\theta.$$

The whole volume is the sum of all these infinitely small volumes. Guided by previous experiences, we write

$$V = \iint_{\text{disk}} f(r \cos \theta, r \sin \theta) r \, dr \, d\theta$$

and, guided by the Leibniz notation, we interpret this to mean that

(26)
$$V = \int_0^{2\pi} \int_0^R f(r \cos \theta, r \sin \theta) r \, dr \, d\theta$$

and
$$V = \int_0^R \int_0^{2\pi} f(r \cos \theta, r \sin \theta) r \, d\theta \, dr.$$

(Remember the convention: outer integral sign goes with outer differential.)

We have now arrived at two equations which have a precise meaning, and we may ask whether they are true. It turns out that they are, and that more is true: we may permit f to take on negative values, and we may replace continuity by piecewise continuity. In order not to get involved in technicalities, we only state the following result.

Theorem 4   LET $f(x, y)$ BE BOUNDED AND PIECEWISE CONTINUOUS, AND EQUAL TO 0 OUTSIDE THE DISK $x^2 + y^2 < R^2$, AND ASSUME THAT $f(r \cos \theta, r \sin \theta)$ IS A PIECEWISE CONTINUOUS FUNCTION OF (r, θ). THEN

(27)
$$\iint_{x^2+y^2<R^2} f(x, y) \, dx \, dy = \int_0^{2\pi} \int_0^R f(r \cos \theta, r \sin \theta) r \, dr \, d\theta$$

$$= \int_0^R \int_0^{2\pi} f(r \cos \theta, r \sin \theta) r \, d\theta \, dr.$$

The proof of this result will be found in an appendix to this chapter (see §7.1).

In applying this theorem, one usually deals with a function that is 0 outside some set D, so that the integral to the left may be replaced by an integral over D. It may happen that the region of integration D is convex in the radial direction, that is, that there are numbers θ_1 and θ_2 such that the ray $x = r \cos \theta$, $y = r \sin \theta$, $r > 0$ intersects D only if $\theta_1 < \theta < \theta_2$. If so, the intersection is a segment between $r = \alpha(\theta)$ and $r = \beta(\theta) > \alpha(\theta)$. In such a case, shown in Fig. 13.26, (27) may be written as

(28)
$$\iint_D f(x, y) \, dx \, dy = \int_{\theta_1}^{\theta_2} \int_{\alpha(\theta)}^{\beta(\theta)} f(r \cos \theta, r \sin \theta) r \, dr \, d\theta.$$

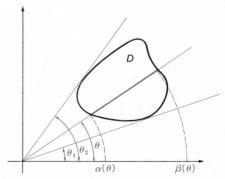

Fig. 13.26

Fig. 13.27

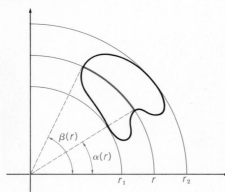

It may also happen that D is convex with respect to θ, so that we may write

(29) $$\iint_D f(x, y)\, dx\, dy = \int_{r_1}^{r_2}\int_{\alpha(r)}^{\beta(r)} f(r \cos \theta, r \sin \theta) r\, d\theta\, dr.$$

The meaning of the term convex and of the numbers r_1, r_2, $\alpha(r)$, $\beta(r)$ is seen from Fig. 13.27.

▶ *Examples* **1.** Assume that D consists of all points with polar coordinates (r, θ) such that

$$\theta_1 < \theta < \theta_2, 0 < r < \phi(\theta)$$

where $\phi(\theta)$ is a positive continuous function. We apply (28) with $f(x, y) = 1$. Since $\alpha(\theta) = 0$, $\beta(\theta) = \phi(\theta)$ and

$$\int_0^{\phi(\theta)} r\, dr = \frac{1}{2} \phi(\theta)^2$$

we obtain that

$$\text{area of } D = \frac{1}{2}\int_{\theta_1}^{\theta_2} \phi(\theta)^2\, d\theta.$$

This is the same formula that we obtained in Chapter 11, §2.6, except for notations.

Fig. 13.28

2. Let D be the region shown in Fig. 13.28. Compute

$$\iint_D e^{x^2+y^2}\, dx\, dy.$$

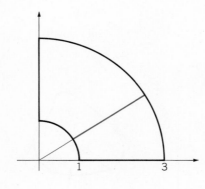

Answer. Clearly θ varies from 0 to $\pi/2$, and r varies from 1 to 3 in the domain considered. Also, $x^2 + y^2 = r^2$. Therefore

$$\iint_D e^{x^2+y^2}\, dx\, dy = \int_0^{\pi/2}\int_1^3 e^{r^2} r\, dr\, d\theta$$

$$= \int_0^{\pi/2}\left[\frac{1}{2}e^{r^2}\right]_{r=1}^{r=3} d\theta = \int_0^{\pi/2} \frac{e^9 - e}{2}\, d\theta = \frac{\pi e(e^8 - 1)}{4}.$$

Without using polar coordinates, one can compute this integral only by techniques of numerical integration.

EXERCISES

53. Pick out five exercises from the set of Exercises 33 to 52 that are especially suited for polar coordinates.

54. In Exercise 45 above, carry out the integration in polar coordinates and check that you get the same result as before.

55. Let D be the region bounded by the spiral $\theta = r$ and the segment $0 \leq x \leq 2\pi$. Compute $\iint_D \arctan (y/x)\, dx\, dy$.

► 56. Let D be the interval $0 < x < 1$, $|y| < 2$. Set up the integral $\iint_D dx\, dy$ in polar coordinates. What is its value?

57. Compute $\iint_D (x^2 + y^2)^2\, dx\, dy$, where D is the interior of one leaf of the four-leafed rose $r = \cos 2\theta$.

58. Compute $\iint_D (1 + x^2 + y^2)^{-1}\, dx\, dy$ where D is the interior of the lemniscate $r^2 = \cos 2\theta$.

59. Let D be the part of the unit circle in the first quadrant. Compute

$$\iint_D \frac{x}{\sqrt{x^2 + y^2}}\, dx\, dy.$$

60. Let D be the ring $1 < x^2 + y^2 < 2$. Compute

$$\iint_D \frac{x^2 y^2}{(x^2 + y^2)^2}\, dx\, dy.$$

61. Let D be the region $\pi < \theta < 5\pi/4$, $\tfrac{1}{4} < x^2 + y^2 < \tfrac{1}{2}$. Compute

$$\iint_D \frac{xy\, dx\, dy}{x^2 + y^2}.$$

We shall now apply double integrals to computing areas of curved surfaces.

§2   Surface Area

2.1 Areas of plane regions

We have already observed that, if the characteristic function $\chi_D(x, y)$ of a bounded set D in the plane is piecewise continuous, then

(1) $$A = \iint_D dx\, dy = \iint_I \chi_D(x, y)\, dx\, dy$$

(where I is some interval containing D) is the area of D. This corresponds to the definition given in Chapter 5, §4.2. Indeed, since the double integral (1) can be computed as an iterated integral, we have

$$A = \int_{-\infty}^{+\infty} \left\{ \int_{-\infty}^{+\infty} \chi_D(x, y)\, dy \right\} dx;$$

we may write $-\infty$ and $+\infty$ as limits of integration since χ_D is 0 outside I. Now if, for a fixed x_0, the intersection of the line $x = x_0$ with D consists of one or several segments, then

$$\int_{-\infty}^{+\infty} \chi_D(x_0, y) \, dy = l(x_0)$$

is the sum of the lengths of these segments. For $\chi_D(x_0, y) = 0$ if (x_0, y) is not a point of D and 1 if y lies in one of the segments considered. Therefore (1) can be written as

$$A = \int_{-\infty}^{+\infty} l(x) \, dx.$$

This is precisely the area formula in Chapter 5, §4.1 (with L the x axis).

Independently of the considerations of Chapter 5, we can convince ourselves that (1) is the right definition of area. Suppose we subdivide I into a mesh of small rectangles (see Fig. 13.29, where the rectangles are squares). The step function $\phi(x, y)$, which equals 1 in those rectangles that lie *completely* in D and is 0 otherwise, satisfies $\phi \leq \chi_D$. Its integral, call it A_1, is the sum of the areas of all small rectangles in D; this is the shaded area in Fig. 13.29. If we add the areas of all rectangles that have *some* points in common with D (shaded in Fig. 13.30), we obtain another number, A_2. This is the integral of a step function $\psi(x, y)$ with $\chi_D \leq \psi$. By Theorem 3 in §1.7, we have $A_1 \leq A \leq A_2$, and we can compute A as accurately as we want, by choosing the mesh fine enough. This corresponds to the intuitive meaning of the statement: the region D has area A.

The definition (1) of area involves a coordinate system. It is intuitively clear, however, that the area of a region does not depend on the position of the coordinate axes used to compute the area. We shall sketch an analytic proof of this in an appendix (§8.2).

Fig. 13.29

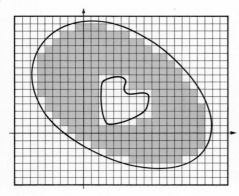

Fig. 13.30

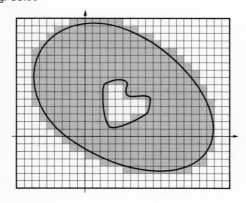

2.2 Areas on inclined planes

Our next aim is to define areas of regions or curved surfaces. As a preparation, let us look at a nonvertical plane and at a reasonable looking set Δ on this plane (Fig. 13.31). Through each point P of Δ, we drop a perpendicular on the plane $z = 0$; let it meet the plane $z = 0$ at the point Q. Then Q is called the **projection** of P on the plane $z = 0$. [In other words, if P has coordinates (x, y, z), its projection Q has coordinates $(x, y, 0)$.] The set D of projections of all points of Δ is called the projection of Δ.

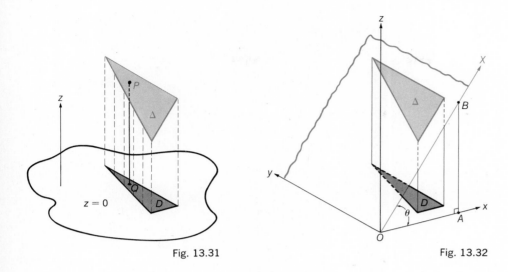

Fig. 13.31   Fig. 13.32

The area of an inclined plane set Δ equals the area of its projection D divided by $\cos\theta$, where θ is the angle between the planes of Δ and D.

To verify this statement, we may assume that the plane of Δ passes through the origin (we may achieve this by translating the coordinate system in the z direction), and that this plane passes through the y axis (this can be achieved by rotating the coordinate system about the z axis). We then have the situation depicted in Fig. 13.32. In the plane of Δ, we introduce the Cartesian coordinate system (X, y). A point in the plane of Δ with coordinates (X, y) has the projection (x, y) with $x = X\cos\theta$, as is seen from the right triangle OAB. (Note that B is the projection of A, $|\overrightarrow{OA}| = x$ and $|\overrightarrow{OB}| = X$.) Thus we obtain Δ from D by keeping all distances in the y direction unchanged and multiplying all distances in the x direction by $1/\cos\theta$. This implies our statement. For a triangle like the one shown in the figure, this is trivial; for every other shape it follows, for instance, by using the area formula of Chapter 5, §4.1 with the y axis as the line L.

2.3 The area formula

Consider now a continuously differentiable function $(x, y) \mapsto z = f(x, y)$ defined in some bounded open set D; we assume that χ_D is piecewise continuous and that the partial derivatives f_x, f_y are bounded. We want to compute the area A of the graph of $z = f(x, y)$, trusting, for the time being, our intuitive ideas about areas.

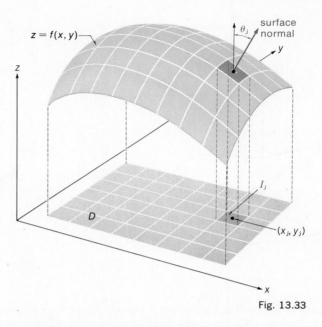

Fig. 13.33

Assume, for the sake of simplicity, that D is an interval (as in Fig. 13.33). We subdivide D into small subintervals. Then our intuitive notion of area tells us that the total area A is the sum of all small areas that project on the small subintervals. Each of these small areas is approximately the area of a plane figure—for a small piece of a surface may be replaced by a piece of its tangent plane. We just learned how to compute areas of inclined plane sets; we conclude that we commit a small error if we replace the area of the portion of our surface lying above a subinterval I_j by the number: (area of I_j)/$\cos \theta_j$, θ_j being the angle between the plane $z = 0$ and the tangent plane to our surface at some point (x_j, y_j) in I_j. Let us compute θ_j. It is the acute angle between a vector normal to the surface at the point with coordinates (x_j, y_j) and $z_j = f(x_j, y_j)$, and a vector in the z direction. We know (see Chapter 12, §2.7) that the vector with the components $(-f_x(x_j, y_j), -f_y(x_j, y_j), 1)$ is normal to the surface at (x_j, y_j, z_j). The unit vector in the z direction has the components $(0, 0, 1)$. Thus (see Chapter 9, §6.2)

$$\cos \theta_j = \frac{1}{\sqrt{1 + f_x{}^2 + f_y{}^2}} \quad \text{computed at } (x_j, y_j).$$

An approximate value of the total area A is

$$\frac{(\text{area of } I_1)}{\cos \theta_1} + \frac{(\text{area of } I_2)}{\cos \theta_2} + \frac{(\text{area of } I_3)}{\cos \theta_3} + \cdots$$

$$= \sqrt{1 + f_x(x_1, y_1)^2 + f_y(x_1, y_1)^2} \cdot \text{area of } I_1$$
$$+ \sqrt{1 + f_x(x_2, y_2)^2 + f_y(x_2, y_2)^2} \cdot \text{area of } I_2$$
$$+ \sqrt{1 + f_x(x_3, y_3)^2 + f_y(x_3, y_3)^2} \cdot \text{area of } I_3 + \cdots.$$

But this is a Riemann sum for a double integral (see §1.8), namely the integral of the function $\sqrt{1 + f_x(x, y)^2 + f_y(x, y)^2}$. We are therefore led to suspect that

$$(2) \qquad A = \iint_D \sqrt{1 + f_x(x, y)^2 + f_y(x, y)^2} \, dx \, dy.$$

This formula gives the right result if $f(x, y) \equiv 0$ (then the integrand is 1) or if $f(x, y)$ is a linear function $f = ax + by + c$ (then the integrand is $\sqrt{1 + a^2 + b^2} = 1/\cos \theta$, θ being the angle between the plane $z = f(x, y)$ and the plane $z = 0$).

Now we change our point of view and adopt formula (2) as the *definition* of the area of the surface $z = f(x, y)$, (x, y) in D.

▶ **Examples** **1.** Find the area of the surface

$$z = \tfrac{2}{3} x \sqrt{x}, \quad 0 < x < 1, \quad 1 < y < 2.$$

Answer. We have $z_x = \sqrt{x}$, $z_y = 0$. By the area formula (2),

$$A = \int_1^2 \int_0^1 \sqrt{1 + z_x^2 + z_y^2} \, dx \, dy = \int_1^2 \int_0^1 \sqrt{1 + x} \, dx \, dy$$

$$= \int_1^2 \left[\frac{2}{3}(1 + x)^{3/2} \right]_{x=0}^{x=1} dy = \frac{2}{3}(2\sqrt{2} - 1).$$

2. Find the area of the surface $z = \cos(\sqrt{3}\, x) + \sin(\sqrt{3}\, y)$, $y > x^2$, $y < 1$.

Answer. We have $z_x = -\sqrt{3} \sin(\sqrt{3}\, x)$, $z_y = \sqrt{3} \cos(\sqrt{3}\, y)$, hence $\sqrt{1 + z_x^2 + z_y^2} = 2$. The inequalities $x^2 < y < 1$ determine a domain (see Fig. 13.34) bounded by a parabola and a horizontal line. The desired area is

$$\int_{-1}^1 \int_{x^2}^1 2dy \, dx = \int_{-1}^1 2(1 - x^2) \, dx = \frac{8}{3}.$$

In general, one should not expect such easy calculations. ◀

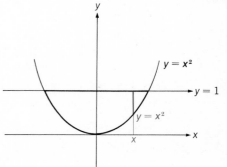

Fig. 13.34

$y = x^2$

$y = 1$

$y = x^2$

EXERCISES

1. Find the area of the surface $z = 9x\sqrt{x}$, $1 < x < 2$, $-1 < y < 0$.
2. Find the area of the surface $z = 4x\sqrt{x} + 3x$, $2 < x < 3$, $1 < y < 2$.
3. Find the area of the surface $z = 9x - y$, $x > 0$, $y > 0$, $x^2 + y^2 < 1$.
4. The plane $3x + 4y - 5z = -6$ intersects an elliptical cylinder parallel to the z axis whose base is $2x^2 + y^2 = 1$. What is the area of that portion of the plane cut off by the cylinder?
5. Find the area of the surface $z = \sqrt{1 - x^2 - y^2}$, $|x| < \frac{1}{2}$, $|y| < |\sqrt{1 - x^2}|$.
6. Find the area of the surface $z = \sqrt{x^2 + y^2}$, $\frac{1}{16} < x^2 + y^2 < \frac{1}{4}$.
7. Find the area of the surface $z = 9x^2$, $7 < y < 8$, $5 < x < 6$.
▶ 8. Find the area of the surface $y = x \tan z$, $1 < x^2 + y^2 < 16$, $-\pi/2 < z < \pi/2$.
9. Find the area above the x, y plane that the cylinder $x^2 + y^2 = 4$ cuts from the sphere $x^2 + y^2 + z^2 = 36$.
10. In formula (2), the surface area is defined for a surface given as the graph of a function $z = f(x, y)$. Write down analogous formulas for the area of a surface given by $x = g(y, z)$ and cut off by a cylinder parallel to the x axis with base D in the y, z plane.
11. Suppose a surface S is defined by $y = h(x, z)$, (x, z) in some region D. Write down a formula for the area of D.
12. Find the area of the surface $y^2 + z^2 = 2x$ cut off by the plane $x = 1$.
13. Find the area above the x, y plane cut off from the cone $x^2 + y^2 = z^2$ by the cylinder $x^2 + y^2 = 2ax$.
14. Find the area of the piece of the paraboloid $x = y^2 + z^2$ contained between the planes $x = 0$ and $x = 12$.
15. Find the area of the ellipsoid $(x/a)^2 + (y/a)^2 + (z/c)^2 = 1$.

2.4 Surfaces of revolution

We recall that a surface of revolution S is obtained by rotating a plane curve C about a straight line l in the plane of the curve (compare Chapter 9, §5.2). Assume that l is the z axis, that C lies in the (x, z) plane, and that C is the graph of the function

$$x \mapsto z = \phi(x), \quad 0 \le a < x < b$$

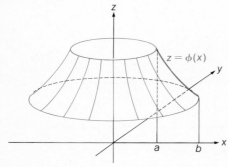

Fig. 13.35

$z = \phi(x)$

(see Fig. 13.35). If C is rotated about the z axis by 360°, it sweeps out a surface S, the graph of

$$(x, y) \mapsto z = \phi(\sqrt{x^2 + y^2}) = \phi(r), \quad a < r < b$$

where r, θ are the polar coordinates of (x, y). By the chain rule,

$$z_x = \phi'(r)\frac{\partial r}{\partial x} = \phi'(r)\frac{x}{r}, \qquad z_y = \phi'(r)\frac{\partial r}{\partial y} = \phi'(r)\frac{y}{r}$$

so that

$$1 + z_x{}^2 + z_y{}^2 = 1 + \phi'(r)^2.$$

By (2) the area of S is

(3)

$$A = \iint\limits_{a^2 < x^2 + y^2 < b^2} \sqrt{1 + \phi'(\sqrt{x^2 + y^2})^2}\; dx\, dy$$

$$= \int_0^{2\pi}\int_a^b \sqrt{1 + \phi'(r)^2}\; r\, dr\, d\theta = 2\pi \int_a^b \sqrt{1 + \phi'(r)^2}\; r\, dr.$$

We may call a dummy variable whatever we please, and we may write

(4)

$$A = 2\pi \int_a^b \sqrt{1 + \phi'(x)^2}\; x\, dx.$$

The same formula gives the area of the surface obtained by rotating the graph of $y = \phi(x)$ about the y axis.

Remark If we rotate the curve C not by 2π but only by the angle ω, $0 < \omega < 2\pi$, then we must replace the factor 2π in (4) by ω. For, in that case, the region of integration (in the x, y plane) in (3) is given by $a < r < b$, $0 < \theta < \omega$.

▶ *Examples*   1. Find the mantle surface of a circular cone of height H and radius R.

Answer. The surface is obtained by rotating about the z axis the segment $z = H - (H/R)x$, $0 < x < R$, called the generator of the cone (see Fig. 13.36). Using (4) with $a = 0$, $b = R$, $\phi(x) = H - (H/R)x$, we obtain

Fig. 13.36

$$A = 2\pi \int_0^R \sqrt{1 + H^2/R^2}\; x\, dx = \frac{2\pi \sqrt{1 + H^2/R^2}\; R^2}{2} = \pi R \sqrt{R^2 + H^2}.$$

Since the length of the generator is $L = \sqrt{R^2 + H^2}$, we have $A = \pi R L$, a well-known formula from elementary geometry.

2. Find the surface area of a sphere of radius R.

Answer. The desired area is twice the area obtained by rotating the circular arc $z = \sqrt{R^2 - x^2}$, $0 < x < R$ about the z axis, or

$$2 \cdot 2\pi \int_0^R \left[1 + \left(\frac{d\sqrt{R^2 - x^2}}{dx}\right)^2\right]^{1/2} x\, dx = 4\pi \int_0^R \frac{Rx\, dx}{\sqrt{R^2 - x^2}} = 4\pi R^2$$

as one is taught in elementary geometry.

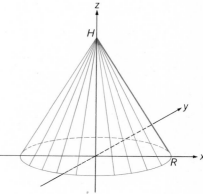

3. What is the area of the part of the sphere $x^2 + y^2 + z^2 = R^2$ which lies over the sector $0 < \theta < \omega < 2\pi$ in the (x, y) plane?

Answer. $2\omega R^2$, by the Remark made above and by Example 3.◄

EXERCISES

16. Let C be the line segment from (x_0, y_0) to (x_1, y_1). Compute the areas of the surfaces of revolution generated by revolving C around the x axis and around the y axis.

17. Let C be the wedge bounded by the rays $\theta = \theta_0$, $\theta = \theta_1$, and the circle $x^2 + y^2 = R^2$. What is the area of the figure of revolution generated by revolving C around the x axis?

18. Repeat Exercise 17 revolving C around the y axis.

19. Find a formula for the area of the surface of revolution generated by revolving the curve $z = \phi(x)$ about the line $z = 1$.

20. Find the area of the surface generated by rotating the parabolic segment $y = x^2$, $y \leq 1$ about the y axis.

21. Repeat Exercise 20, rotating about the x axis.

22. Repeat Exercise 20, rotating about the line $x = 10$.

23. Repeat Exercise 20, rotating about the line $y = -1$.

► 24. Find the area of that part of the unit sphere $x^2 + y^2 + z^2 = 1$ between the planes $z = A$ and $z = B$, where $0 < A < B < 1$.

25. Among all curves of the form $y = \phi(x)$ passing through the points $(0, 1)$ and $(1, 1)$, the one that generates a surface of revolution with the smallest area is the line $y = 1$. Try to prove this.

§3 Triple Integrals

Since we have discussed double integrals rather carefully, we can rapidly extend the theory to functions of three and more variables. No essentially new phenomena appear, so that there will be no need to state formal theorems.

3.1 Piecewise continuous functions

A function $f(x, y, z)$ defined in a three-dimensional interval $x_0 < x < x_1$, $y_0 < y < y_1$, $z_0 < z < z_1$ will be called **piecewise continuous** if it is either continuous in the whole interval or the points at which f fails to be continuous lie on finitely many surfaces. Each of these surfaces must be either the intersection of the interval with a plane parallel to the co-ordinate axes, or representable as a graph of a function $x = \phi_1(y, z)$, *and* as a graph of a function $y = \phi_2(z, x)$ *and* as a graph of a function $z = \phi_3(x, y)$, all three functions being defined and continuous in some closed two-dimensional intervals.

The definition implies that, for all fixed values of x, with perhaps finitely many exceptions, $f(x, y, z)$ is a piecewise continuous function of (x, y); a similar statement is true about y and z.

3.2 Computing triple integrals

If $f(x, y, z)$ is a bounded and piecewise continuous function defined in the interval I: $x_0 < x < x_1$, $y_0 < y < y_1$, $z_0 < z < z_1$, then the integral of f over I, in symbols

$$(1) \qquad \iiint\limits_I f(x, y, z)\, dx\, dy\, dz,$$

is the number that can be computed by integrating $f(x, y, z)$ with respect to x from x_0 to x_1, with respect to y from y_0 to y_1, and with respect to z from z_0 to z_1, *in any order.* For instance, if we decide to integrate in the order x, y, z, then we have

$$(2) \qquad \begin{aligned} \iiint\limits_I f(x, y, z)\, dx\, dy\, dx &= \int_{z_0}^{z_1}\int_{y_0}^{y_1}\int_{x_0}^{x_1} f(x, y, z)\, dx\, dy\, dz \\ &= \int_{z_0}^{z}\left\{\int_{y_0}^{y}\left[\int_{x_0}^{x} f(x, y, z)\, dx\right] dy\right\} dz. \end{aligned}$$

Observe that we again follow the *convention:* the outer integral sign corresponds to the outer differential.

One can also compute (1) by one single and one double integration. Indeed, since

$$\int_{y_0}^{y_1}\int_{x_0}^{x_1} f(x, y, z)\, dx\, dy = \iint\limits_{\substack{y_0 < y < y_1 \\ x_0 < x < x_1}} f(x, y, z)\, dx\, dy$$

we can write instead of (2)

$$(3) \qquad \iiint\limits_I f(x, y, z)\, dx\, dy\, dz = \int_{z_0}^{z_1}\left[\;\iint\limits_{\substack{y_0 < y < y_1 \\ x_0 < x < x_1}} f(x, y, z)\, dx\, dy\right] dz.$$

All single and double integrals occurring in (2) and (3) are integrals of bounded piecewise continuous functions, except perhaps for finitely many values of the variables. The same is true for integrals obtained by changing the order of integration:

$$\int_{x_0}^{x_1}\int_{z_0}^{z_1}\int_{y_0}^{y_1} f(x, y, z)\, dy\, dz\, dx = \int_{y_0}^{y_1}\left[\;\iint\limits_{\substack{x_0 < x < x_1 \\ z_0 < z < z_1}} f(x, y, z)\, dx\, dz\right] dy,$$

and so forth. This statement is analogous to Theorem 1 in §1.5. All orders of integration lead to the same value of (1). This is the analogue of Theorem 2 in §1.5. The proof given for Theorems 1 and 2 in the Supplement (see §§4 and 5) can be modified so as to establish the statement just made.

The definitions of step functions and Riemann sums (see §§1.7 and 1.8) and Theorem 3 of §1.7 extend at once to functions of three variables, and triple integrals have the properties stated in §1.6 for double integrals.

Finally, the triple integral over a bounded region of integration D is defined by the formula

$$\iiint_D f(x, y, z) \, dx \, dy \, dz = \iiint_I \chi_D(x, y, z) f(x, y, z) \, dx \, dy \, dz$$

where I is some interval containing D, provided the characteristic function χ_D is piecewise continuous. In other words, to integrate $f(x, y, z)$ over D, one sets $f = 0$ outside D, and then integrates over some 3-dimensional interval containing D.

▶ **Examples**   **1.** Let I be the cube $0 < x < 1, 0 < y < 1, 0 < z < 1$. Compute

$$\iiint_I xy^2z^3 \, dx \, dy \, dz.$$

Answer.   (Compare Remark 3 in §1.2.) We have

$$\iiint_I xy^2z^3 \, dx \, dy \, dz = \int_0^1\int_0^1\int_0^1 xy^2z^3 \, dx \, dy \, dz$$

$$= \left(\int_0^1 x \, dx\right)\left(\int_0^1 y^2 \, dy\right)\left(\int_0^1 z^3 \, dz\right)$$

$$= \frac{1}{2} \cdot \frac{1}{3} \cdot \frac{1}{4} = \frac{1}{24}.$$

2. Compute

$$\iiint_I xyz \cos(xyz) \, dx \, dy \, dz$$

where I is as in Example 1.

Answer.   The integral equals

$$\int_0^1\int_0^1\int_0^1 xyz \cos(xyz) \, dx \, dy \, dz = \int_0^1\int_0^1 \left[yz \sin(xyz) \right]_{x=0}^{x=1} dy \, dz$$

$$= \int_0^1\int_0^1 yz \sin(yz) \, dy \, dz = \int_0^1 \left[-z \cos(yz) \right]_{y=0}^{y=1} dz$$

$$= \int_0^1 (z - z \cos z) \, dz = \frac{1}{2} - \int_0^1 z \, d \sin z$$

$$= \frac{1}{2} - \left[z \sin z \right]_{z=0}^{z=1} + \int_0^1 \sin z \, dz = \frac{1}{2} - \sin 1 - \left[\cos z \right]_{z=0}^{z=1}$$

$$= \frac{3}{2} - \sin 1 - \cos 1.$$

In the last step, we used integration by parts.

3. Let D be the region determined by the inequalities $x > 0$, $y > 0$, $z < 4$ and $z > x^2 + y^2$. Compute

$$\iiint\limits_D 2x \, dx \, dy \, dz.$$

Solution. D is the region between the planes $x = 0$, $y = 0$, $z = 4$ and the paraboloid $z = x^2 + y^2$. In region D, z varies between 0 and 4. For a fixed z, x and y are positive and vary so that $x^2 + y^2 \leq z$. This means that y varies from $-\sqrt{z}$ to $\sqrt{z}$ and x from $-\sqrt{z - y^2}$ to $\sqrt{z - y^2}$. Hence the integral equals

$$\int_0^4 \iint\limits_{x^2+y^2<z} 2x \, dx \, dy \, dz = \int_0^4 \int_{-\sqrt{z}}^{\sqrt{z}} \int_{-\sqrt{z-y^2}}^{\sqrt{z-y^2}} 2x \, dx \, dy \, dz$$

$$= \int_0^4 \int_{-\sqrt{z}}^{\sqrt{z}} \left[x^2 \right]_{x=-\sqrt{z-y^2}}^{x=\sqrt{z-y^2}} dy \, dz = \int_0^4 \int_0^{\sqrt{z}} (z - y^2) \, dy \, dz$$

$$= \int_0^4 \left[zy - \frac{y^3}{3} \right]_{y=0}^{y=\sqrt{z}} dz = \int_0^4 \frac{2}{3} z^{3/2} \, dz = \left[\frac{4}{15} z^{5/2} \right]_{z=0}^{z=4} = \frac{128}{15}. \blacktriangleleft$$

EXERCISES

In Exercises 1 to 15, integrate the given function f over the given interval.

1. $f = x^2 y^4 z$; $-1 < x < 1$, $0 < y < 2$, $0 < z < 1$.
2. $f = 1/(xyz)$; $1 < x < 2$, $-2 < y < -1$, $2 < z < 3$.
3. $f = (x + y)/z$; $-1 < x < 1$, $-2 < y < 2$, $1 < z < 2$.
4. $f = xyze^{xyz}$; $0 < x < 1$, $0 < y < 1$, $0 < z < 1$.
5. $f = (x + y + z)^2$; $1 < x < 2$, $2 < y < 3$, $3 < z < 4$.
6. $f = uv\sqrt{w}$; $-1 < u < 1$, $0 < v < 2$, $3 < w < 4$.
7. $f = r + 2(s/r)e^t + t^2(1 + 1/r)$; $2 < r < 3$, $2 < s < 3$, $2 < t < 3$.
▶ 8. $f = \log(x^\alpha y^\beta z^\gamma)$; $1 < x < 2$, $2 < y < 3$, $3 < z < 4$.
9. $f = \sqrt{xy + xyz^2}$; $1 < x < 2$, $2 < y < 3$, $3 < z < 4$.
10. $f = (x/y) + (y/z) + (z/x)$; $1 < x < 2$, $1 < y < 2$, $1 < z < 2$.
11. $f = [(x/y) + (y/z)]^3$; $-2 < x < -1$, $-2 < y < -1$, $-2 < z < -1$.
12. $f = r \cos 2t + t \sin 3s$; $0 < r < \pi$, $0 < s < \pi/2$, $0 < t < \pi/4$.
13. $f = \cos(2u - 3v + 4w)$; $\pi/6 < u < \pi/3$, $\pi/4 < v < 3\pi/4$, $\pi/2 < w < 3\pi/4$.
14. $f = x \log(yz) + z^{-1} \log(xy)$; $1 < x < 2$, $1 < y < 2$, $1 < z < 2$.
15. $f = (x + e^{-y} + \cos z)^2$; $0 < x < 1$, $0 < y < 1$, $-\pi/2 < z < \pi/2$.

3.3 Volume

If D is a reasonable region in (x, y, z) space, (that is, if D is contained in some interval and χ_D is piecewise continuous), then

$$\iiint_D dx\, dy\, dz = \iiint_I \chi_D(x, y, z)\, dx\, dy\, dz$$

is the **volume** of D. This definition corresponds to the volume formula given in Chapter 5 (see §4.4) and to our common sense idea of volume. This can be verified by arguments analogous to the ones used for area in §2.1. The volume computed in this way does not depend on the choice of the coordinate system. This is geometrically obvious and can be proved by rephrasing the argument given for areas in the appendix to this chapter (compare §8.2).

Also, the volume has the required additive property: if D is divided into two disjoint sets D_1 and D_2, with volumes V_1 and V_2, the volume of D is $V_1 + V_2$. This follows from the additive property of triple integrals which is analogous to the corresponding property of double integrals (see §1.10).

EXERCISES

▶ 16. Find the volume inside $x^2 + y^2 = 9$, above $z = 0$, and below $x + z = 4$.
17. Find the volume interior to the tetrahedron formed by the plane $x = 0$, $y = 0$, $z = 0$, $6x + 4y + 3z = 12$.
18. Find the volume bounded by the paraboloid $z = x^2 + 2y^2$ and the cylinder $z = 4 - x^2$.
19. Find the volume of an ellipsoid with semiaxes a, b, c.
20. Find the volume common to the two cylinders $x^2 + y^2 = R^2$, $x^2 + z^2 = R^2$.
21. Find the volume bounded by the two paraboloids $z = x^2 + 10y^2$ and $z = 20 - x^2 - 10y^2$.
22. Find the volume of the tetrahedron bounded by the planes $x = 0$, $y = 0$, $z = 0$, and $(x/a) + (y/b) + (z/c) - 1 = 0$.
23. Find the volume of the set defined by $x > 0$, $z > 0$, $0 < y < z$, $x < 4 - y^2$.

3.4 Integration in cylindrical coordinates

In §1.11, we saw that when one wants to compute a double integral using polar coordinates, it is not enough to rewrite the integrand $f(x, y)$ as a function of polar coordinates (r, θ), but one also must multiply the integrand by a factor, namely r (compare Theorem 4 in §1.11). This phenomenon occurs whenever one uses "new" coordinates for integration—the integrand is to be multiplied by a factor, depending on the coordinates used. This factor is called the **Jacobian,** in honor of the mathematician Jacobi.

G. G. J. JACOBI (1804–1857). Among his greatest achievements are the theory of so-called "elliptic" functions, and a new mathematical approach to problems of mechanics that continued the work of Hamilton. The Hamilton-Jacobi theory played an unexpected part in the creation of quantum theory.

Jacobi was an inspiring teacher and established the tradition of combining university teaching with research apprenticeship.

Without going into the general theory, we shall show how to compute triple integrals in cylindrical coordinates. These coordinates, usually denoted by (r, θ, z), have been defined in Chapter 9, §5.1; the definition is shown in Fig. 13.37. We recall that the connection with rectangular coordinates is given by the formulas:

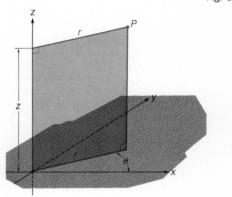

Fig. 13.37

$$(4) \qquad x = r \cos \theta, \qquad y = r \sin \theta, \qquad z = z.$$

Assume now that $f(x, y, z)$ is a bounded piecewise, continuous function which is 0 outside the cylinder $x^2 + y^2 < R^2$, $|z| < H$ (this is no loss of generality) and that the function $(r, \theta, z) \to f(r \cos \theta, r \sin \theta, z)$ is also piecewise continuous. (This could be proved if we used a slightly different definition of piecewise continuity, but we want to avoid technicalities here.) Under these circumstances, we have

$$(5) \qquad \iiint_{\substack{x^2+y^2<R^2 \\ |z|<H}} f(x, y, z)\, dx\, dy\, dz = \int_{-H}^{H} \int_{0}^{2\pi} \int_{0}^{R} f(r \cos \theta, r \sin \theta, z) r\, dr\, d\theta\, dz.$$

The Jacobian is again r; this is not surprising since cylindrical coordinates differ from polar coordinates only by the presence of the third coordinate z.

To prove (5), we note that by equation (3) in §3.2, and since f is 0 outside the cylinder, we have

$$(6) \qquad \iiint_{\substack{x^2+y^2<R^2 \\ |z|<H}} f(x, y, z)\, dx\, dy\, dz = \int_{-H}^{H} \left\{ \iint_{x^2+y^2<R^2} f(x, y, z)\, dx\, dy \right\} dz.$$

But by Theorem 4 in §1.11 we have, for a fixed z,

$$\iint_{x^2+y^2<R^2} f(x, y, z)\, dx\, dy = \int_{0}^{2\pi} \int_{0}^{R} f(r \cos \theta, r \sin \theta) r\, dr\, d\theta.$$

Fig. 13.38

Substituting this into (6), we obtain (5). Of course, one can rewrite (5) in various ways, by changing the order of integration in the right-hand side.

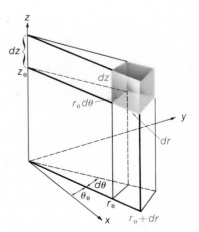

One could also have guessed at (5) by a mathematical myth as depicted in Fig. 13.38. We consider a point with cylindrical coordinates (r_0, θ_0, z_0); the two cylinders $r = r_0$, $r = r_0 + dr$, the two planes $\theta = \theta_0$, $\theta = \theta_0 + d\theta$, and the two planes $z = z_0$, $z_0 + dz$. Here dr, $d\theta$, dz are infinitely small numbers (we may say so, since we are telling a myth). The six surfaces bound an infinitely small box whose volume equals the product of the lengths of the sides; that is, $(dr)(r_0\, d\theta)(dz) = r_0\, dr\, d\theta\, dz$. The triple integral of a function f is the sum of infinitely many terms: value of f at a point times the volume of the small box at this point.

▶**Examples** **1.** Let D be the region defined by the inequalities

(7) $$x^2 + y^2 < 1, \qquad 0 < z < x^2 + y^2.$$

Find $$\iiint\limits_D x^2 y^2 \, dx \, dy \, dz.$$

Answer. We use cylindrical coordinates; the inequalities (7) can be rewritten as $0 < r < 1, 0 < z < r^2$. Hence the triple integral considered equals

$$\int_0^{2\pi} \int_0^1 \int_0^{r^2} (r \cos \theta)^2 (r \sin \theta)^2 r \, dz \, dr \, d\theta$$

$$= \int_0^{2\pi} \int_0^1 \int_0^{r^2} \frac{1}{4} r^5 \sin^2 2\theta \, dz \, dr \, d\theta = \int_0^{2\pi} \int_0^1 \frac{1}{4} r^7 \sin^2 2\theta \, dr \, d\theta$$

$$= \frac{1}{4} \int_0^{2\pi} \sin^2 2\theta \, d\theta \int_0^1 r^7 \, dr = \frac{1}{4} 2\pi \frac{1}{8} = \frac{\pi}{16}. \blacktriangleleft$$

EXERCISES

24. Find the volume cut off the cylinder $r = 4 \cos \theta$ by the sphere $x^2 + y^2 + z^2 = 16$ and the x, y plane.

25. Find the volume cut from the sphere $x^2 + y^2 + z^2 = 36$ by the cylinder $x^2 + y^2 = 9$.

▶ 26. Find the volume above the paraboloid $z = x^2 + y^2$ and below the plane $z = 2y$.

27. If D is the region where $x > 0$, $y > 0$, $0 < z < 1$, $x^2 + y^2 < 4$, compute $\iiint_D (x^2 + y^2)^n \cos z \, dx \, dy \, dz$.

28. If D is the cylindrical shell bounded by the planes $x = 1$, $x = 2$, and by the circular cylinders $y^2 + z^2 = 4, y^2 + z^2 = 9$, compute $\iiint_D e^x \sqrt{y^2 + z^2} \, dx \, dy \, dz$.

29. Two circular cylinders have the same radius R and their axes intersect at right angles. Show that their intersection has volume $16R^3/3$.

30. Find the volume bounded by the x, y plane, the circular cylinders $x^2 + y^2 = 1$, and the parabolic cylinder $x^2 - 1 = z$.

31. If D is the region cut out of the cylinder $x^2 + y^2 = 1$ by the two planes $2x - 3y + z = 0, z = 100$, compute $\iiint_D (5x - 3z) \, dx \, dy \, dz$.

32. Compute $\iiint_D x \, dx \, dy \, dz$ where D is the region of Exercise 30.

33. Compute $\iiint_D [z/(y^2 + z^2)] \, dx \, dy \, dz$ where D is the region $1 < y^2 + z^2 < 2$, $3 \le x \le 4$.

Fig. 13.39

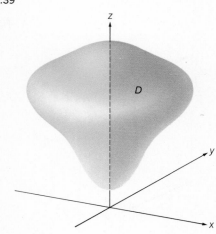

3.5 Bodies of revolution

Let D be a region and l a line, both lying in one plane, D on one side of l. If we rotate D about l by $360°$, it sweeps out a body of revolution B. To compute the volume of B, choose l as the z axis, and let D lie in the right half of the (x, z) plane; see Fig. 13.39. Let $\chi_D(x, z)$ be the characteristic function of D. Then the characteristic function of B is $\chi_D(\sqrt{x^2 + y^2}, z)$. Using cylindrical coordinates, we compute the volume V of B to be (T is some large number)

$$V = \iiint_{x^2+y^2+z^2<T^2} \chi_D(\sqrt{x^2 + y^2}, z) \, dx \, dy \, dz$$

$$= \int_0^{2\pi} \iint_{r^2+z^2<T^2} \chi_D(r, z) r \, dr \, dz \, d\theta = 2\pi \iint_D r \, dr \, dz$$

or, since it does not matter what we call dummy variables,

$$(8) \qquad\qquad V = 2\pi \iint_D x \, dx \, dz.$$

Similarly, if we rotate a region D in the right half of the (x, y) plane about the y axis, the volume is

$$(8') \qquad\qquad V = 2\pi \iint_D x \, dx \, dy$$

and if D is a region in the upper half of the (x, y) plane and we rotate it about the x axis, we obtain the volume

$$(8'') \qquad\qquad V = 2\pi \iint_D y \, dx \, dy.$$

In the latter case, assume that D is the region under the graph of the positive function $y = f(x)$, $a < x < b$. Then

$$V = 2\pi \int_a^b \int_0^{f(x)} y \, dy \, dx = 2\pi \int_a^b \left[\frac{1}{2} y\right]_{y=0}^{y=f(x)} dx = \pi \int_a^b f(x)^2 \, dx.$$

We have already obtained this formula before; see Chapter 5, §4.6.

Remark If D is rotated not by 2π but by an angle ω, $0 < \omega < \pi$, one should replace, in the formula (8), 2π by ω. The proof is left to the reader.

EXERCISES

34. Find the volume obtained by revolving the region $2 < x < 4$, $-x^2 < y < x^3$ about the y axis.
35. Find the volume of the figure generated by revolving the region $|x| < \pi$, $0 < y < \sin x$ around the x axis.
36. Let D be the region bounded by the x axis, the curve $y = e^x$ and the lines $x = a$, $x = b$. Find the volume generated by revolving D around the x axis.
37. Find the volume obtained by revolving the region D in Exercise 36 around the y axis.

Fig. 13.40

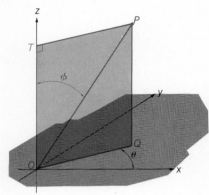

3.6 Integration in spherical coordinates

We recall (see Chapter 9, §5.3, and Fig. 13.40) that spherical coordinates (ρ, θ, ϕ) are connected with Cartesian coordinates by the relations

$$(9) \qquad x = \rho \sin \phi \cos \theta, \qquad y = \rho \sin \phi \sin \theta, \qquad z = \rho \cos \phi$$

and with the cylindrical coordinates by the relations

$$(9') \qquad r = \rho \sin \phi, \qquad \theta = \theta, \qquad z = \rho \cos \phi.$$

The integration formula (valid under appropriate conditions) reads

$$
(10) \quad
\begin{aligned}
&\iiint\limits_{x^2+y^2+z^2<R^2} f(x, y, z)\, dx\, dy\, dz \\
&= \int_0^\pi \int_0^{2\pi} \int_0^R f(\rho \sin \phi \cos \theta,\, \rho \sin \phi \sin \theta,\, \rho \cos \phi)\rho^2 \sin \phi\, d\rho\, d\theta\, d\phi.
\end{aligned}
$$

Thus the Jacobian is $\rho^2 \sin \phi$.

Before proving this, we note that the result could have been guessed from a mathematical myth similar to the one told above, and depicted in Fig. 13.41. The two infinitely close spheres: $\rho = \rho_0$, $\rho = \rho_0 + d\rho$, the two infinitely close planes $\theta = \theta_0$, $\theta = \theta_0 + d\theta$, and the two infinitely close cones $\phi = \phi_0$, $\phi = \phi_0 + d\phi$ bound a "box" of volume $(d\rho)(\rho_0 \sin \phi_0\, d\theta)(\rho_0\, d\phi) = \rho_0{}^2 \sin \phi_0\, d\rho\, d\theta\, d\phi.$

Fig. 13.41

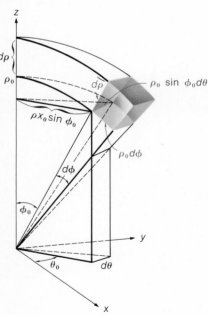

Now we prove (10); we begin by an identity that follows from (5) by assuming that f vanishes outside the sphere of radius R about the origin. Since the inequality $x^2 + y^2 + z^2 < R^2$ can be written in cylindrical coordinates as $z^2 + r^2 < R^2$, we have

$$
(11) \quad
\begin{aligned}
&\iiint\limits_{x^2+y^2+z^2<R^2} f(x, y, z)\, dx\, dy\, dz \\
&= \int_{-R}^R \int_0^{2\pi} \int_0^{R^2-z^2} f(r \cos \theta,\, r \sin \theta,\, z)r\, dr\, d\theta\, dz \\
&= \int_0^{2\pi} \int_{-R}^R \int_0^{R^2-z^2} f(r \cos \theta,\, r \sin \theta,\, z)r\, dr\, dz\, d\theta.
\end{aligned}
$$

The last step is justified since it is legitimate to interchange the order of integration in a multiple integral over an interval. For a fixed θ, consider

$$(12) \qquad \int_{-R}^R \int_0^{R^2-z^2} f(r \cos \theta,\, r \sin \theta,\, z)r\, dr\, dz.$$

We treat z and r as Cartesian coordinates in a (z, r) plane. Then (12) is a double integral over an upper semidisk (see Fig. 13.42). Noting (9′), we interpret (ρ, ϕ) as polar coordinates in the (z, r) plane, so that

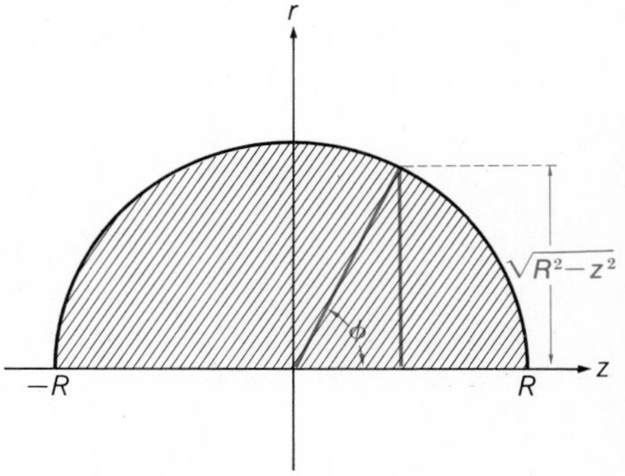

$$\int_{-R}^{R} \int_{0}^{R^2 - z^2} f(r \cos \theta, r \sin \theta, z) r \, dr \, dz$$

$$= \int_{0}^{\pi} \int_{0}^{R} f(\rho \sin \phi \cos \theta, \rho \sin \phi \sin \theta, \rho \cos \phi)(\rho \sin \phi) \rho \, d\rho \, d\phi.$$

If we substitute this into (11), we obtain an identity that differs from (10) only by the order of integrations.

Fig. 13.42

▶ **Example.**   We compute the volume bounded by the sphere of radius R; the problem is tailor-made for spherical coordinates. Thus

$$\iiint\limits_{x^2 + y^2 + z^2 < R^2} dx \, dy \, dz = \int_{0}^{\pi} \int_{0}^{2\pi} \int_{0}^{R} \rho^2 \sin \phi \, d\rho \, d\theta \, d\phi$$

$$= \left(\int_{0}^{\pi} \sin \phi \, d\phi \right) \left(\int_{0}^{2\pi} d\theta \right) \left(\int_{0}^{R} \rho^2 \, d\rho \right)$$

$$= 2 \cdot 2\pi \frac{R^3}{3} = \frac{4\pi R^3}{3}$$

as expected. ◀

EXERCISES

▶ 38. Compute $\iiint_D xyz^2 \, dx \, dy \, dz$ where D is the unit sphere.

39. Find the volume remaining when the cone $3(x^2 + y^2) = z^2$ is removed from the hemisphere $z > 0$, $x^2 + y^2 + z^2 = a^2$.

40. A round hole of radius 1 is bored through the center of a sphere of radius 2. What volume is removed?

41. Compute $\iiint_D (x^2 + y^2 + z^2)^3 \, dx \, dy \, dz$ where D is the region $x > 0$, $y > 0$, $x^2 + y^2 + z^2 < 10$.

42. Compute $\iiint_D xy^2 \, dx \, dy \, dz$ where D is the intersection of the unit sphere, the half-space $z > 0$, and the cone $x^2 + y^2 = z^2$.

43. Compute $\iiint_D x\sqrt{x^2 + y^2 + z^2} \, dx \, dy \, dz$ where D is intersection of the unit sphere and the cylinder $x^2 + y^2 = \frac{1}{2}$.

44. Compute $\iiint_D \arctan (\sqrt{x^2 + y^2}/z) \, dx \, dy \, dz$. D is the region cut out of the upper half of the unit sphere by two circular cones about the z axis with vertex at the origin and vertex angles θ_1 and θ_2.

45. Compute $\iiint_D (x^2 + y^2 + z^2)^{\alpha/2} \, dx \, dy \, dz$ where D is a shell of inner radius ϵ and outer radius R, with center at the origin. For which values of α does this integral approach a finite limit as R becomes arbitrarily large? What is the value of the limit when it exists?

46. For the integral in Exercise 45, for which values of α does the integral approach a finite value as ϵ goes to zero while R remains fixed? What is the value of the limit?

47. Compute $\iiint_D xyz \, dx \, dy \, dz$, where D is the portion of the unit sphere in the "first octant," $x > 0$, $y > 0$, $z > 0$.

3.7 Density

In applications, the integrand in a triple integral is often interpreted as **density.** This means that one thinks of the region of integration, D, as a material body and one assumes that there is a positive continuous function $f(x, y, z)$ such that the mass of every part D_0 of D is given by

$$\iiint\limits_{D_0} f(x, y, z)\, dx\, dy\, dz.$$

The total mass is, of course, the integral over all of D. Since mass cannot be negative, the density function f must be positive or zero. (Negative densities may occur in describing the distribution of electric charges.) The density is measured in units of mass divided by length cubed.

(The image of continuously distributed matter, given by a density function, is very useful in many branches of physics. Yet one should not forget that this is a radical idealization of reality. It neglects the facts that matter consists of molecules, and molecules of atoms, and that the atoms themselves have a complicated internal structure.)

Occasionally one also considers a density function of two variables (measured in units of mass divided by length squared) defined in some region D of the plane $z = 0$. Then we think of D as a material **lamina** which is so thin that its width may be set equal to 0. One can also discuss a continuous distribution of matter along a curved surface (we shall encounter an example in the next subsection) or along a curve. In the latter case, one speaks of a material **wire,** and the density is measured in units of mass divided by length.

▶ *Examples*   *1.* If the density is $f(x, y, z) = x^2 y^4 z^6$, the mass contained in the cube $0 < x < 1$, $0 < y < 1$, $0 < z < 1$ is

$$\int_0^1\int_0^1\int_0^1 x^2 y^4 z^6\, dx\, dy\, dz = \frac{1}{105}.$$

2. Suppose the density of a distribution of matter in the plane $z = 0$ is $f(x, y) = 1$. Then the mass contained in a (reasonable) set D in that plane is equal to the area of D. ◀

EXERCISES

48. In a cylinder of unit height and unit radius, the density is proportional to height above the base and inversely proportional to distance from the axis. At the circumference of the upper boundary, the density is a known quantity, K. Compute the mass of the cylinder.

49. A hemisphere of radius R has density equal to the cube of the distance from the center. What is its mass?

50. A plane lamina has the form of an ellipse with semiaxes a and b, $a > b$. It has planar density equal to the positive distance from the major axis. Compute its mass.

51. A circular wire has linear density equal to $(1 + \sin^2 \phi)$, where ϕ is the angular measure around the wire from a given fixed point on the wire. Compute the mass of the wire.

52. An electrically charged cube of side 1 has a charge density at each point equal to the distance to the bottom face of the cube minus the distance to the left face of the cube. What is the total charge on the cube?

53. A plane lamina has the shape of an isosceles right triangle. The density at each point equals the distance from the right angle. The area of the lamina is 1. What is its mass?

54. A right circular cone has base area 1 and height h. The density at each point equals the distance from the center of the base. What is the total mass?

55. A cube of side 1 is formed in such a way that the faces and the interior have negligible mass in comparison to the edges, which are formed of heavy wire. The linear density along each edge equals the distance from the lower right front corner of the cube. What is the mass of the cube?

3.8 Gravitational attraction

According to Newton's inverse square law, a particle of mass m exerts on another particle, say of mass 1, an attractive force of magnitude $\gamma m / r^2$ where γ is a universal constant and r the distance between the two particles. To simplify writing we may choose the units so that $\gamma = 1$. Then the force exerted by a mass m located at $\mathbf{x} = (x, y, z)$ at a mass 1 located at $\mathbf{X} = (X, Y, Z)$ is

(13) $$\frac{m}{|\mathbf{x} - \mathbf{X}|^3} (\mathbf{x} - \mathbf{X}).$$

This is so since $|\mathbf{x} - \mathbf{X}|$ is the distance between (x, y, z) and (X, Y, Z) and $|\mathbf{x} - \mathbf{X}|^{-1}(\mathbf{x} - \mathbf{X})$ is the unit vector pointing in the direction from (X, Y, Z) to (x, y, z).

Suppose now that we are given N masses, $m_1, m_2, \cdots, m_N$, located at points with position vectors $\mathbf{x}_1, \mathbf{x}_2, \cdots, \mathbf{x}_n$, respectively. The total force exerted by these N particles on a unit mass located at $\mathbf{X}$ is

$$\sum_{j=1}^{N} \frac{m_i}{|\mathbf{x}_j - \mathbf{X}|^3} (\mathbf{x}_j - \mathbf{X}),$$

the sum of the attractive forces of the N particles. This suggests that if we are given a solid D with a density function $f(x, y, z)$, then the force $\mathbf{F}$ exerted by D (on a unit mass at $\mathbf{X}$) can be computed by the formula

$$(14) \qquad \mathbf{F} = \iiint_D \frac{f(\mathbf{x})}{|\mathbf{x} - \mathbf{X}|^3} (\mathbf{x} - \mathbf{X}) \, dx \, dy \, dz.$$

This is a triple integral of a vector-valued function, and is to be interpreted according to the rule stated in Chapter 11, §1.4. The components of the integrand in (13) are

$$\frac{f(x, y, z)(x - X)}{[(x - X)^2 + (y - Y)^2 + (z - Z)^2]^{3/2}},$$

$$\frac{f(x, y, z)(y - Y)}{[(x - X)^2 + (y - Y)^2 + (z - Z)^2]^{3/2}},$$

$$\frac{f(x, y, z)(z - Z)}{[(x - X)^2 + (y - Y)^2 + (z - Z)^2]^{3/2}}.$$

Let $\{\mathbf{e}_1, \mathbf{e}_2, \mathbf{e}_3\}$ be the frame associated with the coordinate system used (see Chapter 9, §1.12). We can write $\mathbf{F} = F\mathbf{e}_1 + G\mathbf{e}_2 + H\mathbf{e}_3$ where

$$F = \iiint_D \frac{f(x, y, z)(x - X)}{[(x - X)^2 + (y - Y)^2 + (z - Z)^2]^{3/2}} \, dx \, dy \, dz,$$

$$(14') \quad G = \iiint_D \frac{f(x, y, z)(y - Y)}{[(x - X)^2 + (y - Y)^2 + (z - Z)^2]^{3/2}} \, dx \, dy \, dz,$$

$$H = \iiint_D \frac{f(x, y, z)(z - Z)}{[(x - X)^2 + (y - Y)^2 + (z - Z)^2]^{3/2}} \, dx \, dy \, dz.$$

Formula (13) may be justified by a mathematical myth: $\mathbf{F}$ is the sum of forces exerted by infinitely small boxes of volume $dx \, dy \, dz$, located at $\mathbf{x}$, and having mass $f(\mathbf{x})(dx \, dy \, dz)$; each such force is, by (13), equal to $f(x)(dx \, dy \, dz)|\mathbf{x} - \mathbf{X}|^{-3}(\mathbf{x} - \mathbf{X})$.

Similar mathematical myths suggest that the gravitational force of a two-dimensional (laminar) distribution of matter should be computed by the formula

$$(15) \quad \mathbf{F} = \iint_D \frac{f(x, y)}{[(x - X)^2 + (y - Y)^2 + Z^2]^{3/2}} (x\mathbf{e}_1 + y\mathbf{e}_2 - \mathbf{X}) \, dx \, dy$$

and that

$$(16) \qquad \mathbf{F} = \int_0^L \frac{f(s)}{|\mathbf{x}(s) - \mathbf{X}|^3} (\mathbf{x}(s) - \mathbf{X}) \, ds$$

is the attractive force of a wire defined by $\mathbf{x} = \mathbf{x}(s)$, $0 \leq s \leq L$, with density function $f(s)$; here s is the arc length (see Chapter 11, §3.12).

If we set $f \equiv$ const in the formulas (14), (15), and (16), we obtain the attractive forces of a solid, lamina, or wire of uniform density.

EXERCISES

56. Evaluate the integrals for the gravitational attraction exerted by a homogeneous circular lamina of density 1 at points on the axis of symmetry.
57. Write down the integral for the attractive force at an arbitrary external point exerted by a solid homogeneous cube with density 1.
58. Compute the gravitational attraction exerted by a circular wire of constant density on a particle located at the center of the circle.
59. Answer the question in Exercise 58, for a square wire.

3.9 Spherical symmetry

We apply the definition given above to the case of a ball (solid sphere) with a spherically symmetrical density function, that is, with a density that depends only on the distance from the center. Let M be the total mass. It will turn out that the *attractive force exerted by such a ball on a mass point outside the ball is the same as that of a mass M concentrated at the origin.*

Let the center of the ball be chosen as the origin of the coordinate system. Then the density is of the form

$$(17) \qquad f(x, y, z) = \phi(\sqrt{x^2 + y^2 + z^2}) = g(\rho).$$

The total mass is best computed by using spherical coordinates. We have [see equation (10) in §3.5]

$$(18) \qquad M = \int_0^\pi \int_0^{2\pi} \int_0^R g(\rho)\rho^2 \sin\phi \, d\rho \, d\theta \, d\phi = 4\pi \int_0^R g(\rho)\rho^2 \, d\rho$$

where R is the radius of the ball.

In order to compute the gravitational attraction $\mathbf{F}$, it is convenient to place the coordinate system so that the particle of unit mass being attracted by the ball lies on the positive z axis, so that it has the position vector

$$\mathbf{X} = Z\mathbf{e}_3.$$

In the notations of §3.8, $X = Y = 0$. If we use spherical coordinates, then the denominator in (14) is

$$(\rho^2 \sin^2\phi \cos^2\theta + \rho^2 \sin^2\phi \sin^2\theta + \rho^2 \cos^2\phi - 2Z\rho \cos\phi + Z^2)^{3/2}$$
$$= (\rho^2 - 2Z\rho \cos\phi + Z^2)^{3/2}.$$

The x component of $\mathbf{F}$ is, by $(14')$,

$$F = \int_0^\pi \int_0^{2\pi} \int_0^R \frac{g(\rho)(\rho \sin \phi \cos \theta)(\rho^2 \sin \phi \, d\rho \, d\theta \, d\phi)}{(\rho^2 - 2Z\rho \cos \phi + Z^2)^{3/2}}$$

$$= \int_0^\pi \int_0^R \frac{g(\rho)\rho^3 \sin^2 \phi}{(\rho^2 - 2Z\rho \cos \phi + Z^2)^{3/2}} \, d\rho \, d\phi \cdot \int_0^{2\pi} \cos \theta \, d\theta = 0$$

since the last factor is 0. The same argument applies to the y component G of the force. Of course, we could have avoided the calculation, since by reasons of symmetry $\mathbf{F}$ must point in the z direction.

To verify the assertion made at the beginning, we must show that the z component H of $\mathbf{F}$ equals $-M/Z^2$, with M given by (18).

By $(14')$, we have

$$H = \int_0^\pi \int_0^{2\pi} \int_0^R \frac{g(\rho)(\rho \cos \phi - Z)\rho^2 \sin \phi \, d\rho \, d\theta \, d\phi}{(\rho^2 - 2Z\rho \cos \phi + Z^2)^{3/2}} = \int_0^R g(\rho)\rho^2 K(\rho) \, d\rho$$

where

(19)
$$K(\rho) = \int_0^\pi \int_0^{2\pi} \frac{(\rho \cos \phi - Z) \sin \phi \, d\theta \, d\phi}{(\rho^2 - 2Z\rho \cos \phi + Z^2)^{3/2}}$$

$$= 2\pi \int_0^\pi \frac{(\rho \cos \phi - Z) \sin \phi \, d\phi}{(\rho^2 - 2Z\rho \cos \phi + Z^2)^{3/2}} .$$

[We can interpret $K(\rho)$ as the z component of the attractive force exerted by an infinitely thin spherical shell of uniform density and of radius ρ on a particle of unit mass located at $(0, 0, Z)$.] In order to compute $K(\rho)$, we introduce instead of ϕ the new variable

$$u = \sqrt{\rho^2 - 2Z\rho \cos \phi + Z^2}.$$

(This is legitimate, since the expression under the radical is not negative.) We have

$$\frac{du}{d\phi} = \frac{Z\rho \sin \phi}{\sqrt{\rho^2 - 2Z\rho \cos \phi + Z^2}} = \frac{Z\rho \sin \phi}{u}$$

which may be written as

$$\sin \phi \, d\phi = \frac{u \, du}{Z\rho} .$$

Then $u = Z - \rho$ for $\phi = 0$ (since $Z > \rho$), $u = Z + \rho$ for $\phi = \pi$ and also

$$\rho^2 - 2Z\rho \cos \phi + Z^2 = u^2, \qquad \rho \cos \phi = \frac{\rho^2 + z^2 - u^2}{2Z} .$$

Substituting this into the single integral in (20), we obtain

$$K(\rho) = 2\pi \int_{Z-\rho}^{Z+\rho} \frac{\left(\dfrac{\rho^2 + Z^2 - u^2}{2Z} - Z \right) \dfrac{u\,du}{Z\rho}}{u^3}$$

$$= 2\pi \int_{Z-\rho}^{Z+\rho} \frac{\rho^2 - Z^2 - u^2}{2Z^2\rho u^2}\,du = -\frac{\pi}{Z^2\rho} \left[\frac{\rho^2 - Z^2}{u} + u \right]_{u=Z-\rho}^{u=Z+\rho}$$

$$= -\frac{4\pi}{Z^2}.$$

Therefore $\qquad H = -\dfrac{4\pi}{Z^2} \int_0^R g(\rho)\rho^2\,d\rho = -\dfrac{M}{Z^2},$

as asserted.

This important result was first obtained by Newton himself. It shows that, in computing the mutual gravitational attraction of astronomical bodies, like the earth, the sun, and the moon, it is permissible to replace these huge bodies by their centers. It seems that Newton did not yet have this result when he first discovered the connection between the inverse square law of gravitation and Kepler's laws of planetary motion, and this could have been one reason for his long delay in publication.

EXERCISES

60. Show that a spherical shell of uniform density exerts no force at any point in its interior.

61. A small hole is bored through the center of the earth from the North to the South Pole. The amount of mass removed is negligible. Show that the gravitational attraction of the earth upon a particle in the hole is *directly* proportional to the particle's distance from the center of the earth. (Use Exercise 60, together with the result in the text. Assume that the earth is spherical and that its density depends only on the distance from its center.) Make a graph of $H(z)$, showing the sharp corner at $z = R$.

4.1 Integrands and regions of integration

It is hardly necessary to point out that everything said above can be repeated, with obvious modifications, for functions of more than three variables. A function $(x_1, x_2, \cdots x_n) \to f(x_1, x_2, \cdots x_n)$ defined in an n-dimensional interval I will be called **piecewise continuous** in I if it is continuous at all points in I, except perhaps at points that belong to finitely many sets $S_1, S_2, \cdots$, such that each S_i is either a set on which one of the coordinates x_k is constant, or a set that can be represented as a graph of a function $x_1 = \phi_1(x_2, x_3, \cdots x_n)$, and as a graph of a function $x_2 = \phi_2(x_1, x_3, \cdots x_n)$, and so forth, each of the functions

$\phi_1, \phi_2, \cdots$ being defined and continuous in some closed $(n-1)$ dimensional interval. In this section, all functions to be integrated are assumed to be bounded and piecewise continuous.

We also agree that all sets over which we shall integrate (regions of integration) will have piecewise continuous characteristic functions.

4.2 Computing the integral

The integral of a function $f(x_1, x_2, \cdots x_n)$ over the n-dimensional interval $I: a_i < x_i < b_i$, $i = 1, 2, \cdots, n$, denoted by

$$(1) \qquad \int \cdots \int_I f(x_1, \cdots, x_n)\, dx_1 \cdots dx_n,$$

is the number computed by successively integrating f with respect to each variable x_i, from a_i to b_i, in any order, for instance,

$$(2) \qquad \int_{a_1}^{b_1}\int_{a_2}^{b_2} \cdots \int_{a_n}^{b_n} f(x_1, x_2, \cdots x_n)\, dx_n\, dx_{n-1} \cdots dx_1.$$

We always adhere to the *convention:* the outer integral sign belongs with the outer differential. However, in order to keep track of the order of integration, one sometimes writes, instead of (2),

$$(2') \qquad \int_{a_1}^{b_1} dx_1 \int_{a_2}^{b_2} dx_2 \cdots \int_{a_n}^{b_n} dx_n f(x_1, \cdots, x_n)$$

or

$$(2'') \qquad \int_{a_1}^{b_1} dx_1 \int_{a_2}^{b_2} dx_2 \cdots dx_{n-1} \int_{a_n}^{b_n} f(x_1, \cdots, x_n)\, dx_n.$$

All properties of double and triple integrals extend the general case. This applies, in particular, to integration over general domains.

▶ *Example*   Integrate the function $(x, y, z, t) \to x$ over the region determined by the inequalities

$$0 < x < 1,\ x < y < 2,\ 3 < z < t^2,\ 3 < t < 6.$$

Answer.   The desired quadruple integral equals

$$\int_0^1 dx \int_x^2 dy \int_3^6 dt \int_3^{t^2} x\, dz = \int_0^1 dx \int_x^2 dy \int_3^6 [xz]_{z=3}^{z=t^2} dt$$

$$= \int_0^1 dx \int_x^2 dy \int_3^6 (xt^2 - 3x)\, dt = \int_0^1 dx \int_x^2 \left[\frac{1}{3} xt^3 - 3xt\right]_{t=3}^{t=6} dy$$

$$= \int_0^1 dx \int_x^2 54x \, dy = \int_0^1 [54xy]_{y=x}^{y=2} dx$$

$$= \int_0^1 (108x - 54x^2) \, dx = 54x^2 - 18x^3 \Big|_0^1 = 36.$$

Note how simple—and how tedious—this is. ◄

EXERCISES

In Exercises 1 to 10, integrate the given function F over the given region D.

1. $F = c_1x_1 + c_2x_2 + \cdots + c_nx_n$; D: $a_1 \le x_1 \le b_1, a_2 \le x_2 \le b_2, \cdots$, $a_n \le x_n \le b_n$.

2. $F = x_1x_2{}^2x_3{}^3 \cdots x_n{}^n$; $D = a \le x_i \le b$ for $i = 1, 2, \cdots n$.

3. $F = x_1{}^{\alpha_1}x_2{}^{\alpha_2} \cdots x_n{}^{\alpha_n}$; $D = a_i \le x_i \le b_i$, $i = 1, \cdots n$. (Specify for which α, a, b the integral is not meaningful.)

► 4. $F = \log(x_1{}^2x_2{}^3x_3{}^2x_4{}^3)$; D: $1 < x_1 < 2 < x_2 < 3 < x_3 < 4 < x_4 < 5$.

5. $F = x_1e^{x_2}(\sin x_3)(\cos x_4)$; D: $0 < x_i < 1$, $i = 1, 2, 3, 4$.

6. $F = tuvw$; D: $t^2 + u^2 \le v^2 + w^2, v^2 + w^2 \le 1$.

7. $F = (x_1 - x_2 + x_3 - x_4)^2$; D: $x_1{}^2 + x_2{}^2 + x_3{}^2 + x_4{}^2 \le 1$.

8. $F = (x_1 - \cos x_2)(x_3 + \sin x_4)$; D: $1 \le x_1 \le 2, 2 \le x_2 \le 3, x_3{}^2 + x_4{}^2 \le 1$.

9. $F = x_1{}^\alpha + 2x_2{}^\alpha + 3x_3{}^\alpha + 4x_4{}^\alpha$; D: $A < x_i < B$, $i = 1, 2, 3, 4$.

10. $F = tx/yz$; D: $-1 \le t \le 1, 1 \le y \le 2, -1 \le x \le 1, 1 \le z \le 2$.

4.3 Volume in n dimensions

The number

$$V = \underset{D}{\iint \cdots \int} dx_1 \, dx_2 \cdots dx_n$$

is called the **n-dimensional volume** of D. If D is the interval $a_1 < x < b_1$, $a_2 < x_2 < b_2, \cdots, a_n < x_n < b_n$, and we set $l_j = b_j - a_j$, then the volume of D is the product of the n sides: $V = l_1l_2 \cdots l_n$.

The set of all points $(x_1, x_2, \cdots, x_n)$ with

$$x_1{}^2 + x_2{}^2 + \cdots + x_n{}^2 < R^2$$

is called the **n-dimensional ball** (solid sphere) of radius R. Let us denote its volume by $\alpha_n(R)$. We know, of course, that

$$\alpha_1(R) = 2R, \qquad \alpha_2(R) = \pi R^2, \qquad \alpha_3(R) = \frac{4}{3}\pi R^3.$$

We have $\alpha_n(R) =$

$$\int_{-R}^{R} dx_n \int_{-\sqrt{R^2-x_n^2}}^{\sqrt{R^2-x_n^2}} dx_{n-1} \int_{-\sqrt{R^2-x_n^2-x_{n-1}^2}}^{\sqrt{R^2-x_n^2-x_{n-1}^2}} dx_{n-2} \cdots \int_{-\sqrt{R^2-x_n^2-\cdots-x_2^2}}^{\sqrt{R^2-x_n^2-\cdots-x_2^2}} dx_1.$$

Hence, writing x instead of x_n, we have

$$\alpha_n(R) = \int_{-R}^{R} \alpha_{n-1}(\sqrt{R^2-x^2})\,dx = 2\int_{0}^{R} \alpha_{n-1}(\sqrt{R^2-x^2})\,dx.$$

Using this recursion formula, we can compute the volumes of n-dimensional balls. For instance

$$\alpha_4(R) = \frac{8\pi}{3}\int_{0}^{R}(R^2-x^2)^{3/2}\,dx.$$

Here we first make the substitution $x = Rt$, then the substitution $t = \cos\phi$, and then use the methods of Chapter 7, §3.5. This gives

$$\alpha_4(R) = \frac{8\pi}{3}\int_{0}^{1}(R^2-R^2t^2)^{3/2}\,R\,dt = \frac{8\pi R^4}{3}\int_{0}^{1}(1-t^2)^{3/2}\,dt$$

$$= \frac{8\pi R^4}{3}\int_{0}^{\pi/2}\sin^4\phi\,d\phi = \frac{8\pi R^4}{3}\cdot\frac{3\pi}{16} = \frac{\pi^2}{2}R^4.$$

The same method works for all n.

EXERCISES

11. Compute $\alpha_5(R)$.
12. Compute $\alpha_6(R)$.
13. The 4-dimensional cone of height h, whose base is a sphere of radius 1 is defined as the set of all x, y, z, t, satisfying $0 < t < h$, $x^2 + y^2 + z^2 > (t/h)^2$. Compute its four-dimensional volume.
14. A "unit simplex" in n-space is defined as the set of points $(x_1, x_2, \cdots x_n)$ satisfying $x_1 > 0$, $x_2 > 0$, $\cdots x_n > 0$, $x_1 + x_2 + \cdots x_n < 1$. Compute the length of the unit 1-simplex, the area of the unit 2-simplex, the volume of the unit 3-simplex, and the 4 dimensional volume of the unit 4-simplex.
15. Try to find a formula for the n-dimensional volume of the unit n-simplex.

—————————————————————— Appendixes to Chapter 13

§5 Centroids

In Chapter 11, §6.4, we arrived at the concept of center of mass by considering the motion of a system of particles. This concept, however, antedates the
science of dynamics. Greek geometers, in particular Archimedes and Pappus
studied centers of mass of curves, surfaces, and solids. These considerations become
quite easy if one uses multiple integrals.

PAPPUS of Alexandria (ca. 340 A.D.) is the author of a historical commentary on Greek geometrical literature known as *Synagogue* (that is, Collection).

5.1 Mass center of a continuous distribution of matter

The **mass center** of N particles with position vectors $x_1, x_2, \cdots x_N$ and masses
$m_1, m_2, \cdots m_N$ has the position vector

$$\mathbf{X} = \frac{1}{\sum_{j=1}^{N} m_j} \sum_{j=1}^{N} m_j \mathbf{x}_j = \frac{1}{M} \sum_{j=1}^{N} m_j \mathbf{x}_j, \qquad \text{where } M = \sum_{j=1}^{N} m_j$$

(see Chapter 11, §6.4). This suggests that the reasonable way of *defining* the position of the mass center of a solid D with variable density $f(x, y, z) = f(\mathbf{x})$ is

$$(1) \qquad \mathbf{X} = \frac{1}{M} \iiint_D f(\mathbf{x})\mathbf{x} \, dx \, dy \, dz \qquad \text{where } M = \iiint_D f(\mathbf{x}) \, dx \, dy \, dz.$$

Needless to say, there is a myth justifying this definition: $\mathbf{X}$ is the mass center of
infinitely many infinitely small boxes with volumes $dx \, dy \, dz$ and masses $f(\mathbf{x}) \, dx \, dy \, dz$.

If D is a region in the plane $z = 0$, with a density function $f(x, y)$ defined in D,
then the mass center of the corresponding material lamina is, by definition, located at

$$(2) \qquad \mathbf{X} = \frac{1}{M} \iint_D f(x, y)(x\mathbf{e}_1 + y\mathbf{e}_2) \, dx \, dy \qquad \text{where } M = \iint_D f(x, y) \, dx \, dy.$$

If we are given a smooth arc of length L, $\mathbf{x} = \mathbf{x}(s)$, $0 < s < L$, with s the arc
length (see Chapter 11, §3.12), and if $f(s)$, $0 < s < L$ is a density function, the
mass center of the corresponding wire is at

$$(3) \qquad \mathbf{X} = \frac{1}{M} \int_0^L f(s)\mathbf{x}(s) \, dr, \qquad \text{where } M = \int_0^L f(s) \, ds.$$

These definitions can be justified in the same way as (1). In all cases M is the total
mass.

Remark In order that the "wires" should have physical meaning we require from
now on that the curve $\mathbf{x} = \mathbf{x}(s)$ be **simple**; this means that different values of

s should correspond to different points $\mathbf{x}(s)$, except that we may have $\mathbf{x}(L) = \mathbf{x}(0)$ if the curve is closed.

EXERCISES

1. Find the center of mass of a right circular cylinder of radius R and height h if the density equals twice the distance from the base.
2. Find the center of mass of a right circular cone of radius R and height h if the density equals the square of the distance from the base.
3. Find the center of mass of a cube of side a whose density equals the square of the distance from the center of one face.
4. Find the center of mass of a plane lamina in the shape of an equilateral triangle of side a, whose density equals the distance from one vertex.
5. Find the center of mass of a bent wire of length 3. The wire travels east for one unit, then north for one unit, then straight up for one unit. Density equals the distance *along* the wire from the starting point (westernmost end).
6. A helical wire lies on a curve $x = \cos t$, $y = \sin t$, $z = t$, $0 \le t \le 6\pi$. The density at each point equals $4t$. Where is the center of mass?

5.2 Centroids

The **centroid** of a reasonable set D in space is the mass center of D for a constant density function. Centroids of plane sets and curves are defined similarly. The formulas for centroids are obtained from the formulas (1), (2), (3) by setting $f \equiv 1$. They read:

$$(4) \qquad \mathbf{X} = \frac{1}{V} \iiint_D \mathbf{x} \, dx \, dy \, dz \quad (V = \iiint_D dx \, dy \, dz = \text{volume of } D)$$

$$(5) \qquad \mathbf{X} = \frac{1}{A} \iint_D (x\mathbf{e}_1 + y\mathbf{e}_2) \, dx \, dy \quad (A = \iint_D dx \, dy = \text{area of } D)$$

$$(6) \qquad \mathbf{X} = \frac{1}{L} \int_0^L \mathbf{x}(s) \, ds \quad (s = \text{arc length, that is, } |x'(s)| = 1).$$

Recalling the definition of arc length (Chapter 11, §3.12), one can rewrite (6) in terms of an arbitrary parameter:

$$(6') \qquad \mathbf{X} = \frac{1}{L} \int_a^b \mathbf{x}(t)|\mathbf{x}'(t)| \, dt \quad (L = \int_a^b |\mathbf{x}'(t)| \, dt = \text{length of the curve}).$$

Since the definitions are stated in vector language, it is clear that they do not depend on the choice of the coordinate system.

5.3 Formulas in components

If we want to compute mass centers, we ought to rewrite the formulas in components. This is, of course, quite easy and we do it explicitly only for the centroids:

(7) $X = \dfrac{1}{V} \iiint\limits_{D} x\, dx\, dy\, dz, \quad Y = \dfrac{1}{V} \iiint\limits_{D} y\, dx\, dy\, dz, \quad Z = \dfrac{1}{V} \iiint\limits_{D} z\, dx\, dy\, dz$

are the coordinates of the centroid of the three-dimensional set D with volume V,

(8) $\qquad\qquad X = \dfrac{1}{A} \iint\limits_{D} x\, dx\, dy, \quad Y = \dfrac{1}{A} \iint\limits_{D} y\, dx\, dy, \qquad Z = 0$

are the coordinates of the centroid of a set D in the plane $z = 0$, with area A,

(9) $\qquad\qquad X = \dfrac{1}{L} \int_{0}^{L} x(s)\, ds, \quad Y = \dfrac{1}{L} \int_{0}^{L} y(s)\, ds, \quad Z = \dfrac{1}{L} \int_{0}^{L} z(s)\, ds$

are the coordinates of the centroid of the curve $x = x(s)$, $y = y(s)$, $z = z(s)$, $0 < s < L$, with $x'(s)^2 + y'(s)^2 + z'(s)^2 = 1$. More generally,

(9′)
$$X = \dfrac{1}{L} \int_{a}^{b} x(t) \sqrt{x'(t)^2 + y'(t)^2 + z'(t)^2}\ dt,$$
$$Y = \dfrac{1}{L} \int_{a}^{b} y(t) \sqrt{x'(t)^2 + y'(t)^2 + z'(t)^2}\ dt,$$
$$Z = \dfrac{1}{L} \int_{a}^{b} z(t) \sqrt{x'(t)^2 + y'(t)^2 + z'(t)^2}\ dt$$

are the coordinates of the curve $x = x(t)$, $y = y(t)$, $z = z(t)$, $a < t < b$, of length L.

EXERCISES

In Exercises 7 to 24, find the centroid of the indicated region. In each case, sketch the region and mark the centroid.

7. The region below the line $y = 2x + 3$ from $x = 1$ to $x = 2$.
▶ 8. The region below the parabola $y = x^2$ from $x = 0$ to $x = 4$.
9. The region below the graph of $x \mapsto 2x^3 - x$ from $x = 1$ to $x = 3$.
10. The region below the graph of $x \mapsto \sin x$ from $x = 0$ to $x = \pi/4$.
11. The region between the graphs of the functions $f_2(x) = x^2$ and $f_1(x) = -x$ from $x = 0$ to $x = 1$.
12. The region between the two parabolas $y = 4 - x^2$ and $y = x^2 - 4$.
13. A right triangle.
14. A trapezoid.
15. The region consisting of all points (x, y) such that $0 < x < 1$ and $y^2 < x$.
16. The set of all points (x, y) with $|y| < e^x$ and $1 < |y| < e^2$.
17. A right circular cone of radius r and height h.
18. The volume cut from one nappe of a cone of vertex angle $60°$ by a sphere of radius 2 whose center is the vertex of the cone.
19. A hemisphere of radius R.
20. The tetrahedron with vertices $(0, 0, 0)$, $(1, 0, 0)$, $(0, 1, 0)$, $(0, 0, 1)$.
21. The intersection of the region $\{0 < z^2 < xy\}$ with the prism (cylinder) parallel to the z axis whose base is the triangle $(0, 0)$, $(4, 0)$, $(4, 4)$.

22. The portion of the sphere $x^2 + y^2 + z^2 \leq R^2$ contained between the half-planes $\theta = -\pi/4$ and $\theta = \pi/4$.
23. The region where $x^2 + y^2 + z^2 \leq 2R^2$ and $Rz > x^2 + y^2$.
24. The portion of the unit sphere cut off between two given parallels of latitude.

In Exercises 25 to 30, find the centroid of the indicated curve. (The curves in Exercises 25 to 29 lie in the x, y plane.)

25. Circle.   27. Equilateral triangle.
▶ 26. Semicircle.   28. $y = 2x^2$, $x^2 \leq 4$.
29. $y = \cosh x$, $0 \leq x \leq 2$.
30. $t \mapsto 3t e_1 + 3t e_2 + 2t e_3$, $0 \leq t \leq 1$.

5.4 Additivity and symmetry

The mass centers defined in §5.1 have properties analogous to those stated in Chapter 11, §6 for mass centers of finitely many particles.

We note first the **additivity property**. Let us, for instance, divide the region D in (1) into two disjoint regions D_1 and D_2. Let M_1 and M_2 be the masses of D_1 and D_2, and let $\mathbf{X}_1$ and $\mathbf{X}_2$ be the corresponding mass centers. Then

$$\mathbf{X} = \frac{1}{M_1 + M_2}(M_1 \mathbf{X}_1 + M_2 \mathbf{X}_2)$$

In other words, $\mathbf{X}$ is the mass center of two particles of mass M_1 and M_2, located at $\mathbf{X}_1$ and $\mathbf{X}_2$, respectively.

To prove (10), we note that by (1) and the additivity property of integrals (discussed in §1.6 for double integrals).

$$M_1 \mathbf{X}_1 + M_2 \mathbf{X}_2 = \iiint\limits_{D_1} f(\mathbf{x})\mathbf{x}\, dx\, dy\, dz + \iiint\limits_{D_2} f(\mathbf{x})\mathbf{x}\, dx\, dy\, dz$$

$$= \iiint\limits_{D} f(\mathbf{x})\mathbf{x}\, dx\, dy\, dz = M\mathbf{X}.$$

A similar argument establishes the additivity property for laminae and wires.

The additivity property can be extended at once to the case where the solid (or the lamina, or the wire) is divided into *several* parts.

Next we note the various **symmetry properties**. If the density $f(x, y, z)$ and the region D are symmetric with respect to a plane, then the mass center (1) lies in that plane. If the density $f(x, y)$ and the region D in the plane $z = 0$ are symmetric with respect to a line l, the mass center (2) lies on this line. The reader should state himself the symmetry property of mass centers of wires.

We prove only one symmetry property. Suppose the plane region D and the density $f(x, y)$ are symmetric with respect to some line l. Choose the coordinate system so that l is the x axis, D is an interval of the form $-a < x < a$, $-b < y < b$ (for we may set $f = 0$ outside the original D), and $f(x, y)$ is an even function of y for every fixed x. Hence the function $y \to yf(x, y)$ is odd and (compare Chapter 5, §5.6)

$$\int_{-b}^{b} yf(x, y)\, dy = 0$$

so that the y component of the mass center is

$$Y = \int_{-a}^{a} \int_{-b}^{b} yf(x, y)\, dy\, dx = 0$$

as asserted.

The additivity and symmetry properties often simplify the calculation of centroids.

▶*Examples* **1.** The centroid of a rectangle is the intersection point of the diagonals.

Proof. Draw two lines, l_1 and l_2 through the intersection point P of the diagonals, parallel to the sides. The rectangle is symmetric with respect to l_1 and to l_2; hence the centroid lies on both lines.

The reader should confirm this by a direct computation.

2. The centroid of a three-dimensional interval (parallelopiped, box) is at the intersection of the main diagonals. In other words: the centroid is the midpoint.

This is proved as the corresponding statement in the example above.

3. Find the centroid of the region shown in Fig. 13.43.

Solution. We divide the region into 4 rectangles as shown in Fig. 13.44. They have areas 24, 8, 6, and 4, respectively, and their centroids are at $(1, 6)$, $(6, \frac{1}{2})$, $(5, \frac{23}{2})$, and $(9, 11)$. The coordinates of the centroid are, therefore,

$$X = \frac{24 \cdot 1 + 8 \cdot 6 + 6 \cdot 5 + 4 \cdot 9}{24 + 8 + 6 + 4} = \frac{23}{7},$$

$$Y = \frac{24 \cdot 6 + 8 \cdot \frac{1}{2} + 6 \cdot \frac{23}{2} + 4 \cdot 11}{24 + 8 + 6 + 4} = \frac{261}{42}.$$

Since we deal with a figure in the plane, there is no need to state explicitly that $Z = 0$. ◀

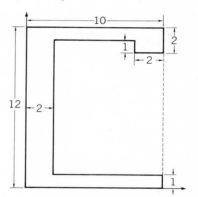

Fig. 13.43

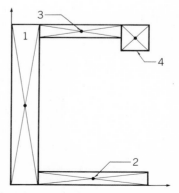

Fig. 13.44

EXERCISES

In Exercises 31 to 42, find the centroids of the indicated regions, solids or curves.

31. The region between the circles $x^2 + y^2 = 1$ and $x^2 + y^2 = 4$.

32. The region consisting of the region in Exercise 33 and the region bounded by the circle $(x - 10)^2 + (y - 10)^2 = 4$.

33. The region consisting of the region in the preceding exercise and the rectangle bounded by the lines $x = -11$, $x = -9$, $y = 1$, and $y = -3$.

34. The region common to the two rectangles: one bounded by the lines $y = 2$, $y = -1$, $x = -10$, and $x = 11$; the other bounded by the lines $y = -12$, $y = 4$, $x = 1$, and $x = 3$.

35. The union of two cubes—the cube of side 2 whose edges are parallel to the axes, and whose lower left front vertex is at the origin; and the cube given by $4 < z < 5$, $\sqrt{2} < x + y < 2\sqrt{2}$, $6\sqrt{2} < x - y < 7\sqrt{2}$.

36. The solid sphere $x^2 + y^2 + z^2 < 10$ from which one removed the cube $0 < x < 1$, $0 < y < 1$, $0 < z < 1$.

37. A regular tetrahedron. (Use symmetry arguments.)

38. The cube $0 < x < 1$, $0 < y < 1$, $0 < z < 1$, in which one has drilled a vertical hole of radius $\frac{1}{8}$ through the point $x = \frac{1}{4}$, $y = \frac{1}{4}$.

39. The drilled cube of Exercise 40, with an additional hole of radius $\frac{1}{8}$ parallel to the x axis through the point $y = \frac{3}{4}$, $z = \frac{1}{2}$.

40. The drilled cube of Exercise 41, to which one has welded a cube $0 < x < \frac{1}{2}$, $0 < y < \frac{1}{2}$, $-\frac{1}{2} < z < 0$.

41. The bent wire of Exercise 5.

42. A wire shaped like the letter A, if the vertex angle is 30° and the crossbar is at the midpoints of the two sides.

43. A semicircle of radius 1 in the x, y plane and a straight segment perpendicular to the x, y plane which has one endpoint at the midpoint of the semicircle.

44. The intersection of a circular cylinder with a plane.

5.5 Pappus' theorems

The following two propositions are due to Pappus, the last of the great Greek mathematicians.

Theorem 1   THE VOLUME V PRODUCED BY ROTATING A PLANE AREA A ABOUT AN AXIS l IN THE SAME PLANE IS

(10) $$V = 2\pi\delta A$$

WHERE δ IS THE DISTANCE FROM THE CENTROID OF THE AREA TO THE AXIS.

Theorem 2   THE AREA A PRODUCED BY ROTATING A PLANE CURVE OF LENGTH L ABOUT AN AXIS l IN THE SAME PLANE IS

(11) $$A = 2\pi\delta L$$

WHERE δ IS THE DISTANCE FROM THE CENTROID OF THE CURVE TO THE AXIS.

Fig. 13.45

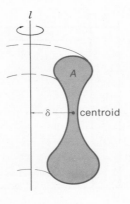

In Theorem 1 (see Fig. 13.45), it is assumed that A is the area of a domain D lying on one side of l. Let D lie in the (x, y) plane, and let l be chosen as the y axis. The x coordinate of the centroid of D, that is, the distance δ, is

$$\delta = X = \frac{1}{A} \iint\limits_{D} x\, dx\, dy$$

as follows from (8). By equation (8') in §3.5, the volume V of the solid B, obtained by rotating D about the y axis, is

$$V = 2\pi \iint\limits_{D} x\, dx\, dy.$$

The two displayed formulas imply (10).

In Theorem 2, it is assumed that the curve, call it C, lies on one side of l. See Fig. 13.46. Let C lie in the (x, y) plane with l chosen as the y axis. We assume that C is the graph of $y = \phi(x)$, $0 \leq a < x < b$. The x coordinate of the centroid of C, that is, the distance δ, is

$$\delta = X = \frac{1}{L} \int_a^b x\sqrt{1 + \phi'(x)^2}\, dx.$$

This is seen by setting $x = t$, $y = \phi(t)$, $z = 0$ in (9′). By equation (4) in §2.4, the area A, of the surface S obtained by rotating D about the y axis, is

$$A = 2\pi \int_a^b x\sqrt{1 + \phi'(x)^2}\, dx.$$

The two displayed formulas imply (11). It is even simpler to verify (11) when C is a segment $x = a$, $\alpha < y < \beta$. It follows that (11) holds whenever C is composed of vertical segments and graphs of continuously differentiable functions $x \to \phi(y)$.

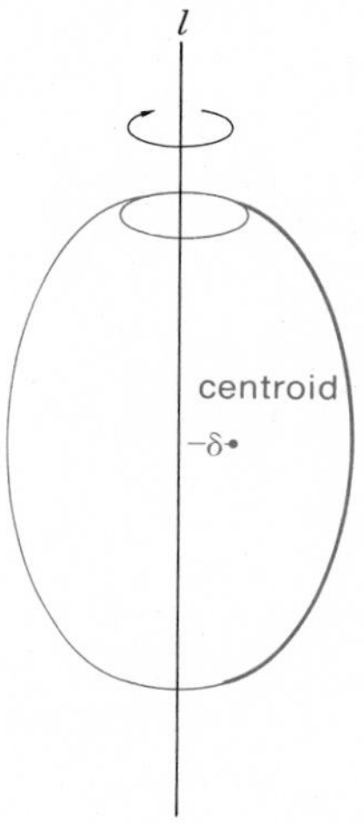

We give below some typical applications of Pappus' Theorems. In some of them the volumes and areas are known from other considerations, and we use this knowledge to find centroids. In others, the centroids are known from symmetry considerations, and we use this to find volumes and areas.

▶ *Examples* **1.** Find the centroid of a semidisk.

Answer. We represent the semidisk as the set D of points (x, y) with $x^2 + y^2 < 1$, $x > 0$. Its area is $A = \pi/2$. The centroid lies on the x axis, by symmetry. Let X be its x coordinate. Rotating D about the y axis, we obtain the unit sphere; its volume is $V = 4\pi/3$. Hence $4\pi/3 = 2\pi X(\pi/2)$, by Pappus' First Theorem, and $X = 4/(3\pi)$.

2. Find the centroid of a semicircle.

Answer. Represent the semicircle as the curve C: $x^2 + y^2 = 1$, $x > 0$. Its length is π. The centroid lies at a point $(X, 0)$, by symmetry. Rotating C about the y axis, we obtain the sphere of area 4π. By Pappus' Second Theorem, $4\pi = 2\pi X\pi$. Hence $X = 2/\pi$.

3. If we rotate a circular disk of radius r about a line at the distance $R > r$ from its center, we obtain a body known as the **solid torus** of radii R and r. The surface obtained by rotating the circumference of radius r about a line at the distance $R > r$ from its center is the surface called **torus** of radii R and r. (The torus looks like an inner tube; it is the boundary surface of the solid torus.)

Pappus' Theorems permit us to determine the volume of the solid torus, and the area of the torus, practically without computing. In both cases $\delta = R$. The volume in question is $2\pi R \cdot \pi r^2 = 2\pi^2 R r^2$. The area is $2\pi R \cdot 2\pi r = 4\pi R r$. ◀

Fig. 13.46

l

centroid

$-\delta$•

EXERCISES

In Exercises 45 to 47, use Pappus' First Theorem to find the volumes of the indicated bodies of revolution.

45. A circular cylinder of radius r and height H.

46. The body obtained by rotating a square of side length 1, with one vertex at the origin and the other at $(0, \sqrt{2})$, about the x axis.

47. The body obtained by rotating the rectangle bounded by the lines $y = a > 0$, $y = b > a$, $x = 0$, and $x = c$, about the x axis.

In Exercises 48 to 50 use Pappus' First Theorem to find the centroid of the indicated plane region.

48. The region bounded by two concentric circles and a line through their center.

49. An equilateral triangle.

50. A right triangle.

51. Use Pappus' Second Theorem to compute the centroid of a semicircle.

52. Use Pappus' Second Theorem to compute the mantle surface of a straight circular cone.

53. Let C be the circle passing through the corners of a given rectangle R. Prove that for all lines l tangent to C, the area generated by revolving R around l is the same.

54. Suppose, in Pappus' Second Theorem, the plane curve is a line segment perpendicular to l. What does the theorem reduce to in this case?

55. Use a symmetry argument together with Pappus' Second Theorem to find the centroid of a quarter circle. (What surface of known area can be obtained by revolving a quarter circle about an axis?)

§6 Green's Theorem

In this section we exhibit, without complete proof, an identity that connects line integrals and double integrals. This identity, known as Green's Theorem, is actually the simplest and most important special case of a very general result which we shall not discuss here.

6.1 Statement of the theorem

Let D be a bounded open set in the plane and let C be an oriented piecewise smooth, simple closed curve

$$x = x(t), \qquad y = y(t), \qquad 0 \le t \le T.$$

(We recall that "simple" means, that for no two different parameter values t, other than 0 and T, do the points $(x(t), y(t))$ coincide.) One calls C the **boundary curve** of D if every point $(x(t), y(t))$ is a boundary point of D, and every boundary point of D can be written as $x(t)$, $y(t)$ for some t. (The concept of a boundary point is defined in Chapter 12, §1.6.) A typical situation is shown in Fig. 13.45.

We say that the boundary curve C of D is **properly oriented** if, "as we go along C in its direction, that is, in the direction of increasing parameter values, the region D is to the left." Here we accept this concept as intuitively clear; a precise definition could be given. In Fig. 13.47, the boundary curve is properly oriented.

GEORGE GREEN (1793–1841) published the theorem that bears his name in a book on the mathematical theory of electricity and magnetism. Green was an essentially self-taught mathematician. He entered Cambridge University at the age of 40, several years after his pioneering work appeared in print.

Fig. 13.47

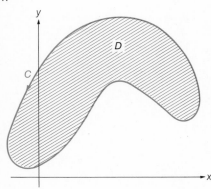

Theorem 1 (Green's Theorem)   LET C BE A PROPERLY ORIENTED (PIECEWISE SMOOTH) BOUNDARY CURVE OF THE OPEN SET D. LET $P(x, y)$, $Q(x, y)$ BE CONTINUOUSLY DIFFERENTIABLE FUNCTIONS (DEFINED IN SOME OPEN SET CONTAINING C AND D). THEN

(1)
$$\iint_D \left(\frac{\partial Q}{\partial x} - \frac{\partial P}{\partial y} \right) dx\, dy = \int_C P\, dx + Q\, dy.$$

One can, of course, replace this by two statements

(2)
$$\int_C P\, dx = -\iint_D \frac{\partial P}{\partial y}\, dx\, dy$$

(3)
$$\int_C Q\, dy = \iint_D \frac{\partial Q}{\partial x}\, dx\, dy.$$

Indeed, if we set $P \equiv 0$ or $Q \equiv 0$ in (1), we obtain (3) or (2), and if we add (2) and (3) we obtain (1).

▶ *Example*   Let D be the disk $x^2 + y^2 < 1$. Then C: $x = \cos t$, $y = \sin t$, $0 \le t \le 2\pi$, is the properly oriented boundary curve of D. Set $P = y$, $Q = 0$. Then,

$$\int_C P\, dx = -\int_0^{2\pi} \sin^2 t\, dt = -\pi = -\iint_D dx\, dy = \iint_D \frac{\partial P}{\partial y}\, dx\, dy. \blacktriangleleft$$

EXERCISES

In Exercises 1 to 10, verify Green's identity by computing

$$\iint_D \left(\frac{\partial Q}{\partial x} - \frac{\partial P}{\partial y} \right) dx\, dy \quad \text{and} \quad \int_C P\, dx + Q\, dy,$$

where C is the properly oriented boundary of D.

1. D is the unit disk $x^2 + y^2 < 1$; $P = Q = 1$.
2. D is the unit square $0 < x < 1$, $0 < y < 1$; $P = x^3 y$, $Q = x - y^2$.
3. D is the triangle $x > 0$, $y > 0$, $x + y < 1$; $P = x^2 - y^2$, $Q = \sqrt{xy}$.
4. D is the region $0 < x < 1$, $0 < y < x^2$; $P = x\sqrt[3]{y}$, $Q = \sqrt{x} + \sqrt{y}$.
5. D is the unit disk; $P = x^2$, $Q = x^2 y^2$.
6. D is the portion of the first quadrant where $1 < x^2 + y^2 < 2$; $P = Q = xy$.
7. D is the square with corners at $(1, 0)$, $(0, 1)$, $(1, 2)$, $(2, 1)$; $P = xe^y + 4$, $Q = 7\cos 4x$.
8. D is the region in the third quadrant where $-5 < x + y < -3$; $P = \log(7 - x)$, $Q = x\cos y$.
9. D is the triangle with vertices (x_0, y_0), (x_1, y_1), (x_2, y_2); $P = x^\alpha y^\beta$, $Q = 0$, where α, β are given positive integers.

10. D is the rectangle $x_0 < x < x_1$, $y_0 < y < y_1$, $P = 0$, $Q = \cos nx \cos my$, where n, m are given integers.
11. D is the region where $y > x^2$ and $x > y^2$; $P = x^{5/4} y^{7/6}$, $Q = x^{10/9} y^{21/8}$.

6.2 Special cases

Green's Theorem (1) implies that

(4)
$$\int_C P\,dx + Q\,dy = 0 \quad \text{if} \quad \frac{\partial P}{\partial y} = \frac{\partial Q}{\partial x}.$$

This is in accordance with the results in Chapter 12, §4.6. Since D is simply connected and $\partial P/\partial y = \partial Q/\partial x$, there is a function f with $P\,dx + Q\,dy = df$, so that

$$\int_C P\,dx + Q\,dy = \int_C df = 0.$$

If we apply (1) to $P = -y$ and to $Q = x$, we obtain interesting formulas for the area

$$A = \iint_D dx\,dy,$$

namely,

(5)
$$A = -\int_C y\,dx$$

(6)
$$A = \int_C x\,dy$$

so that also

(7)
$$A = \frac{1}{2}\int_C x\,dy - y\,dx$$

as is seen by adding (5) and (6).

Suppose now that C can be described in polar coordinates as

$$r = \phi(\theta), \qquad 0 \le \theta \le 2\pi.$$

A parametric representation for C, with the proper orientation, is

$$x = \phi(\theta)\cos\theta, \qquad y = \phi(\theta)\sin\theta, \qquad 0 \le \theta \le 2\pi$$

so that
$$dx = [\phi'(\theta)\cos\theta - \phi(\theta)\sin\theta]\,d\theta,$$
$$dy = [\phi'(\theta)\sin\theta + \phi(\theta)\cos\theta]\,d\theta$$

and by (7) the area A equals

$$\frac{1}{2} \int_0^{2\pi} [-\phi(\theta)\phi'(\theta) \sin\theta\cos\theta + \phi(\theta)^2 \sin^2\theta + \phi(\theta)\phi'(\theta)\cos\theta\sin\theta + \phi(\theta)^2\cos^2\theta]\, d\theta$$

$$= \frac{1}{2} \int_0^{2\pi} \phi(\theta)^2\, d\theta.$$

This proves formula (12) in Chapter 11, §2.6.

EXERCISES

12. Use Green's Theorem to compute the area of the ellipse whose boundary is given by $x = a\cos t$, $y = b\sin t$, $0 \le t \le 2\pi$.
13. Sketch the graph of, and calculate the area bounded by, the *astroid*, $x^{2/3} + y^{2/3} = R^{2/3}$.
14. Sketch the graph of, and calculate the area bounded by, the *cardioid*, $x = 2\cos\theta(1 - \cos\theta) + 1$, $y = 2\sin\theta(1 - \cos\theta)$.
15. Calculate the area bounded by the semicubical parabola $y = x^{3/2}$ and the lines $y = 0$, $x = a$, $x = b$.
16. Calculate the area bounded by $r = \theta$, $0 \le \theta < 2\pi$, and the coordinate axis.
17. Calculate the area under one arch of the cycloid $x = t - \sin t$, $y = 1 - \cos t$.

6.3 Remarks on the proof

We sketch the main ideas in proving Green's Theorem.

Suppose that we divide D into two regions, D_1 and D_2 (as shown in Fig. 13.48) by a piecewise smooth simple curve C_0, whose endpoints divide C into two curves C_1 and C_2. We claim that Green's Theorem holds for D if it holds for D_1 and for D_2.

The (properly oriented) boundary curves of D_1 and D_2 are $C_1 + C_0$ (that is, C_1 followed by C_0) and $(-C_0) + C_2$ (C_0 traversed in the opposite direction followed by C_2). We assume Green's Theorem for D_1 and for D_2. Then

$$-\iint_{D_1} \frac{\partial P}{\partial y}\, dx\, dy = \int_{C_1} P\, dx + \int_{C_0} P\, dx,$$

$$-\iint_{D_2} \frac{\partial P}{\partial y}\, dx\, dy = \int_{-C_0} P\, dx + \int_{C_2} P\, dx = -\int_{C_0} P\, dx + \int_{-C_2} P\, dx.$$

Adding the two equations, we get

$$= -\iint_{D_1} \frac{\partial P}{\partial y}\, dx\, dy - \iint_{D_2} \frac{\partial P}{\partial y}\, dx\, dy = \int_{C_1} P\, dx + \int_{C_2} P\, dx = \int_C P\, dx$$

since $C = C_1 + C_2$. This proves (2). A similar argument holds for (3).

The above argument can be made precise (it is not, since we depended upon looking at the diagram). Also, the argument can be continued, subdividing D_1 and D_2, and so forth. We conclude that in order to establish Green's Theorem it suffices to (1) subdivide D into simple pieces, and (2) prove Green's Theorem for each simple piece.

A subdivision of D is shown in Fig. 13.49. Each piece is either a rectangle or

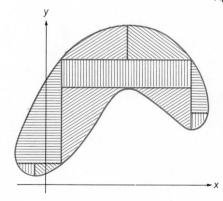

Fig. 13.48

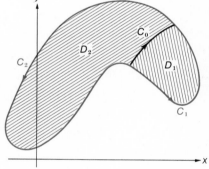

Fig. 13.49

Fig. 13.50

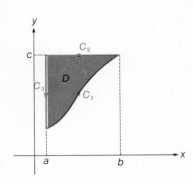

bounded by two segments parallel to the coordinate axes, and by a graph of a monotone smooth function.

We show next how to prove part (2) of Green's Theorem for a typical piece, say for the domain D shown in Fig. 13.50. Its properly oriented boundary is the curve $C = C_1 + C_2 + C_3$ where

C_1: $x = t$, $y = f(t)$, $a \leq t \leq b$, $[f'(t) \geq 0$ for $a < t < b]$,
C_2: $x = b - t$, $y = c$, $0 \leq t \leq b - a$,
C_3: $x = a$, $y = c - t$, $0 \leq t \leq c - f(a)$.

Now, by definition of double integrals and by the fundamental theorem of calculus,

$$(8) \quad - \iint_D \frac{\partial P}{\partial y} \, dx \, dy = - \int_a^b \int_{f(x)}^c \frac{\partial P(x, y)}{\partial y} \, dy \, dx$$

$$= - \int_a^b \{P(x, c) - P[x, f(x)]\} \, dx = \int_a^b P[x, f(x)] \, dx - \int_a^b P(x, c) \, dx.$$

Also, since $dx = 0$ on C_3, we have

$$\int_C P \, dx = \int_{C_1} P \, dx + \int_{C_2} P \, dx = \int_a^b P[t, f(t)] \, dt - \int_0^{b-a} f(b - t, c) \, dt.$$

We make the substitutions $t = x$ and $b - t = x$, respectively, and obtain

$$\int_C P \, dx = \int_a^b P[x, f(x)] \, dx - \int_a^b f(x, c) \, dx$$

which is equal to (8); this proves (2). Relation (3) is proved similarly; one must only write C_1 in the form

$$C_1: \quad x = g(t), \quad y = t, \quad f(a) \leq t \leq c$$

where g is the function inverse to f.

The proof of Green's Theorem for an interval (rectangle with sides parallel to the coordinate axes) is equally simple, and is left to the reader.

EXERCISE

▶ 18. Prove relation (3) for the domain D considered above.

6.4 Extension

Green's Theorem can be extended to regions D bounded by $n > 1$ simple closed curves. A typical case is shown in Fig. 13.51. In this case, the theorem reads

$$\iint_D \left(\frac{\partial Q}{\partial x} - \frac{\partial P}{\partial y} \right) dx \, dy = \sum_{j=1}^n \int_{C_j} P \, dx + Q \, dy.$$

The proof for an "n times connected domain" consists in dividing D into two "simply connected" domains, D_1 and D_2, as in Fig. 13.52, and noting that Green's Theorem holds for D if it holds for D_1 and for D_2.

Fig. 13.51

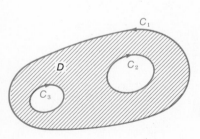

Fig. 13.52

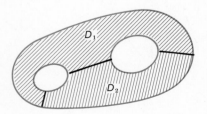

19. Verify Green's Theorem if $P = -y/(x^2 + y^2)$, $Q = x/(x^2 + y^2)$ and D is the ring $0 < \epsilon < x^2 + y^2 < 1$.
20. Verify Green's Theorem if D is defined by the inequalities $x^2 + y^2 < 10$, $x^2 + y^2 > 1$, $(x - 3)^2 + (y - 3)^2 > 1$ and $P = 1$, $Q = 0$.
21. Verify Green's Theorem if D is defined by $3 < |x| < 5$, $3 < |y| < 5$, and $P = x^2 e^y$, $Q = y^2 \cos x$.
22. Verify Green's Theorem if D is defined by $x^2 + y^2 > 1$, $|x| + |y| < 5$ and $P = x + 7$, $Q = y - 2$.

§7 Improper Multiple Integrals

A multiple integral is called improper if either the region of integration, or the integrand (the function to be integrated), or both, are unbounded. We shall consider such integrals only for the case of nonnegative integrands, as we did for simple integrals (see Chapter 5, §6).

7.1 Unbounded integrands

Suppose that $f(x, y)$ is a nonnegative, piecewise continuous but unbounded function defined in some interval I. What meaning can we ascribe to the expression

$$(1) \qquad \iint_I f(x, y) \, dx \, dy?$$

The geometric meaning of the double integral, the volume under the graph of $z = f(x, y)$, suggests the following definition. Let A be a positive number. We denote by f_A the function equal to f whenever $f < A$ and equal to 0 otherwise; that is,

$$(2) \qquad f_A(x, y) = f(x, y) \text{ if } f(x, y) < A, = 0 \text{ if } f(x, y) \geq A.$$

The function f_A is bounded, since $0 \leq f_A(x, y) < A$ in I. If f_A is piecewise continuous, we can form

$$(3) \qquad \iint_I f_A(x, y) \, dx \, dy.$$

This is the volume under the part of the graph of $z = f(x, y)$ where $f \leq A$. Since $f_A \leq f_B$ for $A \leq B$, the number (3) is an increasing function of A. If we let A grow without bounds, then either (3) also grows without bounds, or it remains bounded and approaches a definite limit (see Theorem A in Chapter 5, §9.1). In the latter case, we say that f is **integrable** over I (or that the integral (1) **converges**) and we set

$$(4) \qquad \iint_I f(x, y) \, dx \, dy = \lim_{A \to +\infty} \iint_I f_A(x, y) \, dx \, dy.$$

If (3) does not remain bounded, one says that the integral (1) **diverges** and writes

$$\iint_I f(x, y)\, dx\, dy = +\infty.$$

The **comparison theorem** (see Chapter 5, §6.4) remains valid; if $f \leq g$ in 1, and g is integrable, so is f. No new proof is needed. Also, it can be shown that if f is integrable over I, then

$$\iint_I f(x, y)\, dx\, dy = \int_a^b \left\{ \int_c^d f(x, y)\, dx \right\} dy = \int_c^d \left\{ \int_a^b f(x, y)\, dy \right\} dx$$

where a, b, c, d have the usual meaning, provided the single integrals occurring above exist, as improper integrals. (Actually the last could be proved if we did not insist on working only with piecewise continuous functions.)

Everything said above extends to functions of n variables, for all n. (For $n = 1$, the definition of improper integrals just given turns out to be equivalent to the one given in Chapter 5, §6.) It is clear that one may evaluate improper integrals using polar, cylindrical, or spherical coordinates.

▶ *Examples* **1.** Evaluate the improper integral

$$\iint_{x^2+y^2<1} \frac{dx\, dy}{x^2 + y^2}.$$

First solution. The integrand $(x^2 + y^2)^{-1}$ is unbounded near the origin. For large $A > 0$, the integrand is less than A for $x^2 + y^2 > 1/A$. Hence the integral to be computed is

$$\lim_{A \to +\infty} \iint_{D_A} \frac{dx\, dy}{x^2 + y^2}$$

where D_A is the set of (x, y) with $1/A < x^2 + y^2 < 1$. It is reasonable to set $1/A = \epsilon^2$ and to use polar coordinates:

$$\iint_{D_A} \frac{dx\, dy}{x^2 + y^2} = \int_0^{2\pi} \int_\epsilon^1 \frac{1}{r}\, r\, dr\, d\theta = \int_0^{2\pi} \int_1^\epsilon dr\, d\theta = 2\pi(1 - \epsilon).$$

Hence

$$\iint_{x^2+y^2<1} \frac{dx\, dy}{x^2 + y^2} = 2\pi.$$

Second solution. We use polar coordinates from the beginning:

$$\iint_{x^2+y^2<1} \frac{dx\, dy}{x^2 + y^2} = \int_0^{2\pi} \int_0^1 \frac{1}{r}\, r\, dr\, d\theta = \int_0^{2\pi} \int_0^1 dr\, d\theta = 2\pi.$$

2. Compute

$$\iint_I (xy)^{-1/2}\, dx\, dy$$

where I is the interval $0 < x < 1$, $0 < y < 1$.

Answer. In this case it is convenient to write the integral as an iterated integral:

$$\iint_I (xy)^{-1/2} \, dx \, dy = \int_0^1 \int_0^1 (xy)^{-1/2} \, dx \, dy = \int_0^1 x^{-1/2} \, dx \int_0^1 y^{-1/2} \, dy$$

$$= \left(\int_0^1 x^{-1/2} \, dx \right)^2;$$

the name of a dummy variable is of no importance. Since

$$\int_0^1 x^{-1/2} = 2x^{1/2} \, \Big|_0^1 = 2$$

(this is an improper integral), the original integral equals 4.

3. Evaluate

$$\iiint_D (x^2 + y^2 + z^2)^{-\alpha/2} \, dx \, dy \, dz$$

where D is the unit ball $(x^2 + y^2 + z^2 < 1)$ and α a positive number.

Answer. We use spherical coordinates. The integral equals

$$\int_0^\pi \int_0^{2\pi} \int_0^1 \rho^{-\alpha} \rho^2 \sin \phi \, d\rho \, d\theta \, d\phi = \int_0^\pi \sin \phi \, d\phi \int_0^{2\pi} d\theta \int_0^1 \rho^{2-\alpha} \, d\rho$$

$$= 4\pi \int_0^1 \rho^{2-\alpha} \, d\rho = \begin{cases} 4\pi \dfrac{\rho^{3-\alpha}}{3 - \alpha} \Big|_0^1 = \dfrac{4\pi}{3 - \alpha} & \text{if } \alpha < 3. \\ +\infty \ \text{if } \alpha \geq 3 \end{cases}$$

4. Does the integral

$$\iiint_I \frac{2 - \sin x + e^{yz}}{(x^2 + y^2 + 2z^2)}$$

converge? Here I is some open interval containing the origin.

Answer. Yes. For let M be a number such that $2 - \sin x + e^{yz} \leq M$ in I (it is easy to see that there is such a number). We have

$$0 < \frac{2 - \sin x + e^{yz}}{x^2 + y^2 + 2z^2} \leq \frac{M}{x^2 + y^2 + z^2}.$$

Now apply Example 3 and the comparison theorem. ◄

EXERCISES

1. Evaluate $\iint_D \log \sqrt{x^2 + y^2} \, dx \, dy$ where D is the unit circle.
2. Consider $\iint_D (x^2 + y^2)^{-\alpha/2} \, dx \, dy$ where D is the unit circle. Determine for

which α the integral is finite, compute its value for such α, and compare with Example 3 in the text.

3. Evaluate $\iiint_D |\log \sqrt{x^2 + y^2 + z^2}|\, dx\, dy\, dz$ where D is the unit sphere.

In Exercises 4 to 10, determine whether the integral converges or diverges. In each case, the region D is a circle or sphere around the origin with radius 100.

▶ 4. $\iiint_D [(x - 1)^2 + y^2 + (z - 2)^2]^{-2}\, dx\, dy\, dz.$

5. $\iint_D |\cos xyz|[(x - 2)^2 + (y - 1)^2 + z^2]^{-4}\, dx\, dy\, dz.$

6. $\iiint_D |x|^{-1/2}|y|^{-2/3}|z|^{-3/4}\, dx\, dy\, dz.$

7. $\iint_D \dfrac{(x^2 + y^2 + 1)}{x^2 + |xy|}\, dx\, dy.$

8. $\iint_D |x - 1|^{-1/2}(x + y)^4\, dx\, dy.$

9. $\iint_D \dfrac{e^{x+y}}{\sin\,[(x^2 + y^2)^{1/2}]}\, dx\, dy.$

10. $\iiint_D |x^{-1}y^2z^3|\, dx\, dy\, dz.$

7.2 Integration over unbounded regions

Suppose we want to integrate a nonnegative function $f(x, y)$ over an unbounded region D. We might as well set $f(x, y) = 0$ for (x, y) not in D, and integrate over the whole plane. We write such an integral as

$$(5) \qquad\qquad \int\!\!\!\int_{-\infty}^{+\infty} f(x, y)\, dx\, dy,$$

and interpret it as the volume between the whole (x, y) plane and the surface $z = f(x, y)$.

In order to assign a numerical value to (5), we choose a sequence of bounded regions of integration, $D_1, D_2, D_3, \cdots$ such that (a) each D_j is contained in the next region D_{j+1}, and (b) each disk $x^2 + y^2 < R^2$ is contained in some D_j, and hence also in all following ones. For instance, we can choose for D_j the interval $-j < x < j,\ -j < y < j$. We assume, of course, that all characteristic functions χ_{D_j} are piecewise continuous.

We assume also that f is piecewise continuous and the integrals

$$(6) \qquad\qquad \iint_{D_j} f(x, y)\, dx\, dy$$

converge. (They may be improper integrals.) Since $f > 0$ and condition (a) holds, the numbers (6) form an increasing sequence. We write

$$(7) \qquad\qquad \lim_{j \to +\infty}\ \iint_{D_j} f(x, y)\, dx\, dy = \int\!\!\!\int_{-\infty}^{+\infty} f(x, y)\, dx\, dy.$$

If this is a finite limit, we say that (5) **converges,** or that f is **integrable** over the plane. Otherwise (5) is called **divergent.**

It may seem that the value of (5) just defined depends on the choice of the regions

$D_1, D_2, D_3, \cdots$. This is not so. Indeed, let $\hat{D}_1, \hat{D}_2, \hat{D}_3, \cdots$ be another sequence satisfying conditions (a) and (b). Choose a fixed k; the region $\hat{D}_k$ lies in some disk about the origin; hence also in all D_j, for j sufficiently large. Since $f \geq 0$, the integral

$$\iint\limits_{\hat{D}_k} f \, dx \, dy$$

is not greater than (6), for large j, and hence not greater than the limit (7). The value of (5) obtained by using the sequence $\hat{D}_1, \hat{D}_2, \cdots$ is therefore not greater than that obtained through the sequence $D_1, D_2, \cdots$. An analogous argument shows that the value obtained by using the D_j's is not greater than the value obtained through the $\hat{D}_k$'s. Hence the two values are equal.

The **comparison theorem** is again valid (if $0 \leq f \leq g$, and g is integrable over the whole plane, so is f). Also, integrals over the plane can be computed by iterated integrals and by polar coordinates; for instance,

(8)
$$\iint\limits_{-\infty}^{+\infty} f(x, y) \, dx \, dy = \int_{-\infty}^{+\infty} \left\{ \int_{-\infty}^{+\infty} f(x, y) \, dx \right\} dy$$
$$= \int_{-\infty}^{+\infty} \int_{-\infty}^{+\infty} f(x, y) \, dx \, dy.$$

Finally, everything said above extends to more than two variables.

▶ *Example* Compute

$$\iint\limits_{-\infty}^{+\infty} \frac{dx \, dy}{(1 + x^2 + y^2)^2}.$$

Answer. We choose as D_j the disk $|x^2 + y^2| < j^2$. Using polar coordinates, we have

$$\iint\limits_{-\infty}^{+\infty} \frac{dx \, dy}{(1 + x^2 + y^2)^2} = \lim_{j \to +\infty} \iint\limits_{D_j} \frac{dx \, dy}{(1 + x^2 + y^2)^2} = \lim_{j \to +\infty} \int_0^{2\pi}\int_0^j \frac{r \, dr \, d\theta}{(1 + r^2)^2}$$
$$= \lim_{j \to +\infty} \int_0^{2\pi} d\theta \int_0^j \frac{r \, dr}{(1 + r^2)^2} = \lim_{j \to +\infty} 2\pi \left[-\frac{1}{2} \frac{1}{1 + r^2} \right]_{r=0}^{r=j}$$
$$= \lim_{j \to +\infty} 2\pi \left(\frac{1}{2} - \frac{1}{2(1 + j^2)} \right) = \pi.$$

It is simpler to use polar coordinates at once:

$$\iint\limits_{-\infty}^{+\infty} \frac{dx \, dy}{(1 + x^2 + y^2)^2} = \int_0^{2\pi}\int_0^{+\infty} \frac{r \, dr \, d\theta}{(1 + r^2)^2} = 2\pi \int_0^{+\infty} \frac{r \, dr}{(1 + r^2)^2}$$
$$= 2\pi \left[-\frac{1}{2} \frac{1}{1 + r^2} \right]_0^{+\infty} = \pi.$$

The reader should try to compute the same integral using iterated integration in Cartesian coordinates. This will take longer. ◀

EXERCISES

11. The area under the curve $xy = 1$ for $x \geq 1$ is rotated about the x axis. Show that the resulting solid of revolution has a finite volume but an infinite area. (If we were to build a paint can in the shape of the figure thus constructed, it would hold only a finite amount of paint; yet to paint it would require an infinite amount of paint!)

12. If D is given by $x^2 + y^2 > 1$, for which values of α does $\iint_D (x^2 + y^2)^{\alpha/2} \, dx \, dy$ converge? What is its value?

13. If D is defined by $x^2 + y^2 + z^2 > 1$, for which values of α does the following integral converge? $\iint_D (x^2 + y^2 + z^2)^{\alpha/2} \, dx \, dy \, dz$.

In Exercises 14 to 20, determine which integrals converge and which diverge. D is always the region *exterior* of the unit sphere or circle.

14. $\iint_D |x^3| \, e^{-y^2} \, dx \, dy$.

15. $\iint_D x^{20} y^{10} e^{-\sqrt{|x|+|y|}} \, dx \, dy$.

▶ 16. $\iint_D \dfrac{|\cos xy| \, dx \, dy}{(7 + x^2 + y^2)^3}$.

17. $\iint_D e^{-(x^2+y^2)-1} \, dx \, dy$.

18. $\iint_D e^{-|x|} \, dx \, dy$.

19. $\iint_D e^{-\sqrt{x^2+y^2}} (1 + y^3)^{-1} \, dx \, dy$.

20. $\iint_D e^{-\sqrt{x^2+y^2}} (1 + y^2)^{-1} \, dx \, dy$.

7.3 An important example

We prove the remarkable relation

$$(9) \qquad \int_{-\infty}^{+\infty} e^{-x^2} \, dx = \sqrt{\pi}.$$

This is important in mathematical statistics (see Chapter 6, §8.2).

The proof is based on the double integral

$$T = \iint_{-\infty}^{+\infty} e^{-x^2-y^2} \, dx \, dy.$$

We have, using polar coordinates,

$$T = \int_0^{2\pi}\int_0^{+\infty} e^{-r^2} r \, dr \, d\theta = 2\pi \int_0^{+\infty} \frac{d}{dr}\left(-\frac{e^{-r^2}}{2}\right) dr = \pi\left[-e^{-r^2}\right]_0^{+\infty} = \pi.$$

On the other hand, by (8) and the addition theorem for e^x,

$$T = \iint_{-\infty}^{+\infty} e^{-x^2} e^{-y^2} \, dx \, dy = \int_{-\infty}^{+\infty}\int_{-\infty}^{+\infty} e^{-x^2} e^{-y^2} \, dx \, dy$$

$$= \int_{-\infty}^{+\infty} e^{-x^2} \, dx \int_{-\infty}^{+\infty} e^{-y^2} \, dy = \left(\int_{-\infty}^{+\infty} e^{-x^2} \, dx\right)^2$$

the name of a dummy variable being irrelevant. Now (9) follows.

EXERCISES

21. Prove that $\int_0^{+\infty} (e^{-u}/\sqrt{u}) \, du = \sqrt{\pi}$.

22. Prove that $\int_0^{+\infty} \sqrt{u} e^{-u} \, du = \sqrt{\pi}/2$.

§8 Some Proofs

8.1 Polar coordinates

Here we prove Theorem 4 in §1.11, the formula for integrating in polar coordinates.

Let R be a positive number; we consider in this subsection only such bounded piecewise continuous functions $f(x, y)$ which are 0 for $R > 0$ and for which $(r, \theta) \mapsto f(r \cos \theta, r \sin \theta)$ is a piecewise continuous function of (r, θ) in $0 < r < R$, $0 < \theta < 2\pi$. For such f, set

$$\mathrm{T}[f] = \iint\limits_{x^2 + y^2 < R^2} f(x, y)\, dx\, dy,$$

$$\mathrm{T}_*[f] = \int_0^{2\pi}\!\!\int_0^R f(r \cos \theta, r \sin \theta) r\, dr\, d\theta.$$

We must show that

(1)
$$\mathrm{T}[f] = \mathrm{T}_*[f].$$

Since $\mathrm{T}_*[f]$ is computed by carrying out two single integrations, we have that, if $f_1 \le f_2$, then $\mathrm{T}_*[f_1] \le \mathrm{T}_*[f_2]$. Now if f is given, we can find step functions ϕ and ψ with $\phi \le f \le \psi$, $\mathrm{T}[\phi] \le \mathrm{T}[f] \le \mathrm{T}[\psi]$ and $\mathrm{T}[\psi] - \mathrm{T}[\phi]$ as small as we like (Theorem 3 in §1.7). By the remark just made, we have the inequality $\mathrm{T}_*[\phi] \le \mathrm{T}_*[f] \le \mathrm{T}_*[\psi]$. We conclude that (9) holds in general if it holds for step functions.

Since $\mathrm{T}_*[f]$ is an iterated integral, we have:

$$\mathrm{T}_*[cf] = c\mathrm{T}_*[f], \qquad c \text{ a constant,}$$

and
$$\mathrm{T}_*[f_1 + f_2] = \mathrm{T}_*[f_1] + \mathrm{T}_*[f_2].$$

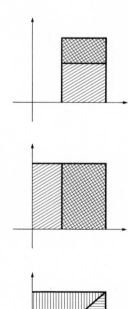

Fig. 13.53

Every step function is a sum of functions of the form $c\chi_I$, I an interval. Hence it suffices to prove (1) when f is the characteristic function of an interval, that is, when $f = 1$ in an interval and $f = 0$ outside the interval. This can be done by a direct calculation.

It pays to note that every interval can be decomposed into intervals lying in a quadrant; an interval lying in a quadrant can be represented as an interval with one side on a coordinate axis from which another such interval has been removed; an interval with one side on an axis can be represented as an interval with two sides on axes from which another such interval has been removed; and that an interval with two sides on the axes can be decomposed into two right triangles with one vertex at the origin and the two legs parallel to the axes (see Fig. 13.53). Hence (1) needs to be proved only for characteristic functions of such triangles. This is easy—see Example 2, Chapter 11, §2.6.

To complete the proof of Theorem 4 in §1.11, we note that the equality

$$\int_0^{2\pi}\!\!\int_0^R f(r \cos \theta, r \sin \theta) r\, dr\, d\theta = \int_0^R\!\!\int_0^{2\pi} f(r \cos \theta, r \sin \theta) r\, d\theta\, dr$$

is a special case of Theorem 2 in §1.5.

8.2 On the definition of area

We noted in §2.1 that the area A of a bounded plane set D can be defined as

$$A = \iint_I \chi_D(x, y) \, dx \, dy,$$

provided the characteristic function χ_D is piecewise continuous; here I is some interval containing D; its choice does not affect the value of A.

Assume that we introduce another Cartesian coordinate system (X, Y) and that the characteristic function of D in the new system, $\hat{\chi}_D(X, Y)$ is still piecewise continuous. Is the area $\hat{A}$, computed by the formula

$$\hat{A} = \iint_{\hat{I}} \hat{\chi}_D(X, Y) \, dX \, dY$$

($\hat{I}$ some interval containing D) equal to A?

Our geometric intuition tells us: yes, of course (see Fig. 13.54). Let us justify the intuition by a proof.

As we noted in §2.1, one can find polygons P_1 and P_2 with sides parallel to the (x, y) axes such that P_1 lies in D and D lies in P_2 and

$$\iint_I \chi_{P_1} \, dx \, dy \leq A \leq \iint_I \chi_{P_2} \, dx \, dy$$

and

$$\iint_I \chi_{P_2} \, dx \, dy - \iint_I \chi_{P_1} \, dx \, dy < \epsilon$$

where ϵ is any given positive number. (Compare Figs. 13.29 and 13.30 where P_1 and P_2 are the shaded areas.) Clearly $\hat{\chi}_{P_1} \leq \hat{\chi}_D \leq \hat{\chi}_{P_2}$, so that

$$\iint_{\hat{I}} \hat{\chi}_{P_1} \, dx \, dy \leq \hat{A} \leq \iint_{I_2} \hat{\chi}_{P_2} \, dx \, dy.$$

We conclude that it suffices to prove that $A = \hat{A}$ for the case when D is a polygon with all sides parallel to one of two perpendicular lines. Since every such polygon can be decomposed into rectangles, it suffices to show that, if D is an arbitrarily situated rectangle with sides a and b lying inside an interval I, then

$$\iint_I \chi_D \, dx \, dy = ab.$$

This is an elementary calculation.

The reader who may be surprised at the simplicity of the argument should note that the existence of the polygons P_1 and P_2 described above follows from the relatively difficult Theorem 3 in §1.7.

Fig. 13.54

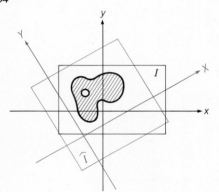

Problems

In Problems 1 to 12, all functions are assumed to be bounded and piecewise continuous in every region D in which they are being integrated. All regions of integration D are assumed to be bounded and to have a piecewise continuous characteristic functions.

1. Suppose that $f(x, y)$ is continuous for $0 < x < 1, 0 < y < 1$ and takes on no negative values. Prove that if

$$\int_0^1 \int_0^1 f(x, y) \, dx \, dy = 0,$$

then $f(x, y)$ is the constant function 0.

2. Let $f(x, y)$ be bounded and continuous everywhere. Set

$$F(x, y) = \int_{y_0}^y \int_{x_0}^x f(u, v) \, du \, dv.$$

What is $F_{xy}(x, y)$?

3. Let $F(x, y)$ have a bounded continuous partial derivative $F_{xy}(x, y) = f(x, y)$. Compute

$$\int_{y_0}^y \int_{x_0}^x f(u, v) \, du \, dv.$$

4. Extend Problems 2 and 3 to triple integrals.

5. Let S be the surface defined by $z = f(x, y)$, (x, y) in D, where $f(x, y)$ is continuously differentiable and D is a "nice" bounded region. Show that the area of S is not less than that of D.

6. Prove, for functions of two variables, Schwarz' inequality and its corollary:

$$\left| \iint_D f(x, y) \, g(x, y) \, dx \, dy \right| \leq \sqrt{\iint_D f(x, y)^2 \, dx \, dy \iint_D g(x, y)^2 \, dx \, dy},$$

$$\sqrt{\iint_D [f(x, y) + g(x, y)]^2 \, dx \, dy} \leq \sqrt{\iint_D f(x, y)^2 \, dx \, dy} + \sqrt{\iint_D g(x, y)^2 \, dx \, dy}.$$

(Hint: Compare Problems 12 and 13 in Chapter 5.)

7. State and prove Schwarz' inequality for functions of 3 variables, and also for functions of $n > 3$ variables.

8. Consider a solid D in 3-space with density function $f(x, y, z)$. The **Newtonian potential** ϕ of this density is defined as

$$\phi(x, y, z) = \iiint_D \frac{f(X, Y, Z) \, dX \, dY \, dZ}{\sqrt{(x - X)^2 + (y - Y)^2 + (z - Z)^2}}$$

or, setting $\mathbf{x} = x\mathbf{e}_1 + y\mathbf{e}_2 + z\mathbf{e}_3$, $\mathbf{X} = X\mathbf{e}_1 + Y\mathbf{e}_2 + Z\mathbf{e}_3$,

$$\phi(\mathbf{x}) = \iiint_D f(\mathbf{X}) \, |\mathbf{X} - \mathbf{x}|^{-1} \, dX \, dY \, dZ.$$

Compute, at a point x, y, z not in or on the boundary of D, $\mathbf{F} = \text{grad } \phi$ and verify that $\mathbf{F}$ is the attractive force exerted by D on a particle of mass 1 located at (x, y, z). [The units are chosen so that the constant of gravitation is 1. You may assume that the distance of (x, y, z) from any point of D is greater than some positive number.]

9. Compute the potential of a sphere $x^2 + y^2 + z^2 < R^2$ with density function $f(x, y, z) = g(\sqrt{x^2 + y^2 + z^2})$. Compare the result with §3.8.

10. Prove that the potential $\phi(x, y, z)$ defined in Problem 8 satisfies at points not in or on the boundary of D the partial differential equation

$$\frac{\partial^2 \phi}{\partial x^2} + \frac{\partial^2 \phi}{\partial y^2} + \frac{\partial^2 \phi}{\partial z^2} = 0.$$

(This is the so-called **Laplace Equation** which plays an important part in mathematical physics.)

11. Let D be a spherical shell $0 < R_1^2 < x^2 + y^2 + z^2 < R_2^2$, with a density function $f(x, y, z) = g(\sqrt{x^2 + y^2 + z^2})$. What gravitational force is exerted on a particle located inside the ball $x^2 + y^2 + z^2 < R_1^2$?

12. The function

$$\phi(x, y) = -\frac{1}{2} \iint\limits_{D} \log \left[(x - X)^2 + (y - Y)^2\right] f(X, Y) \, dX \, dY$$

is called the **logarithmic potential** of the region D with density f. At points not in or on the boundary of D the logarithmic potential satisfies Laplace's Equation

$$\frac{\partial^2 \phi}{\partial x^2} + \frac{\partial^2 \phi}{\partial y^2} = 0.$$

Prove this. [You may assume that the point (x, y) considered has the property: its distance from any point of D is greater than some positive number.]

Supplement

In this book, we have tried to reduce the amount of "pure theory" to a minimum and have developed calculus while bypassing some traditional topics that are of great interest, however, and may be needed by a reader who plans to continue his mathematical studies. These matters are taken up in this Supplement.

We present in §1 and §2 the main theorems on convergence and on continuous functions that have not been proved before. In §§3, 4, and 5, we establish the theorems about multiple integrals which have been stated in the text without proof.

§1 Convergence

All arguments in this Supplement will be based on the least upper bound principle: if a nonempty set of real numbers has an upper bound, it has a least upper bound; if such a set has a lower bound, it has a greatest lower bound (Chapter 1, §6). In this section we prove two theorems on convergence. The first, the Bolzano–Weierstrass Theorem, is the basis of the next sections. The second, Cauchy's Theorem, is quite important, but will not be used later in this Supplement.

1.1 Limit superior and limit inferior

Throughout this section, the letters i, j, and k stand for positive integers. We recall that "for nearly all j" means "for all but finitely many j," that is, "for all j exceeding some integer N."

Let $\{x_j\} = x_1, x_2, x_3, \cdots$ be a sequence of real numbers. A number a^* is called the **limit superior** of the sequence if, for every positive number ϵ, we have: $a^* - \epsilon < x_j$ for infinitely many j, and $a^* + \epsilon > x_j$ for nearly all j. A number a_* is called the **limit inferior** of the sequence $\{x_j\}$ if, for every positive number ϵ, we have: $a_* + \epsilon > x_j$ for infinitely many j, and $a_* - \epsilon < x_j$ for nearly all j.

A sequence $\{x_j\}$ can have at most one limit superior. Because if it had two, say a_1 and $a_2 > a_1$, we could set $\epsilon = (a_2 - a_1)/3 > 0$ and observe that $a_1 + \epsilon > x_j$ for nearly all j and $a_2 - \epsilon < x_j$ for infinitely many j. This is absurd, for we chose ϵ so that $a_1 + \epsilon < a_2 - \epsilon$. One sees similarly that a sequence can have at the most one limit inferior.

The limit inferior and the limit superior of $\{x_j\}$ are usually denoted by lim inf x_j and lim sup x_j.

If a sequence has a limit superior a^* and a limit inferior a_*, then $a^* \geq a_*$. Indeed, we have, for all $\epsilon > 0$ and for nearly all j, $a_* - \epsilon < x_j < a^* + \epsilon$. Hence $a^* - a_* > -2\epsilon$ for all $\epsilon > 0$, that is $a^* \geq a_*$.

If a is both the limit superior and the limit inferior of the sequence $\{x_j\}$, then the sequence **converges** to a. Indeed, for every $\epsilon > 0$, we have: $a - \epsilon < x_j$, $a + \epsilon > x_j$ for nearly all j, that is, $|a - x_j| < \epsilon$ for nearly all j. Conversely, if a sequence has the **limit** a, then a is both limit superior and limit inferior.

1.2 Bounded sequences

We state now the basic existence theorem.

Theorem A   A BOUNDED SEQUENCE OF REAL NUMBERS HAS A LIMIT INFERIOR AND A LIMIT SUPERIOR.

Proof.   Denote the sequence by $x_1, x_2, x_3, \cdots$. By hypothesis, there is a number M such that $|x_j| \leq M$ for all M. For every k, let α_k be the least upper bound of the set of numbers $\{x_k, x_{k+1}, x_{k+2}, \cdots\}$; there is an α_k, since the set is bounded. Since α_k is also an upper bound for the set $\{x_{k+1}, x_{k+2}, \cdots\}$, we have $\alpha_k \geq \alpha_{k+1}$.

Thus $\alpha_1 \geq \alpha_2 \geq \alpha_3 \geq \cdots$, the sequence $\{\alpha_k\}$ is monotone nonincreasing; it is also bounded from below: since $x_j \geq -M$ for all j, we have $\alpha_k \geq -M$. Therefore (see Theorem 3 in Chapter 8, §3.8) the sequence $\{\alpha_k\}$ has a limit α.

(Let us recall briefly how the convergence of $\{\alpha_k\}$ can be proved. Let α be the greatest lower bound of the set numbers $\{\alpha_1, \alpha_2, \alpha_3, \cdots\}$. Assume there is a number $\epsilon > 0$ such that $\alpha_k > \alpha + \epsilon$ for infinitely many k. Since $\alpha_m \geq \alpha_k$ for $m < k$, this would imply that $\alpha_k > \alpha + \epsilon$ for all k and α is not the greatest lower bound of our set. Hence, for every $\epsilon > 0$: $\alpha_k \leq \alpha + \epsilon$ for nearly all k. Also, $\alpha \leq \alpha_k$ for all k. Therefore $\lim_{k \to \infty} \alpha_k = \alpha$.)

We shall show that α is the limit superior of $\{x_j\}$.

Let ϵ be a given positive number. If it were not true that $\alpha + \epsilon > x_j$ for nearly all j, then we would have $\alpha + \epsilon \leq x_j$ for infinitely many j. Hence every set $\{x_k, x_{k+1}, x_{k+2}, \cdots\}$ would contain numbers that are $\geq \alpha + \epsilon$ and all α_k would satisfy $\alpha_k \geq \alpha + \epsilon$. This is absurd, since α is the limit of α_k. If it were not true that $\alpha - \epsilon < x_j$ for infinitely many j, then we would have $\alpha - \epsilon \geq x_j$ for nearly all j. Hence, for k large enough, all numbers of the set $\{\alpha_k, \alpha_{k+1}, \alpha_{k+2}, \cdots\}$ would be $\leq \alpha - \epsilon$ and we would have $\alpha_k \leq \alpha - \epsilon$, which is absurd, since α is the limit of α_k. Thus α satisfies both conditions for a limit superior.

One proves similarly that $\{x_j\}$ has a limit inferior $\beta = \lim_{k \to \infty} \beta_k$ where β_k is the greatest lower bound of the set of numbers $\{x_k, x_{k+1}, x_{k+2}, \cdots\}$.

EXERCISES

1. Prove directly that the limit β defined above exists.
▶ 2. Prove that the limit β defined above is the limit inferior of $\{x_j\}$.

1.3 The Bolzano–Weierstrass Theorem

Let a^* be the limit superior of a sequence $\{x_j\}$. Let $\epsilon > 0$ be given. By the definition, we have $a^* - \epsilon < x_j$ for infinitely many j and $x_j < a^* + \epsilon$ for nearly all j. Hence, $|a^* - x_j| < \epsilon$ for infinitely many j.

It follows that we can find a positive integer j_1 with $|x_{j_1} - a^*| < 1$, another positive integer j_2 with $j_2 > j_1$ and $|x_{j_2} - a^*| < \frac{1}{2}$, and a third $j_3 > j_2$ with $|x_{j_3} - a^*| < \frac{1}{3}$, and so forth. The sequence $\{x_{j_1}, x_{j_2}, x_{j_3}, \cdots\}$ is a subsequence of $\{x_1, x_2, x_3, \cdots\}$ and converges to a^*. One proves similarly that, if the sequence $\{x_j\}$ has a limit inferior a_*, then there is a subsequence of $\{x_j\}$ which converges to a_*.

These considerations, together with Theorem A, imply the following important result.

Theorem B (*Bolzano–Weierstrass*) FROM EVERY BOUNDED SEQUENCE OF REAL NUMBERS, ONE CAN SELECT A CONVERGENT SUBSEQUENCE.

EXERCISE

3. Give an example of a bounded sequence of rational numbers such that no subsequence converges to a rational number.

BERNARD BOLZANO (1781–1848), a philosopher, logician and mathematician, anticipated Weierstrass' rigorous approach to calculus and Cantor's set theory. His main mathematical work, *Paradoxes of Infinity*, appeared posthumously.

Bolzano was a priest and a professor of religious philosophy at the University of Prague. In 1820, the Austrian government dismissed him, since his sermons to the students were considered subversive (they dealt with war and the conflict between individual conscience and obedience to the government).

1.4 The Cauchy convergence criterion

A sequence of numbers $x_1, x_2, x_3, \cdots$ is said to satisfy the **Cauchy condition** if, for every number $\epsilon > 0$, there is an integer N such that: if $j \geq N$ and $k \geq N$, then $|x_i - x_j| < \epsilon$.

Theorem C (Cauchy)   A SEQUENCE OF REAL NUMBERS CONVERGES IF AND ONLY IF IT SATISFIES THE CAUCHY CONDITION.

Proof. First part.   Assume $\{x_j\}$ has the limit a. For a given $\epsilon > 0$, let N be so large that $|x_j - a| < \epsilon/2$ whenever $j \geq N$. (Such an n exists by the definition of the limit.) If $j \geq N$ and $k \geq N$, then $|x_j - x_k| \leq |x_j - a| + |x_k - a| \leq 2(\epsilon/2) = \epsilon$. The Cauchy condition holds.

We disposed of the trivial part of the theorem (convergence implies the Cauchy condition). This holds also in the field of rational numbers. The nontrivial part (Cauchy condition implies convergence) is equivalent to the completeness property of real numbers.

Second part.   Assume that $\{x_j\}$ satisfies the Cauchy condition. For $\epsilon = 1$, the condition implies: there is an N_1 such that for $j > N_1$ we have $|x_{N_1} - x_j| < 1$. We conclude that the sequence is bounded. Indeed, the largest of the numbers: $|x_1|, |x_2|, \cdots, |x_{N_1-1}|, |x_{N_1}| + 1$ is a bound.

By Theorem A, our sequence has a limit superior a^* and a limit inferior a_*. If $a^* \neq a_*$ we have $a^* > a_*$. Set $\epsilon = (a^* - a_*)/3$ and observe that $a^* - \epsilon < x_j$ for infinitely many j, $a_* + \epsilon > x_k$ for infinitely many k. Thus $-x_k > -(a_* - \epsilon)$ for infinitely many k. Hence there exist arbitrarily large j and k such that

$$x_j - x_k \geq (a_* - \epsilon) - (a^* + \epsilon) = a^* - a_* - 2\epsilon = \epsilon.$$

This would violate the Cauchy condition. Hence $a^* = a_*$ and the sequence $\{x_j\}$ has a limit.

EXERCISE

▶ 4. Give an example of a Cauchy sequence of rational numbers that does not converge to a rational number.

1.5 Sequences of points

The Bolzano–Weierstrass and the Cauchy Theorems can be extended to sequences of points. In this Supplement, the word "point" will usually denote a point in a plane. We use a fixed Cartesian coordinate system and we identify a point with its position vector **x**, and with its coordinate pair (x, y). We restrict ourselves to the plane *only* in order to simplify writing. Everything remains valid in spaces of n dimensions, for all n.

A sequence of points $\mathbf{x}_1 = (x_1, y_1)$, $\mathbf{x}_2 = (x_2, y_2)$, $\mathbf{x}_3 = (x_3, y_3), \cdots$, is called **bounded** if the sequence of numbers $|\mathbf{x}_1|, |\mathbf{x}_2|, |\mathbf{x}_3|, \cdots$ is bounded. Since $|\mathbf{x}_j| =$

$(x_j{}^2 + y_j{}^2)^{1/2}$, we have: $|x_j| \leq |\mathbf{x}_j|$, $|y_j| \leq |\mathbf{x}_j|$, $|\mathbf{x}_j| \leq |x_j| + |y_j|$. We conclude that the sequence of points $\{\mathbf{x}_j\}$ is bounded if and only if both sequences of numbers, $\{x_j\}$ and $\{y_j\}$ are.

A sequence of points $\{\mathbf{x}_j\}$ is said to **converge** to $\mathbf{a} = (a, b)$ (or to have the **limit a**) if the sequence of numbers $\{|\mathbf{x}_j - \mathbf{a}|\}$ converges to 0. We conclude, as before, that $\{\mathbf{x}_j\}$ converges to $\mathbf{a}$ if and only if the sequences of numbers, $\{x_j\}$ and $\{y_j\}$ converge to a and b, respectively.

A sequence of points $\{\mathbf{x}_j\}$ is said to satisfy the **Cauchy condition** if, for every number $\epsilon > 0$, there is an integer N such that: if $j \geq N$ and $k \geq N$, then $|\mathbf{x}_j - \mathbf{x}_k| < \epsilon$. We conclude, as before, that this is equivalent to the Cauchy condition for both sequences of numbers: $\{x_j\}$ and $\{y_j\}$.

Theorem D   THE CAUCHY CONVERGENCE THEOREM (THEOREM C) AND THE BOLZANO-WEIERSTRASS THEOREM (THEOREM B) HOLD FOR SEQUENCES OF POINTS.

The proof is contained in the remarks above.

§2 Continuous Functions

2.1 The Boundedness Theorem

Theorem E   LET $f(x)$ BE DEFINED AND CONTINUOUS IN A FINITE CLOSED INTERVAL $[a, b]$, THAT IS, FOR $a \leq x \leq b$. THEN f IS BOUNDED.

Before proving this, we note that the conditions on the interval are essential. The function x is continuous in the infinite interval $-\infty < x < +\infty$, but not bounded in this interval. The function $1/x$ is continuous, but not bounded, in the open finite interval $0 < x < 1$.

Proof.   Assume that f is continuous but unbounded in $[a, b]$. Then, for every $j = 1, 2, 3, \cdots$, there is an x_j with $a \leq x_j \leq b$ and $|f(x_j)| > j$. By Theorem B, there is a subsequence $x_{j_1}, x_{j_2}, x_{j_3}, \cdots$ which converges. Set $\xi = \lim_{i \to \infty} x_{j_i}$. Then $a \leq \xi \leq b$. Since f is continuous, $\lim_{i \to \infty} f(x_{j_i}) = f(\xi)$. Since $|f(x_j)| > j$, we have $|f(\xi)| > j$, for all positive integers j. But this is absurd; hence the assumption that f is unbounded is untenable.

2.2 The Maximum Theorem

Theorem F   LET $f(x)$ BE DEFINED AND CONTINUOUS IN A CLOSED FINITE INTERVAL $[a, b]$. THEN THERE IS A POINT ξ IN THIS INTERVAL WHERE f ATTAINS ITS MAXIMUM (THAT IS, $f(\xi) \geq f(x)$ FOR ALL x, $a \leq x \leq b$).

Proof.   By Theorem E, the set of values taken on by f in $[a, b]$ is bounded; let M be the least upper bound of this set. This means: $M \geq f(x)$ for $a \leq x \leq b$ and, if $N < M$, then there is a point $\hat{x}$ with $a \leq \hat{x} \leq b$ and $f(\hat{x}) > N$. We must show that there is a ξ, $a \leq \xi \leq b$, with $f(\xi) = M$. Assume the contrary. Then $M - f(x) > 0$ for $a \leq x \leq b$; hence $g(x) = 1/[M - f(x)]$ is continuous for $a \leq x \leq b$, Theorem E applies, and there is a number A such that

$$g(x) = \frac{1}{M - f(x)} \leq A \qquad \text{for all } x, a \leq x \leq b.$$

This implies that $M - f(x) \geq 1/A$ or $f(x) \leq M - (1/A)$ in $[a, b]$. But this contradicts the definition of M.

Corollary   UNDER THE HYPOTHESES OF THEOREM F, THERE IS A POINT η WITH $a \leq \eta \leq b$ WHERE f ATTAINS ITS MINIMUM. (THIS MEANS $f(\eta) \leq f(x)$ FOR $a \leq x \leq b$.)

Proof.   Apply Theorem F to the function $x \mapsto -f(x)$.

EXERCISE

1. Prove the corollary to Theorem F directly (that is, not assuming Theorem F).

2.3 Rolle's Theorem and the Mean Value Theorem

Theorem G (Rolle's Theorem in its strong form)   LET $x \mapsto f(x)$ BE CONTINUOUS FOR $a \leq x \leq b$, AND LET $f'(x)$ EXIST FOR $a < x < b$. ASSUME ALSO THAT $f(a) = f(b) = 0$. THEN THERE IS A POINT ξ WITH $a < \xi < b$ AND $f'(\xi) = 0$.

Proof.   By Theorem F and its corollary, there is a point ξ and a point η in $[a, b]$ at which f attains its maximum and minimum, respectively. If $a < \xi < b$, then ξ is a local maximum point and $f'(\xi) = 0$. If $a < \eta < b$, then η is a local minimum point and $f'(\eta) = 0$. If ξ and η are both endpoints of $[a, b]$, then the maximum and minimum of f are both 0. In this case, $f(x) \equiv 0$, and therefore $f'(x) \equiv 0$.

Rolle's Theorem implies the Mean Value Theorem (see Chapter 8, §1.3) and hence the Generalized Mean Value Theorem. Thus these theorems are valid even if the derivatives occurring in their statements are not continuous. Similarly, in Taylor's Theorem there is no need to assume the continuity of the derivative's occurring in the remainder.

In Chapter 4, §8.1, we proved the basic result: *if $f'(x) > 0$ in $[a, b]$, the function f is strictly increasing.* The strong form of the Mean Value Theorem (without assuming continuity of the derivative) gives a new proof of this fact. For, if $a \leq x_1 < x_2 \leq b$, then there is a number ξ between x_1 and x_2 such that $f(x_2) - f(x_1) = f'(\xi)(x_2 - x_1)$. Since $f'(\xi) > 0$, we have $f(x_2) - f(x_1) > 0$.

2.4 Uniform continuity

A function $x \mapsto f(x)$ defined on an interval is called **uniformly continuous** in this interval if, given any positive number $\epsilon > 0$, one can find a positive number δ such that: if x_1 and x_2 are any two points in the interval and $|x_1 - x_2| \leq \delta$, then $|f(x_1) - f(x_2)| \leq \epsilon$.

A uniformly continuous function is always continuous. The converse need not be true. For instance, the function $f(x) = 1/x$ is continuous, but not uniformly continuous, in the interval $0 < x < 1$. Indeed, let n be an integer > 3, and set $x_1 = 1/n$, $x_2 = 2/n$. Then $0 < x < 1$, $0 < x_2 < 1$, $|x_1 - x_2| = 1/n$ and $|f(x_1) - f(x_2)| = n/2$. Choosing n large enough, we can make $|x_1 - x_2|$ as small as we like and $|f(x_1) - f(x_2)|$ as large as we like.

Theorem H   LET $f(x)$ BE DEFINED AND CONTINUOUS IN A CLOSED FINITE INTERVAL $[a, b]$. THEN $f(x)$ IS UNIFORMLY CONTINUOUS IN THIS INTERVAL.

Proof.   Assume f is not uniformly continuous. Then there is a number $\epsilon > 0$ such that for no $\delta > 0$ is it true that: if $|x_1 - x_2| \leq \delta$, then $|f(x_1) - f(x_2)| \leq \epsilon$. In other words: for this $\epsilon > 0$ and for every $\delta > 0$, we can find points x and $\hat{x}$ in our interval such that $|x - \hat{x}| \leq \delta$ and $|f(x) - f(\hat{x})| > \epsilon$. We do this for a sequence of positive numbers $\delta_1, \delta_2, \delta_3, \cdots$ converging to 0 and obtain two sequences of numbers $x_j, \hat{x}_j$ in $[a, b]$ with

$$a \leq x_j \leq b, \qquad a \leq \hat{x}_j \leq b, \qquad \lim_{j \to \infty} |x_j - \hat{x}_j| = 0,$$

$$|f(x_j) - f(\hat{x}_j)| > \epsilon \qquad \text{for all } j.$$

Selecting, if need be, a subsequence, we may assume that the sequence x_j converges (by Theorem B). Set $\xi = \lim_{j \to \infty} x_j$. Then $a \leq \xi \leq b$ and hence, f being continuous in the *closed* interval,

$$\lim_{j \to \infty} f(x_j) = f(\xi).$$

Also

$$\lim_{j \to \infty} |\hat{x}_j - \xi| = \lim_{j \to \infty} |\hat{x}_j - x_j + x_j - \xi| \leq \lim_{j \to \infty} (|\hat{x}_j - x_j| + |x_j - \xi|) = 0,$$

so that $\lim \hat{x}_j = \xi$ and

$$\lim_{j \to \infty} f(\hat{x}_j) = f(\xi).$$

Now, if j is large enough, then $|f(x_j) - f(\xi)| < \epsilon/3$ and $|f(\hat{x}_j) - f(\xi)| < \epsilon/3$. Therefore,

$$|f(x_j) - f(\hat{x}_j)| = |f(x_j) - f(\xi) + f(\xi) - f(\hat{x}_j)| \leq \left(\frac{\epsilon}{3}\right) + \left(\frac{\epsilon}{3}\right) < \epsilon.$$

This contradicts the way the numbers x_j and $\hat{x}_j$ have been selected. The assumption that $f(x)$ is not uniformly continuous is untenable.

EXERCISE

► 2. Give a formal proof of the statement: if $f(x)$ is uniformly continuous in an interval, then $f(x)$ is continuous at every point of the interval.

2.5 Compact sets

We want to extend some of the theorems proved above to functions of several variables. It is advisable to consider functions defined not only on intervals but also on more general sets.

We recall that a set S of points (in the plane) is called **bounded** if it lies in some interval, that is, if the set of all numbers $|\mathbf{x}|$, $\mathbf{x}$ in S, is bounded.

A set S of points (in the plane) is called **closed** if the following condition holds: if the sequence $\{x_j\}$ converges to x, and all x_j belong to S, then x also belongs to S.

A set S of points (in the plane) is called **compact** if, whenever $\{x_j\}$ is a sequence such that all x_j belong to S, one can find a subsequence $\{x_{j_k}\}$ which converges to a point x, also in S.

Theorem I   A SET S OF POINTS (IN THE PLANE) IS COMPACT IF AND ONLY IF S IS BOTH BOUNDED AND CLOSED.

Proof.   If S is not bounded, then there is no number M such that $|x| \leq M$ for all x in S. Hence there are points $x_1, x_2, x_3, \cdots$ in S with $|x_j| > j$. The sequence $\{x_j\}$ has no convergent subsequence. Thus S is not compact.

If S is not closed, there is a sequence $\{x_j\}$ of points in S that converges to a point x not in S. No subsequence of $\{x_j\}$ can converge to a point in S. Thus S is not compact.

If S is bounded, then every sequence of points in S is bounded and hence contains a convergent subsequence (by Bolzano–Weierstrass). If S is also closed, the limit of this subsequence is a point in S. Hence S is compact.

The definitions, and the theorem, extend to sets on the line, in ordinary space, or in n-space, for all n.

▶*Examples   1.* A closed two-dimensional interval, that is, the set of all $x = (x, y)$ with $a \leq x \leq b$, $c \leq y \leq d$, is compact.

Proof.   The interval is clearly bounded. If x_j belongs to the interval for all j, and $\{x_j\}$ converges to x, then we have $a \leq x_j \leq b$, $c \leq y_j \leq d$, so that the limits $x = \lim_{j \to \infty} x_j$ and $y = \lim_{j \to \infty} y_j$ satisfy $a \leq x \leq b$, $c \leq y \leq d$. Hence $x = (x, y)$ belong to the interval.

2. If $S_1, S_2, \cdots S_n$ are finitely many compact sets, and S their union (the set consisting of all points belonging to any one of the sets $S_1, S_2, \cdots S_n$), then S is compact.

Proof.   Let $\{x_j\}$ be an infinite sequence of points of S. Then an infinite subsequence must belong to one of the sets $S_1, S_2, \cdots, S_n$, say, to S_1. Since S_1 is compact, a subsequence of this subsequence converges to a point of S_1, that is, to a point of S.◀

EXERCISES

3. Prove that the union of finitely many closed sets is closed.
▶ 4. Prove that the intersection of any number of closed sets is closed. (The intersection of sets $S_1, S_2, \cdots$ is the set of points that belong to each of the sets $S_1, S_2, \cdots$.)
5. Prove that the intersection of any number of compact sets is compact.

2.6 Continuous functions on compact sets

The definition of maximum value, minimum value, and uniform continuity extend at once to functions of several variables defined on some set.

Theorem J   IF $f(x, y) = f(\mathbf{x})$ IS A CONTINUOUS FUNCTION DEFINED ON A COMPACT
SET S IN THE PLANE, THEN f IS BOUNDED ON S, ATTAINS ON S ITS MAXIMUM AND
MINIMUM VALUES, AND IS UNIFORMLY CONTINUOUS ON S.

This is proved exactly as Theorems E, F, and H. A theorem analogous to J
holds for functions of n variables, for any n.

§3 Differentiation under the Integral Sign

3.1 Statement of the result

We consider a continuous function $f(x, y)$ which has a continuous partial
derivative $f_x(x, y)$ in a closed interval I: $a \leq x \leq b, c \leq y \leq d$. This means that
both the functions $(x, y) \mapsto f(x, y)$ and $(x, y) \mapsto f_x(x, y)$ are continuous in I. We
form the function of one variable

$$(1) \qquad\qquad F(x) = \int_c^d f(x, y)\, dy$$

and claim that

$$(2) \qquad\qquad F'(x) = \int_c^d f_x(x, y)\, dy.$$

This is, except for notations, Theorem 6 in Chapter 12, §2.17.

3.2 The proof

We observe that $f_x(x, y)$ is uniformly continuous in I, by Example 1 in §2.5 and
by Theorem J above. In particular, for every $\epsilon > 0$, there is a number $\delta > 0$ such
that

$$(3) \qquad\qquad |f_x(x + \xi, y) - f_x(x, y)| \leq \epsilon \qquad \text{if } |\xi| \leq \delta$$

Given a point (x, y) in I and a number $h \neq 0$ such that $(x + h, y)$ also lies in I,
there is, by the Mean Value Theorem, a number ξ depending on x, y, and h such
that

$$(4) \qquad\qquad f(x + h, y) = f(x, y) + hf_x(x + \xi, y), \qquad |\xi| < |h|.$$

Therefore

$$\left| \frac{1}{h}\{F(x + h) - F(x)\} - \int_c^d f_x(x, y)\, dy \right|$$

$$= \left| \frac{1}{h}\left\{ \int_c^d f(x + h, y)\, dy - \int_c^d f(x, y)\, dy \right\} - \int_c^d f_x(x, y)\, dy \right|$$

$$= \left| \int_c^d \left\{ \frac{f(x + h, y) - f(x, y)}{h} - f_x(x, y) \right\} dy \right|$$

$$= \left| \int_c^d \{f_x(x + \xi, y) - f_x(x, y)\}\, dy \right|$$

$$\leq \int_c^d |f_x(x + \xi, y) - f_x(x, y)|\, dy \leq (d - c)\epsilon$$

if $|h| \leq \delta$, δ being determined from (3). This means that

$$\lim_{h \to 0} \frac{F(x + h) - F(x)}{h} = \int_c^d F_x(x, y) \, dy,$$

which proves (2).

§4 Riemann Integrals

In the Appendix to Chapter 5 (see §8 of that chapter) we developed the theory of Riemann integrals, making essential use of differentiation. This approach is not suitable for double integrals. Here we give an independent development of Riemann integrals for functions of two variables. Everything we say can be repeated for n variables, including the case $n = 1$.

4.1 Step functions

All functions considered in this section will be defined and bounded in a fixed interval I: $a \leq x \leq b$, $c \leq y \leq d$. All integrals will be taken over I. We described in Chapter 13, §1.7 what is meant by a **subdivision** of I into **subintervals** $I_1, I_2, I_3, \cdots$; such a subdivision is made by drawing several lines parallel to the x axis and several lines parallel to the y axis. We obtain a **refinement** of a subdivision by drawing additional lines—this is illustrated in Fig. S.1. The **mesh** of a subdivision is by definition the largest length of a diagonal of a subinterval.

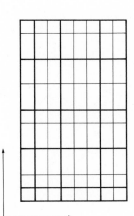

We recall (compare Chapter 13, §1.7) that a step function $\phi(x, y)$ is a function that is constant in each subinterval of a suitable subdivision. The values ϕ takes on the dividing lines are of no importance. If a step function ϕ, belonging to a subdivision of I into N subintervals, takes on the values $z_1, z_2, \cdots, z_N$ in the subintervals $I_1, I_2, \cdots, I_N$, then one *defines* the integral of ϕ (over I) as

(1) $\displaystyle \iint \phi(x, y) \, dx \, dy = z_1 \cdot \text{area of } I_1 + z_2 \cdot \text{area of } I_2 + \cdots + z_N \cdot \text{area of } I_N.$

(There is no need to write I under the integral sign, since I is fixed throughout this section.) We can, if we wish, refine the subdivision and consider ϕ as belonging to the refined subdivision; this does not affect the value of the integral (compare the corresponding statement for functions of one variable in Chapter 5, §8.1).

It is very easy to check that, for step functions ϕ and ψ, we have

(2) $\displaystyle \iint C\phi \, dx \, dy = C \iint \phi \, dx \, dy, \qquad C \text{ a constant,}$

(3) $\displaystyle \iint (\phi + \psi) \, dx \, dy = \iint \phi \, dx \, dy + \iint \psi \, dx \, dy,$

(4) $\displaystyle \iint \phi \, dx \, dy \leq \iint \psi \, dx \, dy \qquad \text{if } \phi \leq \psi;$

compare the corresponding statements for functions of one variable in Chapter 5, §8.1. In this Supplement, "$\phi \leq \psi$" is an abbreviation of the following statement: $\phi(x, y) \leq \psi(x, y)$ for all (x, y) in I. [This also applies to functions other than step functions.]

EXERCISES

1. Prove Statement (2) in §4.1.
► 2. Prove Statement (3) in §4.1.
3. Prove Statement (4) in §4.1.

4.2 Upper and lower sums. Upper and lower integrals

Let ϕ be a step function and $f(x, y)$ any bounded function. If $\phi \geq f$, then the number

$$(5) \qquad S = \iint \phi \, dx \, dy$$

is called an **upper sum** for f (over I). If $\phi \leq f$, then S is called a **lower sum.** The set of upper sums is bounded from below (by $-MA$ where A is the area of I and M a bound for $|f|$). The set of lower sums is bounded from above (by MA). All this follows easily from the monotonicity property (4), compare Chapter 5, §8.2.

The greatest lower bound of all upper sums is called the **upper integral** of f (over I). The least upper bound of all lower sums is called the **lower integral** of f. The upper and lower integrals are denoted by

$$(6) \qquad \overline{\iint} f(x, y) \, dx \, dy \qquad \text{and} \qquad \underline{\iint} f(x, y) \, dx \, dy,$$

respectively. One sees easily that for a step function, the upper and lower integrals are equal to the integral (compare Chapter 5, §8.2).

Lemma A   LET f HAVE (OVER I) THE UPPER AND LOWER INTEGRALS J^* AND J_*, RESPECTIVELY. LET $\epsilon > 0$ BE GIVEN. THEN THERE IS AN UPPER SUM S^* AND A LOWER SUM S_* (FOR f) WITH $J^* \leq S^* \leq J^* + \epsilon$ AND $S_* \leq J_* \leq S_* + \epsilon$.

Proof.   For every upper sum S, we have $S \geq J^*$. If we would have $S > J^* + \epsilon$ for every upper sum, J^* would not have been the greatest lower bound of upper sums. Hence there is an upper sum S^* with $S^* \leq J^* + \epsilon$. The existence of S_* is proved similarly.

4.3 Riemann integrable functions

A (bounded) function $f(x, y)$ is called **Riemann integrable** (over I) if its upper and lower integrals are equal. The common value is called the Riemann integral of f over I and is denoted by

$$(7) \qquad \iint f(x, y) \, dx \, dy.$$

We shall see later that, for piecewise continuous functions, the definition of the integral just given coincides with the one used in Chapter 13, §1.5.

Theorem K   A BOUNDED FUNCTION $f(x, y)$ DEFINED IN AN INTERVAL I IS RIEMANN INTEGRABLE IF AND ONLY IF, FOR EVERY $\epsilon > 0$, THERE IS AN UPPER SUM AND A LOWER SUM (FOR f, OVER I) WHICH DIFFER BY AT MOST ϵ.

Proof. Suppose $J^* = J_*$. Let $\epsilon > 0$ be given. Set $\epsilon_1 = \epsilon/2$. By Lemma A, there is an upper sum S^* and a lower sum S_* with $S^* \leq J^* + \epsilon_1$, $S_* \geq J_* - \epsilon_1$. Hence $S^* - S_* \leq 2\epsilon_1 = \epsilon$.

Suppose next that $J^* \neq J_* > 0$. Then $S^* - S_* \geq J^* - J_* > 0$ for every upper sum S^* and lower sum S_*.

Theorem K shows that the integral (7) is the *only* number that lies between every upper sum and every lower sum.

Using the definition of (7), Properties (2), (3), and (4) of step functions, and Lemma A, it is easy to verify that

(8)
$$\iint Cf \, dx \, dy = C \iint f \, dx \, dy, \qquad C \text{ a constant,}$$

(9)
$$\iint (f + g) \, dx \, dy = \iint f \, dx \, dy + \iint g \, dx \, dy,$$

(10)
$$\iint f \, dx \, dy \leq \iint g \, dx \, dy \qquad \text{if } f \leq g.$$

Here f and g are assumed to be Riemann integrable and it is part of the assertion that Cf and $f + g$ are. We do not carry out the details of the proof, since for piecewise continuous functions one can prove Properties (8), (9), and (10) from the corresponding statements about functions of one variable (see Chapter 13, §1.6).

EXERCISES

In Exercises 4 to 9, f and g are bounded functions defined in the interval I.

▶ 4. If C is a positive constant and S_* and S^* are lower and upper sums for f, show that CS_* and CS^* are lower and upper sums for Cf.

5. Using Exercise 4, show that, if f is integrable over I, so is Cf.

▶ 6. Prove Statement (8) in §4.3.

7. Show that a number is a lower sum (upper sum) for $f + g$ if and only if it can be written as the sum of two lower sums (upper sums) for f and g, respectively.

▶ 8. Prove that, if f and g are integrable, so is $f + g$.

9. Prove Statement (9) in §4.3.

▶ 10. Prove Statement (10) in §4.3.

4.4 Continuous functions

Up to this point, we have followed closely the development sketched in Chapter 5, §8. Now we take a different turn.

Theorem L   A FUNCTION $f(x, y)$ DEFINED AND CONTINUOUS IN A CLOSED INTERVAL IS RIEMANN INTEGRABLE.

Proof. Let $\epsilon > 0$ be given. Let A be the area of I, and set $\epsilon_1 = \epsilon/A$. By Example 1 in §2.5 and by Theorem J in §2.6, the function f is bounded and uniformly continuous. Hence there is a $\delta > 0$ such that: if $|x' - x''| < \delta$, then we have $|f(x') - f(x'')| < \epsilon_1$.

We make a subdivision of I such that the mesh (see §4.1) is less than δ, and we use this subdivision to define two step functions, ϕ and ψ, such that $\phi \leq f \leq \psi$. In every subinterval, we set ϕ equal to the minimum of f, and ψ to the maximum. (The minimum and maximum exist, by Theorem J). Clearly, $\psi - \phi < \epsilon_1$. Hence, using Properties (2), (3), and (4),

$$0 \leq \iint \psi \, dx \, dy - \iint \phi \, dx \, dy = \iint (\psi - \phi) \, dx \, dy \leq \iint \epsilon_1 \, dx \, dy = \epsilon_1 A = \epsilon.$$

We have shown that there are upper and lower sums that differ by less than ϵ. Hence, by Theorem K, f is integrable.

4.5 Covering a curve by a polygon of small area

We want to extend the preceding argument to piecewise continuous functions. This requires a preliminary result.

Lemma B   LET C BE THE GRAPH OF A CONTINUOUS FUNCTION $x \mapsto y = g(x)$, $\alpha \leq x \leq \beta$. LET $\epsilon > 0$ BE A GIVEN NUMBER. ONE CAN FIND FINITELY MANY OPEN TWO-DIMENSIONAL INTERVALS SUCH THAT EVERY POINT OF C LIES IN ONE OF THE INTERVALS, AND THE SUM OF THE AREAS OF THE INTERVALS IS LESS THAN ϵ.

Proof.   Set $\epsilon_1 = \epsilon/[4(\beta - \alpha)]$. By Theorem H in §2.4, the function g is uniformly continuous. Hence there is a number $\delta > 0$ such that: if $|x' - x''| < \delta$, then $|g(x') - g(x'')| < \epsilon_1$. Let N be an integer such that $N\delta > \beta - \alpha$. We divide the interval $[\alpha, \beta]$ into N equal subintervals and denote the endpoints by

$$\alpha = x_0 < x_1 < x_2 < \cdots < x_N = \beta.$$

Each subinterval has length $\delta_1 = (\beta - \alpha)/N < \delta$.

Let I_j denote the closed interval $x_{j-1} \leq x \leq x_{j-1} + \delta_1$, $|g(x_{j-1}) - y| \leq \epsilon_1$. Then I_j has width δ_1, height $2\epsilon_1$ and the midpoint of its left edge is on C above x_{j-1}. Hence the part of C above $[x_{j-1}, x_j]$ lies in I_j, since when x changes by at most δ_1, $y = g(x)$ changes by at most ϵ_1 (see Fig. S.2).

The areas of $I_1, \cdots, I_N$ add up to $2N\delta_1\epsilon_1 = (\beta - \alpha)\epsilon_1 = \epsilon/2 < \epsilon$. Now we put each I_j into an open interval $\hat{I}_j$ whose area exceeds that of I_j by less than $\epsilon/2N$ (see Fig. S.3). The intervals $\hat{I}_1, \cdots, \hat{I}_N$ have the required property.

4.6 Piecewise continuous functions

Lemma C   LET $f(x, y)$ BE PIECEWISE CONTINUOUS IN AN INTERVAL I, AND LET $\epsilon > 0$ BE A GIVEN NUMBER. ONE CAN SUBDIVIDE I INTO SUBINTERVALS, AND DIVIDE ALL SUBINTERVALS INTO TWO CLASSES, SUCH THAT $f(x, y)$ IS CONTINUOUS AT EVERY POINT AND EVERY BOUNDARY POINT OF EVERY INTERVAL OF THE FIRST CLASS, AND THE AREAS OF THE INTERVALS OF THE SECOND CLASS ADD UP TO LESS THAN ϵ.

Proof.   By definition (compare Chapter 13, §1.4), $f(x, y)$ fails to be continuous only on a finite number of sets $C_1, C_2, \cdots, C_r$, each of which is either a curve like the one considered in Lemma B, or a straight segment. (The edges of I are to be included among the C_j, unless f is continuous on these edges.) Set $\epsilon_1 = \epsilon/r$ and

Fig. S.2

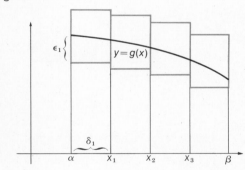

Fig. S.3

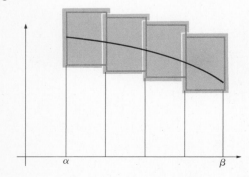

cover each C_j by finitely many open intervals whose areas add up to less than ϵ_1 (see Fig. S.4). If C_j is a curve, this is possible by Lemma B; if C_j is a straight segment, one interval will do.

Now extend each edge of an interval constructed above, and use the vertical and horizontal lines to subdivide I (see Fig. S.5). This gives a subdivision with the required properties, if we assign to the first class all closed subintervals that do not intersect any C_j.

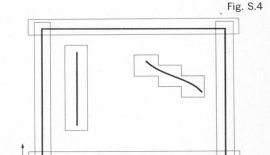

Fig. S.4

Lemma D UNDER THE HYPOTHESIS OF LEMMA C, ONE MAY ALSO REQUIRE THAT IN EVERY CLOSED INTERVAL OF THE FIRST CLASS THE MAXIMUM AND MINIMUM OF f DIFFER BY LESS THAN ϵ. (The maximum and minimum exist by Theorem J in §2.6.)

Proof. First subdivide I as in Lemma C. Denote the set of all points in and on the boundaries of all intervals of the first class by S. This is a compact set (see the examples in §2.5). Hence f is uniformly continuous in S. Therefore there is a $\delta > 0$ such that if $|\mathbf{x'} - \mathbf{x''}| < \delta$, then $|f(\mathbf{x'}) - f(\mathbf{x''})| < \epsilon$.

Now we refine the original subdivision, so as to obtain a subdivision with mesh less than δ. This new subdivision has the required property, if we assign to the first class all those new intervals that are subsets of the old intervals of the first class (see Fig. S.6).

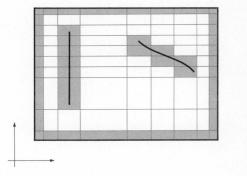

Fig. S.5

Theorem M. A FUNCTION $f(x, y)$ WHICH IS BOUNDED AND PIECEWISE CONTINUOUS IN AN INTERVAL IS RIEMANN INTEGRABLE.

Proof. Let $\epsilon > 0$ be given. We shall construct two step functions, ϕ and ψ, such that

(11) $$\phi \leq f \leq \psi, \qquad \iint \psi \, dx \, dy - \iint \phi \, dx \, dy < \epsilon.$$

By Theorem K, this will prove the assertion.

Let the interval I, where f is defined, have area A, and let M be a bound for $|f|$. Set $\epsilon_1 = \epsilon/(A + 2M)$, and make a subdivision satisfying the conditions of Lemmas C and D, with ϵ replaced by ϵ_1.

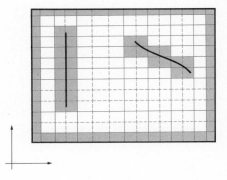

Fig. S.6

In every interval of the first class, we set ϕ and ψ equal to the minimum and maximum of f in that interval, respectively. In every interval of the second class, we set $\phi = -M$, $\psi = M$. Then the first relation (11) holds.

Now we note that $\psi - \phi < \epsilon_1$ in every interval of the first kind, $\psi - \phi = 2M$ in every interval of the second kind. Since the total area of all intervals of the first class is at most A, and the total area of all intervals of the second class is less than ϵ_1, we have

$$\iint \psi \, dx \, dy - \iint \phi \, dx \, dy = \iint (\psi - \phi) \, dx \, dy < \epsilon_1 A + 2M\epsilon_1 = \epsilon$$

as claimed in (11).

EXERCISE

11. Give an example of a bounded nonintegrable function $f(x, y)$ defined for $0 < x < 1, 0 < y < 1$.

§5 Iterated Integrals

We shall join the definition of the double integral given above with that of Chapter 13, §1, and we shall establish some theorems stated there without proof.

5.1 Iterated integrals of continuous functions

Lemma E   LET $f(x, y)$ BE CONTINUOUS IN THE CLOSED INTERVAL FOR I: $a \leq x \leq b$, $c \leq y \leq d$. THEN THE FUNCTION

$$x \rightarrow F(x) = \int_c^d f(x, y)\, dy$$

IS CONTINUOUS FOR $a \leq x \leq b$.

Proof.   Let x_0 be a point in $[a, b]$ and let $\epsilon > 0$ be given. We shall prove that there is a $\delta > 0$ such that if $|x - x_0| < \delta$, then $|F(x) - F(x_0)| < \epsilon$.

Set $\epsilon_1 = \epsilon/(d - c)$. The function f is uniformly continuous in I (by Theorem J) so that there is a $\delta > 0$ such that: if $|\mathbf{x}' - \mathbf{x}''| < \delta$, then $|f(\mathbf{x}') - f(\mathbf{x}'')| < \epsilon_1$. In particular, if $|x - x_0| < \delta$, then $|f(x, y) - f(x_0, y)| < \epsilon_1$ for $c \leq y \leq d$, and therefore

$$|F(x) - F(x_0)| = \left| \int_c^d f(x, y)\, dy - \int_c^d f(x_0, y)\, dy \right| = \left| \int_c^d [f(x, y) - f(x_0, y)]\, dy \right|$$

$$\leq \int_c^d \left| f(x, y) - f(x_0, y) \right|\, dy < \epsilon_1(d - c) = \epsilon$$

as claimed.

The lemma shows that we can form the iterated integral

$$(1) \qquad\qquad \int_a^b \left\{ \int_c^d f(x, y)\, dy \right\} dx.$$

A similar argument shows that we may form the iterated integral

$$(2) \qquad\qquad \int_c^d \left\{ \int_a^b f(x, y)\, dy \right\} dx.$$

We proceed to show that the same is true for bounded piecewise continuous functions.

5.2 Iterated integrals of bounded piecewise continuous functions

We need a preliminary result.

Lemma F   LET $x \mapsto F(x)$ BE A FUNCTION DEFINED NEAR x_0. FOR EVERY $\epsilon > 0$, LET THERE BE TWO FUNCTIONS DEFINED NEAR x_0, $F_1(x)$ AND $F_2(x)$, SUCH THAT $F(x) = F_1(x) + F_2(x)$, $F_1(x)$ IS CONTINUOUS AT x_0 AND $|F_2(x)| < \epsilon$ FOR ALL x FOR WHICH IT IS DEFINED. THEN F IS CONTINUOUS AT x_0.

Proof.   Let $\epsilon > 0$ be given. We shall find a $\delta > 0$ such that if $|x - x_0| < \delta$, then $|F(x) - F(x_0)| < \epsilon$.

Set $\epsilon_1 = \epsilon/3$. We find functions F_1 and F_2, as in the hypothesis of the lemma, but with ϵ replaced by ϵ_1 (so that $|F_2| < \epsilon_1$). Since F_1 is continuous at x_0, there is a $\delta > 0$ such that F_1 and F_2 are defined for $|x - x_0| < \delta$, and if $|x - x_0| < \delta$, then $|F_1(x) - F_1(x_0)| < \epsilon_1$. But in this case,

$$|F(x) - F(x_0)| = |F_1(x) + F_2(x) - F_1(x_0) - F_2(x_0)|$$
$$\leq |F_1(x) - F_1(x_0)| + |F_2(x)| + |F_2(x_0)| < 3\epsilon_1 = \epsilon$$

as claimed.

Theorem N LET $f(x, y)$ BE A BOUNDED PIECEWISE CONTINUOUS FUNCTION DEFINED IN I: $a \leq x \leq b$, $c \leq y \leq d$. THEN THE ITERATED INTEGRALS (1) AND (2) ARE WELL DEFINED.

Proof. We note first that, if we fix the value of the variable x at, say x_0, the function $y \to f(x_0, y)$ fails to be continuous for $c < y < d$ only if the line $x = x_0$ meets one of the curves $C_1, C_2, \cdots, C_n$ on which f is not continuous. Each of these curves is either a graph of a function $x \mapsto g(x)$, $\alpha \leq x \leq \beta$, or a vertical segment. A line $x \mapsto x_0$ intersects a graph of a function in at most one point: Hence if x_0 is not one of the finitely many values that correspond to vertical segments C_j, the function $y \mapsto f(x_0, y)$ is piecewise continuous for $c < y < d$.

Since f is bounded, the function

$$F(x) = \int_c^d f(x, y) \, dy$$

is well defined, for all but finitely many values of x, $a < x < b$. The function F is bounded, because if M is a bound for $|f|$, then

$$|F(x)| \leq \int_c^d |f(x, y)| \, dy \leq M(d - c).$$

It turns out that $F(x)$ is also piecewise continuous. Let x_0 be a point at which $F(x)$ is defined. We shall show that $F(x)$ is continuous at x_0, by verifying the hypothesis of Lemma F.

To simplify matters, assume that the line $x = x_0$ meets only three of the curves C_j: the line $y = c$, the graph of a function $y = g(x)$, $\alpha \leq x \leq \beta$ (with $\alpha < x_0 < \beta$) and the line $y = d$. (See Fig. S.7; if the line $x = x_0$ meets more curves C_j, the argument is the same.) We set $y_0 = g(x_0)$. Let $\epsilon > 0$ be given. We set $\epsilon_1 = \epsilon/5M$. Since $g(x)$ is continuous at x_0, there is a $\delta > 0$ such that $|g(x) - g(x_0)| < \epsilon_1$ if $|x - x_0| \leq \delta$. We may choose δ so small that $g(x)$ is defined for $|x - x_0| \leq \delta$, and that for $|x_1 - x_0| < \delta$, the line $x = x_1$ meets no curves C_j that are not met by $x = x_0$.

Now define, for $|x - x_0| < \delta$,

$$(3) \qquad F_1(x) = \int_{c+\epsilon_1}^{y_0-\epsilon} f(x, y) \, dy + \int_{y_0+\epsilon_1}^{d-\epsilon} f(x, y) \, dy,$$

$$(4) \qquad F_2(x) = \int_c^{c+\epsilon_1} f(x, y) \, dy + \int_{y_0-\epsilon_1}^{y_0+\epsilon_1} f(x, y) \, dy + \int_{d-\epsilon_1}^d f(x, y) \, dy.$$

Since $f(x, y)$ is continuous in the (closed) shaded intervals shown in Fig. S.7, $F_1(x)$ is continuous at x_0, by Lemma E. Since $|f| \leq M$ and the length of the paths

Fig. S.7

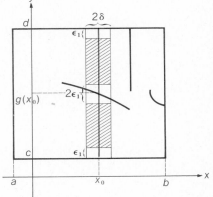

of integration in (4) add up to $4\epsilon_1$, we have $|F_2(x)| \leq 4\epsilon_1 M = 4\epsilon/5 < \epsilon$. Also $F(x) = F_1(x) + F_2(x)$. Hence, by Lemma F, F is continuous at x_0.

It follows that we can form (1); similar reasoning shows that it is legitimate to form (2).

The discussion in this subsection proves Theorem 1 in Chapter 13, §1.5.

5.3 Iterated integrals and double integrals

Theorem O   LET $f(x, y)$ BE BOUNDED AND PIECEWISE CONTINUOUS IN I: $a \leq x \leq b$, $c \leq y \leq d$. THEN

$$(5) \qquad \int_a^b \left\{ \int_c^d f(x, y)\, dy \right\} dx = \iint_I f(x, y)\, dx\, dy = \int_c^d \left\{ \int_a^b f(x, y)\, dx \right\} dy.$$

(*Here the double integral is the Riemann integral defined in §4.3.*)

Thus the two iterated integrals (1) and (2) are equal; this is Theorem 2 in Chapter 13, §1.5. Theorem O also implies that the definition of the integral given in Chapter 13, §1.5 is equivalent (for piecewise continuous functions) to the definition in §4. In view of Theorem K in §4.2, Theorem 3 in Chapter 13, §1.7 is valid.

Proof of Theorem O.   We remark first that the statement of the theorem is true for step functions. This is verified by a completely elementary argument, which the reader can easily supply.

Now let f be bounded and piecewise continuous and let ϕ and ψ be step functions such that

$$\phi(x, y) \leq f(x, y) \leq \psi(x, y).$$

Therefore, by the monotonicity property of ordinary integrals,

$$\int_c^d \phi(x, y)\, dy \leq \int_c^d f(x, y)\, dy \leq \int_c^d \psi(x, y)\, dy, \qquad a < x < b$$

(except for the finitely many values of x for which one of the integrals makes no sense). Hence

$$\int_a^b \left\{ \int_c^d \phi(x, y)\, dy \right\} dx \leq \int_a^b \left\{ \int_c^d f(x, y)\, dy \right\} dx \leq \int_a^b \left\{ \int_c^d \psi(x, y)\, dy \right\} dx.$$

By the remark made above, this is the same as

$$\iint_I \phi\, dx\, dy \leq \int_a^b \left\{ \int_c^d f(x, y)\, dy \right\} dx \leq \iint_I \psi\, dx\, dy.$$

But the only number between every lower and upper sum is the double (Riemann) integral as defined in §4.3. Thus the first equation (5) is verified. The second is proved in exactly the same way.

The arguments of §§3, 4, and 5 can easily be extended to functions of more than two variables.

Problems

1. Find the limit inferior and the limit superior of the following sequences:
   (a) $0, 1, 2, 0, 1, 2, 0, 1, \cdots$.
   (b) $1, 2, 1, 1, 1, 2, 1, 1, 1, 1, 2, 1, \cdots$.
   (c) $1, 1.1, 1, 1.11, 1, 1.111, 1, 1.1111, 1, \cdots$.
2. Find a bounded sequence such that every convergent subsequence converges either to 1, or to 2, or to 3, and there are subsequences converging to these three numbers.
3. Assume that we are given a sequence such that for every integer k there is a subsequence converging to $1/k$. Show that there is a subsequence converging to 0.
4. Find an infinite bounded sequence such that for every number t, $0 \leq t \leq 1$, there is a subsequence converging to t. (This may be a difficult problem.)
5. Suppose that the function $x \mapsto f(x)$ is defined and bounded for $x > A$. For every number $y > A$, denote by $\alpha(y)$ the least upper bound of all numbers $f(x)$, $x \geq y$. Also, denote by $\beta(y)$ the greatest lower bound of all numbers $f(x)$, $x \geq y$. Show that the finite limits

$$\beta = \lim_{y \to +\infty} \beta(y) \quad \text{and} \quad \alpha = \lim_{y \to +\infty} \alpha(y)$$

   exist. We call β and α the **limit inferior** and **limit superior** of f as $x \to +\infty$, and we write

$$\beta = \lim_{x \to +\infty} \inf f(x), \quad \alpha = \lim_{x \to +\infty} \sup f(x).$$

   [Hint: Note that the functions $\alpha(y)$ and $\beta(y)$ are monotone and use Theorem A in Chapter 6, §9.1.]
6. Let $f(x)$ be a bounded function defined for all x near x_0, except perhaps for $x = x_0$. Imitating the procedure of the preceding problem, define

   (*) $\lim_{x \to x_0^-} \inf f(x), \quad \lim_{x \to x_0^+} \inf f(x),$

   (**) $\lim_{x \to x_0^-} \sup f(x), \quad \lim_{x \to x_0^+} \sup f(x).$

   The smaller of the two numbers (*) is called the limit inferior of f at x_0 and is denoted by $\lim \inf_{x \to x_0} f(x)$. The larger of the two numbers (**) is called the limit superior of f at x_0 and is denoted by $\lim \sup_{x \to x_0} f(x)$.
7. Prove that if, under the circumstances described in Problem 6, $\lim \inf_{x \to x_0} f(x) = \lim \sup_{x \to x_0} f(x) = A$, then $\lim_{x \to x_0} f(x) = A$.
8. Assuming Theorem C, that every Cauchy sequence of real numbers converges, prove the least upper bound principle.
9. Prove the following analogue of Theorem C for functions: Let $f(x)$ be defined for all x near x_0, except perhaps for x_0 itself. Assume that for every $\epsilon > 0$ one can find a number $\delta > 0$ such that if $0 < |x' - x_0| < \delta$ and $0 < |x'' - x_0| < \delta$, then $|f(x') - f(x'')| \epsilon$. Then $f(x)$ has a limit at x_0. (This is called the **Cauchy condition** for the existence of limits.)

10. Show that the Cauchy condition is not only sufficient but also necessary for the existence of a limit of a function $f(x)$. (This should be easy.)

11. Let $f_1(x)$, $f_2(x)$, $f_3(x)$, $\cdots$ be a sequence of functions defined in an interval $[a, b]$. The sequence is said to **converge uniformly** in that interval to a function $f(x)$ if for every positive ϵ there is a number N such that $|f_j(x) - f(x)| < \epsilon$ for all $j \geq N$ and all x, $a \leq x \leq b$.

   Show that the sequence $f_j(x) = x^j$, $j = 1, 2, 3, \cdots$, considered in the interval $[0, 1]$, converges to the function $f(x) = 0$ for $x \neq 1$, $f(1) = 1$, but that the convergence is *not* uniform. Show also that the same sequence, considered in an interval $[0, q]$, with $q < 1$, converges uniformly.

12. Show that, if a sequence of continuous functions in an interval converges uniformly, then the limit function is continuous. (This is hard.)

13. Let $\{f_j(x)\}$ be a sequence of continuous functions defined in $[a, b]$. Assume that the sequence converges uniformly to a function $f(x)$ and show that then

$$\lim_{j \to \infty} \int_a^b f_j(x) = \int_a^b f(x)\, dx.$$

14. Assume that $\{f_j(x)\}$ is a sequence of functions defined in $[a, b]$. Assume that, for every $\epsilon > 0$, one can find an integer N such that, if $j \geq N$ and $k \geq N$, then $|f_j(x) - f_k(x)| < \epsilon$ for all x, $a \leq x \leq b$. Prove that the sequence $\{f_j(x)\}$ converges uniformly.

15. State and prove the analogues of the results of Problems 12, 13, and 14 for functions of two variables.

16. A series $\Sigma_{j=0}^{\infty} f_j(x)$ of functions defined in an interval $[a, b]$ is said to converge uniformly if the sequence of partial sums converges uniformly. Show that, if a power series about x_0 has a positive radius of convergence R, and if r is any number such that $0 < r < R$, then the series converges uniformly in the interval $[x_0 - r, x_0 + r]$. (If $R = +\infty$, r can be any positive number.)

17. Let $f(x)$ be the function defined as follows: $f(x) = x$ for $0 \leq x \leq 1$, $f(x) = 2 - x$ for $1 \leq x \leq 2$, $f(x + 2) = f(x)$ for all x. Prove that the series defining the function

$$F(x) = \sum_{n=1}^{\infty} 2^{-n} f(2^n x)$$

converges uniformly in every interval. The sum $F(x)$ is a *continuous function that is not differentiable at any point*. Try to prove this. [Hint: Consider the difference quotients of $F(x)$ at some point x_0. Treat first the case $x_0 = 0$.]

18. If $f(x, y)$ is bounded and Riemann integrable for $a < x < b$, $c < y < d$, then the following statements are true. The functions

$$y \mapsto \underline{\int_a^b} f(x, y)\, dx, \qquad y \mapsto \overline{\int_a^b} f(x, y)\, dx$$

are Riemann integrable for $c < y < d$, the functions

$$x \mapsto \underline{\int_c^d} f(x, y)\, dy, \qquad x \mapsto \overline{\int_c^d} f(x, y)\, dy$$

are Riemann integrable for $a < x < b$, and

$$\iint\limits_{\substack{a<x<b \\ c<y<d}} f(x, y)\, dx\, dy = \int_c^d \left\{ \underline{\int_a^b} f(x, y)\, dx \right\} dy = \int_c^d \left\{ \overline{\int_a^b} f(x, y)\, dx \right\} dy$$

$$= \int_a^b \left\{ \underline{\int_c^d} f(x, y)\, dy \right\} dx = \int_a^b \left\{ \overline{\int_c^d} f(x, y)\, dy \right\} dx.$$

Try to prove these statements. [Hint: Consider step functions $\phi(x, y)$, $\psi(x, y)$ such that $\phi \leq f \leq \psi$. Prove that the functions

$$\Phi(y) = \int_a^b \phi(x, y)\, dx, \qquad \Psi(y) = \int_a^b \psi(x, y)\, dx$$

are step functions, and note that

$$\Phi(y) \leq \underline{\int_a^b} f(x, y)\, dx \leq \overline{\int_a^b} f(x, y)\, dx \leq \Psi(y),$$

except at finitely many points, which do not matter.]

Table 1. Trigonometric functions of numbers 0–1.6

x	$\sin x$	$\tan x$	$\cot x$	$\cos x$	x	$\sin x$	$\tan x$	$\cot x$	$\cos x$
0.00	.00000	.00000	—	1.00000	**0.40**	.38942	.42279	2.3652	.92106
.01	.01000	.01000	99.997	0.99995	.41	.39861	.43463	2.3008	.91712
.02	.02000	.02000	49.993	.99980	.42	.40776	.44657	2.2393	.91309
.03	.03000	.03001	33.323	.99955	.43	.41687	.45862	2.1804	.90897
.04	.03999	.04002	24.987	.99920	.44	.42594	.47078	2.1241	.90475
.05	.04998	.05004	19.983	.99875	.45	.43497	.48306	2.0702	.90045
.06	.05996	.06007	16.647	.99820	.46	.44395	.49545	2.0184	.89605
.07	.06994	.07011	14.262	.99755	.47	.45289	.50797	1.9686	.89157
.08	.07991	.08017	12.473	.99680	.48	.46178	.52061	1.9208	.88699
.09	.08988	.09024	11.081	.99595	.49	.47063	.53339	1.8748	.88233
0.10	.09983	.10033	9.9666	.99500	**0.50**	.47943	.54630	1.8305	.87758
.11	.10978	.11045	9.0542	.99396	.51	.48818	.55936	1.7878	.87274
.12	.11971	.12058	8.2933	.99281	.52	.49688	.57256	1.7465	.86782
.13	.12963	.13074	7.6489	.99156	.53	.50553	.58592	1.7067	.86281
.14	.13954	.14092	7.0961	.99022	.54	.51414	.59943	1.6683	.85771
.15	.14944	.15114	6.6166	.98877	.55	.52269	.61311	1.6310	.85252
.16	.15932	.16138	6.1966	.98723	.56	.53119	.62695	1.5950	.84726
.17	.16918	.17166	5.8256	.98558	.57	.53963	.64097	1.5601	.84190
.18	.17903	.18197	5.4954	.98384	.58	.54802	.65517	1.5263	.83646
.19	.18886	.19232	5.1997	.98200	.59	.55636	.66956	1.4935	.83094
0.20	.19867	.20271	4.9332	.98007	**0.60**	.56464	.68414	1.4617	.82534
.21	.20846	.21314	4.6917	.97803	.61	.57287	.69892	1.4308	.81965
.22	.21823	.22362	4.4719	.97590	.62	.58104	.71391	1.4007	.81388
.23	.22798	.23414	4.2709	.97367	.63	.58914	.72911	1.3715	.80803
.24	.23770	.24472	4.0864	.97134	.64	.59720	.74454	1.3431	.80210
.25	.24740	.25534	3.9163	.96891	.65	.60519	.76020	1.3154	.79608
.26	.25708	.26602	3.7591	.96639	.66	.61312	.77610	1.2885	.78999
.27	.26673	.27676	3.6133	.96377	.67	.62099	.79225	1.2622	.78382
.28	.27636	.28755	3.4776	.96106	.68	.62879	.80866	1.2366	.77757
.29	.28595	.29841	3.3511	.95824	.69	.63654	.82534	1.2116	.77125
0.30	.29552	.30934	3.2327	.95534	**0.70**	.64422	.84229	1.1872	.76484
.31	.30506	.32033	3.1218	.95233	.71	.65183	.85953	1.1634	.75836
.32	.31457	.33139	3.0176	.94924	.72	.65938	.87707	1.1402	.75181
.33	.32404	.34252	2.9195	.94604	.73	.66687	.89492	1.1174	.74517
.34	.33349	.35374	2.8270	.94275	.74	.67429	.91309	1.0952	.73847
.35	.34290	.36503	2.7395	.93937	.75	.68164	.93160	1.0734	.73169
.36	.35227	.37640	2.6567	.93590	.76	.68892	.95045	1.0521	.72484
.37	.36162	.38786	2.5782	.93233	.77	.69614	.96967	1.0313	.71791
.38	.37092	.39941	2.5037	.92866	.78	.70328	.98926	1.0109	.71091
.39	.38019	.41105	2.4328	.92491	.79	.71035	1.0092	.99084	.70385
x	$\sin x$	$\tan x$	$\cot x$	$\cos x$	x	$\sin x$	$\tan x$	$\cot x$	$\cos x$

Table 1, *cont.* Trigonometric functions of numbers 0–1.6

x	sin x	tan x	cot x	cos x		x	sin x	tan x	cot x	cos x
0.80	.71736	1.0296	.97121	.69671		**1.20**	.93204	2.5722	.38878	.36236
.81	.72429	1.0505	.95197	.68950		1.21	.93562	2.6503	.37731	.35302
.82	.73115	1.0717	.93309	.68222		1.22	.93910	2.7328	.36593	.34365
.83	.73793	1.0934	.91455	.67488		1.23	.94249	2.8198	.35463	.33424
.84	.74464	1.1156	.89635	.66746		1.24	.94578	2.9119	.34341	.32480
.85	.75128	1.1383	.87848	.65998		1.25	.94898	3.0096	.33227	.31532
.86	.75784	1.1616	.86091	.65244		1.26	.95209	3.1133	.32121	.30582
.87	.76433	1.1853	.84365	.64483		1.27	.95510	3.2236	.31021	.29628
.88	.77074	1.2097	.82668	.63715		1.28	.95802	3.3413	.29928	.28672
.89	.77707	1.2346	.80998	.62941		1.29	.96084	3.4672	.28842	.27712
0.90	.78333	1.2602	.79355	.62161		**1.30**	.96356	3.6021	.27762	.26750
.91	.78950	1.2864	.77738	.61375		1.31	.96618	3.7471	.26687	.25785
.92	.79560	1.3133	.76146	.60582		1.32	.96872	3.9033	.25619	.24818
.93	.80162	1.3409	.74578	.59783		1.33	.97115	4.0723	.24556	.23848
.94	.80756	1.3692	.73034	.58979		1.34	.97348	4.2556	.23498	.22875
.95	.81342	1.3984	.71511	.58168		1.35	.97572	4.4552	.22446	.21901
.96	.81919	1.4284	.70010	.57352		1.36	.97786	4.6734	.21398	.20924
.97	.82489	1.4592	.68531	.56530		1.37	.97991	4.9131	.20354	.19945
.98	.83050	1.4910	.67071	.55702		1.38	.98185	5.1774	.19315	.18964
.99	.83603	1.5237	.65631	.54869		1.39	.98370	5.4707	.18279	.17981
1.00	.84147	1.5574	.64209	.54030		**1.40**	.98545	5.7979	.17248	.16997
1.01	.84683	1.5922	.62806	.53186		1.41	.98710	6.1654	.16220	.16010
1.02	.85211	1.6281	.61420	.52337		1.42	.98865	6.5811	.15195	.15023
1.03	.85730	1.6652	.60051	.51482		1.43	.99010	7.0555	.14173	.14033
1.04	.86240	1.7036	.58699	.50622		1.44	.99146	7.6018	.13155	.13042
1.05	.86742	1.7433	.57362	.49757		1.45	.99271	8.2381	.12139	.12050
1.06	.87236	1.7844	.56040	.48887		1.46	.99387	8.9886	.11125	.11057
1.07	.87720	1.8270	.54734	.48012		1.47	.99492	9.8874	.10114	.10063
1.08	.88196	1.8712	.53441	.47133		1.48	.99588	10.983	.09105	.09067
1.09	.88663	1.9171	.52162	.46249		1.49	.99674	12.350	.08097	.08071
1.10	.89121	1.9648	.50897	.45360		**1.50**	.99749	14.101	.07091	.07074
1.11	.89570	2.0143	.49644	.44466		1.51	.99815	16.428	.06087	.06076
1.12	.90010	2.0660	.48404	.43568		1.52	.99871	19.670	.05084	.05077
1.13	.90441	2.1198	.47175	.42666		1.53	.99917	24.498	.04082	.04079
1.14	.90863	2.1759	.45959	.41759		1.54	.99953	32.461	.03081	.03079
1.15	.91276	2.2345	.44753	.40849		1.55	.99978	48.078	.02080	.02079
1.16	.91680	2.2958	.43558	.39934		1.56	.99994	92.621	.01080	.01080
1.17	.92075	2.3600	.42373	.39015		1.57	1.00000	1255.8	.00080	.00080
1.18	.92461	2.4273	.41199	.38092		1.58	.99996	−108.65	−.00920	−.00920
1.19	.92837	2.4979	.40034	.37166		1.59	.99982	−52.067	−.01921	−.01920
						1.60	.99957	−34.233	−.02921	−.02920
x	sin x	tan x	cot x	cos x						

Table 2. Trigonometric functions. Arguments in degrees

Degrees	Radians	Sine	Tangent	Cotangent	Cosine		
0	o	o	o	——	1.0000	1.5708	**90**
1	.0175	.0175	.0175	57.290	.9998	1.5533	89
2	.0349	.0349	.0349	28.636	.9994	1.5359	88
3	.0524	.0523	.0524	19.081	.9986	1.5184	87
4	.0698	.0698	.0699	14.301	.9976	1.5010	86
5	.0873	.0872	.0875	11.430	.9962	1.4835	**85**
6	.1047	.1045	.1051	9.5144	.9945	1.4661	84
7	.1222	.1219	.1228	8.1443	.9925	1.4486	83
8	.1396	.1392	.1405	7.1154	.9903	1.4312	82
9	.1571	.1564	.1584	6.3138	.9877	1.4137	81
10	.1745	.1736	.1763	5.6713	.9848	1.3963	**80**
11	.1920	.1908	.1944	5.1446	.9816	1.3788	79
12	.2094	.2079	.2126	4.7046	.9781	1.3614	78
13	.2269	.2250	.2309	4.3315	.9744	1.3439	77
14	.2443	.2419	.2493	4.0108	.9703	1.3265	76
15	.2618	.2588	.2679	3.7321	.9659	1.3090	**75**
16	.2793	.2756	.2867	3.4874	.9613	1.2915	74
17	.2967	.2924	.3057	3.2709	.9563	1.2741	73
18	.3142	.3090	.3249	3.0777	.9511	1.2566	72
19	.3316	.3256	.3443	2.9042	.9455	1.2392	71
20	.3491	.3420	.3640	2.7475	.9397	1.2217	**70**
21	.3665	.3584	.3839	2.6051	.9336	1.2043	69
22	.3840	.3746	.4040	2.4751	.9272	1.1868	68
23	.4014	.3907	.4245	2.3559	.9205	1.1694	67
24	.4189	.4067	.4452	2.2460	.9135	1.1519	66
25	.4363	.4226	.4663	2.1445	.9063	1.1345	**65**
26	.4538	.4384	.4877	2.0503	.8988	1.1170	64
27	.4712	.4540	.5095	1.9626	.8910	1.0996	63
28	.4887	.4695	.5317	1.8807	.8829	1.0821	62
29	.5061	.4848	.5543	1.8040	.8746	1.0647	61
30	.5236	.5000	.5774	1.7321	.8660	1.0472	**60**
31	.5411	.5150	.6009	1.6643	.8572	1.0297	59
32	.5585	.5299	.6249	1.6003	.8480	1.0123	58
33	.5760	.5446	.6494	1.5399	.8387	.9948	57
34	.5934	.5592	.6745	1.4826	.8290	.9774	56
35	.6109	.5736	.7002	1.4281	.8192	.9599	**55**
36	.6283	.5878	.7265	1.3764	.8090	.9425	54
37	.6458	.6018	.7536	1.3270	.7986	.9250	53
38	.6632	.6157	.7813	1.2799	.7880	.9076	52
39	.6807	.6293	.8098	1.2349	.7771	.8901	51
40	.6981	.6428	.8391	1.1918	.7660	.8727	**50**
41	.7156	.6561	.8693	1.1504	.7547	.8552	49
42	.7330	.6691	.9004	1.1106	.7431	.8378	48
43	.7505	.6820	.9325	1.0724	.7314	.8203	47
44	.7679	.6947	.9657	1.0355	.7193	.8029	46
45	.7854	.7071	1.0000	1.0000	.7071	.7854	**45**
		Cosine	Cotangent	Tangent	Sine	Radians	Degrees

Table 3. Natural logarithms of numbers 1.00–10.09

Logarithms of numbers outside this range can be computed by using the functional equation
(log ab = log a + log b) and the following values of the logarithmic function: log .1 =
.6974–3;   log .01 = .3948–5;   log .001 = .0922–7;   log .0001 = .7897–10;   log .00001 =
.4871–12;   log .000 001 = .1845–14.

N	.00	.01	.02	.03	.04	.05	.06	.07	.08	.09
1.0	0.0000	0.0100	0.0198	0.0296	0.0392	0.0488	0.0583	0.0677	0.0770	0.0862
1.1	0.0953	0.1044	0.1133	0.1222	0.1310	0.1398	0.1484	0.1570	0.1655	0.1740
1.2	0.1823	0.1906	0.1989	0.2070	0.2151	0.2231	0.2311	0.2390	0.2469	0.2546
1.3	0.2624	0.2700	0.2776	0.2852	0.2927	0.3001	0.3075	0.3148	0.3221	0.3293
1.4	0.3365	0.3436	0.3507	0.3577	0.3646	0.3716	0.3784	0.3853	0.3920	0.3988
1.5	0.4055	0.4121	0.4187	0.4253	0.4318	0.4383	0.4447	0.4511	0.4574	0.4637
1.6	0.4700	0.4762	0.4824	0.4886	0.4947	0.5008	0.5068	0.5128	0.5188	0.5247
1.7	0.5306	0.5365	0.5423	0.5481	0.5539	0.5596	0.5653	0.5710	0.5766	0.5822
1.8	0.5878	0.5933	0.5988	0.6043	0.6098	0.6152	0.6206	0.6259	0.6313	0.6366
1.9	0.6419	0.6471	0.6523	0.6575	0.6627	0.6678	0.6729	0.6780	0.6831	0.6881
2.0	0.6931	0.6981	0.7031	0.7080	0.7129	0.7178	0.7227	0.7275	0.7324	0.7372
2.1	0.7419	0.7467	0.7514	0.7561	0.7608	0.7655	0.7701	0.7747	0.7793	0.7839
2.2	0.7885	0.7930	0.7975	0.8020	0.8065	0.8109	0.8154	0.8198	0.8242	0.8286
2.3	0.8329	0.8372	0.8416	0.8459	0.8502	0.8544	0.8587	0.8629	0.8671	0.8713
2.4	0.8755	0.8796	0.8838	0.8879	0.8920	0.8961	0.9002	0.9042	0.9083	0.9123
2.5	0.9163	0.9203	0.9243	0.9282	0.9322	0.9361	0.9400	0.9439	0.9478	0.9517
2.6	0.9555	0.9594	0.9632	0.9670	0.9708	0.9746	0.9783	0.9821	0.9858	0.9895
2.7	0.9933	0.9969	1.0006	1.0043	1.0080	1.0116	1.0152	1.0188	1.0225	1.0260
2.8	1.0296	1.0332	1.0367	1.0403	1.0438	1.0473	1.0508	1.0543	1.0578	1.0613
2.9	1.0647	1.0682	1.0716	1.0750	1.0784	1.0818	1.0852	1.0886	1.0919	1.0953
3.0	1.0986	1.1019	1.1053	1.1086	1.1119	1.1151	1.1184	1.1217	1.1249	1.1282
3.1	1.1314	1.1346	1.1378	1.1410	1.1442	1.1474	1.1506	1.1537	1.1569	1.1600
3.2	1.1632	1.1663	1.1694	1.1725	1.1756	1.1787	1.1817	1.1848	1.1878	1.1909
3.3	1.1939	1.1969	1.2000	1.2030	1.2060	1.2090	1.2119	1.2149	1.2179	1.2208
3.4	1.2238	1.2267	1.2296	1.2326	1.2355	1.2384	1.2413	1.2442	1.2470	1.2499
3.5	1.2528	1.2556	1.2585	1.2613	1.2641	1.2669	1.2698	1.2726	1.2754	1.2782
3.6	1.2809	1.2837	1.2865	1.2892	1.2920	1.2947	1.2975	1.3002	1.3029	1.3056
3.7	1.3083	1.3110	1.3137	1.3164	1.3191	1.3218	1.3244	1.3271	1.3297	1.3324
3.8	1.3350	1.3376	1.3403	1.3429	1.3455	1.3481	1.3507	1.3533	1.3558	1.3584
3.9	1.3610	1.3635	1.3661	1.3686	1.3712	1.3737	1.3762	1.3788	1.3813	1.3838
4.0	1.3863	1.3888	1.3913	1.3938	1.3962	1.3987	1.4012	1.4036	1.4061	1.4085
4.1	1.4110	1.4134	1.4159	1.4183	1.4207	1.4231	1.4255	1.4279	1.4303	1.4327
4.2	1.4351	1.4375	1.4398	1.4422	1.4446	1.4469	1.4493	1.4516	1.4540	1.4563
4.3	1.4586	1.4609	1.4633	1.4656	1.4679	1.4702	1.4725	1.4748	1.4770	1.4793
4.4	1.4816	1.4839	1.4861	1.4884	1.4907	1.4929	1.4951	1.4974	1.4996	1.5019
4.5	1.5041	1.5063	1.5085	1.5107	1.5129	1.5151	1.5173	1.5195	1.5217	1.5239
4.6	1.5261	1.5282	1.5304	1.5326	1.5347	1.5369	1.5390	1.5412	1.5433	1.5454
4.7	1.5476	1.5497	1.5518	1.5539	1.5560	1.5581	1.5602	1.5623	1.5644	1.5665
4.8	1.5686	1.5707	1.5728	1.5748	1.5769	1.5790	1.5810	1.5831	1.5851	1.5872
4.9	1.5892	1.5913	1.5933	1.5953	1.5974	1.5994	1.6014	1.6034	1.6054	1.6074
5.0	1.6094	1.6114	1.6134	1.6154	1.6174	1.6194	1.6214	1.6233	1.6253	1.6273
5.1	1.6292	1.6312	1.6332	1.6351	1.6371	1.6390	1.6409	1.6429	1.6448	1.6467
5.2	1.6487	1.6506	1.6525	1.6544	1.6563	1.6582	1.6601	1.6620	1.6639	1.6658
5.3	1.6677	1.6696	1.6715	1.6734	1.6752	1.6771	1.6790	1.6808	1.6827	1.6845
5.4	1.6864	1.6882	1.6901	1.6919	1.6938	1.6956	1.6974	1.6993	1.7011	1.7029
5.5	1.7047	1.7066	1.7084	1.7102	1.7120	1.7138	1.7156	1.7174	1.7192	1.7210
N	.00	.01	.02	.03	.04	.05	.06	.07	.08	.09

Table 3 T5

Table 3, *cont.* Natural logarithms of numbers 1.00–10.09

N	.00	.01	.02	.03	.04	.05	.06	.07	.08	.09
5.5	1.7047	1.7066	1.7084	1.7102	1.7120	1.7138	1.7156	1.7174	1.7192	1.7210
5.6	1.7228	1.7246	1.7263	1.7281	1.7299	1.7317	1.7334	1.7352	1.7370	1.7387
5.7	1.7405	1.7422	1.7440	1.7457	1.7475	1.7492	1.7509	1.7527	1.7544	1.7561
5.8	1.7579	1.7596	1.7613	1.7630	1.7647	1.7664	1.7681	1.7699	1.7716	1.7733
5.9	1.7750	1.7766	1.7783	1.7800	1.7817	1.7834	1.7851	1.7867	1.7884	1.7901
6.0	1.7918	1.7934	1.7951	1.7967	1.7984	1.8001	1.8017	1.8034	1.8050	1.8066
6.1	1.8083	1.8099	1.8116	1.8132	1.8148	1.8165	1.8181	1.8197	1.8213	1.8229
6.2	1.8245	1.8262	1.8278	1.8294	1.8310	1.8326	1.8342	1.8358	1.8374	1.8390
6.3	1.8405	1.8421	1.8437	1.8453	1.8469	1.8485	1.8500	1.8516	1.8532	1.8547
6.4	1.8563	1.8579	1.8594	1.8610	1.8625	1.8641	1.8656	1.8672	1.8687	1.8703
6.5	1.8718	1.8733	1.8749	1.8764	1.8779	1.8795	1.8810	1.8825	1.8840	1.8856
6.6	1.8871	1.8886	1.8901	1.8916	1.8931	1.8946	1.8961	1.8976	1.8991	1.9006
6.7	1.9021	1.9036	1.9051	1.9066	1.9081	1.9095	1.9110	1.9125	1.9140	1.9155
6.8	1.9169	1.9184	1.9199	1.9213	1.9228	1.9242	1.9257	1.9272	1.9286	1.9301
6.9	1.9315	1.9330	1.9344	1.9359	1.9373	1.9387	1.9402	1.9416	1.9430	1.9445
7.0	1.9459	1.9473	1.9488	1.9502	1.9516	1.9530	1.9544	1.9559	1.9573	1.9587
7.1	1.9601	1.9615	1.9629	1.9643	1.9657	1.9671	1.9685	1.9699	1.9713	1.9727
7.2	1.9741	1.9755	1.9769	1.9782	1.9796	1.9810	1.9824	1.9838	1.9851	1.9865
7.3	1.9879	1.9892	1.9906	1.9920	1.9933	1.9947	1.9961	1.9974	1.9988	2.0001
7.4	2.0015	2.0028	2.0042	2.0055	2.0069	2.0082	2.0096	2.0109	2.0122	2.0136
7.5	2.0149	2.0162	2.0176	2.0189	2.0202	2.0215	2.0229	2.0242	2.0255	2.0268
7.6	2.0281	2.0295	2.0308	2.0321	2.0334	2.0347	2.0360	2.0373	2.0386	2.0399
7.7	2.0412	2.0425	2.0438	2.0451	2.0464	2.0477	2.0490	2.0503	2.0516	2.0528
7.8	2.0541	2.0554	2.0567	2.0580	2.0592	2.0605	2.0618	2.0631	2.0643	2.0656
7.9	2.0669	2.0681	2.0694	2.0707	2.0719	2.0732	2.0744	2.0757	2.0769	2.0782
8.0	2.0794	2.0807	2.0819	2.0832	2.0844	2.0857	2.0869	2.0882	2.0894	2.0906
8.1	2.0919	2.0931	2.0943	2.0956	2.0968	2.0980	2.0992	2.1005	2.1017	2.1029
8.2	2.1041	2.1054	2.1066	2.1078	2.1090	2.1102	2.1114	2.1126	2.1138	2.1150
8.3	2.1163	2.1175	2.1187	2.1199	2.1211	2.1223	2.1235	2.1247	2.1258	2.1270
8.4	2.1282	2.1294	2.1306	2.1318	2.1330	2.1342	2.1353	2.1365	2.1377	2.1389
8.5	2.1401	2.1412	2.1424	2.1436	2.1448	2.1459	2.1471	2.1483	2.1494	2.1506
8.6	2.1518	2.1529	2.1541	2.1552	2.1564	2.1576	2.1587	2.1599	2.1610	2.1622
8.7	2.1633	2.1645	2.1656	2.1668	2.1679	2.1691	2.1702	2.1713	2.1725	2.1736
8.8	2.1748	2.1759	2.1770	2.1782	2.1793	2.1804	2.1815	2.1827	2.1838	2.1849
8.9	2.1861	2.1872	2.1883	2.1894	2.1905	2.1917	2.1928	2.1939	2.1950	2.1961
9.0	2.1972	2.1983	2.1994	2.2006	2.2017	2.2028	2.2039	2.2050	2.2061	2.2072
9.1	2.2083	2.2094	2.2105	2.2116	2.2127	2.2138	2.2148	2.2159	2.2170	2.2181
9.2	2.2192	2.2203	2.2214	2.2225	2.2235	2.2246	2.2257	2.2268	2.2279	2.2289
9.3	2.2300	2.2311	2.2322	2.2332	2.2343	2.2354	2.2364	2.2375	2.2386	2.2396
9.4	2.2407	2.2418	2.2428	2.2439	2.2450	2.2460	2.2471	2.2481	2.2492	2.2502
9.5	2.2513	2.2523	2.2534	2.2544	2.2555	2.2565	2.2576	2.2586	2.2597	2.2607
9.6	2.2618	2.2628	2.2638	2.2649	2.2659	2.2670	2.2680	2.2690	2.2701	2.2711
9.7	2.2721	2.2732	2.2742	2.2752	2.2762	2.2773	2.2783	2.2793	2.2803	2.2814
9.8	2.2824	2.2834	2.2844	2.2854	2.2865	2.2875	2.2885	2.2895	2.2905	2.2915
9.9	2.2925	2.2935	2.2946	2.2956	2.2966	2.2976	2.2986	2.2996	2.3006	2.3016
10.0	2.3026	2.3036	2.3046	2.3056	2.3066	2.3076	2.3086	2.3096	2.3106	2.3115
N	.00	.01	.02	.03	.04	.05	.06	.07	.08	.09

Table 4. Mantissas of common logarithms of numbers 1000–9999

N	0	1	2	3	4	5	6	7	8	9
10	0000	0043	0086	0128	0170	0212	0253	0294	0334	0374
11	0414	0453	0492	0531	0569	0607	0645	0682	0719	0755
12	0792	0828	0864	0899	0934	0969	1004	1038	1072	1106
13	1139	1173	1206	1239	1271	1303	1335	1367	1399	1430
14	1461	1492	1523	1553	1584	1614	1644	1673	1703	1732
15	1761	1790	1818	1847	1875	1903	1931	1959	1987	2014
16	2041	2068	2095	2122	2148	2175	2201	2227	2253	2279
17	2304	2330	2355	2380	2405	2430	2455	2480	2504	2529
18	2553	2577	2601	2625	2648	2672	2695	2718	2742	2765
19	2788	2810	2833	2856	2878	2900	2923	2945	2967	2989
20	3010	3032	3054	3075	3096	3118	3139	3160	3181	3201
21	3222	3243	3263	3284	3304	3324	3345	3365	3385	3404
22	3424	3444	3464	3483	3502	3522	3541	3560	3579	3598
23	3617	3636	3655	3674	3692	3711	3729	3747	3766	3784
24	3802	3820	3838	3856	3874	3892	3909	3927	3945	3962
25	3979	3997	4014	4031	4048	4065	4082	4099	4116	4133
26	4150	4166	4183	4200	4216	4232	4249	4265	4281	4298
27	4314	4330	4346	4362	4378	4393	4409	4425	4440	4456
28	4472	4487	4502	4518	4533	4548	4564	4579	4594	4609
29	4624	4639	4654	4669	4683	4698	4713	4728	4742	4757
30	4771	4786	4800	4814	4829	4843	4857	4871	4886	4900
31	4914	4928	4942	4955	4969	4983	4997	5011	5024	5038
32	5051	5065	5079	5092	5105	5119	5132	5145	5159	5172
33	5185	5198	5211	5224	5237	5250	5263	5276	5289	5302
34	5315	5328	5340	5353	5366	5378	5391	5403	5416	5428
35	5441	5453	5465	5478	5490	5502	5514	5527	5539	5551
36	5563	5575	5587	5599	5611	5623	5635	5647	5658	5670
37	5682	5694	5705	5717	5729	5740	5752	5763	5775	5786
38	5798	5809	5821	5832	5843	5855	5866	5877	5888	5899
39	5911	5922	5933	5944	5955	5966	5977	5988	5999	6010
40	6021	6031	6042	6053	6064	6075	6085	6096	6107	6117
41	6128	6138	6149	6160	6170	6180	6191	6201	6212	6222
42	6232	6243	6253	6263	6274	6284	6294	6304	6314	6325
43	6335	6345	6355	6365	6375	6385	6395	6405	6415	6425
44	6435	6444	6454	6464	6474	6484	6493	6503	6513	6522
45	6532	6542	6551	6561	6571	6580	6590	6599	6609	6618
46	6628	6637	6646	6656	6665	6675	6684	6693	6702	6712
47	6721	6730	6739	6749	6758	6767	6776	6785	6794	6803
48	6812	6821	6830	6839	6848	6857	6866	6875	6884	6893
49	6902	6911	6920	6928	6937	6946	6955	6964	6972	6981
50	6990	6998	7007	7016	7024	7033	7042	7050	7059	7067
51	7076	7084	7093	7101	7110	7118	7126	7135	7143	7152
52	7160	7168	7177	7185	7193	7202	7210	7218	7226	7235
53	7243	7251	7259	7267	7275	7284	7292	7300	7308	7316
54	7324	7332	7340	7348	7356	7364	7372	7380	7388	7396

Table 4, *cont.* Mantissas of common logarithms of numbers 1000–9999

N	0	1	2	3	4	5	6	7	8	9
55	7404	7412	7419	7427	7435	7443	7451	7459	7466	7474
56	7482	7490	7497	7505	7513	7520	7528	7536	7543	7551
57	7559	7566	7574	7582	7589	7597	7604	7612	7619	7627
58	7634	7642	7649	7657	7664	7672	7679	7686	7694	7701
59	7709	7716	7723	7731	7738	7745	7752	7760	7767	7774
60	7782	7789	7796	7803	7810	7818	7825	7832	7839	7846
61	7853	7860	7868	7875	7882	7889	7896	7903	7910	7917
62	7924	7931	7938	7945	7952	7959	7966	7973	7980	7987
63	7993	8000	8007	8014	8021	8028	8035	8041	8048	8055
64	8062	8069	8075	8082	8089	8096	8102	8109	8116	8122
65	8129	8136	8142	8149	8156	8162	8169	8176	8182	8189
66	8195	8202	8209	8215	8222	8228	8235	8241	8248	8254
67	8261	8267	8274	8280	8287	8293	8299	8306	8312	8319
68	8325	8331	8338	8344	8351	8357	8363	8370	8376	8382
69	8388	8395	8401	8407	8414	8420	8426	8432	8439	8445
70	8451	8457	8463	8470	8476	8482	8488	8494	8500	8506
71	8513	8519	8525	8531	8537	8543	8549	8555	8561	8567
72	8573	8579	8585	8591	8597	8603	8609	8615	8621	8627
73	8633	8639	8645	8651	8657	8663	8669	8675	8681	8686
74	8692	8698	8704	8710	8716	8722	8727	8733	8739	8745
75	8751	8756	8762	8768	8774	8779	8785	8791	8797	8802
76	8808	8814	8820	8825	8831	8837	8842	8848	8854	8859
77	8865	8871	8876	8882	8887	8893	8899	8904	8910	8915
78	8921	8927	8932	8938	8943	8949	8954	8960	8965	8971
79	8976	8982	8987	8993	8998	9004	9009	9015	9020	9025
80	9031	9036	9042	9047	9053	9058	9063	9069	9074	9079
81	9085	9090	9096	9101	9106	9112	9117	9122	9128	9133
82	9138	9143	9149	9154	9159	9165	9170	9175	9180	9186
83	9191	9196	9201	9206	9212	9217	9222	9227	9232	9238
84	9243	9248	9253	9258	9263	9269	9274	9279	9284	9289
85	9294	9299	9304	9309	9315	9320	9325	9330	9335	9340
86	9345	9350	9355	9360	9365	9370	9375	9380	9385	9390
87	9395	9400	9405	9410	9415	9420	9425	9430	9435	9440
88	9445	9450	9455	9460	9465	9469	9474	9479	9484	9489
89	9494	9499	9504	9509	9513	9518	9523	9528	9533	9538
90	9542	9547	9552	9557	9562	9566	9571	9576	9581	9586
91	9590	9595	9600	9605	9609	9614	9619	9624	9628	9633
92	9638	9643	9647	9652	9657	9661	9666	9671	9675	9680
93	9685	9689	9694	9699	9703	9708	9713	9717	9722	9727
94	9731	9736	9741	9745	9750	9754	9759	9763	9768	9773
95	9777	9782	9786	9791	9795	9800	9805	9809	9814	9818
96	9823	9827	9832	9836	9841	9845	9850	9854	9859	9863
97	9868	9872	9877	9881	9886	9890	9894	9899	9903	9908
98	9912	9917	9921	9926	9930	9934	9939	9943	9948	9952
99	9956	9961	9965	9969	9974	9978	9983	9987	9991	9996

Table 5. The exponential function and its reciprocal in the range 0–6

x	e^x	e^{-x}	x	e^x	e^{-x}
0.0	1.0000	1.0000	**3.0**	20.086	.04979
0.1	1.1052	.90484	3.1	22.198	.04505
0.2	1.2214	.81873	3.2	24.533	.04076
0.3	1.3499	.74082	3.3	27.113	.03688
0.4	1.4918	.67032	3.4	29.964	.03337
0.5	1.6487	.60653	3.5	33.115	.03020
0.6	1.8221	.54881	3.6	36.598	.02732
0.7	2.0138	.49659	3.7	40.447	.02472
0.8	2.2255	.44933	3.8	44.701	.02237
0.9	2.4596	.40657	3.9	49.402	.02024
1.0	2.7183	.36788	**4.0**	54.598	.01832
1.1	3.0042	.33287	4.1	60.340	.01657
1.2	3.3201	.30119	4.2	66.686	.01500
1.3	3.6693	.27253	4.3	73.700	.01357
1.4	4.0552	.24660	4.4	81.451	.01228
1.5	4.4817	.22313	4.5	90.017	.01111
1.6	4.9530	.20190	4.6	99.484	.01005
1.7	5.4739	.18268	4.7	109.95	.00910
1.8	6.0496	.16530	4.8	121.51	.00823
1.9	6.6859	.14957	4.9	134.29	.00745
2.0	7.3891	.13534	**5.0**	148.41	.00674
2.1	8.1662	.12246	5.1	164.02	.00610
2.2	9.0250	.11080	5.2	181.27	.00552
2.3	9.9742	.10026	5.3	200.34	.00499
2.4	11.023	.09072	5.4	221.41	.00452
2.5	12.182	.08208	5.5	244.69	.00409
2.6	13.464	.07427	5.6	270.43	.00370
2.7	14.880	.06721	5.7	298.87	.00335
2.8	16.445	.06081	5.8	330.30	.00303
2.9	18.174	.05502	5.9	365.04	.00274
3.0	20.086	.04979	**6.0**	403.43	.00248

Table 6. The Greek alphabet

Letters		Names	Letters		Names
A	α	alpha	N	ν	nu
B	β	beta	Ξ	ξ	xi
Γ	γ	gamma	O	o	omicron
Δ	δ	delta	Π	π	pi
E	ϵ	epsilon	P	ρ	rho
Z	ζ	zeta	Σ	σ	sigma
H	η	eta	T	τ	tau
Θ	θ	theta	Υ	υ	upsilon
I	ι	iota	Φ	ϕ	phi
K	κ	kappa	X	χ	chi
Λ	λ	lambda	Ψ	ψ	psi
M	μ	mu	Ω	ω	omega

ANSWERS to even numbered unflagged exercises———

Chapter 1

§2.3

2. Postulate **4**.
4. Postulate **2, 4**.
8. Add 76345 to each side of the equation. Use Postulates **3, 4, 5** (not necessarily in that order) to get a contradiction to Postulate **11**.
10. Postulates **1** and **4** do not hold; **5** and **11** make no sense.
12. All but Postulates **4, 5,** and **11**.

§2.4

16. a^{24}.
18. 1.
20. No.
22. No.

§2.5

24. 511/256.
26. Use (5) with $q = a^2$ and $n = 10$.
28. $511 + (3280/6561)$.

§3.1

2. Yes.
4. No.
6. No.
8. Yes.

§3.2

10. $a < b$.
12. None.
14. No.

§3.3

18. $x \leq -1/4$.
20. $x \leq 2/5$.
22. $x \leq 5/3$ or $x > 5$.
26. $x \leq -6/5$ or $x > 1$.
28. $-2 < x \leq -1$.

30. No solution.
32. $x = 1$.
36. $-2 < x < 0$.
38. $-5 < x < -2$.
40. $x < 0$.
42. $0 < x < 1$.
46. $x < -4$ or $-1 < x < 1$.

§3.4

48. $x = 0$.
52. $x \leq 0$.

§3.5

54. 10.
56. $|2| + |-3|$.
58. No. Use $|ab| = |a||b|$.
60. Hint: Use triangle inequality twice.
64. Suppose $|a - b| < r$. Consider the two cases $|a - b| = a - b$ and $|a - b| = -(a - b) = b - a$. Then suppose that $b - r < a < b + r$ and again consider the two cases
$$|a - b| = a - b$$
and $\quad |a - b| = b - a$.
66. -7.
68. The distance between -36 and -54.

§4.5

2. True.
4. True.
6. False.
8. True.
10. True.
12. False.
14. $3/2 = 1.4999\cdots$.
16. $777/33 = 23.5454\cdots$.
18. $29/25 = 1.1599\cdots$.
20. $1/100 = .0099\cdots$.

§5.1

2. There are infinitely many, for example, $.\overline{1}$ or $.1020\overline{1}$.
4. $y < x < z$.

§5.4

6. 1.92.
8. -2.96.
12. 1.024.

§5.5

14. x.
16. x^2.
18. y^2 (for all x, y).

§5.6

20. Yes.
22. Yes.
24. -1.
26. $(1, 11]$.

§6.1

2. Yes.
4. No.
8. No; no.
10. Yes: 0; no; yes.

§6.3

12. The second step amounts to adding q^k to $(1 - q^k)/(1 - q)$.
16. In the second step, "multiply out" $3(k + 1)^2$ and then use the assumption $3k^2 \geq 2k + 1$.
20. In the second step, "multiply out" $(k + 1)^3$ and then use the assumption $k^3 > k^2 + 3$.

§6.4

24. l.u.b. $= .13$; g.l.b. $= .12$.
26. l.u.b. $= 3\sqrt{2}$; g.l.b. $= 0$.

§7.1

2. .05.
6. $0 < a \leq (\sqrt{197} - 14)/2$. (Hint: Use the quadratic formula.)
8. $0 < a \leq .4$.
10. $|(X/Y) - x| \leq .104$.
12. Investigate first $|YZ - y \cdot 1|$ and then $|X(YZ) - x(y \cdot 1)|$. You will need $50b + 15a + 15ab + a^2 + ab \leq .1$. One possible answer is: $b \leq .0004$ and $a \leq .001$.

§7.2

14. Use (6) with $m = 5$: 2.13885.
16. Use (6) with $m = 4$: 2.8843.
20. Use (8) with $m = 3$: 1.704.
22. Use (8) with $m = 4$: 988.0.

§8.2

4. Use **2, 9, 9, 7, 10** to transform $(2 + 3)(a + 1 + b)$ into $[(2 + 3) + (2 + 3)b] + (2 + 3)a$.
6. Since $1/1 = 1^{-1}$, show $1^{-1} = 1$. (Use **11, 12**, and **10**).
8. Since $1/a^{-1} = (a^{-1})^{-1}$, use uniqueness of reciprocal (**12**).
10. Use Exercises 9 and 7.
14. $(a + b)/c = (a + b)c^{-1}$. Then use **7, 9**.
16. $(a + b)^2 = (a + b)(a + b)$. Use Exercise 2 as a start. You will also need the definition of the number 2.
18. Use Example 6, the definition of a^2, and the laws of exponents (§2.4).
20. $(a - b)^3 = (a - b)(a - b)^2$. Use Exercise 16 on $(a - b)^2$. Use Example 6 and Postulates **7** and **9**, great care, and the definition of the number 3.

§8.3

22. 5.5, 5.4$\overline{9}$.
24. 6.6, 6.5$\overline{9}$.

26. .1375, .1374$\overline{9}$.
28. .116.
30. 2.$\overline{009}$.
32. 5.5$\overline{370}$.
34. .$\overline{9801}$.
36. .$\overline{967741935483870}$.
40. 41/99.
42. 9/111.
44. 910/909.

Chapter 2

§1.1

Note: These exercises are to be done without using the equation of a circle or the distance formula, since these have not yet been introduced.
2., 4., 6., 8.

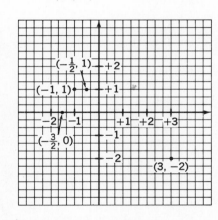

10. $(-2, 3)$ and $(-2, -3)$.
12. $(-3, 4)$.
14. $(13, 5), (13, -5), (-11, 5), (-11, -5)$.

§1.2

18. $(1, \quad 3)$.
22. $(7/2, -3)$.

§1.4

24. 5.
26. $\sqrt{13}$.
28. $\sqrt{2}$.

§2.1

2. The lines intersect at $(0, 1)$ at right angles.

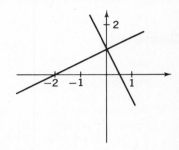

4. Rewrite it as $3x - y - 1 = 0$. Yes.
6. $(40/17, 42/17)$.
8. x axis: $y = 0$; y axis: $x = 0$.

§2.2

10. -6.
12. l_1 descends and l_2 ascends.

§2.3

14. $2x - 3y = 0$.
16. 3.
18. ± 2.

§2.4

22. $4x - y + 5 = 0$.
24. $4x + 9y + 6 = 0$.
26. $3x - 4y - 180 = 0$.
30. -1.
32. $m = 3/2, b = 1/2$.
34. $x + y + 2\sqrt{2} = 0$.
36. $y = (3/2)x$.
38. Find the equation of each side. The altitude from each side must have a slope equal to the negative reciprocal of the slope of the side and must pass through the opposite vertex. The common point is $(27/5, 11/5)$.

§3.1

2. $x^2 + y^2 + 2x - 8y + 8 = 0$.
4. $x^2 + y^2 + 6x - 4y + 11 = 0$.

6. $x^2 + y^2 + 20x - 24y = -235$.
10. Circle of radius 4 and center $(-3, 1)$.
12. Point $(-1, 1/2)$.

§3.2

14. $1/2$.
16. $3x - 4y + 3 = 0$.
20. (i) $\sqrt{2}\,x - 4y - 4\sqrt{2} = 0$.
    (ii) $\sqrt{2}\,x + 4y - 4\sqrt{2} = 0$.

§4.2

2. Focus: $(0, 1/32)$; directrix: $y = -1/32$.
6. $8x = -y^2 + 4y - 20$; vertex is $(-2, 2)$; axis is $y = 2$.
8. Endpoints are $(-2\mid p\mid, p)$ and $(2\mid p\mid, p)$; length is $4\mid p\mid$.
10. Endpoints are $(-1, 4)$ and $(-1, -2)$.

§4.3

12. Focus $(-1, -25/8)$; directrix $y = -23/8$; vertex $(-1, -3)$; axis $x = -1$.

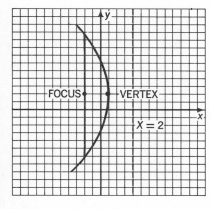

14. Focus $(1/2, 1/2)$; directrix $y = -3/2$; vertex $(1/2, -1/2)$; axis $x = 1/2$.
16. Focus $(-4, -5/4)$; directrix $y = 3/4$; vertex $(-4, -1/4)$; axis $x = -4$.

§4.4

18. $(5/8, 1/4)$.
22. -2 and 18.

§5.3

2. $1/6$.
4. 9.
6. $7/4$.

§6.4

2. No.
4. $\sqrt{15}$.
    The distance between a triple of numbers and a quadruple of numbers is not defined.
6. Center at $(0, 2, 0, 0)$. Radius $= \sqrt{5}$.

Chapter 3

§1.2

2. $0, 20/17, (1 + \sqrt{2})/3$.
4. $-10, 30, 26/125$.
6. $-8/3, 10.03, (3 - 2\sqrt{2})/2$.
8. $-23/3, 13/24, 83/54$.
12. All real numbers *except* for 1 and those in the interval $(-2, -1)$.
14. $z \geq 2$.

§1.4

16. $8, -21, -1, 10$.
18. $22, 1, 7/4, 3$.
22. $1 + \sqrt{z} + [1/(1 + 2)]$, $(1 + \sqrt{z})/(1 + z)$, $1 + (1/\sqrt{1 + z})$, $1/(2 + \sqrt{z})$.
24. $(h \circ f \circ f \circ g \circ g)(s)$, $(fh)(g \circ f \circ g)(s)$.
26. $(-\infty, 1)$.
28. For $-5 \leq u \leq -3$ and $3 \leq u \leq 5$.
30. $[-1, 1]$.

§1.7

32. Even.
34. Even.

38. Only the constant function $f(x) \equiv 0$ is both odd and even.
40. Decreasing.
42. Nondecreasing.
46. Decreasing: $(-\infty, -1/2)$; increasing: $(-1/2, \infty)$.
48. Decreasing: $(-\infty, 0), (0, 1)$, and $(2, \infty)$; increasing: $(1, 2)$.

In solutions 50 to 56, I denotes increasing; D, decreasing; NI, nonincreasing ND, nondecreasing. Note that I implies ND and D implies NI.

50. Neither even nor odd.

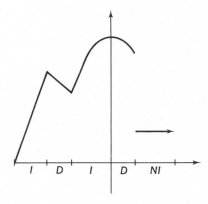

52. Odd.

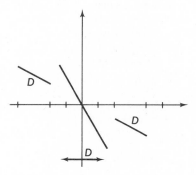

54. Even.

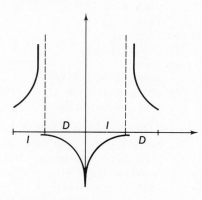

56. Neither even nor odd.

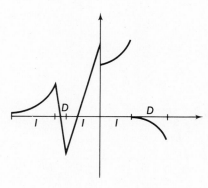

§1.8

58. $t \longmapsto 43{,}200t^2 + 2{,}160t$
60. $s \longmapsto 18{,}000s^2 + 32$ if $s \leq \frac{1}{2}$,
    $s \longmapsto 18{,}000s^2 - 36{,}000s + 18{,}032$ if $s \geq \frac{1}{2}$.
62. $t \longmapsto (450/11)t$ if $t \leq \frac{1}{4}$.
    $t \longmapsto 450/44$ if $t \geq \frac{1}{4}$.

§2.2

2. $s > 1/16$.
6. $u < 0$ or $u > 1$.
8. $(2s - 2 - \sqrt{2})(s - 1 + \frac{1}{2}\sqrt{2})$.

§2.5

10. $h(s) = k(s) \cdot (s^3 - 2s^2 - 4) + (s + 12)$.

12. $A(t) = B(t) \cdot (t^4 - 1)$.
14. The quotient is
    $z^2 + (1 + c)z + c(1 + c)$.
    The remainder is
    $c^2(1 + c) - 1$.
16. $h(u) = k(u) + (3u - 2)$.
18. -2.
20. $1/4$.
22. $x^4 - x^3 - x^2 + x$.
24. $x^7 - \frac{7}{2}x^6 - \frac{29}{4}x^5 - \frac{215}{8}x^4 + 10x^3 - \frac{349}{8}x^2 + \frac{51}{2}x - \frac{9}{2}$.
28. -3 with multiplicity 3; 1 with multiplicity 7; -1 with multiplicity 5.

§2.6

30. $(y^3 - y^2 + 1)/(1 + 2y - y^2)$,
    $(1 + y)/(y + y^2 - y^3)$,
    $(y^4 - 2y^3 - 2y^2 + 3y + 2)^{-1}$.
32. $(z^2 + 2z - 1)/(z^3 + 2z^2)$,
    $(z^4 + z^3 - 3z^2)/(z + 2)$,
    $(1 + z^2 - 3z^4)/(z^2 + 2z^4)$.

§2.7

34. $\dfrac{(x - 1)^{1/2}(1 + x^{1/2}) + x^2}{x(x - 1)^{1/2}}$,

    $\sqrt{x - 1}\,(1 + \sqrt{x})/x^2$,

    $\dfrac{(x - 1)^{1/2} + \left[x^{1/2}(x - 1)^{1/2}\right]}{x}$,

    $\frac{1}{4}3^{1/2} + \frac{1}{2}3^{1/4}$.

§2.10

38. Undefined at 1, -1. Assign the value 0 at $x = 1$.
40. Undefined at 0, 1. Assign the value -1 at $x = 0$.
42. Undefined at 2, -2. Assign the value $1/4$ at $x = 2$.
46. Undefined at 0, 1, -2. Assign the value 0 at $x = 1$.

§3.1

2. Discontinuous at 0.
4. Discontinuous at -3, 0.

6. Discontinuous at -2, 1.
8. $g(s) = 2$ if $s \leq -1/2$;
   s if $-1/2 < s \leq 0$;
   $s^2 + 1$ if $0 < s \leq 8$;
   64 if $s > 8$.
10. $f(x) = x^2$ if $x \leq 0$; x if $x > 0$.

§3.2

12. Yes. $\delta = 1$.
14. No.
18. Yes. $\beta = (5 - b)/2$.
20. Yes. $\delta = (1/\sqrt{B}) - 1$.
22. Yes, $G < A$ near 1 for $A > 3$:
    let $\delta = \frac{1}{2}A - 1$. No, $G > B$
    near 1 for $B < 3$ is false: let
    $B = 5/2$ and consider values
    of z between 1 and $5/4$.
24. (I) Let $A > -1/3 = g(0)$.
        $x^2 - 1/3$ will be less
        than A if $|x| < \sqrt{3A + 1}$.

    (II) Let $B < -1/3 = g(0)$.
        $x^2 - 1/3$ is always
        greater than B, since $x^2 - 1/3 > -1/3$ if $x \neq 0$.

28. $\delta = 1/90$.
30. $\lambda = 1/10$ (Hint: Since t is
    "near 1," assume in this exercise that $|t| < 2$.)

§3.5

34. Let $\psi(x) = x^2 + 2x$. $f = \phi \circ \psi$.
    Use Theorems 2 and 4.
36. Let $H(x) = x^4$.
    $f = \psi/(1 + H \circ \phi)$.
    Use Theorems 1, 2, 3, and 4.
    Note that the denominator is
    never zero.
38. Let $h(x) = x^2$ and
    $g(x) = 1 + \psi(x^2)$.
    $f = h + \phi \circ g$. Use Theorems 1, 2, and 4.

§3.6

40. f has at least two roots.

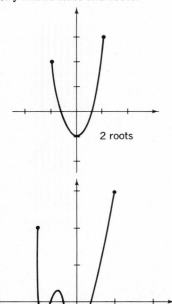

2 roots

4 roots

42.

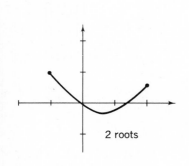

2 roots

§3.7

48. $x \longmapsto (x^{1/3} - 1)^2$.
50. (a) No inverse.
    (b) $x \longmapsto \sqrt{x}$
    (c) $x \longmapsto -\sqrt{x}$.
52.

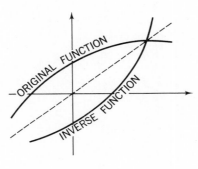

ORIGINAL FUNCTION

INVERSE FUNCTION

54. No inverse.
56.

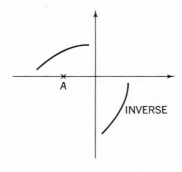

INVERSE

INVERSE→

58.

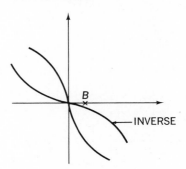

INVERSE

INVERSE→

60. No inverse for A. The inverse for B follows.

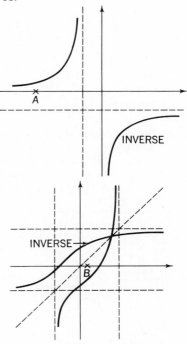

INVERSE

§4.1

4. 5/16.
6. −1.
8. 5/4.
12. Indicated limit does not exist.

§4.2

18. 8.

20. 6.
22. 15.

§4.3

24. (a) Yes. (b) No. (c) Yes.

§5.1

2. −7.
6. No limit.
8. 1.
10. $\lim_{t\to 0^-} H(t) = 0$ (can redefine $H(0) = 0$). $\lim_{t\to 0^+} H(t)$ does not exist. H cannot be made continuous at 0 by altering the value of $H(0)$.

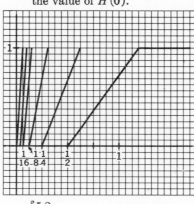

§5.2

14. (a) (b) (c) 19/45.
16. No limit.

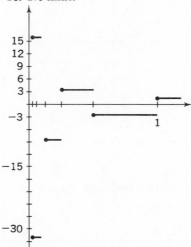

§5.3

20. (a), (b), (c) r/s.
22. (a), (b), (c) 0.
26. (a), (b), (c) $+\infty$.
28. (a) $+\infty$.
    (b) $+\infty$ if $n - m$ is even, $-\infty$ if $n - m$ is odd.
    (c) $+\infty$ if $n - m$ is even, no limit if $n - m$ is odd.

§5.4

30. 32.
34. $+\infty$.
36. $\sqrt{2}/4$.
38. $+\infty$.
40. $+\infty, -\infty, -\infty$.

§6.1

2. Let $c < 0$. Imitate proof in text, being careful about direction of inequalities. Use fact that f satisfies Property II.
4 and 6. Imitate proof in text; watch inequalities.

§6.4

10. Imitate proof in text for fg, having noted that $fg < 0$ near x_0.

§7.1

2. Use the definitions in §5.3 to obtain a positive number b such that $f(b) = B > \gamma$ and a negative number a such that $f(a) = A < \gamma$. Then use the Intermediate Value Theorem for f on the interval (a, b).

Chapter 4

§1.1

2.

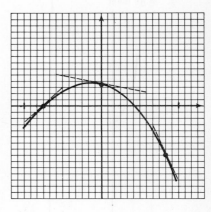

4.

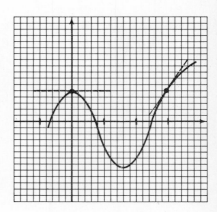

§1.2

6.

h	$\dfrac{f(2 + h) - f(2)}{h}$
1	5
−1	3
1/2	4.5
−1/2	3.5
1/4	4.25
−1/4	3.75

8.

δ	$\dfrac{h(\delta) - h(0)}{\delta}$
1	$1/2$
-1	undefined
.1	$10/11$
$-.1$	$10/9$
.01	$100/101$
$-.01$	$100/99$

§1.5

10. 1.
12. 3/4
14. $12y + x - 86 = 0$.
18. -1.
20. At $x = 0$.

§1.6

24. $-1/2x^{3/2}$.
26. $13/(3x + 2)^2$.
28. $(z - 1)/2z^{3/2}$.

§1.7

30. $2\sqrt{3}$.
32. $1/\sqrt{2s - 1}$.
34. $-1/2\sqrt{1 - u}$.

§1.8

38. $f(x) = 15 + 11(x - 3) + r(x)(x - 3)$ where $r(x) = 2(x - 3)$.
40. $f(y) = 0 + 2(y - \frac{1}{2}) + r(y)(y - \frac{1}{2})$ where $r(y) \equiv 0$.
44. $h(v) = 1 + 3(v - 1) + r(v)(v - 1)$ where $r(v) = v^2 + v - 2$.

§1.9

46. 9.3.
48. 5.999.
50. 1.16.
52. .1999.
54. -1.05.

§1.13

56. 299 ft/sec.
58. Since $v(t) = s'(t) = -60$, the velocity is constant and is always equal to -60. Therefore, the velocity when $t = 1$ is -60 mi/hr.
60. 0 cm/min.
62. $x > 1/3$.

§1.14

64. Approximately -98 cm/sec.
68. $\sqrt{3}/3$.
70. $\sqrt{2}/4$.

§1.15

72. $12s$ where s denotes the length of a side.
74. π.
76. $22.02.

§2.1

2. $h'(s) = 2$,
   $k'(s) = 6s$,
   $(k - h)'(s) = 6s - 2$,
   $(k/h)'(s) = (6s^2 + 6s + 10)/(2s + 1)^2$.
4. $A' = 2t + 1$,
   $B' = 2t - 1$,
   $\frac{1}{2}A - \frac{2}{3}B)'(t) = -7 - 2t/6$,
   $(A/B)'(t) = -2(t^2 + t)/(t^2 - t)^2$.
8. $-1/3$, -86.
10. $(2A - B + 3C)'(x) = 4$,
    $(ABC)'(x) = 3x^2 + 4x - 1$.
12. -14, $1/64$, $-49/64$.

§2.4

14. $\dfrac{10y^{11} - 11y^{10} - 18y^7 + 21y^6 - 1}{(y - 1)^2}$.

16. $\dfrac{4u(u - 3)}{(u^2 + 2u - 3)^2}$.

20. $\dfrac{2t^7 - 8t^6 - 7t^4 - 8t^3 + 39t^2 - 4t - 8}{(t^3 - 1)^2}$.

22. 441.
24. $(5t^4 - 9t^2)(t^3 - 2t + 6) + (3t^2 - 4)(t^5 - 3t^3 - 1)$.
26. $(5w^3 + w^2 - 1)(2w - 10) + (15w^2 + 2w)(w^2 - 10w + 6) + 2/(w + 1)^2$.
28. $3z^2(z^{15} - 3z^{12} + z^6 - 1) \cdot$
    $\cdot (z^6 + 3z + 1)$
    $+ (6z^5 + 3)(z^{15} - 3z^{12} + z^6 - 1) \cdot$
    $\cdot (z^3 + 5)$
    $+ (15z^{14} - 36z^{11} + 6z^5) \cdot$
    $\cdot (z^6 + 3z + 1)(z^3 + 5)$.
30. $z = 1 - 3y$.
32. $2(x^5 - 2x^4 + 3x^3 - x + 1) \cdot$
    $\cdot (5x^4 - 8x^3 + 9x^2 - 1)$.

§2.5

34. $12(z^6 + 3z^2 + z - 3)^{11} \cdot$
    $\cdot (6z^5 + 6z + 1)$.

36. $\dfrac{3x^5 - 5x^4 + x}{\sqrt{x^5 - 2x^4 + x}}$.

40. $15(y^6 + y^3 - 2y + 1)^{14} \cdot$
    $\cdot (6y^5 + 3y^2 - 2)(y^2 + 10)^{10}$
    $+ 20y(y^2 + 1) \cdot$
    $\cdot (y^6 + y^3 - 2y + 1)^{15}$.
42. 5.
44. 40, 40.
46. $3y(y^2 + 1)^2/\sqrt{(y^2 + 1)^3 - 1}$.
48. $8(u^3 - 1)(u^3 + (u^2 - 3)^6)^7 \cdot$
    $\cdot (3u^2 + 12u(u^2 - 3))^5 +$
    $3u^2(u^3 + (u^2 - 3)^6)^8$.
50. 40.

52. $\dfrac{4\sqrt{s}\sqrt{s + \sqrt{s}} + 2\sqrt{s} + 1}{8\sqrt{s}\sqrt{s + \sqrt{s}}\sqrt{s + \sqrt{s + \sqrt{s}}}}$

§2.6

54. 3.
56. $1/2$, $1/5$.
58. $f'(-1)$ does not exist, $f'(3) = 1/14$.
60. $2/3$.
64. $-1/(5x^4 + 3x^2)$.
66. (a), (b), (c):
    $1/(3s^2 - 4s + 1)$.

§2.8

68. $(5/2) z^{3/2} - (4z/3)/(z^2 + 1)^{1/3}$.

70. $(u^{2/5}+3u)(1/3u^{-2/3}-u^{-1/2})$
    $-(-u^{1/3}-2u^{1/2})(\frac{2}{5}u^{-3/5}+3)$
    divided by $(u^{2/5}+3u)^2$.
    (Defined for $u > 0$.)

72. $\frac{3}{2}(x + (1 - x^3)^{1/2})^{1/2} \cdot$
    $\cdot (1 - 3x^2/2\sqrt{1 - x^3})$.
    (Defined for $x < 1$.)

74. $1/2z^{1/2}(z^3 - 1)^{1/3}$
    $-3z^2(z^{1/2} + 2)/(z^3 - 1)^{2/3}$.
    (Defined for $z > 0, z \neq 1$.)

76. $(2/3w^{1/3})(w^{1/2} - w^{1/4})^{1/2}$
    $+ (w^{2/3} + 1)\frac{1}{2}(w^{1/2} - w^{1/4})^{-1/2} \cdot$
    $\cdot (\frac{1}{2}w^{-1/2} - \frac{1}{4}w^{-3/4})$
    (Defined for $w > 1$.)

§3.1

2. $24 - (9/8t^{5/2}) + (72/t^5) (t \neq 0)$.

4. $80s^2(s^2 - 1)^3 + 10(s^2 - 1)^4$.

6. $5(1 - \sqrt{u})^3(5\sqrt{u} - 1)$ divided by $(4u^{3/2})$, $(u > 0)$.

8. The third derivative is
   $240(2y^2+y-6)^4 +$
   $240(4y+1)^2(2y^2+y-6)^3$.

10. $3360w^4 - 3840w^3 + 1440w^2 - 192w + 6$.

12. $\dfrac{-3\sqrt{y} - 2}{16y^{3/2}(1 + \sqrt{y})^{3/2}}$.

14. $21 + 117/256$.

18. $20\frac{1}{4}$.

20. 43.

22. $g^{(k)}(y) =$
    $\dfrac{(-1)^k 1 \cdot 3 \cdot 5 \cdot \cdots \cdot (2k - 1)}{2^k(u + 1)^{(2k+1)/2}}$.

24. $h^{(1)}(z) = 2z + 1 + \frac{1}{2}z^{-1/2}$;
    $h^{(2)}(z) = 2 - \frac{1}{4}z^{-3/2}$;
    $h^{(k)}(z) =$
    $\dfrac{(-1)^{k+1} 3 \cdot 5 \cdot \cdots \cdot (2k - 3)}{2^k z^{(2k+1)/2}}$
    for $k \geq 3$.

§3.3

26. $12, 14, 30$ (all gram cm/sec²).

28. 250 ft lb/sec².

§3.4

30. 10.11 gram cm/sec²; Newton: 10 gram cm/sec².

§4.3

4. Increasing: $x < 1/3$ or $x > 1$; decreasing: $1/3 < x < 1$.

6. Increasing: $-1 < s < 0$ or $s > 1$; decreasing: $s < -1$ or $0 < s < 1$.

8. Increasing: $x < 0$ or $x > 6/7$; decreasing: $0 < x < 6/7$.

§4.4

10. Strict local minimum at $x = 6$, strict local maximum at $x = -1$.

14. Strict local minimum at $u = 0$.

16. Strict local minimum at $y = -1$.

18. Strict local minimum at $t = 0$.

§4.6

20. There is *no* such function (see §4.7).

22. Local maximum: A, C, E, G; local minimum: B, D, F, H; absolute maximum: A; absolute minimum: D, F; derivative is zero: B, E, F, G; derivative fails to exist: A, C, D, H (one-sided derivative).

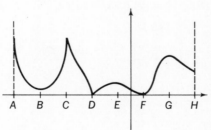

§4.8.

24. Maximum: 0 at $x = 1$; minimum: -2 at $x = -1$.

26. Maximum: $1/2$; minimum: $-1/2$.

28. Increase: $-1 < y < 0, y > 2$; decrease: $0 < y < 2, y < -1$; local maximum at 0; local minima at -1 and 2; absolute minimum at 2; no absolute maximum.

30. Find intervals of increase and decrease and local maxima and minima and analyze this data.

34. $(2 + \sqrt{6})$ by $(4 + 2\sqrt{6})$.

36. 1-inch squares.

38. $\sqrt{2}$ by $\sqrt{2}/2$.

40. (a) Radius of semicircle: $1000/(\pi + 5)$; dimensions of rectangle: $2000/(\pi + 5)$ by $100(10 - \pi)/(\pi + 5)$.

42. $4\frac{1}{6}$ days; $104.16.

46. Upper base $= r$.

48. 1/3 of a mile from the weaker source; $1/(k + 1)$ of a mile from the weaker source.

50. $1500°$ Fahrenheit.

52. $\sqrt{27}$.

§4.10

54. Local minimum: $(2, -5)$; local maximum: $(-2, 27)$; inflec-

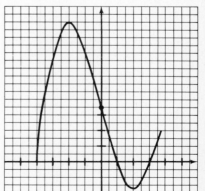

tion point: $(0, 11)$; roots: 1 and $(-1 \pm \sqrt{45})/2$; increasing: $x < -2$ and $x > 2$; decreasing: $-2 < x < 2$; concave: $-\infty < x < 0$; convex: $0 < x < \infty$. (In the drawing all vertical distances are shortened.)

56. Local minima: $(1, -32)$ and $(-2, -5)$; local maximum: $(-1, 0)$; inflection points at $x = (-2 \pm \sqrt{7})/3$; roots: -1 (double root), between 1 and 2, between -2 and -3; increasing: $-2 < x < -1$ and $x > 1$; decreasing: $x < -2$ and $-1 < x < 1$; concave: $(-2 - \sqrt{7})/3 < x < (-2 + \sqrt{7})/3$; convex: $x < (-2 - \sqrt{7})/3$ and $(-2 + \sqrt{7})/3 < x$. (Vertical distances shortened in figure.)

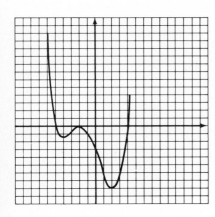

60. Local minimum: $(3, -1)$; local maximum: $(-1, -9)$; no inflection points; roots: 2, 5; increasing: $-\infty < x < -1$ and $3 < x < \infty$; decreasing: $-1 < x < 1$ and $1 < x < 3$; concave: $x < 1$; convex: $x > 1$;
$$f(1^+) = +\infty,$$
$$f(1^-) = -\infty,$$

$$f(+\infty) = +\infty,$$
$$f(-\infty) = -\infty.$$

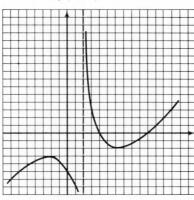

62. Local (and absolute) minimum: $(-1, 0)$; local (and absolute) maximum: $(1, 2)$; double root: -1; inflection points at $t = 0, \pm\sqrt{3}$; increasing: $-1 < x < 1$; decreasing:

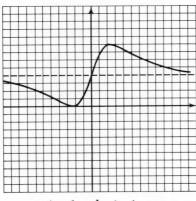

$x < -1$ and $x > 1$; concave: $x < -\sqrt{3}$ and $0 < x < \sqrt{3}$; convex: $-\sqrt{3} < x < 0$ and $x > \sqrt{3}$; $f(\pm\infty) = 1$.

64. No local or absolute maxima or minima; no inflection points; root: $-\frac{3}{2}$; decreasing: $x < \frac{4}{3}$ and $x > \frac{4}{3}$; concave: $x < \frac{4}{3}$; convex: $x > \frac{4}{3}$;
$$h(\tfrac{4}{3}^+) = +\infty,$$
$$h(\tfrac{4}{3}^-) = -\infty,$$
$$h(\pm\infty) = 2/3.$$

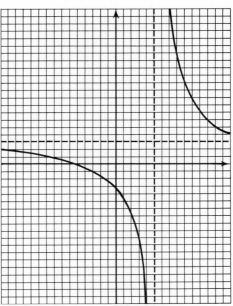

66. Defined only for $t \geq 0$. Local (and absolute) minimum at $(1/9, -14/27)$; no inflection points; roots: 0, 1; decreasing: $0 \leq t \leq 1/9$; increasing: $t > 1/9$; convex: $t \geq 0$.

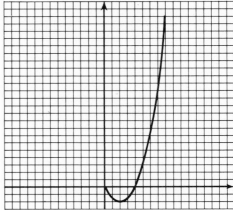

68. Local (and absolute) minimum at $(-1, -1/4)$; inflection point at $(-2, -2/9)$; root: 0; decreasing: $x < -2$ and $x > 1$; increasing: $-2 < x < 1$; concave: $x < -2$;

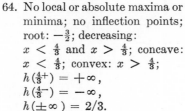

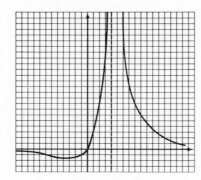

convex: $-2 < x < 1$ and $x > 1$; $g(+\infty) = g(-\infty) = 0$; $g(1^+) = g(1^-) = +\infty$.

70. No local maxima or minima; absolute minimum: 0; inflection point: $(2, \sqrt{3}/3)$; root: 4; $f(-4^+) = +\infty$; f is defined for $-4 < x \le 4$.

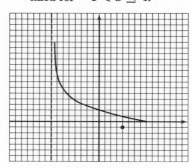

72. h is defined for $x \ge -5$; local maximum: $(-4, 16)$; local (and absolute) minimum: $(0, 0)$; inflection point at $x = -4 + \frac{2}{3}\sqrt{6}$;

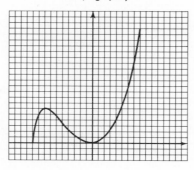

roots: $-5, 0$; increasing: $-5 \le x \le -4$ and $x > 0$; decreasing: $-4 \le x \le 0$; convex: $-4 + \frac{2}{3}\sqrt{6} < x$; concave: $-5 < x < -4 + \frac{2}{3}\sqrt{6}$.

76. g is even; $g(x) \ne 0$ for any x; local maximum at $(0, 3)$; local (and absolute) minima at $(1, 2\sqrt{2})$ and $(-1, 2\sqrt{2})$; inflection points for $x = \pm\sqrt{5}/5$; increasing: $-1 < x < 0$ and $x > 1$; decreasing: $x < -1$ and $0 < x < 1$; concave: $-\sqrt{5}/5 < x < \sqrt{5}/5$; convex: $x < -\sqrt{5}/5$ and $x > \sqrt{5}/5$; $g(+\infty) = g(-\infty) = +\infty$.

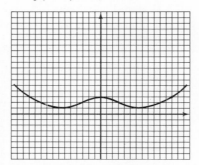

78. h is odd; roots: 0 and $\pm(5/3)^{3/2}$; local minimum at $(1, -2)$; local maximum at $(-1, 2)$; inflection point at $(0, 0)$; increasing: $x < -1$ and $x > 1$; decreasing: $-1 < x < 1$; concave:

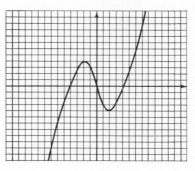

$x < 0$; convex: $x > 0$; $h(+\infty) = +\infty$ and $h(-\infty) = -\infty$.

80. y is odd; root: 0; inflection point: $(0, 0)$; concave: $x < 0$; convex: $x > 0$; no local maxima or minima; $y \to +\infty$ as $x \to +\infty$; $y \to -\infty$ as $x \to -\infty$.

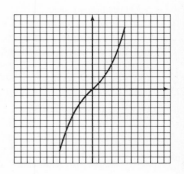

82. No roots; local (and absolute) minimum at $(-1/2, (3/4)^{1/3})$; inflection points at $x = 1, -2$; increasing: $x > -1/2$; decreasing: $x < -1/2$; concave: $x < -2$ and $x > 1$; convex: $-2 < x < 1$; $g(+\infty) = g(-\infty) = +\infty$.

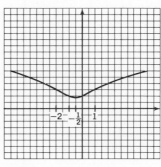

§5.4

2. $\frac{1}{8}x^3 + C$.

4. $\frac{1}{10}(x + 1)^{10} + C$.

6. $\frac{3}{4}u^{4/3} + 2u^{3/2} + C$.

8. $\frac{1}{6}(\sqrt{x} + 1)^6 + C$.

10. $s^4 + \frac{1}{6}s^3 - \frac{1}{2}s^2 + C_1 s + C_2.$

12. $g(s) = \frac{1}{4}s^4 - (1/s) + \frac{3}{4}.$

§5.6

16. $(72/g)$ ft; $(24/g)$ sec.

20. $gT^2/8.$

22. $5\frac{1}{4}$ miles from the starting point.

§5.7

24. $10^{11} c/\sqrt{.0004c^2 + 10^{22}}$ cm/sec.

§5.8

26. Area of "shorter" rectangle is $h \cdot t_0$; of "taller" rectangle, $h \cdot (t_0 + h)$. $A'(t_0) = t_0$. Desired area is $t^2/2$.

28. Area of "shorter" rectangle is $h \cdot \sqrt{t_0}$; of "taller" rectangle, $h \cdot \sqrt{t_0 + h}$. $A'(t_0) = \sqrt{t_0}$. Desired area is $\frac{2}{3}t^{3/2}$.

§6.2

2. (a) Write down the definition of $(cf)'(x_0)$ and use the rule: $\lim (cf) = c \lim f$ (see §4.2).

   (b) Write down the definition of $(f - g)'(x_0)$ and use the rule: $\lim (h - k) = \lim h - \lim k$ (see §4.2).

§7.1

2. 0.

6. $\phi'(0^-) = -2/3, \phi'(0^+) = 2/3.$

8.

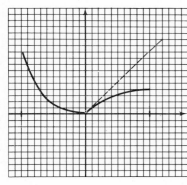

10.

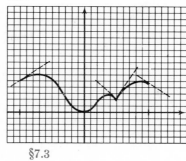

§7.3

12. $f'(t) = t/\sqrt{t^2 - 4}$ for $|t| > 2.$ $f'(2^+) = +\infty$ and $f'(-2^-) = -\infty.$

14. $k'(x) = \frac{1}{3}x^{-2/3} + \frac{2}{3}x^{-1/3}$ for $x \neq 0.$ $f'(0) = +\infty.$ (Use Theorem 2.)

16. For $y < 0,$ $\phi'(y) = \frac{3}{5}(y + 1)^{-2/5};$ for $y > 0, \phi'(y) = \frac{3}{5}(y - 1)^{-3/5};$ $\phi'(0^-) = \frac{3}{5}; \phi'(0^+) = -\frac{3}{5}.$

18.

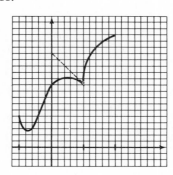

20.

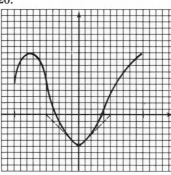

§8.1

2. (a) Assume $f(b) < f(x_0)$ and show this is impossible. (Imitate the proof in the text of the impossibility of having $f(a) > f(x_0)$.)

   (b) Show $f(b) \geq f(x)$ for $x_0 < x < b.$

   (c) Assume $f(b) = f(x_0)$ and show this is impossible.

§8.3

4. Assume, without loss of generality, that $f(0) = 0.$ Show that, given $\epsilon > 0, f(x)/x < 1/\epsilon$ for $x \neq 0,$ x near 0. [Investigate the function $g(x) = f(x) + x/\epsilon.$ Show that $g(0) = 0$ and $g'(x) < 0$ for $x \neq 0.$]

Chapter 5

§1.2

2. 15/8.

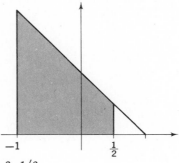

6. 1/6.

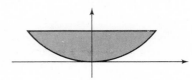

8. 3 1/2.

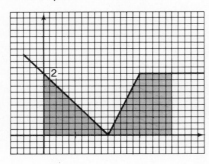

10. $-31/12$.

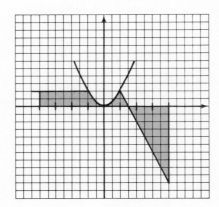

§1.7

12. $7.7 \le \int_{-1}^{3} f(y)\, dy \le 11.0$.

14. $0 \le \int_{-1/2}^{2} \phi(t)\, dt \le 11.4$.

16. $.001625 \le \int_{.1}^{.2} t^2\, dt \le .003125$, exact value: $.002\bar{3}$. (Use $N = 2$.)

18. $.0928 \le \int_{.2}^{.3} (1 - y^2)\, dy \le .0945$ (Use $N = 3$.)

22. $73/120 \le \int_{1/2}^{3/2} \dfrac{2x}{2x + 1}\, dx$
    $\le 83/120$.
    (Use $N = 3$.)

24. $10.931 \le \int_{-1}^{1} f(t)\, dt \le 12.79$.
    [Use $N = 2$ to get an estimate within 2 on $(-1, .9)$ and

$N = 1$ to get an estimate within .5 on $(.9, 1)$.]

§1.9

26. -2.
28. 0.
30. $-7/3$.
32. $-7/2$.
34. Use (3) of Theorem 2.
36. Use (3) of Theorem 2, having compared $1 + x^{22}$ to the constant functions 1 and 2.

§2.1

4. Compute $K(u)$ and $K'(u)$ for each of the cases:
   (i) $-4 \le u \le -1$,
   (ii) $-1 \le u \le 0$,
   (iii) $0 \le u$.
6. Note that
$$\int_a^b f(t)\, dt$$
$$= \int_a^x f(t)\, dt + \int_x^b f(t)\, dt$$
$$= G(x) + H(x)$$
so that $H(x) = c - G(x)$ where $c = \int_a^b f(t)\, dt = $ constant.
8. Note that $L(x) = K(\beta(x))$ and use the chain rule.

§2.2

10. $-7/3$.
12. $1/20$.
14. 3.
18. 2.
22. $(143 + 144 \cdot 2^{2/3})/60$.

§2.3

24. 16.
26. $\frac{6}{5} u^{5/3} - \frac{1}{6} u^6 - \frac{31}{30}$.
28. $2 (t^{3/2} - 1)^5/15$.
30. $x^2 - 1$.

§2.4

34. $v(64) = 173\frac{5}{7}$ ft/sec;
    $s(64) = 2^{11} (235/91)$ ft.

§3.2

2. $47/60$.
4. $\frac{1}{7} u^7 - \frac{1}{9} 7 u^9 - \frac{1}{3} u^3 + C$.
6. $-2\frac{17}{192}$.
8. 1.
10. 820.
12. $\frac{1}{4} (z + 1)^4 - \frac{1}{10} (2z + 3)^5 + C$.
14. $8/3$.
16. $5 (\frac{1}{4} s^4 - \frac{1}{8} s^2 + 2s) + [3 (s^5 + 1)^8/40] + C$.
18. $(13 - 8\sqrt{2})/12$.
20. $x^4/12 - x^2/2 + C$.
22. 6.
24. $3/10$.
26. 3.
30. $(x^3 + 3x^2)/6$.

§3.4

34. 0.
36. $\sqrt{2}$.
38. $2x^{5/2}/5 - (x + 1)^4/2 + C$.
40. $f(x) = \frac{1}{2} x + 1$ and $g(x) = \frac{1}{2} x^2 - \frac{1}{2} x + 1$.
42. ± 1.
44. $\frac{2}{15} (x + 1)^{3/2}(3x - 5) + C$.
48. $2 (3 - 2u)(1 - u)^{-1/2} - 8 (1 - u)^{1/2} + C$.
50. $\frac{3}{4} x^{4/3} (x + 1)^2 - \frac{9}{14} x^{7/3} (x + 1) + \frac{27}{140} x^{10/3} + C$.
52. $2x^2 (x + 4)^{1/2} - \frac{8}{3} x (x + 4)^{3/2} + \frac{16}{15} (x + 4)^{5/2} + C$.
54. $\frac{1}{6} (s^2 + s)(s + 4)^6 - \frac{1}{42} (2s + 1)(s + 4)^7 + \frac{1}{168} (s + 4)^8 + C$.
56. $\frac{1}{22} (3t^2 - 1)(2t + 3)^{11} - \frac{1}{88} t (2t + 3)^{12} + \frac{1}{2288} (2t + 3)^{13} + C$.

§3.6

58. 0.
60. $-1/8 (1 + (x - 1)^4)^2 + C$.
62. $19/3$.
64. $\frac{3}{2} (z^2 + 2z + 3)^{1/3} + C$.
66. $\frac{2}{3} (t^2 + 1/t)^{3/2} + C$.
70. $\frac{1}{6} [(4097)^{3/2} - 2\sqrt{2}]$.
72. $1/5$.

76. 13/144.
78. $\frac{3}{8}(t^{1/3} - 1)^8 +$
$\frac{3}{7}(t^{1/3} - 1)^7 + C.$

§4.2

2. 9/2.
4. 1/6.
6. 928/15.
8. 13/108.
10. 2.
12. 2/3.
18. 3/4.

§4.5

20. 1/3.
22. $\pi R^2 H.$

§4.7

24. $8\pi/15.$
26. $127\pi/7.$
28. $\pi - \pi/a.$ As $a \to \infty$, $V \to \pi.$
30. $\pi/5.$
32. $\sqrt{2}\,\pi/3.$
36. $2\pi.$
38. $56\pi/15.$
40. $8\pi^2.$
42. $5\pi/24.$

§4.8

44. $\frac{1}{27}(85^{3/2} - 40^{3/2}).$
46. 4.
48. $(10\sqrt{7} - 5\sqrt{2})/3.$
50. $\frac{5}{48}(65^{3/2} - 5^{3/2}).$
52. 123/32.
54. Equation of line:

$$f(x) = \frac{d - b}{c - a}(x - a) + b.$$

§4.9.

56. $L(a, b) = \int_a^b [r/\sqrt{r^2 - x^2}]\,dx.$
58. Let
$f(b) = A(a, b)$
$\qquad - \frac{1}{2}r\int_a^b[r/\sqrt{r^2 - x^2}]\,dx.$
Compute $df(b)/db.$ (It will be 0.) Then reason as in the text.
60. $\pi r.$

§5.1

2. About 3136 gram cm/sec^2 ($g = 980$ cm/sec^2).
4. 5.

§5.3

6. 1000 g gram cm or about 980,000 gram cm^2/sec^2.
8. 500 ft lb.
10. 3 cm.

§5.4

12. About 1 mile.
16. (a) $K(t) = \frac{25}{8}t^3;$
    (b) $F(s) = \frac{15}{4}s^{1/5};$
    (c) $W(s) = -\frac{25}{8}s^{6/5};$
    (d) $E = W + K = 0.$

§5.7

20. $\sqrt{2\gamma/R}/2\sqrt{M}.$

§6.3

2. 2.
4. Diverges.
6. 1/100.
10. 1/2.

§6.4

12. Diverges.
14. Make the substitution $x = 2t - 1.$ Then compare with $\int_1^\infty (1/x^2)\,dx.$ Converges.
18. Diverges.
20. Diverges.

§6.5

22. $\pi^2/4.$
24. $\pi/8.$

§8.6

2. Let T be the set of all numbers $\int_a^b \psi(x)\,dx$ for all $\psi \geq f.$ Let $J = \int_a^b f(x)\,dx =$ greatest lower bound of $T.$ For given $\epsilon > 0$, suppose

$$\int_a^b \psi(x)\,dx > J + \epsilon.$$

Show that contradiction follows.
4. Use Lemma C to find step function $\psi \geq g$ with

$$\int_a^b \psi(x)\,dx \geq \int_a^b g(x)\,dx - \epsilon.$$

6. $f(x) = 1/2^n$ for $1 - 2^{-n} < x < 1 - 2^{-n+1},$ $n = 0, 1, 2, \cdots.$

§9.2

2. Let $x \to f(x)$ be a nondecreasing function defined in the interval (a, b), where b is a number and a is either a number or the symbol $-\infty.$ Assume that there is a number M such that $f(x) \geq M$ for $a < x < b.$ Then the finite limit $\alpha = \lim_{x \to a^+} f(x)$ exists and $\alpha \geq M.$ Use the greatest lower bound principle, appropriately reformulating statements 1 and 2 in the text.

Chapter 6

§1.1

2. $1 + \sqrt{5}.$
4. $\sqrt{3}/2.$
6. First solve the algebraic equations $t^2 = u$, $t^3 = u$, that is, find $\sqrt{u}$ and $\sqrt[3]{u}.$ Then solve $t^3 = u^2 + u + 1$, that is, find $\sqrt[3]{u^2 + u + 1}.$ Compute the number

$$\frac{\sqrt{u} + \sqrt[3]{u} + 1}{\sqrt[3]{u^2 + u + 1}}$$

which involves addition and division.

§1.3

8. $2\pi.$
10. $2\pi^2.$

§1.6

12. Use Fig. 6.14:
$\cos(\pi/3) = \sin(\pi/6) = \frac{1}{2}$ and
$\cos(\pi/6) = \sin(\pi/3) = \sqrt{3}/2$.

14. $\pi/12$.

16. $\pi/5$.

18. $20°$.

20. $6°$.

22. $(60/\pi)°$.

24. $1/\sqrt{2}$.

§1.7

26. $\pi/6$.

28. arc sin not defined for $-3\sqrt{2}$.

30. $1/27$.

32. $-3(\operatorname{arc\,cos} x)^2/\sqrt{1 - x^2}$.

34. 2.

§1.8

Remark. In exercises below arc sin may be replaced by $-$arc cos.

38. $\frac{1}{4} \operatorname{arc\,sin} 4u + C$.

40. $\frac{16}{5} \operatorname{arc\,sin} u^{5/2} + C$.

42. $(1/\sqrt{b}) \operatorname{arc\,sin}(\sqrt{b}x/\sqrt{a}) + C$.

44. $\pi/12$.

46. $\pi/6$.

§1.9

48. $-2x \sin x^2$.

50. $\sin x \sin(\cos x)$.

52. $\cos^2\theta - \sin^2\theta = \cos 2\theta$.

56. $-2(1 + s^4)/(1 - s^4)^{3/2}$.

58. $2 - \sqrt{2}$.

62. $(1/2 \cos^2\theta) + C$.

64. $y = -4 \cos\frac{1}{2}x + C_1 x + C_2$.

66. -1.

68. 1.

§1.10

70. $\cos x + \sin x$.

72. $-\cos x + \sin x$.

74. $-3(\cos x + \sin x)$.

§1.11

76. $4 \sin\theta \cos^3\theta - 4 \sin^3\theta \cos\theta$.

78. First show that for $\pi \le \theta \le 2\pi$,
$\cos\frac{1}{2}\theta = -\sqrt{(1 + \cos\theta)/2}$.
Then find the required formula first for $0 \le \theta \le \pi$:
$$\cos\tfrac{1}{4}\theta = \sqrt{\frac{1 + \sqrt{(1 + \cos\theta)/2}}{2}}$$
For $\pi \le \theta \le 2\pi$ one obtains
$$\cos\tfrac{1}{4}\theta = \sqrt{\frac{1 - \sqrt{(1 + \cos\theta)/2}}{2}}.$$

80. $\frac{1}{2}x - \frac{1}{4} \sin 2x + C$.

84. $\frac{1}{6}(1 - \sqrt{2}(2 - \sqrt{2})^{3/2})$.

86.

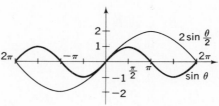

Horizontal tangent: $\cdots -3\pi$, $\pi, 3\pi, \cdots$;
inflection points: $\cdots -4\pi$, $-2\pi, 0, 2\pi, 4\pi, \cdots$;
local (and absolute) maxima: $\cdots -7\pi, -3\pi, \pi, 5\pi, \cdots$;
local (and absolute) minima: $-5\pi, -\pi, 3\pi, 7\pi, \cdots$.

88. Horizontal tangent: $\cdots -3\pi/4$,

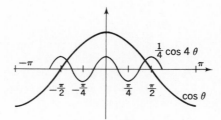

$-\pi/2, -\pi/4, 0, \pi/4, \pi/2, 3\pi/4, \cdots$;
inflection points: $\cdots -3\pi/8$, $-\pi/8, \pi/8, 3\pi/8, \cdots$;
local (and absolute) maxima: $\cdots -\pi/2, 0, \pi/2, \pi, \cdots$;
local (and absolute) minima: $\cdots -3\pi/4, -\pi/4, \pi/4, 3\pi/4, \cdots$.

90.

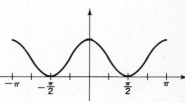

Maxima at $n\pi/2$, n even; minima at $n\pi/2$, n odd; inflection points at $n\pi/4$, n odd.

§1.14

92. $k = \frac{5}{3}$ gram/sec^2.

94. $v(t) = \sqrt{3}[\sin(t\sqrt{3}) - \cos(t\sqrt{3})]$.

96. $s(t) = \cos\frac{1}{2}t + 2\sin\frac{1}{2}t$; amplitude $= \sqrt{5}$.

§2.1

2. $2u/\cos^2(u^2)$.

6. $(2\cos^2 u + u \sin 2u)/\cos^4 u$.

§2.3

8. $\tan 3\theta = \dfrac{3\tan\theta - \tan^3\theta}{1 - 3\tan^3\theta}$.

10. $\sin\theta = 4(s - s^3)(1 - s^2)^{-2}$,
$\cos\theta = (1 - 6s^2 + s^4)(1 - s^2)^{-2}$
where $s = \tan\frac{1}{4}\theta$.

12. We know that $x = \cos\theta$ and

$y = \sin\theta$. Use (13), (16), and (17). Yes.

14. 3/8.

§2.4

16. arc tan 1 — arc tan 0 = $\pi/4$.
18. arc tan $z + [\sqrt{z}/2(1+z)]$.
20. $\dfrac{z-2}{2(z^2+z-1)\sqrt{z-1}}$.
24. Increasing for $u > 0$; concave up for $-1 < u < 1$.
26. Let $2x =$ arc tan u. Then $u =$ tan $(x/2)$. Use (13) and (16).
28. If $u > 0$, use formula for $\cos\frac{1}{2}\theta$ with $\theta = 2$ arc tan u and Exercise 26. The case $u = 0$ is trivial. If $u < 0$, use the facts that arc tan is odd and cos is even.

§2.5

30. $\frac{1}{2}$ arc tan $(2x-1) + C$.
32. arc tan $(2x-3) + C$.
36. $\frac{1}{2}$ arc tan $(x^2+1) + C$.
38. $\frac{1}{2}$ (arc tan $x)^2 + C$.
40. $z = -\frac{1}{2}$(arc tan $u^{-2}) + 1$.

§3.1

2. sec z tan $z/2\sqrt{\sec z}$.
4. $-\cot\sqrt{\theta}\csc^2\sqrt{\theta}/\sqrt{\theta}$.
6. 1.
10.

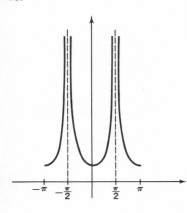

12.

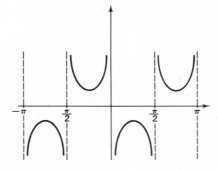

§3.2

14. $2/t\sqrt{t^4-1}$.
16. $(x/2\sqrt{x-1}) + 2x$ arc sec $\sqrt{x}$ where $x > 0$.
18. $\sqrt{\text{arc sin }u^{2/3}}/u^{1/3}\sqrt{1-u^{4/3}}$.
20. $\pi^2/32$.
24. x arc sec $\sqrt{x} - \sqrt{x-1} + C$.
26. Compute $f(1)$ and $f'(x)$ for

$f(x) =$ arc tan $x -$ arc cot $(1/x)$.

§4.1

2. .698.
4. .73567.

§4.3

6. Defined for $t > 1$. $1/t$ log t.
8. Defined for $y > 0$. $3(\log y)^2/y$.
10. Defined for $x > 0$.
$f'(x) = 1/(1+x^2)$ arc tan x.
14. Defined for $v > 1$.

$-\dfrac{\sin(\log\log v)}{v\log v}$.

§4.5

16. 2.046.
18. .754.
20. 1.275.

§4.6

22. 2958.
24. 42.32.

26. .5052.
28. 1.0025.

§4.7

30. $2/3u(\log(u^2))^{2/3}$.
32.

$-\dfrac{1+\log\log s+\log s\log\log s}{(s\log s\log\log s)^2}$.

34. $x = \pm 1$.
36. 2 log $(1+\sqrt{s}) + C$.
38. log $\sqrt{u^2+u^{-2}} + C$.
40. $-\log|\log x| + C$.
42. $\frac{2}{3}(\log z)^{3/2} + C$.
44. $y = \frac{1}{4}x^4$ log $\sqrt{x} - \frac{1}{32}x^4 + \frac{1}{32}$.
46. tan x log (sec x) — tan $x + x + C$.
48. $\frac{1}{2}\tan^2 u + C$.
50. $\frac{1}{2}$ tan $(u^2) + C$.
52. tan (log $u) + C$.
54. $\frac{1}{2}(x^2+1)$ log $(x^2+1) - \frac{1}{2}(x^2+1) + C$.
56. $\frac{2}{3}$ log 2.
60. $\frac{1}{2}\log|1+z| - \frac{1}{4}$ log $(1+z^2) + \frac{1}{2}$ arc tan $z + C$.

§5.1

2. $-5/6$.
4. 81.
6. 3.
8. $0 < x < 1$ or $x > 2$.

§5.4

10. 125 log 5.
12. $(\cos x)\,e^{\sin x}$.
14. 1.
16. $-\frac{1}{2}e^{-2x} + C$.
18. $x > 0$.

§5.5

Remark. We use the notation $e^t = \exp t$.
20. $(1 + 2xe^{x^2})/(x + e^{x^2})$.
22. exp $\sqrt{\log u}/2u\sqrt{\log u}$.
24. $e^z/(1 + e^{2z})$.
26. $(\exp\sqrt{t})(\frac{1}{4} + \frac{3}{4}\sqrt{t})$.

28. $x^{x^2+1}(2 \log x + 1)$.

32. $x^{\sqrt{x}} \cdot \sec^2 (x^{\sqrt{x}}) \cdot$
$\cdot (2 + \log x)/\sqrt{x}$.

34. $2 e^{\sqrt{x}} + C$.

36. $\sin (e^{\theta}) + C$.

38. 18.

42.

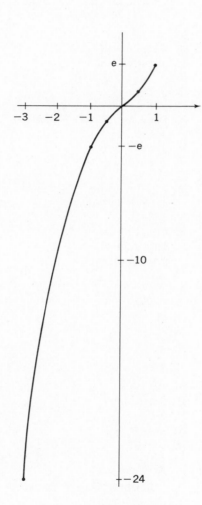

44. Relative minimum at
$(-1, -e^{-3})$, changes in concavity at $-1 - \frac{1}{3}\sqrt{3}$,
$-1 + \frac{1}{3}\sqrt{3}$, 0.

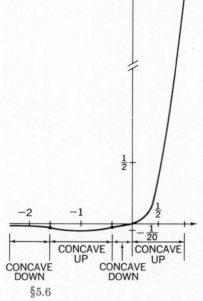

§5.6

48. $\frac{5}{2} \exp (\frac{2}{5}x + \frac{3}{5})$.

50. $\exp (x^3 + \log 2)$.

52. $\exp [(1/x) + 1]$.

§5.7

54. Approximate answers (using Table 5): 2.7, 1.8, 2.2, 13.7.

56. $(x = 1)$ Not defined;
$(x = 2)$ 5.77;
$(x = 3)$ 8.19;
$(x = 4)$ 11.54.

58. $x_0 = 10^5$.

60. $x_0 = e^{10^4}$.

§5.9

62. e.

§5.10

66. 62,500.

§5.11

68. Approximately 32 days.

§5.12

70. Write $\tanh t$ as $\sinh t/\cosh t$ and differentiate.

72. (a) $\cosh (x + y) =$
$\cosh x \cosh y + \sinh x \sinh y$.
To prove this, use (39) and compute.

(b) $\tanh (x + y) =$

$$\frac{\tanh x + \tanh y}{1 + \tanh x \tanh y}.$$

To prove this, write
$\tanh (x + y)$ as
$\sinh (x + y)/\cosh (x + y)$,
use the addition theorems for sinh and cosh, and divide the numerator and denominator by $\cosh x \cosh y$.

76. $(e^x + e^{-x})/(e^x - e^{-x})$.

78. $12 \cosh^3 2t +$
$24 \cosh 2t \sinh^2 2t$.

80. $\frac{1}{4} \sinh 2t + \frac{1}{2}t + C$. [Hint: Use (39).]

82. $t \cosh t - \sinh t + C$. [Hint: Use (39).]

84. $2x \cosh x^2 - [(\sinh \sqrt{x})/2\sqrt{x}]$.

§6.1

2. $\frac{1}{3}e^{2x} - \frac{4}{3}e^{-x}$.

4. $e^{1-x} + 1/\pi \, e^{-x} \sin (\pi e^x)$.

§6.5

6. $N_0 \exp (3/2 \cdot 10^8)$.

10. Use the formula for $G(t)$ just below (9) to obtain
$G(t) = \beta F_0 t \exp (-\beta t)$.

12. $N' + \alpha N = qt^2$; the solution is $N(t) =$
$[N_0 - (2q/\alpha^3)]e^{-\alpha t}$
$+ q(\alpha^2 t^2 - 2\alpha t + 2)/\alpha^3$.

14. $y'(t) + \alpha y(t) = g(t)$; the solution is
$y(t) = y_0 e^{-\alpha t}$
$+ e^{-\alpha t} \int_0^t e^{\alpha \xi} g(\xi) \, d\xi$
or, if $g(t) = bt$,
$y(t) = y_0 e^{-\alpha t}$
$+ (b/\alpha^2)(\alpha t + e^{-\alpha t} - 1)$.

§7.1

2. $2\pi \cosh (\alpha t) + \frac{1}{6}\pi \sqrt{2} \sinh (\alpha t)$
where $\alpha = 3/\sqrt{2}$.

4. $2 \cosh t + \sinh t - 1$.

§7.3

6. $\sec \theta$.

8. $2 + (4t \text{ arc cosh } 2t/\sqrt{4t^2 - 1})$.

10. $\frac{1}{3} \text{ arc sinh } t^3 + C$.

12. $\text{arc cosh } e^x + C$.

14. $z \text{ arc cosh } z - \sqrt{z^2 - 1} + C$.

16. $\frac{2}{3} (\text{arc cosh } x)^{3/2} + C$.

§7.7

18. Follow the hint in the text. For the third, one can also use Exercise 72 of §5.12.

20. Write down the formulas for $\sinh (\alpha + \beta)$ and $\sinh (\alpha - \beta)$ and add.

22. $((1 + x^2)/(1 - x^2))^{1/2}$.

24. Write down the definition and differentiate.

26. $\frac{1}{4} \cosh 2u + C$.

28. $\frac{1}{2} \text{ arc tanh } x^2 + C$.

Chapter 7

§1.5

4. .6949; (8): 1/12, (9): 1/216.

6. 4.85; (8): 13/6; (9): 29/18.

8. 1.593; (8): 3/8; (9): 1/8.

10. 5.01; (8): 4.11; (9): not applicable.

§1.7

12. 8.403; $E \le .0033$; for $n = 2$, $8.\overline{6}$.

14. $.69316$; $E \le .0001$.

16. Exact: $\frac{1}{2} \log 3 \approx .54930$; trapezoidal: .5534, $E \le .0185$; Simpson's: .5494, $E \le .0016$.

18. Exact: $e^{4/5} - 1 \approx 1.2255$; trapezoidal: 1.2296, $E \le .0061$; Simpson's: 1.2255, $E \le .00002$.

20. Exact: $682.\overline{6}$; trapezoidal: 788, $E \le 226.6$; Simpson's: 688, $E \le 10.\overline{6}$.

§2.2

2. 3/8.

---

4. $(5/21) (3x - 8)^{7/5} + C$.

6. 31/800.

8. $-\frac{2}{3}\sqrt{4 - 3x} + C$.

10. $- (1/7) \log | 8 - 7x | + C$.

§2.3

12. $\frac{2}{7}\sqrt{7} \text{ arc tan } [(2x - 1)/\sqrt{7}] + C$.

14. $(2 \text{ arc tan } 3\sqrt{3}) - 2\pi/3$.

16. $\frac{1}{6} \text{ arc tan } \frac{2}{3}$ (approx .098).

§2.4

20. $\dfrac{2x + 1}{1 + (2x + 1)^2}$

$+ \frac{1}{2} \text{ arc tan } (2x + 1) + C$.

22. $\frac{13}{68} + \frac{5}{16} (\text{arc tan } \frac{9}{2} - \text{arc tan } \frac{1}{2}) = .7586$ approx.

24. .0984 (approx).

§2.5

26. $\frac{3}{2} \log | x^2 + 4x + 5 | - 7 \text{ arc tan } (x + 2) + C$.

28. $-\frac{3}{8} \log | 4x^2 + 3x + 1 | + \frac{17}{4}\sqrt{7} \text{ arc tan } [(8x + 3)/\sqrt{7}] + C$.

30. .2461 (approx).

32. $\frac{2}{3} \log | 3x^2 + 3x + 1 | - \frac{10}{3}\sqrt{3} \text{ arc tan } [(6x + 3)/\sqrt{3}] + C$.

§2.8

36. $(12 + \pi)/12 + \log 3\sqrt{2}$.

38. $\log (x^2 | x - 1 |) + 1/(x - 1) + C$.

40. $\pi/3 + \log \sqrt{5/2}$.

44. $2x - \log | x |$
$+ \log | x^2 + 3x + 3 |$

$- (4/\sqrt{3}) \text{ arc tan } \dfrac{2x + 3}{\sqrt{3}}$

$+ C$.

46. $[(1 - 4x)/2(x^2 + 1)] - (2/x)$
$- 2 \text{ arc tan } x + C$.

48. $3 - \sqrt{3}\,\pi$.

50. $\log | x - 2 | - \frac{1}{2} \log | x + 1 |$
$+ \frac{1}{2} \log | x - 1 | + C$.

---

52. $\dfrac{3x}{4 (x^2 + 1)^2} - \dfrac{3x + 8}{8 (x^2 + 1)}$

$- \frac{3}{8} \text{ arc tan } x + C$.

54. $- \dfrac{1}{3x^3} - \dfrac{4x - 2}{2x^2 - 2x + 1}$

$- 4 \text{ arc tan } (2x - 1) + C$.

§3.2

2. $\pi/4$.

4. $x + \log \sqrt{e^{2x} + 1} + \text{arc tan } e^x + C$.

6. $-x + \log (e^x + 1)^2$

$- \dfrac{2}{e^x + 1} + C$

§3.3

8. $(x + 1) + 2\sqrt{x + 1} - 2 \log | x + 1 | - 2 \text{ arc tan } \sqrt{x + 1} + C$.

10. $\frac{2}{3} (x + 4)^{3/2} - 16 (x + 4)^{1/2}$

$- \dfrac{32}{(x + 4)^{1/2}} + C$.

12. $(x + 1) - \frac{5}{3}(x + 1)^{3/5}$
$+ 5 (x + 1)^{1/5}$
$- \frac{5}{2} \log [(x + 1)^{2/5} + 1]$
$- 5 \text{ arc tan } (x + 1)^{1/5} + C$.

§3.4

14. $\frac{1}{16} \log | x |$

$+ \dfrac{\sqrt{(4x - 1)/x}}{4[2 - \sqrt{(4x - 1)/x}]^2} + C$.

16. $\log \left| \dfrac{1 + \sqrt{x(1 - x)}}{1 - x} \right|$

$- 2 \text{ arc tan } y$

$+ \dfrac{2}{\sqrt{3}} \text{ arc tan } \dfrac{2y + 1}{\sqrt{3}}$

$+ C$, where

$y = \sqrt{x - (1/x)}$.

18. $\frac{8}{3}$ arc tan $\sqrt{\dfrac{4+3x}{4-3x}}$

$\qquad - \dfrac{4-3x}{3}\sqrt{\dfrac{4+3x}{4-3x}} + C.$

§3.5

22. $\frac{23}{25}\log|3\cos x + 4\sin x|$
$\quad -\frac{14}{25}x + C.$

24. $\frac{1}{6}\log\left|\dfrac{1+\tan(x/2)}{1-\tan(x/2)}\right|$

$\qquad + \dfrac{1+2\tan(x/2)}{3(1+\tan(x/2))^2} + C.$

26. $\frac{1}{8}\log|\tan(x/2)|$
$\quad -\frac{1}{3}\log|1-\tan^2(x/2)|$
$\quad +\frac{25}{48}\log|4-\tan^2(x/2)| + C.$

28. $\csc x + C.$

30. $\frac{1}{24}\log\left|\dfrac{4+3\tan x}{4-3\tan x}\right| + C.$

32. Write
$a\cos x + b\sin x$
$= r\sin(x+\theta)$
where $r = \sqrt{a^2+b^2}$,
$\sin\theta = a/r$, $\cos\theta = b/r$.
The result is
$-\log|\csc(x+\theta)+\cot(x+\theta)|.$

§3.6

34. $\dfrac{-2}{1+\tanh(x/2)} + C.$

36. $\frac{2}{7}\big[8\log|3-\tanh(x/2)|$
$\quad -3\log|1-\tanh(x/2)|$
$\quad +5\log|1+\tanh(x/2)|$
$\quad -4/(1+\tanh(x/2))\big] + C.$

38. $\displaystyle\int \frac{2y\,dy}{(2+3y-2y^2)(3-y^2)}.$

§3.7

40. $(16-7\sqrt{2})/120.$

42. $(2-x^2)/\sqrt{1-x^2} + C.$

46. arc cos $x - \dfrac{\sqrt{1-x^2}}{x} + C.$

§3.8

48. $(x^2-1)^{1/2} + \frac{1}{3}(x^2-1)^{3/2} + C.$

50. $\frac{1}{2}x\sqrt{x^2-1} - \frac{1}{2}$arc cosh $x + C.$

52. $\sqrt{x^2-1}$ - arc sec $x + C.$

§3.9

56. $\log|\sqrt{1+x^2}+x| + C.$

58. $x/\sqrt{1+x^2} + C.$

§3.10

60. $\dfrac{3\sqrt{3}}{16}\displaystyle\int \frac{\xi^3}{\sqrt{\xi^2-1}}\,d\xi,$

$\quad \xi = \dfrac{2}{\sqrt{3}}\,x,$

form (**VII**).

64. $\dfrac{1}{\sqrt{2}}\displaystyle\int \frac{d\xi}{\sqrt{\xi^2+1}},$

$\quad \xi = 2x-3,$
form (**VIII**).

66. $\displaystyle\int \frac{\sqrt{\xi^2-1}}{8\xi}\,d\xi,$

$\quad \xi = 2x+1,$
form (**VII**).

68. $\displaystyle\int \frac{d\xi}{3\sqrt{\xi^2-1}},$

$\quad \xi = \dfrac{3x-5}{4},$

form (**VII**).

§4.1

2. $(3x^4 - 12x^3 + 34x^2 - 68x + 68)e^x + C.$

4. $(4x^3 - 24x^2 + 94x - 186)\cdot e^{x/2} + C.$

6. $\frac{1}{2}x^2 e^{2x} - \frac{1}{2}xe^{2x} + \frac{1}{2}e^{2x} + \frac{1}{3}x^2 e^{3x}$
$\quad -\frac{2}{9}xe^{3x} + \frac{2}{27}e^{3x} + C.$

8. $\frac{1}{2}(x^4 - 2x^2 + 2)e^{x^2} + C.$

§4.2

12. $2x^{1/2}(\log^3 x - 6\log^2 x + 24\log x - 48) + C.$

14. $-[\log^2(2x)+2\log(2x)+2]/x + C.$

16. $\frac{4}{3}x^3\big[\log^3(3x) - \log^2(3x) + \frac{2}{3}\log(3x) - \frac{2}{9}\big] + C.$

§4.3

18. $(\pi^2 + 18\pi - 6\sqrt{3}\pi - 18)/18$

20. $x^2(\sin x - \cos x) + 2x(\sin x + \cos x) - 2(\sin x - \cos x) + C.$

22. $-x^4\cos x + 4x^3\sin x + 12x^2\cos x - 24x\sin x - 24\cos x + C.$

24. $(\pi/2)^{100}.$ (Do not try to use the formula.)

26. $(x-\pi)^4\sin x + 4(x-\pi)^3\cos x - 12(x-\pi)^2\sin x - 24(x-\pi)\cos x + 24\sin x + C.$

§4.4

28. $\frac{1}{5}\cos^4 x\sin x + \frac{4}{15}\cos^2 x\sin x + \frac{8}{15}\sin x + C.$

30. $\frac{1}{4}\cos^3 x\sin x + \frac{1}{8}x + \frac{1}{16}\sin 2x + C.$

34. $\frac{1}{5}\cos x\sin^4 x + \frac{4}{15}\cos x\sin^2 x + \frac{8}{15}\cos x + C.$

36. $1/384.$

§4.5

38. $(\sqrt{2}-2)/8.$

40. $\frac{1}{4}\sin x + \frac{1}{12}\sin 3x - \frac{1}{20}\sin 5x - \frac{1}{36}\sin 9x + C.$

42. $-\frac{1}{4}\cos 2x + \frac{1}{16}\cos 4x - \frac{1}{32}\cos 8x + C.$

44. $\sin x + \frac{1}{5}\sin 5x + \frac{3}{2}\cos x - \frac{1}{2}\cos 3x + C.$

46. $\frac{1}{8}x + \frac{1}{16}\sin 2x - \frac{1}{64}\sin 8x - \frac{1}{80}\sin 10x + C.$

Chapter 8

§1.1

2. $3/2.$

4. $\sqrt{3}/3.$

8. $216/125.$

10. $\log(e^2 - e).$

§1.4

12. $3\sqrt{3}$.
14. $\pi/2$.
16. 1.
18. $1 - \sqrt{3}$.
20. $(3 + \sqrt{21})/4$.

§1.5

22. $.9; |R(x)| \le .04$.
24. $10 - 1/300;$
    $|R(x)| \le 1/500,000$.
26. $-.02; |R(x)| \le .0003$.
28. $.01; |R(x)| \le 1/10,000$.
30. $1 - .02\pi; |R(x)| \le (.0002)\pi^2$.

§2.1

2. $\xi = 1/2, \eta = 8^{-1/4}$.
4. $\xi = (e - 1)/\sqrt{2(e - 2)}$,
   $$\eta = \left(\frac{2(e - 1)^3}{3(e - 1)^2 - 6(e - 2)}\right)^{1/3}.$$
6. $7 + 195/2744$,
   $|R| \le 1/250,000$.
8. $1.0588, |R| \le .000056$.

§2.2

12. $1 + \frac{4}{3}x + \frac{2}{9}x^2 - \frac{4}{81}x^3 + \frac{5}{243}x^4$.
14. x^2.
16. 0.
18. $(x - 1)^2 - (x - 1)^3 + \frac{11}{12}(x - 1)^4$.
20. $1 + x^5 + \frac{1}{2}x^{10}$.
    [Hint: Find the Taylor polynomial $P(x)$ of degree 2 of e^x. Then $P(x^5)$ is the desired polynomial.]
22. $1 + \frac{4}{3}x^4 + \frac{2}{9}x^8 - \frac{4}{81}x^{12}$.
    [Hint: Find the Taylor polynomial $P(x)$ of degree 3 for $(x + 1)^{4/3}$. Then $P(x^4)$ is the desired polynomial.]

§2.3

24. $455/31104$.
26. $e/120$.
28. $29\sqrt{2}|x|^5/64$.
32. $e/1 \cdot 2 \cdot 3 \cdots (n + 1)$.

§2.4

34. Suppose $j \le n - j$. Write down
    $$\binom{n}{n - j} \quad \text{and} \quad \binom{n}{j}$$
    and see that they are identical.
38. Use Exercise 36 with $n = m - 1$. Then, by Exercise 33, note that
    $$\binom{m}{1} = (m/1)\binom{m - 1}{0}$$
    or
    $$\binom{m - 1}{0} = (1/m)\binom{m}{1}.$$
    Similarly,
    $$\binom{m - 1}{1} = (2/m)\binom{m}{2},$$
    and so forth.

§2.5

40. $1 + 53744/(81 \times 10^5)$;
    $|R| < 5 \times 10^{-9}$.
42. $1.183; |R| < .0014$.
44. $1 + 59/243; |R| < 22/729$.

§2.8

48. $1 + x^8 + x^{16} + \cdots + x^{8m}$;
    $R = x^{8(m+1)}/(1 - x^8)$. Here, m is the smallest integer with $8m \ge n$.
50. $x^2 - x^3 + \cdots + (-1)^{n-2}x^n$;
    $R = (-1)^{n-1}x^{n+1}/(1 + x)$.
52. $-12x^2 + 56x^6 - \cdots + (-1)^m 4m(4m - 1)x^{4m-2}$;
    $$R = \frac{(-1)^{m+1}(4m+4)(4m+3)x^{4m+2}}{1+x^4}.$$
    Here, m is the smallest integer with $4m - 2 \ge n$.
54. Consider two cases: (i) $0 < t < x$ (Note that always $x < 1$) and (ii) $x < t < 0$.
56. $1.344; |R| \le .0486$.

58. $1 + 23/240; |R| \le 1/504$.
60. $74/375; |R| \le 1/15,625$.
62. $191/1536; |R| \le 1/163,840$.
64. $6229/13,440; |R| \le 1/4608$.

§2.10

66. $x^2 - \dfrac{x^6}{1 \cdot 2 \cdot 3} + \dfrac{x^{10}}{1 \cdot 2 \cdot 3 \cdot 4 \cdot 5}$
    $$- \cdots + \frac{(-1)^n x^{2(m+1)}}{1 \cdot 2 \cdot 3 \cdots (2m + 1)};$$
    $$|R| \le \frac{|x|^{2(2m+3)}}{1 \cdot 2 \cdot 3 \cdots (2m + 3)}.$$
    Here, m is the smallest integer with $2m + 2 \ge n$.

68. $x^4 - \dfrac{x^{10}}{1 \cdot 2 \cdot 3} + \dfrac{x^{16}}{1 \cdot 2 \cdot 3 \cdot 4 \cdot 5}$
    $$- \cdots + \frac{(-1)^m x^{6m+4}}{1 \cdot 2 \cdot 3 \cdots (2m + 1)};$$
    $$|R| \le \frac{|x|^{6m+10}}{1 \cdot 2 \cdot 3 \cdots (2n + 3)}.$$
    Here, m is the smallest integer with $6m + 4 \ge n$.

§2.11

70. $73/12$.
72. $7/10$.
74. 747.
76. $-751/784$.

78. $\displaystyle\sum_{k=1}^{5} \sqrt{2k - 1}$

80. $\displaystyle\sum_{k=1}^{4} \frac{1 \cdot 3 \cdots (2k - 1)}{1 \cdot 4 \cdots (3k - 2)}$.

82. $\displaystyle\sum_{k=1}^{5} \frac{(2k - 1)}{3k}$.

84. $\displaystyle\sum_{k=1}^{5} \frac{(2k - 1)^2 (2k)^2}{(2k + 1)(2k + 2)}$.

86. Write out the sum; apply Exercise 35 to each term except the first one. Then

$$\binom{n+1}{0} = \binom{n}{0} = 1.$$

All other terms except the last cancel in pairs.

88. Write out the sum. Rewrite the result in Exercise 35 as

$$\binom{n}{j-1} = \binom{n+1}{j} - \binom{n}{j}.$$

Replace j by $(k+1)$. Apply this to each term in the sum. Everything drops out except

$$\binom{k+n+1}{k+1} \quad \text{and} \quad \binom{k}{k+1}.$$

What is the value of

$$\binom{k}{k+1}?$$

§3.3

4. 2, 0, 0, 2, 6, 12.
6. 1, 1/4, 1/3, 1, 5, 36.
8. 3, 12, 5, 18, 7, 24.
10. 2, 4, 6, 15, 5005.
12. $b_k = 2^{k-2}$, $k = 2, 3, \cdots$.
14. $a_k = (-1)^k(3k+1)$, $k = 0, 1, 2, \cdots$.

16. $z_\lambda = \dfrac{(2\lambda - 4)(2\lambda - 3)}{(2\lambda - 2)}$

$\quad\quad = \dfrac{(\lambda - 2)(2\lambda - 3)}{(\lambda - 1)}$.

§3.4

20. Does not converge.
24. Does not converge.
26. 0.
28. Does not converge.

§3.7

30. 0. Use Theorem 2.
32. 2/3.
34. 0.

36. Show that $|u_r|$ converges to 0.
38. (i) $0 < q < 1$. (I) Let $q < A$. Want $q^{n/(n+1)} < A$. Take logs of both sides and reason carefully. (II) If $0 < q < 1$ and $0 < a < 1$, then $q^a > q$.
   (ii) $q = 1$.
   (iii) $q > 1$. (I) $q^{n/(n+1)} < q$. (II) Trivial for $B < 1$. Let $q > B \geq 1$. Take logs and reason carefully.
40. 2.
42. 0.
46. 0.
48. Converges to 0. (Let $b = 1/a$ and use Example 1.)
50. Diverges.

§3.8

52. Nondecreasing; bounded above by 2/3.
54. Nondecreasing; unbounded.
56. Nonincreasing; bounded below by 0.
58. Nonincreasing; bounded below by 0.
60. Nonincreasing; bounded below by 0.

§4.2

2. 1, 3, 7, 15, 31, 63.
4. 1/2, 7/8, 57/48, 561/384.
8. $\sum_{n=0}^{\infty} x^{2n}/n!$, converges for all x.
10. $\sum_{n=0}^{\infty} a_n(x-1)^n$ where $a_0 = 1$, $a_1 = -1/2$,

$$a_n = \frac{(n+1)(n+3)\cdots(2n-1)}{2\cdot4\cdots n\cdot2^n}$$

   for n even, $n > 0$, and

$$a_n = -\frac{(n+2)(n+4)\cdots(2n-1)}{2\cdot4\cdots(n-1)\cdot2^n}$$

   for n odd, $n > 1$.
12. $\sum_{n=0}^{\infty} b_n(x-1)^n$ where $b_n = -a_n$ for n odd, $b_n = a_n$ for n even, a_n as in Exercise 10.

§4.4

14. Converges to 1.
16. Diverges.
18. Converges.
20. Diverges.
22. Diverges.
24. Diverges.

§4.6

28. Converges to $1/(1-u) + 1/(1-u^2)$.
30. Converges to $\dfrac{1-x^2}{1-x^4}$.
   (Use Theorem 4.)
32. Converges to $\sec^2 x$.
34. Converges to $2/(1-q)^3$.

§4.9

36. 128/99.
38. 1557/4995.
40. 7/12.
42. 39/62.
44. Diverges. (Compare with $\sum 1/k$).
46. Diverges.
50. Diverges.
52. Write $0 \leq a_i - c_i \leq b_i - c_i$. Use Theorem 4, the comparison test, and Theorem 4 again.

§4.10

56. Diverges.
58. Converges.
60. Converges.
62. Converges.
64. Use $s = 2$.
66. Use $s = 3/2$.

§4.11

68. Test fails.
72. Converges.
74. Diverges.
76. Diverges.
78. Test fails.

§4.14

80. Converges absolutely.

82. Converges absolutely.
84. Converges absolutely.
86. Diverges.
90. Converges absolutely.

§4.15

92. (a) Write down the given formula for $\cos x$ twice, once with index j and once with index k. Use the formula for AB stated in Theorem 12 (replacing n by m). Then, replace k by $m - j$ and simplify. Recall that

$$\binom{2m}{2j} = \binom{2m}{2m - 2j}$$

and write down the expression for

$$\frac{1}{(2m)!}\binom{2m}{2m - 2j}.$$

(b) Follows immediately from hint given.

94. Follow hint, noting that, except for the first term, 1, every term appearing in 92 (b) occurs also in 93 (b) but with opposite sign.

§5.3

4. 3/2.
6. 1/e.
8. $R/|p|$.
10. R.
12. Write down series for $\cos x$ and differentiate term by term.
14. Write down series for arc tan t and integrate (from 0 to x) term by term.

§5.5

16. $1 + \frac{2}{3}(x - 1) - \frac{1}{9}(x - 1)^2 + \sum_{n=3}^{\infty} a_n (x - 1)^n$ where

$a_n =$

$$(-1)^{n+1}\frac{1 \cdot 2 \cdot 4 \cdot 7 \cdots (3n - 5)}{n!3^n}.$$

$R = 1$.

18. $\sum_{n=0}^{\infty} \dfrac{n + 1}{2^{n+2}} x^n$. $R = 2$.

20. $\sum_{n=0}^{\infty} a_n (x - \frac{1}{6}\pi)^n$ where

$$a_n = (-1)^{[(n+1)/2]+1}\frac{1}{2 \cdot n!} \text{ for}$$

n odd,

$$a_n = (-1)^{[(n+1)/2]+1}\frac{\sqrt{3}}{2 \cdot n!} \text{ for } n$$

even where $[(n + 1)/2]$ denotes the greatest integer not exceeding $(n + 1)/2$. $R = +\infty$.

22. $\sum_{n=0}^{\infty} (n + 1)x^n$. $R = 1$.
24. $\sum_{n=0}^{\infty} x^{2n}/n!$. $R = +\infty$.

§5.6

28. $|R_{n+1}(x)| \le \dfrac{|x - \frac{1}{6}\pi|^{n+1}}{(n + 1)!} \to 0$

for all x.

30. $|R_{n+1}(x)| \le |x|^{n+1} \to 0$ for $|x| < 1$.

32. $|R_{n+1}(x)| \le$

$$\frac{|\text{polynomial of degree } (n + 1)| \cdot e^{x^2}}{(n + 1)!} \to 0$$

for all x (see Example 3, §3.5).

34. Write out series for $\cos t$, substitute u^2 for t, integrate from 0 to x. Good for all x, since series for $\cos t$ is good for all t.

§5.7

36. $1 - \frac{1}{3}x + \dfrac{1 \cdot 4}{2!3^2} x^2$

$$- \frac{1 \cdot 4 \cdot 7}{3!3^3} x^3 + \frac{1 \cdot 4 \cdot 7 \cdot 10}{4!3^4} x^4$$

$$+ (-1)^n \frac{1 \cdot 4 \cdot 7 \cdots (3n - 2)}{n!3^n} x^n$$

$+ \cdots$. Valid for $|x| < 1$.

38. $1 + \frac{2}{5}x^2 - \dfrac{2 \cdot 3}{2!5^2} x^4$

$$+ \sum_{n=3}^{\infty} (-1)^{n+1} a_n x^{2n},$$

$$a_n = \frac{2 \cdot 3 \cdot 8 \cdot 13 \cdots (5n - 7)}{n!5^n}.$$

Valid for all x.

40. $x + \frac{5}{4}x^2 + \dfrac{5 \cdot 3}{3!2^2} x^3$

$$+ \frac{5 \cdot 3 \cdot 1}{4!2^3} x^4$$

$$- \frac{5 \cdot 3 \cdot 1 \cdot 1}{5!2^4} x^5$$

$$+ \sum_{n=5}^{\infty} (-1)^{n+1} a_n x^{n+1}.$$

$$a_n = \frac{5 \cdot 3 \cdot 1 \cdot 1 \cdot 3 \cdot 5 \cdots (2n-7)}{(n + 2)!2^n}.$$

Valid for $|x| < 1$.

§5.9

42. $x + x^2 + \frac{1}{3}x^3 - \frac{1}{30}x^5 - \frac{7}{360}x^6$. Valid for all x.
44. $1 + 2x + \frac{5}{2}x^2 + \frac{8}{3}x^3 + \frac{65}{24}x^4$. Valid for $|x| < 1$.
46. $1 + 2x + \frac{7}{2}x^2 + 7x^3 + \frac{337}{24}x^4$. Valid for $|x| < \frac{1}{2}$.
48. $x - x^2 + \frac{3}{2}x^3 - \frac{7}{3}x^4$.
50. $1 - x + \frac{1}{2}x^2 - \frac{1}{6}x^3 - \frac{1}{8}x^4$.

§5.10

52. Compute $x/(e^x - 1)$ by the method of §5.9. Then add $x/2$. The result is $\sum_{j=0}^{\infty} b_j x^{2j}$. $B_0 = 1$.
54. $B_2 = -1/360$.

56. $\alpha_1 = 1/6$, $\alpha_2 = 7/360$,
    $\alpha_3 = 31/15$, 120.
58. $a_2 = -1/12$, $a_3 = 0$,
    $a_4 = 1/1440$.

§5.11

62. $1 - \frac{1}{2}x^2 + \frac{5}{24}x^4 - \frac{1}{20}x^6 + \frac{11}{1120}x^8$.
64. $1 + x + \frac{1}{2}x^2 - \frac{1}{6}x^3 - \frac{1}{3}x^4 + \frac{1}{30}x^5$.
66. $1 - \frac{1}{2}x^2 + \frac{1}{2}x^3 - \frac{5}{12}x^4 + \frac{1}{3}x^5$.

§6.1

2. 2.
4. 2.
6. 1/2.
8. 0.
10. $-1/2$.
14. $-\sqrt{2}/2$.

§6.3

16. 0.
18. 1.
20. $+\infty$.
22. 0.
24. 1.
28. 1.
30. $-\infty$.

§6.4

32. 0.
34. 0.
36. 0.
38. 1.
40. $+\infty$.
44. $+\infty$.
46. $+\infty$.

§7.2

2. Partial sums of P (also N) form nondecreasing bounded sequences.
4. Follow method of text. Try grouping positive terms in packets, each of which is larger than 1/3, for convenience. For example, $1/5 + 1/7 + 1/9 > 3/9 = 1/3$.

§7.5

2. Substitute $x - x_0$ for x in the statement and proof.

§9.1

2. 9/4.

§9.2

4. All y.
6. $y = 1/2$.

Chapter 9

§1.2

2. $(-1, 8)$.
6. $(7, -3/2)$.
8. $(4 \cos \frac{1}{9}\pi, 4 \sin \frac{1}{9}\pi)$.

§1.6

10. $P = D = (0, -2)$.

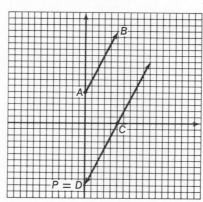

12. $P = (8/5, -4/5)$.

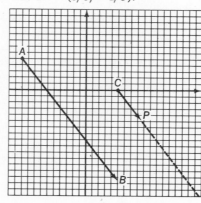

14. $P = (3, -6)$.

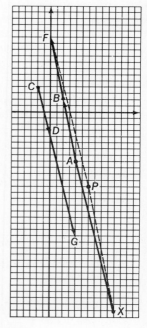

§1.9

16. $\mathbf{a} + \mathbf{b} = (6, -7)$; $\mathbf{a} - \mathbf{b} = (-2, -1)$; $2\mathbf{a} = (4, -8)$; $(-3)\mathbf{b} = (-12, 9)$.
18. $(3, 2)$.
20. $(-3, -3)$.
24. $\alpha_1 = 5$, $\alpha_2 = 5/2$.
26. $\alpha_1 = \alpha_2 = \beta_1 = \frac{1}{2}$, $\beta_2 = -\frac{1}{2}$.

§1.10

28. Use the associative law for numbers.
30. Use the distributive law for numbers.
32. Show that, under this definition, **9** is violated. [Hint: Consider the ordered pair $(1, 1)$ and $\alpha = 1$.]

§1.11

34. Segments defining the vectors lie on the line $y = \frac{4}{3}x$ through O.

36. No.
38. Always independent. Assume **a** and **a** + **b** are dependent. Then they are represented by segments on a line through O. Then so are **a** and **b**.

§1.12

40. No. (Indeed, if **a** $\neq$ **0**, **b** $\neq$ **0**, **c** $\neq$ **0** and **a** + **b** + **c** = **0**, then the three vectors can be represented by the sides of a triangle.)
42. $(19, -23, -12)$.
44. $(5, 1, 15)$.
46. $2, -3, -3/5$.
48. From the first two equations, write **b** in terms of e_1 and e_2. Substitute this expression for **b** in the second and third equations and solve.
50. Use Exercise 39.

§2.2

2. $(1, 1)$.
6. $\mathbf{X} = \frac{1}{3}\mathbf{P} + \frac{2}{3}\mathbf{Q}$,
   $X = (2/3, 11/3)$.
8. $3/2$.

§2.3

10, 12, 14. Let **A** = **O**. Use Theorem 1, §2.2.
16. Let **A** = **O**. Compare the last three sentences of the proof of Example 2.

§3.2

2. $(\sqrt{2}\,\pi/4)$.
4. $(3\sqrt{2}, 3\pi/4)$.
6. $(2, 8\pi/9)$.
8. $(-1, 1)$.
10. $(\sqrt{2}/4, \sqrt{2}/4)$.
12. $(-\sqrt{2}/2, -\sqrt{6}/2)$.
14. $\sqrt{10 - 3\sqrt{2}}$.
16. $a^2 = b^2 + c^2 - 2bc \cos A$.
18. 12.09.

§3.4

20. $3r \sin\theta - 2r \cos\theta = 8$.

24. $r = \theta$.
26. $(2x + 3y)^2 = 1$.
28. $x^2 + y^2 - 2x = 3$.
30. $y = (x^2 + y^2)^{3/2}$.
32.

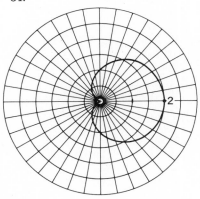

34.

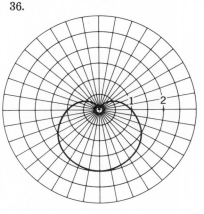

36.

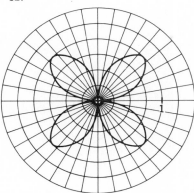

38.

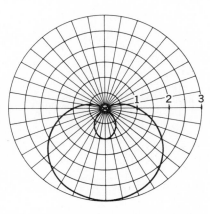

40. Upper half of the graph shown.

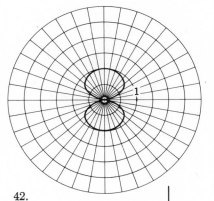

42.

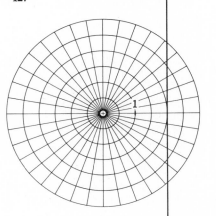

44.

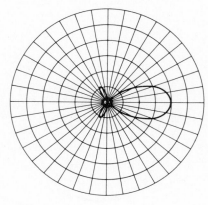

46.

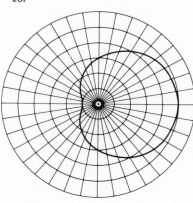

§4.1

4. 6.
6. $(0, 4, 17/3)$, $(17/2, 0, -17/2)$, $(17/5, 17/5, 0)$.

§4.2

8. $x = 3 + 2t, y = 4, z = 5 - 2t$.
10. $x = t, y = 1 - t, z = 0$.
12. $x = -3 + 15t, y = 2 - 13t, z = 0$.
16. No.
18. $(2, 0, 4)$.
20. No.

§4.3

22. $-2y + z - 5 = 0$.
24. $4x + 34y + 7z - 93 = 0$.

26. $x - y = 0$.
30. $u = 1/100$.

§4.5

32. $x + y + z - 10 = 0$.
34. $a = 2$.
36. $\cdots B = C = 0$.
38. $x = t, y = t - 1, z = t - 2$.
40. Parallel.
44. $x = -y, z = 0$.
46. $6x = 7y, y = 2z$.
48. $x = 0, y = 0$.
50. $5x + 6y = 35, y + 5z = 30$.
52. $x = 1, y = 1, z = t$.
54. $x = 0, y = 0, z = t$.
58. $(\frac{1}{2}t - \frac{5}{2}, \frac{1}{2}t - \frac{3}{2}, t)$ for all t.

§4.6

60. Meet at $(45, 45/2, 15)$.
62. Meet at $(2, 3, 4)$.
64. Meet at $(28, 14, 7)$.
66. l lies on σ.
68. Meet at $(2, 1/2, 1/2)$.
70. Parallel.
72. Skew.
74. Skew.
78. Meet at $(5, 2, 1)$.

§4.7

80. $(1, 3, -1)$, 3.
82. $(2, -3, 1/2), \sqrt{2}$.
84. $(0, 0, 4), \sqrt{3}$.
86. No intersection.
88. $z = \pm 2\sqrt{21}$.

§5.2

2. $(2, 3\pi/4, 1)$.
4. $(1, 5\pi/6, 1/4)$.
6. $(\sqrt{5}, \text{arc cos } \frac{1}{5}\sqrt{5}, 3)$.
8. $(5, \text{arc cos } \frac{3}{5}, -\frac{1}{2})$.
10. $r^2 + 2z^2 + 2z - 5 = 0$.
12. $r \cos \theta - 2r \sin \theta + 3z - 5 = 0$.
16. $x^2 + y^2 - z^2 = 0$. See Fig. 9.47, $m = 1$.

§5.3

18. $(\sqrt{3}, \pi/4, \text{arc cos } \frac{1}{3}\sqrt{3})$.

20. $(5\sqrt{10}/2, 7\pi/4, \text{arc cos } \sqrt{\frac{3}{5}})$.
22. $(\sqrt{14}, \text{arc tan } 2, \text{arc cos } \sqrt{\frac{3}{14}})$.
24. $(1, \pi/2, \pi/2)$.
26. $\rho \cos \phi = 1$.
28. $\rho \sin \phi = 3$.

§6.2

2. 1.73.
4. 30.
6. $-1/2$.
8. $58° 54'$ (approx).
10. $120°$.
12. $99° 29'$ (approx).

§6.4

14. ± 3.
16. $60°$.
20. No.

§6.6

22. $2/\sqrt{29}, -3/\sqrt{29}, 4/\sqrt{29}$.
24. $54° 41'$ (approx).
26. $7/\sqrt{206}, -11/\sqrt{206}, 6/\sqrt{206}$.
28. $0°, 20°, 70°$.
30. $r \cos \theta / \sqrt{r^2 + z^2}$, $r \sin \theta / \sqrt{r^2 + z^2}, z / \sqrt{r^2 + z^2}$.
32. $2/\sqrt{30}, -1/\sqrt{30}, 5/\sqrt{30}$.

§6.9

34. $(3/\sqrt{38}, -2/\sqrt{38}, 5/\sqrt{38})$.
36. All are $54° 41'$ (approx).
38. $1/\sqrt{29}$.
40. $137/\sqrt{30}$.
44. $(-1, 1)$, $(-1/3, 1/9)$.
46. $((11 + 7\sqrt{2})/3, (7 + 14\sqrt{2})/3)$ and $((11 - 7\sqrt{2})/3, (7 - 14\sqrt{2})/3)$.

§7.1

2. $(-190, 424, -160, -27)$.
4. No.
8. No.

§7.2

10. There are no such numbers u, v, w.
12. $10x_1 + 5x_2 + 20x_3 + 4x_4 - 20 = 0$.

14. $(18, -8, -8, 10)$.
16. $2x_3 - x_4 = 0$.

§7.3

18. $60°$.
20. Let $\mathbf{a} = (A_1, A_2, \cdots, A_n)$ and $\mathbf{x} = (x_1, x_2, \cdots, x_n)$ and compute $(\mathbf{a}, \mathbf{x})$. Then consult §7.2.
22. $\alpha > 0$: sphere with center $\mathbf{a}$ and radius $\sqrt{\alpha}$; $\alpha = 0$: the point $\mathbf{a}$; $\alpha < 0$: the empty set.

Chapter 10

§1.4

2. $a = 6$, $b = 3$. Center: origin;

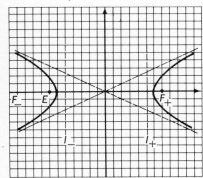

    foci: $(-3\sqrt{5}, 0)$, $(3\sqrt{5}, 0)$; vertices: $(-6, 0)$, $(6, 0)$; asymptotes; $y = \pm\frac{1}{2}x$; axis: x axis; directrices: $x = \pm12\sqrt{5}/5$.
4. $a = 9$. Focus: $(9, 0)$; vertex: $(0, 0)$; axis: x axis; directrix: $x = -9$.

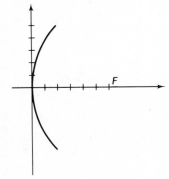

6. $a = 5$, $b = 2$. Center: origin; foci: $(-\sqrt{21}, 0)$, $(\sqrt{21}, 0)$; vertices: $(5, 0)$, $(-5, 0)$, $(2, 0)$, $(-2, 0)$; major axis: x axis; minor axis: y axis; directrices: $x = \pm25/\sqrt{21}$.

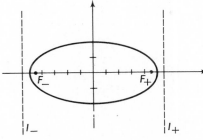

8. $x^2/4 + y^2/3 = 1$. $a = 2$, $b = \sqrt{3}$. Center: origin; foci: $(-1, 0)$, $(1, 0)$; vertices: $(2, 0)$, $(-2, 0)$, $(\sqrt{3}, 0)$ $(-\sqrt{3}, 0)$; major axis: x axis; minor axis: y axis; directrices: $x = \pm4$.

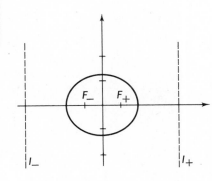

§1.5

10. $x^2 - 4x + 4y^2 - 8 = 0$.
12. $5x^2 + 4y^2 - 40y - 400 = 0$.
14. $4x^2 + y^2 - 8x - 10y + 13 = 0$ and $4x^2 + y^2 - 24x - 2y + 21 = 0$.
18. $3y^2 - 5x^2 - 6y - 27 = 0$.
20. $\frac{1}{64}(y - 8)^2 - \frac{1}{36}(x + 2)^2 = 1$.
22. Foci: $(\pm\sqrt{34}, 0)$; vertices:

$(\pm3, 0)$; eccentricity: $\sqrt{34}/3$; asymptotes: $\pm5/3x$.

§1.6

24. $3x + 20y = 50$.
26. $5x + 6y + 45 = 0$.
28. $x - 2\sqrt{2}\,y = 1$.
30. $7\sqrt{3}\,x + 2y = 28$.
32. $y \pm 15 = \frac{9}{25}(\sqrt{34} \mp 15)$.
36. $x - 2y = 8$ and $91x - 442y = -1352$.

§1.7

38. $r = 5/(1 - \cos\theta)$.
40. $r = 192/(4 - \cos\theta)$.
42. $r = 1/(1 - 2\cos\theta)$.
44. $r = 1/(90 - 900\cos\theta)$.
46. Parabola. Focus $(0, 0)$, vertex $(-1, 0)$.

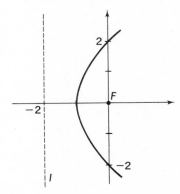

48. $\epsilon = 1/2$, $p = 2$. Ellipse. Foci $(0, 0)$ and $(4/3, 0)$, vertices at $(-2/3, 0)$, $(2, 0)$, $(2/3, 2\sqrt{3}/3)$ and $(2/3, -2\sqrt{3}/3)$.

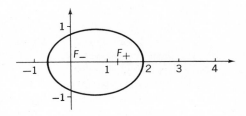

§1.8

52. Ellipse. Foci: $(0, \pm 4\sqrt{2}/3)$;
vertices: $(\pm 2/3, 0)$ and
$(0, \pm 2)$.

54. Hyperbola.
Foci: $(\pm \sqrt{15}/6, 0)$;
vertices: $(\pm 2/3, 0)$.

56. Ellipse.
Foci: $(\pm \sqrt{9{,}999}/10{,}000, 0)$;
vertices: $(\pm 1/100, 0)$ and
$(0, \pm 1/10{,}000)$.

§2.1

2. Right-handed.

§2.3

6. $X = -(6 + \sqrt{3})/2\sqrt{3}$,
$Y = (\sqrt{3} - 2)/2$.

8. $X = (3\sqrt{3} - 2)/2$,
$Y = (3 + 2\sqrt{3})/2$.

10. $X = -\sqrt{2}\cos 15°$,
$Y = -\sqrt{2}\sin 15°$.

12. $(2, 0)$.

§2.4

14. $-56° 18'$, approx.

16. $60° 15'$, approx.

18. $13° 40'$, approx.

20. $37° 52'$, approx.

22. $-26° 34'$, $-39° 38'$, $78° 16'$,
approx.

24. 4.75, approx.

§2.5

28. $y = \frac{1}{3}(a \pm 2^{3/2}a^{1/2})x$; where
$a = 4 - \sqrt{7}$.

30. $y \pm x = \frac{1}{2}\sqrt{2}$.

§2.6

32. $P = (4, 4)$. l has slope $1/2$.
Focus is $(1, 0)$. l_1 has slope
$4/3$; l_2 has slope 0. Substitute
into (8) the values $n = 1/2$,
$m_1 = 4/3$, $m_2 = 0$.

34. Points of intersection:
$x = \pm \frac{5}{3}$, $y = \pm 8\sqrt{2}/3$. Get
slopes of tangents at these
points and show that they are
negative reciprocals of each
other.

§3.1

2. 2.

4. 4.

6. 6.

8. Degree 6: all t except $t = 1$;
degree 5: $t = 1$.

10. (a) 4, (b) 5, (c) 6.

§3.4

12. $X^2 - Y^2 + 2\sqrt{2}\,X = 0$.

14. $3X^2 + Y^2 - 6 = 0$.

16. $5X^2 - 3Y^2 - 10X + 10\sqrt{3}Y - 25 = 0$.

20. $-X^2 + 14Y^2 + 50Y = 0$.

22. $377X^2 + 39Y^2 - 52 = 0$.

§3.5

26. No linear terms.

28. Discriminant $= -15$.
$5\xi^2 - 3\eta^2 - 5 = 0$.

30. Discriminant $= 0$.

32. Discriminant $= -14/25 \neq 0$.
$\frac{1}{5}\xi^2 - \frac{14}{5}\eta^2 + \frac{125}{14} = 0$.

34. No linear terms.

§3.8

36. Hyperbola. Asymptotes $x = -1$, $y = -1$. Vertices $(0, 0)$
and $(-2, -2)$. Foci $(\sqrt{2} - 1, \sqrt{2} - 1)$ and $(-\sqrt{2} - 1, -\sqrt{2} - 1)$.

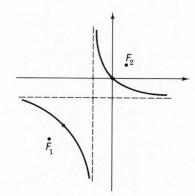

40. Hyperbola.

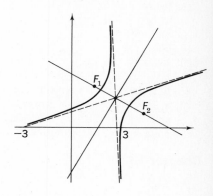

Foci:
$(\sqrt{2} + 4\sqrt{3}/3 - \sqrt{2/3} + 2)$,
$(-\sqrt{2} + 4\sqrt{3}/3, \sqrt{2/3} + 2)$.
Vertices:
$(11\sqrt{3}/6, 3/2)$, $(5\sqrt{3}/6, 5/2)$.
Asymptotes:
$$y = \frac{1 + \sqrt{15}}{2 + \sqrt{5}}x + \frac{4 - 2\sqrt{5}}{2 + \sqrt{5}},$$
$$y = \frac{1 + \sqrt{15}}{2 - \sqrt{5}}x + \frac{4 + 2\sqrt{5}}{2 - \sqrt{5}}.$$

42. Parabola. Vertex at $(\frac{1}{25}, \frac{3}{25})$,
focus at $(\frac{1}{20}, \frac{2}{20})$,
directrix: $2x + y = \frac{1}{5}$.

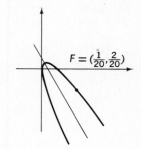

$F = (\frac{1}{20}, \frac{2}{20})$

44. Hyperbola. Foci:
(x_1, y_1) and (x_2, y_2) with
$x_1 = \frac{25}{14}(\sqrt{13} - \frac{1}{5}\sqrt{5})$,
$x_2 = \frac{25}{14}(\frac{1}{5}\sqrt{5} - \sqrt{3})$,
$y_1 = \frac{50}{14}(\sqrt{3} - \frac{1}{5}\sqrt{5})$,
$y_2 = \frac{50}{14}(-\sqrt{3} - \frac{1}{5}\sqrt{5})$.

Vertices:
(0, 0),
$(-\frac{125}{7}\sqrt{5}, -\frac{250}{7}\sqrt{5})$.
Asymptotes:

$$y = \frac{\sqrt{70}+2\sqrt{5}}{\sqrt{5}-2\sqrt{70}}x + \frac{125\sqrt{14}}{14(\sqrt{5}-2\sqrt{70})},$$

$$y = \frac{2\sqrt{5}-\sqrt{70}}{\sqrt{5}+2\sqrt{70}}x - \frac{125\sqrt{14}}{14(\sqrt{5}+2\sqrt{70})}.$$

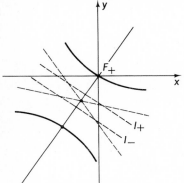

46. Ellipse. Vertices:
$(-2/\sqrt{78}, 10/\sqrt{78})$,
$(2/\sqrt{78}, -10/\sqrt{78})$,
$(-10/\sqrt{754}, -2/\sqrt{754})$,
$(10/\sqrt{754}, 2/\sqrt{754})$.

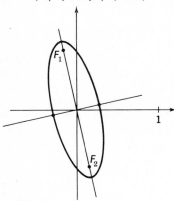

Foci:
$(-2/\sqrt{87}, 10/\sqrt{87})$,
$(2/\sqrt{87}, -10/\sqrt{87})$.
Axes: $y = -5x$, $y = \frac{1}{5}x$.
48. $x^2 - y^2 - 2x + 2y = 0$.
Discriminant $= -1$.

50. $x^2 - 4xy + 4y^2 + 8x - 16y + 16 = 0$.
Discriminant $= 0$.
$\theta = 22\frac{1}{2}°$.
52. $144x^2 - 192xy + 97y^2 - 625 = 0$.
Discriminant $= 9$.
$\theta = \arctan 4/3$.
54. $x^2 + 2xy + y^2 - (6 + 4\sqrt{2})x - (6 - 4\sqrt{2})y + 9 - 4\sqrt{2} = 0$.
Discriminant $= 0$.
$\theta = -45°$.

§4.6

4. Ellipse. Foci $(\pm\sqrt{21}, 0)$.
Vertices $(\pm 5, 0)$ and $(0, \pm 2)$.
6. Hyperbola. Foci at $z = \pm 5\sqrt{5}/3$, $y = 0$. Vertices at $z = \pm 5/3$, $y = 0$. Asymptotes are $y = \pm\frac{1}{2}z$.
8. Ellipse. Vertices $(\pm\frac{5}{8}, 0)$ and $(0, \pm 5/4)$. Foci $(0, \pm 5\sqrt{3}/8)$.
10. Hyperbola. Foci at $z = \pm\sqrt{17}/4$, $x = 0$. Vertices at $z = \pm 2$, $x = 0$. Asymptotes are $x = \pm\frac{1}{8}z$.
12. Hyperbola. Vertices $(\pm\frac{1}{2}, 0)$. Foci $(\pm\sqrt{5}/4, 0)$. Asymptotes are $y = \pm\frac{1}{2}x$.
14. Parabola. Vertex at $z = 1$, $y = 0$. Focus at $z = 63/64$, $y = 0$. Directrix is $z = 65/64$.

§4.7

16. Use (10) and (11):
$x = \frac{1}{2}(1 - \sqrt{3}) + \frac{1}{2}(1 + \sqrt{3})t$,
$y = \frac{1}{2}(1 + \sqrt{3}) - \frac{1}{2}(1 - \sqrt{3})t$,
$z = t$,

and

$x = \frac{1}{2}(1 + \sqrt{3}) - \frac{1}{2}(1 - \sqrt{3})s$,
$y = \frac{1}{2}(1 - \sqrt{3}) + \frac{1}{2}(1 + \sqrt{3})s$,
$z = s$.
18. Use (12) and (13):
$x = \frac{9}{2} + t$,
$y = \frac{9}{2} - t$,
$z = 18t$;

and
$x = -\frac{1}{2} + s$,
$y = \frac{1}{2} + s$,
$z = -2s$.
20. Assume that $0 \le \alpha, \beta < 2\pi$.
(i) Assume $\sin\alpha = 0$. Then $\cos\alpha = 1$. Put these values in (10) with $t = 0$. Show that no value of s in (11) can give the same result. (ii) Similarly, exclude $\cos\alpha = 0$. (iii) Assuming $\sin\alpha \ne 0$, $\cos\alpha \ne 0$, let $s = 0$ in (11). Assume that, for some value t_0, (10) can give the same values. Show that this leads to a contradiction.
22. (i) Dispose of case $A = B = 0$. (ii) Assume $A \ne 0$. Let $s = 0$ in (13). If the lines coincide, for some t_0 we have $Ba = Aa + at_0$, $-Bb = Ab - bt_0$, $0 = 4At_0$. Produce a contradiction. (iii) Assume $B \ne 0$.

§4.10

24. Circle; center: $(0, 0, -2)$; radius 6.
26. Ellipse; foci:
$(1/35, \pm\sqrt{236/245}, 18/35)$;
vertices: $(\frac{1}{35}, \pm 1, \frac{18}{35})$ and $[\frac{1}{35}(1 \pm 6), 0, \frac{1}{35}(18 \pm 3)]$.
28. Parabola:
vertex: $(-1/2, 0, -1/\sqrt{2})$;
focus: $(-1/3, 0, 0)$.
30. (a) Rotate about z axis with $\theta = 90°$; then rotate about (original) y axis with $\theta = 90°$.
(b) Rotate about y axis with $\theta = 90°$; then rotate about original z axis with $\theta = -90°$.

§4.11

32. Ellipsoid.
34. Elliptic cone.
36. Ellipsoid.
38. Hyperboloid of one sheet.
40. Hyperboloid of one sheet.

Chapter 11

§1.3

2. $\mathbf{f}(t) = \big[(t-1)^{1/2} - 1\big]\mathbf{e}_1 + \big[(t-1)^{3/2} - 3(t-1) + 3(t-1)^{1/2}\big]\mathbf{e}_2$.

6. $\mathbf{f}(t) = \frac{1}{2}\sqrt{2}(t-2)^{3/2}\,\mathbf{e}_1 + \frac{1}{4}\sqrt{2}(t-2)^{9/2}\mathbf{e}_2$.

8. $\mathbf{f}(t) = \big[(t+3)^2 + 1\big]\mathbf{e}_1 + \big[\log((t+3)^2 + 1)\big]\mathbf{e}_2$.

10. $\theta(t) = (\pi/6)(t-1)^{1/2},\ r(t) = 1 + \cos\big[(\pi/6)(t-1)^{1/2}\big]$.

12. $\theta(t) = (-5\pi/3)t^{-1/2} + 2\pi/9$, $r(t) = \cos(-5\pi t^{-1/2} + 2\pi/3)$.

§1.5

14. $(-\sin 2t)\mathbf{e}_1 + (\cos t)\mathbf{e}_2$.

16. $\frac{3}{2}\sqrt{t}\,\mathbf{e}_1 - \sec t(t\tan t + 1)\mathbf{e}_2$.

18. $\frac{15}{4}\sqrt{t}\,\mathbf{e}_1$.

20. $-\frac{1}{4}t^{-3/2}\mathbf{e}_1 + e^t\mathbf{e}_2$.

22. $\mathbf{e}_1 + 3\mathbf{e}_2$.

24. $\frac{1}{24}\mathbf{e}_1 - 2\log\frac{5}{4}\mathbf{e}_2$.

28. $\big[\frac{1}{3}(\pi+1)^{3/2} - 1\big]\mathbf{e}_1$.

30. $\mathbf{e}_1 + \big[2\log 2 - \frac{1}{3}\big]\mathbf{e}_2$.

32. $(\frac{1}{12}t^4 + \frac{11}{12}t + 1)\mathbf{e}_1 + (1 - 2t)\mathbf{e}_2$.

34. Use the formula for the integral of a vector-valued function (§1.4). Then use the Fundamental Theorem of Calculus.

36. Let $\mathbf{a} = a_1\mathbf{e}_1 + a_2\mathbf{e}_2$. Use formula for $\lim_{t\to t_0}\mathbf{f}(t)$ in §1.4, the fact that the limit of a sum equals the sum of the limits, and Chapter 3, §4.2, Theorem 2.

38. Write down the formula for the length of a vector.

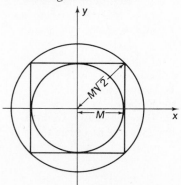

40. Yes. Example:
$\mathbf{f}(t) = (\cos 2\pi t)\mathbf{e}_1 + (\sin 2\pi t)\mathbf{e}_2$.

§1.6

42. $|\alpha(t)\mathbf{f}(t)| = |\alpha(t)||\mathbf{f}(t)|$. Now use Statement 1.

46. Use 2, 3, §1.5 (e) 3, 2.

§2.3

4. (e), different trajectories.

6. (b).

8. (c), the first defines a subarc of the second.

10. (c), the second defines the opposite of the subarc of the first.

§2.4

12. $y = (x^2 - 1)^{3/2}$ for $x \geq 1$.

14. The opposite of
$y = 2x\sqrt{1-x^2},\ 0 \leq x \leq 1$.

16. For $t \geq -1,\ y = (x^{3/2} + 1)^3$, $x \geq 0$. For $t < -1,\ y = (-x^{3/2} + 1)^3,\ x \geq 0$.

18. That subarc of the first quadrant branch of the hyperbola $xy = e$ between $(1, e)$ and $(e, 1)$. The motion is from $(e, 1)$ to $(1, e)$ and back to $(e, 1)$, repeated twice on $0 \leq t \leq 2\pi$.

20. $y = x^2 + x,\ x > \frac{1}{4}$; half of a parabola.

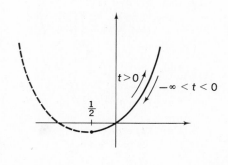

22. $x^2 - y^2 = 1,\ x \geq 0,\ y \geq 0$. One quarter of an equilateral hyperbola.

24. $x^2 + y^2 = 2,\ 0 \leq x \leq 1,\ y > 0$; circular arc.

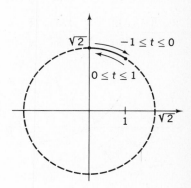

§2.5

28. Hint: Let the center be at the origin. Let P be the point $(a, 0)$, the point from which the end $R = (x, y)$ of the string begins moving. Let T be a point on the circumference. Let t be the angle from the positive x axis to the radius OT. Express x and y in terms of known quantities. Note that arc $PT = at$. The parametric equations are $x = x(t) = a(\cos t + t\sin t)$, $y = y(t) = a(\sin t - t\cos t)$.

30. Hint: Replace b by $-b$ in Exercise 29. The equations are
$x(t) = (a - b)\cos t + b\cos \alpha t$,
$y(t) = (a - b)\sin t - b\sin \alpha t$,
$\alpha = a - (1/b)$.

§2.6

32. Circle with center at $x = 1$, $y = 0$, and radius 1. $A = \pi$.

34. One leaf of lemniscate. $A = 1/2$.

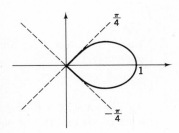

36. Curve is symmetric with respect to both axes. The first quadrant is shown. $A = \sqrt{8(\sqrt{3}-1)} + 2\sqrt{1+\sqrt{3}} - 2\arctan\sqrt{1+\sqrt{3}}$.

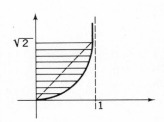

38. Spiral. $A = \frac{28}{3}\pi^3$.

40. $A = (3\pi - 8)/96$.

§3.4

2. $\frac{3}{5}\mathbf{e}_1 - \frac{4}{5}\mathbf{e}_2$.

4. $(5/\sqrt{41})\,\mathbf{e}_1 - (4/\sqrt{41})\,\mathbf{e}_2$.

6. $\dfrac{t}{\sqrt{2t^2+1}}\,\mathbf{e}_1 + \dfrac{\sqrt{t^2+1}}{\sqrt{2t^2+1}}\,\mathbf{e}_2$.

10. $\dfrac{3t^2}{\sqrt{9t^4+4t^2}}\,\mathbf{e}_1 - \dfrac{2t}{\sqrt{9t^4+4t^2}}\,\mathbf{e}_2$.

§3.5

14. $\mathbf{T}(\theta)$ has components
$$\frac{\cos\theta - \theta\sin\theta}{\sqrt{\theta^2+1}}$$

and
$$\frac{\sin\theta + \theta\cos\theta}{\sqrt{\theta^2+1}}.$$

$\tan\Psi = r$.

16. $\mathbf{T}(\theta)$ has components $-1/\sqrt{2}$ and $1/\sqrt{2}$.
$$\tan\Psi = \frac{r(1+\sin 2\theta)}{(\sin\theta - \cos\theta)}.$$

§3.6

18. $(1/\sqrt{5})\mathbf{e}_1 + (2/\sqrt{5})\mathbf{e}_2$.

20. $-\dfrac{3}{\sqrt{11}}\,\mathbf{e}_1 - \dfrac{\sqrt{2}}{\sqrt{11}}\,\mathbf{e}_2$.

22. $-\dfrac{1}{\sqrt{1+\sin^2 t}}\mathbf{e}_1 - \dfrac{\sin t}{\sqrt{1+\sin^2 t}}\mathbf{e}_2$.

24. $-\dfrac{3x^2}{\sqrt{9x^4+1}}\,\mathbf{e}_1 + \dfrac{1}{\sqrt{6x^4+1}}\,\mathbf{e}_2$.

§3.10

26. $\frac{1}{27}(8^{3/2} - 5^{3/2})$.

28. $\sqrt{6} - \sqrt{2}/3$.

30. $\pi/12$.

32. $\int_0^1 \sqrt{9t^4 + 10t^2 + 1}\, dt$.

34. $\int_{\sqrt{\pi/4}}^{\sqrt{\pi/2}} t\sqrt{4(\sin t^2)^2 + 9t^2(\cos t^3)^2}\, dt$.

§3.11

36. $37/3$.

38. 1.

40. $\int_{\pi/6}^{\pi/3}\sqrt{(1+3\cos^2\theta)/4\cos\theta}\, d\theta$.

§3.12

44. $\mathbf{g}(s) = \frac{1}{2}\phi(s)\mathbf{e}_1 + \frac{1}{3}[2\phi(s)+1]^{3/2}\mathbf{e}_2$,
where $\phi(s) = \frac{1}{8}(12\sqrt{2}s - 5^{3/2})^{2/3} - 5$.

46. $\mathbf{g}(s) = \cos(\frac{1}{2}\pi - s)\mathbf{e}_1 + \sin(\frac{1}{2}\pi - s)\mathbf{e}_2,\ 0 \le s \le \pi/2$ (watch orientation).

§3.16

48. $\kappa(x) = 2x^3/(1+x^4)^{3/2}$; $\rho = 1/\kappa$ (since $x > 0$); no inflection points.

50. $\kappa(x) = \cos x$; $\rho = \sec x$ (since $-\pi/2 < x < \pi/2$); no inflection points; local maximum at $(0, 0)$.

52. $\kappa(t) = \dfrac{e^t(\sin t - \cos t)}{(e^{2t} + \sin t)^{3/2}}$;

$\rho(t) = \dfrac{(e^{2t} + \sin t)^{3/2}}{e^t\,|\sin t - \cos t|}$

when $\kappa(t) \ne 0$; inflection points at $(n+\frac{1}{4})\pi$ for every integer n.

54. $\kappa(t) = \dfrac{12t^2(t^2-1)}{(16t^6 + 9t^4 - 6t^2 + 1)^{3/2}}$;

$\rho(t) = \dfrac{(16t^6 + 9t^4 - 6t^2 + 1)^{3/2}}{12t^2\,|t^2-1|}$

for $t \ne \pm 1$; inflection point at $(0, 1)$.

56. $((a^2 - b^2)/a,\ 0)$ and $(0,\ (b^2 - a^2)/b)$.

58. $(11, -10/3)$.

§3.18

60. $-(\sqrt{5}/\sqrt{186})\mathbf{e}_1 + (6\sqrt{5}/\sqrt{186})\mathbf{e}_2 - (1/\sqrt{186})\mathbf{e}_3$.

62. $\dfrac{\cos t - t\sin t}{\sqrt{2-t^2}}\mathbf{e}_1 + \dfrac{\sin t + t\cos t}{\sqrt{2+t^2}}\mathbf{e}_2 + \dfrac{1}{\sqrt{2+t^2}}\mathbf{e}_3$.

64. $4/3$.

66. e.

68. 2.

§4.4

2. $\mathbf{v}(t) = (-\sin 2t)\mathbf{e}_1 + (\sin 2t)\mathbf{e}_2$; speed $= \sqrt{2}\,|\sin 2t|$; $\mathbf{a}(t) = (-2\cos 2t)\mathbf{e}_1 + (2\cos 2t)\mathbf{e}_2$; $\mathbf{a}_N = \mathbf{0}$; $\mathbf{a}_T = (-2\cos 2t)\mathbf{e}_1 + (2\cos 2t)\mathbf{e}_2$.

6. Trajectory: first quadrant branch of $y = 1/x$.
$\mathbf{v}(t) = e^t\mathbf{e}_1 - e^{-t}\mathbf{e}_2$;

speed $= \sqrt{e^{2t} + e^{-2t}}$;

$\mathbf{a}(t) = e^t \mathbf{e}_1 + e^{-t} \mathbf{e}_2$;

$$\mathbf{a}_N = \frac{2e^{-t}}{e^{2t} + e^{-2t}} \mathbf{e}_1 +$$

$$\frac{2e^t}{e^{2t} + e^{-2t}} \mathbf{e}_2;$$

$$\mathbf{a}_T = \frac{e^{3t} - e^{-t}}{e^{2t} + e^{-2t}} \mathbf{e}_1 -$$

$$\frac{e^t - e^{-3t}}{e^{2t} + e^{-2t}} \mathbf{e}_2.$$

8. Trajectory: $y = \log x$;
   $\mathbf{v}(t) = \mathbf{e}_1 + (1/t)\mathbf{e}_2$;

   speed $= \dfrac{\sqrt{t^2 + 1}}{t}$;

   $\mathbf{a}(t) = -(1/t^2)\mathbf{e}_2$;

   $$\mathbf{a}_N = \frac{1}{(t^2 + 1)^{3/2}} \mathbf{e}_1 -$$

   $$\frac{t}{(t^2 + 1)^{3/2}} \mathbf{e}_2;$$

   $$\mathbf{a}_T = -\frac{1}{t(t^2 + 1)} \mathbf{e}_1 -$$

   $$\frac{1}{t^2(t^2 + 1)} \mathbf{e}_2.$$

10. Trajectory: $x = \cos y$; we give
    the values of $\dot{x}, \dot{y}, \ddot{x}, \ddot{y}$:
    $\dot{x} = -\dot{y} \sin y$, $\ddot{x} = -\ddot{y} \sin y$
    $- (\dot{y})^2 \cos y$, $\dot{y} = 1 + \cos t$,
    $\ddot{y} = -\sin t$.

§4.6

12. Put $\delta = \gamma = 0$. Write the
    formula for the speed at time
    t. Note that the projectile
    strikes the ground when
    $y(t) = 0$.

14. (a) $h = \beta^2/2g$; $r = 2\alpha\beta/g$.
    (b) From a sketch, it is clear
    that $\tan \phi = h/(r/2)$.

16. 32 ft, approx.
18. $20\sqrt{2g}$; $4\sqrt{42g}$.

§4.13

22. (a) All terms on the right-
    hand side of the equation
    given in the hint are
    equal. (Use $\alpha = t$.)
    (b) Same remark. (Use $\alpha = 1/n$.)
    (c) Use (a) and (b).

24. (a) Use (17). (b) Suppose
    $\ddot{r} - r\dot{\theta}^2 = cr^k$, c and k con-
    stants.

§5.5

2. Use Galileo's Law for an inclined plane (§4.8) with the
   particle beginning at the point
   $P(\theta = -\pi)$ and sliding to
   Q.

4. $s_1 - s_0 = g\tau^2/2$. This is
   Galileo's law of free fall.

6. The equation says that $s^2 + (ds/dt)^2$ is a constant, say c.
   Write down $dt/ds = 1/s'$ and
   integrate.

§6.4

4. Replace $\mathbf{r}_1, \mathbf{r}_2, \cdots, \mathbf{r}_N$ by $\mathbf{r}_1 + \mathbf{P}, \mathbf{r}_2 + \mathbf{P}, \cdots, \mathbf{r}_N + \mathbf{P}$.

§6.8

8. Let the κ sets be $\{m_1, m_2, \cdots, m_{k_1}\}, \{m_{k_1} + 1, \cdots, m_{k_2}\}, \cdots, \{m_{k_{\kappa-1}}, \cdots, m_{k_\kappa}\}$ (where $k_\kappa = N$). Introduce $\mathbf{R}_1, \mathbf{R}_2, \cdots, \mathbf{R}_\kappa$
   and $M_1, \cdots, M_\kappa$.

10. Let P_3Q intersect P_1P_2 in P.
    Write the position vector of
    P in terms of the position
    vectors of P_1 and P_2. (Com-
    pare Chapter 9.) Then write
    the position vector of Q in
    terms of the position vectors
    of P and P_3.

Chapter 12

§1.1

2. 1.
4. $\sqrt{2x^2 - 1}$.
6. $|\sin t| + |\cos t|$. Note
   periodicity.

§1.3

In Exercises 10 to 28 below,
the functions are defined and
continuous for the indicated
values of the variables.

10. $x \neq 0$ and $y \neq 0$.
12. $x \geq 0$ and $y \geq 0$.
14. $y \neq 0$.
16. $x \neq 0$.
18. $(x, y) \neq (0, 0)$.
20. All (x, y).
22. All (x, y).

§1.4

24. $x \neq 0$; $y \neq 0$; $z \neq 0$.
26. $z \neq 0$.
28. $x + y + z \neq 0$.

§1.6

32. Open.
34. Not open.
38. The set of all (x, y) with
    $xy = 0$, $x \geq 0$, $y \geq 0$.
40. The entire plane.
42. The empty set, and the entire
    plane.

§2.2

2. $\dfrac{\partial f}{\partial x} = 2xy^5$; $\dfrac{\partial f}{\partial y} = 5x^2y^4$.

4. $\dfrac{\partial f}{\partial x} = \dfrac{x}{\sqrt{1 + x^2 + y^2}}$;

   $\dfrac{\partial f}{\partial y} = \dfrac{y}{\sqrt{1 + x^2 + y^2}}$.

6. $\dfrac{\partial f}{\partial x} = 50x^9 - 6y^7,$

$\dfrac{\partial f}{\partial y} = -42xy^6.$

8. $\dfrac{\partial f}{\partial x} = e^{x^2+y^2}[2 - 1/x^2];$

$\dfrac{\partial f}{\partial y} = e^{x^2+y^2}(2y/x).$

12. $\dfrac{\partial f}{\partial x} = \dfrac{4xy^2}{(x^2 + y^4)^2};$

$\dfrac{\partial f}{\partial y} = \dfrac{-4x^2y}{(x^2 + y^2)^2}.$

14. Both equal $\frac{1}{2}$.

16. Both equal $72/169$.

18. $f_x(0, 0) = f_y(0, 0) = 0.$

20. $h_u = g_v = 0.$

22. $[\cos(\cos\sqrt{1 + u^2 + v^2})]$
$\times[-\sin\sqrt{1 + u^2 + v^2}]$
$\times\left[\dfrac{v}{\sqrt{1 + u^2 + v^2}}\right].$

24. $-(\sin x + \sin y).$

§2.7

30. $z = -6 - 2(x-3) + 3(y+2).$

32. $z = 16 + 5(x-4) - 4(y-2).$

34. $z = 2 + 2(x-1) + 2(y-1).$

36. The normal vectors are respectively $(-1, 1, -1)$ and $(5/6, -1/6, -1)$, and their inner product is zero.

§2.11

42. $\dfrac{dz}{dx} = (\log x)(\log x + 2x + 2) + x;$

$\dfrac{dz}{dy} = e^y(2y + y^2 + e^y + 2ye^y).$

46. $g_u(0, 0) = 17;\ g_v(0, 0) = 18.$

48. $F_x = f'(x - 2y)(1)$ and $F_y = f'(x - 2y)(-2)$ by chain rule.

50. $dF/dt > 0$ for all t. Hence $F(t)$ is increasing.

52. No critical points. grad $F = 8\mathbf{e}_1 + \mathbf{e}_2$ and directional derivative $= -7/\sqrt{2}.$

54. $(3, -2)$ is critical point. grad $F = -4(\mathbf{e}_1 + \mathbf{e}_2)$ and directional derivative is $-4\sqrt{2}.$

56. $3x^2 - 6y^2 = 128.$ $\mathbf{e}_1 + 2\mathbf{e}_2$ is tangent direction. The gradient is $24(2\mathbf{e}_1 - \mathbf{e}_2).$

58. $y = -x\cos xy.$ $\mathbf{e}_1 - \mathbf{e}_2$ is tangent direction. The gradient is $\mathbf{e}_1 + \mathbf{e}_2.$

60. $e^x\cos y = 1.$ $\mathbf{e}_2$ is tangent direction. The gradient is $\mathbf{e}_1.$

§2.12

62. $83/20.$

64. $-1.$

66. $1.$

§2.15

70. $f_x = 2z^2 - yz;\ f_y = z^3 - xz;$
$f_z = 4xz + 3yz^2 - xy.$

72. $f_x = \{\exp[x + \sin(zy)]\}z^3;$
$f_y = \{\exp[x + \sin(zy)]\}z^4$
$\quad\times\cos(zy);$
$f_z = \{\exp[x + \sin(zy)]\}$
$\quad\times[3z^2 + y\cos(zy)].$

74. $f_x = -y(1 + z^2)\sin\alpha,$
$f_y = \cos\alpha,$
$f_z = 2yz\sin\alpha,$
where $\alpha = (1 + z^2)x.$

76. $f_x = [yz/(x^2+2y^2+4z^2)]$
$\quad - [2x^2yz/(x^2+2y^2+4z^2)^2]$
$f_y = [xz/(x^2+2y^2+4z^2)]$
$\quad - [4xy^2z/(x^2+2y^2+4z^2)^2]$
$f_z = [xy/(x^2+2y^2+4z^2)]$
$\quad - [8xyz^2/(x^2+2y^2+4z^2)^2].$

78. $\theta_u = ve^{uv+w}/w;$
$\theta_v = ue^{uv+w}/w;$
$\theta_w = e^{uv+w}(w - 1)/w^2.$

80. $x^2/a^2 + y^2/b^2 + z^2/c^2 = 1/a^2 + 9/b^2 + 16/c^2.$ The gradient

has components $(-2/a^2, 6/b^2, 8/c^2).$

84. The first question is meaningless; the origin is a critical point. The answer to the second is: in the direction of the vector $(-2/5, 6/5, -8/5).$

86. $\phi_u = (v^2w^3t)\cos(1 + uv^2w^3t);$
$\phi_v = (2uvw^3t)\cos(1 + uv^2w^3t);$
$\phi_w = (3uv^2w^2t)\cos(1 + uv^2w^3t);$
$\phi_t = (uv^2w^3)\cos(1 + uv^2w^3t).$

§2.17

88. $-\frac{1}{2}(\sin x)(b^2 - a^2) + (\sin b - \sin a).$

90. $2/x.$

92. $1/(2 + x) - 1/(1 + x).$

94. $h_\phi = \displaystyle\int_{-1}^{1}\dfrac{2\phi}{\psi + y^2}\,dy;$

$h_\psi = -\displaystyle\int_{-1}^{1}\dfrac{\phi^2 + y}{(\psi + y^2)^2}\,dy.$

Set $\alpha = 1/\sqrt{\psi}.$ Then
$h_\phi = 4\alpha\phi\arctan\alpha,$
$h_\psi = \phi^2\alpha^3\arctan\alpha$
$\quad - \phi^2\alpha^4/(1 + \alpha^2).$

96. $h'(5) = \log(3/10).$

§3.2

6. $2x^2 + 8x^3y^5 + 8y^{10} + 1.$

8. $\displaystyle\int_0^y e^{t^2}\,dt - \int_0^x e^{t^2}\,dt.$

10. $r^3/3 + r\sin\theta + e^\theta - 1/3.$

§3.3

14. $f_{xxx} = -y^3\cos xy;\ f_{yyy} = -x^3\cos xy;\ f_{xxy} = f_{xyx} = f_{yxx} = -2y\sin xy - xy^2\cos xy;$
$f_{yyx} = f_{yxy} = f_{xyy} = -2x\sin xy - x^2y\cos xy.$

16. $f_{xxxx} = f_{xxyy} = f_{xyxy} = 0.$

§3.5

22. $x = y = z = 1/3.$

24. $(0, 0, -1).$

28. If $c \neq 0$ and $ac = b^2$, then every point on $y = -(b/c)$ is a critical point.

§3.6

32. 6.
34. 0.

§3.9

38. No f exists.
40. $\sin(x \cos y) + \cos(y \sin z)$.

§4.1

2. $19 + e^3 - e^2 + \cos 4 - \cos 9$.
4. $3/8$.
8. $\frac{1}{3} \sum_{j=1}^{n} (t_1^{3j} - t_0^{3j})$.

§4.3

10. $-5/3$.
12. 0.
14. $(D - A) + (5\pi/2)(C - B)$.
20. Both line integrals equal $\sin 2 - 1/\sqrt{2}$.

§4.4

22. $x = y = 0$.
24. $du = 8\cos(zy)\,dx - 8xy\sin(zy)\,dy - 8xy\sin(zy)\,dz$.

26. $du = \dfrac{-dx}{2\sqrt{1 - (x+y)^2}}$
$+ \dfrac{-dy}{2\sqrt{1 - (x+y)^2}}$.

28. $du = [1 + (xze^y)^2]^{-1}[ze^y\,dx + xze^y\,dy + xe^y\,dz]$.

§4.5

30. The integrand is the differential of $\sin x - e^{-y} + z^3/3$. The integral equals $\sin 6 + \sin 1 - 25 + 1/e^2$.
32. 3.
34. 0.

§4.6

36. $f(x, y, z) = xz^2$.
40. $B = C$; $A = D = C - 1$. $C > 0$.

42. Yes, homotopic.
44. Yes, homotopic.
46. Yes, homotopic.
50. (a) Yes.
    (b) Still not simply connected.

§5.2

2. $\frac{1}{2}[(x_0^2 - x_1^2) + (y_0^2 - y_1^2) + (z_0^2 - z_1^2)]$.
4. $\frac{1}{2}[(x_0^2 - x_2^2) + (y_0^2 - y_2^2) + (z_0^2 - z_2^2)]$.
6. 0.

§5.3

8. $w = \gamma m\left[\dfrac{m_1}{r_1} + \cdots + \dfrac{m_n}{r_n}\right]$.

§7.4

2. $\alpha + \beta - \frac{1}{2}[\alpha^2 + 2\alpha\beta + \beta^2]$.
6. 0.
10. $(1/9) - (1/81)[2(x - 1) + 3(y - 1) + 4(z - 1)] + (1/729)[4(x-1)^2 + 9(y-1)^2 + 16(z-1)^2 + 12(x-1)(y-1) + 16(x-1)(z-1) + 24(z-1)(y-1)]$.
12. 1.
14. $(\alpha - 1) + (\beta - 1) - \gamma - \frac{1}{2}[(\alpha - 1)^2 + (\beta - 1)^2 + \gamma^2] + 2(\alpha - 1)\gamma + (\beta - 1)\gamma$.

Chapter 13

§1.3

2. 0.
4. $(e - e^{-1})^2$.
6. -292.44.
10. $(21/2)\log(5/4) + e^{14}(e - 1)^2$.
12. 17.
14. $(1/12)[\cos(36) - \cos(40) + \cos(40 - 0.3) - \cos(36 - 0.3)]$.
16. $\frac{9}{70}(6.7^7 - 3^7) + \frac{37}{70}(5.8^7 - 4.9^7) + \frac{1}{8}(6.7^4 - 3^4)(5.8^4 - 4.9^4)$.
18. $(5/2)(\pi/2 - 1)$.

20. $n \neq -1, -2$: $\dfrac{1}{(n+1)(n+2)}$
$\times [(B + D)^{n+2}$
$- (A + D)^{n+2}$
$- (B + C)^{n+2}$
$+ (A + C)^{n+2}]$;
$n = -2$:
$\log\left[\dfrac{(A + D)(B + C)}{(B + D)(A + C)}\right]$;
$n = -1$:
$(B + D)\log(B + D)$
$- (A + D)\log(A + D)$
$+ (A + C)\log(A + C)$
$- (B + C)\log(B + C)$.

§1.4

24. The points at which $g = 0$ must lie on a finite number of curves as described in the definition of piecewise continuity.
26. No.

§1.8

32. 1128.

§1.10

34. $1/15$.
36. $\pi^2/8$.
38. 5π.
40. $8[1/\sqrt{5} - 1/5 + \arcsin(2/\sqrt{5})]$.
42. For instance,

$$\int_0^1 \int_0^y f\,dx\,dy$$

$$+ \int_{1/2}^{2/3} \int_y^{2-2y} f\,dx\,dy$$

$$+ \int_0^{1/2} \int_{1-y}^{2-2y} f\,dx\,dy.$$

44. $\frac{2}{3}\pi abc$. The volume of half an ellipsoid.
50. $\frac{2}{3}[1/\sqrt{2} + 7/4 - 9/\sqrt{3}]$.

§1.11

54. $\pi R^2 h/3$.
58. $(\pi/2) \log 2$

$$+ 4 \int_0^{\pi/4} \log (\cos \theta)\, d\theta.$$

60. $3\pi/8$.

§2.3

2. $F(3 + 6\sqrt{3}) - F(3 + 6\sqrt{2})$
   where
   $F(u) = \frac{1}{9}(1 + u^2)^{3/2}$
   $- \frac{1}{2}u(1 + u^2)^{1/2}$
   $- \frac{1}{2}\log[u + (1 + u^2)^{1/2}]$.

4. π.
6. $3\sqrt{2}\pi/16$.

10. $\iint_D \sqrt{1 + g_y^2 + g_z^2}\, dy\, dz$.

12. $2\pi[\sqrt{3} - 1/3]$
14. 114π.

§2.4

16. $\frac{1}{2}l\,|\,y_1 + y_2\,|$ and $\frac{1}{2}l\,|\,x_1 + x_2\,|$
    where l is the length of the
    segment. We assume both
    points lie in the first quadrant
18. $\pi R^2 [\cos\theta_0 + \cos\theta_1 + 2\sin\theta_1]$,
    assuming that
    $0 \le \theta_0 \le \theta_1 \le \pi/2$.
20. $(\pi/6)(\sqrt{5} - 1)$.
22. $10\pi \left[\sqrt{5} + \log\left(\frac{1 + \sqrt{5}}{\sqrt{5} - 1}\right)^2 \right]$.

§3.2

2. $\log(3/2)(\log 2)^2$.
4. 1.
6. 0.
8. $\log(4^{\alpha-\beta}27^{\beta-\gamma}256^\gamma)$
   $- (\alpha + \beta + \gamma)$.
10. $9 \log \sqrt{2}$.
12. $\pi^2/4[\pi/2 + 1/3]$.
14. $3(\log 4 - 1) + \log 2 (\log 8 - 2)$.

§3.3

18. 4π.
20. $16 R^3/3$.

22. $abc/6$.

§3.4

24. $128(3\pi - 4)/9$.
28. $(56e\pi/3)(e - 1)$.
30. $\pi(1 - \pi/2)$.
32. 0.

§3.5

34. $1384\pi/5$.
36. $\pi(e^{2b} - e^{2a})/2$.

§3.6

40. $(32 - 14)\sqrt{3}\pi/6$.
42. 0.
44. $\frac{2}{3}\pi(\sin\phi - \phi\cos\phi)\Big|_{\theta_1}^{\theta_2}$.

46. $\alpha > -3; 4\pi R^{\alpha+3}/(\alpha + 3)$.

§3.7

48. πK.
50. $4b^2 a/3$.
52. 0.
54. $\frac{1}{6}\pi[4T(h) - h^2]$
    where

$$T(h) = \int_0^h p(z)^{3/2}\, dz,$$

$$p(z) = \frac{1}{\pi} + \frac{2zh}{\pi} + \left(1 + \frac{1}{\pi^2 h}\right)z^2.$$

The integral T can be evaluated in closed form, but the
result looks very complicated.

§3.8

56. Let the lamina D be the disk
    $x^2 + y^2 < R^2$ in the plane
    $Z = 0$. At $(0, 0, Z)$ we have
    $F = G = 0$ and $H =$
    $2\pi Z[(R^2 + Z^2)^{-1/2} - |Z|^{-1}]$.
58. 0.

§3.9

60. For $|Z| < \rho$ one obtains that
    $K(\rho) = 0$.

§4.2

2. $\dfrac{c_2 c_3 \cdots c_{n+1}}{(n + 1)!}$

   where $c_j = b^j - a^j$,

6. 0.
8. 0.
10. 0.

§4.3

12. $\pi^3 R^6/6$.
14. $1, 1/2, 1/6, 1/24$.

§5.1

2. Midpoint of the axis.
4. $X = 0$, $Y = 3\sqrt{3}a/8$ where
   the origin coincides with the
   vertex and the y axis is the
   bisector of the vertex angle.
6. $X = 0$, $Y = -1/3\pi$, $Z = 4\pi$.

§5.3

10. $X = \dfrac{4 - \pi}{4(\sqrt{2} - 1)}$,

    $Y = \dfrac{4 - \pi}{16\sqrt{2}(\sqrt{2} - 1)}$.

12. $X = 0$, $Y = 0$.
14. $X = (1/3)\left(\dfrac{b_1^3 - b_2^3}{b_1^2 - b_2^2}\right)$,

    $Y = (h/3)\left(\dfrac{b_1 + 2b_2}{b_1 + b_2}\right)$.

16. $X = \dfrac{e^2 + 1}{e^2 - 1}$, $Y = 0$.

18. If the vertex is at the origin
    and the z axis is the axis of
    the cone, $X = Y = 0$, and

    $Z = \dfrac{3}{8(2 - \sqrt{3})}$.

20. $X = Y = Z = 1/4$.
22. $X = \dfrac{3R}{4\sqrt{2}}$, $Y = 0, Z = \dfrac{3R}{8}$.

24. $X = Y = 0$, $Z = 3Rt/16$
with $t =$

$$\frac{2(\beta - \alpha) + \sin 2\alpha - \sin 2\beta}{\cos \alpha - \cos \beta}$$

where β and α are the values of ϕ corresponding to the two latitudes.

28. $X = 0$, $Y = m/L$
where $L =$

$$\frac{\sqrt{5}}{16} + \tfrac{1}{4} \log \frac{1 + \sqrt{5}}{2}$$

and $m =$

$$\frac{3\sqrt{5}}{2^{14}} + \frac{1}{2^8} \log \frac{1 + \sqrt{5}}{2}.$$

30. $X = Y = 3/2$, $Z = 1$.

§5.4

32. $X = Y = 40/7$.
34. $X = 167/178$; $Y = 22/89$.
36. $X = Y = Z$
$= (6 - 80\pi\sqrt{10})/3$.
38. $X = Y$
$= (128 - \pi)/(256 - 4\pi)$,
$Z = 1/2$.
40. $X = (136 - 3\pi)/8a$,
$Y = (34 - \pi)/2a$,

$Z = (30 - \pi)/2a$
where $a = 36 - \pi$.
42. The midpoint of the cross bar.
44. The intersection of the cylinder's axis with the plane.

§5.5

48. If the circles are centered at the origin of the radii a and b, respectively, and the x axis is the line, then $X = 0$,

$$Y = \frac{2(b^3 - a^3)}{3\pi(b^2 - a^2)}.$$

50. If the triangle has sides a and b for the right angle, the centroid is $1/3b$ in from side h and $1/3a$ in from side b.
52. $\pi r \sqrt{h^2 + r^2}$ where r is the radius and h the height.
54. If the line segment has length r, then the theorem says that the area of a circle of radius r is $2\pi(r/2)r$ or πr^2.

§6.1

2. $3/4$.
4. 0.
8. $\cos 3 - \cos 5$.

10. $(1/m)[\sin my_1 - \sin my_0]$
$\times [\cos nx_1 - \cos nx_0]$.

§6.2

12. $ab\pi$.
14. 29π.
16. $4\pi^3/3$.

§6.4

20. 0.
22. 0.

§7.1

2. If $2 > \alpha$, the integral converges to $2\pi/(2 - \alpha)$. If $2 \le \alpha$, the integral diverges.
6. Converges.
8. Converges.
10. Diverges.

§7.2

12. $\alpha < -2$; $-2\pi/(\alpha + 2)$.
14. Converges.
18. Diverges.
20. Converges.

§7.3.

22. $\sqrt{u}e^{-u}\,du = -d(\sqrt{u}\,e^{-u})$
$+ (e^{-u}/2\sqrt{u})\,du$.
Now apply Exercise 21.

SOLUTIONS to flagged exercises

Chapter 1

§2

6. $7 + (-7) = 0$ (Postulate **5**); $(-6) + 6 = 6 + (-6)$ (Postulate **2**); $6 + (-6) = 0$ (Postulate **5**). Therefore, $7 + (-7) = (-6) + 6$.

14. Care must be taken in this exercise and the succeeding ones to use only those laws and rules that have been developed so far.

    Since $a^n b^n = (ab)^n$, we have $a^5 b^5 = (ab)^5$. Using the same law again, we have $a^5 b^5 c^5 = (ab)^5 c^5 = ((ab)c)^5 = (abc)^5$. Therefore, $a^5 b^5 c^5 = 2^5 = 32$.

§3

16. No, the inequality is false for $x = 0$. The inequality is true for $x \neq 0$.

24. We must consider two cases ($x = -3$ is excluded, since division by 0 is meaningless): Case 1: $3 + x > 0$ and $6 - 2x > 2(3 + x)$ *or* Case 2: $3 + x < 0$ and $6 - 2x < 2(3 + x)$. In the first case, we have $x > -3$ and $6 - 2x > 6 + 2x$, that is, $x > -3$ and $x < 0$. Thus, Case 1 gives $-3 < x < 0$. In the second case, we have $x < -3$ and $6 - 2x < 6 + 2x$, that is, $x < -3$ and $x > 0$. This is impossible. The solution is: $-3 < x < 0$.

34. The inequality given here is equivalent to the two inequalities

    $$5x - 2 \leq 10x + 8 \text{ and } 10x + 8 < 2x - 8.$$

    The first of these is true if and only if $x \geq -2$. The second, if and only if $x < -2$. Since these conditions are never satisfied simultaneously, there is no solution to the given inequality.

44. "Quadratic inequalities" are best handled in the same fashion as quadratic equations: gather all terms on the left of the inequality sign, with 0 on the other side; then try to factor the left-hand side (if necessary, use the "quadratic formula" to do this). In this case, we have $x^2 + 3x + 2 > 0$, that is, $(x + 1)(x + 2) > 0$. Since the product of two numbers can be positive only if both are positive or both are negative, we must have *either* $x + 1 > 0$ and $x + 2 > 0$ *or* $x + 1 < 0$ and $x + 2 < 0$. The first alternative gives $x > -1$ and $x > -2$, that is, $x > -1$. The second, $x < -1$ and $x < -2$, that is, $x < -2$. The solution is: either $x > -1$ or $x < -2$ (That is, the inequality is satisfied for all numbers x except those in the interval $[-2, -1]$.).

50. The given inequality is equivalent to the pair of inequalities: (i) $-10^{-j} \leq x - 1$ and (ii) $x - 1 \leq 10^{-j}$. Applying Corollary 2 to (ii) (with $a = x - 1$, $A = 10$, and $n = j$), we deduce that $x - 1 \leq 0$ or $x \leq 1$. Inequality (i) is equivalent to $1 - x \leq 10^{-j}$. Using Corollary 2 again (with $a = 1 - x$, $A = 10$, and $n = j$), we have $1 - x \leq 0$ or $x \geq 1$. Using Exercise 9 of §3.2, we see that 1 is the only value of x satisfying the given inequality.

62. Since $||a| - |b||$ is equal either to $|a| - |b|$ or to $-(|a| - |b|) = |b| - |a|$, we must, in order to establish the given inequality, show that

    (i) $|a - b| \geq |a| - |b|$ and   (ii) $|a - b| \geq |b| - |a|$.

    Statement (i) is Exercise 61. We also get (ii) from Exercise 61 as follows:

    $$|a - b| = |-(a - b)| = |b - a| \geq |b| - |a|.$$

§5

10. Since $1^3 = 1$ while $2^3 = 8$, we know that the first digit is 1. Testing 1.1, we see that $(1.1)^3$ is bigger than 1.3, hence bigger than x^3. So, the second digit is 0. Since $(1.01)^3 > 1.03 > x$, the third digit is also 0. Finally, by guessing and then computing, we find that $(1.003)^3 < x^3$, $(1.004)^3 > x^3$. Therefore, the decimal representation of $\sqrt[3]{x}$ begins with 1.003.

§6

6. The set of all integral perfect squares is the infinite set $\{0, 1, 4, 9, 16, 25, \cdots\}$. It is bounded from below by 0 but it is not bounded from above. Therefore, it is not bounded.

14. First step: For $n = 1$, the given statement reads:

$$1 \cdot 2 = \frac{1 \cdot 2 \cdot 3}{3}$$

which is certainly true. Second step: Assume that

$$(1) \qquad 1 \cdot 2 + 2 \cdot 3 + \cdots + k(k+1) = \frac{k(k+1)(k+2)}{3}$$

for some (arbitrary but fixed) integer $k > 0$. We must use equation (1) (whose truth we have assumed), along with skillful manipulation, to show that the given statement holds for $k + 1$, that is, to show that

$$1 \cdot 2 + 2 \cdot 3 + \cdots + k(k+1) + (k+1)(k+2) = \frac{(k+1)(k+2)(k+3)}{3}.$$

We now proceed to do this. Add the term $(k+1)(k+2)$ to each side of (1), obtaining the equation

$$(2) \qquad 1 \cdot 2 + \cdots + k(k+1) + (k+1)(k+2) = \frac{k(k+1)(k+2)}{3} + (k+1)(k+2).$$

Since the right-hand side of (2) is

$$\frac{k(k+1)(k+2) + 3(k+1)(k+2)}{3} = \frac{(k+1)(k+2)(k+3)}{3},$$

we see that equation (2) is precisely the statement we wished to establish.

18. First step: For $n = 1$, the given statement reads: $4 \geq 1$, which is true. Second step: Assume that $4^k \geq k^2$ for some (arbitrary but fixed) integer $k > 0$. We must show that $4^{k+1} \geq (k+1)^2$. Now, $4^{k+1} = 4^k \cdot 4$. By assumption, $4^k \geq k^2$ so that $4^k \cdot 4 \geq k^2 \cdot 4 = 4k^2$. We now know that $4^{k+1} \geq 4k^2$. We are trying to show that $4^{k+1} \geq (k+1)^2 = k^2 + 2k + 1$. This will follow (Postulate **14**) if we can show that $4k^2 \geq k^2 + 2k + 1$. This last inequality is equivalent to $3k^2 \geq 2k + 1$, the truth of which follows from Exercise 16.

22. The inequality $2z^3 - 1 < 15$ is equivalent to the inequality $z^3 < 8$. If z is positive, we may take the cube root of each side of the inequality (§5.3, Lemma) to obtain $z < 2$. If z is not positive, $z^3 < 8$ is always true. Therefore, S is the set of all rational numbers z with $z < 2$. Clearly, 2 is the least upper bound of S.

§7

4. Using the inequality (3) with $x = 5$ and $y = 8$, we have $|XY - 40| \leq 5b + 8a + ab$. Since we want $|XY - 40| \leq 1$, we can choose a and b small enough so that $5b + 8a + ab \leq 1$. (For then,

by Postulate **14**, we have $|XY - 40| \leq 1$.) If $b \leq 1/15$ and $a \leq 1/24$, we have $5b \leq \frac{1}{3}$ and $8a \leq \frac{1}{3}$ and $ab \leq (1/15) \cdot 24 < \frac{1}{3}$ so that $5b + 8a + ab \leq \frac{1}{3} + \frac{1}{3} + \frac{1}{3} = 1$.

17. We employ inequality (7) of Theorem 2. Since $|x| = |.\overline{717}| < 1$ and $|y| = |.18\overline{19}| < .2$, we have $10^{-m}|x| + 10^{-m}|y| + 10^{-2m} < 1 \cdot 10^{-m} + (1/5)10^{-m} + 10^{-2m}$. Therefore, if we choose m large enough to make $10^{-m} + (10^{-m}/5) + 10^{-2m} \leq .0001 = 10^{-4}$, we shall have the desired result (by Postulate **14**). Observation shows that $m = 5$ suffices. Therefore, we compute $(.71717)(.18191)$ and obtain .1304603947.

§8

2. $(a + b)(c + d) = (a + b)c + (a + b)d$
$$= c(a + b) + d(a + b) = (ca + cb) + (da + db) = (ac + bc) + (ad + bd)$$
$$= ac + (bc + (ad + bd)) = ac + (bc + (bd + ad)) = ac + ((bc + bd) + ad)$$
$$= ac + ((bd + bc) + ad) = ac + (bd + (bc + ad)) = (ac + bd) + (bc + ad).$$

The equalities stated are justified, in order by Postulates **9, 7, 9, 7, 3, 2, 3, 2, 3, 3**.

12. By definition, $1/ab = (ab)^{-1}$ and $(1/a)(1/b) = a^{-1}b^{-1}$. By definition, $(ab)^{-1}$ is that *unique* number whose product with ab is 1. But $a^{-1}b^{-1}$ is also a number whose product with ab is 1 since (using **7, 8, 12, 10, 12**) $(ab)(a^{-1}b^{-1}) = (ab)(b^{-1}a^{-1}) = a(bb^{-1}a^{-1}) = a(1 \cdot a^{-1}) = aa^{-1} = 1$. Therefore $(ab)^{-1} = a^{-1}b^{-1}$ so that $1/ab = (1/a)(1/b)$.

38. As in the last paragraph before the exercises, we note that $.8\overline{1} = .8 + .1(.\overline{1}) = (8/10) + (1/10)(.\overline{1})$. By the rule stated in the text $.\overline{1} = 1/9$. Thus $.8\overline{1} = (8/10) + (1/10)(1/9) = 73/90$.

Chapter 2

§1

16. As a sketch shows, there are two such points, (a, b) and (c, d). If $P = (7, -1)$ and $Q = (10, -1)$, we see that the length of the line segment $\overline{PQ}$ is 3. Using elementary facts about right triangles, we conclude that the distance from Q to (a, b) is 4, so that $(a, b) = (10, 3)$, and the distance from Q to (c, d) is 4, so that $(c, d) = (10, -5)$.

20. Let (a, b) be the coordinates of the new origin in the old system. Using the notation in §1.2, with (x, y) being the coordinates of the old origin in the old system and (X, Y) being the coordinates of the old origin in the new system, we have: $x = y = 0$; $X = -3$; $Y = 2$. Therefore, $a = x - X = 0 - (-3) = 3$ and $b = y - Y = 0 - 2 = -2$. The coordinates of the new origin in the old system are $(3, -2)$.

30. We use formula (4) in §1.4 with $(a_1, b_1, c_1) = (1, 3, 2)$ and $(a_2, b_2, c_2) = (2, 1, 3)$. Then:
$$d = \sqrt{(1-2)^2 + (3-1)^2 + (2-3)^2} = \sqrt{1+4+1} = \sqrt{6}.$$

§2

20. The slope of the given line is $1/3$ [compare (3), §2.2]. Let l be the line whose equation we are asked to find. Then the slope of l must be -3 (§2.3, Theorem 3). We know that l passes through $(6, -3)$. Let (x, y) be any point on l distinct from $(6, -3)$. Then using equation (5) of §2.2, we obtain $(y + 3)/(x - 6) = -3$ or $3x + y - 15 = 0$. This is the equation of a line which does have slope -3 and does pass through $(6, -3)$: it is the equation of l.

28. The line whose equation is given here crosses the x axis when $y = 0$, that is, when $x = 3/2$, and the y axis when $x = 0$, that is, when $y = -4$. We may also solve this problem as follows: Rewrite the equation in the form $(x/a) + (y/b) = 1$. Since $(2x)/3 = x/(3/2)$ and $-y/4 = y/(-4)$, the equation can be written as $x/(3/2) + y/-4 = 1$. By (9), §2.4, we conclude that the x intercept is $3/2$ and the y intercept is -4.

§3

8. We use the method of completing the square: we have $x^2 + y^2 + 4x - 4y + 8 = 0$ or $(x^2 + 4x) + (y^2 - 4y) = -8$. Add to the first parenthesis the square of half the coefficient of x (that is, 2^2 or 4) and add to the second parenthesis the square of half the coefficient of y (that is, $(-2)^2$ or 4). Then add 8 to the right-hand side of the equation to compensate for this addition of two 4's to the left. We now have: $(x^2 + 4x + 4) + (y^2 - 4y + 4) = -8 + 8$ or $(x + 2)^2 + (y - 2)^2 = 0$. This says that the sum of two nonnegative numbers is zero. Therefore, each must be zero, so that $x = -2$ and $y = 2$. The solution set, then, contains but the single point $(-2, 2)$.

18. Let l be the diameter to $(3, 3)$. Since the tangent at $(3, 3)$, $y = x$, has slope 1, l has slope -1. Since $(3, 3)$ lies on l, we may use the point-slope formula [§2.4, (7)] to find the equation of l. We obtain $y - 3 = (-1)(x - 3)$ or $x + y - 6 = 0$. We now know two distinct lines which pass through the center, (a, b), of the circle: l and the line $y = 2x$. Since (a, b) lies on each, we have $a + b - 6 = 0$ and $b = 2a$. Solving this pair of equations simultaneously, we find $a = 2$ and $b = 4$. The radius of the circle is the distance from $(2, 4)$ to $(3, 3)$, which is $\sqrt{2}$. The equation of the circle is $(x - 2)^2 + (y - 4)^2 = 2$.

§4

4. We follow the hint given after Exercise 3.

(1) The distance from a point (x, y) on the parabola to the focus is $\sqrt{(x + 2)^2 + (y + 4)^2}$. The distance from (x, y) to the line $y = 1$ is $1 - y$. Since the vertex and the focus of the parabola lie below the x axis, so does the entire parabola. The distance of a point (x, y) on the parabola from the x axis will be $-y$ and its distance from the line $y = 1$ will be $-y + 1$ or $1 - y$. Equating these two distances, squaring, and simplifying gives $x^2 + 4x + 10y + 19 = 0$.

(2) Translate the parabola so that the vertex is at the origin. The vertex is $(-2, -3/2)$. Let this be the new origin. Using the notation in §1.2, then, $a = -2$ and $b = -3/2$, while $x = a + X$ and $y = b + Y$ [(x, y) denoting old coordinates and (X, Y) new coordinates]. With the vertex now at the origin, we may use (2) in §4.1 with $p = -5/2$ [the new focus is $(0, -5/2)$], to obtain $X^2 = -10Y$ as the equation of the parabola in the new coordinates. Now, put $X = x + 2$ and $Y = y + (3/2)$ to obtain $(x + 2)^2 = -10[y + (3/2)]$ or $x^2 + 4x + 10y + 19 = 0$.

20. The slope of the line tangent to the parabola at $(-10, 100)$ is $2(-10)$ or -20. Let Q have coordinates (x_0, x_0^2). The slope of the line tangent to the parabola at Q is $2x_0$. In order for the two lines to be perpendicular, it is necessary that the product of their slopes be -1. Setting $(2x_0)(-20) = -1$, we have $x_0 = 1/40$. Q is then the point with coordinates $(1/40, 1/1600)$.

§6

5. $\sqrt{(0 - 1)^2 + (1 - 1)^2 + (0 - 2)^2 + (1 + 3)^2} = \sqrt{21}$.

Chapter 3

§1

10. This function is defined for all values of x for which $x - 1 \neq 0$, $x + 2 \neq 0$, and $3x - 2 \geq 0$ (the last inequality is required because the fourth root is defined only for nonnegative numbers). Therefore, we need $x \neq 1$, $x \neq -2$, and $x \geq 2/3$. If $x \geq 2/3$, certainly $x \neq -2$, so that we may sum up by saying that the function is defined for all $x \geq 2/3$ except $x = 1$.

20. $(F + k)(5) = F(5) + k(5) = (25 + 10 - 3) + \frac{1}{4} = 32\frac{1}{4}$.

$((F + k) \circ F)(2) = (F + k)(F(2)) = (F + k)(5) = 32\frac{1}{4}$.

$(F \circ (Fk))(-1) = F((Fk)(-1)) = F((Fk)(-1)) = F(F(-1) \cdot k(-1)) = F((-4) \cdot (-1/2)) = F(2) = 5$.

$(F + (k \circ k))(3) = F(3) + k(k(3)) = 12 + k(1/2) = 12 + (-2) = 10$.

36. This function is neither odd nor even. It is not odd since, for example, $g(-1)=0$ while $-g(1)=-\sqrt{2}$; and it is not even since, using the same example, $g(-1)=0$ while $g(1)=\sqrt{2}$.

44. The domain of this function may be divided into intervals [namely, $(-\infty, 0]$ and $(0, \infty)$], on each of which the function is increasing. Thus on $(-\infty, 0]$, if $z_1 < z_2$, then $z_1 + 1 < z_2 + 1$ so that $f(z_1) < f(z_2)$. And on $(0, \infty)$, if $z_1 < z_2$, then $z_1^2 < z_2^2$ (since z_1 and z_2 are positive) so that $f(z_1) < f(z_2)$. However, as it stands (that is, considered on its entire domain) the function is "none of these" since, for example, $0 < 1/2$ but $1 = f(0) > f(1/2) = 1/4$ while, as we saw above, the function is increasing on $(-\infty, 0]$.

§2

4. The graph of $x \mapsto x^2 + 3zx + 4$ will intersect the line $y = 1$ (that is, the graph of $x \mapsto 1$) at those points at which $x^2 + 3zx + 4 = 1$, that is, at those points at which $x^2 + 3zx + 3 = 0$. Our problem is then reduced to finding those values of z for which $x \mapsto x^2 + 3zx + 3$ has two distinct roots. For this latter function, $D = 4(1)(3) - (3z)^2 = 12 - 9z^2$. We want to know when $D < 0$, so that, we solve the inequality $12 - 9z^2 < 0$. We have: $9z^2 > 12$, that is, $z^2 > 4/3$, so that $z < -2\sqrt{3}/3$ or $z > 2\sqrt{3}/3$.

26. Since -2 is a double root and 2 is a single root, $(y-2) \cdot (y+2)^2 = y^3 + 2y^2 - 4y - 8$ is a factor of the given polynomial. We now divide $y^5 - 7y^3 + 2y^2 + 12y - 8$ by $y^3 + 2y^2 - 4y - 8$. There is no remainder; the quotient is $y^2 - 2y + 1 = (y-1)^2$ so that 1 is a double root.

36. (a) $(f \circ g \circ h \circ f)(x) = (f \circ g \circ h)((1 + \sqrt{x})^{1/2}) = (f \circ g)(4(1 + \sqrt{x})^{-1/2}) = f(\frac{1}{2}(1 + \sqrt{x})^{1/4}) = (1 + \frac{1}{2}\sqrt{2}(1 + \sqrt{x})^{1/8})^{1/2}$.

(b) For $x = 225 = 15^2$, we obtain $\sqrt{2}$.

44. Since

$$f(x) = \frac{(x-3)(x-3)}{(x+3)(x-3)(x-3)},$$

it is clear that f is undefined at $x = 3$ and $x = -3$. If $x \neq 3$ then $x - 3 \neq 0$ so that we may divide numerator and denominator by $(x-3)$, obtaining $g(x) = 1/(x+3)$. If $x \neq 3$, $f(x)$ and $g(x)$ are identical—that is, they assign the same value to every real number x. But if $x = 3$, $g(x) = 1/6$ whereas $f(x)$ is undefined. As explained in the text, it is natural to *define* $f(3)$ to be $1/6$. However, no "natural" value may be assigned to f at $x = -3$, for, as x approaches -3, $f(x)$, takes on arbitrarily large positive and negative values.

§3

16. We shall show that $f(x) < A$ if $|x| < \sqrt{A-1}$. The inequality $x^2 + 1 < A$ is equivalent to the inequality $x^2 < A - 1$ which, since $A > 1$, is equivalent to the pair of inequalities $-\sqrt{A-1} < x < \sqrt{A-1}$. This, in turn, is equivalent to $|x| = |x - 0| < \sqrt{A-1}$. Therefore, if $\alpha = \sqrt{A-1}$ and if $|x| < \alpha$, we *do* have $f(x) < A$. Since $f(x) = x^2 + 1 \geq 1$ for all x, $f(x)$ can never be less than a number A which is itself ≤ 1.

Remark. In the following solutions we use the symbol $\leftrightarrow$ as an abbreviation for the words "is equivalent to".

26. We shall verify Properties (I) and (II) of the definition of continuity: (I) Let $A > 2 = h(-1)$. If $t > -1$, $h(t) = -2 < 2 < A$. Assume $t < -1$. Then, $h(t) = t^2 + 1 < A \leftrightarrow t^2 < A - 1 \leftrightarrow -\sqrt{A-1} < t < \sqrt{A-1} \leftrightarrow -\sqrt{A-1} + 1 < t + 1 < \sqrt{A-1} + 1$. Since $t < -1$, $|t+1| = -(t+1)$. The left-hand side of the last inequality then gives $\sqrt{A-1} - 1 > |t+1|$. Since $A > 2$, $\sqrt{A-1} - 1 > 0$. Put $\delta = \sqrt{A-1} - 1$. Then if $|t+1| < \delta$, that is, if t lies in an interval of length 2δ with midpoint at -1, we have $h(t) < A$. (II) Let $B < 2 = h(-1)$. If

$t < -1, h(t) = t^2 + 1 > 2 > B$. Assume $t > -1$. Then $h(t) = -2t > B \leftrightarrow t < -\frac{1}{2}B \leftrightarrow t + 1 < 1 - \frac{1}{2}B$. Since $t > -1$, $|t+1| = t+1$. Therefore, put $\delta = 1 - \frac{1}{2}B > 0$. Then, if $|t+1| < \delta$, we have $h(t) > B$.

32. Let $\epsilon > 0$ be given. We must produce a number δ (which will here—and in general—depend on ϵ) such that, for all x with $|x - 1| < \delta$, we shall have $|f(x) - f(1)| < \epsilon$, that is, $|f(x) - 1| < \epsilon$. Now, $|f(x) - 1| < \epsilon \leftrightarrow |(2x+3)/5 - 1| < \epsilon \leftrightarrow |(2x/5) - (2/5)| < \epsilon \leftrightarrow (2/5)|x - 1| < \epsilon \leftrightarrow |x - 1| < 5\epsilon/2$. Therefore, if we choose $\delta = 5\epsilon/2$, it will be true that if $|x - 1| < \delta$, then $|f(x) - f(1)| < \epsilon$.

44. We imitate the proof of Theorem 6: $f(x) = x^9(1 - 100/x^5 + 3/x^6 - 2/x^9)$. Choose $|x|$ so large that $|-100/x^5|$, $|3/x^6|$, and $|-2/x^9|$ will each be less than $1/9$. A little experimentation shows that $|x| = 4$ will suffice. Let $k = 4$. Then, using the last paragraph of the proof, we see that if $x > 4$, we have $f(x) > 0$, and if $x < 4$, we have $f(x) > 0$, and if $x < 4$, we have $f(x) < 0$. This means that f has a root in $[-4, 4]$. Since neither -4 nor 4 is a root (test), f has a root in $(-4, 4)$.

46. It can be shown that the given function is continuous and monotone decreasing so that it does have an inverse. However, we are asked only to *find* the inverse function. Let $y = f(t) = (3t + 1)/t = 3 + (1/t)$. Then $y - 3 = 1/t$ or $t = 1/(y - 3)$. Thus $t = g(y) = 1/(y - 3)$ is the desired inverse function (where y is now the independent variable). It is more common, however, to reinstate t as the independent variable, so that we would say "the inverse function is $t \mapsto 1/(t - 3)$." The preceding method is quite general: once we know that a function $y = f(x)$ has an inverse, we find that inverse by solving the equation $y = f(x)$ for x in terms of y. It is then customary, after this has been done, to switch x and y so that x is the independent variable.

§4

2. The function h is defined and continuous at $t = 4$ and $h(4) = 1/17$. Therefore, $\lim_{t \to 4} h(t) = h(4) = 1/17$.

10. Let $f(x) = (x\sqrt{x} - x + \sqrt{x} - 1)/(x - 1)$. The function f is not defined at $x = 1$. We try to extend the definition of f to 1 in such a way that f will be continuous at 1. As usual, this is done by attempting to replace f by a function g such that $f(x) = g(x)$ when $x \neq 1$ and such that g is defined and continuous at 1. We then *define* $f(1)$ to be $g(1)$. We may write

$$f(x) = (\sqrt{x} - 1)(x + 1)/(\sqrt{x} - 1)(\sqrt{x} + 1).$$

(This factorization is permissible, since we are interested in values of x close to 1, and hence positive, so that $\sqrt{x}$ *is* defined.) If $x \neq 1$, then $\sqrt{x} - 1 \neq 0$ so that $f(x) = g(x) = (x + 1)/(\sqrt{x} + 1)$. g is defined and continuous at 1 and $g(1) = 1$. Define $f(1) = 1$. Then f is continuous at 1 so that $\lim_{x \to 1} f(x) = f(1) = 1$.

14. Since both functions $y \mapsto y^2 + y$ and $y \mapsto 12y^3$ are continuous everywhere, and both have the value $4/9$ at $y = 1/3$, we conclude that H can be made continuous at $1/3$ by redefining $H(1/3) = 4/9$. Thus

$$\lim_{y \to 1/3} H(y) = H(1/3) = 4/9.$$

16. We are asked to find $\lim_{x \to 0} \sqrt{fg}(x)$. If we put $h(x) = \sqrt{x}$, then the function $\sqrt{fg}$ is the function $h \circ fg$. We know that $\lim_{x \to 0}(fg)(x) = \lim_{x \to 0} f(x) \cdot \lim_{x \to 0} g(x) = 12 \cdot 3 = 36$ (Theorem 1). Therefore, we could use Theorem 2 to find $\lim_{x \to 0}(h \circ fg)(x)$ if we knew $\lim_{x \to 36} h(x)$. This is easy: h is continuous at 36 and $h(36) = 6$. Therefore, $\lim_{x \to 36} h(x) = 6$. We now employ Theorem 2. (The function f mentioned there is the function fg of this exercise; the function g mentioned there is the function h here.). Since $\lim_{x \to 0}(fg)(x) = 36$ and $\lim_{x \to 36} h(x) = 6$, $\lim_{x \to 0}(h \circ fg)(x) = 6$.

26. (a) In general, to find limits at infinity of rational functions (quotients of polynomials), the follow-

ing "trick" is useful. Divide numerator and denominator by the term of highest power in the denominator. Thus:

$$\lim_{x \to +\infty} \frac{x^3}{1 + x^2} = \lim_{x \to +\infty} \frac{x^3/x^2}{(1/x^2) + (x^2/x^2)} = \lim_{x \to +\infty} \frac{x}{(1/x^2) + 1}$$

Clearly, as $x \to +\infty$, the denominator approaches 1 while the numerator becomes arbitrarily large. Statement (a) is correct.

(b) $\lim_{x \to -\infty} \dfrac{x^3}{1 + x^2} = \lim_{x \to -\infty} \dfrac{x}{(1/x^2) + 1} = -\infty.$

(c) Since $x^3/(1 + x^2)$ is continuous at 2 and takes on the value 8/5 for $x = 2$, we have

$$\lim_{x \to 2} x^3/(1 + x^2) = 8/5.$$

But then $\lim_{x \to 2^+} x^3/(1 + x^2) = 8/5$. (Note that the statement "$\lim_{x \to x_0} f(x) = \alpha$" implies *both* statements "$\lim_{x \to x_0^+} f(x) = \alpha$" and "$\lim_{x \to x_0^-} f(x) = \alpha$," whereas neither of the two latter statements alone implies the former.)

§5

4. As defined, the function A assigns to 3 the value $3 + 1 = 4$. However, as $u \to 3^+$, that is, as u approaches 3 through values larger than 3, the function A is evaluated by the rule $u^2 - 1$. Clearly, as $u \to 3^+$, $A(u)$ is heading for $(3)^2 - 1 = 8$. But, as defined $A(3) = 4$. However, we may redefine A so that $a(3) = 8$. Then A, when considered as defined for $3 \geq u$, is continuous at 3. Hence, $\lim_{u \to 3^+} A(u) = 8$.

12. If $x \neq 1$ but x is near 1, $(x + 1)/|x - 1| > 0$. Further, $\lim_{x \to 1} |x - 1|/(x + 1) = 0$. Therefore, $\lim_{x \to 1} (x + 1)/|x - 1| = +\infty$. But the existence of a limit as $x \to x_0$ implies the existence of a limit, in fact the same limit, as $x \to x_0^+$ and as $x \to x_0^-$. Consequently, all three limits in this exercise are $+\infty$.

18. Using the rule stated just before the examples,

$$\lim_{y \to \infty} \frac{1 + y + 2y^2}{10 + 5y - 4y^2} = \lim_{t \to 0} \frac{1 + 1/t + 2/t^2}{10 + 5/t - 4/t^2} = \lim_{t \to 0} \frac{t^2 + t - 2}{10t^2 + 5t - 4} = \frac{1}{2}.$$

Thus $\frac{1}{2}$ is also the answer to (a) and (b).

24. Let $p(x) = rx^n + r_{n-1}x^{n-1} + \cdots + r_1 x + r_0$ and let $q(x) = sx^m + s_{m-1}x^{m-1} + \cdots + s_1 x + s_0$. Then

$$\lim_{x \to \infty} \frac{p(x)}{q(x)} = \lim_{t \to 0} \frac{p(1/t)}{q(1/t)} = \lim_{t \to 0} \frac{(r/t^n) + (r_{n-1}/t^{n-1}) + \cdots + (r_1/t) + r_0}{(s/t^m) + (s_{m-1}/t^{m-1}) + \cdots + (s_1/t) + s_0}.$$

Multiply numerator and denominator of the last expression by t^{m-n}. Then:

$$\lim_{x \to \infty} \frac{p(x)}{q(x)} = \lim_{t \to 0} \frac{rt^{m-n} + r_{n-1}t^{m-n+1} + \cdots + r_1 t^{m-1} + r_0 t^m}{s + s_{m-1}t + \cdots + s_1 t^{m-1} + s_0 t^m}.$$

Since $m - n > 0$, t appears to a positive power in each term in the numerator. Therefore, each term in the numerator approaches 0 as $t \to 0$. Since the denominator approaches s, the indicated limit is $0/s = 0$. The answer to all three parts of this exercise is 0.

32. We first compute

$$\lim_{x \to 1^-} 1/(x^2 + x - 2) = \lim_{x \to 1^-} 1/(x + 2)(x - 1).$$

If x is less than 1, but near 1, $x + 2$ is positive and $x - 1$ is negative. Therefore, $1/(x^2 + x - 2)$ is negative. Since $x^2 + x - 2 \to 0$ as $x \to 1^-$, $\lim_{x \to 1^-} 1/(x^2 + x - 2) = -\infty$. Therefore,

$$\lim_{x \to 1^-} (2x + 1)^{100}/(x^2 + x - 2) = \lim_{x \to 1^-} (2x + 1)^{100} \lim_{x \to 1^-} 1/(x^2 + x - 2) = (3^{100})(-\infty) = -\infty.$$

§6

8. Repeat the proof in the text up to and including the fifth line. Then let y_0 be a number such that $\alpha \le y_0 \le \beta$. Let B be a number such that $g(y_0) > B$. We must show that $g > B$ near y_0. We may assume that $B \ge a$ (since g takes on no values less than a). Set $g(y_0) = B_0$. Then $a \le B_0 \le b$ and $f(B_0) = y_0$. Set $f(B) = y_1$. Then $y_1 \ge \alpha$. Since $B_0 > B$ ($B_0 = g(y_0) > B$) and f is increasing, we have $y_0 = f(B_0) > f(B) = y_1$. If y is a number such that $y_1 < y < y_0$, then by the Intermediate Value Theorem, $y = f(C)$ for some C with $B < C < B_0$. Since $y = f(C), g(y) = C$. Hence $g(y) > B$ for $y_1 < y < y_0$. On the other hand, since g is increasing, $g(y) > B$ if $y > y_0$. Thus, $g > B$ near y_0 and Property II is verified.

Chapter 4

§1

16. Let $f(t) = t/(t^2 - 1)$. Then

$$f'(2) = \lim_{h \to 0} \frac{\dfrac{2+h}{(2+h)^2 - 1} - \dfrac{2}{3}}{h} = \lim_{h \to 0} \frac{-2h - 5}{3h^2 + 12h + 9} = -\frac{5}{9}.$$

The equation of the tangent is $s - 2/3 = -(5/9)(t - 2)$ or $9s + 5t - 16 = 0$.

22.
$$g'(x) = \lim_{h \to 0} \frac{\dfrac{1}{2(x+h) - 3} - \dfrac{1}{2x - 3}}{h} = \lim_{h \to 0} \frac{(2x - 3) - (2(x+h) - 3)}{h(2x - 3)(2(x+h) - 3)}$$

$$= \lim_{h \to 0} \frac{-2h}{h(2x - 3)(2(x+h) - 3)} = \lim_{h \to 0} \frac{-2}{(2x - 3)(2(x+h) - 3)}$$

$$= -\frac{2}{(2x - 3)^2}.$$

36. The symbol $(ds/dt)_{t=-1}$ means "the derivative (of the given function) evaluated at $t = -1$." Therefore, we compute ds/dt:

$$\lim_{h \to 0} \frac{\{1/[(t+h)^2 + (t+h) + 1]\} - \{1/(t^2 + t + 1)\}}{h}$$

$$= \lim_{h \to 0} \frac{t^2 + t + 1 - [(t+h)^2 + (t+h) + 1]}{h(t^2 + t + 1)[(t+h)^2 + (t+h) + 1]}$$

$$= \lim_{h \to 0} \frac{-2ht - h^2 - h}{h(t^2 + t + 1)[(t+h)^2 + (t+h) + 1]}$$

$$= \lim_{h \to 0} \frac{-2t - h - 1}{(t^2 + t + 1)[(t+h)^2 + (t+h) + 1]}.$$

Since this last function is continuous at $h = 0$, we have $(-2t - 1)/(t^2 + t + 1)^2$ as the desired limit. Therefore,

$$(ds/dt)_{t=-1} = \frac{-2(-1) - 1}{((-1)^2 - 1 + 1)^2} = 1/1 = 1.$$

42. Let $f(t) = 1/t$. Then (compare §1.6) $f'(t) = -1/t^2$. Therefore, $f'(1) = -1$. Since $f(1) = 1$, we have (using formula (8) with $x_0 = 1$) $f(t) = 1 - 1(t - 1) + r(t)(t - 1)$. That is

$$1/t = 2 - t + r(t)(t - 1),$$

$$r(t) = \frac{(1/t) + t - 2}{t - 1} = t - \frac{1}{t}.$$

Thus, $r(t)$ is continuous at 1 and $r(1) = 0$.

66. Using the formula $v(t) = -2\beta t$, we have $-128 = -2\beta t$ or $t = 64/\beta$ sec when the velocity is -128 ft/sec. How far has the brick fallen at this time? Using the formula $s = -\beta t^2$, we have

$$s = -\beta \cdot 4096/\beta^2 \quad \text{or} \quad s = -4096\beta \sim -256 \text{ ft.}$$

This means that the position of the brick is 256 feet *below* the point from which it was dropped—that is, it has fallen 256 ft.

§2

6. By rule (5), $(fg)' = f'g + fg'$. Therefore, $(fg)'(0) = (f'g + fg')(0) = (f'g)(0) + (fg')(0) = f'(0)g(0) + f(0)g'(0) = 1 + (-3) = -2$. And by rule (7), $(g/f)' = (g'f - gf')/f^2$, so that $(g/f)'(0) = (g'f - gf')/f^2(0) = [(g'f)(0) - (gf')(0)]/f^2(0) = [g'(0)f(0) - g(0)f'(0)]/(f(0))^2 = -3 - 1/1 = -4$.

18. Let $y = f(x)$. To find the slope of the desired line, we compute $f'(0)$. Using the quotient rule (7), we have

$$f'(x) = \frac{2x(x^3 + 1) - 3x^2(x^2 + 1)}{(x^3 + 1)^2},$$

so that $f'(0) = 0$. The equation of the line is $y - 1 = 0(x - 0)$ or $y = 1$.

38. We may write $f(t) = (t^3 + 2t - t^{1/2})^{1/2}$. We use the chain rule with $f = g \circ h$ where $g(u) = u^{1/2}$ and $h(t) = t^3 + 2t - t^{1/2}$. Then

$$f'(t) = g'(h(t)) \cdot h'(t) = 1/2(t^3 + 2t - t^{1/2})^{-1/2} \cdot (3t^2 + 2 - 1/2t^{-1/2})$$
$$= (3t^2 + 2) - (1/2\sqrt{t})/2\sqrt{t^3 + 2t - \sqrt{t}}.$$

62. (a) $(dg/dy)_{y=2}$ may also be written as $g'(2)$. Putting $y_0 = 2$, we see from the statement of Theorem 7 that we must find the number x_0 such that $f(x_0) = 2$, that is, we must solve the equation $x^4 + 1 = 2$. Clearly, $x = \pm 1$ are the only (real) solutions to this equation, and, since the domain on which the function is increasing is $x > 0$, we can eliminate $x = -1$ so that the desired number x_0 is 1. Then, using the theorem, we have $g'(2) = 1/f'(1) = 1/4$ since $f'(x) = 4x^3$. We sum up what we did: $g'(2) = g'(f(1)) = 1/f'(1) = 1/4$. Similarly,

$$(dg/dy)_{y=82} = g'(82) = g'(f(3)) = 1/f'(3) = 1/108.$$

(b) Reasoning as above,

$$(dh/dy)_{y=2} = h'(2) = h'(f(-1)) = 1/f'(-1) = -1/4,$$

$$(dh/dy)_{y=82} = h'(82) = h'(f(-3)) = 1/f'(-3) = -1/108.$$

§3

16. We make use of the chain rule here. (If we put $F(x) = x^3$, we have $G = F \circ H$.) Thus,

$$G'(x) = 3(H(x))^2 \cdot H'(x).$$

Then, using the rule for differentiating the product of two functions, we have

$$G''(x) = (6(H(x)) \cdot H'(x)) \cdot H'(x) + 3(H(x))^2 \cdot H''(x).$$

Therefore, $G''(1) = 6 \cdot 2 \cdot \frac{1}{3} \cdot \frac{1}{3} + 3(2)^2 \cdot \frac{1}{4} = 4\frac{1}{3}$.

32. Using the relativistic equation of this section, we have

$$10 \text{ gram cm/sec}^2 = m_0(5 \text{ cm/sec}^2)/\sqrt{(1 - .64)^3} = 5m_0 \text{ cm/sec}^2/.216$$

or $m_0 = .432$ grams. Newtonian physics would predict a mass of 2 grams.

§4

2. This function is undefined at $t = -1$. If we consider f only on the interval $(-\infty, -1)$, we have $f'(t) = 3/(t + 1)^2 > 0$. Therefore, f is increasing on $(-\infty, -1)$. Similarly, if we consider f only on the interval $(-1, \infty)$, $f'(t) > 0$ and f is increasing.

12. $f'(x) = 100x^{99} + 1000x^{999} = 100x^{99}(1 + 10x^{900})$. Since $x^{900} \geq 0$ for all x, we can have $f'(x) = 0$ only if $x = 0$. Therefore, 0 is the only critical point. $f''(x) = 9900x^{98} + 999,000x^{998}$, so that $f''(0) = 0$. Theorem 5 (the "second derivative test") is, therefore, of no help here. However, if we note that $f'(x) < 0$ if $x < 0$ and $f'(x) > 0$ if $x > 0$, we may use the first line of the table immediately preceding Theorem 5 (with $x_0 = 0$) to deduce that f has a strict local minimum at $x = 0$. [The procedure just used, that is, investigating the first derivative to the left and to the right of a critical point, and then applying the facts in the table, is often called the "first derivative test."]

32. Let x be the length of a side of the base and let h be the height. The problem here is to find a function f that gives surface area in terms of x and then to "minimize f." Since V is fixed and $V = x^2h$, we have $h = V/x^2$. (Obviously, $x = 0$ is meaningless here.) The surface area of the box is

$$2x^2 + 4xh = 2x^2 + 4V/x.$$

Let $f(x) = 2x^2 + 4V/x$. We have now expressed surface area as a function of x. $f'(x) = 4x - 4V/x^2$. $f'(x) = 0$ if and only if $x = V/x^2$ or $x = 3\sqrt{V}$. Since $f''(x) = 4 + 8V/x^3$, $f''(3\sqrt{V}) = 12 > 0$, so that f has a local minimum when $x = 3\sqrt{V}$. To check that this also gives the absolute minimum, we note that, when $0 < x < 3\sqrt{V}$, $f'(x) < 0$ and f is decreasing and when $x > 3\sqrt{V}$, $f'(x) > 0$ and f is increasing. Therefore, the dimension should be $x = 3\sqrt{V}$ and $h = 3\sqrt{V}$, that is, the box should be a cube.

44. Let (x, y) be a point on the curve. Then $y = \sqrt{x^2 + 6x + 10}$. The distance from $(1, 0)$ to a point (x, y) is $d = \sqrt{(x - 1)^2 + (y - 0)^2} = \sqrt{x^2 - 2x + 1 + y^2}$. We seek to "minimize d." If we set $f = d^2$, we can just as well minimize f.

$$f(x) = x^2 - 2x + 1 + y^2 = x^2 - 2x + 1 + (x^2 + 6x + 10) = 2x^2 + 4x + 11.$$

Then $f'(x) = 4x + 4$. The only critical point is $x = -1$ and $f''(-1) = 4 > 0$, so that -1 gives a local minimum. This is clearly an absolute minimum since f is a parabola. Since the function d takes the value $\sqrt{4 + 5} = 3$ at $x = -1$, the shortest distance is 3.

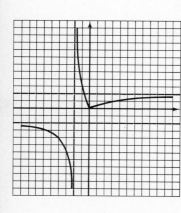

58. For $x \geq 0$, $f(x) = x/(x + 1)$ and $f'(x) = 1/(x + 1)^2 > 0$ so that f increases for $x \geq 0$.

$$f''(x) = -2/(x + 1)^3 < 0$$

so that f is concave for $x \geq 0$. There are no local maxima or minima and no inflection points for $x \geq 0$. $f(+\infty) = 1$. For $x < 0$, $f(x) = -x/(x + 1)$. There is a break at $x = -1$. We have

$$f(-1^+) = +\infty \quad \text{and} \quad f(-1^-) = -\infty.$$

$f'(x) = -1/(x+1)^2 < 0$ if $x \neq -1$ so that f decreases for $x < 0$. $f''(x) = 2/(x+1)^3$ which is positive for $x > -1$ and negative for $x < -1$ so that f is convex for $-1 < x < 0$ and concave for $-\infty < x < -1$. $f(-\infty) = -1$.

74. Let $f(x) = 8/x^3 - 6/x$. f is not defined for $x = 0$. For $x \neq 0$, $f(x) = 2(4 - 3x^2)/x^3$ so that f has roots at $\pm 2\sqrt{3}/3$. $f'(x) = 6(x^2 - 4)/x^4$ so that $f'(x) = 0$ for $x = \pm 2$. $f''(x) = 12(8 - x^2)/x^5$. $f''(2) > 0$ so that $(2, -2)$ is a local minimum and $f''(-2) < 0$ so that $(-2, 2)$ is a local maximum. $f''(x) = 0$ for $x = \pm 2\sqrt{2}$ and f'' changes sign in going through these points so each one gives an inflection point. $f(0^+) = +\infty$, $f(0^-) = -\infty$, $f(+\infty) = f(-\infty) = 0$.

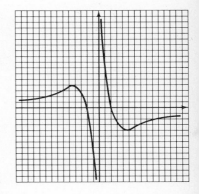

§5

14. Since $A''(v) = 3v^2$, we know that $A'(v) = (3v^3/3) + C = v^3 + C$. Since $A'(0)$ is to be equal to 1, we must have $0^3 + C = 1$, that is, $C = 1$. Therefore, $A'(v) = v^3 + 1$. Then $A(v) = (v^4/4) + v + k$, k being constant. Since we want $A(0) = -1$, we need $k = -1$. The desired function is $A(v) = (v^4/4) + v - 1$.

18. If the ball reaches a maximum height of 50 ft, the velocity will be zero at that value of t for which $s(t) = 50$. Let t_1 be this particular value. Then $t_1 = v_0/g$ or $v_0 = gt_1$ (since $v(t_1) = -gt_1 + v_0 = 0$). Now, $s(t_1) = -\frac{1}{2}gt_1^2 + v_0t_1 = 50$. Replacing v_0 by gt_1 we obtain $-\frac{1}{2}gt_1^2 + gt_1^2 = 50$ or $t_1 = 10/\sqrt{g}$. Since $v_0 = gt_1$, $v_0 = 10\sqrt{g} \approx 1.77$ ft/sec. This is the minimal velocity required.

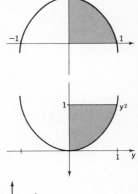

§7

4. $f'(3^-) = 1$ because, as y approaches 3 from the left, $f(y) = y + 1$ and the derivative is 1. $f'(3^+) = 4$ because, as y approaches 3 from the right, $f(y) = y^2 - 2y + 1$ and the derivative is $2y - 2$ which, at $y = 3$, has the value 4.

Chapter 5

§1

4. The parabola $1 - y^2$ is precisely the same shape as the parabola y^2, except that it is inverted and translated up by 1. Thus the desired area is the same as that shown at right: and this area is the same as the area of the square whose upper right-hand vertex is $(1, 1)$ minus the area under y^2 from 0 to 1. Therefore, the desired integral is $1 - \int_0^1 y^2 \, dy = 1 - \frac{1}{3} = \frac{2}{3}$.

20. Since $1/x$ is decreasing on $[1, 2]$, we compute $[(b-a)/N][f(a) - f(b)] = 1/N \cdot (1 - \frac{1}{2}) = 1/2N$. If $N = 2$, $1/2N$ will not exceed $1/4$. Therefore, divide $[1, 2]$ into 2 equal subintervals, as shown. $\phi(x)$, the step function under the curve, is equal to $2/3$ on $(1, 3/2)$ and $1/2$ on $(3/2, 2)$. Thus, $\int_1^2 \phi(x) \, dx = (1/2)(2/3) + (1/2)(1/2) = 7/12$. $\psi(x)$, the step function above the curve, is equal to 1 on $(1, 3/2)$ and $2/3$ on $(3/2, 2)$. Thus, $\int_1^2 \psi(x) \, dx = (1/2)(1) + (1/2)(2/3) = 10/12$. Therefore, $7/12 \leq \int_1^2 1/x \, dx \leq 10/12$. (In Chapter 6, we shall show that $\int_1^2 1/x \, dx = \log 2 \sim .6931$.)

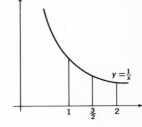

§2

2. If $x \leq 2$, $G(x) = -\int_x^2 \phi(r) \, dr = -\frac{1}{2}(2 - x)(2 - x) = -2 + 2x - \frac{1}{2}x^2$. (Use formula for area of triangle to compute integral.) Then $G'(x) = 2 - x = \phi(x)$. If $x > 2$, $G(x) = \int_x^2 \phi(r) \, dr = \frac{1}{2}(x - 2)(4 + 2x) = x^2 - 4$. (Use formula for area of trapezoid.) Then $G'(x) = 2x = \phi(x)$. That G is continuous is easily seen by graphing $-2 + 2x - \frac{1}{2}x^2$ for $x \leq 2$ and $x^2 - 4$ for $x > 2$ on the same axes.

16. The desired area is given by $\int_2^5 \sqrt{z - 1} \, dz$. The problem, then, is to find a primitive function of the integrand, $\sqrt{z - 1}$, and use the fundamental theorem. $\frac{2}{3}(z - 1)^{3/2}$ is an antiderivative. Therefore, the area $= \frac{2}{3}(5 - 1)^{3/2} - \frac{2}{3}(2 - 1)^{3/2} = \frac{2}{3}8 - \frac{2}{3}1 = 14/3$.

20. We note that the graph of $v \to v\sqrt{v^2 + 9}$ goes through the origin and is positive for $0 < v \leq 4$.

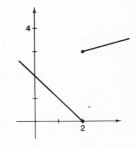

Hence, the area is $\int_0^4 v \sqrt{v^2 + 9}\, dv$. Since $1/3 \cdot (v^2 + 9)^{3/2}$ is a primitive function of $v\sqrt{v^2 + 9}$,

$$\int_0^4 v\sqrt{v^2 + 9}\, dv = \tfrac{1}{3}(v^2 + 9)^{3/2}\Big|_0^4 = \tfrac{125}{3} - \tfrac{27}{3} = 32\tfrac{2}{3}.$$

32. $h(x) = \int_0^x (s^3 + 1)\, dx$. Then $h(x) = (s^4/4) - s\,|_0^x = (x^4/4) - x$. Now, $f(x)$ is a primitive for $h(x)$, that is, $f'(x) = h(x)$. So, $f'(x) = (x^4/4) - x$ and $f(x) = (x^5/20) - (x^2/2) + C$. Since $f(0) = -1$ we have $C = -1$. Therefore, $f(x) = (x^5/20) - (x^2/2) - 1$ and $f(-1) = -(1/20) - (1/2) - 1 = -31/20$.

§3

28. We first compute the number

$$\int_{-1}^1 t^3\, dt = (t^4/4)\,|_{-1}^1 = (1)^4/4 - (-1)^4/4 = (1/4) - (1/4) = 0.$$

Thus,

$$\int_{-1}^1 \{x + \int_{-1}^1 t^3\, dt\}dx = \int_{-1}^1 x\, dx = (x^2/2)\,|_{-1}^1 = (1)^2/2 - (-1)^2/2 = (1/2) - (1/2) = 0.$$

32. Since

$$d(2\sqrt{x} + x^4 - 1) = f(x)dx, \qquad f(x) = d(2\sqrt{x} + x^4 - 1)/dx.$$

Hence, $f(x) = 1/\sqrt{x} + 4x^3$ and $f(1) = 1 + 4 = 5$.

46. In general, when integrating by parts a product of two functions, the function chosen as f in the formula should be one for which df will be "easier to handle" than f. Clearly, in this case, if we choose $f(z) = (2z - 3)$, then $df = 2\, dz$, an obvious "gain." Therefore, we choose $f(z) = 2z - 3$ and $dg(z) = \sqrt{z - 3}\, dz$. We have:

$$\int (2z - 3)\sqrt{z - 3}\, dz = \int (2z - 3)\, d\,\tfrac{2}{3}(z - 3)^{3/2} = \frac{2}{3}\int (2z - 3)\, d(z - 3)^{3/2}.$$

This is now in the correct form for applying the formula $\int f\, dg = fg - \int g\, df$. We obtain:

$$\tfrac{2}{3}\int (2z - 3)\, d(z - 3)^{3/2} = \tfrac{2}{3}\{(2z - 3)(z - 3)^{3/2} - \int (z - 3)^{3/2}\, d(2z - 3)\}$$

$$= \tfrac{2}{3}\{(2z - 3)(z - 3)^{3/2} - \int (z - 3)^{3/2} 2\, dz\} = \tfrac{2}{3}(2z - 3)(z - 3)^{3/2} - \tfrac{4}{3}\int (z - 3)^{3/2}\, dz$$

$$= \tfrac{2}{3}(2z - 3)(z - 3)^{3/2} - \tfrac{4}{3}\cdot\tfrac{2}{5}(z - 3)^{5/2} + C = \tfrac{2}{5}(z - 3)^{3/2}(2z - 1) + C.$$

68. We try the substitution $t = \psi(y) = (y^{3/2} + 1)$, motivated by the fact that then $\psi'(y)$ will involve $y^{1/2}$. ψ has a continuous derivative, $\tfrac{3}{2}y^{1/2}$, defined on $0 \le y \le 1$. Since $\psi'(y) = \tfrac{3}{2}y^{1/2}$, we have $y^{1/2} = \tfrac{2}{3}\psi'(y)$ so that,

$$\int_0^1 y^{1/2}(y^{3/2} + 1)^{1/2}\, dy = \frac{2}{3}\int_0^1 (\psi(y))^{1/2}\psi'(y)\, dy = \frac{2}{3}\int_1^2 t^{1/2}\, dt = \tfrac{2}{3}\cdot\tfrac{2}{3}t^{3/2}\Big|_1^2 = \tfrac{4}{9}(2\sqrt{2} - 1)$$

74. We do this exercise using the Leibniz notation. The "obvious" substitution seems to be

$$t = \psi(z) = z + 1.$$

$dt/dz = 1$ is continuous on $[0, 1]$. However, $dt = dz$ and this would appear to leave the z in the numerator of the integrand unaccounted for. As a matter of fact, this substitution *is* useful since, if $t = z + 1$, then $z = t - 1$ and we have:

$$\int_0^1 \frac{z}{(z + 1)^4}\, dz = \int_1^2 \frac{t - 1}{t^4}\, dt = \int_1^2 \frac{t}{t^4}\, dt - \int_1^2 \frac{1}{t^4}\, dt = \int_1^2 t^{-3}\, dt - \int_1^2 t^{-4}\, dt = -\frac{1}{2t^2}\Big|_1^2 + \frac{1}{3t^3}\Big|_1^2$$

$$= -\tfrac{1}{8} + \tfrac{1}{2} + \tfrac{1}{24} - \tfrac{1}{3} = \tfrac{1}{12}.$$

§4

14. The curves intersect at $(-3, 9)$, $(0, 0)$, and $(4, 16)$. The desired area is:

$$\int_{-3}^{0} (x^3 - 12x - x^2)\, dx + \int_{0}^{4} (x^2 - x^3 + 12x)\, dx = (\tfrac{1}{4}x^4 - 6x^2 - \tfrac{1}{3}x^3)\Big|_{-3}^{0} + (\tfrac{1}{3}x^3 - \tfrac{1}{4}x^4 + 6x^2)\Big|_{0}^{4}$$

$$= -(81/4 - 54 + 9) + (64/3 - 64 + 96) = 78\tfrac{1}{12}.$$

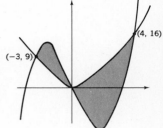

16. The base of the solid is shown in the figure. For any value of x between 0 and 1, the cross section is a square whose side has length equal to x. Therefore, $A(x) = x^2$ and

$$V = \int_{0}^{1} A(x)\, dx = \int_{0}^{1} x^2\, dx = \tfrac{1}{3}.$$

34. The tangent line at $(1, 1)$ has as its equation $y - 1 = \tfrac{2}{3}(x - 1)$ or $x = \tfrac{3}{2}(y - \tfrac{1}{3})$ and crosses the y axis at $y = \tfrac{1}{3}$. The desired region is shaded. The desired volume V may be obtained by computing V_1, the volume obtained by rotating the region under $x = y^{3/2}$ (that is, $y = x^{2/3}$) from $y = 0$ and $y = 1$ around the y axis, and V_2, the volume obtained by rotating the region under $x = \tfrac{3}{2}(y - \tfrac{1}{3})$ from $y = \tfrac{1}{3}$ to $y = 1$ around the y axis, and then noting that $V = V_1 - V_2$.

$$V_1 = \pi \int_{0}^{1} (y^{3/2})^2\, dy = \pi y^4/4 \Big|_{0}^{1} = \pi/4$$

and

$$V_2 = \pi \int_{1/3}^{1} [\tfrac{3}{2}(y - \tfrac{1}{3})]^2\, dy = \tfrac{9}{4}\pi (y - \tfrac{1}{3})^3/3 \Big|_{1/3}^{1} = 8\pi/36 - 0 = 2\pi/9.$$

Thus, $V = \pi/4 - 2\pi/9 = \pi/36$.

§5

14. Since $v(t) = -gt$, $K(t) = \tfrac{1}{2}mv(t)^2 = \tfrac{1}{2}mg^2t^2$. Therefore, $K'(t) = mg^2 t$ and $K''(t) = mg^2$, a constant.

18. Note that g appears in formula (13). However, $g = \gamma M/R^2$ [compare (9)] where M is the mass of the earth and R is the radius of the earth. Therefore, (13) may be restated as $V = \sqrt{2\gamma(M/R^2)R} = \sqrt{2\gamma}\cdot\sqrt{M/R}$ where V is now the escape velocity from a body of mass M and radius R. In this case, $V = \sqrt{13.34 \times 10^{-8}}\,\sqrt{2.65 \times 10^{20}} = \sqrt{35.35} \times 10^6$ or approximately 6×10^6 cm/sec.

§6

8. The integrand becomes infinite at $v = 1$. Therefore, we write

$$\int_{0}^{\sqrt{2}} v(v^2 - 1)^{-4/5}\, dv = \int_{0}^{1} v(v^2 - 1)^{-4/5}\, dv + \int_{1}^{\sqrt{2}} v(v^2 - 1)^{-4/5}\, dv$$

$$= \lim_{\epsilon \to 0^+} \int_{0}^{1-\epsilon} v(v^2 - 1)^{-4/5}\, dv + \lim_{\epsilon \to 0^+} \int_{1+\epsilon}^{\sqrt{2}} v(v^2 - 1)^{-4/5}\, dv$$

$$= \lim_{\epsilon \to 0^+} \tfrac{5}{2}(v^2 - 1)^{1/5}\Big|_{0}^{1-\epsilon} + \lim_{\epsilon \to 0^+} \tfrac{5}{2}(v^2 - 1)^{1/5}\Big|_{1+\epsilon}^{\sqrt{2}}$$

$$= \lim_{\epsilon \to 0^+} \tfrac{5}{2}[(\epsilon^2 - 2\epsilon)^{1/5} + 1] + \lim_{\epsilon \to 0^+} \tfrac{5}{2}[1 - (\epsilon^2 + 2\epsilon)^{1/5}] = 5.$$

16. Note that

$$\int_0^1 (x-1)^{2/3}\,dx = \lim_{\epsilon\to0^+}\int_0^{1-\epsilon}(x-1)^{2/3}\,dx = \lim_{\epsilon\to0^+} 3\,(x-1)^{1/3}\Big|_0^{1-\epsilon} = \lim_{\epsilon\to0^+}\,(-3\epsilon^{1/3}+3) = 3.$$

But, for $0 < x < 1$, $\sqrt{x} < 1$, so that $\sqrt{x}/(x-1)^{2/3} < 1/(x-1)^{2/3}$. Therefore, the given integral converges.

Chapter 6

§1

36. $$\frac{d\,(\text{arc cos } u)\,(\text{arc cos } 2u)}{du} = \text{arc cos } u\,\frac{d\,\text{arc cos } 2u}{du} + \text{arc cos } 2u\,\frac{d\,\text{arc cos } u}{du}$$

$$= \text{arc cos } u\cdot\frac{-1}{\sqrt{1-4u^2}}\cdot 2 + \text{arc cos } 2u\cdot\frac{-1}{\sqrt{1-u^2}}$$

$$= -\frac{2\,\text{arc cos } u}{\sqrt{1-4u^2}} - \frac{\text{arc cos } 2u}{\sqrt{1-u^2}}.$$

54. Set $t = 2z+1$. Then $u = (\cos^2 t)(\sin t^2)$ and

$$\frac{du}{dt} = \frac{d\cos^2 t}{dt}\sin t^2 + \cos^2 t\,\frac{d\sin t^2}{dt} = \frac{d\cos^2 t}{d\cos t}\frac{d\cos t}{dt}\sin t^2 + \cos^2 t\,\frac{d\sin (t^2)}{d(t^2)}\frac{dt^2}{dt}$$

$$= 2\cos t\,(-\sin t)\sin t^2 + \cos^2 t\cos t^2\cdot 2t = 2\cos t\,(t\cos t\cos t^2 - \sin t\sin t^2).$$

Also,

$$\frac{du}{dz} = \frac{du}{dt}\frac{dt}{dz} = 4\cos\,(2z+1)\big[(2z+1)\cos\,(2z+1)\cos\,(2z+1)^2 - \sin\,(2z+1)\sin\,(2z+1)^2\big].$$

60. Let $u = \sqrt{x}$. Then $du = (1/2\sqrt{x})\,dx$. So,

$$\int (\cos\sqrt{x})\,(dx/\sqrt{x}) = 2\int \cos u\,du = 2\sin u + C = 2\sin\sqrt{x} + C.$$

82. $\cos^2 x \sin^2 x = (\tfrac{1}{2}+\tfrac{1}{2}\cos 2x)(\tfrac{1}{2}-\tfrac{1}{2}\cos 2x) = \tfrac{1}{4}-\tfrac{1}{4}\cos^2 2x$.

But $\cos^2 2x = \tfrac{1}{2}+\tfrac{1}{2}\cos 4x$; thus $\cos^2 x \sin^2 x = \tfrac{1}{8}-\tfrac{1}{8}\cos 4x$ and the desired integral is

$$\int (\tfrac{1}{8}-\tfrac{1}{8}\cos 4x)\,dx = \tfrac{1}{8}x - \tfrac{1}{32}\sin 4x + C.$$

§2

4. Since $d\tan x/dx = 1+\tan^2 x$, we have $d^2\tan x/dx^2 = 2\tan x\cdot(1+\tan^2 x) = 2\tan x + 2\tan^3 x$, and $d^3\tan x/dx^3 = 2(1+\tan^2 x) + 6\tan^2 x(1+\tan^2 x) = 6\tan^4 x + 8\tan^2 x + 2$. For $x = 9\pi/4$, we have $6(\tan\tfrac{9}{4}\pi)^4 + 8(\tan\tfrac{9}{4}\pi)^2 + 2 = 6\cdot 1 + 8\cdot 1 + 2 = 16$.

22. $d x\,\text{arc tan }(1/\sqrt{x})/dx = x\cdot\dfrac{-\tfrac{1}{2}x^{-3/2}}{1+1/x} + \text{arc tan }(1/\sqrt{x}) = -\dfrac{\sqrt{x}}{2\,(x+1)} + \text{arc tan }(1/\sqrt{x}).$

Then

$$d x\,\text{arc tan }(1/\sqrt{x})/dx^2 = \frac{-(x+1)x^{-1/2} + 2x^{1/2}}{4\,(x+1)^2} - \frac{1}{2\,(x+1)\sqrt{x}} = -\frac{x+3}{4\,(x+1)^2\sqrt{x}}.$$

34. Let $u = \phi(t) = 2t^3$. Then, $\phi'(t) = 6t^2 > 0$, $\phi'(t)$ is continuous, and $du = 6t^2\, dt$.

$$\int_1^2 \frac{t^2\, dt}{1 + 4t^6} = \frac{1}{6}\int_2^{16} \frac{du}{1 + u^2} = \tfrac{1}{6} \text{ arc tan } u \Big|_2^1 = \tfrac{1}{6} \text{ arc tan } 16 - \tfrac{1}{6} \text{ arc tan } 2.$$

Using the table of trigonometric functions, we note that $\tan 86° = 14.301$ and $\tan 87° = 19.081$. Interpolating, we find that $16 \approx \tan 86\tfrac{1}{3}° = \tan(259\pi/540)$, or arc tan $16 \approx 259\pi/540$. Similarly, $2 \approx \tan 63\tfrac{1}{2}° = \tan(127\pi/360)$, or arc tan $2 \approx 127\pi/360$. Then, $\tfrac{1}{6}$ arc tan $16 - \tfrac{1}{6}$ arc tan $2 \approx 137\pi/6480$.

§3

8. Since $\sec^2 x = d \tan x/dx$, we have $\int \sec^2 x \tan^2 x\, dx = \int \tan^2 x\, d(\tan x) = (\tan^3 x/3) + C$.

22. $\displaystyle\int_1^{\sqrt{2}} x \text{ arc sec } x\, dx = \int_1^{\sqrt{2}} \text{ arc sec } x\, d(x^2/2) = (x^2/2) \text{ arc sec } x \Big|_1^{\sqrt{2}} - \int_1^{\sqrt{2}} \frac{x}{2\sqrt{x^2 - 1}}\, dx.$

This last integral is improper. To evaluate it, we note that

$$\int_1^{\sqrt{2}} \frac{x}{2\sqrt{x^2 - 1}}\, dx = \lim_{\epsilon \to 0^+} \int_{1 + \epsilon}^{\sqrt{2}} \frac{x}{2\sqrt{x^2 - 1}}\, dx = \lim_{\epsilon \to 0^+} \frac{\sqrt{x^2 - 1}}{2}\Big|_{1+\epsilon}^{\sqrt{2}} = \lim_{\epsilon \to 0^+}\left(\frac{1}{2} - \frac{\sqrt{\epsilon^2 + 2\epsilon}}{2}\right) = \frac{1}{2}.$$

Therefore

$$\int_1^{\sqrt{2}} x \text{ arc sec } x\, dx = (x^2/2) \text{ arc sec } x \Big|_1^{\sqrt{2}} - 1/2$$

$$= \text{ arc sec } \sqrt{2} - (1/2) \text{ arc sec } 1 - 1/2 = \pi/4 - 0 - 1/2 = (\pi - 2)/4.$$

§4

12. Although arc tan is defined for all values, log is defined only for $y > 0$. Therefore, this function is defined only for $y > 0$. The derivative is

$$\frac{2y \log y}{1 + y^4} + \frac{\text{arc tan } y^2}{y}.$$

52. Since $(1/u)\, du = d(\log u)$, and since $1 + \tan^2(\log u) = \sec^2(\log u)$, our integral equals

$$\int \sec^2(\log u)\, d(\log u) = \tan(\log u) + C.$$

58. Following the hint, we set

$$\frac{1}{x^2 + x} = \frac{a}{x} + \frac{b}{x + 1}$$

and solve for a and b. (This is called the "method of partial fractions" and will be discussed in Chapter 7, §2.) Combining the two fractions on the right we have

$$\frac{1}{x^2 + x} = \frac{ax + a + bx}{x^2 + x}.$$

If these two fractions are to be equal for all x (for which $x^2 + x \neq 0$), the numerators must be equal for all such x. Thus, setting $x = 0$, we see that $1 = a$. Next, setting $x = 1$, we have $1 =$

$x + 1 + bx$. Thus, $b = -1$. Therefore, $1/(x^2 + x) = 1/x - 1/(x + 1)$. The integration is now simple:

$$\int \frac{dx}{x^2 + x} = \int \frac{dx}{x} - \int \frac{dx}{x + 1} = \log |x| - \log |x + 1| + C = \log \left| \frac{x}{x + 1} \right| + C.$$

§5

30. Since $z = (x^2 + 1)^{\log x}$, we have $z = \exp \left[(\log x)\log (x^2 + 1) \right]$. Therefore,

$$dz/dx = \exp \left[(\log x)\log (x^2 + 1) \right] \left[\frac{\log (x^2 + 1)}{x} + \frac{2x \log x}{x^2 + 1} \right]$$

$$= (x^2 + 1)^{\log x} \left[\frac{\log (x^2 + 1)}{x} + \frac{2x \log x}{x^2 + 1} \right].$$

40. In general, if the integrand consists of a polynomial times e^x, integration by parts works if the polynomial is chosen as f and $e^x\, dx$ as dg. Thus

$$\int x^2 e^x\, dx = x^2 e^x - \int 2xe^x\, dx.$$

We again use integration by parts, this time with x as f and $e^x\, dx$ as dg, to obtain

$$\int x^2 e^x\, dx = x^2 e^x - 2 \left(xe^x - \int e^x\, dx \right) = x^2 e^x - 2xe^x + 2e^x + C.$$

46. We try $f(x) = e^{4x}$. It is true that $f'(x) = 4e^{4x} = 4f(x)$, but $f(1) \neq e$. Note, however, that if $f'(x) = \alpha f(x)$ and if $g(x) = Cf(x)$, then $g'(x) = \alpha g(x)$. Thus, if we replace f by Cf, we shall not lose the property that the derivative equals four times the function. Since $f(1) = e^4$, we use $C = 1/e^3$. That is, we replace e^{4x} by $(1/e^3)e^{4x} = e^{4x-3}$ and now use f to denote the function $f(x) = e^{4x-3}$. Then $f'(x) = 4e^{4x-3} = 4f(x)$ and $f(1) = e^{4-3} = e$ as required. (Note: This is a particular case of the general method suggested just before Exercise 49. Let us assume that the required f is of the form $f(x) = e^{g(x)}$. Then, $f'(x) = g'(x)e^{g(x)} = g'(x)f(x) = 4f(x)$. Therefore, $g(x) = 4x + C$. Now, $f(x) = e^{4x+C}$ and $f(1) = e$ so that $C = -3$. We now verify that $f(x) = e^{4x-3}$ does have the required properties.)

64. As in the proof of Theorem 8, we compute

$$\lim_{x \to 0} \log (1 - x)^{(x+1)/x} = \lim_{x \to 0} \frac{(x + 1) \log (1 - x)}{x} = \lim_{x \to 0} \left[\log (1 - x) + (1/x) \log (1 - x) \right]$$

$$= \lim_{x \to 0} \log (1 - x) + \lim_{x \to 0} \left[(1/x) \log (1 - x) \right] = 0 + \lim_{x \to 0} \left[(1/x) \log (1 - x) \right].$$

Using the same device as in text, we have, upon setting $t = -x$,

$$\lim_{x \to 0} \log \left[(1/x)(1 - x) \right] = \lim_{t \to 0} \left[(-1/t) \log (1 - t) \right] = -\lim_{t \to 0} \frac{\log (1 + t) - \log 1}{t}$$

$$= -(d \log t/dt)_{t=1} = -1/1 = -1.$$

Thus,

$$\lim_{x \to 0} (1 - x)^{(x+1)/x} = \lim_{x \to 0} \exp \{ \log \left[(1 - x)^{(x+1)/x} \right] \} = \exp \lim_{x \to 0} \log \left[(1 - x)^{(x+1)/x} \right] = e^{-1} = 1/e.$$

74. $\dfrac{d\,\cosh^{1/2} z^2}{dz} = \tfrac{1}{2}\cosh^{-1/2} z^2\,\dfrac{d\,\cosh z^2}{dz} = \tfrac{1}{2}\cosh^{-1/2} z^2\,\sinh z^2\,\dfrac{dz^2}{dz}$

$$= \dfrac{z\,\sinh z^2}{\sqrt{\cosh z^2}}$$

§6

8. First, we note that $r_{\text{cr}} = 4 \cdot 10^{10}/2 \cdot 10^8 = 2 \cdot 10^2$. Therefore,

$$r = \tfrac{1}{2} r_{\text{cr}} = 10^2 \quad \text{and} \quad \beta = \beta_0/r = 4 \cdot 10^{10}/10^2 = 4 \cdot 10^8.$$

Then,

$$N = N_0 \exp\left(-2 \cdot 10^8 t\right) + \exp\left(-2 \cdot 10^8 t\right) \int_0^t \exp\left(2 \cdot 10^8 \xi\right) \cdot 100\,d\xi$$

$$= N_0 \exp\left(-2 \cdot 10^8 t\right) + 100 \exp\left(-2 \cdot 10^8 t\right) \left[\dfrac{\exp\left(2 \cdot 10^8 \xi\right)}{2 \cdot 10^8}\,\Big|_0^t\right]$$

$$= N_0 \exp\left(-2 \cdot 10^8\right) + \tfrac{1}{2} \cdot 10^6 - \dfrac{\exp\left(-2 \cdot 10^8 t\right)}{2 \cdot 10^6}\,.$$

The limit is the sum of the first two terms.

§7

30. (a) Use (20) and (9) to obtain:

$$\text{arc sech } x = \text{arc cosh } \frac{1}{x} = \log\left(\frac{1}{x} + \sqrt{\frac{1}{x^2} - 1}\right) = \log\frac{1 + \sqrt{1 - x^2}}{x}\,.$$

We write x rather than $|x|$ since arc sech x is defined only for $0 < x \le 1$.
(b) Use (20) and (7) and note that arc csch x is defined for $x > 0$ and $x < 0$.

$$\text{arc csch } x = \text{arc sinh } \frac{1}{x} = \log\left(\frac{1}{x^2} + \sqrt{1 + \frac{1}{x^2}}\right)$$

(c) Use (19) and (15) and note that arc coth x is defined for $|x| > 1$.

Chapter 7

§1.5

2. We have $x_0 = 0$, $x_1 = 1/3$, $x_2 = 2/3$, $x_3 = 1$, $x_4 = 4/3$, $x_5 = 5/3$, $x_n = x_6 = 2$. $h = 1/3$. It is convenient to make a table of these x values and the corresponding y values:

x	0	1/3	2/3	1	4/3	5/3	2
y	0	28/81	70/71	2	364/81	760/81	18

We may first look at the error estimates. ($x^4 + x$ is monotone on $[0, 2]$.) From (8), we have $E \le |18 - 0|\,(1/3) = 6$.
    From (9), $E \le (1/12)(2 - 0)(48)(1/9) = 8/9$, since $|f''(x)| = |12x^2| \le 48 = M_2$.

We may now apply the trapezoidal rule:

$$\int_0^2 (x^4 + x) \cdot dx \approx 1/6 \ (0 + 56/81 + 140/81 + 4 + 728/81$$

$$+ \ 1520/81 + 18) = 4226/486 \approx 8.69.$$

Of course, it is simple to integrate $x^4 + x$ directly. The exact answer is 8.4.

§2

18. Since $x^2 + 4 = 4\big[(x/2)^2 + 1 \big] = 4 \, (t^2 + 1)$ where $t = x/2$ and $dt = dx/2$, we have:

$$\int \frac{dx}{(x^2 + 4)^3} = \int \frac{2 \, dt}{64 \, (t^2 + 1)^3} = \frac{1}{32} \int \frac{dt}{(t^2 + 1)^3}$$

$$= \frac{1}{32} \left[\frac{1}{4} \frac{t}{(1 + t^2)^2} + (1 - \tfrac{1}{4}) \int \frac{dt}{(1 + t^2)^2} \right]$$

$$= \frac{1}{128} \frac{t}{(1 + t^2)^2} + \frac{3}{128} \left[\frac{1}{2} \frac{t}{1 + t^2} + (1 - \tfrac{1}{2}) \int \frac{dt}{1 + t^2} \right]$$

$$= \frac{1}{128} \frac{t}{(1 + t^2)^2} + \frac{3}{256} \frac{t}{1 + t^2} + \frac{3}{256} \text{ arc tan } t + C$$

$$= \frac{1}{16} \frac{x}{(4 + x^2)^2} + \frac{3}{128} \frac{x}{4 + x^2} + \frac{3}{256} \text{ arc tan } \frac{x}{2} + C.$$

34. Since the derivative of $x^2 - 2x + 2$ is $2x - 2$, we want to have $(2x - 2)$ in the numerator. We have $2 - x$ in the numerator. Now $2 - x = -x + 2 = -\tfrac{1}{2}(2x - 2) + 1$. Therefore,

$$\int \frac{2 - x}{(x^2 - 2x + 2)^3} \, dx = \int \frac{-\tfrac{1}{2}(2x - 2) + 1}{(x^2 - 2x + 2)^3} \, dx$$

$$= -\frac{1}{2} \int \frac{2x - 2}{(x^2 - 2x + 2)^3} \, dx + \int \frac{1}{(x^2 - 2x + 2)^3} \, dx$$

$$= -\frac{1}{2} \int \frac{d \, (x^2 - 2x + 2)}{(x^2 - 2x + 2)^3} + \int \frac{dx}{(x^2 - 2x + 2)^3} \, .$$

The first integral is equal to $\tfrac{1}{4}(x^2 - 2x + 2)^2$. The second can be handled by the methods of the two previous sections, as follows: $x^2 - 2x + 2 = x^2 - 2x + 1 + 1 = (x - 1)^2 + 1 = t^2 + 1$ where $t = x - 1$ and $dt = dx$. Then by (4), the second integral is

$$\int \frac{dt}{(t^2 + 1)^3} = \frac{1}{4} \frac{t}{(t^2 + 1)^2} + \frac{3}{4} \int \frac{dt}{(t^2 + 1)^2}$$

$$= \frac{1}{4} \frac{t}{(t^2 + 1)^2} + \frac{3}{4} \left[\frac{1}{2} \frac{t}{t^2 + 1} + \frac{1}{2} \int \frac{dt}{t^2 + 1} \right]$$

$$= \frac{1}{4} \frac{t}{(t^2 + 1)^2} + \frac{3}{8} \frac{t}{t^2 + 1} + \frac{3}{8} \text{ arc tan } t + k.$$

Finally,

$$\int \frac{2-x}{(x^2-2x+2)^3}\,dx = \frac{1}{4}\frac{x}{(x^2-2x+2)^2} + \frac{3}{8}\frac{x-1}{x^2-2x+2} + \frac{3}{8}\text{arc tan }(x-1) + C.$$

42. $\dfrac{x^4+3x^3+4x^2+4x+1}{(x^2+x+1)^3}$

$$= \frac{Ax+B}{(x^2+x+1)} + \frac{Cx+D}{(x^2+x+1)^2} + \frac{Ex+F}{(x^2+x+1)^3}$$

$$= \frac{(Ax+B)(x^2+x+1)^2 + (Cx+D)(x^2+x+1) + (Ex+F)}{(x^2+x+1)^3}$$

$$= \frac{Ax^5+(2A+B)x^4+(3A+2B+C)x^3+(2A+3B+C+D)x^2+(A+2B+C+D+E)x+(B+D+F)}{(x^2+x+1)^3}$$

Therefore, $A = 0$; $2A + B = 1$, $B = 1$; $3A + 2B + C = 3$, $C = 1$; $2A + 3B + C + D = 4$, $D = 0$; $A + 2B + C + D + E = 4$, $E = 1$; $B + D + F = 1$, $F = 0$. Therefore, the integrand is $[1/(x^2 + x + 1)] + [x/(x^2 + x + 1)^2] + [x/(x^2 + x + 1)^3]$, and we have

$$\int \frac{dx}{x^2+x+1} + \int \frac{x}{(x^2+x+1)^2}\,dx + \int \frac{x}{(x^2+x+1)^3}\,dx$$

$$= \frac{2}{\sqrt{3}}\text{arc tan }\frac{2x+1}{\sqrt{3}} - \frac{1}{2(x^2+x+1)} - \frac{1}{4(x^2+x+1)^2} + C.$$

§3

20. Using the general method, we have, with $y = \tan(x/2)$,

$$\frac{1-\sin x}{1+\sin x} = \frac{1-[2y/(1+y^2)]}{1+[2y/(1+y^2)]} = \frac{y^2-2y+1}{y^2+2y+1}.$$

Since

$$dx = \frac{2}{1+y^2}\,dy,$$

the integral becomes

$$2\int \frac{y^2-2y+1}{(y+1)^2(y^2+1)}\,dy.$$

Setting

$$\frac{y^2-2y+1}{(y+1)^2(y^2+1)} = \frac{A}{y+1} + \frac{B}{(y+1)^2} + \frac{Cy+D}{y^2+1}$$

and solving, we find $A = C = 0$, $B = 2$, $D = -1$. Accordingly, we integrate

$$2\int \frac{2}{(y+1)^2}\,dy - 2\int \frac{1}{y^2+1}\,dy$$

and obtain

$$-\frac{4}{y+1} - 2\arctan y + C \quad \text{or} \quad -\frac{4}{1+\tan (x/2)} - x + C.$$

44. Putting $x = \cos y$, so that $1 - x^2 = \sin^2 y$, and $dx = -\sin y \, dy$, we have

$$\int x^3 (1 - x^2)^{1/4} \, dx = -\int \cos^3 y \sin^{3/2} y \, dy$$

$$= -\int (1 - \sin^2 y)(\sin^{3/2} y) \, d(\sin y) = -\int (1 - u^2) u^{3/2} \, du \quad (\text{where } u = \sin y)$$

$$= -\tfrac{2}{5} u^{5/2} + \tfrac{2}{9} u^{9/2} + C = -\tfrac{2}{5}(\sin y)^{5/2} + \tfrac{2}{9}(\sin y)^{9/2} + C$$

$$= -\tfrac{2}{5}(1 - x^2)^{5/4} + \tfrac{2}{9}(1 - x^2)^{9/2} + C.$$

56. We shall first find an indefinite integral of $\sqrt{x^2 + 1}/x^4$. Using Method (a),

$$\int \frac{\sqrt{x^2 + 1}}{x^4} \, dx = \int \frac{\sec y}{\tan^4 y} \sec^2 y \, dy = \int \frac{\sec^3 y}{\tan^4 y} \, dy = \int \frac{\cos y}{\sin^4 y} \, dy.$$

Since $\cos y \, dy = d(\sin y)$, we obtain $-1/3 \sin^3 y$. Now if $y = \arctan x$, we have $\sin y = x/\sqrt{1 + x^2}$ so that

$$\int_1^{\sqrt{3}} \frac{\sqrt{x^2 + 1}}{x^4} \, dx = -\frac{(x^2 + 1)^{3/2}}{3x^3} \bigg|_1^{\sqrt{3}} = \frac{6\sqrt{6} - 8}{9\sqrt{3}}.$$

62. Since $\alpha\gamma - \beta^2 = -1 < 0$, we set

$$\xi = \frac{3x + 4}{\sqrt{16 - 15}} = 3x + 4.$$

Then,

$$3x^2 + 8x + 5 = \tfrac{1}{3}\big[(3x + 4)^2 - 1\big] = \tfrac{1}{3}(\xi^2 - 1) \quad \text{and} \quad d\xi = 3 \, dx.$$

Therefore,

$$\int \frac{dx}{(3x^2 + 8x + 5)^{3/2}} = \int \frac{\tfrac{1}{3} d\xi}{\big[\tfrac{1}{3}(\xi^2 - 1)\big]^{3/2}} = \sqrt{3} \int \frac{d\xi}{(\xi^2 - 1)^{3/2}}$$

which is of form (VII).

§4

10. Using (2),

$$\int x^{1/3} \log^2 x \, dx = \frac{x^{4/3} \log^2 x}{4/3} - \frac{2}{4/3} \int x^{1/3} \log x \, dx = \tfrac{3}{4} x^{4/3} \log^2 x - \tfrac{3}{2} \int x^{1/3} \log x \, dx.$$

The remaining integral can be integrated by parts, with $u = \log x$ and $dv = x^{1/3} \, dx$. We have

$$\int x^{1/3} \log x \, dx = \tfrac{3}{4} x^{4/3} \log x - \int \tfrac{3}{4} x^{4/3} \frac{dx}{x} = \tfrac{3}{4} x^{4/3} \log x - \tfrac{9}{16} x^{4/3} + C.$$

Therefore,

$$\int x^{1/3} \log^2 x \, dx = \tfrac{3}{4} x^{4/3} \log^2 x - \tfrac{9}{8} x^{4/3} \log x + \tfrac{27}{32} x^{4/3} + C.$$

32. Using (6), we have

$$\int_{\pi/4}^{\pi/3} \cos^3 x \sin^3 x \, dx = -\tfrac{1}{6} \cos^4 x \sin^2 x \Big|_{\pi/4}^{\pi/3} + \frac{1}{3} \int_{\pi/4}^{\pi/3} \cos^3 x \sin x \, dx.$$

In the last integral, let $u = \cos x$. Then $du = -\sin x \, dx$, and we have

$$\int_{\pi/4}^{\pi/3} \cos^3 x \sin x \, dx = -\int_{\sqrt{2}/2}^{1/2} u^3 \, du = -\frac{u^4}{4} \Big|_{\sqrt{2}/2}^{1/2} = \frac{3}{64}.$$

Therefore,

$$\int_{\pi/4}^{\pi/3} \cos^3 x \sin^3 x \, dx = -\frac{1}{6}\left(\frac{1}{16}\right)\left(\frac{3}{4}\right) + \frac{1}{6}\left(\frac{1}{4}\right)\left(\frac{1}{2}\right) + \frac{1}{64} = \frac{11}{384}.$$

Chapter 8

§1

6. Since $f(1) = f(-1) = 3$, we want all numbers ξ strictly between -1 and 1 such that $f'(\xi) = 0$. Now, $f'(x) = 10x^9 + 4x^3 = x^3(10x^6 + 4)$, so that the only number ξ for which $f'(\xi) = 0$ is $\xi = 0$.

§2

10. We have $f(x) = e^{-x}$, $x = .01$, $x_0 = 0$. Also $f'(x) = f'''(x) = -e^{-x}$ and $f''(x) = e^{-x}$. Using (6), we have

$$e^{-.01} = e^0 - e^0(.01) + \tfrac{1}{2}e^0(.01)^2 + R(x) = .99005 + R(x).$$

$|f'''(t)| = |-e^{-t}| = e^{-t}$ is decreasing. Hence $|f'''(t)| < |f'''(0)| = 1$ for $0 < t < .01$. Therefore, $|R| \leq 1(.01)^3/6 < .00000017$.

30. $|f^{(7)}(x)| = |\sin x|$. In the interval, $3\pi/8 \leq x \leq 5\pi/8$, $|\sin x| = \sin x$ achieves its largest value at $x = \pi/2$. Thus, $|f^{(7)}(x)| \leq \sin(\pi/2) = 1$ for $3\pi/8 \leq x \leq 5\pi/8$. Therefore,

$$|R_7(x)| \leq \frac{|x - (\pi/2)|^7}{1\cdot2\cdot3\cdot4\cdot5\cdot6\cdot7} \leq \frac{\pi/2}{1\cdot2\cdot3\cdot4\cdot5\cdot6\cdot7\cdot8}.$$

Since, in the interval considered, we have $|x - (\pi/2)| \leq \pi/8 < \tfrac{1}{2}$, it follows that

$$|R_7(x)| < 1/(1\cdot2\cdot3\cdot4\cdot5\cdot6\cdot7) = 1/10080.$$

36.
$$2^n = (1 + 1)^n = \binom{n}{0} + \binom{n}{1} + \binom{n}{2} + \cdots + \binom{n}{n}.$$

46. By (20) and (21),

$$(.8)^{.01} = (1 + (-.2))^{.01} = 1 + (.01)(-.2) + \frac{(.01)(.01 - 1)}{1\cdot2}(-.2)^2$$

$$+ \frac{(.01)(.01 - 1)(.01 - 2)}{1\cdot2\cdot3}(-.2)^3 + R = .997775732 + R$$

where

$$R = \frac{(.01)(.01 - 1)(.01 - 2)(.01 - 3)}{1\cdot2\cdot3\cdot4}(1 + \xi)^{.01-4}(-.2)^4 = .0000013134(1 + \xi)^{-3.99}$$

with some ξ, $-.2 < \xi < 0$. Hence $.8 < 1 + \xi < 1$, $(.8)^{-3.99} > (1 + \xi)^{-3.99} > 1$. Therefore, $(1 + \xi)^{-3.99} < 1/(.8)^{3.99} < 1/(.8)^4 = 1/.4096 < 3$, and finally $|R| < .0000039402$.

§3

2. $a_1 = (-1)^1(1-1)^3 = 0$, $a_2 = (-1)^2(2-1)^3 = 1$, $a_3 = (-1)^3(3-1)^3 = -8$,
     $a_4 = (-1)^4(4-1)^3 = 27$, $a_5 = (-1)^5(5-1)^3 = -64$, $a_6 = (-1)^6(6-1)^3 = 125$.

18. The general term of this sequence is $a_n = (n-1)/n$, $n = 1, 2, \cdots$. The limit appears to be 1. We verify this: (I) Let A be any number such that $1 < A$. Obviously, nearly all terms of the sequence are less than A; in fact *all* terms are less than A since $(n-1)/n < 1 < A$ for all n. (II) Let B be any number such that $1 > B$. We want to show that, if n is large enough, $(n-1)/n > B$. But this will be true if $n - 1 > nB$, that is, if $n(1-B) > 1$ or $n > 1/(1-B)$. (Since $B < 1$, $1 - B$ is positive.) Choose $N \geq 1/(1-B)$. Then, for all a_n with $n > N$, $a_n > B$.

22. This sequence does not converge. This is obvious since, instead of having the terms come as close as we like to *one* fixed number, they oscillate between two different numbers. The proof: Suppose there were a limit α. (i) If $\alpha \geq \frac{1}{2}$, (II) is violated since 0 is a number B such that $\alpha > B$ whereas it is not true that nearly all terms of the sequence are $> B = 0$. (ii) If $\alpha \leq -\frac{1}{2}$, (I) is violated with $A = 0$. (iii) If $-\frac{1}{2} < \alpha < \frac{1}{2}$, let $A = \frac{1}{2}(\alpha + \frac{1}{2})$. Then $\alpha < A$, but infinitely many terms are $> A$.

44. $\lim v_k = \lim [2k/(5k+1)] + \lim [\sin k/(5k+1)]$. $\lim [2k/(5k+1)] = \lim [2/(5+k^{-1})] = \frac{2}{5}$. To handle $\lim [\sin k/(5k+1)]$, let $b_k = 1/(5k+1)$. Then $\lim b_k = 0$. Since $|\sin k/(5k+1)| \leq b_k$, we also have $\lim [\sin k/(5k+1)] = 0$, as one sees at once. Therefore, $\lim v_k = \frac{2}{5} + 0 = \frac{2}{5}$.

§4

6. Let the series be $\sum_{j=1}^{\infty} a_j$ and let S_n be the nth partial sum. Then $S_1 = 1/(1+1) = \frac{1}{2}$ so that $a_1 = \frac{1}{2}$. $S_2 = 2/(2+1) = \frac{2}{3} = a_1 + a_2 = \frac{1}{2} + a_2$. Therefore, $a_2 = \frac{1}{6}$. In general, $S_n = S_{n-1} + a_n$, that is, $n/(n+1) = [(n-1)/n] + a_n$ or $a_n = 1/n(n+1)$. The series is $\sum_{j=1}^{\infty} 1/j(j+1) = \frac{1}{2} + \frac{1}{6} + \frac{1}{12} + \cdots$.

26. It will be helpful to write out the first few terms of the series:

$$\sum_{k=0}^{\infty} (-1)^k t^{k/2} = 1 - t^{1/2} + t - t^{3/2} + t^2 - t^{5/2} + \cdots.$$

This is a geometric series with $q = -\sqrt{t}$. Thus

$$\sum_{k=0}^{\infty} (-1)^k t^{k/2} = \sum_{k=0}^{\infty} (-\sqrt{t})^k = \frac{1}{1 + \sqrt{t}}.$$

48. Since $\sin k < 1$, $3^k - \sin k > 3^k - 1$ and $1/(3^k - \sin k) < 1/(3^k - 1) \leq 1/2^k$. (The last inequality is obvious, since $3^k - 1 \geq 2^k$ for $k = 1, 2, \cdots$, but it can be easily proved by induction.) The series $\sum_{k=1}^{\infty} 1/2^k = \sum_{k=1}^{\infty} (1/2)^k$ converges. By Theorem 4, so does the given series.

54. Let $f(t) = 1/t \log^2 t$. Then f is continuous, nonnegative and decreasing for $t \geq 2$. Further,

$$\int_2^{+\infty} f(t)\, dt = \lim_{A \to \infty} \int_2^A \frac{dt}{t \log^2 t} = \lim_{A \to \infty} \left[\frac{-1}{\log t} \right]_2^A = \lim_{A \to \infty} \left[\frac{1}{2} - \frac{1}{\log A} \right] = \frac{1}{2} < \infty.$$

Since the improper integral exists, the series $\sum_{m=2}^{\infty} f(n) = \sum_{m=2}^{\infty} 1/(n \log^2 n)$ converges.

70. $\dfrac{a_{n+1}}{a_n} = \dfrac{1 \cdot 6 \cdot 11 \cdots (5n-4)(5n+1)}{2 \cdot 6 \cdot 10 \cdots (4n-2)(4n+2)} \cdot \dfrac{2 \cdot 6 \cdot 10 \cdots (4n-2)}{1 \cdot 6 \cdot 11 \cdots (5n-4)} = \dfrac{5n+1}{4n+2} \to \dfrac{5}{4} > 1.$

The series diverges.

88. Since $(2n^2+1)/(n^3+3) > 1/n$ [Reason: $2n^3 + n > n^3 + 3$ for $n > 1$] the series does not converge absolutely. Let $p_n = (2n^2+1)/(n^3+3)$. Then $p_n > p_{n+1}$ for n large enough. [Reason:

$(2n^2 + 1)((n + 1)^3 + 3) > (n^3 + 3)(2(n + 1)^2 + 1)$ for n large enough, as can be seen by multiplying out both sides of this inequality and recalling that the terms of highest power predominate as $n \to \infty$.] And

$$\lim_{n \to \infty} p_n = \lim_{n \to \infty} \frac{2 + (1/n^2)}{n + (3/n^2)} = 0.$$

By Theorem 9, the series converges conditionally.

§5

2. Using the ratio test, we find that

$$\lim_{n \to \infty} \left| \frac{2(n + 1) - 1}{3(n + 1) + 1} x^{n+1} \div \frac{2n - 1}{3n + 1} x^n \right| = \lim_{n \to \infty} \left| \frac{2n + 1}{3n + 4} \cdot \frac{3n + 1}{2n - 1} x \right|$$

$$= |x| \lim_{n \to \infty} \frac{6n^2 + 5n + 1}{6n^2 + 5n - 4} = |x| \cdot 1 = |x|.$$

Therefore, by the ratio test, the series converges for $|x| < 1$, diverges for $|x| > 1$, that is, $R = 1$.

26. Using Exercise 15 and §2.3 (14), we have

$$|f^{(n+1)}(x)| = \frac{1 \cdot 3 \cdot 5 \cdots (2n - 1)}{2^n} |x|^{-(2n+1)/2}$$

which, since $1 \le x < 2$, is less than $[1 \cdot 3 \cdot 5 \cdots (2n - 1)]/2^n$. Therefore,

$$|R_{n+1}| \le \frac{1 \cdot 3 \cdot 5 \cdots (2n - 1)}{2^n (n + 1)!} |x - 1|^{n+1}.$$

Since $1 \le x < 2$, $|x - 1|^{n+1} < 1$, and $|x - 1|^{n+1} \to 0$ as $n \to \infty$. On the other hand,

$$\frac{1 \cdot 3 \cdot 5 \cdots (2n - 1)}{2^n (n + 1)!} = \frac{1 \cdot 3 \cdot 5 \cdot 7 \cdots (2n - 3) \cdot (2n - 1)}{2 \cdot 4 \cdot 6 \cdot 8 \cdots (2n - 2) \cdot (2n)(2n + 2)} < 1.$$

Therefore, $|R_{n+1}| \to 0$ as $n \to \infty$. Since the radius of convergence is 1, we know that the series converges for $|x - 1| < 1$ or $0 < x < 2$.

60. Since

$$\frac{1}{1 - t} = 1 + t + t^2 + t^3 + t^4 + \cdots,$$

we have

$$\frac{1}{1 - x^2} = 1 + x^2 + x^4 + x^6 + x^8 + \cdots$$

and

$$\frac{1}{1 - x^2} - 1 = x^2 + x^4 + x^6 + x^8 + \cdots.$$

Therefore,

$$\sin\left(\frac{1}{1 - x^2} - 1\right) = (x^2 + x^4 + x^6 + \cdots) - \frac{(x^2 + x^4 + x^6 + \cdots)^3}{3!} + \frac{(x^2 + x^4 + x^6 + \cdots)^5}{5!} - \cdots$$

$$= x^2 + x^4 + x^6 - \frac{x^6}{6} + x^8 - \frac{3x^8}{6} + x^{10} - \frac{5x^{10}}{5} + \frac{x^{10}}{120} + \cdots$$

$$= x^2 + x^4 + \tfrac{5}{6}x^6 + \tfrac{1}{2}x^8 - \tfrac{33}{40}x^{10} + \cdots$$

[Note that only the first term in $(x^2 + x^4 + x^6 + \cdots)^5$ was needed.]

§6

12. Applying l'Hospital's rule twice, we have

$$\lim_{x \to 0} \frac{(\arc\tan x)^2}{\log (x^2 + 1)} = \lim_{x \to 0} \frac{2 (\arc\tan x) \cdot [1/(x^2+1)]}{2x/(x^2+1)}$$

$$= \lim_{x \to 0} \frac{\arc\tan x}{x} = \lim_{x \to 0} \frac{1/(x^2+1)}{1} = 1.$$

26. This is of the form 0/0. We have

$$\lim_{x \to +\infty} \frac{\arc\tan (2/x)}{1/x} = \lim_{x \to +\infty} \frac{(-2/x^2)/[1 + (4/x^2)]}{-1/x^2} = \lim_{x \to +\infty} \frac{2x^2}{x^2 + 4}.$$

This is now of the form $+\infty/+\infty$ and we try again:

$$\lim_{x \to +\infty} \frac{2x^2}{x^2 + 4} = \lim_{x \to \infty} \frac{4x}{2x} = 2.$$

42. This is a case of 0^0, which is not covered by the examples. However, we proceed as in Cases 4 and 5:

$$\lim_{x \to 0} \log (\sin x)^x = \lim_{x \to 0} x \log \sin x = \lim_{x \to 0} \frac{\log \sin x}{1/x}.$$

This is now a case of ∞/∞, which we can handle.

$$\lim_{x \to 0} \frac{\log \sin x}{1/x} = \lim_{x \to 0} \frac{\cot x}{-1/x^2} = \lim_{x \to 0} \frac{-x^2}{\tan x}.$$

This is still indeterminate, of form 0/0. We try again:

$$\lim_{x \to 0} \frac{-x^2}{\tan x} = \lim_{x \to 0} \frac{-2x}{\sec^2 x} = \frac{0}{1} = 0.$$

Therefore, $\lim_{x \to 0} (\sin x)^x = e^0 = 1.$

Chapter 9

§1

4. According to the discussion in §1.1, we see that R must lie on the line l through P and Q and that $|\overrightarrow{PQ}|$ must equal $|\overrightarrow{QR}|$. The equation of l is $y = \frac{4}{3}x - \frac{1}{3}$, and $|\overrightarrow{PQ}| = 5$. Let R have coordinates (x_0, y_0). Then, since $|\overrightarrow{QR}|$ is to be 5, we must solve the equation $\sqrt{(x_0 - 4)^2 + (y_0 - 5)^2} = 5$. Since R lies on l, replace y_0 by $\frac{4}{3}x_0 - \frac{1}{3}$, square both sides, and simplify, to obtain $x_0^2 - 8x_0 + 7 = 0$. Therefore, $x_0 = 7$ or 1. We reject $x_0 = 1$ (since then $R = P$ and we know $\overrightarrow{PQ} \sim \overrightarrow{QP}$ is false). Therefore, R has coordinates $(7, 9)$.

22. $3 (\alpha_1, \alpha_2) + 2 (3, 0) = (3\alpha_1 + 6, 3\alpha_2) = (3\alpha_1 + 6) \mathbf{e}_1 + 3\alpha_2 \mathbf{e}_2 = 9\mathbf{e}_1 + 12\mathbf{e}_2$. By uniqueness of components, $3\alpha_1 + 6 = 9$ and $3\alpha_2 = 12$. Therefore, $\alpha_1 = 1$ and $\alpha_2 = 4$.

52. To prove that $\{\mathbf{f}_1, \mathbf{f}_2, \mathbf{f}_3\}$ is a basis we must show that (I) every vector $\mathbf{w}$ can be written as $\mathbf{w} = a\mathbf{f}_1 + b\mathbf{f}_2 + c\mathbf{f}_3$, and (II) this can be done in one way only.

If $\alpha = 1$, then $\{2\mathbf{e}_1, \mathbf{e}_2, \mathbf{v}\}$ is not a basis, for in this case $\mathbf{0} = 0 (2\mathbf{e}_1) + 0\mathbf{v} = 2\mathbf{e}_1 - 2\mathbf{v}$.

If $a \neq 1$, then $\{2\mathbf{e}_1, \mathbf{e}_2, \mathbf{v}\}$ is a basis. Indeed, we have

$$\mathbf{e}_3 = \frac{1}{1 - \alpha} (\mathbf{v} - \alpha\mathbf{e}_1).$$

Given a vector $\mathbf{w} = A\mathbf{e}_1 + B\mathbf{e}_2 + C\mathbf{e}_3$ we can write

$$\mathbf{w} = A \cdot 2 \cdot \tfrac{1}{2}\mathbf{e}_1 + B\mathbf{e}_2 + C\frac{1}{1-\alpha}\mathbf{v} - C\frac{\alpha}{1-\alpha}\mathbf{e}_1$$

$$= \left(\tfrac{1}{2}A - \frac{C}{2}\frac{\alpha}{1-\alpha}\right)2\mathbf{e}_1 + B\mathbf{e}_2 + \frac{C}{1-\alpha}\mathbf{v}$$

$$= a(2\mathbf{e}_1) + b\mathbf{e}_2 + c\mathbf{v}$$

where $a = \tfrac{1}{2}A - (C\alpha/2 - 2\alpha)$, $b = B$, $c = C/(1+\alpha)$. Condition I is satisfied.

Next, assume that $a(2\mathbf{e}_1) + b\mathbf{e}_2 + c\mathbf{v} = \hat{a}(2\mathbf{e}_1) + \hat{b}\mathbf{e}_2 + \hat{c}\mathbf{v}_1$ that is

$$\begin{aligned}
\mathbf{0} &= (a - \hat{a})(2\mathbf{e}_1) + (b - \hat{b})\mathbf{e}_2 + (c - \hat{c})\mathbf{v} \\
&= 2(a - \hat{a})\mathbf{e}_1 + (b - \hat{b})\mathbf{e}_2 + (c - \hat{c})\alpha\mathbf{e}_1 + (c - \hat{c})(1 - \alpha)\mathbf{e}_3 \\
&= (2a - 2\hat{a} + \alpha c - \alpha\hat{c})\mathbf{e}_1 + (b - \hat{b})\mathbf{e}_2 + (c - \hat{c})(1 - \alpha)\mathbf{e}_3.
\end{aligned}$$

Since $\{\mathbf{e}_1, \mathbf{e}_2, \mathbf{e}_3\}$ is a basis, we have $(c - \hat{c})(1 - \alpha) = 0$, $b - \hat{b} = 0$, $2a - 2\hat{a} + \alpha c - \alpha\hat{c} = 0$ and therefore, assuming $\alpha \neq 1$, $c = \hat{c}$, $b = \hat{b}$, $\alpha = \alpha$. Thus, condition II is satisfied.

§2

4. We use Theorem 1. A quick sketch shows that $|\overrightarrow{PX}| + |\overrightarrow{PQ}| = |\overrightarrow{QX}|$. Since $|\overrightarrow{PX}| = \tfrac{1}{4}|\overrightarrow{PQ}|$, we have $|\overrightarrow{QX}| = 5|\overrightarrow{PX}|$. Therefore, $|\overrightarrow{PX}|/|\overrightarrow{QX}| = \tfrac{1}{5} = |\beta/\alpha|$ or $|\beta| = \tfrac{1}{5}|\alpha|$. Since X lies in I (Fig. 9.22), $\beta < 0$. Hence, $|\beta| = -\beta$. Therefore $\alpha > 0$ (since $\alpha + \beta = 1$) and $-\beta = \alpha/5$. Since also $\alpha + \beta = 1$, we have $\alpha = \tfrac{5}{4}$, $\beta = -\tfrac{1}{4}$. $\mathbf{X} = \tfrac{5}{4}\mathbf{P} - \tfrac{1}{4}\mathbf{Q}$ and $X = (\tfrac{7}{4}, -\tfrac{1}{2})$.

§3

22. Since $x = r\cos\theta$, $y = r\sin\theta$, and $x^2 + y^2 = r^2$, we have $(r^2)^{3/2} = r(\cos\theta + \sin\theta)$ or $r^2 = \cos\theta + \sin\theta$.

§4

2. Let $X = (x_1, x_2, x_3)$. Let $n = 4$, $m = 5$. Then

$$\mathbf{X} = [m/(m + n)]\mathbf{P} + [n/(m + n)]\mathbf{Q} = \tfrac{5}{9}\mathbf{P} + \tfrac{4}{9}\mathbf{Q}$$

is the position vector of X. Therefore,

$$\begin{aligned}
x_1 &= \tfrac{5}{9}(1) + \tfrac{4}{9}(2) = \tfrac{13}{9}, \\
x_2 &= \tfrac{5}{9}(-3) + \tfrac{4}{9}(5) = \tfrac{5}{9}, \\
x_3 &= \tfrac{5}{9}(0) + \tfrac{4}{9}(-7) = -\tfrac{28}{9}.
\end{aligned}$$

Thus the point with coordinates $(13/9, 5/9, -28/9)$ is the desired point.

14. The line l of Exercise 8 will meet the yz plane ($x = 0$) when $x = 0$, that is, when $3 + 2t = 0$ or $t = -3/2$. For $t = -3/2$, $y = 4$ and $z = 5 - 2(-3/2) = 8$. Therefore, l meets the yz plane at $(0, 4, 8)$. l does not meet the xz plane ($y = 0$) since every point on l has 4 as its y coordinate. When $z = 0$, $5 - 2t = 0$ or $t = 5/2$. For $t = 5/2$, $x = 8$ and $y = 4$. Therefore, l meets the xy plane ($z = 0$) at $(8, 4, 0)$.

28. We note first that P is not on l, for if it were, $4 = t$ would require $-1 = 2(4)$. We find two points on l: put $t = 0$, then $(0, 0, 0)$ is on l; put $t = 1$, then $(1, 2, 3)$ is on l. This exercise is now of the type of Example 1. Let $Ax + By + Cz + D = 0$ be the desired plane. Then, since $(0, 0, 0)$ lies on this plane, we have $A(0) + B(0) + C(0) + D = 0$, that is, $D = 0$. Since $(1, 2, 3)$ and $(4, -1, 0)$

are also on the plane, we have $A + 2B + 3C + D = 0$ and $4A - B + 0 \cdot C + D = 0$, that is, $A + 2B + 3C = 0$ and $4A - B = 0$. From this it is clear that $B = 4A$ and $A = -\frac{1}{3}C$, so that $B = -\frac{4}{3}C$. Choose $C = 3$. Then, $A = -1$, $B = -4$, $C = 3$, and $D = 0$, and the plane is $-x - 4y + 3z = 0$.

42. Putting $A = 2$, $B = -6$, $C = 7$ and $A_1 = 6$, $B_1 = 1$, $C_1 = 4$, we see that none of the numbers (15) is zero, which means that we can solve for any one of the three variables in terms of the other two. This means that we are free to let any one of the variables be equal to t. We choose $y = t$. The two equations become $2x + 7z = 5 + 6t$ and $6x + 4z = 1 - t$. Solving these simultaneously, we find $z = \frac{19}{17}t + \frac{14}{17}$ and $x = -\frac{31}{34}t - \frac{13}{34}$.

56. We rewrite the first two equations as $x - y = -z$ and $2x - 3y = -z$. Therefore $x - y = 2x - 3y$ or $x = 2y$. The first and third equations may be written as $y - z = x$ and $10y - 5z = x$. Therefore, $y - z = 10y - 5z$ or $z = \frac{9}{4}y$. Put $y = t$. Then $x = 2t$ and $z = \frac{9}{4}t$. A parametric representation, then, is $x = 2t$, $y = t$, $z = \frac{9}{4}t$.

76. The parametric equations of l_1 are: $x = t$, $y = 1 - 2t$, $z = 0$. We substitute these values for x, y, z into the equations of l_2 and obtain $1 - t = -1 + 3t = 1 - t$. These equations are satisfied if and only if $t = 1/2$. Therefore, l_1 intersects l_2 in precisely one point: $(1/2, 0, 0)$.

§5

14. This curve is symmetric about the z axis. Therefore, it is enough to consider the curve $C: z = x^3$, $0 \le x$. C is the right half of the curve shown in Fig. 3.3 on p. 91. The equation of S is $z = r^3$ or $z = (x^2 + y^2)^{3/2}$.

§6

18. Let $\mathbf{a} = a_1\mathbf{e}_1 + a_2\mathbf{e}_2 + a_3\mathbf{e}_3$. Then, $a_1 + a_2 = (\mathbf{a}, \mathbf{e}_1) + (\mathbf{a}, \mathbf{e}_2) = (\mathbf{a}, \mathbf{e}_1 + \mathbf{e}_2) = 2$, $a_1 - a_2 = (\mathbf{a}, \mathbf{e}_1) - (\mathbf{a}, \mathbf{e}_2) = (\mathbf{a}, \mathbf{e}_1 - \mathbf{e}_2) = 3$, and $a_3 = (\mathbf{a}, \mathbf{e}_3) = 0$. Solving $a_1 + a_2 = 2$ and $a_1 - a_2 = 3$, we find $a_1 = \frac{5}{2}$ and $a_2 = -\frac{1}{2}$. Therefore, $\mathbf{a} = \frac{5}{2}\mathbf{e}_1 - \frac{1}{2}\mathbf{e}_2$.

42. Using Theorem 6, we have

$$\delta = \frac{\left| 2v - 2v + 2v^2 - 6 \right|}{\sqrt{4 + 1 + 4}} = \frac{\left| 2v^2 - 6 \right|}{3} = 1000.$$

Therefore, $2v^2 - 6 = \pm 3000$. This gives $2v^2 = 3006$ or $2v^2 = -2994$. The second is impossible. Since $v > 0$, the first gives $v = \sqrt{1530}$.

§7

6. Call these vectors $\mathbf{f}_1, \mathbf{f}_2, \mathbf{f}_3, \mathbf{f}_4$. Then $\mathbf{f}_1 = \mathbf{e}_1 + \mathbf{e}_2 + \mathbf{e}_3$, $\mathbf{f}_2 = \mathbf{e}_2$, $\mathbf{f}_3 = \mathbf{e}_1 + \mathbf{e}_3$, $\mathbf{f}_4 = \mathbf{e}_4$. Note that $\mathbf{f}_1 - \mathbf{f}_3 = \mathbf{e}_2$. But then $\mathbf{e}_2$ is a vector that can be written in the form $\alpha_1\mathbf{f}_1 + \alpha_2\mathbf{f}_2 + \alpha_3\mathbf{f}_3 + \alpha_4\mathbf{f}_4$ in *two* distinct ways, namely, $\alpha_1 = \alpha_3 = \alpha_4 = 0$, $\alpha_2 = 1$ and also $\alpha_1 = 1$, $\alpha_3 = -1$, $\alpha_2 = \alpha_4 = 0$. This violates the condition for a basis.

Chapter 10

§1

16. We first introduce new coordinates by a translation so that the foci will be symmetric with respect to the origin: let $\xi = x - 3$ and $\eta = y$. In this *new* coordinate system, the foci are at $(-3, 0)$ and $(3, 0)$. Thus, $c = 3$. Since $\epsilon = 3/2$ and since $\epsilon = c/a$, $a = 2$. Therefore, since $c = \sqrt{a^2 + b^2}$ for a hyperbola, $b = \sqrt{5}$. The equation, in the new coordinate system, is $\xi^2/4 - \eta^2/5 = 1$. Returning to the original coordinates, we have $\frac{1}{4}(x - 3)^2 - \frac{1}{5}y^2 = 1$ or $5x^2 - 4y^2 - 30x + 25 = 0$.

34. Since the standard form of the parabola is $y^2 = 4ax$, we have $a = 5/2$. By (19), the tangent at (x_0, y_0) has equation $yy_0 = 5(x + x_0)$ or $y = (5/y_0)x + (5/y_0)x_0$ $(y_0 \neq 0$, otherwise the tangent is the y axis). The slope is thus $5/y_0$. Set this equal to 2 to get $y_0 = 5/2$. Then $x_0 = 5/8$. The tangent at $(5/8, 5/2)$ is $y - 5/2 = 2(x - 5/8)$ or $4y - 8x = 5$.

50. Divide numerator and denominator by 20 to reduce to form (24). We have

$$r = \frac{\frac{1}{20}}{1 - \frac{5}{4}\cos\theta}.$$

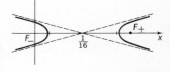

Therefore, $\epsilon = 5/4$ and $p = 1/25$. Since $\epsilon > 1$, the conic is a hyperbola. One focus is at the origin and its corresponding directrix is $x = -1/25 = -a/\epsilon$. Therefore, $\epsilon = 25a$ or since $\epsilon = 5/4$, $a = 1/20$. Therefore, $c = a\epsilon = 1/16$ and $b = \sqrt{c^2 - a^2} = 3/80$.

§2

4. We note that, for $\theta = 150°$, we have $\cos\theta = -\sqrt{3}/2$ and $\sin\theta = 1/2$. Then the given equations may be written as: $\mathbf{f}_1 = (\cos\theta)\mathbf{e}_1 + (\sin\theta)\mathbf{e}_2$ and $\mathbf{f}_2 = (-\sin\theta)\mathbf{e}_1 + (\cos\theta)\mathbf{e}_2$. By the sentence following (1) and by Theorem 1, $\{\mathbf{f}_1, \mathbf{f}_2\}$ is right-handed.

26. The slope of l_1 is $m_1 = 1$. The slope of l_2 is $m_2 = 7$. Since $1 + 7 = 8 \neq 0$, we may use (8) to obtain $n^2 - \frac{3}{2}n - 1 = 0$. Solving this for n, we obtain $n = 2$ or $n = -\frac{1}{2}$. The lines l_1 and l_2 intersect when $x + 1 = 7x - 2$, that is, when $x = 1/2$. The point of intersection is $(\frac{1}{2}, \frac{3}{2})$. The equations of the bisectors are $y - \frac{3}{2} = 2(x - \frac{1}{2})$ or $2y = 4x + 1$ and $y - \frac{3}{2} = -\frac{1}{2}(x - \frac{1}{2})$ or $4y + 2x = 7$.

§3

18. Since $A \neq C$, $\tan 2\theta = 2B/(A - C) = 4/3$. Since the rotation angle θ can always be considered to be between $0°$ and $90°$ [by the remarks following (4)], we can consider 2θ to be an acute angle in a right triangle. Then the leg opposite 2θ has length 4 and the leg adjacent to 2θ has length 3. The hypotenuse is 5. Then $\sin 2\theta = 4/5$ and $\cos 2\theta = 3/5$. Hence, $\sin\theta = \sqrt{\frac{1}{2}(1 - \cos 2\theta)} = 1/\sqrt{5}$ and $\cos\theta = \sqrt{\frac{1}{2}(1 + \cos 2\theta)} = 2/\sqrt{5}$. The formulas, then, are $x = \frac{1}{5}\sqrt{5}(2X - Y)$, $y = \frac{1}{5}\sqrt{5}(X + 2Y)$. The given equation becomes: $5X^2 - \sqrt{5}X + \frac{1}{5}\sqrt{5}Y = 0$.

24. The discriminant is $-1 \neq 0$. We now use the formula for the new coordinates given just before the example in the text, remembering that X and Y now play the roles of x and y, and ξ and η the roles of X and Y. We have: $X = \xi - \sqrt{2}$ and $Y = \eta$. The equation $X^2 - Y^2 + 2\sqrt{2}X = 0$ becomes $\xi^2 - \eta^2 - 2 = 0$.

38. In Exercise 14, we have reduced the equation to $X^2/2 + Y^2/6 = 1$. To get the larger constant under the "first variable," rotate by 90, putting $u = Y$ and $v = -X$ to obtain $u^2/6 + v^2/2 = 1$. In u, v coordinates, we have $a = \sqrt{6}$, $b = \sqrt{2}$, $c = 2$; foci at $(\pm 2, 0)$; vertices at $(\pm\sqrt{6}, 0)$ and $(0, \pm\sqrt{2})$. To return to x, y coordinates, we have (from Exercise 14) $x = \frac{1}{2}\sqrt{2}(X - Y) = \frac{1}{2}\sqrt{2}(-v - u)$ and $y = \frac{1}{2}\sqrt{2}(X + Y) = \frac{1}{2}\sqrt{2}(u - v)$. Thus, in x, y coordinates, we have foci at $(-\sqrt{2}, \sqrt{2})$ and $(\sqrt{2}, -\sqrt{2})$ and vertices at $(-\sqrt{3}, \sqrt{3})$, $(\sqrt{3}, -\sqrt{3})$, $(-1, -1)$, and $(1, 1)$.

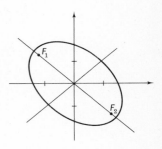

§4

2. When $z = 2$, the equation of the quadric surface becomes $4x^2 + y^2 = 5/9$. This is the ellipse $\frac{x^2}{\frac{5}{36}} + \frac{y^2}{\frac{5}{9}} = 1$. The major axis is the y axis. The vertices are $(\pm\sqrt{5}/6, 0)$ and $(0, \pm\sqrt{5}/3)$. The foci are $(0, \pm\sqrt{15}/6)$.

Chapter 11

§1

4. We must express $\mathbf{f}(t)$ as $\mathbf{f}(t) = x(t)\mathbf{e}_1 + y(t)\mathbf{e}_2$ where the functions x and y satisfy the equation of the curve, that is, $y(t) = \sqrt{x(t)^2 + 1}$ for each $t \geq \frac{1}{4}$. Since the distance of the point from the

origin is proportional to t, we also have $\sqrt{x(t)^2 + y(t)^2} = kt$. Thus,

$$\sqrt{x(\tfrac{1}{4})^2 + y(\tfrac{1}{4})^2} = \tfrac{1}{4}k$$

or $\sqrt{0 + 1} = k/4$, that is, $k = 4$. Therefore,

$$\sqrt{x(t)^2 + y(t)^2} = 4t.$$

Replacing $y(t)$ by $\sqrt{x(t)^2 + 1}$ in the last equation, and squaring both sides, we obtain $2x(t)^2 + 1 = 16t^2$ or $x(t) = \pm\sqrt{8t^2 - \tfrac{1}{2}}$. We may drop the minus sign, since the point is always in the first quadrant. Therefore, $x(t) = \sqrt{8t^2 - \tfrac{1}{2}}$ and $y(t) = \sqrt{x(t)^2 + 1} = \sqrt{8t^2 + \tfrac{1}{2}}$.

26. Using the formula for the integral of a vector-valued function given in §1.4, we have

$$\int_1^2 \left(\frac{1}{t}\mathbf{e}_1 + \frac{1}{t+1}\mathbf{e}_2\right) dt = \left(\int_1^2 \frac{dt}{t}\right)\mathbf{e}_1 + \left(\int_1^2 \frac{dt}{t+1}\right)\mathbf{e}_2$$

$$= (\log 2 - \log 1)\mathbf{e}_1 + (\log 3 - \log 2)\mathbf{e}_2 = (\log 2)\mathbf{e}_1 + (\log \tfrac{3}{2})\mathbf{e}_2.$$

44. We first note that the limit of a sum of vector-valued functions is the sum of the limits of these functions. This follows easily from the corresponding statement for number-valued functions. By Statement 4, we must show that

$$\lim_{h \to 0} \frac{1}{h}\{(\mathbf{f} + \mathbf{g})(t_0 + h) - (\mathbf{f} + \mathbf{g})(t_0)\} = \mathbf{a} + \mathbf{b}.$$

By our first remark, this limit is equal to

$$\lim_{h \to 0} \frac{1}{h}\{\mathbf{f}(t_0 + h) - \mathbf{f}(t_0)\} + \lim_{h \to 0} \frac{1}{h}\{\mathbf{g}(t_0 + h) - \mathbf{g}(t_0)\}.$$

By hypothesis, $\mathbf{f}'(t_0) = \mathbf{a}$ and $\mathbf{g}'(t_0) = \mathbf{b}$. By statement 4, then, we may replace the two limits above by $\mathbf{a}$ and $\mathbf{b}$ respectively, which is the desired result.

§2

2. If the two functions are to give the same trajectory, for each t between 2 and 4, there must be a u between $\tfrac{1}{4}$ and $\tfrac{1}{2}$ such that $t/(t+1) = 1/(u+1)$ and $(t+1)/(t-1) = (1+u)/(1-u)$. Solving these, we find in both cases $u = 1/t$. Therefore (compare §2.2) let $\psi(\xi) = 1/\xi$. ψ is decreasing on $\alpha = \tfrac{1}{4} \le \xi \le \tfrac{1}{2} = \beta$. $\psi(\tfrac{1}{2}) = 2 = a$. $\psi(\tfrac{1}{4}) = 4 = b$. If we call the first function in this exercise $\mathbf{f}$ and the second $\mathbf{h}$, we do have

$$\mathbf{f}(\psi(\xi)) = \frac{1/\xi}{(1/\xi) + 1}\mathbf{e}_1 + \frac{(1/\xi) + 1}{(1/\xi) - 1}\mathbf{e}_2 = \frac{1}{\xi + 1}\mathbf{e}_1 + \frac{1 + \xi}{1 - \xi}\mathbf{e}_2 = \mathbf{h}(\xi).$$

Therefore we have case (b): opposite curves.

26. Let Q have coordinates $(\cos\theta, \sin\theta)$ for some θ, $0 \le \theta < \pi$. The line through $P = (-1, 0)$ and Q has a parametric representation

(*) $$x = -1 + (1 + \cos\theta)t, \qquad y = (\sin\theta)t$$

The point R corresponds to a value t such that $t > 1$ (then Q is between P and R) and

$$1 = (x - \cos\theta)^2 + (y - \sin\theta)^2 = (t - 1)^2[(1 + \cos\theta)^2 + \sin^2\theta]$$

$$= (t - 1)^2 \, 2(1 + \cos\theta) = (t - 1)^2 \, 4\cos^2 \tfrac{1}{2}\theta.$$

Thus $t = 1 + \tfrac{1}{2}\sec\tfrac{1}{2}\theta$. Substitute this value in (*), and note that $\sin\theta = 2\sin\tfrac{1}{2}\theta\cos\tfrac{1}{2}\theta$. We obtain, for the coordinates of R, the representation

$$x = \cos\theta + \cos\tfrac{1}{2}\theta, \qquad y = \sin\theta + \sin\tfrac{1}{2}\theta, \qquad 0 \le \theta < \pi.$$

§3

8. Since we want $\mathbf{T}(y)$, let y play the role of t so that $x = (y + 2)^{3/5}$.

$$\mathbf{f}(y) = (y + 2)^{3/5}\mathbf{e}_1 + y\mathbf{e}_2.$$

$$\mathbf{T}(y) = \frac{\tfrac{3}{5}(y+2)^{-2/5}\mathbf{e}_1 + \mathbf{e}_2}{\sqrt{[9/25(y+2)^{4/5}]+1}} = \frac{3}{\sqrt{9 + 25(y+2)^{4/5}}}\mathbf{e}_1 + \frac{5(y+2)^{2/5}}{\sqrt{9 + 25(y+2)^{4/5}}}\mathbf{e}_2$$

12. The components of $\mathbf{T}(\theta)$ are

$$\frac{(-\sin\theta)\cos\theta - (1+\cos\theta)\sin\theta}{\sqrt{(1+\cos\theta)^2 + (-\sin\theta)^2}} \quad \text{and} \quad \frac{(-\sin\theta)(\sin\theta) + (1+\cos\theta)\cos\theta}{\sqrt{(1+\cos\theta)^2 + (-\sin\theta)^2}}.$$

These can be simplified to

$$\frac{-(\sin\theta + \sin 2\theta)}{\sqrt{2}\sqrt{1+\cos\theta}} \quad \text{and} \quad \frac{\cos\theta + \cos 2\theta}{\sqrt{2}\sqrt{1+\cos\theta}}.$$

$$\tan\Psi = \frac{r}{dr/d\theta} = -\frac{r}{\sin\theta}.$$

42. We have $y = \tfrac{3}{2}x - \tfrac{1}{2}$. Let $x(t) = t$ and $y(t) = \tfrac{3}{2}t - \tfrac{1}{2}$. Then $\mathbf{f}(t) = t\mathbf{e}_1 + (\tfrac{3}{2}t - \tfrac{1}{2})\mathbf{e}_2$, $0 < t$, represents the curve. Using (16),

$$s(\xi) = \int_0^\xi \sqrt{1 + \tfrac{9}{4}}\, dt = (\sqrt{13}/2)\xi.$$

Therefore,

$$\xi = (2/\sqrt{13})s = \phi(s) \quad \text{and} \quad \mathbf{g}(s) = \mathbf{f}(\phi(s)) = (2/\sqrt{13})s\mathbf{e}_1 + [(3/\sqrt{13})s - \tfrac{1}{2}]\mathbf{e}_2$$

is the parametrization in terms of arc length.

§4

4. We have $x^2/a^2 + y^2/b^2 = 1$. The trajectory is an ellipse. $\mathbf{v}(t) = (-a\sin t)\mathbf{e}_1 + (b\cos t)\mathbf{e}_2$. The speed is the length of the velocity vector, that is, $\sqrt{a^2\sin^2 t + b^2\cos^2 t}$.

$\mathbf{a}(t) = (-a\cos t)\mathbf{e}_1 - (b\sin t)\mathbf{e}_2$. To compute the normal and tangential components of acceleration, we shall neet $\mathbf{T}(t)$, κ, and $\mathbf{N}(t)$.

$\mathbf{T}(t) = -(a\sin t/|\mathbf{v}|)\mathbf{e}_1 + (b\cos t/|\mathbf{v}|)\mathbf{e}_2$; $\kappa = (ab\sin^2 t + ab\cos^2 t)/|\mathbf{v}|^3 = ab/|\mathbf{v}|^3$, and $\mathbf{N}(t) = -(b\cos t/|\mathbf{v}|)\mathbf{e}_1 - (a\sin t/|\mathbf{v}|)\mathbf{e}_2$. Then,

$$\mathbf{a}_N = \kappa(ds/dt)^2\mathbf{N} = ab/|\mathbf{v}|^3.$$

$$|\mathbf{v}|^2\mathbf{N} = \frac{-ab^2\cos t}{a^2\sin^2 t + b^2\cos^2 t}\mathbf{e}_1 - \frac{a^2 b\sin t}{a^2\sin^2 t + b^2\cos^2 t}\mathbf{e}_2.$$

And

$$\mathbf{a}_T = (d^2s/dt^2)\mathbf{T} = \frac{a^2\sin t\cos t - b^2\sin t\cos t}{|\mathbf{v}|}\mathbf{T}$$

$$= \frac{-a(a^2-b^2)\sin^2 t\cos t}{a^2\sin^2 t + b^2\cos^2 t}\mathbf{e}_1 + \frac{b(a^2-b^2)\sin t\cos^2 t}{a^2\sin^2 t + b^2\cos^2 t}\mathbf{e}_2.$$

20.

Since $\tan \theta = \beta/\alpha = 2$, we know that $\beta = 2\alpha$.

$$\mathbf{r}(t) = \alpha t \mathbf{e}_1 + (-\tfrac{1}{2}gt^2 + \beta t + 16)\mathbf{e}_2 = \alpha t \mathbf{e}_1 + (-\tfrac{1}{2}gt^2 + 2\alpha t + 16)\mathbf{e}_2.$$

The projectile strikes the ground after 2 sec, that is, when $t = 2$, $y(t) = 0$ or $-2g + 4\alpha + 16 = 0$ and $\alpha = \tfrac{1}{2}g - 4$. The horizontal distance traveled is $x(2) = 2\alpha = {}'g - 8$ or approximately 24 ft.

§6

2. Yes, it is possible.

Suppose that two components of the double star have masses m_1, m_2, periods T_1, T_2 (about their common center of mass) and move along ellipses with major semi-axes a_1, a_2. According to the observation in the text, the first component moves as if it were a planet of an immovable sun with mass

$$M_1 = m_2/[1 + (m_1/m_2)]^2 = m_2^3/(m_1 + m_2)^2.$$

Similarly, the second component moves like a planet of an immovable sun with mass $M_2 = m_1^3/(m_1 + m_2)^2$. By the results of §4.12, we have

$$\frac{m_2^3}{(m_1 + m_2)^2} = \frac{4\pi}{\gamma}\frac{a_1^2}{T_1^2}, \qquad \frac{m_1^3}{(m_1 + m_2)^2} = \frac{4\pi}{\gamma}\frac{a_2^3}{T_2^2}.$$

Thus

$$\frac{m_1}{m_2} = \frac{a_2}{a_1}\left(\frac{T_1}{T_2}\right)^{2/3}.$$

6. The answer to the question in Exercise 5 is "yes". Consider a man sitting on a chair in his room. His body is subject to the weight $\mathbf{w}$ and to the reaction of the chair, $-\mathbf{w}$ (in view of Newton's Third Law). The total force is $\mathbf{w} + (-\mathbf{w}) = \mathbf{0}$, the acceleration is $\mathbf{0}$. The pressure between the body and the chair is perceived by the man as his weight.

Assume now that the man, and the chair, are both falling freely (as in an elevator with a cut cable, or in a spacecraft in orbit). The man's body experiences the acceleration $(1/m)\mathbf{w}$, where m is the mass of the body. Hence, there is no other force except $\mathbf{w}$ acting on the man's body. The chair offers no resistance. The man feels "weightless".

Chapter 12

§1

8. $[e^m \cos n]^2 - [e^m \sin n]^2 = e^{2m}[\cos^2 n - \sin^2 n] = e^{2m} \cos 2n.$
$2[e^m \cos n] \cdot [e^m \sin n] = e^{2m}[2(\cos n)(\sin n)] = e^{2m} \sin 2n.$

30. Open. If (x_1, y_1) lies in our set, then $x_1 + y_1 = -\alpha$ and $\alpha > 0$. If (x_0, y_0) has distance less than $\alpha/2$ from (x_1, y_1), then $x_0 > x_1 + (\alpha/2)$, $y_0 > y_1 + (\alpha/2)$, so that

$$x + y < x_1 + y_1 + \alpha < 0.$$

Thus all points near (x_1, y_1) lie in the set considered.

36. If (x_1, y_1) lies in E, then for some k, (x_1, y_1) lies in A_k. Since A_k is open, there is an interval containing (x_1, y_1) which lies wholly in A_k and hence wholly in E. Thus E is open. If A_k is the interval $|x| < 1/k$, $|y| < 1/k$, then only $(0, 0)$ lies in F, that is, in all the A_k. Thus there is no interval about $(0, 0)$ which lies in all the A_k.

§2

10. $\dfrac{\partial f}{\partial x} = \dfrac{1}{x \tan y} \dfrac{\partial}{\partial x} (x \tan y) = \dfrac{\tan y}{x \tan y} = \dfrac{1}{x} \cdot \dfrac{\partial f}{\partial y} = \dfrac{1}{x \tan y} \dfrac{\partial}{\partial y} (x \tan y) = \dfrac{x \sec^2 y}{x \tan y} = \dfrac{2}{\sin 2y}.$

Or, note first that $f = \log x + \log \tan y$.

26. Since $v = \int_a^{a^2+b^2} (de^x/dx)\, dx$, we have that $v = e^{a^2} + b^2 - e^a$, $v_a = 2ae^{a^2+b^2} - e^a$, and for $a = 1$, $b = 2$, $v_a = 2e^5 - e$.

28. We have $\phi = \frac{1}{2} \log (a^2 + b^2)$ and

$$\phi_a = \frac{1}{2} \frac{1}{a^2 + b^2} 2a = \frac{a}{a^2 + b^2}, \quad \phi_b = \frac{b}{a^2 + b^2},$$

so that $a\phi_a + b\phi_b = (a^2 + b^2)/(a^2 + b^2) = 1$.

38. Consider a point (x_0, y_0, z_0) on the ellipsoid, with $z_0 > 0$. The upper half of the ellipsoid is the graph of the function

$$(x, y) \mapsto z = c\sqrt{1 - (x/a)^2 - (y/b)^2}.$$

We compute that $z_x = -c^2 x/a^2 z$, $z_y = -c^2 y/b^2 z$. The equation of the tangent plane at (x_0, y_0, z_0) reads

$$z = z_0 + z_x(x_0, y_0)(x - x_0) + z_y(x_0, y_0)(y - y_0) = z_0 - \frac{c^2 x_0 (x - x_0)}{a^2} - \frac{c^2 y_0 (y - y_0)}{b^2}$$

$$= c^2 \left(\frac{z_0^2}{c^2} + \frac{x_0^2}{a^2} + \frac{y_0^2}{b^2} \right) - c^2 \frac{x_0 x}{a^2} - c^2 \frac{y_0 y}{b^2} = c^2 \left(1 - \frac{x_0 x}{a^2} - \frac{y_0 y}{b^2} \right)$$

or

$$\frac{x_0 x}{a^2} + \frac{y_0 y}{b^2} + \frac{z_0 z}{c^2} = 1.$$

The cases $z_0 < 0$ and $z_0 = 0$ are treated similarly. (In the latter case, interchange y and z; consider the function $(x, z) \mapsto y$.)

40. As for the ellipsoid, replace x^2, y^2, z^2 by $x_0 x$, $y_0 y$, $z_0 z$, respectively. The tangent planes are

$$\frac{x x_0}{a^2} + \frac{y y_0}{b^2} - \frac{z z_0}{c^2} = 1, \qquad \frac{x x_0}{a^2} - \frac{y y_0}{b^2} - \frac{z z_0}{c^2} = 1.$$

This is verified as is Exercise 38.

44. $\dfrac{dz}{dt} = \dfrac{\partial z}{\partial x} \dfrac{dx}{dt} + \dfrac{\partial z}{\partial y} \dfrac{dy}{dt} = y x^{y-1} \cos t + x^y (\log x)(-\sin t) = [y \cos t - x(\log x) \sin t] x^{y-1}$

$= [\cos^2 t - \sin^2 t (\log (\sin t))](\sin t)^{\cos t - 1}.$

Or $z(t) = (\sin t)^{\cos t}$ and hence $\log z = \cos t \log (\sin t)$. Thus

$$\frac{dz}{dt} = z(t)[(-\sin t) \log (\sin t) + \cos^2 t/\sin t] = (\sin t)^{\cos t - 1} [\cos^2 t - \sin^2 t (\log (\sin t))].$$

68. We differentiate both sides of the given equation with respect to x, treating y as a function of x, and obtain:

$$\frac{2x + 2y\,(dy/dx)}{x^2 + y^2} = \frac{x\,(dy/dx) - y}{x^2 + y^2} \quad \text{or} \quad \frac{dy}{dx} = \frac{y + 2x}{x - 2y}.$$

At $(1, 0)$ we get $(dy/dx) = 2$.

82. Since $f_x = 6xy^3z^4$, $f_y = 9x^2y^2z^4$, $f_z = 12x^2y^3z^3$ we see that $f_x = f_y = f_z = 0$ if and only if one of the coordinates x, y, z is 0. Every point on a coordinate plane is critical, every other point non-critical.

§3

2. $(\partial/\partial x)f = (\partial/\partial x)\sum a_j(x)b_j(y) = \sum (\partial/\partial x)(a_j(x)b_j(y)) = \sum a'_j(x)b_j(y)$. Similarly, $\partial f_x/\partial y = \sum a'_j(x)b'_j(y)$. As above, $\partial f/\partial y = \sum a_j(x)b'_j(y)$ and $\partial f_y/\partial x = \sum a'_j(x)b'_j(y)$.

4. Let $g = f_x$. g satisfies the conditions of Theorem 1. Thus $g_{xy} = g_{yx}$ or $f_{xxy} = f_{xyx}$. Since $f_{xy} = f_{yx}$, $f_{yxx} = f_{xyx}$.

12. First check f_{yx} and f_{xy} for equality. $f_{xy} = (-1/y^2)e^{x/y} + (-x/y^3)e^{x/y}$ and $f_{yx} = (-1/y^2)e^{x/y} + (-x/y^3)e^{x/y}$. Thus such an f exists. Integrate f_x with respect to x and obtain $f(x, y) = e^{x/y} + g(y)$. We need $f_y(x, y) = (-x/y^2)e^{x/y} + g'(y) = (-x/y^2)e^{x/y}$, or $g'(y) = 0$. Thus $f(x, y) = e^{x/y} + c$ and since $f(1, 1) = 0$, $c = -e$.

18. We differentiate the equation $f(x, y) = f(-x, y) = -f(x, -y)$ twice: $(\partial/\partial x)$ followed by $(\partial/\partial y)$, and apply the chain rule to the second and third members. Thus we have $f_x(x, y) = f_x(-x, y)(-1) = -f_x(x, -y)$ and then $f_{xy}(x, y) = -f_{xy}(-x, y) = -f_{xy}(x, -y)(-1)$. Similarly, from the above, we have $f_{xy}(x, y) = -f_{xy}(-x, -y)$ and thus $f_{xyx}(x, y) = -f_{xyx}(-x, -y)(-1)$.

20. Let the box have dimensions x by y by z. Then $V = xyz$ and $S = 2[xy + yz + xz]$. Thus $z = [S/2 - xy]/(y + x)$ and

$$V_x = \frac{y}{y + x}[S/2 - xy] - \frac{xy^2}{y + x} - \frac{xy}{(y + x)^2}[S/2 - xy] = \left(\frac{y}{y + x}\right)^2[S/2 - 2xy - x^2];$$

$$V_y = \left(\frac{x}{y + x}\right)^2[S/2 - 2xy - y^2].$$

The only critical point (with $x > 0$, $y > 0$) is $x = y = \sqrt{S/6}$ to which corresponds $z = \sqrt{S/6}$ Theorem 3 shows that this is a maximum.

26. (a) $U_x = 3x^2 - 3y$; $U_y = 3y^2 - 3x$; $U_{xx} = 6x$; $U_{xy} = -3$; $U_{yy} = 6y$. Set $U_x = 0 = U_y$ and obtain $(0, 0)$ and $(1, 1)$ as critical points. At $(0, 0)$, $U_{xx}U_{yy} - U_{xy}^2 = -9$. Thus $(0, 0)$ is a saddle point. At $(1, 1)$, $U_{xx}U_{yy} - U_{xy}^2 = 27$ and $U_{xx} + U_{yy} = 12$. Thus $(1, 1)$ is a strict minimum. (b) No critical points occur interior to the triangle, so we investigate the behavior of U on the boundary. $U(0, y) = y^3$; $U(x, 0) = x^3$ have no critical points in the interior of their domains. $U(x, 1 - x) = x^3 + (1 - x)^3 - 3x(1 - x) = 6x^2 - 6x + 1$ also has no critical points for $|x| < 1$. $U(0, 0) = 0$; $U(0, 1) = 1$; $U(1, 0) = 1$. Thus the minimum is 0 and the maximum 1.

30. $f_a = \cos b \sin(t^2)$; $f_{ab} = -\sin b \sin(t^2)$; $f_{abt} = -2t \sin b \cos(t^2)$. At the point $(1, \pi/2, \sqrt{\pi})$ the derivative is $2\sqrt{\pi}$.

36. Note first that if a, b, c are positive integers with $a + b + c = 3$, then

$$\frac{\partial^3 H}{\partial u^a \partial v^b \partial w^c} = 3^a 4^b 5^c t(t - 1)(t - 2)(3u + 4v + 5w)^{t-3}.$$

Next

$H_t = (3u + 4v + 5w)^t \log(3u + 4v + 5w);$
$H_{tt} = (3u + 4v + 5w)^t [\log(3u + 4v + 5w)]^2;$
$H_{ttt} = (3u + 4v + 5w)^t [\log(3u + 4v + 5w)]^3;$
$H_{ut} = 3(3u + 4v + 5w)^{t-1} + 3t(3u + 4v + 5w)^{t-1}[\log(3u + 4v + 5w)];$
$H_{utt} = [6 + 3t[\log(3u + 4v + 5w)]](3u + 4v + 5w)^{t-1}\log(3u + 4v + 5w);$

and so forth.

42. $f_{yx} = (-\sin y)(\cos y) \exp(x \cos y) - 1/z$,

while

$f_{xy} = (-\sin y) \exp(x \cos y) - x(\sin y)(\cos y) \exp(x \cos y) - 1/z$.

Since $f_{xy} \neq f_{yx}$, there is no f with the given partial derivatives.

§4

6. $dx_1 = a_1 \, dt, \cdots, dx_n = a_n \, dt$. The integral becomes

$$\int_{t_0}^{t_1} [a_1^2 + \cdots + a_n^2] t \, dt = (1/2)[a_1^2 + \cdots + a_n^2](t_1^2 - t_0^2).$$

16. A simple sketch shows that C is the lower half of the circle of radius 1 centered at $(0, 1)$ and traversed in the counterclockwise direction. To parameterize C use polar coordinates with center at $(0, 1)$. Thus

$$C : x = \cos\theta, \, y = 1 + \sin\theta, \, -\pi \leq \theta \leq 0.$$

Hence, on C, $P = x + y = 1 + \cos\theta + \sin\theta$, $Q = x - y = -1 + \cos\theta - \sin\theta$, $dx = -\sin\theta \, d\theta$, $dy = \cos\theta \, d\theta$, and

$$\int_C P \, dx + Q \, dy = \int_{-\pi}^{0} (-\sin\theta - \sin\theta \cos\theta - \sin^2\theta - \cos\theta + \cos^2\theta - \sin\theta \cos\theta) \, d\theta$$

$$= \int_{-\pi}^{0} (-\sin\theta - \cos\theta + \cos 2\theta - \sin 2\theta) \, d\theta$$

$$= \cos\theta - \sin\theta + \tfrac{1}{2}\sin 2\theta + \tfrac{1}{2}\cos 2\theta \,\Big|_{-\pi}^{0} = 2.$$

18. The first integral $= \int_0^{4\pi} (\cos\theta \sin\theta)^2 (1 - \cos\theta) \, d\theta = (1/8)\int_0^{4\pi} (1 - \cos 4\theta)(1 - \cos\theta) \, d\theta = (1/8)[\theta - (\sin 4\theta)/4 - \sin\theta + (\sin 3\theta)/6 + (\sin 5\theta)/10]_0^{4\pi} = \pi/2$. This is also the value of the second integral, as can be seen by a direct calculation as above. The equality of the two integrals could be obtained easier by observing that setting $\theta = t^2$ one recognizes C_1 and C_2 as different parameterizations of the same path.

30. Since $\cos x \, dx + e^{-y} \, dy + z^2 \, dz = df$, where $f(x, y, z) = \sin x - e^{-y} + \tfrac{1}{3}z^3$, and $f(6, 0, -2) = \sin 6 - \tfrac{8}{3}$, $f(-1, 2, 4) = -\sin 1 - e^{-2} + \tfrac{256}{3}$, the desired integral equals $\sin 6 + \sin 1 + e^{-2} - 88$.

38. We need an f with $f_x = 1 + (x + y)^{-1}$, $f_y = (x + y)^{-1}$, $f_z = 1$. The integrability conditions are satisfied. Knowing f_x we conclude that $f(x, y, z) = \log|x + y| + g(y, z)$. The condition on f_y gives $g_y = 0$. Thus $g(y, z) = h(z)$. The condition on f_z gives $h'(z) = 1$ or $h(z) = z + c$, c a constant. Thus $f(x, y, z) = \log|x + y| + c$. This solution is valid in any domain not containing a segment of the line $x + y = 0$.

48. Yes. Any two curves joining the same points can be continuously deformed one into the other, without leaving D. No. $C_1 : x = \tfrac{3}{2}\cos t$; $y = \tfrac{3}{2}\sin t$; $0 \leq t \leq \pi$ and $C_2 : x = \tfrac{3}{2}\cos t$; $y = -\tfrac{3}{2}\sin t$; $0 \leq t \leq \pi$ both join $(\tfrac{3}{2}, 0)$ to $(-\tfrac{3}{2}, 0)$ in D and cannot be deformed into each other without leaving D.

§5

10. $x'(t) = \alpha$; $y'(t) = -gt + \beta$. The kinetic energy of the projectile is $\tfrac{1}{2}m(\alpha^2 + (gt)^2 + \beta^2 - 2gt\beta)$. The potential energy is $mgy = -\tfrac{1}{2}mg^2t^2 + \beta gtm + mg\delta$. The total energy is $(m/2)(\alpha^2 + \beta^2 + 2g\delta)$.

§7

4. $f_u = 3(u - v)^2$; $f_v = -3(u - v)^2$; $f_{uu} = 6(u - v)$; $f_{vv} = 6(u - v)$; $f_{uv} = -6(u - v)$; $f_{uuu} = 6$;

$f_{vvv} = f_{uuv} = -6; f_{vvu} = 6$. Thus the third Taylor polynomial at $u = 2$, $v = -1$ is

$$27[1 + (u - 2) - (v + 1)] + 9[(u - 2)^2 - 2(u - 2)(v + 1) + (v + 1)^2]$$
$$+ [(u - 2)^3 - 3(u - 2)^2(v + 1) + 3(u - 2)(v + 1)^2 - (v + 1)^3].$$

Since $(u - v)^3$ is itself a third degree polynomial, calculus was not really needed. We could have written $(u - v)^3 = [(u - 2) - (v + 1) + 3]^3$ and expanded.

8. When $|u| < 1$, we have $(1 + u)^{-1} = 1 - u + u^2 - u^3 + \cdots$. If $|x| < 2$, $|y| < \frac{1}{2}$, then $|xy| < 1$ and

$$(1 + xy)^{-1} = 1 - xy + x^2y^2 - x^3y^3 \cdots.$$

If the series is cut off after $(x, y)^{2n}$, the remainder term is $R_{2n+1} = (xy)^{2n+1}(1 - xy)^{-1}$ and it tends to zero for $|xy| < 1$.

Chapter 13

§1

8. $\displaystyle\int_{0.25}^{0.30} \int_{0.1}^{0.2} ye^{xy}\, dx\, dy = \int_{0.25}^{0.30} \left[e^{xy}\right]_{x=0.1}^{x=0.2} dy = \int_{0.25}^{0.30} \left[e^{0.2y} - e^{0.1y}\right] dy = \left[e^{0.2y}/0.2 - e^{0.1y}/0.1\right|_{0.25}^{0.30}$

$$= 5(e^{0.06} - e^{0.05}) - 10(e^{0.03} - e^{0.025}).$$

22. $\displaystyle\int_{x_0}^{x_1} \int_{y_0}^{y_1} e^{ax} \sin by\, dy\, dx = \left\{\int_{x_0}^{x_1} e^{ax}\, dx\right\}\left\{\int_{y_0}^{y_1} \sin by\, dy\right\}$

$$= 1/(ab)[e^{ax_1} - e^{ax_0}][\cos (by_0) - \cos (by_1)].$$

28. For some interval I, take the lines that divide I into subintervals on which ψ is constant and those that divide I into subintervals where ϕ is constant. The resulting subdivision gives subintervals on which both ϕ and ψ, and hence ω, are constant.

30. $\displaystyle S = \sum_{i,j=1}^{4} (i^3j - 4j^2) = [-3 + 4 + 23 + 60] + 2[-7 + 0 + 19 + 56]$

$$+ 3[-11 - 4 + 15 + 52] + 4[-15 - 8 + 11 + 48] = 520.$$

$$\int_0^4 \int_0^4 [x^3y - 4y^2]\, dy\, dx = 170.66 \cdots.$$

In this case the Riemann sum is very far from the integral.

46. The sides of our triangle are the lines $6x = 7y$, $3x = 5y$, $3x - 4y = 3$. A sketch shows that the desired integral equals

$$\int_{-6}^{0} \int_{7y/6}^{(4y+3)/3} e^{x+y}\, dx\, dy + \int_{0}^{3} \int_{5y/3}^{(4y+3)/3} e^{x+y}\, dx\, dy = \int_{-6}^{0} \left[e^{(7y+3)/3} - e^{13y/6}\right] dy + \int_{0}^{3} \left[e^{(7y+3)/3} - e^{8y/3}\right] dy$$

$$= -(3/7)e^{-13} + (6/13)e^{-6} - (47/21)e^8 + 86/39.$$

48. $f(x, y) = -f(-x, -y)$. Since the function and region are both symmetric about the origin, $\iint = 0$.

52. $\displaystyle\int_0^1 \int_{\sqrt{x}}^{x^2} x^ny^m\, dy\, dx = [1/(m + 1)]\int_0^1 \left[x^{2m+n+2} - x^{n+(m+1)/2}\right] dx$

$$= \frac{-3m + 5}{(m + 1)(2m + n + 3)(2n + m + 1)}$$

(Recall that $m \geq 1$, $n \geq 1$.)

56. The value of the integral is the area of D, that is 4. It is four times the area of the triangle with vertices $(0, 0)$, $(0, 1)$ and $(1, 2)$, hence equal to

$$4 \int_0^{\arctan 2} \int_0^{\sec \theta} r \, dr \, d\theta = 2 \int_0^{\arctan 2} \sec^2 \theta \, d\theta = 2 \tan (\text{arc tan } 2) = 4.$$

§2

8. We use polar coordinates and rewrite the equation of the surface as

$$z = \text{arc tan } (y/x) = \theta, \, -\pi/2 < \theta < \pi/2, \, 1 < r < 4.$$

Since $z_x = -\sin \theta/r$, $z_y = \cos \theta/r$, the area equals

$$\int_{-\pi/2}^{\pi/2} \int_1^4 \sqrt{1 + r^2} \, dr \, d\theta = \frac{\pi}{2} \left(4\sqrt{17} - \sqrt{2} + \log \frac{4 + \sqrt{17}}{1 + \sqrt{2}} \right).$$

24. Use polar coordinates.

$$z_x = -x/z; \qquad z_y = -y/z; \qquad \sqrt{1 + z_x^2 + z_y^2} = \sqrt{\frac{z^2 + x^2 + y^2}{z^2}} = 1/z.$$

Thus

$$A = 2 \int_0^{2\pi} \int_A^B (1 - r^2)^{-1/2} \, dr \, d\theta = 4\pi [\text{arc sin } B - \text{arc sin } A].$$

§3

16. $V = \int_{-3}^3 \int_{-\sqrt{9-y^2}}^{+\sqrt{9+y^2}} \int_0^{4-x} dz \, dx \, dy = \int_{-3}^3 (4x - x^2/2) \Big|_{-\sqrt{9-y^2}}^{+\sqrt{9-y^2}} dy$

$$= 8 \int_{-3}^3 \sqrt{9 - y^2} \, dy = 4 \left[y\sqrt{9 - y^2} + 9 \text{ arc sin } (y/3) \right] \Big|_{-3}^3 = 36\pi.$$

26. For all points $(x, y, z) = (r \cos \theta, r \sin \theta, z)$ in the volume considered $x^2 + y^2 < 2y$, that is $y > 0$ (or $\sin \theta > 0$) and $x^2 + (y - 1)^2 < 1$ which shows that (x, y) lies in the circle of radius 1 and center $(0, 1)$. Thus $|\theta| < \pi/2$. We conclude that our volume is described by $r^2 < z < 2r \sin \theta$, $0 < \theta < \pi/2$. Hence

$$V = \int_0^{\pi/2} \int_0^{2\sin\theta} \int_{r^2}^{2r\sin\theta} r \, dz \, dr \, d\theta = \int_0^{\pi/2} \int_0^{2\sin\theta} (2r^2 \sin \theta - r^3) \, dr \, d\theta$$

$$= \int_0^{\pi/2} \left[(16/3) \sin^4 \theta - 4 \sin^4 \theta \right] d\theta$$

$$= (4/3) \left[\frac{\sin^3 \theta \cos \theta}{-4} + (3/4)[\theta/2 - (1/4) \sin 2\theta] \right] \Big|_0^{\pi/2} = \pi/4.$$

38. $xyz^2 = \rho^4 (\cos \phi \sin \phi)^2 \cos \theta \sin \theta$; the desired integral is

$$\int_0^\pi \int_0^{2\pi} \int_0^1 \rho^6 \left(\frac{\sin 2\phi}{2} \right)^2 \frac{\sin 2\theta}{2} \, d\rho \, d\theta \, d\phi = 0 \quad \text{since} \quad \int_0^{2\pi} \sin 2\theta \, d\theta = 0.$$

§4

4. $F = 2 \log x_1 + 3 \log x_2 + 2 \log x_3 + 3 \log x_4$. Thus

$$\int_1^2 \int_2^3 \int_3^4 \int_4^5 F \, dx_1 \cdots dx_4$$

$$= 2 \int_1^2 \log x_1 \, dx_1 + 3 \int_2^3 \log x_2 \, dx_2 + 2 \int_3^4 \log x_3 \, dx_3 + 3 \int_4^5 \log x_4 \, dx_4$$

$$= -10 + \log \, (5^{15} 3^3 / 4^5).$$

§5

8. $A = \displaystyle\int_0^4 x^2 \, dx = 64/3.$

$$X = (3/64) \int_0^4 \int_0^{x^2} x \, dy \, dx = (3/64) \int_0^4 x^3 \, dx = 3.$$

$$Y = (3/64) \int_0^4 \int_0^{x^2} y \, dy \, dx = (3/128) \int_0^4 x^4 \, dx = 24/5.$$

26. Let the semicircle be given by $x^2 + y^2 = R^2, y > 0$. Then $X = 0$ by symmetry (see §5.4 below) and

$$Y = (1/\pi R) \int_0^\pi R \sin \theta R \, d\theta = \frac{2R}{\pi}.$$

46. The centroid is $(0, 1/\sqrt{2})$ and is $1/\sqrt{2}$ distance from the axis. Thus $V = 2\pi \, (1/\sqrt{2})1 = \sqrt{2}\pi.$

§6

6. The double integral in Green's theorem is, in the case considered,

$$\iint_D (y - x) \, dx \, dy = \int_0^{\pi/2} \int_0^{\sqrt{2}} r^2 \, (\cos \theta - \sin \theta) \, dr \, d\theta = 0.$$

Also $P \, dx + Q \, dy = xy \, (dx + dy) = 0$ on the boundary curve of D except for the arc $x = \sqrt{2} \cos \theta$, $y = \sqrt{2} \sin \theta, 0 \leq \theta \leq \pi/2$. The line integral in Green's theorem is therefore

$$\int_0^{\pi/2} 2 \cos \theta \sin \theta \, (-\sqrt{2} \sin \theta + \sqrt{2} \cos \theta) \, d\theta = 0.$$

18. Repeat above argument in the text with $C_3: x = g(t), y = t, f(a) \leq t \leq c$.

$$\iint_D Q_x \, dx \, dy = \int_{f(a)}^c \int_a^{g(y)} Q_x \, dx \, dy = \int_{f(a)}^c [Q(g(y), y) - Q(a, y)] \, dy.$$

Since $dy = 0$ on C_2,

$$\int_C Q \, dy = \int_{f(a)}^c Q(g(y), y) \, dy + \int_c^{f(a)} Q(a, y) \, dy.$$

§7

4. Let D be broken up into a small sphere about $(1, 0, 2)$ and the remaining space. The convergence of the integral is in doubt in the small sphere of radius r. In this sphere, use spherical coordinates with origin in the center, that is, write $Z = 2 + \rho \cos \phi$; $y = \rho \sin \theta \sin \phi$; $x = \rho \sin \theta \sin \phi + 1$. The integral becomes

$$\lim_{\epsilon \to 0^+} \int_0^{2\pi} \int_0^\pi \int_\epsilon^r \rho^{-2} \sin \phi \, d\rho \, d\phi \, d\theta = \lim_{\epsilon \to 0^+} 4\pi[1/\epsilon - 1/r] = +\infty.$$

Thus the integral diverges.

16. We use the comparison theorem. For $x^2 + y^2 > 1$, the integrand is not greater than $\mid x^2 + y^2 \mid^{-3}$, and

$$\iint_{x^2+y^2>1} \mid x^2 + y^2 \mid^{-3} dx \, dy = \int_0^{2\pi} \int_1^{+\infty} r^{-5} \, dr \, d\theta = \pi/2.$$

Hence the integral in the exercise converges.

SUPPLEMENT

§1

2. Let ϵ be a given positive number. If $\beta - \epsilon$ were not less than x_j for nearly all j, then for each set $\{x_k, x_{k+1}, \cdots\}$ there would be a number $\le \beta - \epsilon$. Hence $\beta_k \le \beta - \epsilon$ for each k, which contradicts the definition of β as the limit of $\{\beta_k\}$. Thus $\beta - \epsilon < x_j$ for nearly all j. If $\beta + \epsilon$ not $> x_j$ for infinitely many j, then for nearly all j, $x_j \ge \beta + \epsilon$. Hence $\beta_k \ge \beta + \epsilon$ for all large k, which contradicts the definition of β. Thus $\beta + \epsilon > x$; for infinitely many j.

4. Take the decimal representation of an irrational number, $x = .\alpha_1\alpha_2\alpha_3\cdots$, say $\pi = 3.1415926536 \ldots$, and define $a_n = .\alpha_1\alpha_2\cdots\alpha_n$. a_n are clearly rational and converge to x, an irrational. The sequence is Cauchy by Theorem C.

§2

2. Let x_0 be any point in the interval. Let $\epsilon > 0$ be given. Since f is uniformly continuous, there is a $\delta > 0$ such that, if x and $\hat{x}$ are points in the interval and $\mid x - \hat{x} \mid < \delta$ then $\mid f(x) - f(\hat{x}) \mid < \epsilon$. Thus if $\hat{x}$ is a given point, f is continuous at $\hat{x}$.

4. Let S be the intersection of some collection of closed sets $\{S_\alpha\}$ where α runs through some indexing set A. Let $\{x_n\}$ be a sequence in S that converges to some x. $\{x_n\}$ is then a sequence in each S_α which converges to x. Since each S_α is closed, x must lie in each S_α and hence in S.

§4

2. Let I be subdivided so that on each subinterval I_j, ϕ, and ψ take on the respective constant values z_j and w_j. Thus

$$\iint (\phi + \psi) \, dx \, dy = (z_1 + w_1)A(I_1) + (z_2 + w_2)A(I_2) + \cdots + (z_n + w_n)A(I_n)$$

$$= [z_1A(I_1) + z_2A(I_2) + \cdots + z_nA(I_n)] + [w_1A(I_1) + w_2A(I_2) + \cdots w_nA(I_n)]$$

$$= \iint \phi \, dx \, dy + \iint \psi \, dx \, dy.$$

4. Since S^* is an upper sum for f, $S^* = \iint \phi \, dx \, dy$ where ϕ is a step function and $\phi \ge f$. But if $C \ge 0$, $C\phi \ge Cf$ and $\iint C\phi \, dx \, dy = C\iint \phi \, dx \, dy = CS^*$ is an upper sum for Cf. S_* is treated in the same manner.

6. Since f is integrable, Cf is integrable by Exercise 5. $C\int\int f\, dx\, dy = CS$ where S is the glb of upper sums S^* and lub of lower sums S_*. If $C \geq 0$, CS is the glb of CS^* which are all the upper sums of Cf; CS is the lub of CS_*, all the lower sums of Cf. Thus $CS = \int\int Cf\, dx\, dy$. If C is negative, the argument is similar.

8. Let $S^*{}_g$, $S^*{}_f$, S_{g*}, S_{f*}, be respective upper and lower sums for g and f. Then $S^* = S^*{}_g + S^*{}_f$ is an upper sum for $f + g$ and $S_* = S_{g*} + S_{f*}$ is a lower sum by Exercise 7.

$$|\, S^* - S_* \,| \leq |\, S^*{}_g - S_{g*} \,| + |\, S^*{}_f - S_{f*} \,| < \epsilon/2 + \epsilon/2 = \epsilon$$

for appropriate upper and lower sums of f and g, since they are integrable. Hence $f + g$ is integrable.

10. If $f \leq g$, let $\epsilon > 0$ be given. There are step functions ϕ and ψ such that $\phi \leq f \leq g \leq \psi$ and

$$\int\int f\, dx\, dy - \int\int \phi\, dx\, dy < \epsilon \qquad \text{and} \qquad \int\int \psi\, dx\, dy - \int\int g\, dx\, dy < \epsilon.$$

By (4) we conclude that $\int\int \phi\, dx\, dy \leq \int\int \psi\, dx\, dy$. Hence $\int\int f\, dx\, dy < \int\int g\, dx\, dy + 2\epsilon$. Since ϵ was arbitrary, the conclusion follows.

Index